BIOLOGY

today and tomorrow

WITH PHYSIOLOGY

SECOND EDITION

CECIE STARR
CHRISTINE A. EVERS
LISA STARR

THOMSON
BROOKS/COLE

Australia • Canada • Mexico • Singapore
Spain • United Kingdom • United States

PUBLISHER Jack C. Carey

VICE-PRESIDENT, EDITOR-IN-CHIEF Michelle Julet

SENIOR DEVELOPMENT EDITOR Peggy Williams

ASSOCIATE DEVELOPMENT EDITOR Suzannah Alexander

EDITORIAL ASSISTANT Kristina Razmara

TECHNOLOGY PROJECT MANAGER Keli Amann

MARKETING MANAGER Stacy Best, Kara Kindstrom

MARKETING ASSISTANT Brian Smith

MARKETING COMMUNICATIONS MANAGER Nathaniel Bergson-Michelson

PROJECT MANAGER, EDITORIAL PRODUCTION Andy Marinkovich

CREATIVE DIRECTOR Rob Hugel

PRINT BUYER Karen Hunt

PERMISSIONS EDITOR Joohee Lee, Sarah Harkrader

PRODUCTION SERVICE Grace Davidson & Associates

TEXT AND COVER DESIGN Gary Head

PHOTO RESEARCHER Myrna Engler

COPY EDITOR Christy Goldfinch

ILLUSTRATORS Gary Head, ScEYEnce Studios, Lisa Starr

COMPOSITOR Lachina Publishing Services

TEXT AND COVER PRINTER QuebecorWorld—Versailles

COVER PHOTOGRAPHER Jay Barnes

A female crab flower spider (Misumena vatia) *raises its legs and shows its fangs in a defensive posture. Such spiders exploit the mutually beneficial relationship between flowers and their pollinators. A bee or fly that is lured into range by the flower's color and scent is snatched by the spider's elongated front legs and subdued by a venomous bite from the fangs.*

Printed in the United States of America
3 4 5 6 7 09 08 07

Library of Congress Control Number: 2005936176

ISBN-13: 978-0-495-01654-0
ISBN-10: 0-495-01654-3 (softcover)

ISBN-13: 978-0-495-10285-4
ISBN-10: 0-495-10285-7 (softcover)

ISBN-13: 978-0-495-10919-8
ISBN-10: 0-495-10919-3 (hardcover)

For more information about our products, contact us at:
Thomson Learning Academic Resource Center

1-800-423-0563

For permission to use material from this text or product, submit a request online at http://www.thomsonrights.com.
Any additional questions about permissions can be submitted by e-mail to thomsonrights@thomson.com.

BOOKS IN THE BROOKS/COLE BIOLOGY SERIES

Biology: The Unity and Diversity of Life, Eleventh, Starr/Taggart
Engage Online for Biology: The Unity and Diversity of Life
Biology: Concepts and Applications, Sixth, Starr
Basic Concepts in Biology, Sixth, Starr
Biology Today and Tomorrow, Second, Starr
Biology, Seventh, Solomon/Berg/Martin
Human Biology, Sixth, Starr/McMillan
Biology: A Human Emphasis, Sixth, Starr
Human Physiology, Fifth, Sherwood
Fundamentals of Physiology, Second, Sherwood
Human Physiology, Fourth, Rhoades/Pflanzer

Laboratory Manual for Biology, Fourth, Perry/Morton/Perry
Laboratory Manual for Human Biology, Morton/Perry/Perry
Photo Atlas for Biology, Perry/Morton
Photo Atlas for Anatomy and Physiology, Morton/Perry
Photo Atlas for Botany, Perry/Morton
Virtual Biology Laboratory, Beneski/Waber
Introduction to Cell and Molecular Biology, Wolfe
Molecular and Cellular Biology, Wolfe
Biotechnology: An Introduction, Second, Barnum

Introduction to Microbiology, Third, Ingraham/Ingraham
Microbiology: An Introduction, Batzing
Genetics: The Continuity of Life, Fairbanks/Anderson
Human Heredity, Seventh, Cummings
Current Perspectives in Genetics, Second, Cummings
Gene Discovery Lab, Benfey

Animal Physiology, Sherwood, Kleindorf, Yarcey
Invertebrate Zoology, Seventh, Ruppert/Fox/Barnes
Mammalogy, Fourth, Vaughan/Ryan/Czaplewski
Biology of Fishes, Second, Bond
Vertebrate Dissection, Ninth, Homberger/Walker

Plant Biology, Second, Rost/Barbour/Stocking/Murphy
Plant Physiology, Fourth, Salisbury/Ross
Introductory Botany, Berg

General Ecology, Second, Krohne
Essentials of Ecology, Third, Miller
Terrestrial Ecosystems, Second, Aber/Melillo
Living in the Environment, Fourteenth, Miller
Environmental Science, Tenth, Miller
Sustaining the Earth, Seventh, Miller
Case Studies in Environmental Science, Second, Underwood
Environmental Ethics, Third, Des Jardins
Watersheds 3—Ten Cases in Environmental Ethics, Third, Newton/Dillingham

Problem-Based Learning Activities for General Biology, Allen/Duch
The Pocket Guide to Critical Thinking, Second, Epstein

Thomson Higher Education
10 Davis Drive
Belmont, CA 94002-3098
USA

Asia (including India)
Thomson Learning
5 Shenton Way
#01-01 UIC Building
Singapore 068808

Australia/New Zealand
Thomson Learning Australia
102 Dodds Street
Southbank, Victoria 3006
Australia

Canada
Thomson Nelson
1120 Birchmount Road
Toronto, Ontario M1K 5G4

UK/Europe/Middle East/Africa
Thomson Learning
High Holborn House
50/51 Bedford Row
London WC1R 4LR
United Kingdom

CONTENTS IN BRIEF

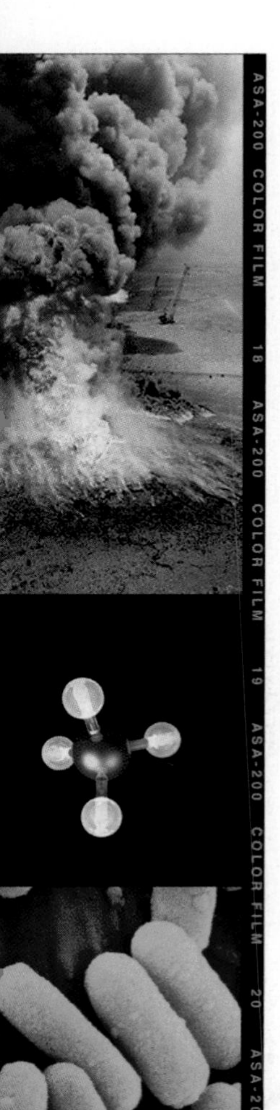

DETAILED CONTENTS

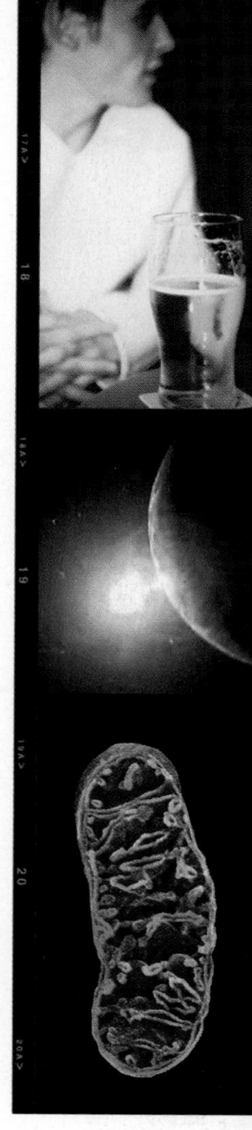

Unit Two Genetics

7 How Cells Reproduce

8 Observing Patterns in Inherited Traits

9 DNA Structure and Function

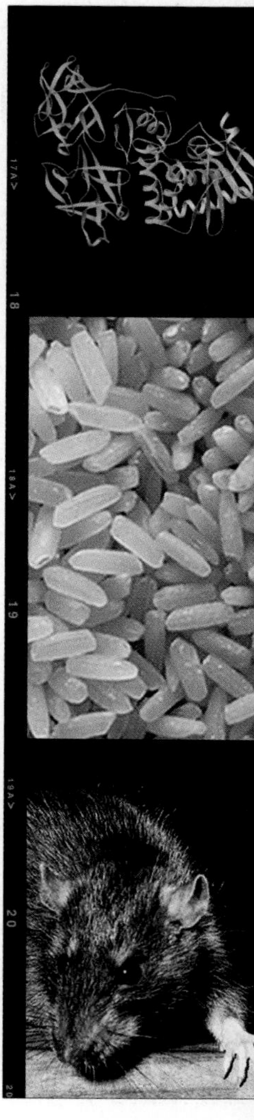

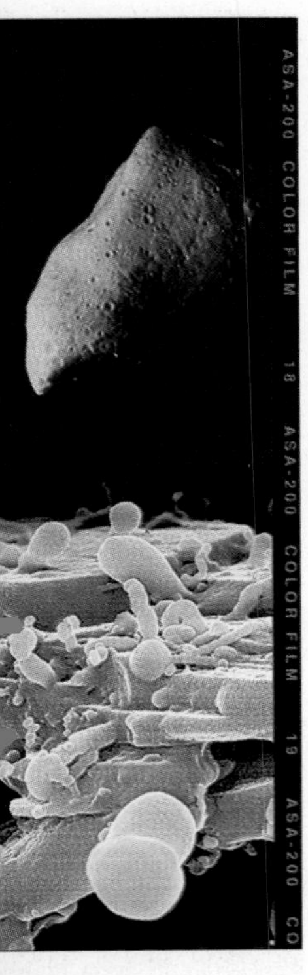
ASA-200 COLOR FILM 18
ASA-200 COLOR FILM 19
ASA-200 CO

ASA-200 COLOR FILM 18
ASA-200 COLOR FILM 19
ASA-200 COLOR FILM 20
ASA-200 COLOR FILM

Unit Five How Animals Work

20 Animal Tissues and Organ Systems

21 How Animals Move

22 Circulation and Respiration

23 Immunity

24 Digestion, Nutrition, and Excretion

25 Neural Control and the Senses

26 Endocrine Controls

27 Reproduction and Development

Unit Six Ecology

28 Population Ecology

29 Community Structure and Biodiversity

30 Ecosystems

31 The Biosphere

32 Behavioral Ecology

Preface

Biology opens a window on processes of nature that are so familiar that we might not even think about them. It reveals connections, sometimes subtle and sometimes not, between us and everything else on Earth. This is profound stuff that many nonmajors students are ill-prepared to grasp. Many enter college without adequate preparation in science, and often with the mistaken belief that they are not smart enough to learn biology. Should textbook writers reinforce their apprehension by squeezing all the juice out of a story, by producing a pedagogical husk? We think not. We find biology fascinating and eagerly devour news of recent discoveries. We believe that, with the proper guidance, students too can become lifelong learners who understand and are excited by the science of life.

This book is briefer than our other titles, but in it we strive to share our enthusiasm for biology with easy-to-follow writing on relevant topics. For this edition, we have shortened and streamlined many sections. We did not water down the science, but instead emphasize processes and relationships over detail.

Promote Critical Thinking Like all textbooks at this level, we walk students through examples of problem solving and experiments throughout the book. Many historical and theoretical experiments have been integrated into the text. Each chapter ends with several *Critical Thinking* questions, some illustrated. Some are more challenging than others, but all invite students to think outside the memorization box. We return again and again in chapter introductions and in the text to examples of how science is carried out. Introductory students are not scientists, however, so we do not expect them to learn the language and processes of science by intuition alone; we help them build these skills in a paced way.

Make It Brief, With Clear Explanations To keep the book length manageable, we were selective about which topics to include but were not stingy with clear explanations. If something is worth reading, why reduce it to a factoid? Factoids invite mind-numbing memorization, which does not promote critical thinking about the world and our place in it.

For instance, you can safely bet that most nonmajors simply do not want to memorize each catalytic step of crassulacean acid metabolism. They *do* want to learn about the biological basis of sex, and many female students want to know what will be going on inside them if and when they get pregnant. Good explanations can help them make their own informed decisions on many biology-related issues, including STDs, fertility drugs, prenatal diagnoses of genetic disorders, gene therapies, and abortions.

Our choices for which topics to condense, expand, or delete were not arbitrary. They reflect three decades of feedback from teachers throughout the world.

Make It Relevant Most students taking this course will not become biologists, but biological research will affect their lives in direct and often controversial ways. *What they learn today will have impact on how they make decisions tomorrow*—in the voting booth as well as in their personal lives.

Each chapter starts with an IMPACTS/ISSUES essay on a topic of current interest related to its content. For instance, the microevolution chapter opens with how the use of warfarin has favored "super rats." Essays are expanded in custom videoclips and in sidebars on relevant pages within chapters, as shown on the facing page. We return to the essay topics in exercises on the student website and in the PowerLecture, an all-in-one PowerPoint tool for instructors. We also ask the students, HOW WOULD YOU VOTE? on an application related to the essay's topic. The exercise invites them to read a selection of articles, pro and con, before voting.

12 PROCESSES OF EVOLUTION

IMPACTS, ISSUES

Rise of the Super Rats

Slipping in and out of the pages of human history are rats—*Rattus*—the most notorious of mammalian pests. One kind of rat or another has distributed pathogens and parasites that cause bubonic plague, typhus, and other deadly infectious diseases. The death toll from fleas that bit infected rats and then bit people has exceeded the death toll in all wars combined.

The rats themselves are far more successful. By one estimate, there is one rat for every person in urban and suburban centers of the United States. In addition to spreading diseases, rats chew their way through walls and wires of homes and cities. In any given year, they cause economic losses approaching 19 billion dollars.

For years, people have been fighting back with traps, ratproof storage facilities, and various poisons. During the 1950s, they started using baits laced with warfarin. This compound interferes with blood clotting. Rats ate the baits, then died within days after bleeding internally or losing blood through cuts or scrapes. Warfarin was extremely effective. Compared with other rat poisons, it had a lot less impact on harmless species.

In 1958, however, a Scottish researcher reported that warfarin did not work against some rats. Similar reports from other European countries followed. About twenty years later, 10 percent of the urban rats caught in the United States were warfarin resistant. *What happened?*

To find out, researchers compared warfarin-resistant rat populations with still-vulnerable rats. They traced the difference to a gene on one of the rat chromosomes. At that gene locus, a dominant allele was common in warfarin-resistant rat populations but very rare among the vulnerable ones. "What happened" was evolution by natural selection. Warfarin was exerting selective pressure on populations of rats. The previously rare dominant allele proved to be adaptive. The lucky rats that inherited the allele survived and produced more offspring. The unlucky ones that inherited the recessive allele had no built-in defense, and died. Over time, the dominant allele's frequency increased in all rat populations exposed to the poison.

Of course, selection pressures can and often do change. When warfarin resistance increased in rat populations, people stopped using warfarin. And guess what: The dominant allele's frequency declined. Now the latest worry is the evolution of "super rats," which even more potent rodenticides can't seem to kill.

When you hear someone question whether life evolves, remember this: With respect to life, **evolution** simply means that heritable change is occurring in some line of descent. The actual mechanisms that can bring about such change are the focus of this chapter. Later chapters highlight how these mechanisms have contributed to the evolution of new species.

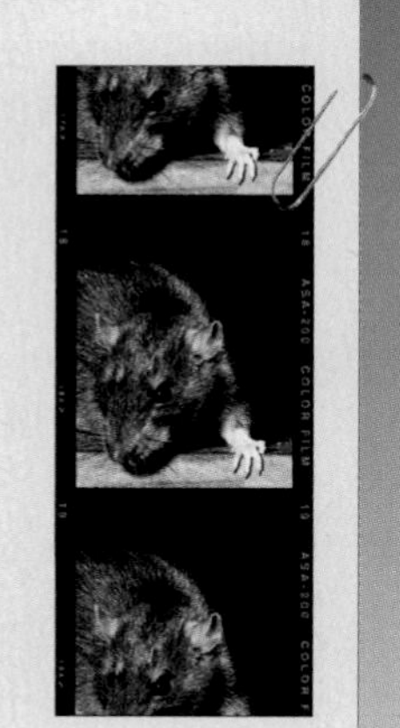

How Would You Vote? *Antibiotic-resistant strains of bacteria are becoming dangerously pervasive. Standard animal husbandry practice includes the repeated dosing of healthy animals with antibiotics—the same ones prescribed to people. Should this practice stop? See BiologyNow for details, then vote online.*

Key Concepts

EVOLUTIONARY VIEWS EMERGE
The world distribution of species, similarities and differences in body form, and the fossil record gave early evidence of evolution—of changes in lines of descent. Charles Darwin and Alfred Wallace had an idea of how those changes occur.

VARIATION AND ADAPTATION
An adaptation is a heritable aspect of form, function, behavior, or development that promotes survival and reproduction. It enhances the fit between the individual and prevailing conditions in its environment.

MICROEVOLUTIONARY PROCESSES
An individual does not evolve. A *population* evolves, which means its shared pool of alleles changes. Over generations, any allele may increase in its frequency among individuals, or it may become rare or lost. Mutation, genetic drift, natural selection, and gene flow change allele frequencies in a population. These processes of microevolution change the observable characteristics that define a population and, more broadly, a species.

Links to Earlier Concepts

Section 1.4 sketched out the key premises of the theory of natural selection. Here you will read about evidence that led to its formulation. You may wish to refresh your memory of protein structure (2.8), basic terms of genetics (8.1), chromosomes and crossing over (8.4, 8.5), DNA replication and repair (9.3), and gene mutation (10.5). This chapter puts the chromosomal basis of inheritance (8.4, 8.7) and continuous variation in populations (8.3) into the broader context of evolution.

IMPACTS, ISSUES

Warfarin-resistant rats led to the development of anticoagulants that are more toxic and that persist longer in the environment. They are weakening and killing owls, hawks, coyotes, and other predators that have eaten poisoned rats. Between 1985 and 1999, for example, the number of barn owls with anticoagulants in their blood rose from 5 to 36 percent. And that was just one study in Great Britain.

Students throughout the country are already voting online, and they are accessing campuswide, statewide, and nationwide tallies. This interactive approach to issues reinforces the premise that an individual's actions can make a difference.

Make It Easy To Follow On each chapter's opening page is a preview of key concepts, each with a simple title. We repeat these titles at the bottom of appropriate pages as reminders of the chapter's conceptual organization.

New to this edition are LINKS TO EARLIER CONCEPTS that can help students follow the big connections within and between the chapters. The opening page lists the key sections in earlier chapters that students should be familiar with before they start the chapter. For instance, before reading about neural function, a student may wish to scan an earlier chapter's section on active transport. We repeat linking icons in the chapter's page margins as reminders of these and other relevant sections. This feature demonstrates how the concepts in the book are not separate topics, but are as closely interconnected as biology itself.

A conversational writing style eases students into the story. They can stay focused on that story without worrying about highlighting something they might be tested on. Why? We already highlight key points for them, in blue boxes at the end of each section. These HIGHLIGHTED KEY POINTS function as a *running in-text summary*. All chapters end with a *section-by-section summary* that reinforces what they have learned.

Students also can stay on track with ANIMATED DIAGRAMS. They can walk through the text's step-by-step art as a preview of major concepts, then check out the steps online. These visual learning devices are available in narrated, animated form, on *BiologyNow* and on the PowerLecture DVD. Exposing students to the material in a variety of modalities accommodates diverse learning styles and reinforces understanding.

Offer Easy-to-Use Media Tools Integrated into each end-of-chapter summary are associated media tools that will help students focus their study and understand the key concepts. With this new edition, taking advantage of these online assets is easier than ever. Students register their 1PASS ACCESS CODE at *http://1pass.thomson.com*, then log in to access all resources outlined below. An access code is packaged in each new copy of the book.

BIOLOGYNOW™ tests which topics students have not yet mastered and creates customized learning plans to focus their study and review. Responses to the diagnostic pretest questions activate a personalized learning plan. Answer incorrectly, and the plan lists the relevant text sections, figures, chapter videos, and animations that the student may review.

A *Post-Test* can be used as a self-assessment tool or submitted to an instructor. Because BiologyNow is built into iLrn, Thomson Learning's course management system, the answers and results can be fed directly into an electronic gradebook. BiologyNow can be integrated with WebCT and Blackboard, so students may log in through these systems as well.

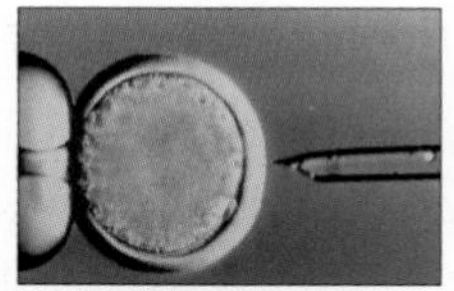

1 Removal of the nucleus from a sheep egg during a cloning. At the center is an unfertilized sheep egg. A pipette has positioned the egg, which a microneedle is about to penetrate.

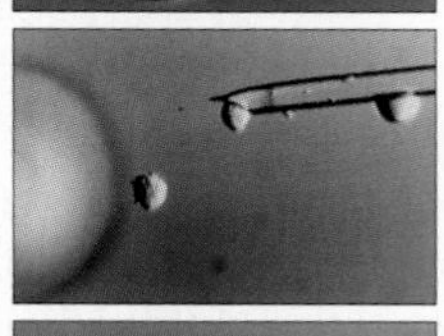

2 The microneedle has now emptied the sheep egg of its own nucleus, which holds the cell's DNA.

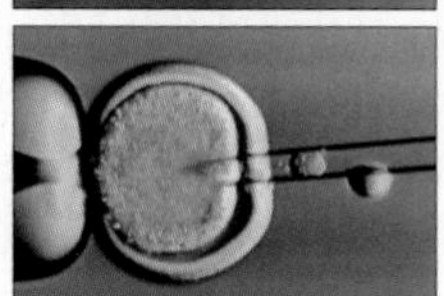

3 A nucleus from a donor cell is about to enter the enucleated egg. An electric spark will stimulate the egg to enter mitotic cell divisions. At an early stage, the ball of cells will be implanted in the womb of a surrogate mother sheep.

Figure 9.8 *Animated!* Nuclear transfer of sheep cells. In this series of photos, a microneedle replaces the nucleus of a sheep egg with a nucleus from an adult sheep cell. Newer methods involve direct transfer of nuclear DNA that has been treated with mitotic cell extracts to condense the chromosomes.

Interactive flashcards with audio pronunciation use the definitions from this edition's highly revised glossary. All the boldface terms within the text are presented in this resource.

The *How Do I Prepare?* section of BiologyNow has tutorials in math, chemistry, and graphing, as well as a review of basic study skills.

1PASS grants access to INFOTRAC COLLEGE EDITION™, an exclusive online searchable database of more than 5,000 periodicals and close to 18 million articles. The articles are in full-text form and can be located easily and quickly with a key word search. For each chapter, the *How Would You Vote* exercise in BiologyNow references specific InfoTrac articles and websites. As students vote on each issue, the website provides a running tally by campus, state, and the country. Instructors can assign the exercises from iLrn, or students can access them through the website at *http://biology.brookscole.com/btt2*.

An online ISSUES AND RESOURCES INTEGRATOR correlates chapter sections with applications, videos, InfoTrac articles, and websites. This guide is updated each semester.

VMENTOR is an online tutorial service available from 6 AM to 12 PM Monday through Saturday. Students may interact with a live tutor in a virtual classroom by voice communication, whiteboard, and text messaging.

New college editions of the book can also include access to the AUDIOBOOK. Students can either listen to narration online or download MP3 files for use on a portable MP3 player.

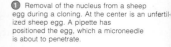

Acknowledgments

No short list can convey our thanks to the extended team that made this book not only possible, but also excellent. John Jackson and Walt Judd deserve special thanks for their detailed evaluations, perspective, and continued commitment to excellence. Dave Rintoul and his students helped us fine-tune some important last-minute details. Through focus-group participation, reviews, and class testing, he and the other instructors listed below helped us transform our first edition into the polished, cohesive textbook you see here, one that particularly addresses the needs of students who have been previously underserved.

With Jack Carey, Susan Badger, Sean Wakely, Kathie Head, Michelle Julet, and Michael Johnson, Thomson Learning proved why it is one of the world's foremost publishers; we doubt that any authors get finer support anywhere. Keli Amann managed production of all student and instructor media tools. Peggy Williams again brought her tenacity, intelligence, and humor to the developmental editing. Andy Marinkovich calmly kept us on track, and Grace Davidson made this book happen despite several hurricanes, both literal and figurative. Gary Head created great designs and graphics; Steve Bolinger did so for the media tools. Myrna Engler, Suzannah Alexander, Kristina Razmara, Karen Hunt, Stacy Best—the list goes on. Thank you to each one of you; this book would not exist without your help.

CECIE STARR, CHRIS EVERS, AND LISA STARR *November 2005*

General Advisors

DANIEL J. FAIRBANKS *Brigham Young University*
JOHN D. JACKSON *North Hennipin Community College*
WALTER JUDD *University of Florida*
E. WILLIAM WISCHUSEN *Louisiana State University*

Focus Groups, Reviewers, and Class Testing

JO DALE H. ALES *Baton Rouge Community College*
DIANNE ANDERSON *San Diego City College*
KENNETH D. ANDREWS *East Central University*
BERT ATSMA *Union County College*
ANDREW STEPHEN BALDWIN *Mesa Community College*
GREGORIO B. BEGONIA *Jackson State University*
DAVID BELT *Johnson County Community College*
PAUL A. BILLETER *College of Southern Maryland*
ANDREW BLAUSTEIN *Oregon State University*
JUDY BLUEMER *Morton College*
MARK G. BOLLONE *Tallahassee Community College*
SUSAN BOWER *Pasadena City College*
WILLIAM BROWER *Del Mar College*
STEVEN G. BRUMBAUGH *Greenriver Community College*
WILBERT BUTLER, JR. *Tallahassee Community College*
THOMAS R. CAMPBELL *Los Angeles Pierce College*
J. MICHELLE CAWTHORN *Georgia Southern University*
THOMAS T. CHEN *Santa Monica College*
THOMAS F. CHUBB *Villanova University*
JAY L. COMEAUX *Louisiana State University*
EDWARD A. DEGRAUW *Portland Community College*
JEAN DESAIX *University of North Carolina*
BRIAN DINGMANN *University of Minnesota, Crookston*
ERNEST F. DUBRUL *University of Toledo*
PARAMJIT DUGGAL *Metropolitan Community College, Maple Woods*
PAMELA K. ELF *University of Minnesota, Crookston*
ROBERT C. EVANS *Rutgers University-Camden*
W. LOGAN FINK *Pensacola Junior College*
APRIL A. FONG *Portland Community College*
EDISON R. FOWLKS *Hampton University*
TERESA LANE FULCHER *Pellissippi State Technical Community College*
RIC A. GARCIA *Clemson University*
MICHELLE GEARY *West Valley College*
MARY GOBBETT *University of Indianapolis*
MELVIN H. GREEN *University of California, San Diego*
CARLA GUTHRIDGE *Cameron University*
GALE HAIGH *McNeese State University*
JERRIE R. HANIBLE *Southeastern Louisiana University*
LINDA HARRIS-YOUNG *Motlow State Community College*
DAVID J. HENSON *Eastern Arizona College*
JULIANA HINTON *McNeese State University*
SONGQIAO (SARA) HUANG *Los Angeles Valley College*
DONNA HUFFMAN *Calhoun Community College*
KIMBERLY HURD *Bakersfield College*
DESIRÉE JACKSON *Texas Southern University*
ROSS S. JOHNSON *Virginia State University*
GLENN H. KAGEYAMA *California State Polytechnic University, Pomona*
RITA A. KARPIE *Brevard Community College*
JUDITH KELLY *Henry Ford Community College*
PAULINE LIZOTTE *Valencia Community College*
JUAN LOPEZ-BAUTISTA *University of Alabama, Tuscaloosa*
DAVID LORING *Johnson County Community College*
DANIEL MARK *Metropolitan Community College, Maple Woods*
LINDA MARTIN-MORRIS *University of Washington*
DENNIS MCDERMOT *Lake Erie College*
RON MOLLICK *Christopher Newport University*
JILL NISSEN *Montgomery College*
ALEXANDER E. OLVIDO *Virginia State University*
JOSHUA M. PARKER *Community College of Southern Nevada*
ROBERT P. PATTERSON *North Carolina State University*
JOHN S. PETERS *College of Charleston*
HAROLD PLETT *Fullerton College*
JENNIE M. PLUNKETT *San Jacinto College*
JAMES V. PRICE *Utah Valley State College*
JERRY W. PURCELL *San Antonio College*
KIRSTEN RAINES *San Jacinto College*
WENDA RIBEIRO *Thomas Nelson Community College*
PELE EVE RICH *Mt. San Jacinto College*
DAVID A. RINTOUL *Kansas State University*
FRANK A. ROMANO III *Jacksonville State University*
DAVID ROSEN *Diablo Valley College*
ETTA V. RUPPERT *Clemson University*
ERIK P. SCULLY *Towson University*
MIKE SHIFFLER *University of Alabama*
MARCIA SHOFNER *University of Maryland*
GREG A. SIEVERT *Emporia State University*
LINDA D. SMITH-STATON *Pellissippi State Technical Community College*
LARRY L. ST. CLAIR *Brigham Young University*
JACQUELINE J. STEVENS *Jackson State University*
JUDY STEWART *Community College of Southern Nevada*
ALICE B. TEMPLET *Nicholls State University*
MICHAEL T. TOLIVER *Eureka College*
JEANETTE H. TUCKER *Northern Virginia Community College*
HEATHER D. VANCE-CHALCRAFT *East Carolina University*
PATRICIA M. WALSH *University of Delaware*
JENNIFER WARNER *University of North Carolina at Charlotte*
MICHAEL WENZEL *California State University, Sacramento*
LISA A. WERNER *Pima Community College*
EMILY C. WHITELEY *Catawba Valley Community College*
MICHAEL WINDELSPECHT *Appalachian State University*
MARK WITMER *Ithaca College*
MARK WYGODA *McNeese State University*
MARTIN D. ZAHN *Thomas Nelson Community College*
ROSEMARY ZARAGOZA *University of Chicago*

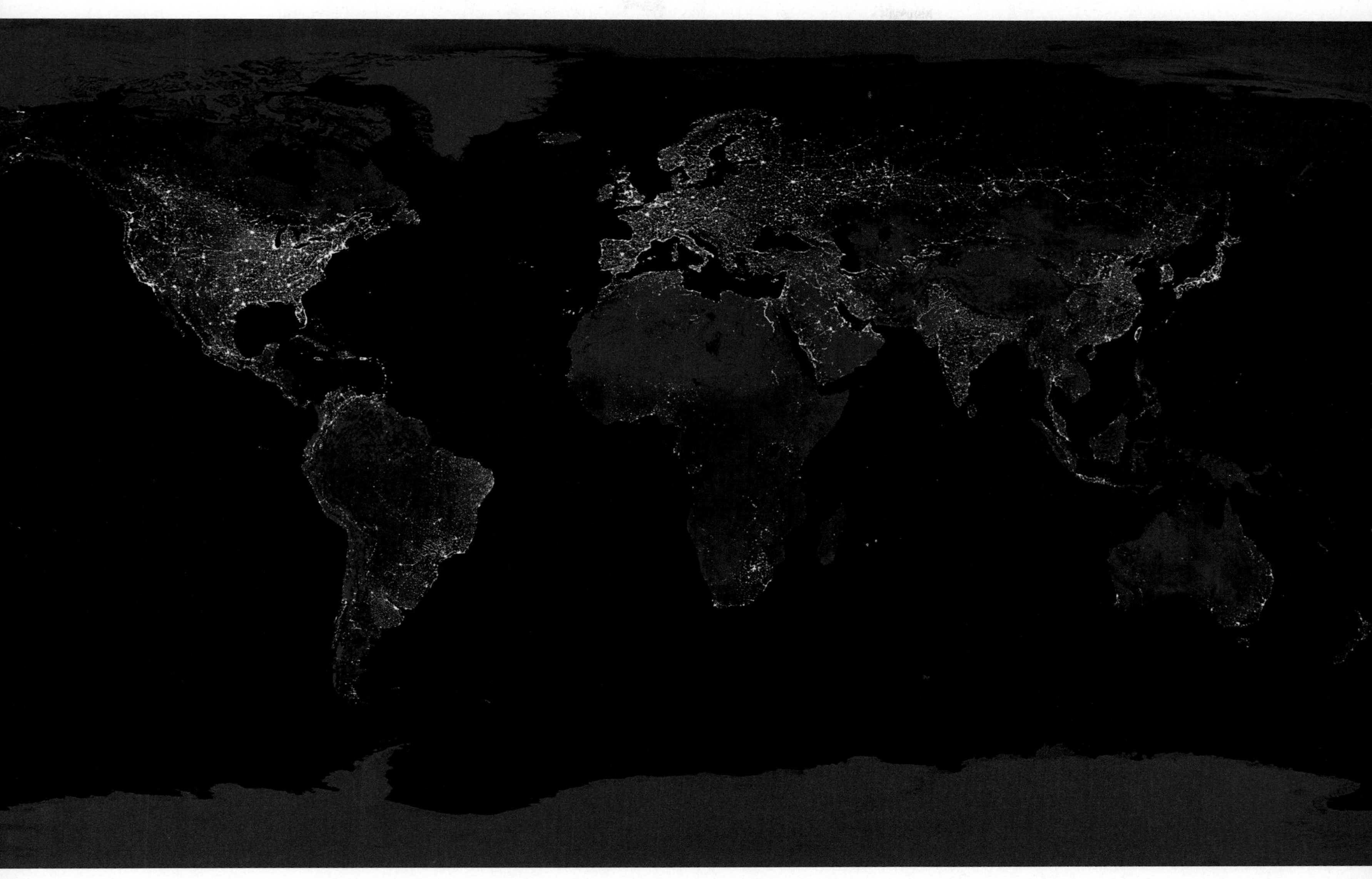

Current configurations of the Earth's oceans and land masses—the geologic stage upon which life's drama continues to unfold. This composite satellite image reveals global energy use at night by the human population. Just as biological science does, it invites you to think more deeply about the world of life—and about our impact upon it.

1 INVITATION TO BIOLOGY

IMPACTS, ISSUES

What Am I Doing Here?

Leaf through a newspaper on any given Sunday and you may get an uneasy feeling that the world is out of control. There are stories about the Middle East, where great civilizations have come and gone. You won't find much on the amazing coral reefs in the surrounding seas, especially at the northern end of the Red Sea. Now the news is about oil and politics, bioterrorists and war.

Think back on the 1991 Persian Gulf conflict, when thick smoke from oil fires blocked out sunlight, and black rain fell. Iraqis released 460 million gallons of crude oil into the Gulf. Uncounted numbers of reef organisms and thousands of birds died.

Today Kuwaitis wonder if the oil fires have caused their increased cancer rates. They join New Yorkers who worry about developing lung problems from breathing dense, noxious dust that filled the air after the terrorist attack on the World Trade Center.

Nature, too, seems to have it in for us. The AIDS pandemic is unraveling African societies. Forests are burning fiercely. Storms, droughts, and heat waves are monstrous, polar ice caps and once-vast glaciers are melting, and the whole atmosphere is warming up.

It's enough to make you throw down the paper and long for the good old days, when things were simpler.

Of course, read up on the good old days and you'll find they weren't so good. Bioterrorists were around in 1346, when soldiers catapulted the corpses of bubonic plague victims into a walled city under siege. Infected people and rats fled the city and helped fuel the Black Death, a plague that left 25 million dead in Europe.

Nature was busy with us back then, too. In 1918, the Spanish flu raced around the world and left between 30 and 40 million people dead.

Then, as now, many felt helpless in a world that seemed out of control. Perhaps it is a human condition to think we may control something if we understand it. For millions of years, we and our ancestors have been trying to make sense of our natural world and how we fit into it. We observe it, come up with ideas, then test the ideas. This is the nature of science. But the more pieces of the puzzle we fit together, the bigger the puzzle seems to be.

You could walk away from this challenge and simply not think about it. You could let others tell you what to think. Or you could choose to develop your own understanding. Maybe you are interested in the pieces of the puzzle that affect your health, your food, or your children, should you choose to reproduce. Maybe you just find life fascinating. No matter what your focus might be, you can deepen your perspective. You can learn ways to sharpen how you interpret the natural world, including human nature. This is the gift of biology, the scientific study of life.

How Would You Vote? *The seas of the Middle East support some of the world's most spectacular coral reef ecosystems. Should the United States provide funding to help preserve the reefs? See BiologyNow for details, then vote online.*

Key Concepts

LIFE'S UNDERLYING UNITY
Life's organization extends from the molecular level to the biosphere. Shared features at the molecular level are the basis of life's unity.

LIFE'S DIVERSITY
Life shows spectacular diversity as well as unity. Each one of millions of kinds of organisms, past and present, has unique traits that help define it as a separate species.

EXPLAINING UNITY IN DIVERSITY
Evolutionary theories, especially the theory of natural selection, help us see a profound connection between life's underlying unity and its diversity.

HOW WE KNOW
Biologists find out about life by observing, asking questions, formulating and testing hypotheses, and reporting results in ways that others can test.

Links to Earlier Concepts

This book parallels nature's levels of organization, from atoms to the biosphere. Learning about the structure and function of atoms and molecules primes you to understand the structure of cells. Learning about protein synthesis, active transport, and other processes that keep a single cell alive helps you understand how multicelled organisms survive, because each of their cells uses the same processes. Knowing what it takes to survive helps you sense why and how organisms interact with one another and the environment. At the start of each chapter, we will be reminding you of structural and functional connections. Within chapters, you will come across keychain icons with references to relevant sections in earlier chapters.

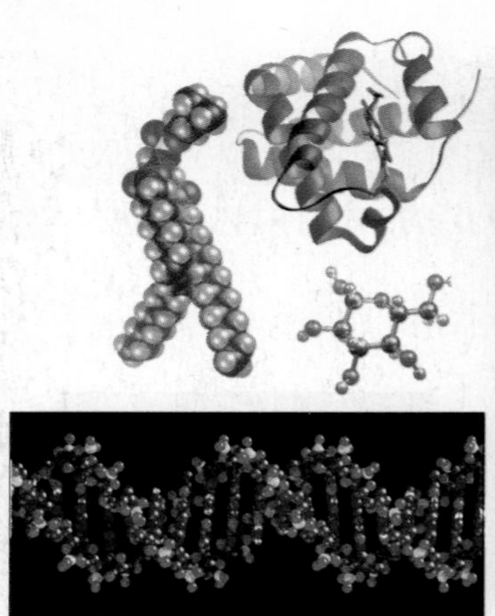

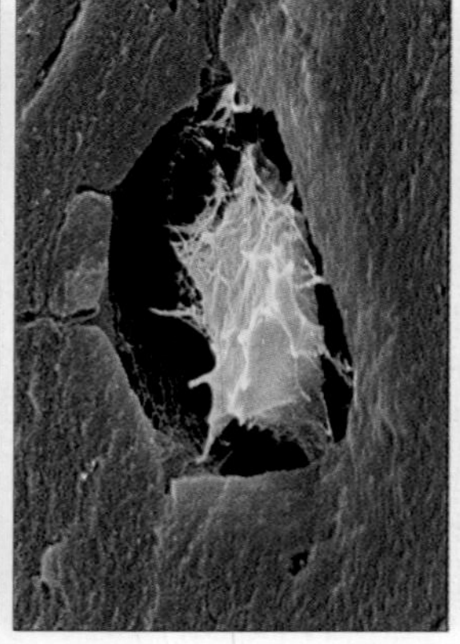
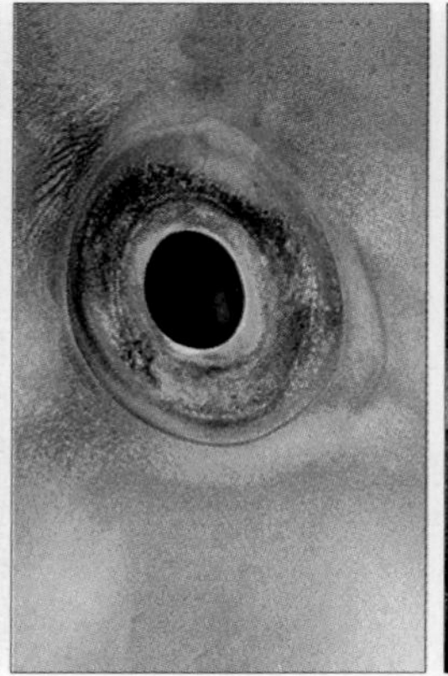
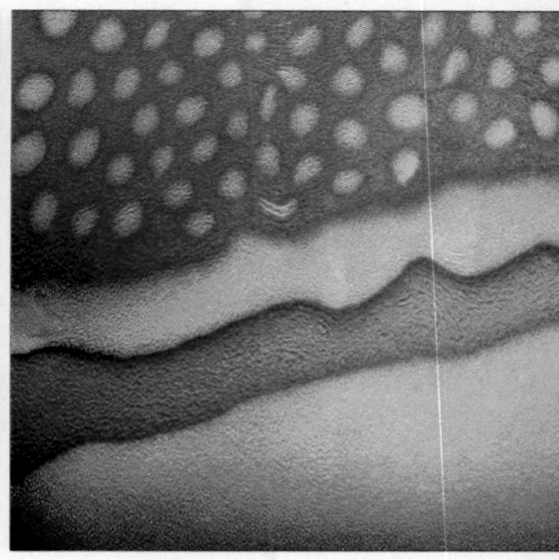
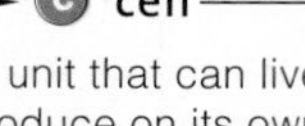

b molecule
Two or more joined atoms of the same or different elements. The "molecules of life" are complex carbohydrates, lipids, proteins, DNA, and RNA. In today's world, only living cells make them.

c cell
Smallest unit that can live and reproduce on its own or as part of a multicellular organism. It has an outer membrane, DNA, and other components.

d tissue
Organized cells and substances that interact in a specialized activity. Many cells (white) made this bone tissue from their own secretions.

e organ
Two or more tissues interacting in some task. A parrotfish eye, for example, is a sensory organ used in vision.

f organ system
Organs interacting physically, chemically, or both in some task. Parrotfish skin is an organ system with tissue layers, organs such as glands, and other parts.

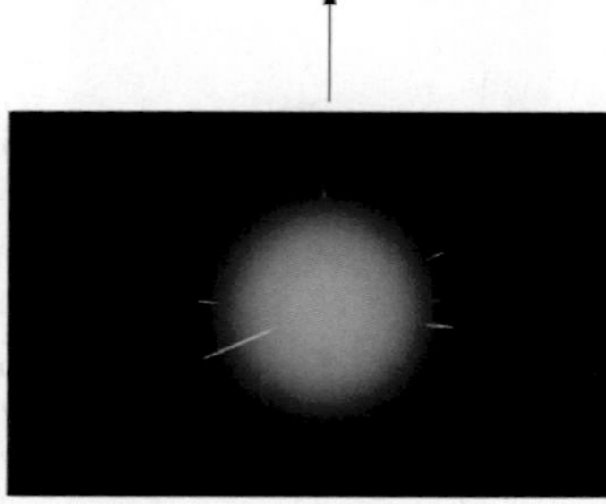

a atom
Smallest unit of an element that still retains the element's properties. Electrons, protons, and neutrons are its building blocks. This hydrogen atom's electron zips around a proton in a spherical volume of space.

Figure 1.1 ***Animated!*** Increasingly complex levels of organization in nature, extending from subatomic particles to the biosphere.

1.1 Life's Levels of Organization

Biologists are interested in all forms of life, past and present. They study organisms down to their atoms and up to their interactions on a global scale. In doing so, they have discovered a great pattern of organization.

FROM SMALL TO SMALLER

Imagine yourself on the deck of a sailing ship, about to journey around the world. The distant horizon of a vast ocean beckons, and suddenly you sense that you are just one tiny part of the great scheme of things.

Now imagine one of your red blood cells can think. It realizes it is only a small part of the great scheme of your body. A string of 375 cells like itself would fit across a pin's head—and you have trillions of cells. One of the fat molecules at the red blood cell's surface is thinking about how small it is. A string of 1,200,000 molecules like itself would stretch across that pinhead.

Now a hydrogen atom in the molecule is pondering the great scheme of a fat molecule. Hydrogen atoms are so small that it would take 53,908,355 of them to reach across the pinhead! With that lone atom, you have reached the starting point of life's great pattern of organization.

FROM SMALLER TO VAST

That pattern starts at the level of atoms, which are the basic building blocks of matter. Like nonliving things, all organisms are made of atoms. Figure 1.1*a* depicts a hydrogen atom. At the next level, atoms combine into molecules (Figure 1.1*b*). Complex carbohydrates, fats and other lipids, proteins, DNA, and RNA are called the "molecules of life." In nature, only living cells can make them.

The pattern reaches the threshold of life as these molecules assemble into cells (Figure 1.1*c*). A **cell** is the smallest living unit, having the capacity to survive and reproduce on its own, given raw materials, energy

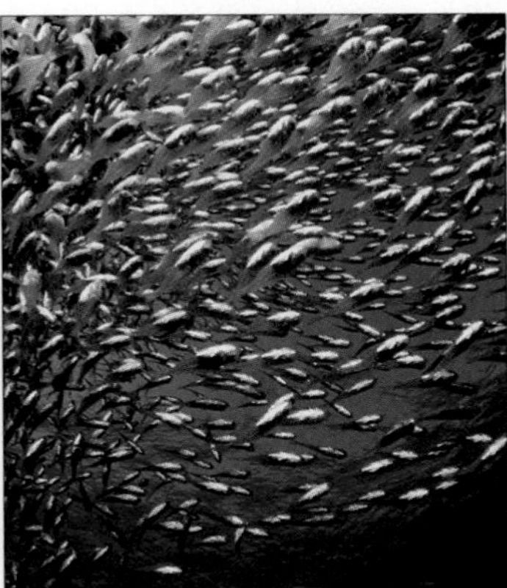

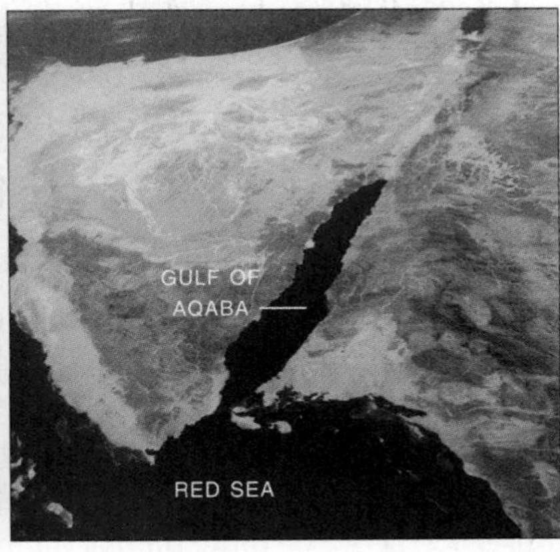

g multicelled organism
Individual made of different types of cells. Cells of most organisms, including this Red Sea parrotfish, are organized as tissues, organs, and organ systems.

h population
Group of single-celled or multicelled individuals of the same species occupying a specified area. This is a fish population in the Red Sea.

i community
All populations of all species occupying a specified area. This is part of a coral reef in the Gulf of Aqaba at the northern end of the Red Sea.

j ecosystem
A community that is interacting with its physical environment. It has inputs and outputs of energy and materials. Reef ecosystems flourish in warm, clear seawater throughout the Middle East.

k the biosphere
All regions of the Earth's waters, crust, and atmosphere that hold organisms. In the vast universe, Earth is a rare planet. Without its abundance of free-flowing water, there would be no life.

inputs, and suitable conditions in its environment. In multicelled species, cells are organized into tissues, organs, and organ systems. Figure 1.1*d*–g defines these levels of organization. Often there are trillions of cells, with many specialized types that interact directly and indirectly in ways that keep the whole body alive. In size, multicelled organisms range from microscopic to giant, as are redwoods and whales.

The **population** is the next level of organization. Each population is a group of the same kind of single-celled or multicelled organisms in some specified area. Fields of red poppies in a valley or schools of fish in a lake are such groups (Figure 1.1*h*).

At the next level of organization, the **community** includes all populations of all species in some specified area (Figure 1.1*i*). An example could be all of the kelps, corals, shrimps, fishes, sea anenomes, plankton, crabs, and everything else living in an underwater cave in the Red Sea. Another example would be the populations of tiny organisms that live, reproduce, and die quickly in a drop of water on the cupped petals of a flower.

The next level is the **ecosystem**, or a community interacting with its environment (Figure 1.1*j*). Finally, the **biosphere** is the highest level of life. It encompasses all regions of Earth's crust, waters, and atmosphere in which organisms live.

Bear in mind, life is more than the sum of its parts. At each successive level of organization, new properties emerge that are not inherent in any part by itself, as when living cells emerge from lifeless molecules.

This book is a journey through the globe-spanning organization of life. So take a moment to study Figure 1.1. You can use it as a road map of where each part fits into the great scheme of things.

Nature shows levels of organization, from the simple to the increasingly complex.

Life's unique characteristics originate at the molecular level. They extend through individual cells, populations, communities, ecosystems, and the biosphere.

IMPACTS, ISSUES

Nearly half of the oil exports from the Persian Gulf are shipped through the Gulf of Oman, where satellite imaging revealed this vast slick formed by an oil spill. Such water pollution is damaging reefs all over the world.

1.2 Overview of Life's Unity

"Life" is not easy to define. It is too big, and it has been changing for 3.9 billion years! Even so, you can frame a definition in terms of its unity and diversity. Here is the unity part: All living things grow and reproduce with the help of DNA, energy, and raw materials. They sense and respond to their environment. But details of their traits differ among many millions of kinds of organisms. That is the diversity part—variation in traits.

DNA, THE BASIS OF INHERITANCE

You will never, ever find a rock made of nucleic acids, proteins, and complex carbohydrates and lipids. In the natural world, only living cells make these molecules. The signature molecule of life is the nucleic acid called DNA. No chunk of granite or quartz has it.

DNA holds information for building proteins from smaller molecules: the amino acids. By analogy, if you follow suitable instructions and invest enough energy in the task, you might organize a pile of a few kinds of ceramic tiles (representing amino acids) into diverse patterns (representing proteins), as in Figure 1.2.

Why are the proteins so important? Many kinds are structural materials, and others are enzymes. Enzymes are the main worker molecules in cells. They build, split, and rearrange the molecules of life in ways that keep a cell alive. Without enzymes, the information held in DNA could not be used. There would be no new organisms.

In nature, each organism inherits its DNA—and its traits—from parents. *Inheritance* means an acquisition of traits after parents transmit their DNA to offspring. Think about it. Why do baby storks look like storks and not like pelicans? Because storks inherit stork DNA, not pelican DNA.

Reproduction refers to actual mechanisms by which parents transmit DNA to offspring. For slugs, humans, trees, and other multicelled organisms, information in DNA guides *development*—the transformation of the first cell of a new individual into a multicelled adult, typically with many different tissues and organs.

ENERGY, THE BASIS OF METABOLISM

DNA is only part of the picture. Becoming alive and maintaining life requires energy—the capacity to do work. Each normal living cell has ways to obtain and convert energy from its surroundings. By the process called **metabolism**, a cell acquires and uses energy to maintain itself, grow, and make more cells.

Where does the energy come from? Nearly all of it flows from the sun into the world of life, starting with **producers**. Producers are plants and other organisms that make their own food from simple raw materials. All other organisms are **consumers**. They cannot make their own food; they must eat other organisms.

When, say, zebras browse on plants, some energy stored in plant tissues is transferred to them. Later on, energy is transferred to a lion as it devours the zebra. It gets transferred again as decomposers go to work, acquiring energy from the remains of zebras, lions, and other organisms.

Decomposers are consumers, mostly bacteria and fungi that break down sugars and other molecules to simpler materials. Some of the breakdown products are cycled back to producers as raw materials. Over time, energy that plants originally captured from the sun returns to the environment.

ENERGY AND LIFE'S ORGANIZATION

Energy happens to flow in one direction, from the environment, through producers, and then consumers. Eventually, it ends up being lost as heat (Figure 1.3). Energy transfers maintain life's organization, both in

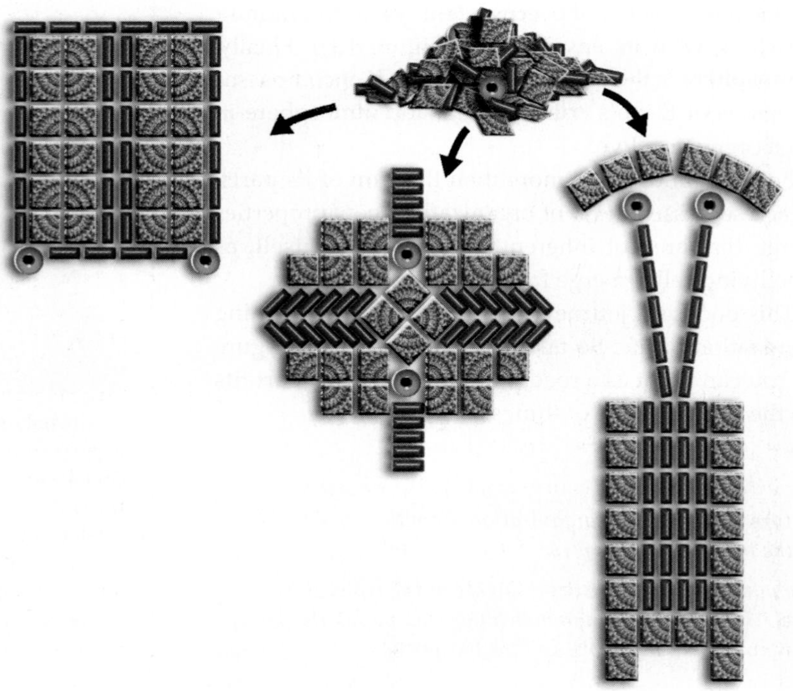

Figure 1.2 Examples of objects built from the same materials according to different assembly instructions.

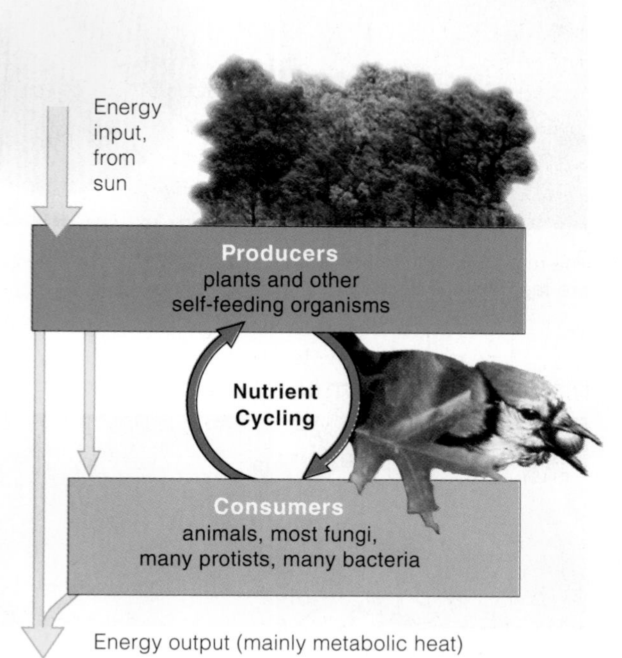

Figure 1.3 ***Animated!*** The one-way flow of energy and the cycling of materials in the world of life.

Figure 1.4 A roaring response to signals from pain receptors, activated by a lion cub flirting with disaster.

individual organisms, and as all organisms interact throughout the biosphere. Later on, you will see how life's interconnectedness relates to problems such as food shortages, the AIDS pandemic, global warming, acid rain, cholera, and rapid losses in biodiversity.

LIFE'S RESPONSIVENESS TO CHANGE

Organisms keep their internal operating conditions within a range that their cells can tolerate. This state, called **homeostasis**, is a defining feature of life.

Only living things respond to the environment. Yet even a rock seems responsive, as when it yields to gravity and tumbles down a hill, or changes its shape slowly under the repeated batterings of wind and rain. The difference is this: Unlike rocks, living things make responses to offset changes in their environment. Such compensatory changes maintain homeostasis.

How? The answer begins with receptors. **Receptors** are structures that detect stimuli (singular, stimulus). A **stimulus** is a specific kind of energy, such as sunlight energy, chemical energy (as when a substance becomes more concentrated outside a cell than on the inside), or the mechanical energy of a bite (Figure 1.4).

Different kinds of receptors respond to different kinds of stimuli. Their responses trigger changes in cell activities. As a simple example, after you finish eating a piece of fruit, sugars leave your gut and enter your bloodstream. Think of blood and fluid around cells as an *internal* environment. The sugar from the fruit you ate changes that internal environment when it enters your bloodstream. A change in blood sugar level is a deviation from homeostasis that your body works to reverse. Cells in your pancreas (an organ) detect extra sugar in your blood and respond by secreting insulin. Many kinds of cells in your body have receptors for this hormone. When their receptors bind insulin, cells absorb sugar, removing it from the blood. After enough cells do so, the blood sugar level returns to normal. Thus, homeostasis is maintained.

Organisms reproduce based on instructions in DNA, which they inherit from their parents.

All living organisms reproduce, grow, and stay alive by way of metabolism—by ongoing energy conversions and energy transfers at the cellular level.

Organisms interact through a one-way flow of energy and a cycling of materials through the biosphere. Collectively, their interdependencies have global impact.

Organisms sense and respond to changing conditions in controlled ways. Their responses help them maintain tolerable conditions in their internal environment.

IMPACTS, ISSUES

Satellite image of a phytoplankton bloom in the Persian Gulf. These communities of mostly microscopic producers start aquatic food webs. Sewage, runoff from croplands, and other nutrient inputs fan the blooms (huge increases in population sizes). When clouds of smoke block out the sun and oil slicks coat the water, entire ecosystems are disrupted, starting with catastrophic losses of the producers.

1.3 If So Much Unity, Why So Many Species?

Although unity pervades the world of life, so does diversity. Organisms differ enormously in body form, in the functions of their body parts, and in behavior.

Superimposed on life's unity—its common heritage—is tremendous diversity. Millions of different kinds of organisms, or **species**, live on Earth. Many millions more lived during the past 3.9 billion years, but their lineages disappeared; about 99.9 percent of all species have become extinct.

For centuries, scholars have tried to make sense of diversity. One, Carolus Linnaeus, devised a scheme for classifying organisms by assigning a two-part name to each newly identified species. The first part designates the genus (plural, genera). Each **genus** is one or more species grouped on the basis of a number of unique shared traits. The second part of the name refers to the particular species within the genus.

For instance, *Scarus gibbus* is a humphead parrotfish. We abbreviate a genus name once it's been spelled out in a document. Thus, we might call another species in the same genus, the midnight parrotfish, *S. coelestinus*.

Biologists are still working out how to organize the genera. Most favor a classification system that groups all organisms into three sets, which are the domains Archaea, Bacteria, and Eukarya. Eukarya encompasses protists, fungi, plants, and animals (Figures 1.5 and 1.6).

All **archaea** and **bacteria** are single cells, and all are *prokaryotic*, meaning they do not contain a nucleus (a membrane-enclosed sac that keeps DNA separated from the rest of the cell's interior.) Prokaryotes can be producers or consumers. Of all groups of organisms, theirs show the most metabolic diversity.

Archaea survive in boiling ocean water, freezing desert rocks, sulfur-rich lakes, and other habitats as harsh as those thought to have prevailed when life originated. Bacteria are sometimes called eubacteria, which means "true bacteria," to distinguish them from archaea. They are far more common than archaeans, and they live throughout the world in diverse habitats.

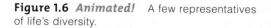

IMPACTS, ISSUES

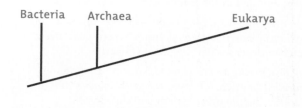

Species names make sense—when you know a bit of Latin. Some *Scarus* species live in the Persian Gulf. Figure 1.1g shows *S. gibbus*, a humphead parrotfish (the Latin *gibbus* means hump). *S. ferrugineus* is the rusty parrotfish; its species name means rust-colored. One of its Caribbean relatives, the midnight parrotfish (*S. coelestinus*), lives up to its name—"celestial" in Latin. Worldwide, there are 80 *Scarus* species.

Bacteria

Archaea

Eukarya

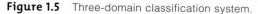

Figure 1.5 Three-domain classification system.

Bacteria

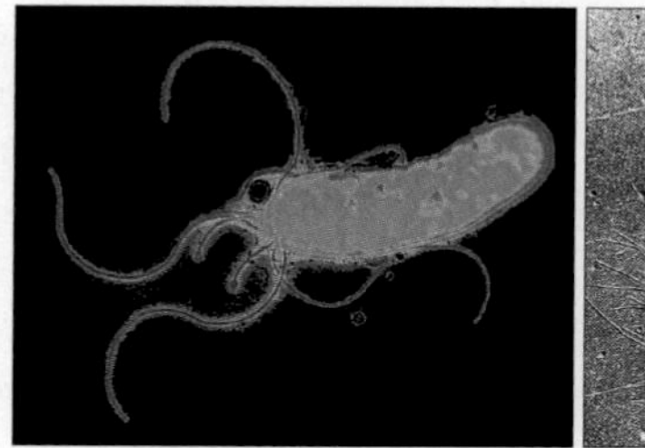

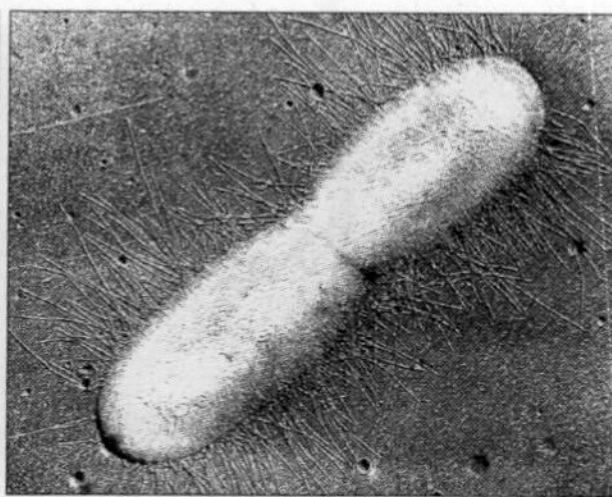

The most common prokaryotes; collectively, these single cells are the most metabolically diverse species on Earth.

Archaea

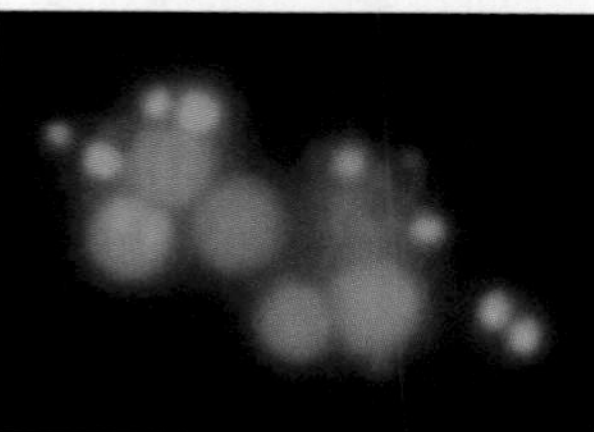

These prokaryotes are evolutionarily closer to eukaryotes than to bacteria. *Left*, a colony of methane-producing cells. *Right*, two species from a hydrothermal vent on the seafloor.

Figure 1.6 ***Animated!*** A few representatives of life's diversity.

Plants, fungi, animals, and protists are *eukaryotic*, which means their cells start out life with a nucleus. Eukaryotes are generally larger and far more complex than prokaryotes. Differences among protist lineages are so great that this group might be subdivided into several separate groups, which would result in a major reorganization within the domain.

Plants are multicelled, photosynthetic producers. They can make their own food by using simple raw materials and sunlight as an energy source.

Eukarya

Protists Single-celled and multicelled eukaryotic species that range from the microscopic to giant seaweeds. Many biologists are now viewing the protists as many major lineages.

Plants Generally, photosynthetic, multicelled eukaryotes, many with roots, stems, and leaves. Plants are the primary producers for ecosystems on land. Redwoods and flowering plants are examples.

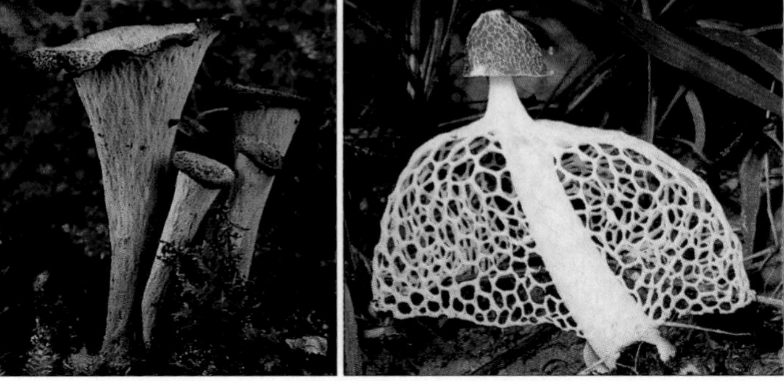

Fungi Single-celled and multicelled eukaryotes; decomposers, parasites, and pathogens. Without decomposers, communities would become buried in their own wastes.

Animals Multicelled eukaryotes that ingest tissues or juices of other organisms. Like this basilisk lizard, most actively move about during at least part of their life.

Most **fungi**, such as the mushrooms sold in grocery stores, are multicelled consumers with a distinctive way of feeding. They secrete enzymes that digest food outside the fungal body, and then their individual cells absorb the digested nutrients. **Animals** are multicelled consumers that ingest the tissues of other organisms. Different animals are herbivores (grazers), carnivores (meat eaters), scavengers, and parasites. All develop by a series of embryonic stages, and all actively move about during their life.

Pulling this information together, are you getting a sense of what it means when someone says that life shows unity *and* diversity?

To make the study of life's diversity more manageable, we group organisms related by descent from a shared ancestor.

We recognize three domains—Archaea, Bacteria, and Eukarya (protists, fungi, plants, and animals).

1.4 An Evolutionary View of Diversity

How can organisms be so much alike and still show tremendous diversity? A theory of evolution by way of natural selection is one explanation.

Your traits make you and 6.4 billion other individuals members of the human population. Traits are different aspects of an organism's form, function, or behavior. For example, height and hair color are human traits. Like most traits of any natural population, there are different versions of them among individuals.

Variation in most traits arises through **mutations**, or changes in DNA. Some mutations lead to new traits that make an individual better able to find food, a mate, and so on. We call these *adaptive* traits.

An adaptive trait tends to become more common over generations, because it gives individuals a better chance to live and bear more offspring than individuals that do not have it. When different forms of a trait are becoming more or less common, evolution is under way. To biologists, **evolution** simply means heritable change in a line of descent. Mutations, the source of new traits, provide the variation that serves as the raw material for evolution.

"Diversity" refers to variations in traits that have accumulated in lines of descent. For now, think about what a great naturalist, Charles Darwin, discovered about evolution:

WILD ROCK DOVE

First, populations tend to increase in size, past the capacity of their environment to sustain them, so their members must compete for resources (food, shelter). Second, individuals of natural populations differ from one another in the details of their shared traits. Most variation has a heritable basis.

Third, when individuals differ in their ability to survive and reproduce, the traits that help them do so tend to become more common in the population over generations. This process is called **natural selection**.

Pigeon breeding is *artificial* selection. One form of a trait is favored over others in an artificial environment under contrived, manipulated conditions. Take a look at the pigeons in Figure 1.7. They differ in size, feather color, and other traits. Suppose pigeon breeders allow only the pigeons with the darkest and curliest-tipped feathers to mate. In time, pigeons with black, curly-tipped feathers make up the entire captive population.

Darwin saw that breeding practices could be an easily understood model for natural selection, a favoring of some forms of a given trait over others in nature.

Just as breeders are "selective agents" promoting reproduction of particular captive pigeons, different agents operate across the range of variation in the wild. Pigeon-eating peregrine falcons are among them (Figure 1.7). Swifter or better camouflaged pigeons are more likely to avoid peregrine falcons and live long enough to reproduce, compared with not-so-swift or more conspicuous pigeons.

Traits are variations in form, function, or behavior that arise as a result of mutations in DNA. Some traits are more adaptive than others to prevailing conditions.

Natural selection is an outcome of differences in survival and reproduction among individuals of a population that vary in one or more heritable traits. The process of evolution, or change in lines of descent, gives rise to life's diversity.

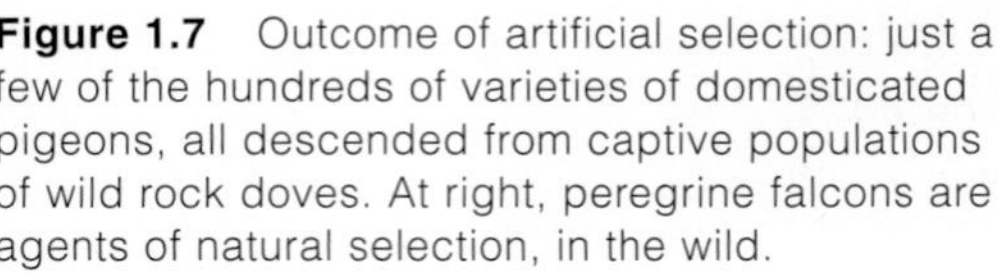

Figure 1.7 Outcome of artificial selection: just a few of the hundreds of varieties of domesticated pigeons, all descended from captive populations of wild rock doves. At right, peregrine falcons are agents of natural selection, in the wild.

1.5 The Nature of Biological Inquiry

The preceding sections introduced some big concepts. Consider approaching this or any other collection of facts with a critical attitude: "Why should I accept that they have merit?" The answer requires a look at how biologists make inferences about observations, then test their inferences against actual experience.

OBSERVATIONS, HYPOTHESES, AND TESTS

To get a sense of how science works, you might start with practices that are common in scientific research:

1. Observe some aspect of nature, carefully check what others have found out about it, then frame a question or identify a problem related to your observation.

2. Think of a **hypothesis**, a proposal that explains the cause of something that is observable, or that makes a generalization about it.

3. Using that hypothesis as a guide, make a **prediction**—a statement of what you should find in the natural world if you were to go out looking for it. This is often called the if–then process. The "if" part of a prediction is the actual hypothesis. *If* gravity does not pull objects toward the Earth, *then* it should be possible to observe apples falling up, not down, from a tree.

4. Devise one or more ways to **test** the prediction, as by making observations, conducting an experiment, or formulating a model. **Models** are theoretical, detailed descriptions of partially-understood processes.

5. If tests do not confirm your prediction, check your results. Rethink your tests, and the prediction. If the tests and prediction still stand, start again by revising your hypothesis.

6. Repeat the tests or devise new ones—the more the better, because hypotheses that withstand many tests are likely to have a higher probability of being useful.

7. Objectively analyze and report test results, and the conclusions you drew from them.

You might hear someone refer to these practices as the scientific method, as if all scientists march to the drumbeat of an absolute, fixed procedure. They do not. Many observe, describe, and report on some aspect of nature, then leave the hypothesizing to others. Some scientists stumble onto information that they are not looking for. Of course, it is not always a matter of luck. Chance seems to favor a mind that has already been prepared, by education and experience, to recognize what new information means. The scientific method is not a single way scientists do their research. It is a critical attitude about being shown rather than told, and taking a logical approach to problem solving.

ABOUT THE WORD "THEORY"

So what exactly constitutes the truth? Or proof of it? Most scientists carefully avoid the word "truth" when discussing science, preferring instead to use "accurate" in reference to data. In science, there is only evidence that supports a hypothesis.

For example, the hypothesis that the sun comes up every day has been supported by observable evidence for all of recorded history. Scientists no longer attempt to disprove that hypothesis, for the simple reason that after thousands of years of observation, no one has yet seen otherwise. The hypothesis is consistent with all of the data that has been collected, and it is now used to make different predictions. When a hypothesis meets these criteria, it has become a **scientific theory**.

You may hear someone apply the word "theory" to a speculative idea, as in the expression "only a theory." But a scientific theory differs from speculation, because it offers valid, testable predictions. Untestable ideas do not belong in science.

The predictive power of a scientific theory has been tested many times, and in many ways, and researchers have yet to find evidence that disproves it. However, a theory cannot be tested under every circumstance, so scientists do not ever assume that a theory is infallible. For example, based on the theory that the sun comes up every day, the prediction that the sun will come up again tomorrow seems like a good one. Yet, a scientist sees tomorrow as an opportunity for that prediction to fail. Like every other theory, the theory that the sun comes up every day will always be open to revision.

After more than a century of many thousands of tests, Darwin's theory of natural selection still stands, with only minor revision. It explains diverse issues, such as how river dams can alter ecosystems and why certain cancers can run in families. Such a well-tested theory is as close to the "truth" as scientists venture. As for any theory, we can say it has a high probability of not being wrong. Even so, the door is always open for scientists to look for data that might disprove it.

A scientific approach to studying nature is based on asking questions, formulating hypotheses, making predictions, testing the predictions, and objectively reporting results.

A scientific theory is a well-tested concept that can be used to explain causes of a broad range of related phenomena. Every scientific theory remains open to testing and revision.

IMPACTS, ISSUES

The Persian Gulf is far saltier than most seas, and it's about to get saltier. Several new dams on the Tigris and Euphrates rivers are reducing the flow of fresh water into the Gulf to a trickle. Shown here, the Ataturk Dam on the Euphrates River, where it flows through Turkey. How will increased salinity affect life in the Gulf? It could become an agent of natural selection. It could favor individuals among the Gulf's inhabitants that are better at tolerating higher salt levels.

1.6 The Power of Experimental Tests

Natural processes are often interrelated. Researchers unravel how they work together by studying a single aspect at a time. They design experiments to observe the function, cause, or effect of that aspect in isolation.

Hypothesis
Olestra® causes intestinal cramps.

Prediction
People who eat potato chips made with Olestra will be more likely to get intestinal cramps than those who eat potato chips made without Olestra.

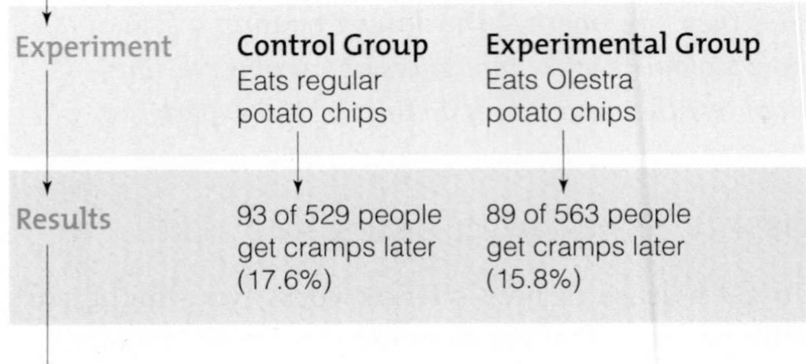

Conclusion
Percentages are about equal. People who eat potato chips made with Olestra are just as likely to get intestinal cramps as those who eat potato chips made without Olestra. These results do not support the hypothesis.

Figure 1.8 ***Animated!*** Example of a typical sequence of steps taken in a scientific experiment.

AN ASSUMPTION OF CAUSE AND EFFECT

A scientific experiment starts with this premise: *Any aspect of nature has an underlying material cause that can be tested by controlled experiments.* This is how science is different from faith. Each has its place, but a scientific hypothesis is testable in the natural world, in stringent ways that might well disprove it.

Most aspects of nature are outcomes of interacting variables. A **variable** is a feature of an object or event that may differ over time or among the representatives of that object or event. Researchers design experiments that test one variable at a time.

They define a **control group**, which is a standard for comparison with one or more **experimental groups**. There are two kinds of control groups. One is *identical* to an experimental group, but is tested differently. For instance, consider an experiment with two identical groups of mice. The control group is given water, and the experimental group receives water with a drug in it. (In this experiment, the variable is the drug.) The other type of control group *differs* from an experimental group, but is tested the same way. For example, two different species of mice might be given the same drug. In this experiment, one of the species is the control group, and the other species is the experimental group.

EXAMPLE OF AN EXPERIMENTAL DESIGN

Here is a real-life example of an experiment. In 1996, the FDA approved Olestra®, a synthetic fat replacement made from sugar and vegetable oil, as a food additive. Potato chips were the first Olestra-laced food product on the market in the United States. Controversy raged soon after. Some people who got stomach cramps after eating Olestra chips suggested this hypothesis: Olestra causes cramps. Researchers designed an experiment to test the hypothesis. Their prediction was this: If Olestra causes cramps, then people who eat Olestra are more likely to get cramps than people who do not.

To test this prediction, they used a Chicago theater as the "laboratory." They asked more than 1,100 people between ages thirteen and thirty-eight to watch a movie and eat their fill of potato chips. Each person got an unmarked bag that held thirteen ounces of chips. The individuals who received a bag of Olestra-laced potato chips were the experimental group. Individuals who got a bag of regular chips were the control group.

Afterward, the researchers contacted the subjects and tabulated any reports of gastrointestinal cramps. Of 563 people in the experimental group, 89 (or 15.8 percent) complained of problems. But so did 93 of 529 people (17.6 percent) in the control group, who had munched on regular chips! The experiment yielded no evidence that eating Olestra-laced potato chips, at least in one sitting, causes gastrointestinal problems. Figure 1.8 summarizes this experiment.

EXAMPLE OF A FIELD EXPERIMENT

Some biologists perform their experiments in natural settings. This one was performed by Durrell Kapan, an evolutionary biologist, in the rain forests of Ecuador.

Many toxic or bad-tasting species are vividly colored, often with distinctive patterning. After eating a few of them and suffering the consequences, predators learn to avoid individuals that display particular visual cues.

Some butterflies that are good-tasting mimic others that are bad-tasting. **Mimicry** is a case of looking like something else and confusing predators (or prey). The naturalist Fritz Müller wondered if mimicry between different species of bad-tasting butterflies would benefit both species. Young birds sample butterflies before learning to avoid eating bad-tasting ones, so if both butterfly species are sampled, then each would lose fewer individuals to predatory birds.

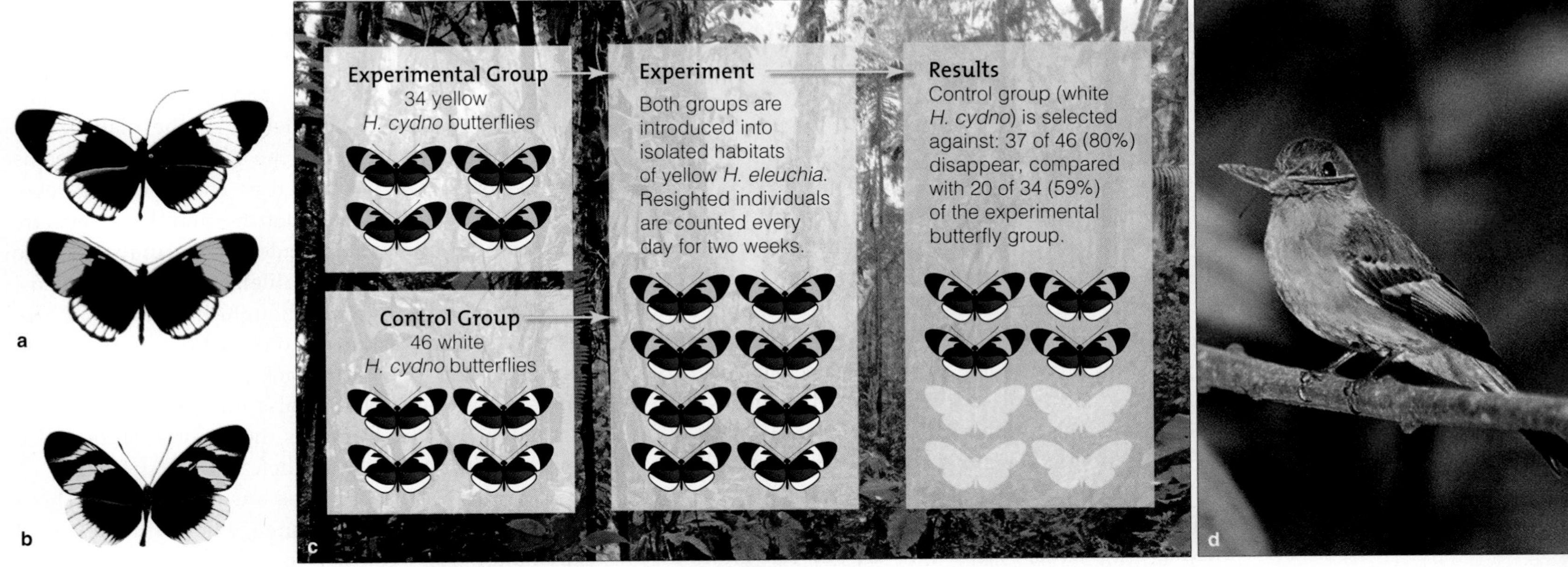

a
b
Experimental Group
34 yellow
H. cydno butterflies
Control Group
46 white
H. cydno butterflies
Experiment
Both groups are introduced into isolated habitats of yellow H. eleuchia. Resighted individuals are counted every day for two weeks.
Results
Control group (white H. cydno) is selected against: 37 of 46 (80%) disappear, compared with 20 of 34 (59%) of the experimental butterfly group.
c
d

Figure 1.9 *Heliconius* butterflies. (**a**) The two forms of *H. cydno* and (**b**) yellow *H. eleuchia*. (**c**) Kapan's experiment. Both forms of *H. cydno* butterflies were captured and transferred to habitats of *H. eleuchia*. Local predatory birds (**d**), familiar with bad-tasting yellow *H. eleuchia*, avoided the yellow *H. cydno* butterflies but ate most of the white ones.

Kapan tested Müller's hypothesis by performing an experiment on some bad-tasting butterflies in Ecuador. There are two forms of *Heliconius cydno*, a bad-tasting species of butterfly. One has yellow markings on its wings; the other has white (Figure 1.9*a*).

The yellow form of *H. cydno* resembles a different unpalatable species, *H. eleuchia* (Figure 1.9*b*), that lives in another part of the forest. Kapan decided to test the hypothesis that this resemblance would benefit both species. Kapan predicted: If birds have learned to avoid *H. eleuchia* in their part of the forest, then they also will avoid imported yellow *H. cydno* butterflies.

He captured both forms of *H. cydno*. The form with white markings became the control group. The form with the yellow markings was the experimental group. He released both of the groups into isolated *H. eleuchia* habitats. For two weeks, the approximate life span of the butterflies, he counted the survivors.

Kapan found that butterflies of the control group were less likely to survive in the new forest habitat. Because they did not display the visual cue recognized by the local birds—yellow markings—they were more likely to be eaten. The experimental group did better, as the test results in Figure 1.9*c* indicate. We can expect that predatory birds recognized a familiar cue for bad taste—yellow markings—and avoided them. Besides confirming Kapan's prediction, these test results also provide evidence of natural selection in action.

In natural settings, experimenters rarely observe *all* individuals of a group. By necessity, they select subsets or samples of populations, events, and other aspects of nature. But data taken from a sample might differ from data taken from the whole. The difference is called **sampling error**. Results are less likely to be distorted (sampling error can be minimized) with large samples.

BIAS IN REPORTING RESULTS

Experimenters risk interpreting their results in terms of what they want to prove. That is why scientists design experiments to yield *quantitative* results, actual counts or some other data that can be measured and collected objectively. Data give other scientists an opportunity to confirm tests and check conclusions.

This last point gets us back to the value of thinking critically. Scientists expect one another to put aside pride or bias and test ideas in ways that may prove them wrong. If someone will not do so, others will—because science is a competitive community. It is also cooperative. Scientists share ideas, knowing it is just as useful to expose errors as to applaud insights.

Experiments simplify observations in nature by restricting a researcher's focus to one variable at a time. A variable is any feature of an object or event that may differ over time or among representatives of the object or event.

Tests are based on the premise that any aspect of nature has one or more underlying causes. Scientific hypotheses can be tested in ways that might disprove them.

1.7 The Scope and Limits of Science

Beyond the realm of scientific inquiry, some events are unexplained. Why do we exist, and for what purpose? Why do we have to die at a particular moment? Such questions lead to *subjective* answers, which come from within as an integrated outcome of all the experiences and mental connections that shape our consciousness. People differ enormously in this regard. That is why subjective answers do not readily lend themselves to scientific analysis and experiments.

This is not to say subjective answers are without value. No human society can function for long unless its individuals share a commitment to standards for making judgments, even if they are subjective. Moral, aesthetic, philosophical, and economic standards vary from one society to the next. But they all guide people in deciding what is important and good, and what is not. All attempt to give meaning to what we do.

Every so often, scientists stir up controversy when they explain something that was thought to be beyond natural explanation, or belonging to the supernatural. This is often the case when a society's moral codes are interwoven with traditional interpretations of the past. Exploring a long-standing view of the natural world from a scientific perspective might be misinterpreted as questioning morality.

As one example, centuries ago in Europe, Nikolaus Copernicus observed the motion of the planets and concluded that Earth circles the sun. Today this seems obvious. Back then, it was heresy. The prevailing belief was that the Creator made Earth—and, by extension, humans—the immovable center of the universe. One respected scholar, Galileo Galilei, studied Copernicus' model of the solar system, thought it was a good one, and said so. He was forced to retract his statement, on his knees, and put Earth back as the fixed center of things. (Word has it that he also muttered, "Even so, it *does* move.") Later, Darwin's theory of evolution also ran up against prevailing belief.

Today, as then, society has sets of standards. Those standards might be questioned when a new, logical explanation runs counter to prevailing beliefs. This does not mean scientists who raise questions are less moral, less lawful, or less spiritual than anyone else. It only means an extra standard guides their work. Their ideas about nature must be testable in the external world, in ways that can be repeated.

The external world, not internal conviction, is the testing ground for the theories generated in science.

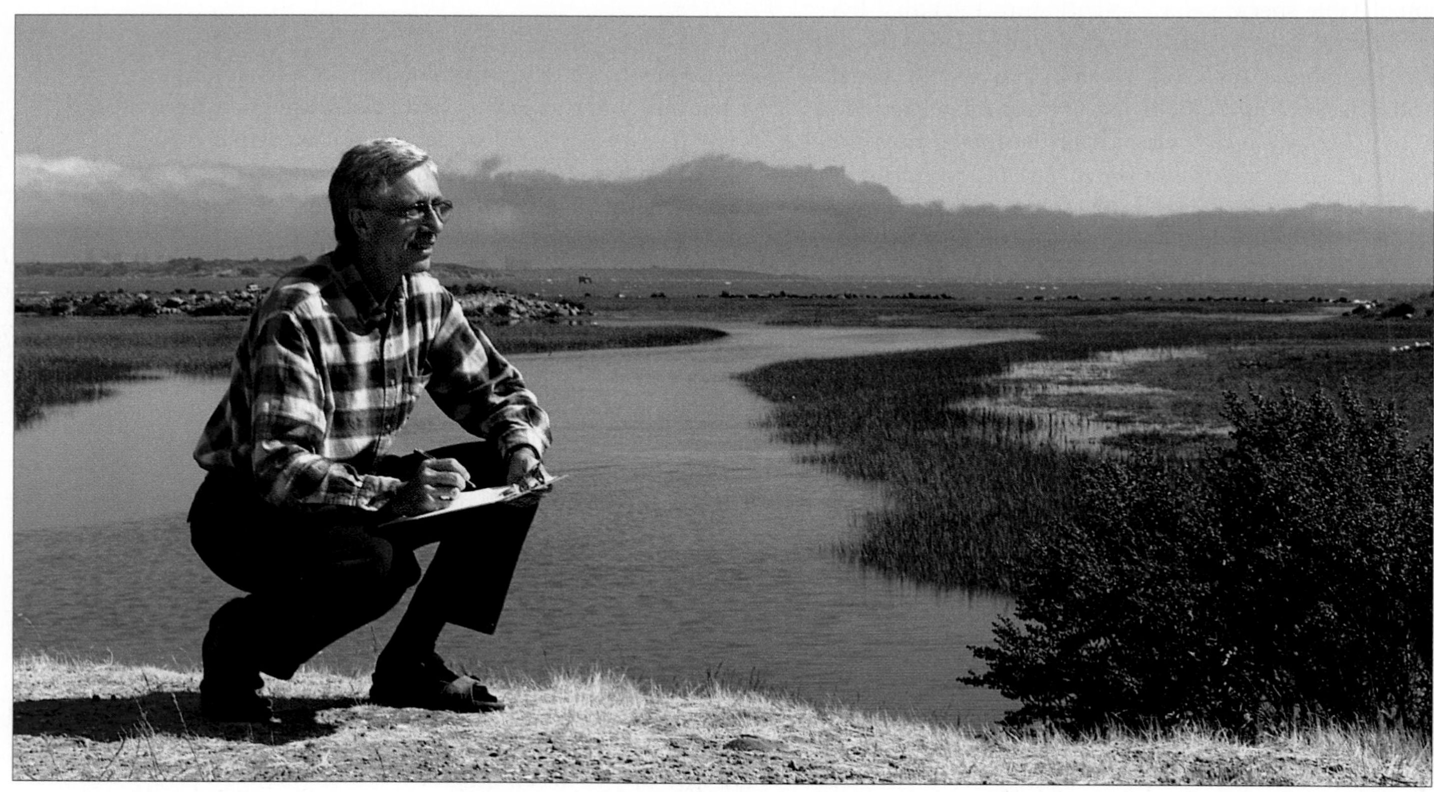

Summary

Section 1.1 Life shows many levels of organization. All things, living and nonliving, are made of atoms. Atoms join to form molecules, and molecules of life assemble into cells. Cells are the smallest living units. An organism may be a single cell or multicellular. In most multicelled species, cells are organized as tissues and organs.

A population consists of individuals of the same species in a specified area. A community consists of all populations occupying the same area. An ecosystem is a community and its environment. The biosphere includes all regions of Earth's atmosphere, waters, and land where we find living organisms.

BiologyNow
Explore levels of biological organization with the interaction on BiologyNow.

Section 1.2 Life shows unity. All organisms inherit DNA instructions for building proteins and pass them on to offspring. All organisms require energy and raw materials from the environment to grow and reproduce. All sense changes in the surroundings and respond to them in controlled ways (Table 1.1).

BiologyNow
Using instructions with the animation on BiologyNow, view how different objects are assembled from the same materials. Also view energy flow and materials cycling.

Section 1.3 Life shows tremendous diversity. Millions of species exist. Each is unique in some aspects of its body plan, functions, and behavior. Species that are related by descent from a common ancestor are grouped into three domains: Archaea, Bacteria, and Eukarya. Protists, plants, fungi, and animals are eukaryotes.

BiologyNow
Explore the characteristics of the three domains of life with the interaction on BiologyNow.

Section 1.4 Mutations change DNA and give rise to new variations of heritable traits. Natural selection occurs if a variation affects survival and reproduction. A population is evolving when adaptive forms of traits are becoming more common than less adaptive forms. Evolution means heritable change in a line of descent.

BiologyNow
Learn more about natural selection and evolution with Infotrac readings on BiologyNow.
Also, see the Infotrac article entitled "Will We Keep Evolving?," Ian Tattersall, Time, *April 2000.*

Section 1.5 Scientific methods are varied, but all are based on a logical approach to explaining nature. Scientists observe some aspect of nature, then develop a hypothesis about what might have caused it. They use the hypothesis to make predictions that can be tested by making more observations, building models, or doing experiments.

Scientists analyze test results, draw conclusions, and share this information with other scientists. A hypothesis that does not hold up under repeated testing is modified or discarded. A scientific theory is a long-standing hypothesis that has been supported by many tests.

Section 1.6 Most aspects of nature are complex, an outcome of many interacting variables. Researchers test the predictive power of hypotheses by performing experiments. An experiment allows a scientist to observe what happens to an experimental group when one variable is different. In every scientific experiment, an experimental group is compared with a control group. Scientists share their results so others can confirm their tests and verify their conclusions.

Section 1.7 Science cannot answer all questions. It deals only with aspects of nature that lend themselves to systematic observation, hypotheses, predictions, and experimental testing.

Table 1.1 Summary of Life's Characteristics
Shared characteristics that reflect life's unity
1. Organisms consist of one or more cells.
2. Organisms are constructed of the same kinds of atoms and molecules according to the same laws of energy.
3. Organisms engage in metabolism; they acquire and use energy and materials to survive and reproduce.
4. Organisms sense and make controlled responses to conditions in their internal and external environments.
5. Heritable information encoded in DNA gives organisms a capacity to grow and reproduce. That information also guides the development of complex multicelled organisms.
6. Characteristics that define a population can change over the generations; the population can evolve.
Foundations for life's diversity
1. Mutations (heritable changes in the structure of DNA) give rise to variation in heritable traits, including most details of body form, functioning, and behavior.
2. Diversity is the sum total of variations that have accumulated in different lines of descent over the past 3.9 billion years, by way of natural selection and other processes of evolution.

Self-Quiz

Answers in Appendix I

1. The smallest unit of life is the ______ .

2. ______ is the acquisition of traits after parents transmit their DNA to offspring.
   a. metabolism
   b. reproduction
   c. homeostasis
   d. inheritance

3. ______ is the capacity of cells to extract energy from sources in the environment, and use it to live, grow, and reproduce.

4. ______ is a state in which the internal environment is being maintained within a tolerable range.

5. A trait is ______ if it improves an organism's ability to survive and reproduce in a given environment.

6. Differences in heritable traits arise through ______ .

7. Researchers assign all species to one of three ______ .

8. ______ secure energy from their surroundings.
   a. producers
   b. consumers
   c. decomposers
   d. all of the above

9. DNA ______ .
   a. contains instructions for building proteins
   b. undergoes mutation
   c. is transmitted from parents to offspring
   d. all of the above

10. A control group is ______ .
   a. a standard against which experimental groups can be compared
   b. identical to experimental groups except for one variable only
   c. both a and b

11. Match the terms with the most suitable descriptions.

____ adaptive trait	a. statement of what you can expect to observe in nature
____ natural selection	b. proposed explanation; an educated guess
____ scientific theory	c. improves chances of surviving and reproducing
____ hypothesis	d. long-standing, much tested hypothesis used to interpret other hypotheses
____ prediction	e. outcome of differences in survival, reproduction among individuals of a population that differ in details of their traits

Additional questions are available on **BiologyNow™**

Critical Thinking

1. Witnesses in a court of law are asked to "swear to tell the truth, the whole truth, and nothing but the truth." What are some of the problems inherent in this question? Can you think of a better alternative?

2. Many popular magazines publish articles on exercise, diet, and other health-related topics. Authors disagree on what is the "best" diet or the "best" supplements. What kinds of evidence do you think the articles should include so that you can decide whether to accept what they say?

3. The Olestra® potato chip experiment in Section 1.6 was a *double-blind* study: Neither the subjects of the experiment nor the researchers knew which potato chips were in which bag until after all the subjects had reported. Why do researchers perform double-blind experiments? What do think are some challenges in designing and carrying out such an experiment?

4. In 1988 Dr. Randolph Byrd and his colleagues made a study of 393 patients admitted to San Francisco General Hospital's Coronary Care Unit. In this experiment, "born-again" Christian volunteers were asked to pray daily for a patient's rapid recovery and also for the prevention of complications and death. None of the patients knew if he or she was being prayed for or not, and no volunteers or patients knew each other.

How each patient fared was classified by Byrd as "good," "intermediate," or "bad." Byrd reported that the patients who had been prayed for fared better than those who had not. His was the first experiment to document, in a scientific fashion, results in support of the prediction that prayer can have beneficial effects on the medical outcome of seriously ill patients.

Publication of these results elicited a storm of criticism from scientists who cited bias in Byrd's experimental design. For instance, Byrd himself classified the patients before observations of them were finished. Think about how bias might play a role in interpreting medical data. Why do you think this study got such a dramatic response from the scientific community?

5. A scientific theory makes a general conclusion from particular instances. The assumption is that an outcome that has been observed to happen with regularity will happen again. However, there is no way to anticipate all of the variables that may affect an outcome, as this parable illustrates.

Once there was an intelligent turkey. It lived in a pen, attended by a kind master, and had nothing to do but reflect upon the world's wonders and its regularities. The turkey observed some major regularities. Every morning began with the sky becoming light, followed by the clop, clop, clop of the master's footsteps, which was followed by the appearance of delicious food. Other things varied, but food always followed footsteps.

The sequence of events was so predictable that it became the entire basis of the turkey's theory of the goodness of the world.

One morning, after a hundred confirmations of the theory, the turkey listened for the clop, clop, clop of its master's footseps, heard them, and had its head chopped off.

The unfortunate turkey did not understand that explanations about the world only have a high or low probability of not being wrong. Some people take this uncertainty to mean that "facts are irrelevant—facts change." Should we stop doing scientific research? Why or why not?

IMPACTS, ISSUES

Science or Supernatural?

About 2,000 years ago, in the mountains of Greece, the oracle of Delphi delivered prophecies from Apollo after inhaling sweet-smelling fumes that had collected in the sunken floor of her temple. Her prophecies tended to be rambling and cryptic. Why? She was babbling in a hydrocarbon-induced trance. Geologists recently found intersecting, earthquake-prone faults under the temple. Whenever the faults slipped, methane, ethane, and ethylene seeped out from the depths. All three gases are mild narcotics and hallucinogens.

methane

ethane

ethylene

Methane, ethane, and ethylene are nothing more than a few carbon and hydrogen atoms; hence the name hydrocarbons. Thanks to scientific inquiry, we now know more about them than the ancient Greeks did. For example, we know that methane was present when Earth first formed. We know that it is released into the atmosphere when volcanoes erupt, when we burn wood, peat, or fossil fuels, and when termites and cattle pass gas. We know it is one of the greenhouse gases contributing to global warming.

A legacy of the buried organic remains and wastes of countless marine organisms, methane also seeps from the ocean floor. There, the high pressure and low temperature of deep water freeze the gas as it bubbles up. A thousand billion tons of icy methane crystals lie at the bottom of the seas around the world. What will happen if the ocean warms up enough to melt those crystals is anyone's guess.

In short, knowledge about lifeless substances like methane can tell you a lot about life—past, present, and future—from Greek history, to current topics of health and disease, to physical and chemical conditions that influence the biosphere.

How Would You Vote? *Undersea methane deposits might be developed as a vast supply of energy, but the environmental costs are unknown. Should we continue to move toward exploiting this resource? See BiologyNow for details, then vote online.*

Key Concepts

ATOMS BOND

All substances are made of one or more elements. Atoms are the smallest units of an element that still retain the element's properties. They are composed of protons, neutrons, and electrons. Bonds between atoms form molecules. Whether and how one atom will bond with another depends on the arrangements of their electrons.

WATER AND PH

Water's unique characteristics, including temperature-stabilizing effects, cohesion, and solvent properties, make life possible on Earth. Living things adapt to changing amounts of hydrogen ions and other substances dissolved in water.

BIOLOGICAL MOLECULES

Living cells combine small carbon-containing subunits into complex carbohydrates and lipids, proteins, and nucleic acids. All of these large molecules of life have a backbone of carbon atoms. Functional groups bonded to the backbone help determine the properties of a molecule.

Links to Earlier Concepts

With this chapter, we start at life's basic level of organization, so take a moment to review the simple chart in Section 1.1. You will come across an early example of how the body's built-in mechanisms return the internal environment to a homeostatic state when conditions shift beyond ranges that cells can tolerate (1.2).

Here again, you will be considering one of the consequences of mutation in DNA (1.4), this time with sickle-cell anemia as the example.

LINK TO SECTION 1.1

2.1 Atoms and Their Interactions

A line of a million atoms would fit on the period ending this sentence. About ninety-two types of atoms make up everything in the world around you.

ATOMS AND ISOTOPES

Think about an apple. An apple, like all other matter, takes up space and responds to gravity. Now imagine using anything in your garage to demolish the apple completely: crush it, burn it, pour acid on it, plug it into an electrical socket, blow it up. When you finish, the smallest remaining bit of apple will be an atom.

Atoms are fundamental forms of matter—they can't be degraded to anything else, at least not by ordinary means. But if you use a little nuclear physics, you can separate atoms into smaller particles called protons, neutrons, and electrons. A **proton** (p^+) has a positive *charge*, or defined amount of electricity. Protons and neutrons are in the nucleus, or core region, of an atom. Zipping about that nucleus are the negatively charged **electrons** (e^-). The positive charge of a proton and the negative charge of an electron balance each other; an atom that has the same number of electrons as protons has no net electrical charge.

All matter consists of different kinds of atoms called **elements**. Figure 2.1 lists the most common elements in humans and in Earth's crust. Each element has its own *unique* number of protons in its atoms. Hydrogen atoms have one proton, carbon atoms have six.

Atoms of the same element do not necessarily have the same number of neutrons. Atoms with the same number of protons but a different number of neutrons are **isotopes**. Carbon has three isotopes, nitrogen has two, and so on. We identify isotopes by superscript numbers on standardized symbols for elements. For example, the isotopes of carbon are ^{12}C (or carbon 12, the most common isotope of carbon, with six protons and six neutrons), ^{13}C (with six protons and seven neutrons), and ^{14}C (six protons, eight neutrons).

Many isotopes are *radio*isotopes, which means that the atoms are unstable, or radioactive. An atom that is radioactive spontaneously emits energy and particles as its nucleus disintegrates. This process, radioactive decay, transforms one element into a different, more stable element at a known rate. For instance, in about 5,700 years, about half of the atoms in a sample of ^{14}C will decay into ^{14}N (nitrogen). Measuring proportions of radioactive elements can tell us how old the Earth is (very old!). It tells us when organisms, now fossilized, originated and later vanished.

Living systems use different isotopes of an element in the same way: Carbon is carbon, regardless of how many neutrons it has. We can substitute a radioactive isotope for a stable atom in a molecule, and then feed it into a living system as a *tracer*, or probe. The energy from radioactive decay acts like a shipping label that researchers can track as it passes through the system.

For example, PET (positron-emission tomography) uses tracers to form images of body tissues. Clinicians attach a radioisotope to a molecule such as glucose and inject it into a patient. All active body cells take up glucose, radioactive or not. Some cells take up more tracer (thus becoming more radioactive) than others, indicating differences in cell activity. These differences can be detected and converted to color-coded scans, as you will see in Chapter 25.

Figure 2.1 You are far more than the sum of your chemical parts, which, commercially speaking, are worth just $118.63, retail price. Proportions of the most common elements making up humans and most other living things are shown here. There are a lot more *trace* elements, each less than 0.01 percent of body weight.

human body	
Oxygen	61.0%
Carbon	23.0
Hydrogen	10.0
Nitrogen	2.6
Calcium	1.4
Phosphorus	1.1
Potassium	0.2
Sulfur	0.2

Earth's crust	
Oxygen	46.0%
Silicon	27.0
Aluminum	8.2
Iron	6.3
Calcium	5.0
Magnesium	2.9
Sodium	2.3
Potassium	1.5

ELECTRONS AND ENERGY LEVELS

Simple physics is great for explaining the motion of an apple falling off a tree. An electron is so tiny that it belongs to a world where simple physics just does not apply. Different forces bring about its motion. Where it can go depends on how many other electrons are buzzing around an atom's nucleus. All electrons in an atom occupy orbitals, which are volumes of space in which the electrons are likely to be at any instant.

Think of an atom as an apartment building with a nucleus in the basement, and many floors of rooms to rent to electrons. Each floor of that apartment building

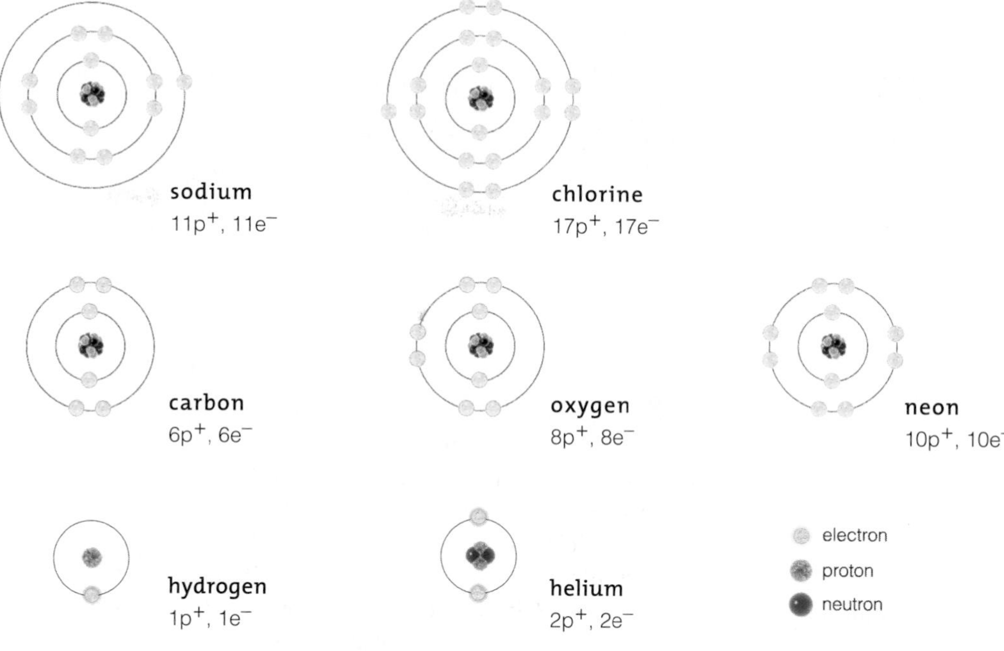

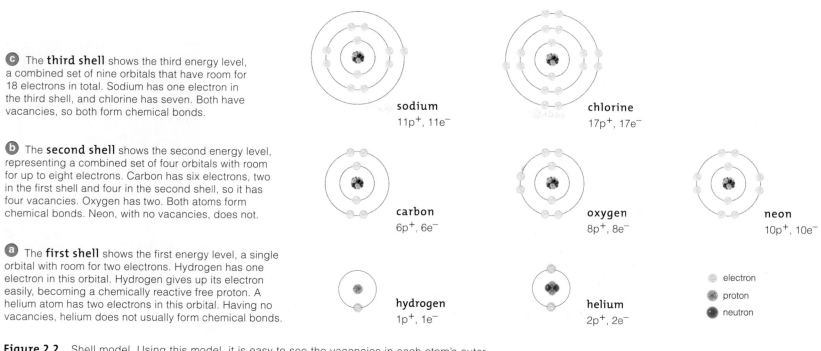

c The **third shell** shows the third energy level, a combined set of nine orbitals that have room for 18 electrons in total. Sodium has one electron in the third shell, and chlorine has seven. Both have vacancies, so both form chemical bonds.

b The **second shell** shows the second energy level, representing a combined set of four orbitals with room for up to eight electrons. Carbon has six electrons, two in the first shell and four in the second shell, so it has four vacancies. Oxygen has two. Both atoms form chemical bonds. Neon, with no vacancies, does not.

a The **first shell** shows the first energy level, a single orbital with room for two electrons. Hydrogen has one electron in this orbital. Hydrogen gives up its electron easily, becoming a chemically reactive free proton. A helium atom has two electrons in this orbital. Having no vacancies, helium does not usually form chemical bonds.

Figure 2.2 Shell model. Using this model, it is easy to see the vacancies in each atom's outer orbitals. Each circle represents all of the orbitals on one energy level. Larger circles correspond to higher energy levels. This model is highly simplified; a more realistic rendering would show the electrons as fuzzy clouds of probability density about ten thousand times bigger than the nucleus.

is one energy level. Each orbital is like one room that can accommodate up to two electrons. There is only one room on the first floor (one orbital at the lowest energy level), and it fills first. Hydrogen, the simplest atom, has a single electron in that room. Helium has two. Additional electrons rent second-floor rooms in larger atoms. When the second floor rooms are filled, more electrons can rent third-floor rooms, and so on. *Electrons fill orbitals at successively higher energy levels.*

The farther an electron is from the basement (the nucleus), the greater its energy. An electron in a first-floor room cannot move to the second or third floor, let alone the penthouse, unless a boost of energy gets it there. Suppose it absorbs the right amount of energy, say, from sunlight, to get excited about moving up. Move it does. If nothing fills that lower room, though, the electron will quickly go back to it, emitting extra energy as it does. Later, you will see how some cells in plants and in your eyes can capture that energy.

In the **shell model**, nested "shells" correspond to energy levels (Figure 2.2). They offer us an easy way to check for electron vacancies in various atoms. Bear in mind, atoms do not look like these flat diagrams. The shells are not three-dimensional volumes of space, and they do not show the individual electron orbitals.

Atoms that have vacancies in the outermost "shell" tend to give up, acquire, or share electrons with other atoms. This kind of electron-swapping between atoms is known as **chemical bonding** (Section 2.2). Atoms with zero vacancies rarely bond with other atoms. By contrast, the most common atoms in organisms—such as oxygen, carbon, hydrogen, nitrogen, and calcium—have vacancies in orbitals at their outermost energy level. And they do bond with other atoms.

A union between orbitals of two or more atoms—a *chemical bond*—is the glue that makes molecules out of atoms. Some molecules consist of atoms of just one element. Molecular nitrogen (N_2), with two nitrogen atoms, is like this. So is molecular oxygen (O_2).

Molecules with unvarying proportions of two or more elements are **compounds**. A water molecule, for instance, always has one oxygen and two hydrogen atoms. The ones in a Siberian lake, a bathtub, a leaf, or anywhere else have exactly twice as many hydrogen as oxygen atoms. In a **mixture**, two or more molecules intermingle without bonding. For example, you can make a mixture by swirling water and sugar together. The proportions of elements in a mixture can vary.

Electrons occupy orbitals, or defined volumes of space around an atom's nucleus. Successive orbitals correspond to levels of energy, which become higher with distance from the nucleus.

One or at most two electrons can occupy any orbital. Atoms that have vacancies in their highest-level orbitals can interact with other atoms.

A molecule is two or more atoms joined in a chemical bond. Atoms of two or more elements are bonded together in compounds. A mixture consists of intermingled molecules.

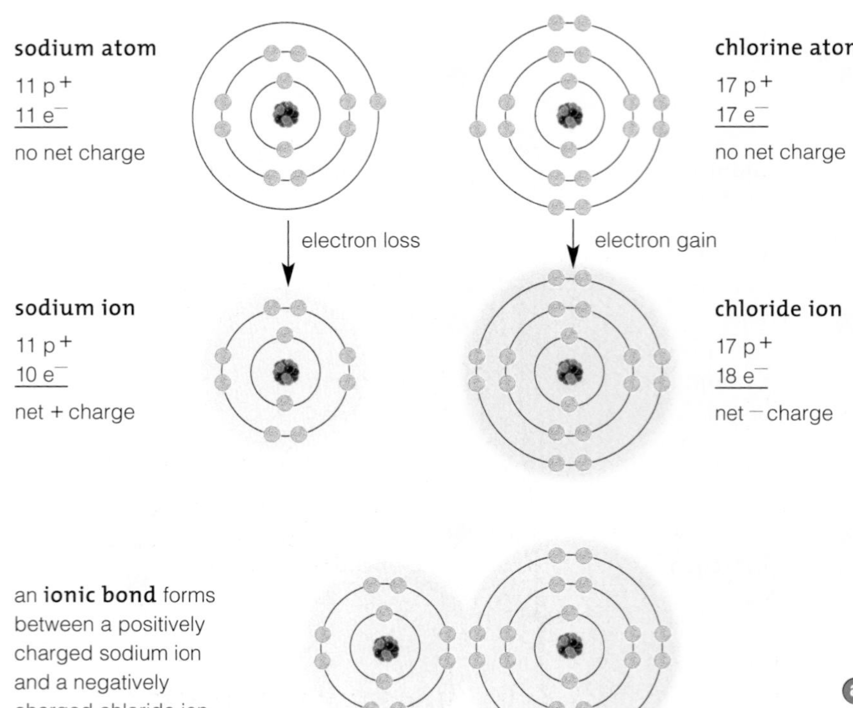

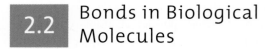

ⓐ Ionic bonds. A sodium atom loses the lone electron in its outermost shell, becoming a positively charged sodium ion (Na^+). A chlorine atom gains an electron in its outermost shell, becoming a negatively charged chloride ion (Cl^-). The two ions form an ionic bond: They stay together by the mutual attraction of opposite charges. Each crystal of table salt, or NaCl, is a cubic lattice of ionically bonded sodium and chloride ions.

Figure 2.3 ***Animated!*** Important bonds in biological molecules.

2.2 Bonds in Biological Molecules

The distinctive properties of biological molecules start with the properties of bonds that unite their atoms.

ION FORMATION AND IONIC BONDING

An electron, recall, has a negative charge equal to a proton's positive charge. When an atom has as many electrons as protons, these charges balance each other, so the atom will have a net charge of zero.

Atoms with more electrons than protons carry a net negative charge, and atoms with more protons than electrons carry a net positive charge. An atom that has either a positive or negative charge is known as an **ion**. Ions form when atoms gain or lose electrons.

Example: An uncharged chlorine atom has seven electrons in the third shell, which is most stable when filled with eight electrons. Chlorine tends to grab an electron from other places to fill the vacancy in its third shell. That extra electron will make it a chloride ion (Cl^-), with a net negative charge. A sodium atom tends to lose the lone electron in its third shell. If it does, it becomes a sodium ion with a net positive charge (Na^+), and two full shells.

Opposite charges attract. If a positively charged ion encounters a negatively charged ion, they may stay close together. A close association of ions is an **ionic bond**. Figure 2.3*a* shows crystals of table salt, or NaCl. In such crystals, ionic bonds constrain the ions in an orderly arrangement.

COVALENT BONDING

In an ionic bond, an atom that has gained electrons is associated with an atom that has lost electrons. What if both atoms tend to gain (or to lose) electrons? By *sharing* electrons, each atom's vacancy becomes partly filled with an electron from the other atom. Two atoms that are sharing a pair of electrons are connected by a **covalent bond**. A covalent bond is much stronger than an ionic bond (Figure 2.3*b*).

We depict chemical bonding patterns with structural formulas to show how atoms are physically arranged in a molecule. A single line that connects two atoms in a structural formula represents two shared electrons, or one covalent bond. Molecular hydrogen, with one covalent bond, is written H—H.

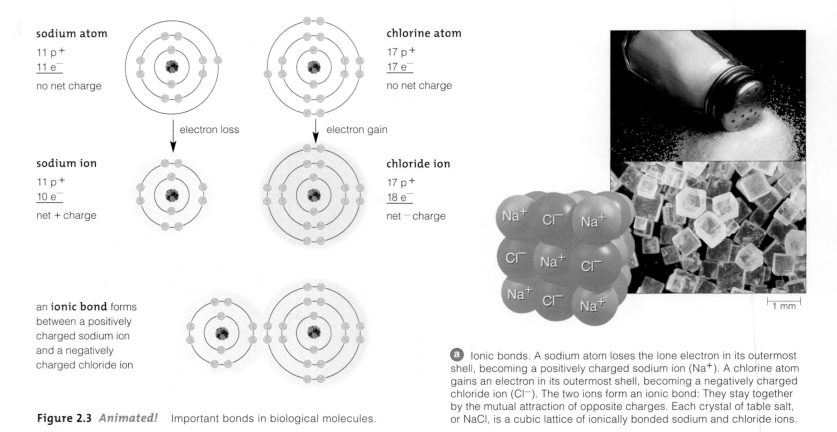

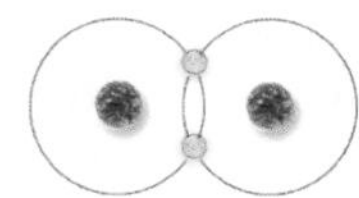

molecular hydrogen (H—H)

Two hydrogen atoms, each with one proton, share two electrons in a single covalent bond that is nonpolar.

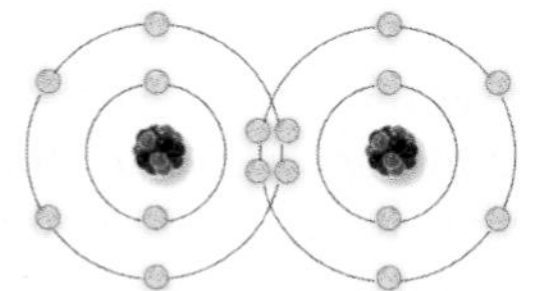

molecular oxygen (O=O)

Two oxygen atoms, each with eight protons, share four electrons in a double covalent bond, also nonpolar.

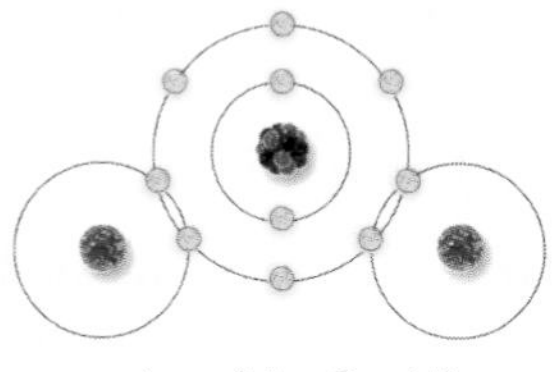

water (H—O—H)

Two hydrogen atoms each share an electron with oxygen. The resulting two covalent bonds form a water molecule. These bonds are polar. The oxygen exerts a greater pull on the shared electrons, so it bears a slight negative charge. Each of the hydrogens bears a slight positive charge.

b Covalent bonds. Two atoms with unpaired electrons in their outermost shell become more stable by sharing electrons. Two electrons are shared in each covalent bond. When the electrons are shared equally, the covalent bond is nonpolar. If one atom exerts a slightly greater pull on the shared electrons, the covalent bond is polar.

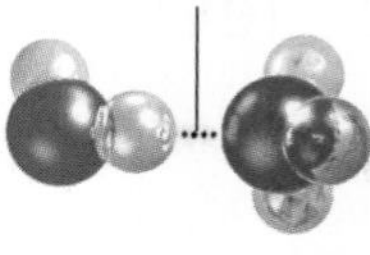

Two molecules interacting in one hydrogen (H) bond.

H bonds helping to hold part of two large molecules together.

Many H bonds hold DNA's two strands together. Individually, each H bond is weak, but collectively they stabilize DNA's large structure.

c Hydrogen bonds. Such bonds can form at a hydrogen atom that is already covalently bonded in a molecule. The atom's slight positive charge weakly attracts an atom with a slight negative charge that is already covalently bonded to something else. As shown above, this can happen between one of the hydrogen atoms of a water molecule and the nitrogen atom of an ammonia molecule.

Two atoms can share two electron pairs in a *double* covalent bond. Molecular oxygen (O=O) is like this. In a *triple* covalent bond, two atoms share three pairs, as they do in molecular nitrogen (N≡N). Each time you breathe in, a tremendous number of gaseous O_2 and N_2 molecules flows into your lungs.

In a *nonpolar* covalent bond, atoms share electrons equally, so there is no difference in charge between the "ends" of the bond. Molecular hydrogen (H_2) has such symmetry, as does methane (CH_4).

A covalent bond is *polar* if the atoms do not share electrons equally. One atom pulls the electrons a little more toward its end of the bond, so that part of the molecule bears a slight negative charge. This leaves a slight positive charge on the other end of the bond. A water molecule (H—O—H) has two polar covalent bonds; the oxygen carries a slight negative charge, and the hydrogens each carry a slight positive charge. We return to the polarity of water in the next section.

HYDROGEN BONDING

A hydrogen atom taking part in a polar covalent bond bears a slight positive charge, so it attracts negatively charged atoms. When the negatively charged atom is already taking part in a different covalent bond, the interaction between it and the hydrogen atom is called a **hydrogen bond** (Figure 2.3*c*). Hydrogen bonds are individually weak, so they break easily. But they also form easily, and often.

Many hydrogen bonds working together become a strong force that plays a very important role in the structure, and so the function, of biological molecules. Hydrogen bonding between different parts of a very large molecule hold it in a particular shape. Hydrogen bonds hold the two nucleotide strands of large DNA molecules together, for example.

Ions form when atoms acquire a net charge by gaining or losing electrons. Two ions of opposite charge attract each other. They can associate in an ionic bond.

In a covalent bond, atoms share electrons. When atoms share a pair of electrons equally, the bond is nonpolar. When the sharing is not equal, the bond is polar—slightly positive at one end, slightly negative at the other.

In a hydrogen bond, a covalently bound hydrogen atom bearing a slight positive charge is attracted to a covalently bound atom bearing a slight negative charge.

LINK TO SECTION 1.2

2.3 Water's Life-Giving Properties

Life originated in water. Organisms either live in it or they carry water around inside cells and tissue spaces. The properties of water that are so essential to life exist because of its unique hydrogen bonding behavior.

POLARITY OF THE WATER MOLECULE

Figure 2.4*a* shows the structure of one water molecule. In it, two hydrogen atoms have formed polar covalent bonds with an oxygen atom. The molecule has no net charge. Even so, the oxygen pulls the shared electrons more than the hydrogen atoms do. Thus, the molecule of water has an "end" with a slight negative charge, and two "ends" with slight positive charges.

The polarity of each water molecule attracts other water molecules. Hydrogen bonds form between them (Figure 2.4*b*). Water also forms hydrogen bonds with sugars and other polar molecules. That is why polar molecules are **hydrophilic** (water-loving) substances.

That same polarity repels oils and other nonpolar molecules, which are **hydrophobic** (water-dreading) substances. Shake a bottle filled with water and salad oil, then set it on a table. Soon, new hydrogen bonds replace the ones that the shaking broke. The reunited water molecules push out oil molecules, which cluster as oil droplets or as an oily film at the water's surface.

The same kinds of interactions occur at the thin, oily membrane between the water inside and outside cells. Membrane organization, and life itself, begins with hydrophilic and hydrophobic interactions. You'll be reading about membrane structure in Chapter 3.

WATER'S TEMPERATURE-STABILIZING EFFECTS

All molecules jiggle nonstop, and they move faster as they absorb heat. **Temperature** measures the energy of their motion. Extensive hydrogen bonding in water restricts the jiggles of individual water molecules. In other words, the hydrogen bonds absorb some of the energy of motion of water molecules. The result is that more heat is needed to raise the temperature of water

IMPACTS, ISSUES

Partly because of rising levels of methane and other greenhouse gases that are contributing to global warming, the Arctic ice cap is melting. At current rates, it will be gone in fifty years. So will the polar bears. Already their season for hunting seals is shorter, bears are thinner, and they are giving birth to fewer cubs.

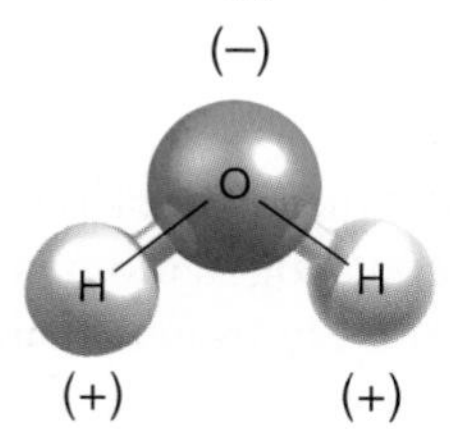

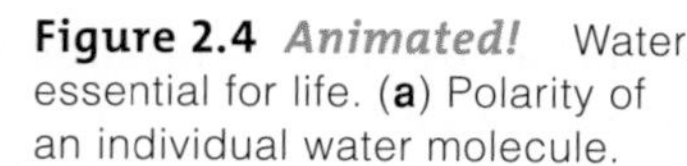

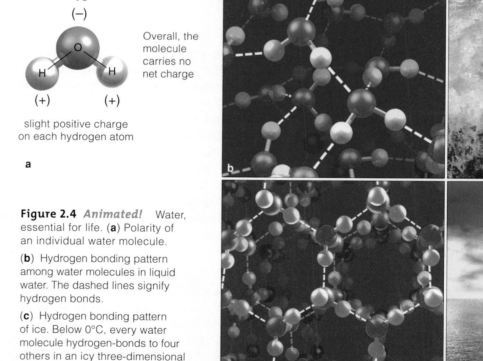

Figure 2.4 ***Animated!*** Water, essential for life. (**a**) Polarity of an individual water molecule.

(**b**) Hydrogen bonding pattern among water molecules in liquid water. The dashed lines signify hydrogen bonds.

(**c**) Hydrogen bonding pattern of ice. Below 0°C, every water molecule hydrogen-bonds to four others in an icy three-dimensional array. Water molecules are farther apart in ice than they would be in liquid water, because they vibrate less and so more hydrogen bonds can form. Ice floats because it has fewer molecules per unit volume (it is less dense) than liquid water.

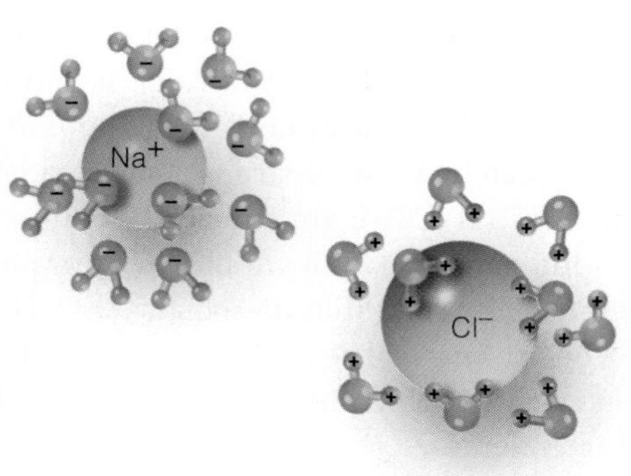

Figure 2.5 Two spheres of hydration.

Figure 2.6 Examples of water's cohesion. (**a**) When a pebble hits liquid water and forces molecules away from the surface, the individual water molecules don't fly off every which way. They stay together in droplets. Why? Countless hydrogen bonds exert a continuous inward pull on individual molecules at the surface.

(**b**) And just how does water rise to the very top of trees? Cohesion, and evaporation from leaves, pulls it upward.

b

compared to other fluids. So, water is a heat reservoir, meaning that its temperature stays fairly stable even when the temperature of the environment does not.

When water temperature is below its boiling point, hydrogen bonds form as fast as they break. Energy input (heat) increases molecular motion so much that the hydrogen bonds stay broken, and molecules at the water's surface escape into air. By this process, called **evaporation**, heat energy converts liquid water to a gas. Energy input overcomes the attraction between molecules of water, which break free.

The surface temperature of water decreases during evaporation. Evaporative water loss (sweat is about 99 percent water) helps you and some other mammals cool off on hot, dry days.

Below 0°C, water molecules do not move enough to break their hydrogen bonds, and they become locked in the latticelike bonding pattern of ice (Figure 2.4*c*). Ice is less dense than water. During winter freezes, ice sheets may form near the surface of ponds, lakes, and streams. These "ice blankets" insulate the liquid water beneath them, helping to protect many fishes, frogs, and other aquatic organisms from freezing.

WATER'S SOLVENT PROPERTIES

Water is an excellent solvent, meaning ions and polar molecules easily dissolve in it. A dissolved substance is called a **solute**. In general, a substance is *dissolved* after water molecules cluster around ions or molecules of it and keep them dispersed in fluid.

Water molecules that cluster around a solute form a "sphere of hydration." Such spheres form around any solute in cellular fluids, a maple tree's sap, blood, the fluid in your gut, and every other fluid associated with life. Spheres of hydration keep ions dispersed in fluid. Watch it happen as you pour table salt (NaCl) into a cup of water. After a while, the crystals dissolve as they separate into sodium ions (Na^+) and chloride ions (Cl^-). Every Na^+ attracts the oxygen ends of many water molecules, and every Cl^- attracts the hydrogen ends of many water molecules (Figure 2.5).

WATER'S COHESION

Another life-sustaining property of water is cohesion. **Cohesion** means that molecules resist separating. You see its effect as surface tension when you toss a pebble into a pond or some other body of water (Figure 2.6*a*). The water bounces, waves, sprays, and bubbles, but it does not fly apart into molecules. Hydrogen bonds are exerting a continuous, inward pull on the individual water molecules. This pull is so strong that they stick together in drops and pools rather than spreading out in a thin film like other liquids do.

Cohesion is in play inside organisms, too. Plants, for example, absorb nutrient-laden water while they grow. Very narrow columns of liquid water rise inside pipelines of vascular tissues, which extend from roots to leaves and other plant parts. Water evaporates from leaves as molecules break free and diffuse into the air (Figure 2.6*b*). The cohesive force of hydrogen bonds pulls replacements into the leaf cells, in ways you will read about in Section 18.7.

Being polar, water molecules hydrogen-bond to one another and to other polar (hydrophilic) substances. They tend to repel nonpolar (hydrophobic) substances.

Extensive hydrogen bonding between water molecules gives water unique properties that make life possible.

Water has cohesion, temperature-stabilizing effects, and a capacity to dissolve many substances.

LINK TO SECTION 1.2

2.4 Acids and Bases

Ions are dissolved in fluids inside and outside a cell. Among the most influential are hydrogen ions. They have major effects because they are chemically active.

THE PH SCALE

At any instant in liquid water, some water molecules are split into hydrogen (H^+) and hydroxide (OH^-) ions. Hydrogen ions are the basis of the **pH scale**. The scale is a way to measure the relative amount of hydrogen ions in solutions such as seawater, blood, or sap. The greater the H^+ concentration, the lower the pH. Pure water (not rainwater or tap water) always has as many H^+ as OH^- ions. This state is neutrality, or pH 7.0 (Figure 2.7).

A one unit decrease from neutrality corresponds to a tenfold increase in H^+ concentration, and an increase by one unit corresponds to a tenfold decrease in H^+ concentration. One way to get a sense of the range is to taste dissolved baking soda (pH 9), water (pH 7), and lemon juice (pH 2).

Acids lose hydrogen ions in solution. **Bases** acquire them. *Acidic* solutions, such as lemon juice, gastric fluid, and coffee, contain more H^+ than OH^-; the pH of these solutions is below 7. *Basic*, or alkaline, solutions like egg white, baking soda, and seawater contain more OH^- than H^+ and have a pH greater than 7.

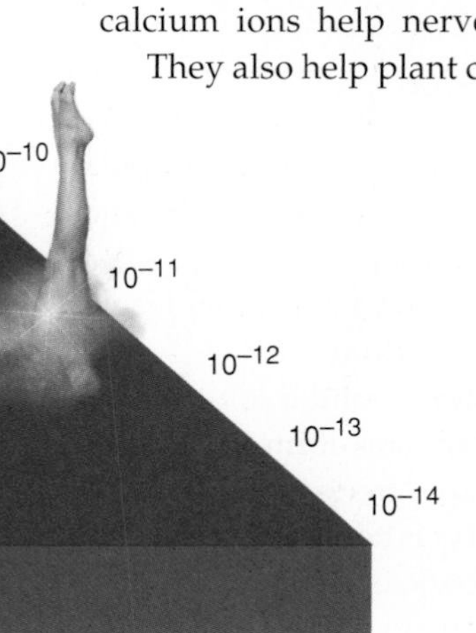

Figure 2.7 ***Animated!*** The pH scale, representing concentrations of hydrogen ions in a solution. Also shown are approximate pH values of some liquids. This pH scale ranges from 0 (most acidic) to 14 (the most basic). A change of 1 on the scale means a tenfold change in H^+ concentration.

Nearly all of life's chemistry occurs near pH 7. Most of your body's internal environment (tissue fluids and blood) is between pH 7.3 and 7.5. Seawater is more basic than the body fluids of the organisms living in it.

Acids and bases can be weak or strong. Weak acids, such as carbonic acid (H_2CO_3), are stingy H^+ donors. Strong acids readily give up H^+ in water. An example is the hydrochloric acid (HCl) that dissociates into H^+ and Cl^- inside your stomach. The hydrogen ions make gastric fluid very acidic, which in turn activates local protein-digesting enzymes.

HCl splashing up out of the stomach can cause *acid indigestion*. Antacids taken for this condition, including milk of magnesia, release OH^- ions that combine with H^+ to reduce the pH of stomach contents.

High concentrations of strong acids or bases can disrupt ecosystems, and make it impossible for cells to survive. Read the labels on containers of ammonia, drain cleaner, and other products commonly stored in households. Many cause severe *chemical burns*. So does sulfuric acid in car batteries. Fossil fuel burning and nitrogen fertilizers release strong acids that lower the pH of rainwater (Figure 2.8). Some regions are quite sensitive to this *acid rain*. Alterations in the pH of soil and water can harm organisms. We return to this topic in Section 30.7.

SALTS AND WATER

A **salt** is any compound that dissolves easily in water and releases ions *other than* H^+ and OH^-. It commonly forms when an acid interacts with a base. For example:

$$\underset{\text{HYDROCHLORIC ACID}}{\text{HCl (acid)}} + \underset{\text{SODIUM HYDROXIDE}}{\text{NaOH (base)}} \rightleftharpoons \underset{\text{SODIUM CHLORIDE}}{\text{NaCl (salt)}} + H_2O$$

When dissolved in water, the salt product of this reaction (NaCl) dissociates into sodium ions (Na^+) and chloride ions (Cl^-). Many salt ions are important components of cellular processes. For example, sodium, potassium, and calcium ions help nerve and muscle cells function. They also help plant cells take up water from soil.

BUFFERS AGAINST SHIFTS IN PH

Cells must respond quickly to even slight shifts in pH, because enzymes and many other essential biological molecules can function properly only within a narrow range of pH. A slight deviation from that range halts cellular processes.

Most body fluids maintain a stable pH with buffers. A **buffer system** is a dynamic chemical partnership between a weak acid or base and its salt. These two related chemicals work in an equilibrium to counter slight shifts in pH. For example, if a basic substance is added to a buffered fluid, the weak acid donates some H^+ ions to the solution. Excess OH^- ions from the base combine with H^+ ions from the acid, forming water.

Carbon dioxide, a by-product of many reactions, combines with water in the blood to compose a buffer system of carbonic acid and bicarbonate ions. The acid can combine with free OH^-, forming bicarbonate (salt) ions and water:

$$\underset{}{OH^-} + \underset{\text{CARBONIC ACID}}{H_2CO_3} \longrightarrow \underset{\text{BICARBONATE (SALT)}}{HCO_3^-} + \underset{\text{WATER}}{H_2O}$$

The carbonate ions can combine with free H^+, forming carbonic acid:

$$\underset{\text{BICARBONATE}}{HCO_3^-} + H^+ \longrightarrow \underset{\text{CARBONIC ACID}}{H_2CO_3}$$

Together, these reactions keep the blood pH between 7.3 and 7.5, but only up to a point. A buffer can absorb only so many excess ions. Even slightly more than that limit will cause the pH to swing widely.

The effects of buffer system failure in a biological system can be devastating. In *acute respiratory acidosis*, carbon dioxide accumulation leads to excess carbonic acid in the blood. The resulting drop in blood pH may cause a person may fall into a *coma*, a dangerous state of unconsciousness.

Coma may also occur in *alkalosis*, which is a rise in blood pH. Nausea, vomiting, and headache are among the other outcomes. If blood pH increases to just 7.8, prolonged muscle spasms called *tetany* may occur.

Ions in fluids on the inside and outside of cells have key roles in cell function. Acidic substances release hydrogen ions when dissolved, and basic substances accept them. Salts release ions other than H^+ and OH^-.

Acid–base interactions help maintain pH, which reflects the H^+ concentration in a fluid. Buffer systems keep the pH of body fluids at levels suitable for life.

Figure 2.8 Sulfur dioxide emissions from a coal-burning power plant. Some airborne pollutants, including sulfur dioxide, dissolve in water vapor to form acidic solutions that are a component of acid rain.

2.5 Molecules of Life—From Structure to Function

Under present-day conditions on Earth, only living cells can make complex carbohydrates, lipids, proteins, and nucleic acids. These are the molecules of life, and their structures hold clues to how each kind functions.

WHAT IS AN ORGANIC COMPOUND?

The molecules of life are **organic compounds**, which are compounds with covalent bonds between carbon and hydrogen atoms. The term is a holdover from a time when chemists defined "organic" substances as those made by animals and vegetables, as opposed to "inorganic" substances from other sources, like rock. Now, we can make organic compounds in laboratories. And we found out that these compounds were present on Earth even before life originated.

Vitamins, aspirin, sugars, and caffeine are organic compounds. They and most other organic compounds have at least one **functional group**: an atom (other than hydrogen) or cluster of atoms bonded to carbon. Hydrocarbons are the exception; they have carbon and hydrogen atoms only. Methane, ethane, and ethylene (shown in the chapter opener) are examples. We use a color code to depict the atoms shown at right.

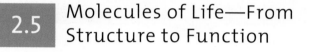

- carbon (C)
- oxygen (O)
- hydrogen (H)
- nitrogen (N)
- calcium (C)
- phosphorus (P)
- potassium (K)
- sulfur (S)
- sodium (Na)
- chlorine (Cl)

IT ALL STARTS WITH CARBON'S BONDING BEHAVIOR

Living organisms consist mainly of oxygen, hydrogen, and carbon. Most of the oxygen and hydrogen are in the form of water. Put water aside, and carbon makes up more than half of what is left.

Carbon's importance in life arises from its versatile bonding behavior. *Each carbon atom can covalently bond with as many as four other atoms.* Such bonds, in which two atoms share one, two, or three pairs of electrons, are relatively stable. Covalent bonds often join carbon atoms as "backbones" to which hydrogen, oxygen, and other elements are attached. The three-dimensional shapes of organic compounds start with these bonds.

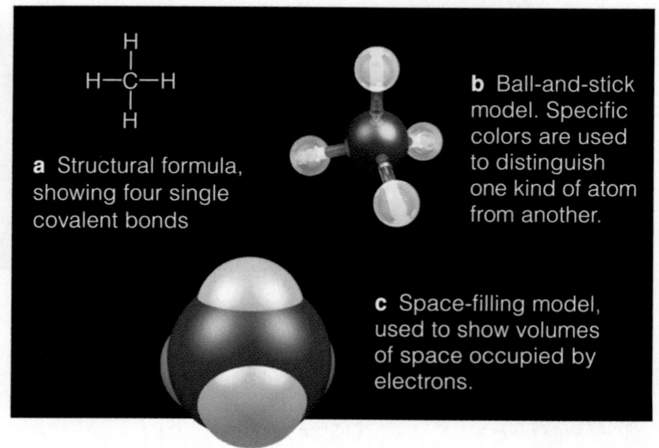

H
H—C—H
H
a Structural formula, showing four single covalent bonds
b Ball-and-stick model. Specific colors are used to distinguish one kind of atom from another.
c Space-filling model, used to show volumes of space occupied by electrons.

Figure 2.9 Molecular models for methane (CH_4).

Figure 2.9 shows some different ways to represent methane, the simplest organic compound. A colorless, odorless gas, methane is one carbon atom covalently bonded to four hydrogen atoms (CH_4). The ball-and-stick model conveys its bond angles and the way the atoms are distributed. The space-filling model is better to show relative sizes and surfaces.

Figure 2.10*a* is a ball-and-stick model for glucose, another organic compound. Hydrogen and oxygen are covalently bonded to a backbone of six carbon atoms. This backbone usually coils back on itself, with two carbons joined to form a ring structure (Figure 2.10*b*).

Other models look simpler. A flat structural model might show carbons but not hydrogen atoms bonded to it (Figure 2.10*c*). Some icons for a ring show only atoms that are *not* carbon or hydrogen.

We use these depictions of structure to help us visualize how molecules function. For instance, virus particles can infect a cell by attaching to its membrane proteins. Like Lego blocks, the virus surface proteins have ridges, clefts, and charged regions that match up to clefts, ridges, and charged regions of proteins at cell membrane surfaces.

IMPACTS, ISSUES

In ocean sediments and polar regions, methane is trapped in the form of methane hydrates. They represent Earth's largest carbon reservoir, twice as large as all the known reserves of fossil fuels. Could we harvest that reservoir to help meet our energy demands? Maybe. But releasing so much gas might alter the worldwide climate. About 183 million years ago, a catastrophic burp of methane from the ocean floor may have wiped out most deep-sea organisms.

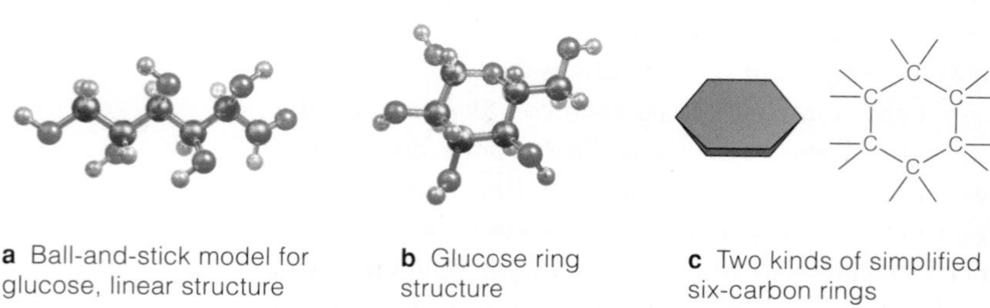

c Two kinds of simplified six-carbon rings

Figure 2.10 Some ways of representing organic compounds.

FUNCTIONAL GROUPS

A functional group is an atom or cluster of atoms that imparts reactive properties to a molecule. How much can one functional group do? Consider a seemingly tiny difference—reversed functional groups—between the sex hormones shown in Figure 2.11. Early on, a vertebrate embryo appears neither male nor female; it has a set of tubes that will develop into a reproductive system. In the presence of testosterone, the tubes will become male sex organs, and male traits will develop. In the absence of testosterone, however, the tubes will become female sex organs that secrete estrogens. Those estrogens will guide the formation of female traits.

Figure 2.12 lists some major functional groups. The number, kind, and arrangement of such groups dictate features of carbohydrates, lipids, proteins, and nucleic acids. For example, hydrocarbons (organic molecules of only carbon and hydrogen atoms) are hydrophobic, or nonpolar. Fatty acids have chains of hydrocarbons, which is why lipids with fatty acid tails don't dissolve in water. Sugars, a class of compounds called alcohols, contain one or more polar *hydroxyl* (—OH) groups. Molecules of water quickly form hydrogen bonds with these groups; that is why sugars dissolve fast in water.

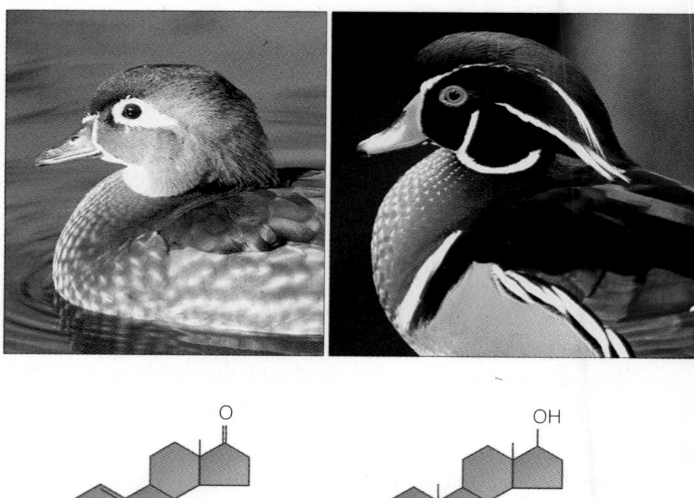

Figure 2.11 Observable differences in traits between female and male wood ducks (*Aix sponsa*), brought about by estrogen and testosterone. These two sex hormones have the same carbon ring structure. They differ only in the position of functional groups attached to the ring.

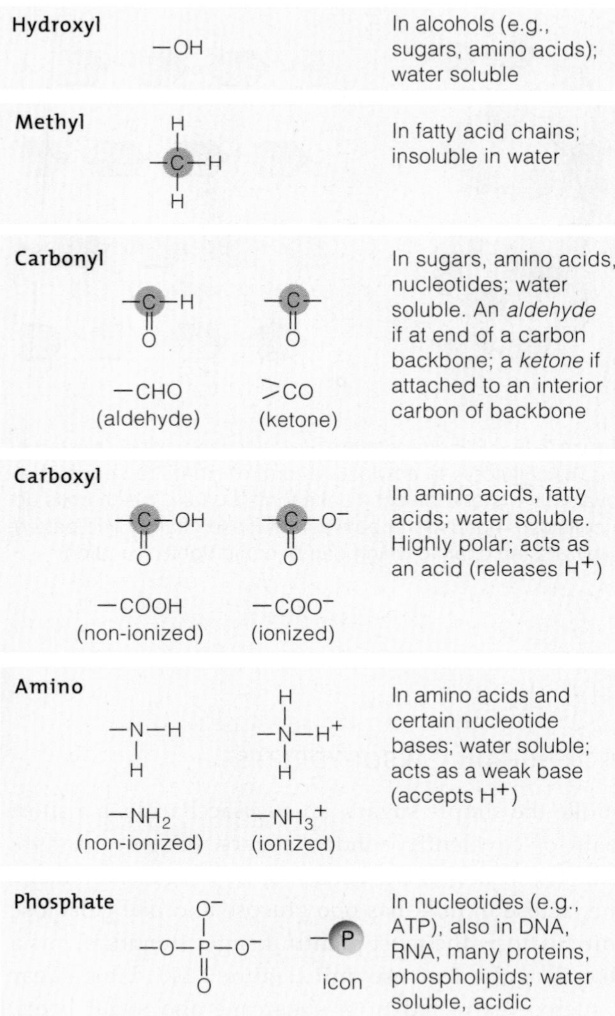

Hydroxyl
—OH
In alcohols (e.g., sugars, amino acids); water soluble
Methyl
In fatty acid chains; insoluble in water
Carbonyl
—CHO
(aldehyde)
>CO
(ketone)
In sugars, amino acids, nucleotides; water soluble. An aldehyde if at end of a carbon backbone; a ketone if attached to an interior carbon of backbone
Carboxyl
—COOH
(non-ionized)
—COO⁻
(ionized)
In amino acids, fatty acids; water soluble. Highly polar; acts as an acid (releases H+)
Amino
—NH2
(non-ionized)
—NH3+
(ionized)
In amino acids and certain nucleotide bases; water soluble; acts as a weak base (accepts H+)
Phosphate
icon
In nucleotides (e.g., ATP), also in DNA, RNA, many proteins, phospholipids; water soluble, acidic

Figure 2.12 ***Animated!*** Common functional groups in biological molecules, with examples of their locations.

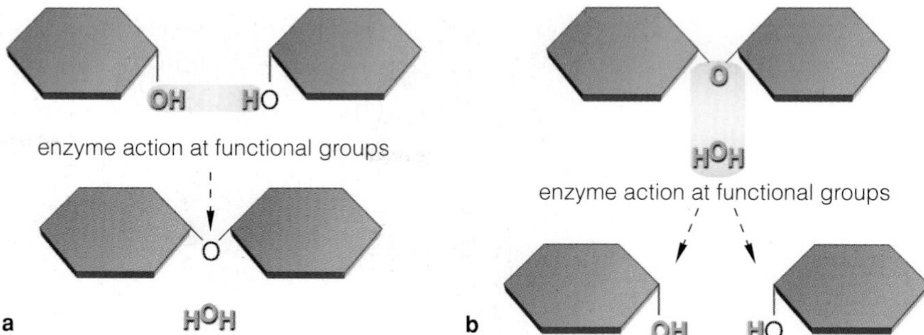

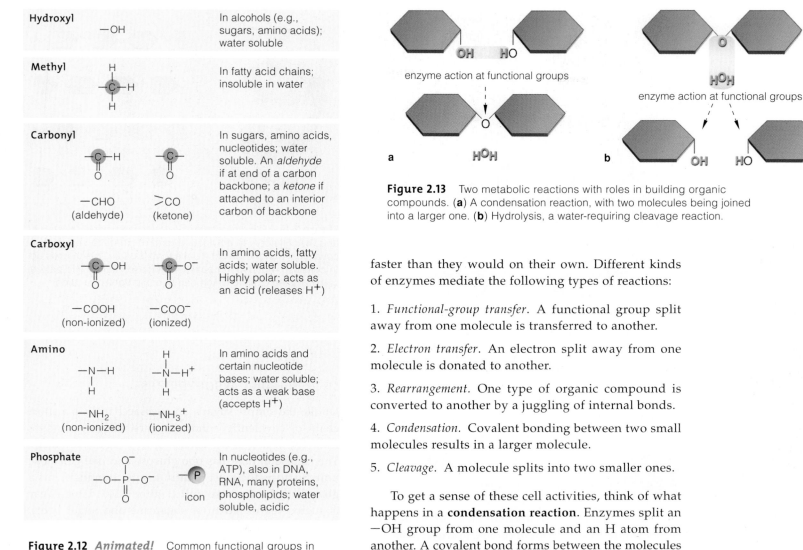

Figure 2.13 Two metabolic reactions with roles in building organic compounds. (**a**) A condensation reaction, with two molecules being joined into a larger one. (**b**) Hydrolysis, a water-requiring cleavage reaction.

HOW DO CELLS BUILD ORGANIC COMPOUNDS?

Cells build big molecules mainly from four families of small organic compounds: simple sugars, fatty acids, amino acids, and nucleotides. These compounds are called monomers when many of the same kind bond together to form a larger molecule. Molecules with many repeating monomer units are called polymers (*mono–*, one; *poly–*, many). Starch, for example, is a polymer of glucose units. Plant cells make starch when glucose is plentiful, then break it apart when they require glucose units as a source of instant energy or to build something else.

So how do cells actually do the construction work? The **reactions** by which a cell builds, rearranges, or splits apart all organic compounds require energy and **enzymes**: proteins that make substances react much faster than they would on their own. Different kinds of enzymes mediate the following types of reactions:

1. *Functional-group transfer.* A functional group split away from one molecule is transferred to another.
2. *Electron transfer.* An electron split away from one molecule is donated to another.
3. *Rearrangement.* One type of organic compound is converted to another by a juggling of internal bonds.
4. *Condensation.* Covalent bonding between two small molecules results in a larger molecule.
5. *Cleavage.* A molecule splits into two smaller ones.

To get a sense of these cell activities, think of what happens in a **condensation reaction**. Enzymes split an —OH group from one molecule and an H atom from another. A covalent bond forms between the molecules at their exposed sites. The discarded atoms often form water, or H_2O (Figure 2.13*a*). Polymers such as starch form by way of repeated condensation reactions.

Another example: **Hydrolysis**, a type of cleavage reaction, is like condensation reversed (Figure 2.13*b*). Enzymes split molecules, then attach an —OH group and a hydrogen atom (both from a water molecule) to each of the exposed sites. With hydrolysis, cells cleave polymers into monomers when these are required for building blocks or for energy.

Complex carbohydrates, lipids, proteins, and nucleic acids are the molecules of life.

Organic compounds have diverse three-dimensional shapes and functions that start with their carbon backbone and with functional groups covalently bonded to it.

Enzyme-mediated reactions build the molecules of life mainly from smaller organic compounds—simple sugars, fatty acids, amino acids, and nucleotides.

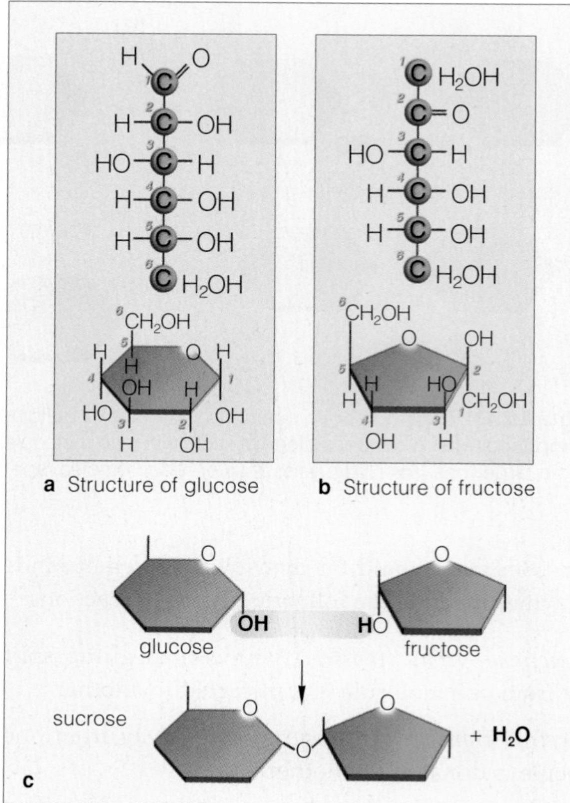

Figure 2.14 ***Animated!*** (**a**,**b**) Straight-chain and ring forms of glucose and fructose. The carbon atoms of these simple sugars are numbered in sequence, starting at the one closest to the aldehyde or ketone group. (**c**) Condensation of two monosaccharides into a disaccharide.

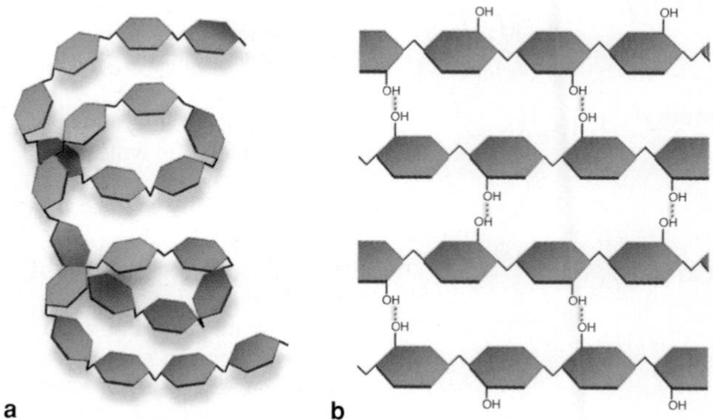

Figure 2.15 Bonding patterns for glucose units in (**a**) starch and (**b**) cellulose. In amylose, a form of starch, a series of covalently bonded glucose units form a chain, which coils up. In cellulose, bonds form between glucose chains. The pattern stabilizes the chains, which can become tightly bundled.

2.6 The Truly Abundant Carbohydrates

Which biological molecules are most plentiful in nature? ***Carbohydrates****. Most consist of carbon, hydrogen, and oxygen in a 1:2:1 ratio* $(CH_2O)_n$*. Cells use carbohydrates as structural materials and transportable or storable forms of energy. Monosaccharides, oligosaccharides, and polysaccharides are the main classes.*

THE SIMPLE SUGARS

"Saccharide" is from a Greek word that means sugar. The *mono*saccharides (one sugar unit) are the simplest carbohydrates. They all have at least two —OH groups bonded to their carbon backbone and one aldehyde or ketone group. Most dissolve easily in water. Common sugars have a backbone of five or six carbon atoms and tend to form a ring structure when dissolved.

Ribose and deoxyribose are the sugar monomers of RNA and DNA, respectively; each has five carbon atoms. Glucose has six carbons (Figure 2.14*a*). Cells use glucose as an instant energy source, as a structural unit, or as a precursor (parent molecule) to make other compounds, including vitamin C (a sugar acid) and glycerol (an alcohol with three —OH groups).

IMPACTS, ISSUES

Are you one of the methane producers? Cattle, sheep, and a lot of people are. Bacteria in their gut feed on partly digested carbohydrates and release methane, along with other gases. Researchers are working with diets and vaccines to retard methanogenic bacteria in the gut. Why bother? Huge livestock herds release methane; they may be a big factor in global warming.

SHORT-CHAIN CARBOHYDRATES

Unlike the simple sugars, an *oligo*saccharide is a short chain of covalently bonded sugars. (*Oligo–* means a few.) The *di*saccharides consist of two sugar monomers. The lactose in milk has one glucose and one galactose unit. Sucrose, the most plentiful sugar in nature, has a glucose and a fructose unit (Figure 2.14). Table sugar is sucrose extracted from sugarcane and sugar beets. Glycolipids are lipids with oligosaccharide side chains. In later chapters, you will encounter glycolipids with essential roles in self-recognition and immunity.

COMPLEX CARBOHYDRATES

The "complex" carbohydrates, or *poly*saccharides, are straight or branched chains of many sugar monomers —often hundreds or thousands of them. Different kinds consist of one or more types of monomers. The most common polysaccharides are cellulose, glycogen, and starch. Even though all three are made of glucose, they differ a lot in their properties. Why? The answer starts with differences in bonding patterns between their glucose units, which are joined together in chains.

In starch, the covalent bonding pattern of glucose units makes the chain coil like a spiral staircase, as in Figure 2.15*a*. Many —OH groups project out from a coiled starch chain (Figure 2.16*a*). These unprotected groups are easily accessible to hydrolysis enzymes. Thus, starch can be broken apart into its glucose units quickly. Plant cells store much of their glucose in the form of starch.

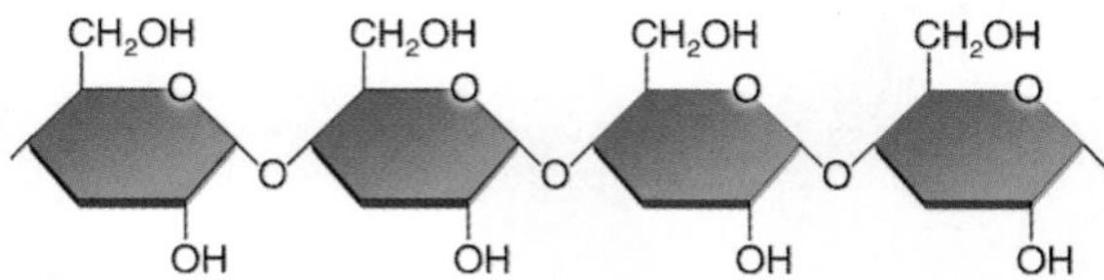

a Structure of amylose, a soluble form of starch. Cells inside tree leaves briefly store excess glucose monomers as starch grains in their chloroplasts, which are tiny, membrane-bound sacs that specialize in photosynthesis.

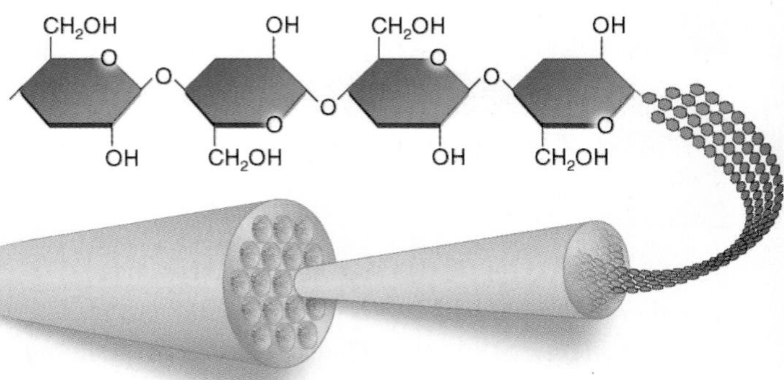

b Structure of cellulose. In cellulose fibers, chains of glucose units stretch side by side and hydrogen-bond at –OH groups. The many hydrogen bonds stabilize the chains in tight bundles that form long fibers. Few organisms produce enzymes that can digest this insoluble material. Cellulose is a structural component of plants and plant products, such as cotton dresses and wood.

c Glycogen. Animal cells build this polysaccharide when the body has excess glucose. It is especially abundant in the liver and muscles of highly active animals, including fishes and people.

Figure 2.16 Molecular structure of starch (**a**), cellulose (**b**), and glycogen (**c**), and their typical locations in a few organisms. All three carbohydrates consist only of glucose units.

In cellulose, many side-by-side chains of glucose monomers hydrogen-bond to each other at their —OH groups (Figure 2.15*b*). This bonding pattern stabilizes the chains into a tight bundle that resists hydrolysis. Long fibers of cellulose are a structural component of plant cell walls (Figure 2.16*b*). Like steel rods inside reinforced concrete, cellulose fibers in cell walls are tough, insoluble, and resistant to weight loads and mechanical stress, such as strong winds against stems.

In animals, glycogen is the sugar-storage equivalent of starch in plants (Figure 2.16*c*). Muscle and liver cells store a lot of it. When the sugar level in blood falls, liver cells hydrolyze glycogen into glucose monomers that enter the blood. Exercise strenuously, but briefly, and muscle cells tap glycogen for a burst of energy.

A different polysaccharide, chitin, is distinctive in that it has nitrogen-containing groups attached to its glucose units. Chitin strengthens the external skeletons and other hard body parts of a number of animals, such as crabs, earthworms, insects, spiders, and ticks of the sort shown in Figure 2.17. Chitin also stiffens the cell walls of many kinds of fungi.

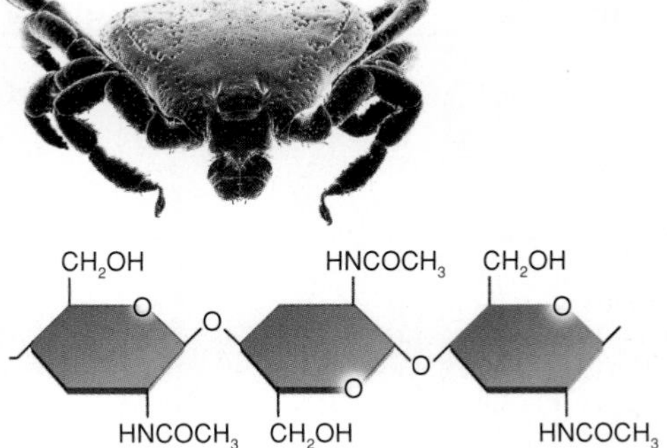

Figure 2.17 (*Top*) A tick's body covering is a protective cuticle reinforced with a polysaccharide called chitin (*right*). You may "hear" chitin when big spider legs clack across an aluminum oil pan on a garage floor.

The simple sugars (such as glucose), oligosaccharides, and polysaccharides (such as starch) are carbohydrates. Each cell requires carbohydrates as structural materials, and as storable or transportable packets of energy.

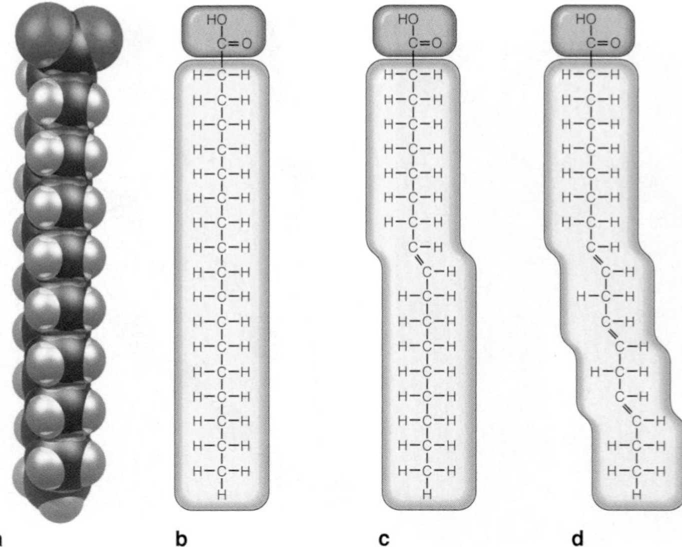

Figure 2.18 Three fatty acids. (**a**,**b**) Space-filling model and structural formula for stearic acid. The carbon backbone is fully saturated with hydrogen atoms. (**c**) Oleic acid, with a double bond in its backbone, is an unsaturated fatty acid. (**d**) Linoleic acid, also unsaturated, has three double bonds in its backbone.

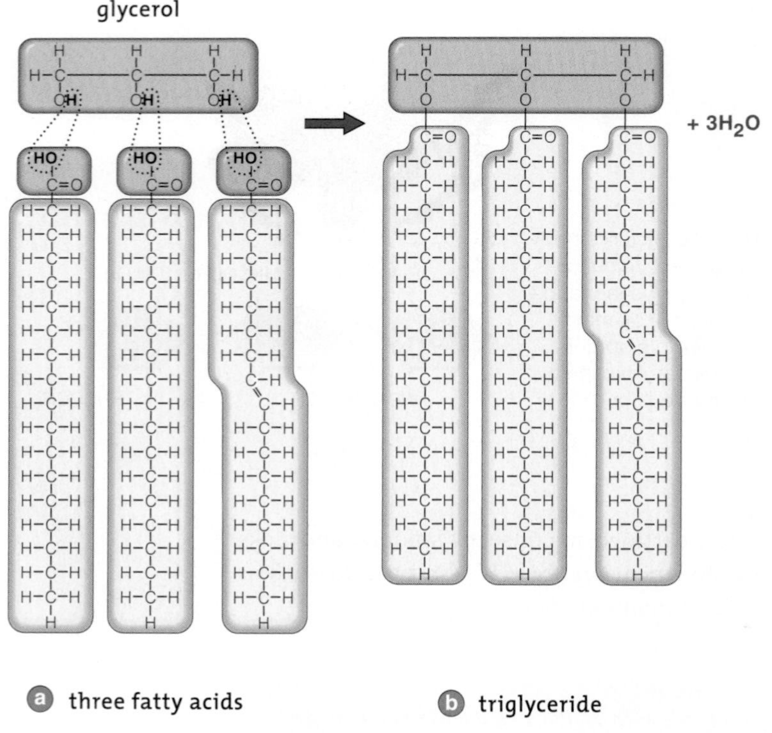

Figure 2.19 ***Animated!*** Condensation of (**a**) three fatty acids and one glycerol molecule into (**b**) a triglyceride.

2.7 Greasy, Fatty—Must Be Lipids

If something is greasy or oily to the touch, it may be a lipid or have lipid parts. Cells use different kinds of lipids for energy, as structural materials, and as signaling molecules. Fats, phospholipids, and waxes all have fatty acid tails. Sterols have a backbone of four carbon rings.

FATS AND FATTY ACIDS

Lipids are nonpolar hydrocarbons. They don't dissolve much in water, but they readily dissolve in nonpolar substances, as butter does in warm cream sauce.

The lipids called **fats** have one, two, or three fatty acids attached to a glycerol molecule. A **fatty acid** has a backbone of as many as thirty-six carbon atoms, a carboxyl group at one end, and hydrogen atoms at most or all of the remaining carbons (Figure 2.18).

Unsaturated fatty acids have one or more double bonds. *Saturated* fatty acids have single bonds only. Saturated fatty acids are tightly packed in animal fats, and so are solid at room temperature. Most plant fats stay liquid at room temperature, as "vegetable oils." The double bonds in unsaturated fatty acid tails form kinks that tend to prevent them from packing tightly.

"Neutral" fats, such as butter, lard, and vegetable oils, are **triglycerides**. Each triglyceride has three fatty acids stretching out like flexible tails from a glycerol unit (Figure 2.19). Triglycerides are the most abundant lipids in your body and its richest reservoir of energy. Gram for gram, the breakdown of lipids yields more than twice the energy of carbohydrate breakdown. All vertebrates store triglycerides as droplets in fat cells that make up adipose tissue.

Layers or patches of adipose tissue also insulate the body and cushion some of its parts. For instance, as in many other kinds of animals, a thick layer of triglycerides beneath the skin of Emperor penguins

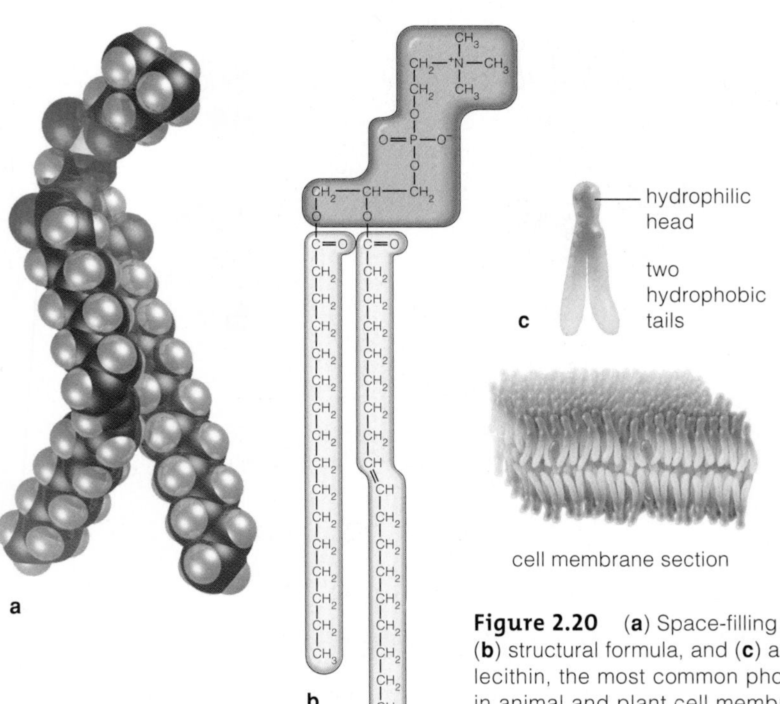

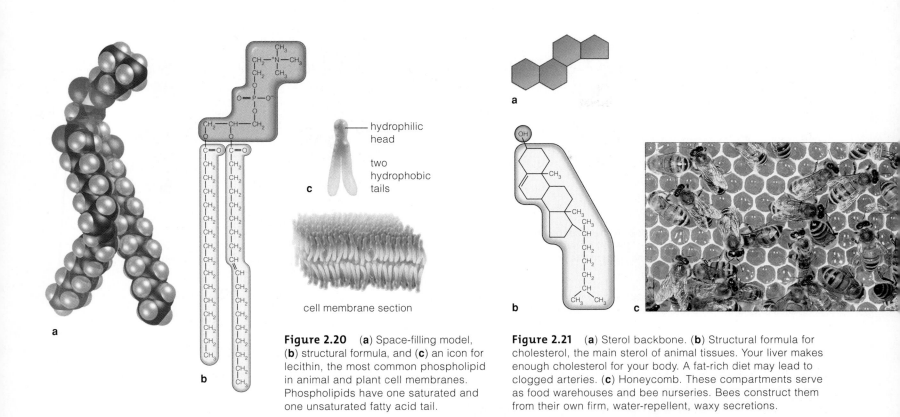

Figure 2.20 (**a**) Space-filling model, (**b**) structural formula, and (**c**) an icon for lecithin, the most common phospholipid in animal and plant cell membranes. Phospholipids have one saturated and one unsaturated fatty acid tail.

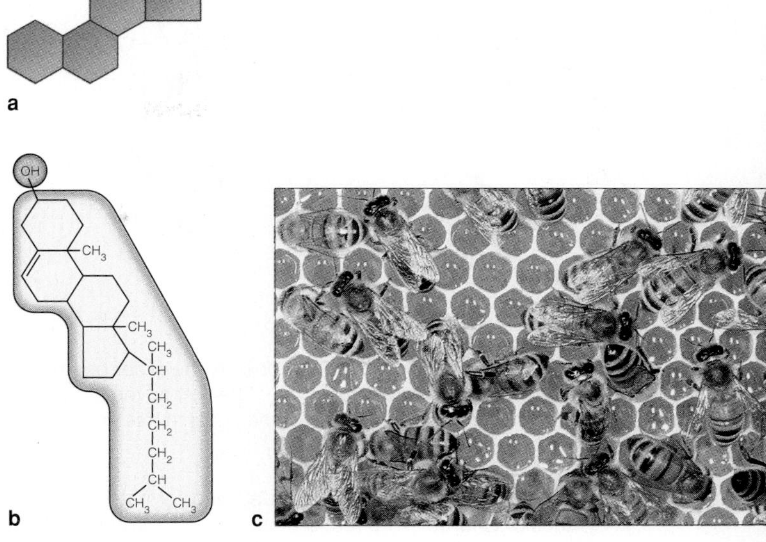

Figure 2.21 (**a**) Sterol backbone. (**b**) Structural formula for cholesterol, the main sterol of animal tissues. Your liver makes enough cholesterol for your body. A fat-rich diet may lead to clogged arteries. (**c**) Honeycomb. These compartments serve as food warehouses and bee nurseries. Bees construct them from their own firm, water-repellent, waxy secretions.

insulates their bodies so they can keep warm in the extremely cold Antarctic winter months.

PHOSPHOLIPIDS

A **phospholipid** has a glycerol backbone, a hydrophilic head with a phosphate group, and two nonpolar fatty acid tails (Figure 2.20). Phospholipids are the principal component of cell membranes, which have two layers of lipids. Heads of one layer are dissolved in the cell's fluid interior; heads of the other layer are dissolved in the surroundings. Sandwiched between the two layers are all of the fatty acid tails, which are hydrophobic. Section 3.3 takes a look at membrane functions.

CHOLESTEROL AND OTHER STEROLS

Sterols are among the many lipids with no fatty acids. They differ in the number, position, and type of their functional groups, but all have a rigid backbone of four carbon rings (Figure 2.21*a*).

Sterols are key components of every eukaryotic cell membrane. The most common type in animal tissues is cholesterol (Figure 2.21*b*). Cholesterol is remodeled into compounds as diverse as steroids, bile salts, and vitamin D. Bile salts assist fat digestion in the small intestine. The sex hormones, including the two shown in Figure 2.11, are steroids. Sex hormones have roles in development and function of sexual organs, secondary sexual traits, and adult sexual behavior.

WAXES

Waxes have long-chain fatty acids tightly packed and linked to long-chain alcohols or carbon rings. All have a firm consistency; all repel water. Surfaces of plants have a cuticle that contains waxes and another lipid, cutin. A plant cuticle restricts water loss and thwarts some parasites.

Waxes also protect, lubricate, and lend pliability to skin and hair. Birds secrete waxes, fats, and fatty acids that waterproof their feathers. Bees use beeswax for honeycomb, which houses each new bee generation as well as honey (Figure 2.21*c*).

Being largely hydrocarbon, lipids can dissolve in nonpolar substances, but they resist dissolving in water.

Triglycerides, or neutral fats, have a glycerol head and three fatty acid tails. They are a major energy reservoir. Phospholipids are the main components of cell membranes.

Sterols such as cholesterol serve as membrane components and precursors of steroid hormones and other compounds. Waxes are firm yet pliable components of water-repelling and lubricating substances.

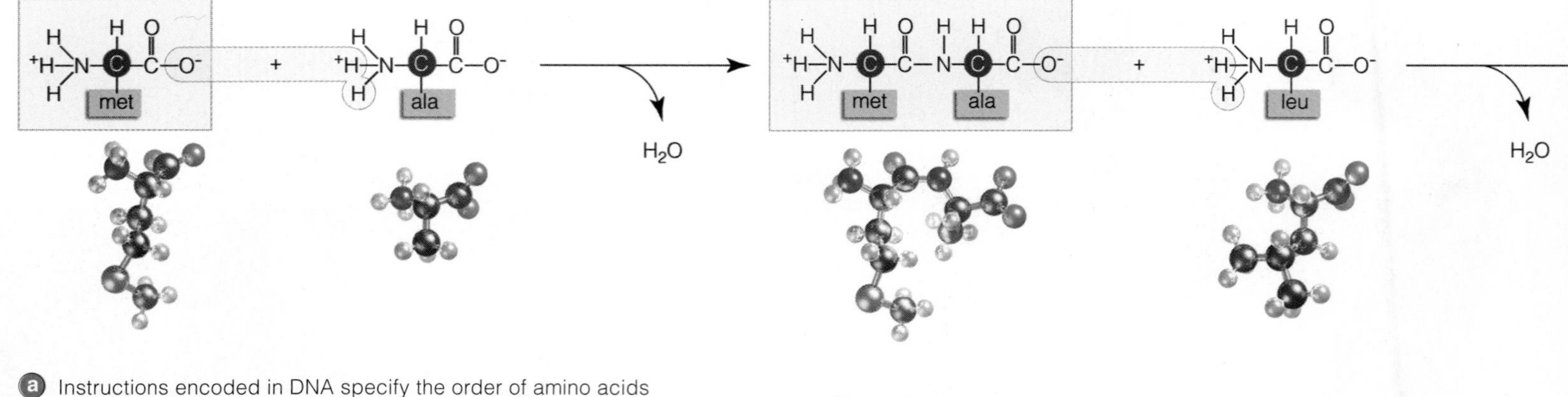

ⓐ Instructions encoded in DNA specify the order of amino acids to be joined in a polypeptide chain. The first amino acid is usually methionine (met). Alanine (ala) comes next in this one.

ⓑ In a condensation reaction, a peptide bond forms between the methionine and alanine. Leucine (leu) is next.

Figure 2.22 ***Animated!*** Peptide bond formation during protein synthesis. Chapter 10 offers a closer look at protein synthesis.

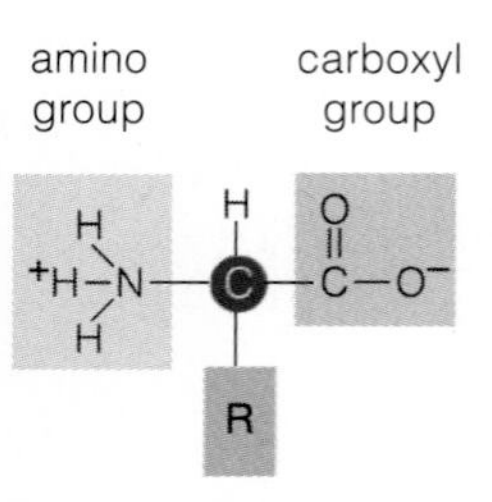

R group (20 kinds, each with distinct properties)

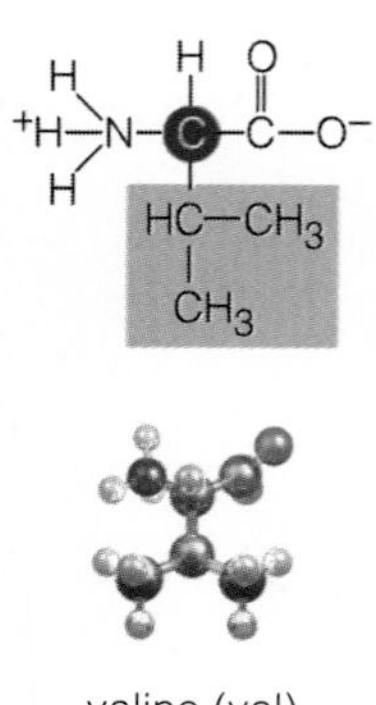

valine (val)

Figure 2.23 Generalized formula and one representative of the twenty common amino acids. The *green* boxes highlight the R groups.

2.8 Proteins—Diversity in Structure and Function

Of all large biological molecules, proteins are the most diverse. Some speed reactions; others are the stuff of spider webs, feathers, bones, hair, and other body parts. Nutritious types abound in seeds and eggs. Many move substances, help cells communicate, or defend against pathogens. Amazingly, cells build thousands of different proteins from only twenty kinds of amino acids!

WHAT IS AN AMINO ACID?

An **amino acid** is a small organic compound with an amino group ($-NH_3^+$), a carboxyl group ($-COO^-$, the acid part), a hydrogen atom, and one or more atoms called its R group. In most cases, these components are attached to the same carbon atom. The R group is a functional group that defines a particular amino acid and gives it special properties. Figures 2.22 and 2.23 show a few kinds that you will encounter later on.

LEVELS OF PROTEIN STRUCTURE

When a cell constructs a protein, it strings amino acids together, one after the other. Instructions coded in DNA specify the order in which any of the twenty kinds of amino acids will occur in a given protein. A *peptide* bond forms when a condensation reaction joins the amino group of one amino acid and the carboxyl group of the next. Each **polypeptide chain** consists of many amino acids. Its carbon backbone (Figure 2.24*a*) contains nitrogen atoms in this regular pattern:

—N—C—C—N—C—C—

The sequence of amino acids in a polypeptide chain is its *primary* structure. Hydrogen bonding between different amino acids causes different regions of each polypeptide chain to twist, bend, loop, or fold. This is *secondary* structure. Some stretches of amino acids coil into a helical shape, a bit like a spiral staircase; other regions form sheets or loops (Figure 2.24*b*). Bear in mind, the primary structure of each type of protein is unique, but similar patterns of coils, sheets, and loops occur in most of them.

Much as an overly twisted rubber band coils back on itself, the coils, sheets, and loops of a protein fold up even more, into compact domains. A "domain" is a part of a protein that forms a functional unit. This is the third level of organization, or *tertiary* structure, of a protein. Tertiary structure is what makes a protein a working molecule (Figure 2.24*c*). For instance, barrel-shaped domains of some proteins function as tunnels through membranes.

Many proteins are two or more polypeptide chains that are bonded together or closely associated with one another. This is a protein's fourth level of organization, its *quaternary* structure. Most enzymes and many other proteins are globular, with multiple polypeptide chains folded into rounded shapes. Hemoglobin, described shortly, is an example.

Protein structure does not stop here. Enzymes often attach short, linear, or branched oligosaccharides to a

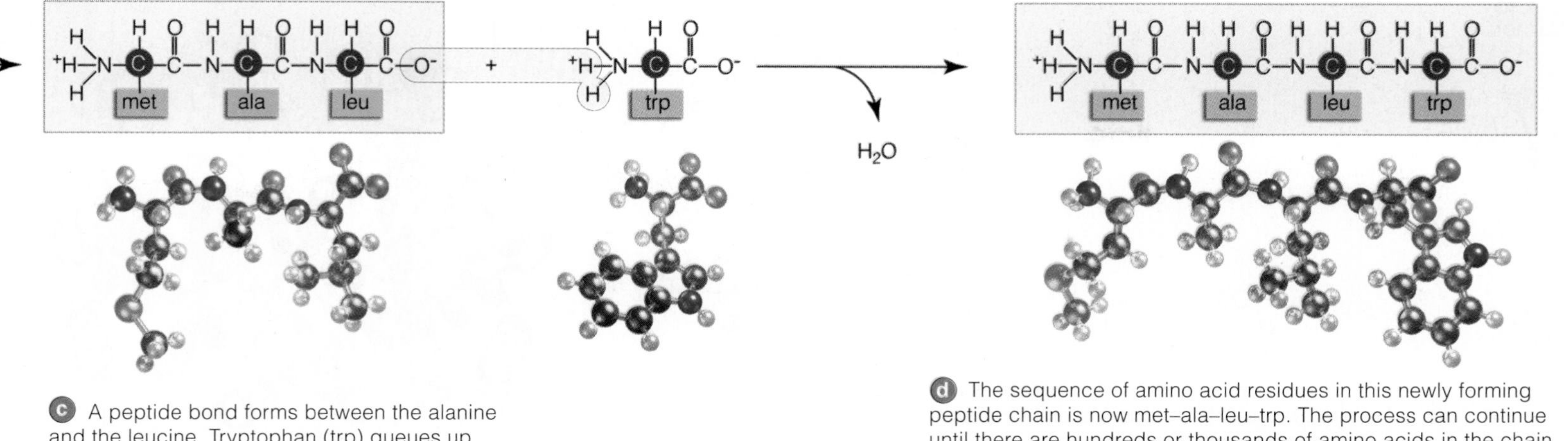

c A peptide bond forms between the alanine and the leucine. Tryptophan (trp) queues up.

d The sequence of amino acid residues in this newly forming peptide chain is now met–ala–leu–trp. The process can continue until there are hundreds or thousands of amino acids in the chain.

Figure 2.24 Protein structure. (**a**) Primary structure of a protein is its sequence of amino acids. (**b**) Hydrogen bonds (dotted lines) that connect different parts of a polypeptide chain form a protein's secondary structure, which can be coiled or sheetlike. (**c**) Coils and sheets pack together into stable, functional domains, the protein's third level of structural organization.

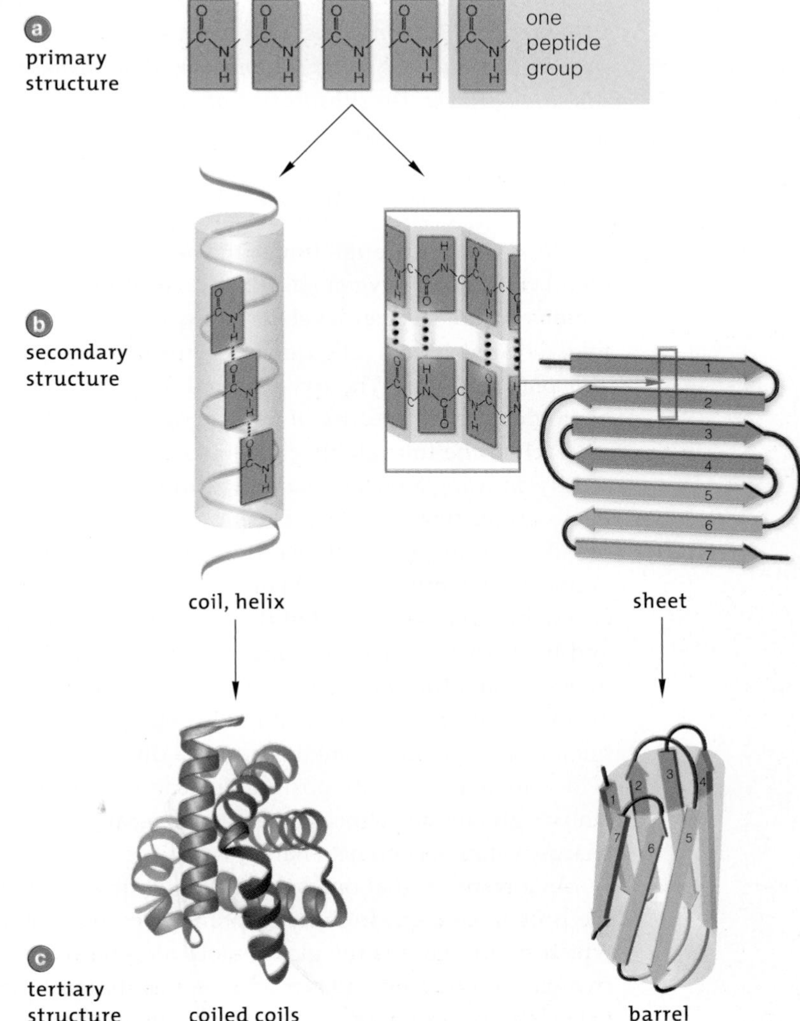

new polypeptide chain, making a *glyco*protein. Many glycoproteins occur at the cell surface or are secreted from cells. Lipids also get attached to many proteins. The cholesterol, triglycerides, and phospholipids that your body absorbs after a meal are transported about as components of *lipo*proteins.

Finally, some proteins array themselves by many thousands into a much larger structure. Many such conglomerate proteins are fibrous; their polypeptide chains are organized as strands or sheets. Some of them contribute to the cell's shape and organization. Others help cells and cell parts move.

A protein has primary structure, a sequence of amino acids covalently bonded as a polypeptide chain.

Local regions of a polypeptide chain become twisted and folded into helical coils, sheetlike arrays, and loops. These arrangements are the protein's secondary structure.

A polypeptide chain or parts of it become organized as structurally stable, compact, functional domains. Such domains are a protein's tertiary structure.

Many proteins show quaternary structure; they consist of two or more polypeptide chains.

LINK TO SECTION 1.4

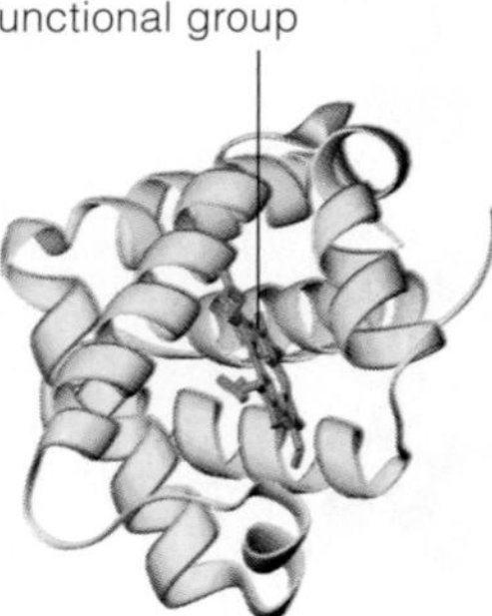

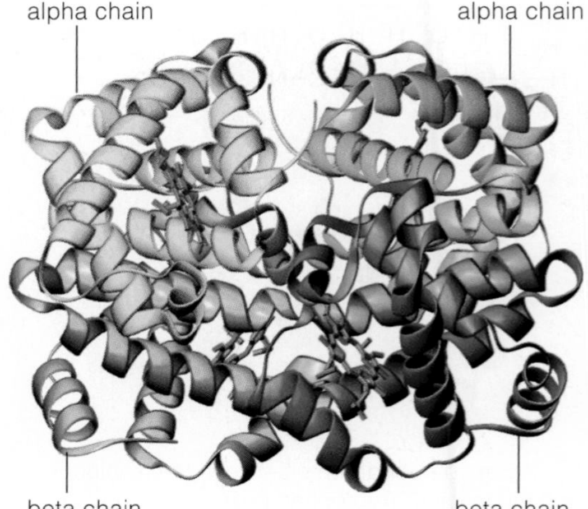

Figure 2.25 (**a**) Globin. This polypeptide chain cradles heme, a functional group that contains an iron atom.

(**b**) Hemoglobin, an oxygen-transport protein in red blood cells. This is one of the proteins with quaternary structure. It consists of four globin molecules (two alphas and two betas) held together by hydrogen bonds.

2.9 Why Is Protein Structure So Important?

Cells are good at making proteins that are just what their DNA specifies. But sometimes cells make mistakes or nature sabotages the specifications, as with ultraviolet radiation attacks. The substitution of just one wrong amino acid may have far-reaching consequences.

JUST ONE WRONG AMINO ACID . . .

As blood moves through lungs, the hemoglobin in red blood cells binds oxygen gas, then gives it up in body regions where oxygen levels are low. After oxygen is released, red blood cells quickly return to the lungs and pick up more. The oxygen-binding properties of hemoglobin exist because of its structure.

Each of the four globin chains in hemoglobin folds tightly, forming a pocket that holds an iron-containing heme group (Figure 2.25*a*). During its life span, a red blood cell transports billions of oxygen molecules, all bound to 250 million or so hemoglobin molecules.

Globin comes in two slightly different forms, alpha and beta. Two of each form make up one hemoglobin molecule in adult humans (Figure 2.25*b*). Glutamate is normally the sixth amino acid in the beta globin chain, but a DNA mutation sometimes puts a different amino acid, valine, in the sixth position instead (Figure 2.26). Unlike glutamate, which carries an overall negative charge, valine has no net charge.

As a result of that one substitution, a tiny patch of the protein changes from hydrophilic to hydrophobic, which in turn causes the globin molecule's behavior to change slightly. Hemoglobin with this mutation in its beta chain is designated HbS, or sickle hemoglobin.

HbS molecules can form large, stable, rod-shaped aggregates. Red blood cells containing these aggregates become distorted into a sickle shape. Sickled cells clog tiny blood vessels, and this disrupts blood circulation.

Every human inherits two genes for beta globin, one from each of the two parents. (Genes are units of DNA that encode proteins.) Cells use both genes when they make beta globin. If one is normal and the other has the valine mutation, a person can make enough normal hemoglobin to lead a relatively normal life. But someone who inherits two mutant genes can only make the mutant HbS hemoglobin. The outcome is sickle-cell anemia, a severe genetic disorder. Figure 2.26*d* lists the far-reaching effects of sickle-cell anemia on tissues and organs.

PROTEINS UNDONE—DENATURATION

The shape of a protein defines its biological activity. A globin molecule cradles heme, an enzyme speeds some reaction, a receptor transduces an energy signal. These proteins and others cannot function unless they stay coiled, folded, and packed in a precise way. Their shape depends on many hydrogen bonds and other interactions—which heat, shifts in pH, or detergents can disrupt. At such times, polypeptide chains unwind and change shape in an event called **denaturation**.

Consider albumin, a protein in the white of an egg. When you cook eggs, the heat does not disrupt the covalent bonds of albumin's primary structure, but it does destroy the weaker hydrogen bonds that maintain the protein's shape, and so the protein unfolds. When a translucent egg white turns opaque, we know that the

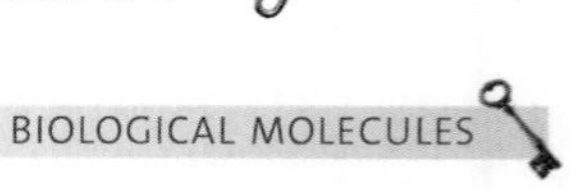

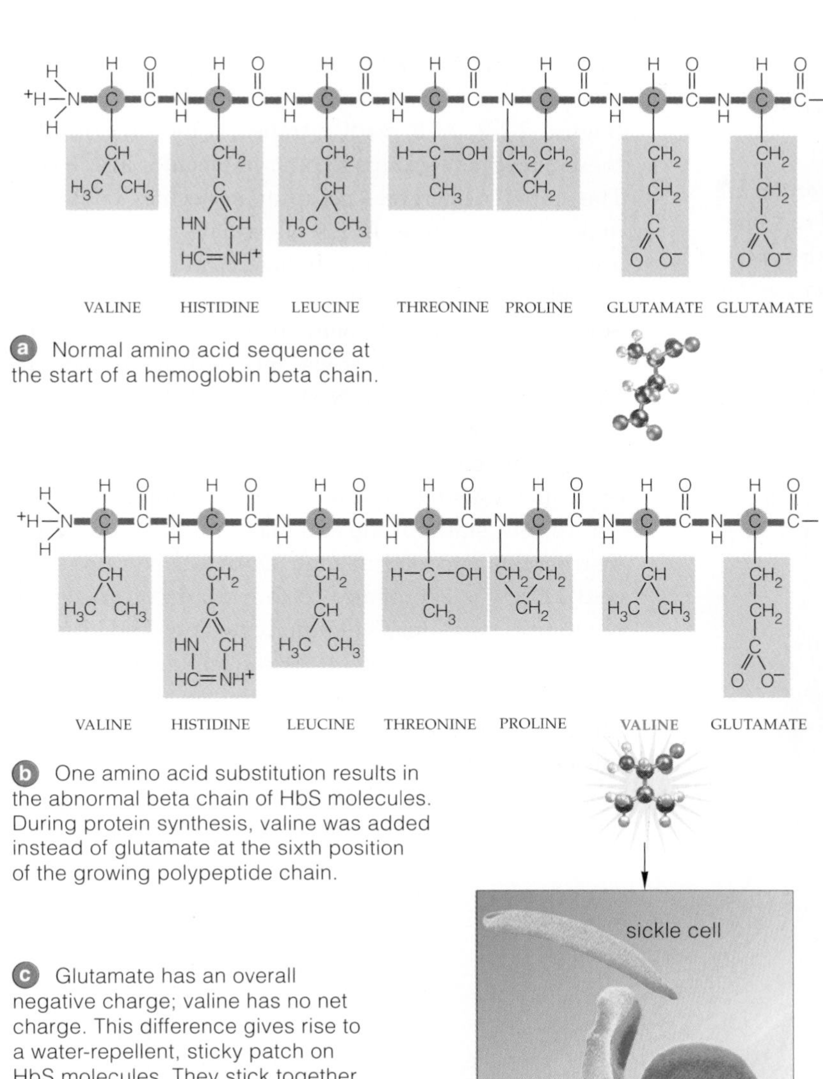

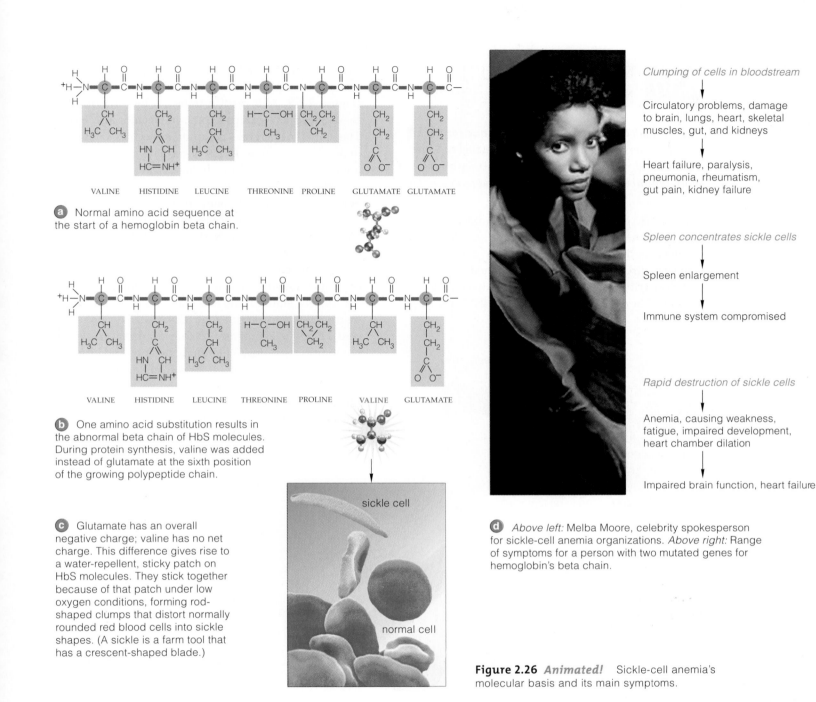

(a) Normal amino acid sequence at the start of a hemoglobin beta chain.

(b) One amino acid substitution results in the abnormal beta chain of HbS molecules. During protein synthesis, valine was added instead of glutamate at the sixth position of the growing polypeptide chain.

(c) Glutamate has an overall negative charge; valine has no net charge. This difference gives rise to a water-repellent, sticky patch on HbS molecules. They stick together because of that patch under low oxygen conditions, forming rod-shaped clumps that distort normally rounded red blood cells into sickle shapes. (A sickle is a farm tool that has a crescent-shaped blade.)

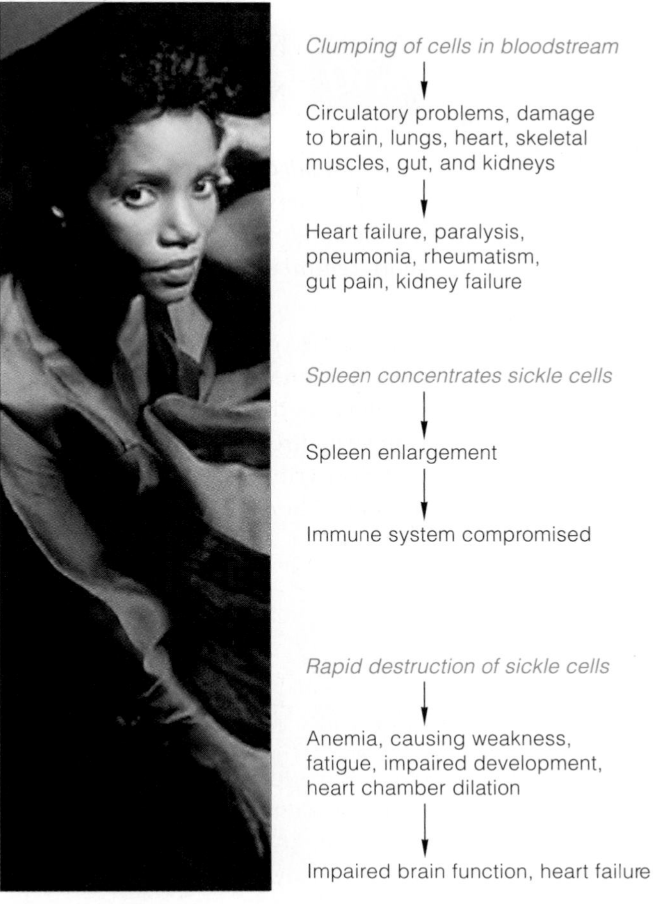

(d) *Above left:* Melba Moore, celebrity spokesperson for sickle-cell anemia organizations. *Above right:* Range of symptoms for a person with two mutated genes for hemoglobin's beta chain.

Figure 2.26 ***Animated!*** Sickle-cell anemia's molecular basis and its main symptoms.

albumin is denatured. For a few proteins, denaturation is reversible if normal conditions return, but albumin is not one of them. There is no way to uncook an egg.

So what is the big take-home lesson? Hemoglobin, hormones, enzymes, transporters—these are the kinds of proteins that help us survive. Twists and folds in their polypeptide chains form anchors, or membrane-spanning barrels, or jaws that can grip enemy agents in the body. Mutation can alter the chains enough to block or enhance an anchoring, transport, or defensive function. Sometimes the consequences are awful. Yet changes in the sequences and functional domains also give rise to variation in traits—the raw material of evolution. *Learn about protein structure and function, and you are on your way to comprehending life in its richly normal and abnormal expressions.*

The structure of proteins dictates function. Mutations that alter a protein's structure can also alter its function, with potentially drastic consequences for an individual.

2.10 Nucleotides and the Nucleic Acids

*Certain small organic compounds called **nucleotides** are energy carriers, enzyme helpers, and messengers. Some are the building blocks for DNA and RNA. They are, in short, central to metabolism, survival, and reproduction.*

Nucleotides consist of one sugar, one nitrogen-containing base, and one or more phosphate groups. Deoxyribose or ribose is the sugar. Both sugars have a five-carbon ring structure, but ribose has an attached oxygen atom and deoxyribose does not. The bases have a single or double carbon ring structure (Figures 2.27 and 2.28).

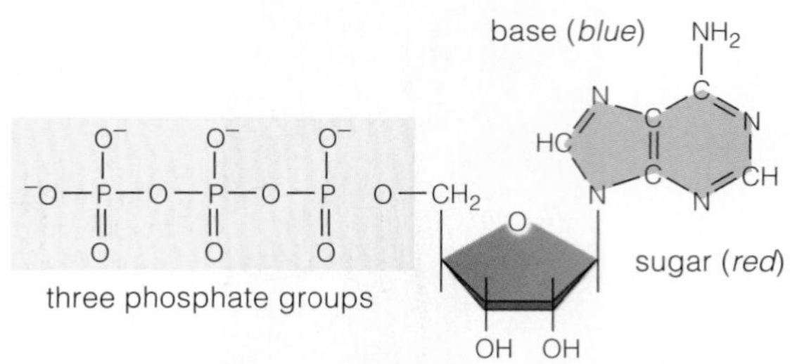

Figure 2.27 Structural formula for ATP.

The nucleotide **ATP** (adenosine triphosphate) has a row of three phosphate groups attached to its sugar (Figure 2.27). ATP readily transfers the outermost of these phosphate groups to other molecules inside cells. The transfer energizes acceptor molecules enough to make them reactive. ATP's phosphate-group transfers are crucial aspects of metabolism.

Other nucleotides have different metabolic roles. Some are subunits of **coenzymes**, or enzyme helpers. Coenzymes move electrons and hydrogen from one reaction site to another. NAD^+ and FAD are examples.

Still other nucleotides act as chemical messengers within and between cells. Later, you will read about one called cAMP (cyclic adenosine monophosphate).

All cells start out life and maintain themselves with instructions encoded in their **DNA** (deoxyribonucleic acid). DNA is a polymer of four kinds of nucleotide monomers: adenine, guanine, cytosine, and thymine (Figure 2.28*a*). It is one of the **nucleic acids** that store and retrieve heritable information in all cells. In all nucleic acids, a covalent bond joins the sugar of one nucleotide monomer and the phosphate group of the next (Figure 2.28*b*).

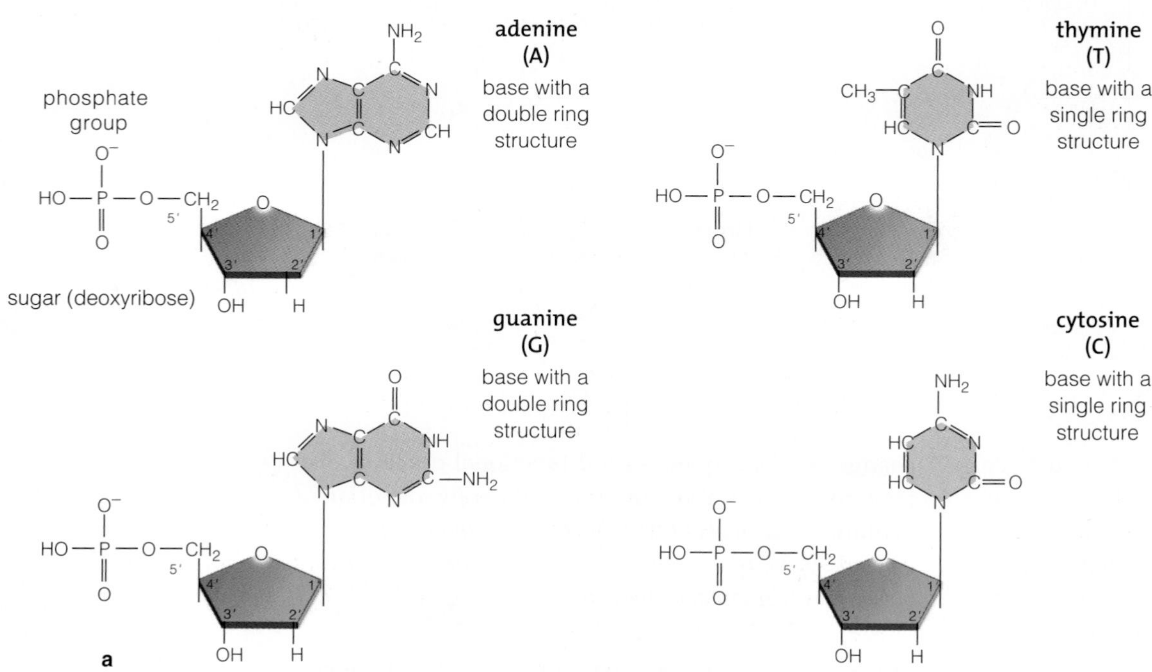

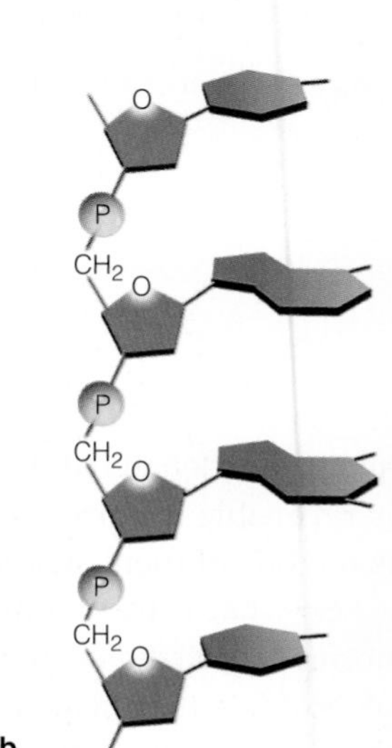

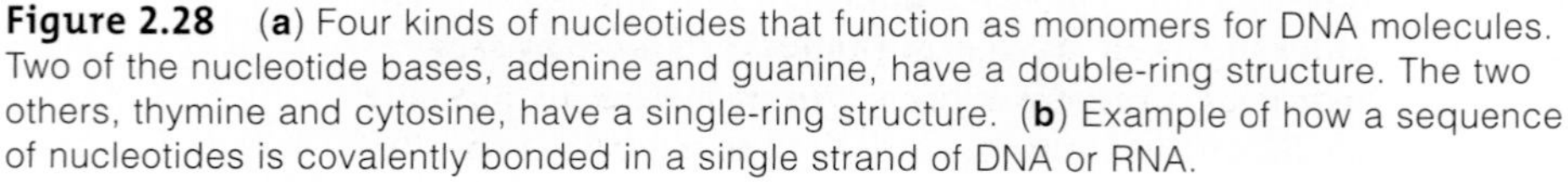

Figure 2.28 (**a**) Four kinds of nucleotides that function as monomers for DNA molecules. Two of the nucleotide bases, adenine and guanine, have a double-ring structure. The two others, thymine and cytosine, have a single-ring structure. (**b**) Example of how a sequence of nucleotides is covalently bonded in a single strand of DNA or RNA.

Figure 2.29 shows how hydrogen bonds between bases join two strands of DNA along the length of a molecule. Think of every "base pairing" as one rung of a ladder, and the two sugar–phosphate backbones as the ladder's two posts. The ladder twists and turns in a regular pattern, forming a double helical coil.

The sequence of bases in DNA encodes heritable information about how to build all of the proteins that give each cell the potential to grow, maintain itself, and reproduce. Part of the sequence is unique for each species. Some parts are identical, or nearly so, among many species. We will return to DNA in Chapter 9.

Like DNA, the **RNAs** (ribonucleic acids) consist of four sugar–phosphate monomers—adenine, guanine, cytosine, and uracil (not thymine). Unlike DNA, most RNAs are single stranded. In eukaryotes, one type of RNA carries DNA's protein-building instructions from the nucleus to the cytoplasm. There, other RNAs build proteins using the instructions.

Ever since life originated, RNA has been central in all processes of retrieving and using the hereditary information stored in DNA. We return to RNA and its role in protein synthesis in Chapter 10.

Small organic compounds called nucleotides serve as coenzymes, subunits of nucleic acids, energy carriers, and chemical messengers.

The nucleic acid DNA consists of two nucleotide strands joined by hydrogen bonds and twisted as a double helix. Its nucleotide sequence contains information about how to build all of a cell's proteins.

RNA is a single-stranded nucleic acid. Different RNAs have roles in the processes by which a cell retrieves and uses genetic information in DNA to build proteins.

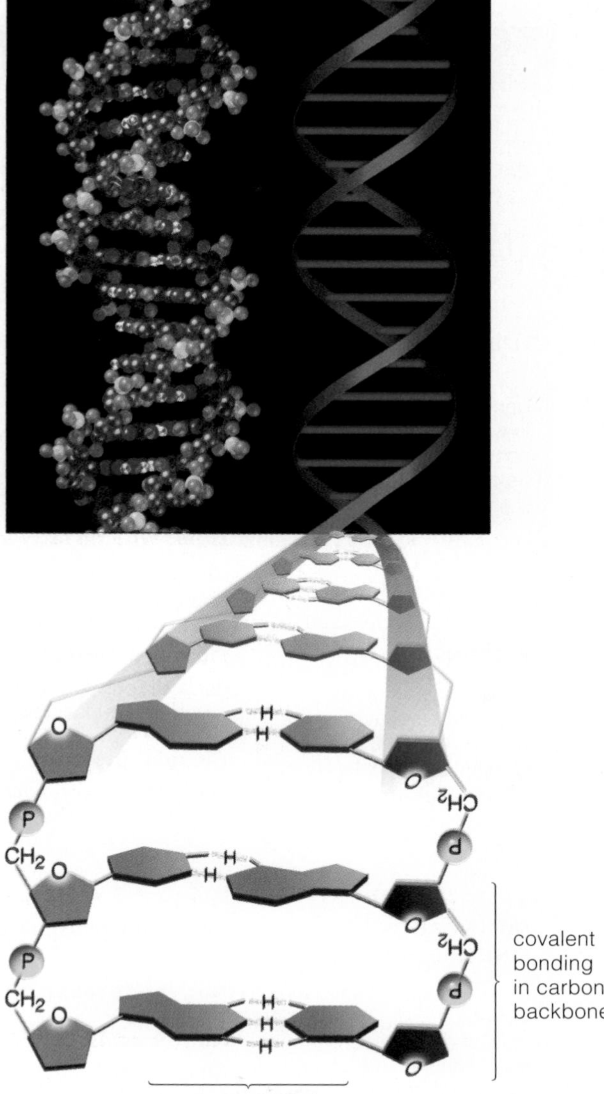

Figure 2.29 Models for the DNA molecule, with its two strands of nucleotides hydrogen-bonded at their bases and twisted into a double helix.

Summary

Section 2.1 All substances consist of elements, each of which consists of one kind of atom. Atoms are fundamental forms of matter that have mass, take up space, and cannot be broken down by ordinary means like heat.

Each atom consists of positively charged protons, negatively charged electrons, and (except for hydrogen) uncharged neutrons. Protons and neutrons occupy the core region, or nucleus, of the atom.

Section 2.2 Whether a given atom will interact with others depends on the number and arrangement of its electrons. Atoms with electron vacancies in the highest energy level (outermost shell) tend to donate, accept, or share electrons with other atoms. When they do so, they become united in a chemical bond.

Atoms that lose or gain electrons are ions. In an ionic bond, ions of opposite charge attract each other and stay together. In a covalent bond, atoms share one or more pairs of electrons. Such a bond is polar when electrons are not shared equally and nonpolar when sharing is equal. Hydrogen bonds can form between a hydrogen atom and a different atom when both are already taking part in polar covalent bonds.

Biology Now

Compare the types of bonds in biological molecules with the animation on Biology Now.

Section 2.3 Polar covalent bonds join two hydrogen atoms to one oxygen atom in each water molecule. The water molecule's polarity invites extensive hydrogen bonding between water molecules in liquid water and ice. Extensive hydrogen bonding is the basis of water's

Table 2.1 Summary of the Main Organic Compounds in Living Things

Category	Main Subcategories	Some Examples and Their Functions	
CARBOHYDRATES . . . contain an aldehyde or a ketone group, and one or more hydroxyl groups	**Monosaccharides** (simple sugars)	Glucose	Energy source
	Oligosaccharides (short-chain carbohydrates)	Sucrose (a disaccharide)	Most common form of sugar; the form transported through plants
	Polysaccharides (complex carbohydrates)	Starch, glycogen Cellulose	Energy storage Structural roles
LIPIDS . . . are mainly hydrocarbon; generally do not dissolve in water but do dissolve in nonpolar substances, such as other lipids	**Lipids with fatty acids**		
	Glycerides: Glycerol backbone with one, two, or three fatty acid tails	Fats (e.g., butter), oils (e.g., corn oil)	Energy storage
	Phospholipids: Glycerol backbone, phosphate group, one other polar group, and (often) two fatty acids	Phosphatidylcholine	Key component of cell membranes
	Waxes: Alcohol with long-chain fatty acid tails	Waxes in cutin	Conservation of water in plants
	Lipids with no fatty acids		
	Sterols: Four carbon rings; the number, position, and type of functional groups differ among sterols	Cholesterol	Component of animal cell membranes; precursor of many steroids and vitamin D
PROTEINS . . . are one or more polypeptide chains, each with as many as several thousand covalently linked amino acids	**Fibrous proteins**		
	Long strands or sheets of polypeptide chains; often tough, water-insoluble	Keratin Collagen	Structural component of hair, nails Structural component of bone
	Globular proteins		
	One or more polypeptide chains folded into globular shapes; many roles in cell activities	Enzymes Hemoglobin Insulin Antibodies	Great increase in rates of reactions Oxygen transport Control of glucose metabolism Tissue defense
NUCLEIC ACIDS (AND NUCLEOTIDES) . . . are chains of units (or individual units) that each consist of a five-carbon sugar, phosphate, and a nitrogen-containing base	**Adenosine phosphates**	ATP cAMP (Section 20.8)	Energy carrier Messenger in hormone regulation
	Nucleotide coenzymes	NAD^+, $NADP^+$, FAD	Transfer of electrons, protons (H^+) from one reaction site to another
	Nucleic acids		
	Chains of thousands to millions of nucleotides	DNA, RNAs	Storage, transmission, translation of genetic information

unique properties: its ability to resist temperature changes, display internal cohesion, and easily dissolve polar or ionic substances. These properties of water make life possible.

BiologyNow

Learn how water's structure gives rise to its unique properties with animation on BiologyNow.

Section 2.4 The pH of a solution indicates its H^+ (hydrogen ion) concentration. Acids have a pH that is less than 7.0; bases have a pH greater than 7.0. Buffer systems maintain pH of body fluids; most biological processes operate only within a narrow range of pH.

BiologyNow

Discover the pH of common substances with the interaction on BiologyNow.

Section 2.5 Organic compounds consist primarily of carbon and hydrogen atoms. Carbon atoms bond covalently with up to four other atoms, often forming long chains or rings. Functional groups attached to the carbon backbone contribute to an organic compound's chemical properties. Enzyme-driven reactions construct carbohydrates, proteins, lipids, and nucleic acids from smaller organic subunits (Table 2.1).

BiologyNow

Investigate the properties of functional groups with the interaction on BiologyNow.

Section 2.6 Carbohydrates include simple sugars, oligosaccharides, and polysaccharides. Living cells use carbohydrates as instant energy sources, transportable or storage forms of energy, and structural materials.

BiologyNow

See how sucrose is formed with animation on BiologyNow.

Section 2.7 Lipids are greasy or oily compounds that tend not to dissolve in water but dissolve easily in nonpolar compounds, such as other lipids. Neutral fats (triglycerides), phospholipids, waxes, and sterols are lipids. Cells use lipids as major sources of energy and as structural materials.

Biology Now
See a triglyceride form with animation on BiologyNow.

Sections 2.8, 2.9 A sequence of amino acids (a polypeptide chain) is a protein's primary structure. At higher levels of structural organization, polypeptide chains twist, coil, and bend into functional domains. Many proteins consist of two or more chains. Changes in a protein's structure may alter its function.

Biology Now
Explore amino acid structure and learn about peptide bond formation with the animation on BiologyNow.

Section 2.10 ATP and other nucleotides have vital roles in metabolism. DNA and RNA are nucleic acids, each composed of four kinds of nucleotide subunits. The sequence of nucleotide bases in DNA encodes instructions for how to construct all of a cell's proteins. Different kinds of RNA molecules access and use this information to build proteins.

Self-Quiz

Answers in Appendix I

1. Is this statement true or false: Every atom consists of protons, neutrons, and electrons.

2. Atoms share electrons unequally in a(n) ______ bond.
   a. ionic c. polar covalent
   b. nonpolar covalent d. hydrogen

3. Liquid water shows ______ .
   a. polarity d. cohesion
   b. hydrogen-bonding capacity e. b through d
   c. notable heat resistance f. all of the above

4. Hydrogen ions (H^+) are ______ .
   a. the basis of pH values c. dissolved in blood
   b. targets of certain buffers d. all of the above

5. Name the molecules of life and the families of small organic compounds from which they are built.

6. Which of the following is a class of molecules that encompasses all of the other molecules listed?
   a. triglycerides c. waxes e. lipids
   b. fatty acids d. sterols f. phospholipids

7. Each carbon atom can share pairs of electrons with as many as ______ other atoms.
   a. one b. two c. three d. four

8. ______ is a simple sugar (a monosaccharide).
   a. Glucose c. Ribose e. both a and b
   b. Sucrose d. Chitin f. both a and c

9. Unlike saturated fats, the fatty acid tails of unsaturated fats incorporate one or more ______ .
   a. single covalent bonds b. double covalent bonds

10. ______ are to proteins as ______ are to nucleic acids.
    a. Sugars; lipids c. Amino acids; hydrogen bonds
    b. Sugars; proteins d. Amino acids; nucleotides

11. A denatured protein has lost its ______ .
    a. hydrogen bonds c. function
    b. shape d. all of the above

Figure 2.30 Lacking most basic tools and practices of modern scientists, medieval alchemists intertwined their scientific endeavors with mysticism and astrology.

12. Match the terms with their most suitable description.

____	salt	a. long chain of amino acids
____	protein	b. atomic nucleus component
____	trace element	c. releases ions other than H^+ and OH^- when dissolved in water
____	hydrophilic	d. dissolves easily in water
____	covalent bond	e. double strands of nucleotides
____	proton	f. electrons shared between atoms
____	DNA	g. makes up less than 0.001 percent of body weight

Additional questions are available on **Biology Now™**

Critical Thinking

1. By weight, oxygen is the most abundant element in organisms, ocean water, and Earth's crust. Predict which element is the most abundant in the whole universe.

2. Medieval scientists and philosophers called alchemists (Figure 2.30) were predecessors of modern-day chemists. Many of them tried to transform lead (82 protons) into gold (79 protons). Explain why they never could have possibly succeeded.

3. David, an inquisitive three-year-old, poked his fingers into warm water in a metal pan on the stove and did not sense anything hot. Then he touched the pan itself and got a nasty burn. Explain why.

4. In the following list, identify which is the carbohydrate, the fatty acid, the amino acid, and the polypeptide:
   a. $^+NH_3$—CHR—COO^- c. $(glycine)_{20}$
   b. $C_6H_{12}O_6$ d. $CH_3(CH_2)_{16}COOH$

5. A clerk in a health-food store tells you "natural" vitamin C extracts from rose hips are better than synthetic tablets of this vitamin. Given what you know about the structure of organic compounds, what would be your response? How would you design an experiment to test whether a natural and synthetic version of a vitamin differ?

6. It seems there are "good" and "bad" unsaturated fats. The double bond in *cis* fatty acid tails gives them a rigid kink. A *trans* fatty acid tail is not kinked; it can straighten.
   Some *trans* fatty acids occur naturally in beef. But most form by processes that solidify vegetable oils for margarine and shortening, synthetic substances that are widely used in prepared foods (such as cookies) and in french fries and other fast-food products. *Trans* fatty acids are linked to high levels of LDL, a "bad" form of cholesterol that can set the stage for a heart attack. Speculate on why your body might have an easier time dealing with *cis* fatty acids than *trans* fatty acids.

3 HOW CELLS ARE PUT TOGETHER

IMPACTS, ISSUES

Animalcules and Cells Fill'd With Juices

Do you ever think of yourself as a fraction of a kilometer tall? Probably not. Yet that is how we measure cells—in micrometers, or thousandths of a millimeter, which is a thousandth of a meter, which is a thousandth of a kilometer. The bacterial cells atop the pin at right are a few micrometers "tall."

Before those cells were fixed on the head of a pin so someone could take their picture, they were living members of the most ancient lineage, the prokaryotes. Prokaryotes are the simplest of cells, structurally. Their DNA is not housed in a nucleus. Your cells, by contrast, are eukaryotic. A nucleus evolved in their single-celled ancestors, along with other internal compartments that keep metabolic activities organized.

Nearly all cells are invisible to the naked eye. No one knew about them until the seventeenth century, when the first microscopes were made. The instruments were not very complicated. Galileo Galilei, for instance, used two glass lenses inside a cylinder, but the arrangement was good enough to reveal details of an insect's eyes.

At midcentury, Robert Hooke focused a microscope on thinly sliced cork. He named the small compartments he saw *cellulae*, meaning small rooms—thus the origin of the biological term "cell." Actually, they were dead plant cell walls, but Hooke did not think of them as dead because neither he nor anyone else knew cells could be alive. He observed cells "fill'd with juices" in green plant tissues, but did not realize they were alive, either.

Given the simplicity of their instruments, it is truly amazing that the pioneers in microscopy observed as much as they did. Antoni van Leeuwenhoek, a Dutch shopkeeper, had exceptional skill constructing lenses. By the late 1600s, he was looking at sperm, protists, some bacteria, and "many very small animalcules, the motions of which were very pleasing to behold," in scrapings of tartar from his teeth.

In the 1820s, much improved lenses brought cells into sharper focus. Robert Brown, a botanist, was the first to identify a plant cell nucleus. Later, the botanist Matthias Schleiden wondered whether each plant cell develops as an independent unit even though it is part of the plant. In 1839, the zoologist Theodor Schwann reported that cells have a life of their own even when they are part of a body. The physiologist Rudolf Virchow decided that every cell must come from one that already exists.

Microscopic analysis yielded three generalizations, which together constitute the *cell theory*. First, every organism consists of one or more cells. Second, the cell is the smallest unit of organization that still displays the properties of life. Third, the continuity of life arises directly from the growth and division of single cells.

This chapter introduces the defining features of prokaryotic and eukaryotic cells. It is not meant for memorization, but rather it is an invitation into otherwise invisible worlds.

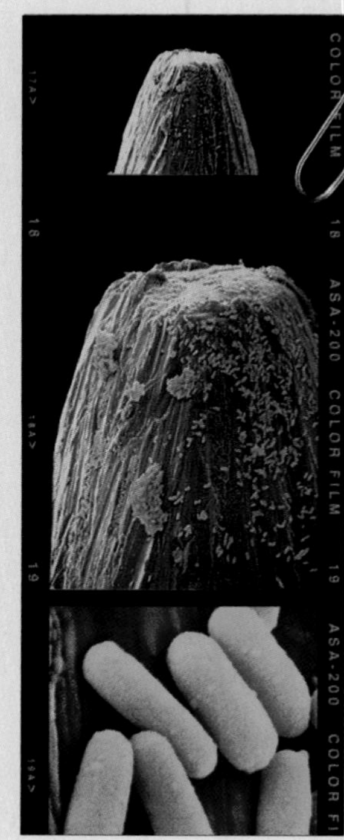

How Would You Vote? *Researchers are in the process of modifying prokaryotes in efforts to find out just how little complexity is required for life. Should this research continue? See BiologyNow for details, then vote online.*

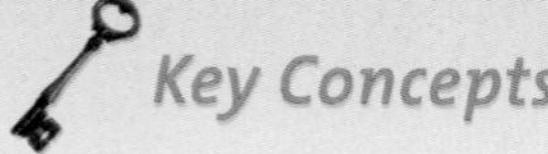

Key Concepts

BASIC CELL FEATURES

Nearly all cells are microscopic in size. They all start out life with a plasma membrane, a semifluid interior called cytoplasm, and an inner region of DNA. The plasma membrane controls the flow of specific substances into and out of cells. Two layers of phospholipid molecules are the structural basis of cell membranes. Proteins in this bilayer and those attached to its surfaces carry out many different membrane functions.

PROKARYOTIC CELLS

Compared to eukaryotic cells, prokaryotes have little internal complexity, and no nucleus. However, when taken as a group, they are the most metabolically diverse organisms. The archaea and bacteria are the only prokaryotes, but all species have prokaryotic ancestors.

EUKARYOTIC CELLS

Eukaryotic cells contain organelles, membrane-enclosed compartments that divide the cell interior into functional regions for specialized tasks. A major organelle, the nucleus, keeps the cell's DNA organized and away from cytoplasmic machinery.

Links to Earlier Concepts

Look back on your road map through the levels of organization in nature (Section 1.1). This chapter arrives at the level of living cells. You will start to see how lipids (2.7) are structurally organized into cell membranes, where DNA and RNA (2.10) reside inside cells, and where carbohydrates (2.6) are built and broken apart. You will also expand your view of how cell structure and function depend on protein structure and function (2.8–2.9). You will see how the principles of natural selection introduced in Section 1.4 are being used to make hypotheses (1.5) about how eukaryotic cells evolved.

3.1 What Is "A Cell"?

Inside your body and at its moist surfaces, trillions of cells live in interdependency. In northern forests, four-celled structures called pollen grains drift down from pine trees. In scummy pondwater, free-living single cells called bacteria and amoebas move about. How are these cells alike, and how do they differ?

Let's begin with three points of the **cell theory**. First, every organism consists of one or more cells. Second, a **cell** is the smallest unit with the properties of life; it has the capacity for metabolism, controlled responses to the environment, growth, and reproduction. Third, under conditions now prevailing in nature, only living cells give rise to new cells.

Cells differ in size, shape, and activities. Yet all are similar in three respects. They all start out life with a plasma membrane, a region of DNA, and cytoplasm (Figure 3.1).

1. **Plasma membrane**. This thin, outermost membrane separates each cell from its outside environment. The membrane lets internal metabolic reactions proceed in controllable ways, but does not isolate the cell interior. It is like a house with many windows and doors that do not open for just anyone. Water and gases cross in and out freely, but the flow of other substances across the plasma membrane is controlled.

2. **Nucleus** or **nucleoid**. In every eukaryotic cell, DNA occupies a membrane-enclosed sac called the nucleus. Inside prokaryotic cells, DNA occupies a region with no membrane. This region is called the nucleoid.

3. **Cytoplasm**. Cytoplasm is everything in between the plasma membrane and the region of DNA. It consists of a semifluid matrix and other components, including **ribosomes** (structures on which proteins are built).

Structurally, prokaryotes are the simplest of cells; nothing separates their DNA from the cytoplasm. Only the archaea and bacteria are prokaryotic cells. All other organisms—from amoebas and peach trees to puffball mushrooms and elephants—are eukaryotic. Eukaryotes have internal membranes dividing the cytoplasm into functional compartments. The compartment called the nucleus is their key defining feature.

> *All living cells have an outer plasma membrane, cytoplasm, and an internal region that contains their DNA.*
>
> *Bacteria and archaea are prokaryotic cells. Unlike eukaryotic cells, they do not have an abundance of organelles, and they do not have a nucleus.*
>
> *All other cells are eukaryotic. They have internal membranes that form organelles, particularly a nucleus.*

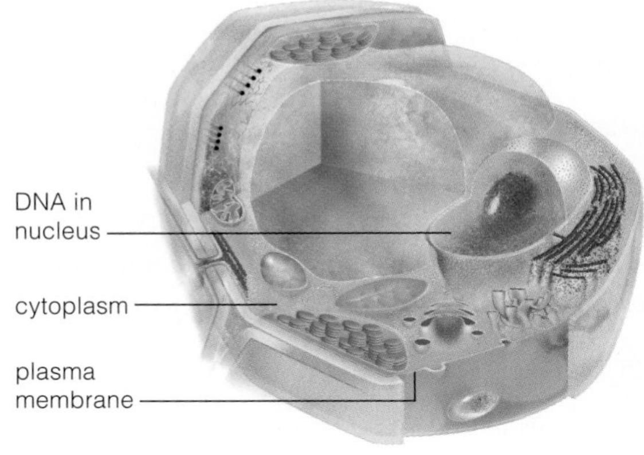

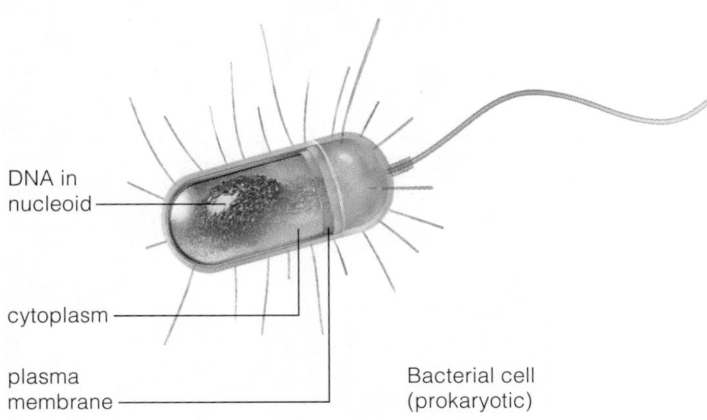

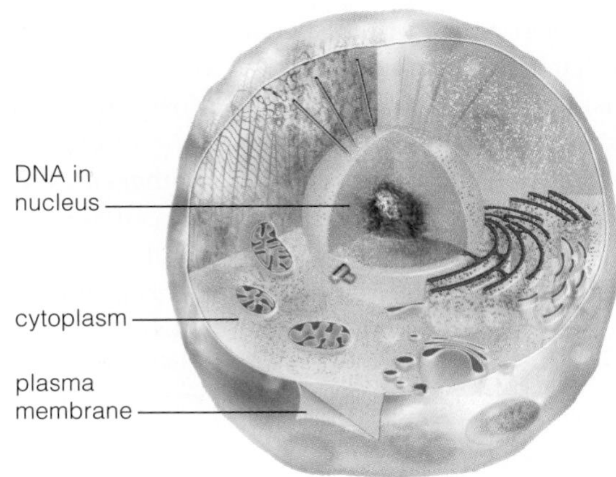

Figure 3.1 Overview of the general organization of prokaryotic cells and eukaryotic cells. The three cells are not drawn to the same scale.

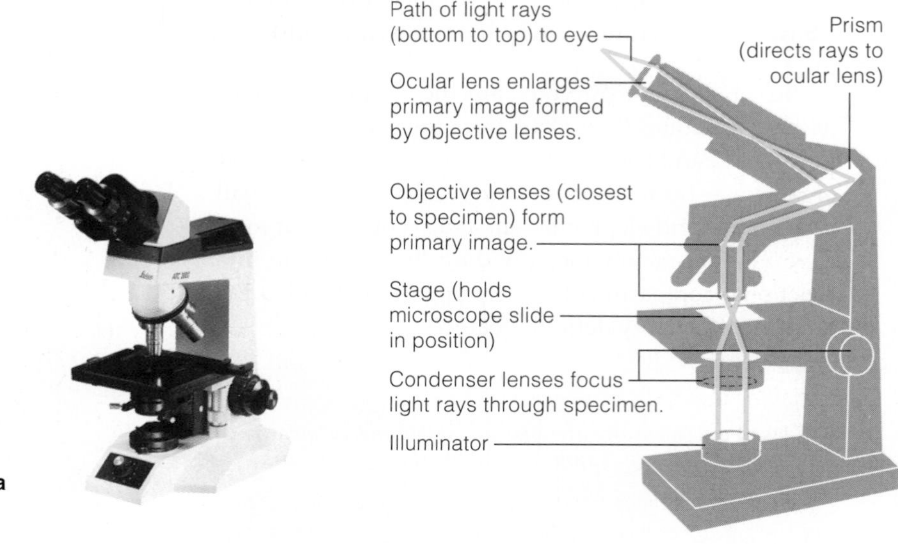

Figure 3.2 ***Animated!*** Different types of microscopes. (**a**) Compound light microscope. (**b**) Electron microscope.

LINK TO SECTION 2.3

3.2 Most Cells Are *Really Small*

Few cells can be seen without the aid of a microscope, because physical constraints restrict their size.

TYPES OF MICROSCOPES

Like the earlier instruments, some microscopes still use visible light to illuminate objects. Light passes through a cell or other specimen, and then curved glass lenses inside the microscope bend the light and focus it into an enlarged image of the specimen (Figure 3.2). Only cells that are thin enough for light to pass through will be visible with a light microscope. Most cells will look uniformly dense unless they are *stained*, or exposed to dyes that some of their parts but not others soak up.

With the best light microscope, you can distinguish details or objects 200 billionths of a millimeter across (Figure 3.3*a,b*). Electron microscopes bring into focus objects 100,000 times smaller than that.

Electron microscopes use magnetic lenses to bend beams of electrons. A *transmission* electron microscope projects electrons through a sample and focuses them into an image of the specimen's internal components. In *scanning* electron microscopes, a beam of electrons bounces off a specimen, and is then converted into an image of the specimen's surface. Images of specimens magnified under a microscope are called micrographs. Those taken with electron microscopes have fantastic detail (Figure 3.3*c,d*).

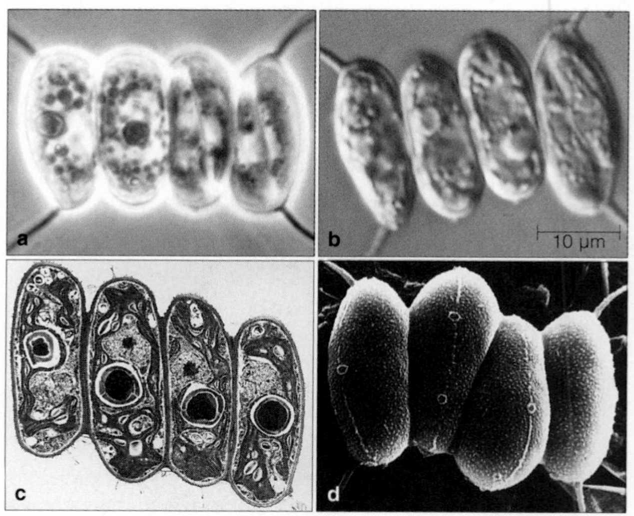

Figure 3.3 How different microscopes reveal different aspects of the same organism, a type of green algae. All four images are at the same magnification. (**a**,**b**) Light micrographs. (**c**) Transmission electron micrograph. (**d**) Scanning electron micrograph. A horizontal bar on or near a micrograph provides a size reference. A micrometer (µm) is a millionth of a meter.

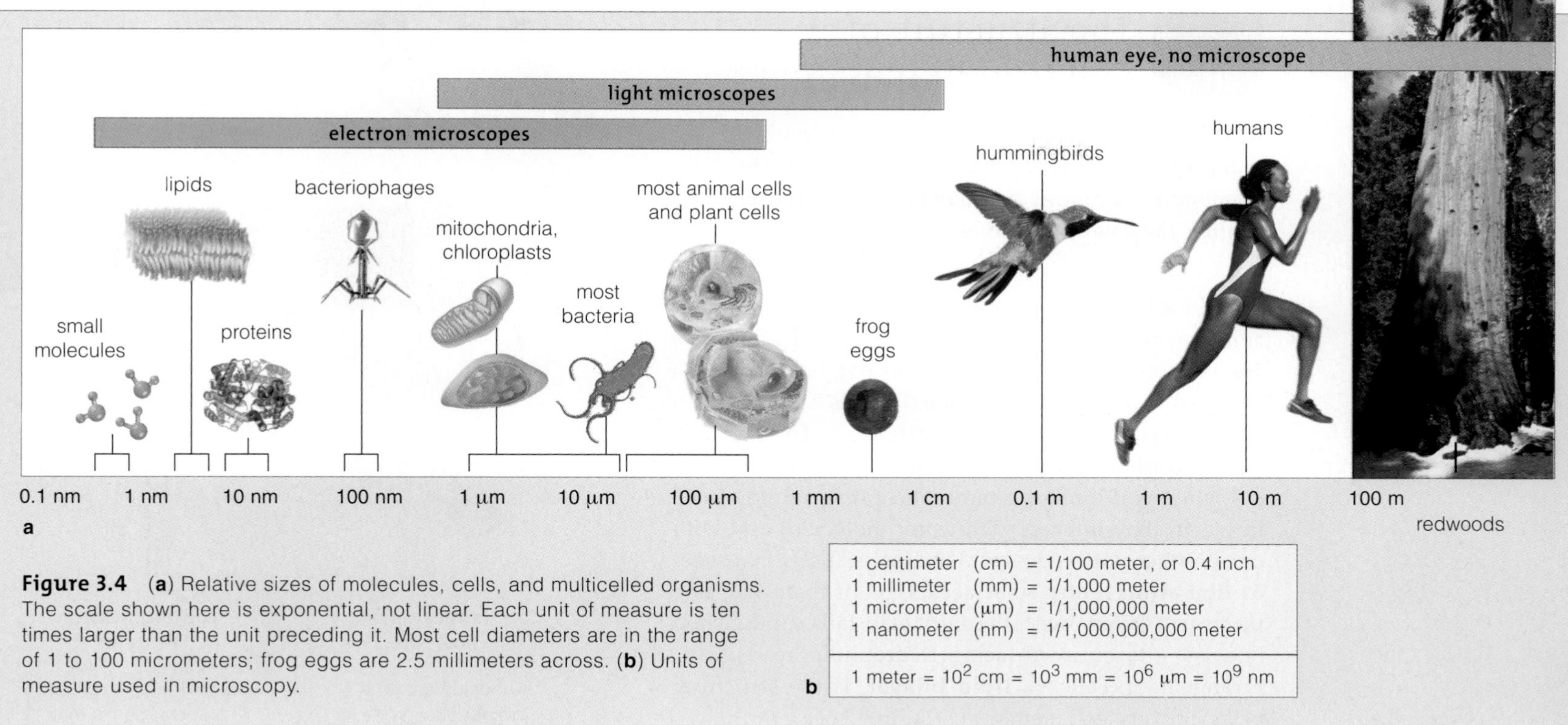

Figure 3.4 (**a**) Relative sizes of molecules, cells, and multicelled organisms. The scale shown here is exponential, not linear. Each unit of measure is ten times larger than the unit preceding it. Most cell diameters are in the range of 1 to 100 micrometers; frog eggs are 2.5 millimeters across. (**b**) Units of measure used in microscopy.

1 centimeter (cm)	= 1/100 meter, or 0.4 inch
1 millimeter (mm)	= 1/1,000 meter
1 micrometer (µm)	= 1/1,000,000 meter
1 nanometer (nm)	= 1/1,000,000,000 meter

1 meter = 10^2 cm = 10^3 mm = 10^6 µm = 10^9 nm

WHY AREN'T ALL CELLS BIG?

A few types of cells can be seen without a microscope. These cells include yolks of bird eggs, cells in the red part of watermelons, and fish eggs. These cells can get big because they are not doing much, metabolically, at maturity. Most metabolically active cells are too tiny to be seen by the unaided eye (Figure 3.4).

So why aren't all cells big? A physical relationship called the **surface-to-volume ratio** constrains increases in cell size. By this relationship, any object's volume increases with the cube of its diameter, but the surface area increases only with the square.

Apply this to a round cell. As Figure 3.5 shows, *if a cell expands in diameter during growth, then its volume will increase faster than its surface area will*. Suppose you make a round cell grow four times wider than normal. Its volume increases by 64 times, but the surface area increases by only 16 times. This means that each unit of plasma membrane must service four times as much cytoplasm as before. If a cell's diameter becomes too great, the inward flow of nutrients and the outward flow of wastes will not be able to keep up with its metabolic activity.

A large, round cell also would have trouble moving materials through its cytoplasm. Random motions of molecules can distribute materials through tiny cells. If a cell is not tiny, you can expect it to be long or thin, or to have outfoldings or infoldings that increase its surface relative to its volume. *The smaller or narrower or more frilly-surfaced the cell, the more efficiently materials cross its surface and become distributed through the interior.*

Surface-to-volume constraints also shape the body plans of multicelled species. For example, small cells attach end to end in strandlike algae, so that each one interacts directly with its surroundings.

diameter (cm):	0.5	1.0	1.5
surface area (cm^2):	0.79	3.14	7.07
volume (cm^3):	0.06	0.52	1.77
surface-to-volume ratio:	13.17:1	6.04:1	3.99:1

Figure 3.5 Example of the surface-to-volume ratio. This physical relationship between increases in volume and surface area restricts the size and the shape of cells.

Most cells are microscopically small. Different types of microscopes reveal cells and their internal structures.

The surface-to-volume ratio constrains increases in cell size. It also influences cell shape and the body plans of multicelled organisms.

LINKS TO SECTIONS 2.7, 2.9

3.3 The Structure of Cell Membranes

All cell membranes consist of a lipid bilayer in which many different kinds of proteins are embedded. The membrane is a continuous boundary that selectively controls the flow of substances across it.

Think back on the phospholipids, the most abundant components of cell membranes (Section 2.7). Each has one phosphate-containing head and two fatty acid tails attached to a glycerol backbone (Figures 2.20 and 3.6*a*). The head is hydrophilic; water attracts it. The two tails are hydrophobic; water repels them.

When you immerse many phospholipid molecules in water, they interact with water molecules and with one another until they spontaneously cluster in a sheet or film at the water's surface. Some of them will line up in two layers, with the fatty acid tails sandwiched between the outward-facing hydrophilic heads. This arrangement, called a **lipid bilayer**, is the structural basis of every cell membrane (Figure 3.6*b*,*c*).

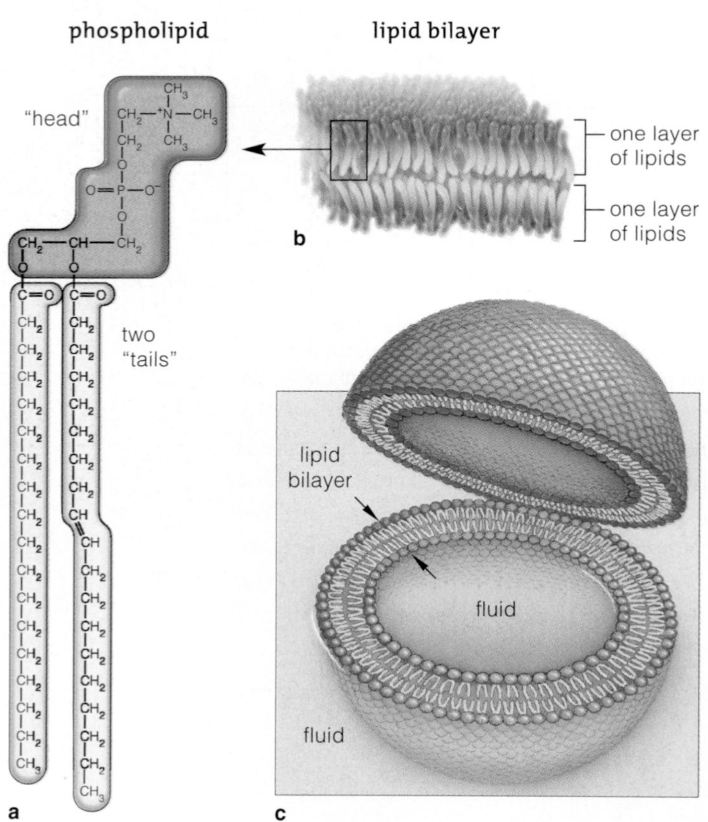

Figure 3.6 Lipid bilayer organization of cell membranes. (**a**) One of the phospholipids, the most abundant membrane components. (**b**) These lipids and others are arranged as two layers. Their hydrophobic tails are sandwiched between their hydrophilic heads in the bilayer. (**c**) Arrangement of a lipid bilayer that forms a cell.

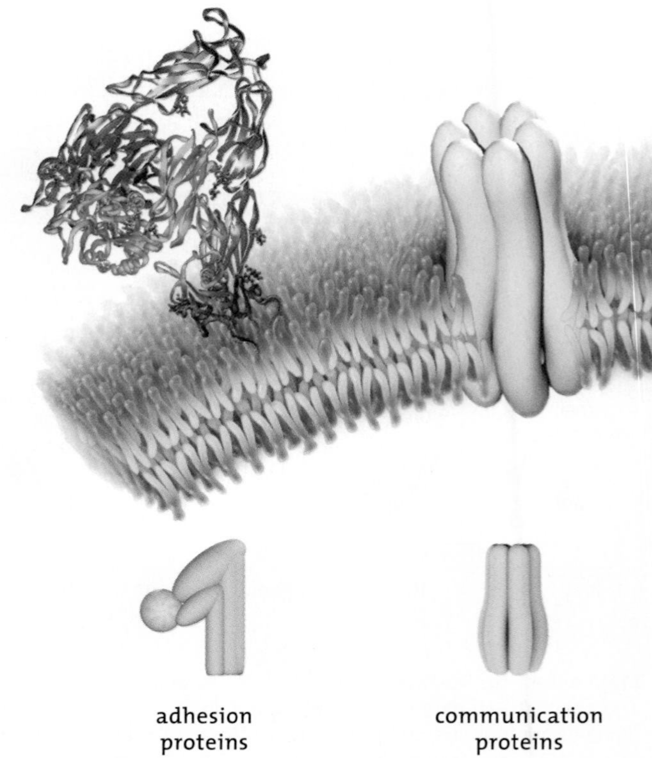

Adhesion proteins project outward from plasma membranes of multicelled species. They help cells stick to one another, or to protein matrixes that are part of tissues.

Communication proteins of adjoining cells match up, forming cytoplasm-to-cytoplasm channels between cells. Signals can flow through these channels. Part of a gap junction between heart muscle cells is shown.

Figure 3.7 ***Animated!*** Part of a plasma membrane, which consists of a lipid bilayer and more proteins than you see here. Later chapters use the icons below these models.

The **fluid mosaic model** describes the organization of membranes (Figure 3.7). A cell membrane is a mixed composition—a *mosaic*—of phospholipids, glycolipids, sterols, and proteins. Membrane phospholipids have different kinds of heads and tails that vary in length and saturation. Unsaturated fatty acids have at least one double covalent bonds in their carbon backbone, and fully saturated fatty acids have none (Section 2.7).

Also by this model, a membrane is *fluid* because of the motions and interactions of its components. Most phospholipids drift sideways, spin on their long axes, and flex their tails, so they do not pack together into a solid layer. Most membrane phospholipids have at least one kinked (unsaturated) fatty acid tail.

Hydrogen bonds and other weak interactions help proteins associate with the phospholipids. Many of the

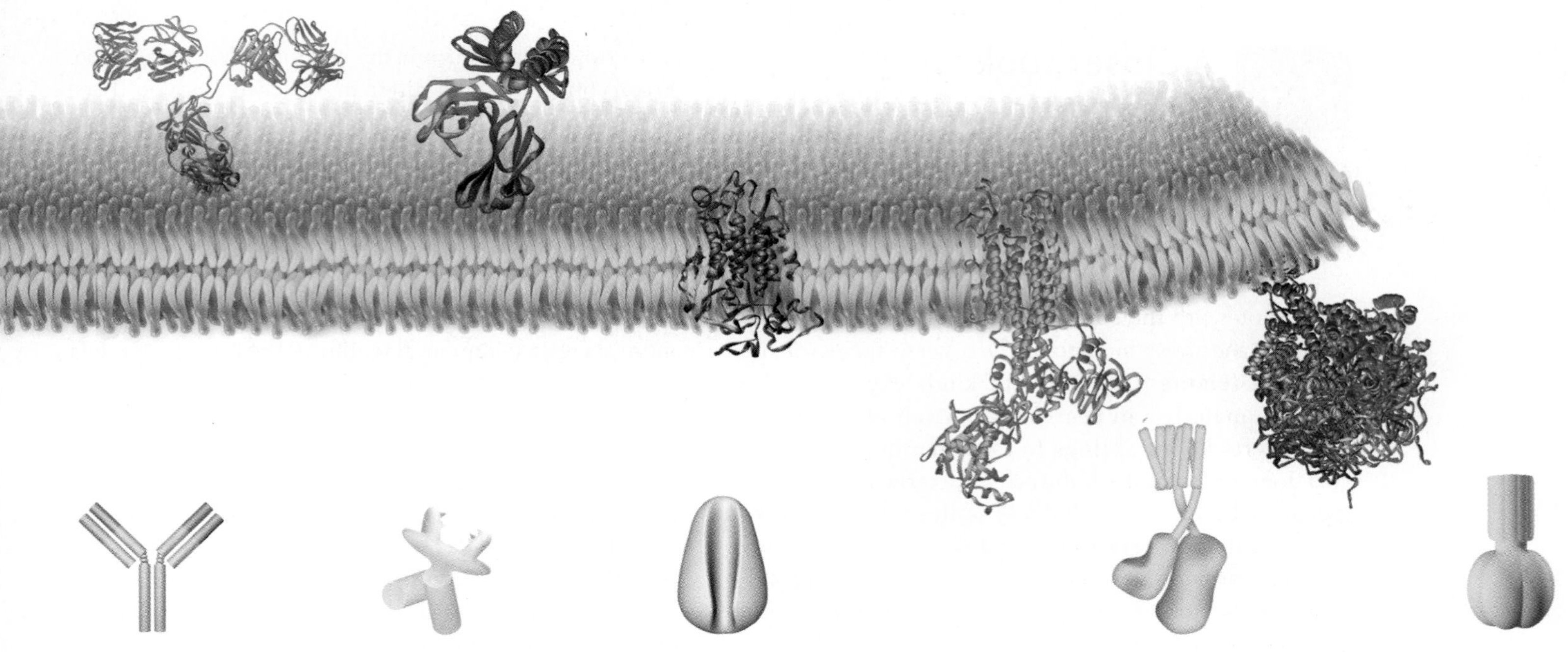
receptor proteins
Receptors are docks for outside substances that serve as signaling molecules. When a receptor binds one of these molecules, the cell will change its activities.
recognition proteins
A plasma membrane's recognition proteins are like molecular fingerprints. They identify each cell as *self* (belonging to one's own body or tissue) or *nonself* (foreign to the body).
passive transporters
All passive transporters are channels across a membrane. Solutes move freely through the channel to the side where they are less concentrated. Some passive transporters are open all the time, but others have molecular gates that open and close in controlled ways.
active transporters
Active transporters are channels across a membrane. They pump solutes through the channel to the side where they are more concentrated. Pumping action requires an energy input.
These two examples are a calcium pump (*blue*) and an ATP synthase (*multicolored*).

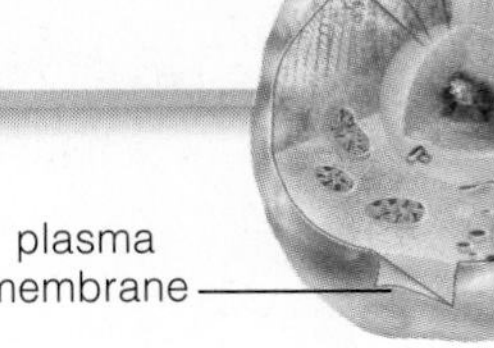
plasma membrane

membrane proteins span the bilayer, with hydrophilic parts extending beyond both of its surfaces. Others are anchored to underlying cell structures.

The lipid bilayer functions mainly as a barrier to water-soluble substances. Proteins carry out nearly all other membrane tasks. Many proteins are receptors for signals. Transport proteins move certain solutes across the bilayer. Some transport proteins require an energy input; others do not. Still other proteins are enzymes that mediate events at the membrane.

Especially among multicelled species, the plasma membrane bristles with diverse proteins. Some kinds identify the cell as belonging to the body, and others help defend it against attacks. Special proteins also help cells communicate with one another or stick together in tissues. Figure 3.7 shows some important categories of membrane proteins and gives a brief description of their functions.

The fluid mosaic model is a good starting point for thinking about cell membranes. But keep in mind that different membranes have different types of molecules. Even one bilayer's two surfaces are not exactly alike. For example, some carbohydrate side chains attached to proteins and lipids project from the cell, but not into it. Differences among plasma membranes and internal membranes reflect the roles they play in cells.

All cell membranes consist of two layers of lipids—mainly phospholipids—and diverse proteins. Hydrophobic parts of the lipids are sandwiched between hydrophilic parts, which face the cytoplasmic fluid or extracellular fluid.

All cell membranes have protein receptors, transporters, and enzymes. The plasma membrane also incorporates adhesion, communication, and recognition proteins.

LINK TO SECTION 1.3

3.4 A Closer Look at Prokaryotic Cells

The word prokaryote means "before the nucleus." The name reminds us that bacteria and archaea originated before cells with a nucleus evolved.

Prokaryotes are the smallest known cells. As a group they are the most metabolically diverse forms of life on Earth (Figure 3.8). Different kinds exploit energy and raw materials in nearly every environment, from dry deserts to hot springs to mountain ice. There are two domains of prokaryotic cells: Bacteria and Archaea (Sections 1.3 and 14.1). Cells of both groups are alike in outward appearance, size, and where they live.

Most prokaryotes have a semirigid or rigid cell wall that surrounds the plasma membrane, supports the cell, and imparts shape to it. Prokaryotic cell walls differ from eukaryotic cell walls, but all of them are permeable to dissolved substances, which are free to move to and away from the cell's plasma membrane. Eukaryotic cell walls differ in structure, as you'll see in Section 3.8. Sticky polysaccharides often surround the prokaryotic cell wall. These help the cell attach to interesting surfaces such as river rocks, teeth, and the vagina. Many disease-causing (or pathogenic) bacteria have a jellylike protective capsule of polysaccharides around their wall.

Projecting from many prokaryotic cells are one or more **flagella** (singular, flagellum). Prokaryotic flagella are motile structures that differ from similar-looking flagella found on some kinds of eukaryotic cells; they rotate like propellers rather than waving side to side. Prokaryotic cells use the flagella to move through fluid environments (Figure 3.8*d*). Other surface projections include pili (singular, pilus). These protein filaments help the cells attach to surfaces and to one another.

Like eukaryotic cells, bacteria and archaea depend on a plasma membrane to selectively control the flow of substances into and out of the cytoplasm. Their lipid bilayer, too, bristles with diverse transporters,

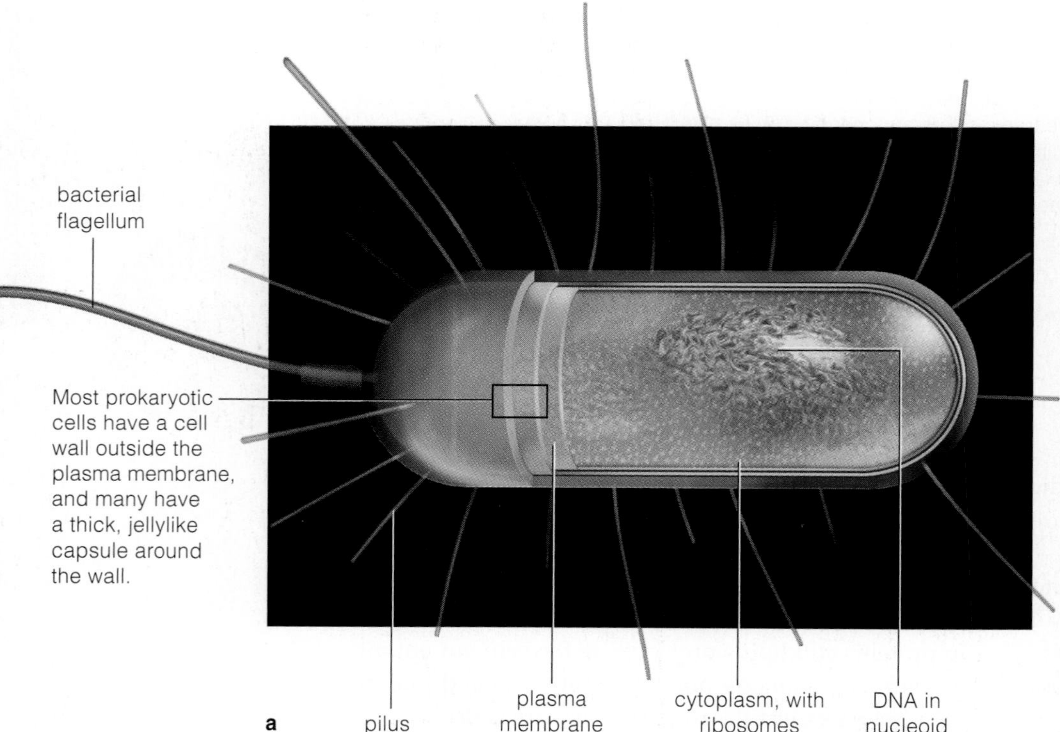

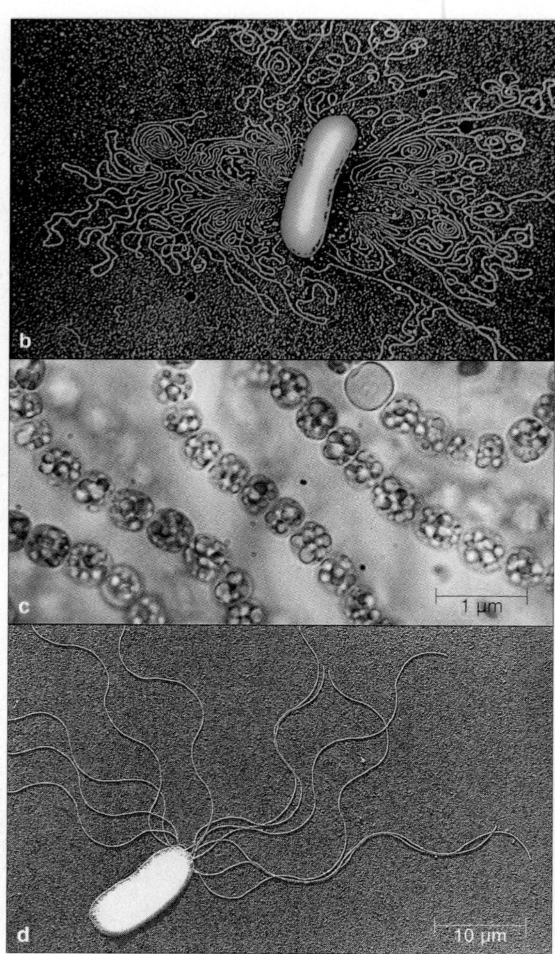

Figure 3.8 (**a**) Generalized sketch of a typical prokaryotic cell.

(**b**) Micrograph of *Escherichia coli.* Researchers manipulated this bacterial cell to release its single, circular molecule of DNA, which looks like tangled string.

(**c**) Cells of various bacterial species are shaped like balls, rods, or corkscrews. Ball-shaped cells of *Nostoc*, a photosynthetic bacterium, stick together in a thick, jellylike sheath of their own secretions.

(**d**) Like this *Pseudomonas marginalis* cell, many species have one or more bacterial flagella that propel the cell in fluid environments.

receptors, and channels. It also incorporates built-in machinery for reactions. For example, photosynthesis proceeds at the plasma membrane of many species of bacteria. Organized arrays of proteins capture light energy and convert it to chemical energy in the form of ATP, which is used to build sugars.

Given their size and simplicity, prokaryotes do not require a complex internal framework. Just under the plasma membrane, arrays of protein filaments in the cytoplasm compose a simple internal "skeleton." This protein scaffold helps shape their sturdy cell wall.

The prokaryotic cytoplasm also contains ribosomes on which polypeptide chains are built, and one DNA molecule in the form of a circle.

Prokaryotes are the simplest cells, but as a group they show the most metabolic diversity.

Bacteria and archaea are different groups of prokaryotic cells. Their DNA is not housed inside a nucleus. Most have a permeable cell wall around their plasma membrane that structurally supports and imparts shape to the cell.

Table 3.1 Common Features of Eukaryotic Cells

Organelles and Their Main Functions	
Nucleus	Localizing and protecting the cell's DNA
Endoplasmic reticulum (ER)	Routing and modifying new polypeptide chains; synthesizing lipids; other tasks
Golgi body	Modifying polypeptide chains into mature proteins; sorting and shipping proteins and lipids for secretion or for use inside cell
Vesicles	Transporting or storing a variety of substances; digesting substances and structures in the cell; other functions
Mitochondria	Making many ATP molecules with energy released from glucose breakdown
Non-Membranous Structures and Their Functions	
Ribosomes	Assembling polypeptide chains
Cytoskeleton	Imparting overall shape and internal organization to the cell; moving the cell and its internal structures

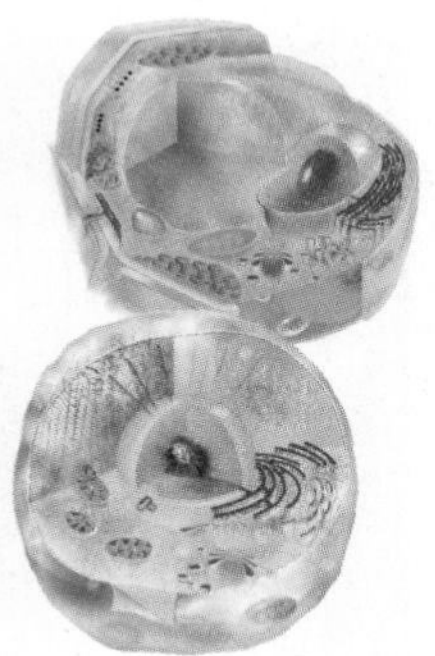

3.5 A Closer Look at Eukaryotic Cells

All cells engage in biosynthesis and energy production, but eukaryotic cells compartmentalize the tasks. Their interior is divided into organelles, each responsible for a specialized function. Ribosomes, a cytoskeleton, and other cell structures support those functions.

Like the prokaryotes, eukaryotic cells have ribosomes in the cytoplasm. Unlike them, eukaryotic cells have an intricate internal skeleton of proteins that is called a cytoskeleton. They also start out life with **organelles**: internal compartments such as the nucleus. *Eu–* means true; *karyon*, meaning kernel, refers to a nucleus. Table 3.1 lists the organelles found in most eukaryotic cells. Some specialized cells may contain additional kinds of organelles and other structures.

What advantages do compartments offer? Each has a semipermeable membrane that selectively controls the types and amounts of substances crossing it. This control maintains the microenvironment inside sacs. Compartments concentrate substances for reactions, isolate toxic or disruptive chemicals, and export others. For instance, sacs called lysosomes contain enzymes that would digest the entire cell if they escaped.

Compartmentalization is quite an organizational challenge for cells. Just as your body's organ systems interact to keep your body running, organelles also must interact to keep a cell running. Substances move from one organelle to another, and to and from the plasma membrane, in controlled ways. As you will see later, diffusion, transport proteins, and motor proteins on tiny tracks are the movers.

Some substances travel throughout the cytoplasm along a pathway that involves a series of organelles. One series acts as a *secretory* pathway. It moves newly formed polypeptide chains from ribosomes through the organelles known as endoplasmic reticulum, then through Golgi bodies, then to the plasma membrane. Another series, the *endocytic* pathway, moves outside substances into the cell.

Substances that move along either pathway travel in tiny vesicles that bud from organelle membranes and the plasma membrane. Proteins on these vesicles help them deliver their cargo to the right destination.

All eukaryotic cells start out life with a nucleus and other organelles, as well as ribosomes and a cytoskeleton.

Specialized cells typically incorporate additional kinds of organelles and other structures.

Organelles physically separate chemical reactions. Organelles assemble, store, and move substances to and from the plasma membrane, or to specific destinations in the cytoplasm.

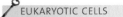

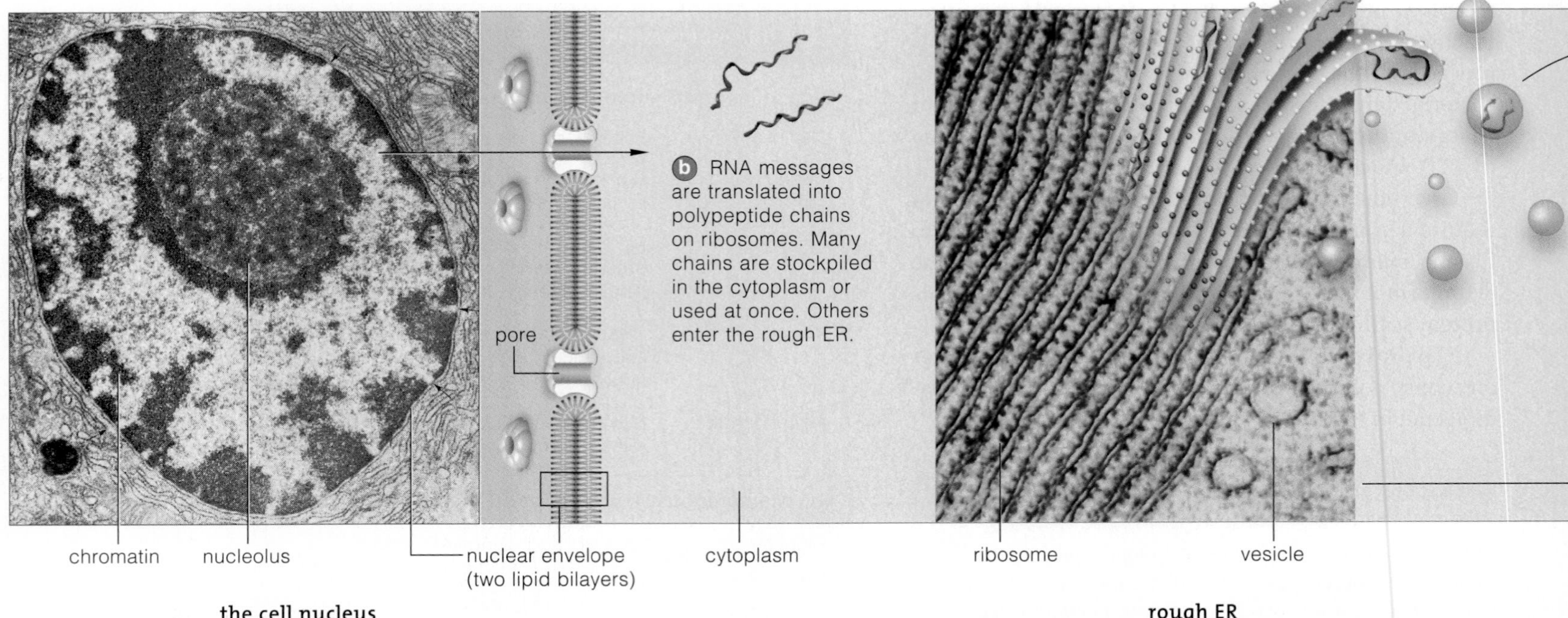

(a) DNA instructions for making proteins are transcribed in the nucleus and moved to the cytoplasm. RNAs are the messengers and protein builders.

(c) Flattened sacs of rough ER form one continuous channel between the nucleus and smooth ER. Polypeptide chains that enter the channel undergo modification. They will be inserted into organelle membranes or secreted from the cell.

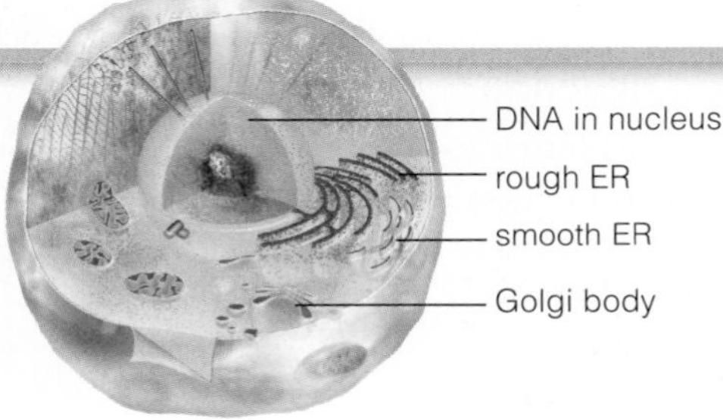

Figure 3.9 ***Animated!*** (**a**) The nucleus, which houses eukaryotic DNA. (**b–d**) Main components of the endomembrane system. Here, many proteins are processed, lipids are assembled, and both products are sorted and shipped to destinations inside the cell or to the plasma membrane for export.

THE NUCLEUS

Unlike prokaryotes, eukaryotic cells have their genetic material distributed among a number of linear DNA molecules of different lengths. The term **chromosome** refers to one double-stranded DNA molecule together with many attached protein molecules. Human body cells have forty-six chromosomes in their nucleus; frog body cells have twenty-six. **Chromatin** is the name for all of the DNA and associated proteins in a nucleus.

The nucleus has two functions. First, it isolates a cell's DNA from potentially damaging reactions in the cytoplasm. This structural separation keeps the DNA molecules organized and makes it easier for the cell to copy them before it divides. Second, it controls access to DNA through the receptors, transport proteins, and pores at its surface. This functional separation helps the cell to control the timing of protein synthesis.

Enclosing the semifluid interior of the nucleus is a **nuclear envelope**: two lipid bilayers studded with pores and transport proteins (Figure 3.9*a*). Each nucleus also has at least one nucleolus (plural, nucleoli). This is a construction site where subunits of ribosomes are built from RNA and proteins. The subunits pass through the nuclear envelope's pores to the cytoplasm (Figure 3.9*b*). Ribosome subunits link up briefly when they assemble amino acids into polypeptide chains (Section 2.8).

THE ENDOMEMBRANE SYSTEM

Many new polypeptide chains enter the flattened sacs and tubes of the **endomembrane system**: the ER, Golgi bodies, and vesicles. Every protein destined for export or for insertion into cell membranes passes through the endomembrane system.

Endoplasmic reticulum, or **ER**, is a dynamic channel of flattened membrane sacs and tubes that starts at the nuclear envelope and extends through the cytoplasm.

Rough ER is studded with ribosomes attached to its outer surface (Figure 3.9*c*). Newly formed polypeptide chains enter it or become inserted into its membrane if (and only if) they contain a special signal sequence of amino acids. Enzymes in the rough ER channel modify polypeptide chains into final form. You can see a lot of rough ER in cells that make, store, and secrete proteins.

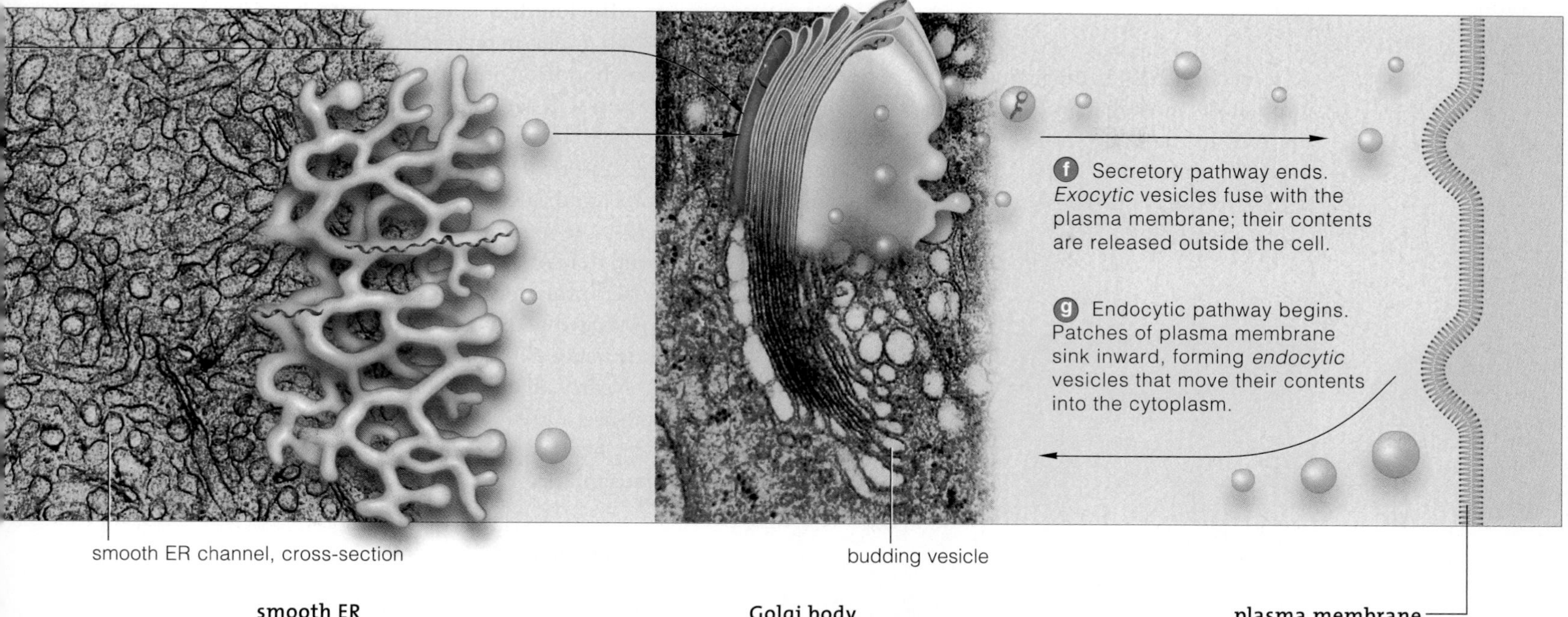
f Secretory pathway ends. *Exocytic* vesicles fuse with the plasma membrane; their contents are released outside the cell.
g Endocytic pathway begins. Patches of plasma membrane sink inward, forming *endocytic* vesicles that move their contents into the cytoplasm.
smooth ER channel, cross-section
budding vesicle
smooth ER
Golgi body
plasma membrane

d Some proteins in the channel continue on, to smooth ER. Many become smooth ER enzymes or membrane proteins. The enzymes make lipids, inactivate toxins, and mediate other tasks.

e A Golgi body receives, processes substances that arrive in vesicles from the ER. New vesicles transport the processed substances to the plasma membrane or other parts of the cell.

h Exocytic vesicles release cell products and wastes to the outside. Endocytic vesicles move nutrients, water, and other substances into the cytoplasm from outside (Section 4.7).

For example, ER-rich gland cells inside your pancreas synthesize and secrete enzymes that end up in your small intestine and help digest meals.

Smooth ER does not have ribosomes (Figure 3.9*d*). It makes lipid molecules that become part of cell membranes. It also takes part in fatty acid breakdown and degrades some kinds of toxins. In muscle cells, a type of smooth ER called sarcoplasmic reticulum has a role in contraction.

Patches of ER membrane bulge and break away as vesicles that deliver polypeptide chains and lipids to **Golgi bodies**. Golgi bodies are folded, flattened sacs that look like stacks of pancakes (Figure 3.9*e*). They attach sugar side chains to some proteins and lipids, and split others. Finished products are packaged into vesicles and then shipped to other regions of the cell, the plasma membrane, or the outside (Figure 3.9*f*).

Vesicles that bud away from the Golgi membranes include the organelles of intracellular digestion. For instance, **lysosomes** contain enzymes that break down carbohydrates, proteins, nucleic acids, and some lipids. Other types of vesicles deliver ingested bacteria, cell parts, and other materials to lysosomes for destruction. Vesicles called **peroxisomes** hold enzymes that digest fatty acids and amino acids. An important function of peroxisomes is the breakdown of hydrogen peroxide, a toxic by-product of fatty acid metabolism. As you will see in the next chapter, enzymes convert hydrogen peroxide to water and oxygen, or use it for reactions that also break down alcohol and other toxins. Drink alcohol, and the peroxisomes of liver and kidney cells normally will degrade nearly half of it.

Some vesicles fuse to form large vacuoles. One of these is the **central vacuole** of a mature plant cell. Ions, amino acids, sugars, and toxic substances accumulate in the fluid-filled interior of a central vacuole, which expands and forces the pliable cell wall to enlarge.

The nucleus keeps eukaryotic DNA molecules organized and separated from the metabolic machinery of the cytoplasm.

In the ER and Golgi bodies of the endomembrane system, polypeptide chains are processed and lipids are assembled.

Vesicles help integrate cell activities. Different kinds store substances or transport them around the cell. Some break down nutrients and toxins.

MITOCHONDRIA

ATP (Section 2.10) is the main energy carrier in cells. It jump-starts most metabolic reactions. How do cells get the ATP in the first place? One way is to break down organic compounds in the ATP-forming reactions of aerobic respiration. These reactions run to completion in **mitochondria** (singular, mitochondrion). A much-folded inner membrane divides the interior of each mitochondrion into two compartments (Figure 3.10). As you will see in Sections 5.2 and 6.3, hydrogen ions released from the breakdown of organic compounds are shuttled across the inner membrane, into the inner mitochondrial compartment. Hydrogen ion flow back across the membrane to the outer compartment drives formation of ATP. Oxygen keeps the reactions going by accepting electrons. Each time you breathe in, you are securing oxygen mainly for the mitochondria in your many trillions of cells.

Nearly all eukaryotic cells have mitochondria. Cells that use a great deal of energy, such as those of your liver, heart, and skeletal muscles, may each contain a thousand or more mitochondria.

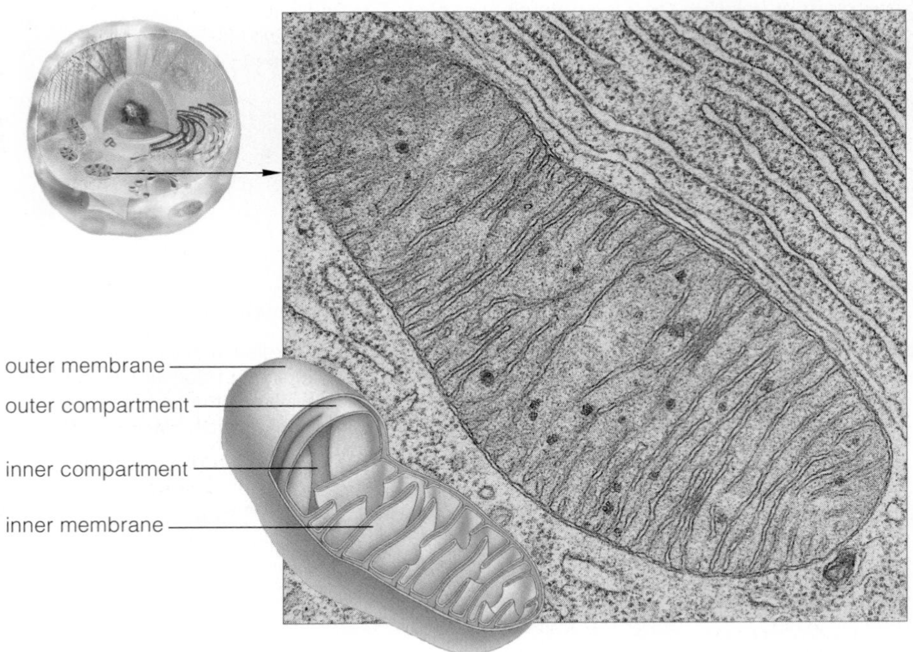

Figure 3.10 Typical mitochondrion. This organelle specializes in producing ATP.

CHLOROPLASTS

Many plant cells have plastids, which are organelles of photosynthesis, storage, or both. Only photosynthetic eukaryotic cells contain the important plastids called **chloroplasts**. In these organelles, energy from the sun drives the formation of ATP and NADPH, which are then used to form organic compounds.

A chloroplast has two outer membranes around a semifluid interior, the stroma, which bathes an inner membrane. Often this single membrane is folded back on itself into a series of stacked, interconnected disks (Figure 3.11). Each stack is a **thylakoid**. Embedded in the thylakoid membrane are light-trapping pigments, including chlorophylls, as well as enzymes and other proteins with roles in photosynthesis. Glucose, then sucrose, starch, and other organic compounds are built from carbon dioxide and water in the stroma.

Considered individually, chloroplasts are a lot like photosynthetic bacteria. Like mitochondria, they may have evolved by way of endosymbiosis, as described in the next section.

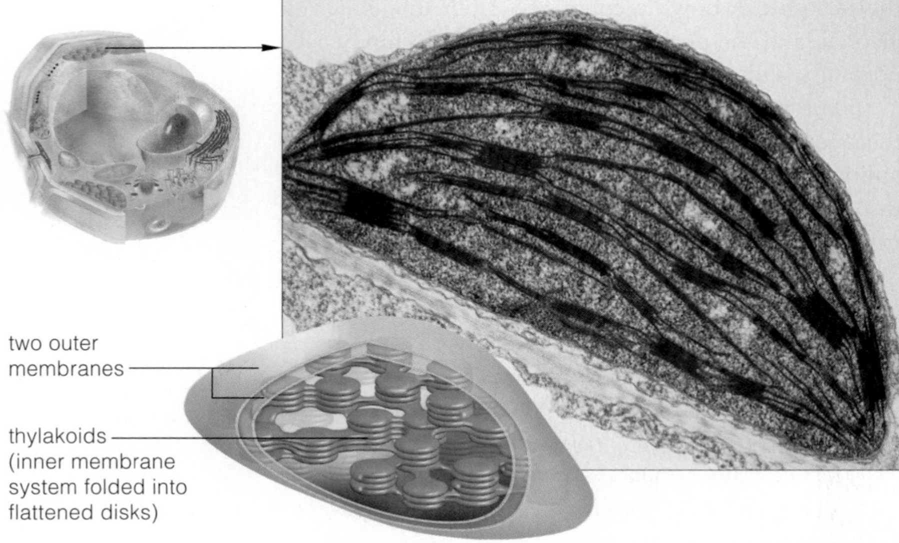

Figure 3.11 Typical chloroplast, the key defining feature of photosynthetic eukaryotic cells.

SUMMARY OF MAJOR ORGANELLES

Before continuing with eukaryotic cell structures, take a quick look at Figures 3.12 and 3.13. They summarize the organelles we have considered so far.

The organelles called mitochondria are the ATP-producing powerhouses of all eukaryotic cells.

Reactions that release energy from organic compounds occur in mitochondria. The reactions require oxygen.

Photosynthetic eukaryotic cells have chloroplasts, organelles that specialize in making sugars and other carbohydrates.

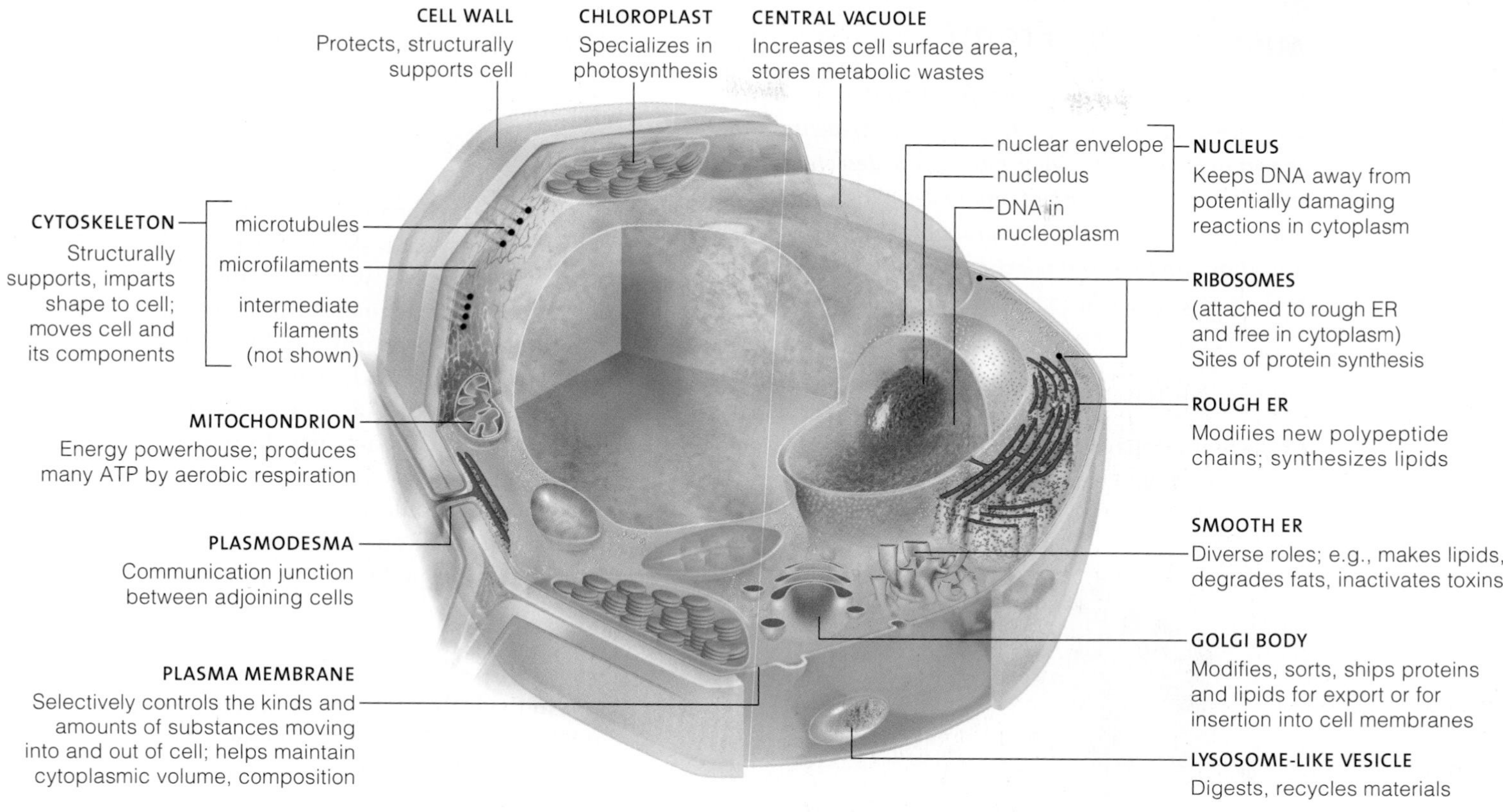
CELL WALL
Protects, structurally supports cell
CHLOROPLAST
Specializes in photosynthesis
CENTRAL VACUOLE
Increases cell surface area, stores metabolic wastes
nuclear envelope
nucleolus
DNA in nucleoplasm
NUCLEUS
Keeps DNA away from potentially damaging reactions in cytoplasm
CYTOSKELETON
Structurally supports, imparts shape to cell; moves cell and its components
microtubules
microfilaments
intermediate filaments (not shown)
RIBOSOMES
(attached to rough ER and free in cytoplasm) Sites of protein synthesis
MITOCHONDRION
Energy powerhouse; produces many ATP by aerobic respiration
ROUGH ER
Modifies new polypeptide chains; synthesizes lipids
PLASMODESMA
Communication junction between adjoining cells
SMOOTH ER
Diverse roles; e.g., makes lipids, degrades fats, inactivates toxins
GOLGI BODY
Modifies, sorts, ships proteins and lipids for export or for insertion into cell membranes
PLASMA MEMBRANE
Selectively controls the kinds and amounts of substances moving into and out of cell; helps maintain cytoplasmic volume, composition
LYSOSOME-LIKE VESICLE
Digests, recycles materials

Figure 3.12 Typical organelles and structures of plant cells.

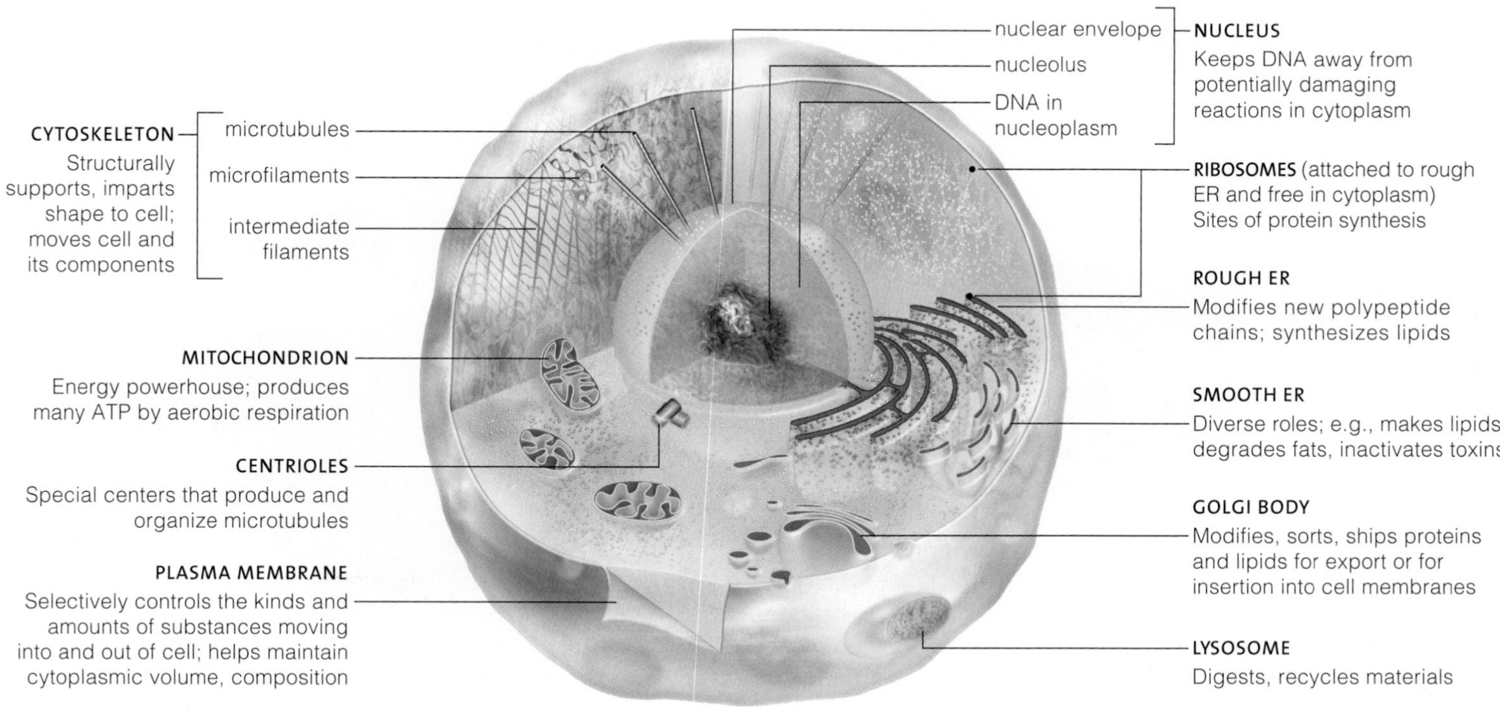
nuclear envelope
nucleolus
DNA in nucleoplasm
NUCLEUS
Keeps DNA away from potentially damaging reactions in cytoplasm
CYTOSKELETON
Structurally supports, imparts shape to cell; moves cell and its components
microtubules
microfilaments
intermediate filaments
RIBOSOMES (attached to rough ER and free in cytoplasm) Sites of protein synthesis
ROUGH ER
Modifies new polypeptide chains; synthesizes lipids
MITOCHONDRION
Energy powerhouse; produces many ATP by aerobic respiration
SMOOTH ER
Diverse roles; e.g., makes lipids, degrades fats, inactivates toxins
CENTRIOLES
Special centers that produce and organize microtubules
GOLGI BODY
Modifies, sorts, ships proteins and lipids for export or for insertion into cell membranes
PLASMA MEMBRANE
Selectively controls the kinds and amounts of substances moving into and out of cell; helps maintain cytoplasmic volume, composition
LYSOSOME
Digests, recycles materials

Figure 3.13 Typical organelles and structures of animal cells.

LINKS TO SECTIONS 1.4, 1.5

3.6 Where Did Organelles Come From?

Thanks to globe-hopping microfossil hunters, we have evidence of early life, including the fossil treasures shown in Figure 3.14. Today, most of the descendant species have a profusion of organelles. Where did the organelles come from? Some probably evolved by gene mutations and natural selection. For others, researchers make a case for evolution by way of endosymbiosis.

ORIGIN OF THE NUCLEUS AND ER

Prokaryotic cells, recall, do not have an abundance of organelles. Some have infoldings of plasma membrane that hold enzymes and other metabolic agents (Figure 3.15). By applying the theory of natural selection, we can hypothesize that such infoldings may also have evolved among the ancestors of eukaryotic cells. The infoldings became narrow channels in which nutrients and other substances were concentrated.

What selective advantage did the infoldings offer? More membrane meant more metabolic machinery for reactions, transport tasks, and waste disposal. Did ER channels evolve this way? Did they also protect the cell's metabolic machinery from uninvited guests?

By the latter scenario, infoldings that surrounded the DNA became a nucleus that helped protect the cell from foreign DNA (Figure 3.15). Bacteria and simple eukaryotic cells called yeasts do transfer bits of DNA between themselves. Early eukaryotic cells equipped with a nucleus could copy and use their messages of inheritance, free from competition with a potentially disruptive hodgepodge of foreign DNA.

THEORY OF ENDOSYMBIOSIS

Early in the history of life, the insertion of one cell into another led to partnerships among a variety of ancient prokaryotes. Whether they began as undigested meals or as parasites, uninvited guests evolved into what are now mitochondria, chloroplasts, and other organelles. This is one premise of a theory of **endosymbiosis**, as proposed by Lynn Margulis and others. (*Endo–* means within; *symbiosis* means living together.) One species (a guest) lives out its life inside another species (the host). The close interaction benefits one or both.

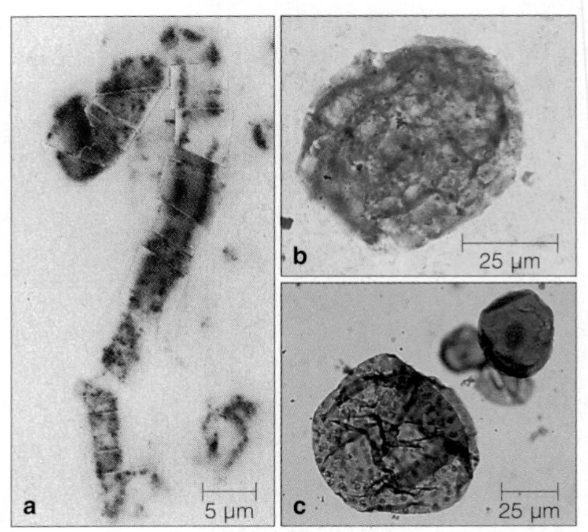

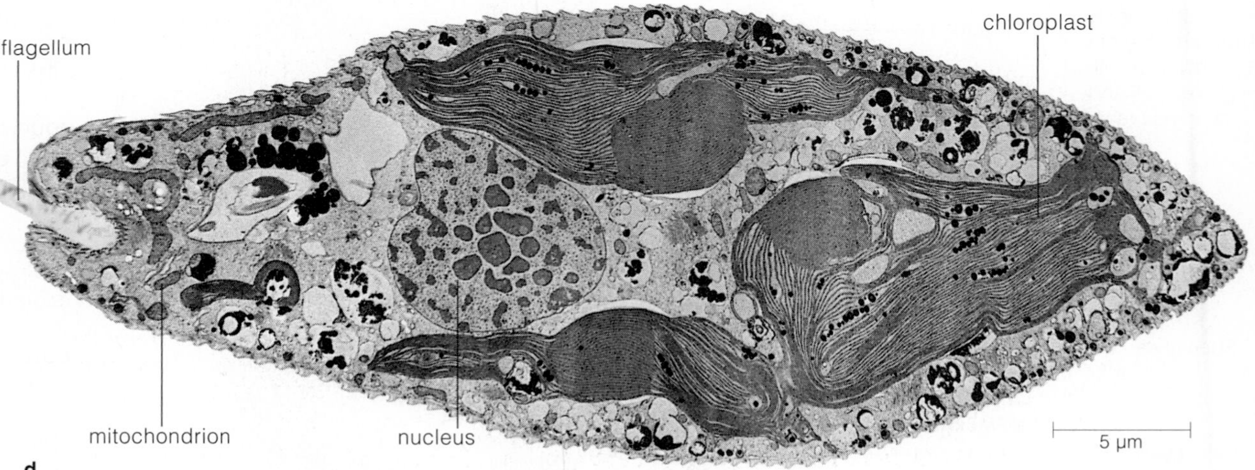

Figure 3.14 A few fossilized cells: From Australia, (**a**) a strand of what may be walled prokaryotic cells 3.5 billion years old and (**b**) one of the early eukaryotes—a protist that is 1.4 billion years old. From Siberia, (**c**) aquatic eukaryotic cells about 850 million years old. (**d**) A fine example of diverse organelles—hallmarks of eukaryotic cells: *Euglena*, a single-celled protist, sliced lengthwise. It has a long flagellum, which could not completely fit in the image area at this magnification.

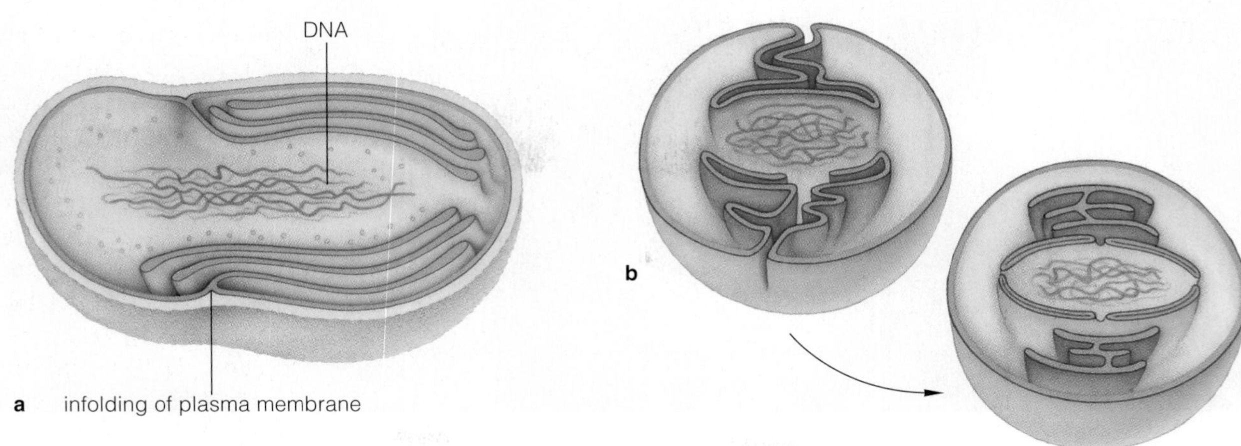

Figure 3.15 (**a**) Sketch of a bacterial cell (*Nitrobacter*) that lives in soil. Permanent infoldings of the plasma membrane penetrate the cytoplasm. (**b**) Model for the origin of the nuclear envelope and endoplasmic reticulum. In prokaryotic ancestors of eukaryotic cells, infoldings of the plasma membrane may have evolved into these important organelles.

According to this theory, eukaryotic cells evolved after the oxygen-producing pathway of photosynthesis emerged and changed the atmosphere. By 2.1 billion years ago, some populations of prokaryotic cells had adapted to the presence of oxygen and were extracting energy from glucose by way of aerobic respiration.

Some of the ancestors of eukaryotic cells certainly preyed on the aerobic bacteria. Others were perhaps hosts for parasitic infection by them. Suppose that a meal refused to be digested, or that a host persisted despite an ongoing parasitic infection. Endosymbiotic interactions would have begun at that time.

Internalized cells thrived in the sheltered, nutrient-rich environment of the host cytoplasm. The uninvited guests no longer had to spend energy on acquiring or making raw materials. The DNA regions that specified unneeded materials eventually mutated or were lost.

How did the hosts benefit? In time, they started to use some of the ATP made by their guests for growth and increased or diversified activities. Eventually they came to depend upon it. The two kinds of organisms became incapable of independent life. The guests had evolved into mitochondria.

Indeed, mitochondria of eukaryotic cells are a lot like bacteria in size and structure. Each has its own DNA and divides independently of cell division. The inner membrane of a mitochondrion also resembles the plasma membrane of a bacterial cell.

Chloroplasts, too, may have endosymbiotic origins. They would have provided a selective advantage to an aerobically respiring host by releasing needed oxygen. Chloroplasts do resemble some bacteria in the details of their metabolism and genes. They differ from one another in shape and light-absorbing pigments, like photosynthetic bacteria do. Their DNA self-replicates, and they divide independently of the cell's division.

However they arose, the hybrid cells had a nucleus, an endomembrane system, mitochondria, and in some cases, chloroplasts. They were all eukaryotic cells, the first protists. They had efficient metabolic systems, and they evolved fast. In no time at all, evolutionarily speaking, their descendants evolved into plants, fungi, and animals.

EVIDENCE OF ENDOSYMBIOSIS

Is such a theory far-fetched? A chance discovery in Jeon Kwang's laboratory suggests otherwise. In 1966, a type of rod-shaped bacteria had infected his culture of single-celled *Amoeba*. Some infected cells died right away. Others grew more slowly, and were smaller and vulnerable to starving to death. Kwang maintained the infected culture.

Five years later, amoebas stuffed with bacteria were flourishing, and actually had become symbiotic with them. Researchers discovered the symbiosis because antibiotics specific for prokaryotic cells killed infected amoebas. Also, uninfected amoebas stripped of their nucleus died after receiving a nucleus from an infected cell. Yet, more than 90 percent survived the transplant when a few bacteria tagged along with the nucleus. Later studies revealed that the infected amoebas had lost their capacity to make an essential enzyme, and were dependent on the bacteria to make it for them.

The nucleus and ER may have evolved through infoldings of the plasma membrane. Mitochondria, chloroplasts, and other organelles might have evolved by endosymbiosis. Such ideas may help explain how you and all other eukaryotes evolved from simple prokaryotic beginnings.

IMPACTS, ISSUES

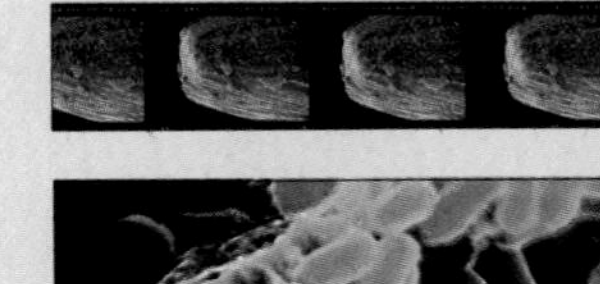

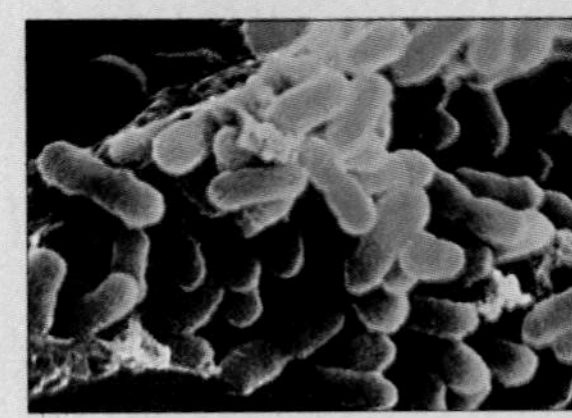

What did the ancestors of mitochondria look like? Gene studies offer clues. *Reclinomonas americana* is the protist with the structurally simplest mitochondria. Genes of *R. americana* mitochondria resemble the genes of *Rickettsia prowazekii*, the bacteria that cause typhus. Like mitochondria, parasitic *R. prowazekii* divide only inside eukaryotic cells.

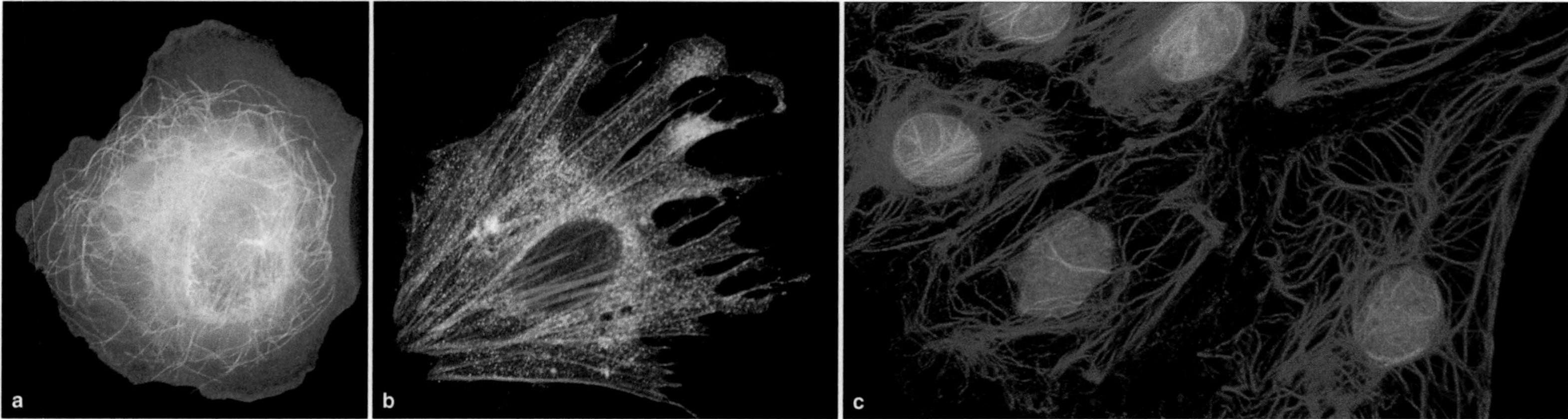

Figure 3.16 The beautiful cytoskeleton. (**a**) Microtubules (*gold*) and actin filaments (*red*) in an epithelial cell. (**b**) Actin filaments (*red*) and associated proteins (*green, blue*) help new epithelial cells such as this one migrate in tissues as human embryos are developing. (**c**) Intermediate filaments (*red*) of cultured kangaroo rat cells. The *blue* organelle inside each cell is the nucleus.

3.7 The Dynamic Cytoskeleton

Like you, all eukaryotic cells have an internal structural framework—a skeleton. Unlike your skeleton, theirs has elements that are not permanently rigid; they assemble and disassemble at different times.

In between the nucleus and plasma membrane of all eukaryotic cells is a **cytoskeleton**—an interconnected system of many protein filaments. Parts of the system reinforce, organize, and move cell structures and often the whole cell. Different parts of the cytoskeleton are shown in the micrographs in Figure 3.16. Some of the cytoskeletal elements are permanent; others form at certain times in the cell's life.

Two major cytoskeletal elements, microtubules and microfilaments, occur in all eukaryotic cells. Another type, intermediate filaments, strengthens some animal cells. Prokaryotic cells do not have a cytoskeleton, but some have microfilament-like proteins that reinforce the cell body.

Microtubules are long, hollow cylinders of tubulin subunits (Figure 3.17). They organize the cell interior and form a dynamic framework that moves structures such as chromosomes to specific locations. Controls govern which microtubules grow or fall apart at any given time. Microtubules growing in one direction—say, the forward end of a prowling amoeba—might get a protein cap that keeps them intact. Those at the cell's trailing end aren't used, aren't capped, and fall apart.

Figure 3.17 Poisonous plants, (**a**) the autumn crocus, and (**b**) western yew. Both of these plants make toxic substances that block microtubule assembly (*above*).

Colchicine, made by the autumn crocus (*Colchicum autumnale*), blocks microtubule assembly in the cells of animals that eat the plant (Figure 3.17*a*). Animal cells cannot divide without microtubules. The western yew (*Taxus brevifolia*) makes taxol, a different microtubule poison (Figure 3.17*b*). Doctors use a synthetic version of taxol to stop the uncontrolled cell divisions of some kinds of cancers.

Microfilaments strengthen or reconfigure a cell's shape (Figure 3.16*a,b*). Crosslinked, bundled, and gel-like arrays of them form a **cell cortex** that underlies and reinforces the plasma membrane. Microfilaments also anchor proteins, and they help muscles contract. An animal cell divides when microfilaments around its midsection contract, pinching it in two.

Think of the bustle at a train station during a busy holiday season, and you get an idea of what goes on in cells. Microtubules and microfilaments in cells are like dynamically assembled tracks. Motor proteins are the freight engines (Figure 3.18).

Intermediate filaments are the most stable parts of some cytoskeletons (Figure 3.16*c*). They strengthen and help maintain structures. For example, lamins anchor the actin and myosin of contractile units inside muscle cells. Other types anchor cells in tissues.

One or at most two kinds of intermediate filaments occur in certain animal cells. Researchers use them to identify the type of cell. *Cell typing* is a useful tool in diagnosing the tissue origin of diverse cancers.

MOVING ALONG WITH MOTOR PROTEINS

All eukaryotic cells contain similar microtubules and microfilaments. In spite of the uniformity, both kinds of elements serve diverse functions. How? *Other* proteins associate with them. Among the accessory proteins are **motor proteins**. These move cell parts in a sustained, directional way when repeatedly energized by ATP.

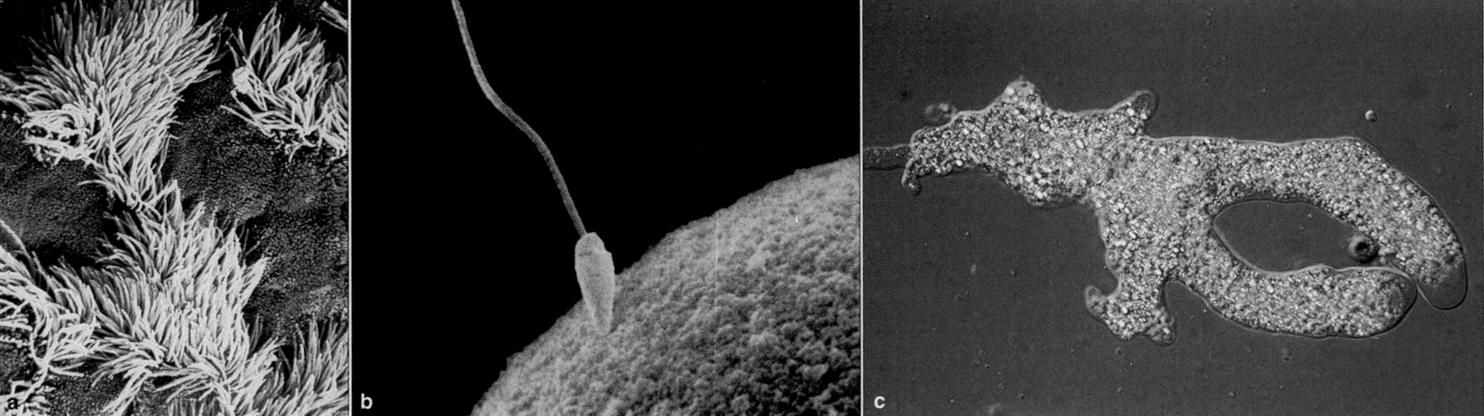

Figure 3.19 Cilia, a flagellum, and false feet. (**a**) Cilia (*gold*) projecting from the surface of cells that line an airway to human lungs. (**b**) Human sperm, with flagellum, about to penetrate an egg. (**c**) A predatory amoeba (*Chaos carolinense*), extending two pseudopods around a single-celled green alga (*Pandorina*).

Some motor proteins move chromosomes. Others slide one microtubule over another. Some chug along tracks in nerve cells that extend from your spine to your toes. Many engines are organized in series, each moving some vesicle partway along the track before giving it up to the next in line. Motor proteins in plant cells drag the chloroplasts into new light-intercepting positions as the sun's angle changes overhead. Motor proteins also contribute to animal muscle movement, as you will see in Chapter 21.

CILIA, FLAGELLA, AND FALSE FEET

Besides moving their internal parts, many eukaryotic cells use microtubules and motor proteins to move the whole cell or to stretch out part of it. Consider flagella and **cilia** (cilium). These motile structures project from the cell surface (Figure 3.19*a,b*).

Compared to cilia, flagella usually are longer and less profuse. Sperm and many other cells swim with the help of their whiplike flagella. In many multicelled organisms, ciliated cells continually stir air or fluid. Many thousands of ciliated cells line airways in your chest. Their coordinated beating propels bacteria and airborne particles away from the lungs.

Flagella and cilia move by a sliding mechanism that involves a motor protein called dynein. Paired microtubules extend the same distance into the tip of a cilium or flagellum. When ATP energizes them, short dynein arms projecting from each microtubule grab the next microtubule, tilt in a short, downward stroke, and then release. Repeated action of the dynein arms forces the microtubules to slide past each other. They don't slide far, but each *bends* a bit so that the sliding motion becomes converted to a bending motion.

Amoebas and some other types of eukaryotic cells form **pseudopods**, or "false feet" (Figure 3.19*c*). These temporary, irregular lobes project from the cell and are used for locomotion and prey capture. Microfilaments that elongate in the lobe also push it forward. Motor proteins on the microfilament tracks drag the plasma membrane with them.

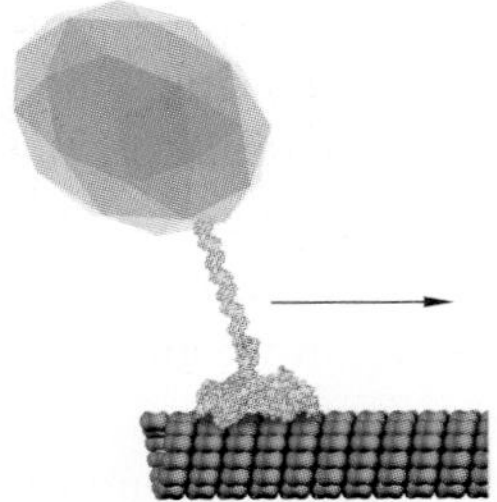

Figure 3.18 ***Animated!*** A motor protein, kinesin, on a microtubule. Kinesin inches along the length of the microtubule. If the microtubule is anchored near the cell's center, the kinesin moves its freight away from the center.

A cytoskeleton of protein filaments is the basis of eukaryotic cell shape, internal structure, and movement.

Microtubules help organize the eukaryotic cell and move its parts. Microfilaments form networks that reinforce the cell surface. Intermediate filaments strengthen and maintain the shape of some animal cells.

When energized by ATP, motor proteins move along tracks of microtubules and microfilaments. They deliver specific cell components that are bound to them to new locations or into new positions.

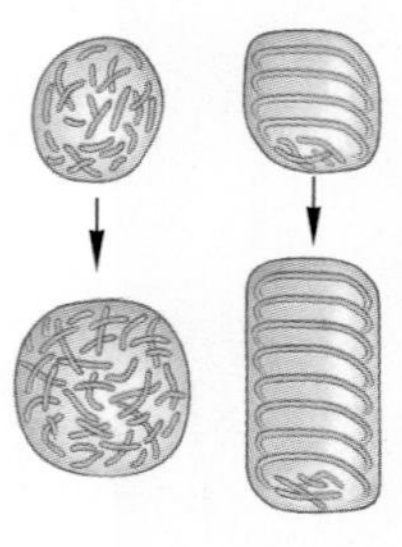

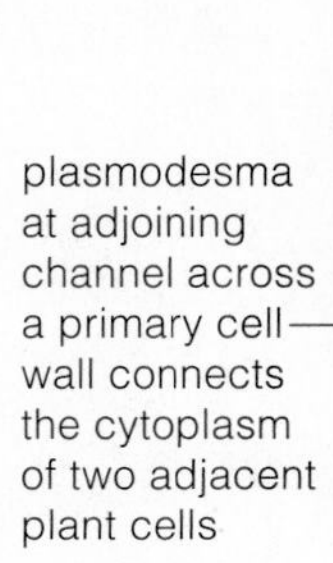

a Randomly oriented cellulose strands in a growing primary wall let a cell expand in all directions. Cross-oriented strands let it lengthen only.

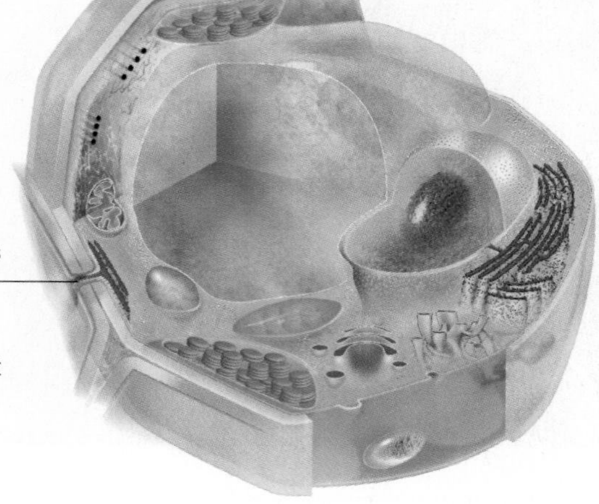

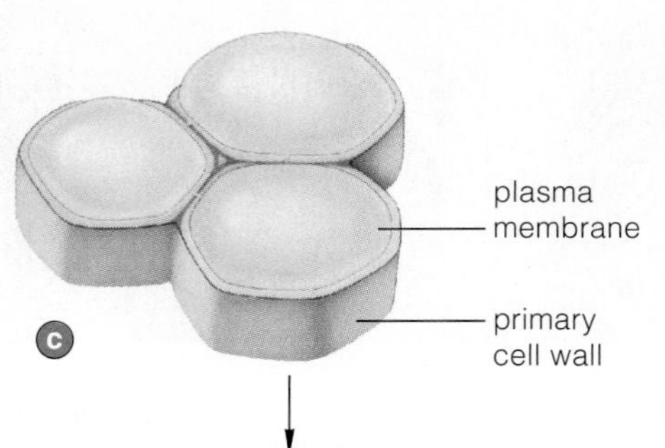

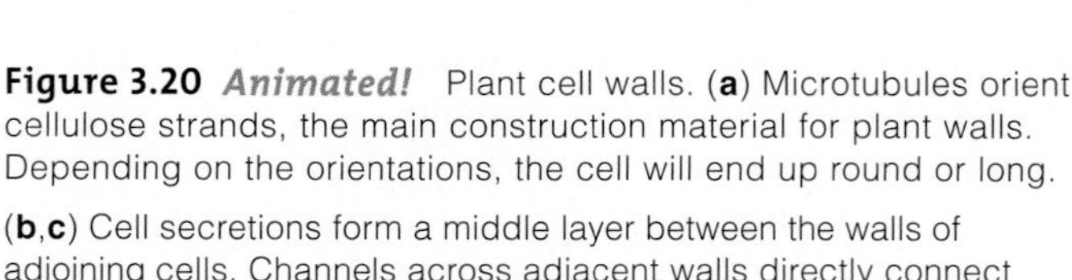

Figure 3.20 ***Animated!*** Plant cell walls. (**a**) Microtubules orient cellulose strands, the main construction material for plant walls. Depending on the orientations, the cell will end up round or long.

(**b**,**c**) Cell secretions form a middle layer between the walls of adjoining cells. Channels across adjacent walls directly connect the cytoplasm of adjacent cells.

(**d**) In many plant cells, more layers are deposited on the inside of the primary wall. They strengthen the wall and maintain its shape. When the cell dies, the stiffened walls remain. (**e**) This happens in water-conducting pipelines that thread through most plant tissues. Interconnected, stiffened walls of dead cells form the tubes.

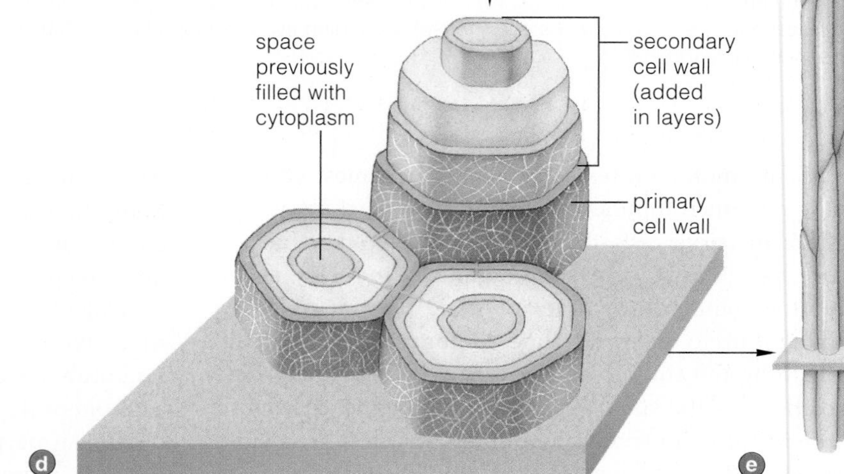

LINK TO SECTION 2.6

3.8 Cell Surface Specializations

Our survey of eukaryotic cells concludes with a look at cell walls and other specialized surface structures. Some structures are made of cell secretions. Others are clusters of proteins that connect neighboring cells.

EUKARYOTIC CELL WALLS

Single-celled eukaryotic species are directly exposed to the environment. Many have a **cell wall**, a structural component that encloses the plasma membrane. A cell wall protects and physically supports the cell. It is porous, so water and solutes easily move to and from the plasma membrane. A cell would die without these exchanges. Many single-celled protists have a cell wall around their plasma membrane. So do plant cells and many types of fungal cells.

For example, in the growing parts of multicelled plants, young cells secrete molecules of pectin and other glue-like polysaccharides, as well as cellulose. All of these materials are components of a plant cell's **primary wall** (Figure 3.20). The sticky cell wall glues abutting cells together. Thin and pliable, the wall permits the cell to enlarge.

Cells that have only a thin primary wall retain the capacity to divide or change shape as they grow and develop. Many plant cell types stop enlarging when they are mature. Mature cells secrete material on the primary wall's inner surface to form a rigid **secondary wall** that reinforces a cell's shape (Figure 3.20*d*). These secondary wall deposits are extensive, and contribute to structural support of the plant.

In woody plants, up to 25 percent of the secondary wall is made of lignin. This organic compound makes plant parts more waterproof, less susceptible to plant-attacking organisms, and stronger.

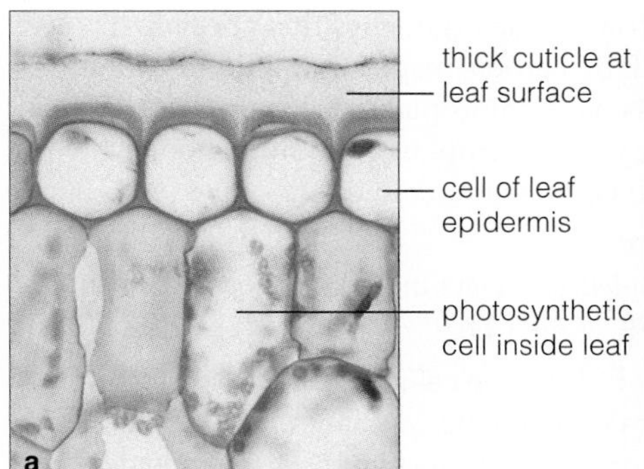

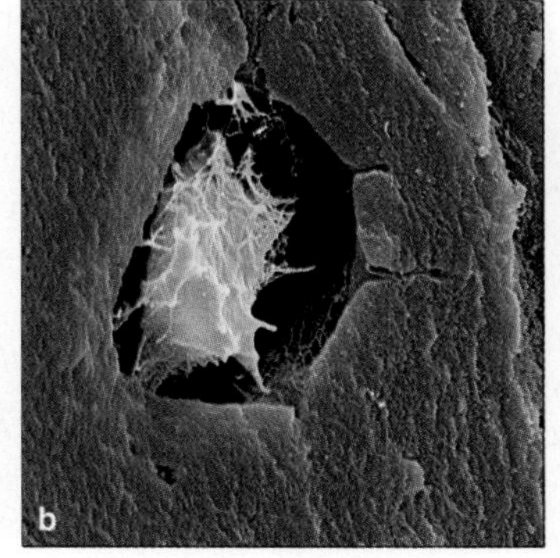

Figure 3.21 (**a**) Section through a plant cuticle, an outer surface layer of cell secretions. (**b**) A living cell inside bone tissue, the stuff of vertebrate skeletons.

At plant surfaces exposed to outside air, waxes and other cell secretions build up and form a cuticle. This protective, semitransparent surface limits water losses from plants (Figure 3.21*a*).

MATRIXES BETWEEN ANIMAL CELLS

Animal cells have no walls, but intervening between many of them are matrixes of cell secretions and other materials absorbed from the surroundings. For instance, the cartilage at the knobby ends of leg bones consists of scattered cells and protein fibers formed from their secretions, all embedded in polysaccharides. In bone tissue, the living cells form an extensive matrix (Figure 3.21*b* and Section 21.1).

CELL JUNCTIONS

Even when a wall or some other structure imprisons a cell in its own secretions, the cell still has contact with the outside world through its plasma membrane. In multicelled species, membrane components extend into adjoining cells and the surroundings. Among the components are **cell junctions**: molecular structures through which a cell sends and receives signals and materials. Cell junctions also cement a cell to other cells and to underlying membranes.

In plants, for instance, many tiny channels cross the primary walls of adjacent living cells and interconnect their cytoplasm. Figure 3.20*b* shows one. Each channel is a plasmodesma (plural, plasmodesmata). Substances flow quickly from cell to cell across these junctions.

Three types of cell-to-cell junctions are common in most animal tissues (Figure 3.22). *Tight* junctions link cells of most tissues, including epithelia that line outer surfaces, internal cavities, and organs. Tight junctions seal abutting cells so water-soluble substances cannot pass between them. That is why gastric fluid does not leak across the stomach lining and damage internal tissues. *Adhering* junctions occur in skin, the heart, and other organs that stretch. *Gap* junctions that link the cytoplasm of certain adjoining cells are open channels for a rapid flow of small molecules, most notably in heart muscle.

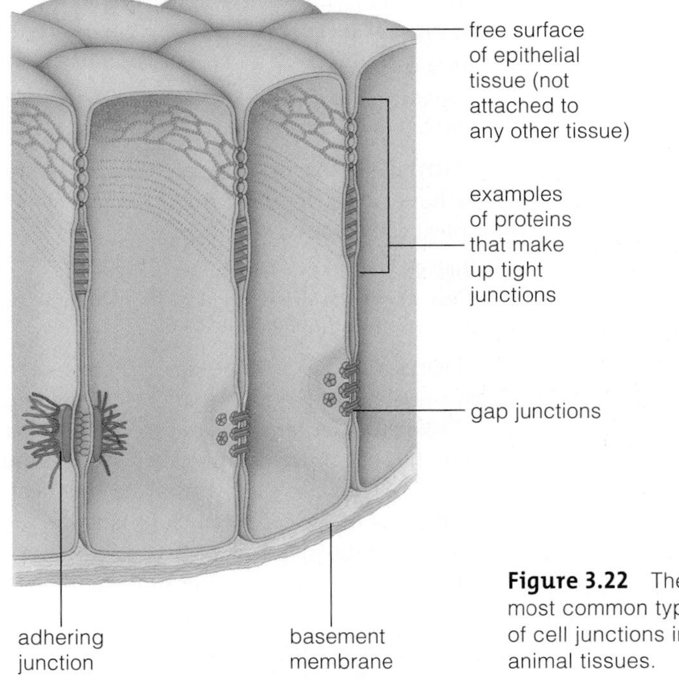

Figure 3.22 The most common types of cell junctions in animal tissues.

A variety of protist, plant, and fungal cells have a porous wall that surrounds the plasma membrane.

Young plant cells have a thin primary wall pliable enough to permit expansion. Some mature cells also deposit a lignin-reinforced secondary wall that affords structural support.

Animal cells have no walls, but they and many other cells often secrete substances that help form matrixes of tissues. Junctions form between cells of multicelled organisms.

Summary

Section 3.1 The cell is the smallest unit that still displays the properties of life. All cells start out life with an outer plasma membrane, cytoplasm, and a nucleus or nucleoid area that contains DNA.

Section 3.2 Most cells are microscopically small because a surface-to-volume ratio constrains their size. Different microscopes reveal cell shapes and structures.

BiologyNow
Learn how microscopes function with the animation on BiologyNow.

Section 3.3 Cell membranes consist mainly of lipids and proteins. A lipid bilayer gives a membrane its fluid properties and prevents water-soluble substances from freely crossing it. Proteins embedded in the bilayer or positioned at its surfaces carry out many functions.

BiologyNow
Explore the structure of cell membranes with the animation on BiologyNow.

Section 3.4 Bacteria and archaea are the prokaryotes. They have no nucleus and are the smallest, structurally simplest cells known (Table 3.2).

BiologyNow
View the structure of a typical prokaryote with the animation on BiologyNow.

Sections 3.5, 3.6 All eukaryotic cells start out life with a nucleus and other organelles, ribosomes, and a cytoskeleton. Organelles are membranous sacs that divide the cell's interior into functional compartments (Table 3.2). They physically separate chemical reactions from the rest of the cell. A nucleus isolates and protects the cell's genetic material.

In the ER and Golgi bodies of the endomembrane system, new polypeptide chains are modified and lipids are assembled. Many proteins and lipids become part of membranes or packaged inside vesicles that function in transport, storage, and other cell activities.

Mitochondria are the organelles that specialize in forming ATP by aerobic respiration. Photosynthetic eukaryotic cells have chloroplasts, which make sugars and carbohydrates. Chloroplasts and mitochondria may have evolved by endosymbiosis.

BiologyNow
See how organelles interact in a secretory pathway with the animation on BiologyNow.

Section 3.7 Eukaryotic cells have a cytoskeleton of microtubules, microfilaments, and intermediate filaments. The cytoskeleton supports and imparts shape; it also moves internal structures, motile structures, and often the whole cell.

BiologyNow
Learn about the role of motor proteins with the animation on BiologyNow.

Section 3.8 Many bacterial, protist, fungal, and plant cells have an outer, supporting wall. Animal cells do not have walls. In multicelled organisms, cell junctions form structural and functional connections between cells.

BiologyNow
Explore the structure of plant cell walls and cell junctions with the animation on BiologyNow.

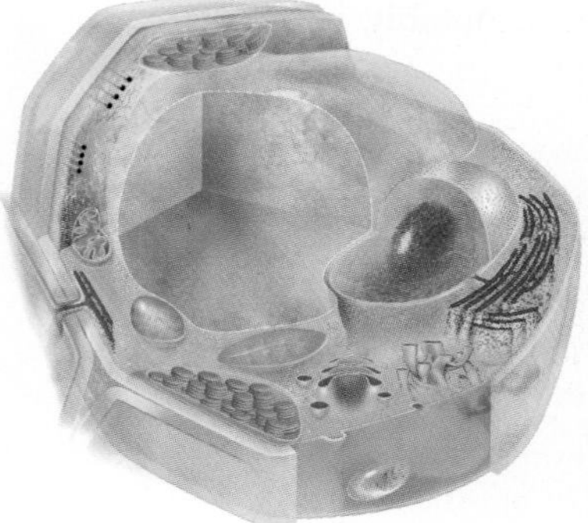

Self-Quiz

Answers in Appendix I

1. Cell membranes consist mainly of a ______ .
   a. carbohydrate bilayer and proteins
   b. protein bilayer and phospholipids
   c. lipid bilayer and proteins
   d. none of the above

2. Organelles ______ .
   a. are membrane-bound compartments
   b. are typical of eukaryotic cells, not prokaryotic cells
   c. separate chemical reactions in time and space
   d. All of the above are features of organelles.

3. Cells of many protists, plants, and fungi, but not animals, commonly have ______ .
   a. mitochondria　　c. ribosomes
   b. a plasma membrane　　d. a cell wall

4. Is this statement true or false: The plasma membrane is the outermost component of all cells. Explain your answer.

5. Unlike eukaryotic cells, prokaryotic cells ______ .
   a. lack a plasma membrane　　c. have no nucleus
   b. have RNA, not DNA　　d. all of the above

6. Identify as many of the components of the cells shown at left as you can.

7. Match each cell component with its function.

____ mitochondrion	a. protein synthesis
____ chloroplast	b. initial modification of new polypeptide chains
____ ribosome	c. modification of new proteins; sorting, shipping tasks
____ rough ER	d. photosynthesis
____ Golgi body	e. formation of many molecules of ATP

Additional questions are available on **BiologyNow™**

Critical Thinking

1. In a classic episode of *Star Trek*, a stupendous amoeba engulfs an entire starship. Spock blows the amoeba to bits before it reproduces and eats the universe. Think of at least one problem a biologist would identify with this scenario.

2. Your professor shows you an electron micrograph of a cell with many mitochondria and Golgi bodies. You notice the cell also contains a great deal of rough endoplasmic reticulum. What kinds of cellular activities would require such an abundance of the three kinds of organelles?

3. Peggy, a graduate student, discovers a single-celled organism swimming in a freshwater pond. Which cellular features might convince her that the cell is a bacterium? What features would suggest it is a protist?

4. *Kartagener syndrome* is a genetic disorder caused by a mutated form of the protein dynein. Affected people have chronically irritated sinuses, and mucus builds up in the airways to their lungs. Bacteria form huge populations in the thick mucus. Their metabolic by-products and the inflammation they trigger combine to damage tissues.

Males affected by the syndrome can make sperm but are infertile (Figure 3.23). Some have become fathers with the help of a procedure that injects sperm cells directly into eggs. Explain how an abnormal dynein molecule could cause the observed effects.

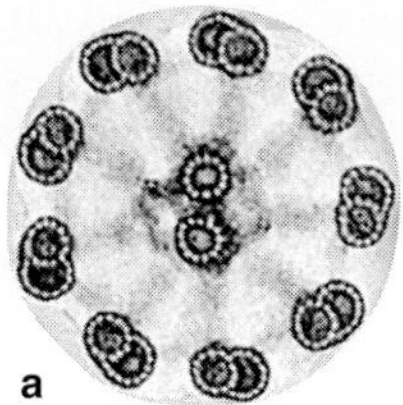

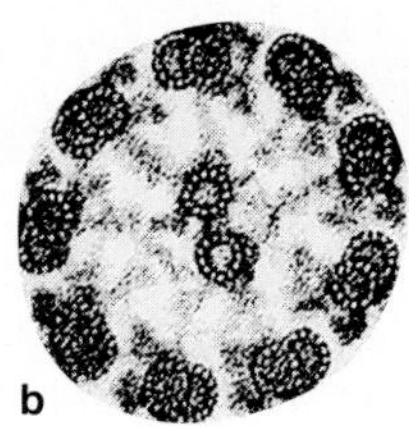

Figure 3.23 Cross-section through the flagellum of a sperm cell from (**a**) a male affected by Kartagener syndrome and (**b**) an unaffected male. Notice the dynein arms that extend from the paired microtubules.

Table 3.2 Summary of Typical Components of Prokaryotic and Eukaryotic Cells

		Prokaryotic	Eukaryotic			
Cell Component	Function	Bacteria, Archaea	Protists	Fungi	Plants	Animals
Cell wall	Protection, structural support	✓*	✓*	✓	✓	*None*
Plasma membrane	Control of substances moving into and out of cell	✓	✓	✓	✓	✓
Nucleus	Physical separation and organization of DNA	*None*	✓	✓	✓	✓
DNA	Encoding of hereditary information	✓	✓	✓	✓	✓
RNA	Transcription, translation of DNA messages into polypeptide chains of specific proteins	✓	✓	✓	✓	✓
Nucleolus	Assembly of subunits of ribosomes	*None*	✓	✓	✓	✓
Ribosome	Protein synthesis	✓	✓	✓	✓	✓
Endoplasmic reticulum (ER)	Initial modification of many of the newly forming polypeptide chains of proteins; lipid synthesis	*None*	✓	✓	✓	✓
Golgi body	Final modification of proteins, lipids; sorting and packaging them for use inside cell or for export	*None*	✓	✓	✓	✓
Lysosome	Intracellular digestion	*None*	✓	✓*	✓*	✓
Mitochondrion	ATP formation	**	✓	✓	✓	✓
Photosynthetic pigments	Light–energy conversion	✓*	✓*	*None*	✓	*None*
Chloroplast	Photosynthesis; some starch storage	*None*	✓*	*None*	✓	*None*
Central vacuole	Increasing cell surface area; storage	*None*	*None*	✓*	✓	*None*
Bacterial flagellum	Locomotion through fluid surroundings	✓*	*None*	*None*	*None*	*None*
Eukaryotic flagellum or cilium	Locomotion through or motion within fluid surroundings	*None*	✓*	✓*	✓*	✓
Complex cytoskeleton	Cell shape; internal organization; basis of cell movement and, in many cells, locomotion	*Rudimentary****	✓*	✓*	✓*	✓

* Known to be present in cells of at least some groups.
** Many groups use oxygen-requiring (aerobic) pathways of ATP formation, but mitochondria are not involved.
*** Protein filaments form a simple scaffold that helps support the cell wall in at least some species.

4 HOW CELLS WORK

IMPACTS, ISSUES

Beer, Enzymes, and Your Liver

The next time someone asks you to have a drink, stop for a moment to think about the challenge confronting cells that keep a drinker alive—especially a heavy drinker. It makes little difference whether one swallows 12 ounces of beer, 5 ounces of wine, or 1–1/2 ounces of eighty-proof vodka. Each of these drinks has the same amount of "alcohol" or, more precisely, ethanol.

Ethanol—CH_3CH_2OH—has both water-soluble and fat-soluble components, so it is absorbed quickly by the stomach and small intestine. From there it enters the bloodstream. Most tissues in the body contain enzymes (catalase, cytochrome P-450, and alcohol dehydrogenase) that convert ethanol to acetaldehyde. Acetaldehyde can be toxic if it accumulates. In healthy people, another enzyme converts it to nontoxic acetic acid.

Almost all of the alcohol you drink ends up in the liver. But its metabolism there causes a shift in mitochondrial reactions that can be toxic to liver cells. Given the liver's central role in alcohol metabolism, heavy drinkers gamble with alcohol-induced liver diseases. Over time, there are fewer and fewer liver cells for detoxification.

What are some of the possible outcomes? Alcoholic hepatitis is a disease characterized by inflammation and destruction of liver tissue. A different disease, alcoholic cirrhosis, permanently scars the liver. In time, the liver just stops working, with devastating effects.

The liver is the largest gland in the human body, and its activity affects everything else. You would have a very hard time digesting and absorbing food without it. Your cells would have a hard time synthesizing and taking up carbohydrates, lipids, and proteins, and staying alive.

There is more to think about. The liver rids the body of toxic compounds other than acetaldehyde. It also makes plasma proteins essential for blood clotting, immunity, and maintaining fluid volume. This brings us to a self-destructive behavior called binge drinking, the idea being to consume large amounts of alcohol in a brief period.

Binge drinking is now the most serious drug problem on college campuses in the United States, though very few students recognize their own drinking problem. Consider this finding from one survey: Almost half of 17,600 students at 140 colleges and universities are binge drinkers, meaning they consumed at least four alcoholic drinks in one day, in the two weeks prior to the survey.

Binge drinking does more than damage the liver. Put aside the related 500,000 injuries from accidents, the 70,000 cases of date rape, the 400,000 cases of (whoops) unprotected sex among students in an average year. Binge drinking can kill you. Drink too much, too fast, and you can abruptly end the beating of your heart.

With this example, we turn to metabolism, the cell's capacity to acquire energy and use it to build, degrade, store, and release substances in controlled ways. At times, the activities of your cells may be the last thing you want to think about. But they help define who you are and how you live, liver and all.

How Would You Vote? *There aren't enough liver donors for all the people waiting for liver transplants. Should a person's life-style be a factor in deciding who gets one? See BiologyNow for details, then vote online.*

Key Concepts

HOW CELLS USE ENERGY
Energy flows into the web of life, mainly from the sun. Cells tap into this one-way flow by energy-acquiring processes, starting with photosynthesis. They convert inputs of energy to forms that can be used to drive metabolic reactions.

HOW ENZYMES WORK
Without enzymes, substances would not react fast enough to maintain life. By controlling enzyme action, cells control the inputs and outputs of substances that keep them alive.

MEMBRANES AND METABOLISM
Cells have built-in mechanisms for increasing or decreasing water and solute concentrations across the plasma membrane and internal cell membranes. These adjustments are essential for metabolic reactions and metabolic pathways.

Links to Earlier Concepts

Reflect again on the road map for life's organization (Section 1.1). Here you will gain insight into how all organisms tap into a grand, one-way flow of energy to maintain that organization (1.2). You will start thinking about how cells use electrons and protons (2.1) in ways that help make ATP (2.10). Remember what you learned about electron arrangements and hydrogen bonding (2.2)? Here you will see how the biological function of enzymes follows from their structure (2.8), and how electron arrangements contribute to cofactor activity. Remember the structure of cell membranes (3.3, 3.5) and motor proteins (3.7) as you learn more about their function.

4.1 Inputs and Outputs of Energy

We know, almost without thinking about it, that skydivers fall down instead of up, that a hot pan eventually cools off. Our sense of time moving irrevocably forward comes from these observations—which are actually observations of the tendency of energy to disperse irrevocably.

THE ONE-WAY FLOW OF ENERGY

The dictionary defines **energy** as "the capacity to do work." What exactly does that mean? Let's start with the energy part. Energy exists in many different forms. Motion, light, sound, chemical bonds, electricity, heat, and magnetism are some examples.

Energy cannot be created or destroyed. This is the **first law of thermodynamics**. Any system taken as a whole, including the universe, has a given amount of energy. That energy may be distributed in different forms.

Energy disperses spontaneously. This is the **second law of thermodynamics**. For example, a hot pan gives off heat. We will never see all of that heat flow back from the air into the pan.

Each form of energy is useful to some degree for doing work. Work may occur as energy is transferred between systems or bodies. To a biologist, this means that every organism harvests energy from the environment, then uses it for growth and maintenance. Every such event involves energy transfers, and often the conversion of one form of energy into another.

Energy transfers are never entirely efficient. This is another aspect of the second law. With every transfer, some energy dissipates, and by doing so becomes less useful for doing work. The energy of a hot pan may boil water, but only before it disperses. If you think about it, this has to be a one-way trend: Energy that is available to do work in the universe is decreasing.

Can life be one glorious pocket of resistance to this depressing flow? Energy becomes concentrated in each new organism as molecules get organized. Even so, the second law applies to life on Earth. Life's main energy source is the sun, which has been losing energy ever since it formed 5 billion years ago. Photosynthetic cells intercept light energy from the sun and convert it to chemical bond energy in sugars, starches, and other compounds. Organisms that eat plants get the stored chemical energy by breaking and rearranging chemical bonds. With each conversion, however, a bit of energy escapes into the environment, usually as heat. *Overall, energy flows in one direction.* Organisms maintain their organization only by being continually supplied with energy lost from someplace else (Figure 4.1).

ENERGY LOST
Energy continuously flows from the sun.
ENERGY GAINED
Sunlight energy reaches environments on Earth. Producers secure some and convert it to stored forms of energy. They and all other organisms convert stored energy to forms that can drive cellular work.

Figure 4.1 A one-way flow of energy into the world of life compensates for the one-way flow of energy out of it.

ATP—THE CELL'S ENERGY CURRENCY

All cells stay alive by *coupling* energy inputs to energy outputs, mainly with adenosine triphosphate, or **ATP**. This nucleotide consists of a five-carbon sugar (ribose), a base (adenine), and three phosphate groups (Figure 4.2*a*). ATP readily gives up a phosphate group to other molecules and primes them to react. Such phosphate-group transfers are known as **phosphorylations**.

ATP is the currency in a cell's economy. Cells earn it by investing in energy-requiring reactions. They spend it in energy-releasing reactions that keep them alive. We use a cartoon coin to symbolize ATP.

Because ATP is the main energy carrier for so many reactions, you may infer, correctly, that cells have ways to renew it. When ATP gives up a phosphate group, ADP (adenosine diphosphate) forms. ATP forms again when ADP binds to inorganic phosphate (P_i) or to a phosphate group that was split away from a different molecule. Regenerating ATP by this **ATP/ADP cycle** helps drive most metabolic reactions (Figure 4.2*b*).

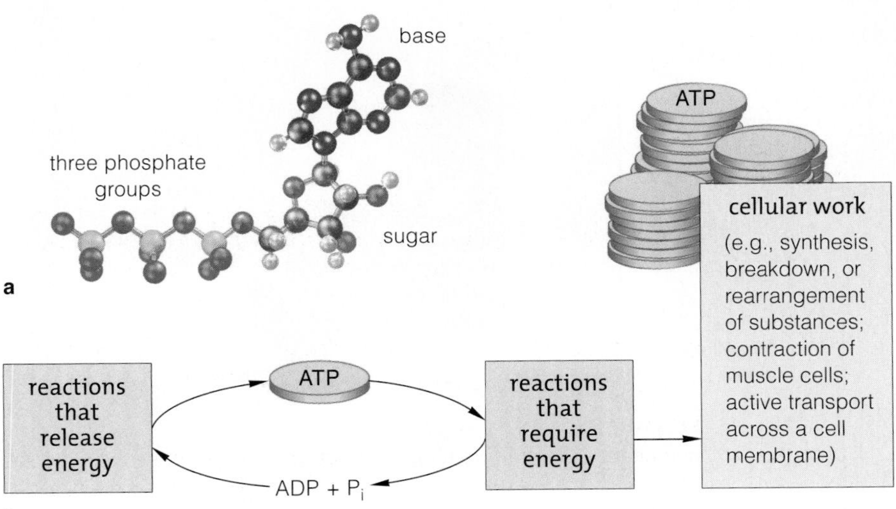

Figure 4.2 (**a**) Ball-and-stick model for the energy carrier ATP. (**b**) ATP couples energy-releasing reactions with energy-requiring ones. In the ATP/ADP cycle, recurring phosphate-group transfers turn ATP into ADP, and back again to ATP.

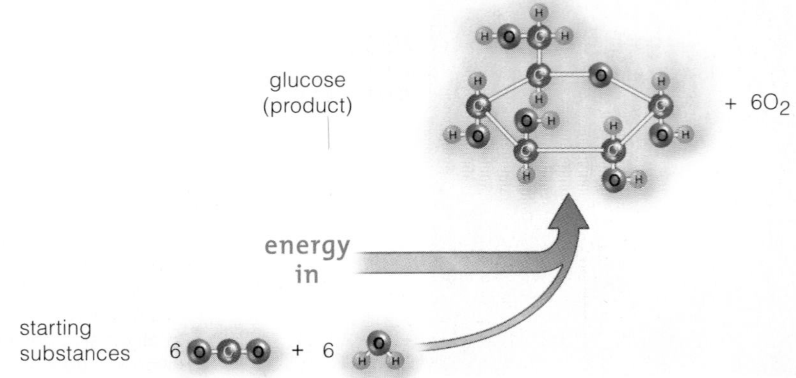

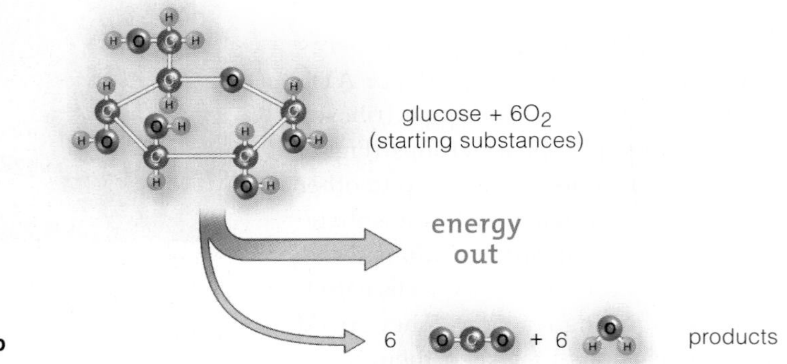

Figure 4.3 ***Animated!*** Up and down the energy hills. (**a**) Endergonic reactions require a net input of energy. (**b**) Exergonic reactions end with a net release of energy.

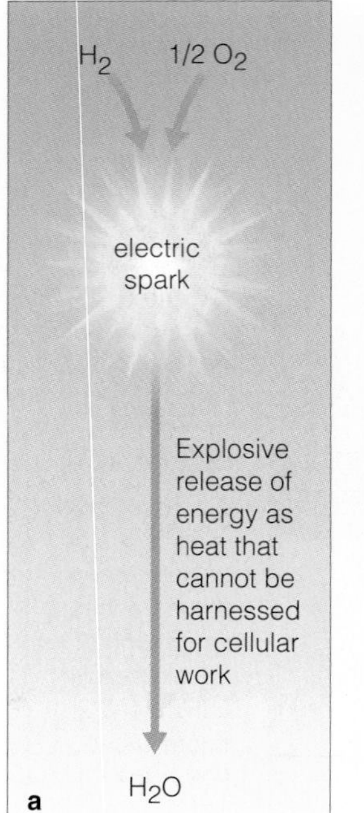

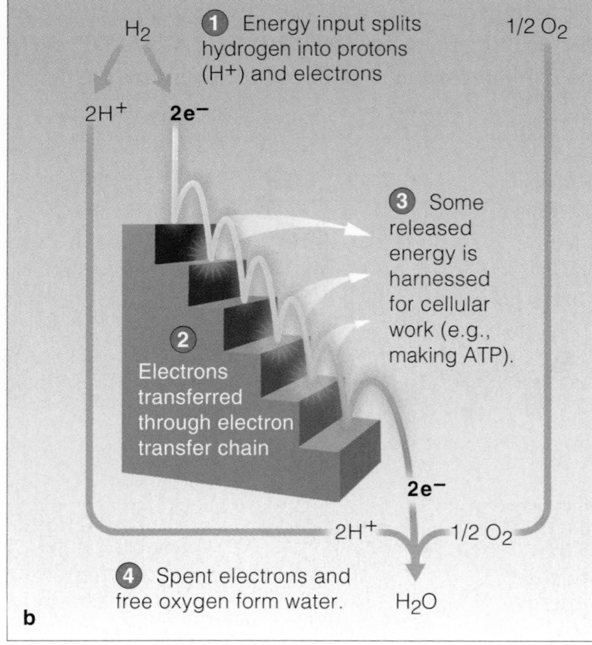

Figure 4.4 ***Animated!*** Uncontrolled versus controlled energy release. (**a**) Free hydrogen and oxygen exposed to an electric spark react and release energy all at once. (**b**) Electron transfer chains allow the same reaction to proceed in smaller, more manageable steps that can handle the released energy.

UP AND DOWN THE ENERGY HILLS

Cells store and retrieve energy when they convert one molecule to another. Inherent in each chemical bond is a certain amount of energy. The difference between the energy of all bonds that compose a reaction's starting compounds and the energy of all bonds that compose its products is the basis of energy flow in metabolism.

By investing in a reaction in which the bonds of the products have more energy than those of the starting materials, cells store energy in the reaction's products. This kind of reaction takes a net input of energy, and is called *endergonic* (meaning energy in). Photosynthetic cells, for example, use light energy to drive formation of ATP, and then use the energy from the ATP to make glucose from carbon dioxide and water. Unlike either light or ATP, glucose is stable and storable.

We think of glucose as a high-energy compound because cells can harvest energy inherent in its bonds to build other molecules. It takes a small investment of energy to break glucose's bonds, but the formation of more stable reaction products (CO_2 and H_2O) releases a lot more energy than the amount invested. It is as if glucose is at the top of an "energy hill," and carbon dioxide and water are at the bottom (Figure 4.3). Such reactions that end in a net release of energy are called *exergonic* (meaning energy out).

Running exergonic reactions would be pointless if cells could not capture released energy. For example, a spark ignites a mixture of hydrogen and oxygen gases. Water forms with an explosive burst of energy (Figure 4.4*a*). Cells cannot capture this kind of disorganized, intense release of energy.

Cells harvest energy in manageable increments by organized, stepwise reactions. In **oxidation–reduction reactions**, one molecule gives up electrons (it becomes oxidized) to another (which becomes reduced). Redox reactions occur in **electron transfer chains**, arrays of enzymes and other molecules that pass electrons from one to the next. Think of the electrons bouncing down a staircase (Figure 4.4*b*). Energy they lose at each step is captured by cells for ATP synthesis and other tasks.

Energy is the capacity to do work. It flows in one direction, from more usable to less usable forms. Organisms maintain their organization by using energy lost from someplace else.

ATP is the main energy carrier in all living cells. It couples energy-releasing and energy-requiring reactions. ATP primes molecules to react by transferring a phosphate group to them.

The sun is the primary source of energy for the web of life. Cells store and retrieve energy by chemical reactions. They harvest energy gradually with organized, stepwise reactions.

4.2 Inputs and Outputs of Substances

How cells get energy is only one aspect of metabolism. Another is the accumulation, conversion, and disposal of materials by energy-driven reactions. Most of these reactions are part of stepwise metabolic pathways.

WHAT ARE METABOLIC PATHWAYS?

Cells use and store thousands of substances in orderly, enzyme-mediated reaction sequences called **metabolic pathways**. In *biosynthetic* pathways, small molecules become assembled into larger molecules. Biosynthetic pathways require energy inputs. Photosynthesis is the main biosynthetic pathway. *Degradative* pathways are exergonic, overall. They break down large molecules to smaller products for a net energy release. In aerobic respiration, the main degradative pathway, glucose is converted to CO_2 and H_2O.

Not all metabolic pathways are linear, a straight line from reactants to products. In cyclic pathways, the last step regenerates a reactant that is the cycle's point of entry. In branched pathways, reactants or intermediates are sent to two or more different reaction sequences.

THE NATURE OF METABOLIC REACTIONS

For any metabolic reaction, the starting substances are called *reactants*. Substances formed during a sequence of reactions are *intermediates*, and those left at the end are the *products*. ATP and other *energy carriers* activate enzymes and other molecules by making phosphate-group transfers. *Enzymes* are catalysts: They can speed specific reactions enormously. *Cofactors* are metal ions and coenzymes such as NAD^+. They help enzymes by transferring functional groups, atoms, and electrons between reaction sites. *Transport proteins* move solutes within or across membranes. Because they can change local concentrations of reactants, intermediates, and products, transport proteins and controls over them can influence the timing and direction of metabolism.

Metabolic reactions do not run only forward, from reactants to products. Most also run in reverse, with products being converted back into reactants at least to some extent. Such reversible reactions tend toward **chemical equilibrium**, when the reaction rate is about the same in both directions (Figure 4.5).

The numbers of reactant and product molecules are not usually equal at chemical equilibrium. Imagine it is like a party where people drift between two rooms. The number in each room stays the same—say, thirty in one and ten in the other—even as individuals move back and forth. Why bother to think about this? *Each cell can bring about big changes in its activities simply by controlling a few steps of reversible metabolic pathways.*

For instance, most kinds of cells use a lot of ATP in order to function properly. They continuously convert glucose into pyruvate and ATP by a pathway of nine enzyme-mediated reactions called glycolysis (Section 6.2). When glucose supplies become too low, cells can reverse this pathway and build glucose from pyruvate and other non-carbohydrate molecules. How? Six steps of glycolysis are reversible and the other three are bypassed. Energy input from ATP drives the bypass reactions in the energetically unfavorable (or uphill) direction. If cells did not have this reverse pathway, they would not be able to compensate for episodes of starvation or intense exercise, when the blood glucose level can become dangerously low.

Figure 4.5 Chemical equilibrium. With a high concentration of reactant molecules (represented as wishful frogs), a reaction runs most strongly in the forward direction, to products (the princes). When the concentration of product molecules is high, it runs most strongly in reverse. At equilibrium, the rates of the forward and reverse reactions are the same.

Metabolic pathways are orderly, enzyme-mediated reaction sequences, some biosynthetic, others degradative.

All cells increase, maintain, and decrease supplies of substances by thousands of reactions. They rapidly shift activities by controlling key steps of reversible reactions.

Metabolic reactions start with reactants, typically form intermediates, and end with products. They use energy carriers (ATP), enzymes, cofactors, and transport proteins.

LINKS TO SECTIONS 2.1, 2.5, 2.9

4.3 How Enzymes Make Substances React

What would happen if you left a cupful of glucose out in the open? Not much. Years would pass before you would see evidence of its conversion to carbon dioxide and water. Yet that same conversion takes only a few seconds in your body. Enzymes make the difference.

Enzymes are catalysts that can make reactions occur hundreds to millions of times faster than they would on their own. Enzyme molecules can work again and again; a reaction does not use them up or alter them. They recognize only certain reactants, which are called their **substrates**. Except for a few RNAs, all enzymes are proteins.

LOWERING THE ENERGY HILL

Reactions do not proceed until the reactants have a minimum amount of internal energy—the activation energy. Visualize activation energy as a barrier—a hill or a brick wall (Figure 4.6). Enzymes lower it. How? *Compared with the surrounding environment, they offer a stable microenvironment, more favorable for reaction.*

Enzyme molecules are far larger than substrates. Each has one or more **active sites**: pockets or crevices where substrates bind and where reactions occur. Part of the substrate molecule is complementary in shape, size, charge, and solubility to the active site. This fit is why each enzyme can recognize and bind its substrate among the thousands of substances inside cells.

Think back on the main types of enzyme-mediated reactions (Section 2.5). With *functional group transfers*, one molecule gives up a functional group to another. With *electron transfers*, electrons are removed from one molecule and donated to another. With *rearrangements*, a juggling of internal bonds converts one molecule to another kind. In *condensation*, two or more molecules are covalently bound into a larger molecule. Finally, with *cleavage* reactions, a larger molecule is split into smaller ones.

When we talk about activation energy, *we really are talking about the energy it takes to align reactive chemical groups, destabilize electric charges, and break bonds.* These events put a substrate into its **transition state**. In this state, substrate bonds are at the breaking point, and the reaction can run easily to product.

The binding between an enzyme and its substrate is weak and temporary (that is why the reaction does not change the enzyme). However, energy released when

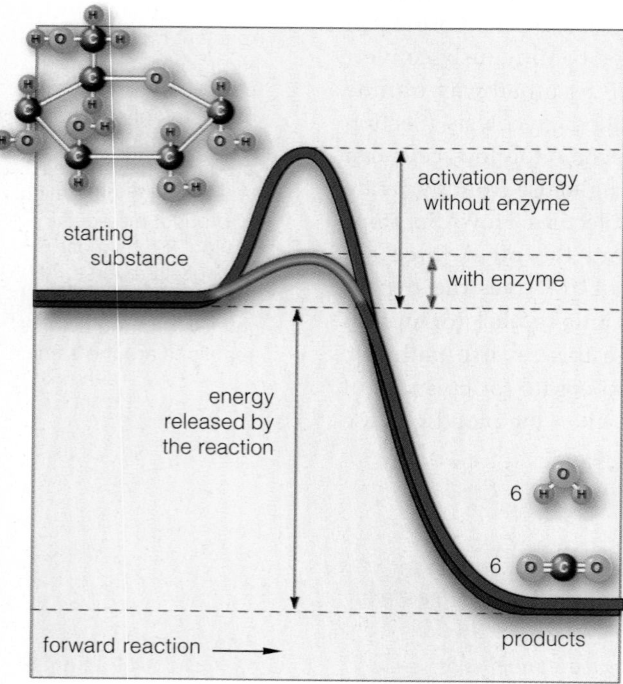

Figure 4.6 ***Animated!*** Activation energy: the minimum amount of internal energy that reactants must have before a reaction will run to products. An enzyme enhances the reaction rate by lowering the activation energy; by doing so it lowers the energy hill.

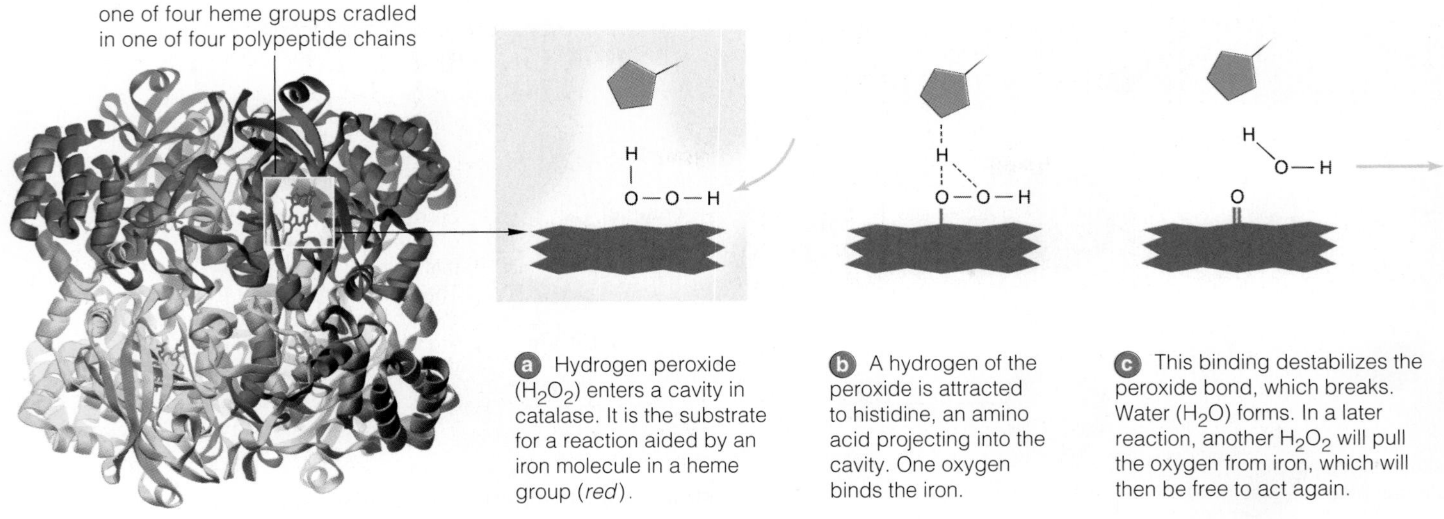

a Hydrogen peroxide (H_2O_2) enters a cavity in catalase. It is the substrate for a reaction aided by an iron molecule in a heme group (*red*).

b A hydrogen of the peroxide is attracted to histidine, an amino acid projecting into the cavity. One oxygen binds the iron.

c This binding destabilizes the peroxide bond, which breaks. Water (H_2O) forms. In a later reaction, another H_2O_2 will pull the oxygen from iron, which will then be free to act again.

Figure 4.7 How catalase works. This enzyme has four polypeptide chains and four heme groups.

the weak bonds form stabilizes the transition state. This "binding energy" keeps enzyme and substrate together long enough for the reaction to be completed.

Four mechanisms work alone or in combination to move substrates to their transition state:

Helping substrates get together. When substrates are at low concentrations, they rarely react. Binding at an active site is an effective local boost in concentration, by as much as ten millionfold.

Orienting substrates in positions favoring reaction. On their own, substrates collide from random directions. By contrast, the weak but extensive bonds at an active site keep substrates close together.

Shutting out water. Because of its capacity to form hydrogen bonds so easily, water can interfere with the breaking and formation of chemical bonds during reactions. Some active sites are heavily populated with nonpolar amino acids. These repel water and so keep it away from the reaction.

Inducing changes in enzyme shape. Weak interactions with an enzyme may make a substrate change shape. Sometimes a substrate does not fit into an active site very well. The enzyme then stretches or squeezes the substrate into a particular shape, often next to another molecule. By optimizing its fit with the substrate, the enzyme can move the substrate to its transition state.

Substrates are not exactly complementary to active sites in an enzyme. Why? If an active site were to bind substrate too tightly, the complex would be too stable. Enzyme and substrate would stick together, and that would be a chemical, not enzymatic, reaction.

HELP FROM COFACTORS

LINK TO SECTION 2.2

Nearly a third of all enzymatic reactions use cofactors. **Cofactors** are non-protein compounds and metal ions that associate with certain enzymes and are required for their function. Organic cofactors such as vitamins are coenzymes, and many of them contain metal ions. Metals donate and accept electrons easily. Proximity to a metal ion can alter the arrangement of electrons in a substrate, and so bring on its transition state. That is what happens in a catalase, an enzyme with four heme cofactors. The iron atom in each heme helps catalase break down hydrogen peroxide to water and oxygen (Figure 4.7). Catalase, an **antioxidant**, also neutralizes other strong oxidizers such as *free radicals*, substances with unpaired electrons. Free radicals are dangerously reactive by-products of aerobic respiration, attacking DNA, lipids, and other biological molecules. We make less catalase as we age, so free radicals accumulate.

Activation energy is the minimum amount of energy required to convert reactants into products. Enzymes greatly increase reaction rates by lowering activation energy.

Together with cofactors, enzymes drive their substrates to a transition state, when the reaction can most easily run to completion. This happens in the enzyme's active site.

In the active site, substrates move to the transition state by mechanisms that concentrate and orient them, that exclude water, and that induce an optimal fit with the active site.

IMPACTS, ISSUES

One of the mutant forms of the alcohol-degrading enzyme called aldehyde dehydrogenase is not good at breaking down acetaldehyde. It is quite common in individuals of many Asian populations, and it gets them flushed, dizzy, and nauseous soon after they drink alcohol.

NAD^+ assists this enzyme, but the body has only so many NAD^+ molecules. When this coenzyme is helping to break down alcohol, it can't enter the body's energy-producing reactions. It is diverted into fat-synthesizing reactions. That is why a fatty liver is the first stage of alcoholic liver disease.

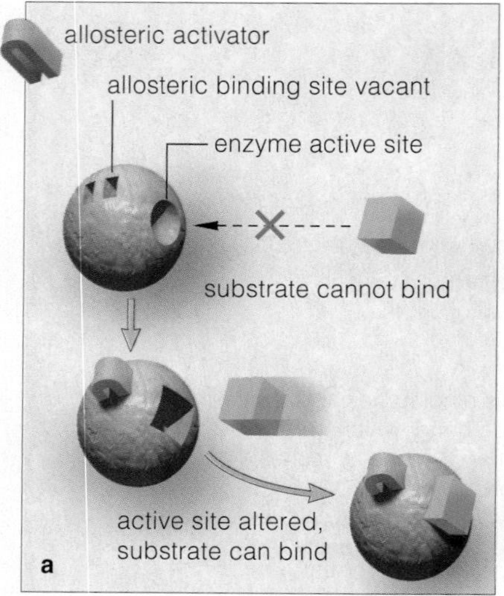

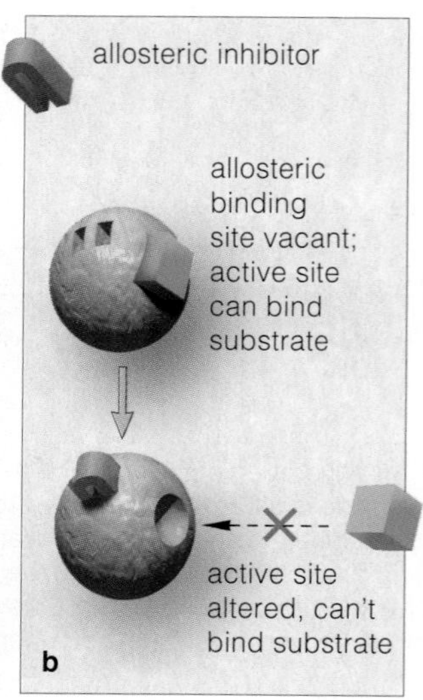

Figure 4.8 Allosteric control over enzyme activity. (**a**) An active site is unblocked when an activator binds to a vacant allosteric site. (**b**) An active site is blocked when an inhibitor binds to a vacant allosteric site.

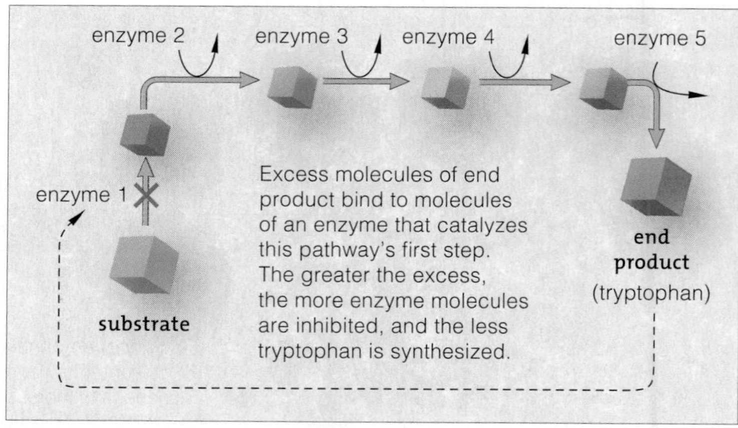

Figure 4.9 Feedback inhibition of a metabolic pathway. Five kinds of enzymes act in sequence to convert a substrate to tryptophan.

LINK TO SECTION 2.4

HOW IS ENZYME ACTIVITY CONTROLLED?

Controls over enzymes can raise, lower, and maintain concentrations of substances. Some controls can adjust the rate of enzyme synthesis. Others activate or inhibit the enzymes that are already built.

In some cases, a molecule that acts as an activator or inhibitor reversibly binds to its own *allosteric* site, not the active site, on the enzyme (*allo–*, other; *steric*, structure). Binding alters the enzyme's shape in a way that either hides or exposes the active site (Figure 4.8). Allosteric enzymes are part of many feedback loops.

Feedback loops adjust the concentrations of many substances (Figure 4.9). Picture a bacterial cell making tryptophan and the other amino acids—the building blocks for proteins. When the cell has made enough proteins, tryptophan synthesis will continue until its accumulation causes **feedback inhibition**. This means some result of a specific activity *shuts down the activity*. Here, tryptophan stops tryptophan synthesis. Unused tryptophan binds to an allosteric site on an enzyme in the tryptophan synthesis pathway. Binding blocks the enzyme's active site, so less tryptophan is made. More of the active sites are available when few tryptophan molecules are around, and the synthesis rate increases.

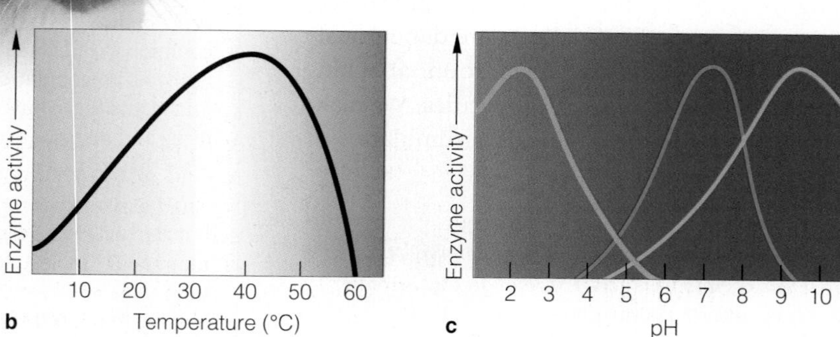

Figure 4.10 Enzymes and the environment. (**a**) Temperature outside the body affects the fur color of Siamese cats. Epidermal cells that give rise to the cat's fur produce a brownish-black pigment, melanin. Tyrosinase, an enzyme in the melanin production pathway, is heat-sensitive in the Siamese. It becomes inactive in warmer parts of the cat's body, which end up with less melanin, and lighter fur. Put this cat in booties for a few weeks and its warm feet will turn white. (**b**) How increases in temperature affect one enzyme's activity. (**c**) How rising pH values affect three kinds of enzymes.

EFFECTS OF TEMPERATURE, PH, AND SALINITY

Temperature is a measure of molecular motion. As it rises, it boosts reaction rates both by increasing the chance that a substrate will bump into an enzyme, and by raising a substrate's internal energy. Remember, the more energy a reactant has, the closer it is to jumping the activation energy barrier. Outside a certain range of temperature, hydrogen bonds break. Enzyme shape changes, and the active site no longer binds substrate. The reaction rate falls sharply (Figure 4.10*a,b*). Such declines typically occur in fevers above 44°C (112°F), which are usually deadly.

Enzyme function is also affected by pH and salt concentration, because they influence both hydrogen bonding and reaction chemistry. Most enzymes have a narrow range of pH and salinity under which they function properly (Figure 4.10*c*). This range is found where the enzyme occurs in nature. For instance, in the human body, most enzymes work best when the pH is between 6 and 8.

Enzyme action adjusts the concentrations and kinds of substances available in cells. Controls over enzymes can enhance or inhibit their activity.

Enzymes work best when the cellular environment stays within limited ranges of temperature, pH, and salinity. The actual ranges differ from one type of enzyme to the next.

4.4 Diffusion and Metabolism

What determines whether a substance will move one way or another across a cell membrane? Part of the answer has to do with diffusion.

Think about the water bathing the surfaces of a cell membrane. Plenty of substances are dissolved in it, but the kinds and amounts close to the two membrane surfaces differ. The membrane set up the difference and busily maintains it. How? Its structure lets some substances but not others cross it in certain ways, at certain times. This is called **selective permeability**.

Lipids of a membrane's bilayer are mostly nonpolar, so they let small, nonpolar molecules such as O_2 and CO_2 slip across. Water molecules are polar, but some slip through gaps that open when hydrophobic tails of many lipids flex and bend. A bilayer is not permeable to ions or large, polar molecules such as glucose; these cross only with the help of proteins. Water frequently crosses with them (Figure 4.11).

Membrane barriers and crossings are vital, because metabolism depends on the cell's capacity to increase, decrease, and maintain the concentrations of ions and molecules required for reactions. As well, they supply cells or organelles with raw materials, remove wastes, and collectively maintain the cell's volume and pH.

LINKS TO SECTIONS 2.3, 3.3

WHAT IS A CONCENTRATION GRADIENT?

A **concentration gradient** is a difference in the number per unit volume of molecules (or ions) of a substance between two regions. In the absence of other forces, molecules move into an adjoining region where they are not as concentrated. Why? They randomly collide many millions of times a second. This goes on in both regions; it just happens more often in a bigger crowd. In any given interval, more molecules are moving to the less crowded region than are moving out of it.

Diffusion is the net movement of like molecules or ions down a concentration gradient. It contributes to the movement of substances through the cytoplasm and across membranes. In multicelled species, it moves substances between body parts and between the body and its environment. For example, when oxygen builds up in leaf cells, it may diffuse into air inside the leaf, then into air outside, where its concentration is lower.

A substance tends to diffuse in a direction set by its *own* concentration gradient, not by gradients of other solutes. You can see this by dropping dye into water. Dye molecules diffuse into the region where they are less concentrated, and water molecules move into the region where *they* are less concentrated (Figure 4.12).

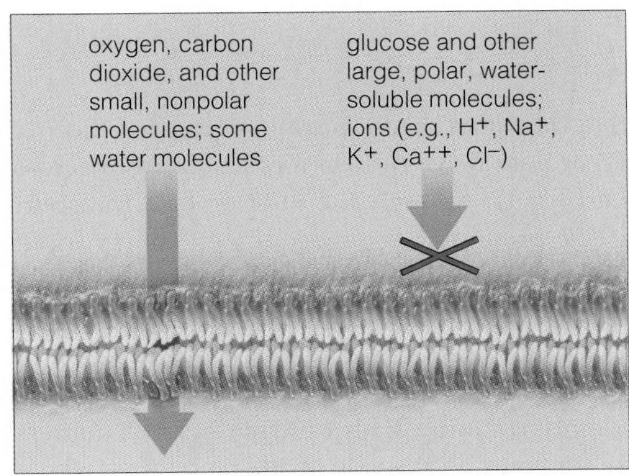

Figure 4.11 ***Animated!*** Selective permeability of cell membranes. Small, nonpolar molecules and some water molecules cross the lipid bilayer. Ions and large, polar, water-soluble molecules and the water dissolving them cross with the help of transport proteins.

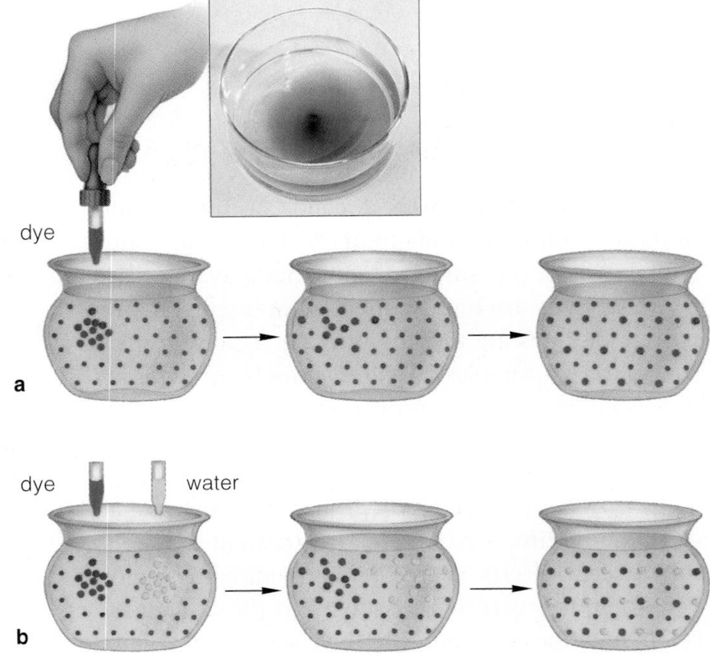

Figure 4.12 ***Animated!*** Two examples of diffusion. (**a**) A drop of dye enters a bowl of water. Gradually, the dye molecules become evenly dispersed through the molecules of water.

(**b**) The same thing occurs with the water molecules. Here, dye (*red*) and water (*yellow*) are added to the same bowl. Each of the substances will show a net movement down its own concentration gradient.

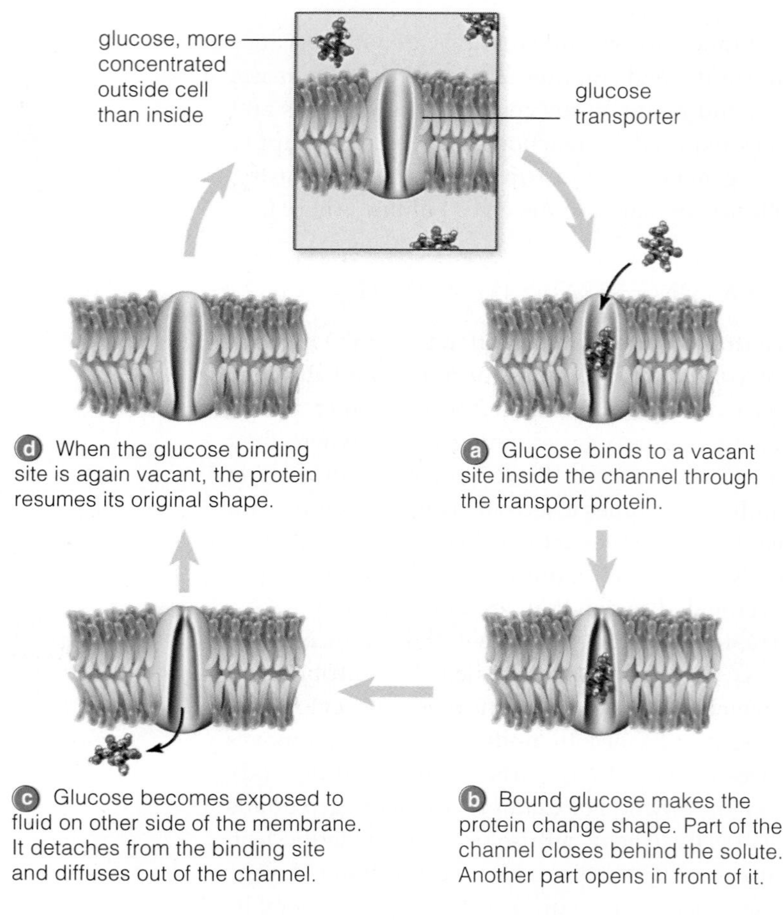

glucose, more concentrated outside cell than inside
glucose transporter
d When the glucose binding site is again vacant, the protein resumes its original shape.
a Glucose binds to a vacant site inside the channel through the transport protein.
c Glucose becomes exposed to fluid on other side of the membrane. It detaches from the binding site and diffuses out of the channel.
b Bound glucose makes the protein change shape. Part of the channel closes behind the solute. Another part opens in front of it.

Figure 4.13 ***Animated!*** Passive transport. This model shows one of the glucose transporters that span the plasma membrane. Glucose crosses in both directions. The *net* movement of this solute is down its concentration gradient until its concentrations are equal on both sides of the membrane.

LINK TO SECTION 3.3

WHAT DETERMINES DIFFUSION RATES?

How fast a solute diffuses depends on the steepness of its concentration gradient, its size, the temperature, and electric or pressure gradients that may be present.

First, rates are high with steep gradients, because more molecules are moving out of a region of greater concentration compared with the number moving into it. Second, more heat energy in warmer regions makes molecules move faster and collide more often. Third, smaller molecules diffuse faster than large ones do.

Fourth, an electric gradient may alter the rate and direction of diffusion. An **electric gradient** is simply a difference in electric charge between adjoining regions. For example, every ion dissolved in fluids bathing a cell membrane contributes to the electric charge at one side or the other. Opposite charges attract. So the fluid with more negative charge overall exerts the greatest pull on positively charged substances, such as sodium ions. Many cell activities, including information flow through nervous systems, involve the combined force of electric and concentration gradients.

Fifth, as you will see shortly, diffusion also may be affected by a **pressure gradient**. This is a difference in the pressure being exerted in two adjoining regions.

Diffusion is the net movement of molecules or ions of a substance into an adjoining region where they are not as concentrated.

The force of a concentration gradient can drive the directional movement of a substance across membranes. The gradient's steepness, temperature, molecular size, and electric and pressure gradients affect diffusion rates.

4.5 Working With and Against Diffusion

Large, polar molecules and ions cannot cross a lipid bilayer without the help of transport proteins.

Many solutes diffuse across cell membranes through a channel or tunnel inside transport proteins. When one solute molecule or ion enters the channel and weakly binds to the protein, the protein's shape changes. The channel closes behind the bound solute and opens in front of it. This exposes the solute to the opposite side of the membrane. There, the solute is released as the binding site reverts to its former state.

PASSIVE TRANSPORT

In **passive transport**, a concentration gradient, electric gradient, or both drive diffusion of a substance across a cell membrane, through the interior of a transport protein. The protein does not require an energy input to assist the movement. That is why passive transport is also known as facilitated diffusion.

With a concentration gradient, the *net* direction of movement depends on how many molecules of solute are randomly colliding with transporters. Encounters are more frequent on the side of the membrane where the solute concentration is greatest. So the solute's *net* movement tends to be in the direction where it is less concentrated (Figure 4.13).

If nothing else were going on, passive transport would continue until concentrations were equal across the membrane. But other processes affect the outcome. For instance, a glucose transport protein helps glucose from blood enter cells, which use it for biosynthesis and for quick energy. This happens even if the blood glucose level is low. Why? As fast as molecules diffuse into the cells, others are used up. By using up glucose, then, cells maintain a gradient that favors the uptake of *more* glucose.

ACTIVE TRANSPORT

Are solute concentrations ever equal on both sides of a membrane? Only in a dead cell. Living cells never stop expending energy to pump solutes into and out of their interior. With **active transport**, energy-driven protein motors help move a specific solute across the cell membrane *against* its concentration gradient.

Only specific solutes can bind to functional groups that line the interior channel of an active transporter. When they do so, the transporter accepts a phosphate group from ATP. The phosphate-group transfer alters the shape of the transporter in a way that releases the solute on the other side of the membrane.

Figure 4.14 depicts a **calcium pump**. This active transporter helps keep the concentration of calcium in a cell at least a thousand times lower than it is outside. Calcium ions make muscles contract. They will keep contracting until most of the calcium ions are cleared away from the muscle cells by staggering numbers of calcium pumps.

A different protein, the **sodium–potassium pump**, is a cotransporter. When ATP activates it, this pump binds sodium ions (Na^+) on one side of the membrane and releases them on the other side. The release favors the binding of potassium ions (K^+) at a different site in the tunnel through the transporter, which reverts to its original shape after K^+ is released to the other side.

These and other active transport systems maintain many key gradients across membranes. The gradients are vital to many processes. We return to mechanisms of active transport in later chapters.

Transport proteins bind solutes on one side of a membrane and reversibly change shape. The change moves the solute through a tunnel in the interior of the transport protein, to the other side of the membrane.

In passive transport, a solute diffuses through a transporter, and its net movement is down its concentration gradient.

In active transport, the net movement of one type of solute is uphill, against its concentration gradient. The transporter must be activated by an energy input from ATP to counter the energy inherent in the gradient.

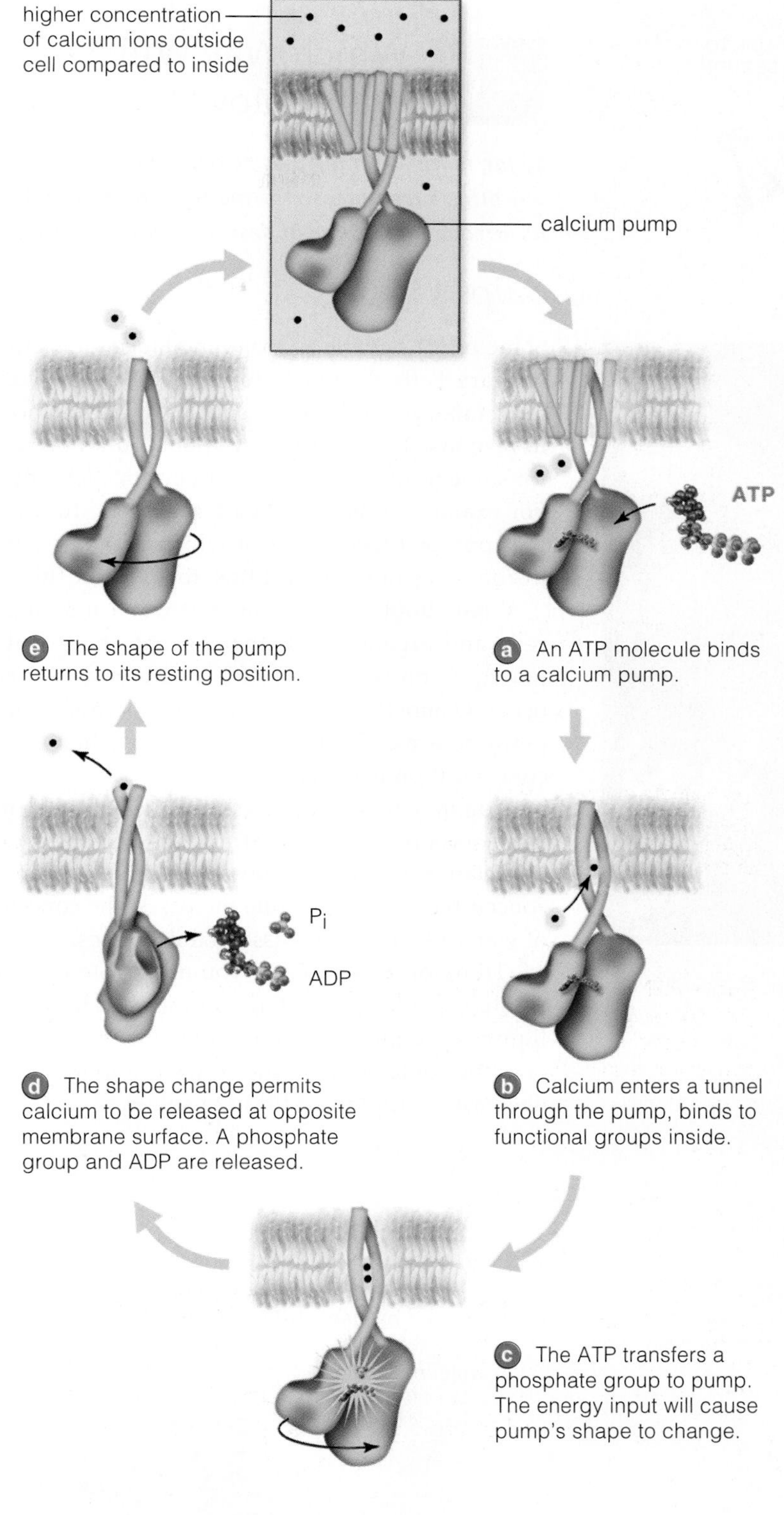

higher concentration of calcium ions outside cell compared to inside
calcium pump
ATP
a An ATP molecule binds to a calcium pump.
b Calcium enters a tunnel through the pump, binds to functional groups inside.
c The ATP transfers a phosphate group to pump. The energy input will cause pump's shape to change.
Pi
ADP
d The shape change permits calcium to be released at opposite membrane surface. A phosphate group and ADP are released.
e The shape of the pump returns to its resting position.

Figure 4.14 ***Animated!*** Active transport. This example uses a calcium pump that spans the plasma membrane. The sketch shows its channel for calcium ions. ATP transfers a phosphate group to the pump, thus providing energy that can drive the movement of calcium *against* a concentration gradient across the cell membrane.

LINK TO SECTION 2.3

4.6 Which Way Will Water Move?

By far, more water diffuses across cell membranes than any other substance, so the main factors that influence its directional movement deserve special attention.

OSMOSIS

Something as gentle as a running faucet or as mighty as Niagara Falls demonstrates **bulk flow**, the concerted movement of a mass of molecules. It occurs most often in response to a pressure gradient. Bulk flow accounts for some movement of water in multicelled organisms. For example, a beating heart generates fluid pressure that pumps blood, most of which is water. Sap flows inside tubes in trees, and this, too, is bulk flow.

What about the movement of water into and out of cells and organelles? If the concentration of water is not equal on both sides of a membrane, osmosis will occur. **Osmosis** is diffusion of water across a selectively permeable membrane, into a region where the water concentration is lower.

You may be wondering: How can water be more or less concentrated? Concentration is nothing more than the relative amounts of substances in a solution. As the concentration of any solute increases, the concentration of water in which it is dissolved decreases.

Think of it like this. If you add protein or another solute to a glass of water, you increase the volume of liquid. That greater volume still holds the same number of water molecules, so the concentration of water has decreased with the addition of solute.

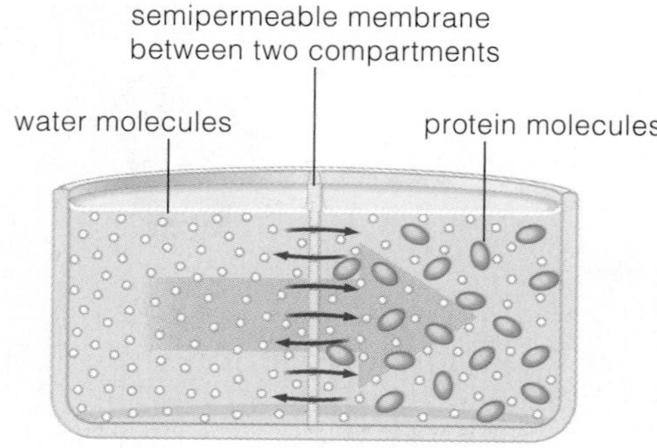

Figure 4.15 Solute concentration gradients and osmosis. A membrane divides this container. Water but not proteins can cross it. Pour 1 liter of water in the left compartment and 1 liter of a protein-rich solution in the right one. The proteins occupy some of the space in the right one. The net diffusion of water in this case is from left to right (large *gray* arrow).

Now suppose you divide the interior of another glass of water with a selectively permeable membrane that permits water but not protein to diffuse across it. Adding protein to the water on only one side forms a gradient across the membrane. That gradient is a force that pushes water molecules across the membrane into the protein solution (Figure 4.15). A gradient remains as long as there is a difference in water concentration on either side of a membrane.

EFFECTS OF TONICITY

Suppose you decide to test the statement that water tends to move toward a region where solutes are more concentrated. You make three sacs from a membrane that water but not sucrose can cross. You fill each sac with a solution that is 2 percent sucrose, then immerse one in a liter of water. You immerse another sac in a solution that is 10 percent sucrose. And you immerse the third sac in a solution that is 2 percent sucrose.

In each experiment, tonicity dictates the extent and direction of water movement. *Tonicity* refers to relative solute concentrations of two fluids on opposite sides of a membrane. When the two fluids have different solute concentrations, the one with fewer solutes in it is **hypotonic** with respect to the other. The fluid with more solutes is **hypertonic**. Two fluids that have the same solute concentration are **isotonic**. Water diffuses from a hypotonic fluid to a hypertonic fluid.

Normally, the fluid inside of your cells is isotonic with respect to the fluid outside. If the fluid outside becomes hypotonic, too much water will diffuse into those cells and make them burst. If it gets hypertonic, water will diffuse out, and the cells will shrivel.

Most cells have built-in mechanisms that adjust to changes in tonicity. Red blood cells do not. Figure 4.16 shows what happens to them when tonicity changes.

EFFECTS OF FLUID PRESSURE

The transport of solutes across the plasma membrane keeps animal cells from bursting. Cells of plants and many protists, fungi, and bacteria avoid bursting with the help of pressure on their cell walls.

Pressure differences as well as solute concentrations affect osmosis. Figure 4.17 shows how water diffuses from a hypotonic solution into a hypertonic solution until the concentration of solutes is the same on both sides of the membrane. The *volume* of the hypertonic solution increases (because its solutes cannot cross the membrane). **Hydrostatic pressure** is the force that any volume of fluid exerts against a wall, a membrane, or some structure enclosing it. (In plants, this pressure is called *turgor*.) Hydrostatic pressure that has built up

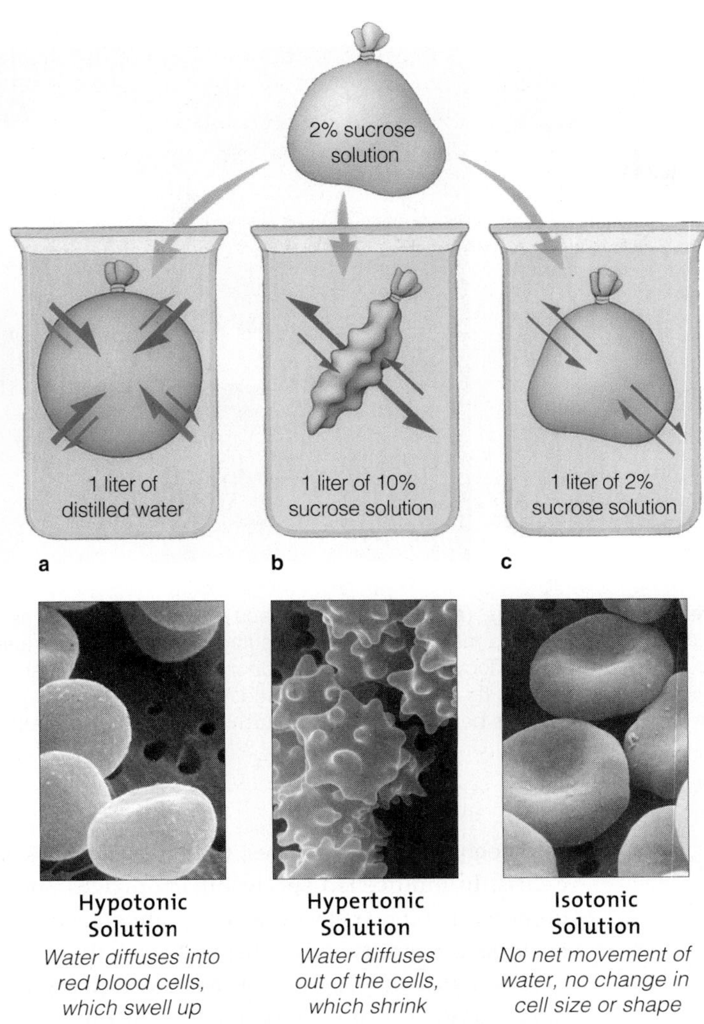

Figure 4.16 ***Animated!*** Tonicity and the direction of water movement. (**a–c**) Inside each of three containers, arrow widths signify the direction and the relative amounts of flow. The micrographs below each sketch show the shape of a human red blood cell that is immersed in fluids of higher, lower, or equal concentrations of solutes. The solutions inside and outside red blood cells are normally isotonic. This type of cell has no way to adjust to drastic change in solute levels in its fluid surroundings.

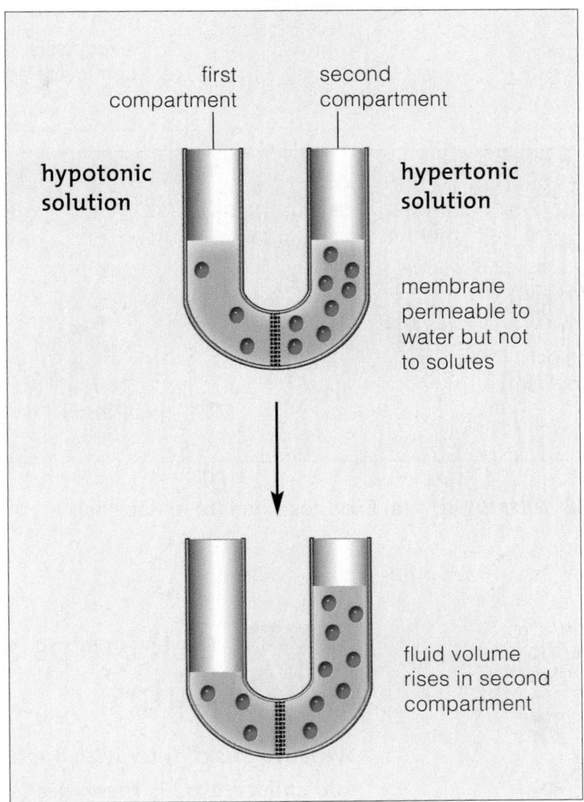

Figure 4.17 ***Animated!*** Experiment showing an increase in fluid volume as an outcome of osmosis. A selectively permeable membrane separates two compartments. Eventually, the solute concentration will be the same on both sides of the membrane, but the fluid volume in the second compartment will be greater because there are more solute molecules in it.

in a cell can counteract any further inward diffusion of water. The amount of hydrostatic pressure that must be applied in order to halt the osmosis of water into a hypotonic solution is called **osmotic pressure**.

As a young plant cell grows, many vesicles coalesce into one large central vacuole. Water diffuses into the vacuole and puts more fluid pressure on the pliable primary cell wall. The wall expands, so the cell volume increases. The continued expansion of the wall (and of the cell) ends when internal fluid pressure is enough to counteract the uptake of more water.

Plant cells are vulnerable to water losses, which can occur when soil dries out or becomes too salty. Water stops diffusing in and starts diffusing out, so internal fluid pressure falls, cytoplasm shrinks, and the plant wilts. In later chapters, you will see how fluid pressure affects the distribution of water and solutes inside the body of plants and animals.

Osmosis is a net diffusion of water between two solutions that differ in solute concentration and are separated by a selectively permeable membrane. The greater the number of molecules and ions dissolved in a solution, the lower its water concentration.

Water tends to move osmotically to regions of greater solute concentration (from hypotonic to hypertonic solutions). There is no net diffusion between isotonic solutions.

Fluid pressure that a solution exerts against a membrane or wall also influences the osmotic movement of water.

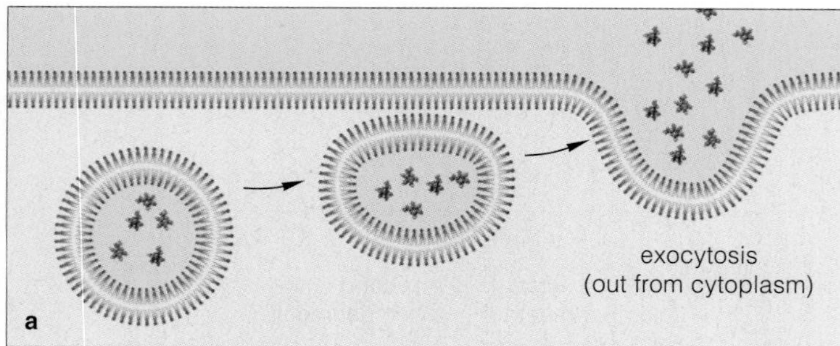

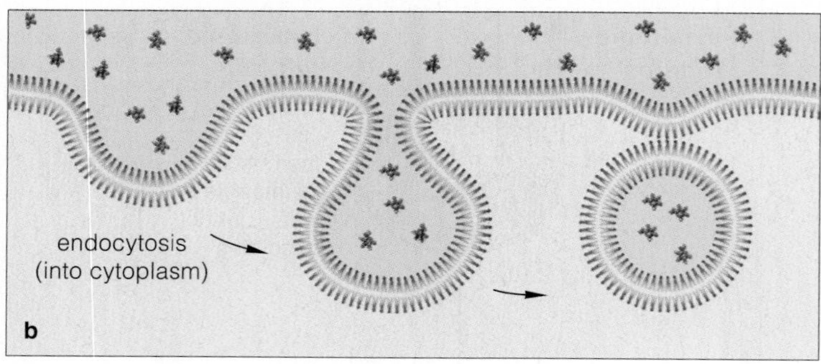

Figure 4.18 ***Animated!*** (**a**) Exocytosis and (**b**) endocytosis.

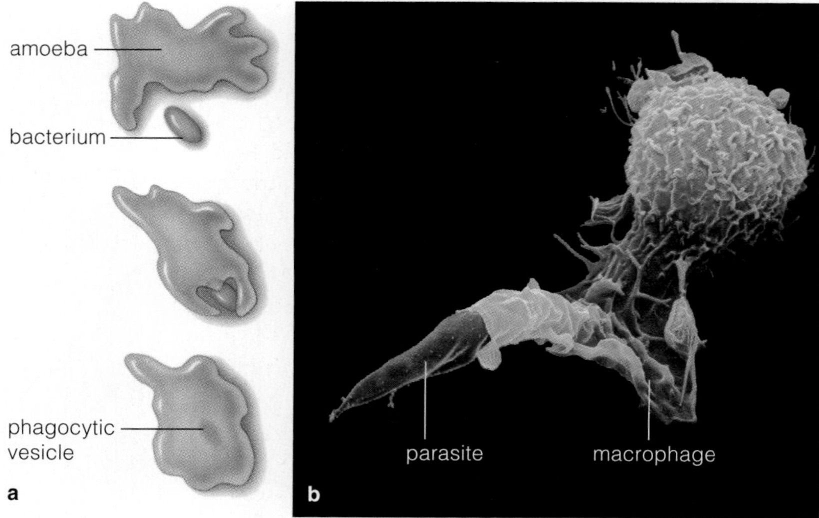

Figure 4.19 Phagocytosis. (**a**) Lobes of an amoeba's cytoplasm extend around a bacterial cell. The plasma membrane extensions flow over the cell and fuse, forming a vesicle. This endocytic vesicle moves deeper into the cytoplasm and fuses with lysosomes. Its contents are digested. (**b**) Macrophage engulfing *Leishmania mexicana.* This parasitic protozoan causes a dangerous disease.

LINKS TO SECTIONS 3.5, 3.7

4.7 Cell Burps and Gulps

We leave this chapter with another look at exocytosis and endocytosis. By these mechanisms, vesicles move substances to and from the plasma membrane. Vesicles help the cell take in and expel bulk amounts of items that are more than transport proteins can handle.

Think back on the membrane traffic to and from a cell surface (Figure 3.9). By **exocytosis**, a vesicle moves to the surface, and the protein-studded lipid bilayer of its membrane fuses with the plasma membrane. As this exocytic vesicle is losing its identity, its contents are released to the surroundings (Figure 4.18*a*).

There are three pathways of **endocytosis**, but all take up substances near the cell surface. A small patch of plasma membrane balloons inward and pinches off inside the cytoplasm, forming an endocytic vesicle that moves its contents to some organelle or stores them in a cytoplasmic region (Figure 4.18*b*).

With *receptor-mediated* endocytosis, receptors at the membrane bind to molecules of a hormone, vitamin, mineral, or another substance. A tiny pit forms in the plasma membrane beneath the receptors. The pit sinks into the cytoplasm and closes back on itself, and in this way it becomes a vesicle.

Phagocytosis ("cell eating") is a common endocytic pathway, but it is an active form of it. Phagocytes such as amoebas engulf microbes, food particles, or cellular debris. In multicelled species, macrophages and some other white blood cells do this to pathogenic viruses or bacteria, cancerous body cells, and other threats.

Phagocytosis also involves receptors. Receptors that bind a target cause microfilaments to form a mesh just beneath the phagocyte's plasma membrane. When the microfilaments contract, they squeeze some cytoplasm toward the margins of the cell, forming a bulging lobe called a pseudopod (Figure 4.19). Pseudopods flow all around the target and form a vesicle. This sinks into the cytoplasm and fuses with lysosomes, the organelles of intracellular digestion. Lysosomes digest trapped items into fragments and smaller, reusable molecules.

Bulk-phase endocytosis is not as selective. A vesicle forms around a small volume of the extracellular fluid no matter what kinds of substances are dissolved in it. Bulk-phase endocytosis constantly removes patches of plasma membrane, balancing the steady additions that arrive in the form of exocytic vesicles.

IMPACTS, ISSUES

Alcohol's effect on the body's phagocytes is one reason alcoholics are especially prone to infectious diseases. Alcohol interferes with the production of these white blood cells, and it also hampers the way they normally follow chemical gradients to infected tissues.

Whereas transport proteins in plasma membranes deal only with ions and small molecules, exocytosis and endocytosis move large packets of materials across a plasma membrane.

By exocytosis, a cytoplasmic vesicle fuses with the plasma membrane, and its contents are released outside the cell.

By endocytosis, a small patch of the plasma membrane sinks inward and seals back on itself, forming a vesicle inside the cytoplasm. Membrane receptors often mediate this process.

Summary

Section 4.1 Cells require energy for metabolism, or chemical work to grow and maintain themselves.

Energy flows in one direction, from usable to less usable forms. Life maintains its complex organization by being resupplied with energy lost from someplace else. Sunlight is the primary energy source for the web of life.

ATP, the main energy carrier, couples reactions that release energy with reactions that require it. It primes molecules to react through phosphate-group transfers.

Electron transfers, or oxidation–reduction reactions, often proceed in series at cell membranes.

BiologyNow

Learn about energy changes in chemical reactions and the role of ATP, and compare controlled and uncontrolled energy release with the animations on BiologyNow.

Section 4.2 Metabolic pathways are stepwise, enzyme-mediated reaction sequences. Photosynthesis and other energy-requiring, *biosynthetic* pathways build larger molecules from smaller ones. Energy-releasing, *degradative* pathways such as aerobic respiration break down large molecules to smaller products. Table 4.1 lists the participants.

Cells increase, maintain, and lower concentrations of substances by coordinating thousands of reactions. They rapidly shift rates of metabolism by controlling a few steps of reversible pathways.

Section 4.3 Enzymes are catalysts; they enormously enhance rates of specific reactions. Pockets or cavities in these big molecules create favorable microenvironments for the reaction; these are the active sites.

Activation energy for a reaction is the energy it takes to align reactive chemical groups, destabilize electric charges, and break bonds.

Enzymes lower the activation energy by boosting concentrations of substrates in the active site, orienting substrates in positions that favor reaction, shutting out most or all water, and often changing substrate shape.

Cofactors (metal ions, coenzymes, or both) help an enzyme catalyze a reaction.

Controls over enzyme action influence the kinds and amounts of substances available. Enzymes function best within a limited range of temperature, pH, and salinity.

BiologyNow

Investigate how enzymes facilitate reactions, and mechanisms of enzyme control with the animations and interactions on BiologyNow.

Section 4.4 Molecules or ions of a substance tend to move from regions of higher to lower concentration. Diffusion is movement in response to a concentration gradient. Its rates are influenced by the steepness of the concentration gradient and by temperature, molecular size, and gradients in electrical charge and pressure.

Built-in cellular mechanisms work with and against gradients to move solutes across membranes.

Table 4.1 Participants in Metabolic Reactions

Reactant	Substance that enters a metabolic reaction or pathway; also called an enzyme's substrate
Intermediate	Substance that forms in a reaction or pathway, between the reactants and the end products
Product	Substance left at end of a reaction or pathway
Enzyme	A protein that greatly enhances reaction rates; a few RNAs also do this
Cofactor	Coenzyme (such as NAD^+) or metal ion; assists enzymes or taxis electrons, hydrogen, or functional groups between reaction sites
Energy carrier	Mainly ATP; couples energy-releasing reactions with energy-requiring ones
Transport protein	Protein that passively assists or actively pumps specific solutes across a cell membrane

Molecular oxygen, carbon dioxide, and other small nonpolar molecules diffuse across a membrane's lipid bilayer. Ions and large, polar molecules such as glucose cross it with the help of transport proteins. Some water moves through proteins and some crosses the bilayer.

BiologyNow

Investigate diffusion across membranes with the interaction on BiologyNow.

Section 4.5 Transport proteins move water-soluble substances across a membrane by reversible changes in

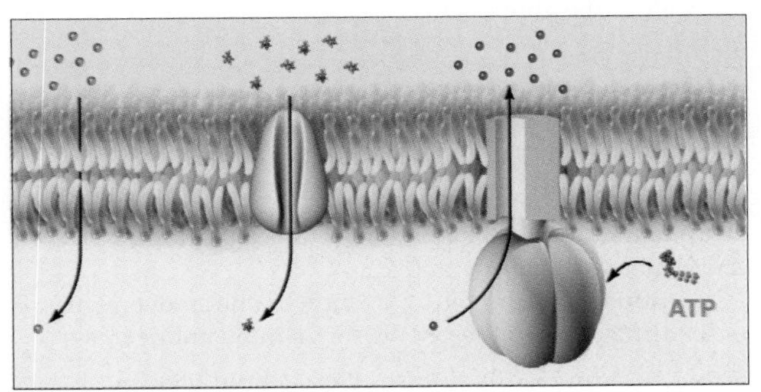

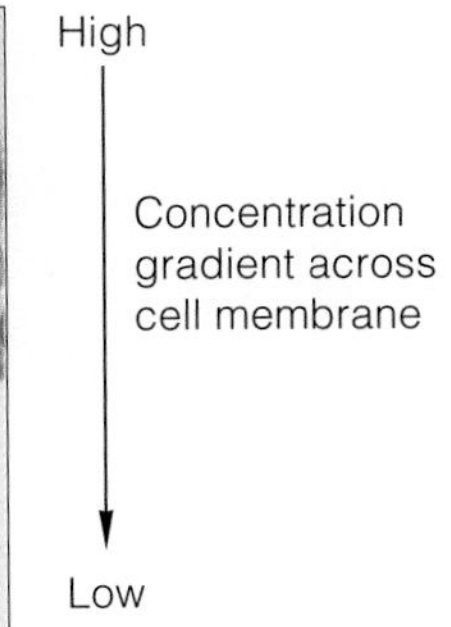

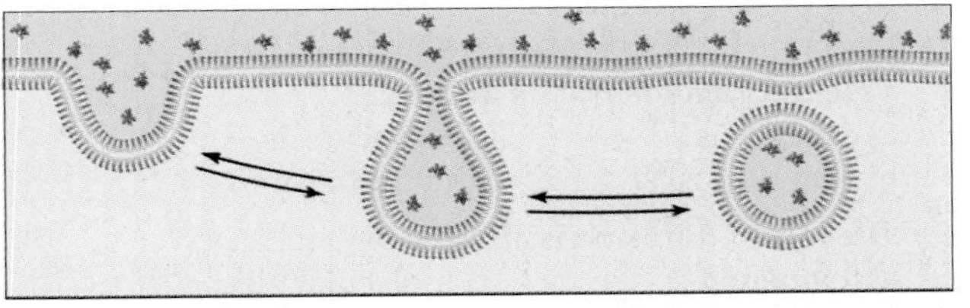

Figure 4.20 Summary of membrane crossing mechanisms.

shape. Passive transport does not require energy inputs. Active transport requires an energy input from ATP to move a solute against its concentration gradient.

BiologyNow

Compare the processes of passive and active transport, using the animation on BiologyNow.

Section 4.6 Osmosis is the diffusion of water across a selectively permeable membrane. The water molecules move down the water concentration gradient, which is influenced by solute concentrations and pressure.

BiologyNow

Explore the effects of osmosis with the interaction and animation on BiologyNow.

Section 4.7 By exocytosis, a cytoplasmic vesicle fuses with the plasma membrane, and its contents are released outside (Figure 4.20). By endocytosis, a patch of plasma membrane forms a vesicle that sinks into the cytoplasm.

BiologyNow

Use the animation on BiologyNow to discover how membrane components are cycled.

Self-Quiz

Answers in Appendix I

1. ______ is life's primary source of energy.
   a. Food b. Water c. Sunlight d. ATP

2. If we liken a chemical reaction to an energy hill, then an ______ reaction is an uphill run.
   a. endergonic
   b. exergonic
   c. ATP-assisted
   d. both a and c

3. Which of the following statements is correct? A metabolic pathway ______ .
   a. has an orderly sequence of reaction steps
   b. is mediated by enzymes
   c. may be biosynthetic or degradative, overall
   d. all of the above

4. An enzyme ______ .
   a. is usually a protein
   b. lowers the activation energy of a reaction
   c. is destroyed by the reaction it catalyzes
   d. a and b

5. Immerse a living cell in a hypotonic solution, and water will tend to ______ .
   a. diffuse into the cell
   b. diffuse out of the cell
   c. show no net movement
   d. move in by endocytosis

6. ______ can readily diffuse across a lipid bilayer.
   a. Glucose
   b. Oxygen
   c. Carbon dioxide
   d. b and c

7. Sodium ions cross a membrane at transport proteins that receive an energy boost. This is a case of ______ .
   a. passive transport
   b. active transport
   c. facilitated diffusion
   d. a and c

8. Vesicle formation occurs in ______ .
   a. exocytosis
   b. phagocytosis
   c. endocytosis
   d. all of the above

9. The rate of diffusion is affected by ______ .
   a. temperature
   b. electrical gradients
   c. molecular size
   d. all of the above

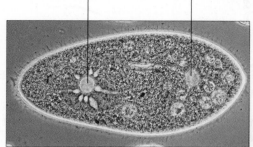

Figure 4.21 *Paramecium.*

10. Match the substance with its suitable description.

____	coenzyme or metal ion	a. reactant
____	adjusts gradients at membrane	b. enzyme
____	substance entering a reaction	c. cofactor
____	substance formed during a reaction	d. intermediate
____	substance at end of reaction	e. product
____	enhances reaction rate	f. energy carrier
____	mainly ATP	g. transport protein

Additional questions are available on **BiologyNow™**

Critical Thinking

1. Often, beginning physics students are taught the basic concepts of thermodynamics with two phrases: a) You can't win; and b) You can't break even. Explain these phrases in terms of the two laws of thermodynamics.

2. Water moves osmotically into *Paramecium*, a single-celled aquatic protist. If unchecked, the influx would bloat the cell and rupture its plasma membrane, killing the cell. An energy-requiring mechanism that involves contractile vacuoles expels excess water (Figure 4.21). Water enters the vacuole's tubelike extensions and collects inside. A full vacuole contracts and squirts water out of the cell through a pore. Are *Paramecium's* surroundings hypotonic, hypertonic, or isotonic?

3. Some nutritional supplements include plant enzymes. Explain why it is not likely that plant enzymes will aid your digestion.

4. Why do you think applying lemon juice to sliced apples keep them from turning brown?

5. Explain why hydrogen peroxide bubbles when you dribble it on an open cut but does not bubble on skin that is unbroken.

6. Most of the cultivated fields in California are heavily irrigated. Over the years, most of the imported water has evaporated from the soil, leaving behind solutes. What problems will the altered soil cause plants?

7. Imagine you're a juvenile shrimp in an estuary, where fresh water draining from the land mixes with saltwater from the sea. Many people own homes around a lake, and they want boat access to the sea. They ask their city for permission to build a canal to your estuary. If they succeed, what is likely to happen to you?

5 WHERE IT STARTS—PHOTOSYNTHESIS

IMPACTS, ISSUES

Sunlight and Survival

Think about the last piece of apple, lettuce, chicken, pizza, or any other food you put in your mouth. Where did it come from? Look beyond the refrigerator, the market or restaurant, and the farm. Look to plants, the starting point for nearly all of the food—the carbon-based compounds—you eat.

All plants are **autotrophs**, or "self-nourishing" organisms. Autotrophs use energy and carbon from nonliving sources to build their own food, the organic compounds that sustain them and, ultimately, every other living thing. Most bacteria, many protists, and all fungi and animals are not autotrophs. Like you, they are **heterotrophs**, which feed on autotrophs, one another, and organic wastes and remains.

Plants are a type of *photo*autotroph. By the process of **photosynthesis**, they use energy from sunlight and carbon from carbon dioxide to make sugar. Each year, plants make 220 billion tons of sugar.

That is a LOT of sugar.

They also release a LOT of oxygen in the process.

It wasn't always this way. The first prokaryotic cells probably were *chemo*autotrophs. Like some existing archaea, they built organic compounds using energy extracted from simple inorganic chemicals such as hydrogen and ammonia. Both gases were part of the nasty brew that was Earth's early atmosphere.

Things did not change much for about a billion years, then light-sensitive molecules evolved in a few lineages. The immense supply of energy from the sun began to be tapped by these first photoautotrophs. Not long after this, the photosynthetic machinery in their descendants became remodeled in a new way. Water molecules, an abundant resource, were used as a source of electrons for the reactions. Oxygen atoms released from uncountable water molecules diffused out of the photosynthetic cells into the environment. The atmosphere changed forever.

Free oxygen (O_2) reacts fast with metals, including enzyme cofactors. In these reactions, toxic oxygen radicals form. The new abundance of oxygen in the environment put selection pressure on all populations. Lineages that could not neutralize the oxygen radicals became extinct or marginalized in muddy sediments, deep water, or other anaerobic (oxygen-free) habitats.

As you will read at the chapter's end, detoxification pathways evolved in some lineages. One pathway, aerobic respiration, allowed cells to use the reactive properties of oxygen with splendid metabolic results.

As you read this chapter on photosynthesis, keep in mind that its emergence, and its continuity, are a big reason why *you* can exist, and read this book, and think about what it takes to stay alive.

How Would You Vote? *The oxygen in Earth's atmosphere is a sure sign that photosynthetic organisms flourish here. New technology allows astronomers in search of alien life to measure the oxygen content of atmospheres surrounding planets too distant to visit. Should public funds be used to continue this research? See BiologyNow for details, then vote online.*

Key Concepts

CATCHING THE RAINBOW

Energy enters the world of life when chlorophyll and other pigments absorb energy in the sun's rays. Plants, some bacteria, and many protists capture sunlight energy to make glucose and other carbohydrates in a process called photosynthesis. In plant cells and protists, photosynthesis occurs in two stages, in chloroplasts.

MAKING ATP AND NADPH

In the first stage of photosynthesis, sunlight energy becomes converted to chemical bond energy in ATP. NADPH forms and free oxygen escapes into the air.

MAKING SUGARS

The second stage is the "synthesis" part of photosynthesis. Enzymes assemble sugars from atoms of carbon and oxygen (delivered by carbon dioxide). These synthesis reactions use energy and hydrogen atoms from the ATP and NADPH that formed in the first stage of photosynthesis.

Links to Earlier Concepts

Before diving into photosynthesis, you may want to look at Section 2.1 to review electron energy levels, and particularly how photons and electrons interact. You will be using your knowledge of carbohydrate structure (2.6), chloroplasts (3.5), active transport proteins (4.5), and concentration gradients (4.4).

Remember the concepts of energy flow and the underlying organization of life (4.1 and 4.2)? They will help you understand how energy flows through photosynthesis reactions. This chapter also offers you a prime example of how cells harvest energy by using electron transfer chains (4.1).

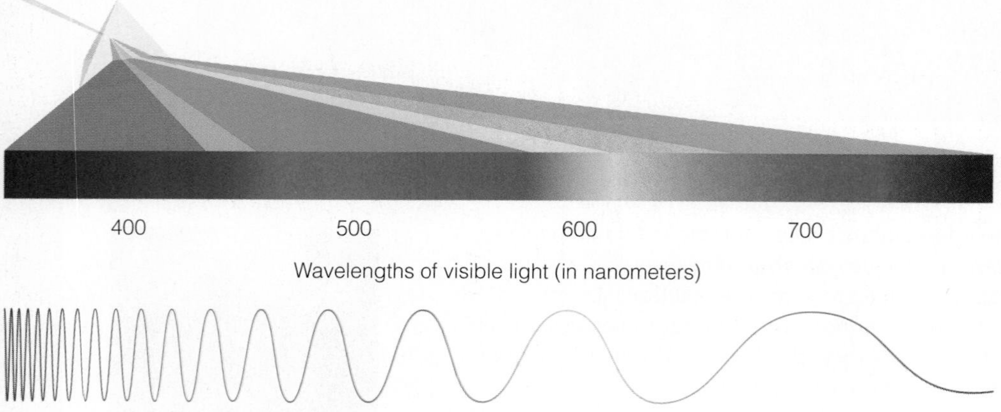

Figure 5.1 Light. Radiant energy from the sun undulates across space in waves. The horizontal distance between the crests of two successive waves is a wavelength. Visible light is a very small part of the electromagnetic spectrum, which includes all radiant energy.

Although light travels in waves, its energy is organized in packets called photons. The shorter the wavelength of a photon, the higher its energy. Photosynthesizers use photons with wavelengths between 380 and 750 nanometers. Like many other organisms, we can see these wavelengths of visible light as different colors. Radiant energy with shorter wavelengths, including ultraviolet (UV) light, is so energetic it can alter or break the chemical bonds of DNA and proteins.

Figure 5.2 Pigments in the changing leaves of autumn. In green leaves, photosynthetic cells are making chlorophylls, which mask carotenoids, xanthophylls (a kind of anthocyanin), and other accessory pigments.

In autumn, chlorophyll synthesis lags behind its breakdown in many species of deciduous trees. Accessory pigments then show through, giving the leaves characteristic red, orange, and yellow colors.

LINKS TO SECTIONS 2.1, 3.5

5.1 The Rainbow Catchers

Plants do something you'll never do. Using the metabolic process of photosynthesis, they make their own food from no more than light, water, and carbon dioxide.

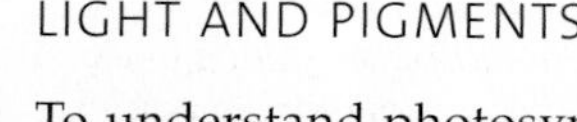

LIGHT AND PIGMENTS

To understand photosynthesis, you will need to know about some properties of sunlight energy. That energy undulates across space like waves moving across a sea. The distance between each wave is called **wavelength**. Wavelengths of light are measured in nanometers, or billionths of a meter. The shorter the wavelength, the greater the energy.

Our story starts with energy in the sun's rays—not all of it, just light of wavelengths between 380 and 750 nanometers (Figure 5.1). These are the wavelengths of visible light, the only ones that drive photosynthesis. Humans and many other organisms see the different wavelengths of visible light as different colors.

Pigments are molecules that capture light energy. They are the key molecular bridges from sunlight to photosynthesis. Each kind of pigment can absorb only specific wavelengths of light. All other wavelengths are not absorbed, but instead are transmitted through the molecule or reflected by it.

Reflected light gives each pigment its characteristic color. For example, the primary pigment in almost all photoautotrophs, **chlorophyll *a***, absorbs mainly red and blue-violet light. It reflects all other light—which is mainly green. That is why plants that have a lot of chlorophyll *a* appear intensely green.

Accessory pigments capture other wavelengths of light. **Chlorophyll *b*** reflects green and yellow light, so it is yellow-green. Red, orange, and yellow **carotenoids** impart color to some flowers, fruits, and vegetables. **Xanthophylls** may be yellow, brown, purple, or blue. **Anthocyanins** color some red or purple flowers and foods, including cherries, blueberries, and rhubarb. Accessory pigments are masked by green chlorophylls in the leaves of many plants until autumn (Figure 5.2).

Phycobilins, which are red or blue-green, are the signature pigments of red algae and cyanobacteria. A few prokaryotes of ancient lineages still use unusual pigments like the intensely purple bacteriorhodopsin.

When you were reading about chemical bonds in Section 2.2, you might have wondered why you would ever need to know that information. Here is one case. A pigment has at least one region of atoms in which single covalent bonds alternate with double covalent bonds. It acts as an antenna that receives light energy. The light-catching portion of a chlorophyll is its ring structure, which holds a magnesium atom (Figure 5.3).

Energy inputs, remember, boost electrons to higher energy levels. Energy is absorbed by a pigment only if it matches the energy required to boost an electron of the pigment's antenna region to a higher energy level. A boosted electron quickly returns to a lower energy level, emitting its extra energy. Photosynthesizers can harvest the surplus energy of an excited electron by converting it to chemical bond energy of ATP.

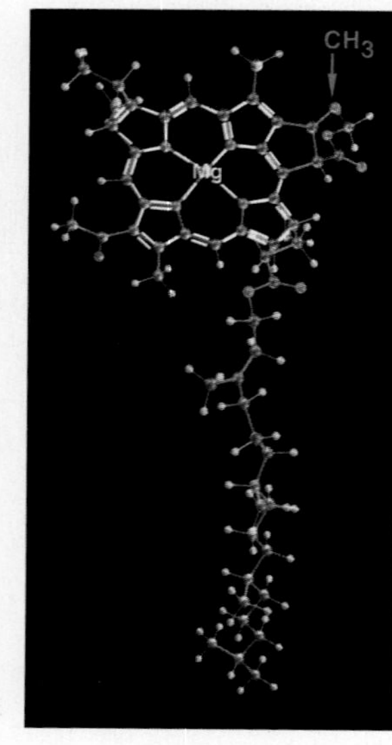

Figure 5.3 Model for chlorophyll *a*. Chlorophyll *b* differs by only a single functional group (COO^- instead of the CH_3 shown here).

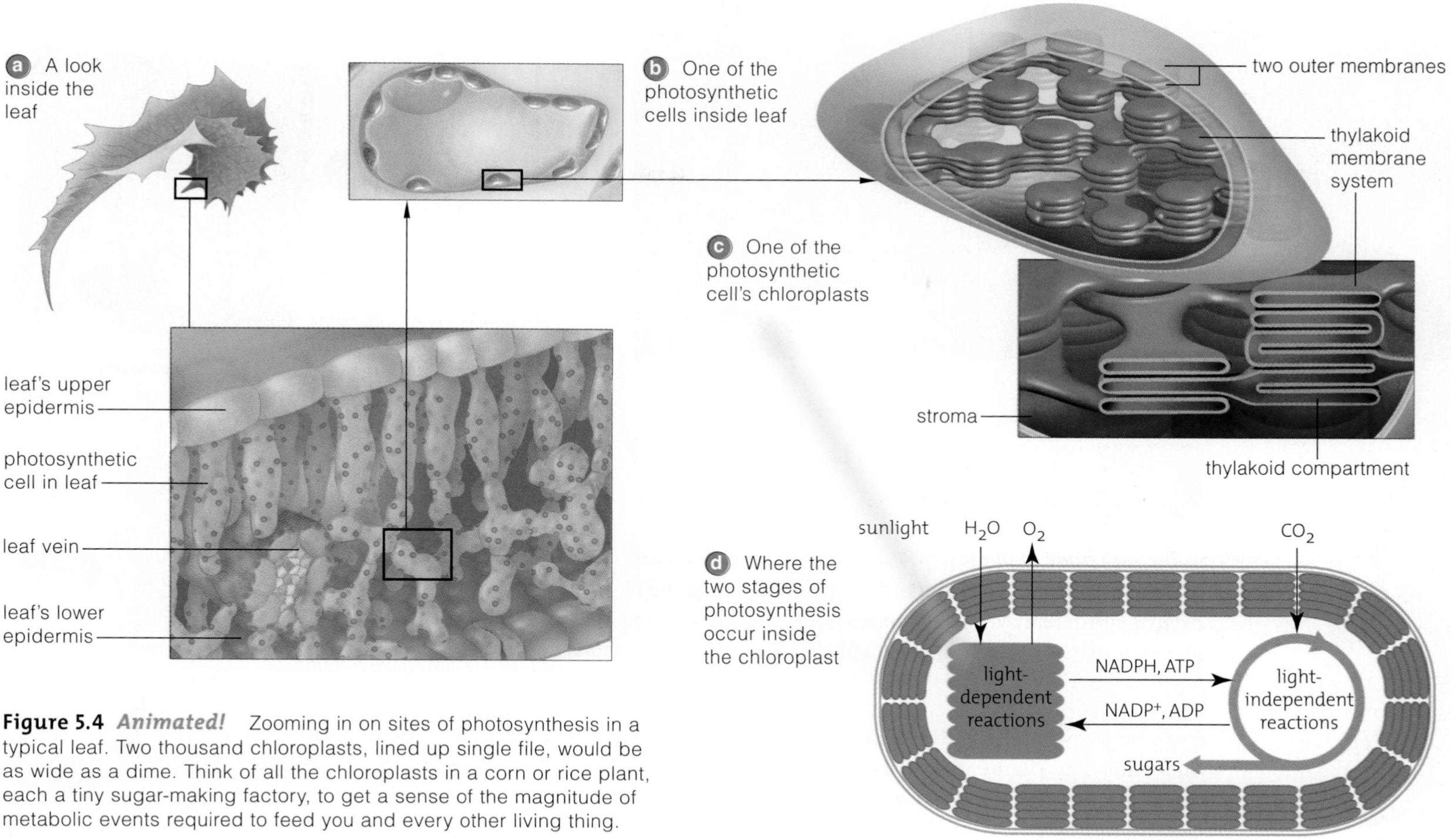

a A look inside the leaf

b One of the photosynthetic cells inside leaf

c One of the photosynthetic cell's chloroplasts

two outer membranes

thylakoid membrane system

stroma

thylakoid compartment

leaf's upper epidermis

photosynthetic cell in leaf

leaf vein

leaf's lower epidermis

d Where the two stages of photosynthesis occur inside the chloroplast

sunlight H_2O O_2 CO_2

light-dependent reactions

NADPH, ATP

$NADP^+$, ADP

light-independent reactions

sugars

Figure 5.4 ***Animated!*** Zooming in on sites of photosynthesis in a typical leaf. Two thousand chloroplasts, lined up single file, would be as wide as a dime. Think of all the chloroplasts in a corn or rice plant, each a tiny sugar-making factory, to get a sense of the magnitude of metabolic events required to feed you and every other living thing.

TWO STAGES OF REACTIONS

Photosynthesis proceeds in two reaction stages. In the first stage—the **light-dependent reactions**—sunlight energy is converted to chemical bond energy of ATP. Water molecules are split, and typically the coenzyme $NADP^+$ picks up the released hydrogen and electrons, thus becoming NADPH. The oxygen atoms released from water molecules escape into the surroundings.

The second stage, the **light-independent reactions**, runs on energy donated by the ATP. Glucose and other carbohydrates are made with carbon and oxygen atoms (secured from carbon dioxide and water), and with the hydrogen and electrons delivered by NADPH.

Photosynthesis is often summarized this way:

$$\underset{\text{water}}{12H_2O} + \underset{\text{carbon dioxide}}{6CO_2} \xrightarrow[\text{enzymes}]{\text{light energy}} \underset{\text{oxygen}}{6O_2} + \underset{\text{glucose}}{C_6H_{12}O_6} + \underset{\text{water}}{6H_2O}$$

Let's focus on what goes on inside **chloroplasts**, the organelles of photosynthesis in all plants and certain protists. A chloroplast has a semifluid interior, called the **stroma**, enclosed by two outer membranes (Figure 5.4). A third membrane—the **thylakoid membrane**—forms a compartment inside the stroma. It folds back on itself into what often look like stacks of pancakes (thylakoids) connected by channels. The space inside the folds is one continuous compartment. The light-dependent reactions occur at the thylakoid membrane, and sugars are built in the stroma.

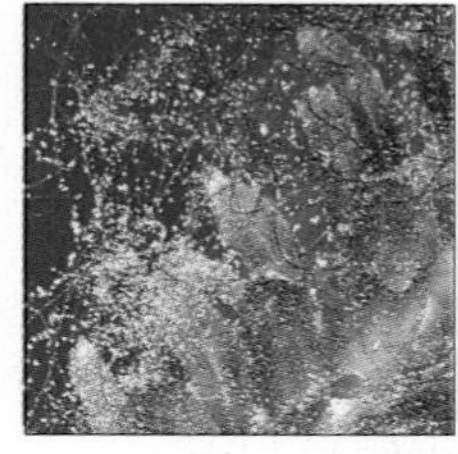

Oxygen bubbling away from leaves of *Elodea*.

Chlorophylls and accessory pigments absorb wavelengths of visible light. They are the key molecular bridge between sunlight and photosynthesis.

In the first stage of photosynthesis, sunlight energy drives formation of ATP and NADPH, and oxygen is released. In chloroplasts, the reactions proceed at an inner membrane system of interconnected, flattened sacs called thylakoids.

The second stage occurs in the stroma. Energy from ATP drives the synthesis of sugars. Carbon dioxide provides carbon and oxygen atoms for the reactions, and NADPH delivers electrons and hydrogen atoms.

The net reaction of photosynthesis is the conversion of water and carbon dioxide to glucose and oxygen.

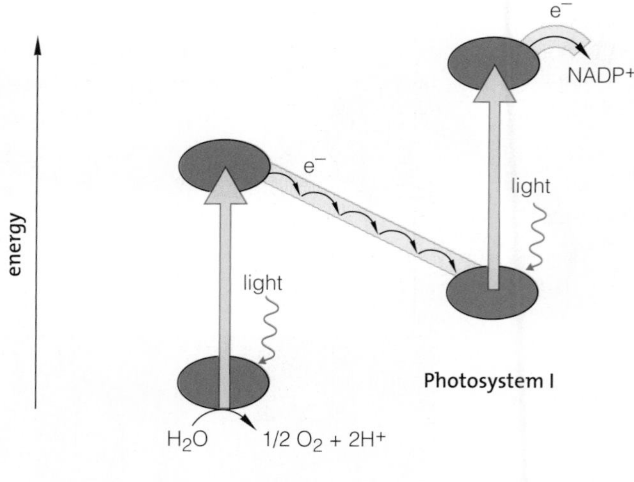

LINKS TO SECTIONS 4.1, 4.2, 4.4, 4.5

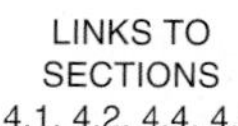

5.2 Light-Dependent Reactions

In the first stage of photosynthesis, sunlight energy is used to drive ATP formation. Water molecules are split, and their oxygen diffuses away. NADP+ combines with the remaining electrons and hydrogen to form NADPH.

Figure 5.6 Energy flow in photosynthesis reactions.

(**a**) The noncyclic pathway. Electrons that photosystem II pulls from water move through an electron transfer chain, to photosystem I. Another transfer chain delivers them to NADP+.

(**b**) The cyclic pathway. Electrons from photosystem I move through a transfer chain, and then return to photosystem I.

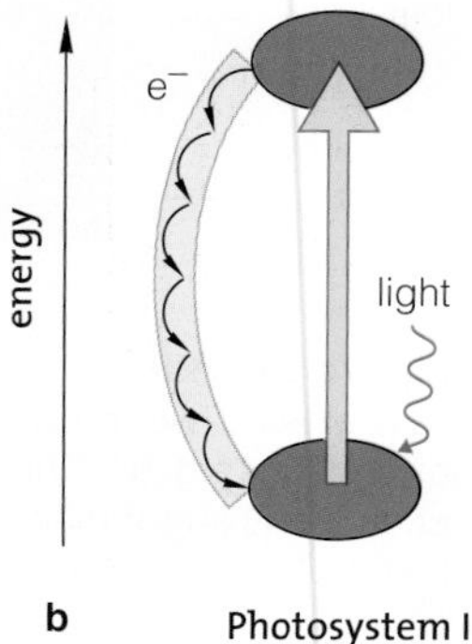

Imagine a ray of light illuminating a single pigment molecule. The pigment absorbs the light energy, and one of its electrons jumps to a higher energy level. The electron quickly loses its extra energy as light or heat when it drops back to its unexcited state.

In the thylakoid membrane, however, energy that excited electrons give up is kept in play. Embedded in the membrane are many hundreds of light-harvesting complexes (Figure 5.5). These are circular clusterings of pigments (chlorophylls, carotenoids, and so on) and other proteins. When the pigments in light-harvesting complexes absorb photon (light) energy, they do not waste it. Electrons of these pigments hold on to energy by passing it back and forth, like a volleyball.

Energy released from one cluster gets passed to another, which passes it on to another, and so on until it arrives at a photosystem. **Photosystems**, or reaction centers, are arrays of hundreds of pigments and other molecules. Chloroplasts have two types, called I and II. Each photosystem is surrounded by hundreds of light-harvesting complexes.

Look back on Figure 5.3, which shows the structure of chlorophyll. Two molecules of chlorophyll *a* are at the center of a photosystem. Their flat rings face each other so closely that both rings act together as one king-sized antenna. When light-harvesting neighbors pass energy to a photosystem, electrons come right off of that special pair of chlorophylls.

The freed electrons immediately enter an electron transfer chain positioned next to the photosystem in the thylakoid membrane. (An electron transfer chain, remember, is an ordered array of enzymes and other molecules that transfer electrons in sequence.) As they transfer electrons, molecules of this chain also transfer hydrogen ions (H^+) across the thylakoid membrane, from the stroma into the inner compartment. They do this again and again, so that concentration and electric gradients form across the thylakoid membrane. The combined force of those gradients attracts the H^+ ions back toward the stroma.

But H^+ cannot diffuse across the membrane's lipid bilayer. It can cross only through channels inside **ATP synthases**, a type of active transport protein. The flow of hydrogen ions through ATP synthases powers the attachment of phosphate groups to ADP molecules in the stroma. In this way, ATP forms.

As long as electrons flow through transfer chains, the cell can keep on producing ATP. But photosystems themselves do not have unlimited electrons to donate. Where do they come from? Photosystem II replaces its lost electrons by pulling them from water molecules—which *are* essentially unlimited. When photosystem II takes their electrons, water molecules dissociate into hydrogen ions and molecular oxygen. The oxygen gas diffuses out of the chloroplast, then out of the cell and into the air. Hydrogen ions stay and contribute to the gradients that drive ATP formation.

Light-Harvesting Complex

Figure 5.5
A ringlike array of pigment molecules can intercept rays of sunlight coming from any direction.

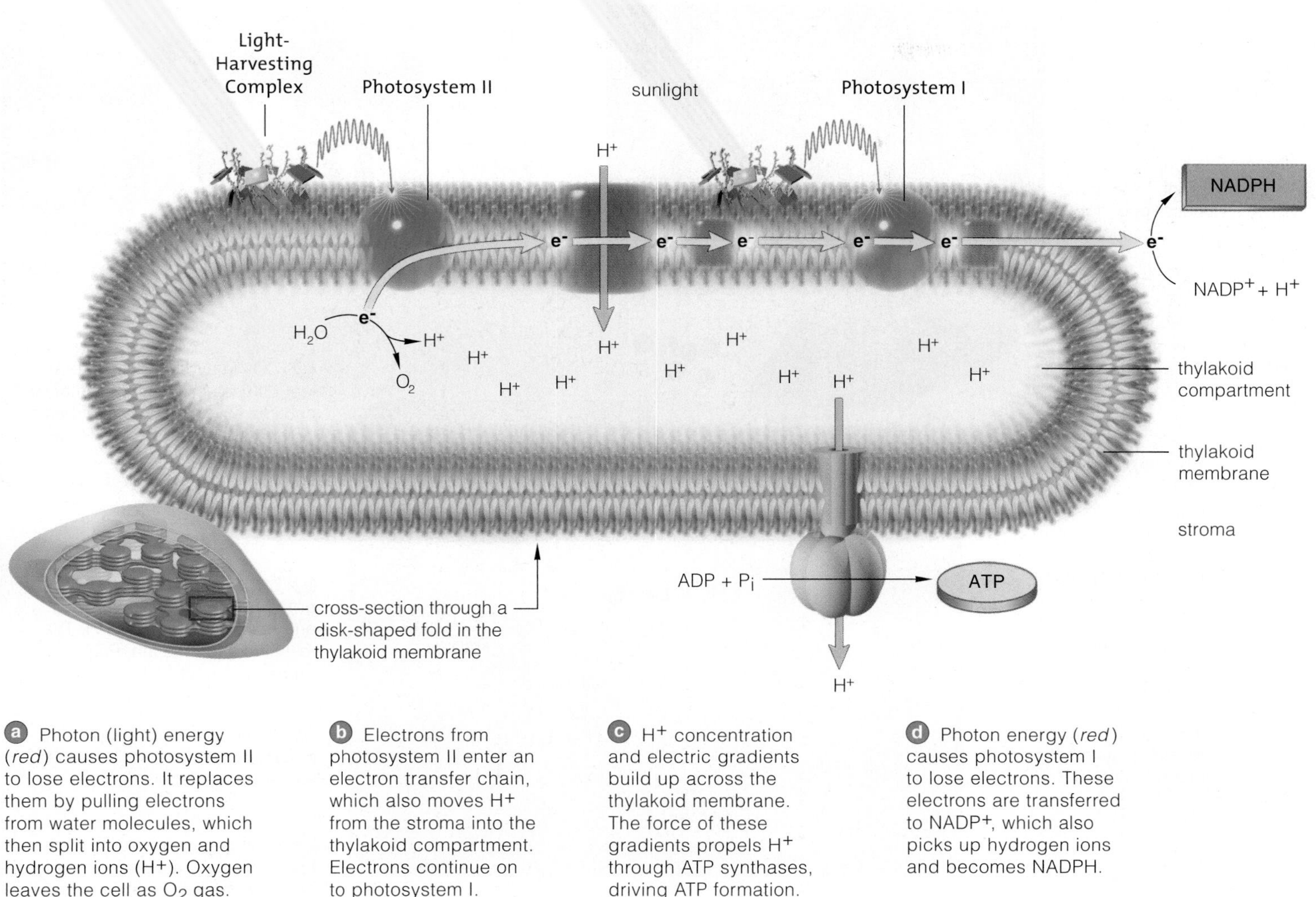

a Photon (light) energy (*red*) causes photosystem II to lose electrons. It replaces them by pulling electrons from water molecules, which then split into oxygen and hydrogen ions (H^+). Oxygen leaves the cell as O_2 gas.

b Electrons from photosystem II enter an electron transfer chain, which also moves H^+ from the stroma into the thylakoid compartment. Electrons continue on to photosystem I.

c H^+ concentration and electric gradients build up across the thylakoid membrane. The force of these gradients propels H^+ through ATP synthases, driving ATP formation.

d Photon energy (*red*) causes photosystem I to lose electrons. These electrons are transferred to $NADP^+$, which also picks up hydrogen ions and becomes NADPH.

Figure 5.7 ***Animated!*** How ATP and NADPH form in the first stage of photosynthesis. The drawing represents a cross-section through one of the disk-shaped folds of the thylakoid membrane.

So where do the electrons end up? After passing through the electron transfer chain, they continue to a type I photosystem. Here, light-harvesting complexes volley energy to a special pair of chlorophylls at the photosystem's reaction center, causing them to release electrons. An intermediary molecule transfers them to $NADP^+$—which attracts hydrogen ions at the same time. In this way, NADPH forms (Figures 5.6*a* and 5.7).

If too much NADPH forms, it accumulates in the stroma, and the photosystem II pathway backs up. In such a case, photosystem I may run independently of photosystem II so the cell can continue to make ATP. In a cyclic pathway, light energy frees electrons from the special chlorophyll pair in photosystem I. The released electrons enter an electron transfer chain that moves hydrogen ions into the thylakoid compartment. The resulting H^+ gradient drives ATP formation. Electrons are recycled back to photosystem I at the end of this transfer chain, and so no NADPH forms. Electron flow in the cyclic pathway is shown in Figure 5.6*b*.

Today, both the cyclic and noncyclic pathways of electron flow operate in all photosynthetic eukaryotes. Which one dominates at a given time depends on the organism's metabolic demands for ATP and NADPH.

In the light-dependent reactions, sunlight energy drives the formation of ATP, NADPH, or both. ATP alone forms in a cyclic pathway that starts and ends with one type of photosystem.

Both ATP and NADPH form by a noncyclic pathway in which electrons flow from split water molecules, through two types of photosystems, and finally to $NADP^+$. This is the oxygen-releasing pathway in photosynthesis.

Leaves of basswood, a typical C3 plant. Far right, basswood leaf cross section.

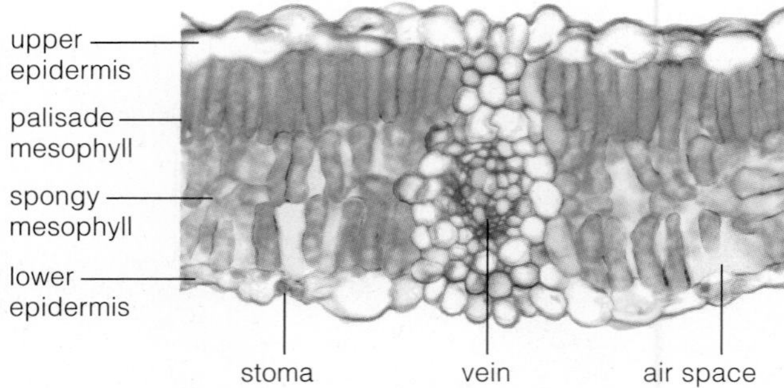

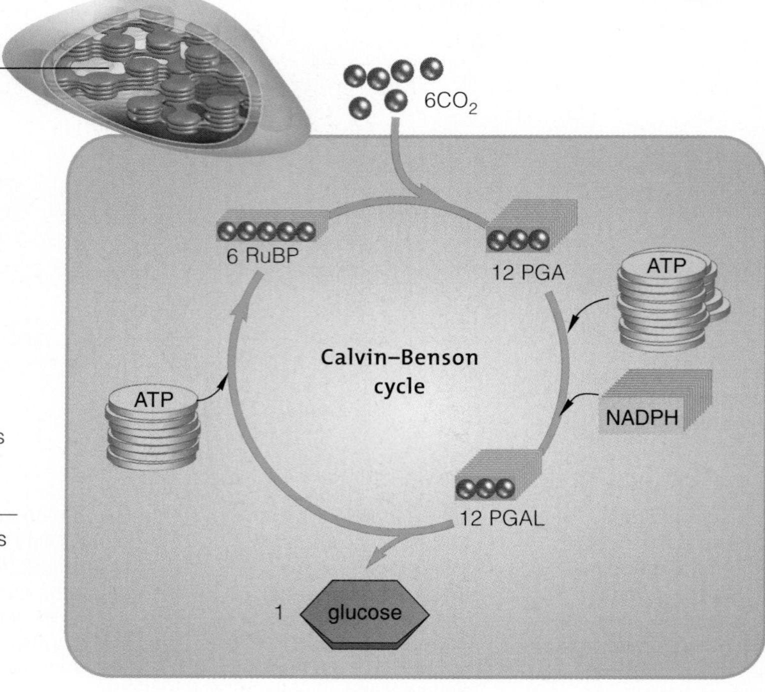

a CO_2 in air spaces inside a leaf diffuses into a photosynthetic cell. Six times, rubisco attaches a carbon atom from CO_2 to RuBP. The resulting intermediate splits each time into two PGAs, for a total of 12 PGAs.

b Each PGA gets a phosphate group from ATP, plus hydrogen and electrons from NADPH. The resulting intermediate, PGAL, is primed for reaction.

c Two of the twelve PGAL molecules combine to form one six-carbon molecule of glucose.

d Glucose enters reactions that form carbohydrate end products of photosynthesis—mainly sucrose, starch, and cellulose.

e Ten of the PGAL get phosphate groups from ATP. In terms of energy, this primes them for an uphill run—for synthesis reactions that regenerate RuBP.

f It takes six turns of the Calvin–Benson cycle (six carbon atoms) to make one glucose molecule in this mesophyll cell.

Figure 5.8 ***Animated!*** Light-independent reactions of photosynthesis. The sketch is a summary of all six turns of the Calvin–Benson cycle and its product, one glucose molecule. *Brown* circles signify carbon atoms.

IMPACTS, ISSUES

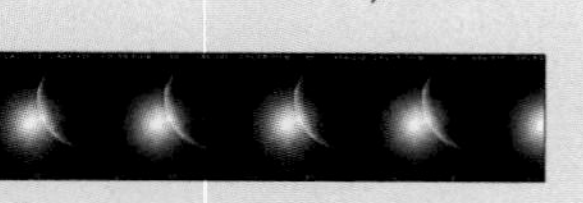

How do phytoplankton get the carbon to make sugars? Many contain enzymes that release CO_2 from bicarbonate dissolved in the water. One diatom acts like a C4 plant; it fixes carbon by the C4 pathway in its cytoplasm and then by the Calvin–Benson cycle in its chloroplasts.

5.3 Light-Independent Reactions

The chloroplast is a sugar factory, and the Calvin–Benson cycle is its machinery. These cyclic, light-independent reactions are the "synthesis" part of photosynthesis.

THE SUGAR FACTORY

Sugars form in the **Calvin–Benson cycle**, a pathway that runs inside the stroma of chloroplasts (Figure 5.8). The cycle also is called light-*independent* because light is not needed to run the reaction. It uses the ATP and NADPH that formed in the light-dependent reactions. Plants get their carbon and oxygen building blocks from carbon dioxide (CO_2) in the air. Algae of aquatic habitats get them from CO_2 dissolved in water.

The enzyme **rubisco** joins a carbon from CO_2 to five-carbon ribulose bisphosphate (RuBP). The resulting unstable intermediate is the molecule that starts the Calvin–Benson cycle. It splits at once into two stable molecules of phosphoglycerate (PGA), each having a backbone of three carbons. The process of securing carbon from the environment by incorporating it into a stable organic compound is called **carbon fixation**.

Each PGA gets a phosphate group from ATP, and hydrogen and electrons from NADPH. For every six CO_2 molecules fixed, twelve phosphoglyceraldehydes (PGAL) form. Ten PGAL become rearranged in a way that regenerates RuBP. The other two combine to make a six-carbon glucose with a phosphate group attached.

Most of the glucose is converted at once to sucrose or starch by other pathways that conclude the light-independent reactions. Sucrose is a transportable form of carbohydrate in plants; starch is the main storage form. Plants use both as energy sources, and also as building blocks for other complex molecules.

DIFFERENT PLANTS, DIFFERENT PATHWAYS

If sunlight intensity, air temperature, rainfall, and soil composition never varied, photosynthesis might be the same in all plants. But environments differ, and so do details of photosynthesis. For example, you see such differences on hot days when water is scarce.

A waxy cuticle stops water loss from leaves. Water and gases move into and out of leaves only across tiny openings called **stomata** (singular, stoma), as in Figure 5.9. Stomata close on hot, dry days, so water and O_2 do not exit leaves and CO_2 does not enter them. Plants cannot make as much sugar when their photosynthetic cells are exposed to too much O_2 and too little CO_2.

That's why beans, sunflowers, and other **C3 plants** don't grow well without irrigation in hot, dry climates. Their name is a nod to the *three*-carbon PGA, the first stable intermediate to form during the Calvin–Benson cycle. When oxygen builds up in C3 leaves, rubisco binds oxygen, not CO_2. This alternate reaction yields CO_2, so the plant loses carbon instead of fixing it.

The C4 cycle evolved independently over millions of years, in many lineages. In **C4 plants**, *four*-carbon oxaloacetate forms first in reactions that run through two cell types. In mesophyll cells, the C4 cycle will fix carbon no matter how much O_2 there is. The cycle deposits CO_2 directly into bundle-sheath cells where it enters the Calvin–Benson cycle (Figure 5.9). The CO_2 level near rubisco is kept high enough to block the competing reaction. C4 plants use more ATP than C3 plants do because they have to fix carbon twice, but they can make more sugar on dry days.

CAM plants open stomata at night and fix carbon by repeated turns of a C4 cycle, and then the Calvin–Benson cycle runs on the next day (Figure 5.10). CAM stands for crassulacean acid metabolism. These plants include cacti and other succulents, which have juicy, water-storing tissues and thick surface layers. Both are adaptations to hot, dry climates. Some CAM plants survive prolonged droughts by closing their stomata even at night. They fix CO_2 released during aerobic respiration, which is enough to support slow growth.

> *Driven by ATP energy, the light-independent reactions make sugars with hydrogen from NADPH, and with carbon from carbon dioxide.*
>
> *C3 plants, C4 plants, and CAM plants respond differently to hot, dry conditions.*

Leaves of corn, a typical C4 plant

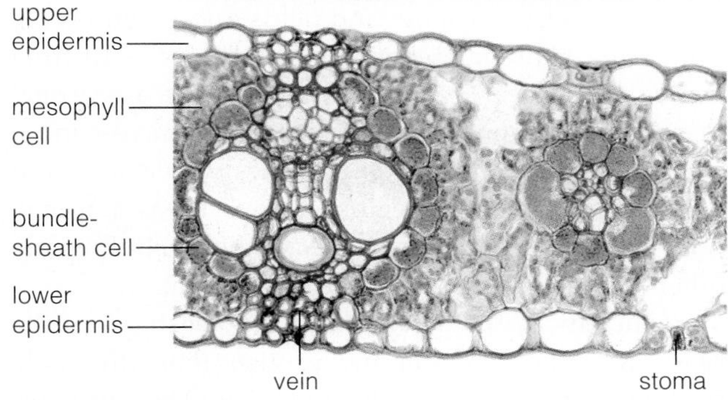

a Corn leaf, cross-section

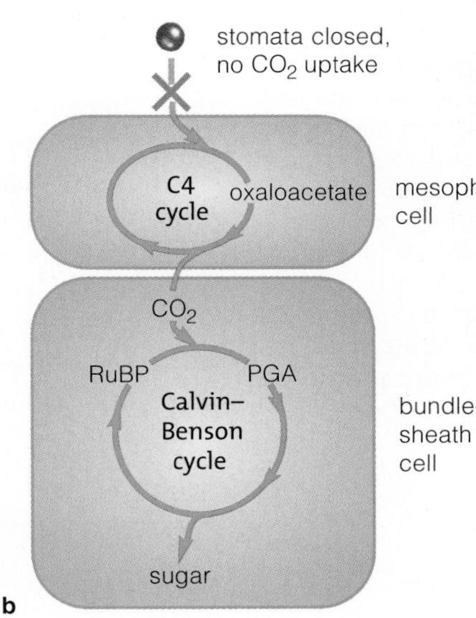

b

Figure 5.9 (**a**) Typical C4 leaves. (**b**) How C4 plants fix carbon in hot, dry weather, when there is too little CO_2 and too much O_2 inside leaves. A C4 cycle is common in grasses, corn, and other plants that evolved in the tropics. In their mesophyll cells, CO_2 gets fixed by an enzyme that ignores O_2. That reaction releases carbon dioxide in adjoining bundle sheath cells, where the Calvin–Benson cycle runs.

By contrast, many C3 plants evolved in moist temperate zones. The basswood tree (Figure 5.8) is one of them. On hot, dry days, it does not grow as well as a C4 plant. With stomata closed, the oxygen level rises, and rubisco uses O_2 instead of CO_2 in a different reaction. The plant loses carbon instead of fixing it.

Flat, fleshy leaves of a beavertail cactus, one of the CAM plants.

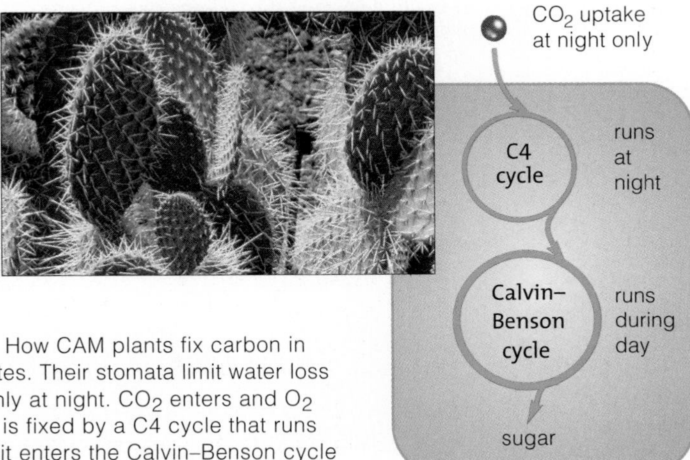

Figure 5.10 How CAM plants fix carbon in hot, dry climates. Their stomata limit water loss by opening only at night. CO_2 enters and O_2 departs. CO_2 is fixed by a C4 cycle that runs at night, then it enters the Calvin–Benson cycle during the day in the same cell.

5.4 Pastures of The Seas

We conclude this chapter with a story to reinforce how some photosynthesizers fit into the world of living things.

IMPACTS, ISSUES

About 20 percent of the world's photosynthesis takes place inside the glass-like walls of diatoms. Princeton researchers found out that these elaborate walls help diatoms concentrate CO_2 and so increase the rate of the light-independent reactions. Collectively, photosynthesizers use a lot of rubisco for the reactions. Rubisco makes up 30 to 40 percent of the soluble protein in plant cells. It may be the world's most abundant enzyme!

Each spring, the renewed growth of photoautotrophs is evident as trees leaf out and fields turn green. At the same time, innumerable single-celled species drifting through the ocean's surface waters make a seasonal response. You cannot see them without a microscope; a row of a million cells of one species would be about one inch long. In some regions, a cup of seawater may hold 24 million cells of one species, not including any other aquatic species suspended in the cup.

Collectively, these cells form the "pastures of the seas." Most are bacteria and protists that ultimately feed nearly all other marine organisms. Their primary productivity is the start of vast aquatic food webs.

Imagine zooming in on just one small patch of "pasture" in an Antarctic sea. There, tiny shrimplike crustaceans are feeding on tinier photosynthesizers. Dense concentrations of these crustaceans—krill—are food for other animals, including fishes, penguins, seabirds, and immense blue whales. A single, mature whale is straining four tons of krill from the water today, as it has been doing for many months. And before they themselves were eaten, those four tons of krill had munched through 1,200 tons of the pasture!

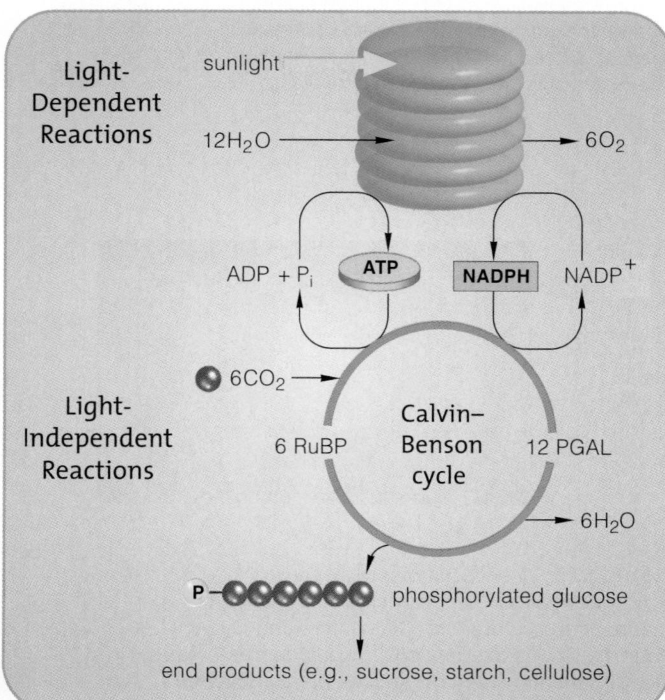

Figure 5.11 Summary of the key reactants, intermediates, and products of photosynthesis.

Collectively, tiny marine photoautotrophs have a tremendous impact on global climate. They use about 45 billion metric tons of carbon from CO_2 each year to build organic compounds by photosynthesis.

That is a LOT of photosynthesis.

Think about all the carbon dioxide that we humans release into the atmosphere. We do this, for example, by burning fossil fuels and vast tracts of forests to clear land for farming, and even by making cement. Without help from the little marine photoautotrophs, atmospheric carbon dioxide would accumulate very quickly and accelerate global warming (Section 30.6)

Although drastic global change is a real possibility, we continue to dump many tons of industrial wastes, sewage, and fertilizers in runoff from croplands into the ocean every day. Pollutants dramatically alter the chemical composition of seawater. How will marine photoautotrophs function in this chemical brew?

The flow of energy and the cycling of carbon through the biosphere start with photoautotrophs.

Summary

Section 5.1 Photosynthesis is the main metabolic pathway by which carbon and energy enter the web of life. Photoautotrophs get carbon from carbon dioxide and energy from the sun or inorganic compounds. Animals and other heterotrophs are not self-nourishing; they get carbon and energy from organic compounds made by autotrophs or other heterotrophs.

Photosynthesis runs on energy that pigments harvest from visible light. Pigments capture light of only certain wavelengths. Other wavelengths are reflected as the characteristic color of each pigment.

Chlorophyll *a*, the main photosynthetic pigment, absorbs mainly violet and red light. Accessory pigments capture additional wavelengths of light.

Photosynthesis has two stages: light-dependent and light-independent reactions. This equation and Figure 5.11 summarize the process:

$$\underset{\text{water}}{12H_2O} + \underset{\text{carbon dioxide}}{6CO_2} \xrightarrow[\text{enzymes}]{\text{light energy}} \underset{\text{oxygen}}{6O_2} + \underset{\text{glucose}}{C_6H_{12}O_6} + \underset{\text{water}}{6H_2O}$$

In chloroplasts, light-dependent reactions occur at a thylakoid membrane that forms a single compartment in a semifluid interior (stroma). The light-independent reactions occur in the stroma.

BiologyNow

View the sites where photosynthesis takes place with the animation on BiologyNow.

Section 5.2 Pigments clustered in the thylakoid membrane absorb light energy and pass it to many photosystems. Light-dependent reactions start when electrons released from photosystems enter electron transfer chains.

The released electrons can move through a noncyclic or a cyclic pathway. In the noncyclic reactions, light energy prompts photosystem II to lose electrons, which then move through an electron transfer chain and enter photosystem I. Light energy prompts photosystem I to lose electrons which, along with hydrogen ions, convert $NADP^+$ to NADPH.

In the cyclic reactions, electrons from photosystem I enter a different electron transfer chain, then are recycled back to the same photosystem. NADPH does not form.

In both pathways, electron flow through transfer chains causes H^+ to accumulate in the thylakoid compartment, forming concentration and electric gradients across the thylakoid membrane. H^+ flows in response to the gradients, through ATP synthases. The flow causes phosphate ATP to form in the stroma.

Electrons lost from photosystem I are replaced by electrons lost from photosystem II. Photosystem II replaces its lost electrons by pulling electrons from water molecules; the H^+ and O_2 also are released.

BiologyNow

Review the pathways by which light energy is used to form ATP with the animation on BiologyNow.

Section 5.3 The light-independent reactions are the synthesis part of photosynthesis. They occur in the stroma. The enzyme rubisco attaches carbon from CO_2 to RuBP to start the Calvin–Benson cycle. In this cyclic pathway, energy from ATP, carbon and oxygen from CO_2, and hydrogen and electrons from NADPH are used to make glucose, which quickly enters reactions that form the products of photosynthesis (mainly glucose, sucrose, and starch). It takes six turns of the Calvin–Benson cycle to fix the six CO_2 molecules required to make one six-carbon glucose molecule.

On hot dry days, plants shut stomata and oxygen can build up inside leaves. In C3 plants, this causes rubisco to attach oxygen (rather than carbon) to RuBP. The plant then loses carbon instead of fixing it. C4 plants raise the CO_2 level near rubisco by fixing carbon twice, in two cell types. CAM plants capture light energy during the day, and then fix carbon at night.

BiologyNow

Take a spin around the Calvin–Benson cycle with the animation on BiologyNow.

Read the InfoTrac article "Light of our Lives," Norman Miller, Geographical, *January 2001.*

Section 5.4 Photoautotrophs and chemoautotrophs produce the food that sustains themselves and all other organisms. Many different aerobic and anaerobic kinds influence the global cycling of oxygen, carbon, nitrogen, and other substances. Aerobic photoautotrophs live in mind-boggling numbers in the seas as well as on land.

Self-Quiz

Answers in Appendix I

1. Photosynthetic autotrophs use ______ from the air as a carbon source and ______ as their energy source.

2. Light-*dependent* reactions in plants occur at the ______ .
   a. thylakoid membrane c. stroma
   b. plasma membrane d. cytoplasm

3. In the light-*dependent* reactions, ______ .
   a. carbon dioxide is fixed c. CO_2 accepts electrons
   b. ATP and NADPH form d. sugars form

4. What accumulates inside the thylakoid compartment during the light-*dependent* reactions?
   a. glucose c. hydrogen ions
   b. RuBP d. carbon dioxide

5. Light-*independent* reactions proceed in the ______ .
   a. cytoplasm b. plasma membrane c. stroma

6. The Calvin–Benson cycle starts when ______ .
   a. light is available
   b. carbon dioxide is attached to RuBP
   c. electrons leave photosystem II

7. Match each event with its most suitable description.
   ____ NADPH formation a. rubisco required
   ____ CO_2 fixation b. ATP, NADPH required
   ____ PGAL formation c. electrons cycled back to photosystem I
   ____ ATP formation only d. water molecules split

Additional questions are available on **BiologyNow™**

Critical Thinking

1. Krishna exposes plants to a radioisotope of carbon ($^{14}CO_2$), which they absorb. In which compound will he see the labeled carbon appear first if the plants are C3? C4?

2. Approximately 400 years ago, Jan Baptista van Helmont experimented on the nature of photosynthesis. He wanted to know where growing plants get the materials necessary for increases in size. He planted a tree seedling weighing 5 pounds in a barrel filled with 200 pounds of soil. He watered the tree regularly. Five years passed. Then Van Helmont weighed the tree and the soil. The tree weighed 169 pounds, 3 ounces. The soil weighed 199 pounds, 14 ounces. Because the tree gained so much weight and the soil lost so little, he concluded the tree had gained weight after absorbing the water he had added to the barrel. Given what you know about the composition of biological molecules, why was he misguided? Explain what really happened.

3. The green alga in Figure 5.12*a* lives in seawater. Its main pigments absorb red light. Its accessory pigments help harvest energy in sunlit waters, and some shield it against ultraviolet radiation. Other green algae live in ponds, in lakes, on rocks, even in snow. The red alga in Figure 5.12*b* grows on tropical reefs in clear, warm water. Its phycobilin absorbs green and blue-green wavelengths that penetrate deep water. Some of its relatives live in deep, dimly lit waters; their pigments are nearly black.

   If wavelengths are such a vital source of energy, why aren't all pigments black? (*Hint*: Photoautotrophs first evolved in the seas.)

Figure 5.12 (**a**) A green alga (*Codium*) from shallow coastal waters. (**b**) Red alga from a tropical reef.

6 HOW CELLS RELEASE CHEMICAL ENERGY

IMPACTS, ISSUES

When Mitochondria Spin Their Wheels

In the early 1960s, Swedish physician Rolf Luft reflected on some odd symptoms of a young patient. The woman felt weak and too hot all the time. Even on the coldest winter days, she could not stop sweating and her skin was flushed. She was thin in spite of a huge appetite. Luft inferred that his patient's symptoms pointed to a metabolic disorder. Tests designed to detect her rate of metabolism showed that her oxygen consumption was the highest ever recorded!

Microscopic examination of a tissue sample from the patient's muscles revealed mitochondria, the cell's ATP-producing powerhouses. But there were far too many of them. The mitochondria were working hard, but producing very little ATP for their efforts.

Her disorder, now called *Luft's syndrome*, was the first to be linked directly to a defective organelle. By analogy, someone with this mitochondrial disorder functions like a city with half of its power plants shut down. Muscles, the brain, and other body parts with the highest energy demands are hurt the most.

More than a hundred other mitochondrial disorders are now known. One, a genetic disorder called *Friedreich's ataxia*, runs in families. Like siblings Leah and Joshua at *left*, affected people develop weak muscles, ataxia (loss of coordination), and visual problems. Many die young because of heart problems.

Leah and Joshua

A mutant gene and its abnormal protein product give rise to Friedreich's ataxia. The abnormal protein causes iron to accumulate inside mitochondria. Iron is required for electron transfers that drive ATP formation, but too much invites an accumulation of **free radicals**—unbound molecular fragments with an unstable number of electrons. These highly reactive fragments can attack all of the molecules of life.

Type 1 diabetes, atherosclerosis, amyotrophic lateral sclerosis (Lou Gehrig's disease), Parkinson's, Alzheimer's, and Huntington's diseases—defective mitochondria contribute to every one of these problems. So when you consider mitochondria in this chapter, do not assume they are too remote from your interests. Without them, you would not make enough ATP even to read about how they do it.

You already read about the way ATP molecules form during photosynthesis. Turn now to how all organisms make ATP by tapping into the chemical bond energy of organic compounds—especially glucose.

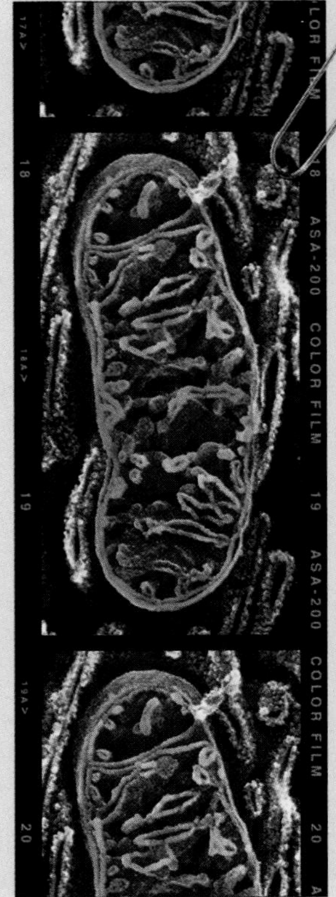

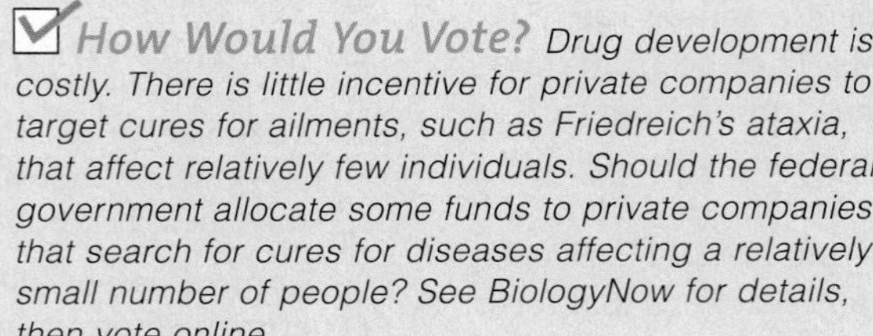

How Would You Vote? *Drug development is costly. There is little incentive for private companies to target cures for ailments, such as Friedreich's ataxia, that affect relatively few individuals. Should the federal government allocate some funds to private companies that search for cures for diseases affecting a relatively small number of people? See BiologyNow for details, then vote online.*

Key Concepts

ENERGY-RELEASING PATHWAYS

All organisms release chemical bond energy of organic molecules to drive ATP formation. The main pathways start in the cytoplasm with glycolysis. Products of glycolysis can enter the aerobic respiration pathway, which uses oxygen, or anaerobic pathways such as alcoholic and lactate fermentation, which do not use oxygen.

YIELDS AND TALLIES

Only aerobic respiration ends in mitochondria. Fermentation's net energy yield per glucose molecule is always smaller than that of aerobic respiration. Molecules other than glucose can enter the aerobic pathway as alternative energy sources.

EVOLUTIONARY CONNECTIONS

Aerobic respiration and photosynthesis are connected on a global scale. Photosynthesis uses the carbon dioxide released by aerobic respiration, which uses oxygen released by photosynthesis. This connection can be traced to the evolution of novel metabolic pathways.

Links to Earlier Concepts

This chapter expands the picture of life's dependence on energy flow (Section 4.1) by showing how all organisms tap energy stored in glucose and convert it to different, transportable forms, especially ATP (4.2). Review the structure of glucose and other carbohydrates (2.6) and lipids (2.7), as well as the introduction to membrane function (3.3), mitochondria (3.5), and mechanisms of active transport (4.5). You will see another example of controlled energy release at electron transfer chains (4.1, 5.2), and you will explore global connections between photosynthesis (5.2, 5.3) and aerobic respiration.

6.1 Overview of Energy-Releasing Pathways

LINKS TO SECTIONS 4.1, 4.2

Organisms stay alive by taking in energy. Regardless of its source, energy must be converted to a form that can drive thousands of diverse life-sustaining reactions. The chemical bond energy of ATP serves that purpose.

The first energy-releasing metabolic pathways were operating billions of years before Earth's oxygen-rich atmosphere evolved, so we can expect that they were *anaerobic*; the reactions did not use free oxygen. Many prokaryotes still live in oxygen-free habitats, making ATP by **fermentation**, breaking down carbohydrates in pathways that do not use oxygen. Many eukaryotic cells still use fermentation.

Cells of nearly all species use **aerobic respiration**, a breakdown pathway that uses oxygen to extract energy very efficiently from glucose and other carbohydrates. Every time you inhale, you provide your aerobically-respiring cells with a fresh supply of oxygen.

Both energy-releasing pathways start with the same set of reactions in the cytoplasm. During these initial reactions, called **glycolysis**, enzymes rearrange a six-carbon glucose molecule into two pyruvate molecules. **Pyruvate** is an organic compound with a three-carbon backbone. Once glycolysis ends, the energy-releasing pathways differ. Only the aerobic pathway continues in mitochondria, where oxygen accepts electrons at the end of electron transfer chains.

By contrast, the anaerobic pathways start and end in the cytoplasm, and the final electron acceptor is not oxygen; it is some other substance.

Of energy-releasing pathways, aerobic respiration gets the most bang for the buck, typically yielding thirty-six ATP for each molecule of glucose. Compare this with two ATP per glucose yielded by anaerobic pathways (Figure 6.1). If you were a single cell, you would not require much ATP; you could depend on fermentation for all of your energy needs. Being far larger and highly active, however, you require the high energy yield of aerobic respiration. Aerobic respiration can be summarized this way:

$$\underset{\text{GLUCOSE}}{C_6H_{12}O_6} + \underset{\text{OXYGEN}}{6O_2} \longrightarrow \underset{\text{CARBON DIOXIDE}}{6CO_2} + \underset{\text{WATER}}{6H_2O} + \text{ATP}$$

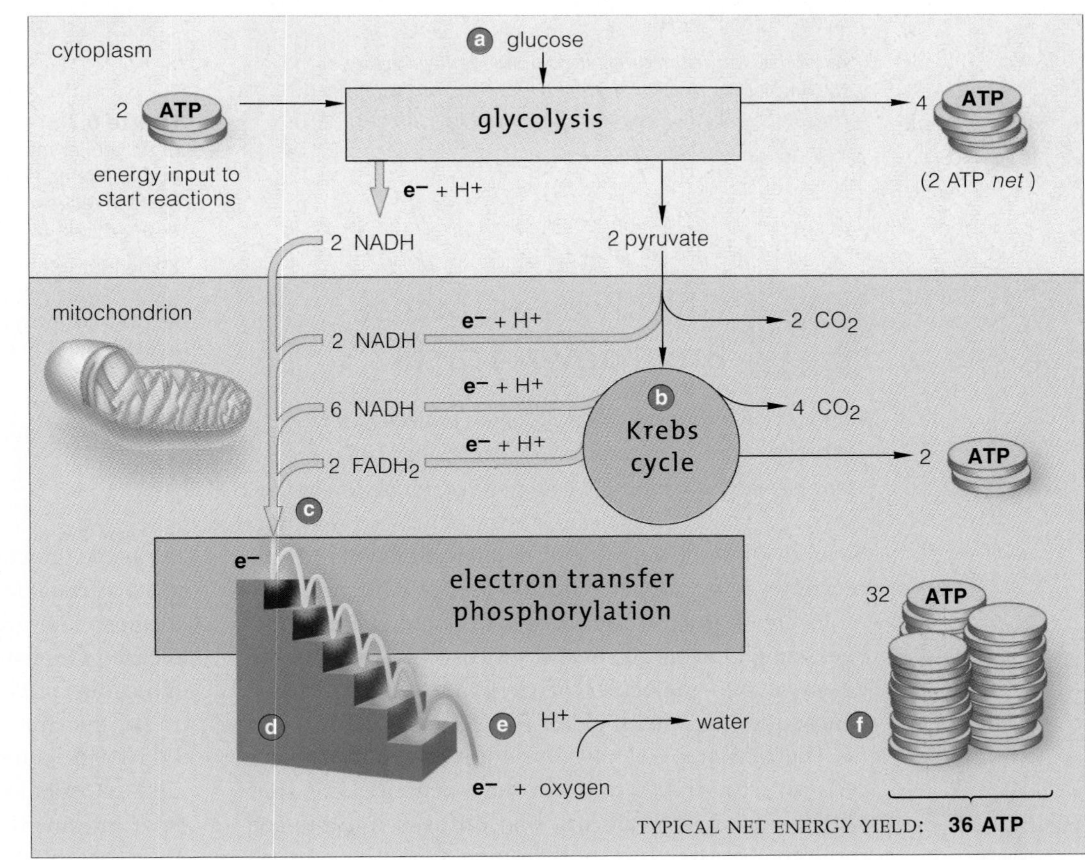

Figure 6.1 *Animated!* Overview of the three stages of aerobic respiration. Reactions start in the cytoplasm, and end inside mitochondria.

(**a**) The first stage is glycolysis. Enzymes partly break down glucose to pyruvate.

(**b**) In the second stage, the pyruvate is broken down to carbon dioxide.

(**c**) Coenzymes NAD^+ and FAD pick up electrons and hydrogen to become NADH and $FADH_2$.

(**d**) In the final stage, the NADH and $FADH_2$ give up electrons to electron transfer chains. The flow of electrons through the transfer chains results in H^+ gradients that drive ATP formation.

(**e**) At the end of the transfer chains, oxygen accepts electrons and hydrogen, forming water.

(**f**) From start to finish, a typical net energy yield from a single glucose molecule is thirty-six ATP.

LINKS TO SECTIONS 2.6, 5.3

The summary equation tells you which substances are present at the pathway's start and end. In between, however, are three reaction stages.

Again, glycolysis is the first stage. The second is a cyclic pathway called the **Krebs cycle** and a few steps preceding it. In these reactions, enzymes break down pyruvate to carbon dioxide and water. Electrons and hydrogen atoms released by the reactions are picked up by the coenzymes $\mathbf{NAD^+}$ and **FAD**, which become NADH and $FADH_2$.

Few ATP form during glycolysis or the Krebs cycle. The big energy payoff is the third stage, as NADH and $FADH_2$ give up their cargo of electrons and hydrogen to electron transfer chains—the machinery of **electron transfer phosphorylation**. Operation of these chains establishes hydrogen ion gradients, which in turn drive formation of ATP. Oxygen inside the mitochondrion accepts the electrons and hydrogen at the end of the chains, thereby forming water.

Nearly all metabolic reactions run on energy released from glucose and other organic compounds. The main energy-releasing pathways start in the cytoplasm with glycolysis, a series of reactions that break down glucose to pyruvate.

Anaerobic pathways have a small net energy yield, typically two ATP per glucose.

Aerobic respiration, an oxygen-dependent pathway, runs to completion in mitochondria. From start (glycolysis) to finish, it typically has a net energy yield of thirty-six ATP.

glucose

GLYCOLYSIS

pyruvate

to second stage of aerobic respiration or to a different energy-releasing pathway

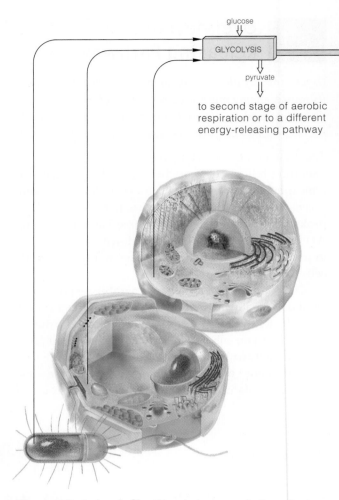

Figure 6.2 ***Animated!*** Glycolysis occurs in the cytoplasm of all prokaryotic and eukaryotic cells. Glucose is the reactant in this example. Two pyruvate, two NADH, and four ATP form. Cells invest two ATP to start the reactions, however, so the *net* energy yield is two ATP.

Depending on the type of cell and environmental conditions, the pyruvate may enter the second set of reactions of the aerobic pathway, including the Krebs cycle. Or it may be used in other reactions, such as those of fermentation.

6.2 Glycolysis—Glucose Breakdown Starts

Let's track what happens to a glucose molecule in the first stage of aerobic respiration. Remember, the same steps happen in anaerobic energy-releasing pathways.

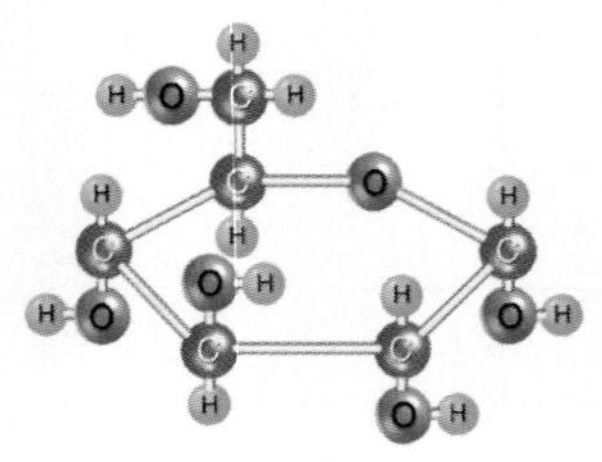

glucose

Several kinds of sugars can enter the enzyme-driven reactions of glycolysis. Glucose is one of them. Each molecule of glucose has six carbon atoms that form its backbone. During glycolysis, glucose is partly broken down to two molecules of pyruvate, a three-carbon compound, for a net yield of two ATP (Figure 6.2).

The first steps of glycolysis are *energy-requiring*—cells invest two ATP to begin the reactions. Let's start with a molecule of glucose that diffused into the cell through a passive transport protein. One ATP donates a phosphate group to the glucose. Phosphorylation traps the glucose inside the cell, because the resulting molecule (glucose–6–phosphate) cannot go through the glucose transporter. Glucose–6–phosphate then accepts a phosphate group from a second ATP. The resulting 6-carbon intermediate splits at once into two 3-carbon molecules of PGAL (Section 5.3).

In the first *energy-releasing* step of glycolysis, both PGAL are converted to intermediates that each give up a phosphate to ADP, which results in two ATP. Two later intermediates do the same thing. Thus, four ATP form by **substrate-level phosphorylation**. We define this metabolic event as a direct transfer of a phosphate group from a reaction substrate to another molecule—

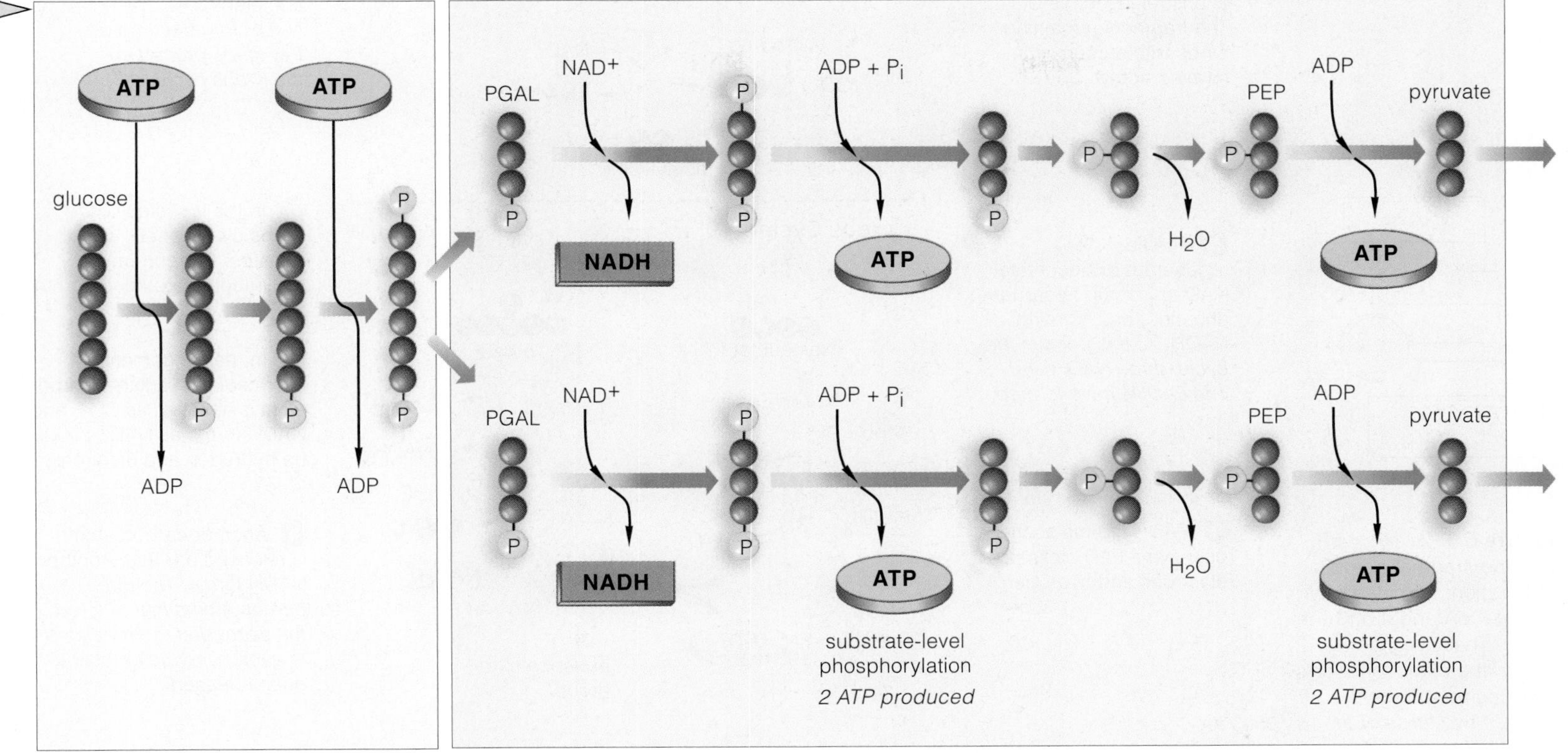

a Track the six carbon atoms (*brown* circles) of glucose. Glycolysis starts with an energy investment of two ATP.

One ATP transfers a phosphate group to glucose. Then, the phosphorylated glucose accepts a phosphate group from another ATP. The intermediate splits at once into two, each three-carbon PGAL molecules.

b Two NADH form when each PGAL gives up two electrons and a hydrogen atom to NAD^+.

Two intermediates each transfer a phosphate group to ADP. *Two ATP have now formed by substrate-level phosphorylation.* The original energy investment of two ATP has been recovered.

c Two intermediates form and are rearranged. Each one gives up a hydrogen atom and an —OH group. These combine as water. Two molecules called PEP form by these reactions.

d Each PEP transfers a phosphate group to ADP. *Once again, two ATP have formed by substrate-level phosphorylation.*

In sum, glycolysis has a net energy yield of two ATP for each glucose molecule. Two NADH also form during the reactions, and two molecules of pyruvate are the end products.

in this case, to ADP. Meanwhile, the two PGAL give up their electrons and hydrogens to two NAD^+, which become NADH. Later these two NADH will give up their electrons and hydrogens at another reaction site, and so revert to NAD^+. In other words, like all other coenzymes, NAD^+ is reusable.

Four ATP have now formed. Remember two ATP were invested to start the reactions. The *net* yield of glycolysis is two ATP and two NADH.

To summarize, glycolysis converts the bond energy of glucose (a storable, stable form of energy) to the bond energy of ATP (a transportable form of energy). The electrons and hydrogen from glucose that are picked up by NAD^+ enter the next stage of reactions. And so can the end products of glycolysis—the two pyruvate molecules.

Glycolysis is a series of reactions that partially break down glucose or some other carbohydrate to two molecules of pyruvate, thereby releasing energy.

Two NADH and four ATP form. However, when we subtract the two ATP required to start the reactions, the net energy yield of glycolysis is two ATP from one glucose molecule.

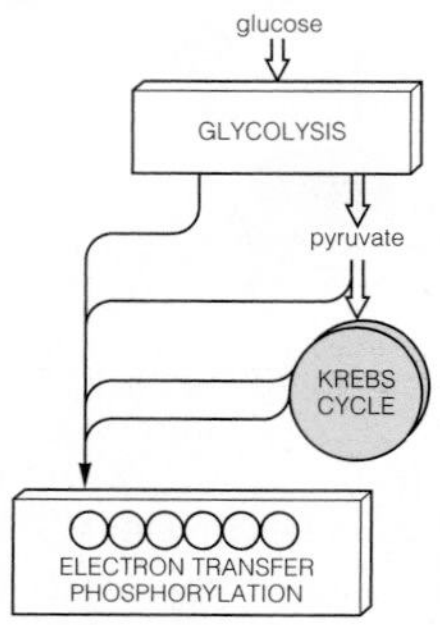

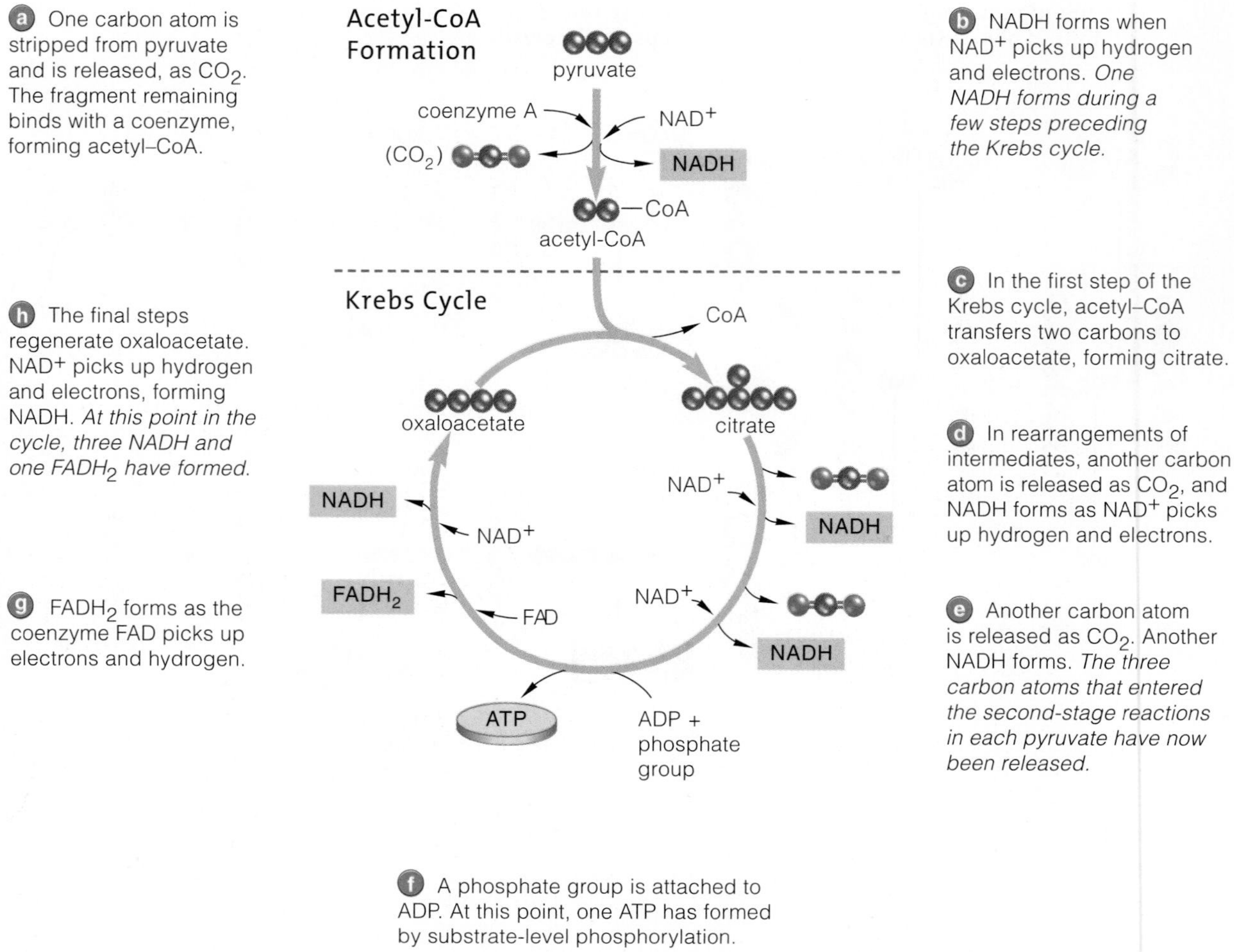

Acetyl-CoA Formation

ⓐ One carbon atom is stripped from pyruvate and is released, as CO_2. The fragment remaining binds with a coenzyme, forming acetyl–CoA.

ⓑ NADH forms when NAD^+ picks up hydrogen and electrons. *One NADH forms during a few steps preceding the Krebs cycle.*

Krebs Cycle

ⓒ In the first step of the Krebs cycle, acetyl–CoA transfers two carbons to oxaloacetate, forming citrate.

ⓓ In rearrangements of intermediates, another carbon atom is released as CO_2, and NADH forms as NAD^+ picks up hydrogen and electrons.

ⓔ Another carbon atom is released as CO_2. Another NADH forms. *The three carbon atoms that entered the second-stage reactions in each pyruvate have now been released.*

ⓕ A phosphate group is attached to ADP. At this point, one ATP has formed by substrate-level phosphorylation.

ⓖ $FADH_2$ forms as the coenzyme FAD picks up electrons and hydrogen.

ⓗ The final steps regenerate oxaloacetate. NAD^+ picks up hydrogen and electrons, forming NADH. *At this point in the cycle, three NADH and one $FADH_2$ have formed.*

Figure 6.3 ***Animated!*** Second stage of aerobic respiration: formation of acetyl–CoA and the Krebs cycle. These reactions occur in a mitochondrion's inner compartment. *It takes two turns of the cycle to break down two pyruvates from glycolysis.* A total of two ATP, eight NADH, two $FADH_2$, and six CO_2 molecules form.

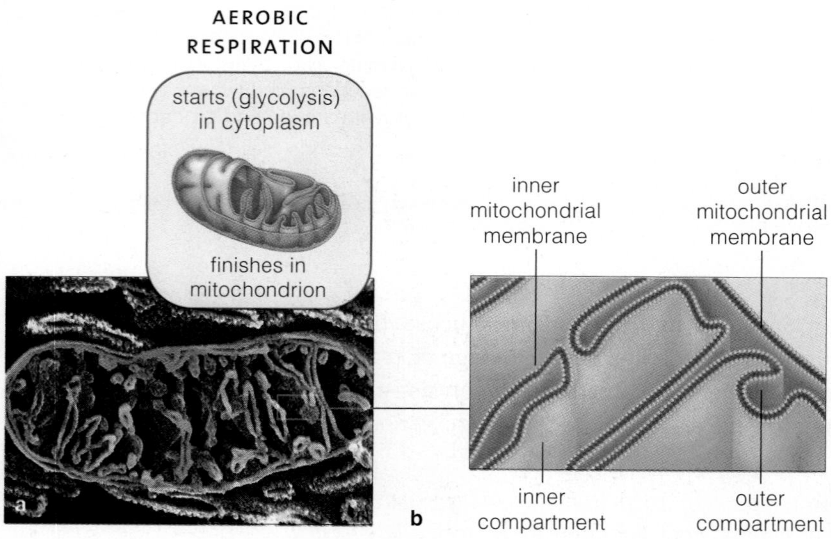

Figure 6.4 ***Animated!*** Functional zones inside the mitochondrion. (**a**) Micrograph of a mitochondrion, sliced lengthwise. Most eukaryotic species contain these organelles. (**b**) An inner membrane divides a mitochondrion's interior into two compartments. The second and third stages of aerobic respiration take place at this membrane.

6.3 Second and Third Stages of Aerobic Respiration

The second and third stages of the aerobic pathway unfold inside mitochondria. The two pyruvate molecules formed by glucose breakdown in glycolysis enter a series of reactions that result in the release of carbon dioxide and, later, the formation of many ATP molecules.

Suppose two pyruvate molecules formed by glycolysis leave the cytoplasm and enter a **mitochondrion** (plural, mitochondria). In this organelle, the second and third stages of the aerobic pathway are completed. Figures 6.3 and 6.4 show the reactions and where they occur.

ACETYL–COA FORMATION AND THE KREBS CYCLE

During the second stage, what remains of the original glucose molecule is broken down completely to water and carbon dioxide. Six carbon atoms, three from each

YIELDS AND TALLIES

Ⓐ Electrons and hydrogen from NADH and $FADH_2$ that formed during the first and second stages enter electron transfer chains.

Ⓑ As electrons are being transferred through these chains, H^+ ions are moved across the inner membrane, into the outer compartment.

Ⓒ As H^+ accumulates in the outer compartment, chemical and electric gradients form across the inner membrane.

Ⓓ Hydrogen ions follow the gradients through the interior of ATP synthases, driving ATP formation from ADP and phosphate (P_i).

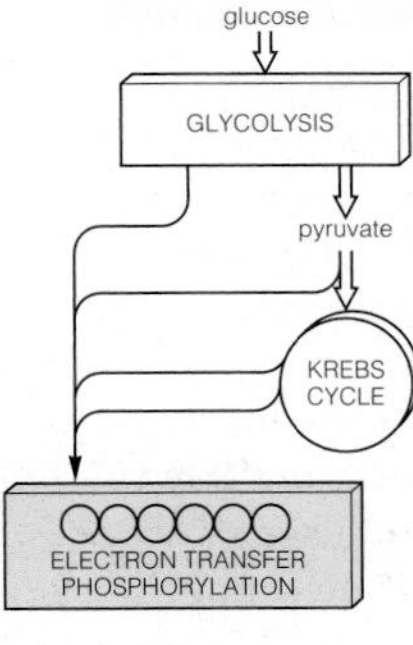

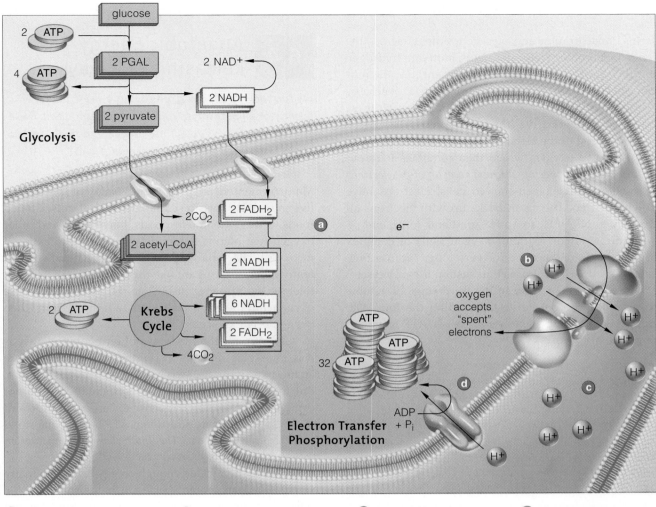

Figure 6.5 ***Animated!*** Summary of electron transfers in aerobic respiration.

pyruvate, enter these reactions. Six carbon atoms depart, as six molecules of carbon dioxide that escape into the surroundings (Figure 6.3).

In the preparatory reactions, an enzyme removes one carbon atom from each pyruvate and attaches it to oxygen, forming CO_2. Each two-carbon fragment then combines with a coenzyme to form acetyl–CoA, a type of cofactor that is the starting material for the Krebs cycle. Two NADH also form.

Each acetyl–CoA molecule begins the Krebs cycle by transferring its two-carbon acetyl group to four-carbon oxaloacetate, which becomes citrate—a form of citric acid. (That is why the Krebs cycle is also known as the citric acid cycle.) Intermediates are rearranged to become oxaloacetate. Cells have limited amounts of oxaloacetate, and so must regenerate it to keep the Krebs cycle running. Two ATP form by substrate-level phosphorylation. Six more NADH and two $FADH_2$ also form. With their cargo of electrons and hydrogen, these coenzymes—*eight NADH and two* $FADH_2$—are the big potential payoff for the cell.

ELECTRON TRANSFER PHOSPHORYLATION

In the third stage, ATP formation goes into high gear. It begins when coenzymes donate electrons to electron transfer chains in the inner mitochondrial membrane (Figure 6.5). The flow of electrons through the chains ultimately causes the attachment of phosphate to ADP molecules—thus the name of these reactions, *electron transfer phosphorylation*.

As electrons are passed through the transfer chains, hydrogen ions in the inner compartment are shuttled into the outer one. The accumulation of H^+ ions there

LINKS TO SECTIONS 3.3, 3.5, 4.5

IMPACTS, ISSUES

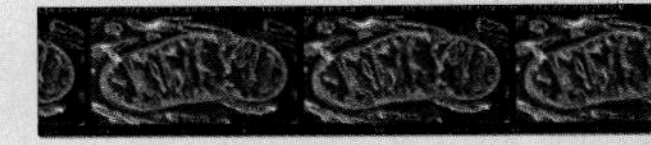

The enzyme pyruvate dehydrogenase speeds conversion of pyruvate to acetyl–CoA. When mutated, this enzyme disrupts the formation of citrate in the Krebs cycle. A fetus with this severe mutation will die before birth.

forms concentration and electric gradients across the inner membrane. These gradients motivate hydrogen ion flow back to the inner compartment, but the ions can only cross the membrane through ATP synthases (Figure 6.5). The flow of hydrogen ions through these transport proteins drives the formation of ATP.

At the end of the electron transfer chains, electrons are passed to oxygen, which then combines with H^+ to form water. *Oxygen is the final acceptor of the electrons from glucose.* In oxygen-starved cells, electrons have nowhere to go. The chain backs up with electrons all the way to NADPH, so no H^+ gradients form, and no ATP forms, either. Without oxygen, cells of complex organisms do not survive for very long, because they cannot produce enough ATP to sustain life processes.

Thirty-two ATP form in the third stage of aerobic respiration. The net harvest, including the four ATP from the preceding stages, is thirty-six ATP from one molecule of glucose.

IMPACTS, ISSUES

What if there is a leak between the inner and outer mitochondrial compartments? Here is one hypothesis: The reactions run smoothly until electron transfer phosphorylation. The H^+ gradient used in ATP formation cannot form because the inner membrane is too leaky. Luft's disease results. We can't test this idea. No one besides the woman described in the chapter introduction has been diagnosed with this condition, and she is no longer living.

In the second stage of aerobic respiration, two molecules of pyruvate from glycolysis are converted to acetyl–CoA, which enters the Krebs cycle. Two ATP and ten coenzymes form. All of pyruvate's carbons depart, in the form of carbon dioxide.

In aerobic respiration's third stage, coenzymes deliver their electrons and hydrogen to electron transfer chains. Electron flow through the chains sets up gradients across the inner membrane. These drive the synthesis of thirty-two ATP.

Aerobic respiration typically nets thirty-six ATP molecules from each glucose molecule metabolized.

6.4 Anaerobic Energy-Releasing Pathways

We turn now to the use of glucose as a substrate for fermentation pathways. Unlike aerobic respiration, these anaerobic pathways do not use oxygen as a final acceptor of electrons. Their final steps regenerate NAD^+.

Many fermenters are protists and bacterial species that live in marshes, ocean sediments, animal guts, canned foods, sewage treatment ponds, bogs, mud, and other oxygen-free places. Some die when they are exposed to oxygen. Bacterial species that cause botulism and many other diseases are like this. Some fermenters can tolerate the presence of oxygen. Still other kinds use oxygen, but they also can switch to fermentation when oxygen becomes scarce.

Glycolysis is the first stage of fermentation, just as it is in aerobic respiration (Figures 6.6 and 6.7). Here again, two pyruvate, two NADH, and two ATP form. But unlike aerobic respiration, fermentation does not completely break down glucose to carbon dioxide and water. It produces no additional ATP beyond the small yield from glycolysis. *The final steps in fermentation pathways regenerate the coenzyme NAD^+.*

Fermentation yields enough energy to sustain many single-celled anaerobic species. It helps some aerobic cells when the oxygen level is stressfully low. It is not enough to sustain large, multicelled organisms, this being why you will never see an anaerobic elephant.

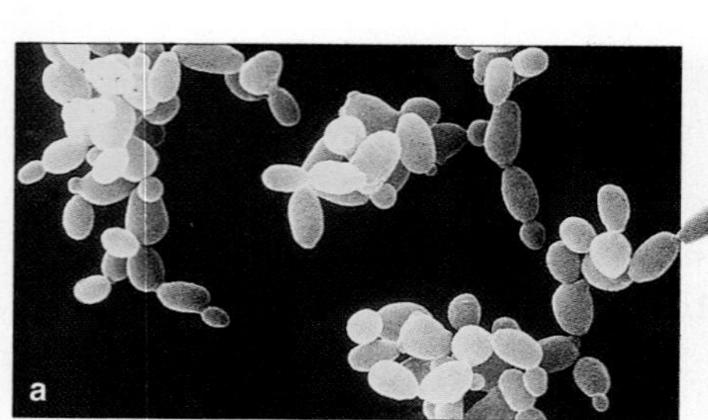

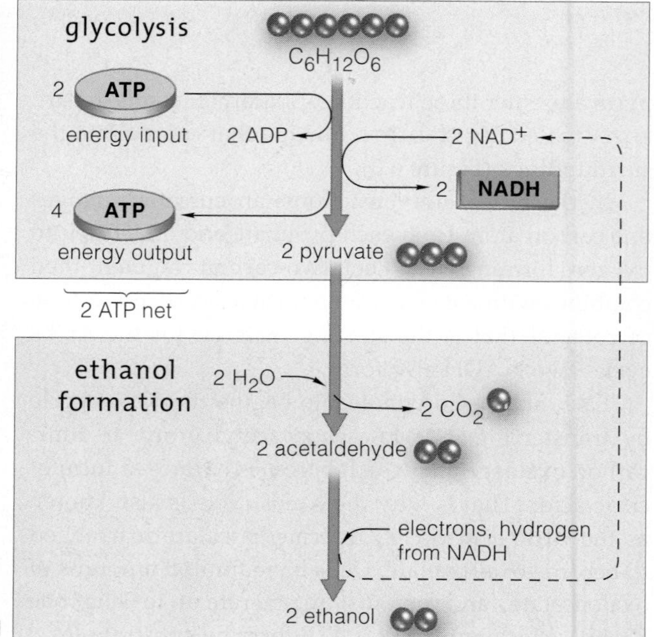

Figure 6.6 ***Animated!*** Alcoholic fermentation. (**a**) Yeasts, which are single-celled fungi, use this anaerobic pathway to make ATP.

(**b**) A vintner examining a fermentation product of *Saccharomyces*. This yeast lives on sugar-rich tissues of ripe grapes.

(**c**) Carbon dioxide released from cells of *S. cerevisiae* makes bread dough rise.

(**d**) Alcoholic fermentation. An intermediate, acetaldehyde, functions as this pathway's final electron acceptor. The end product of the reactions is ethanol (ethyl alcohol).

ALCOHOLIC FERMENTATION

In **alcoholic fermentation**, the three-carbon backbones of the two pyruvate molecules from glycolysis are split into two molecules of acetaldehyde and two of carbon dioxide. Next, the acetaldehydes accept electrons and hydrogen from NADH, thus becoming the alcohol called ethanol (Figure 6.6).

Some single-celled fungi called yeasts are famous because they use this pathway. Bakers mix one type, *Saccharomyces cerevisiae*, into dough. Fermenting yeast cells release carbon dioxide, and the dough expands (rises) as the gas forms bubbles in it. Oven heat forces the bubbles out of spaces they had occupied in the dough, and the alcohol product evaporates away.

Some wild and cultivated strains of *Saccharomyces* are used to produce wine. Crushed grapes are left in vats along with the yeast, which converts sugar in the juice to ethanol. Ethanol is toxic to microbes. When a fermenting brew's ethanol content nears 10 percent, yeast cells start dying and fermentation ends.

LACTATE FERMENTATION

In **lactate fermentation**, NADH gives up electrons and hydrogen to two pyruvate molecules from glycolysis. The transfer converts each pyruvate to lactate, a three-carbon compound (Figure 6.7).

Lactobacillus and other kinds of bacteria use lactate fermentation. Their fermenting action can spoil food, yet some species have commercial uses. For instance, bacteria that break down the glucose in milk give us dairy products like cheeses, yogurt, and buttermilk. Lactate fermentation is used to cure meats and pickle fruits and vegetables, such as sauerkraut. Lactate is an acid so it gives these foods a sour taste.

In humans, rabbits, and many other animals, some types of cells use lactate fermentation. When demands for energy are intense but brief—say, in a short race—some muscle cells use this pathway instead of aerobic respiration. Lactate fermentation works quickly but not for long. It does not produce enough ATP from glucose to sustain prolonged activity.

In fermentation pathways, an intermediate that forms during the reactions functions as the final acceptor of electrons originally derived from glucose.

Alcoholic fermentation and lactate fermentation both have a net energy yield of two ATP for each glucose molecule metabolized. That ATP formed in glycolysis. The rest of the reactions regenerate NAD+, the coenzyme that keeps both of these pathways operational.

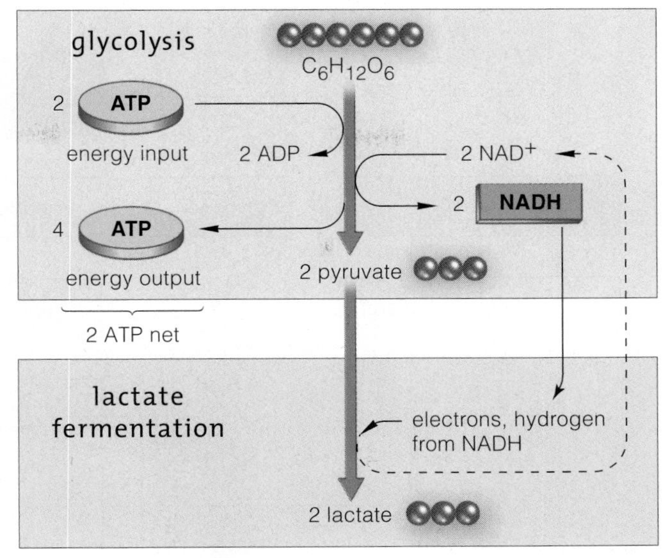

Figure 6.7 Lactate fermentation. In this anaerobic pathway, electrons originally stripped from glucose end up in lactate, the reaction product. The reactions have a net energy yield of two ATP (from glycolysis).

IMPACTS, ISSUES

Many people affected by mitochondrial disease are hit by *lactic acidosis*; they make far too much lactate. Some drugs used to treat AIDS also cause this condition by blocking mitochondrial function. High lactate levels make the blood dangerously acidic and cause muscle weakness, abdominal pain, nausea, fatigue, and, in extreme cases, death.

6.5 Alternative Energy Sources in the Body

LINKS TO SECTIONS 2.6, 2.7

So far, you've looked at what happens after glucose molecules enter an energy-releasing pathway. Now start thinking about what cells do when they have too much or too little glucose.

THE FATE OF GLUCOSE AT MEALTIME AND IN BETWEEN MEALS

What happens to glucose at mealtime? While you and all other mammals are eating, glucose and other small organic molecules are being absorbed across the gut lining, and your blood is transporting them through the body. The rising glucose concentration in blood prompts an organ, the pancreas, to secrete insulin. This hormone stimulates cells to take up glucose faster.

Cells trap the incoming glucose by converting it to glucose–6–phosphate. This is the first intermediate of glycolysis, formed by a phosphate group transfer from ATP (Figure 6.2). When glucose is phosphorylated this way, it cannot be transported back out of the cell.

When glucose intake exceeds cellular demands for energy, ATP-producing machinery goes into high gear. Unless a cell is using ATP rapidly, the concentration of ATP in its cytoplasm rises. Then, glucose–6–phosphate from glycolysis is diverted into a different pathway,

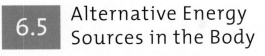

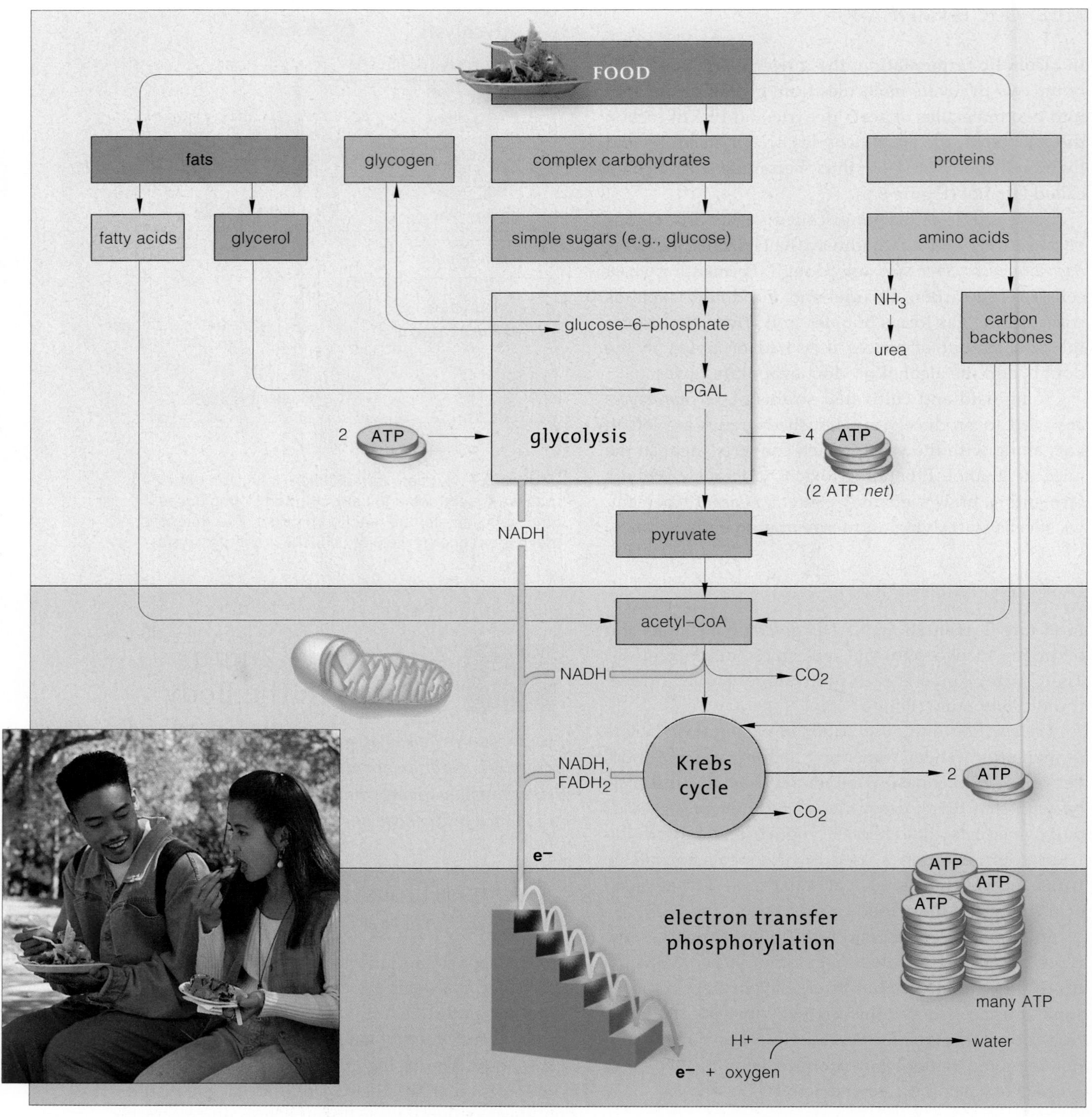

Figure 6.8 Reaction sites where a variety of organic compounds can enter the different stages of aerobic respiration. These compounds are alternative energy sources in the human body.

In humans and other mammals, complex carbohydrates, fats, and proteins cannot enter the aerobic pathway directly. The digestive system, and then individual cells, must first break apart these molecules into simpler subunits that can be degraded.

and ends up being converted to glycogen, a storage polysaccharide (Section 2.6). The glucose–to–glycogen pathway is the dominant one in muscle and liver cells, which have the largest stores of glycogen in the body.

Between meals, the blood level of glucose declines. If the decline were not countered, that would be bad news for the brain, your body's glucose hog. At any time, your brain is taking up more than two-thirds of the freely circulating glucose. Why? Its many hundreds of millions of neurons (or nerve cells) utilize this sugar as their primary energy source. Glucose-starved brain cells don't function, and so dizziness, weakness, and confusion occur.

In response to a decline in blood sugar, pancreas cells secrete glucagon, a hormone that prompts liver cells to convert stored glycogen into glucose and send it back to the blood. Only liver cells do this; muscle cells will not give it up. Blood glucose levels rise, and brain cells keep functioning. *Hormones control whether your body's cells use free glucose as an energy source or tuck it away.*

Don't let this explanation lead you to believe cells squirrel away gigantic amounts of glycogen. In adult humans, glycogen makes up only about 1 percent of the body's total energy reserves, the energy equivalent of two cups of cooked pasta. If you stop eating, you will completely deplete your liver's stash of glycogen in less than twelve hours.

Of the total energy reserves in, say, a typical adult who eats well, 78 percent (about 10,000 kilocalories) is concentrated in body fat and 21 percent in proteins.

ENERGY FROM FATS

How does a human body access its fat reservoir? A fat molecule, recall, has a glycerol head and one, two, or three fatty acid tails (Section 3.4). The body stores most fats as triglycerides, which have three tails each. Triglycerides accumulate in fat cells of adipose tissue. This tissue is strategically located beneath the skin of buttocks and other body regions.

When the blood glucose level falls, triglycerides are tapped as an energy alternative. Enzymes in fat cells cleave bonds between glycerol and fatty acids, which both enter the blood. Enzymes in the liver convert the glycerol to PGAL. And PGAL, recall, is one of the key intermediates for glycolysis (Figure 6.2). Nearly all cells of your body take up circulating fatty acids, and enzymes inside them cleave the fatty acid backbones. The fragments become converted to acetyl–CoA, which can enter the Krebs cycle.

Compared to glucose, a fatty acid tail has far more carbon-bound hydrogen atoms, so it yields more ATP. Between meals or during sustained exercise, fatty acid conversions supply about half of the ATP that muscle, liver, and kidney cells require.

What happens if you eat too many carbohydrates? Exceed the glycogen-storing capacity of your liver and muscle cells, and the excess gets converted to fats. *Too much glucose ends up as excess fat.*

ENERGY FROM PROTEINS

Some enzymes in your digestive system split dietary proteins into their amino acid subunits, which can be absorbed into the bloodstream. Cells use amino acids to make new proteins and other nitrogen-containing compounds. Even so, if you eat more protein than your body needs, amino acids will be broken down too. Their $-NH_3^+$ group is removed, and then becomes ammonia (NH_3). Depending on the type of amino acid, the carbon backbones are split to form acetyl–CoA, pyruvate, or an intermediate of the Krebs cycle. Your cells can divert any of these organic compounds into the Krebs cycle (Figure 6.8).

As you can see, maintaining and accessing energy reserves is complicated business. Controlling the use of glucose is special because it is the fuel of choice for the brain. However, providing all of your cells with energy starts with the kinds of food you eat.

In humans and other mammals, the entrance of glucose or other organic compounds into an energy-releasing pathway depends on the kinds and proportions of carbohydrates, fats, and proteins in the diet as well as on the type of cell.

IMPACTS, ISSUES

CPT deficiency is a mitochondrial disorder that interferes with fatty acid metabolism. Normally, prolonged exercise, fasting, and other events that can deplete glycogen trigger fatty acid breakdown. When the body does not have enough of the enzyme known as CPT, cells break down the proteins in muscle cells instead, which causes painful muscle cramps.

6.6 Connections With Photosynthesis

LINKS TO SECTIONS 5.2, 5.3

In this unit you read about photosynthesis and aerobic respiration—the main pathways by which cells trap, store, and release energy. Take a moment to consider how these two processes became interconnected.

All organisms require ongoing supplies of energy and organic compounds to survive, grow, and reproduce. Autotrophs get them from the physical environment. They grow, their populations grow, and they become concentrated stores of food tempting to heterotrophs.

Autotrophs are more than carbon-rich food for the biosphere. Early practitioners of the noncyclic pathway of photosynthesis filled the atmosphere with oxygen. Before then, the iron and other metals abundant on all surfaces both above and below the seas was in reduced form. All of the oxygen released by early autotrophs immediately became bound to (oxidized) these metals. Over time, all of the exposed iron rusted, as evidenced by global bands of red iron deposits below the seafloor. Eventually, there was little exposed metal that was not

Figure 6.9 Bubbling mud pit at Yellowstone National Park. Anaerobic archaeans live deep in the superheated water of this hot spring. They use the hydrogen sulfide gas bubbling from underlying magma as their energy source. A metabolic by-product is sulfuric acid, which dissolves adjacent rocks into mud.

oxidized. Oxygen, bubbling from the now tremendous populations of photoautotrophs, gushed unimpeded into the atmosphere.

In a wink of geologic time, maybe a few hundred thousand years, oxygen levels rose in the seas and the sky. It was too fast for most anaerobic species to adapt, and they perished in a mass extinction. Others clung to life in marginalized anaerobic habitats. We find their descendants in hydrothermal vents where molten rock and superheated water spew out from fissures in the seafloor. These archaeans get hydrogen and electrons from hydrogen sulfide gas in mineral-rich volcanic water (Figure 6.9). Other anaerobes that persist in soil get energy from nitrogen-rich wastes and the remains of other organisms.

Some ancient species survived because they had metabolic pathways to detoxify oxygen. These later became remodeled to *use* oxygen as the final electron acceptor for reactions that release energy from carbon compounds. *As a direct outcome of selection pressure by a by-product of photosynthesis, aerobic respiration evolved.*

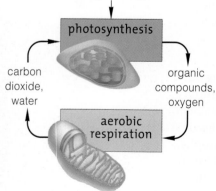

sunlight energy
photosynthesis
carbon dioxide, water
organic compounds, oxygen
aerobic respiration

With the evolution of aerobic respiration, the flow of carbon, and oxygen through the web of life came full circle. Why? Products of aerobic respiration—carbon dioxide and water—are precisely the raw materials that photoautotrophs use.

Such processes might seem far removed from your daily interests. But the food that nourishes you and most other heterotrophs would not be produced without them. We will return to this point in later chapters. It affords perspective on how we as a species are having a global impact on our primary sources of food and oxygen.

Aerobic respiration evolved as a direct outcome of selection pressure by oxygen, a by-product of photosynthesis.

Products of aerobic respiration, carbon dioxide and water, are starting materials for photosynthesis.

http://biology.brookscole.com/BTT2

Summary

Section 6.1 All organisms, including photosynthetic ones, access the chemical bond energy of glucose and other organic compounds, then use it to make ATP. They do so because ATP is a transportable form of energy that can jump-start nearly all metabolic reactions. They break apart compounds and release electrons and hydrogen, both of which have roles in ATP formation.

Glycolysis, the partial breakdown of one glucose to two pyruvate molecules, takes place in the cytoplasm of all cells. It is the first stage of all the main energy-releasing pathways, and it does not use free oxygen.

Anaerobic pathways end in the cytoplasm. The oxygen-requiring pathway called aerobic respiration ends in mitochondria. It harvests far more ATP from glucose than anaerobic pathways do.

BiologyNow
Get an overview of aerobic respiration with the animation on BiologyNow.

Section 6.2 It takes two ATP molecules to jump-start glycolysis. Two molecules of the coenzyme NAD^+ accept electrons and hydrogen originally from glucose, forming 2 NADH. Two pyruvate and four ATP form. The net energy yield is 2 ATP (because 2 ATP were invested up front). The end products of glycolysis are two molecules of pyruvate.

BiologyNow
Take a step-by-step journey through glycolysis with the animation on BiologyNow.

Section 6.3 After glycolysis, aerobic respiration goes through two more reaction stages, in mitochondria.

The second stage consists of acetyl–CoA formation and the Krebs cycle. It takes two full turns of these cyclic reactions to break down both of the pyruvates from glycolysis of one glucose molecule. In total, the second stage of aerobic respiration results in the formation of six CO_2, two ATP, eight NADH, and two $FADH_2$.

The third stage, electron transfer phosphorylation, proceeds at the inner mitochondrial membrane. There, electron transfer chains accept electrons and hydrogen from NADH and $FADH_2$. Electron flow through the transfer chains causes H^+ to accumulate in the inner compartment, so concentration and electric gradients build up across the inner membrane. H^+ follows its gradients back to the outer mitochondrial compartment, through the interior of ATP synthases. The flow drives attachment of phosphate to ADP, forming many ATP. Free oxygen picks up the electrons at the end of the transfer chains and combines with H^+, forming water.

Aerobic respiration has a typical net energy yield of 36 ATP for each glucose molecule metabolized.

BiologyNow
Explore a mitochondrion and observe the reactions inside. Study how each step in aerobic respiration contributes to energy harvests with animations on BiologyNow.

Section 6.4 Fermentation pathways do not require oxygen, and they take place only in the cytoplasm. They start with the formation of two ATP during glycolysis. The remaining reactions regenerate NAD^+ but produce no additional ATP. The net energy yield is only the two ATP that formed in glycolysis.

In alcoholic fermentation, pyruvate is first converted to acetaldehyde and carbon dioxide. When NADH transfers electrons and hydrogen to acetaldehyde, two ethanol molecules form and the NAD^+ is regenerated.

In lactate fermentation, NAD^+ is regenerated when electrons and hydrogen are transferred from NADH to pyruvate, which then becomes lactate.

BiologyNow

Compare alcoholic and lactate fermentation with the animation on BiologyNow.

Section 6.5 In the human body, simple sugars from carbohydrates, glycerol and fatty acids from fats, and carbon backbones of amino acids from proteins enter the aerobic pathway as alternative energy sources.

BiologyNow

Follow the breakdown of different organic molecules with the interaction on BiologyNow.

Section 6.6 Aerobic respiration arose after the noncyclic pathway of photosynthesis had enriched the atmosphere with oxygen. The two pathways are linked on a global scale. Photosynthesizers use water and carbon dioxide products of aerobic respiration when building organic compounds, and most organisms now depend on the oxygen released by photosynthesis.

Self-Quiz

Answers in Appendix I

1. Glycolysis starts and ends in the ______ .
   a. nucleus
   b. mitochondrion
   c. plasma membrane
   d. cytoplasm

2. Which of the following molecules does *not* form during glycolysis?
   a. NADH
   b. pyruvate
   c. $FADH_2$
   d. ATP

3. Aerobic respiration is completed in the ______ .
   a. nucleus
   b. mitochondrion
   c. plasma membrane
   d. cytoplasm

4. In the third stage of aerobic respiration, ______ is the final acceptor of electrons that originally were part of a glucose molecule.
   a. water
   b. hydrogen
   c. oxygen
   d. NADH

5. ______ engage in lactate fermentation.
   a. *Lactobacillus* cells
   b. Some muscle cells
   c. Sulfate-reducing bacteria
   d. a and b

6. In alcoholic fermentation, ______ is the final acceptor of electrons stripped from glucose.
   a. oxygen
   b. pyruvate
   c. acetaldehyde
   d. sulfate

7. The fermentation pathways produce no more ATP beyond the small yield from glycolysis, but the remaining reactions ______ .
   a. regenerate FAD
   b. regenerate NAD^+
   c. dump electrons on an inorganic substance (not oxygen)

8. In certain organisms and under certain conditions, ______ can be used as an energy alternative to glucose.
   a. fatty acids
   b. glycerol
   c. amino acids
   d. all of the above

9. Match the event with its most suitable description.
   ____ glycolysis
   ____ fermentation
   ____ Krebs cycle
   ____ electron transfer phosphorylation
   a. ATP, NADH, $FADH_2$, and CO_2 form
   b. glucose to two pyruvate
   c. NAD^+ regenerated, two ATP net
   d. H^+ flows through ATP synthases

Additional questions are available on **BiologyNow™**

Critical Thinking

1. Living cells of your body never use their nucleic acids as alternative energy sources. Suggest why.

2. Suppose you start a body-building program. You are already eating plenty of carbohydrates. Now a qualified nutritionist recommends that you start a protein-rich diet that includes protein supplements. Speculate on how extra dietary proteins will be put to use, and in which tissues.

3. Each year, Canada geese lift off in precise formation from their northern breeding grounds to spend the winter months in warmer climates (Figure 6.10). Then they make the return trip in spring. As is the case for other migratory birds, their flight muscle cells are efficient at using fatty acids as an energy source. Remember, the carbon backbone of fatty acids can be cleaved into small fragments that can be converted to acetyl–CoA for entry into the Krebs cycle.

   Suppose a lesser Canada goose from Alaska's Point Barrow has been steadily flapping along for about three thousand kilometers and is approaching Klamath Falls, Oregon. It looks down and notices a rabbit sprinting like the wind from a coyote with a taste for rabbit. With a stunning burst of speed, the rabbit reaches the safety of its burrow.

   Which energy-releasing pathway predominated in muscle cells in the rabbit's legs? Why was the Canada goose relying on a different pathway for most of its journey? And why wouldn't the pathway of choice in goose flight muscle cells be much good for a rabbit making a mad dash from its enemy?

Figure 6.10 Canada geese.

4. At very high altitudes, oxygen levels are low and climbers can suffer from altitude sickness. They are short of breath, weak, dizzy, and confused. Curiously, the early symptoms of *cyanide poisoning* are similar. This potent poison binds tightly to cytochrome, the last component in the mitochondrial electron transfer chain. When cyanide is bound to it, cytochrome can't transfer electrons to the next component of the chain. Explain why cytochrome shutdown might cause the same symptoms as altitude sickness.

7 HOW CELLS REPRODUCE

IMPACTS, ISSUES

Henrietta's Immortal Cells

Each human starts out as a fertilized egg. By the time of birth, the body consists of about a trillion cells, each one descended from that single egg. Even in an adult, billions of cells still divide every day and replace their damaged or worn-out predecessors.

In 1951, George and Margaret Gey of Johns Hopkins University were trying to develop a way to keep human cells dividing outside the body. An "immortal" cell line could help researchers study basic life processes as well as cancer and other diseases. Using cells to study cancer would be a better alternative than experimenting with patients and risking their already vulnerable lives.

For almost thirty years, the Geys tried to grow human cells. But they could not stop the cellular descendants from dying within a few weeks. Mary Kubicek, the Geys' assistant, tried again and again to establish a lineage of cultured human cancer cells. She was about to give up, but she prepared one last sample. She named the cells *HeLa*, a code consisting of the first two letters of the donor patient's first and last names. These HeLa cells started to divide, over and over, every twenty-four hours.

Henrietta Lacks

Sadly, cancer cells in the patient were dividing just as fast. Eight months after she had been diagnosed with cervical cancer, Henrietta Lacks, a young mother from Baltimore, was dead.

Although Henrietta passed away, her cells lived on in the Geys' laboratory. HeLa cells were shipped to research laboratories all over the world. The Geys used them to identify which strains of polio virus were harmful. The cells were also used to test the resulting polio vaccine. Other researchers used HeLa cells to investigate cancer, viral growth, protein synthesis, the effects of radiation, and more. Some HeLa cells even traveled into space.

Henrietta was only thirty-one when runaway cell divisions killed her. Decades later, her legacy continues to help humans through her cellular descendants that are still dividing day after day.

Understanding cell division—and, ultimately, how new individuals are put together in the image of their parents—starts with answers to three questions. *First*, what kind of information guides inheritance? *Second*, how is information copied in a parent cell before being distributed to each daughter cell? *Third*, what kinds of mechanisms actually parcel out the information to daughter cells?

It will take more than one chapter to survey the nature of cell reproduction and other mechanisms of inheritance. In this chapter, we introduce the structures and mechanisms that cells use to reproduce.

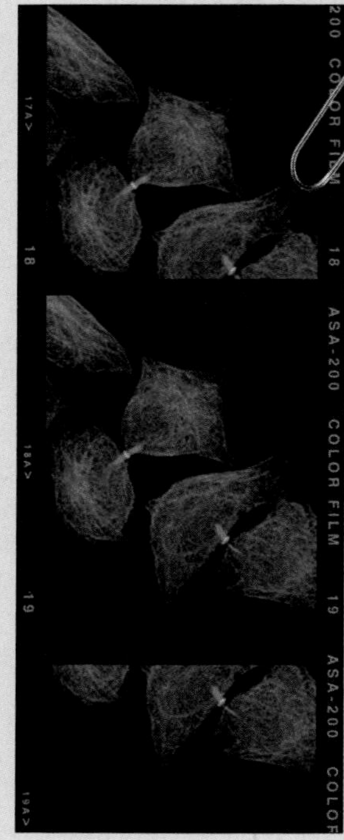

How Would You Vote? *It is illegal to sell your organs, but you can sell your cells, including eggs, sperm, and blood cells. HeLa cells are still being sold all over the world. Should the family of Henrietta Lacks share in the profits? See BiologyNow for details, then vote online.*

Key Concepts

WHAT DIVIDES, AND WHEN
Cells of each species have a characteristic number of chromosomes, which differ in length and shape. Division mechanisms parcel out chromosomes and cytoplasm to daughter cells.

MITOSIS AND CYTOPLASMIC DIVISION
Mitosis, a nuclear division mechanism of eukaryotic cells, maintains the parental chromosome number. The cytoplasm divides by a separate process.

MEIOSIS AND SEXUAL REPRODUCTION
Meiosis, another nuclear division mechanism, divides the parental chromosome number by half. It occurs only in cells that give rise to gametes or spores. Later, fertilization restores the parental chromosome number.

CANCER
Built-in mechanisms control the timing and rate of cell division. On rare occasions, surveillance mechanisms fail, and cell division becomes uncontrollable. Tumor formation is the outcome.

Links to Earlier Concepts

Before reading this chapter, quickly review the introduction to the nucleus of eukaryotic cells (Section 3.5).

You will be also drawing upon your knowledge of microtubules, their assembly and disassembly, and the motor proteins associated with them (3.7).

A quick look back at plant cell walls (3.8) will help you see why plant cells do not divide by pinching their cytoplasm into two parcels, as animal cells do.

The sections on the structure of DNA (2.10), and the nature of inheritance and diversity (1.2, 1.4) gave you the framework to understand how the processes of asexual and sexual reproduction contribute to the evolution of a species.

7.1 Overview of Cell Division Mechanisms

The continuity of life depends on reproduction. By this process, parents produce a new generation of cells or multicelled individuals like themselves. Cell division is the bridge between generations.

A dividing cell faces a challenge. Each of its daughters must get information encoded in the parental DNA and enough cytoplasm to start up its own operation. DNA "tells" a cell how to build proteins. Some proteins are structural materials; others are enzymes that speed up construction of organic compounds. If a cell does not inherit all of the information it needs to build proteins, it will not be able to grow or function properly.

Each daughter cell also inherits a blob of the parent cell's cytoplasm. That blob contains enough enzymes, organelles, and other metabolic machinery for the new cell to begin functioning.

MITOSIS AND MEIOSIS

Because their DNA is packaged into a single nucleus, eukaryotic cells cannot simply divide in half. They do split their cytoplasm into daughter cells, but not until *after* the DNA is copied and repackaged in two nuclei. Nuclear division occurs by either **mitosis** or **meiosis**. As you will see, although these two processes have a lot in common, their outcomes differ.

In multicelled eukaryotes, mitosis and cytoplasmic division occur only in *somatic* (body) cells. These two processes are the foundation of increases in body size, the replacement of dead cells, and tissue repair. Many organisms also reproduce by mitosis, making clones, or exact copies, of themselves.

Meiosis, a different mechanism of nuclear division, precedes the formation of gametes or spores. Meiosis is the basis of sexual reproduction. Gametes, which are mature reproductive cells (including eggs and sperm), form by the meiotic divisions of *germ* cells, which are immature reproductive cells. Spores form by meiosis in the life cycle of many protists, plants and fungi.

What about the prokaryotes—archaea and bacteria? These cells reproduce asexually by prokaryotic fission, a different mechanism. We will consider prokaryotic fission later, in Section 14.2.

KEY POINTS ABOUT CHROMOSOME STRUCTURE

Every eukaryotic cell has a characteristic number of DNA molecules, each with many attached proteins. Together, a molecule of DNA and its proteins are one **chromosome**. A cell duplicates its chromosomes before the nucleus divides. A chromosome and its duplicate stay attached to each other as **sister chromatids** until late in the nuclear division process:

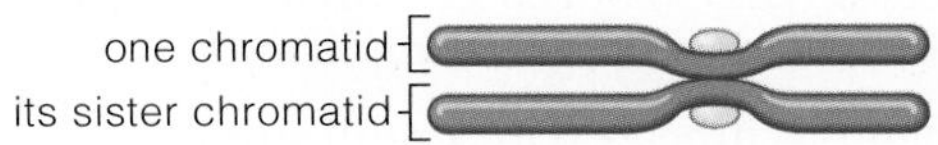

During the early stages of either mitosis or meiosis, each duplicated chromosome coils back on itself again and again, into a highly condensed form (Figure 7.1). This orderly coiling arises from interactions between DNA and proteins associated with it. At regular intervals, DNA "spools" twice around proteins called histones. Each spool of DNA and histones is a **nucleosome**, the chromosome's smallest unit of structural organization.

While each duplicated chromosome is condensing, it constricts where the sister chromatids are attached to each other. This constriction is called a **centromere**. Its location is different for each kind of chromosome.

So what is the point of all this organization? Tight packing helps keep chromosomes from getting tangled while they are moved and sorted into parcels *during* nuclear division. *Between* divisions, the packing keeps enzymes from accessing the cell's genetic information. Loosening specific regions allows a cell to use only the hereditary information that it requires at that time.

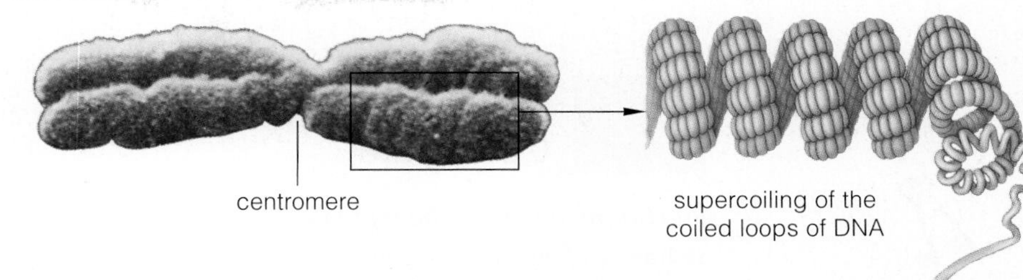

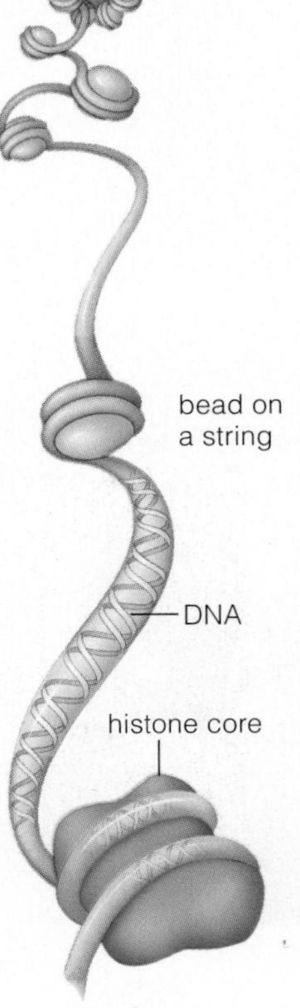

Figure 7.1 ***Animated!*** Scanning electron micrograph of a duplicated human chromosome. Interacting proteins coil the loops of coiled DNA. At a deeper level of organization, proteins and DNA interact in a cylindrical fiber. The smallest unit of organization of a chromosome is the nucleosome: part of a DNA molecule looped twice around a core of histones.

When a cell divides, each daughter cell gets chromosomes and some cytoplasm. Nuclear division and cytoplasmic division are two separate processes in eukaryotic cells.

One nuclear division mechanism, mitosis, is the basis of body growth, cell replacements, tissue repair, and asexual reproduction in eukaryotes.

Meiosis, another nuclear division mechanism, is the basis of gamete and spore formation in sexual reproduction.

LINKS TO SECTIONS 2.10, 3.5

7.2 Introducing the Cell Cycle

Start thinking about the reproduction of a cell as a cyclic sequence of events. This cell cycle is not the same as a life cycle, which is a series of stages that individuals of each species pass through during their lifetime.

A **cell cycle** is a series of events from one cell division to the next (Figure 7.2). It starts when a new daughter cell forms by way of mitosis and cytoplasmic division. It ends when that cell divides. Interphase, mitosis, and cytoplasmic division constitute one turn of the cycle.

During **interphase**, the cell increases in size, roughly doubles the number of components in its cytoplasm, and duplicates its DNA. For most cells, interphase is the longest portion of the cell cycle. Biologists divide it into three stages:

G1 Interval (*G*ap 1) of cell growth and functioning before the onset of DNA replication

S Time of **S**ynthesis (DNA replication)

G2 Second interval (*G*ap 2), after DNA replication, when the cell prepares for division

control or out of control?

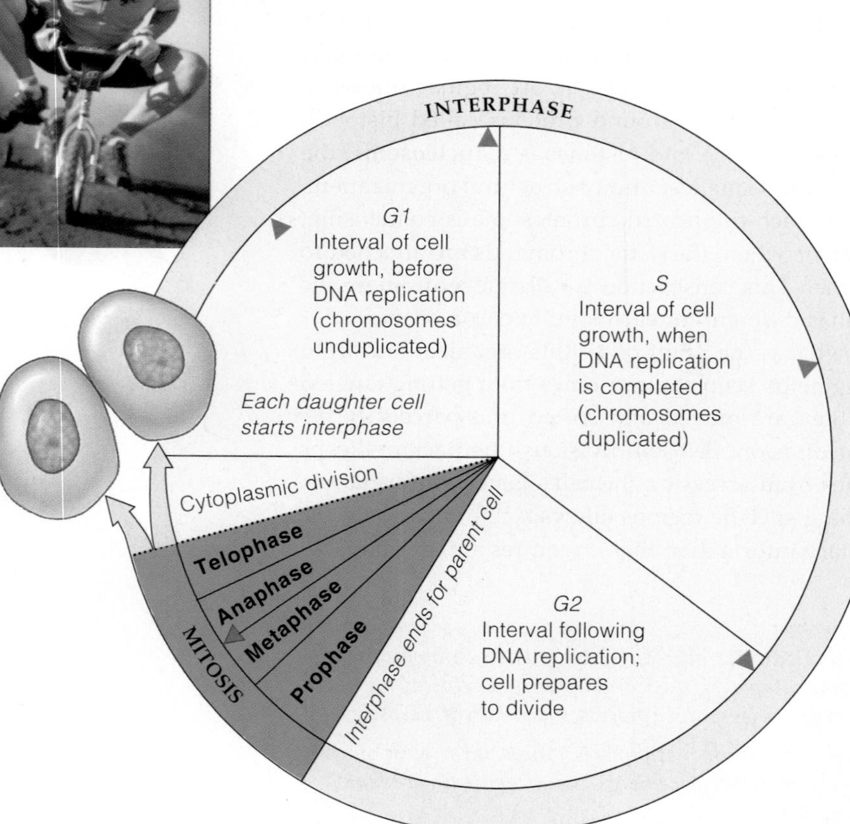

Figure 7.2 ***Animated!*** Eukaryotic cell cycle, generalized. The length of each interval differs among different cell types. *Red* triangles indicate checkpoints in the cycle.

G1, S, and G2 are code names for some events that are just amazing, considering how much DNA is stuffed into a nucleus. If you could stretch out all the DNA molecules from one of your somatic cells in a line, that line would be longer than you are tall, about 7.5 feet. DNA from one salamander cell would stretch 540 feet!

The wonder is, enzymes and other proteins in cells *selectively* access, activate, and silence information in all that DNA. They also make base-by-base copies of every DNA molecule before they divide. Most of this work is completed in interphase.

G1, S, and G2 of interphase have distinct patterns of biosynthesis. G1 is a growth phase. Cells remain in G1, sometimes for years, until they reach adequate cell size and the external environment is appropriate for replication. Cells destined to divide enter S, when they copy their DNA as well as the proteins attached to it. During G2, they make proteins that drive mitosis.

The length of the cycle is about the same for cells of the same type. However, it differs from one cell type to another. For example, every second, 2 to 3 million red blood cells form to replace worn-out ones circulating in your bloodstream. It takes 12 hours for the parent of a red blood cell to divide. Cells in the growing root tip of a bean plant divide every 19 hours, and those in the regenerating limb of an axolotl (a type of amphibian) divide every 53 hours. The cells in a sea urchin embryo can divide every 2 hours.

Once S begins, events will usually proceed at about the same rate and continue through mitosis. Because that rate is the same for all cells of a species, you might wonder if the cell cycle has built-in molecular brakes. It does. The red triangles in Figure 7.2 indicate where some of these controls exert their influence over the cell cycle. Apply the brakes that are supposed to work in G1, and the cycle stalls in G1. Lift the brakes, and the cell cycle will run to completion. Said another way, *control mechanisms govern the rate of cell division.*

Imagine a car losing its brakes just as it starts down a steep mountain road. As you will read later on in the chapter, cancer begins as controls over the cell cycle are lost. Stressful conditions often interrupt the cell cycle. For instance, when deprived of any vital nutrient, free-living cells called amoebas will remain in interphase until that nutrient is restored.

Interphase, mitosis, and cytoplasmic division constitute one turn of the cell cycle.

In interphase, a new cell increases its size, roughly doubles the number of its cytoplasmic components, and duplicates its chromosomes. The cycle ends after the cell undergoes mitosis and then divides.

7.3 Mitosis Maintains the Chromosome Number

Mitosis is a nuclear division mechanism that maintains the parental chromosome number in daughter cells.

LINK TO SECTION 3.7

After G2, a cell divides by mitosis to produce identical daughter cells, each with the same number and kind of chromosomes as the parent. The **chromosome number** is the sum of all chromosomes in a cell of a given type. The body cells of gorillas have 48, those of human cells have 46, and pea plant cells have 14 (Figure 7.3).

Actually, your cells have a **diploid** number ($2n$) of chromosomes; there are two of each type. Those 46 are like volumes of two sets of books numbered from 1 to 23. You have two volumes of, say, chromosome 22—*a pair of them*. Except for the different sex chromosomes (X and Y), chromosomes of a pair have the same length and shape, and carry the same hereditary instructions.

Think of them as two sets of books on how to build a house. Your father gave you one set. Your mother had her own ideas about wiring, plumbing, and so on. She gave you an alternate edition that has the same topics but says slightly different things about many of them.

With mitosis and cytoplasmic division, one diploid parent cell produces two diploid daughter cells (Figure 7.4). This does not mean that each daughter cell merely gets forty-six or forty-eight or fourteen chromosomes. If only the total mattered, then one cell might get, say, two pairs of chromosomes 22 and none of chromosome 9. But then neither cell could function like its parent *without two of each type of chromosome.*

Cells divide all of their chromosomes evenly into daughter cells by using a dynamic structure called the **mitotic spindle**. This network of microtubules grows from opposite sides of the cell during mitosis. Some of the microtubules overlap in the middle of the cell, and others attach to duplicated chromosomes.

Remember, all the chromosomes of a eukaryotic cell duplicate *before* mitosis starts. The spindle determines where each sister chromatid will end up before the cell divides. Microtubules extending from one side of the cell connect to one chromatid of every chromosome. Microtubules from the other side of the cell connect to its sister chromatid. The spindle then pulls the sister chromatids of each duplicated chromosome apart, to opposite poles of the cell.

Thus, each side of the cell gets a mass of DNA that includes one copy of each chromosome. When the cell divides, those masses of DNA will become the nuclei of the daughter cells.

Mitosis proceeds in four stages: *prophase*, *metaphase*, *anaphase*, and *telophase*. These are discussed next.

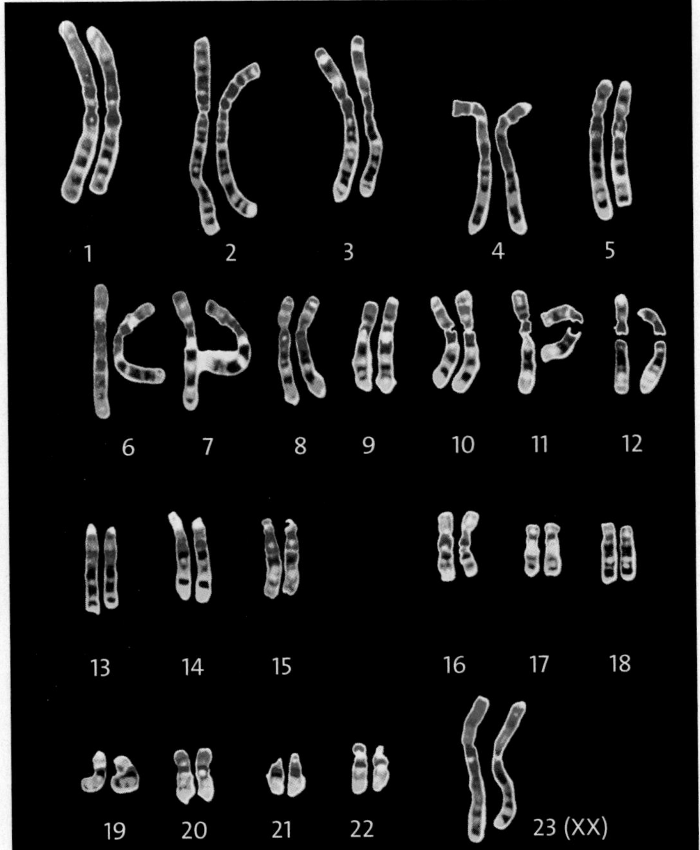

Figure 7.3 From a diploid cell of a human female, twenty-three pairs of chromosomes. The last two are a pair of sex chromosomes. Females have an XX pairing, and males have an XY pairing. The X and the Y chromosome differ in length, shape, and which genetic information they carry, but they pair up during mitosis nonetheless.

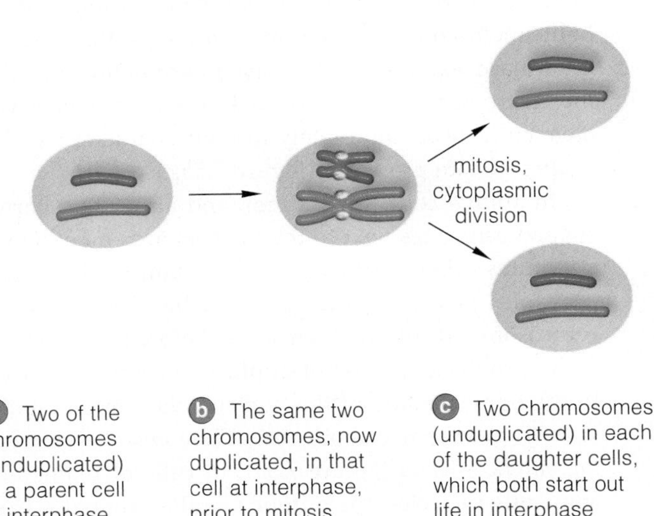

mitosis, cytoplasmic division
a Two of the chromosomes (unduplicated) in a parent cell at interphase
b The same two chromosomes, now duplicated, in that cell at interphase, prior to mitosis
c Two chromosomes (unduplicated) in each of the daughter cells, which both start out life in interphase

Figure 7.4 Preview of how mitosis maintains a parental chromosome number, one generation to the next. For clarity, the spindle is not shown here.

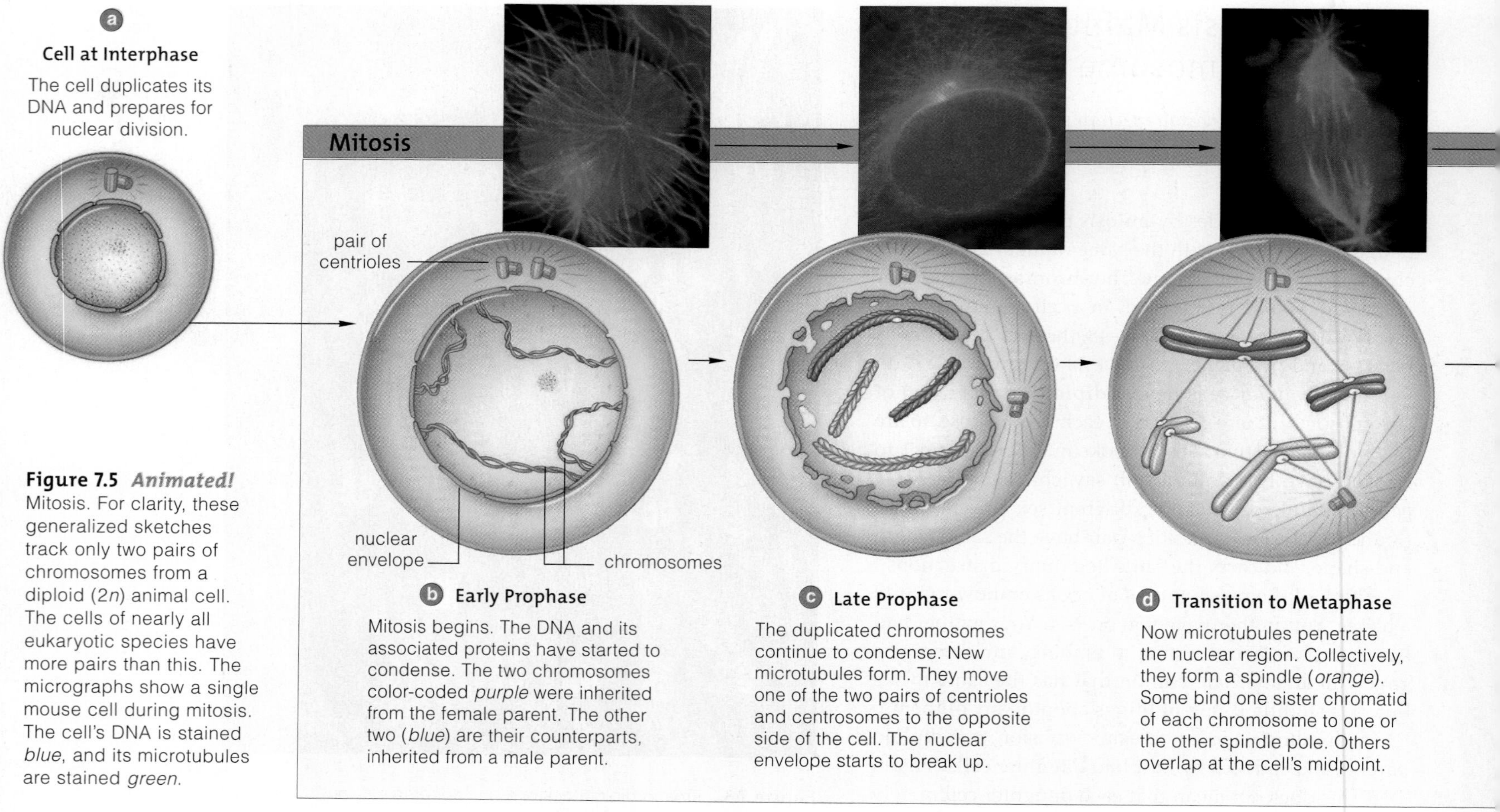

Cell at Interphase

The cell duplicates its DNA and prepares for nuclear division.

Mitosis

pair of centrioles

nuclear envelope

chromosomes

b Early Prophase

Mitosis begins. The DNA and its associated proteins have started to condense. The two chromosomes color-coded *purple* were inherited from the female parent. The other two (*blue*) are their counterparts, inherited from a male parent.

c Late Prophase

The duplicated chromosomes continue to condense. New microtubules form. They move one of the two pairs of centrioles and centrosomes to the opposite side of the cell. The nuclear envelope starts to break up.

d Transition to Metaphase

Now microtubules penetrate the nuclear region. Collectively, they form a spindle (*orange*). Some bind one sister chromatid of each chromosome to one or the other spindle pole. Others overlap at the cell's midpoint.

Figure 7.5 ***Animated!*** Mitosis. For clarity, these generalized sketches track only two pairs of chromosomes from a diploid ($2n$) animal cell. The cells of nearly all eukaryotic species have more pairs than this. The micrographs show a single mouse cell during mitosis. The cell's DNA is stained *blue*, and its microtubules are stained *green*.

PROPHASE: MITOSIS BEGINS

We can tell that a cell is in prophase, the first stage of mitosis, when its chromosomes become visible under a light microscope. Each chromosome was duplicated in interphase; each is now two sister chromatids joined at the centromere (Figure 7.5*a*). All the chromosomes will have condensed completely in their characteristic "X" shapes by late prophase (Figure 7.5*b*,*c*).

In animal cells, some fungi, and algae, two barrel-shaped centrioles next to the nucleus are embedded in structures called centrosomes. In certain kinds of cells, centrioles give rise to flagella or cilia. Centrioles and centrosomes duplicate themselves before prophase.

In prophase, one set of duplicated centrioles moves to the opposite side of the nucleus, then microtubules start to grow out of both sets. *These microtubules form the mitotic spindle* (Figure 7.5*d*). In cells of plants and some other species, the mitotic spindle originates from structures that are similar to centrosomes. In all cells, the areas from which spindle microtubules grow are called spindle poles.

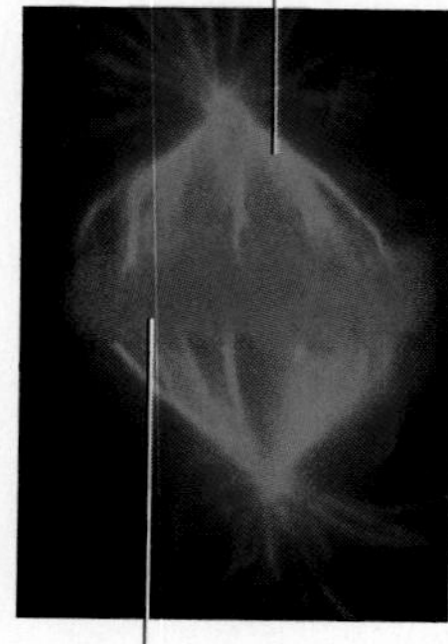

TRANSITION TO METAPHASE

During the transition from prophase to metaphase, the nuclear envelope breaks up into many tiny, flattened vesicles. Spindle microtubules are now free to interact with the chromosomes and one another. Microtubules growing from one spindle pole bind one chromatid of each chromosome, and microtubules from the other spindle pole bind its sister.

The opposing sets of microtubules engage in a tug-of-war, growing and shrinking until they are the same length. Then, all of the duplicated chromosomes will be aligned at the spindle's equator, midway between poles. This is metaphase (Figure 7.5*e*). The alignment is crucial for the next stage.

FROM ANAPHASE THROUGH TELOPHASE

At anaphase, sister chromatids of each chromosome separate from each other and move toward opposite spindle poles. Two mechanisms bring this about. First, the microtubules attached to a chromatid shorten and

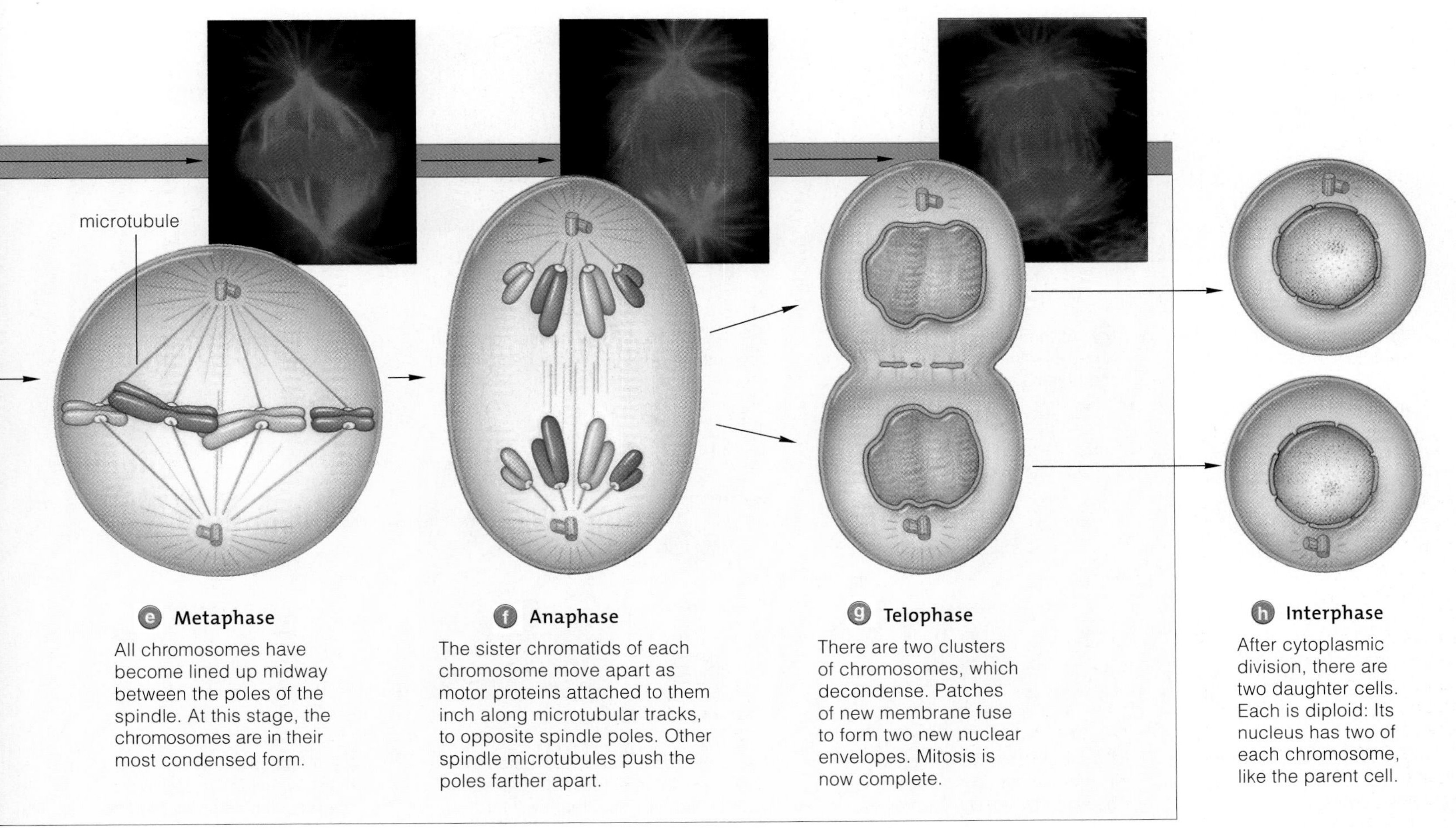

e Metaphase

All chromosomes have become lined up midway between the poles of the spindle. At this stage, the chromosomes are in their most condensed form.

f Anaphase

The sister chromatids of each chromosome move apart as motor proteins attached to them inch along microtubular tracks, to opposite spindle poles. Other spindle microtubules push the poles farther apart.

g Telophase

There are two clusters of chromosomes, which decondense. Patches of new membrane fuse to form two new nuclear envelopes. Mitosis is now complete.

h Interphase

After cytoplasmic division, there are two daughter cells. Each is diploid: Its nucleus has two of each chromosome, like the parent cell.

pull it toward a spindle pole. As the spindle shortens, ATP-driven motor proteins drag the chromatids along microtubules toward the poles.

At the same time, the spindle poles themselves are being pushed farther apart. Some of the microtubules extending from spindle poles do not bind chromatids; they overlap midway between spindle poles. Motor proteins ratchet these microtubules past each other, away from the center of the cell. As they do, they push the spindle poles apart.

Sister chromatids, recall, are identical. Once they detach from each other at anaphase, each is a separate chromosome in its own right (Figure 7.5*f*). Telophase starts when the two sets of chromosomes are clustered at opposite spindle poles. Then, all the chromosomes decondense (Figure 7.5*g*).

Vesicles derived from the old nuclear envelope fuse and form patches of membrane around the clusters. Patch joins with patch until a new nuclear envelope encloses each set of chromosomes. Thus, by the end of telophase, two nuclei have formed.

If the parent cell was diploid, each cluster has two chromosomes of each type. With mitosis, remember, a new nucleus has the same chromosome number as the parent nucleus. Once the two nuclei form, telophase is over—and so is mitosis (Figure 7.5*h*).

Prior to mitosis, chromosomes are duplicated, so each chromosome consists of two sister chromatids.

In prophase, chromosomes condense. The mitotic spindle tethers sister chromatids of duplicated chromosomes to opposite spindle poles.

At metaphase, all duplicated chromosomes are aligned midway between the spindle poles.

At anaphase, microtubules move the sister chromatids of each chromosome apart, to opposite spindle poles.

At telophase, a new nuclear envelope forms around each of two clusters of decondensing chromosomes.

Thus two daughter nuclei have formed. Each has the same chromosome number as the parent cell's nucleus.

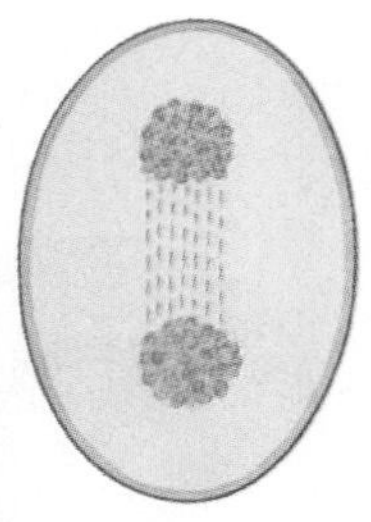

1 Mitosis is over, and the spindle is disassembling.

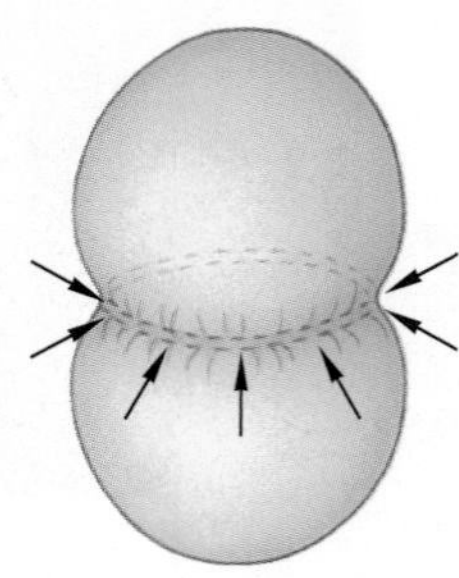

2 At the former spindle equator, a ring of microfilaments attached to the plasma membrane contracts.

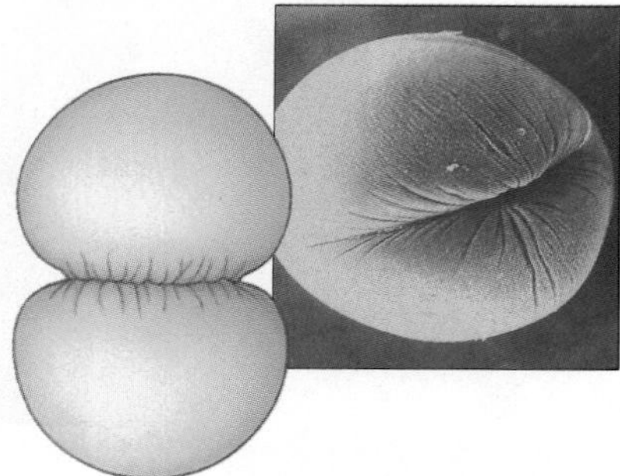

3 As the microfilament ring shrinks in diameter, it pulls the cell surface inward.

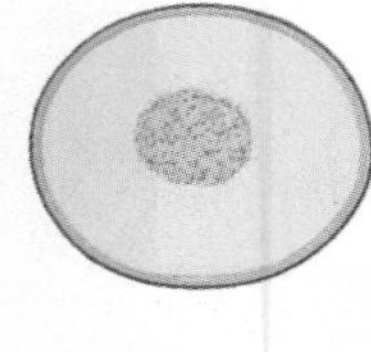

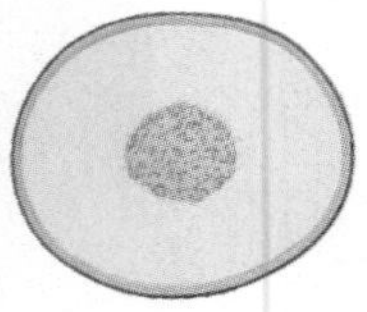

4 Contractions continue; the cell is pinched in two.

a Animal cell division

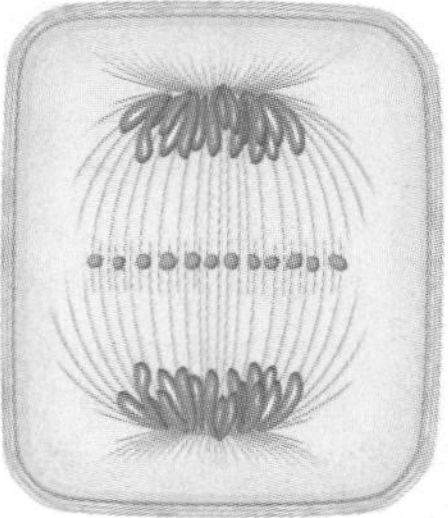

1 As mitosis ends, vesicles cluster at the spindle equator. They contain materials for a new primary cell wall.

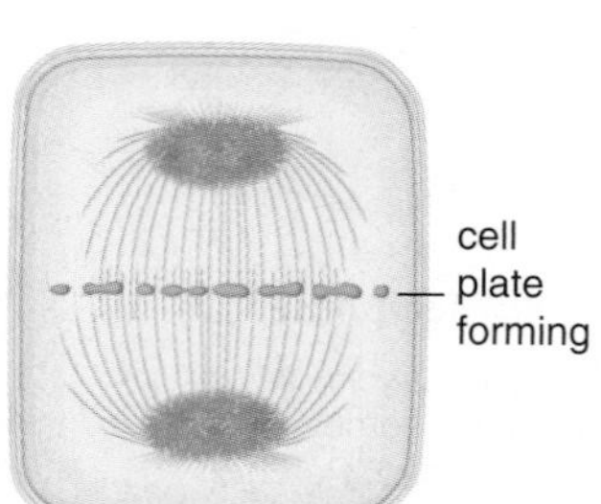

2 Vesicle membranes fuse. The wall material is sandwiched between two new membranes that lengthen along the plane of a newly forming cell plate.

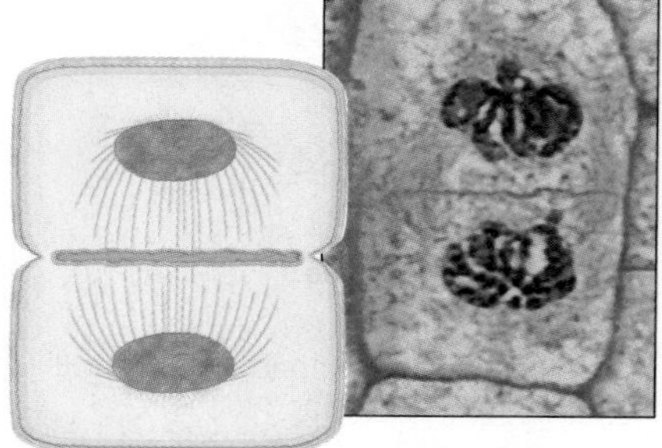

3 Cellulose is deposited inside the sandwich. In time, these deposits will form two cell walls. Others will form the middle lamella between the walls and cement them together.

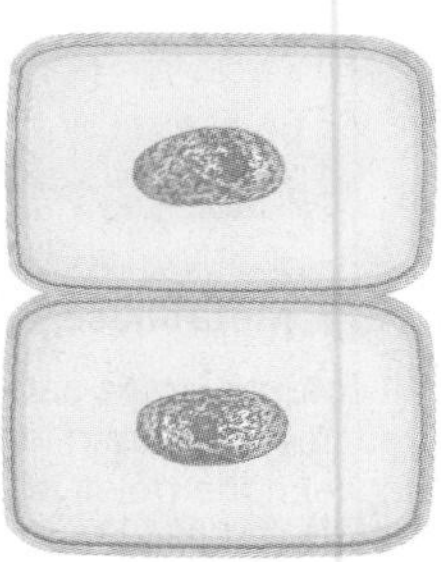

4 A cell plate grows at its margins until it fuses with the parent cell plasma membrane. The primary wall of growing plant cells is still thin. New material is deposited on it.

b Plant cell division

Figure 7.6 ***Animated!*** Cytoplasmic division of an animal cell (**a**) and a plant cell (**b**).

LINK TO
SECTION
3.8

animal cell pinching in two

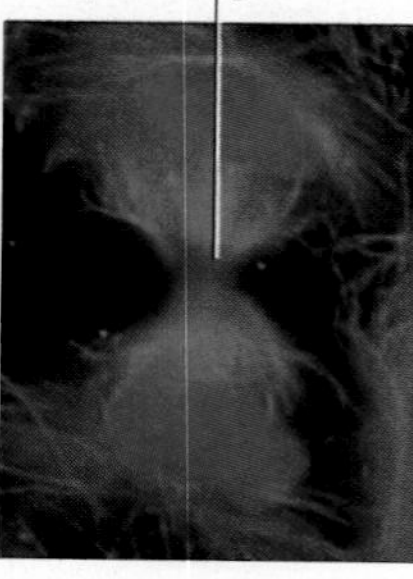

7.4 Division of the Cytoplasm

In most cell types, the cytoplasm usually divides at some time between late anaphase and the end of telophase. The mechanism of cytoplasmic division—or, formally, cytokinesis—differs among species.

After meiosis, an animal cell's cytoplasm can pinch in two by a process called **cleavage**. Midway between the cell's two poles, a patch of plasma membrane sinks in to form a thin indentation (Figure 7.6*a*). This *cleavage furrow* is the first visible sign that the cytoplasm is in the process of dividing. The furrow advances until it surrounds the cell. As it does so, it deepens along the plane corresponding to the former spindle's midpoint.

What is happening? Under the plasma membrane, a thin band of actin filaments is wrapped around the cell's midsection. When energized by ATP, the actin filaments contract and slide past one another, in a way that shrinks the band's diameter. The shrinking band is attached to the plasma membrane, and thus drags the membrane inward until the cytoplasm is pinched in two. Two daughter cells form, each with a nucleus and cytoplasm enclosed in a plasma membrane.

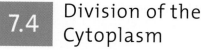

The contractile force of actin filaments is not strong enough to pinch through the stiff cell walls of plant cells. Plant cell cytoplasm divides instead by **cell plate formation** (Figure 7.6*b*). Vesicles full of wall-building materials fuse to form a disk-shaped structure known as the cell plate. In time, cellulose deposits at the plate are thick enough to form a cross-wall through the cell. New plasma membrane extends across both sides of it.

The new wall grows until it bridges the cytoplasm and partitions the parent cell.

LINK TO SECTION 1.2

After mitosis, a separate mechanism partitions the cytoplasm into two daughter cells, each with a nucleus.

A contractile ring mechanism partitions animal cells. Actin filaments form a band around the cell's midsection; as the band contracts the cytoplasm is pinched in two.

Cell plate formation partitions plant cells. Vesicles deposit material to form a cross-wall, which connects to the parent cell wall.

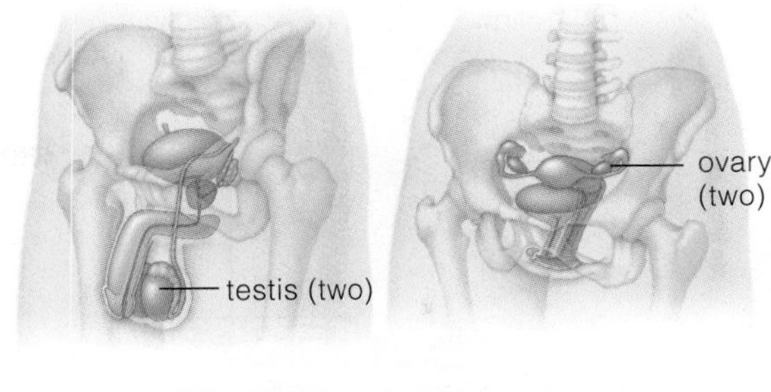

Figure 7.7 Examples of reproductive organs.

7.5 Meiosis and Sexual Reproduction

Meiosis is a nuclear division mechanism that is central to sexual reproduction. It reduces the parental number of chromosomes by half for forthcoming gametes.

ASEXUAL VERSUS SEXUAL REPRODUCTION

When an orchid or aphid reproduces all by itself, what sort of offspring does it get? By the process of **asexual reproduction**, all offspring inherit the same DNA from a single parent. That DNA contains all the heritable information necessary to make a new individual. Rare mutations aside, then, asexually produced individuals are *clones*, genetically identical copies of the parent.

Inheritance gets far more interesting with **sexual reproduction**, a process involving meiosis, formation of gametes, and fertilization—a union of two gametes. Unlike mitosis, meiosis sorts chromosomes into parcels *twice* prior to cytoplasmic division. Meiosis is the first step leading to the formation of gametes.

As you know, the chromosome number is the total number of chromosomes in a single cell. If that cell is diploid ($2n$), it has a *pair* of each type of chromosome. Human body cells have 46 chromosomes, or 23 pairs. Thus, we can say that human somatic cells are diploid, expressed as $2n=46$. So are the human germ cells that give rise to gametes. After meiosis, each gamete ends up with 23 chromosomes—one of each type. Meiosis has reduced the parental chromosome number by half, to a **haploid** number (n). Human gametes are haploid; they have 23 chromosomes, expressed as $n=23$.

In most multicelled eukaryotes, cells that form in specialized reproductive structures or organs give rise to gametes (Figure 7.7). Fertilization occurs as male and female gametes fuse to form a new cell that is the start of a new individual. In most sexual reproducers, this cell is diploid, with pairs of chromosomes. One of the chromosomes of each pair came from the female gamete, and one from the male gamete.

Except for a pair of nonidentical sex chromosomes, paired chromosomes have the same length, shape, and complement of genetic information. The pairs, which zipper together during meiosis, are called **homologous chromosomes** (*hom–* means alike).

A **gene** is a section of a chromosome that contains instructions, encoded in the DNA, for building some structural or functional part of an organism. Genes on homologous chromosomes match up. In other words, each pair of chromosomes has pairs of genes (Figure 7.8). If all pairs of all genes were identical, then sexual reproduction would produce only clones. Imagine if that were true—then you, everyone you know, and the entire human population would be clones.

Fortunately for us, paired genes on homologous chromosomes usually are *not* identical. Most genes are a bit different from their partners on the homologous chromosome. Why? Gene mutations make permanent changes in the molecular structure of DNA. Different mutations accumulate in the DNA of different lines of descent, so it is unlikely that two parents would have identical genes unless they were closely related.

Alleles are different molecular forms of the same gene. The differences affect thousands of traits. This is one reason why individuals of a sexually reproducing species do not all look alike. *With sexual reproduction, offspring inherit new combinations of alleles, which lead to variations in the details of their traits.*

This section introduces the cellular basis of sexual reproduction. More importantly, it starts you thinking about far-reaching effects of gene shufflings at certain stages of the process. The process introduces variations in traits among offspring that are typically acted upon by agents of natural selection. Thus, *variation in traits is a foundation for evolution.*

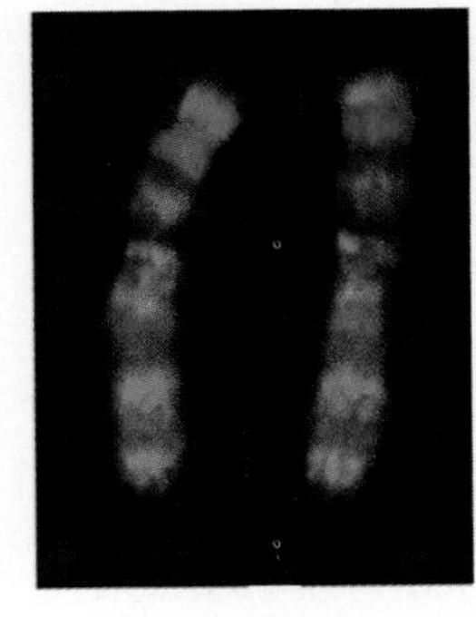

Figure 7.8 Homologous chromosomes. Any gene on one chromosome may be slightly different from its partner on the other chromosome, but at this magnification both of the chromosomes appear to be identical.

Meiosis I

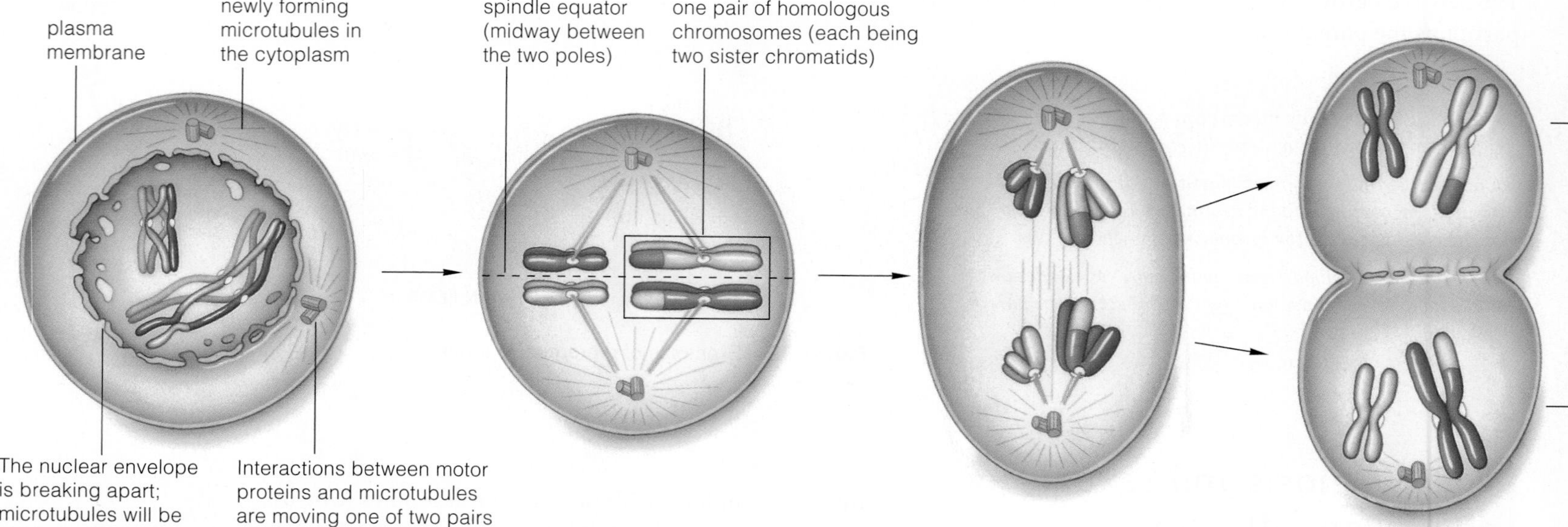

a Prophase I
Chromosomes have been duplicated by the end of interphase I. Now they start to condense. Each pairs with its homologue, and the two usually swap segments. This event, called *crossing over*, is indicated by the break in color on the pair of larger chromosomes. Newly forming microtubules of the spindle become attached to each sister chromatid.

b Metaphase I
Motor proteins projecting from the microtubules move the chromosomes and spindle poles apart. Chromosomes are tugged into position midway between the spindle poles. The spindle becomes fully formed by the dynamic interactions among motor proteins, microtubules, and chromosomes.

c Anaphase I
Some microtubules extend from the spindle poles and overlap at its equator. These *lengthen* and push the poles apart. Other microtubules extending from the poles *shorten* and pull each chromosome away from its homologous partner. These motions move the homologous partners to opposite poles.

d Telophase I
The cytoplasm of the cell divides at some point. There are now two haploid (n) cells with one of each type of chromosome that was present in the parent ($2n$) cell. *All chromosomes are still in the duplicated state.*

Figure 7.9 ***Animated!*** Meiosis in one type of animal cell. This nuclear division mechanism reduces the parental chromosome number in immature reproductive cells by half, to the haploid number, for forthcoming gametes. To keep things simple, we track only two pairs of homologous chromosomes. Maternal chromosomes are shaded *purple* and paternal chromosomes *blue.*

TWO DIVISIONS, NOT ONE

Meiosis is like mitosis in some ways, but the outcome is different. As in mitosis, a germ cell duplicates its DNA in interphase. The two DNA molecules and their proteins stay attached at the centromere, the notably constricted region along their length. For as long as they remain attached, they are sister chromatids:

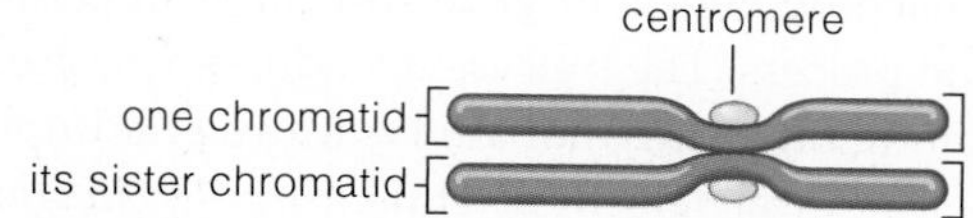

As in mitosis, the microtubules of a spindle apparatus parcel chromosomes into new nuclei by a process with several distinct stages.

With meiosis, however, *chromosomes go through two consecutive divisions that end with the formation of four haploid nuclei.* Each of these two nuclear divisions can be compared to mitosis, but interphase does not occur between divisions. In meiosis, the nucleus of a germ cell divides twice after one cycle of replication. These two divisions are called meiosis I and meiosis II:

interphase (*DNA replication before meiosis I*)	MEIOSIS I	no interphase (*no DNA replication before meiosis II*)	MEIOSIS II
	PROPHASE I		PROPHASE II
	METAPHASE I		METAPHASE II
	ANAPHASE I		ANAPHASE II
	TELOPHASE I		TELOPHASE II

In meiosis I, each duplicated chromosome aligns with its partner, *homologue to homologue.* Then, after the two

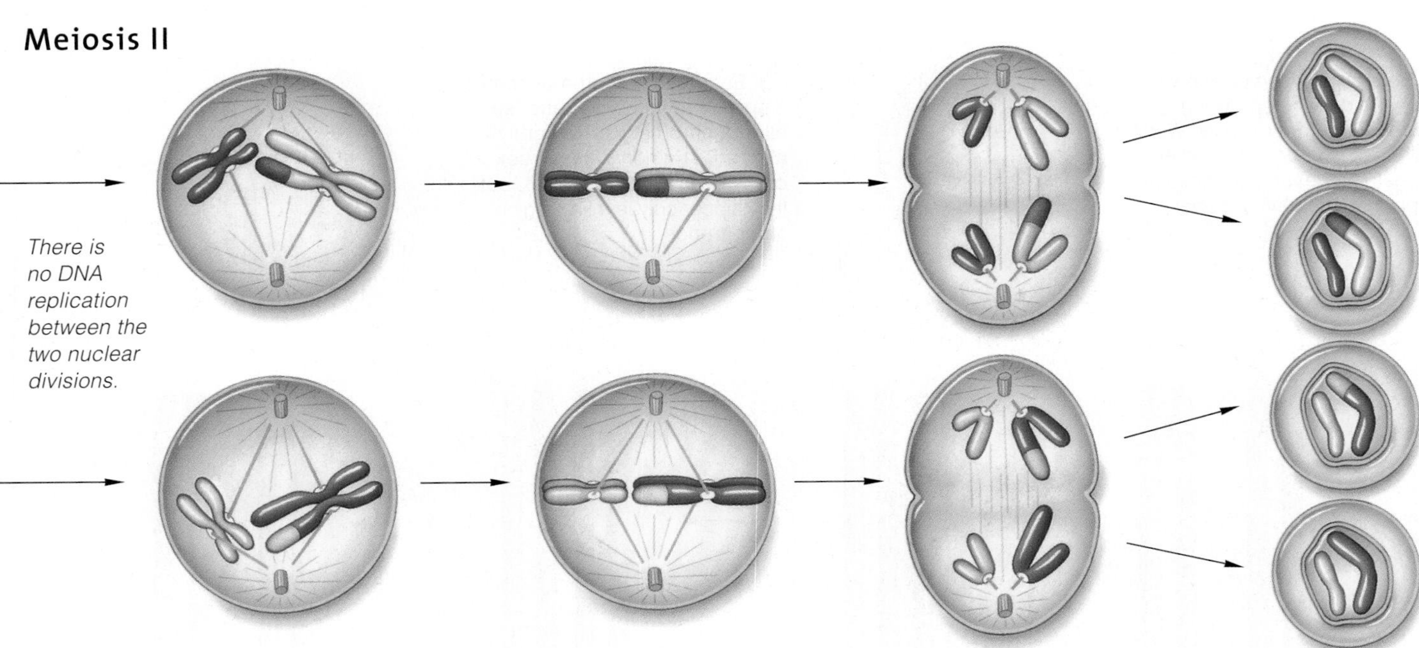

e Prophase II
Microtubules have moved one member of the centriole pair to the opposite spindle pole in each of two daughter cells. During prophase II, microtubules attach to the chromosomes and begin to drag them toward the middle of the cell.

f Metaphase II
The microtubules, motor proteins, and duplicated chromosomes interact, which positions all of the duplicated chromosomes midway between the two spindle poles.

g Anaphase II
The attachment between the sister chromatids of each chromosome breaks, and the two are moved to opposite spindle poles. Each former "sister" is now a chromosome on its own.

h Telophase II
By the time telophase II is finished, there are four daughter nuclei. When cytoplasmic division is over, each daughter cell is haploid (n). All chromosomes are in the unduplicated state.

chromosomes of every pair have lined up with each other, they are moved apart:

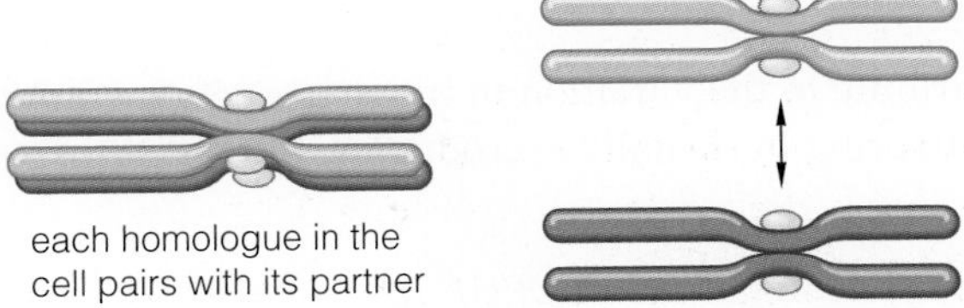

The cytoplasm typically divides after each homologue has separated from its partner. Two haploid daughter cells form by this first cytoplasmic division.

Later, during meiosis II, *the two sister chromatids of each chromosome are separated from each other:*

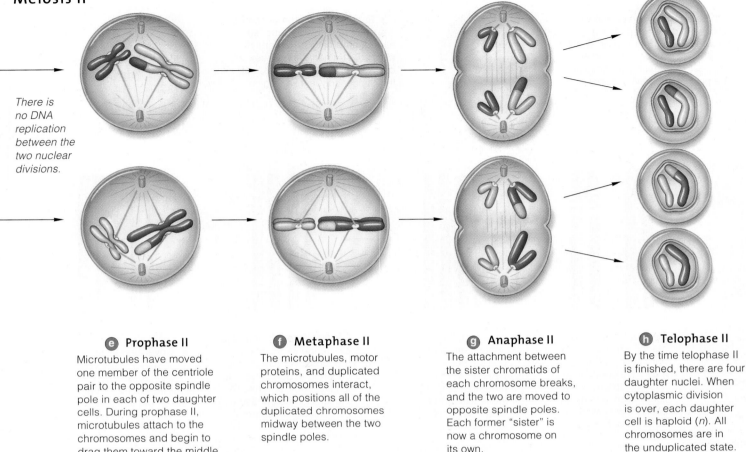

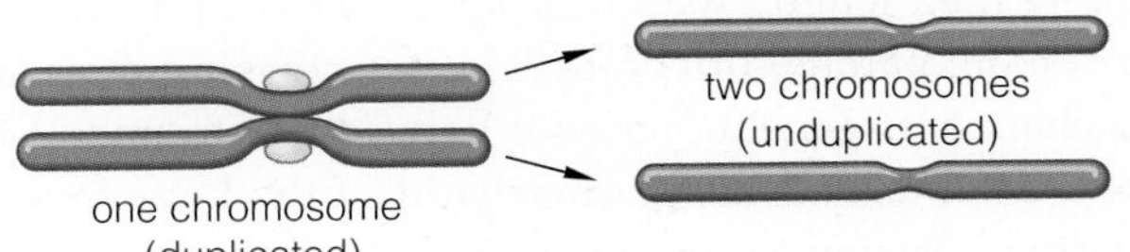

Each sister chromatid has now become a chromosome in its own right. There are now four identical parcels of chromosomes, each containing one chromosome of each type. Nuclear envelopes form around the parcels, enclosing them as four nuclei. Typically the cytoplasm divides once more, so the outcome is four haploid (n) cells. Thus, two meiotic divisions reduced the parental chromosome number of a germ cell by half. Figure 7.9 shows the chromosomal movements in the context of the stages of meiosis.

Asexual reproduction by way of mitotic cell division results in genetically identical copies of a parent.

Sexual reproduction introduces variation in traits by bestowing novel combinations of alleles on offspring.

Meiosis is a nuclear division mechanism that reduces the chromosome number of a parental cell by half—to the haploid number (n). Cytoplasmic division typically follows.

ⓐ Both chromosomes shown here were duplicated during interphase, before meiosis. When prophase I is under way, sister chromatids of each chromosome are positioned close together.

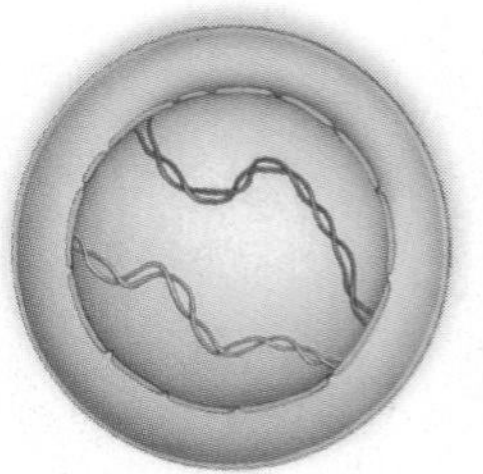

ⓑ Each chromosome becomes zippered to its homologue, so all four chromatids are tightly aligned. The nonidentical sex chromosomes also pair up, but they zipper only in a tiny region at their ends.

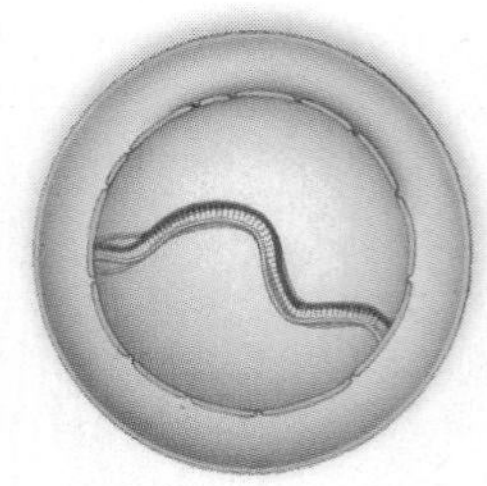

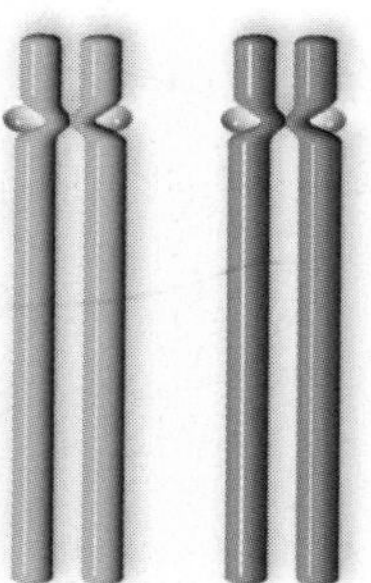

ⓒ For the sake of clarity, we show the chromosomes as if they were condensed. Actually, they are in a tightly aligned form, but not condensed, during prophase I.

ⓓ The intimate contact encourages one crossover (and usually more) to happen at various intervals along the length of nonsister chromatids.

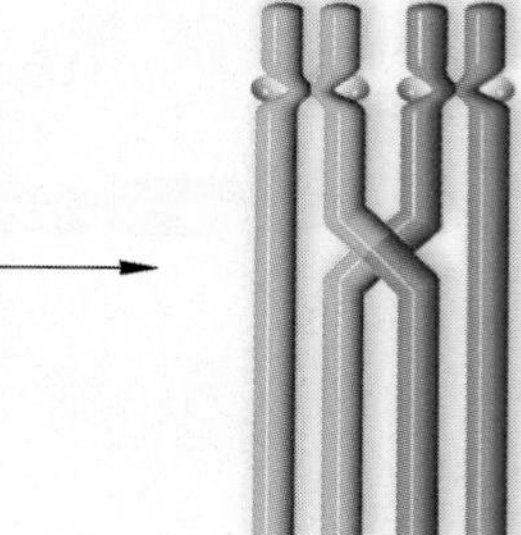

ⓔ Nonsister chromatids exchange segments at the crossover sites. They continue to condense into thicker, rodlike forms. By the start of metaphase I, they will be unzippered from each other.

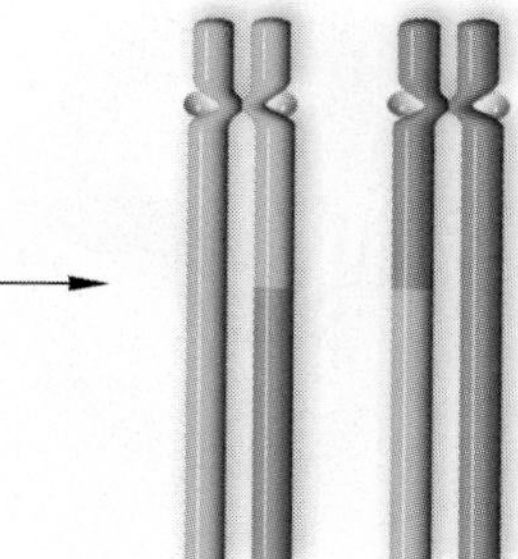

ⓕ Crossing over breaks up old combinations of alleles and puts new ones together in the cell's pairs of homologous chromosomes.

Figure 7.10 ***Animated!*** Key events of prophase I, the first stage of meiosis. For the sake of clarity, we show only one pair of homologous chromosomes and one crossover event. Typically, more than one crossover occurs. *Blue* signifies the paternal chromosome; *purple* signifies its maternal homologue.

LINK TO SECTION 1.4

7.6 How Meiosis Puts Variation in Traits

One function of meiosis is the reduction of the parental chromosome number by half. However, two other functions are just as important. They both give sexual reproduction an evolutionary advantage over asexual reproduction.

Figure 7.9 mentioned briefly that pairs of homologous chromosomes swap pieces during prophase I. It also showed how a homologous chromosome lines up with its partner during metaphase I. Both events introduce new combinations of alleles into gametes that form at some point after meiosis. Together with chromosome shufflings that occur during fertilization, these events contribute to the variation in traits among generations of offspring in sexually reproducing species.

CROSSING OVER IN PROPHASE I

Figure 7.10*a* is a simple sketch of a pair of duplicated chromosomes, early in prophase I of meiosis. Notice how they are in long, thin, threadlike form. All of the chromosomes in a germ cell stretch out like this, and then each is drawn close to its homologue. Molecular interactions stitch homologues together point by point along their length, with little space between them. The intimate, parallel orientation favors **crossing over**, an exchange of genetic material between a chromatid of one chromosome and a chromatid of the homologous partner. During a crossing over event, two "nonsister" chromatids break at exactly the same locations along their length. The chromatids exchange corresponding segments at the breaks; they swap genes.

Gene swapping would be pointless if genes never varied. But remember, each gene can come in slightly different forms: alleles. You can predict that a number of the alleles on one chromosome will *not* be identical to their partner alleles on the homologous chromosome. Every crossover event is an opportunity to exchange different versions of heritable information.

We will look at the implications of crossing over in later chapters. For now, just remember this: *Crossing over leads to recombinations among genes of homologous chromosomes, and eventually to variation in traits among gametes, then offspring.*

METAPHASE I ALIGNMENTS

Major shufflings of intact chromosomes start during the transition from prophase I to metaphase I. Suppose this is happening right now in one of your germ cells. Crossovers have already made genetic mosaics of the chromosomes, but put this aside to simplify tracking. Just call the twenty-three chromosomes you inherited from your mother the *maternal* chromosomes, and the twenty-three homologues from your father the *paternal* chromosomes.

At metaphase I, the microtubules from both spindle poles have the duplicated chromosomes lined up at the spindle equator (Figure 7.9*b*). Have they attached all maternal chromosomes to one pole and all paternal chromosomes to the other? Maybe, but probably not. When the microtubules were growing, they connected to the first chromosome they contacted. Because the attachment was random, there is no particular pattern to the metaphase I positions of maternal and paternal chromosomes. Carry this thought one step further. During anaphase I, when a duplicated chromosome is moved away from its homologous partner, *either one* of them can end up at either spindle pole.

Think of the possibilities while tracking just three pairs of homologues. By metaphase I, these three pairs may be arranged in any one of four possible positions (Figure 7.11). This means that eight combinations (2^3) are possible for forthcoming gametes.

Cells that give rise to human gametes have twenty-three pairs of homologous chromosomes, not three. So each time a human sperm or egg forms, you can expect a total of *8,388,608* (or 2^{23}) possible combinations of maternal and paternal chromosomes!

Moreover, in each sperm or egg, many hundreds of alleles inherited from the mother might not "say" the exact same thing about hundreds of different traits as alleles inherited from the father. Are you beginning to get an idea of why such fascinating combinations of traits show up the way they do among the generations of your own family tree?

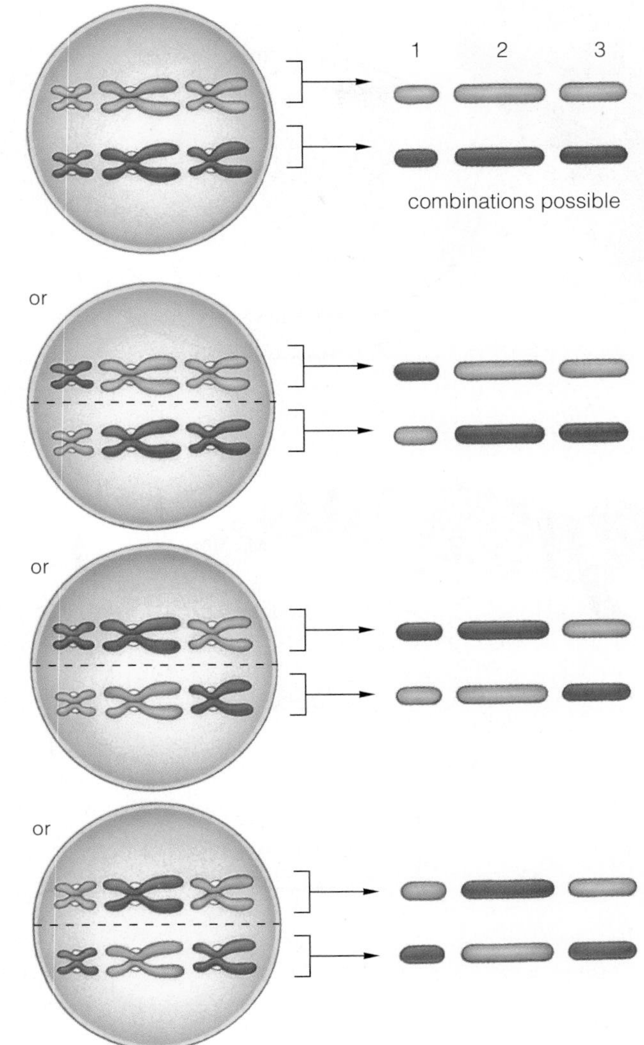

Figure 7.11 ***Animated!*** Possible outcomes for random alignment of three pairs of homologous chromosomes at metaphase I. The three types of chromosomes are labeled 1, 2, and 3. With four alignments, eight combinations of maternal (*purple*) and paternal chromosomes (*blue*) are possible in gametes.

Crossing over, an interaction between a pair of homologous chromosomes, breaks up old combinations of alleles and puts new ones together during prophase I of meiosis.

The random attachment and subsequent positioning of each pair of maternal and paternal chromosomes at metaphase I lead to different combinations of maternal and paternal traits in each new generation.

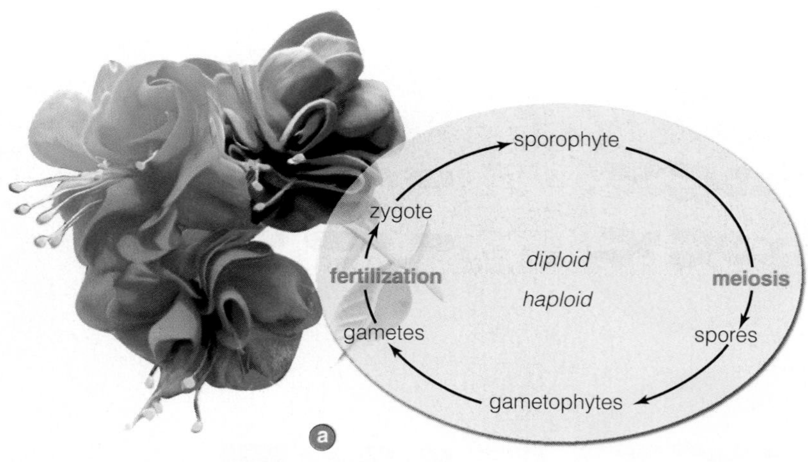

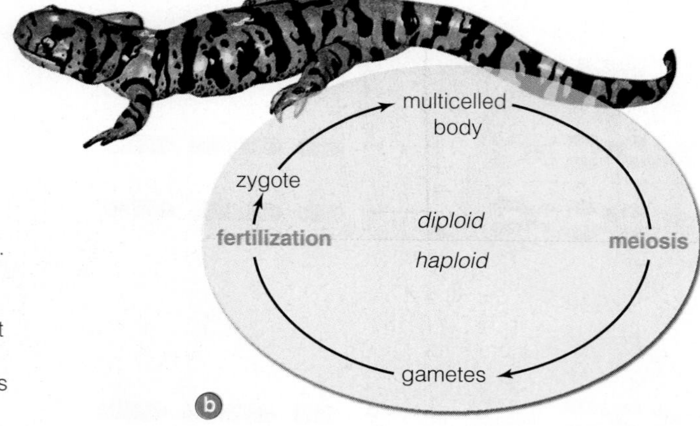

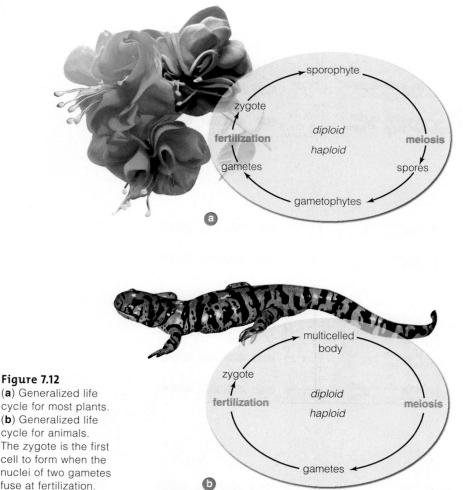

Figure 7.12
(**a**) Generalized life cycle for most plants. (**b**) Generalized life cycle for animals. The zygote is the first cell to form when the nuclei of two gametes fuse at fertilization.

7.7 From Gametes to Offspring

What happens to the gametes that form after meiosis? Later chapters have specific examples. Here, just focus on where they fit into the life cycles of plants and animals.

Gametes are not all the same in their details. Human sperm have one tail, opossum sperm have two, and roundworm sperm have none. Crayfish sperm look like pinwheels. Most eggs are microscopic, and yet an ostrich egg inside its shell can weigh five pounds. A flowering plant's male gamete is just a sperm nucleus.

GAMETE FORMATION IN PLANTS

Plants on land face seasonal changes, and fertilization must coincide with spring rains and other conditions that favor growth of offspring. That is why most plant life cycles alternate between the production of spores (by sporophytes) and the production of gametes (by gametophytes). Figure 7.12*a* summarizes this cycle. A *sporophyte* is the plant body that bears reproductive structures. Trees and all other plants with roots, stems, and leaves are typical sporophytes. Most are diploid.

Spores form by meiosis within sporophyte tissues. A **spore** is a haploid resting structure. When spores germinate, or resume growth, they undergo mitosis to form haploid *gametophytes*, gamete-producing bodies. Gametophytes form, for example, inside stamens and carpels, the reproductive structures of flowers. During sexual reproduction, two haploid gametes will fuse to form a diploid zygote, which by mitotic cell divisions will give rise to another sporophyte. We will return to the topic of plant reproduction in later chapters.

GAMETE FORMATION IN ANIMALS

Diploid germ cells give rise to the gametes of animals (Figure 7.12*b*). In the male reproductive system, a germ cell develops into a primary spermatocyte. This large, immature cell enters meiosis. Four haploid cells result and develop into spermatids (Figure 7.13). These cells undergo changes, such as the formation of a tail, and become **sperm**, a type of mature male gamete.

In female animals, a germ cell becomes an **oocyte**, or immature egg. Unlike sperm, an oocyte stockpiles many cytoplasmic components, and its four daughter cells differ in size and function (Figure 7.14).

As an oocyte divides after meiosis I, one daughter cell called the secondary oocyte gets nearly all of the cytoplasm. The other cell, a first polar body, is small. Later, both of these haploid cells enter meiosis II, then cytoplasmic division. One of the secondary oocyte's daughter cells develops into a second polar body. The other gets most of the cytoplasm and develops into a gamete. The mature female gamete is called an ovum (plural, ova) or, more often, an **egg**. The three polar bodies that formed do not function as gametes; they are not rich in nutrients or plump with cytoplasm. In time they will degenerate. But their formation assures that the egg will have a haploid chromosome number. Also, by getting most of the cytoplasm, the egg holds enough metabolic machinery to support the early cell divisions of the new individual.

MORE SHUFFLINGS AT FERTILIZATION

The chromosome number characteristic of the parents is restored at **fertilization**, a time when a female and male gametes unite and their haploid nuclei fuse. If meiosis did not precede fertilization, the chromosome number would double in each generation. Doublings disrupt cell function. Why? Duplicate information on

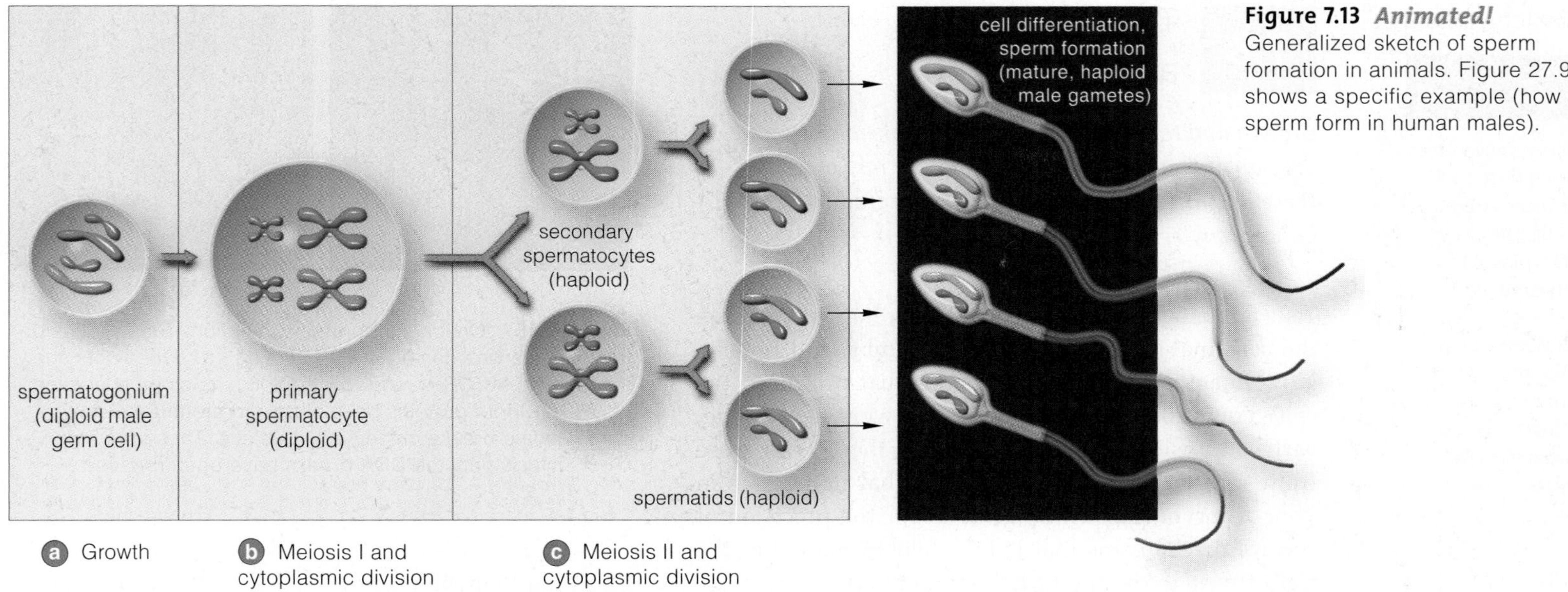

Figure 7.13 ***Animated!*** Generalized sketch of sperm formation in animals. Figure 27.9 shows a specific example (how sperm form in human males).

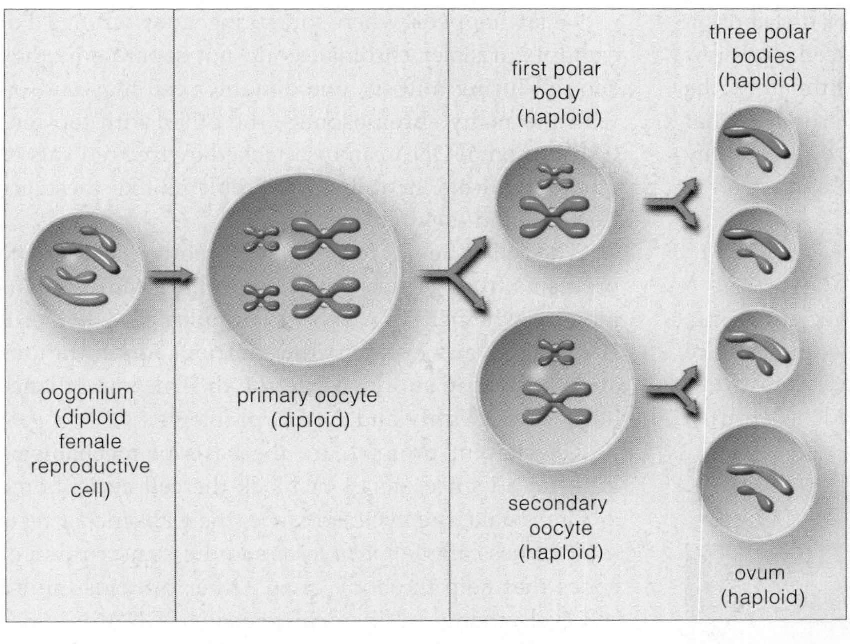

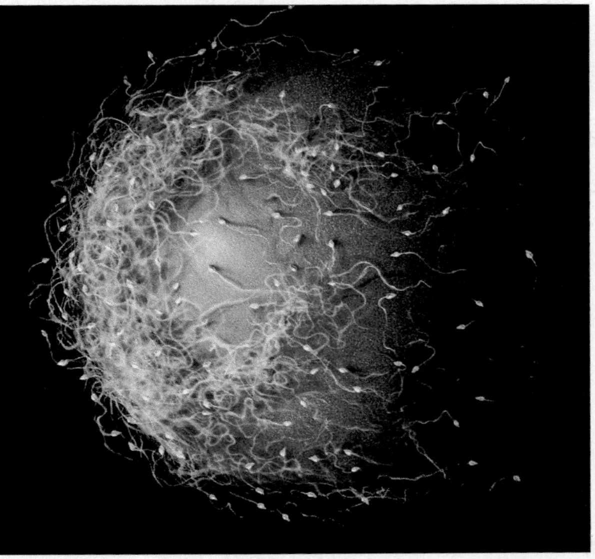

Figure 7.14 ***Animated!*** Animal egg formation. Eggs are far larger than sperm and larger than the three polar bodies. The painting above, based on a scanning electron micrograph, depicts human sperm surrounding an ovum.

extra chromosomes becomes converted into duplicate structural and functional parts of the cell, changing the finely tuned balance of its operation.

Fertilization also adds to variation among offspring. Reflect on the possibilities for humans alone. During prophase I, every human chromosome undergoes an average of two or three crossovers. In addition to the crossovers, random positioning of pairs of paternal and maternal chromosomes at metaphase I results in one of millions of possible chromosome combinations in each gamete. And of all male and female gametes that form, *which* two actually get together is a matter of chance. The sheer number of combinations that can exist at fertilization is staggering!

Distribution of random mixes of chromosomes into gametes and fertilization further contribute to the variation in traits that occur in offspring produced by sexual reproduction.

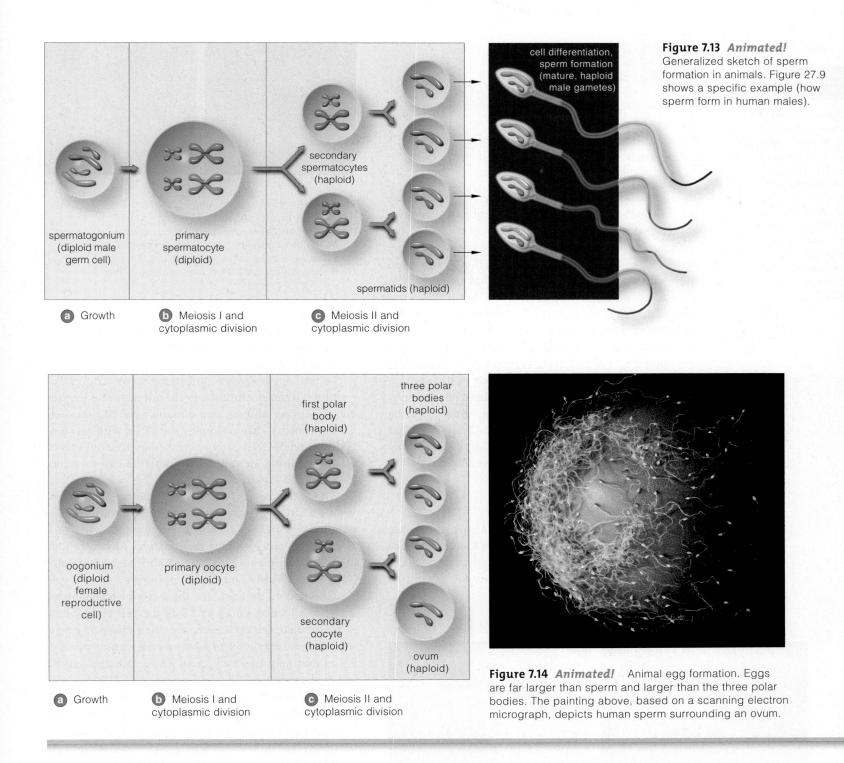

IMPACTS, ISSUES

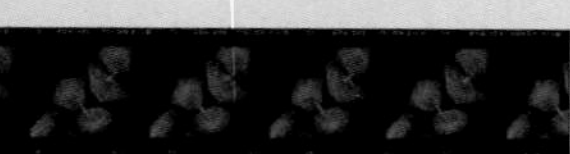

The filmstrip shows HeLa cells derived from Lack's cervical cancer. Like other cancer cells, they have a distorted plasma membrane and altered cytoskeleton. Cervical cancer develops most often after infection by HPV, the human papillomavirus. HeLa cells contain the DNA of this virus.

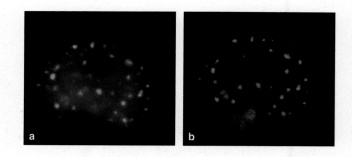

7.8 The Cell Cycle and Cancer

Growth and reproduction depend on controls over cell division and programmed cell death. On rare occasions, these controls fail in a somatic cell or reproductive cell. Cancer may be the outcome.

Take a moment to look closely at your hands. Visualize the cells making up your palms, thumbs, and fingers. Now imagine the mitotic divisions that produced the cell generations before them as you were developing, early on, inside your mother. All of those cells arose from tiny clumps of identical cells that budded from your embryonic body. Be grateful for the precision of control mechanisms that led to their formation at the right times, in the right places (Figure 7.15).

Health and survival depend on the proper timing and completion of cell cycle events. Genetic disorders arise from mistakes during duplication or distribution of even one chromosome. Uncontrolled cell divisions can destroy surrounding tissues and, ultimately, the individual. Such losses often begin in body cells that have lost control over the cell cycle. More rarely, they start in germ cells that give rise to sperm and eggs.

THE CELL CYCLE REVISITED

Millions of cells in your skin, bone marrow, gut lining, liver, and elsewhere in your body divide and replace their worn-out, dead, and dying predecessors every second of every day. The cells do not divide arbitrarily.

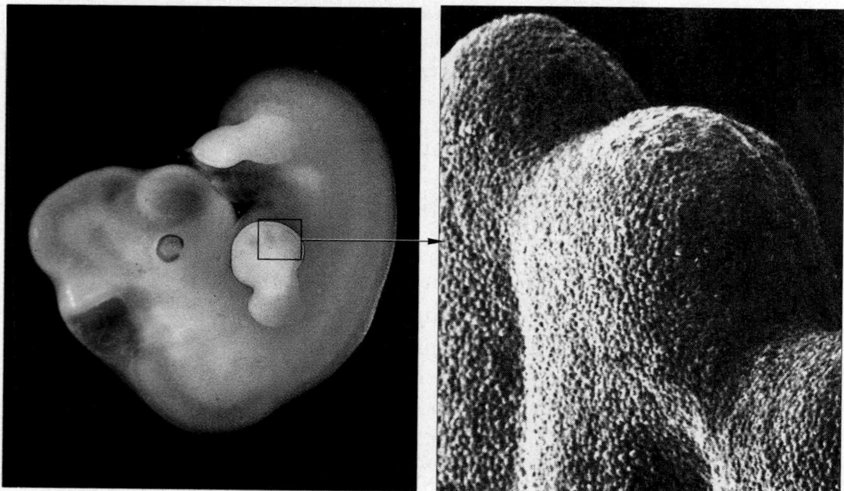

Figure 7.15 The paddlelike structure of a human embryo that develops into a hand by mitosis, cytoplasmic divisions, and programmed cell death. The scanning electron micrograph on the right shows individual cells.

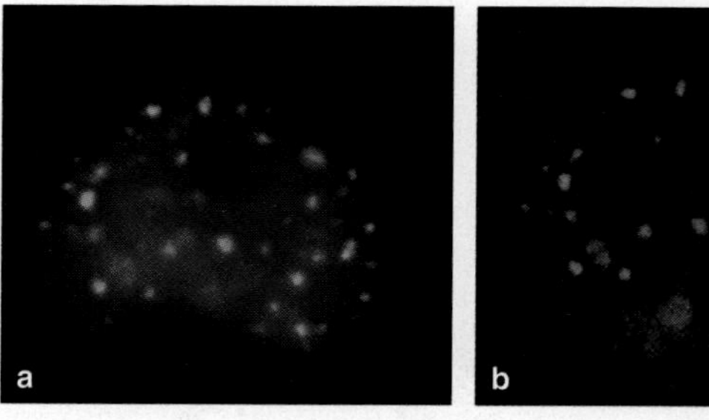

Figure 7.16 Checkpoint proteins in action. The DNA inside this nucleus was damaged by radiation. (**a**) *Green* dots show the location of *53BP1*, and (**b**) *red* dots pinpoint the location of *BRCA1*. Both proteins have clustered around the same chromosome breaks in the same nucleus. These proteins block mitosis until the DNA breaks have been repaired.

Many mechanisms control cell growth, replication of DNA, and division. Some of them put the machinery of division to rest.

What happens when something goes wrong? For example, if sister chromatids do not separate as they should during mitosis, one daughter cell may end up with too many chromosomes, the other with too few. Chromosomal DNA can be attacked by free radicals or other chemicals, or radiation. Problems like these are frequent and inevitable.

The cell cycle has built-in checkpoints that keep the problems from getting out of control. Certain proteins monitor whether DNA gets fully replicated, whether it is damaged, and even whether nutrient concentrations are sufficient to support cell growth. This surveillance helps cells identify and correct problems.

Checkpoint proteins are the basis of mechanisms that can advance, delay, or block the cell cycle. Some of them make the cycle advance; their absence arrests it. The ones called *growth factors* invite transcription of genes that help the body grow. Other proteins inhibit cell cycle changes. When chromosomal DNA becomes damaged, several checkpoint gene products interrupt mitosis until it is repaired (Figure 7.16). If the DNA remains broken or incomplete, a cascade of signaling events induces cell death.

Sometimes a checkpoint gene mutates so that its protein product no longer functions properly. When all checkpoint mechanisms fail, the cell loses control over its replication cycle. In some cases it gets stuck in mitosis, dividing again and again with no interphase. In other cases, damaged chromosomes are replicated or cells do not die as they are supposed to, because signals calling for cell death are disabled. A growing mass of defective cellular descendants forms a tumor.

Moles and other tumors are **neoplasms**, abnormal masses of cells that lost controls over their cell cycle.

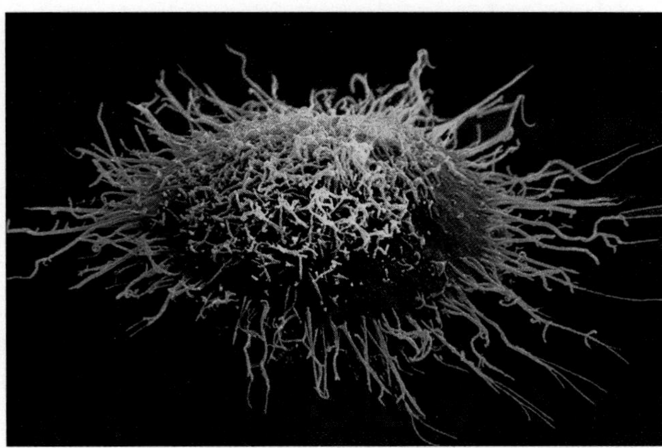

Cervical cancer cell, the kind that killed Henrietta Lacks.

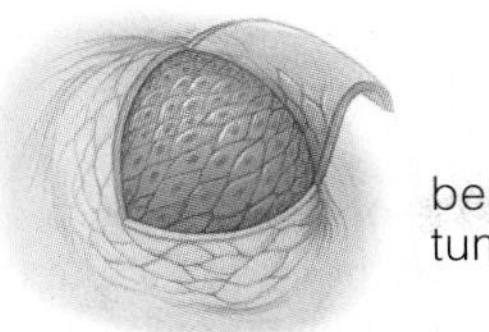

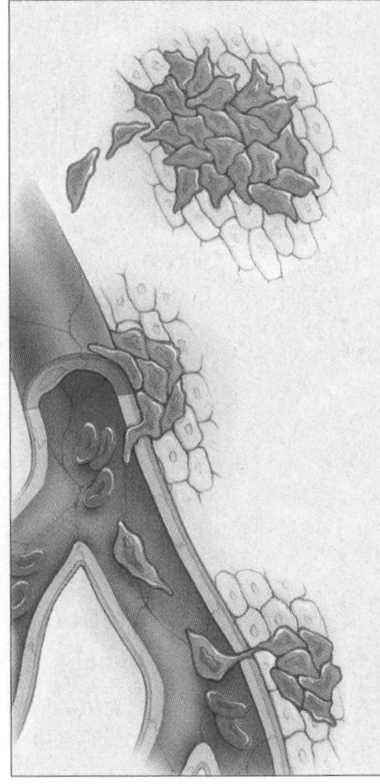

a Cancer cells break away from their home tissue.
b The metastasizing cells become attached to the wall of a blood vessel or lymph vessel. They secrete digestive enzymes onto it. Then they cross the wall at the breach.
c Cancer cells creep or tumble along inside blood vessels, then leave the bloodstream the same way they got in. They start new tumors in different tissues.

Figure 7.17 *Animated!* Comparison of benign and malignant tumors. Benign tumors typically are slow-growing and stay put in their home tissue. Cells of a malignant tumor migrate abnormally through the body and establish colonies even in distant tissues.

Ordinary skin moles and other *benign*, noncancerous neoplasms grow very slowly, and their cells retain the surface recognition proteins that are supposed to keep cells inside a home tissue. Unless a benign neoplasm grows too large or becomes irritating, it does not pose a threat to the body.

CHARACTERISTICS OF CANCER

Cancers are abnormally growing and dividing cells of a *malignant* neoplasm. The cells disrupt surrounding tissues, both physically and metabolically. Cancer cells are grossly disfigured. They can break loose from their home tissues, slip into and out of blood vessels and lymph vessels, and invade other tissues where they do not belong (Figure 7.17).

All cancer cells display four characteristics. First, they grow and divide abnormally. Controls that limit overcrowding in tissues are lost and cell populations reach high densities. The number of tiny blood vessels that service the growing cell mass increases.

Second, the cytoskeleton and plasma membrane of cancer cells are badly altered. The plasma membrane becomes leaky and has abnormal or lost proteins. The cytoskeleton shrinks, becomes disorganized, or both. Enzyme action can shift, as in an amplified reliance on ATP formation by glycolysis.

Third, cancer cells have a weakened or nonexistent capacity for adhesion. Proteins that function in cell-cell recognition are lost or altered, so cells do not stay put. They break away and colonies of them start forming in distant tissues. This process of abnormal cell migration and tissue invasion is called *metastasis*.

Fourth, cancer cells have lethal effects. Unless they can be eradicated by surgery, chemotherapy, or other procedures, their uncontrollable divisions will put an individual on a painful road to death.

Each year in the developed countries alone, 15 to 20 percent of all deaths result from cancer. This is not just a human problem. Cancers are known to occur in most of the animal species studied to date.

Cancer is a multistep process. But researchers have already identified many of the mutated forms of genes that contribute to it. They also are working to identify drugs that specifically target and destroy cancer cells or stop them from dividing.

HeLa cells, for instance, were used in early tests of taxol, the microtubule poison that stops spindles from forming. With this kind of research, we may one day have drugs that can put the brakes on cancer cells.

Built-in checkpoints can advance, delay, or block the cell cycle in response to environmental or internal conditions.

Loss of controls over cell division results in abnormal masses of cells called neoplasms. Malignant neoplasms are cancers; cells in them reach high densities and can metastasize.

IMPACTS, ISSUES

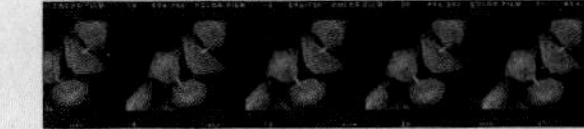

In laboratories all over the world, cells of human nerves, muscles, skin, and lungs are dividing away in culture. Immortal cell lines have been derived from more than seventy different human tissues, from both normal and cancerous cells. They are used to test new drugs, study cell function, and also to produce vaccines. In the United States, for example, cultures of cells called fibroblasts make most of the vaccine used against rabies.

http://biology.brookscole.com/BTT2

Summary

Section 7.1 By processes of reproduction, parents produce a new generation of individuals like themselves. Cell division is the bridge between generations. When a cell divides, its daughter cells each receive a required number of DNA molecules and some cytoplasm.

A eukaryotic chromosome is a molecule of DNA together with its associated proteins. When duplicated, the chromosome consists of two sister chromatids. Until late in mitosis (or meiosis), the two sister chromatids remain attached at their centromere.

Only eukaryotic cells undergo mitosis, meiosis, or both. These nuclear division mechanisms partition the chromosomes of a parent cell into daughter nuclei. A separate mechanism divides the cytoplasm.

Mitosis is the basis of body growth of multicelled organisms, cell replacements, and tissue repair. Many eukaryotic species also reproduce asexually by mitosis. Meiosis, the basis of sexual reproduction, precedes the formation of gametes or spores.

BiologyNow

Explore the structure of a chromosome with animation on BiologyNow.

Section 7.2 The cell cycle begins when a new cell forms, and it ends when the cell reproduces by nuclear and cytoplasmic division. A cell carries out most of its functions in interphase: it increases in mass, and also duplicates each of its chromosomes.

BiologyNow

Investigate the stages of the cell cycle with the interaction on BiologyNow.

Section 7.3 The sum of all chromosomes in cells of a given type is the chromosome number. Human body cells have a diploid chromosome number of 46, or two copies of 23 types of chromosome. Mitosis maintains the chromosome number, one generation to the next.

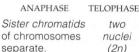

Mitosis proceeds through prophase, metaphase, anaphase, and telophase. A mitotic spindle separates the sister chromatids of each chromosome into two daughter nuclei (Figure 7.18).

Section 7.4 Cytoplasmic division mechanisms differ. Animal cells undergo cleavage. A microfilament ring under the plasma membrane contracts, pinching the cytoplasm in two. In plant cells, a cross-wall forms in the cytoplasm and divides it.

Section 7.5 Asexual reproduction yields clones, or offspring that are genetically identical to a single parent. Sexual reproduction involves two parents that engage in meiosis, gamete formation, and fertilization. It leads to the variation in traits among offspring.

Alleles are slightly different molecular forms of the same gene. These differences affect traits. Meiosis and fertilization mix alleles in each generation of offspring. The offspring of most sexual reproducers inherit pairs of chromosomes, one from a maternal and one from a paternal parent. Except in individuals that have inherited nonidentical sex chromosomes (e.g., X with Y), the pairs are homologous (alike); each pair of chromosomes interacts during meiosis.

Mitosis

PROPHASE

METAPHASE
Chromosomes align at spindle equator.

ANAPHASE
Sister chromatids of chromosomes separate.

TELOPHASE
two nuclei (2n)

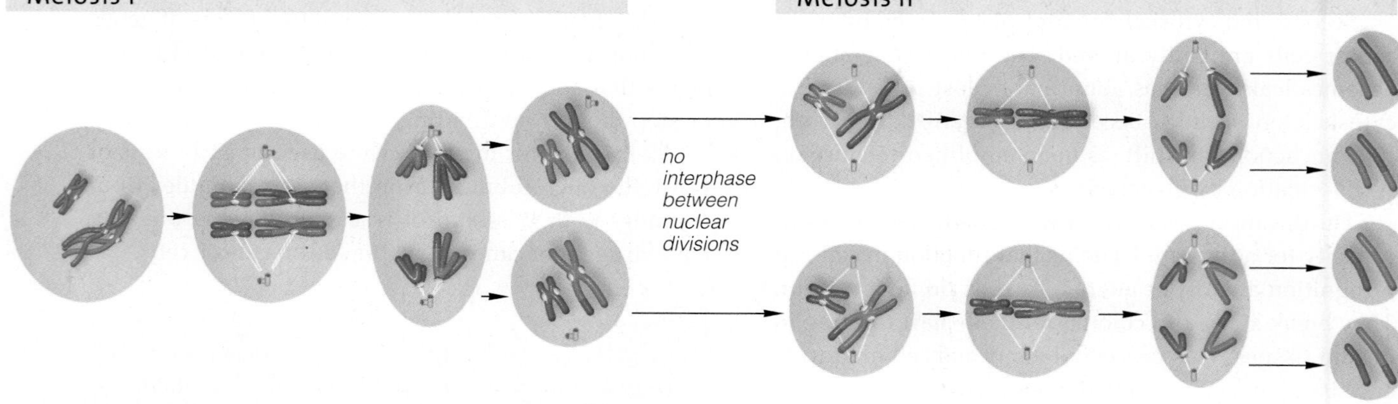

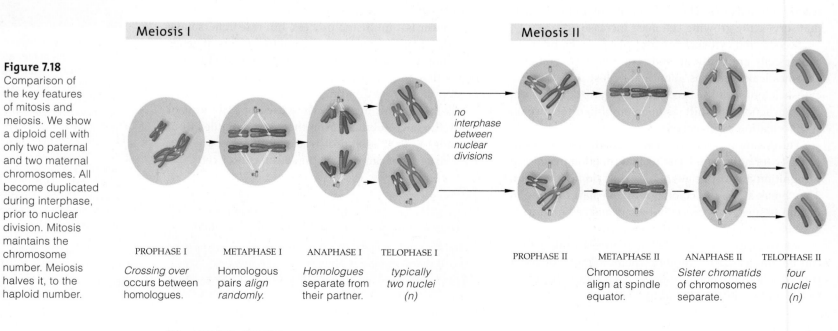

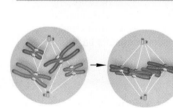

Figure 7.18 Comparison of the key features of mitosis and meiosis. We show a diploid cell with only two paternal and two maternal chromosomes. All become duplicated during interphase, prior to nuclear division. Mitosis maintains the chromosome number. Meiosis halves it, to the haploid number.

Meiosis is central to sexual reproduction. With two nuclear divisions, it halves the chromosome number. The fusion of two gametes at fertilization restores the chromosome number (Figure 7.18).

Section 7.6 Like mitosis, meiosis uses a spindle to move and assort chromosomes. But meiosis occurs only in cells that give rise to gametes or spores. It leads to new combinations of traits in offspring.

*Non*sister chromatids of homologous chromosomes undergo crossing over during prophase I. They break at corresponding sites and exchange segments, so each gets novel combinations of alleles.

During metaphase I, and again at metaphase II, sister chromatids are randomly attached to one spindle pole or the other, so each may end up in any gamete.

BiologyNow
Investigate how crossing over and metaphase I alignments affect allele combinations, and learn how gametes form with the animations on BiologyNow.

Section 7.7 Life cycles vary. In animals, haploid eggs and sperm form. In plants, haploid spores, then gamete-producing bodies form. Gametes combine at fertilization, restoring the diploid chromosome number and further shuffling genetic information.

Section 7.8 Cancer is a multistep process involving altered cells that grow and divide abnormally. Such cells have lost their ability to regulate the cell cycle. Cancer cells have a weakened ability to adhere in tissues, and sometimes they migrate to new tissues.

Self-Quiz

Answers in Appendix I

1. Mitosis and cytoplasmic division function in ______ .
   a. asexual reproduction of single-celled eukaryotes
   b. growth, tissue repair, often asexual reproduction
   c. gamete formation in prokaryotes
   d. both a and b

2. A duplicated chromosome has ______ chromatid(s).
   a. one b. two c. three d. four

3. A somatic cell having two of each type of chromosome has a(n) ______ chromosome number.
   a. diploid b. haploid c. tetraploid d. abnormal

4. After mitosis, the chromosome number of a daughter cell is ______ the parent cell's.
   a. the same as c. rearranged compared to
   b. one-half d. doubled compared to

5. Only ______ is not a stage of mitosis.
   a. prophase c. metaphase
   b. interphase d. anaphase

6. Match each stage of mitosis with the events listed.
   ____ metaphase a. sister chromatids move apart
   ____ prophase b. chromosomes start to condense
   ____ telophase c. daughter nuclei form
   ____ anaphase d. all duplicated chromosomes are aligned at the spindle equator

7. Sexual reproduction requires ______ .
   a. meiosis c. gamete formation
   b. fertilization d. all of the above

8. Generally, a pair of homologous chromosomes ______ .
   a. carry the same genes c. interact at meiosis
   b. are the same length, shape d. all of the above

9. Meiosis ______ the parental chromosome number.
   a. doubles c. maintains
   b. reduces d. corrupts

10. Meiosis is a division mechanism that produces ______ .
   a. two cells c. eight cells
   b. two nuclei d. four nuclei

11. Pairs of duplicated, homologous chromosomes end up at opposite spindle poles during ______ .
   a. prophase I c. anaphase I
   b. prophase II d. anaphase II

12. Sister chromatids of each duplicated chromosome end up at opposite spindle poles during ______ .
   a. prophase I c. anaphase I
   b. prophase II d. anaphase II

13. Match each term with its description.
   ____ chromosome number a. different molecular forms of the same gene
   ____ alleles b. none between meiosis I, II
   ____ metaphase c. all chromosomes aligned at spindle equator
   ____ interphase d. all chromosomes in a cell of a given type

Additional questions are available on **BiologyNow™**

Critical Thinking

1. Pacific yews (*Taxus brevifolius*) are among the slowest growing trees, which makes them vulnerable to extinction. People started stripping their bark and killing them when they heard that *taxol*, a chemical extracted from the bark, may work against breast and ovarian cancer. It takes bark from about six trees to treat one patient. Do some research to find out why taxol has potential as an anticancer drug and what has been done to protect the trees.

2. X-rays emitted from some radioisotopes damage DNA, especially in cells undergoing DNA replication. Humans exposed to high levels of x-rays face *radiation poisoning*. Hair loss and a damaged gut lining are early symptoms. Speculate why. Also speculate on why radiation exposure is used as a therapy to treat some cancers.

3. Why can we expect meiosis to give rise to genetic differences between parent cells and their daughter cells in fewer generations than mitosis?

4. Aphids reproduce asexually or sexually at different times of year. How might their reproductive flexibility be an adaptation that allows them to avoid predators?

5. The bdelloid rotifer lineage started at least 40 million years ago (Figure 7.19). About 360 known species of these tiny animals are found in many aquatic habitats worldwide. They show tremendous genetic diversity. Why do you think scientists were surprised to discover that all of the bdelloid rotifers are female?

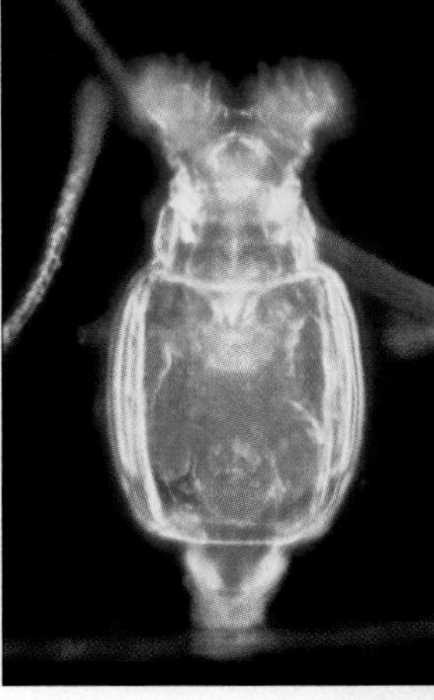

Figure 7.19
Bdelloid rotifer

8 OBSERVING PATTERNS IN INHERITED TRAITS

IMPACTS, ISSUES

Menacing Mucus

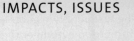

Cystic fibrosis (CF) is a debilitating, frequently fatal genetic disorder. In 1989, researchers identified the mutated gene that causes it. In 2001, the American College of Obstetricians and Gynecologists suggested that all prospective parents be screened for mutated versions of the gene. The suggestion led to the first mass screening for carriers of a genetic disorder.

The gene encodes a membrane transport protein called CFTR. It helps chloride and water move into and out of cells that secrete mucus or sweat. More than 10 million people in the U.S. inherited one normal and one abnormal copy of the CFTR gene. Some of them suffer from sinus problems, but no other symptoms develop. Most do not know they carry the abnormal gene.

CF develops in anyone who inherits a mutant form of the gene from both parents. Thick, dry mucus clogs bronchial airways to their lungs (shown in the filmstrip) and makes it hard to breathe. The mucus is too thick for the ciliated cells lining the airways to sweep out, and bacteria thrive in it.

Daily routines of posture changes and thumps on the chest and back help clear the mucus, and antibiotics help control infections. Yet even with lung transplants, most patients can expect to succumb to lung failure or airway infections before their thirtieth birthday.

The severity of CF and the prevalence of carriers in the general population persuaded doctors to screen hundreds of thousands of prospective parents for the mutated gene. Many people worried that laboratory errors or confusion over the test results might prompt some parents to abort normal fetuses, but the testing continued. Since 2003, screening for CTFR mutations has been offered to all prospective parents in the U.S.

So here we are, working our way through ethical consequences of understanding human genetics. It started long before DNA, genes, or chromosomes were known, in a small garden, with a monk named Gregor Mendel. By breeding many generations of pea plants in monastery garden plots, Mendel discovered evidence of how parents bestow units of hereditary information upon their offspring.

This chapter starts out with the methods and some representative results of Mendel's experiments. His pioneering work remains a classic example of how a scientific approach can pry open important secrets about the natural world. To this day, it serves as the foundation for modern genetics.

How Would You Vote? *Advances in genetics have led to our ability to detect mutant genes that cause medical disorders in human embryos and fetuses. Should society encourage women to give birth only if their child will not develop severe medical problems? How severe? See BiologyNow for details, then vote online.*

Key Concepts

MENDEL'S PATTERNS

Gregor Mendel gathered the first indirect, experimental evidence of the genetic basis of inheritance. His meticulous work tracking traits in many generations of pea plants gave him clues that heritable traits are specified in units. These units are distributed into one gamete or another according to certain patterns. The units were later identified as genes.

BEYOND MENDEL

Not all traits have clearly dominant or recessive forms. One allele of a pair may be fully or partly dominant over its partner, or codominant with it. Two or more gene pairs often influence the same trait, and some single genes influence many traits. Environmental factors can also influence the final form of a trait.

CHROMOSOMES AND INHERITANCE

Human autosomes and sex chromosomes undergo expected gene shufflings by crossing over. Many genetic disorders arise from rare changes in the chromosome number or structure.

Links to Earlier Concepts

Before starting this chapter, review the definitions for genes, alleles, chromosomes, and diploid and haploid (Sections 7.1 and 7.3).

This chapter will show you how crossing over and metaphase I alignments during meiosis (7.5, 7.6) determine patterns of inheritance.

As you read, you may wish to refer back to the introductions to experimental design (1.5, 1.6), natural selection (1.4), protein structure (2.8), and pigments (5.1).

8.1 Tracking Traits With Hybrid Crosses

LINKS TO SECTIONS 7.1, 7.3

We turn now to recurring inheritance patterns among humans and other sexually reproducing species.

TERMS USED IN MODERN GENETICS

In Mendel's time, no one knew about genes, meiosis, or chromosomes. As we follow his thinking, we can clarify the picture by substituting some modern terms used in inheritance studies, including those in Figure 8.1.

1. **Genes** are units of DNA information about traits, transmitted from parents to offspring. Each gene has a specific location (locus) on a chromosome.

2. Cells with a diploid chromosome number ($2n$) have pairs of genes, on pairs of homologous chromosomes.

3. Mutation alters a gene's molecular structure. It may cause a trait to change, as when a gene for flower color specifies purple and a mutated form specifies white. Different molecular forms of the same gene are **alleles**.

4. When offspring inherit a pair of *identical* alleles for a trait generation after generation, we say they belong to a true-breeding lineage. Offspring of a cross between two individuals that breed true for different forms of a trait are **hybrids**; each one has inherited *nonidentical* alleles for the trait.

5. A *homozygous* condition occurs when two alleles on homologous chromosomes are identical. A *heterozygous* condition occurs when they are not identical.

6. An allele is *dominant* when its effect on a trait masks that of any *recessive* allele paired with it. We use capital letters to signify dominant alleles and lowercase letters for recessive ones. *A* and *a* are examples.

7. Pulling this all together, a **homozygous dominant** individual has a pair of dominant alleles (*AA*) for the trait. A **homozygous recessive** individual has a pair of recessive alleles (*aa*). A **heterozygous** individual has a pair of nonidentical alleles (*Aa*).

8. *Gene expression* is the process by which information coded in a gene is converted to structural or functional parts of a cell. Expressed genes determine traits.

9. Two terms help keep the distinction clear between genes and the traits they specify. *Genotype* refers to the particular alleles that an individual carries. *Phenotype* refers to an individual's observable traits.

10. F_1 stands for first-generation offspring, and F_2 for second-generation offspring.

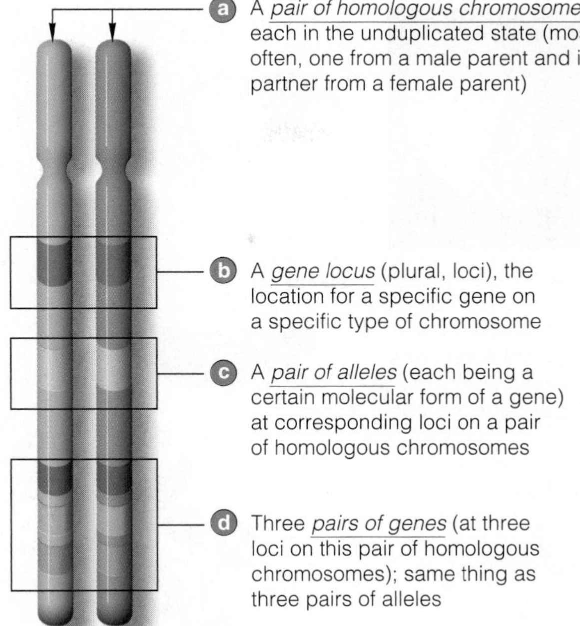

a A pair of homologous chromosomes, each in the unduplicated state (most often, one from a male parent and its partner from a female parent)
b A gene locus (plural, loci), the location for a specific gene on a specific type of chromosome
c A pair of alleles (each being a certain molecular form of a gene) at corresponding loci on a pair of homologous chromosomes
d Three pairs of genes (at three loci on this pair of homologous chromosomes); same thing as three pairs of alleles

Figure 8.1 ***Animated!*** A few genetic terms. Garden pea plants and other species with a diploid chromosome number have pairs of genes, on pairs of homologous chromosomes. Genes come in slightly different molecular forms called alleles. Different alleles specify different versions of the same trait. An allele at any given location on a chromosome may or may not be identical to its partner on the homologous chromosome.

MENDEL'S EXPERIMENTAL APPROACH

Reflect on cystic fibrosis, and a question comes to mind: How did we come to know such amazing things about genes? It all started with Gregor Mendel (Figure 8.2). He guessed that sperm and eggs carry distinct "units" of information about heritable traits. After tracking certain traits of pea plants generation after generation, Mendel found indirect but *observable* evidence of how parents transmit genes to offspring.

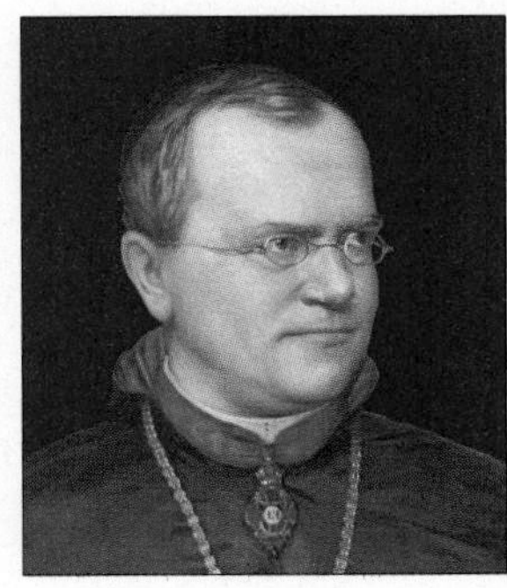

Figure 8.2 Gregor Mendel (1822–1884). He is considered to be the founder of modern genetics.

Mendel spent most of his adult life in a monastery in a city that is now part of the Czech Republic. Having grown up on a farm, he was familiar with agricultural principles and how to apply them. He kept abreast of the breeding experiments and developments described in the literature. Shortly after entering the monastery, he took courses in mathematics, physics, and botany at the University of Vienna. At the time, very few scholars showed interest in plant breeding *and* mathematics.

Shortly after his university training, Mendel began experiments with the garden pea plant, *Pisum sativum*. This plant is self-fertilizing. Sperm form in the same flower as eggs and can fertilize them. Some pea plants breed true for certain traits. This means that successive generations will be just like the parents in one or more

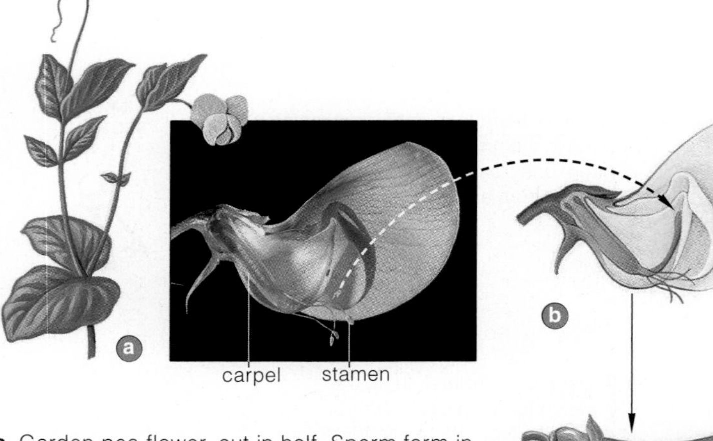

a Garden pea flower, cut in half. Sperm form in pollen grains, which originate in male floral parts (stamens). Eggs develop, fertilization takes place, and seeds mature in female floral parts (carpels).

b Pollen from a plant that breeds true for purple flowers is brushed onto a floral bud of a plant that breeds true for white flowers. The white flower had its stamens snipped off. This is one way to guarantee a plant will not self-fertilize.

c Later, seeds develop inside pods of the cross-fertilized plant. An embryo within each seed develops into a mature pea plant.

d Each new plant's flower color is indirect but observable evidence that hereditary material has been transmitted from the parent plants.

Figure 8.3 ***Animated!*** Garden pea plant (*Pisum sativum*), which can self-fertilize or cross-fertilize. Experimenters can control the transfer of its hereditary material from one flower to another.

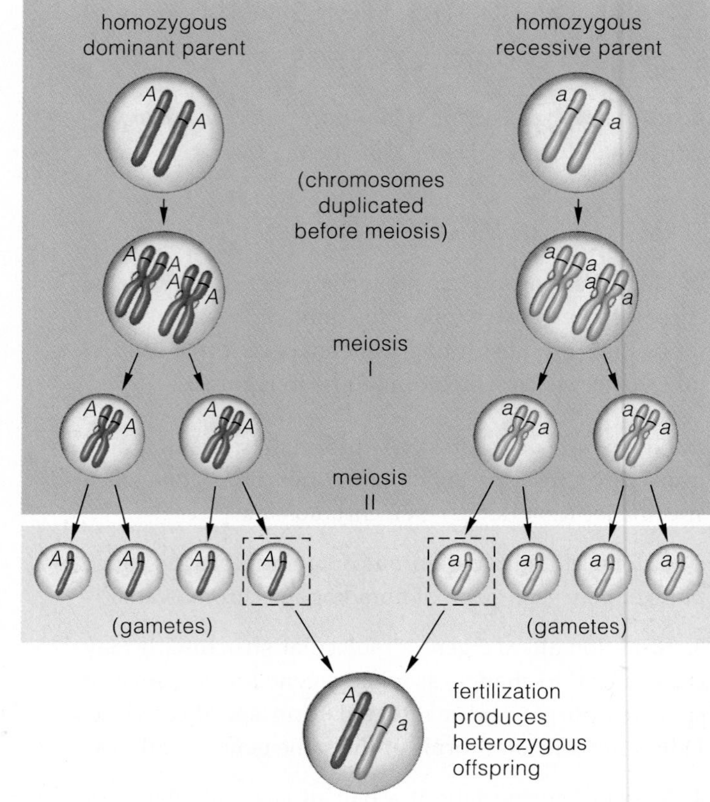

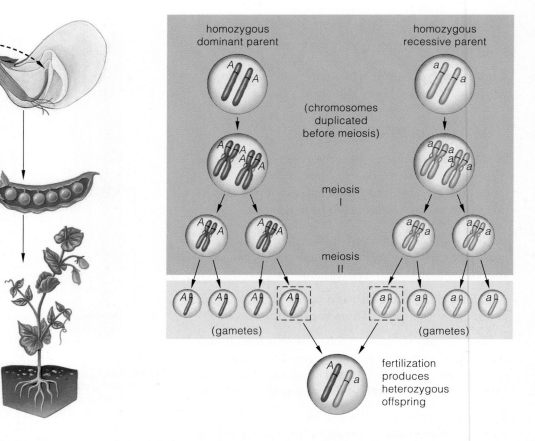

Figure 8.4 One gene of a pair separates from the other gene in a monohybrid cross. Two parents that breed true for two versions of a trait produce only heterozygous offspring.

LINKS TO SECTIONS 1.5, 1.6, 7.5

traits, as when all plants grown from the seeds of white-flowered parent plants also have white flowers.

Pea plants may be cross-fertilized by transferring pollen from one plant to another. Mendel did this with plants that bred true for *different* versions of the same trait—say, purple or white flowers, as in Figure 8.3. He thought that clearly observable differences would help him track a given trait over many generations. If there were patterns to a trait's inheritance, *then the patterns might tell him something about heredity itself.*

MENDEL'S THEORY OF SEGREGATION

Mendel hypothesized that pea plants inherit two "units" (genes) of information for a trait, one from each parent. He tested his idea with **monohybrid experiments**. In such tests, parents that breed true for two different forms of a trait are bred together, or crossed ($AA \times aa = Aa$). The experiment itself is the self-fertilization of the F_1 heterozygotes, or "monohybrids" ($Aa \times Aa$).

For one set of experiments, he crossed plants that bred true for purple flowers with plants that bred true for white flowers. All F_1 offspring had purple flowers.

Let's express this test cross in modern terms. Pea plant cells have pairs of homologous chromosomes. Assume one parent is homozygous dominant for flower color (AA), and the other parent is homozygous recessive (aa). After meiosis, all gametes of the AA parent have one A allele, and all those of the aa parent have one a allele (Figure 8.4). If any of these sperm fertilize any of these eggs, only one kind of zygote can result: Aa. All F_1 offspring will have purple flowers.

When the F_1 plants were allowed to self-fertilize, most of the F_2 offspring had purple flowers, but some of them had white flowers. After tracking offspring of his experiments with thousands of plants, Mendel was able to calculate that about three out of every four F_2 plants had purple flowers, and one out of every four had white flowers (Figure 8.5).

The ratio hinted that fertilization is a chance event having a number of possible outcomes. Mendel knew about probability, which applies to chance events *and so could help him predict the possible outcomes of genetic crosses.* **Probability** simply means this: The chance that each outcome of an event will occur is proportional to the number of ways in which that outcome can occur.

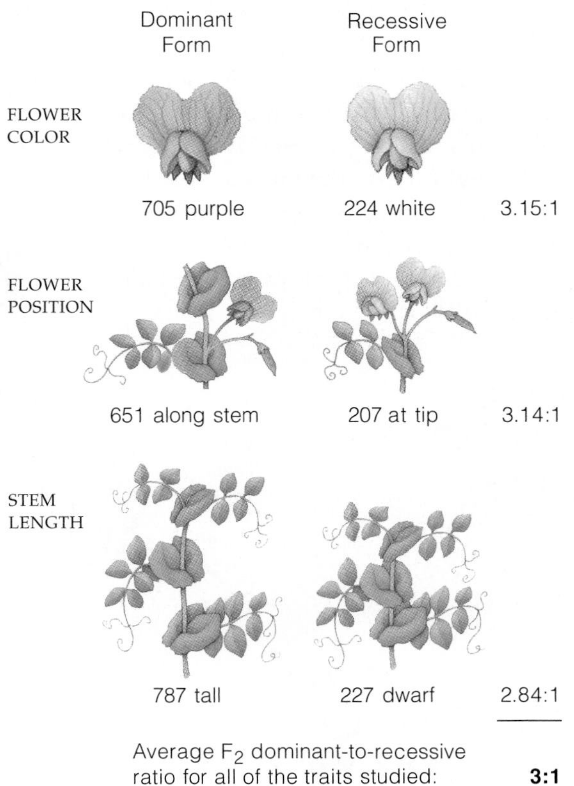

Figure 8.5 Monohybrid experiments with pea plants. Mendel's counts of F_2 offspring having dominant or recessive hereditary "units" (alleles). The 3:1 phenotypic ratio held for other traits, including pea pod shape and color.

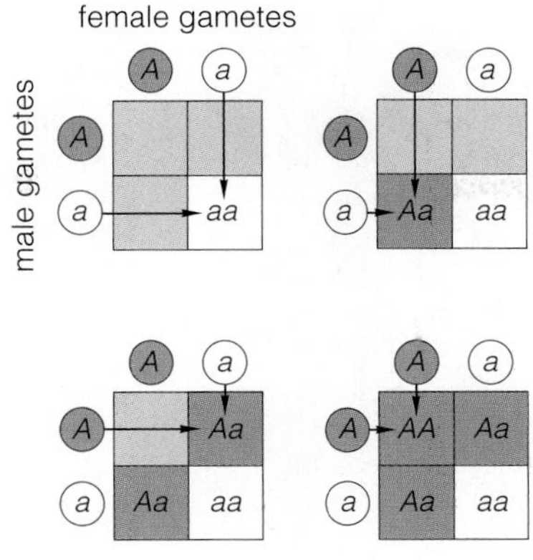

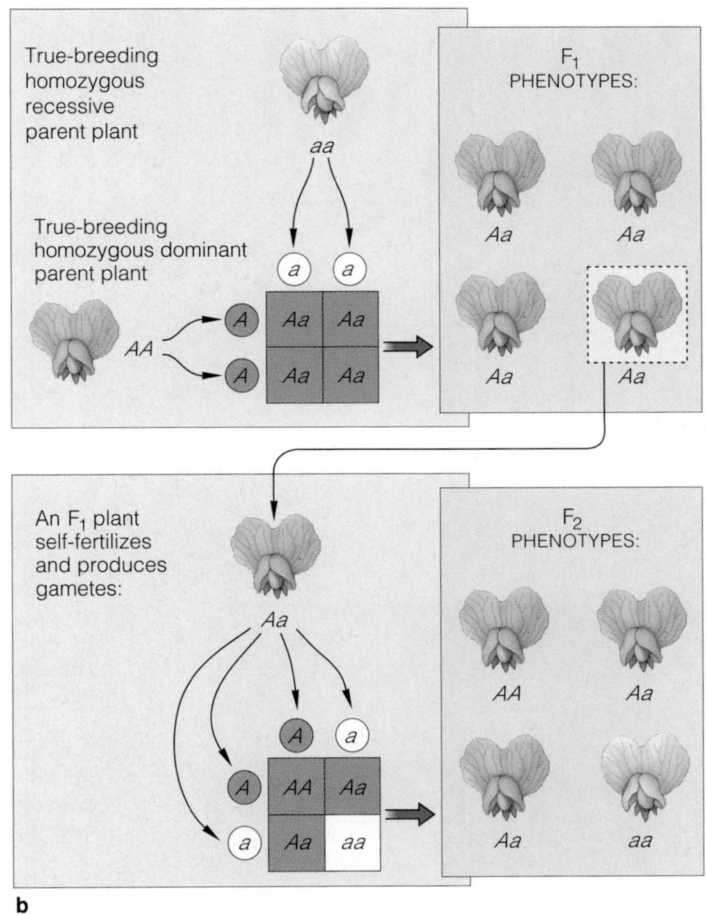

Figure 8.6 ***Animated!*** (**a**) Punnett-square method of predicting the probable outcomes of genetic crosses. Circles signify gametes. *Italics* indicate dominant or recessive alleles. The possible genotypes among offspring are written inside the squares.

(**b**) Results of one of Mendel's monohybrid experiments. The ratio of dominant to recessive phenotypes in the F_2 (second-generation) plants was 3:1.

A **Punnett-square method**, explained and applied in Figure 8.6, shows the possibilities. If half of a plant's sperm (or eggs) are *a* and half *A*, we can expect four outcomes with each fertilization:

POSSIBLE EVENT	PROBABLE OUTCOME
sperm *A* meets egg *A*	1/4 *AA* offspring
sperm *A* meets egg *a*	1/4 *Aa*
sperm *a* meets egg *A*	1/4 *Aa*
sperm *a* meets egg *a*	1/4 *aa*

Each F_2 plant has 3 chances in 4 of inheriting at least one dominant allele (purple flowers). It has 1 chance in 4 of inheriting two recessive alleles (white flowers). That is a probable phenotypic ratio of 3:1.

Mendel's observed ratios were not *exactly* 3:1, but he knew about sampling error. To understand, flip a coin several times. A coin is as likely to end up heads as tails. But often it will end up heads, or tails, several times in a row. If you flip the coin only a few times, the observed ratio of heads to tails might differ a lot from the predicted ratio of 1:1. Flip it many times, and you are far more likely to approach the predicted ratio. Mendel minimized the sampling error in his results by counting many offspring.

Results from his monohybrid experiments became the basis of Mendel's theory of **segregation**, which we state here in modern terms:

MENDEL'S THEORY OF SEGREGATION *Diploid cells have pairs of genes, on pairs of homologous chromosomes. The two genes of each pair are separated from each other during meiosis, so they end up in different gametes.*

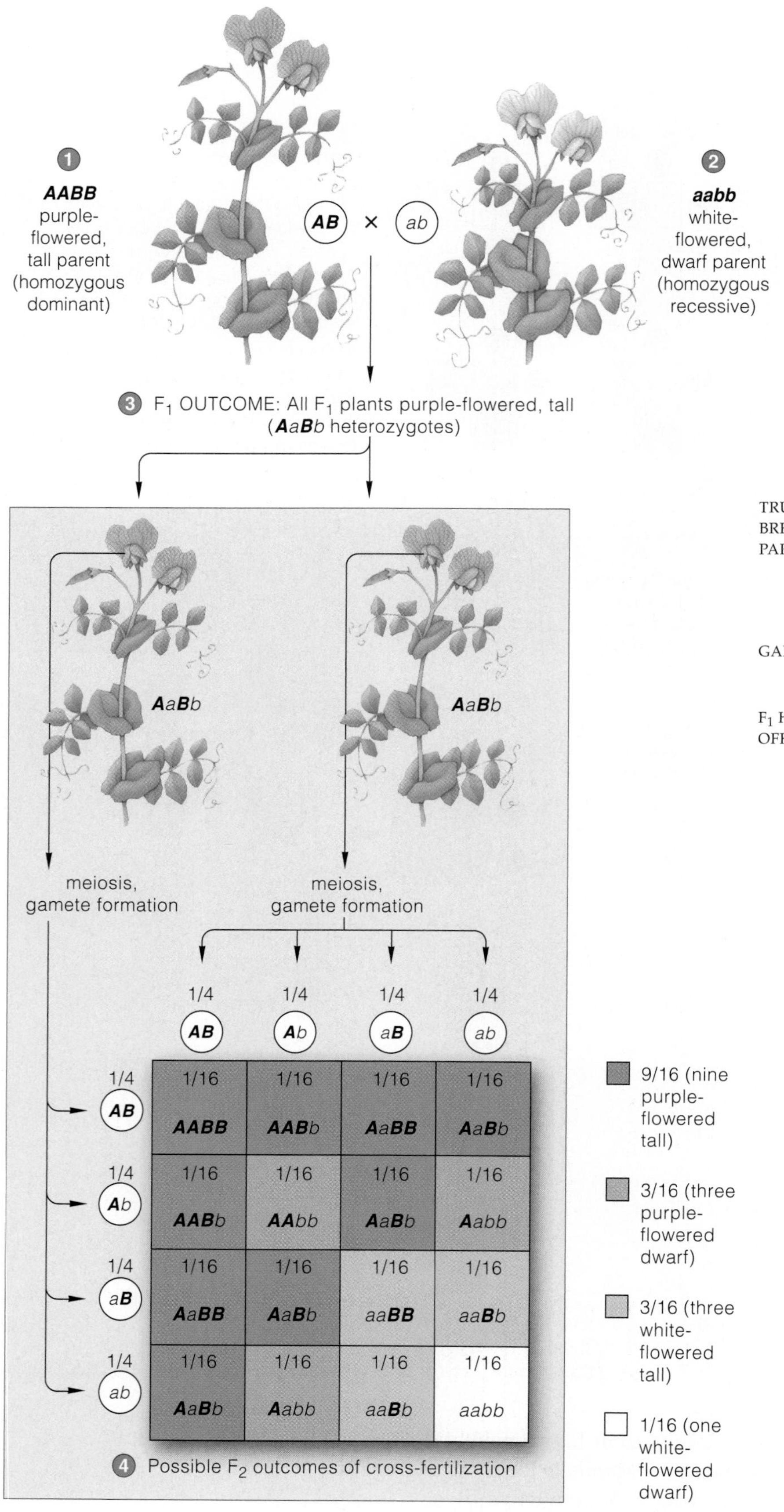

1 **AABB** purple-flowered, tall parent (homozygous dominant)

2 **aabb** white-flowered, dwarf parent (homozygous recessive)

AB × ab

3 F_1 OUTCOME: All F_1 plants purple-flowered, tall (***A**a**B**b* heterozygotes)

AaBb *AaBb*

meiosis, gamete formation

meiosis, gamete formation

	1/4 **AB**	1/4 **A**b	1/4 a**B**	1/4 ab
1/4 **AB**	1/16 **AABB**	1/16 **AAB**b	1/16 **A**a**BB**	1/16 **A**a**B**b
1/4 **A**b	1/16 **AAB**b	1/16 **AA**bb	1/16 **A**a**B**b	1/16 **A**abb
1/4 a**B**	1/16 **A**a**BB**	1/16 **A**a**B**b	1/16 aa**BB**	1/16 aa**B**b
1/4 ab	1/16 **A**a**B**b	1/16 **A**abb	1/16 aa**B**b	1/16 aabb

9/16 (nine purple-flowered tall)

3/16 (three purple-flowered dwarf)

3/16 (three white-flowered tall)

1/16 (one white-flowered dwarf)

4 Possible F_2 outcomes of cross-fertilization

Figure 8.7 ***Animated!*** A dihybrid cross. Here, two traits are tracked: flower color and plant height. *A* and *a* stand for dominant and recessive alleles for flower color. *B* and *b* stand for dominant and recessive alleles for height. The Punnett square shows all of the possible F_2 combinations.

MENDEL'S THEORY OF INDEPENDENT ASSORTMENT

In another set of experiments, Mendel showed how *two* pairs of genes are sorted into gametes. For this he used **dihybrids**, which are heterozygous for two genes (*AaBb*). A **dihybrid experiment** is a cross between two identical dihybrid parents (*AaBb* × *AaBb*).

We can duplicate one of Mendel's dihybrid crosses. First, he bred dihybrids for flower color (alleles *A* or *a*) and for height (*B* or *b*):

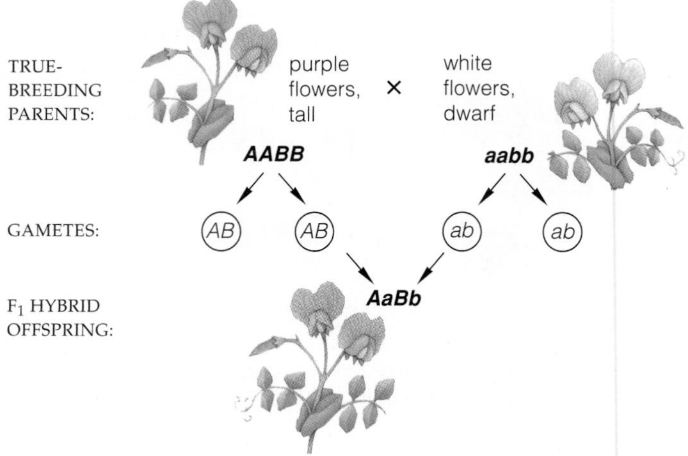

All F_1 offspring from this cross are dihybrids for flower color and height, all purple-flowered and tall (*AaBb*). Mendel then made dihybrid crosses by crossing these F_1 plants. The same ratio of phenotypes appeared in the F_2 offspring of the dihybrid crosses, time after time (Figure 8.7). To understand why Mendel's results were so predictable, think about how gene pairs are sorted into gametes during meiosis.

How two gene pairs (*Aa* and *Bb*) sort into gametes depends partly on whether the two genes are on the same chromosome. Let's assume they are not. Say one pair of homologous chromosomes carries the *A* and *a* alleles and that another pair carries the *B* and *b* alleles.

Now think about how these chromosomes become positioned at the spindle's equator during metaphase I of meiosis (Section 7.5). The chromosome with the *A* allele is positioned to move to either one of the spindle poles. Its homologue (carrying the *a* allele) will move to the opposite spindle pole. The same is true for the chromosomes that carry the *B* and *b* alleles. Following meiosis, there are four possible combinations of alleles in gametes: *AB*, *Ab*, *aB*, and *ab* (Figure 8.8).

Given the random metaphase I alignments, several combinations of alleles can occur at fertilization. Simple multiplication (four sperm types × four egg types) tells us that sixteen combinations of genotypes are possible among the F_2 offspring of a dihybrid cross.

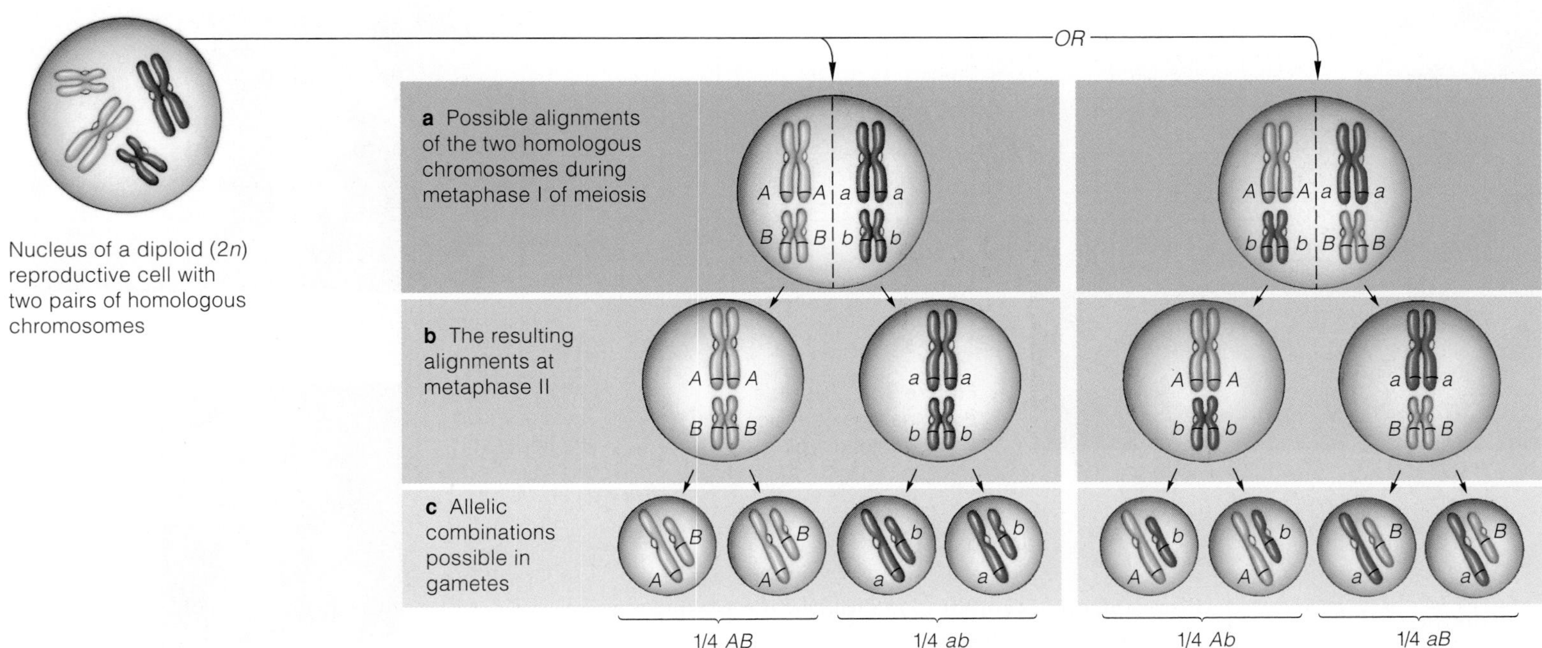

Figure 8.8 An example of independent assortment during meiosis. Either chromosome of a pair may become attached to either pole of the spindle during meiosis. When just two pairs are tracked, two different metaphase I lineups are possible.

Each of these genotypes will result in a particular phenotype. Nine of the sixteen will be tall and purple-flowered, three will be dwarf purple-flowered, three will be tall white-flowered, and 1 will be dwarf white-flowered. Mendel saw this predicted phenotypic ratio, 9:3:3:1, among the F_2 offspring of his dihybrid crosses.

Mendel could predict the outcome of his dihybrid crosses, but without any knowledge of chromosomes, genes, or meiosis, he could not have understood the underlying reason for their predictability. He did not know that genes for a pea plant's traits are distributed on seven pairs of homologous chromosomes.

Mendel hypothesized, correctly, that the two "units of inheritance" which determine flower color in pea plants were sorting into offspring independently of the two "units" which determine height. Neither he, nor anyone else at the time, knew what those units were.

Mendel's hypothesis is now called the **theory of independent assortment**, stated here in modern terms: During meiosis, the genes on each pair of homologous chromosomes are assorted into gametes independently of how all other pairs are assorted.

Independent assortment and segregation give rise to genetic variation. In a monohybrid cross for one gene pair, three genotypes are possible: *AA*, *Aa*, and *aa*. With more pairs, there are more possible combinations. If the parents differ at n genes, each with two alleles, 3^n combinations are possible in offspring. If they differ in twenty genes, that number is nearly 3.5 billion!

The theory does not apply to *all* gene combinations, however, as you will see in the next sections.

MENDEL'S THEORY OF INDEPENDENT ASSORTMENT

In meiosis, genes on pairs on homologous chromosomes become sorted into gametes independently of how gene pairs on other chromosomes are sorted.

8.2 Not-So-Straightforward Phenotypes

Mendel focused on traits that have clearly dominant and recessive forms. However, expression of genes for most traits is not as straightforward.

ABO BLOOD TYPES—A CASE OF CODOMINANCE

In *codominance*, neither of the alleles of a gene pair in heterozygotes is dominant or recessive: both alleles are expressed fully and at the same time. The two parental forms of the trait appear together as a new phenotype.

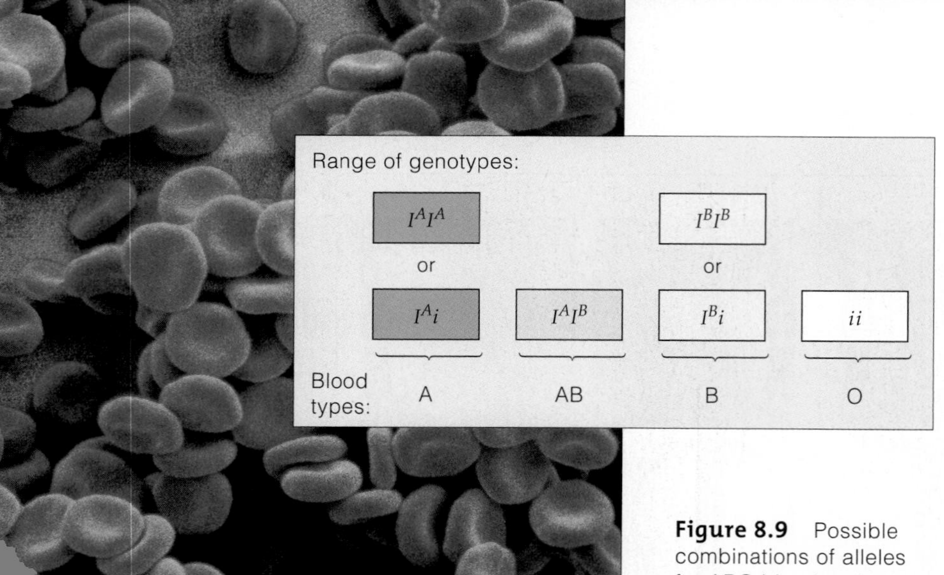

Figure 8.9 Possible combinations of alleles for ABO blood typing.

All F_1 offspring heterozygous for flower color:

Cross two of the F_1 plants, and the F_2 offspring will show three phenotypes in a 1:2:1 ratio:

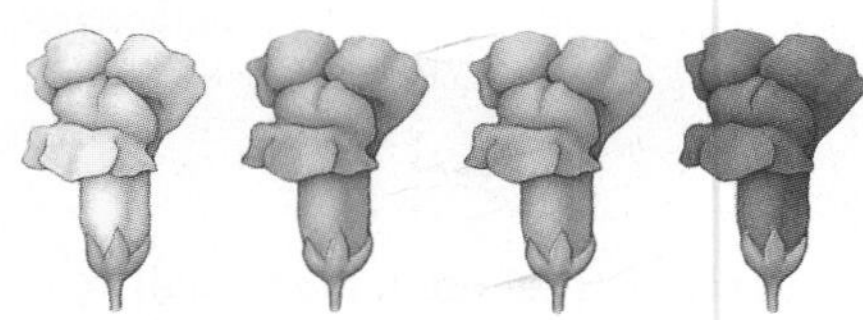

Figure 8.10 Incomplete dominance in heterozygous (pink) snapdragons, in which an allele that affects red pigment is paired with a "white" allele.

LINKS TO SECTIONS 2.6, 3.3, 5.1

For example, one type of glycolipid in the plasma membrane imparts an identity to your red blood cells. This glycolipid occurs in different forms. An analytical method called *ABO blood typing* reveals which of the forms a person has.

An enzyme dictates the glycolipid's final structure. Humans have three alleles that code for this enzyme. Two of them, I^A and I^B, are codominant when paired. The third, i, is recessive; a pairing with I^A or I^B masks its effect. Together, they are a **multiple allele system**, defined as the presence of three or more alleles of a single gene among the individuals of a population.

Which alleles do you have? If you have I^AI^A or I^Ai, your blood is type A. If you have I^BI^B or I^Bi, it is type B. If you have codominant alleles I^AI^B, it is AB—you have both versions of the enzyme. If you have ii, your blood cells do not have the glycolipid identifier, and so your blood type is O (Figure 8.9).

Receiving incompatible blood cells in a transfusion is dangerous. The immune system attacks cells with unfamiliar markers, causing them to clump or burst—with potentially lethal consequences. This transfusion reaction is the reason for blood typing. Type O blood is compatible with all other blood types, so people who have it are called universal blood donors. If you have AB blood, you can receive a transfusion of any blood type; you are called a universal recipient.

INCOMPLETE DOMINANCE

In *incomplete* dominance, one allele of a pair is partly dominant over its partner. A heterozygote displays a third phenotype, one that is a blend of the different homozygote phenotypes. For instance, a cross between true-breeding red and white snapdragons will yield *pink*-flowered F_1 offspring. Cross two of the F_1 plants and the offspring will have red, white, or *pink* flowers (Figure 8.10). Why pink? Red snapdragon plants have two alleles for red pigment, so their flowers are red. White snapdragon plants also have two alleles for the red pigment, but neither one is functional. Flowers of these plants have no red pigment, so they are white. Heterozygote snapdragon plants have one "red" allele and one "white" allele. They make half the amount of red pigment that the red homozygotes make, so their flowers appear pink.

WHEN PRODUCTS OF TWO OR MORE GENE PAIRS INTERACT

Traits also arise from interactions among products of two or more gene pairs. One gene product may mask or alter the expression of another gene product, and some expected phenotypes may not appear at all. For example, several gene pairs govern the hair color of

BLACK LABRADOR

YELLOW LABRADOR

CHOCOLATE LABRADOR

	EB	Eb	eB	ee
EB	EEBB	EEBb	EeBB	EeBb
Eb	EEBb	EEBb	EeBB	EeBb
eB	EeBB	EeBb	eeBB	eeBb
ee	EeBB	EEeb	eeBB	eeBb

Figure 8.11 Coat color among Labrador retrievers, an example of differences in phenotype arising from interactions among gene pairs. Alleles *B* and *b* encode melanin. Alleles *E* and *e* govern how much of the melanin is deposited in hair.

Labrador retrievers. Their hair can be black, yellow, or brown depending on interactions between products of gene pairs. One set of alleles encodes melanin, a dark pigment. Allele *B* (black) has a stronger effect and is dominant to *b* (brown). The alleles of a different gene control how much melanin is deposited in hair. Allele *E* permits deposition. Allele *e* does not, so dogs with two recessive alleles (*ee*) have yellow hair (Figure 8.11).

Alleles at another locus (*C*) may override those two. They encode the first enzyme in a melanin-producing pathway. A *CC* or *Cc* individual makes the functional enzyme, but an individual with two recessive alleles (*cc*) does not. *Albinism*, the absence of melanin, results.

SINGLE GENES WITH A WIDE REACH

Alleles at a single locus may affect two or more traits. **Pleiotropy** occurs when the activity of one gene has multiple phenotypic effects. Cystic fibrosis (box, *right*) and sickle-cell anemia (Section 2.9) are two examples.

Figure 8.12 Flo Hyman, at left, captain of the United States volleyball team that won an Olympic silver medal in 1984. Two years later, during a game in Japan, she slid silently to the floor and died. A dime-sized weak spot in the wall of her aorta had burst. At least two college basketball stars have also died abruptly as a result of Marfan syndrome.

Marfan syndrome is another. This genetic disorder begins with certain mutations in the gene for fibrillin. Fibrillin is one of the proteins in connective tissue, the most abundant vertebrate tissue. Long, thin fibers of fibrillin, loose or cross-linked with the protein elastin, impart elasticity to the connective tissues of the heart, skin, blood vessels, skeleton, tendons, and around the eyes. Normal fibrillin fibers passively recoil after being stretched, as by the beating heart.

A mutated fibrillin gene can result in altered fibrillin. Connective tissues that lack normal fibrillin fibers form throughout the body. The outcome is Marfan syndrome, which affects 1 in 5,000 people throughout the world—men and women of all ethnicities.

Marfan mutations particularly affect the aorta, the largest blood vessel that transports blood away from the heart. Muscle cells in the aorta's thick wall do not function properly, and the aortic wall itself is not as elastic as it should be. Being under pressure, the wall eventually widens, becoming thinner and leaky. Cells infiltrate and multiply in its epithelial lining. Calcium deposits can accumulate inside. Inflamed, weakened, and thinned, the aorta may rupture suddenly during strenuous exercise.

Precisely because of its far-reaching effects, Marfan syndrome can be difficult to diagnose. Many affected people are not aware they have it. Until recent medical advances, it killed most of them before the age of fifty. Olympian Flo Hyman was one (Figure 8.12).

In codominance, an allele is expressed independently of its partner on the homologous chromosome. Neither one masks the other.

In incomplete dominance, both alleles affect phenotype.

Two or more genes may affect a single phenotypic trait.

With pleiotropy, alleles at one locus affect multiple traits.

IMPACTS, ISSUES

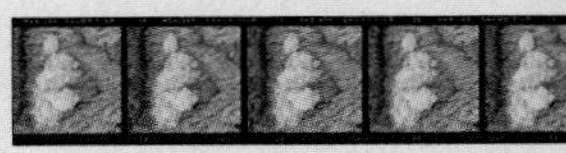

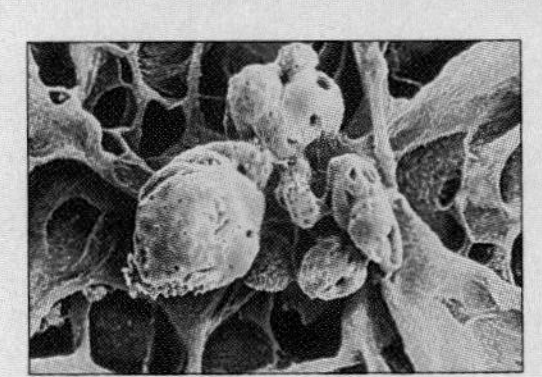

The mutant gene that causes CF has pleiotropic effects. Lung infections are the greatest threat, but CF also increases the risk of gallstones, such as those shown above. Too much chloride is lost in sweat, shifting the body's salt–water balance and making the heart beat irregularly. Also, mucus thickens. This blocks secretion of digestive enzymes from the pancreas and causes nutritional problems. It also blocks ducts in the male reproductive tract. Because sperm cannot be ejaculated, sterility is the outcome.

LINKS TO SECTIONS 4.3, 5.1

8.3 Complex Variations in Traits

For most populations or species, individuals show rich variation for many of the same traits. This variation arises from gene mutations, gene interactions, and variations in environmental conditions.

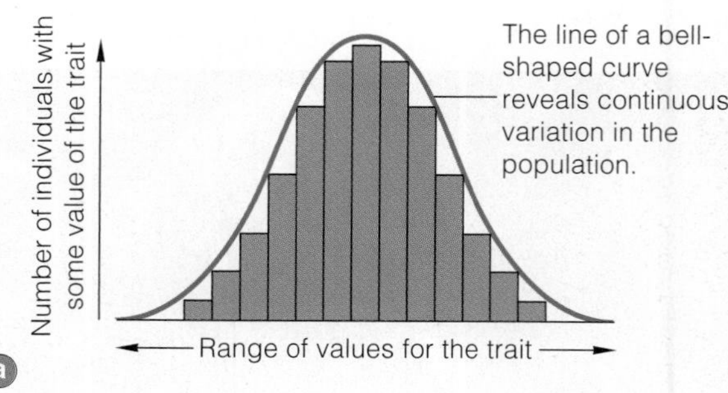

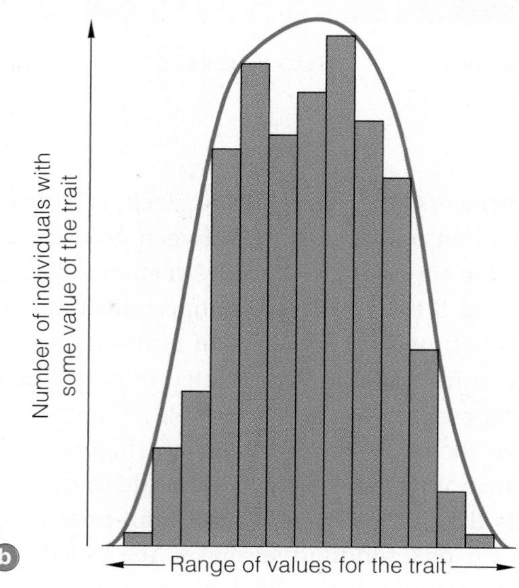

Figure 8.14 ***Animated!*** Continuous variation. The number of individuals in each category is graphed against the range of measured values of a trait. (**a**) Bell-shaped curve of continuous variation in an idealized population. (**b**) A real-life example of a bell-curve for continuous variation in a trait, a graph of height distribution among the females at far right (Figure 8.15*b*).

REGARDING THE UNEXPECTED PHENOTYPE

As Mendel demonstrated, phenotypic effects of one or more gene pairs often show up in predictable ratios. However, phenotypic effects of genes for some traits are quite difficult to predict.

As an example, *camptodactyly* is a rare abnormality that affects the shape and movement of fingers. Some people who carry the mutant allele for this heritable trait have immobile, bent fingers on both hands. Other people who carry the same allele have immobile, bent fingers only on one hand. Still others are not affected in any obvious way.

What causes this variation? Remember that organic compounds are synthesized in a sequence of metabolic reactions. Different enzymes mediate each of the steps. Any gene that specifies an enzyme can mutate in any number of ways. A mutant gene may result in, say, a defective version of an enzyme that fails to function, or one that does not function as effectively as the normal enzyme. Environmental conditions also can affect the performance of a crucial enzyme. Any of these factors may introduce unpredictable variation in phenotypes.

CONTINUOUS VARIATION IN POPULATIONS

Generally, individuals of each population show a range of small differences in many traits. This characteristic of populations is called **continuous variation**. It is primarily an outcome of the genes that affect a trait, and environmental factors that alter their expression. The more genes and environmental factors there are that affect a particular trait, the smaller are the differences in the range of phenotypes that result.

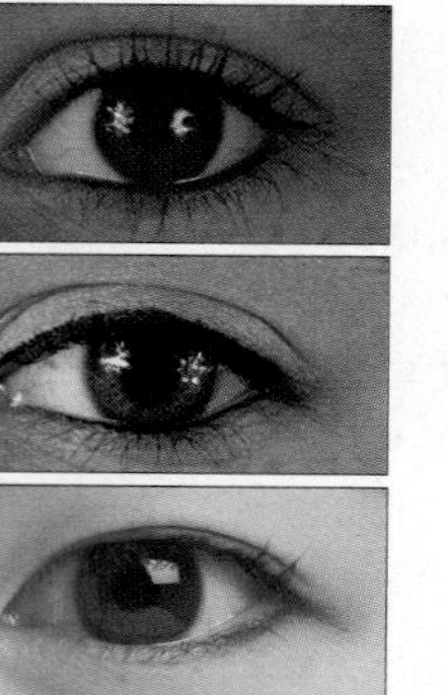

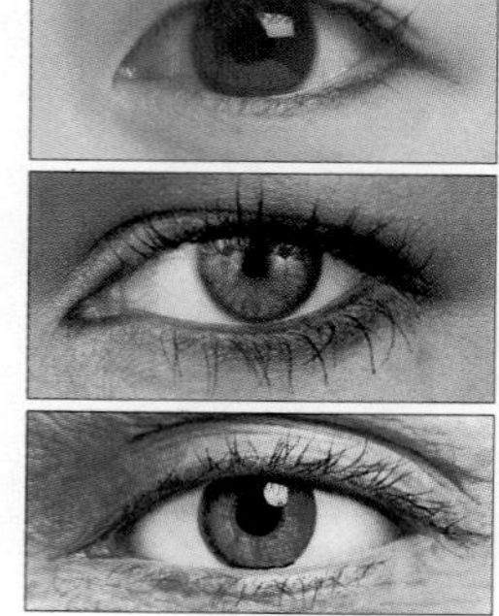

Figure 8.13 Part of the range of continuous variation in human eye color. Products of different gene pairs interact in producing and distributing melanin, which helps color the eye's iris. Different combinations of alleles result in small color differences. The frequency distribution for the eye-color trait is continuous over a range from black to light blue.

Eye color is one example of continuous variation. It arises from variable amounts of pigment deposited in the doughnut-shaped structure beneath the cornea, the iris. The color of the iris is the cumulative outcome of several gene products, some of which help synthesize or distribute melanin. Dark irises that are almost black have dense deposits of melanin. This pigment absorbs most wavelengths of light, so very little incoming light is reflected out of dark eyes. Melanin deposits are not as extensive in brown eyes. Light brown or hazel eyes have even less melanin (Figure 8.13).

Green, gray, or blue eyes do not have green, gray, or blue pigments. Their iris has some melanin, but not much. Thus, many or most of the wavelengths of light that do enter the eyeball are reflected out.

5'3" 5'4" 5'5" 5'6" 5'7" 5'8" 5'9" 5'10" 5'11" 6'0" 6'1" 6'2" 6'3" 6'4" 6'5"
Height (feet/inches)
a

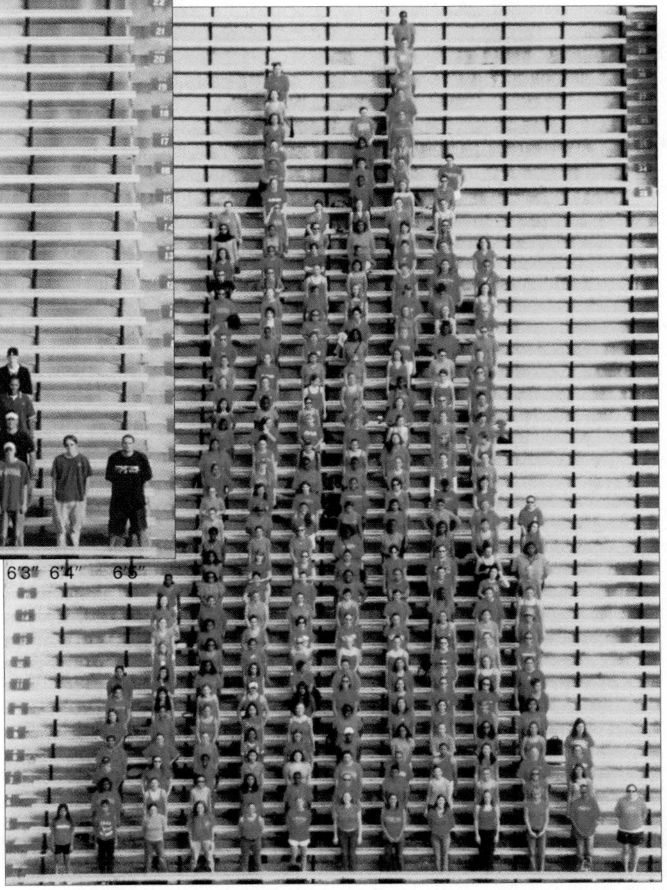

4'11" 5'0" 5'1" 5'2" 5'3" 5'4" 5'5" 5'6" 5'7" 5'8" 5'9" 5'10" 5'11"
Height (feet/inches)
b

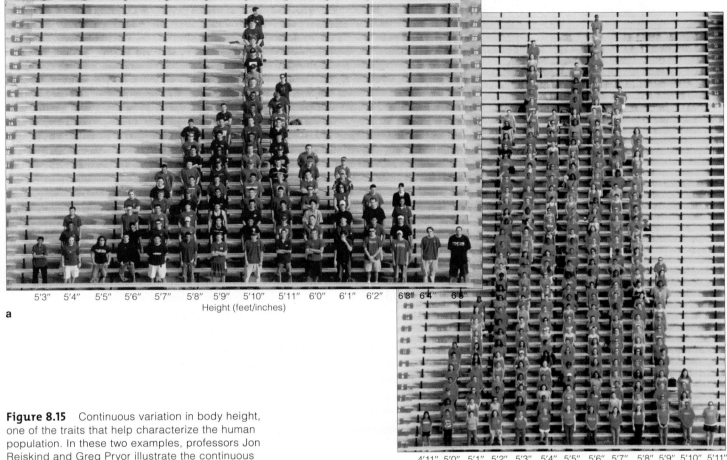

Figure 8.15 Continuous variation in body height, one of the traits that help characterize the human population. In these two examples, professors Jon Reiskind and Greg Pryor illustrate the continuous variation in height of their biology students at the University of Florida. They divided all of the students into two groups, (**a**) males and (**b**) females. For both groups, they divided the range of possible heights incrementally, measured the students, and assigned each to the appropriate category.

How can you describe the continuous variation of a trait within a group? You might start by dividing the full range of of phenotypes for the trait—say, height—into measurable categories, such as number of inches. Then, you would count how many individuals in your group fall into each category. Finally, you would graph the data as a bar chart, such as the one in Figure 8.14*a*. This type of graph shows how different phenotypes are distributed in a group.

In Figure 8.14, the shortest bars indicate categories having the fewest individuals. The tallest bar signifies the category with the most individuals. In this case, a graph line skirting the top of all of the bars is a **bell curve**. Such bell-shaped curves are typical of any trait that shows continuous variation. Figure 8.14*b* is a bell curve that was graphed from real-life measurements of the height of many biology students at the University of Florida (Figure 8.15*b*).

ENVIRONMENTAL EFFECTS ON PHENOTYPE

LINKS TO SECTIONS 2.8, 2.9

We have mentioned, in passing, that the environment often contributes to variable gene expression among a population's individuals. Now consider a few cases.

Like a Siamese cat, the Himalayan rabbit has light hair on some of its body, and dark hair elsewhere. The Himalayan rabbit is homozygous for the c^h allele of one of the enzymes involved in melanin production. The c^h allele specifies a form of the enzyme that stops working when the temperature of the skin rises above a certain point.

Above about 33°C (91°F) the cells that give rise to a Himalayan rabbit's hairs cannot make melanin, and the resulting hairs are light. This happens in body regions that are massive enough to conserve a fair amount of metabolic heat. Ears and other slender extremities are cooler because they tend to lose metabolic heat faster.

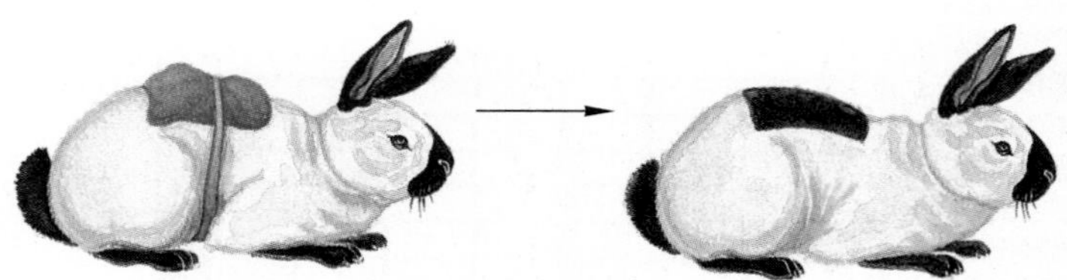

Figure 8.16 ***Animated!*** Observable effect of an environmental factor that alters gene expression. A Himalayan rabbit normally has black hair only on its long ears, nose, tail, and legs—the cooler parts of its body. In one experiment, a patch of a rabbit's white fur was removed and an icepack placed over the hairless patch. Where the colder temperature had been maintained, the hairs that grew back were black.

Figure 8.16 shows one experiment demonstrating the effect of temperature on the c^h allele.

We also can identify environmental effects on the genes that govern phenotypes in plants. Consider one experiment using yarrow. This plant can grow from cuttings, so it is a useful experimental organism. Why? All cuttings from the same plant will have the same genotype, so experimenters can discount genes as the basis for the differences that appear among them.

In this classic study, cuttings (clones) from several yarrow plants were transplanted to locations at three different elevations. Researchers periodically observed the growth of the yarow plants in their new habitats. They saw that cuttings from a single plant often grew differently at the different altitudes. For example, two cuttings from one plant grew quite tall at the lowest elevation and the highest elevation, but a third cutting remained short at mid-elevation (Figure 8.17). Though the plants were genetically identical, their phenotypes differed depending on their environment.

Similarly, plant one kind of hydrangea in a garden and it may produce pink blossoms or blue blossoms. Soil acidity affects the function of gene products that dictate the color of hydrangea flowers.

What about humans? One of our genes codes for a transporter protein that moves serotonin across the plasma membrane of brain cells. Serotonin has several important effects, one of which is to counter anxiety and depression when traumatic events challenge us. For a long time, researchers have known that some people handle stress without getting too upset, while others spiral into a deep and lasting depression.

Mutation of the gene for the serotonin transporter compromises responses to stress. It is as if some of us are bicycling through life without an emotional helmet. Only when we take a fall does the phenotypic effect—depression—appear. Other genes also affect emotional states, but mutation of this one reduces our capacity to snap out of it when bad things happen.

Figure 8.17 Experiment demonstrating the impact of environmental conditions on gene expression in yarrow (*Achillea millefolium*). Cuttings from the same parent plant were grown at three different elevations.

Enzymes mediate each step of most metabolic pathways. Mutations, interactions among genes, and environmental conditions may affect one or more steps. The outcome is variation in phenotypes.

For most traits, individuals of a population or species show continuous variation—a range of small differences.

Variation in traits arises not only from gene mutations and interactions, but also in response to variations in environmental conditions that each individual faces.

8.4 The Chromosomal Basis of Inheritance

You already know about the structure of chromosomes and what happens to them during meiosis. Now start correlating them with human inheritance patterns.

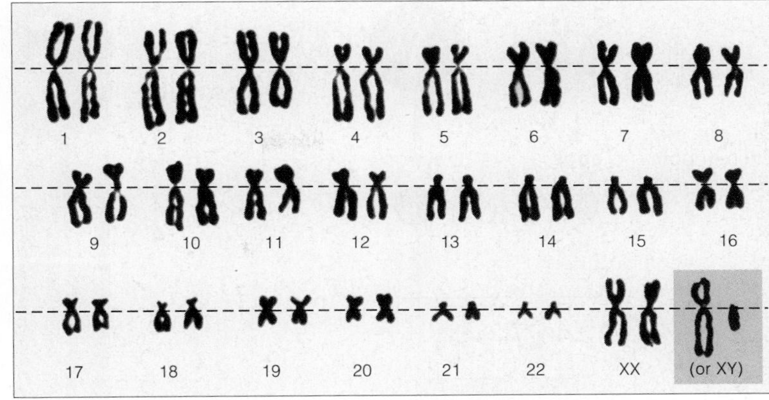

Figure 8.18 ***Animated!*** Karyotyping, a diagnostic tool that reveals an image of a single cell's diploid complement of chromosomes. This human karyotype shows 22 pairs of autosomes and both pairs of sex chromosomes—XX *or* XY.

A REST STOP ON OUR CONCEPTUAL ROAD

Before driving on into the land of human inheritance, take a few minutes to review this road map. It might give you perspective on key concepts.

A *gene*, again, is a unit of DNA information about a heritable trait. Genes are distributed among a number of chromosomes in eukaryotic cells. A gene has its own normal location, or locus, on one type of chromosome.

A cell with a diploid chromosome number ($2n$) has *pairs of homologous chromosomes*. Each chromosome of a pair is identical to its homologue in length, shape, and gene sequence. The exception is a pair of nonidentical sex chromosomes, such as X and Y.

A pair of genes on homologous chromosomes may or may not be the same, because genes mutate. For any locus, all of the slightly different molecular forms of a gene are called *alleles*.

A *wild-type* allele is the most common form of a gene in a natural population. Any less common form of the gene is a *mutant* allele.

All genes on the same chromosome are physically joined. They can become separated when *crossing over* occurs during prophase I of meiosis. By this process, the nonsister chromatids of homologous chromosomes exchange corresponding segments. Crossing over is a route to new combinations of alleles that did not exist in a parental cell.

Variable alignments of homologous chromosomes at metaphase I of meiosis are the basis of *independent assortment*, which also sorts nonparental combinations of alleles into gametes and offspring.

KARYOTYPING

Sometimes the *structure* of a chromosome can change at mitosis or meiosis. A parental chromosome *number* can change also, with variable consequences. How do we know about these changes? A diagnostic tool called *karyotyping* helps us analyze an individual's diploid complement of chromosomes (Figure 8.18).

With this procedure, a sample of cells taken from an individual is put in a growth medium that stimulates mitotic division. The medium also contains colchicine, a microtubule poison that interferes with dynamics of the mitotic spindle (Section 3.7). Thus, the cells enter mitosis, but colchicine prevents them from dividing, so their cell cycle becomes arrested at metaphase.

The cells and the medium are transferred to a tube. Then, the cells are separated from the liquid medium with a *centrifuge*. Like a washing machine spin cycle, a centrifuge whirls tubes around a center post. The force of spinning moves the cells away from the center of rotation, and they collect at the bottom of the tube in a clump. The culture medium is removed from the tube, and then a hypotonic solution (Section 4.6) is added. The cells swell up, so the chromosomes inside of them move apart. The cells are spread on a microscope slide and stained so the chromosomes become visible under a microscope (Section 3.2).

The microscope reveals metaphase chromosomes in every cell. A micrograph of one cell is cut apart and reassembled so that the images of all the chromosomes are aligned at the centromeres and arranged according to size and shape, and length. This is the individual's **karyotype**, which is compared with a normal standard for diagnostic purposes.

AUTOSOMES AND SEX CHROMOSOMES

Like many other species, normal humans are either male or female, genetically speaking anyway. Also as in many other species, human body cells are diploid ($2n$), having pairs of homologous chromosomes. With one exception, each chromosome of a pair is identical to its homologue. One member of that nonidentical pair is a unique sex chromosome that occurs either in males or females, but not both.

LINKS TO SECTIONS 3.2, 3.7, 4.6, 7.1, 7.3, 7.5, 7.6

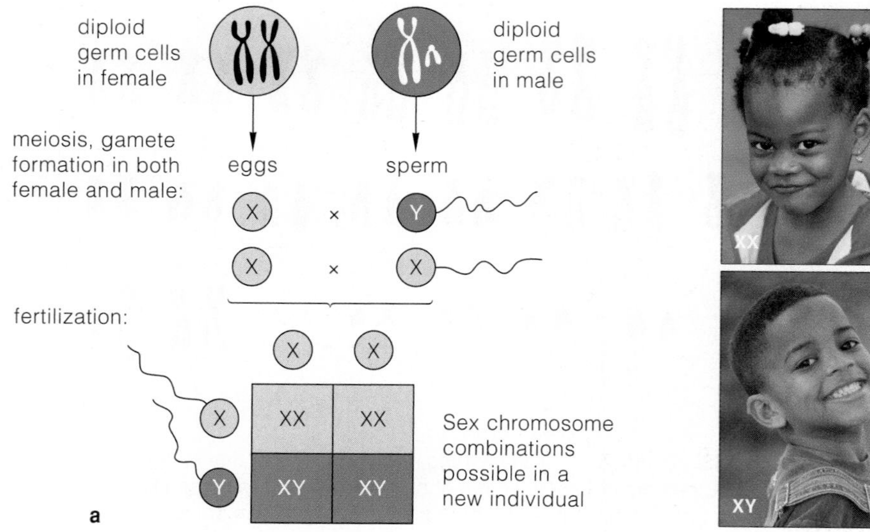

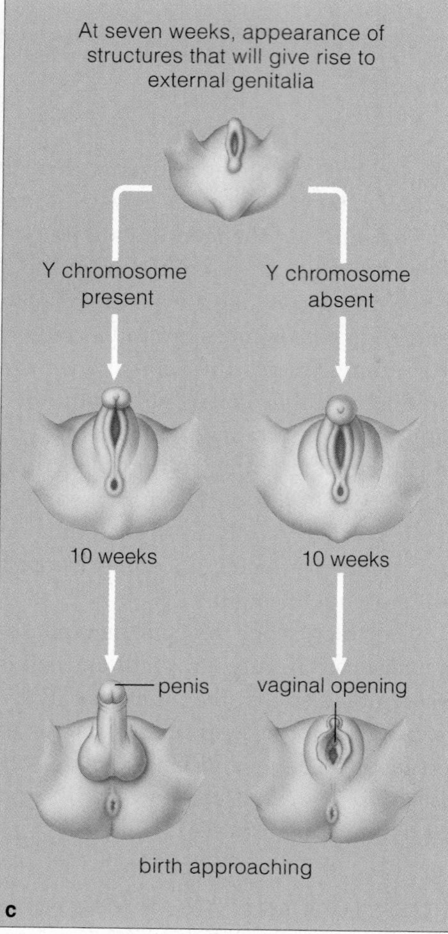

Figure 8.19 ***Animated!*** (**a**) Pattern of sex determination in humans. (**b**) An early human embryo appears neither male nor female. Then, ducts and other structures that develop into male *or* female reproductive organs form. In an XX embryo, ovaries form *in the absence of the SRY gene on the Y chromosome.* In an XY embryo, the gene product causes testes to form. Testes prompt the development of other male traits. (**c**) External reproductive organs in human embryos.

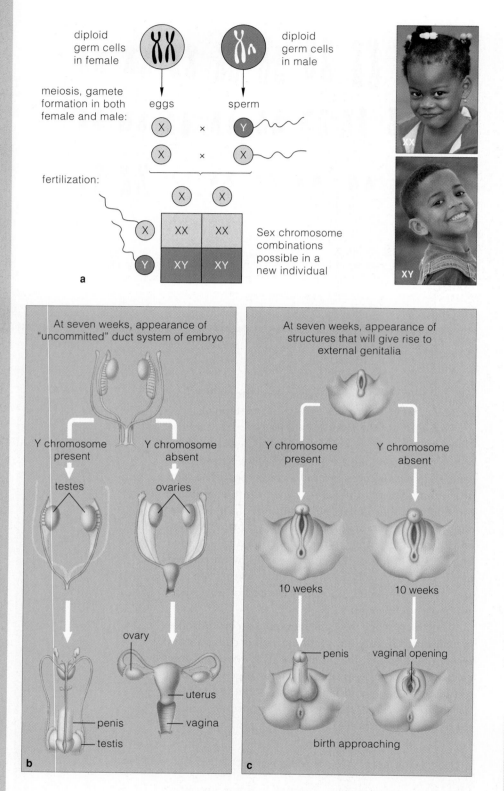

For instance, a diploid cell in a human female has two X chromosomes (XX). A diploid cell in a human male has one X and one Y chromosome (XY). Human X and Y chromosomes differ. In outward appearance, the considerably shorter Y chromosome appears to be a remnant of the X. The two also differ in which genes they carry. In prophase I, the X and Y chromosomes associate briefly in one small region. That small region is enough to let them interact as a homologous pair of chromosomes during meiosis.

Human X and Y chromosomes fall into the category of **sex chromosomes**, those that determine whether an individual is male or female. All other chromosomes are **autosomes**, which are the same in both sexes.

Each normal egg produced by a human female has one X chromosome. Half the sperm formed in a human male carry an X chromosome, and half carry a Y. If an X-bearing sperm fertilizes an egg, the new individual will become female. However, if the sperm carries a Y chromosome, then the new individual will become a male. The Punnett square in Figure 8.19*a* depicts sex determination in humans.

SRY, one of 330 genes on a human Y chromosome, is the master gene for sex determination in males. *SRY* expression in XY embryos triggers formation of testes, which are primary male reproductive organs. Testes make testosterone, a sex hormone that is responsible for the emergence of male sexual traits. An XX embryo has no *SRY* gene, so the primary female reproductive organs, ovaries, form instead (Figure 8.19*b*,*c*). Ovaries make estrogens and other sex hormones that cause the development of female sexual traits.

The human X chromosome carries 2,062 genes. Like other chromosomes, it carries some genes associated with sexual traits, such as the distribution of body fat and hair. But most of its genes deal with *nonsexual* traits, such as blood-clotting functions. Such genes can be expressed in males as well as in females. Remember, males also carry one X chromosome.

Diploid cells have pairs of genes, on pairs of homologous chromosomes. At each gene locus, the alleles (alternative forms of a gene) may be either identical or nonidentical.

Abnormal events at meiosis or mitosis can change the structure and number of chromosomes.

Autosomes are pairs of chromosomes that are the same in males and females of a species. One other pair, the sex chromosomes, differ between males and females.

The SRY gene on the human Y chromosome dictates that a new individual will develop into a male. In the absence of the Y chromosome (and the gene), a female develops.

8.5 Impact of Crossing Over on Inheritance

LINK TO SECTION 7.6

Crossing over between homologous chromosomes is one of the major pattern-busting events in inheritance.

We now know there are many genes on each type of autosome and sex chromosome. All the genes on one chromosome are called a linkage group. For example, the fruit fly (*Drosophila melanogaster*) has four linkage groups corresponding to its four pairs of homologous chromosomes. Indian corn (*Zea mays*) has ten linkage groups, corresponding to its ten pairs. Humans have twenty-three linkage groups, and so on.

If linked genes stayed linked all through meiosis, then there would be no surprising mixes of parental traits. You could expect parental phenotypes among, say, F_2 offspring of dihybrid crosses to show up in a predictable ratio. As early experiments with fruit flies showed, however, that ratio was predictably incorrect for many pairs of genes on the same chromosome. For example, in one experiment, 17 percent of F_2 offspring of dihybrid crosses had a combination of alleles that was found in neither parent.

Plenty of genes on the same chromosome do not stay linked during meiosis, but some do so more often than others. Why? Genes that are close together on the chromosome are separated less frequently by crossing over than genes that are far apart. Think of two genes that are on the same chromosome. *The probability that a crossover will disrupt their linkage is proportional to the distance between them.*

If genes *A* and *B* are twice as far apart as genes *C* and *D*, we would expect crossing over to disrupt the linkage between *A* and *B* much more often:

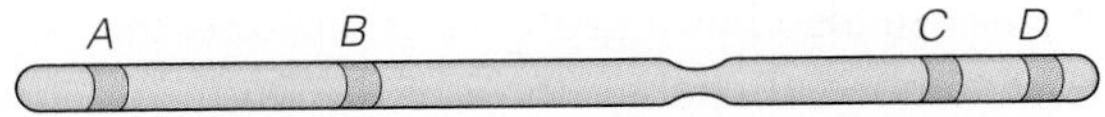

Two genes are very closely linked when the distance between them is small. Their combinations of alleles nearly always end up inside the same gamete. Linkage is more vulnerable to crossing over when the distance between them is greater (Figure 8.20). When two gene loci are far apart, crossing over is so frequent that the genes sort independently of each other into gametes.

All of the human gene linkages were identified by tracking phenotypes in families over the generations. One thing is clear from these studies: Crossovers are not rare. For most eukaryotes, meiosis cannot even be completed properly until at least one crossover occurs between each pair of homologous chromosomes.

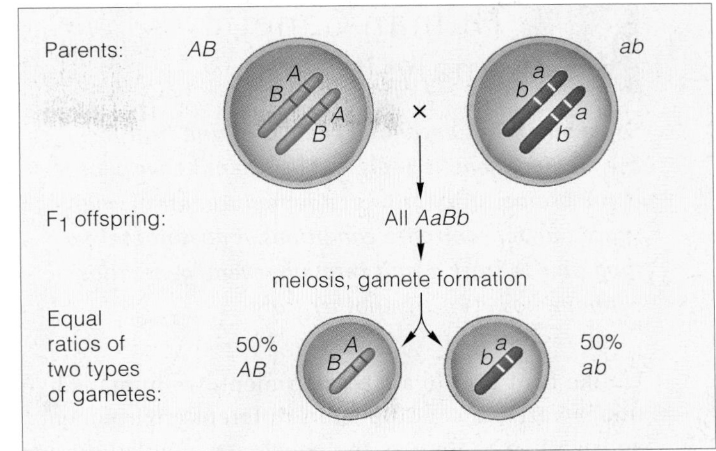

a Full linkage between two genes—no crossing over. Half the gametes have one parental genotype, and half have the other. Genes that are close together on a chromosome frequently stay together in gametes.

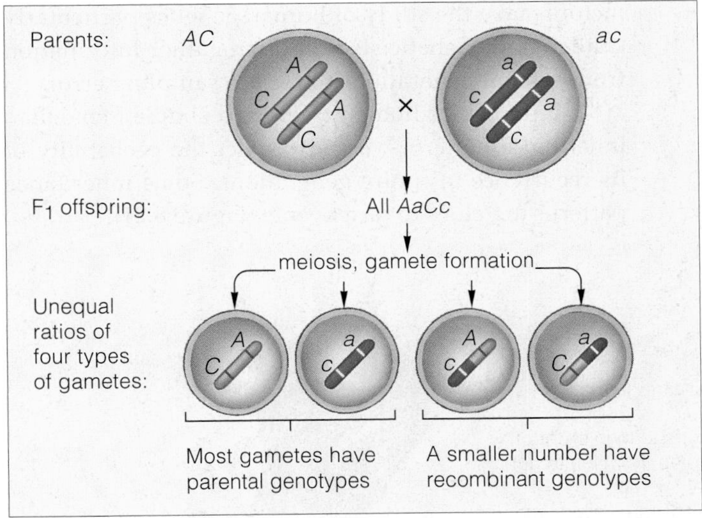

b Incomplete linkage; crossing over affected the outcome. Genes that are far apart on a chromosome are more frequently separated by crossing over than adjacent genes.

Figure 8.20 Crossover events between two gene loci influence the outcome of dihybrid crosses.

All of the genes at different locations along the length of a chromosome belong to the same linkage group. They do not all assort independently at meiosis.

Crossing over between homologous chromosomes disrupts gene linkages and results in nonparental combinations of alleles in chromosomes.

The farther apart two genes are on a chromosome, the greater will be the frequency of crossing over between them.

LINK TO SECTION 1.4

8.6 Human Genetic Analysis

Some organisms, including pea plants and fruit flies, are ideal for genetic analysis. They do not have a lot of chromosomes. They grow and reproduce fast in small spaces, under controlled conditions. It does not take a long time to track a trait through several generations. Humans, however, are another story.

Unlike fruit flies in a lab experiment, we humans live under variable conditions, in different environments, and we live as long as the geneticists who study our traits. Most of us select our own mates and reproduce if and when we want to. Most of our families are not large, which means there may not be enough offspring for geneticists to infer a pattern of inheritance. These factors make the study of human genetics particularly challenging. Geneticists will often gather information from several generations to reduce sampling error.

If a particular human trait has a simple Mendelian inheritance pattern, we can predict the probability of its recurrence in future generations. Some inheritance patterns are clues to past events (Figure 8.21).

Figure 8.21 An intriguing pattern of inheritance. Eight percent of the men in Central Asia carry nearly identical Y chromosomes, which implies descent from a shared ancestor. If so, then 16 million males living between northeastern China and Afghanistan—close to 1 of every 200 men alive today—belong to a lineage that probably started with the warrior and notorious womanizer Genghis Khan. His abundant offspring ruled an empire that stretched from China to Vienna.

Such information is often displayed in **pedigrees**, or charts of genetic connections among individuals. Standardized methods, definitions, and symbols that represent kinds of individuals are used to construct the charts (Figure 8.22*a*). Those who analyze pedigrees rely on their knowledge of probability and Mendelian inheritance patterns, which may yield clues to a trait. For example, as you will see, clues to many disorders point to a dominant or recessive allele, and even to its location on a certain autosome or sex chromosome.

Some traits are deviations from the average. Such a genetic *abnormality* is only a rare or uncommon version of a trait, as when a person is born with six fingers on each hand instead of five (Figure 8.22*b*). Whether this condition is disfiguring or interesting is a matter of opinion. On the other hand, a **genetic disorder** is an inherited condition that, sooner or later, will result in medical problems that can be quite severe. A genetic abnormality or disorder is characterized by a specific set of symptoms called a **syndrome**.

A disease also has a group of symptoms that arises from an abnormal change in how the body functions. However, a **disease** is an illness caused by infectious, dietary, or environmental factors, and does not result from a heritable mutation. It is appropriate to use the term *genetic* disease if such factors modify previously workable genes in a way that disrupts body functions.

Alleles that underlie severe genetic disorders are rare in populations because they put their bearers at

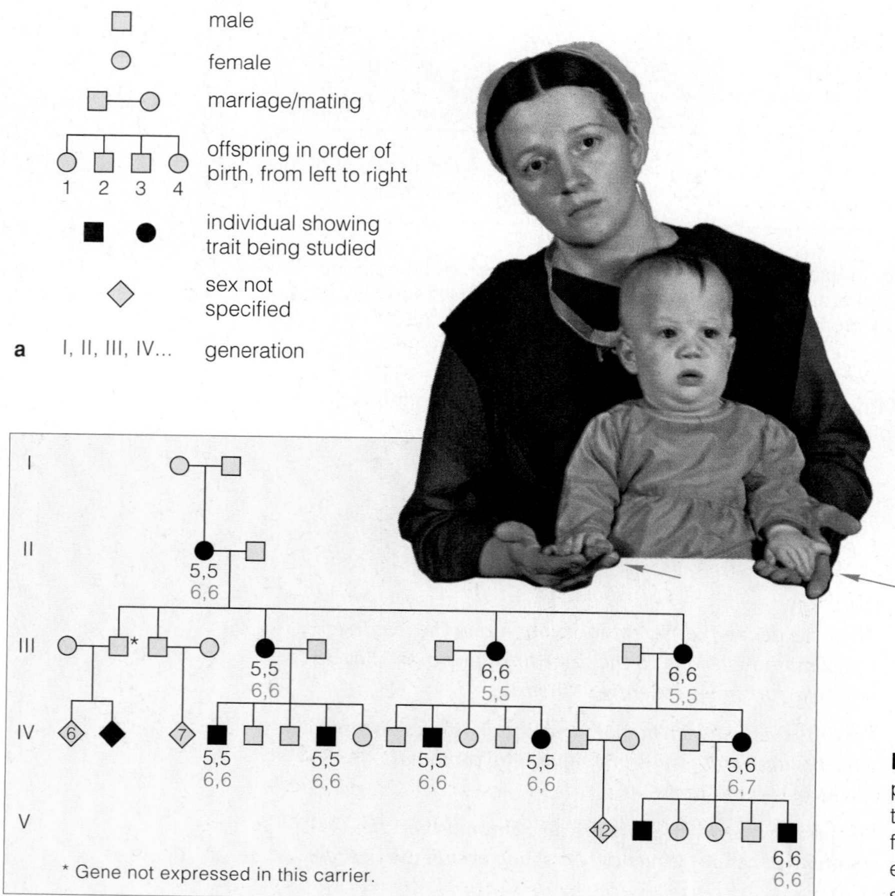

Figure 8.22 ***Animated!*** (**a**) Some standardized symbols used in pedigrees. (**b**) A pedigree for *polydactyly*, characterized by extra toes, fingers, or both. *Black* numerals signify the known number of fingers on each hand; *blue* numerals signify the number of toes on each foot. Polydactyly is one of the symptoms of Ellis–van Creveld syndrome (Section 12.7).

risk. They do not completely disappear, because rare mutations reintroduce them. Also, in heterozygotes, a normal allele that is paired with a harmful one might compensate for it. Heterozygotes can transmit mutated alleles to offspring, with effects of the sort described in the next section.

Pedigree analysis can reveal inheritance patterns. From such patterns, geneticists may determine the probability that certain genes will be passed on to children.

A genetic abnormality is a rare or less common version of an inherited trait. A genetic disorder is an inherited condition that results in mild to severe medical problems.

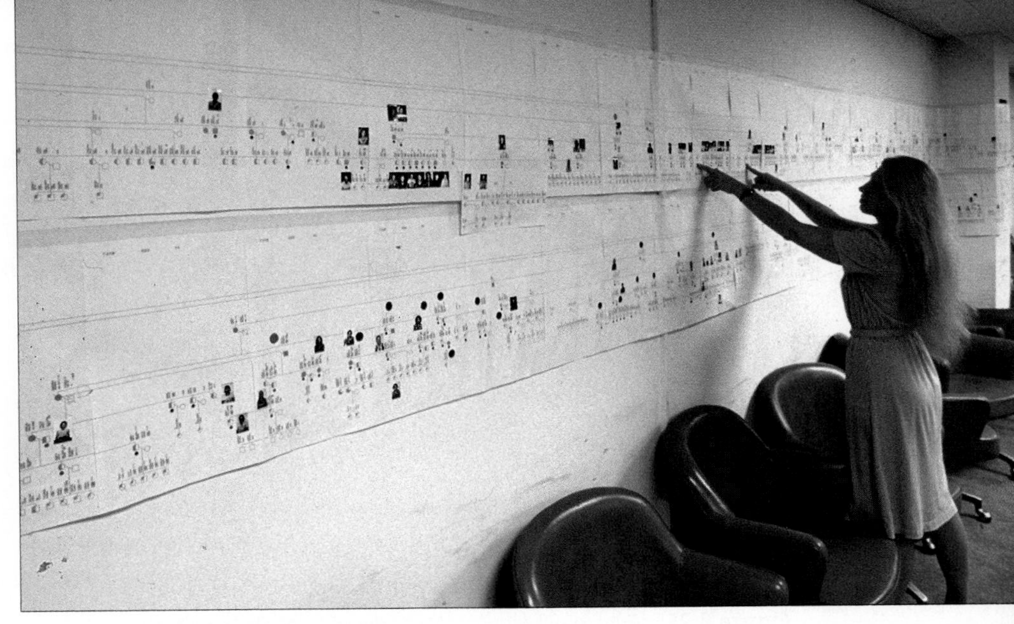

Figure 8.23 Pedigree for *Huntington disease*, a progressive degeneration of the nervous system. Researcher Nancy Wexler and her team constructed this extended family tree for nearly 10,000 Venezuelans. Their analysis of unaffected and affected individuals revealed that a dominant allele on human chromosome 4 is the culprit. Wexler has a special interest in the disease; it runs in her family.

8.7 Examples of Human Inheritance Patterns

Some human phenotypes arise from dominant or recessive alleles on an autosome or X chromosome, inherited in simple Mendelian patterns. Others arise from changes in chromosome structure or number.

AUTOSOMAL DOMINANT INHERITANCE

Two clues point to an autosomal dominant allele for a trait. First, the trait typically appears each generation; the allele is usually expressed even in heterozygotes. Second, if one parent is heterozygous and the other is homozygous recessive, any child of theirs will have a 50 percent chance of being heterozygous.

A few dominant alleles persist in populations even when they cause severe genetic disorders. Spontaneous mutations reintroduce some. For others, expression of a dominant allele may not interfere with reproduction, or people reproduce before symptoms are severe.

For instance, with *Huntington disease*, the nervous system deteriorates, muscle movements become more and more involuntary, and death follows (Figure 8.23). Symptoms may not start until after age thirty. Many affected individuals have already reproduced by then, which puts children at risk of inheriting the disorder. Affected people usually die in their forties or fifties.

The mutation that causes this disorder changes one of the proteins necessary for normal development of brain cells. It is one of the *expansion* mutations, which occurs as multiple repeats of the same DNA segment. The repeats disrupt gene function.

Achondroplasia is an autosomal dominant disorder that affects about 1 in 10,000 people. Commonly, the homozygous dominant condition results in stillbirth, yet heterozygotes can still reproduce. Cartilage parts of the skeleton form abnormally in achondroplasiacs. Adults have abnormally short arms and legs relative to other body parts. They are less than 4 feet, 4 inches tall. Figure 8.24 shows three such adults.

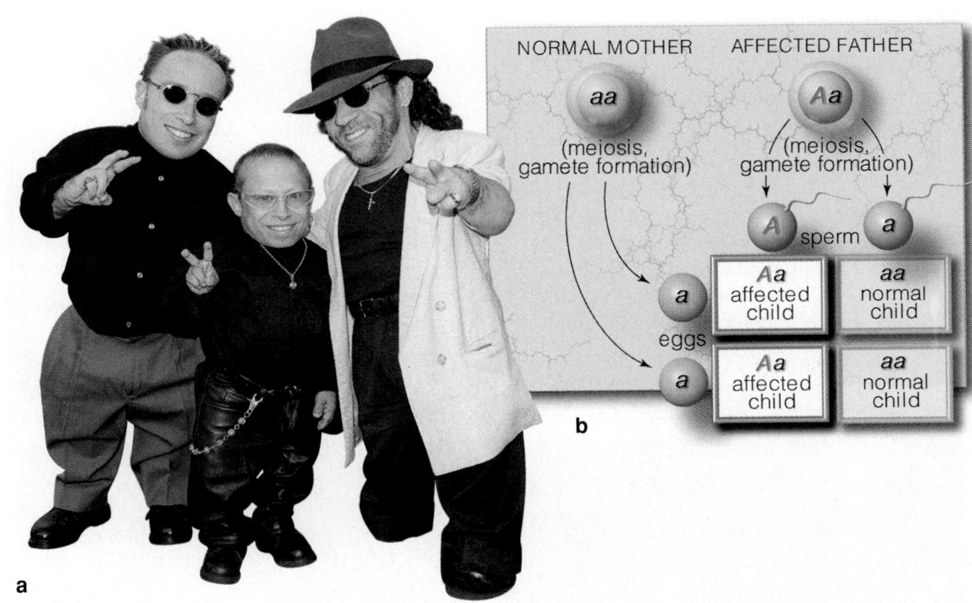

Figure 8.24 ***Animated!*** Autosomal dominance. (**a**) Three males affected by achondroplasia. Verne Troyer (center) stands only two feet, eight inches tall. (**b**) One pattern for autosomal dominant inheritance. A dominant allele (*red*) is fully expressed in the carriers.

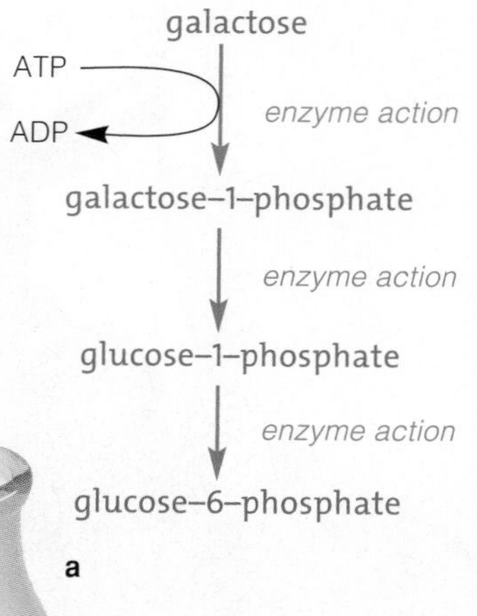

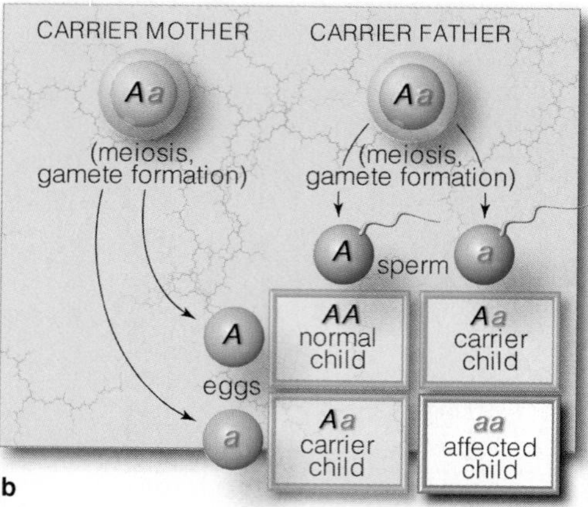

Figure 8.25 Autosomal recessive inheritance. (**a**) How galactose is converted to a form that can enter glycolysis. Mutation in a gene for any of the enzymes in the conversion pathway can give rise to galactosemia. (**b**) One pattern for autosomal recessive inheritance. In this case, both parents are heterozygous carriers of the recessive allele (coded *red*).

AUTOSOMAL RECESSIVE INHERITANCE

For some traits, inheritance patterns reveal two clues that point to a recessive allele on an autosome. First, if both parents are heterozygous for the recessive allele, each child of theirs has a 25 percent chance of being homozygous recessive, and also a 50 percent chance of being heterozygous. Second, any child of homozygous recessive parents will be homozygous recessive also.

About 1 in 100,000 newborns is homozygous for a recessive allele that causes *galactosemia*. Those affected do not have working copies of one of the enzymes that digest lactose, and a reaction intermediate builds up to toxic levels. Lactose normally becomes converted to glucose and galactose, then to glucose–1–phosphate (which is broken down by glycolysis or converted to glycogen). The complete conversion by the pathway is blocked in galactosemics (Figure 8.25).

Galactose rises to levels that may be detected in urine. The excess results in malnutrition, diarrhea, and vomiting. It also damages the eyes, liver, and brain. When untreated, galactosemics typically die early in life. If they are quickly placed on a restricted diet that excludes dairy products, they grow up symptom-free.

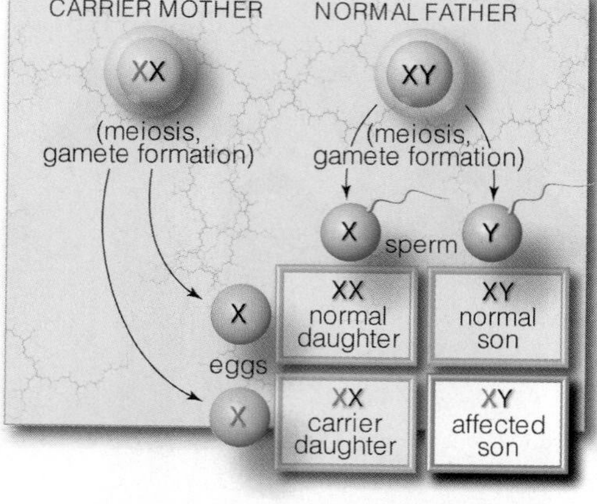

Figure 8.26 ***Animated!*** One pattern for X-linked inheritance. In this case, the mother carries the recessive allele on one of her X chromosomes (coded *red*).

X-LINKED RECESSIVE INHERITANCE

An X-linked gene is found on the X chromosome. With X-linked genetic disorders, females are not affected as often as males, because a dominant allele on a female's other X chromosome can mask a recessive one (Figure 8.26). A son will not inherit an X-linked allele from his father, but a daughter will. When she does, each of her sons has a 50 percent chance of inheriting it. Recessive

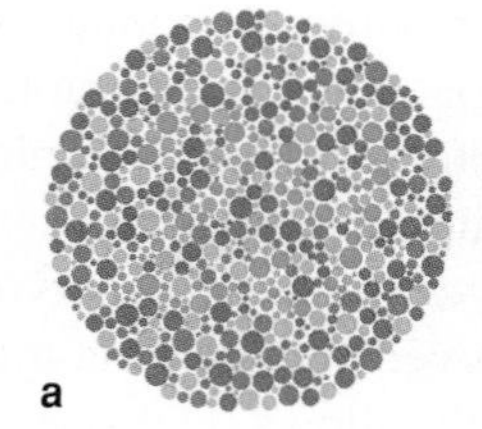

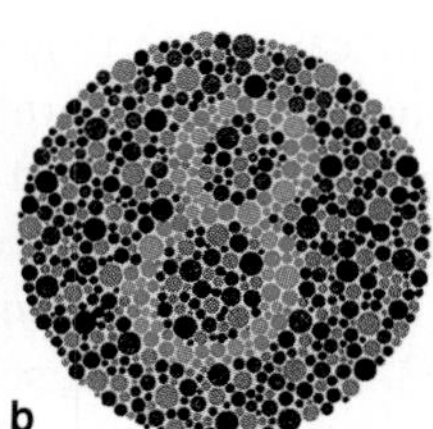

Figure 8.27 *Left,* what red-green color blindness means, using ripe red cherries on a green-leafed tree as an example. In this case, perception of blues and yellows is not affected, but the affected person probably would be at a loss during harvest time in a cherry orchard.

Above, two of many Ishihara plates, which are standardized tests for different forms of color blindness. (**a**) You may have one form of red-green color blindness if you see the numeral "7" instead of "29" in this circle, (**b**) You may have another form if you see a "3" instead of this "8."

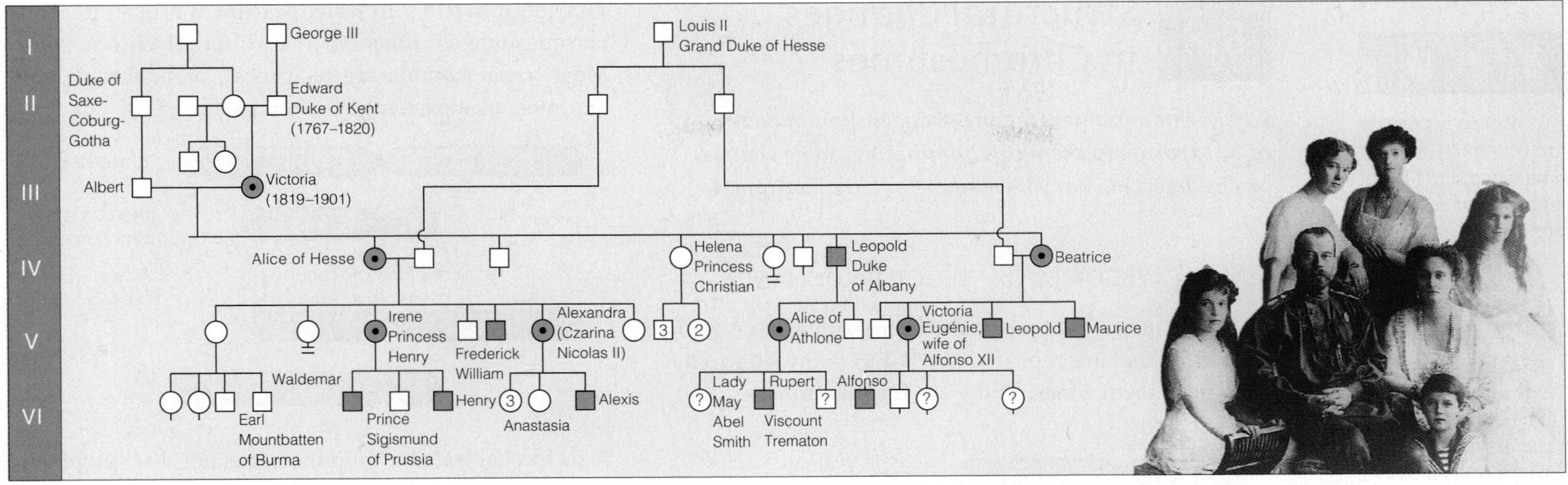

Figure 8.28 X-linked recessive inheritance in royal families. *Left*, a partial pedigree, or a chart of genetic connections, among descendants of Queen Victoria of England. It focuses on carriers and affected males who inherited the X-linked allele for hemophilia A (*white* circles and squares). At one time, the recessive allele was present in eighteen of Victoria's sixty-nine descendants, who sometimes intermarried. *Right*, the last Russian royal family. Czarina Alexandra was a carrier for hemophilia. Through her obsession with the vulnerability of her son Alexis, a hemophiliac, she became involved in political intrigue that helped trigger the Russian Revolution of 1917.

alleles on the human X chromosome cause more than 300 genetic disorders, even though this chromosome holds only 4 percent of all human genes.

Color blindness is an inability to distinguish among some or all colors. It can result from several common recessive disorders associated with X-linked genes. Mutations in these genes change the light-absorbing capacity of sensory receptors in the eyes. Normally, humans can spot the difference between 150 colors. A person who is red–green color blind sees fewer than 25 colors; some or all receptors that respond to visible light of red and green wavelengths are weakened or absent. Other people who are color blind confuse red and green colors, and still others see only shades of gray instead of green (Figure 8.27). The trait is more common in men, but heterozygous women also show some symptoms. (Can you explain why?) People with color blindness are identified with standardized tests such as those in Figure 8.27.

Duchenne muscular dystrophy (DMD) is one of a group of X-linked recessive disorders characterized by rapid degeneration of muscles, starting early in life. About 1 of 3,500 boys is affected. The recessive allele encodes a structural protein called dystrophin, which provides structural support to muscle cells. It anchors much of the cell cortex to the plasma membrane. In cases where dystrophin is abnormal or absent, the cell cortex weakens, and muscle cells die. The debris left behind in tissues triggers chronic inflammation. Most cases of DMD are diagnosed between the ages of three and seven. There is no way to stop the progression of the disorder. When the affected boy is about twelve years old, he will start to use a wheelchair. His heart muscles will start to break down. Even with medical care, he will probably die before age twenty-five, most often as a result of respiratory failure.

Hemophilia A, a blood-clotting disorder, is a well-studied example of an X-linked recessive inheritance pattern. Most people have a clotting mechanism that quickly stops the bleeding from minor injuries. Some proteins involved in clotting are products of genes on the X chromosome. Blood clotting is altered in people with a mutated form of one of these genes. Clotting time is close to normal in heterozygous females, but bleeding is prolonged in males that have the disorder. Affected males bruise easily, and the internal bleeding causes problems in muscles and joints. Hemophilia A now affects about 1 in 7,000 males, but the frequency was much higher among the royal families of Europe and Russia in the nineteenth century. This is probably because the common practice of inbreeding kept the harmful allele in the family tree (Figure 8.28).

The map in Appendix IV shows where the mutations that cause these disorders occur in the human genome.

Genetic analyses of family pedigrees have revealed simple Mendelian inheritance patterns for certain traits, as well as for many genetic disorders that arise from expression of alleles on an autosome or X chromosome.

IMPACTS, ISSUES

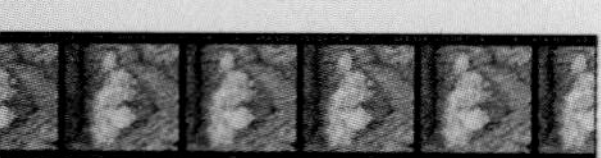

The CFTR gene resides on chromosome 7. The most common cause of CF is a deletion from the gene region. The mutation makes the gene's protein product fold abnormally after it is synthesized. When the protein enters the ER for processing, it is destroyed. The mutant protein never reaches the plasma membrane, and transport channels never do form.

8.8 Structural Changes in Chromosomes

Rarely, chromosome structure changes spontaneously or by exposure to chemicals or radiation. Some changes can be detected. Many have severe or lethal outcomes.

MAJOR CATEGORIES OF STRUCTURAL CHANGE

DUPLICATION Even normal chromosomes have gene sequences that are repeated several to many hundreds or thousands of times. These are **duplications**:

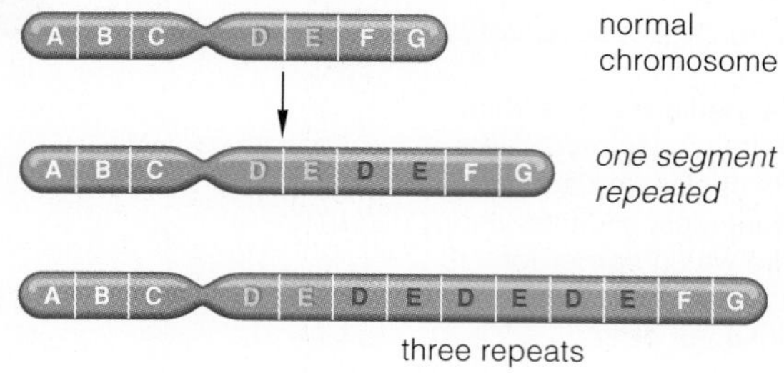

Although no genetic information has been lost with a duplication, the presence of extra genes often disrupts the body's fine balance of gene expression, resulting in neural problems and physical abnormalities.

DELETION A **deletion** is the loss of part or all of a chromosome:

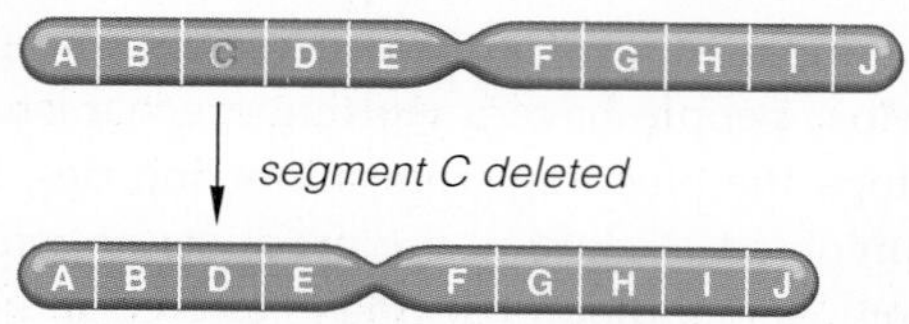

Most deletions result in the loss of several genes. They usually cause serious disorders, and are often lethal.

INVERSION With an **inversion**, part of the sequence of DNA within the chromosome becomes oriented in the reverse direction, with no molecular loss:

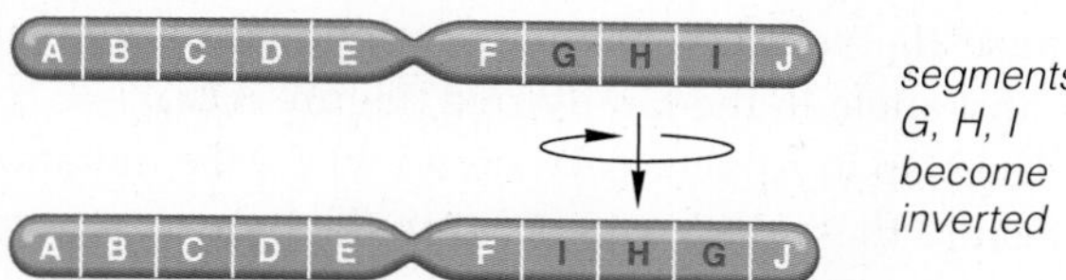

If an inversion does not disrupt a crucial gene region, it may cause no medical problems. However, because an inverted chromosome mispairs during meiosis, it can cause chromosome deletions in offspring.

TRANSLOCATION In **translocation**, a broken part of a chromosome is attached to a different chromosome. Most translocations are reciprocal, or balanced; both chromosomes exchange broken parts:

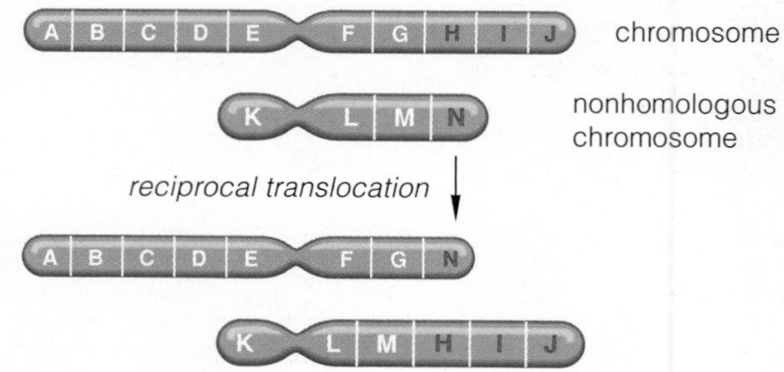

A reciprocal translocation that does not disrupt genetic information may have no adverse affect on its bearer. Many people do not realize they carry a translocation until one of their children is born with a severe genetic disorder. The two translocated chromosomes assort independently into gametes during meiosis, so each gamete has a 50 percent chance of receiving only one of them. The outcome is an unbalanced translocation, the effects of which range from very severe to lethal.

Any alteration in chromosome structure may have grave medical effects on its bearer. For example, even a small deletion in one part of chromosome 5 results in an abnormally shaped larynx and mental impairment. When affected infants cry, they sound like kittens. The name of this disorder is *cri-du-chat*, which translates to meow in French (Figure 8.29).

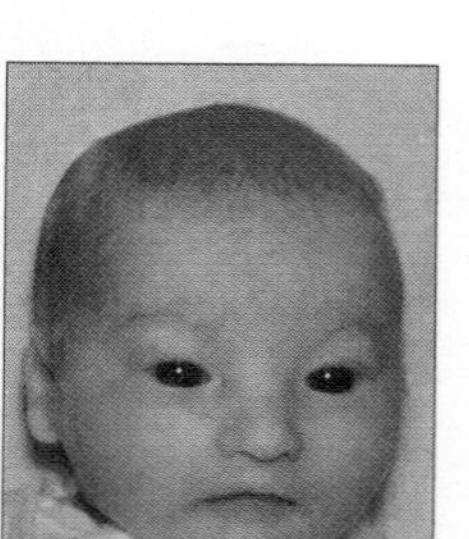

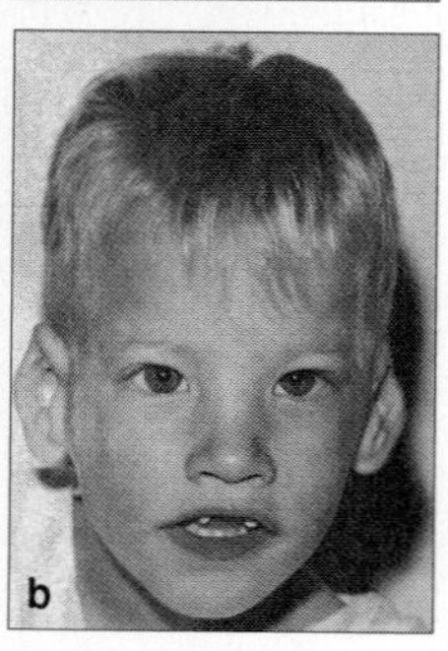

Figure 8.29 A case of cri-du-chat syndrome. (**a**) This infant's ears are low on the side of the head relative to his eyes. (**b**) Same boy, four years later. The characteristic high-pitched monotone of cri-du-chat children may persist into their adulthood.

DOES CHROMOSOME STRUCTURE EVOLVE?

Chromosome structure alterations have had important impact on the evolution of every species. For example, certain duplications might allow one copy of a gene to mutate while a different copy carries out its original function. The globin genes of primates appear to have evolved like this. Several alternative forms of primate globin have different capacities to bind oxygen under a wide range of cellular conditions.

Such changes in chromosome structure might have contributed to the differences between closely related primate species, such as apes and humans. Eighteen of twenty-three pairs of human chromosomes are almost identical with those of chimpanzees and gorillas. The other five differ only by inversions and translocations.

A chromosome or part of it may become duplicated, deleted, inverted, or moved to a new location.

Most changes in chromosome structure are harmful.

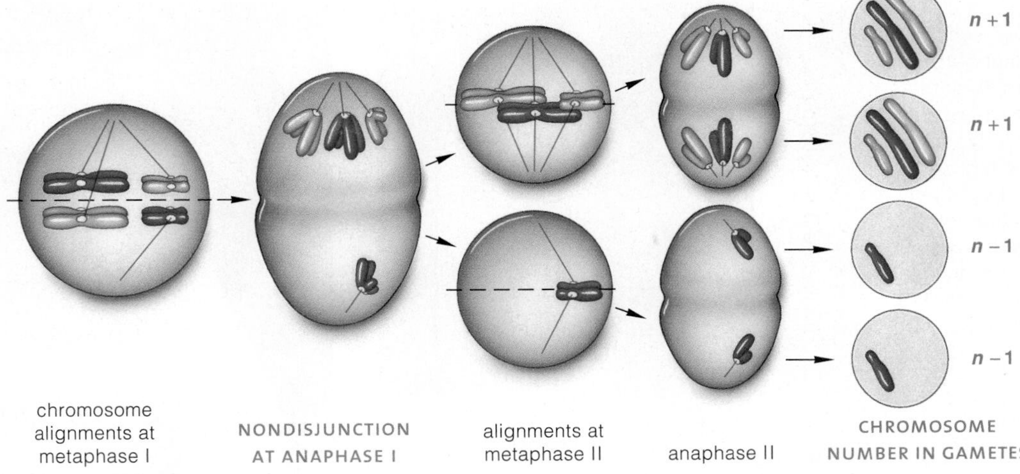

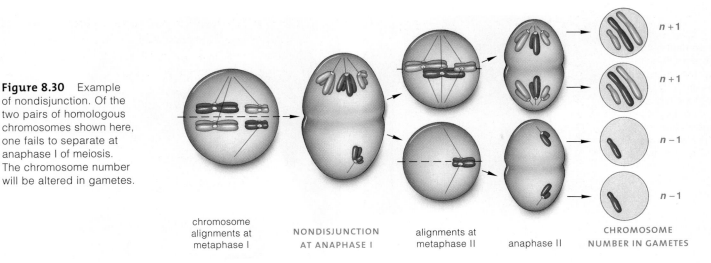

Figure 8.30 Example of nondisjunction. Of the two pairs of homologous chromosomes shown here, one fails to separate at anaphase I of meiosis. The chromosome number will be altered in gametes.

8.9 Change in the Number of Chromosomes

Occasionally, abnormal events occur before or during cell division, and gametes and new individuals end up with the wrong chromosome number. Consequences range from minor to lethal changes in form and function.

LINK TO SECTION 7.5

In **aneuploidy**, cells have one too many or one too few chromosomes. Autosomal aneuploidy is usually fatal for humans and has been linked to most miscarriages. In **polyploidy**, cells have three or more of each type of chromosome. Half of all species of flowering plants, some insects, fishes, and other animals are polyploid (Section 13.8). Polyploidy is lethal for humans. All but about 1 percent of human polyploids die before birth; rare newborns die soon afterward.

Nearly all such changes in chromosome number arise through **nondisjunction**, whereby one or more pairs of chromosomes do not separate as they should either during mitosis or meiosis (Figure 8.30). Trisomy or monosomy, both of which are types of aneuploidy, may be the result. Suppose a normal gamete fuses with an $n + 1$ gamete (one extra chromosome). The new individual will be trisomic ($2n + 1$), with three of one type of chromosome and two of every other type. If a normal gamete fuses with an $n - 1$ gamete, the new individual will be monosomic ($2n - 1$).

AN AUTOSOMAL CHANGE AND DOWN SYNDROME

A few trisomics are born alive, but only trisomy 21 individuals reach adulthood. A newborn with three chromosomes 21 will develop *Down syndrome*. This autosomal disorder is the most frequent type of altered chromosome number in humans. Occurring once in every 800 to 1,000 births, it affects more than 350,000 people in the United States alone. Figure 8.31 shows a karyotype for a trisomic 21 female. About 95 percent of all cases arise by nondisjunction during meiosis.

Affected individuals have upward-slanting eyes, a fold of skin that starts at the inner corner of each eye, a deep crease across each palm and foot sole, one (not two) horizontal creases on the fifth finger, and slightly flattened facial features. Not all individuals develop every symptom. Some newborns show only a few. We even see some of these defining features in the human population at large.

That said, trisomic 21 individuals have moderate to severe mental impairments and heart defects. Their skeleton develops abnormally, so older children have

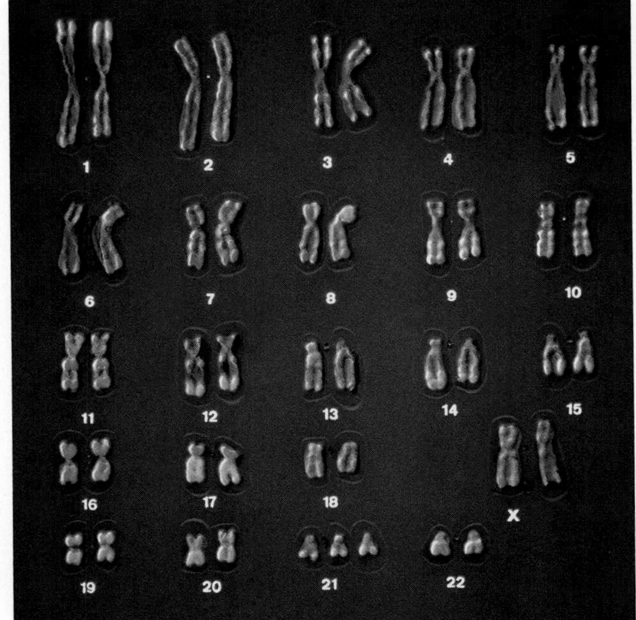

Figure 8.31 Karyotype revealing the trisomic 21 condition of a human female.

Figure 8.32 Down syndrome. (a) About 80 percent of trisomy 21 babies are born to mothers who are not yet 35 years old. These women are in the age categories with the highest fertility rates, and they simply have more babies than older women. The risk of having a trisomic 21 baby actually rises dramatically with the mother's age. (b) A boy with trisomy 21 and his brothers.

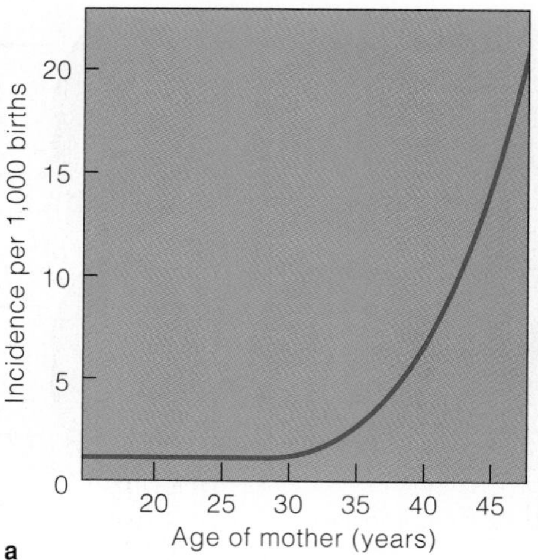

shortened body parts, loose joints, and poorly aligned bones of the hips, fingers, and toes. Skeletal muscles and reflexes are weaker than normal. Their speech and other motor skills develop slowly. With medical care, they live fifty-five years, on average.

The incidence of nondisjunction rises as mothers become older (Figure 8.32*a*). It may originate with the father, but less often. Trisomy 21 is one of hundreds of conditions that are detectable with prenatal diagnosis (Section 8.10). With early special training and medical intervention, affected individuals still can take part in normal activities. As a group, they tend to be cheerful and sociable people (Figure 8.32*b*).

CHANGES IN THE SEX CHROMOSOME NUMBER

Nondisjunction also causes alterations in the number of X and Y chromosomes, with a frequency of about 1 in 400 live births. Most often they lead to difficulties in learning and motor skills, such as a speech delay, but problems may be so subtle that the underlying cause is never diagnosed.

FEMALE SEX CHROMOSOME ABNORMALITIES *Turner syndrome* arises by inheritance of an X chromosome and no corresponding X or Y chromosome. On average, 1 in 2,500 to 10,000 newborn girls have this condition. Nondisjunction originating with the father accounts for 75 percent of Turner syndrome cases. Compared to other sex chromosome abnormalities, cases are few. Why? Probably at least 98 percent of XO zygotes abort spontaneously early in pregnancy.

Figure 8.33 One young girl with Turner syndrome.

Despite the near lethality, XO survivors are not as disadvantaged as the other aneuploids. They grow up well proportioned but not tall—four feet, eight inches, on average (Figure 8.33). Most do not have functional ovaries and so cannot produce enough sex hormones to mature sexually. Secondary sexual traits, including breast enlargement, are affected. Any eggs that form in the ovaries disintegrate before two years of age.

Another example: A few females inherit three, four, or five X chromosomes. This *XXX condition* occurs at a frequency of about 1 in 1,000 live births. Most adults are an inch or so taller and more slender than average. They are fertile. Except for slight learning difficulties, most fall within the normal range of social behavior.

MALE SEX CHROMOSOME ABNORMALITIES One in 500 to 2,000 males has inherited one Y and two or more X chromosomes, mainly by nondisjunction. They have an XXY or, rarely, XXXY, XXXXY, or XY/XXY mosaic genotype. (Mosaic individuals have different numbers of chromosomes in different populations of their body cells.) About 67 percent of affected people inherited the extra chromosome from their mother.

The resulting *Klinefelter syndrome* develops after the onset of puberty. XXY males tend to be overweight and tall. The testes and the prostate gland usually are smaller than average. Many XXY males are within the normal range of intelligence. Some have short-term memory loss and other learning disabilities. Affected individuals make less testosterone and more estrogen than normal, with feminizing outcomes. Their sperm counts are low and hair is sparse, the voice is pitched high, and the breasts are a bit enlarged. Testosterone injections at puberty can reverse the feminized traits.

About 1 in 500 to 1,000 males has one X and two Y chromosomes. Those with this *XYY condition* tend to be taller than average, with mild mental impairment, but most are otherwise normal. At one time, XYY males were thought to be genetically predisposed to a life of crime. This misguided opinion originated with a sampling error—a study of too few cases in narrowly selected groups, including prison inmates—and flaws in experimental design (the same researchers gathered karyotypes *and* personal histories). A false report that a murderer of young nurses was XYY did not help.

In 1976 a Danish geneticist reported on a study of 4,139 tall males, twenty-six years old, who reported to their draft board. Besides giving results of physical examinations and intelligence testing, the records held clues to social and economic status, education, and any criminal convictions. Twelve males were XYY, which meant there were more than 4,000 males in the control group. The only finding was that mentally impaired, tall males who engage in criminal activity are likely to get caught—irrespective of karyotype.

The majority of XXY, XXX, and XYY children may not even be properly diagnosed. Some are dismissed unfairly as being underachievers.

Nondisjunction in germ cells, gametes, or early embryonic cells changes the number of autosomes or number of sex chromosomes. The change affects development and the resulting phenotypes.

Nondisjunction at meiosis causes most sex chromosome abnormalities, which typically lead to subtle difficulties with learning, and with speech and other motor skills.

8.10 Some Prospects in Human Genetics

With the first news of pregnancy, parents-to-be typically wonder if their baby will be normal. Quite naturally, they want their baby to be free of genetic disorders, and most babies are. What are the options when they are not?

BIOETHICAL QUESTIONS

It is a fact that we humans do not approach heritable disorders and diseases the same way. We will attack diseases with antibiotics, surgery, and other weapons. But how do we attack a heritable "enemy" that can be transmitted to offspring? Should we institute regional, national, or global programs to identify people who may carry harmful alleles? Do we tell them they are "defective" and might bestow some disorder on their children? Who decides which alleles are bad? Should society bear the cost of treating all genetic disorders before and after birth? If so, should society have a say in whether an affected embryo will be born at all, or aborted? An **abortion** is the expulsion of a pre-term embryo or fetus from the uterus. As you most likely have learned by now, such questions are just the tip of an ethical iceberg.

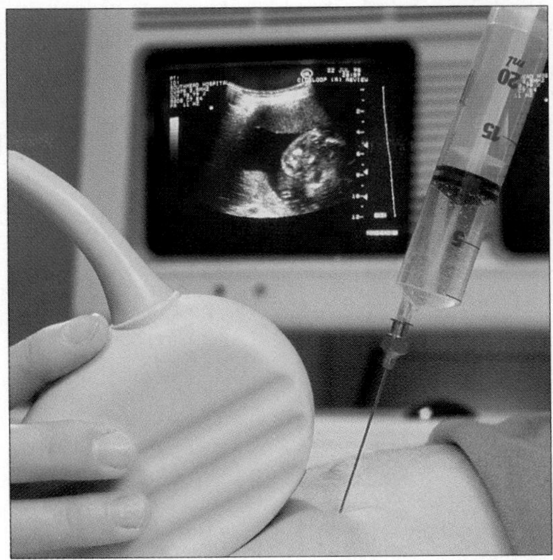

Figure 8.34 Amniocentesis, a prenatal diagnostic tool. A pregnant woman's doctor holds an ultrasound emitter against her abdomen while drawing a sample of amniotic fluid into a syringe. He monitors the path of the needle with an ultrasound screen, in the background. Then he directs the needle into the amniotic sac that holds the developing fetus and withdraws twenty milliliters or so of amniotic fluid. The fluid holds fetal cells and wastes that can be analyzed for genetic disorders.

CHOICES AVAILABLE

GENETIC COUNSELING *Genetic counseling* starts with the diagnosis of parental genotypes, pedigrees, and genetic testing for known metabolic disorders. Using information gained from these tests, genetic counselors can predict the risk for disorders in future children.

If a previous child or a close relative has a genetic disorder, prospective parents may seek counseling. If one of them is a carrier for the disorder, the counselor will advise them of the chances that a child of theirs might inherit the disorder.

Even if neither prospective parent is a carrier for any genetic disorder, there is a 3 percent risk that any child will be born with a birth defect. The risk increases with the woman's age at the time of pregnancy.

PRENATAL DIAGNOSIS A genetic counselor describes available diagnostic procedures to prospective parents. Methods of *prenatal diagnosis* are used to determine whether an embryo or fetus has a genetic abnormality. (*Prenatal* means before birth. *Embryo* is the stage of prenatal human development until eight weeks after fertilization. After eight weeks, *fetus* is the correct term.)

Suppose a forty-five-year-old woman gets pregnant and worries about Down syndrome. She might choose to have *amniocentesis,* usually performed eight to twelve weeks after conception (Figure 8.34). A tiny sample of fluid inside the amnion, a membranous sac enclosing the fetus, is withdrawn. Some of the cells that the fetus has shed are suspended in the sample. These cells are analyzed for many severe genetic disorders, such as

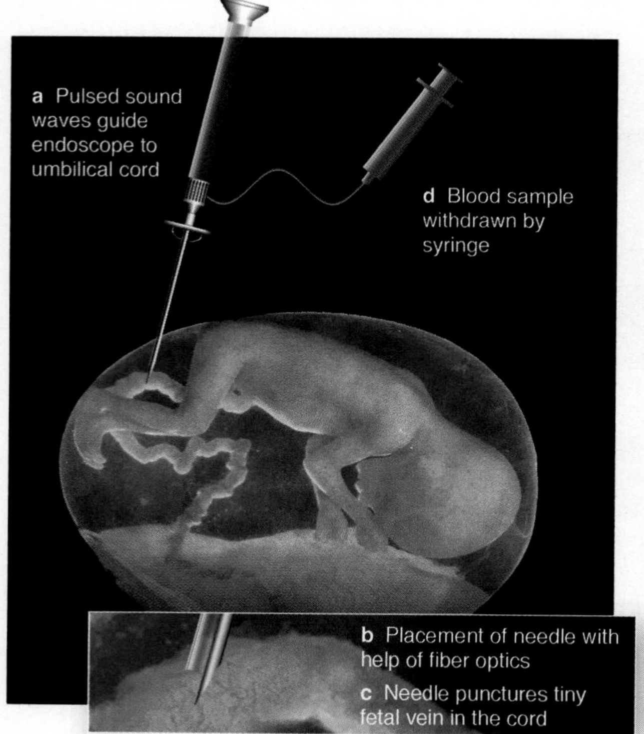

Figure 8.35 Fetoscopy for prenatal diagnosis.

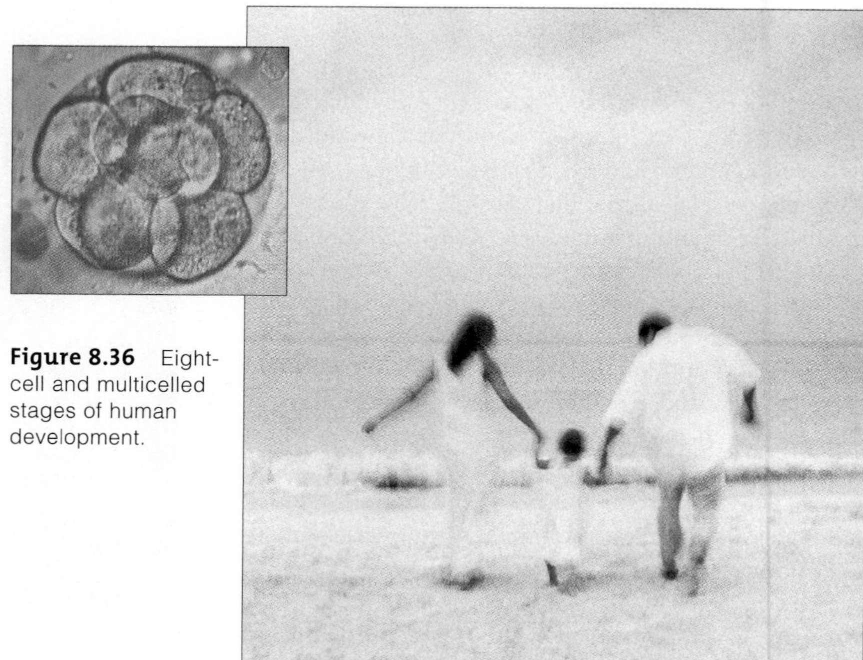

Figure 8.36 Eight-cell and multicelled stages of human development.

sickle-cell anemia, Down syndrome, Turner syndrome, Kleinefelter syndrome, hemophilia, cystic fibrosis, and Duchenne muscular dystrophy.

Chorionic villi sampling (CVS) is another procedure. A clinician withdraws a few cells from the chorion, a fluid-filled, membranous sac that encloses the amnion. CVS can be performed earlier than amniocentesis can, as early as the eighth week of pregnancy.

Directly visualizing the developing fetus is possible with *fetoscopy*. An endoscope, a fiber-optic device, uses pulsed sound waves to scan the uterus and locate parts of the fetus, umbilical cord, or placenta (Figure 8.35). Fetoscopy also is used to diagnose blood cell disorders, such as sickle-cell anemia and hemophilia.

All three procedures may puncture a fetus or cause infection. If a punctured amnion does not reseal itself quickly, too much fluid leaks out, and this endangers the fetus. Amniocentesis raises the risk of miscarriage by 1 to 2 percent. After CVS, 0.3 percent of newborns have missing or deformed fingers and toes. Fetoscopy raises the risk of a miscarriage by 2 to 10 percent.

PREIMPLANTATION DIAGNOSIS This procedure relies on **in-vitro fertilization**. Sperm and eggs taken from prospective parents are mixed in a petri dish. One (or more) eggs may become fertilized. If so, mitotic cell divisions form a ball of eight cells within forty-eight hours (Figure 8.36).

All the cells in the ball have the same genes, and each is not yet committed to being specialized. Doctors carefully remove one of the cells and analyze its genes. If the cell has no detectable genetic defects, the ball is inserted into the mother's uterus, where it may develop into an embryo. Couples who are at risk of passing on cystic fibrosis, muscular dystrophy, and other disorders often have opted for the procedure. Many *"test-tube" babies* have been born in good health.

PHENOTYPIC TREATMENTS The medical symptoms of some genetic disorders may be treated by controlling the diet, adjusting environmental factors, surgery, or hormone replacement therapy.

Dietary control works in *phenylketonuria*, or PKU. In this case, a gene specifies an enzyme that converts the amino acid phenylalanine to tyrosine. Someone who is homozygous recessive for a mutant form of the gene does not make this enzyme. Phenylalanine accumulates, and becomes diverted into other metabolic pathways. The result can be impaired brain functioning. Affected people can lead normal lives by restricting their dietary phenylalanine intake. For example, they can avoid soft drinks and other food products that contain aspartame, a sweetener that contains phenylalanine.

REGARDING ABORTION What happens when prenatal diagnosis reveals a severe problem? Do the prospective parents opt for an induced abortion? We can only say here that they must weigh awareness of the severity of the disorder against their ethical and religious beliefs. Worse, they must play out their personal tragedy on a larger stage dominated by a nationwide battle between highly vocal "pro-life" and "pro-choice" factions.

Summary

Section 8.1 Genes are heritable units of information about traits. Each gene has its own locus on one type of chromosome. Different forms of the same gene are called alleles. Diploid cells have two copies of every gene, usually one inherited from each of two parents, on homologous chromosomes.

An individual with two dominant alleles for a trait (*AA*) is homozygous dominant. A homozygous recessive has two recessive alleles (*aa*). A heterozygote has two nonidentical alleles (*Aa*) for a trait, and the dominant one may mask effects of the partner allele.

Mendel's experiments with pea plants revealed patterns of inheritance. Monohybrid experiments led to the theory of segregation, as stated here in modern terms: Pairs of genes on homologous chromosomes separate during meiosis and end up in different gametes.

Dihybrid experiments led to a theory of independent assortment, as stated here in modern terms: Pairs of genes on homologous chromosomes sort into gametes independently of those on different chromosomes. The random alignment of homologous chromosomes during metaphase I of meiosis is the basis of this outcome.

BiologyNow

Learn how Mendel crossed garden pea plants, and definitions of important genetic terms. Carry out monohybrid crosses and observe the results of a dihybrid cross with interactions on BiologyNow.

Section 8.2 Some alleles are not fully dominant or are codominant. Products of gene pairs often interact to influence the same trait. A single gene may have effects on two or more traits. Products of two or more pairs of genes may interact to affect a single trait.

BiologyNow

Explore patterns of non-Mendelian inheritance with interactions on BiologyNow.

Section 8.3 Many gene interactions give rise to continuous variation for a trait among individuals of a population. Environmental factors influence gene expression and may contribute to that variation.

BiologyNow

Plot the continuous distribution of height for a class with the interaction on BiologyNow.

Section 8.4 Human somatic cells are diploid ($2n$), with twenty-three pairs of homologous chromosomes. One pair are sex chromosomes—XX in females, XY in males. All other chromosomes are autosomes, which are the same in both sexes.

BiologyNow

See how sex is determined in humans with the interaction on BiologyNow.

Section 8.5 Genes on the same chromosome do not all assort independently during meiosis. Crossing over results in new (nonparental) combinations of alleles in chromosomes. The farther apart two genes are on the same chromosome, the more frequently they will undergo crossing over.

Section 8.6 Geneticists often use pedigrees, or charts of genetic connections in a lineage over time, to estimate probabilities that offspring will inherit a given trait.

Section 8.7 Dominant or recessive alleles on either an autosome or X chromosome can be tracked when they are inherited in simple Mendelian patterns.

BiologyNow

Investigate autosomal and X-linked inheritance with the interactions on BiologyNow.

Section 8.8 On rare occasions, a chromosome's structure can change. Part of it may be duplicated, deleted, inverted, or translocated (moved to a new location on a different chromosome). Most alterations are harmful or lethal.

Section 8.9 Rarely, chromosome number can change, as by nondisjunction during meiosis. Aneuploids have one extra or one less chromosome. About half of all flowering plants, some insects, fish, and other animals are polyploid (with three or more of each chromosome).

Section 8.10 Genetic counselors review the options available for potential parents at risk of having children with a genetic disorder.

BiologyNow

Examine a human pedigree and also observe amniocentesis with animation on BiologyNow.

Self-Quiz

Answers in Appendix I

1. Alleles are ______ .
   a. different molecular forms of a gene
   b. different phenotypes
   c. self-fertilizing, true-breeding homozygotes

2. A heterozygote has a ______ for the trait being studied.
   a. pair of identical alleles
   b. pair of nonidentical alleles
   c. haploid condition, in genetic terms
   d. a and c

3. Crosses between F_1 pea plants resulting from the cross *AABB* × *aabb* lead to F_2 phenotypic ratios close to ______ .
   a. 1:2:1 b. 3:1 c. 1:1:1:1 d. 9:3:3:1

4. The probability of a crossover occurring between two genes on the same chromosome is ______ .
   a. unrelated to the distance between them
   b. increased if they are close together
   c. increased if they are far apart

5. Chromosome structure can be altered by a ______ .
   a. deletion c. inversion e. all of the above
   b. duplication d. translocation

6. Genes on the same chromosome are said to be ______ .
   a. linked c. homologous e. all of the above
   b. alleles d. autosomes

7. Most genes for human traits are located on _______ .
   a. the X chromosome
   b. the Y chromosome
   c. autosomes

8. Nondisjunction at meiosis can result in _______ .
   a. an inversion c. a translocation
   b. crossing over d. aneuploidy

9. Turner syndrome (XO) is an example of _______ .
   a. pleiotropy c. aneuploidy
   b. polyploidy d. gene linkage

10. Match each example with the suitable description.
   ____ dihybrid cross a. *bb*
   ____ monohybrid cross b. *AaBb* × *AaBb*
   ____ homozygous condition c. *Aa*
   ____ heterozygous condition d. *Aa* × *Aa*

11. Match the chromosome terms appropriately.
   ____ crossing over a. number and defining features of an individual's metaphase chromosomes
   ____ deletion b. segment of a chromosome moves to a nonhomologous chromosome
   ____ nondisjunction c. disrupts gene linkages
   ____ translocation d. can result in gametes with wrong chromosome number
   ____ karyotype e. a chromosome segment lost
   ____ linkage group f. all genes on a chromosome

Additional questions are available on **Biology Now™**

Figure 8.39 Two albino organisms.

Genetics Problems

Answers in Appendix II

1. A certain recessive allele *c* is responsible for *albinism*, an inability to produce or deposit melanin, a brownish-black pigment, in body tissues. Humans and a number of other organisms can have this phenotype. Figure 8.39 shows two stunning examples. In each of the following cases of albinism, what are the possible genotypes of the father, the mother, and their children?
   a. Both parents have normal phenotypes; some of their children are albino and others are unaffected.
   b. Both parents are albino and have albino children.
   c. The woman is unaffected, the man is albino, and they have one albino child and three unaffected children.

2. One gene has alleles *A* and *a*. Another has alleles *B* and *b*. For each genotype, what type(s) of gametes will form? Assume that independent assortment occurs.
   a. *AABB* c. *Aabb*
   b. *AaBB* d. *AaBb*

3. Refer to Problem 2. What will be the genotypes of offspring from the following matings? Indicate the frequencies of each genotype among them.
   a. *AABB* × *aaBB* c. *AaBb* × *aabb*
   b. *AaBB* × *AABb* d. *AaBb* × *AaBb*

4. Certain dominant alleles are so essential for normal development that an individual who is homozygous recessive for a mutant recessive form cannot survive. Such recessive, *lethal alleles* can be perpetuated in the population by heterozygotes.

Figure 8.40 Manx cat, a breed that has no tail.

Consider the Manx allele (M^L) in cats. Homozygous cats (M^LM^L) die when they are still embryos inside the mother cat. In heterozygotes (M^LM), the spine develops abnormally. The cats end up with no tail (Figure 8.40).

Two M^LM cats mate. What is the probability that any one of their *surviving* kittens will be heterozygous?

5. Human females are XX and males are XY.
   a. Does a male inherit the X from his mother or father?
   b. With respect to X-linked alleles, how many different types of gametes can a male produce?
   c. If a female is homozygous for an X-linked allele, how many types of gametes can she produce with respect to that allele?
   d. If a female is heterozygous for an X-linked allele, how many types of gametes might she produce with respect to that allele?

6. A mutant allele that causes Marfan syndrome follows a pattern of autosomal dominant inheritance. What is the chance that any child will inherit it if one parent does not carry the allele and the other is heterozygous for it?

7. A dominant allele *W* confers black fur on guinea pigs. A guinea pig that is homozygous recessive (*ww*) has white fur. Fred would like to know whether his pet black-furred guinea pig is homozygous dominant (*WW*) or heterozygous (*Ww*). How might he determine his pet's genotype?

8. The somatic cells of individuals with Down syndrome usually have an extra chromosome 21; they contain forty-seven chromosomes.
   a. At which stages of meiosis I and II could a mistake alter the chromosome number?
   b. A few individuals have forty-six chromosomes, including two normal-appearing chromosomes 21 and a longer-than-normal chromosome 14. Explain how this chromosome abnormality may have arisen.

9. In the human population, mutation of two different genes on the X chromosome causes two types of X-linked hemophilia (types A and B). In a few known cases, a woman is heterozygous for both mutant alleles (one on each of her two X chromosomes). All of her sons should have either hemophilia A or B.

However, on very rare occasions, such a woman gives birth to a son who does not have hemophilia, and his one X chromosome does not have either mutant allele. Explain how such an X chromosome could arise.

9 DNA STRUCTURE AND FUNCTION

IMPACTS, ISSUES

Here, Kitty, Kitty, Kitty, Kitty, Kitty

In 1997, geneticist Ian Wilmut made headlines when he created an exact genetic copy—a **clone**—of a fully grown animal. His team used a technique called nuclear transfer. They slipped a nucleus from a sheep udder cell into an unfertilized egg that had its own nucleus gently sucked out beforehand. The hybrid egg developed into an embryo, then into a lamb, which they named Dolly.

Dolly was the first of a series of cloned animals. Researchers all over the world since have cloned cows, horses, rabbits, pigs, mules, mice, and cats, but the process remains far from routine. Not many clonings end successfully. Most nuclear transfer attempts fail to produce an embryo, and embryos that do form tend to die before birth or shortly after. Researchers made 692 nuclear transfer attempts to get one cloned guar, an endangered wild ox. The calf promptly died.

Nuclear transfer clones that survive are prone to health or development problems. Many age quickly or become unusually overweight as they age. By age five, Dolly was just like a ten-year-old sheep, fat and arthritic. Other clones are too large from birth or have enlarged organs. Cloned mice develop lung and liver problems, and almost all die prematurely. Cloned pigs have heart problems, they limp, and one never did develop a tail or, worse still, an anus.

Structural and functional problems occur in animal clones because most of the DNA in an adult cell nucleus is inactive. It has to be reprogrammed to function like the DNA of an egg, switched back on. This sounds easier than it is; most animal cloning attempts still rely heavily on trial-and-error methods.

Why do scientists keep trying to clone animals? The potential benefits are enormous. Animals with certain desirable traits could be mass-produced. Replacement cells or organs may soon cure people with degenerative diseases that are now incurable. Endangered animals may be saved from extinction; extinct animals may be brought back to life. Commercial pet cloning may be the most financially viable practical application of cloning technology yet. Several companies now are banking pet DNA for future cloning; some are making cloned cats, like the kitten in the filmstrip, for paying pet owners.

Controversies over animal cloning continue to rage. Perfecting the methods to make healthy animal clones brings us closer to cloning humans, both technically and ethically. Think about it. If cloning a cat for a grieving pet owner is acceptable, would it be also acceptable to clone a lost child for a grieving parent?

Think about these issues as you read through the rest of the chapters in this unit of the book. You will learn how cells replicate and repair their DNA, how DNA is converted to physical parts of a cell, and how researchers use these events to make clones.

How Would You Vote? *Sickly or deformed clones are considered unfortunate but acceptable casualties of animal cloning research, but the technique also may soon provide benefits to human patients. Should animal cloning be banned? See BiologyNow for details, then vote online.*

Key Concepts

THE DNA CHASE
DNA carries instructions for making all of the structural and functional materials that cells—and bodies—use in order to function. The simplicity of its structure enables DNA to encode virtually unlimited amounts of this genetic information.

HOW DNA IS REPLICATED
Before a cell divides, enzymes and other proteins make a copy of, or replicate, the cell's DNA. During replication, double-stranded DNA molecules unwind. Each strand serves as a guide for assembling a new strand.

CLONING
What does it mean when we say that DNA holds heritable information? The answer hits home when the messages encoded in its base sequences are tapped to make an exact copy of an adult animal.

Links to Earlier Concepts

This chapter builds on your understanding of hydrogen bonding (Section 2.2), condensation reactions (2.5) and DNA structure and function (2.10). Reminding yourself about the alpha helix structure of proteins (2.8) will help you visualize the helical structure of DNA.

What you learned about chromosomes, mitosis, and meiosis (7.1, 7.3, 7.5) will let you put the molecular basis of DNA replication into a larger perspective.

Your knowledge of the cell cycle and controls over it (7.2, 7.8) will help you understand some of the technical difficulties inherent in cloning procedures.

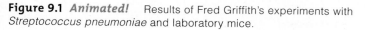

LINK TO SECTION 2.10

9.1 The Hunt for Fame, Fortune, and DNA

With this chapter, we turn to investigations that led to our understanding of DNA. That story is more than a march through details of its structure and function. It also reveals the scientific discovery process.

EARLY AND PUZZLING CLUES

In the 1800s, Johann Miescher, a physician, wanted to identify the composition of the nucleus. He collected white blood cells from pus-filled bandages and sperm from fish. These cells do not contain much cytoplasm, so it is easy to isolate the contents of their nuclei. He described the substance he found as large and acidic, containing nitrogen and phosphorus. This substance would later be called **deoxyribonucleic acid**, or **DNA**.

Now fast-forward to 1928. An army medical officer, Frederick Griffith, was trying to make a vaccine for pneumonia. He did not succeed, but he did isolate two forms of *Streptococcus pneumoniae*, a bacterium that can cause pneumonia. Colonies of one form had a rough appearance; colonies of the other were smooth. Griffith called the forms *R* and *S*. He inadvertently contributed to our understanding of heredity by using the different forms in four experiments (Figure 9.1).

First, he injected mice with live *R* cells. The mice did not develop pneumonia. *The R form was harmless.*

Second, he injected other mice with live *S* cells. The mice died. Blood samples from them teemed with live *S* cells. *The S form was pathogenic; it caused the disease.*

Third, he killed *S* cells by exposing them to high temperature. *Mice injected with dead S cells did not die.*

Fourth, he mixed live *R* cells with heat-killed *S* cells and injected them into mice. The mice died—and blood samples drawn from them teemed with live S cells!

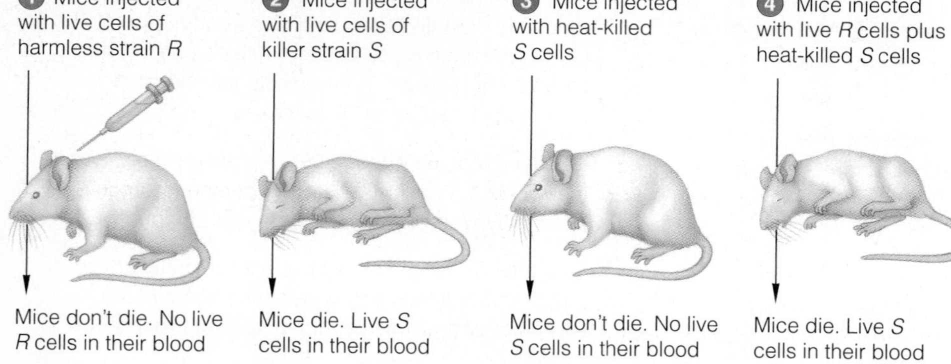

1 Mice injected with live cells of harmless strain R
2 Mice injected with live cells of killer strain S
3 Mice injected with heat-killed S cells
4 Mice injected with live R cells plus heat-killed S cells
Mice don't die. No live R cells in their blood
Mice die. Live S cells in their blood
Mice don't die. No live S cells in their blood
Mice die. Live S cells in their blood

Figure 9.1 ***Animated!*** Results of Fred Griffith's experiments with *Streptococcus pneumoniae* and laboratory mice.

What went on in the fourth experiment? Maybe heat-killed *S* cells in the mix were not really dead. But if that were so, then mice injected with heat-killed *S* cells alone (in experiment 3) would have died. Maybe the harmless *R* cells had mutated into a killer form. But if that were so, then mice injected with the *R* cells alone (experiment 1) would have died.

The simplest explanation was that heat killed the *S* cells, but it did not destroy their hereditary material—including whatever part of it that specified "infectious." Somehow, that material had been transferred from the dead *S* cells to living *R* cells, which put it to use.

The transformation was permanent and heritable. Even after a few hundred generations, the transformed *R* cell descendants were still infectious.

What *was* the hereditary material that caused the transformation? Scientists were looking in earnest, but most were thinking "proteins." Why? The molecules of inheritance had to be just as diverse as the information they encode. Virtually unlimited mixes of twenty kinds of amino acids are possible in proteins. By comparison, other molecules seemed too simple.

Griffith's results intrigued Oswald Avery, who in 1916 began transforming the *R* strain with fluid from *S* strain cultures. Something in that fluid was hereditary material. *R* cells were still transformed by *S* cell fluid after Avery added protein-digesting enzymes to it, so protein could not be the hereditary material. However, *R* cells were *not* transformed by *S* cell fluid to which Avery had added a DNA-digesting enzyme.

DNA was looking good.

CONFIRMATION OF DNA FUNCTION

By the 1950s, researchers trying to identify hereditary material were experimenting with viruses that infect bacteria, called **bacteriophages**. Like all other viruses, these infectious particles carry hereditary information about how to build new virus particles. After viruses infect a host cell, the cell starts making virus particles instead of its own components.

As researchers knew, some bacteriophages consist only of DNA and protein. Micrographs had revealed that the bacteriophage's protein coat stays on the *outer surface* of infected cells. The viruses had to be injecting genetic material into host cells, but was that material DNA, protein, or both? Figure 9.2 describes two of the landmark experiments that proved DNA is the carrier of heritable material, not protein.

Then, in 1951, a scientist named Linus Pauling and his colleagues deduced the structure of the alpha helix, a coiled motif that is part of the secondary structure of many proteins (Section 2.8). People suddenly realized that proteins had defined, organized structures.

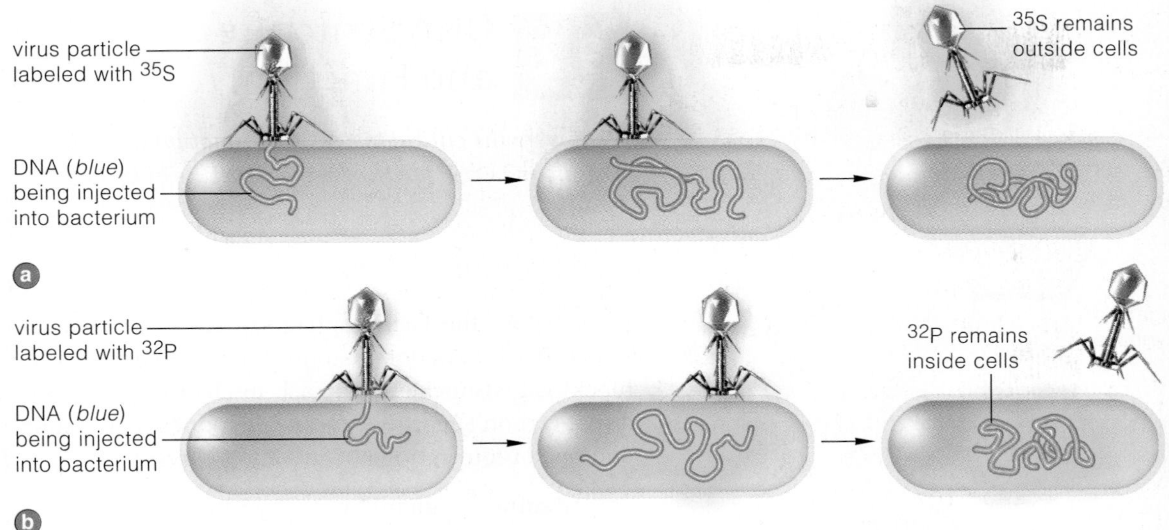

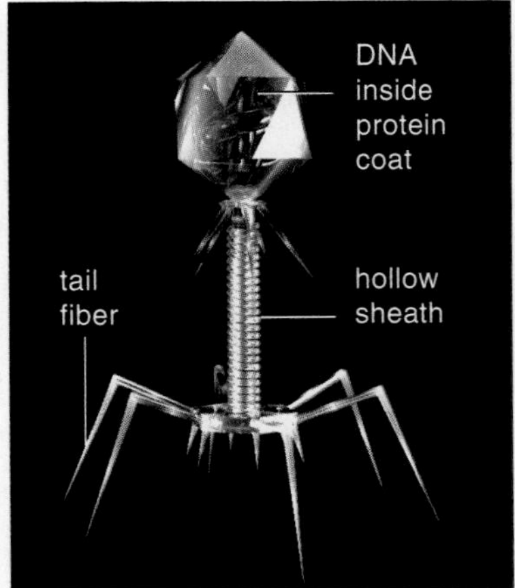

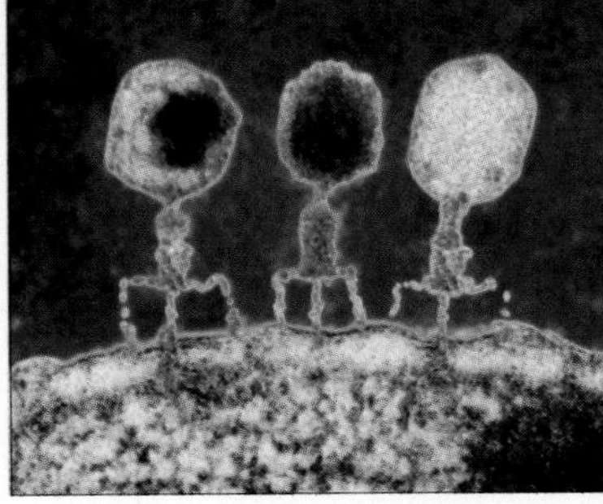

micrograph of virus particles injecting DNA into an *E. coli* cell

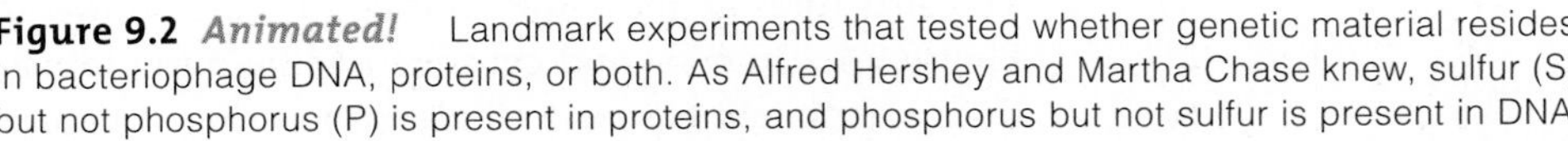

Figure 9.2 ***Animated!*** Landmark experiments that tested whether genetic material resides in bacteriophage DNA, proteins, or both. As Alfred Hershey and Martha Chase knew, sulfur (S) but not phosphorus (P) is present in proteins, and phosphorus but not sulfur is present in DNA.

(**a**) In one experiment, bacteria were infected with virus particles labeled with a radioisotope of sulfur (^{35}S). The sulfur had labeled the viral proteins. The mixture was whirled in a kitchen blender, which dislodged the viruses from the bacterial cells. Radioactive sulfur was detected mainly in the solution, not inside the cells. The viruses had not injected protein into the bacteria.

(**b**) In another experiment, bacteria were infected with virus particles labeled with a radioisotope of phosphorus (^{32}P). The phosphorus had labeled the viral DNA. When the viruses were dislodged from the bacteria, the radioactive phosphorus was detected mainly inside the bacterial cells. The viruses had injected DNA into the cells—evidence that DNA is the genetic material of this virus.

The discovery was electrifying. If someone could pry open the secrets of proteins, then why not DNA? And if DNA's structural details could be worked out, then wouldn't they be clues to its functions? *Someone would go down in history as having dissected the secret of life!*

ENTER WATSON AND CRICK

Having a shot at fame and fortune quickens the pulse of men and women in any profession, and scientists are no exception. However, science is a community effort. Individuals share not only what they find but also what they do not understand. Even if an experiment does not yield an expected result, it may turn up something that others can use or raise questions others can answer. Unexpected results, too, may be clues to secrets of the natural world.

And so scientists all over the world started sifting through results and scrambling after the prize. Among them were James Watson, a postdoctoral student from Indiana University, and Francis Crick, a researcher at Cambridge University. They spent hours arguing over everything they had read about DNA's size, shape, and bonding requirements. They fiddled with cardboard cutouts. They badgered chemists into helping them to identify bonds they might have overlooked. They built models from bits of metal connected by suitably angled "bonds" of wire.

In 1953, they built a model that fit all the pertinent biochemical rules and insights they had gleaned from other sources (Figure 9.3). They discovered the double-helix structure of the DNA molecule. As you will see in the next section, the striking simplicity of this structure is how life shows such unity at the level of molecules, and such diversity at the level of organisms.

After many decades of ingenious experiments, we now know that the information required to pass parental traits to offspring is encoded in the structure of DNA.

Figure 9.3 In 1953, James Watson and Francis Crick with their newly unveiled model of the DNA molecule.

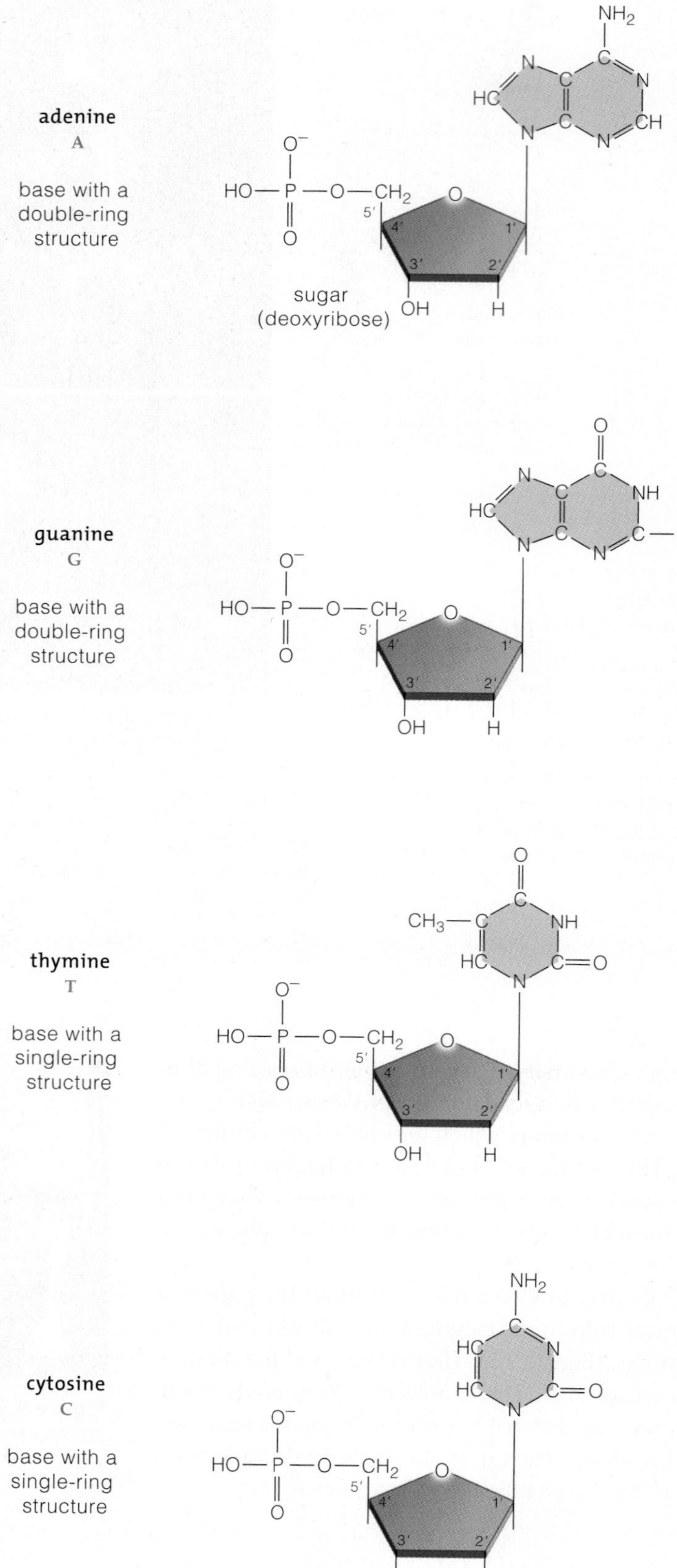

Figure 9.4 All chromosomes in a cell contain DNA. What does DNA contain? Four kinds of nucleotides: A, G, T, and C. Each nucleotide is named after the base structure it carries.

9.2 DNA Structure and Function

What were the critical pieces of information that led Watson and Crick to discover the structure of DNA?

DNA'S BUILDING BLOCKS

Long before the bacteriophage research, biochemists knew that DNA contains only four kinds of building blocks called nucleotides. Each **nucleotide** consists of a five-carbon sugar (deoxyribose), a phosphate group, and one of four nitrogen-containing bases:

adenine	**guanine**	**thymine**	**cytosine**
A	G	T	C

The structures of these four nucleotides are shown in Figure 9.4. Thymine and cytosine are the pyrimidines; their component base is a single carbon ring structure. Adenine and guanine are purines, which have larger, bulkier bases with two carbon rings.

By 1949, the biochemist Erwin Chargaff had shared with the scientific community two crucial insights into the composition of DNA. First, the amount of adenine relative to guanine differs from one species to the next. Second, the amounts of thymine and adenine in a DNA molecule are exactly the same, and so are the amounts of cytosine and guanine. We may show this as:

$$A = T \quad \text{and} \quad G = C$$

The symmetrical proportions had to mean something important. In fact, it was a tantalizing clue to how the nucleotides are arranged in a DNA molecule.

FAME AND GLORY

In 1951, Rosalind Franklin came to King's College in London to set up an x-ray crystallography laboratory. In this technique, x-rays are directed at a crystalline sample of molecules. Atoms in the molecules scatter the x-rays in a pattern that can be captured on film as an **x-ray diffraction image**. An x-ray diffraction image consists only of dots and streaks; it is not a picture of a molecule. But researchers can use the image to find clues about a molecule's structure, such as its size and shape, and the spacing between repeating patterns.

Many scientists tried to use x-ray crystallography to solve the structure of DNA, but DNA is a very large molecule, difficult to crystallize, and its shape changes depending on how it is prepared. No one yet had been able to make a clear x-ray diffraction image of DNA. Franklin was an expert at preparing samples for x-ray crystallography because of her work on the structure of coal and other forms of carbon. Coal is complex and unorganized (as are also large biological molecules like DNA). Franklin had devised a new mathematical way to interpret x-ray diffraction images, and, like Pauling, had built three-dimensional molecular models.

Franklin's assignment at King's College: investigate the structure of DNA. No one told her Maurice Wilkins was already working on that same problem, just down the hall. Even the graduate student who was assigned to assist her failed to mention it. Wilkins did not know what Franklin's assignment was; he assumed she was a technician hired to do his x-ray crystallography work because he didn't know how to do it himself. And so a clash began. To Franklin, Wilkins seemed inexplicably irritable. To Wilkins, Franklin displayed an appalling lack of the deference that technicians of the era usually showed researchers.

Wilkins had a prized cache of DNA which he gave to his "technician" for x-ray crystallography studies. Franklin's meticulous preparation of that sample gave her the first clear x-ray diffraction image of DNA as it exists inside cells (Figure 9.5). Franklin gave a research presentation on what she had learned from it in 1952. DNA, she said, had two chains twisted into a double helix, with phosphate groups forming the backbone on the outside of the helix and the bases arranged in some as yet unknown way on the inside. She presented her calculations of DNA's diameter, the distance between the two chains, the pitch (angle) of the helix, and the distance between bases. She had calculated that each turn of the helix contains ten bases (Figure 9.6).

With his crystallography background, Crick would have recognized the significance of her image and her calculations—*if* he had been there. (A pair of chains in opposing directions would be the same even if flipped 180°.) Watson was in the audience but did not realize the full implications of her data.

Franklin began preparing a report on her findings. Meanwhile, Crick read a copy of her presentation, and Wilkins reviewed Franklin's best diffraction image with Watson. Watson and Crick now had all the information

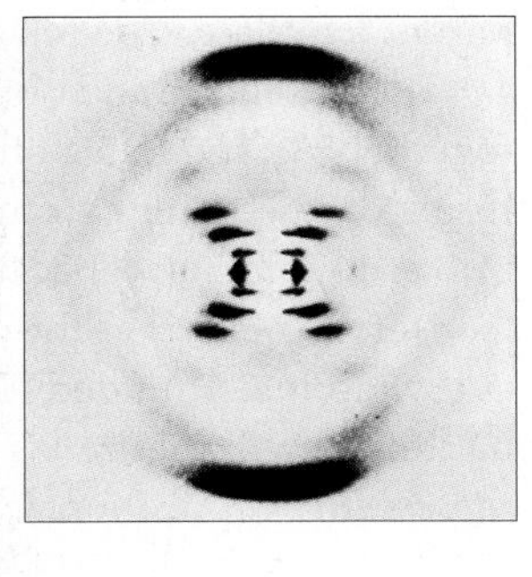

Figure 9.5 Rosalind Franklin and her famous x-ray diffraction image of DNA. Franklin died at 37, of ovarian cancer that likely resulted from her intensive work with x-rays. Because the Nobel Prize is not given posthumously, she did not share in the 1962 honor that went to Watson, Crick, and Wilkins for discovery of the structure of DNA.

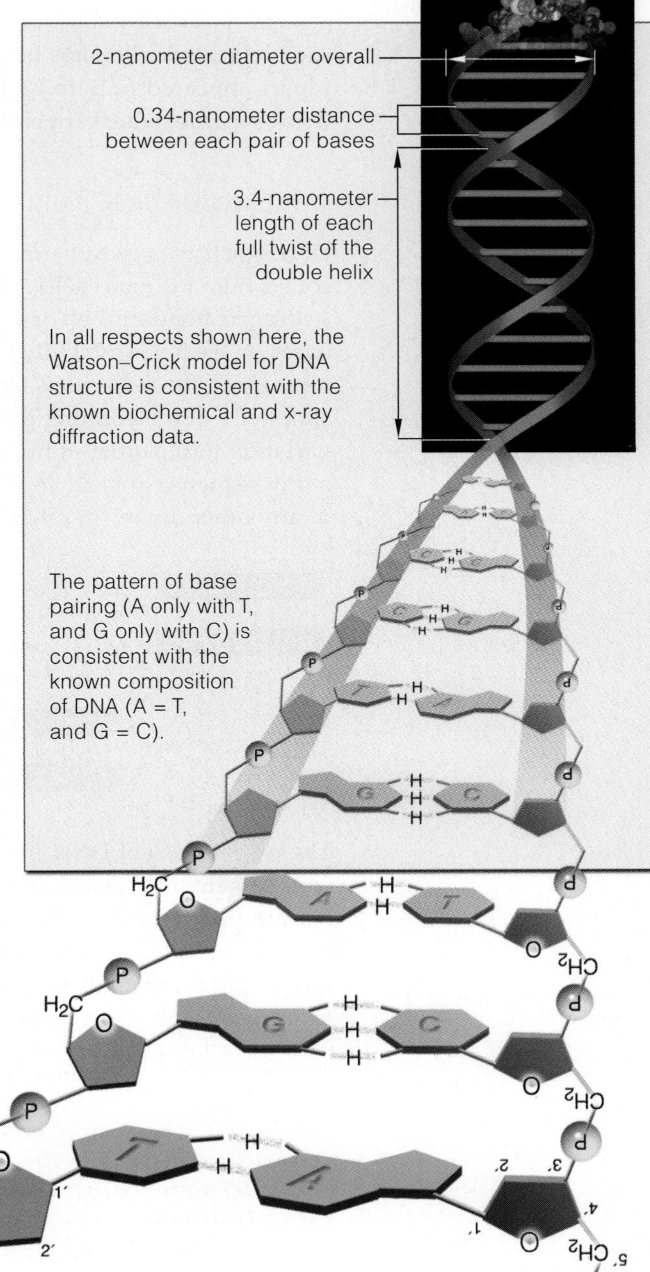

2-nanometer diameter overall
0.34-nanometer distance between each pair of bases
3.4-nanometer length of each full twist of the double helix
In all respects shown here, the Watson–Crick model for DNA structure is consistent with the known biochemical and x-ray diffraction data.
The pattern of base pairing (A only with T, and G only with C) is consistent with the known composition of DNA (A = T, and G = C).

Figure 9.6 ***Animated!*** Composite of three different models for the DNA double helix. Two sugar–phosphate backbones run in opposite directions. Think of one strand as being upside down from the other.

LINKS TO SECTIONS 2.2, 2.5, 7.3, 7.8, 8.8, 8.9

they needed to build their model of the DNA helix—one with two sugar–phosphate external chains running in opposite directions and paired bases inside.

Pauling, who had earlier discovered the alpha helix of proteins, already had been thinking "helix." So had everyone else—including Wilkins, Watson, and Crick. Later, Watson wrote, "We thought, why not try it on DNA? We were worried that *Pauling* would say, why not try it on DNA? Certainly he was a clever man. He was a hero of mine. But we beat him at his own game. I still can't figure out why."

Pauling, like most other people, had thought the nucleotide bases had to be on the outside of the helix, so they would be accessible to the enzymes that copy DNA. Franklin's data showed otherwise.

Franklin's report on her findings appeared third in a series of articles about the structure of DNA in the same magazine. Perhaps because of its placement, her report appeared only to be in support of Watson and Crick's article, which appeared first in the series.

PATTERNS OF BASE PAIRING

DNA consists of two chains, or strands, of nucleotides, coiled into a double helix. The helix is maintained by hydrogen bonding between the internally positioned bases on each strand. Only two kinds of base pairings form along the molecule's length: A—T and G—C. The simplicity of the bonding pattern belies almost infinite variation in the order of nucleotide bases. For instance, a tiny sequence of bases in DNA from a rose, a human, or any other organism might be:

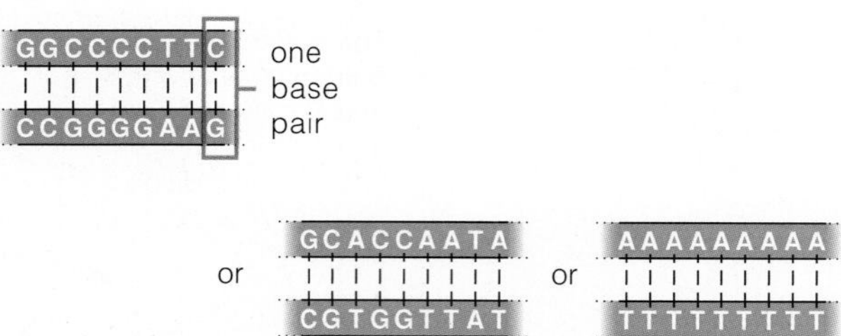

The two strands of DNA match up; they are said to be complementary to each other. Although the molecular bonding pattern (A to T, and G to C) is the same in every molecule of DNA, the actual sequence of bases differs for every species. *This molecular variation is the foundation for the diversity of life.*

The pattern of base pairing between the two strands in DNA is constant for all species—A with T, and G with C. However, each species has a number of unique sequences of base pairs along the length of their DNA molecules.

9.3 DNA Replication and Repair

The discovery of DNA structure was a turning point in studies of inheritance. Suddenly it became clear how cells duplicate their DNA before they divide.

HOW IS A DNA MOLECULE DUPLICATED?

Until Watson and Crick presented their model, no one could explain **DNA replication**, or how the molecule of inheritance is duplicated before a cell divides. This is how it goes.

DNA replication uses a team of molecular workers. Replication enzymes become active along the entire length of the DNA molecule at the same time. First, enzymes pry apart the hydrogen bonds between the two nucleotide strands of a DNA molecule in several places along its length. The double helix opens up and unwinds in both directions from the unzipped sites, exposing stretches of nucleotide bases.

Then, a new strand of DNA is assembled on each parent strand according to the base-pairing rule (A to T, and G to C). **DNA polymerases** catalyze formation of two new DNA strands from free nucleotides, using the two parent strands as templates. Cells have stores of free nucleotides that can pair up with the exposed bases. The free nucleotides themselves provide energy for strand assembly. DNA polymerase splits off two of the three phosphate groups on each, and this releases energy that drives the attachments.

Last, **DNA ligases** seal any tiny gaps in the new DNA. One continuous, complementary strand of DNA winds up with the parent strand that was its template. The other new DNA strand winds up with the other parent template. The result is two DNA helixes.

The parent strands of DNA remain intact, but not together. As soon as a stretch of a new partner strand forms on the parent strand, the two twist up together into a double helix. Because the parent DNA is saved (conserved) during the replication process, half of every double-stranded DNA molecule is "old" and half is "new" (Figure 9.7). The process is called *semi-conservative* replication.

MONITORING AND FIXING DNA

Errors occasionally are introduced into a new DNA molecule so that it acquires extra bases, loses bases, or no longer is complementary to the parent molecule. Some replication errors occur after the parent DNA is damaged by radiation or chemicals, but most arise simply because DNA polymerases make mistakes.

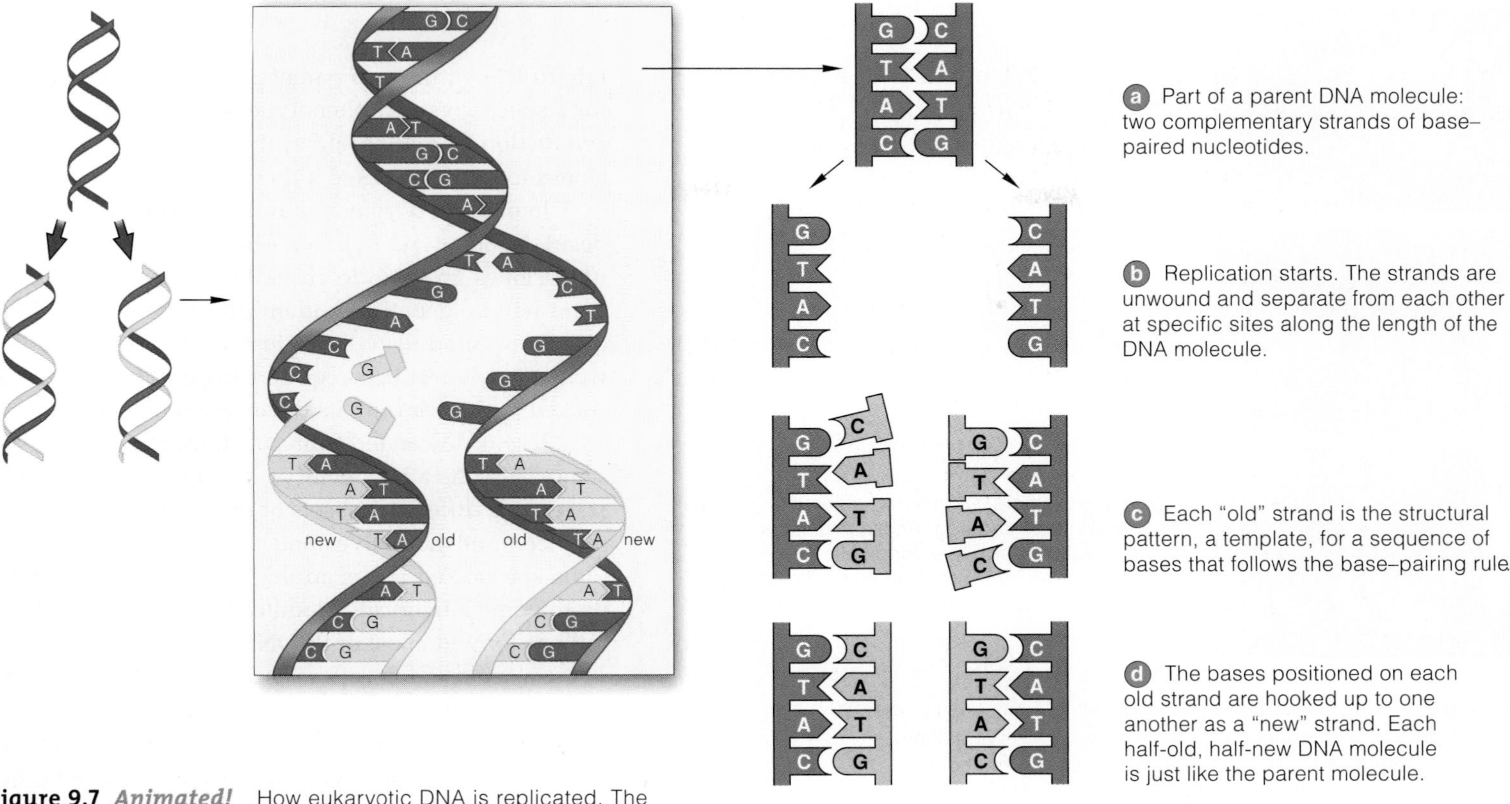

new
old
old
new
a Part of a parent DNA molecule: two complementary strands of base–paired nucleotides.
b Replication starts. The strands are unwound and separate from each other at specific sites along the length of the DNA molecule.
c Each "old" strand is the structural pattern, a template, for a sequence of bases that follows the base–pairing rule.
d The bases positioned on each old strand are hooked up to one another as a "new" strand. Each half-old, half-new DNA molecule is just like the parent molecule.

Figure 9.7 ***Animated!*** How eukaryotic DNA is replicated. The original two-stranded DNA molecule is *blue*. Each parent strand remains intact. A new strand (*yellow*) is assembled on each one. Hence the name, semiconservative replication.

Luckily, DNA proofreading mechanisms fix most errors as they occur. DNA polymerases themselves proofread new base pairings and correct mismatches. Other enzymes recognize and repair damaged DNA. If an error is not fixed, replication may be arrested, and the cell cycle stops. On the rare occasion that all of these processes fail, the cell acquires a mutation—a permanent alteration in its DNA sequence.

A mutation may result in the altered or diminished function of a somatic cell, and so disrupt how all of its progeny operate. This is how cancer begins. Mutations in germ cell DNA can give rise to genetic disorders in offspring. However, mutations can benefit a species: they introduce variation in traits that serve as the raw material of evolution.

DNA is replicated prior to cell division. Enzymes unwind its two strands. Each strand remains intact throughout the process—it is conserved—and enzymes assemble a new, complementary strand on each one.

DNA proofreading and repair systems fix nearly all errors and breaks, and thus help maintain the integrity of a cell's genetic information. Unrepaired errors are mutations.

9.4 Using DNA To Clone Mammals

LINKS TO SECTIONS 7.2, 7.8

Here we return to a topic introduced at the start of this chapter. Researchers can now use DNA from an adult mammal to bypass sexual reproduction, which gives rise to mixes of traits from two parents. Adult DNA cloning produces an exact genetic copy of a single adult.

"Cloning" can be a confusing word. It is a laboratory method of making multiple copies of DNA fragments. It also applies to natural and manipulated interventions in reproduction or development.

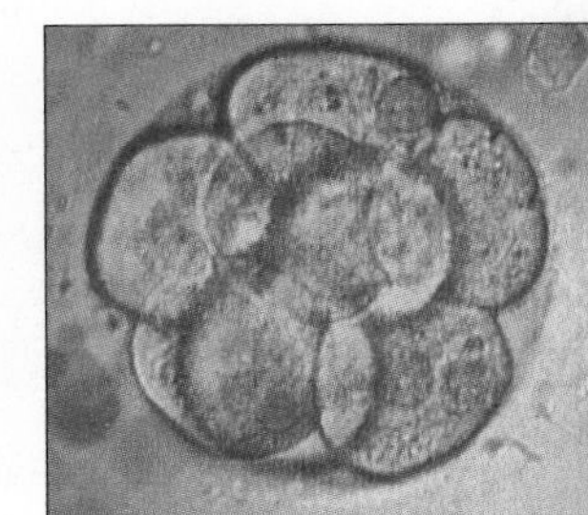

Embryo cloning occurs all the time in nature. For instance, soon after fertilized human eggs start dividing, a tiny ball of cells forms. Once in every seventy-five or so pregnancies, the ball splits, and the two parts grow and develop into identical twins. A laboratory method, "artificial twinning," simulates what happens in nature. For example, balls of cells grown from fertilized cattle eggs in petri dishes can be divided into identical-twin embryos. These are implanted into surrogate mothers, which give birth to cloned calves. Embryo cloning has been practiced for decades. However, embryo clones

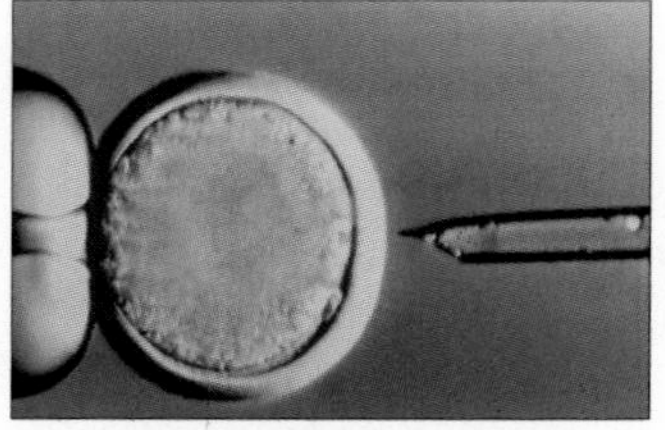

1 Removal of the nucleus from a sheep egg during a cloning. At the center is an unfertilized sheep egg. A pipette has positioned the egg, which a microneedle is about to penetrate.

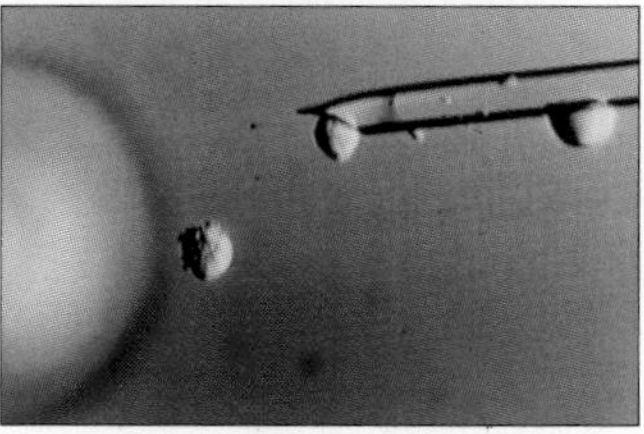

2 The microneedle has now emptied the sheep egg of its own nucleus, which holds the cell's DNA.

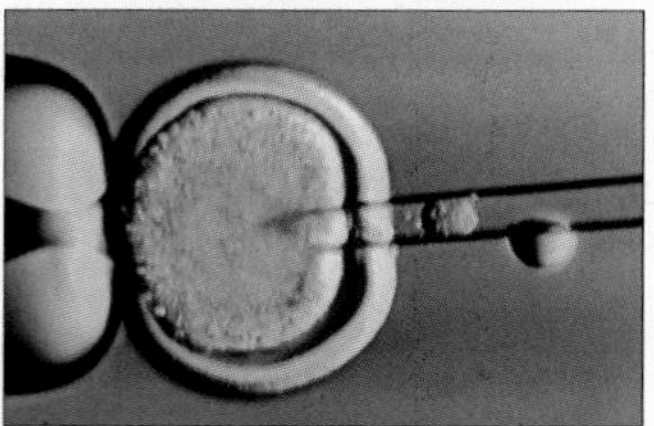

3 A nucleus from a donor cell is about to enter the enucleated egg. An electric spark will stimulate the egg to enter mitotic cell divisions. At an early stage, the ball of cells will be implanted in the womb of a surrogate mother sheep.

Figure 9.8 ***Animated!*** Nuclear transfer of sheep cells. In this series of photos, a microneedle replaces the nucleus of a sheep egg with a nucleus from an adult sheep cell. Newer methods involve direct transfer of nuclear DNA that has been treated with mitotic cell extracts to condense the chromosomes.

inherit DNA from two parents. If breeders are looking for a specific valued phenotype, such as plentiful milk production or mating vigor, they have to wait for the clones to grow up to see if they display the trait.

Cloning a cell from an adult animal that displays a desired phenotype is faster. There is no need to wait until clones grow up to check their phenotype—all of them will be genetically identical to the parent. But a cell from an adult will not just start dividing as if it were a newly fertilized egg starting out life. It must be tricked into rewinding its developmental clock.

All cells descended from a fertilized egg inherit the same DNA. As a new embryo develops, different cells start using different subsets of their DNA. As they do, they commit themselves and all of their descendants to be specialized in the adult. This is a one-way path. By the time a liver cell, blood cell, or other specialized cell has formed, most of its DNA is unused, turned off.

To clone an adult cell, scientists must first transform a specialized cell into an unspecialized cell by turning some of its unused DNA back on. Turning off parts of a cell's DNA is a normal process of development, but turning it back on is a different story. The DNA must be reprogrammed into directing the development of a complete individual.

Nuclear transfer is one way to do this. A researcher replaces the nucleus of an unfertilized egg with the nucleus or DNA from from an animal cell (Figure 9.8). The egg's cytoplasm provides the transplanted DNA with the appropriate environment to begin dividing. About 1 of 10 times, a cluster of embryonic cells forms that can be implanted into a surrogate mother. About 1 of every 10 implantations results in a live birth.

The technique is advancing rapidly. For instance, pet owners who have had their cats cloned report that their clones are healthy, lively, and uncannily like the DNA donor (Figure 9.9). Still, nuclear transfer is far from routine. For example, dogs are notoriously hard to clone, and for some as yet unknown reason cloned frogs remain tadpoles until they die.

The real issue, of course, is that humans—like cats, dogs, and sheep—are mammals. Not long ago, cloning mammals seemed fraught with technical problems. It still is, but as the techniques improve, making a genetic copy of a human no longer seems like science fiction.

We will return to the topic of manipulating DNA in Chapter 11.

Figure 9.9 Gallery of cloned cats. (**a**) Rainbow, genetic donor for cc (**b**), the first cloned cat. Their coats differ in color because cc was cloned from a single cell from Rainbow. In that cell, the black color allele was active, but not the orange.

(**c**) DNA donor Tahini, a Bengal cat, and (**d**) Tabouli and Baba Ganoush, two of her clones. Their eye color differs; a Bengal cat's eye color changes as it matures. The three cats are household pets of the founder of Genetic Savings and Clone, a pet cloning company.

With nuclear transfer cloning techniques, DNA of a cell taken from an adult donor can be reprogrammed to direct the development of a whole organism—one that is genetically identical to the donor.

Summary

Section 9.1 Experimental tests on bacteriophages and bacteria provided the first solid evidence that DNA is the hereditary material of all cells.

Biology Now
Learn about experiments that revealed the function of DNA with the animation on BiologyNow.

Section 9.2 DNA consists only of nucleotides, each having a five-carbon sugar (deoxyribose), a phosphate group, and one of four kinds of nitrogen-containing bases: adenine, thymine, guanine, or cytosine.

A DNA molecule consists of two nucleotide strands coiled together into a double helix. Bases of one strand hydrogen-bond with bases of the other.

Bases of the two DNA strands pair in a constant way. Adenine pairs with thymine (A–T), and guanine with cytosine (G–C). Which base follows another along a strand varies among species. The DNA of each species incorporates some number of unique stretches of base pairs that set it apart from the DNA of all other species.

Biology Now
Investigate the structure of DNA with the animation on BiologyNow.

Section 9.3 A cell replicates its DNA before dividing. In DNA replication, enzymes unwind a DNA molecule's double helix at several sites along its length. Then, DNA polymerases use each strand of the unwound DNA as a template to assemble new, complementary strands of DNA from free nucleotides. Two double-stranded DNA molecules result.

Proofreading mechanisms fix most base-pairing mistakes and strand breaks. These mechanisms help keep a cell's genetic information intact and accurate. Repair enzymes fix base mismatches and damaged sites in the DNA. An unrepaired error becomes a mutation.

Biology Now
See how a DNA molecule is replicated with the animation on BiologyNow.

Section 9.4 Embryo splitting and nuclear transfers are two methods that produce clones, or individuals that are genetically identical. Cloning adult animals involves reprogramming a cell's DNA to direct the development of a new individual.

Self-Quiz

Answers in Appendix I

1. Which is *not* a nucleotide base in DNA?
   a. adenine c. uracil e. cytosine
   b. guanine d. thymine f. All are in DNA.
2. What are the base-pairing rules for DNA?
   a. A–G, T–C c. A–U, C–G
   b. A–C, T–G d. A–T, G–C
3. One species' DNA differs from others in its ______ .
   a. sugars c. base sequence
   b. phosphates d. all of the above
4. When DNA replication begins, ______ .
   a. the two DNA strands unwind from each other
   b. the two DNA strands condense for base transfers
   c. two DNA molecules bond
   d. old strands move to find new strands
5. DNA replication requires ______ .
   a. free nucleotides c. many enzymes
   b. new hydrogen bonds d. all of the above
6. Match the terms appropriately.
   ____ bacteriophage
   ____ clone
   ____ nucleotide
   ____ DNA ligase
   ____ DNA polymerase
   a. nitrogen-containing base bonded to a sugar and one or more phosphate groups
   b. adds nucleotides to a growing DNA strand
   c. only DNA and protein
   d. seals breaks in DNA
   e. carbon copy of donor

Additional questions are available on **Biology Now™**

Critical Thinking

1. Matthew Meselson and Frank Stahl's experiments supported the semiconservative model of replication. These researchers obtained "heavy" DNA by growing *Escherichia coli* with ^{15}N, a radioactive isotope of nitrogen. They also prepared "light" DNA by growing *E. coli* in the presence of ^{14}N, the more common isotope. An available technique helped them identify which of the replicated molecules were heavy, light, or hybrid (one heavy strand and one light). Use pencils of two colors, one for heavy strands and one for light. Starting with a DNA molecule having two heavy strands, arrange them so that they show the formation of daughter molecules after replication in a ^{14}N-containing medium. Show the four DNA molecules that would form if the daughter molecules were replicated a second time in the ^{14}N medium. Would the resulting DNA molecules be heavy, light, or mixed?

2. During DNA replication, the template DNA forms a Y-shaped *replication fork* where it is unwinding. New DNA is assembled on both strands of the template at the same time. Look back at Figures 9.6 and 9.7. DNA polymerase only synthesizes DNA in the 5′ to 3′ direction (*right*). This means nucleotides can be added to each growing strand of DNA *only on its 3′ end*. Which way do the polymerases move with respect to the fork as they are copying each of the parent strands near it? What has to happen as the DNA template unwinds more and more?

3. Mutations are permanent changes in base sequences of DNA and the original source of genetic variation—the raw material of evolution. Yet how can mutations accumulate, given that cells have repair systems that fix structurally altered or discontinuous DNA strands during replication?

4. In 1999, scientists discovered a woolly mammoth that had been frozen in glacial ice for the past 20,000 years. They thawed it very carefully so they could use its DNA to clone a woolly mammoth. It turns out there was not enough preserved material to work with. They plan to try again the next time a frozen woolly mammoth comes along. Speculate on the pros and cons, both technical and ethical, of cloning an extinct animal.

10 GENE EXPRESSION AND CONTROL

IMPACTS, ISSUES

Ricin and Your Ribosomes

Ricin is very poisonous. The castor oil plant (*Ricinus communis*) has ricin in all of its tissues, but its oil is an ingredient in many plastics, paints, cosmetics, textiles, and adhesives. The oil, and the ricin, is concentrated in the plant's seeds (beans, *below*). The ricin is discarded when the oil is extracted.

A dose of ricin as small as a grain of salt can kill you; only plutonium and botulism toxin are more deadly. When Germany unleashed mustard gas against allied troops during World War I, both the United States and England investigated whether ricin could be used as a weapon. The research was shelved when the war ended.

Fast-forward to 1969, at the height of the Cold War between Russia and the West. A Bulgarian writer, Georgi Markov, had defected to England. As he strolled down a busy London street, an assassin jammed the tip of an umbrella into Markov's leg. The tip held a tiny ball laced with ricin. Markov died in agony three days later.

Ricin is on stage once again. During the 2001 war in Afghanistan, ricin turned up in some caves hastily abandoned by terrorists. In 2003, police acting on an intelligence tip stormed a London apartment, where they collected laboratory glassware, bomb circuitry, castor oil beans and other raw materials, and recipes for producing ricin, cyanide, botulism toxin, and other natural poisons.

In 2004, traces of ricin were found in a United States Senate mailroom and State Department building, and in an envelope addressed to the White House. In 2005, a Florida man with castor oil beans and an AK-47 assault rifle was arrested. Jars of banana baby food laced with ground castor oil beans also made the news in 2005, as well as an Arizona man intending to settle a vendetta with his homemade ricin.

Ricin exerts its deadly effects by inactivating ribosomes, the cell's machinery for making proteins. Ricin itself is a protein. One of its two polypeptide chains, shown in the filmstrip, helps ricin insert itself into cells. The other chain wrecks part of the ribosome where amino acids are joined together. It pulls adenine subunits from the ribosome's protein-building RNA molecule. As the ribosome's three-dimensional shape unravels, protein synthesis stops. There is no antidote.

It is possible to go on with your life without knowing what a ribosome is or what it does. It is also possible to recognize that protein synthesis is not a topic invented to torture biology students. It is something that is worth appreciating for how it keeps us alive.

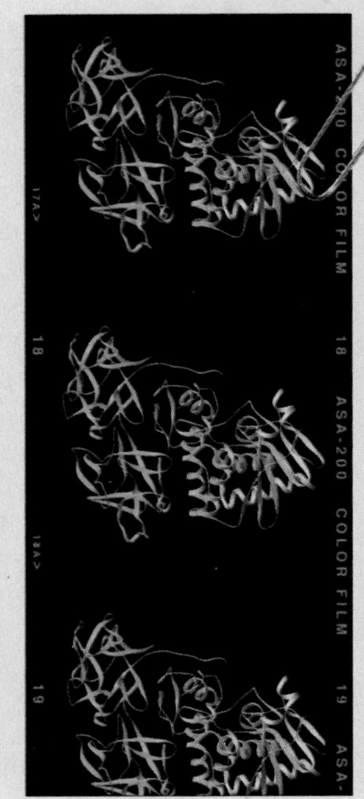

How Would You Vote? *Scientists are developing a ricin vaccine even though a large-scale terrorist attack using ricin is unlikely, because it is difficult to disperse in air. Should we carry out mass immunizations with a ricin vaccine? See BiologyNow for details, then vote online.*

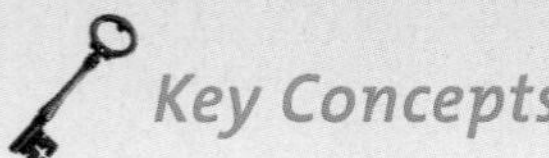

Key Concepts

PROTEIN SYNTHESIS
DNA is a cell's master copy of genetic information, archived safely in the nucleus. RNAs are disposable copies of useful parts of the DNA archive. Three kinds of RNAs participate in building proteins. mRNA carries protein-building messages from DNA, and tRNA and rRNA translate those messages into chains of amino acids.

MUTATION
A mutation, or permanent alteration in the nucleotide sequence of DNA, may change a protein's structure and function. When the protein is an essential part of cell architecture or metabolism, the outcome may be a disruption of cellular function.

GENE CONTROLS
Controls govern when, how, and to what extent each cell uses its DNA. All cells counter changes in their external environment by adjusting transcription rates. In addition, eukaryotic cells have a long-term program of development that is based on selective gene expression.

Links to Earlier Concepts

Once again you will encounter the nucleic acids DNA and RNA (Sections 2.10, 9.2). Transcription has features in common with DNA replication, so you may wish to review Section 9.3 before you start.

In this chapter, we return again to protein structure and function (2.8), this time in the context of RNA translation. Remember the impact that mutations can have on protein structure (2.9, 8.2) as you learn about the actual mechanisms of mutation.

Your knowledge of the chromosomal basis of inheritance (Section 8.4) will help you understand gene controls.

10.1 Making and Controlling the Cell's Proteins

Two cellular processes convert DNA's protein-building information into a new protein molecule. First, part of the DNA serves as a template for making a strand of RNA. Then the RNA is used to make a protein.

If DNA is the book of protein-building information in cells, its alphabet has only four letters: the nucleotides adenine, thymine, guanine, and cytosine. How does a cell make all of its proteins from this simple alphabet? The answer starts with the order, or sequence, of the four nucleotide bases in a DNA molecule.

Genetic information is coded in the sequence itself as sets of three bases. These "genetic words" occur one after the other along the length of a chromosome. Like sentences, a linear series of genetic words can form a meaningful parcel of information. The parcel is a **gene**, a nucleotide sequence that a cell can convert to an RNA (ribonucleic acid) or protein product.

Most gene products are proteins. It takes two steps, transcription and translation, to convert a gene into a protein. In the first step, **transcription**, a gene serves as a template for making RNA. You know that the DNA double helix unwinds entirely when a cell replicates its DNA. At other times, DNA unwinds just enough to expose one gene for transcription. Then, the nucleotide base sequence of the gene is copied as a strand of RNA.

In the second step of protein synthesis, **translation**, the genetic information carried by a strand of RNA is decoded, or translated, into a sequence of amino acids that is assembled into a new polypeptide chain.

In short, DNA guides the synthesis of RNA, then RNA guides the synthesis of proteins:

$$\text{DNA} \xrightarrow{\textit{transcription}} \text{RNA} \xrightarrow{\textit{translation}} \text{PROTEIN}$$

A cell can regulate protein synthesis by using various control mechanisms that affect timing and duration of transcription and translation. Some of these controls are triggered if the concentrations of nutrients or other substances change. Others are activated by signaling molecules that call for changes in cell activities. These kinds of controls are the basis for homeostasis, and also for embryonic development in eukaryotic organisms.

The nucleotide base sequence along one of DNA's two strands guides RNA synthesis. RNA guides protein synthesis.

Controls over transcription and translation govern which genes are expressed, at what times, and in what amounts.

10.2 How Is RNA Transcribed From DNA?

LINKS TO SECTIONS 2.10, 9.2

In transcription, the first step in protein synthesis, the nucleotide sequence in a DNA region becomes a template for assembling a strand of ribonucleic acid, or RNA.

THREE CLASSES OF RNA

Transcription of most genes produces **messenger RNA**, or **mRNA**—the only class of RNA that carries *protein-building* instructions. Transcription of some other genes produces **ribosomal RNA**, or **rRNA**, a component of ribosomes. A ribosome is a large molecular structure upon which polypeptide chains are built. Transcription of other genes produces **transfer RNA**, or **tRNA**, which delivers amino acids one at a time to a ribosome. All three types of RNA participate in protein synthesis.

THE NATURE OF TRANSCRIPTION

An RNA molecule is almost, but not quite, like a short single strand of DNA. Like DNA, RNA contains four types of nucleotides. Each nucleotide has a five-carbon sugar, a ribose (not DNA's deoxyribose), a phosphate group, and a base. Three of the nucleotides—adenine, cytosine, and guanine—are the same in RNA and DNA. But in RNA, the fourth nucleotide is uracil, not thymine (Figure 10.1*a,b*). Like thymine, uracil can pair with adenine. This means a new RNA strand can be built according to the same base-pairing rules as DNA (Figure 10.1*c*).

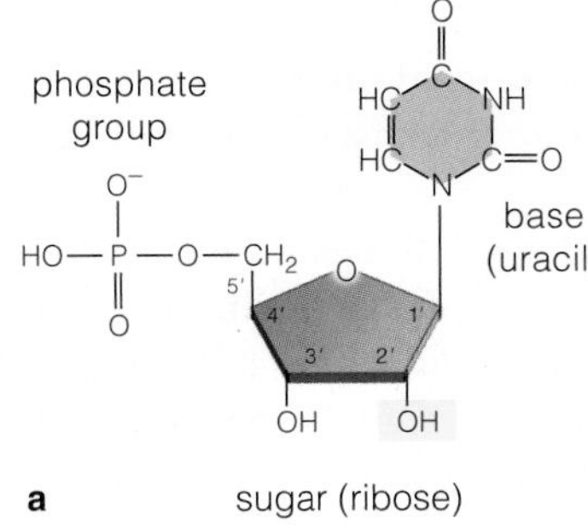

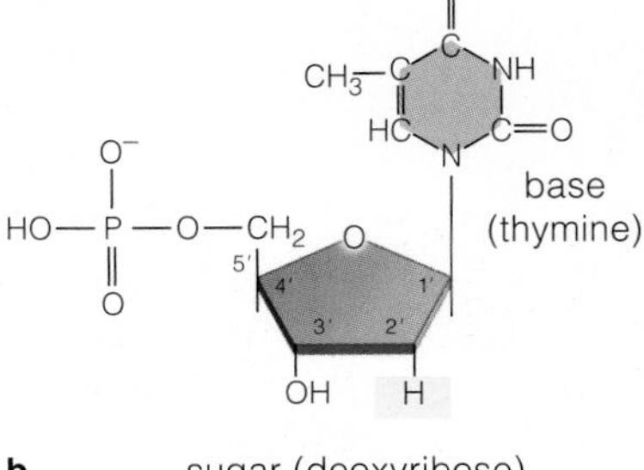

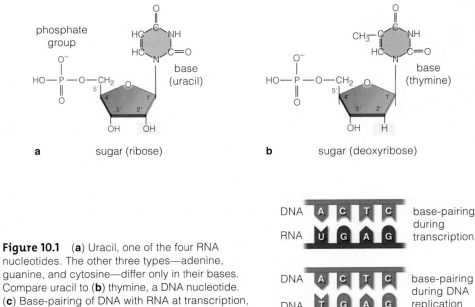

Figure 10.1 (**a**) Uracil, one of the four RNA nucleotides. The other three types—adenine, guanine, and cytosine—differ only in their bases. Compare uracil to (**b**) thymine, a DNA nucleotide. (**c**) Base-pairing of DNA with RNA at transcription, compared to base-pairing during DNA replication.

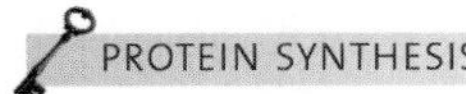

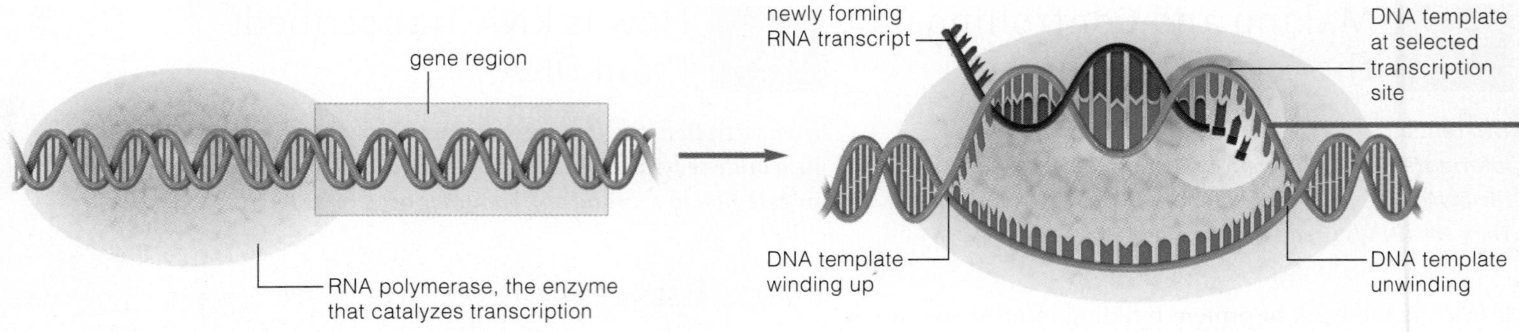

a RNA polymerase binds the DNA at a promoter near a gene. The polymerase will begin linking RNA nucleotides (adenine, cytosine, guanine, and uracil) into a strand of RNA, in the order specified by the gene.

b All through transcription, the DNA double helix becomes unwound in front of the RNA polymerase. The newly forming RNA strand briefly winds up with its DNA template. As the RNA strand lengthens, it unwinds from the DNA (and the two DNA strands wind up again).

Figure 10.2 ***Animated!*** Transcription. (**a**) RNA polymerase binds to a promoter near a gene region in the DNA. (**b**) The nucleotide base sequence on one of DNA's two strands (not both) is used as a template for assembling RNA nucleotides. (**c**) The nucleotide base sequence of the transcribed RNA is complementary to that of the DNA template. (**d**) The result of transcription is a new molecule of RNA.

LINK TO SECTION 9.3

The process of transcription *differs* from the process of DNA replication in three ways.

First, transcription produces one single-stranded molecule of RNA, not two DNA double helixes.

Second, the enzyme called **RNA polymerase**, not DNA polymerase, adds RNA nucleotides one at a time to a growing RNA molecule.

Third, a small, gene-containing part of one DNA strand is the template for transcription, not the whole DNA molecule.

A gene is at most a few thousand nucleotides long, out of millions of nucleotides on a chromosome. How does RNA polymerase know where to start? Just like a sentence, a gene contains signals that tell its readers where to begin and end. The initial capital letter of a sentence means "start," and a period at the end means "stop." The genetic signals for "start" and "stop" are special nucleotide sequences at the beginning and end of a gene. They are recognized by RNA polymerase.

Transcription is initiated when RNA polymerase binds a **promoter**, a special nucleotide sequence near a gene in the DNA. The polymerase swiftly moves along the DNA until it finds the signal for the start of the gene. Using the nucleotide sequence of the gene as a template, the polymerase joins one RNA nucleotide after another into a strand of RNA (Figure 10.2). When it arrives at the signal for the end of the gene, the new RNA is released and transcription stops.

FINISHING UP RNA TRANSCRIPTS

In eukaryotes, most new RNA transcripts are not in a final form; they must be modified before they become functional molecules. Just as a dressmaker may snip off some loose threads or add bows on a dress before it leaves the shop, so do eukaryotic cells tailor their "pre-RNA" before it leaves the nucleus. All types of RNA undergo such modification.

For example, most eukaryotic genes contain one or more **introns**, nucleotide sequences that are removed from a pre-RNA molecule. Introns intervene between **exons**, which are sequences that remain in the RNA. Introns are transcribed into RNA along with exons, but they are snipped out before the RNA leaves the nucleus in mature form. Think of it this way: *Ex*ons are *ex*ported from the nucleus, and *in*trons stay *in* the nucleus, where they are degraded.

Exons can be snipped apart and spliced together in more than one way. Alternative splicing lets cells use the same gene to make proteins that differ slightly in

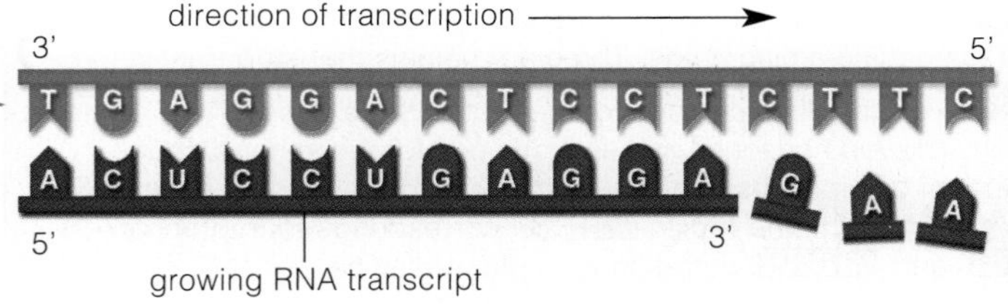

(c) RNA polymerase links RNA nucleotides, one at a time, into a strand of RNA. The order of nucleotides added to the growing strand is determined by base-pairing (U–A, G–C) with the DNA template. Many other proteins assist this process.

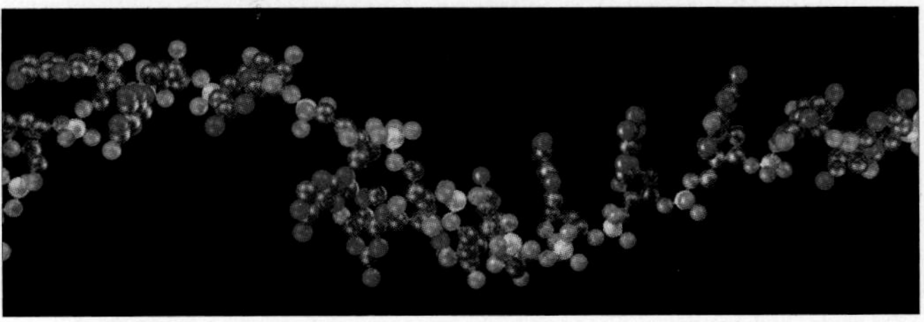

(d) At the end of the gene region, the new RNA transcript is released from its DNA template. Shown below it is a model for a transcribed strand of RNA.

form and function. Such splicing at introns is a way to maximize the information-storing capacity of a DNA molecule, hence its capacity to make diverse proteins.

Additional modification events further alter pre-mRNA, the transcripts that become mRNAs. After all the splicing events are complete, a modified guanine "cap" is attached to the start of a pre-mRNA (Figure 10.3). Later, in the cytoplasm, the cap will help bind the mRNA to a ribosome.

Then, a tail of 100 to 300 adenines is attached to the end of the pre-mRNA. Degradation enzymes begin to disassemble an mRNA immediately upon its arrival in the cytoplasm, beginning at the tail. The length of the tail paces enzyme access to the sequence part of the mRNA, and thus determines how long the mRNA will function. The tail keeps the protein-building message intact for as long as the cell needs more of that protein. Figure 10.3 summarizes mRNA processing.

In transcription, a gene sequence on one of the two strands of a DNA molecule serves as a template for RNA polymerase to assemble a single strand of RNA from free nucleotides.

In eukaryotes, each new RNA transcript, or pre-RNA, undergoes modification before leaving the nucleus.

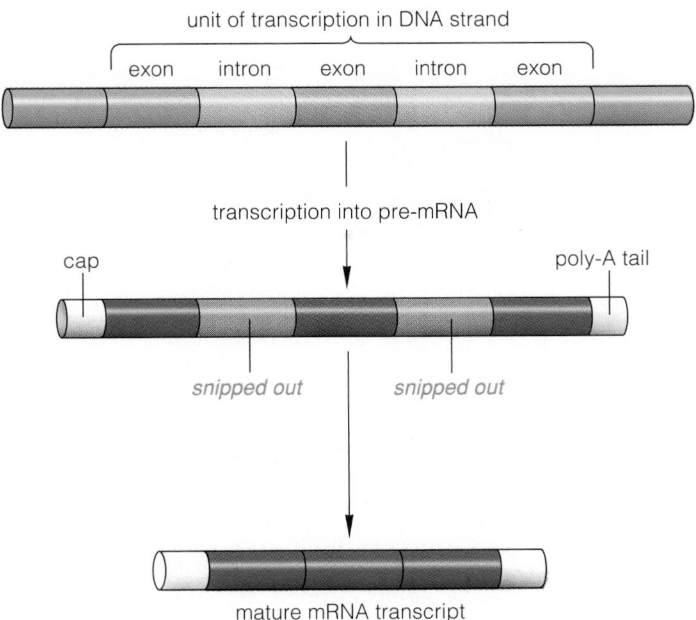

Figure 10.3 ***Animated!*** How pre-mRNA transcripts are processed into final form. Inside the nucleus of eukaryotes, some or all introns are removed, and the transcript gets a modified guanine cap and a tail. The tail is a sequence of many adenines (hence the name, poly-A tail).

first base	second base: U	C	A	G	third base
U	phenylalanine	serine	tyrosine	cysteine	U
U	phenylalanine	serine	tyrosine	cysteine	C
U	leucine	serine	STOP	STOP	A
U	leucine	serine	STOP	tryptophan	G
C	leucine	proline	histidine	arginine	U
C	leucine	proline	histidine	arginine	C
C	leucine	proline	glutamine	arginine	A
C	leucine	proline	glutamine	arginine	G
A	isoleucine	threonine	asparagine	serine	U
A	isoleucine	threonine	asparagine	serine	C
A	isoleucine	threonine	lysine	arginine	A
A	methionine (START)	threonine	lysine	arginine	G
G	valine	alanine	aspartate	glycine	U
G	valine	alanine	aspartate	glycine	C
G	valine	alanine	glutamate	glycine	A
G	valine	alanine	glutamate	glycine	G

Figure 10.4 ***Animated!*** The genetic code. Codons in mRNA are nucleotide bases "read" in blocks of three. Sixty-one of the base triplets correspond to specific amino acids. Three are signals that stop translation.

The *left* vertical column lists choices for the first of three bases in an mRNA codon. The *top* horizontal row lists choices for the second codon. The *right* vertical column lists choices for the third. To give three examples, reading from left to right, the triplet UGG corresponds to tryptophan. Both UUU and UUC correspond to phenylalanine.

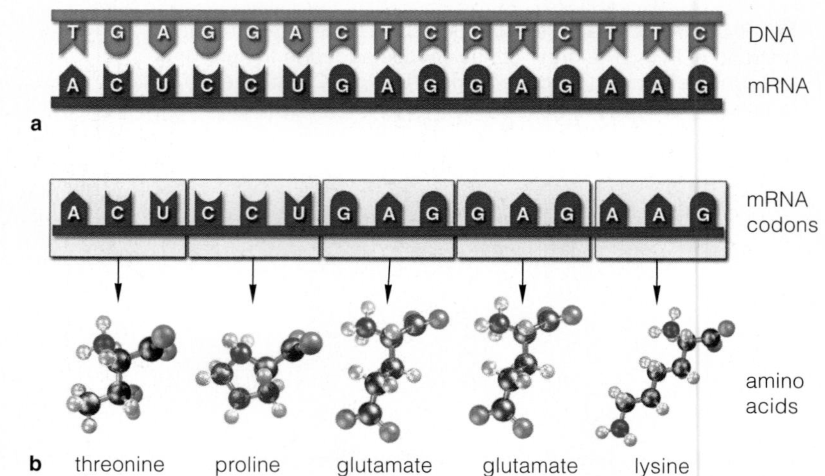

Figure 10.5 Example of the correspondence between genes and proteins. (**a**) An mRNA transcript of a gene region of DNA. Three nucleotide bases, equaling one codon, specify one amino acid. This series of codons (base triplets) specifies the sequence of amino acids shown in (**b**).

LINKS TO SECTIONS 2.8, 3.6

10.3 Deciphering mRNA

Like a strand of DNA, an mRNA transcript is a linear sequence of nucleotide bases. A protein-building message is encoded in consecutive groups of three bases.

Like this sentence, an mRNA is a linear sequence of words—genetic words spelled with a four nucleotide alphabet—that is readable if you can understand the language in which it is written. Researchers Marshall Nirenberg, Philip Leder, Severo Ochoa, and Gobind Korana learned the language of mRNA; they deduced the correspondence between genes and proteins.

An mRNA codes for a sequence of amino acids. To translate its nucleotide base sequence into an amino acid sequence, you have to know that mRNA bases are read *three at a time*, as triplets. Base triplets in mRNA were given the name **codons**. The order of the three bases in each codon determines which amino acid it specifies. There are sixty-four codons (Figure 10.4), many more than necessary to specify all twenty kinds of amino acids. Most amino acids are encoded by more than one codon. For instance, both GAA and GAG call for the amino acid glutamate.

The order of codons in an mRNA determines the order of amino acids in the corresponding polypeptide chain (Figure 10.5). Some codons signal the beginning and the end of the coding region. In most species, the first AUG in a transcript is the START signal for protein translation. AUG codes for methionine, so methionine is the first amino acid of most new polypeptide chains. UAA, UAG, and UGA do not call for an amino acid. They are STOP signals that block further additions of amino acids to a new chain.

The set of sixty-four different codons is the **genetic code**. For every *organism*, protein synthesis adheres to this code. One *organelle*, the mitochondrion of many species, makes its own proteins by using a few unique codons. These mitochondrial codons were a clue that led to the theory of how eukaryotic organelles arose

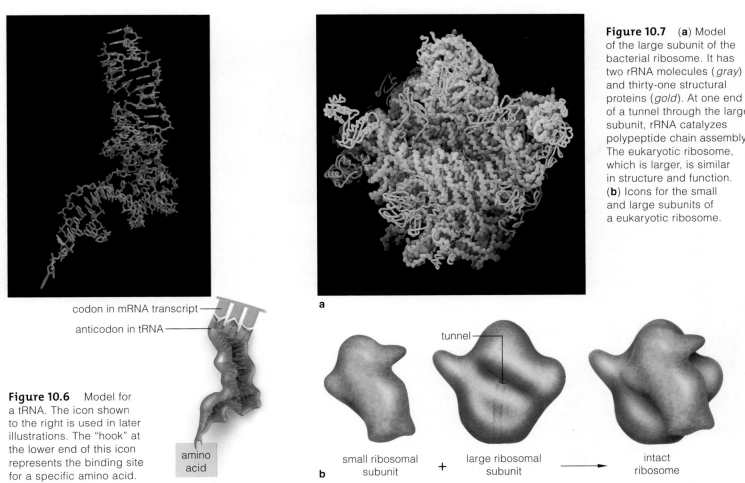

codon in mRNA transcript

anticodon in tRNA

amino acid

Figure 10.6 Model for a tRNA. The icon shown to the right is used in later illustrations. The "hook" at the lower end of this icon represents the binding site for a specific amino acid.

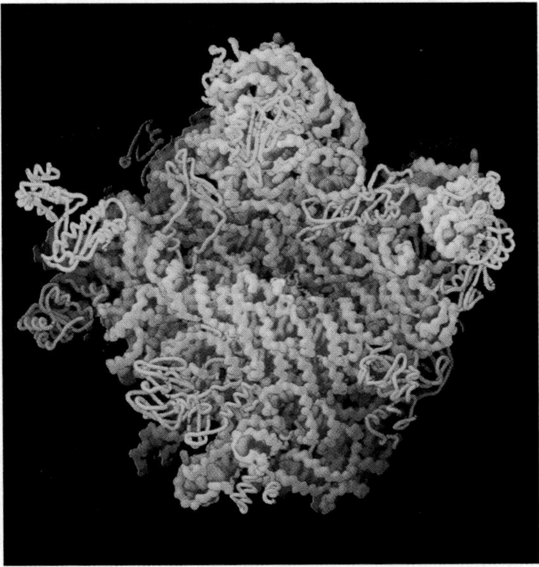

a

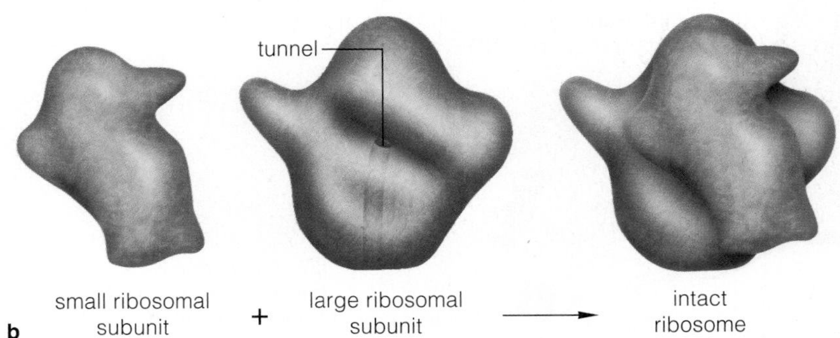

b

Figure 10.7 (**a**) Model of the large subunit of the bacterial ribosome. It has two rRNA molecules (*gray*) and thirty-one structural proteins (*gold*). At one end of a tunnel through the large subunit, rRNA catalyzes polypeptide chain assembly. The eukaryotic ribosome, which is larger, is similar in structure and function. (**b**) Icons for the small and large subunits of a eukaryotic ribosome.

by endosymbiosis. Mitochondria, remember, probably evolved from foreigners that stayed inside their host cells (Section 3.6).

A team of tRNAs and rRNAs translates the genetic code carried by an mRNA into a polypeptide chain. The cytoplasm of every cell contains free amino acids and tRNA molecules. A tRNA has an attachment site for one amino acid, and it also has an **anticodon**, or a base triplet that is complementary to an mRNA codon (Figure 10.6). tRNAs link the genetic code and amino acids. During translation, tRNA anticodons base-pair with codons in an mRNA, in order. The amino acid cargoes of the tRNAs thus become positioned in the order specified by the codons in the mRNA.

There are sixty-four codons but not as many kinds of tRNAs, so the tRNAs must be able to match up with more than one codon. According to base-pairing rules, adenine pairs with uracil, and cytosine with guanine. How can an anticodon pair with more than one codon? Base-pairing rules are not so strict for the third base in a codon–anticodon pair. For example, AUU, AUC, and AUA all specify isoleucine. All three codons can base-pair with the tRNA that carries isoleucine. Freedom in codon–anticodon pairing is known as "wobble."

Interactions between tRNAs and mRNA take place at ribosomes. Ribosomes have two different subunits (Figure 10.7). Each is built from rRNA and structural proteins in the nucleus. In the cytoplasm, a large and a small ribosomal subunit join into an intact, functional ribosome when an mRNA is to be translated.

The genetic code is a set of sixty-four codons, which are nucleotide bases in mRNA that are read in sets of three. Different amino acids are specified by different codons.

mRNA carries DNA's protein-building instructions from the nucleus into the cytoplasm.

tRNAs deliver amino acids to ribosomes. Their anticodons base-pair with codons in the order specified by mRNA.

Polypeptide chains are built on ribosomes, each consisting of a large and small subunit made of rRNA and proteins.

IMPACTS, ISSUES

In spite of its dangers, ricin shows promise as a cancer-fighting toxin. By one technique, ricin's ribosome-inactivating polypeptide chain is attached to cancer-specific antibodies or other weapons. Once inside cancer cells, ricin kills them by blocking protein synthesis. One concern: The highly toxic ricin dribbles out of dying cancer cells.

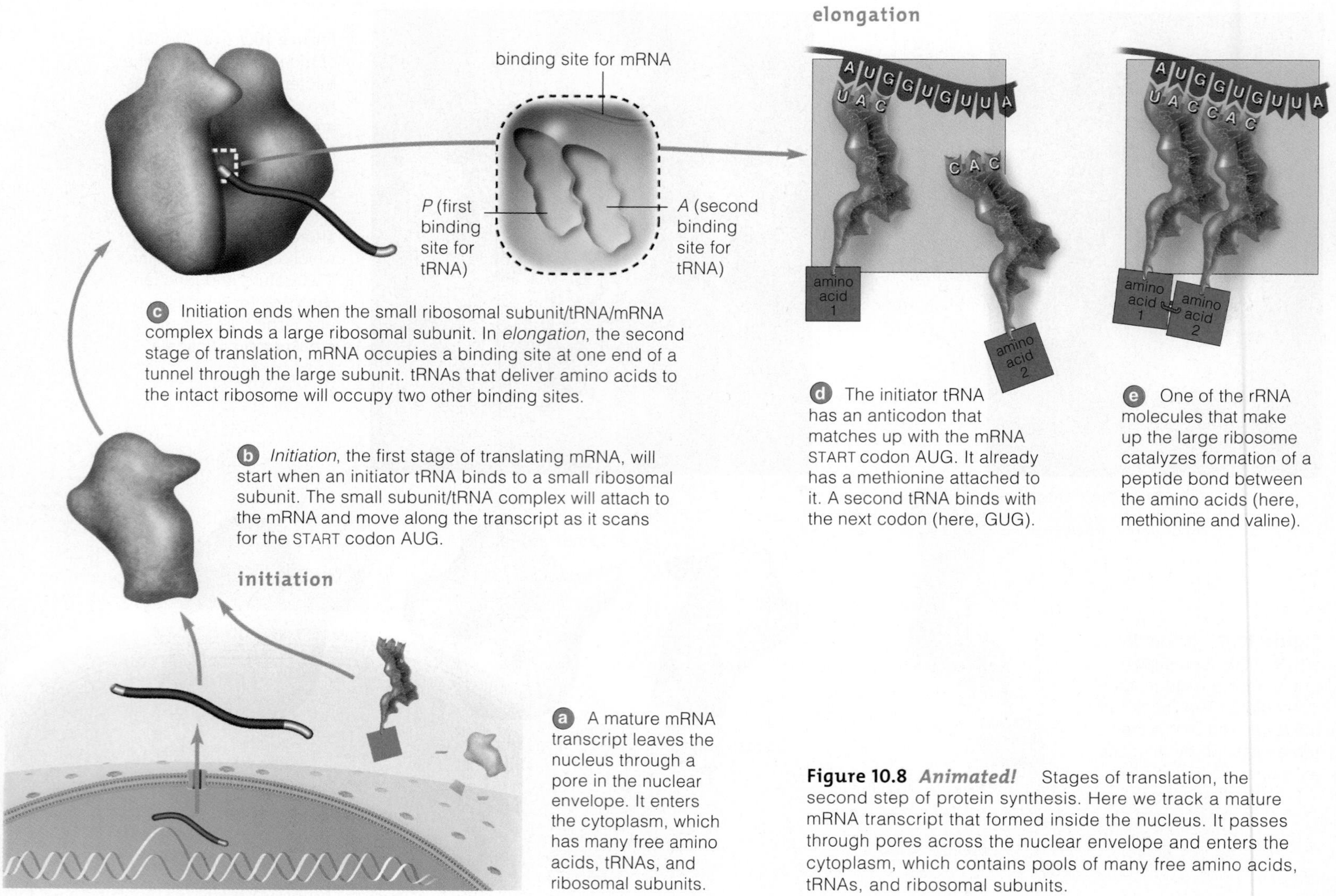

elongation

binding site for mRNA

P (first binding site for tRNA)

A (second binding site for tRNA)

amino acid 1

amino acid 2

c Initiation ends when the small ribosomal subunit/tRNA/mRNA complex binds a large ribosomal subunit. In *elongation*, the second stage of translation, mRNA occupies a binding site at one end of a tunnel through the large subunit. tRNAs that deliver amino acids to the intact ribosome will occupy two other binding sites.

d The initiator tRNA has an anticodon that matches up with the mRNA START codon AUG. It already has a methionine attached to it. A second tRNA binds with the next codon (here, GUG).

e One of the rRNA molecules that make up the large ribosome catalyzes formation of a peptide bond between the amino acids (here, methionine and valine).

b *Initiation*, the first stage of translating mRNA, will start when an initiator tRNA binds to a small ribosomal subunit. The small subunit/tRNA complex will attach to the mRNA and move along the transcript as it scans for the START codon AUG.

initiation

a A mature mRNA transcript leaves the nucleus through a pore in the nuclear envelope. It enters the cytoplasm, which has many free amino acids, tRNAs, and ribosomal subunits.

Figure 10.8 ***Animated!*** Stages of translation, the second step of protein synthesis. Here we track a mature mRNA transcript that formed inside the nucleus. It passes through pores across the nuclear envelope and enters the cytoplasm, which contains pools of many free amino acids, tRNAs, and ribosomal subunits.

LINKS TO SECTIONS 2.8, 3.5

10.4 From mRNA To Protein

DNA's hereditary information is stored intact, safely, in one place. Think of mRNAs as disposable intermediaries that deliver messages from DNA to ribosomes, which translate them into polypeptide chains.

Translation is the process of converting the genetic information carried by an mRNA into a polypeptide chain. It proceeds in the cell cytoplasm through three stages: initiation, elongation, and termination. In the first stage, *initiation*, an intact ribosome, an mRNA, and a special initiation tRNA form a cluster called an initiation complex (Figure 10.8*b*,*c*).

In the *elongation* stage of translation, a polypeptide chain is assembled as the mRNA passes between the two ribosomal subunits, like a thread moving through the eye of a needle. tRNAs bring amino acids to the ribosome, then bind the mRNA in the order specified by its codons. Part of an rRNA molecule in the large ribosomal subunit has enzymatic activity. It catalyzes the formation of peptide bonds between amino acids (Figure 10.8*d*–*f*). You already saw how peptide bonds form during protein synthesis in Figure 2.22. Peptide bonds continue to form in a series between the amino acid that was most recently attached and the next one delivered to the intact ribosome (Figure 10.8*g*,*h*).

During the last stage of translation, *termination*, the ribosome reaches the mRNA's STOP codon. No tRNA has an anticodon that corresponds to a STOP codon, so

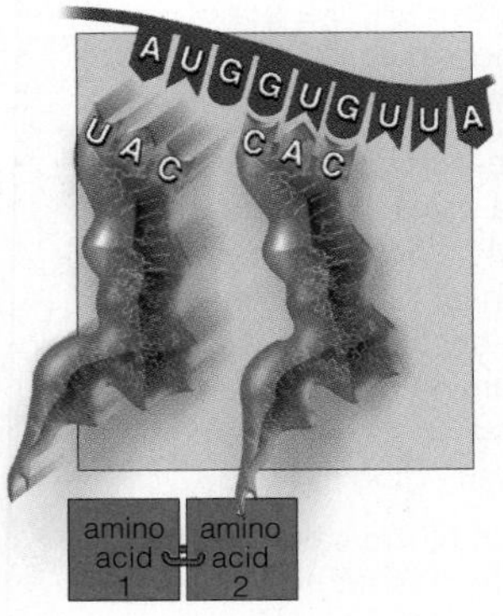

f The first tRNA is released, and the ribosome moves to the next codon position.

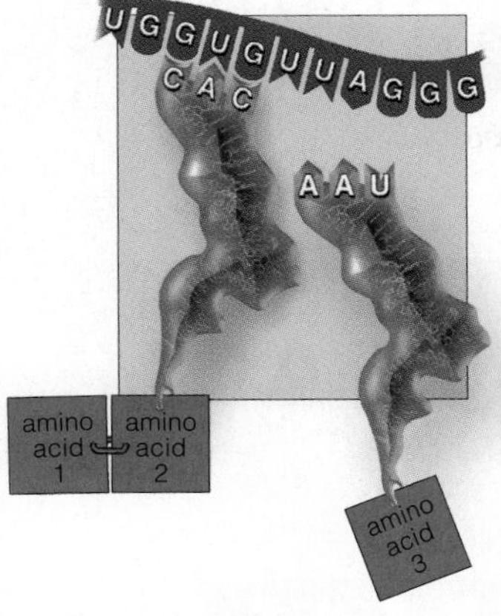

g A third tRNA binds with the next codon (here it is UUA). The ribosome catalyzes peptide bond formation between amino acids 2 and 3.

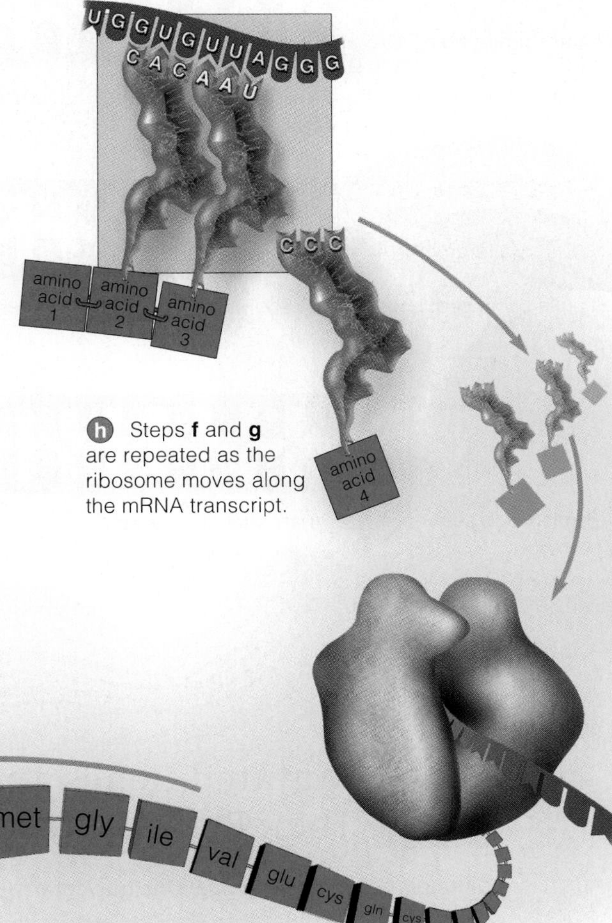

h Steps **f** and **g** are repeated as the ribosome moves along the mRNA transcript.

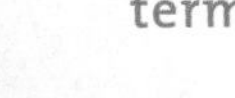

termination

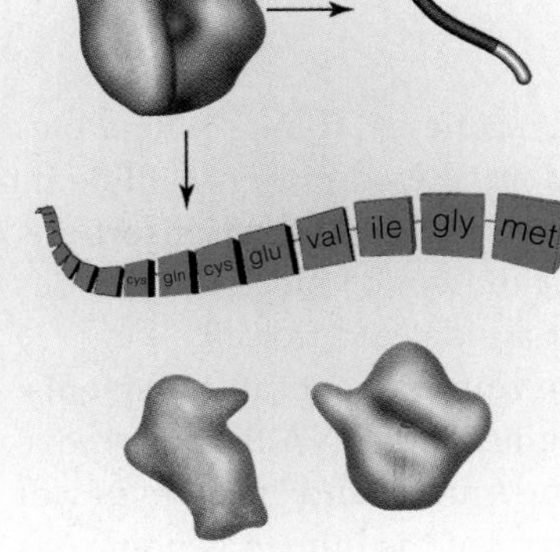

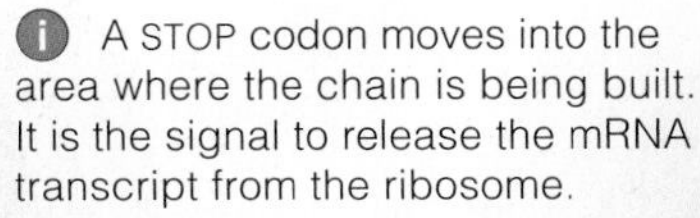

i A STOP codon moves into the area where the chain is being built. It is the signal to release the mRNA transcript from the ribosome.

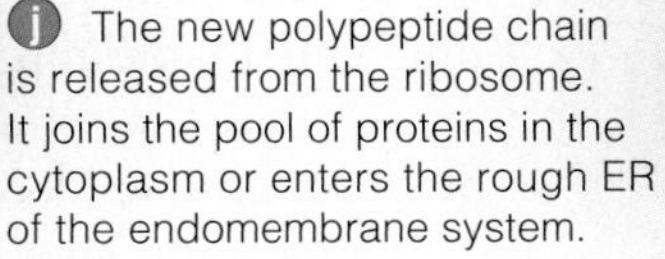

j The new polypeptide chain is released from the ribosome. It joins the pool of proteins in the cytoplasm or enters the rough ER of the endomembrane system.

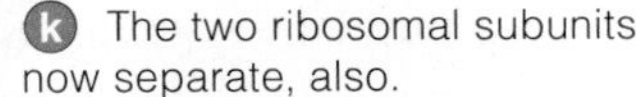

k The two ribosomal subunits now separate, also.

translation ends. Proteins called release factors bind to the ribosome, triggering enzyme activity that detaches both the mRNA and the polypeptide chain from the ribosome (Figure 10.8*i–k*).

Unfertilized eggs and other cells that will rapidly synthesize many copies of different proteins usually stockpile mRNA transcripts in their cytoplasm. In cells that are quickly using or secreting proteins, you often observe polysomes. A polysome is a cluster of many ribosomes, all translating the same mRNA transcript at the same time. Ribosomes move down the mRNA like many trains chugging along the same track. Under the electron microscope, they look like a string of pearls.

Many newly forming polypeptide chains carry out their functions in the cytoplasm. Many others have a special sequence of amino acids that allows them to enter the ribosome-studded, flattened sacs of rough ER (Figure 3.9). In the organelles of the endomembrane system, the polypeptides will take on final form before shipment to their ultimate destinations.

Translation is initiated when a small ribosomal subunit and an initiator tRNA arrive at an mRNA's start codon. A large ribosomal subunit binds to them.

tRNAs deliver amino acids to a ribosome in the order dictated by the linear sequence of mRNA codons. A polypeptide chain lengthens as peptide bonds form between the amino acids.

Translation ends when a STOP codon triggers events that cause the polypeptide chain and the mRNA to detach from the ribosome.

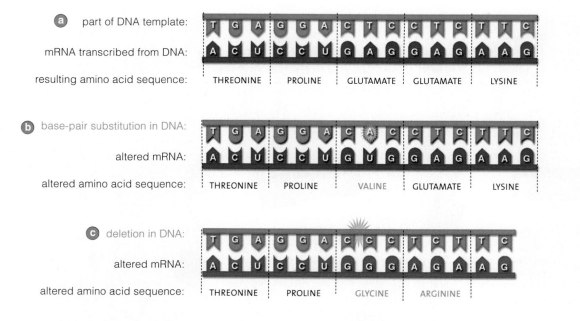

a part of DNA template:

mRNA transcribed from DNA:

resulting amino acid sequence:

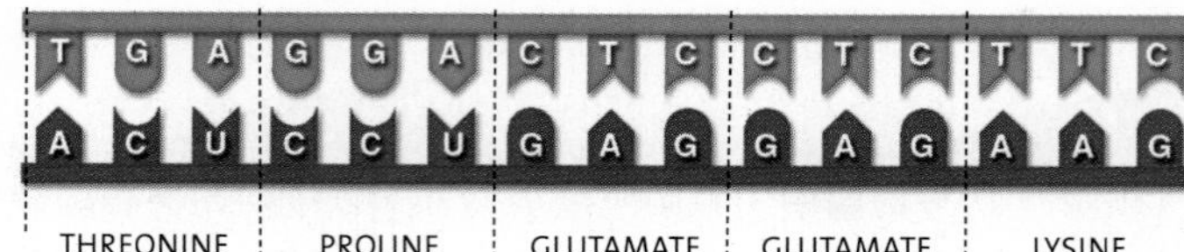

b base-pair substitution in DNA:

altered mRNA:

altered amino acid sequence:

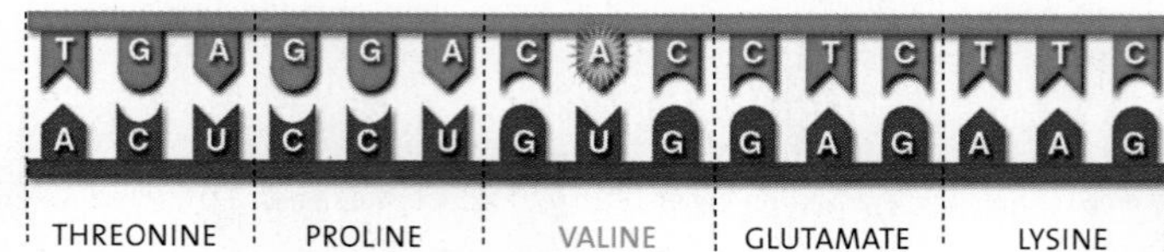

c deletion in DNA:

altered mRNA:

altered amino acid sequence:

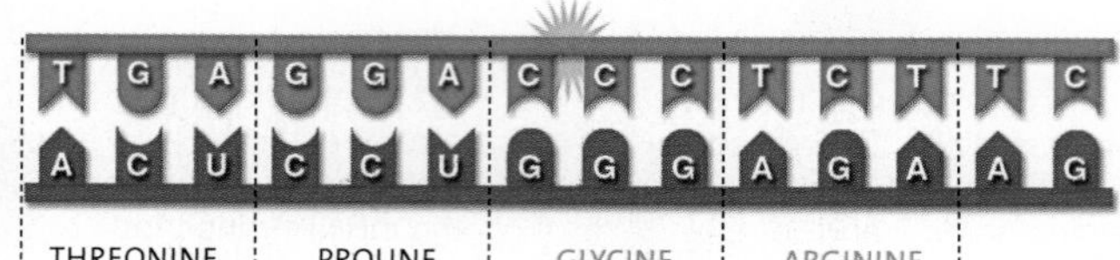

Figure 10.9 ***Animated!*** Mutation. (**a**) Part of a gene, mRNA, and amino acid sequence of the hemoglobin beta chain.

(**b**) A base-pair substitution replaces a thymine with adenine in the DNA. When the altered mRNA transcript is translated, valine replaces glutamate as the sixth amino acid of the new polypeptide chain. The outcome is sickle hemoglobin.

(**c**) Deletion of the same thymine causes a frameshift mutation. The reading frame for the rest of the mRNA shifts, and a different protein product forms. This mutation causes thalassemia, a different type of red blood cell disorder.

LINKS TO SECTIONS 2.8, 2.9, 8.2, 9.3

10.5 Mutated Genes and Their Protein Products

Gene sequences can change. One nucleotide might be substituted for another in the DNA sequence, an extra nucleotide may become inserted, or a nucleotide may be lost. Such changes are mutations.

IMPACTS, ISSUES

Each year, about a billion pounds of castor oil are manufactured for use in plastics, cosmetics, paints, and other products. The leftover castor meal is rich in the highly toxic ricin. Some researchers are trying to develop ricin-free castor plants by selectively mutating genes required for ricin synthesis. Castor meal left over from ricin-free plants could be used as nutritious feed for cattle, chickens, and other domestic animals.

COMMON MUTATIONS

In the gene mutations called **base-pair substitutions**, a base is copied incorrectly during DNA replication. A substitution causes a codon to change, and may result in a protein that incorporates a wrong amino acid or terminates when it is incomplete. There is room for error; remember, more than one codon may specify the same amino acid. For example, if one codon changes from UCU to UCC, it probably would not have dire effects because both codons specify serine. However, other mutations give rise to proteins with altered or lost functions. In the example shown in Figure 10.9*b*, an adenine replaced a certain thymine in the gene for beta hemoglobin. The resulting mutant gene's product has a single amino acid substitution that causes sickle-cell anemia (Section 2.9).

Figure 10.9*c* illustrates a different mutation, one in which a nucleotide, thymine, was *deleted*. Recall, base sequences are read in blocks of three. A deletion is one of the *frameshift* mutations, because it shifts the "three-bases-at-a-time" reading frame. A frameshift garbles the genetic message, and an altered protein will be its outcome. Frameshift mutations result from **insertions** or **deletions**, in which one, two, or more nucleotides are inserted into DNA or are deleted from it.

Other mutations arise from transposable elements, or **transposons**, small segments of DNA that can move around in a genome. They are found in all types of organisms; about 40 percent of the human genome has been transposed. Geneticist Barbara McClintock found that some kinds of transposons move spontaneously to a new location, or even to a different chromosome. When transposons disrupt a gene, the gene's product may become altered or cease to function entirely. The unpredictability of transposons is a source of curious variations in traits. Figure 10.10 gives an example.

HOW DO MUTATIONS ARISE?

Given the very swift pace of DNA replication (about twenty nucleotide bases per second in eukaryotes, and a thousand per second in bacteria), it is not surprising that mutations happen spontaneously while DNA is being replicated. All DNA polymerases make mistakes at a predictable rate, but fix most of their own errors. Uncorrected errors may be perpetuated as mutations.

Mutations can arise if a cell is exposed to certain hazards in the environment. For example, x-rays and other forms of **ionizing radiation** have enough energy to break chromosomes into pieces (Figure 10.11). Later,

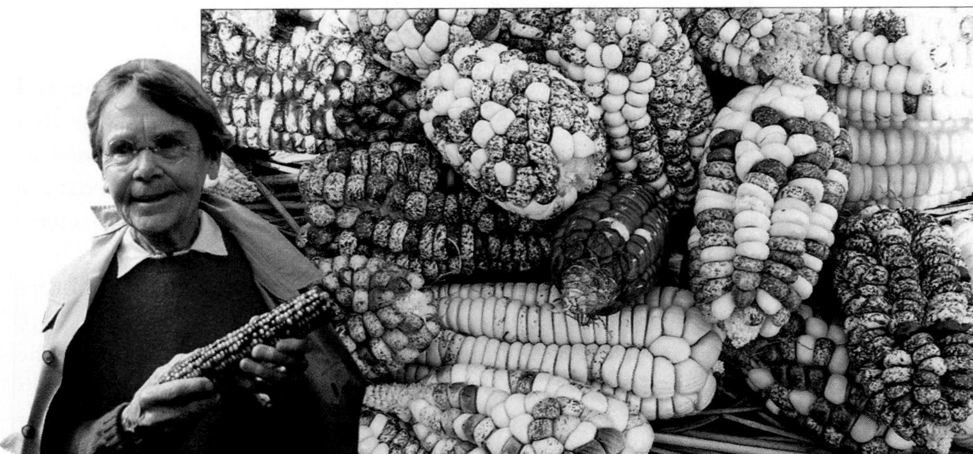

Figure 10.10 Barbara McClintock, who won a Nobel Prize for discovering that transposons slip into and out of different locations in DNA. The curiously nonuniform coloration of kernels in strains of Indian corn (*Zea mays*) sent her on the road to discovery. Several genes govern pigment formation and deposition in corn kernels. Mutations in these genes produce yellow, white, red, orange, blue, and purple kernels. Unstable mutations caused by transposons jumping in and out of these genes can cause streaks or spots in *individual* kernels.

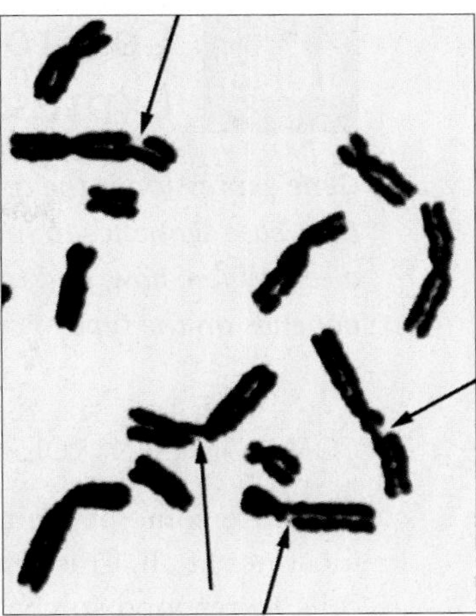

Figure 10.11 Chromosomes from a human cell that was exposed to ionizing radiation. The broken pieces will be lost during DNA replication. The extent of chromosome damage in an exposed cell depends on how much radiation it absorbed.

during DNA replication, broken pieces may get lost. Ionizing radiation can also damage DNA indirectly when it penetrates living tissue, leaving a destructive trail of reactive free radicals in its wake. Free radicals also alter DNA. Doctors and dentists always use the lowest possible doses of x-rays in order to minimize damage to their patients' DNA.

Ultraviolet (UV) light is one form of **nonionizing radiation** that can boost electrons of nucleotide bases to a higher energy level. Destabilized electrons change the bonding properties of thymine and cytosine. For instance, exposure to UV light can cause two adjacent thymine bases on the same strand of DNA to bond covalently to each other. This bond is called a thymine dimer, and it wrinkles the DNA. During replication, DNA polymerase skips the wrinkled region, and thus mutations may occur. Thymine dimers are the major source of mutations that cause certain kinds of cancer, and so they are the primary reason that exposing your unprotected skin to sunlight can cause skin cancer. At least seven gene products interact as a combined DNA repair mechanism to remove them.

Natural and synthetic chemicals also accelerate the rate of mutations. Alkylating agents are one example. These chemicals transfer small hydrocarbons (such as methyl groups) to nucleotide bases in DNA. Alkylated nucleotides halt DNA replication, or mispair during it. Either event can result in a mutation. Alkylated DNA also disintegrates during enzymatic attempts at repair. Chemicals in cigarette smoke alkylate DNA.

THE PROOF IS IN THE PROTEIN

Over evolutionary time, mutations are rare. When one arises in a somatic cell of an organism that reproduces solely by sexual means, its good or bad effects will not endure beyond the life of the organism. The mutation will not be passed on to offspring.

However, if a mutation arises in the DNA of a germ cell or gamete, it may enter the evolutionary arena. The same can happen if a mutation arises in an organism or cell that reproduces asexually.

Either way, nature's test is this: An altered protein specified by a heritable mutation may have harmful, neutral, or beneficial effects on the individual's ability to function in its prevailing environment. The effects of uncountable mutations in millions of species have had spectacular evolutionary consequences in terms of diversity. That is a topic of later chapters.

A mutation is an alteration involving one or more nucleotide bases in a gene. The most common are base-pair substitutions, insertions, and deletions.

DNA polymerases make errors, but also repair most of them. Unrepaired errors are perpetuated as mutations.

Exposure to harmful radiation and certain chemicals in the environment also can cause mutations in DNA.

A protein specified by a mutated gene may have harmful, neutral, or beneficial effects.

LINK TO SECTION 7.1

10.6 Controls Over Gene Expression

Gene expression is the conversion of the information in a DNA base sequence to functional and structural parts of a cell. When, how, and to what extent a gene is expressed depends on the type of cell and its chemical environment.

COMMON CONTROLS

Cells use some of their genes a lot, some a little, and most not at all. This regulation of gene activity allows cells to respond to changing conditions, either inside or outside the cell.

Diverse regulatory proteins adjust gene expression before, during, or after transcription and translation. They include signaling molecules such as hormones, which are small chemicals that can initiate changes in cell activities. With **negative control**, some regulatory proteins slow or stop gene expression; with **positive control**, others promote or enhance it.

Some DNA base sequences are sites of control over transcription. Gene expression is initiated, enhanced, or blocked depending on what binds to the sequence. A promoter, introduced in Section 10.2, is a sequence that marks the location of a gene for RNA polymerase to begin transcription. An *enhancer* is a binding site for proteins that increase the level of transcription.

Chemical modification is another form of control. For example, methyl groups physically block an RNA polymerase from accessing DNA. Thus, methylation prevents transcription. When the methyl groups are removed, the genes become active again. Attachment of functional groups to histones is another chemical control (Section 7.1 and Figure 10.12).

The signaling mechanisms are easier to understand in the specific context of how plants and animals work (Section 17.5, and Units Four and Five). Here, we just sample the kinds of events they set in motion.

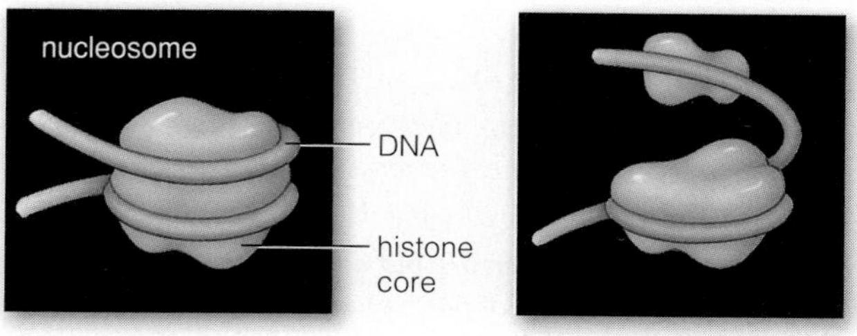

Figure 10.12 How loosening of the DNA–histone packaging in chromosomes exposes genes for transcription. Attachment of a functional group to a histone makes it loosen its grip on the DNA that is wound around it. Enzymes that are associated with transcription attach or detach the groups.

BACTERIAL CONTROL OF THE LACTOSE OPERON

Think about the dot of the letter "i." About a thousand bacterial cells would stretch side by side across that dot. Each of those microscopic cells depends just as much on gene controls as you do! When nutrients are plentiful and other conditions favor growth, bacteria divide quickly.

In bacteria, transcription occurs in the cytoplasm. Translation begins even before mRNA transcripts are fully assembled. Genes for the enzyme in a metabolic pathway often are organized along the chromosome in the same sequence as the reaction steps. All the genes in the sequence may be transcribed as one continuous strand of RNA.

With this bit of background, consider an example of how one type of bacteria adjusts transcription rates to availability of the nutrient lactose. *Escherichia coli* lives in the mammalian gut, and it dines on nutrients traveling past. Mammalian infants generally are fed milk. Milk does not contain glucose, the sugar of choice for *E. coli*. It contains lactose, a different sugar.

Once mammals are weaned, milk intake typically dwindles or stops. *E. coli* cells still take advantage of lactose if and when it becomes available by regulating the expression of three genes for lactose-metabolizing enzymes. One promoter precedes all three genes, and there are operators are on both sides of the promoter. The **operators** are binding sites for a *repressor*, a type of regulatory protein that can prevent transcription. The promoter, operators, and genes together are called the **lac operon** (Figure 10.13).

The genes for lactose-metabolizing enzymes are not used when they are not needed. In the absence of lactose, a repressor molecule binds to both operators. The binding causes the DNA region that contains the promoter to twist into a loop, as in Figure 10.13*b*. RNA polymerase cannot bind to a looped-up promoter, and so cannot transcribe the genes.

The three genes for lactose-metabolizing enzymes are used when lactose is in the gut. *E. coli* cells convert some lactose to allolactose, which binds the repressor and changes its shape. The altered repressor molecule can no longer bind to the operators. The looped DNA unwinds, and RNA polymerase transcribes the genes.

Unlike *E. coli*, humans develop *lactose intolerance*. Cells in the lining of our small intestine make and then secrete lactase, a lactose-digesting enzyme, into the gut. As many people age, however, concentrations of lactase decline. Lactose accumulates in the large intestine (colon), where it fosters population explosions of resident bacteria. As the bacteria digest the lactose, they produce carbon dioxide gas, which accumulates, distends the colon, and causes pain.

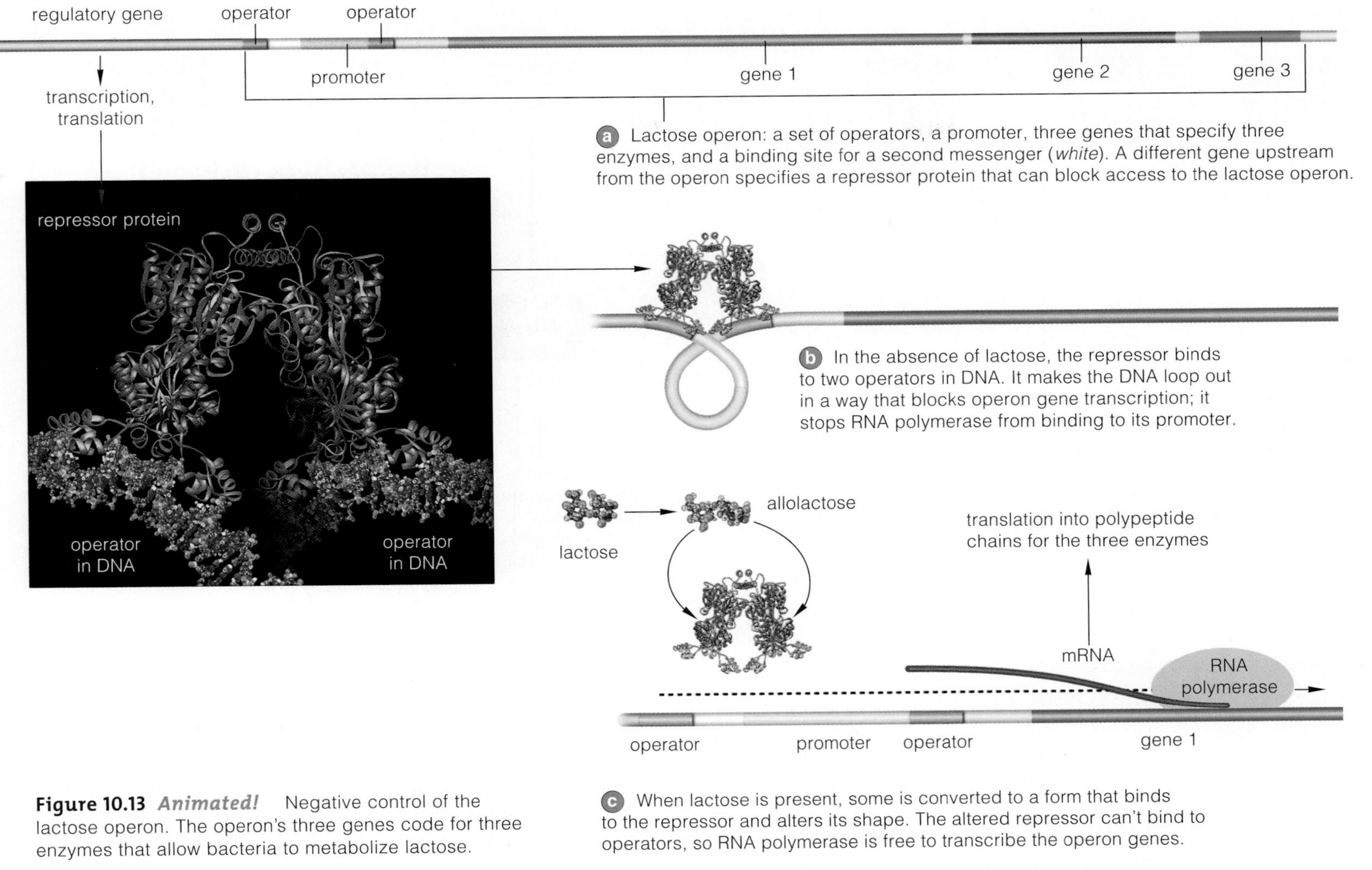

regulatory gene
operator
operator
transcription, translation
promoter
gene 1
gene 2
gene 3
a Lactose operon: a set of operators, a promoter, three genes that specify three enzymes, and a binding site for a second messenger (white). A different gene upstream from the operon specifies a repressor protein that can block access to the lactose operon.
repressor protein
operator in DNA
operator in DNA
b In the absence of lactose, the repressor binds to two operators in DNA. It makes the DNA loop out in a way that blocks operon gene transcription; it stops RNA polymerase from binding to its promoter.
lactose
allolactose
translation into polypeptide chains for the three enzymes
mRNA
RNA polymerase
operator
promoter
operator
gene 1
c When lactose is present, some is converted to a form that binds to the repressor and alters its shape. The altered repressor can't bind to operators, so RNA polymerase is free to transcribe the operon genes.

Figure 10.13 ***Animated!*** Negative control of the lactose operon. The operon's three genes code for three enzymes that allow bacteria to metabolize lactose.

EUKARYOTIC GENE CONTROLS

Later in the book, you will be reading about how you and other complex organisms developed from a single cell. For now, tentatively accept this premise: All cells of your body started out life with the same genes, because every one arose by mitotic cell divisions from the same fertilized egg. All of your cells transcribe many of the same genes, because most aspects of their structure and basic housekeeping activities are similar.

In other ways, however, *nearly all of your body cells became specialized in composition, structure, and function*. This process of **cell differentiation** occurs during the development of all multicelled organisms. Differences arise among cells that use different subsets of genes. Specialized tissues and organs are the result.

For example, nearly all of your cells continually transcribe the genes for glycolysis enzymes, but only immature red blood cells can transcribe the genes for hemoglobin. Only your liver cells transcribe the genes for enzymes that neutralize certain toxins. When your eyes first formed, their cells began to access the genes necessary for synthesizing crystallin. No other cells activate the genes for this protein, which helps make the transparent fibers of the lens in each eye.

By some estimates, cells rarely use more than 5 to 10 percent of their genes at a time. Gene controls keep the rest of them switched off. Which genes are active at any time depends on the type of organism, the stage of growth and development it is passing through, and the controls operating at different stages.

Controls over gene expression allow cells to respond to changing environmental conditions.

Gene expression controls also govern cell differentiation in complex eukaryotes.

Cells become specialized when they start using a unique fraction of their genes.

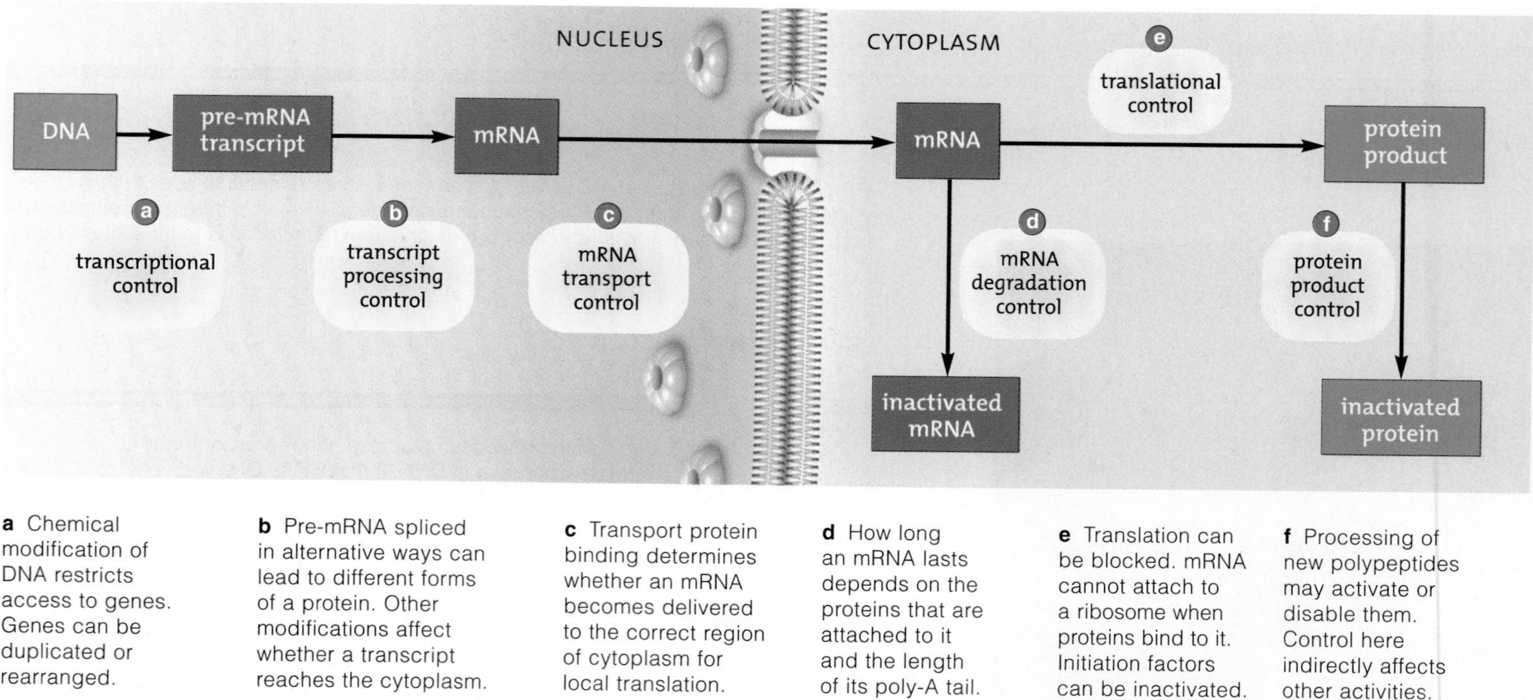

a Chemical modification of DNA restricts access to genes. Genes can be duplicated or rearranged.

b Pre-mRNA spliced in alternative ways can lead to different forms of a protein. Other modifications affect whether a transcript reaches the cytoplasm.

c Transport protein binding determines whether an mRNA becomes delivered to the correct region of cytoplasm for local translation.

d How long an mRNA lasts depends on the proteins that are attached to it and the length of its poly-A tail.

e Translation can be blocked. mRNA cannot attach to a ribosome when proteins bind to it. Initiation factors can be inactivated.

f Processing of new polypeptides may activate or disable them. Control here indirectly affects other activities.

Figure 10.14 ***Animated!*** Controls over eukaryotic gene expression.

We conclude this chapter with examples of two of the eukaryotic gene controls outlined in Figure 10.14.

HOMEOTIC GENES AND BODY PLANS In most species of eukaryotes, the products of **homeotic genes** guide formation of organs and limbs by turning on other genes in local areas of a developing embryo. Different homeotic genes are transcribed in specific parts of a developing embryo, so their products accumulate in local tissue regions. Body parts form as the products interact with one another and with other regulatory proteins to switch on genes in certain cells of the body.

Homeotic genes were discovered through mutations that cause cells in a *Drosophila* embryo to develop into a body part that should be somewhere else (Figure 10.15). For instance, the *antennapedia* gene is normally transcribed where a thorax and legs should form. In all other regions, cells do not transcribe *antennapedia*. Figure 10.15*b* shows what happens when a mutation allows this gene to be wrongly transcribed in the body region that is supposed to be a head.

We know of more than a hundred homeotic genes. They control transcription by the same mechanisms in all eukaryotes, and many are interchangeable among species as evolutionarily distant as yeasts and humans. They probably evolved in the most ancient eukaryotic cells. Every homeotic gene product has a domain of about sixty amino acids that regulates transcription of other genes by binding to certain control elements in DNA. These domains differ among species only by *conservative* substitutions, in which one amino acid has replaced another with similar chemical properties. We will return to this topic in Chapter 13.

X CHROMOSOME INACTIVATION Female humans and female calico cats both have two X chromosomes in their

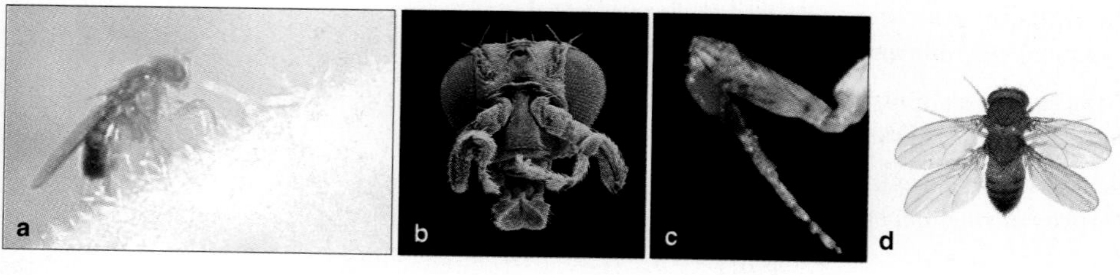

Figure 10.15 Gene controls that govern where body parts develop in fruit flies. (**a**) Normal fly. (**b**) A mutation that affects transcription of the homeotic gene *antennapedia* results in legs on the head. (**c**) Expression of a transcriptional control gene can induce an eye to form not only on a fruit fly's head, but also on its legs, wings, and antennae. (**d**) A mutation in one of this fly's homeotic genes caused it to develop two thoraxes.

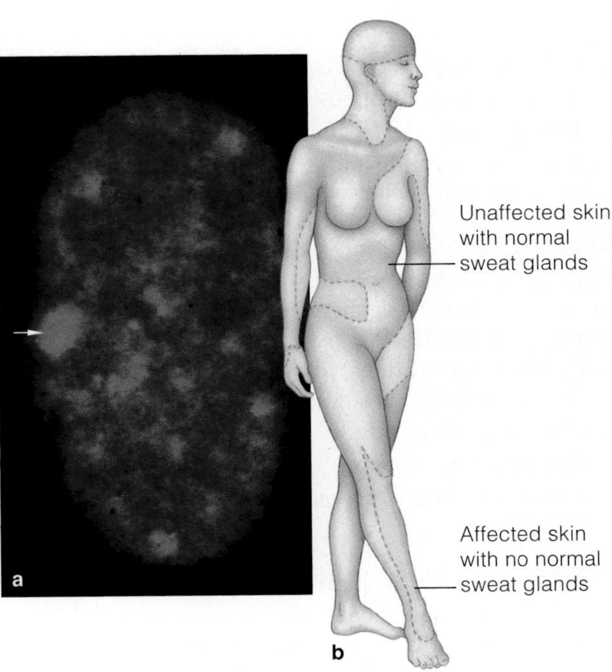

Figure 10.16 (**a**) A condensed X chromosome, also called a Barr body (*arrow*), in the somatic cell nucleus of a human female. The X chromosome in cells of human males is not condensed this way. (**b**) Mosaic tissue effect that occurs in anhidrotic ectodermal dysplasia.

Figure 10.17 ***Animated!*** Why is this female cat "calico"? In her body cells, one of the two X chromosomes has a dominant allele for the brownish-black pigment melanin. Expression of the allele on the other X chromosome results in orange fur. When this cat was an embryo, one X chromosome was inactivated at random in each cell.

Patches of different colors reflect which allele was inactivated in cells that formed a given tissue region. White patches are an outcome of an interaction that involves a different gene, the product of which blocks melanin synthesis.

diploid cells. One is stretched long and thin. The other is scrunched up, even during interphase. That scrunching is not a chromosome abnormality. It is a programmed shutdown of all but about three dozen genes on *one* of two homologous X chromosomes. This **X chromosome inactivation** happens in female embryos of all placental mammals and marsupials. Figure 10.16*a* shows a single condensed X chromosome in a nucleus at interphase. We call this condensed chromosome a **Barr body**.

One X chromosome gets inactivated when embryos are still a tiny ball of cells. In placental mammals, the shutdown is random, in that *either* chromosome could become condensed. The maternal X chromosome may be inactivated in one cell; the paternal or the maternal X chromosome may be inactivated in a cell next to it. Once that random molecular decision occurs in a cell, all of that cell's descendants make the same decision as they go on dividing to form tissues. What is the outcome? *A fully developed female has patches of tissue in which genes of the maternal X chromosome are expressed, and patches of tissue in which the genes of the paternal X chromosome are expressed*. She has become a "mosaic" for X chromosome expression (Figure 10.17).

When alleles on two homologous X chromosomes are not identical, differences occur among patches of tissues throughout the body. The mosaic effect occurs in *anhidrotic ectodermal dysplasia*, an X-linked disorder characterized by abnormalities in the skin and in the structures derived from it, including teeth, hair, nails, and sweat glands. Human females with this disorder are heterozygous for a recessive allele that causes an absence of sweat glands. Their mosaic tissues can be observed because sweat glands are absent in patches of skin (Figure 10.16*b*). Where the glands are absent, the recessive allele is on the X chromosome that was not shut down.

LINK TO SECTION 8.4

In mammals, the X-chromosome shutdown has an important function. Remember, males have one X and one Y chromosome. This means females have twice as many X chromosome genes. Inactivating one of their X chromosomes balances gene expression between the sexes. The normal development of female mammalian embryos depends on this type of control mechanism, which is called **dosage compensation**.

Controls operate before, during, and after transcription and translation. A variety of control elements interact with one another and with DNA, RNA, and gene products.

When, how, and to what extent a gene is expressed depends on the type of cell and its functions, on the cell's chemical environment, and on signals for change.

Diverse controls over eukaryotic genes come into play at different stages of growth and development. Homeotic genes and dosage compensation are two examples.

Summary

Section 10.1 All enzymes and other proteins consist of polypeptide chains. Each chain, a linear sequence of amino acids, corresponds to a nucleotide sequence in a gene. The path from a gene to a protein has two steps: transcription and translation (Figure 10.18).

Section 10.2 Translation requires messenger RNA, transfer RNA, and ribosomal RNA. In transcription, enzymes unwind the two strands of the DNA double helix in a gene region. RNA polymerases covalently bond RNA nucleotides into a new RNA transcript, in an order complementary to the exposed nucleotides on the DNA template. Adenine, guanine, cytosine, and uracil are the RNA nucleotides.

Each new RNA molecule becomes modified before it leaves the nucleus. Its noncoding regions (exons) are removed and the remaining coding regions (introns) become spliced together as a continuous transcript.

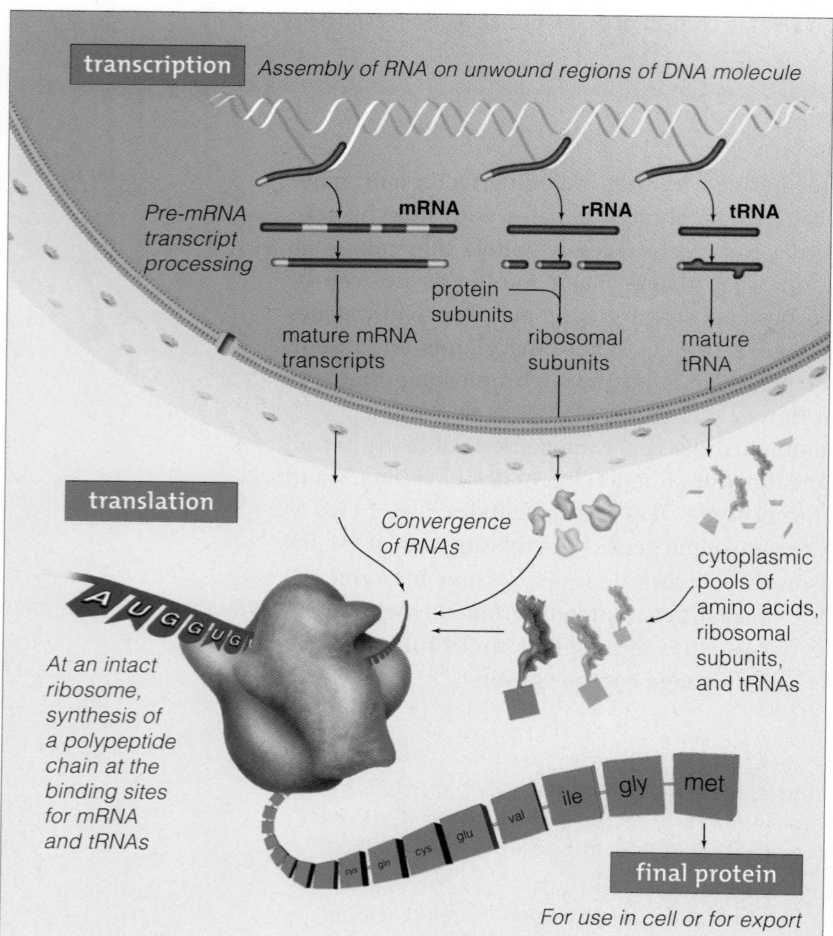

Figure 10.18 Summary of protein synthesis in eukaryotes. DNA is transcribed into RNA in the nucleus. RNA is translated into protein in the cytoplasm. Prokaryotic cells do not have a nucleus; transcription and translation occur in their cytoplasm.

One end of a new mRNA is capped, and the other gets a tail that helps control how long it will last in the cell.

BiologyNow
Learn how genes are transcribed and transcripts processed with animation on BiologyNow.

Section 10.3 Only messenger RNA (mRNA) carries DNA's protein-building information from the nucleus to the cytoplasm. Its genetic message is written in codons, sets of three nucleotides that specify amino acids. Sixty-four codons, a few of which act as START or STOP signals for translation, constitute a highly conserved genetic code.

BiologyNow
Explore the genetic code with the interaction on BiologyNow.

Section 10.4 Translation proceeds through three stages. In initiation, ribosomal subunits, an mRNA, and an initiator tRNA join as an initiation complex. During the elongation stage, tRNAs deliver amino acids to the intact ribosome in the order dictated by codons in the mRNA. The ribosome links the amino acids into a polypeptide chain. At termination, a STOP codon and other factors trigger release of the mRNA and the polypeptide chain, and cause the ribosome's subunits to separate from each other.

BiologyNow
Observe the translation of an mRNA transcript with the animation on BiologyNow.

Section 10.5 A mutation is a permanent alteration in a DNA nucleotide sequence. Mutations introduce changes in protein structure, and thus protein function. These changes may lead to small or large differences in traits among individuals of a population.

BiologyNow
Investigate the effects of mutation with animation on BiologyNow.

Section 10.6 Regulatory elements interact with RNA, DNA, protein products, and one another before, during, and after transcription and translation. They govern whether, when, and how a gene is expressed.

All cells respond fast to short-term shifts in nutrients and other environmental factors. Regulatory proteins adjust gene transcription rates in response. Control of bacterial operons is an example.

In eukaryotic cells, other controls also come into play during development. Selective gene expression is the basis of differentiation. All cells of multicelled organisms inherit the same genes. During development, different cell lineages begin to use different subsets of those genes. As they do, the cells diverge in structure and function, and eventually they form specialized tissues and organs.

BiologyNow
Review how eukaryotic gene controls influence development with animation on BiologyNow.

Self-Quiz

Answers in Appendix I

1. DNA contains many different genes that are transcribed into different ______ .
   a. proteins
   b. mRNAs only
   c. mRNAs, tRNAs, rRNAs
   d. all of the above

2. An RNA molecule is typically ______ .
   a. a double helix
   b. single-stranded
   c. double-stranded
   d. triple-stranded

3. An mRNA molecule is produced by ______ .
   a. replication
   b. duplication
   c. transcription
   d. translation

4. Each codon calls for a specific ______ .
   a. protein
   b. polypeptide
   c. amino acid
   d. carbohydrate

5. Anticodons pair with ______ .
   a. mRNA codons
   b. DNA codons
   c. RNA anticodons
   d. amino acids

6. Cell differentiation ______ .
   a. occurs in all complex multicelled organisms
   b. requires unique genes in different cells
   c. involves selective gene expression
   d. both a and c
   e. all of the above

7. Control mechanisms adjust gene expression in response to changing ______ .
   a. nutrient availability
   b. solute concentrations
   c. signals from other cells
   d. all of the above

8. An operon typically governs ______ .
   a. prokaryotic genes
   b. eukaryotic genes
   c. genes of all types
   d. DNA replication

9. X chromosome inactivation is an example of ______ .
   a. a chromosome abnormality
   b. dosage compensation
   c. calico cats
   d. homeotic gene control

10. Homeotic genes ______ .
    a. are binding sites that flank a bacterial operon
    b. fill in body plan details of developing embryo
    c. control X chromosome inactivation
    d. both a and c

11. A cell with a Barr body is ______ .
    a. prokaryotic
    b. from a male mammal
    c. from a female mammal
    d. infected by the Barr virus

12. Match the terms with the most suitable description.

____ alkylating agent	a. coding region of mRNA transcript
____ chain elongation	b. base triplet for amino acid
____ exons	c. second stage of translation
____ genetic code	d. base triplet; pairs with codon
____ anticodon	e. an environmental agent that induces mutation in DNA
____ introns	f. set of 64 codons for mRNA
____ codon	g. removed from a pre-mRNA transcript before translation

Additional questions are available on **Biology Now™**

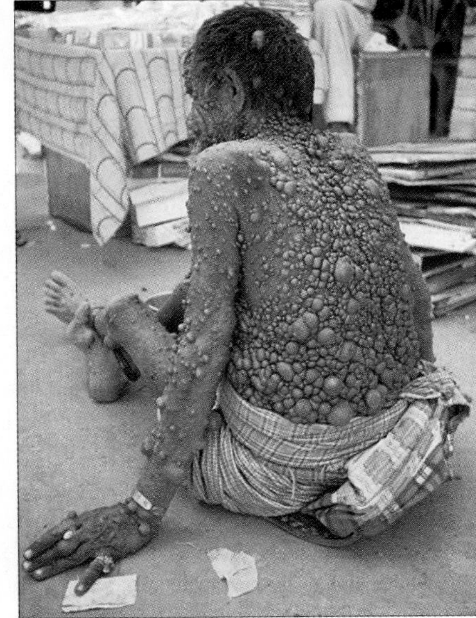

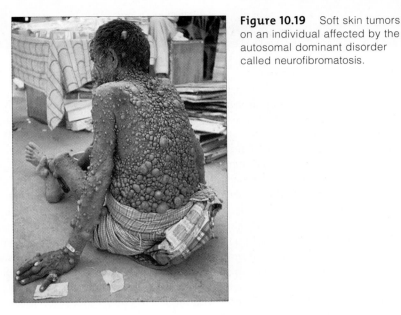

Figure 10.19 Soft skin tumors on an individual affected by the autosomal dominant disorder called neurofibromatosis.

Critical Thinking

1. Using Figure 10.4, translate this base sequence in part of an mRNA transcript into an amino acid sequence:

5′—GGUUUCUUCAAGAGA—3′

2. *Antisense drugs* may help us fight cancer and viral diseases, including SARS. They are short mRNA strands that are complementary to mRNAs associated with these illnesses. Speculate on how these drugs work.

3. Cigarette smoke is mostly carbon dioxide, nitrogen, and oxygen. The rest contains at least fifty-five different chemicals identified as carcinogenic, or cancer-causing, by the International Agency for Research on Cancer (IARC). When these carcinogens enter the bloodstream, enzymes convert them to a series of chemical intermediates that are easier to excrete. Some of them bind irreversibly to DNA. Propose one mechanism by which smoking cigarettes causes cancer.

4. *Neurofibromatosis* is a human autosomal dominant disorder caused by mutations in the *NF1* gene. Soft, fibrous tumors form in the peripheral nervous system and skin (Figure 10.19). Because the mutant allele is dominant, an affected child usually has an affected parent. Yet in 1991, a boy developed neurofibromatosis even though his parents did not. The *NF1* gene from his father contained a transposon. Neither his father or mother had a transposon in any of the copies of their own *NF1* genes. Explain the cause of neurofibromatosis in the boy and how it arose.

5. On the right are pictured Rainbow the calico cat and cc, her clone. They were introduced in Section 9.4. One cell donated by Rainbow was cloned to form cc. Despite being an exact genetic copy of her donor, cc is not calico. Explain why cc's coat does not match Rainbow's. What other coat colors would you expect to see in different single-cell clones of Rainbow?

11 STUDYING AND MANIPULATING GENOMES

IMPACTS, ISSUES

Golden Rice or Frankenfood?

Not long ago, the World Health Organization made an estimate that 124 million undernourished children around the world have vitamin A deficiencies. These children may become permanently blind and develop other disorders. Geneticists who wanted to help them transferred genes from daffodils into rice plants. The plants began making beta-carotene, a yellow pigment converted in the body to vitamin A. The pigment is stored in the seeds of the plants—the grains of "golden rice." Researchers predict that including these golden grains in a child's diet will prevent vitamin A deficiency

Rice is the main foodstuff for 3 billion people in impoverished countries. Getting beta-carotene into rice grains seemed the least costly way to deliver vitamin A to those who need it the most, but doing so was beyond the scope of conventional breeding practices. Rice plants just will not breed with daffodil plants.

Golden rice is a genetically modified food. Its creation was an extension of thousands of years of artificial selection and cross-breeding practices that we have been using to coax new breeds of plants and animals from wild ancestral stocks. Meatier turkeys, seedless watermelons, big juicy corn kernels from puny hard ones—the list goes on.

No one wants children to suffer or die. But many people oppose the idea of genetically modified foods, including golden rice. Maybe the unsettling part about the more recent human-directed activities is that the pace has picked up, hugely. We are getting quite good at tinkering with genetics. We do this both for pure research and also for useful, practical applications.

Some say we must never alter the DNA of anything. The concern is that we as a species simply do not have the wisdom to bring about genetic changes without causing irreparable harm. One is reminded of our very human tendency to leap before we look.

And yet, we dream of the impossible. Something about the human experience gave us a capacity to imagine wings of our own making, and that capacity carried us to the frontiers of space. Someone else dreamed of turning plain rice into more nutritious food that might keep some children from going blind.

In this unit, you began with the cell division mechanisms by which parents pass on DNA to new generations. You moved to the chromosomal and molecular basis of inheritance, then to gene controls that guide life's continuity. The sequence parallels the history of genetics. Now, you have arrived at the point in time when geneticists hold molecular keys to the kingdom of inheritance. What they are unlocking is already having an impact on life in the biosphere.

How Would You Vote? *Nutritional labeling is required on all packaged food in the United States, but genetically modified food products may be sold without labeling. Should food distributors be required to label all products made from genetically modified plants or livestock? See BiologyNow for details, then vote online.*

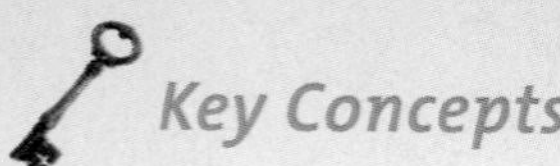

Key Concepts

DNA TECHNOLOGIES

With DNA technologies, we can identify, isolate, copy, and otherwise analyze and manipulate genes of any species. These tools have allowed researchers to decipher the linear order of all 3 billion nucleotide bases that make up the complete set of human chromosomes.

GENETIC ENGINEERING

In genetic engineering, copies of normal or modified genes are inserted into individual organisms for research and for practical applications, including medicine, agriculture, and gene therapy. Many ethical and social issues remain unresolved.

Links to Earlier Concepts

This chapter continues the story of fame and fortune that began with the search for the structure of DNA (Section 9.1). It builds on earlier discussions about the molecular structure of DNA (2.10, 9.2), DNA replication and repair (9.3), and mutations (10.5). You may wish to quickly review the nature of mRNA transcription and processing (10.2) and controls over gene transcription (10.6). You will also come across more uses for radioisotopes (2.1), and will see why inheritance patterns (7.6, 8.7) and the lactose operon (10.6) are not necessarily of obscure interest.

11.1 A Molecular Toolkit

Just as a carpenter has a toolkit for manipulating wood and metal, a molecular biologist also has a toolkit—a set of techniques for manipulating and studying DNA.

LINKS TO SECTIONS 2.10, 9.2, 9.3, 10.6

THE SCISSORS: RESTRICTION ENZYMES

In the 1950s, excitement over the discovery of DNA's structure gave way to frustration. No one could figure out how to determine the order of the nucleotides in DNA—the molecule was just too big. Identifying one particular base within the context of billions of others was an enormous technical challenge. DNA had to be divided into manageable, identifiable sections before its sequence could be determined. But how?

Then, in 1970, Hamilton Smith and his colleagues were studying *Haemophilus influenzae*, a bacteria. The bacteria were protecting themselves from infection by cutting up invading viral DNA before it could insert itself into the bacterial chromosome. This process was not random degradation; certain enzymes were making the cuts at specific nucleotide sequences.

In time, hundreds of strains of bacteria and a few eukaryotic cells yielded thousands of other enzymes like those Smith discovered. Each of these **restriction enzymes** cuts any double-stranded DNA wherever a particular nucleotide sequence occurs. For example, the enzyme EcoRI cuts between the G and the A of the sequence GAATTC. EcoRI and many other restriction enzymes make staggered cuts. These cuts leave single-stranded tails, or "sticky ends," on the cut fragments. Each tail can base-pair with the tail of any other DNA fragment that was cut by the same enzyme, because their sticky ends match up. DNA ligase appropriated from replication and repair processes is used to "glue" restriction fragments together. The result is a molecule of recombinant DNA (Figure 11.1).

CLONING VECTORS

Bacteria, recall, have one circular chromosome. Many also have **plasmids**, much smaller circles of DNA with a few extra genes (Figure 11.2*a*). Bacteria divide often, making huge populations of genetically identical cells. Before each division, replication enzymes copy both the chromosomal DNA *and* the plasmid DNA. Researchers discovered that foreign DNA inserted into a plasmid gets replicated and distributed into each daughter cell along with the plasmid. Bacteria could thus be used to copy a small fragment of DNA again and again.

A plasmid that accepts foreign DNA and slips into a host bacteria, yeast, or another kind of cell is called a **cloning vector**. Cloning vectors typically contain several unique restriction enzyme recognition sites that can be used for cloning (Figure 11.2*b*).

A cell that takes up a cloning vector with foreign DNA in it may found a huge population of descendant cells. Each will contain an identical copy of the vector and a "cloned" copy of the foreign DNA. DNA cloning

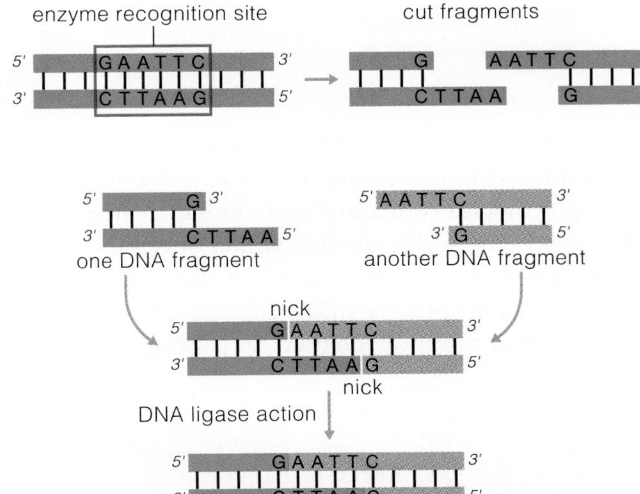

Figure 11.1 Restriction fragments are spliced together to form a recombinant DNA molecule.

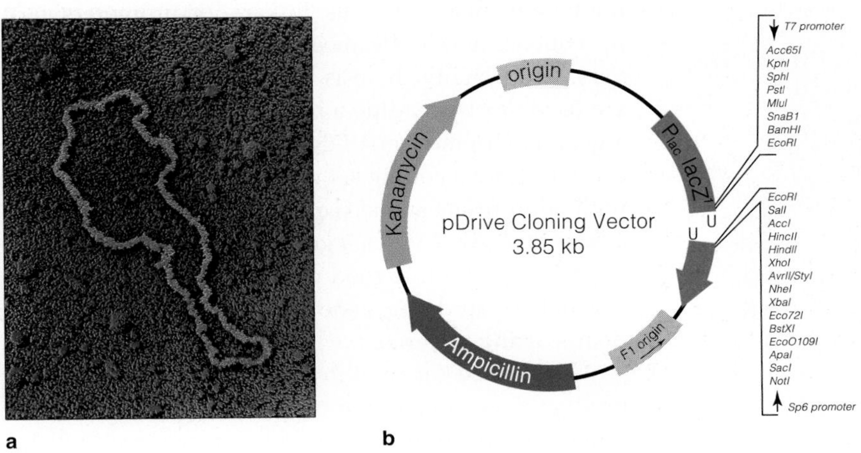

Figure 11.2 Cloning vectors. (**a**) Micrograph of a plasmid. (**b**) A commercially available cloning vector, with its useful restriction enzyme sites listed at right. This vector includes antibiotic resistance genes (*blue*, *purple*) and part of the bacterial lactose operon (*red*). These genes help researchers identify cells that take up recombinant DNA molecules.

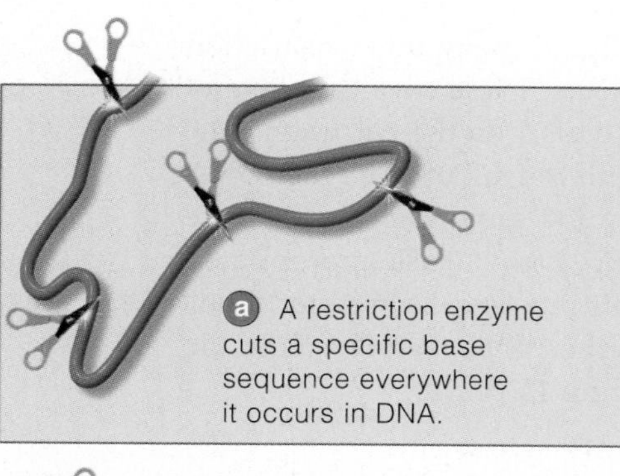
a A restriction enzyme cuts a specific base sequence everywhere it occurs in DNA.

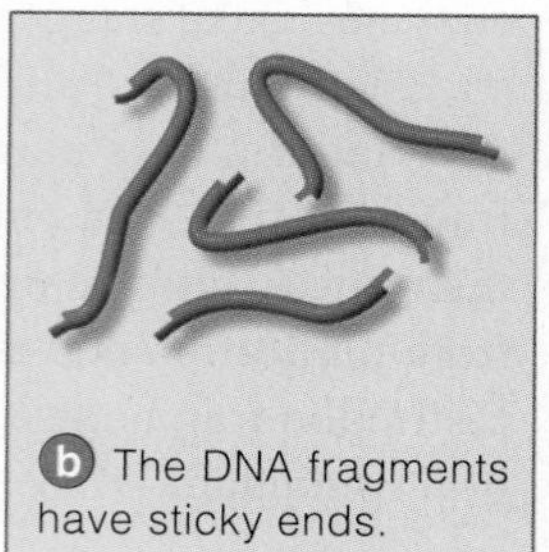
b The DNA fragments have sticky ends.

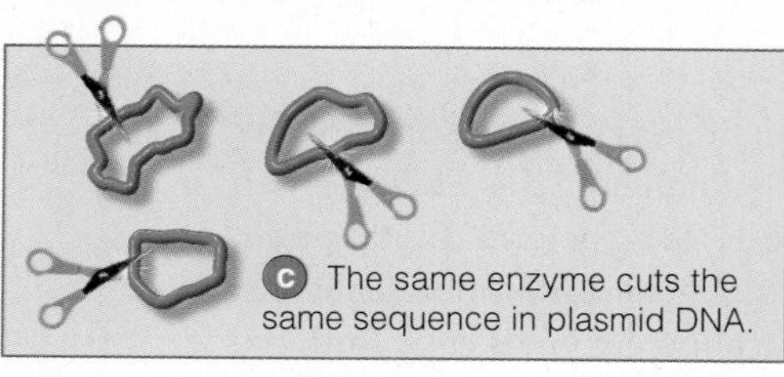
c The same enzyme cuts the same sequence in plasmid DNA.

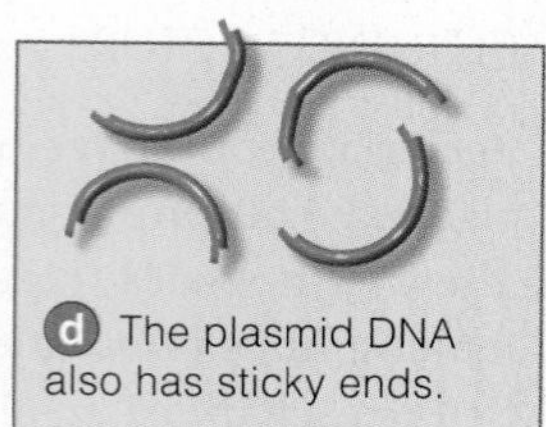
d The plasmid DNA also has sticky ends.

e The DNA fragments and plasmid DNA are mixed with DNA ligase.

f The result? A collection of recombinant plasmids that incorporate foreign DNA fragments.

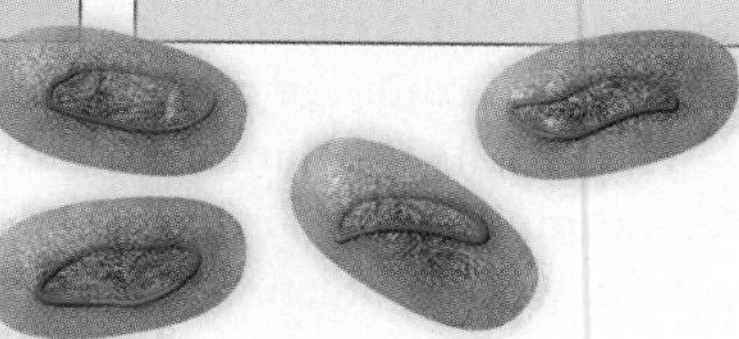
g Host cells that can divide rapidly take up the recombinant plasmids.

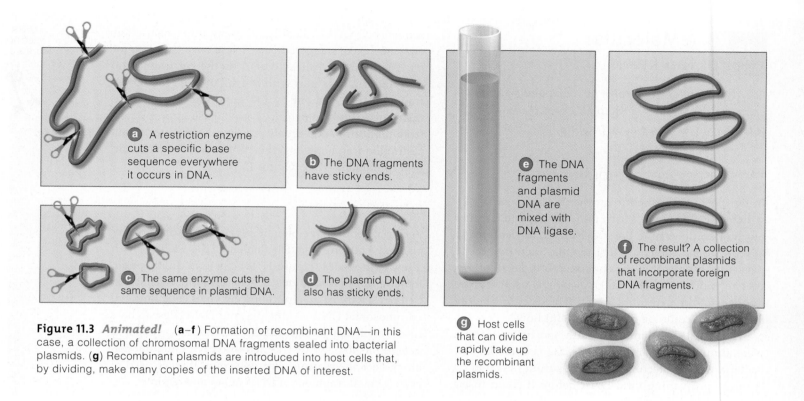

Figure 11.3 ***Animated!*** (**a**–**f**) Formation of recombinant DNA—in this case, a collection of chromosomal DNA fragments sealed into bacterial plasmids. (**g**) Recombinant plasmids are introduced into host cells that, by dividing, make many copies of the inserted DNA of interest.

LINK TO SECTION 10.2

is one tool in the molecular toolkit. It helps scientists to isolate and identify manageable sections of huge DNA molecules, and also to obtain enough of each bit to be able to study it (Figure 11.3).

cDNA CLONING

Figuring out where a gene is in chromosomal DNA is not by any means obvious. Eukaryotic promoters vary in sequence and in distance from a gene. Splice signals vary, and usually there is no way to tell which ones are used just by reading a nucleotide sequence. Unlike a gene in chromosomal DNA, an mRNA conveniently contains only a coding sequence, minus introns, along with appropriate signal sequences for its translation in eukaryotes. Also, when a gene is active, its transcribed mRNA is present in good quantity. For these reasons, researchers studying gene expression use mRNA as their starting material.

Genes used for protein expression in research and product development are typically cloned, but mRNA cannot be cloned directly. Restriction enzymes will cut only double-stranded DNA, so an mRNA must be first converted into double-stranded DNA before it can be spliced into a vector. A retroviral replication enzyme, **reverse transcriptase**, is used to assemble a strand of **cDNA**, or *complementary* DNA, on an mRNA template (Figure 11.4). A hybrid molecule, one strand of mRNA base-paired with one strand of cDNA, is the outcome. DNA polymerase added to the mixture then strips the RNA away as it copies the first strand of cDNA into a second DNA strand. The result is a double-stranded DNA copy of the original mRNA. Because it is double-stranded, that copy may be used for cloning.

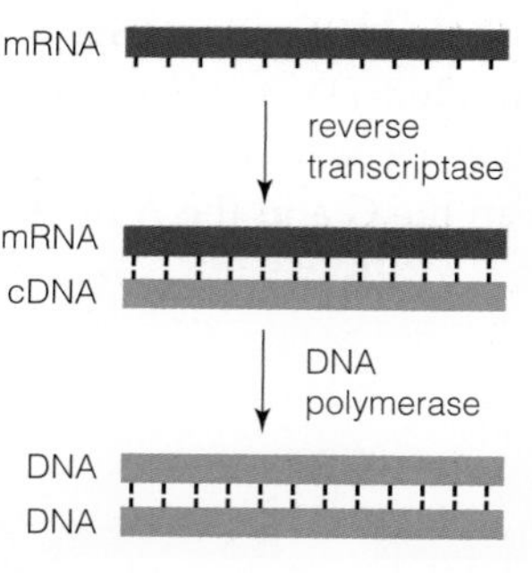
mRNA
reverse transcriptase
mRNA
cDNA
DNA polymerase
DNA
DNA

Figure 11.4 cDNA. Reverse transcriptase assembles DNA on an mRNA template. An RNA–cDNA hybrid molecule is the result.

Then, DNA polymerase removes the mRNA as it assembles a new DNA strand on the cDNA. The result is a double-stranded DNA "copy" of an mRNA template.

In DNA cloning, restriction enzymes cut DNA isolated from any species. DNA ligase seals the fragments into plasmids. RNA is copied into double-stranded DNA before cloning.

A plasmid cloning vector can slip into bacteria, yeast, or other host cell. Host cells divide rapidly, thereby making multiple, identical copies of the foreign DNA.

11.2 Haystacks to Needles

To study one gene, researchers must first find it among all the others in an organism's DNA. It is a bit like searching for a needle in a haystack. Once found, the gene is copied many times to make enough material for experiments.

ISOLATING GENES

A **library** is a collection of cells that host fragments of DNA from a particular organism. Collectively, all cells in a *genomic library* contain the organism's entire set of genetic material, or **genome**. A *cDNA library* is derived from mRNA, and so it represents only genes that are being expressed.

A clone that contains a particular gene of interest occurs in a library along with thousands or millions of clones that contain other genes. A probe can be used to find a single clone of interest among many others. A **probe** is a short fragment of DNA, complementary in sequence to a specific gene of interest, tagged with a detectable label such as a radioisotope or pigment.

How do researchers make a suitable probe? If they already know the desired gene sequence, they can use it to design and make a short chain of nucleotides. Or they can use some DNA from a closely related gene.

Probes distinguish one DNA sequence from all of the others in a library of clones. A probe will base-pair only with DNA that contains a complementary DNA sequence. In other words, a probe sticks to the gene of interest, but it does not stick to any other DNA. Base-pairing between any two strands of nucleic acids from different sources is called **nucleic acid hybridization**.

Researchers can pinpoint a DNA fragment that has the gene of interest, and also the clone that contains it, by detecting the label on a hybridized probe. Once the clone is found, it can be isolated from the other clones in the library. That clone will be grown in a nutrient medium to form a gigantic population of genetically identical cells, from which researchers can extract the cloned DNA in sufficient quantity for experiments. An example of how researchers isolate a bacterial clone by using a radioactive probe is shown in Figure 11.5.

PCR

Researchers can isolate and mass-produce a particular DNA fragment without cloning, by using a technique called the polymerase chain reaction, or PCR. **PCR** is a hot–and–cold cycled reaction that uses a heat-tolerant polymerase to replicate a targeted DNA fragment by the billion. It can amplify, or increase the quantity of,

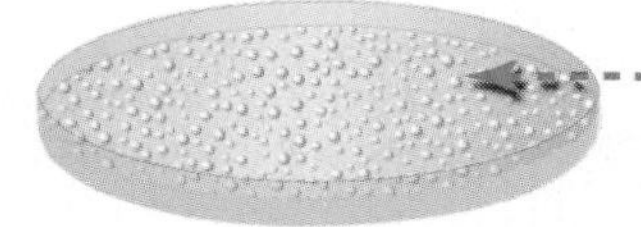

a Individual cells from a library of bacterial cell clones are spread over the surface of a solid growth medium in a dish. The bacterial cells divide repeatedly to form colonies, clusters of millions of genetically identical daughter cells.

b A piece of special paper pressed onto the surface of the growth medium will bind some cells from each colony.

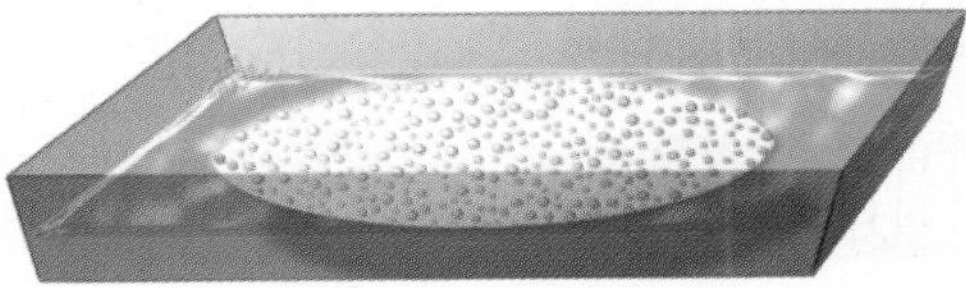

c The paper is soaked in a solution that ruptures the cells and releases their DNA. The DNA clings to the paper in spots mirroring the distribution of the colonies.

d A probe is added to the liquid bathing the paper. The probe hybridizes with (sticks to) only the spots of DNA that contain the targeted sequence.

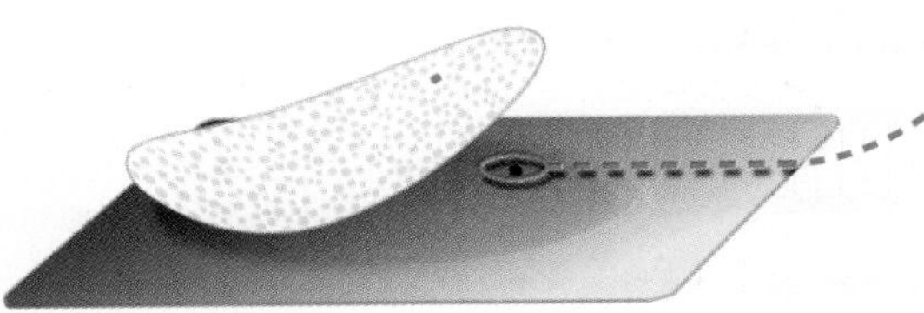

e The bound probe makes a radioactive spot that will darken x-ray film. The position of the spot on the film relative to all the other spots of DNA is used to find the original colony in the dish. Cells from that colony alone are grown to isolate the cloned gene of interest.

Figure 11.5 ***Animated!*** Example of nucleic acid hybridization. Here, a radioactive probe helps identify a clone, in this case a bacterial colony, that contains a targeted gene.

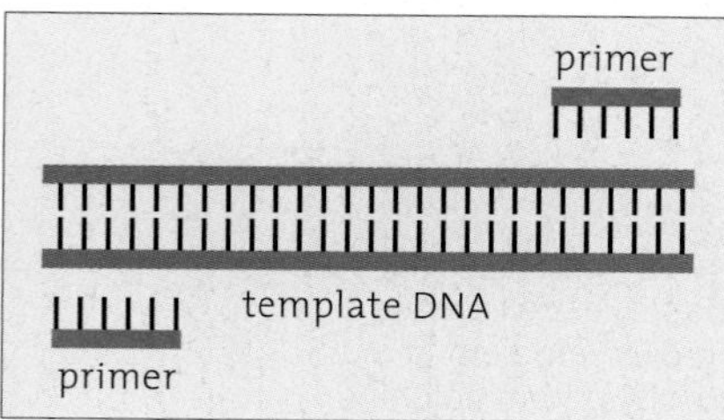

a Primers, free nucleotides, and DNA template are mixed with heat-tolerant DNA polymerase.

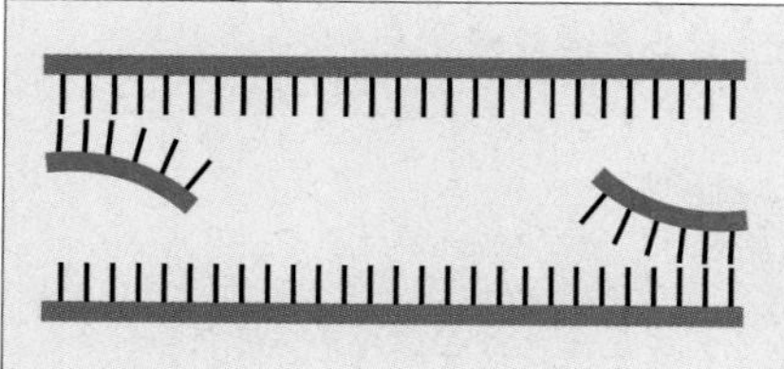

b When the mixture is heated, the DNA denatures. When it is cooled, some primers hydrogen-bond to the DNA template.

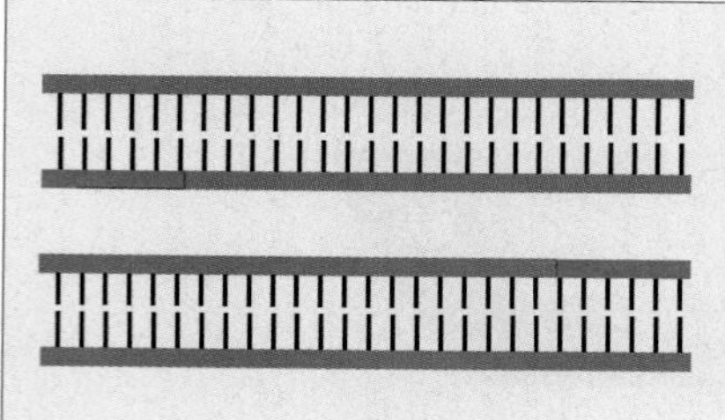

c *Taq* polymerase uses the primers to initiate synthesis. It assembles a copy of the DNA template from free nucleotides. The first round of PCR is completed.

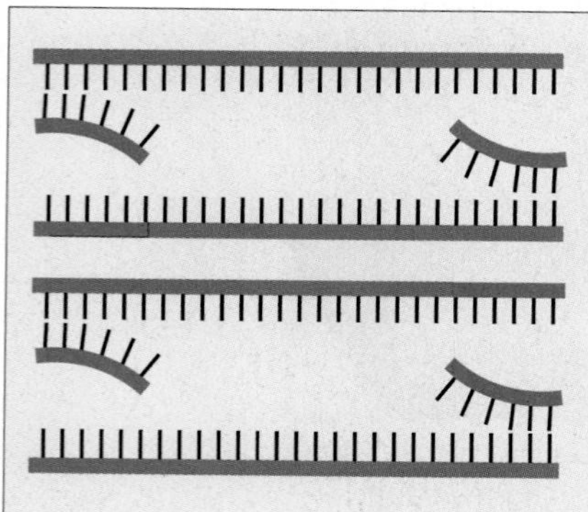

d The mixture is heated again. This denatures all the DNA into single strands. When the mixture is cooled, some of the primers hybridize with the DNA.

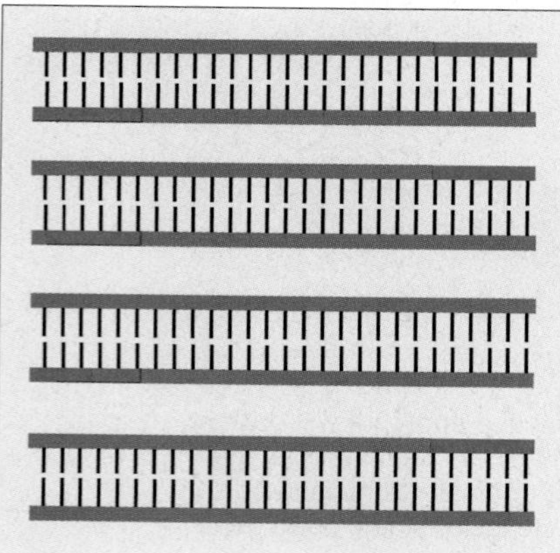

e *Taq* polymerase uses the primers to initiate synthesis and then copies the DNA. The second round of PCR is complete. Each successive round of synthesis can double the number of DNA molecules.

Figure 11.6 ***Animated!*** Two rounds of the polymerase chain reaction, or PCR. A bacterium, *Thermus aquaticus*, is the source of the *Taq* polymerase. Thirty or more cycles of PCR can yield a billionfold increase in the number of starting DNA template molecules.

an improbably tiny amount of DNA. PCR transforms a needle in a haystack, one DNA fragment mixed with millions of others, into a huge stack of needles with a little hay in it (Figure 11.6).

The starting material for PCR can be any sample that contains at least one molecule or fragment of a target DNA. It can be DNA extracted from a mixture of ten million different clones, a single sperm cell, a drop of blood at a crime scene, a mummy—essentially anything that has DNA in it.

In a PCR reaction, researchers mix the starting DNA with free nucleotides, DNA polymerase, and primers. **Primers** are synthetic single strands of DNA, usually between ten and thirty bases long. They are designed to base-pair only with nucleotide sequences on either end of the DNA sequence to be amplified.

The researchers then expose the mixture to cycles of high and low temperatures. At high temperature, the two strands of a DNA double helix separate into single strands. During a high-temperature PCR cycle, every DNA molecule unwinds and becomes single-stranded. Then, the mixture is allowed to cool. Like molecular sleuths, some of the primers find the template needle in the haystack mixture, and hybridize with it.

The high temperature needed to separate strands of DNA would destroy most DNA polymerases (Section 2.9). The special polymerase used for PCR reactions is harvested from *Thermus aquaticus*, a species of bacteria that thrives in superheated spring water (Chapter 14). The *Taq* polymerase is necessarily heat-tolerant.

Like all other DNA polymerases, *Taq* polymerase recognizes primers hybridized with template DNA as places to initiate synthesis. Synthesis proceeds along the template until the temperature cycles up and the DNA separates into single strands again. The newly synthesized DNA is a copy of the target sequence, so the number of DNA template molecules has doubled.

When the temperature cycles down again, primers rehybridize to new template DNA molecules, and the reactions start all over. With each successive round of temperature cycling, the number of copies of targeted DNA doubles. After thirty PCR cycles, the number of template molecules will have been amplified by about a billionfold.

Probes may be used to help identify a clone that hosts a DNA fragment of interest among many clones in a library. That clone is cultured, and its DNA is harvested.

The polymerase chain reaction (PCR) is a method of quickly and exponentially amplifying even a tiny amount of DNA. Compared to cloning methods, PCR more rapidly amplifies a target DNA sequence.

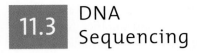

11.3 DNA Sequencing

Sequencing reveals the order of nucleotide bases in DNA. This technique uses DNA polymerase to partially replicate a DNA template. Manual sequencing methods have been replaced largely by automated techniques.

Once a targeted DNA molecule has been isolated in sufficient quantity, the order of its nucleotide bases may be determined with **DNA sequencing**—another technique in the molecular toolkit.

Researchers mix the DNA template to be sequenced with nucleotides, primer, and DNA polymerase. The reaction mixture also includes nucleotides modified to terminate DNA synthesis when added to the end of a growing strand of DNA. Also, each of the four kinds of modified nucleotide bases (G, A, T, and C) is tagged to fluoresce a different color under laser light.

As in DNA replication, the polymerase constructs a new strand of DNA by stringing nucleotides together according to the template DNA sequence. Each time, the polymerase randomly attaches either a standard *or* a modified nucleotide to the end of the new strand of DNA. If it attaches one of the modified nucleotides, synthesis of that nucleotide chain ends. After enough time, there are new strands of DNA that end at each base in the template sequence. The mixture now holds millions of DNA fragments of different lengths, each of which is labeled with one of four fluorescent pigments.

The fragments are separated by **gel electrophoresis**. With this technique, an electric field pulls the DNA fragments through a polymer matrix, a gel that is a little like Jello®. This matrix hinders the movement of larger molecules more than that of smaller ones, as a dense forest would hinder the movement of elephants more than tigers. So different sized fragments migrate at different rates through the gel according to length—larger fragments move more slowly than smaller ones. Fragments of the same length migrate through the gel at the same rate, and form distinct bands.

As the bands reach the end of the gel, they pass through a laser beam. Remember, each DNA fragment has a modified nucleotide attached to its end, one that fluoresces a certain color under laser light. Thus, each band fluoresces with the color of one kind of modified nucleotide. The order of the colored bands indicates the DNA sequence (Figure 11.7).

DNA sequencing reveals the order of nucleotides in DNA. The technique uses gel electrophoresis, which separates fragments of DNA according to their length.

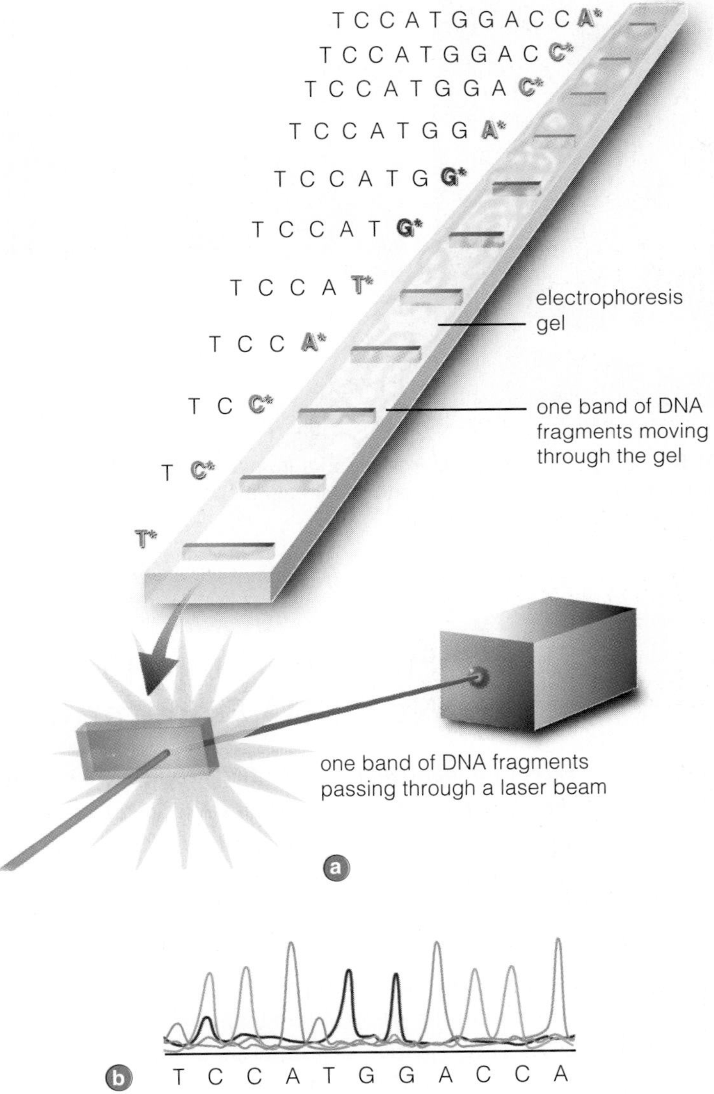

Figure 11.7 ***Animated!*** DNA sequencing. (**a**) DNA fragments are synthesized using a template and fluorescent nucleotides, then separated by gel electrophoresis. Fragments of the same length migrate through the gel at the same rate, and so form distinct bands. (**b**) The order of the fluorescent bands that appear in the gel indicates the template DNA sequence.

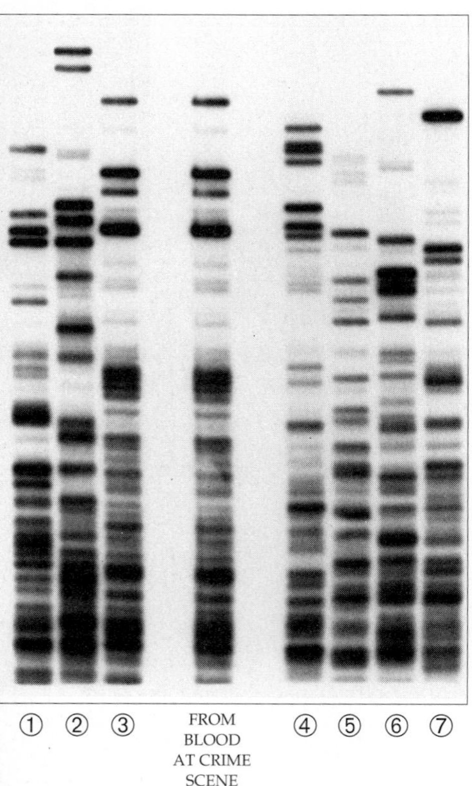

Figure 11.8 Comparison of the DNA fingerprint from a bloodstain left at a crime scene and from blood samples of seven suspects (the circled numbers). Can you point out which of the seven is a match?

LINKS TO SECTIONS 7.6, 8.7

11.4 First Just Fingerprints, Now DNA Fingerprints

Except for identical twins, no two people have exactly the same base sequence in their DNA. Scientists can distinguish one person from another on the basis of differences in those sequences.

Each human has a unique set of fingerprints. Like all other sexually reproducing species, each also has a **DNA fingerprint**—a unique array of DNA sequences inherited in a Mendelian pattern from parents. About 99 percent of the DNA in all humans is the same, but the other 1 percent is unique to each individual. Most unique stretches of DNA are sprinkled throughout the human genome as **tandem repeats**—many copies of the identical base sequences, positioned one after the next along the length of a chromosome.

For example, one person's DNA might contain four repeats of the nucleotides TTTTC in a certain location. Another person's DNA might have TTTTC repeated fifteen times in the same location. One person might have five repeats of CGG, and another might have fifty. These repetitive sequences spontaneously slip into DNA during replication, and their numbers grow or shrink over generations. The mutation rate is high around tandem repeat regions.

DNA fingerprinting reveals differences in the tandem repeats among individuals. A restriction enzyme cuts genomic DNA into fragments, the sizes of which are unique to an individual, because they reflect regions of tandem repeats. Thus, the genetic differences between people can be revealed by gel electrophoresis.

As in DNA sequencing, the fragments form bands on an electrophoresis gel according to their length. The banding pattern of a person's genomic DNA fragments on an electrophoresis gel is his or her DNA fingerprint. For all practical purposes, it is identical only between identical twins, because the odds that two unrelated people would share an identical DNA fingerprint are about one in three trillion.

PCR can be used to amplify tandem-repeat regions. Differences in the sizes of amplified DNA fragments are again detected with gel electrophoresis. One drop of blood or semen, or a few cells from a hair follicle at a crime scene or on a suspect's clothing yield enough DNA to amplify with PCR.

DNA fingerprints help forensic scientists identify criminals, victims, and innocent suspects. Figure 11.8 shows some DNA fragments of genomic DNA that were separated by gel electrophoresis. Those samples of DNA were taken from seven people and from a bloodstain that was left at a crime scene. One of the DNA fingerprints matched.

Defense attorneys initially challenged the use of DNA fingerprinting as evidence in court. However, today the procedure has been firmly established as accurate and unambiguous. Now, DNA fingerprinting is routinely submitted as evidence in disputes over paternity, and it is being widely used to convict the guilty as well as exonerate the innocent. To date, such evidence has helped secure the release of more than 140 innocent people from prison.

DNA fingerprint analysis also has confirmed that human bones exhumed from a shallow pit in Siberia belonged to five individuals of the Russian imperial family, all shot to death in secrecy in 1918. It also was used to identify the remains of those who died in the World Trade Center on September 11, 2001.

Tandem repeats—many copies of short DNA sequences in a chromosome—differ from one person to the next. They are the basis of DNA fingerprint comparisons.

11.5 Tinkering With the Molecules of Life

The structural and comparative analysis of genomes is yielding information about evolutionary trends as well as potential therapies for genetic diseases.

EMERGENCE OF MOLECULAR BIOLOGY

In 1953, James Watson and Francis Crick unveiled their model of the DNA double helix and ignited a global blaze of optimism about genetic research. The very book of life seemed to open up for scrutiny. In reality, it dangled just beyond reach. Major scientific breakthroughs are not very often accompanied by the simultaneous discovery of the tools to study them. New methods of doing research had to be invented before that book could become readable.

In 1972, Paul Berg and his associates were the first to make recombinant DNA by fusing fragments of DNA from one species into the genetic material from a different species. Their technique allowed them to isolate and replicate manageable subsets of DNA from any organism they wanted to study. The discipline of molecular biology was born, and suddenly everybody was worried about it.

Although scientists knew that DNA itself was not toxic, they could not predict with absolute certainty what would happen every time they fused genetic material from different organisms. Would they create new super-pathogens by accident? Could DNA from normally harmless organisms be fused to create a new form of life? What if their creation escaped into the environment and transformed other organisms?

In a remarkably immediate and responsible display of self-regulation, scientists reached a consensus on safety guidelines for doing DNA research. Adopted at once by the National Institutes of Health (NIH), the guidelines listed laboratory procedural precautions. They covered the design and use of host organisms that could survive only under the narrow range of conditions that occur in the laboratory.

A golden age of recombinant DNA research soon followed. The emphasis had shifted from the chemical and physical properties of DNA to its exact molecular structure. In 1977, Allan Maxam, Walter Gilbert, and Fred Sanger developed a method for determining the nucleotide sequence of cloned DNA fragments. The tools for reading the book of life, opened more than twenty years earlier, were now available for everyone to use.

DNA sequencing was cool, a visually rewarding, data-rich technique that soon entranced more than a few scientists (Figure 11.9). Tremendous amounts of sequence data started to amass, from diverse species. Computer technology was simultaneously advancing, but it was barely keeping pace with the demand for sequence data analysis and storage. In 1982, the NIH allocated 3 million dollars to fund the first large-scale DNA sequence database in the United States, one that was accessible to the public.

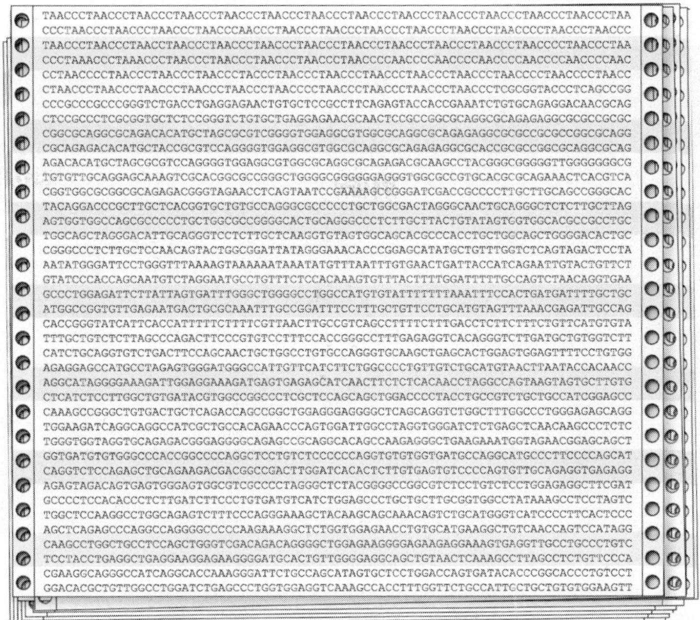

Figure 11.9 DNA sequence data. Here, a few thousand bases of the DNA sequence of the human genome.

LINK TO SECTION 9.1

THE HUMAN GENOME PROJECT

Around 1986, it seemed that everyone was arguing about sequencing the human genome. Many insisted that knowing the sequence would offer tremendous support for medicine and pure research. Others said that such a project would divert funding from other research that had greater urgency as well as a greater likelihood of success.

At the time, sequencing three billion bases seemed like an impossible task. Sequencing has a maximum resolution of about 500 bases of sequence data per run, so at least 6 million sequencing runs would be necessary. With the techniques available at the time, this would have taken at least fifty years.

But techniques were getting better every year, and more bases were being sequenced and analyzed in less time. Automated (robotic) sequencing had just been invented, as had PCR, the polymerase chain reaction. Although both techniques were still cumbersome, expensive, and far from standardized, many sensed their potential.

Figure 11.10 A few of the supercomputers used to sequence the human genome, at Celera Genomics in Maryland.

LINK TO SECTION 10.5

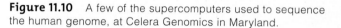

Waiting for faster methods seemed to be the most efficient means of sequencing the human genome, but exactly when was the technology going to be fast *enough*? Who would decide? It was during this heated debate in 1987 that several independent organizations launched their own efforts to sequence the human genome. Among them was Genome, a company started by Walter Gilbert. Gilbert announced that his company not only was going to sequence the human genome, but the company was also going to patent it.

In 1988, the NIH effectively annexed the Human Genome Project by hiring James Watson as its head and providing genome researchers with 200 million dollars per year. A consortium formed between the NIH and other institutions working on the different versions of the project.

Amid ongoing squabbles over patent issues, the bulk of the Human Genome Project continued as such in the United States until 1998, when scientist Craig Venter started Celera Genomics (Figure 11.10). Venter declared that *his* new company would be the first to complete the sequence. His challenge prompted the United States government to propel its sequencing efforts into high gear.

Sequencing of the human genome was officially completed in 2003—fifty years after the discovery of the structure of DNA. Around 99 percent of the coding regions in human DNA have been sequenced with a high degree of accuracy. To date, more than 21,000 genes have been identified. This does not mean that we know what all of those genes encode; it only means we know they are definitely genes.

IMPACTS, ISSUES

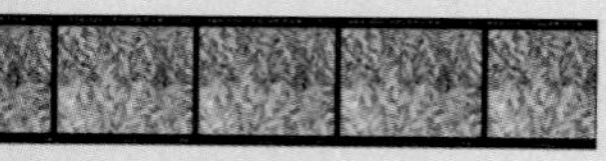

In 2002, an international research team finished a draft sequence of the rice genome. The data were made available on the Internet free of charge. Rice is the staple food for more than half the world's population. It was the first crop plant to have its genome fully sequenced. This work will be used to design new rice varieties with more nutritional value, bigger crop yields, and more resistance to drought and pests.

What do we do with this vast amount of data? The next step is to investigate questions about precisely what the sequence means: where all of the genes are, where they are not, what the genes encode, and what the control mechanisms are and how they operate.

GENOMICS

Research into genomes of humans and other species has converged into a new research field—**genomics**. The *structural* genomics branch deals with the actual mapping and sequencing of genomes of individuals. The *comparative* genomics branch is concerned with finding evolutionary relationships among groups of organisms. Comparative genomics researchers analyze similarities and differences among genomes.

Comparative genomics has practical applications as well as potential for research. The basic premise is that the genomes of all existing species are derived from common ancestors. For instance, pathogens share some conserved genes with their human hosts even though the lineages diverged long ago. Shared gene sequences, how they are organized, and where they differ might be clues to where our immune defenses against these pathogens are strongest or the most vulnerable.

Genomics has potential for **human gene therapy**—the transfer of one or more normal or modified genes into a person's body cells to correct a genetic defect or boost resistance to a disease. For example, some gene therapies deliver modified cells into a patient's tissue. Others use stripped-down viruses as vectors that can inject genes into a person's cells. In many cases, these therapies can alleviate the patient's symptoms even if the modified cells produce just a small amount of the gene's product.

However, a gene is not easy to manipulate within the context of a living individual, even if we know its sequence and where it lies in the genome. No one, for example, can predict where a virus-injected gene will end up in a person's chromosomes. The danger is that the insertion might disrupt other genes, particularly those which control cell division and growth (Section 11.7 gives an example). One-for-one gene swaps using genetic recombination methods are possible, but like other gene therapies they are still experimental.

Fifty years after the structure of DNA was discovered, the human genome sequence was completed.

We are applying knowledge about the genomes of humans and other species to reconstruct evolutionary histories and to design gene therapies.

11.6 Practical Genetics

Genetic engineering changes the genetic makeup of an organism, often with the intent to alter its phenotype. A gene may be transferred between species, or modified and reinserted into an organism of the same species.

DESIGNER PLANTS

As crop production expands to keep pace with human population growth, it puts unavoidable pressure on ecosystems everywhere. Irrigation leaves mineral and salt residues in soils. Tilled soil erodes, taking topsoil with it. Runoff clogs rivers, and fertilizer in it causes algae to grow so much that fish suffocate. Pesticides harm humans, other animals, and beneficial insects. Pressured to produce more food at lower cost and with less damage to the environment, some farmers are turning to genetically engineered crop plants.

Cotton plants with a built-in insecticide gene kill only the insects that eat it, so farmers that grow them do not have to use as many pesticides. Genetically modified wheat has double the yield per acre. Certain transgenic tomato plants survive in salty soils that wither other plants; they also absorb and store excess salt in their leaves, thus purifying saline soil for future crops. **Transgenic** simply refers to an organism into which DNA from another species has been inserted, as in Figure 11.11.

The cotton plants in Figure 11.11*a* were genetically engineered for resistance to a relatively short-lived herbicide. Spraying fields with this herbicide kills all of the weeds but not the engineered cotton plants. The practice means farmers can use fewer toxic chemicals and less of them. They also do not have to till the soil as much to control weeds, so there is less runoff.

Aspen tree seedlings in which a lignin biosynthesis pathway has been modified still make lignin, but not as much (Figure 11.11*b*). Wood from lignin-deficient trees might make it easier to manufacture paper and clean-burning fuels such as ethanol.

Engineering plants starts with cloning vectors that carry genes into plant cells. *Agrobacterium tumefaciens* is a bacterial species that infects many plants, including beans, peas, potatoes, and other vital crops. Its plasmid genes cause tumor formation on these plants; hence the name Ti plasmid (*T*umor-*i*nducing). The Ti plasmid is used as a vector for transferring new or modified genes into plants (Figure 11.12). Researchers excise the tumor-inducing genes, then insert a desired gene into the Ti plasmid. Plant cells cultured with the modified

Figure 11.11 Transgenic plants. (**a**) Cotton plant (*left*), and cotton plant with a gene for herbicide resistance (*right*). Both were sprayed with weed killer.

(**b**) When biosynthesis for lignin was partially blocked in these genetically engineered aspen seedlings, cellulose production increased. Root, stem, and leaf growth were greatly enhanced. Lignin is a tough polymer that strengthens plant cell walls. Unmodified seedling is on the left.

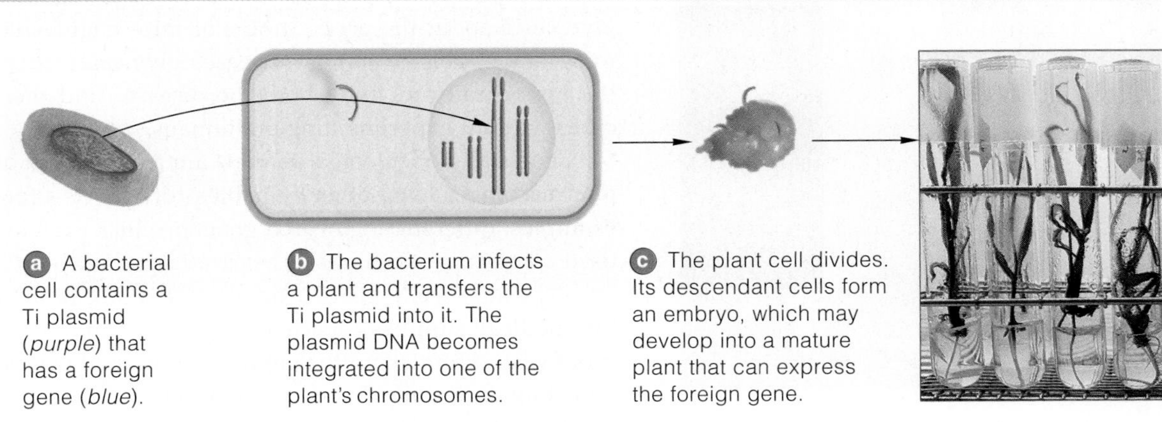

ⓐ A bacterial cell contains a Ti plasmid (*purple*) that has a foreign gene (*blue*).
ⓑ The bacterium infects a plant and transfers the Ti plasmid into it. The plasmid DNA becomes integrated into one of the plant's chromosomes.
ⓒ The plant cell divides. Its descendant cells form an embryo, which may develop into a mature plant that can express the foreign gene.

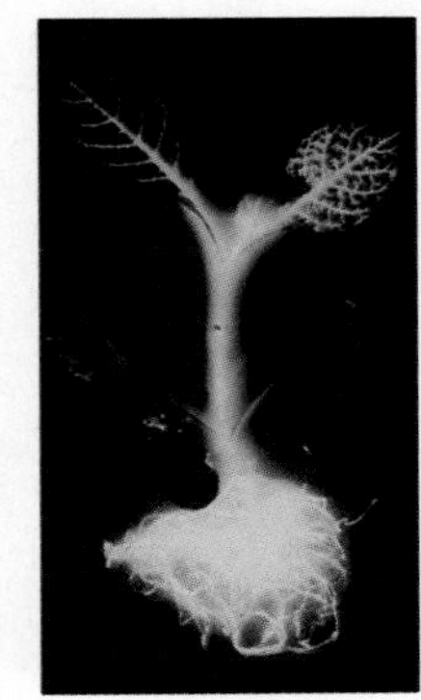

A young plant expressing a fluorescent gene product

Figure 11.12 ***Animated!*** Gene transfer from a bacterium, *Agrobacterium tumefaciens*, to a plant cell by way of a Ti plasmid.

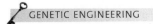

IMPACTS, ISSUES

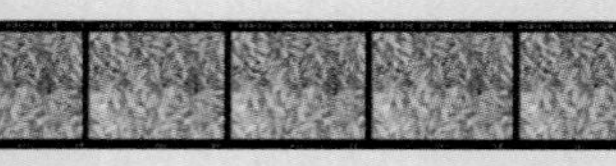

A new version of golden rice contains DNA from two different sources: a soil bacterium and maize. A mixture of these DNAs was introduced into rice embryos. Once inside, the foreign genes directed the production of two enzymes that catalyze the synthesis of beta-carotene in endosperm, a nutritive tissue to be used by the embryo seedling. The new golden rice has 20 times more beta-carotene than the original version.

plasmid take it up. Whole plants may be regenerated. Modified *A. tumefaciens* bacteria can deliver genes into some food crop plants, including wheat, corn, and rice. Researchers can also transfer genes into plants by way of electric shocks, chemicals, or blasts of microscopic particles coated with DNA.

To date, sixty-five genetically modified plants are approved by the USDA for general crop use in the United States, and hundreds more are in the works. Among them are berries, apples, and melons that ripen later, disease-resistant plums, papaya, grapes, carrots, and walnuts, and reduced-nicotine tobacco.

GENETICALLY ENGINEERED BACTERIA

Decoding the genetic scripts of species opens the door to great possibilities. Consider genetically engineered bacteria that produce medically valued proteins. Huge populations churn out vast quantities of desired gene products in stainless steel vats.

Diabetics, who must get daily insulin injections, are among the beneficiaries of these bacteria. At one time, insulin for their injections was extracted from pigs and cattle. In 1982, synthetic genes for human insulin were transferred into *E. coli* cells. The cells became the first large-scale, cost-effective "factory" for production of a medically important protein. Some other engineered products of bacteria are various blood-clotting factors, taxol, and human growth hormone.

Modified bacteria also hold potential for industry and for cleaning up environmental messes. Many can break down crude oil and other organic wastes into less harmful compounds, and cycle them as nutrients through ecosystems. When sprayed on oil spills, they may avert ecological disasters of the sort that occurred in the Persian Gulf (Chapter 1). Others absorb excess phosphates and heavy metals.

Plasmid-containing species are being used in basic research, agriculture, and gene therapies. Deciphering bacterial genes is also helping researchers reconstruct the evolutionary history of life. The next two sections give a few more examples of this application.

What about the "bad" bunch, the pathogenic fungi, bacteria, and viruses? Natural selection favors mutated genes that improve a pathogen's chances of evading a host's defenses. Mutation is frequent among pathogens that reproduce fast, so designing new antibiotics and other defenses against their altered gene products is a constant challenge. If we understand their genes, we may be able to keep up with them. The story of HIV, a rapidly mutating virus that causes AIDS, hints at the magnitude of the problem (Section 23.8).

BARNYARD BIOTECH

The first mammals enlisted for experiments in genetic engineering were lab mice. Transgenic mice appeared on the research scene in 1982 when scientists built a plasmid containing a gene for rat somatotropin (growth hormone). They injected the recombinant DNA into fertilized mouse eggs, and then implanted the eggs into female mice. One-third of the resulting offspring grew much larger than their littermates. The rat gene had become integrated into their DNA and was being expressed, with impressive results.

Now, transgenic animals are commonly used for medical research. The function and regulation of many gene products have been discovered using "knockout mice," in which targeted genes are inactivated by using recombination techniques. Defects in the resulting mice give clues about the gene. Strains of mice engineered to be susceptible to human diseases give researchers the opportunity to study both the diseases and their cures without experimenting on humans.

Genetically engineered animals are also sources of pharmacological and other valuable proteins. As a few examples, different engineered goats produce proteins used to treat cystic fibrosis, heart attacks, and nerve gas exposure. Some goats produce spider silk protein for medical supplies, space equipment, and bullet-proof vests. Human antithrombin, used to treat people with blood clotting disorders, is another product of engineered goats (Figure 11.13*a*). Rabbits make human IL-2 (interleukin-2) a protein that stimulates immune cell divisions. Cattle produce human collagen that can be used for surgical repair of cartilage, bone, and skin.

a

b

Figure 11.13 Genetically modified animals. (**a**) Mira, a goat transgenic for human antithrombin III, an anticlotting factor. (**b**) Inquisitive transgenic pig at the Virginia Tech Swine Research facility.

Figure 11.14 Featherless chicken developed by traditional cross-breeding methods in Israel. Such chickens survive in hot deserts where cooling systems are not an option.

Genetic engineering has also given us dairy goats with heart-healthy milk, pigs whose low phosphorus manure is easier on the environment, freeze-resistant salmon, extra-hefty sheep, low-fat pigs, and mad cow disease-resistant cows. Allergen-free cats are next.

Tinkering with the genetics of animals for the sake of human convenience does raise some serious ethical issues, particularly so because failed experiments can have gruesome results. However, is transgenic animal research not simply an extension of thousands of years of acceptable barnyard breeding practice (Figure 11.14)? The techniques have changed, but not the intent. Like our ancestors, we continue to have a vested interest in improving our livestock.

Transgenic plants help farmers grow crops more efficiently.

Transgenic animals are widely used in medical research. Some are sources of medically valued proteins. Food animals are being engineered to be more nutritious, more disease resistant, or easier to raise.

11.7 Weighing the Benefits and Risks

We as a society continue to work our way through the ethical implications of DNA research even while we are applying the new techniques to medicine, industry, agriculture, and environmental remediation.

Over 15,500 serious genetic disorders affect between 3 and 5 percent of all newborns, and they cause 20 to 30 percent of all infant deaths per year. They account for about 50 percent of the mentally impaired and nearly 25 percent of all hospital admissions.

Rhys Evans (Figure 11.15) was born with a severe immune deficiency known as SCID-X1, which stems from mutations in a gene called *IL2RG*. Children who are affected by this disorder can live only in germ-free isolation tents, because they cannot fight infections.

Figure 11.15 Gene therapy. Rhys Evans was born with a gene mutation that causes SCID-X1. His immune system never could fight infections. A gene transfer freed him from life in a germ-free isolation tent.

In 1998, a viral vector was used to insert unmutated copies of *IL2RG* into cells taken from the bone marrow of eleven boys with SCID-X1. Each child's modified cells were infused back into his bone marrow. Months afterward, ten of the children left their isolation tents for good. Their immune systems had been repaired by the gene therapy. Since then, several other SCID-X1 patients, including Rhys Evans, have improved thanks to other gene therapy trials.

Three children from the initial experiment in 1998 have since developed leukemia, and one died. Their illness surprised researchers, who had predicted that any cancer related to the therapy would be extremely rare. Our understanding of how the human genome works clearly lags behind our ability to modify it.

Modifying the human genome has profound ethical implications even beyond unexpected risks. To many, correcting genetic disorders with gene therapy seems like a socially acceptable goal. Now take this idea one step further. Is it acceptable to change some genes of a normal human in order to alter or enhance traits?

The philosophy of selecting desirable human traits, called *eugenics*, is not new, but it is still controversial. Who, for example, decides which forms of a trait are the most desirable? Realistically, cures for most severe but rare genetic disorders will not be pursued because the financial payback will not be enough to cover the research. Eugenics, however, might just turn a profit. How much might potential parents pay to ensure that their children will be tall or blue-eyed or fair-skinned? Would it be okay to engineer "superhumans" with breathtaking strength or intelligence? How about an injection that would help you lose that extra weight, and keep it off permanently? The borderline between interesting and abhorrent is not the same for everyone.

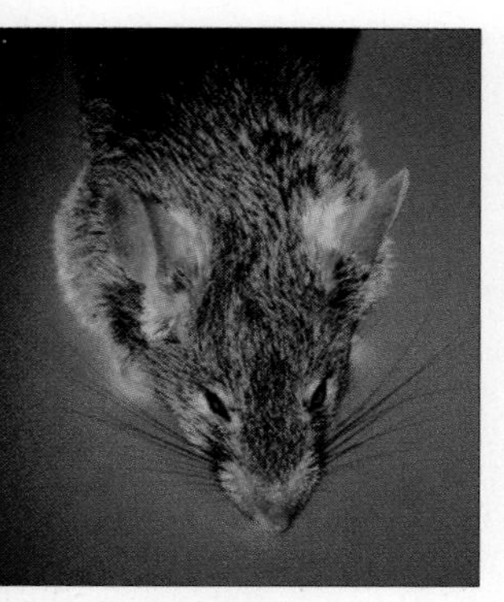
transgenic mouse expressing green fluorescent protein

In a survey conducted not long ago in the United States, more than 40 percent of those interviewed said it would be fine to use gene therapy to make smarter or better looking babies. In one poll of British parents, 18 percent would use genetic enhancement to prevent their children from being aggressive, and 10 percent would use it to keep them from being homosexual.

KNOCKOUT CELLS AND ORGAN FACTORIES

Each year, about 75,000 people are on waiting lists for an organ transplant, but human donors are in short supply. There is talk of harvesting organs from pigs, because pig organs function very much like ours do. Transferring an organ from one species into another is called **xenotransplantation**.

The human immune system battles anything that it recognizes as "nonself." It rejects a pig organ at once. Researchers have successfully engineered pigs lacking immunity-inducing molecules on the surface of their cells. The human immune system might not recognize and attack organs from these pigs. Tissues and organs from such animals could benefit millions of people, including those who suffer from diabetes, Parkinson's disease, kidney failure, AIDS, and hepatitis.

Critics of xenotransplantation are concerned that, among other things, pig to human transplants would put selective pressure on pig viruses to cross species and infect humans. These concerns are not unfounded. In 1918, an influenza pandemic killed twenty million people worldwide. It originated with a strain of avian flu that mutated to become infectious in humans.

REGARDING "FRANKENFOOD"

Genetically engineered foods are widespread in the United States. In 2003, 81 percent of soybean crops and 40 percent of corn crops were modified to withstand weedkillers. Such crops are no longer fed only to farm animals. At least 70 percent of all processed foods in grocery stores—like tofu, cereal, soy sauce, vegetable oils, breads, frozen pizzas, hot dogs, soft drinks, and beer—may contain some ingredients from genetically engineered plants.

Read the research and form your own opinions. The alternative is to be swayed by media hype (the term "Frankenfood," for instance), or by potentially biased reports from other groups that might have a different agenda (such as herbicide manufacturers).

Our ability to manipulate genomes is outpacing attempts to work through some of its bioethical implications.

Summary

Section 11.1 Recombinant DNA technology uses restriction enzymes that cut DNA into fragments. The fragments may be spliced into cloning vectors by using DNA ligase. Recombinant plasmids are taken up by host cells, such as bacteria, to make multiple, identical copies of the foreign DNA. mRNA can be cloned after it is copied into double-stranded DNA.

Section 11.2 A gene library is a mixed collection of host cells that have taken up cloned DNA. A particular gene can be isolated from a library by using a probe, a short stretch of DNA that can base-pair with the gene and that is traceable with a radioactive or pigment label. Probes help researchers identify one particular clone among millions of others. Base-pairing between strands of DNA or RNA from different sources is called nucleic acid hybridization.

The polymerase chain reaction (PCR) is a way to rapidly copy particular pieces of DNA. A DNA template is mixed with nucleotides, primers, and a heat-resistant DNA polymerase. Each cycle of PCR proceeds through a series of temperature changes that doubles the number of DNA molecules. After thirty cycles, PCR amplification can result in a billionfold increase in the number of DNA molecules.

BiologyNow
Learn how researchers isolate and copy genes with the interaction on BiologyNow.

Section 11.3 DNA sequencing reveals the order of nucleotides in DNA fragments. As DNA polymerase is copying a template DNA, progressively longer fragments stop growing when one of four kinds of modified, fluorescent nucleotides becomes attached.

Electrophoresis separates the resulting labeled DNA fragments into bands according to length. The order of the colored bands as they migrate through the gel reflects which fluorescent base was added to the end of each fragment, and so indicates the template DNA nucleotide sequence.

BiologyNow
Investigate DNA sequencing with the animation on BiologyNow.

Section 11.4 Tandem repeats are multiple copies of a short DNA sequence that follow one another along a chromosome. The number and distribution of tandem repeats, unique in each person, can be revealed by gel electrophoresis; they form a DNA fingerprint.

BiologyNow
Observe the process of DNA fingerprinting with the animation on BiologyNow.

Section 11.5 Discovery of DNA's double helical structure sparked interest in deciphering its genetic messages. A global race to complete the sequence of the human genome spurred rapid development of new techniques to study and manipulate DNA. The entire human genome has been sequenced and is now being analyzed in detail.

Genomics, the study of human and other genomes, is shedding light on evolutionary relationships and has practical uses. Human gene therapy transfers normal or modified genes into body cells with the intent to correct genetic defects.

Section 11.6 Genetic engineering is the directed modification of the genetic makeup of an organism, often with intent to modify its phenotype. Researchers insert normal or modified genes from one organism into another of the same or different species. Gene therapies also reinsert altered genes into individuals.

Genetically engineered bacteria produce medically valued proteins. Transgenic crop plants help farmers produce food more efficiently. Genetic engineering of animals allows commercial production of human proteins, as well as research into genetic disorders.

BiologyNow
See how the Ti plasmid is used to genetically engineer plants with the animation on BiologyNow.

Section 11.7 Gene therapy and the modification of animals for xenotransplantation are examples of developing technologies. As with any new technology, potential benefits must be weighed against potential risks, including ecological and social repercussions.

Self-Quiz

Answers in Appendix I

1. _______ is the transfer of normal genes into body cells to correct a genetic defect.
   a. Reverse transcription   c. PCR
   b. Nucleic acid hybridization   d. Gene therapy

2. DNA is cut at specific sites by _______ .
   a. DNA polymerase   c. restriction enzymes
   b. DNA probes   d. reverse transcriptase

3. Fill in the blank: A _______ is a small circle of bacterial DNA that is not part of the bacterial chromosome.

4. By reverse transcription, a molecule of _______ is assembled on a(n) _______ template.
   a. mRNA; DNA   c. DNA; ribosome
   b. cDNA; mRNA   d. protein; mRNA

5. By gel electrophoresis, fragments of DNA can be separated according to _______ .
   a. sequence   b. length   c. species

6. DNA sequencing uses _______ .
   a. standard and modified nucleotides
   b. primers and DNA polymerase
   c. gel electrophoresis and a laser beam
   d. all of the above

7. _______ can be used to insert genes into human cells.
   a. PCR   c. Xenotransplantation
   b. Modified viruses   d. cDNA

8. Match the terms with the most suitable description.
   ____ DNA fingerprint   a. selecting "desirable" traits
   ____ Ti plasmid   b. uses heat stable polymerase
   ____ PCR   c. used in some gene transfers
   ____ nucleic acid hybridization   d. a person's unique collection of tandem repeats
   ____ eugenics   e. base pairing of nucleotide sequences from different DNA or RNA sources

Additional questions are available on **BiologyNow™**

Critical Thinking

1. Lunardi's Market put out a bin of nice tomatoes having vine-ripened redness, flavor, and texture. A sign identified them as genetically engineered produce. Most shoppers selected unmodified tomatoes in the adjacent bin even though those tomatoes were pale pink, mealy-textured, and tasteless. Which tomatoes would you pick? Why?

2. Scientists at Oregon Health Sciences University produced ANDi, the first transgenic primate, by inserting a jellyfish gene into the fertilized egg of a rhesus monkey. (The gene encodes a protein that fluoresces green.) The long-term goal of this project is not to make fluorescent green monkeys, but the transfer of human genes into organisms with genomes most like ours. Transgenic primates could be studied to gain insight into human genetic disorders, which might lead to cures for those who are affected and vaccines for those at risk. Would it be ethical to transfer a monkey gene into a human embryo to cure a genetic defect? Or to bestow immunity against a potentially fatal disease such as AIDS?

3. Animal viruses can mutate so that they infect humans, occasionally with disastrous results. In 1918, an influenza pandemic that apparently originated with a strain of avian flu left in its wake 50 million people dead worldwide. Researchers recently isolated samples of that virus, the influenza A-H1N1 strain, from bodies of infected people that were fortuitously preserved in Alaskan permafrost since 1918. From the samples, the scientists reconstructed the DNA sequence of the entire viral genome, and from that sequence they reconstructed the actual virus. 39,000 times more infectious than recent influenza strains, the reconstructed A-H1N1 virus is 100 percent lethal in mice.

Understanding how the A-H1N1 strain works in a living context can help us defend ourselves against others that may be like it. For example, researchers are using the reconstructed virus to discover which of its mutations made it so infectious and deadly in humans. This research is of immediate importance. A deadly new strain of avian influenza in Asia has some mutations in common with the A-H1N1 strain. Perhaps most pertinent, researchers can test the effectiveness of antiviral drugs and vaccines on the reconstructed virus, and develop new ones.

Critics of the A-H1N1 reconstruction are concerned that if this virus escapes the containment facilities (this is not without precedent), it might cause another pandemic. Worse, terrorists could use the published DNA sequence and methods to make the virus for nefarious purposes. Do you think this research makes us more or less safe?

12 PROCESSES OF EVOLUTION

IMPACTS, ISSUES

Rise of the Super Rats

Slipping in and out of the pages of human history are rats—*Rattus*—the most notorious of mammalian pests. One kind of rat or another has distributed pathogens and parasites that cause bubonic plague, typhus, and other deadly infectious diseases. The death toll from fleas that bit infected rats and then bit people has exceeded the death toll in all wars combined.

The rats themselves are far more successful. By one estimate, there is one rat for every person in urban and suburban centers of the United States. In addition to spreading diseases, rats chew their way through walls and wires of homes and cities. In any given year, they cause economic losses approaching 19 billion dollars.

For years, people have been fighting back with traps, ratproof storage facilities, and various poisons. During the 1950s, they started using baits laced with warfarin. This compound interferes with blood clotting. Rats ate the baits, then died within days after bleeding internally or losing blood through cuts or scrapes. Warfarin was extremely effective. Compared with other rat poisons, it had a lot less impact on harmless species.

In 1958, however, a Scottish researcher reported that warfarin did not work against some rats. Similar reports from other European countries followed. About twenty years later, 10 percent of the urban rats caught in the United States were warfarin resistant. *What happened?*

To find out, researchers compared warfarin-resistant rat populations with still-vulnerable rats. They traced the difference to a gene on one of the rat chromosomes. At that gene locus, a dominant allele was common in warfarin-resistant rat populations but very rare among the vulnerable ones. "What happened" was evolution by natural selection. Warfarin was exerting selective pressure on populations of rats. The previously rare dominant allele proved to be adaptive. The lucky rats that inherited the allele survived and produced more offspring. The unlucky ones that inherited the recessive allele had no built-in defense, and died. Over time, the dominant allele's frequency increased in all rat populations exposed to the poison.

Of course, selection pressures can and often do change. When warfarin resistance increased in rat populations, people stopped using warfarin. And guess what: The dominant allele's frequency declined. Now the latest worry is the evolution of "super rats," which even more potent rodenticides can't seem to kill.

When you hear someone question whether life evolves, remember this: With respect to life, **evolution** simply means that heritable change is occurring in some line of descent. The actual mechanisms that can bring about such change are the focus of this chapter. Later chapters highlight how these mechanisms have contributed to the evolution of new species.

How Would You Vote? *Antibiotic-resistant strains of bacteria are becoming dangerously pervasive. Standard animal husbandry practice includes the repeated dosing of healthy animals with antibiotics—the same ones prescribed to people. Should this practice stop? See BiologyNow for details, then vote online.*

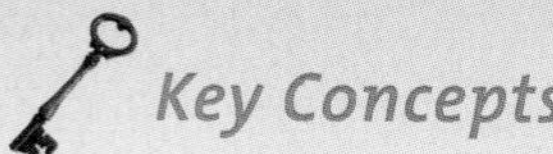

Key Concepts

EVOLUTIONARY VIEWS EMERGE

The world distribution of species, similarities and differences in body form, and the fossil record gave early evidence of evolution—of changes in lines of descent. Charles Darwin and Alfred Wallace had an idea of how those changes occur.

VARIATION AND ADAPTATION

An adaptation is a heritable aspect of form, function, behavior, or development that promotes survival and reproduction. It enhances the fit between the individual and prevailing conditions in its environment.

MICROEVOLUTIONARY PROCESSES

An individual does not evolve. A *population* evolves, which means its shared pool of alleles changes. Over generations, any allele may increase in its frequency among individuals, or it may become rare or lost. Mutation, genetic drift, natural selection, and gene flow change allele frequencies in a population. These processes of microevolution change the observable characteristics that define a population and, more broadly, a species.

Links to Earlier Concepts

Section 1.4 sketched out the key premises of the theory of natural selection. Here you will read about evidence that led to its formulation. You may wish to refresh your memory of protein structure (2.8), basic terms of genetics (8.1), chromosomes and crossing over (8.4, 8.5), DNA replication and repair (9.3), and gene mutation (10.5). This chapter puts the chromosomal basis of inheritance (8.4, 8.7) and continuous variation in populations (8.3) into the broader context of evolution.

12.1 Early Beliefs, Confounding Discoveries

Prevailing beliefs can influence how we interpret clues to natural processes and their observable outcomes.

QUESTIONS FROM BIOGEOGRAPHY

At one time Europeans viewed nature as a great Chain of Being extending from the "lowest" forms of life to humans, and on to spiritual beings. Each kind of being, or **species** as it was called, was one separate link in the chain. All the links had been designed and forged at the same time at one center of creation. They had not changed since. Once all the links were discovered and described, the meaning of life would be revealed.

Then Europeans embarked on their globe-spanning explorations and discovered the world is a lot bigger than Europe. Tens of thousands of unique plants and animals were brought back home to be catalogued as more links in the chain. But a few naturalists moved beyond cataloguing the species and started identifying *patterns* in where species live, and how species might or might not be related. These people were pioneers in **biogeography**, the study of patterns in the geographic distribution of species and communities. They were the first to ponder the relationship between ecology and evolution.

Some patterns were intriguing. For example, many plants and animals are found only on islands in the middle of the ocean and in other remote places. Many species that are strikingly similar live far apart, and vast expanses of open ocean or impassable mountain ranges keep them separated, in isolation.

Consider the three long-necked, long-legged, and flightless birds in Figure 12.1*a–c*. They live on different continents. Why are they alike? The two plants shown in Figure 12.1*d,e* have such differences in reproductive parts that they cannot be close relatives. They live on different continents. Yet both have spines, tiny leaves, and short fleshy stems. Why are *they* so much alike?

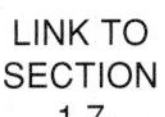

LINK TO SECTION 1.7

Curiously, the flightless birds all live in the same kind of environment about the same distance from the equator. They sprint about in flat, open grasslands of dry climates, and they raise their long necks to keep an eye on predators in the distance. Both plant species also live about the same distance from the equator in the same kind of environment—a desert—where water is seasonally scarce. Both store water in their fleshy stems. Their stems have a very thick, water-conserving cuticle. Both species have rows of sharp spines that deter thirsty, hungry animals. If all birds and plants arose in one place, then how did such similar kinds end up in such distant, remote places?

QUESTIONS FROM COMPARATIVE MORPHOLOGY

Questions about similarities and differences in body plans among groups of organisms led to another field of inquiry, now called **comparative morphology**. For example, the bones of a human arm, a whale flipper, and a bat wing differ in size, shape, and function. As the next chapter explains, they have similar positions in the body and the same tissues arranged in much the same patterns. These parts develop in much the same way in embryos. The scholars who deduced all of this wondered: Why are some animals that are so different in some features so much alike in others?

Figure 12.1 Species that resemble one another, strikingly so, even though they are native to distant geographic realms. (**a**) South American rhea, (**b**) Australian emu, and (**c**) African ostrich. All three types of birds live in similar habitats. They are unlike most birds in several traits, most notably in their long, muscular legs and their inability to fly. (**d**) A spiny cactus native to the hot deserts of the American Southwest. (**e**) A spiny spurge native to southwestern Africa.

By one hypothesis, basic body plans are so perfect there was no need to come up with a new one for each organism at the time of its creation. Yet if that were so, then how could there be parts having no function? For example, some ancient whales had ankle bones, but whales do not walk (Figure 12.2). Why the bones? Our coccyx is like some tailbones in many other mammals. We do not have a tail. Why do we have parts of one?

QUESTIONS ABOUT FOSSILS

About the same time, geologists were mapping layers of rock exposed by erosion or quarrying. As you will read later on, they added to the confusion when they found fossils in the same kinds of layers in different parts of the world. **Fossils** came to be recognized as stone-hard evidence of earlier forms of life. Figure 12.2 shows an example.

A puzzle: Many deep layers held fossils of simple marine life. Some layers above them contained fossils that were similar, but more intricate. In higher layers, the fossils looked like modern species. What did these sequences in complexity among fossils of a given type mean? Were they evidence of lines of descent?

Taken as a whole, the findings from biogeography, comparative morphology, and geology did not fit with prevailing beliefs. Scholars floated novel hypotheses. If a simultaneous dispersal of all species from a center of creation was unlikely, *then perhaps species originated in more than one place*. If species had not been created in a perfect state—and fossil sequences and "useless" body parts among species implied that they had not—*then perhaps species had become modified over time.*

SQUEEZING NEW EVIDENCE INTO OLD BELIEFS

Georges Cuvier was among those trying to make sense of the growing evidence for change. For years he had compared fossils with living organisms. He was aware of the abrupt changes in the fossil record and was the first to recognize that they indicated times of mass extinctions. By Cuvier's hypothesis, a single time of creation had populated the entire world, which was an unchanging stage for the human drama. Monstrous earthquakes, floods, and other major catastrophes did happen, and many people died. Each time, survivors repopulated the world. By Cuvier's reckoning, there were no new species. Naturalists simply had not yet found all of the fossils. His hypothesis was favored for a long time. It even became elevated to the rank of theory, one that later became known as **catastrophism**.

Jean Lamarck had a different idea: The environment modifies traits of individuals in ways that are passed on to offspring. Outside pressures and internal needs cause permanent changes in body form and function, and offspring inherit the acquired changes. Life, created long ago in a simple state, gradually improved—up the Chain of Being. Apply his idea to modern giraffes. Say they had a short-necked ancestor. Pressed by the need to find food, it kept stretching its neck to reach leaves beyond the reach of other animals. Stretching made the neck lengthen permanently. The longer neck was inherited by offspring that stretched their necks, too. So generations of animals stretching to reach ever loftier leaves led to the modern giraffe.

The environment *is* a factor in evolution. However, Lamarck's idea, like others proposed at the time, has not been supported by rigorous tests.

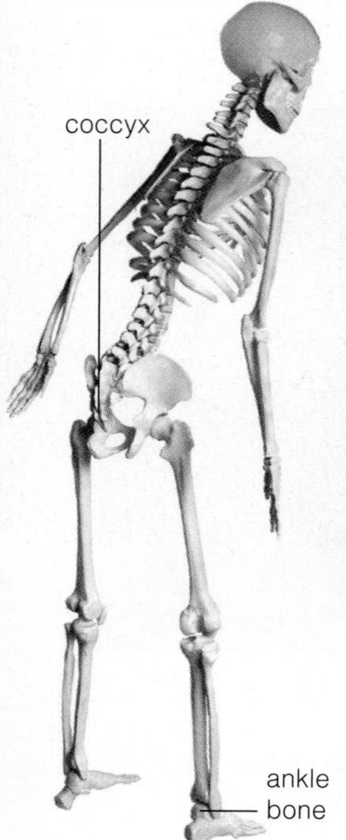

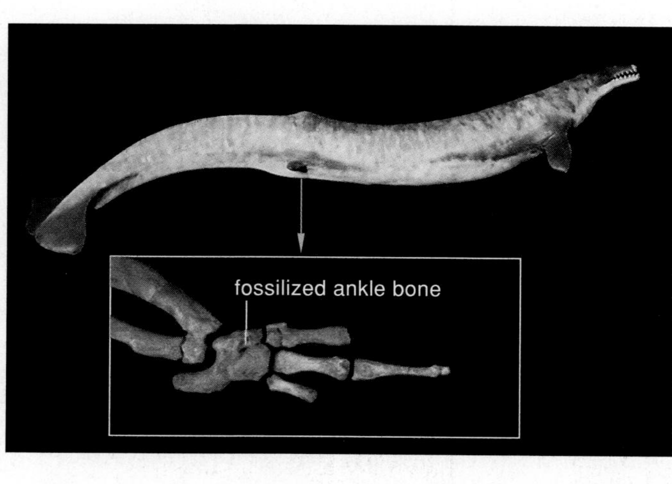

Figure 12.2 Body parts that have no apparent function. *Above*, reconstruction of an ancient whale (*Basilosaurus*), with a head as long as a sofa. This marine predator was fully aquatic, so it did not use its hindlimbs to support body weight as you do—yet it had ankle bones. We use our ankles, but not our coccyx bones.

VOYAGE OF THE *BEAGLE*

In 1831, in the midst of the confusion, Charles Darwin was twenty-two years old and wondering what to do with his life. Ever since he was eight, he had wanted to hunt, fish, collect shells, or simply watch insects and birds—anything but sit in school. Later, at his father's insistence, he tried to study medicine in college. The crude, painful procedures used on patients at that time sickened him, so he packed for Cambridge. He earned a degree in theology, but nature still beckoned him. A botanist, John Henslow, sensed Darwin's real interests. Henslow arranged for him to be appointed as ship's naturalist aboard a ship, the *Beagle*.

The *Beagle* (Figure 12.3) was about to embark on a five-year voyage that would take Darwin around the world. During the Atlantic crossing, he collected and studied marine life. At stops along coasts and islands, he observed diverse species in environments ranging from shores to mountains. He read Henslow's parting gift, a book called *Principles of Geology* by Charles Lyell. He started circling the question of evolving life, now on the minds of many respected scholars.

Lyell and other geologists were proposing a radical hypothesis. Catastrophes, they said, had no more effect on Earth history than did subtle processes of change. They studied how long it takes for rain, the pounding surf, and other forces of nature to sculpt the surface. They considered how long it takes layers of sediments to be deposited in beds of rivers and seas. If deposition occurred just as slowly in the past, then it must have taken millions of years for thick stacks to form. And what about catastrophes? Massive floods, more than a hundred huge earthquakes, and twenty or so volcanic eruptions strike in a typical year. Catastrophes, they said, are just not that unusual.

Their view of gradual, uniformly repetitive change became a theory of **uniformity**. It directly challenged a prevailing view that the Earth was only about 6,000 years old. Traditionalists thought people had recorded everything that happened in those thousands of years. In all that time, no one had ever mentioned seeing a species evolve. By Lyell's calculations, geologic forces must have taken millions of years to sculpt the present landscape. *Was that not enough time for species to evolve in diverse ways?*

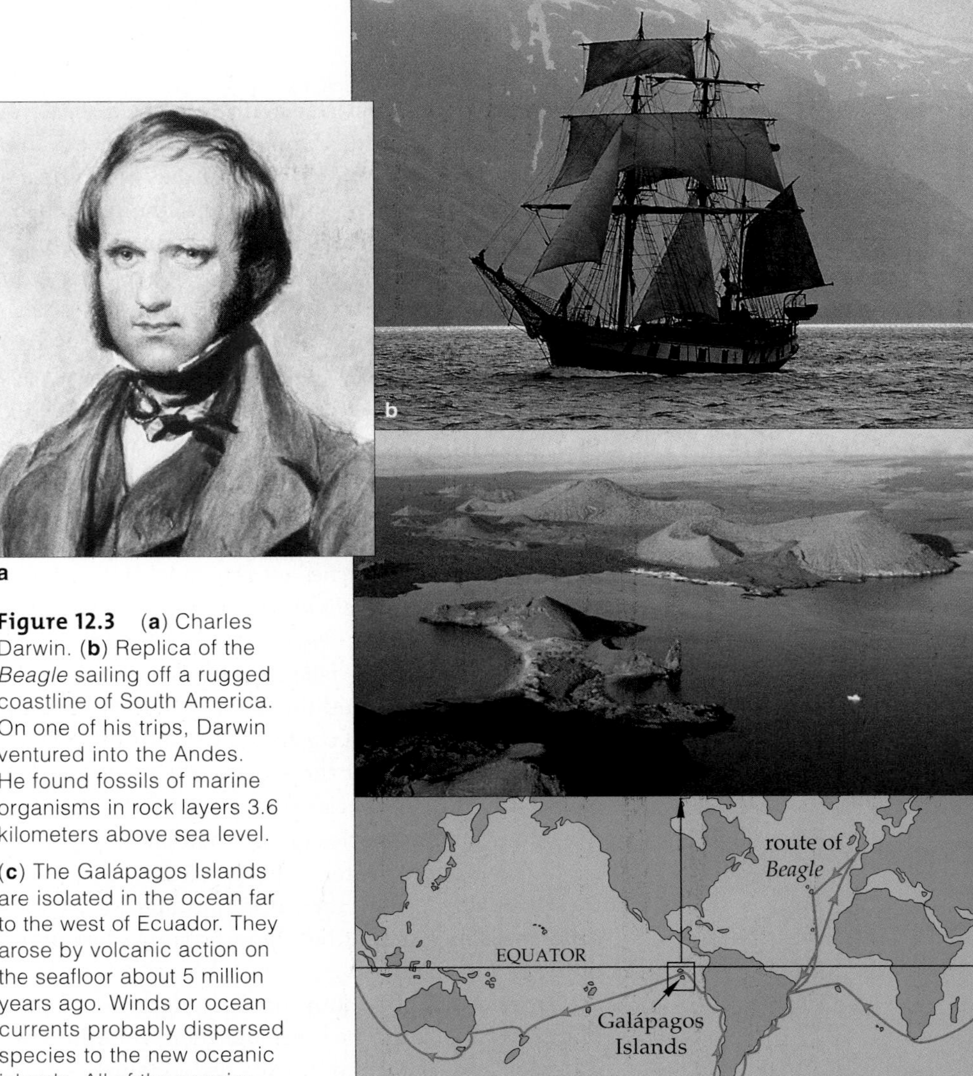

Figure 12.3 (**a**) Charles Darwin. (**b**) Replica of the *Beagle* sailing off a rugged coastline of South America. On one of his trips, Darwin ventured into the Andes. He found fossils of marine organisms in rock layers 3.6 kilometers above sea level.

(**c**) The Galápagos Islands are isolated in the ocean far to the west of Ecuador. They arose by volcanic action on the seafloor about 5 million years ago. Winds or ocean currents probably dispersed species to the new oceanic islands. All of the species native to the islands are their descendants.

OLD BONES AND ARMADILLOS

When Darwin returned to England, he talked to other naturalists about possible evidence that life evolves. By carefully studying detailed notes from his voyage, he came up with possibilities.

In Argentina, for example, he had observed fossils of glyptodonts, now extinct. Of all animals on Earth, only living armadillos are like them (Figure 12.4). Of all places on Earth, armadillos live only in the same places where glyptodonts had lived. If the two kinds of animals had been created at the same time, lived in the same place, and were so much alike, why is only one still alive? Would it not be reasonable to assume that glyptodonts were early ancestors of armadillos? Many of their shared traits might have been retained over many thousands of generations. Other traits may have been modified in the armadillo branch of a family tree. *Descent with modification* seemed possible. What, then, could be the driving force for evolution?

Figure 12.4 (**a**) From Texas, a modern armadillo, about a foot long excluding the tail. (**b**) A Pleistocene glyptodont, which was about as big as a Volkswagen Beetle, and now extinct. Glyptodonts shared unusual traits and a restricted distribution with the existing armadillos. Yet the two kinds of animals are widely separated in time. Their similarities were a clue that helped Darwin develop a theory of evolution by natural selection.

A KEY INSIGHT — VARIATION IN TRAITS

While Darwin assessed his notes, an influential essay by Thomas Malthus, a clergyman and economist, made him consider a topic of social interest. Malthus had

LINK TO SECTION 1.4

correlated population size with famine, disease, and war. Humans, Malthus claimed, run out of food, living space, and other resources because they reproduce too much. The larger a population gets, the more people reproduce. Resources dwindle, and struggles intensify. Many people starve, get sick, or engage in war and other forms of competition for remaining resources.

Darwin deduced that *any* population has a capacity to produce more individuals than the environment can support. Even one sea star can release 2,500,000 eggs per year, but the seas do not fill up with sea stars.

Darwin also reflected on species he had observed during his voyage. Individuals of those species were not alike in their details. They varied in size, color, and other traits. *It dawned on Darwin that variations in traits affect an individual's ability to secure resources in the environment—and thus to survive and reproduce.*

Did the Galápagos Islands reveal evidence of this? Between the islands and the South American coast are 900 kilometers of open ocean. The islands offer diverse habitats on their rocky shores, deserts, and mountains. Nearly all of their species live nowhere else, although they do resemble species on the mainland.

Did wind and water deliver colonizing species to the Galápagos? If so, then different species living on different islands might be island-hopping descendants of the first colonizers.

Darwin thought about the Galápagos finch species (Figure 12.5). Could he correlate their variations with different environmental challenges? Imagine yourself in his place, watching a population of large-billed, seed-cracking birds. A few birds with a stronger bill crack seeds too tough for the others. If most of the seeds that form during a given season have hard coats, then the strong-billed birds will have a competitive edge and a better chance of surviving and producing offspring. If the trait is heritable, then the advantage will help their descendants, also. If environmental factors continue to "select" the most adaptive version of this trait, then the population may end up being mostly strong-billed birds. *And a population is evolving when forms of heritable traits change over the generations.*

Figure 12.5 Four of thirteen finch species on the Galápagos Islands. (**a**) *Geospiza conirostris* and (**b**) *G. scandens*. Both eat cactus flowers and fruits. Other finches eat cactus seeds. (**c**) *Certhidea olivacea*, a tree-dwelling finch, uses its slender bill to probe for food. (**d**) *Camarhynchus pallidus* eats wood-boring insects such as termites. It learns to break cactus spines and twigs to suitable lengths, then hold the "tools" and use them to probe bark for insects. Compare Figure 13.18, which shows bills of the spectacularly diverse Hawaiian honeycreepers.

Let's trace Darwin's observations and conclusions about a key evolutionary process, using modern terms:

1. *Observation:* Natural populations have an inherent reproductive capacity to increase in numbers through successive generations.
2. *Observation:* No population can indefinitely grow in size, because its individuals run out of food, living space, and other resources.
3. *Inference:* Sooner or later, individuals will wind up competing for dwindling resources.
4. *Observation:* Individuals share a pool of heritable information about traits, encoded in genes.
5. *Observation:* Mutations have given rise to alleles, or slightly different molecular forms of genes, which are a source of differences in phenotypic details.
6. *Inferences:* Some phenotypes are better than others at helping an individual survive and reproduce. They increase an individual's **fitness**—its success in a given environment, as measured by its relative contribution to future generations. Over successive generations, the alleles for such phenotypes increase in frequency.
7. *Conclusions:* **Natural selection** is an outcome of the differences in reproduction among individuals of a population that vary in details of their shared traits. Environmental agents of selection act on the range of variation, and the population may evolve as a result.

Anticipating that his view would be controversial, Darwin searched for flaws in his reasoning. He took too long. More than ten years after he wrote up but did not publish his theory, another respected naturalist—Alfred Wallace—came up with the same theory. At a scientific meeting in 1858, the theory was presented and attributed to both. Wallace was still in the field, so he knew nothing about the meeting, and Darwin did not attend it. The next year, Darwin finally published *On the Origin of Species*, which has detailed evidence in support of the theory.

You may have heard that Darwin's book fanned an intellectual firestorm, but most scholars were quick to accept evolution. It was, however, fiercely debated that evolution occurs by natural selection. Decades passed before evidence from genetics led to the widespread acceptance of the theory of natural selection.

As Darwin and Wallace perceived, natural selection is the outcome of differences in survival and reproduction among individuals of a population that show variation in their shared traits. It can lead to increased fitness—that is, to increased reproductive success in a given environment.

12.2 The Nature of Adaptation

Observable traits are not always easy to correlate with specific conditions in an organism's environment.

"Adaptation" is one of those words that have different meanings in different contexts. An individual plant or animal often can quickly adjust its form, function, and behavior. Junipers in inhospitably windy places grow less tall than junipers of the same species that grow in more sheltered places. A sudden clap of thunder may make you lurch the first time you hear it, but then you may get used to the sound over time and ignore it. These are examples of *short-term* adaptations, because they last only as long as the individual does.

Over the long term, an **adaptation** is some heritable aspect of form, function, behavior, or development that improves the odds for surviving and reproducing in a given environment. It is an *outcome* of microevolution —natural selection especially—an enhancement of the fit between the individual and prevailing conditions.

Figure 12.6 (**a**) Severe, rapid wilting of one commercial tomato plant (*Lycopersicon esculentum*) that absorbed salty water. (**b**) Galápagos tomato plant, *L. cheesmanii*, which stores most absorbed salts in its leaves, not in its fruits.

SALT-TOLERANT TOMATOES

As an example of long-term adaptation, compare how tomato species handle salty water. Tomatoes evolved in Ecuador, Peru, and the Galápagos Islands. The type found most often in markets, *Lycopersicon esculentum*, has eight close relatives in the wild. If you mix ten grams of table salt with sixty milliliters of water, then pour it into the soil around *L. esculentum* roots, the plant will wilt severely in less than thirty minutes (Figure 12.6*a*). Even when the soil has only 2,500 parts per million of salt, this species grows poorly. Yet the Galápagos tomato (*L. cheesmanii*) survives in seawater-washed soils (Figure 12.6*b*). We know its salt tolerance is a heritable adaptation. How? Crosses between the wild and commercial species produce a small, edible hybrid. This hybrid tolerates irrigation water that is two parts fresh and one part salty. It is gaining interest in places where fresh water is scarce and where salts have built up in croplands.

It may take modification of only a few traits to get new salt-tolerant plants. Revving up a single gene for a sodium/hydrogen ion transporter lets tomato plants use salty water and still bear edible fruits.

ADAPTATION TO WHAT?

You can safely assume a polar bear (*Ursus maritimis*) is adapted to the Arctic, and that its form and function would be a flop in a desert. But it is not always easy to ascertain a direct relationship between an adaptation and the environment. A prevailing environment might be very different from the one in which a trait evolved.

Consider the llama. It is native to the cloud-piercing peaks of the Andes range in western South America (Figure 12.7*a*). It lives 4,800 meters (16,000 feet) above sea level. Compared to humans at lower elevations, its lungs have more air sacs and blood vessels. The llama heart has larger chambers, so it pumps larger volumes of blood. Llamas do not have to make extra blood cells, as people do when they move permanently from the lowlands to high elevations. (Extra cells make blood "stickier," so the heart has to pump harder.) But the most publicized adaptation is this: Llama hemoglobin is better than ours at latching on to oxygen. It picks up oxygen in the lungs far more efficiently.

Superficially, at least, the oxygen-binding affinity of llama hemoglobin appears to be an adaptation to thin air at high altitudes. But is it? Apparently not.

Llamas belong to the same family as the dromedary camels (Figure 12.7*b*). Their shared camelid ancestors evolved in the Eocene grasslands and deserts of North America. Later, the ancestors of the camels and llamas went their separate ways. Camel forerunners moved into Asia's low-elevation grasslands and deserts by a land bridge, which later submerged when the sea level rose. Llama forerunners moved a different way—down the Isthmus of Panama, and on into South America.

Intriguingly, a dromedary camel's hemoglobin also shows a high oxygen-binding capacity. So if the trait arose in a shared ancestor, then in what respect was it adaptive at *low* elevations if it is also adaptive at *high* elevations? We know that camels and llamas did not just happen to evolve to be similar. They are very close kin,

Figure 12.7 The Llama (**a**) and the camel (**b**) share a common ancestor.

LINKS TO SECTIONS 8.3, 8.4, 8.5, 10.5

and their most recent ancestors lived in very different environments with different oxygen concentrations.

Who knows why the trait was originally favored? Eocene climates were alternately warm and cool, and hemoglobin's oxygen-binding capacity does go down as temperatures go up. Did it prove adaptive during a long-term shift in climate? Or were its effects neutral at first? What if a small ancestral population became homozygous for the mutant gene simply by chance?

Use all of these "what-ifs" as a reminder to think critically about the connections between an organism's form and function. Identifying connections takes more than intuition.

A long-term adaptation is any heritable aspect of form, function, behavior, or development that contributes to the fit between the individual and its environment.

An adaptive trait improves the odds of surviving and reproducing, or at least it did so under conditions that prevailed when genes coding for the trait first evolved.

12.3 Individuals Don't Evolve, Populations Do

Evolution starts with changes in the gene pool of a population. Mutation is the source of new alleles, and sexual reproduction spreads them around.

VARIATION IN POPULATIONS

As Darwin and Wallace both perceived, *individuals do not evolve; populations do*. By definition, a **population** is a group of individuals of the same species occupying a given area. To understand how a population evolves, start with variation in the features that characterize it.

Individuals of a population share the same body plan. Some birds have two eyes, two wings, feathers, three toes forward and one toe back, and so on. These are *morphological* traits (*morpho–*, meaning form). Cells and body parts of all individuals work much the same way in metabolism, growth, and reproduction. These are *physiological* traits, relating to how the body works in its environment. Individuals respond the same way to basic stimuli, as when babies instinctively imitate adult facial expressions. These are *behavioral* traits.

However, the individuals of a population also show variation in the details of their shared traits. You know this just by thinking about the variations in the color and patterning of pigeon feathers or butterfly wings or snail shells (Figure 12.8). Almost every trait of any species may vary, sometimes dramatically.

Many traits, such as those of Gregor Mendel's pea plants, show *qualitative* differences. They have two or more distinct forms (morphs), such as purple or white flowers. Having just two is called dimorphism; having three or more is called polymorphism. Do not forget that for many traits, individuals of a population show *quantitative* differences, a range of incrementally small variations in a specified trait (Section 8.3).

THE "GENE POOL"

Genes encode information about heritable traits. All of a population's individuals inherit the same number and kinds of genes (except for a pair of nonidentical sex chromosomes). Together, they and their offspring represent a **gene pool**—a pool of genetic resources.

For sexual reproducers, nearly all of the genes in the shared pool have two or more slightly different molecular forms, or **alleles**. An individual might or might not inherit identical alleles for any trait. This is the source of variations in *phenotype*—differences in details of shared traits. Whether you have red, black, brown, or blond hair depends upon which alleles you inherited from your two parents.

You read about the inheritance of alleles in earlier chapters. Here we summarize the key events involved:

1. Gene mutation (produces new alleles)

2. Crossing over events in meiosis I (introduce novel combinations of alleles in chromosomes)

3. Independent assortment at meiosis I (puts mixes of maternal and paternal chromosomes in gametes)

4. Fertilization (combines alleles from two parents)

5. Change in chromosome number or structure (loss, duplication, or repositioning of genes)

Only mutation *creates* new alleles. Other events shuffle *existing* alleles into different combinations—but what a shuffle! There are $10^{116,446,000}$ possible combinations of human alleles. Not even 10^{10} people are alive today. Unless you have an identical twin, it is very unlikely that any other person with your exact genetic makeup has ever lived, or ever will.

STABILITY AND CHANGE IN ALLELE FREQUENCIES

Researchers typically track **allele frequencies**, or the abundance of certain alleles in a population. They start from a theoretical reference point known as **genetic equilibrium**. A population at that point is *not* evolving with respect to the allele frequencies being studied.

You will read more about genetic equilibrium in Section 12.4. For now, know that five conditions are necessary for genetic equilibrium to occur, and those conditions will almost never prevail at the same time. *Mutations* are rare but inevitable, and *natural selection, gene flow,* and *genetic drift* can drive a population out of genetic equilibrium. **Microevolution** refers to the small-scale changes in allele frequencies that originate from those four processes, as discussed next.

Figure 12.8 Fabulous polymorphism evident among individuals of the snail populations on Caribbean islands.

MUTATIONS REVISITED

Mutations, again, are heritable changes in DNA that usually give rise to altered gene products. They are the original source of all alleles. We cannot predict when or in which individual they will occur. Yet within each species, mutations accumulate at a predictable rate. In humans, that rate is around one mutation for every replication cycle—or about one in three billion bases. That means one out of every four gametes, on average, will carry a mutation.

Mutations may give rise to structural, functional, or behavioral alterations that diminish an individual's chances of surviving and reproducing. Even one small biochemical change can be devastating. For instance, skin, bones, tendons, lungs, blood vessels, and many of the other vertebrate organs incorporate collagen. When the collagen gene mutates in a way that affects collagen function, drastic changes in the body occur. A mutation that severely changes a phenotype usually causes death; it is called a **lethal mutation**.

With respect to natural selection, a mutation can be neutral if it has no effect on survival or reproduction. It neither helps nor hurts the individual. If you carry a mutant gene that keeps your earlobes attached to your head instead of swinging freely, this in itself should not stop you from surviving and reproducing as well as anybody else. Because a neutral mutation has no effect on survival and reproduction, natural selection does not alter its frequency in the population.

Every so often, a mutation proves useful. A mutant gene product that affects growth may make a corn plant grow larger or faster and so give it the best access to sunlight and nutrients. A neutral mutation may prove helpful after a condition in the environment changes. Even when it bestows only a small advantage, chance events or natural selection might preserve the mutant gene in the DNA and favor its representation in the next generation.

Mutations are so rare, they usually have little or no immediate effect on the population's allele frequencies. But both beneficial and neutral mutations have been accumulating in different lineages for billions of years. Through all that time, they have functioned as the raw material for evolutionary change, for the staggering range of biodiversity, past and present. Actually, the reason you do not look like a bacterium or an avocado or an earthworm—or even your neighbors down the street—started with mutations that arose at different times, in different lines of descent.

Each population or species is characterized by certain morphological, physiological, and behavioral traits. The details of these traits vary among its individuals, and most of the variation has a heritable basis.

Different combinations of alleles among individuals of a population give rise to variations in phenotype—that is, to differences in the details of their shared structural, functional, and behavioral traits.

In sexually reproducing species, a population's individuals share a pool of genetic resources—a gene pool.

Mutation alone creates new alleles. Whether an allele proves neutral, beneficial, or harmful depends on the mutation and on prevailing environmental conditions.

Figure 12.9 ***Animated!*** Proportions of butterflies with dark-blue, medium-blue, and white wings in a hypothetical population that remains at genetic equilibrium.

LINK TO SECTION 8.1

12.4 When Is A Population *Not* Evolving?

How do researchers know whether or not a population is evolving? They can start by tracking deviations from the baseline of genetic equilibrium.

Gene pools can be stable only if five conditions are met. There is no mutation, the population is infinitely large, the population is reproductively isolated from all other populations of the species, mating is random, and all individuals survive and produce the same number of offspring. As you can imagine, these conditions rarely, if ever, occur at once in nature. Nonetheless, they serve as a hypothetical starting point.

Allele frequencies for any gene in a shared pool will stay stable unless the population is evolving. Godfrey Hardy and Wilhelm Weinberg developed a formula that tracks allele frequencies in a population of any sexually reproducing species. Their formula can be used to check whether the population is in genetic equilibrium.

Butterflies are sexually reproducing organisms with pairs of genes on homologous chromosomes. Start by assuming the population is at genetic equilibrium for a single pair of alleles that influence wing color. Allele *A* is associated with dark-blue wings, *a* with white, and *Aa* with medium-blue. The proportions of genotypes for wing color at genetic equilibrium are:

$$p^2\,(AA) + 2pq(Aa) + q^2\,(aa) = 1.0$$

where p and q are frequencies of alleles *A* and *a*. This formula is what is known as the *Hardy–Weinberg rule.*

The frequencies of *A* and *a* must add up to 1.0. For example, if *A* occupies half of all the loci for this gene in the population, then *a* must occupy the other half (0.5 + 0.5 = 1.0). If *A* occupies 90 percent of all the loci, then *a* must occupy 10 percent (0.9 + 0.1 = 1.0).

No matter what the proportions,

$$p + q = 1.0$$

At meiosis, alleles of each gene pair segregate and end up in separate gametes. So the proportion of gametes having the *A* allele is p, and the proportion having the *a* allele is q. Check this Punnett square for genotypes that are possible in the next generation (*AA*, *Aa*, and *aa*):

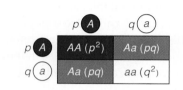

	p (A)	q (a)
p (A)	*AA* (p^2)	*Aa* (pq)
q (a)	*Aa* (pq)	*aa* (q^2)

The frequencies add up to 1: $p^2 + 2pq + q^2 = 1.0$.

Let's say the population contains 1,000 butterflies, and each one produces two gametes:

490 *AA* individuals produce 980 *A* gametes
420 *Aa* individuals produce 420 *A* and 420 *a* gametes
90 *aa* individuals produce 180 *a* gametes

The frequency of *A* and *a* among the 2,000 gametes is

$$A = \frac{980 + 420}{2{,}000 \text{ alleles}} = \frac{1{,}400}{2{,}000} = 0.7 = p$$

$$a = \frac{180 + 420}{2{,}000 \text{ alleles}} = \frac{600}{2{,}000} = 0.3 = q$$

At fertilization, the gametes combine at random and give rise to the next generation. If the population stays constant at 1,000 individuals, you now have 490 *AA*, 420 *Aa*, and 90 *aa* individuals. Because the frequencies of alleles for dark-blue, medium-blue, and white wings are the same as they were in the original gametes, they will give rise to the same phenotypic frequencies you observed in the second generation (Figure 12.9).

As long as the five conditions hold, the pattern will also hold. When they do not, one or more evolutionary forces are at work, and allele frequencies will change.

When traits do not appear in proportions predicted from the Hardy–Weinberg rule, one or more conditions stated in the rule are not holding. The hunt can begin for one or more specific evolutionary forces driving the change.

12.5 Natural Selection Revisited

Natural selection operates on the variation in traits that characterize a population. It affects which individuals survive and reproduce under existing environmental conditions, one generation to the next.

DIRECTIONAL SELECTION

In cases of **directional selection**, allele frequencies that give rise to a range of variation in phenotype tend to shift in a consistent direction. The shift is a response to directional change in the environment or to one or more new conditions in it. A new mutation that proves beneficial also may cause the directional shift. Either way, forms at one end of the phenotypic range become more common than midrange forms (Figure 12.10).

ROCK POCKET MICE Directional selection is at work among rock pocket mice (*Chaetodipus intermedius*) of the Sonoran Desert in Arizona. Of more than eighty genes known to affect coat color in mice, researchers found one gene that governs a difference between two populations of this mouse species.

Rock pocket mice are small mammals that spend the day in underground burrows. At night they forage for seeds, scampering over rock outcroppings in the desert. Individuals of the largest population of mice live among tawny-colored granite rocks. Their tawny fur camouflages them from predators (Figure 12.11*a*).

A smaller population of pocket mice lives in the same region, but these mice scamper over dark basalt rocks of ancient lava flows. The individuals of this population have dark fur, and they are camouflaged from predators on the dark rocks (Figure 12.11*b*). We can expect that local night-flying predatory birds are selective agents that affect fur color. For instance, owls have an easier time seeing mice with fur that does not match the rock color.

Researchers used genetic data on laboratory mice to formulate a hypothesis about coat color differences in the two populations of wild mice. They predicted that a gene mutation causes the difference.

DNA was collected from dark-colored pocket mice at one lava flow, and from light-colored mice living on granite outcroppings. Analysis of both sets of DNA showed that the DNA sequence of the coat color gene in all dark-fur mice differed by four nucleotides from that of their light-furred neighbors. In the population of dark mice, the allele frequencies had evolved in a consistent direction as a result of selection pressure, so dark fur became more common.

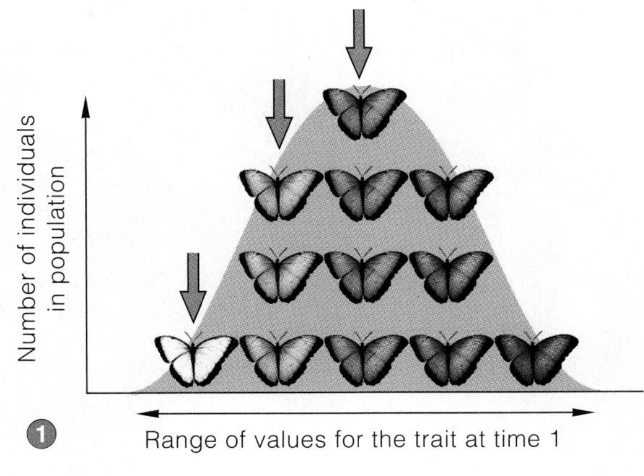

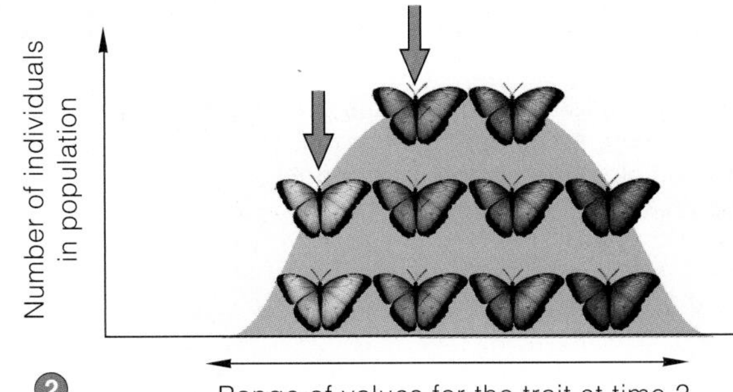

Number of individuals in population

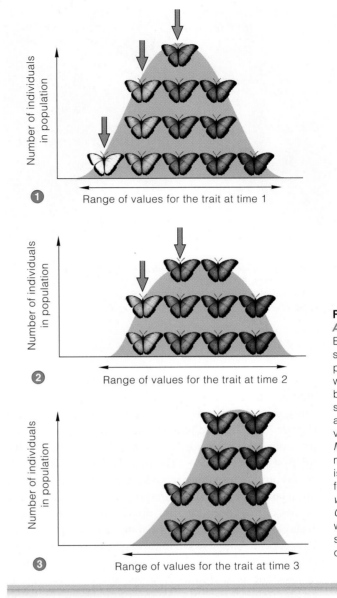

3 Range of values for the trait at time 3

Figure 12.10 ***Animated!*** Example of directional selection, using the phenotypic variation within a population of butterflies. The bell-shaped curve spans a range of continuous variation in wing color. *Medium-blue*, the most common form, is between extreme forms of the trait—*white* and *dark purple*. *Orange* arrows signify which forms are being selected against over time.

Figure 12.11 (**a**,**b**) The two color morphs of rock pocket mice. (**c**) A lava basalt area in the Sonoran Desert of Arizona.

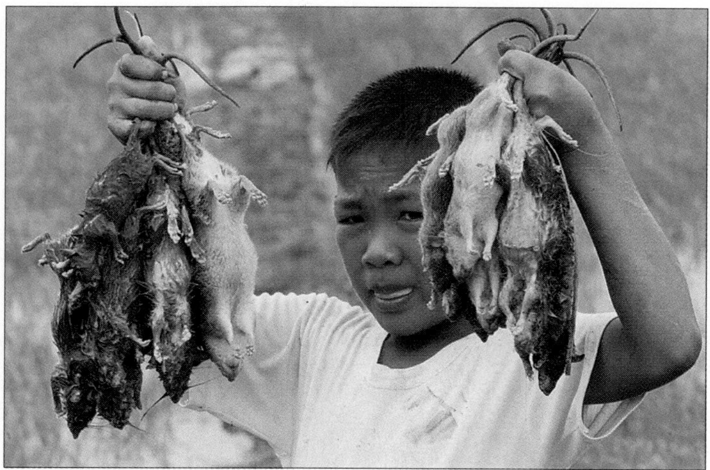

Figure 12.12 The super-rat threat. Rats infesting the Philippines have destroyed more than 20 percent of that country's 80,000 hectares of rice fields. Rice is the main food source for most people of Southeast Asia.

IMPACTS, ISSUES

Warfarin-resistant rats led to the development of anticoagulants that are more toxic and that persist longer in the environment. They are weakening and killing owls, hawks, coyotes, and other predators that have eaten poisoned rats. Between 1985 and 1999, for example, the number of barn owls with anticoagulants in their blood rose from 5 to 36 percent. And that was just one study in Great Britain.

PESTICIDE RESISTANCE The overuse of pesticides can cause directional selection, as it did for the super rats (Figure 12.12). Typically, a heritable aspect of body form, physiology, or behavior helps a few individuals survive the first pesticide doses. The most resistant individuals are favored. If resistance has a heritable basis, it becomes more common in each generation. The chemicals act as agents of selection, favoring the most resistant forms! Today, 450 species of pests resist one or more pesticides. Worse, pesticides also kill natural predators of the pests. When freed from their natural constraints, populations of resistant individuals overgrow, and crop damage is greater than before. This outcome of directional selection is called *pest resurgence*.

Pesticide use is decreasing where crops have been genetically engineered to resist pests. However, these plants are exerting selection pressure on populations of pests. Traits that help any pests survive the engineered defenses tend to be favored.

ANTIBIOTIC RESISTANCE When your grandparents were young, tuberculosis, pneumonia, and scarlet fever caused one-fourth of the annual deaths in the United States. Since the 1940s, we have been relying on natural and synthetic antibiotics to fight such bacterial diseases. Natural **antibiotics** are toxins that some microorganisms in soil release to kill bacterial competitors for nutrients (Section 1.6). Streptomycins, for instance, block protein synthesis in target cells. Penicillins disrupt formation of covalent bonds that hold a bacterial cell wall together, and by weakening the wall they cause the cell to rupture.

Antibiotics also have been overprescribed, often for simple infections that most people can fight on their own. Standard farming practices have traditionally included preventative doses of antibiotics for all farm animals, whether or not they are sick.

The overabundance of antibiotics in the human arena and the agricultural environment has put great selection pressure on bacterial populations. Remember how quickly bacteria grow and divide; their evolution proceeds at a much accelerated rate compared with ours. The intestinal bacteria *Escherichia coli* can divide once every 17 minutes, so an average two-week course of antibiotics potentially can exert selective pressure on over a thousand generations.

The genetic variation in bacterial gene pools allows some cells to survive this antibacterial onslaught even as others do not. Resistant strains are becoming the norm, particularly in hospitals, and because of this are becoming a greater threat to human health each year. Even as doctors and researchers scramble to find new antibiotics, this trend does not bode well for the many millions of people who contract cholera, tuberculosis, and other bacterial diseases each year.

SELECTION AGAINST OR IN FAVOR OF EXTREME PHENOTYPES

Figure 12.13 shows outcomes of two more modes of natural selection. One works against phenotypes at the fringes of a range of variation. The other favors them.

STABILIZING SELECTION With **stabilizing selection**, intermediate forms of a trait in a population are favored and alleles for the extreme forms are not. This mode of selection tends to preserve the intermediate phenotypes in the population (Figure 12.13).

As an example, prospects are not good for human babies who weigh far more or far less than average at birth (Figure 12.14). Newborns weighing less than 5.51 pounds, if they survive, tend to suffer from high blood pressure, diabetes, and heart disease as adults.

A different example is stabilizing selection on the body mass of sociable weavers of the African savanna. These birds cooperate in constructing and using large communal nests in regions where trees and other good nesting sites are scarce. For seven years, researchers captured, measured, tagged, released, and recaptured birds during the breeding seasons. Their studies show that body mass in these birds is a trade-off between the risks of starvation and predation. Birds that have intermediate mass have the selective advantage. Lean birds do not store enough fat to avoid starvation, and we can expect that the fat ones are more attractive to predators and not as good at escaping.

Number of individuals in population

1 Range of values for wing-color trait at time 1

Number of individuals in population

1 Range of values for wing-color trait at time 1

Number of individuals in population

2 Range of values for wing-color trait at time 2

Number of individuals in population

2 Range of values for wing-color trait at time 2

Number of individuals in population

3 Range of values for wing-color trait at time 3

Number of individuals in population

3 Range of values for wing-color trait at time 3

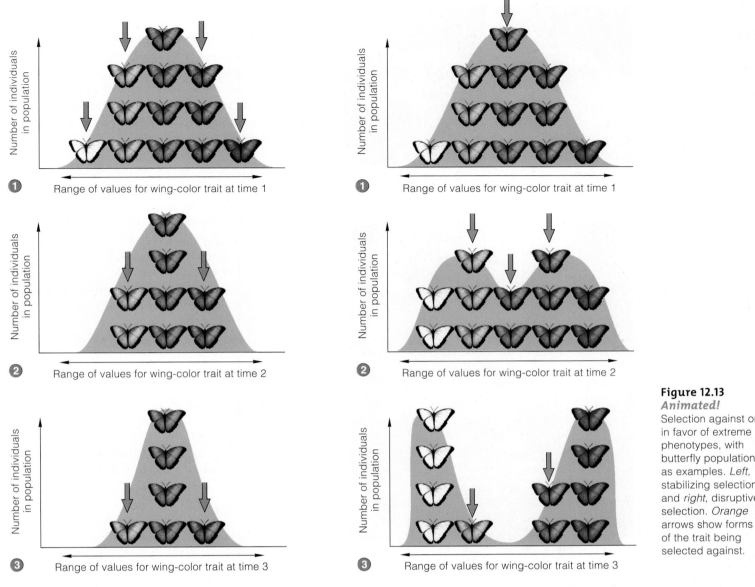

Figure 12.13 ***Animated!*** Selection against or in favor of extreme phenotypes, with butterfly populations as examples. *Left*, stabilizing selection and *right*, disruptive selection. *Orange* arrows show forms of the trait being selected against.

DISRUPTIVE SELECTION With **disruptive selection**, forms at both ends of the range of variation are favored and intermediate forms are selected against.

Consider the black-bellied seedcracker (*Pyrenestes ostrinus*) of Cameroon, in West Africa. Both females and males of these African finches have large or small bills—but none of the birds have medium-sized bills. The pattern holds throughout the geographic range of this species. It is like everyone in Texas being four feet *or* six feet tall, with no one in between.

If the pattern is unrelated to geography or gender, what causes it? Seed-cracking ability must directly influence survival. If only two bill sizes persist, then disruptive selection might be eliminating birds with intermediate-size bills. Which factors affect feeding

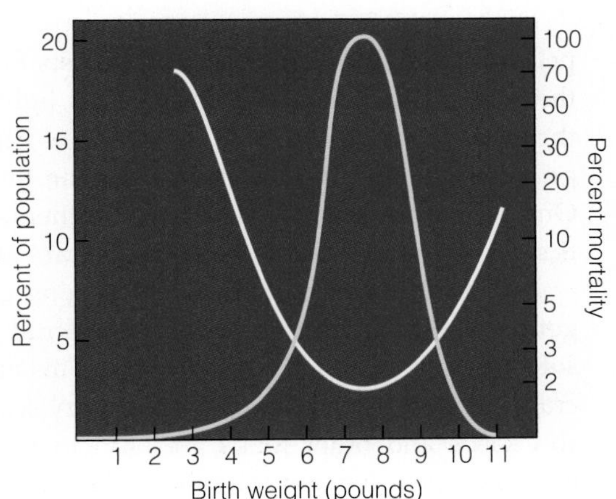

Figure 12.14 Weight distribution for 13,730 human newborns (*yellow* curve) correlated with death rate (*white* curve).

Figure 12.15 Disruptive selection in African finch populations. Selection pressures favor birds with bills that are either 12 *or* 15 millimeters wide. The size difference is correlated with competition for scarce food resources during the local dry season.

Figure 12.16 Outcome of sexual selection. This male bird of paradise (*Paradisaea raggiana*) is engaged in a flashy courtship display. He caught the eye (and, perhaps, sexual interest) of the smaller, less colorful female. The males compete fiercely for females, which are the selective agents. (Why do you suppose drab-colored females have been favored?)

performance? Cameroon's swamp forests flood during the wet season, and fires sparked by lightning burn throughout the dry season. Two kinds of fire-resistant, grasslike plants called sedges dominate these forests. One sedge has hard seeds and the other, soft. When finches reproduce, hard *and* soft seeds are abundant.

All birds prefer soft seeds for as long as they can get them. Birds with small bills are better at picking up soft, tiny seeds; and birds with large bills are better at cracking larger, hard ones. When the dry season peaks, soft seeds and other foods are hard to find, and the birds compete fiercely for seeds. The scarcity of both types of seeds during recurring periods of drought has had a disruptive effect on bill size in the seedcracker population. Birds with intermediate sizes are selected against, and now all bills are either 12 *or* 15 millimeters wide (Figure 12.15).

In these seedcrackers, bills of a particular size have a genetic basis. In experimental crosses between two birds with the two optimal bill sizes, all offspring had a bill of one size or the other.

With directional selection, allele frequencies underlying a range of variation tend to shift in a consistent direction in response to directional change in the environment.

With stabilizing selection, intermediate phenotypes are favored and extreme phenotypes at both ends of the range of variation are eliminated.

With disruptive selection, intermediate forms of traits are selected against; extreme forms in the range of variation are favored.

12.6 Maintaining Variation in a Population

Natural selection theory helps explain diverse aspects of nature, including male–female differences as well as a balance between sickle-cell anemia and malaria.

SEXUAL SELECTION

The individuals of many sexually reproducing species show a distinct male or female phenotype, or **sexual dimorphism** (*dimorphos,* having two forms). Often the males are larger and flashier than females. Courtship rituals and male aggression are common.

These adaptations take energy and time away from survival activities. Why do they persist? The answer is **sexual selection**: a form of natural selection in which the genetic winners are the ones that can outreproduce others of the population. The most adaptive traits help individuals defeat same-sex rivals for mates or are the ones most attractive to the opposite sex.

By choosing mates, one gender acts as an agent of selection on its own species. For example, the females of some species shop among a clustering of males, which differ in appearance and courtship behavior. The selected males, as well as the females making the selection, pass on their alleles to the next generation.

Flashy structures and behaviors are correlated with species in which males have little or nothing to do with raising offspring. The female chooses a male by observing signs of his health and vigor, which might improve her odds of producing healthy and vigorous offspring (Figure 12.16).

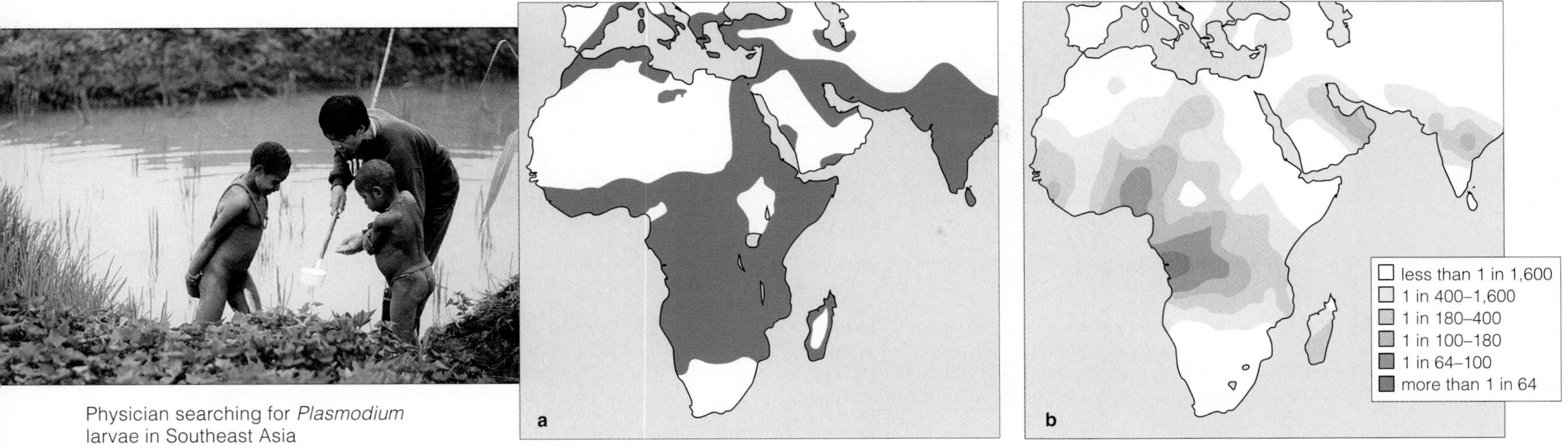

Physician searching for *Plasmodium* larvae in Southeast Asia

Figure 12.17 (**a**) Distribution of malaria cases in Africa, Asia, and the Middle East in the 1920s, before the start of programs to control mosquitoes, the vector for *Plasmodium.* (**b**) Distribution and frequency of the allele that causes sickle-cell anemia. Notice the close correlation between the maps.

SICKLE-CELL ANEMIA—LESSER OF TWO EVILS?

With *balancing* selection, two or more alleles of a gene are being maintained at relatively high frequencies—at least one percent—in the population. Their persistence is called **balanced polymorphism** (*polymorphos*, having many forms). Allele frequencies might shift, but often they return to the same values over the long term. We see this balance when conditions favor heterozygotes. In some way, their nonidentical alleles for a trait give them higher fitness compared to homozygotes, which, recall, have identical alleles for the trait.

Consider the human heterozygotes that carry an allele for Hb^S. The Hb^S allele, remember, codes for a mutant version of hemoglobin, an oxygen-transporting protein in the blood. Homozygotes (Hb^S/Hb^S) develop *sickle-cell anemia*, a severe disorder (Section 2.9).

The frequency of Hb^S is highest in the tropical and subtropical parts of Asia and Africa. Often, Hb^S/Hb^S homozygotes will die in their teens or early twenties. Yet, in these same regions, heterozygotes (Hb^A/Hb^S) make up nearly a third of the human population! Why is this combination of alleles being maintained at such high frequency?

The balancing act is most pronounced in areas that have the highest incidence of *malaria* (Figure 12.17 and Section 14.3). A mosquito transmits the parasitic agent of malaria, *Plasmodium*, to human hosts. The protozoan multiplies in the liver and, later, in red blood cells. The infected cells rupture and release new parasites during severe, recurring bouts of infection.

It turns out that Hb^A/Hb^S heterozygotes are more likely to survive an infection of malaria than people who produce only the normal hemoglobin (Hb^A/Hb^A homozygotes). There are several potential reasons for this. For one example, in heterozygotes, infected cells sickle under normal environmental conditions. Their abnormal shape brings them to the attention of the host's immune system, which will destroy them along with their parasitic cargo. In Hb^A/Hb^A homozygotes, infected cells do not sickle, and so the parasite remains hidden from the immune system.

Also, heterozygotes have one normal hemoglobin allele. Although they are not completely healthy, they produce enough normal hemoglobin to support body functions. That is why heterozygotes are more likely to survive long enough to reproduce, compared to very sick Hb^S/Hb^S homozygotes.

So the persistence of the "harmful" Hb^S allele is a matter of relative evils. Natural selection favors the Hb^A/Hb^S combination in malaria-ridden areas because heterozygotes show more resistance to the disease. In such environments, the combination has more survival value than either homozygotic condition (Hb^S/Hb^S or Hb^A/Hb^A). Malaria has been a selective force for many thousands of years in tropical and subtropical areas of Asia, the Middle East, and Africa.

With sexual selection, some version of a trait simply gives the individual an advantage in reproductive success. Sexual dimorphism is one outcome of sexual selection.

Balanced polymorphism is a state in which natural selection is maintaining two or more alleles over the generations at frequencies greater than one percent.

LINKS TO SECTIONS 2.9, 8.1, 8.7

IMPACTS, ISSUES

Rat populations show balanced polymorphism as a result of selection for a dominant allele that confers resistance to warfarin. This allele also promotes increases in the amount of vitamin K that all mammals require. In the absence of warfarin, rats with the dominant allele are less fit than those without it. In the presence of warfarin, the relative fitnesses are reversed.

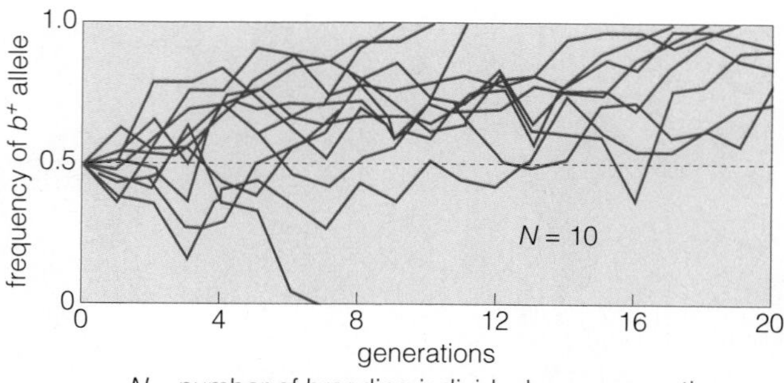

(a) The size of twelve populations of beetles was maintained at 10 breeding individuals per generation for twenty generations. Allele b^+ was lost and b became fixed in one population. Notice that alleles can be fixed or lost even in the absence of selection.

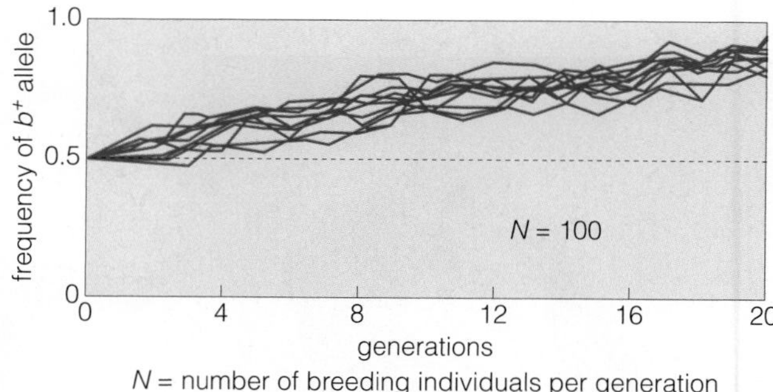

(b) The size of twelve populations was maintained at 100 individuals per generation for twenty generations. Allele b did not become fixed. Drift was far less in each generation than it was in the small populations tracked in (**a**).

Figure 12.18 ***Animated!*** The effect of genetic drift on allele frequencies in small and large populations. The starting frequency of the mutant allele (b^+) was 0.5.

LINK TO SECTION 8.1

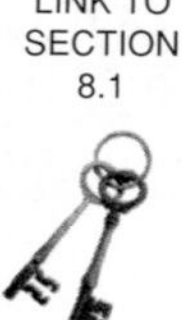

12.7 Genetic Drift—The Chance Changes

Random changes in allele frequencies drift toward a loss of genetic diversity. The drifting has greatest impact on small populations.

CHANCE EVENTS AND POPULATION SIZE

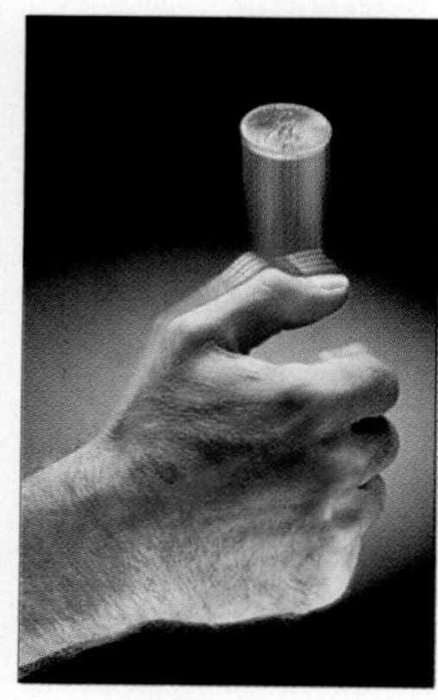

Genetic drift is a random change in allele frequencies over time, brought about by chance alone. Researchers measure it in terms of probability rules. *Probability* is the chance that something will happen relative to the number of times it could happen (Section 8.1). We can measure relative frequency as a fraction on a scale of 0 to 1—or 0 to 100 percent of the time. For instance, if ten million people enter a drawing for a month-long vacation in Paris, all expenses paid, each has an equal chance of winning: 1/10,000,000, or a very improbable 0.00001 percent.

By one probability rule, you are less likely to see an event occurring with expected probability if that event happens only rarely. For example, each time you flip a coin, there is a 50 percent chance it will land heads up. With 10 flips, the proportion of times that heads land up will probably deviate greatly from 50 percent. With 1,000 flips, large deviations are less likely.

We can apply the same rule to populations. Because population sizes are not infinite, there will be random changes in allele frequencies. These random changes tend to have minor impact on large populations, but greatly increase the odds that an allele will become dramatically more or less prevalent if a population is very small.

Steven Rich and his coworkers used small and large populations of the flour beetle (*Tribolium castaneum*) to study genetic drift. They started with beetles that bred true for the allele b^+ (the superscript plus signifies the wild-type allele) and other beetles that bred true for the mutant allele b. They hybridized individuals from both groups to get a population of F_1 heterozygotes (b^+b), which they divided into sets of twelve. Different sets consisted of 10, 20, 50, and 100 randomly selected male and female beetles, and the subpopulation sizes were maintained for twenty generations.

Figure 12.18 shows two of the test results. Drift was greatest in the sets of 10 beetles and least in the sets of 100 beetles. Notice the loss of b^+ from one of the small populations (one graph line ends at 0 in Figure 12.18*a*). Only allele b remained. When all of the individuals of a population have become homozygous for one allele only at a locus, we say that **fixation** has occurred.

Thus, *random change in allele frequencies leads to the homozygous condition and a loss of genetic diversity over time.* This is genetic drift's outcome in all populations; it simply happens faster in small populations. Once alleles from the parent population have become fixed, their frequencies will not change again unless new alleles are introduced.

BOTTLENECKS AND THE FOUNDER EFFECT

Genetic drift is pronounced when a few individuals rebuild a population or start a new one. This happens after a **bottleneck**, a drastic reduction in population size brought about by selection pressure or a calamity. Suppose contagious disease, habitat loss, or hunting nearly wipes out a population. Even when a moderate number of individuals survive the bottleneck, allele frequencies will have been altered at random.

By the 1890s, northern elephant seals were on the edge of extinction, with only twenty known survivors. Hunting restrictions implemented since have allowed the population to recover to about 170,000 individuals. Each is homozygous for all of the genes analyzed yet.

Genetic outcomes can be similarly unpredictable after a few individuals leave a population and establish a new colony elsewhere. This form of bottlenecking is called the **founder effect**. The allele frequencies in the founders may differ quite dramatically from those in the original population. Genetic diversity can be much reduced relative to the original gene pool, as when a lone seed founds a population on a remote island in the middle of the ocean (Figure 12.19).

GENETIC DRIFT AND INBRED POPULATIONS

Genetic drift is pronounced in an inbred population. **Inbreeding** is mating between close relatives, which share many identical alleles. Inbreeding can lower the overall genetic diversity of a population by increasing the frequency of the homozygous state.

Most human societies forbid or discourage incest (inbreeding between parents and children or between siblings). But inbreeding among other close relatives is common in small communities that are geographically or culturally isolated from a larger population. The Old Order Amish of Pennsylvania, for instance, are a moderately inbred group having distinct genotypes. One outcome of Amish inbreeding is a relatively high frequency of the recessive allele that causes *Ellis–van Creveld syndrome*. Affected individuals have extra toes, fingers, or both, and short limbs. Section 8.6 shows one individual. The allele might have been rare when the few founders immigrated to Pennsylvania. Now, about 1 in 8 of the population are heterozygous for the allele, and 1 in 200 are homozygous for it.

Genetic drift is the random change in allele frequencies over the generations, brought about by chance alone. The magnitude of its effect is greatest in small populations, such as a population that makes it through a bottleneck.

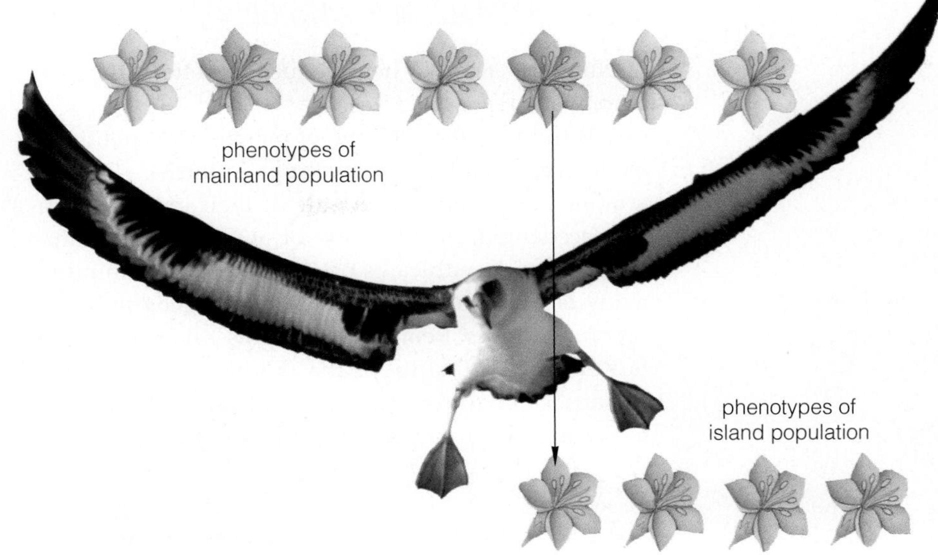

Figure 12.19 Founder effect. This wandering albatross carries seeds, stuck to its feathers, from the mainland to a remote island. By chance, most of the seeds carry an allele for orange flowers that are rare in the original population. Without further gene flow or selection for color, genetic drift will fix the allele on the island.

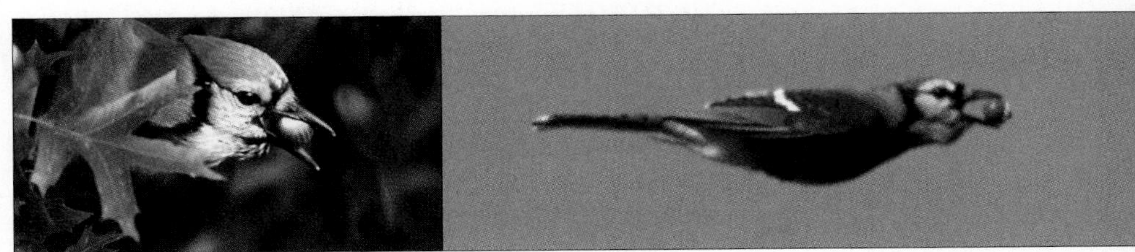

Figure 12.20 Blue jay, a mover of acorns that helps keep genes flowing between separate oak populations.

12.8 Gene Flow—Keeping Populations Alike

Individuals, and their alleles, move into and out of populations, and this physical flow counters changes introduced by other microevolutionary processes.

Individuals of the same species do not always stay put. A population will lose alleles whenever an individual permanently leaves it, an act called *emigration*. It gains alleles when new individuals permanently move in, an act called *immigration*. In both cases, **gene flow**—a physical flow of alleles between populations—occurs. Gene flow tends to counter the effects of mutation, natural selection, and genetic drift. It keeps separated populations genetically similar.

Think of the acorns that blue jays disperse when they gather nuts for the winter. Each fall the jays visit acorn-bearing oak trees repeatedly, then bury acorns in the soil of home territories that may be as much as a mile away (Figure 12.20). Alleles flowing in with the "immigrant acorns" help reduce genetic differences among stands of oak trees.

Or think of the millions of people from politically explosive, economically bankrupt countries who seek a more stable home. The scale of their emigrations is unprecedented, but the flow of genes is not. Human history is rich with cases of gene flow that minimized many of the genetic differences among geographically separate groups. Remember Genghis Khan? His genes flowed from China to Vienna (Figure 8.21). Similarly, the armies of Alexander the Great brought the genes for green eyes from Greece all the way to India.

Gene flow is the physical movement of alleles into and out of a population, through immigration and emigration. It tends to counter the effects of mutation, natural selection, and genetic drift.

Summary

Section 12.1 Awareness of evolution, or changes in lines of descent, emerged long ago from biogeography, comparative morphology, and geology. Charles Darwin and Alfred Wallace proposed a theory that natural selection results in evolution. Here are its key points:

Populations grow in size until resources dwindle and individuals compete for them. When individuals have forms of traits that make them more competitive, they tend to have more offspring. Over generations, those forms increase in frequency. Nature "selects" variations in traits that are more effective at helping individuals survive and reproduce.

Section 12.2 Long-term, heritable adaptations are aspects of form, function, behavior, or development that improve the chance of surviving and reproducing, or at least did so under conditions that prevailed when genes for the trait first emerged.

Section 12.3 Individuals of a population generally have the same number and kinds of genes for the same traits. Mutations are the source of *new* alleles (different molecular forms of genes). Individuals who inherit different allele combinations vary in details of one or more traits. An allele at any locus may become more or less common relative to other kinds or may be lost.

Microevolution refers to changes in allele frequencies of a population brought about by mutation, natural selection, genetic drift, and gene flow (Table 12.1).

Section 12.4 Genetic equilibrium is a state in which a population is not evolving. According to the Hardy–Weinberg rule, this occurs only if there is no mutation, the population is infinitely large and isolated from all other populations of the species, there is no natural selection, mating is random, and all individuals survive and reproduce the same number of offspring. Deviations from the theoretical baseline indicate that microevolution is occurring.

BiologyNow
Investigate gene frequencies and genetic equilibrium with the interaction on BiologyNow.

Section 12.5 Natural selection is an outcome of differences in survival and reproduction among individuals of a population that vary in forms of a trait. Selection pressures operating over the range of phenotypic variation shift or maintain allele frequencies in the population's gene pool.

Directional selection favors the forms at one end of a phenotypic range. Stabilizing selection favors intermediate forms. With disruptive selection, forms at both ends of a range of variation are favored over intermediate forms.

BiologyNow
View the animation of directional selection, and the animation of disruptive and stabilizing selection. Both on BiologyNow.

Section 12.6 Sexual selection, by females or males, leads to forms of traits that favor reproductive success. Persistence in phenotypic differences between males and females (sexual dimorphism) is one outcome.

Selection may result in balanced polymorphism, with nonidentical alleles for a trait being maintained over time at frequencies greater than 1 percent.

Section 12.7 Genetic drift is a random change in allele frequencies over time due to chance alone. The random changes tend to lead to the homozygous condition and loss of genetic diversity through the generations. The effect of genetic drift is greatest in small populations, such as ones that pass through a bottleneck or arise from a small group of founders.

BiologyNow
Learn more about genetic drift with the interaction on BiologyNow.

Table 12.1 Summary of Microevolutionary Processes

Mutation	A heritable change in DNA
Natural selection	Changes in relative frequencies of alleles for a heritable trait in a population; outcome of differences in survival and reproduction among individuals that vary in a trait
Genetic drift	Random fluctuation in allele frequencies over generations due to chance occurrences alone
Gene flow	Individuals, and their alleles, move into and out of populations; the physical flow counters the effects of the other microevolutionary processes

Section 12.8 Gene flow shifts allele frequencies by physically moving alleles into a population (by way of immigration) and out of it (by emigration). It tends to counter mutation, natural selection, and genetic drift and keep populations of the species alike.

Self-Quiz

Answers in Appendix I

1. Individuals don't evolve, ______ do.

2. Biologists define evolution as ______ .
   a. the origin of a species
   b. heritable change in a line of descent
   c. acquiring traits during the individual's lifetime
   d. both a and b

3. ______ is the original source of new alleles.
   a. Mutation
   b. Natural selection
   c. Genetic drift
   c. Gene flow
   d. All are original sources of new alleles

4. Natural selection is an outcome of ______ .
   a. differences in forms of traits
   b. differences in survival and reproduction among individuals that differ in one or more traits
   c. both a and b

5. Directional selection ______ .
   a. eliminates common forms of alleles
   b. shifts allele frequencies in a consistent direction
   c. does not favor intermediate forms of a trait
   d. works against adaptive traits

6. Disruptive selection ______ .
   a. eliminates uncommon forms of alleles
   b. shifts allele frequencies in a consistent direction
   c. does not favor intermediate forms of a trait
   d. both b and c

7. ______ tends to counter changes that occur in the allele frequencies among populations of a species.
   a. Genetic drift
   b. Gene flow
   c. Mutation
   d. Natural selection

8. Match the evolution concepts.

____	gene flow	a. source of new alleles
____	natural selection	b. changes in a population's allele frequencies due to chance alone
____	mutation	c. allele frequencies change owing to immigration, emigration, or both
____	genetic drift	d. outcome of differences in survival, reproduction among individuals that vary in forms of shared traits

Additional questions are available on **BiologyNow™**

Critical Thinking

1. Martha is studying a population of tropical birds. The males have brightly colored tail feathers and the females do not. She suspects the difference is maintained by sexual selection. Design an experiment to test her hypothesis.

Figure 12.21 Two designer dogs: the Great Dane (*legs, left*) and the chihuahua (*right*).

2. A few families in a remote region of Kentucky show a frequency of *methemoglobinemia*, an autosomal recessive disorder. Skin of affected individuals appears bright blue. Homozygous recessives lack an enzyme that maintains hemoglobin in its normal molecular form. Without it, a blue form of hemoglobin accumulates in blood and shows through the skin. Formulate a hypothesis to explain why the blue-offspring trait recurs among a cluster of families but is rare in the human population at large.

3. About 50,000 years ago, humans began domesticating wild dogs. By 14,000 years ago, they started to favor new varieties (breeds) by way of artificial selection. Individual dogs having desirable forms of traits were selected from each new litter and, later, encouraged to breed. Those with undesired forms of traits were passed over.

After favoring the pick of the litter for hundreds or thousands of generations, we ended up with sheep-herding border collies, badger-hunting dachshunds, bird-fetching retrievers, and sled-pulling huskies. At some point we began to delight in odd, extraordinary forms.

In practically no time at all, evolutionarily speaking, we picked our way through the pool of variant dog alleles and came up with such extreme breeds as chihuahuas and Great Danes (Figure 12.21).

Sometimes the canine designs exceeded the limits of biological common sense. How long would a tiny, nearly hairless, nearly defenseless, finicky-eating chihuahua last in the wild? Not long. What about English bulldogs, bred for a stubby snout and compressed face? Breeders thought these traits would let the dogs get a better grip on the nose of a bull. (Why they wanted dogs to bite bulls is a story in itself.) So now the roof of the bulldog mouth is ridiculously wide and often flabby, so bulldogs have trouble breathing. Sometimes they get so short of breath that they pass out.

Explain what aspects of artificial selection processes are like, and unlike, natural selection processes.

Figure 13.2 A slice through time—Butterloch Canyon, Italy, once at the bottom of a sea. Its sedimentary rock layers formed over hundreds of millions of years. Here, scientists Cindy Looy and Mark Sephton climb to reach the Permian–Triassic boundary layer, where they will look for fossilized fungal spores.

FOSSILS IN SEDIMENTARY ROCK LAYERS

Stratified (stacked) layers of sedimentary rock are rich sources of fossils. They formed long ago from deposits of volcanic ash, silt, and other materials. Sand and silt piled up after rivers transported them from land to the sea, just as da Vinci had suspected. Sandstones formed from sand, and shales from silt. The deposits were not continuous. For instance, the sea level changed during ice ages. Tremendous volumes of water became locked in glaciers, rivers dried up, and depositions ended in some regions. When the climate warmed and glaciers melted, depositions resumed.

The formation of sedimentary rock layers is called **stratification**. The deepest layer was the first to form; those closest to the surface were the last. Most formed horizontally, as in Figure 13.2, because particles tend to settle in response to gravity. You may see tilted or ruptured layers, as along a road that was cut out of a mountainside. Major crustal movements or upheavals disturbed them much later in time.

Most fossils are in sedimentary rock. Understand how rock layers form, and you know that the fossils in them formed at specific times in the past. Specifically, *the older the rock layer, the older the fossils.*

INTERPRETING THE FOSSIL RECORD

We have fossils for more than 250,000 known species. Judging from the current range of biodiversity, there must have been many, many millions more. Yet the fossil record will never be complete. Why is this so?

The odds are against finding evidence of an extinct, ancient species. At least one individual had to die and be gently buried before it decomposed or something ate it. The burial site had to have escaped obliteration by erosion, lava flows, or other geologic assaults. The fossil had to end up in a place where someone could actually find it, as in a sedimentary layer exposed by a river that had slowly carved out a canyon.

Most ancient species did not lend themselves well to preservation. For example, unlike bony fishes and hard-shelled mollusks, the jellyfishes and soft worms do not show up as much in the fossil record. Yet they probably were just as common, or more so.

Also think about population density and body size. One plant population might release millions of spores in a single season. The earliest humans lived in small bands and raised few offspring. What are the odds of finding even one fossilized human bone compared to spores of plants that lived at the same time?

Finally, imagine one line of descent, a **lineage**, that vanished when its habitat on a remote volcanic island sank into the sea. Or imagine two lineages, one lasting only briefly and the other for billions of years. Which is more likely to be represented in the fossil record?

IMPACTS, ISSUES

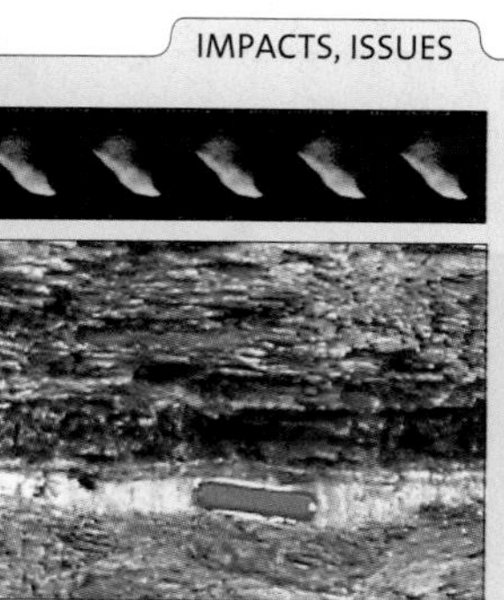

Iridium is rare on Earth, not in asteroids. A thin, dark layer of it dates to the Cretaceous–Tertiary (K–T) boundary, a sign that the Earth took a huge hit. Iridium-rich dust blew skyward and then drifted all over the globe. It marks the K–T mass extinction.

Fossils are physical evidence of organisms that lived in the distant past. Stratified layers of sedimentary rock that are rich in fossils are a historical record of life. The deepest layers generally contain the oldest fossils.

The fossil record is incomplete. Major geologic events obliterated much of it. Our sampling of the record is slanted toward species that had large bodies, hard parts, dense populations, and broad distribution. It also is slanted toward species that persisted for a long time.

Even so, the fossil record is now substantial enough to reconstruct patterns and trends in the history of life.

13.2 Dating Pieces of the Puzzle

How do we assign fossils to a place in time? In other words, how do we know how old they really are?

RADIOMETRIC DATING

People have been coming across fossils for as long as they have been digging up rocks. For a long time they could assign only *relative* ages to their fossil treasures, not absolute ones. For instance, a fossilized mollusk in a rock layer was said to be younger than a fossil below it and older than a fossil above it, and so on.

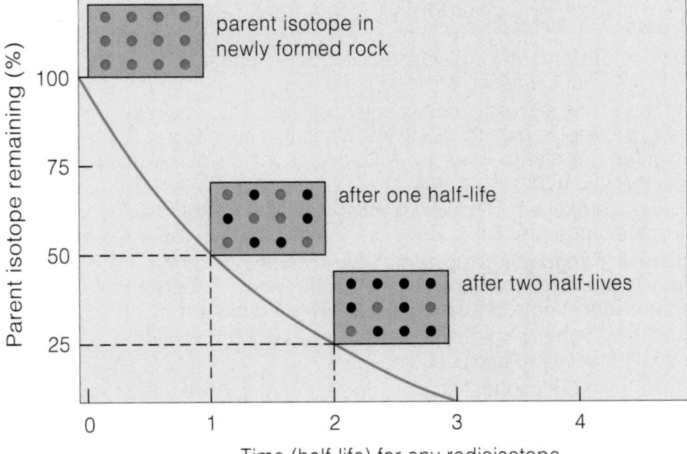

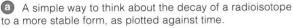

(a) A simple way to think about the decay of a radioisotope to a more stable form, as plotted against time.

(b) Long ago, trace amounts of ^{14}C carbon and a lot more ^{12}C were incorporated in tissues of a living mollusk. The carbon was part of the organic compounds making up the tissues of its prey, which it captured in the sea. As long as the mollusk was living, the proportion of ^{14}C to ^{12}C in its tissues remained the same.

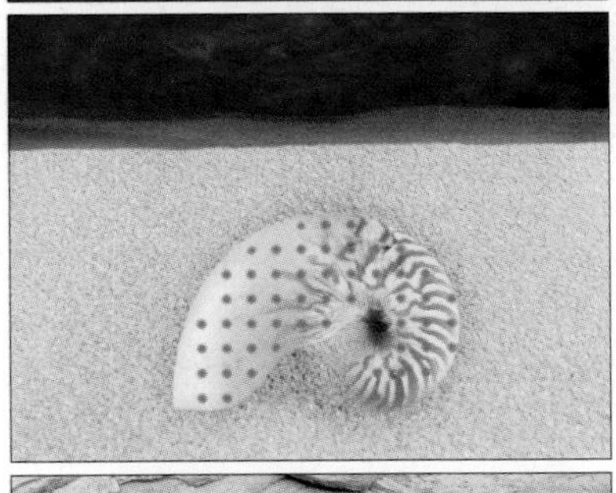

(c) When the mollusk died, it stopped gaining carbon. Over time, the proportion of ^{14}C to ^{12}C in its remains fell because of the radioactive decay of the ^{14}C. Half of the ^{14}C had decayed in 5,370 years, half of what was left was gone after another 5,370 years, and so on.

(d) Fossil hunters find the fossil. They measure its $^{14}C/^{12}C$ ratio to determine half-life reductions since death. The ratio for this fossil turns out to be one-eighth of the $^{14}C/^{12}C$ ratio in living organisms. Thus the mollusk lived about 16,000 years ago.

Figure 13.3 ***Animated!*** (**a**) The decay of radioisotopes at a fixed rate to more stable forms. The half-life of each kind of radioisotope is the time it takes for 50 percent of a sample to decay. After two half-lives, 75 percent of the sample has decayed, and so on.

(**b–d**) Radiometric dating of a fossil. Carbon 14 (^{14}C) forms in the atmosphere. There, it combines with free oxygen, the result being carbon dioxide. Along with far greater quantities of the more stable carbon isotope (^{12}C), trace amounts of carbon 14 enter food webs by way of photosynthesis. All living organisms incorporate carbon in their body tissues.

Things changed when the nature of radioisotope decay was discovered. Remember, a radioisotope is an atom of an element with an unstable nucleus (Section 2.1). The nucleus decays spontaneously, losing energy and subatomic particles until the atom becomes a more stable element. In **radiometric dating**, the proportions of a radioisotope to its daughter elements in a sample can be used to calculate the sample's age. This works because, like the ticking of a perfect clock, the rate of decay for each isotope is constant.

We cannot predict the exact instant of decay of any one atom, but a predictable number of a radioisotope's atoms in any sample will decay in a characteristic time span. No change in pressure, temperature, or chemical state alters that rate.

The time it takes for half of a radioisotope's atoms to decay is called its **half-life** (Figure 13.3*a*). For instance, an atom of uranium 238 will decay into thorium 234, which decays into something else, and so on through intermediate isotopes until it becomes lead, the final, stable daughter element for this decay series. The half-life of uranium 238 is 4.5 billion years.

Radiometric dating can be used to reveal the age of rocks. By measuring the uranium 238–to–lead ratio in the oldest known rocks, geologists figured out that Earth formed more than 4.6 billion years ago.

Volcanic rocks can be dated, but most fossils occur in sedimentary rock. Dating a sedimentary rock yields the date its component materials formed, not the date they actually sedimented. The only way to date very old fossils in sedimentary rock is to find their position relative to volcanic rocks in the same area. Fossils that retain some carbon can be dated directly by measuring their ratio of carbon 12 to carbon 14 (Figure 13.3*b–d*).

LINK TO SECTION 2.1

PLACING FOSSILS IN GEOLOGIC TIME

Early geologists carefully counted backward through layer upon layer of sedimentary rock. They used their findings to construct a chronology of Earth history—a **geologic time scale**. By comparing information from around the world, they found four abrupt transitions in fossil sequences and used them as boundaries for four great intervals.

Eon	Era	Period	Epoch	Millions of Years Ago	Major Geologic and Biological Events That Occurred Millions of Years Ago (mya)
PHANEROZOIC	**CENOZOIC**	**QUATERNARY**	Recent	0.01	
			Pleistocene	1.8	1.8 mya to present. Major glaciations. Modern humans evolve. The most recent *extinction crisis* is under way.
		TERTIARY	Pliocene	5.3	65–1.8 mya. Major crustal movements, collisions, mountain building. Tropics, subtropics extend poleward. When climate cools, dry woodlands, grasslands emerge. *Adaptive radiations* of flowering plants, insects, birds, mammals.
			Miocene	22.8	
			Oligocene	33.7	
			Eocene	55.5	
			Paleocene	65	65 mya. Asteroid impact; *mass extinction* of all dinosaurs and many marine organisms.
	MESOZOIC	**CRETACEOUS**	Late	99	99–65 mya. Pangea breakup continues, inland seas form. Adaptive radiations of marine invertebrates, fishes, insects, and dinosaurs. Origin of angiosperms (flowering plants).
			Early	145	145–99 mya. Pangea starts to break up. Marine communities flourish. *Adaptive radiations* of dinosaurs. 145 mya. Asteroid impact? Mass extinction of many species in seas, some on land. Mammals, some dinosaurs survive.
		JURASSIC		213	
		TRIASSIC		248	248–213 mya. *Adaptive radiations* of marine invertebrates, fishes, dinosaurs. Gymnosperms dominate land plants. Origin of mammals. 248 mya. *Mass extinction.* Ninety percent of all known families lost.
	PALEOZOIC	**PERMIAN**		286	286–248 mya. Supercontinent Pangea and world ocean form. On land, *adaptive radiations* of reptiles and gymnosperms.
		CARBONIFEROUS		360	360–286 mya. Recurring ice ages. On land, *adaptive radiations* of insects, amphibians. Spore-bearing plants dominate; cone-bearing gymnosperms present. Origin of reptiles. 360 mya. *Mass extinction* of many marine invertebrates, most fishes.
		DEVONIAN		410	410–360 mya. Major crustal movements. Ice ages. *Mass extinction* of many marine species. Vast swamps form. Origin of vascular plants. *Adaptive radiation* of fishes continues. Origin of amphibians.
		SILURIAN		440	440–410 mya. Major crustal movements. *Adaptive radiations* of marine invertebrates, early fishes.
		ORDOVICIAN		505	505–440 mya. All land masses near equator. Simple marine communities flourish until origin of animals with hard parts.
		CAMBRIAN		544	544–505 mya. Supercontinent breaks up. Ice age. *Mass extinction.*
PROTEROZOIC				2,500	2,500–544 mya. Oxygen accumulates in atmosphere. Origin of aerobic metabolism. Origin of eukaryotic cells. Divergences lead to eukaryotic cells, then protists, fungi, plants, animals.
ARCHEAN AND EARLIER					3,800–2,500 mya. Origin of photosynthetic prokaryotic cells. 4,600-3,800 mya. Origin of Earth's crust, first atmosphere, first seas. Chemical, molecular evolution leads to origin of life (from proto-cells to anaerobic prokaryotic cells).

Figure 13.4 ***Animated!*** Geologic time scale. *Red* boundaries mark times of the greatest mass extinctions. If these spans were to the same scale, the Archean and Proterozoic portions would extend downward, spill off the page, and extend across the room. Compare Figure 13.5.

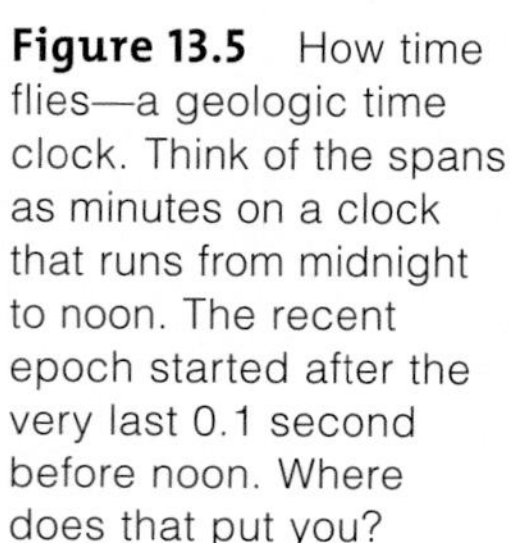

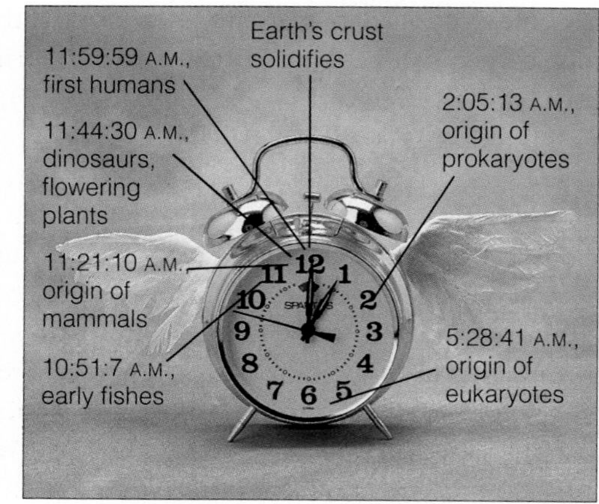

Figure 13.5 How time flies—a geologic time clock. Think of the spans as minutes on a clock that runs from midnight to noon. The recent epoch started after the very last 0.1 second before noon. Where does that put you?

The current geologic time scale now correlates with **macroevolution**, or major patterns, trends, and rates of change among lineages. Life originated in one of the intervals, the Archean eon (Figure 13.4).

The predictable property of radioisotope decay allows scientists to accurately estimate the age of volcanic rocks and carbon-containing fossils.

A geologic time scale shows the chronology of Earth history. Intervals of life history on the scale were initially organized using evidence from the fossil record. Dates assigned to the intervals later were determined by radiometric dating.

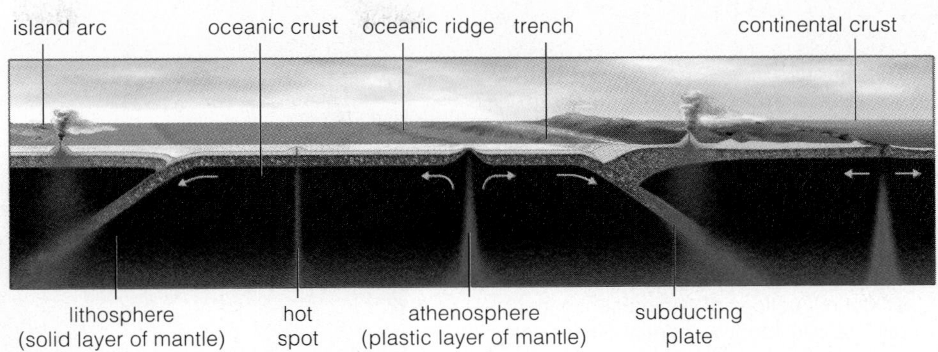

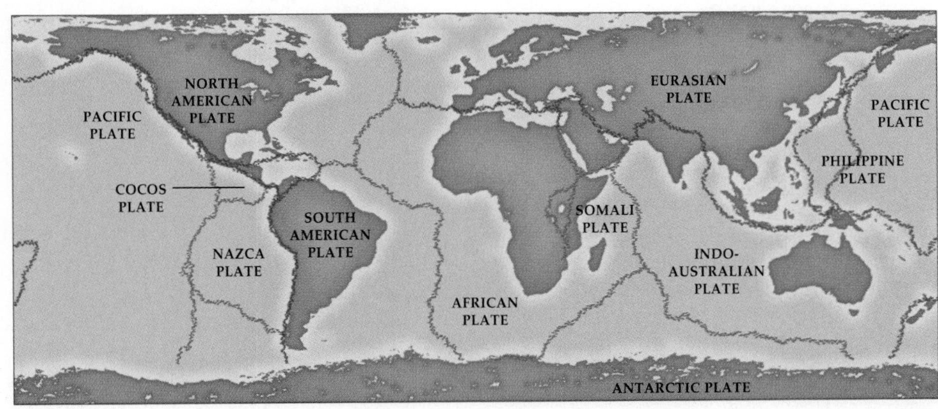

Figure 13.6 Some forces of geologic change.

(**a**) Immense, rigid parts of Earth's crust split, drift apart, and collide very slowly, at almost imperceptible rates. Huge plumes of molten material drive the movement. The plumes well up from the interior, spread laterally under the crust, and rupture it at deep mid-oceanic ridges. At these ridges, molten material seeps out, cools, and slowly forces the seafloor away from the rupture, which displaces plates from the ridges. The leading edge of one plate plows under an adjoining plate and uplifts it.

Besides these forces, superplumes ruptured the crust at what are now called "hot spots" in the mantle. The Hawaiian Archipelago has been forming this way. Continents also can rupture in their interior. Deep rifting and splitting are happening now in Missouri, at Lake Baikal in Russia, and in eastern Africa.

(**b**) Present configuration of Earth's crustal plates. *Red* lines indicate where spreading is occurring. *Blue* lines show where one plate is being driven under another. The Cascades, Andes, and other great mountain ranges paralleling the coasts of continents formed this way.

13.3 Evidence From Biogeography

By clinking their hammers against the rocks, geologists realized that Earth had not stayed put beneath our feet.

When early geologists were beginning to map vertical stacks of sedimentary rock, the theory of uniformity prevailed. The geologists knew that mountain building and erosion had repeatedly altered the surface of the Earth. Eventually, however, it became clear that those recurring geologic events were only part of a bigger picture. Like life, Earth had changed irreversibly.

AN OUTRAGEOUS HYPOTHESIS

For instance, the Atlantic coasts of South America and Africa seemed to "fit" like jigsaw puzzle pieces. Were all continents once part of a bigger one that had split into fragments and drifted apart? One model for the proposed supercontinent—Pangea—took into account the world distribution of fossils and existing species. It also took into account glacial deposits, which held clues to ancient climate zones.

Most scientists did not accept the continental drift theory. To them, continents drifting about on their own across the Earth's mantle was an outrageous idea; they preferred to think that continents did not move.

But more evidence kept piling up. Iron-rich rocks are molten when they form. Their iron particles orient north–south in response to the Earth's magnetic poles and stay that way after the rocks cool and harden. Yet the tiny iron compasses in the 200 million year old rocks of North and South America did *not* point pole to pole. So scientists came up with a hypothetical map. They made a north–south alignment work by joining North America and western Europe—and by orienting them in a way that did not look like current maps.

Deep-sea probes also showed that the seafloor has been spreading from mid-oceanic ridges (Figure 13.6). Molten rock spewing from a ridge flows sideways in both directions, and then hardens into new crust. The spreading crust forces older crust into deep trenches elsewhere. All of the ridges and trenches are actually edges of relatively thin but enormous plates, a bit like

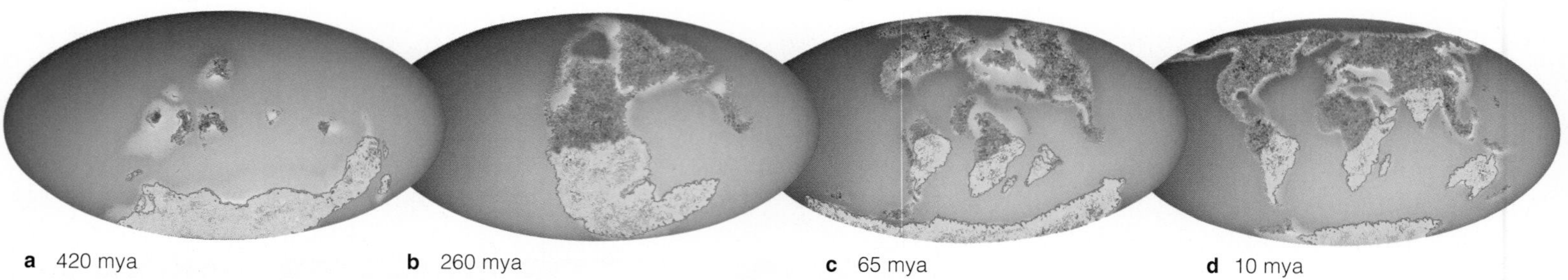

Figure 13.7 ***Animated!*** A series of reconstructions of drifting continents. (**a**) The early supercontinent Gondwana (*yellow*). (**b**) Later, all major land masses collided and formed the supercontinent Pangea. (**c**) Positions of fragments that drifted apart after Pangea split apart 65 million years ago, and (**d**) their positions 10 million years ago.

About 260 million years ago, seed ferns and other plants lived nowhere except on the area of Pangea that had once been Gondwana. So did mammal-like reptiles named therapsids. (**e**) Fossilized leaf of one of the seed ferns, *Glossopteris*. (**f**) *Lystrosaurus*, a therapsid about 1 meter (3 feet) long. This tusked herbivore fed on fibrous plants in dry floodplains.

pieces of a gigantic cracked eggshell. The movement is imperceptibly slow.

Together, these findings put continental drift into a broader explanation of Earth's crustal movements, the **plate tectonics theory**. Researchers soon found ways to use the new theory's predictive power.

For instance, by this theory, 420 million years ago the supercontinent **Gondwana** comprised most of the land on Earth (Figure 13.7*a*). By 260 million years ago, Gondwana had joined with all of the other major land masses to become the supercontinent **Pangea** (Figure 13.7*b*). Pangea fragmented about 65 million years ago. By 10 million years ago, the parts of Pangea that once were Gondwana had formed Antarctica, India, Africa, Australia, and South America. The rest of Pangea had become North America, Europe, Siberia, and all other land in the northern hemisphere (Figure 13.7*c,d*).

Fossils of *Glossopteris*, a seed fern, and *Lystrosaurus*, a reptile (Figure 13.7*e,f*), were found in India, Africa, Australia, and South America, but not found in North America or Europe. Scientists hypothesized that both species had evolved together on Gondwana, and they predicted that the same fossils would also be found in Antarctica. In support of the hypothesis and the plate tectonics theory, Antarctic explorers found the fossils.

A BIG CONNECTION

Let's review. In the remote past, crustal movements put huge land masses on collision courses. In time the masses converged to form supercontinents, which split at deep rifts and formed new ocean basins. Gondwana drifted south from the tropics, across the south pole, then north until it crunched into other land masses.

A new supercontinent, Pangea, now extended from pole to pole, and a single world ocean lapped against its coasts. All the while, the erosive forces of water and wind resculpted the land. Asteroids and meteorites hit the crust. The impacts and aftermaths had long-term effects on global temperature and climate.

Such changes on land and in the ocean and atmosphere influenced life's evolution. Imagine early life in shallow, warm waters along continents. Shorelines vanished as continents collided and wiped out many lineages. But even as old habitats vanished, new ones opened up for survivors—and evolution took off in new directions.

Over the past 3.8 billion years, gradual and catastrophic events changed the Earth's crust, the atmosphere, and the ocean—and thus influenced the evolution of life.

13.4 More Evidence From Comparative Morphology

To biologists, remember, evolution means heritable changes in lines of descent. Comparisons of body forms and structures of major groups of organisms yield clues to evolutionary trends.

Comparative morphology is the study of body forms and structures of major groups of organisms, such as vertebrates and flowering plants. Often, comparison reveals similarities in one or more body parts, which hint at inheritance from a shared ancestor. Such body parts are called **homologous structures** (*homo–* means the same). In these cases, genetic similarities persist even when the structures have different functions in different kinds of organisms.

MORPHOLOGICAL DIVERGENCE

Populations of a species diverge genetically when gene flow ends between them (Chapter 12). In time, some of the morphological traits that help define these species commonly diverge, also. Change from the body form of a common ancestor is a major macroevolutionary pattern called **morphological divergence**.

Even if the same body part of two related species become dramatically different, some other aspects of the species may remain alike. A careful look beyond unique modifications may reveal the shared heritage.

For example, all the vertebrates on land descended from the first amphibians. Divergences led to what we generally call reptiles, then to birds and mammals. A kind of "stem reptile" probably was ancestral to these groups. Fossilized, five-toed limb bones tell us that the ancestral species crouched low to the ground (Figure 13.8*a*). Its descendants diversified into many habitats on land. When environmental conditions changed, a few of the descendants that had become adapted to land returned to the seas.

The five-toed limb was evolutionary clay that was modeled into different kinds of limbs with different functions (Figure 13.8*b–g*). It was modified for flight in extinct reptiles called pterosaurs, most birds, and in bats. In lineages that led to penguins and porpoises, it became a flipper that was used for swimming. Among moles, it became stubby, useful for burrowing, and in elephants, it became strong and pillarlike. In horses, it became a long, one-toed limb suitable for fast running. The five-toed limb was also modified in humans to become the arm and hand. A thumb that was the basis of stronger and more precise movements evolved in opposition to four fingers on the hand.

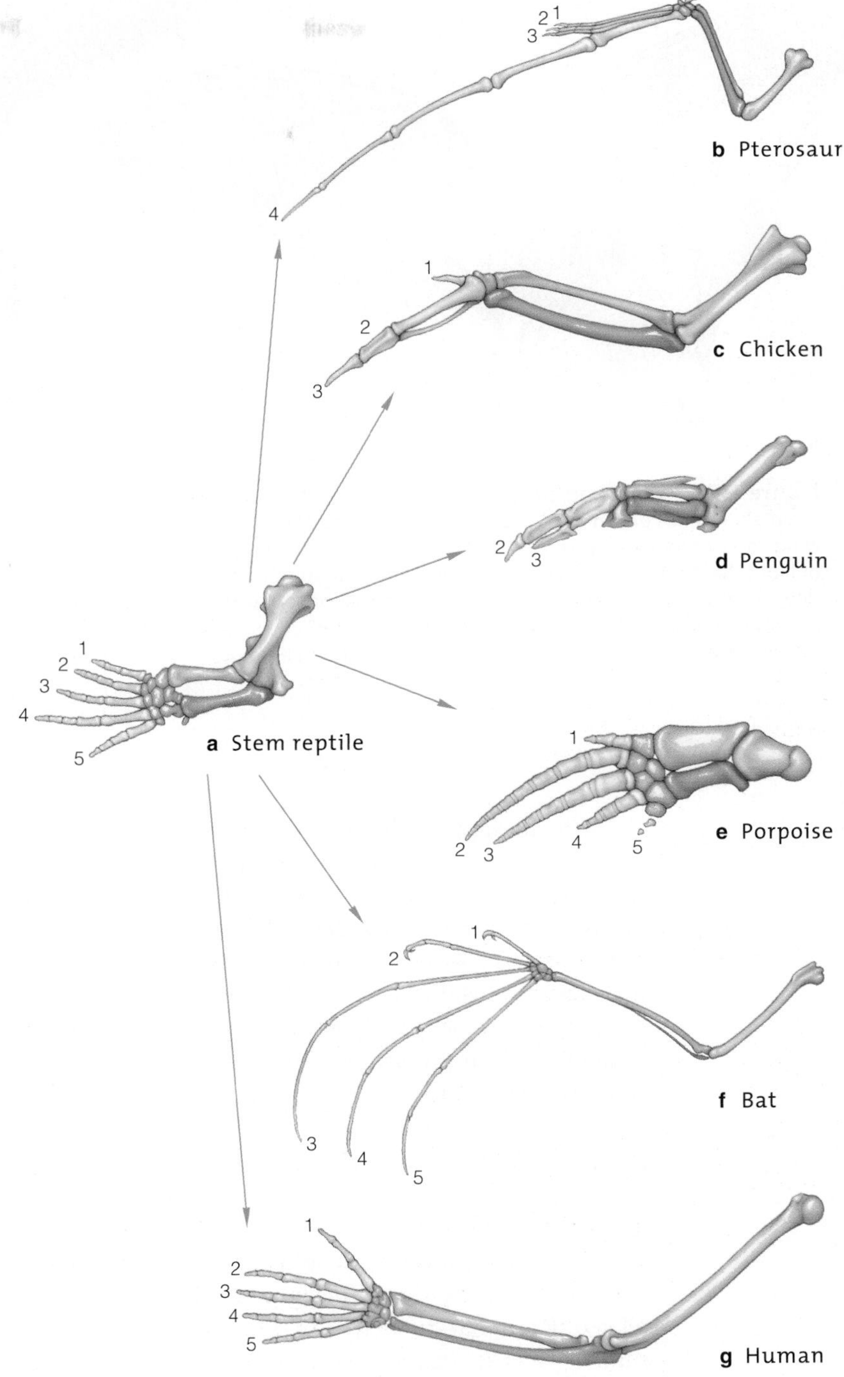

Figure 13.8 ***Animated!*** Morphological divergence among vertebrate forelimbs, starting with bones of a stem reptile (a cotylosaur). Similarities in the number and position of skeletal elements were preserved when diverse forms evolved. Some bones were lost over time (compare the numbers 1 through 5). The drawings are not to the same scale.

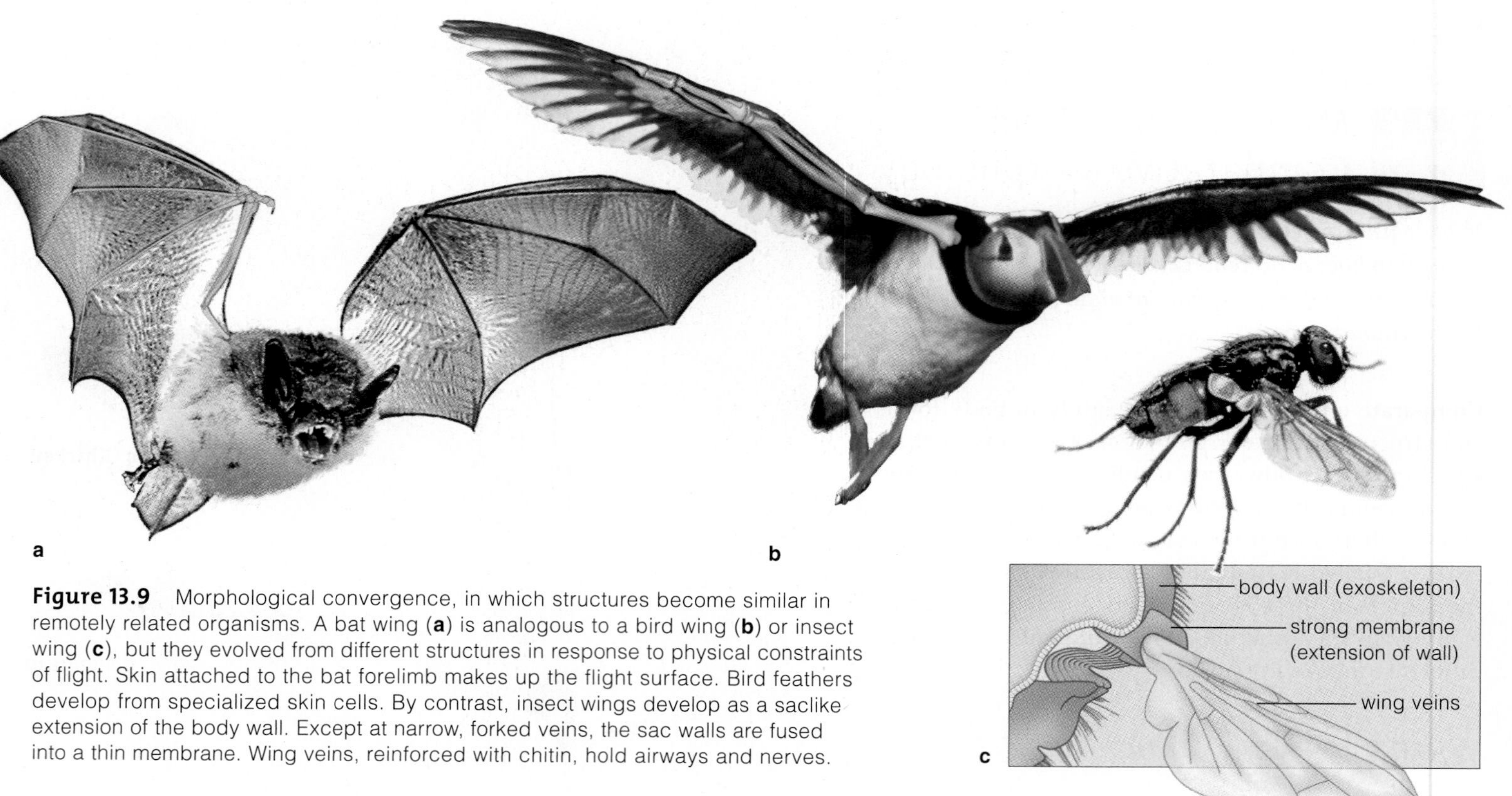

Figure 13.9 Morphological convergence, in which structures become similar in remotely related organisms. A bat wing (**a**) is analogous to a bird wing (**b**) or insect wing (**c**), but they evolved from different structures in response to physical constraints of flight. Skin attached to the bat forelimb makes up the flight surface. Bird feathers develop from specialized skin cells. By contrast, insect wings develop as a saclike extension of the body wall. Except at narrow, forked veins, the sac walls are fused into a thin membrane. Wing veins, reinforced with chitin, hold airways and nerves.

Even though forelimbs are not the same in size, shape, or function from one group of vertebrates to the next, they clearly are alike in the structure and positioning of their bony elements. They also are alike in the patterns of nerves, blood vessels, and muscles that develop inside them. In addition, comparisons of early embryos of different vertebrates reveal strong resemblances in patterns of bone development. Such similarities point to a shared ancestor.

MORPHOLOGICAL CONVERGENCE

Body parts with similar form or function in different lineages are not always homologous. Sometimes they evolved *independently* in lineages that are not closely related. Parts that differed at first may have evolved in similar ways as organisms became subjected to similar environmental pressures. **Morphological convergence** refers to cases where dissimilar body parts evolved in similar ways in evolutionarily distant lineages.

For instance, you just read about the homologous forelimbs of birds and bats. Internal structures aside, however, the flight surfaces of bird and bat wings are *not* homologous. The flight surface of the bird wing evolved as a sweep of feathers, all derived from skin. The forelimb structurally supports the wing (Figure 13.9*b*). By contrast, the flight surface of the bat wing is a thin membrane, an extension of the skin itself. The wing is attached to and reinforced by the bony digits of the bat forelimb (Figure 13.9*b*). The insect wing also resembles bird and bat wings in function—flight—but it is not homologous with either of them. The insect wing develops as an extension of the outer body wall. It has no underlying bony elements for support; it is instead reinforced with chitin (Figure 13.9*c*).

The differences between bat, bird, and insect wings are evidence that each of these animal groups adapted independently to the same physical constraints that govern how a wing can function. The flight surfaces of all three are **analogous structures**. These are structural adaptations that are the result of similar challenges in the environment, but they are derived from dissimilar body parts of distantly-related lineages. They are not modifications of the comparable body parts of closely related lineages.

With morphological divergence, comparable body parts became modified in different ways in different lines of descent from a common ancestor.

Such divergences resulted in homologous structures. Even if these body parts differ in size, shape, or function, they have an underlying similarity because of shared ancestry.

In morphological convergence, dissimilar body parts became similar in independent lineages that are not closely related.

Such body parts are analogous structures. They converged in form as an outcome of similar pressures.

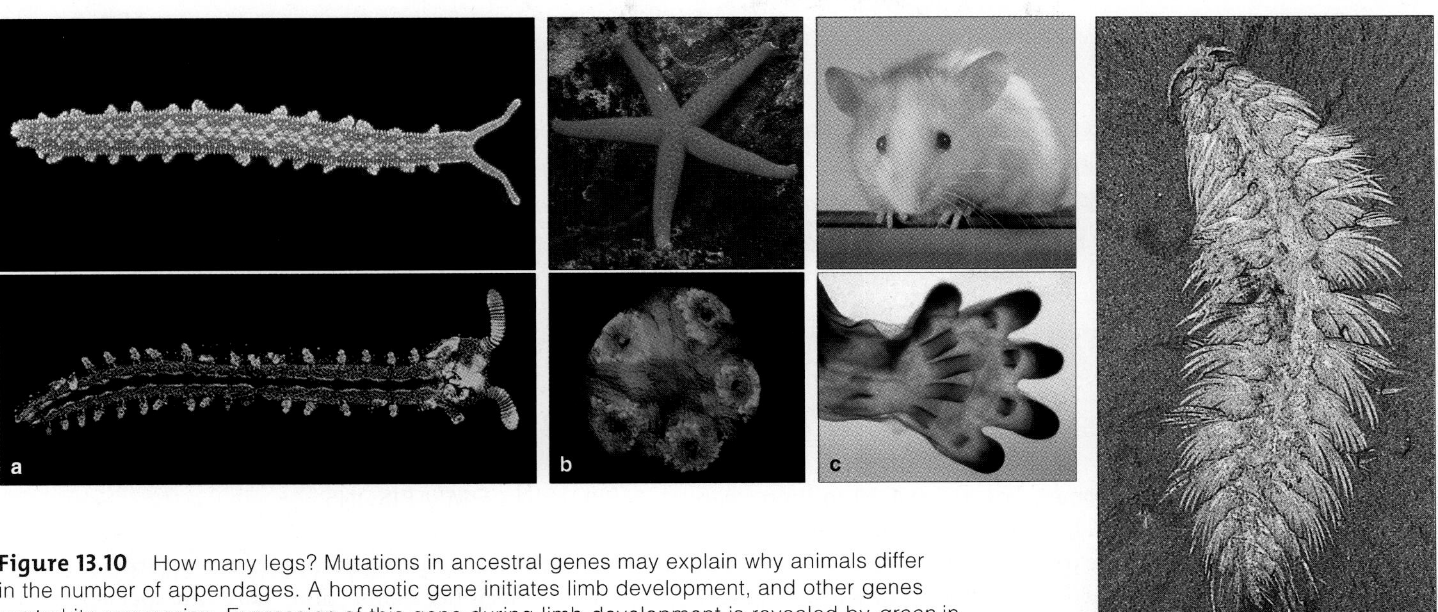

Figure 13.10 How many legs? Mutations in ancestral genes may explain why animals differ in the number of appendages. A homeotic gene initiates limb development, and other genes control its expression. Expression of this gene during limb development is revealed by *green* in (**a**) a velvet walking worm, and (**b**) a sea star; and *blue* in (**c**) a mouse foot. (**d**) Cambrian legs.

13.5 Evidence From Patterns of Development

Comparing the patterns of embryonic development often yields evidence of evolutionary relationships.

Multicelled embryos of plants and animals develop step-by-step from a fertilized egg, and there are built-in constraints on how the body plan develops. Most mutations and changes in chromosome structure are selected against because they will disrupt an inherited program of development. Even so, a mutation having neutral or beneficial effects can move a lineage past one of the constraints.

Mutations in master genes can do this. Recall from Section 10.6 that homeotic genes guide the formation of tissues and organs in orderly patterns. A homeotic gene mutation can disrupt the body plan, sometimes drastically so. Such disruptions typically lead to huge problems, but occasionally an altered body plan may prove to be advantageous.

For example, a limb bud will form anywhere in an embryo that a particular homeotic gene is expressed. This gene's product is a signal for clusters of dividing embryonic cells to "stick out from the body." Where expression of this gene is suppressed, limb buds do not form. What is most remarkable is that expression of *the same gene* can initiate limb bud development in organisms as diverse as crabs, mice, beetles, sea stars, butterflies, fishes, and humans. Evolution of layers of controls over this homeotic gene may have resulted in successively fewer limbs (Figure 13.10).

Modifications in genes that influence growth rates might have caused the major proportional differences between chimpanzee and human skull bones (Figure 13.11). For humans, the facial bones and skull bones around the brain increase in size at fairly consistent rates, from infant to adult. The rate of growth is faster for chimp facial bones, so the proportions of an infant skull and an adult skull differ significantly.

LINK TO SECTION 10.6

In addition, transposons may have caused variation among lineages. As Section 10.5 explains, these short DNA segments spontaneously and repeatedly slip into

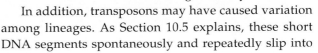

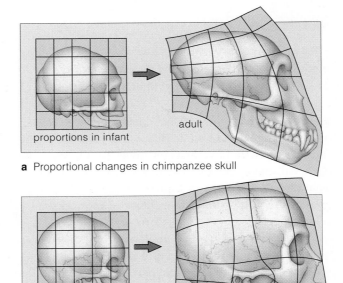

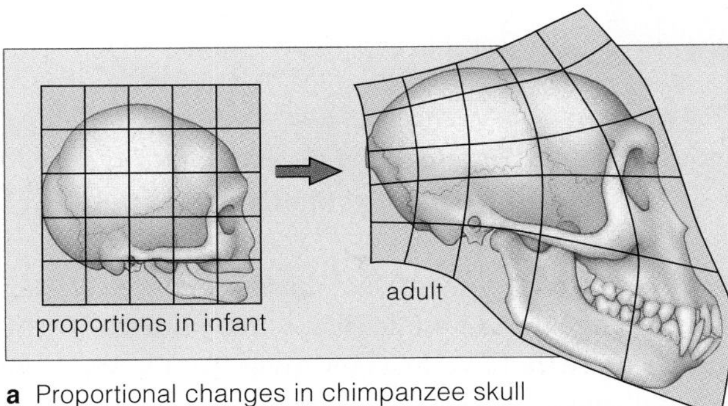

Figure 13.11 ***Animated!*** Differences between two primates, a possible outcome of mutations that changed the timing of steps in the body's development. The skulls are depicted as paintings on a rubber sheet that is divided into a grid. Stretching both sheets deforms the grids in a way that corresponds to the differences in growth patterns between these primates.

$^{+}NH_3$-gly asp val glu lys gly lys lys ile phe ile met lys cys ser gln cys his thr val glu lys gly gly lys his lys thr gly pro asn leu his gly leu phe gly arg lys thr gly gln ala pro gly tyr ser

$^{+}NH_3$-ala ser phe ser glu ala pro pro gly asn pro asp ala gly ala lys ile phe lys thr lys cys ala gln cys his thr val asp ala gly ala gly his lys gln gly pro asn leu his gly leu phe gly arg gln ser gly thr thr ala gly tyr ser

$^{+}NH_3$-thr glu phe lys ala gly ser ala lys lys gly ala thr leu phe lys thr arg cys leu gln cys his thr val glu lys gly gly pro his lys val gly pro asn leu his gly ile phe gly arg his ser gly gln ala glu gly tyr ser

LINKS TO SECTIONS 11.2, 11.3

new places in genomes. For at least 30 million years, only primate DNA has carried transposons called *Alu* elements. We can expect that they were pivotal in the evolution of primates. By current estimates, 99 percent of the DNA sequences of human genes are identical to chimpanzee genes. The remaining 1 percent accounts for the species differences. Uniquely positioned *Alu* elements may be one factor.

Similarities in patterns of development are often clues to an evolutionary relationship among plant and animal lineages.

Heritable changes that alter key steps in a developmental program may be enough to bring about major differences between adult forms of related lineages. Transposons and single gene mutations can bring about such changes.

13.6 Evidence From DNA, RNA, and Proteins

All species are a mix of ancestral and novel traits, including biochemical ones. The kinds and numbers of traits they do or do not share are clues to relationships.

Each species has its own DNA base sequence. We can also expect genes to mutate in any line of descent. The more recently that two lineages diverged, the less time each will have had to accumulate unique mutations. That is why the RNA and proteins of closely related species are more similar than those of more distantly related ones.

Identifying biochemical similarities and differences among species is rapid and precise, thanks to DNA sequencing (Section 11.3). Extensive sequence data for many genomes and proteins are continually compiled into internationally accessible databases. With these data, we know that 31 percent of the 6,000 yeast cell genes have counterparts in the human genome. So do 40 percent of the 19,023 roundworm genes, and also 50 percent of the 13,601 fruit fly genes.

PROTEIN COMPARISONS

Similarities in amino acid sequences of proteins can be used to decipher connections between species. If two species have many proteins with similar or identical amino acid sequences, the species probably are closely related. When the sequences differ considerably, many mutations have accumulated, which indicates that a long time passed since the species shared an ancestor.

A few essential genes have evolved very little; they are highly *conserved* even across very diverse species. Conserved genes also are clues about ancient ancestral connections. One such gene encodes cytochrome *c*, a protein that is a crucial component of electron transfer chains in species that range from aerobic bacteria to humans. In humans, its primary structure consists of only 104 amino acids. Figure 13.12 shows the striking similarity between the entire amino acid sequence of cytochrome *c* from a yeast, a plant, and an animal.

NUCLEIC ACID COMPARISONS

Most phenotypic differences between any two species boil down to differences in the nucleotide sequences of their respective genomes. If you find the number of bases that differ between different genomes, you can estimate the relative evolutionary distance between the different species. DNA from nuclei, mitochondria, and chloroplasts, and particularly DNA regions that encode ribosomal RNA, are compared across different species for this kind of estimation.

Mitochondrial DNA (mtDNA), which mutates fast, is used to compare individuals of eukaryotic species. In sexually reproducing species, it is inherited intact from one parent—usually the mother. Thus, changes between maternally related individuals probably were caused by mutations, not by genetic recombinations.

DNA sequencing is now the most common method of comparing DNA among different species because it yields specific information, but two other methods are still in use. Nucleic acid hybridization (Section 11.2) can be used to compare DNA of two different species. DNA from both species is mixed in a way that double-

r ala ala asn lys asn lys gly ile ile trp gly glu asp thr leu met glu tyr leu glu asn pro lys lys tyr ile pro gly thr lys met ile phe val gly ile lys lys lys glu glu arg ala asp leu ile ala tyr leu lys lys ala thr asn glu-COO^-
er ala ala asn lys asn lys ala val glu trp glu glu asn thr leu tyr asp tyr leu leu asn pro lys lys tyr ile pro gly thr lys met val phe pro gly leu lys lys pro gln asp arg ala asp leu ile ala tyr leu lys lys ala thr ser ser-COO^-
r asp ala asn ile lys lys asn val leu trp asp glu asn asn met ser glu tyr leu thr asn pro lys lys tyr ile pro gly thr lys met ala phe gly gly leu lys lys glu lys asp arg asn asp leu ile thr tyr leu lys lys ala cys glu-COO^-

Figure 13.12 ***Animated!*** Comparison of the amino acid sequence of the cytochrome *c* protein from yeast (*top row*), wheat (*middle*), and primate (*bottom*). *Gold* highlights amino acids that are identical in all three. The probability that such a resemblance resulted by chance is extremely low. Cytochrome *c* is a vital component of electron transfer chains.

stranded hybrid molecules form. The more base pairs that match, the more hydrogen bonds form between the two strands. The amount of heat that is needed to separate the strands of a hybrid DNA molecule can be used as a measure of their similarity. Why? More heat is needed to disrupt more hydrogen bonding.

Also, restriction fragments of DNA from different species are separated by gel electrophoresis. The band patterns that result can be compared just like DNA fingerprints (Section 11.4).

These comparative studies either reinforce or invite modification of ideas about evolutionary relationships that originated with morphological comparisons.

MOLECULAR CLOCKS

Some researchers estimate the timing of divergence by comparing the numbers of neutral mutations in genes that have been highly conserved in different lineages. Because the mutations have little or no effect on the individual's survival or reproduction, we can expect that neutral mutations have accumulated in conserved genes at a fairly constant rate.

The accumulation of neutral mutations to the DNA of a lineage has been likened to the predictable ticks of a **molecular clock**. Turn the hands of such a clock back, so that the total number of ticks unwind down through past geologic intervals. Where the last tick stops, that is the approximate time when molecular, ecological, and geographic events put the lineage on its own unique evolutionary road. Thus, the number of differences in DNA base sequences or amino acid sequences between species can be plotted as a series of branches on an evolutionary tree. The branches reflect relative times of divergences among species.

Biochemical similarity is greatest among the most closely related species and smallest among the most remote.

13.7 Reproductive Isolation, Maybe New Species

LINKS TO SECTIONS 8.3, 12.7, 12.8

Speciation, a macroevolutionary event, begins with reproductive isolation and ends with a new species.

Species is a Latin word that means "kind," as in "one kind of plant." This generic definition does not help much when we are trying to figure out whether, say, a population of plants in one place belongs to the same species as a population of plants somewhere else. You may see variations in traits between the populations and variations within them, because plants can inherit diverse combinations of alleles. Also individuals of a species may vary in form because gene expression can be affected by living in different environments (Figure 13.13). In other words, we may not be able to identify a biological species on the basis of appearance alone.

Actually, individuals of *most* species vary greatly in details of form. So perhaps we should also look for a basic function that unites populations of a species and isolates them from populations of other species.

Reproduction is such a basic, defining function. It is the core of the **biological species concept**. Ernst Mayr, an evolutionary biologist, phrased it this way: Species are interbreeding natural populations that have been reproductively isolated from other such groups. In his view, it does not matter how much phenotypes vary. Populations belong to the same species as long as their members share traits that allow them to interbreed and produce fertile offspring. Being able to contribute to a shared gene pool is qualification for membership.

Mayr's species concept does not apply to asexually reproducing organisms, such as bacteria. It is a useful framework for studying the vast majority of species, which reproduce sexually. A new species arises when it achieves reproductive isolation. This isolation does not happen deliberately. *Any structural, functional, or behavioral difference that favors reproductive isolation is simply a by-product of genetic change.*

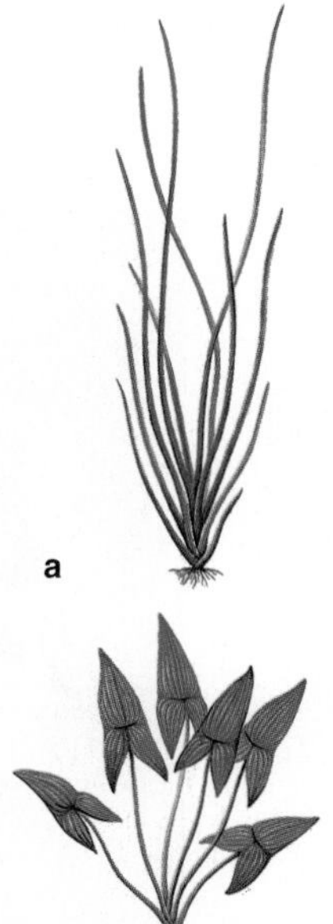

Figure 13.13 Differences in form between two plants of the same species (*Sagittaria sagittifolia*) growing (**a**) in water and (**b**) on land. Leaf shapes are responses to different conditions in the environment, not differences in genes.

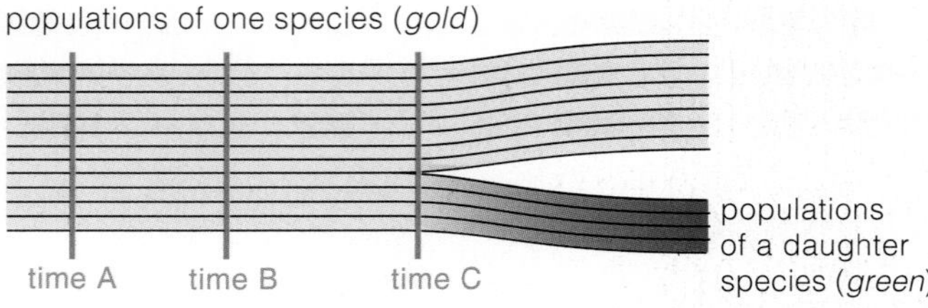

Figure 13.14 One way to think about genetic divergence. Each horizontal band represents a different population.

LINK TO SECTION 12.8

REPRODUCTIVE ISOLATING MECHANISMS

Gene flow alone can counter genetic changes between populations of a species. **Gene flow**, remember, is the movement of alleles into and out of a population by immigration and emigration. This microevolutionary process helps maintain diversity in a shared reservoir of alleles. **Genetic divergence** occurs after gene flow has stopped entirely (Figure 13.14). Then, mutation, natural selection, and genetic drift all begin to operate independently in each isolated population (Section 12.8). Their gene pools change. Eventually, the genetic divergence may lead to new species.

The process by which a new species arises is called **speciation**. It may happen gradually, or within a few generations, as commonly occurs among plants.

Regardless of differences in the duration or details, **reproductive isolating mechanisms** begin after gene flow has ended. These heritable aspects of body form, function, or behavior prevent interbreeding between divergent populations. Some mechanisms prevent the successful pollination or mating between individuals in the different populations. Others block fertilization or embryonic development after mating or pollination occurs. Still others cause hybrid offspring to be weak or infertile.

Mechanical Isolation. Sexual body parts of a species may not fit, mechanically, with the sexual body parts of a different species. Figure 13.15*a* shows the precise fit between a zebra orchid and its preferred pollinator, a single species of wasp.

Two sage species can keep pollen to themselves by attracting different insect pollinators. Pollen-bearing stamens of white sage extend from a nectar cup, above petals that constitute a sturdy platform. Bumblebees and large carpenter bees brush against the stamens when they land on the platform. Small honey bees that land on it are too small to reach the stamens, and so do not pick up pollen. The flower of black sage is too

Figure 13.15 *Animated!* (**a**) Mechanical isolation. Few pollinating insects fit as well as this wasp on a zebra orchid. Petals form a landing platform below the stamens. The scent of this flower stimulates male wasps to attempt mating with it. They are, of course, unsuccessful, but their actions do pollinate the flower.

(**b**) Temporal isolation. *Magicicada septendecim*, a periodical cicada that matures underground and emerges to reproduce every seventeen years. Its populations often overlap the habitats of a sibling species (*M. tredecim*), which reproduces every thirteen years. Adults live only a few weeks. Birds consider them a taste thrill and frenziedly gorge on them to the point of vomiting, which makes one suspect why periodical cicadas stay underground for so long.

(**c**) Behavioral isolation. Courtship displays precede sex among many kinds of birds, including these albatrosses. Individuals recognize tactile, visual, and acoustical signals, such as a prancing dance followed by back arching, skyward pointing bill and exposed throat, and wing spreading.

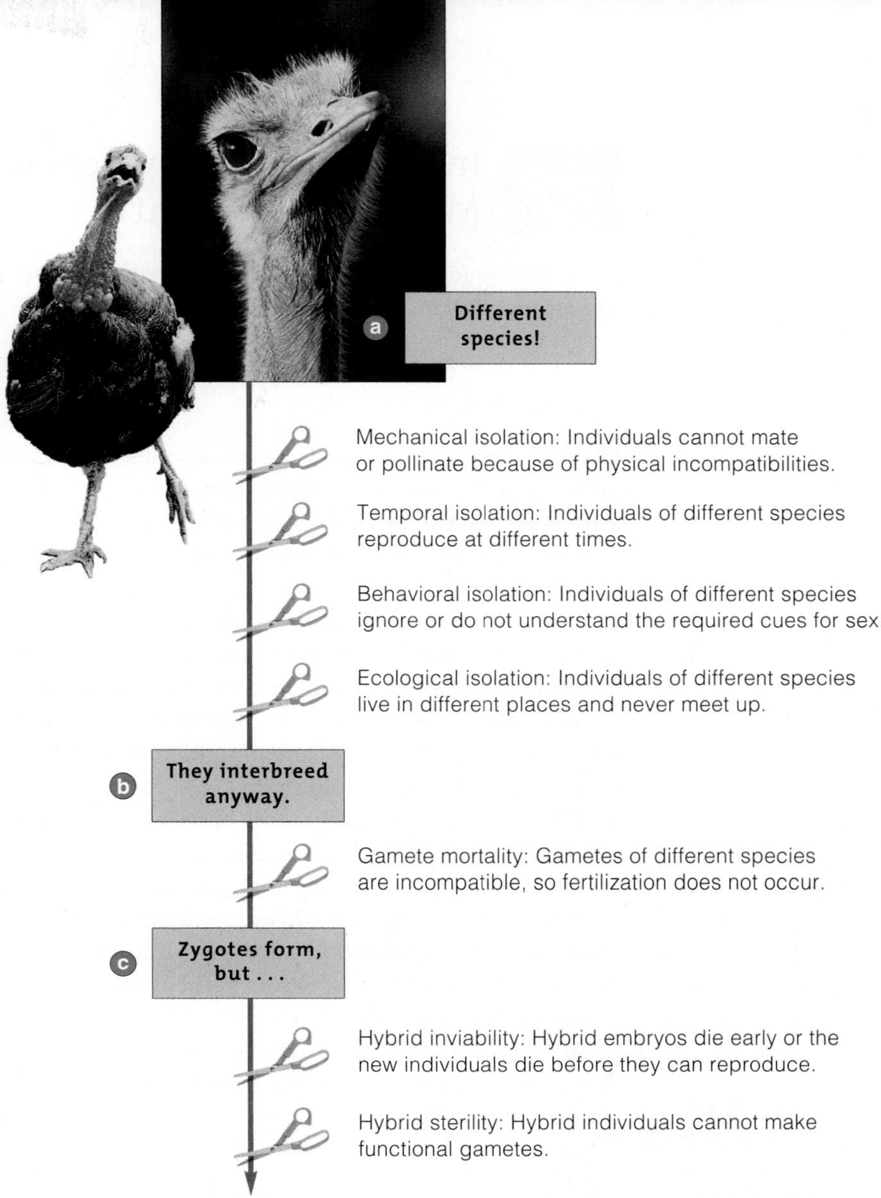

Different species!
Mechanical isolation: Individuals cannot mate or pollinate because of physical incompatibilities.
Temporal isolation: Individuals of different species reproduce at different times.
Behavioral isolation: Individuals of different species ignore or do not understand the required cues for sex.
Ecological isolation: Individuals of different species live in different places and never meet up.
They interbreed anyway.
Gamete mortality: Gametes of different species are incompatible, so fertilization does not occur.
Zygotes form, but . . .
Hybrid inviability: Hybrid embryos die early or the new individuals die before they can reproduce.
Hybrid sterility: Hybrid individuals cannot make functional gametes.
No offspring, sterile offspring, weak offspring that die before reproducing, or sterility of later generations

Figure 13.16 ***Animated!*** When certain reproductive isolating mechanisms prevent interbreeding. There are barriers to (**a**) getting together, mating, or pollination, (**b**) successful fertilization, and (**c**) survival, fitness, or fertility of hybrid embryos or offspring.

small to support the bigger bees, so this plant can be pollinated only by smaller bee species.

Temporal isolation. Diverging populations cannot interbreed when they differ in timing of reproduction. Three species of periodical cicadas often live in the same habitats in the eastern United States. They differ in size and other traits, but all mature underground. They only emerge to reproduce every 17 years (Figure 13.15*b*). Each cicada species has a *sibling* species that emerges to reproduce every 13 years. Sibling species are identical, or nearly so, but they are reproductively isolated. Each cicada species and its sibling cannot get together except once every 221 years!

Behavioral isolation. Behavioral differences bar gene flow between related species in the same vicinity. For instance, before male and female birds copulate, they often engage in courtship displays (Figure 13.15*c*). A female is genetically prewired to recognize distinctive singing, head bobbing, wing spreading, and prancing by a male of her species as an overture to sex. Females of different species usually ignore his behavior.

Ecological isolation. Populations occupying different microenvironments in a habitat might be ecologically isolated. In the open forests of seasonally dry foothills of the Sierra Nevada are two manzanita species. One lives at elevations between 600 and 1,850 meters, the other between 750 and 3,350 meters. Where the ranges overlap, these species hybridize only rarely. Like other manzanita shrubs, they depend on mechanisms that conserve water all through the dry season. One species is adapted to sheltered sites, where water stress is not intense. The other lives in drier, more exposed sites on rocky hillsides, so cross-pollination is unlikely.

Gamete mortality. Gametes of different species may have evolved incompatibilities at the molecular level. For example, when the pollen of one species lands on a plant of a different species, it usually cannot recognize the molecular signals that are supposed to trigger its growth down through the plant's tissues, to the egg.

Postzygote isolating mechanisms may act while an embryo develops. Unsuitable interactions among some genes or gene products cause early death, sterility, or weak hybrids with low survival rates. Some hybrids are sturdy but sterile. Mules, the offspring of a female horse and male donkey, are examples of such hybrids.

Figure 13.16 summarizes the points at which the different reproductive isolation mechanisms can cut a connection between species.

A species is one or more populations of individuals that share a unique common ancestor, that share a gene pool under natural conditions and produce fertile offspring, and that are reproductively isolated from other species.

Speciation is the process by which a daughter species forms from a population or subpopulation of a parent species. The process varies in its details and duration.

Species originate by genetic divergence. Differences build up in the gene pools of diverging populations as mutation, natural selection, and genetic drift operate independently in each one.

Reproductive isolating mechanisms may evolve simply as by-products of genetic changes. They are heritable traits that, one way or another, prevent interbreeding.

13.8 Interpreting the Evidence: Models for Speciation

Three models for speciation differ in their basic premise of how populations become reproductively isolated.

GEOGRAPHIC ISOLATION

Genetic changes that lead to a new species often arise after a *physical separation* occurs between populations. This is called **allopatric speciation**, and it is the most common speciation route. Some physical barrier that arises stops the gene flow between two populations or subpopulations of one species. (*Allo*– means different; *patria* can be taken to mean homeland.) Reproductive isolating mechanisms evolve in the two populations. In time, speciation is complete. Interbreeding will no longer be possible even if the daughter species come into contact with one another.

How effective any geographic barrier is at blocking gene flow depends on the organisms' means of travel, how fast they travel, and also whether they are forced or behaviorally inclined to disperse. Some measurable distance often separates populations of a species, so gene flow among them is an intermittent trickle, not a steady stream. Barriers can arise abruptly and shut off the trickles. In the 1800s, a strong earthquake buckled part of the Midwest. The Mississippi River changed its course, cutting through habitats of insects that could not swim or fly. Gene flow across the river stopped in an instant of geologic time.

The fossil record suggests that geographic isolation also occurs over great spans of time. This happened after glaciers advanced down into North America and Europe during the ice ages and cut off populations of plants and animals from one another. After the glaciers retreated, descendants of the separate populations met up. Although related by descent, some were no longer reproductively compatible. They had evolved into new species. Genetic divergence was not as great between other populations; their descendants still interbreed. Reproductive isolation in their case was incomplete, so speciation did not follow.

Also, remember how Earth's crust is fractured into gigantic plates? Slow but colossal movements of the plates have altered configurations of land masses. As Central America was forming, part of an ancient ocean basin was uplifted and became a land bridge—called the Isthmus of Panama. Some camelids crossed that bridge into South America. Geographic separation led to new species, llamas and vicunas (Figure 13.17).

THE INVITING ARCHIPELAGOS

Archipelagos are island chains at some distance away from a continent. Some are so close to the mainland that gene flow is more or less unimpeded, and there is little if any speciation. The Florida Keys are a case in point. Other archipelagos are so isolated that they can

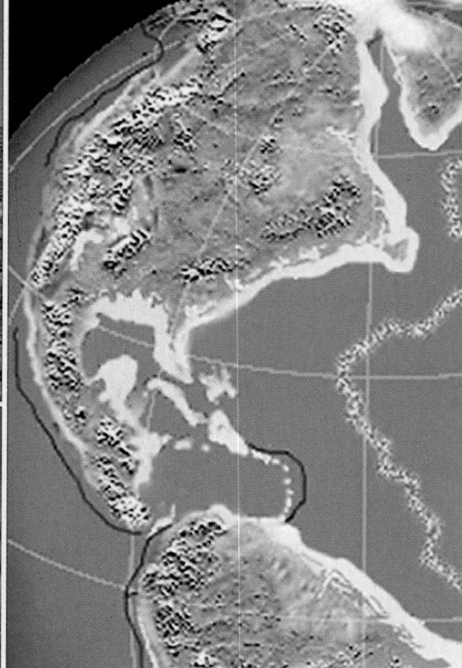

Late Eocene paleomap, before a land bridge formed between North and South America. At that time, North America and Eurasia were connected by a land bridge.

Figure 13.17 Allopatric speciations. The earliest camelids, no bigger than a jackrabbit, evolved in the Eocene grasslands and deserts of North America. By the end of the Miocene, they included the now-extinct *Procamelus*. The fossil record and comparative studies indicate that *Procamelus* may have been the common ancestral stock for llamas (**a**), vicunas (**b**), and camels (**c**). One descendant lineage that dispersed into Africa and Asia evolved into modern camels. A different lineage, ancestral to the llamas and vicunas, dispersed into South America after gradual movements of Earth's crust formed a land bridge between the two continents.

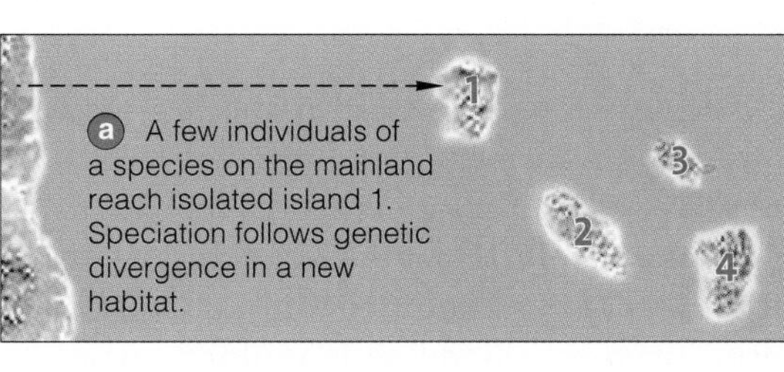

1
2
3
4
a A few individuals of a species on the mainland reach isolated island 1. Speciation follows genetic divergence in a new habitat.

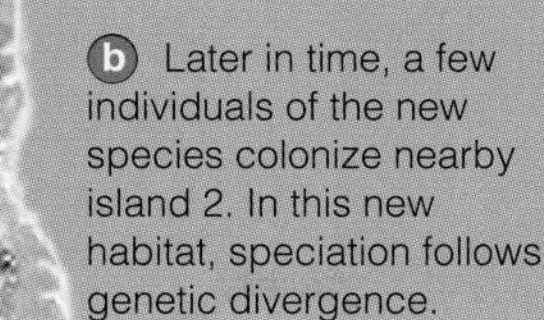

b Later in time, a few individuals of the new species colonize nearby island 2. In this new habitat, speciation follows genetic divergence.

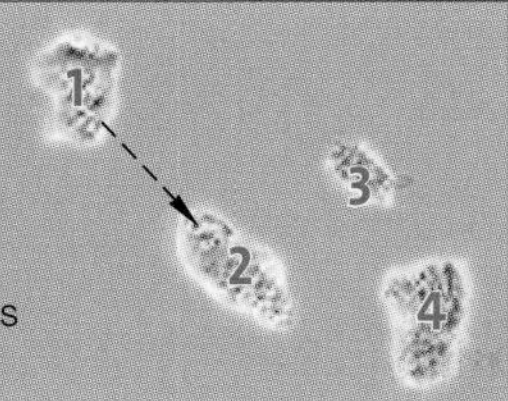

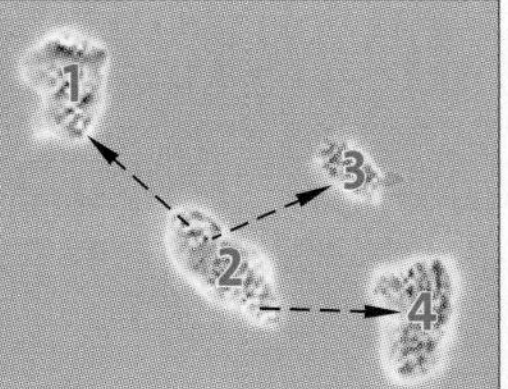

Akepa (*Loxops coccineus*)

Akekee (*L. caeruleirostris*)

Nihoa finch (*Telespiza ultima*)

Palila (*Loxioides bailleui*)

Maui parrotbill (*Pseudonestor xanthrophrys*)

Apapane (*Himatione sanguinea*)

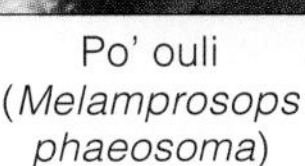

Po' ouli (*Melamprosops phaeosoma*)

Alauahio (*Paroreomyza montana*)

Kauai Amakihi (*Hemignathus kauaiensis*)

Akiapolaau (*H. munroi*)

Akohekohe (*Palmeria dolei*)

Iiwi (*Vestiaria coccinea*)

Figure 13.18 ***Animated!*** (**a**–**c**) Sketch of allopatric speciation on an isolated archipelago. (**d**) Representatives of more than twenty species of Hawaiian honeycreepers, an example of a burst of speciations on an isolated archipelago. Honeycreeper bills are adapted to diverse food sources, such as insects, seeds, fruits, and floral nectar cups.

The shared ancestor of all of Hawaii's honeycreepers probably looked like this housefinch (*Carpodacus*)

serve as model laboratories for the study of evolution. Foremost among them are the Hawaiian Archipelago, nearly 4,000 kilometers west of the California coast, and the Galápagos Islands, 900 kilometers or so from the Ecuadoran coast. These islands are the very tops of volcanoes that, in time, broke the surface of the sea. We can therefore expect that their fiery surfaces were initially barren, devoid of life.

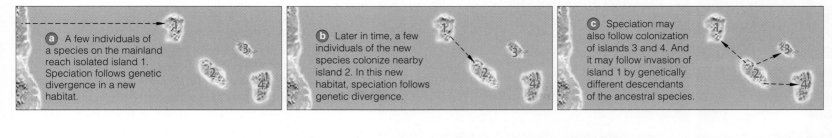

In one view, winds or ocean currents carry a few individuals of some mainland species to such islands (Figure 13.18*a–c*). Descendants colonize other islands in the chain. Habitats and selection pressures differ within and between islands, so allopatric speciation proceeds by way of divergences. Later, new species may even invade islands that were colonized by their ancestors. Distances between islands in archipelagos are enough to favor divergence but not enough to stop the occasional colonizers.

Flurries of speciation in the Hawaiian Archipelago appear to be the result of a few colonizers in a species-poor place. The youngest island, Hawaii, formed less than a million years ago. Here alone we see diverse habitats, ranging from cooled lava beds, rain forests, and alpine grasslands to high, snow-capped volcanoes. When ancestors of Hawaiian honeycreepers arrived, they found a buffet of fruits, seeds, nectars, and tasty insects, and not many competitors for them. The near absence of competition from other species triggered rapid allopatric speciations into the vacant adaptive zones. Figure 13.18*d* hints at the variation that arose among Hawaiian honeycreepers. Like thousands of other animal and plant species, they are unique to this island. As another example of speciation potential, the Hawaiian Islands combined make up less than 2 percent of the world's land masses. Yet they are home to 40 percent of all fruit fly (*Drosophila*) species.

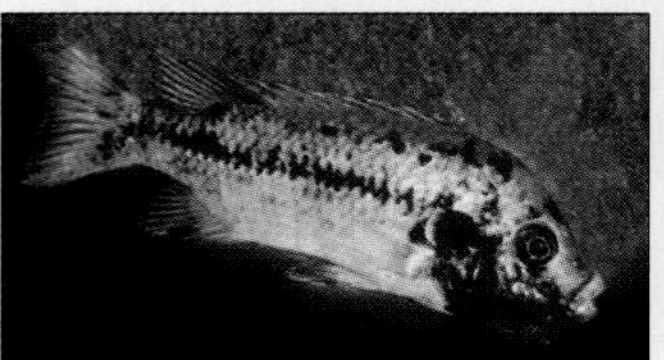

Figure 13.19 A small, isolated crater lake in Cameroon, West Africa, where different species of cichlids may have originated by way of sympatric speciation.

LINKS TO SECTIONS 8.8, 8.9

SYMPATRIC SPECIATION

By the model for **sympatric speciation**, a species may form *within* the home range of an existing species, in the absence of a physical barrier. (*Sym–* means together with, as in "together with others in the homeland.")

EVIDENCE FROM CICHLIDS IN AFRICA In Cameroon, West Africa, many species of freshwater fishes called cichlids may have arisen by sympatric speciation. The fish live in lakes that formed in the collapsed cones of small volcanoes (Figure 13.19). The cichlids probably colonized the lakes before volcanic action severed the inflow from a nearby river system.

Nuclear DNA and mitochondrial DNA from eleven cichlid species in Barombi Mbo, one of the small crater lakes, were compared to the DNA from cichlid species in nearby lakes and rivers. Comparisons of the DNA sequences showed that cichlid species in Barombi Mbo are more closely related to one another than to any neighboring species. All of the Barombi Mbo cichlids are descended from the same ancestral species—and speciation must have occurred *within* this lake.

What could have caused the divergences that led to speciation? The lake is only 2.5 kilometers across, with no obvious barriers. Water condition throughout the lake is uniform. Even the shoreline is uniform. Also, cichlids are very good swimmers, and so individuals of different species often meet.

However, the Barombi Mbo cichlid species do show some *ecological* separation, in that feeding preferences localize the different species in different regions of the lake. Yet all the species breed close to the lake bottom, in sympatry. Perhaps small-scale ecological separation was enough to foster sexual selection among potential mates. Sexual selection may have led to reproductive isolation, then speciation.

POLYPLOIDY'S IMPACT Reproductive isolation might happen within a few generations through **polyploidy**, in which individuals inherit three or even more sets of chromosomes characteristic of their species. Either a somatic cell fails to divide mitotically after its DNA is duplicated, or nondisjunction at meiosis results in an unreduced chromosome number (Section 8.9).

Plant speciation is rapid when polyploids produce fertile offspring by self-fertilizing or cross-fertilizing with an identical polyploid. The ancestor of common bread wheat may have been a species of wild wheat that spontaneously hybridized about 11,000 years ago with another wild species (Figure 13.20). Much later, a spontaneous chromosome doubling and an additional hybridization resulted in *Triticum aestivum*, a common bread wheat with a chromosome number of 42.

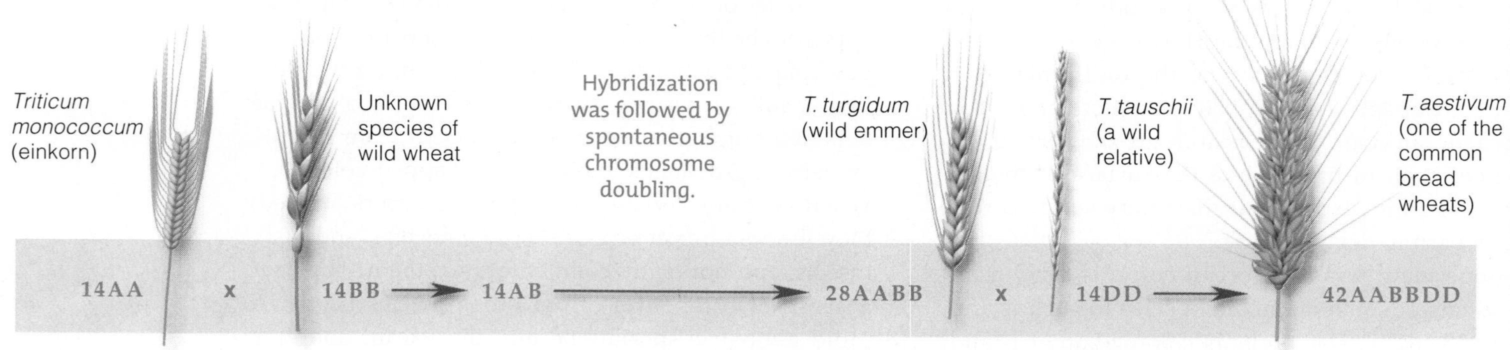

Triticum monococcum (einkorn)
Unknown species of wild wheat
Hybridization was followed by spontaneous chromosome doubling.
T. turgidum (wild emmer)
T. tauschii (a wild relative)
T. aestivum (one of the common bread wheats)
14AA
x
14BB
14AB
28AABB
x
14DD
42AABBDD
a Einkorn has a diploid chromosome number of 14 (two sets of 7). It probably hybridized with another wild wheat species that had the same number of chromosomes.
b About 8,000 years ago, wild emmer originated from a hybrid wheat plant in which the chromosome number had doubled. Wild emmer is tetraploid; it has two sets of 14 chromosomes.
c A wild relative of wheat that had a diploid chromosome number of 14 probably hybridized with wild emmer. Common bread wheats have a chromosome number of 42.

Figure 13.20 ***Animated!*** Presumed sympatric speciation in wheat. Wheat grains 11,000 years old and diploid wild wheats have been found in the Near East.

Figure 13.21 Parapatric speciation. (**a**) Giant velvet worm and (**b**) blind velvet worm. (**c**) These rare species of velvet walking worms live in overlapping regions of northeastern Tasmania. Hybrid offspring are sterile, which may be the main reason these two species maintain separate identities in the absence of an obvious physical barrier between their habitats.

About 95 percent of fern species and 30–70 percent of flowering plants are polyploid species. So are a few conifers, mollusks, insects, and other arthropods, as well as fishes, amphibians, and reptiles.

PARAPATRIC SPECIATION

By the model for **parapatric speciation**, populations that are maintaining contact along a common border become distinct species. (*Para–* means near, as in "near another homeland.") This happens when populations face different selection pressures across a single area. Hybrids form in a contact zone between populations. They are restricted to this zone because they are less fit than nonhybrid individuals (Figure 13.21).

By the allopatric speciation model, some type of physical barrier intervenes between populations or subpopulations of a species and prevents gene flow among them. Gene flow ends, and genetic divergences give rise to daughter species.

By the sympatric speciation model, daughter species arise from a group of individuals within an existing population. Polyploid flowering plants probably formed this way.

By the parapatric speciation model, populations in contact along a common border evolve into distinct species.

13.9 Patterns of Speciation and Extinctions

All species, past and present, are related by descent. They share genetic connections through lineages that extend back in time to the molecular origin of the first prototypic cells. Later chapters focus on evidence that supports this view. Here, simply start thinking about ways to interpret the large-scale histories of species.

BRANCHING AND UNBRANCHED EVOLUTION

The fossil record reveals two patterns of evolutionary change, one branching, the other unbranched. The first is called **cladogenesis** (from *klados*, branch; and *genesis*, origin). In this pattern, a lineage splits, the populations become genetically isolated, and they diverge. This is the pattern of speciation described earlier.

In **anagenesis**, one species can evolve into another species without branching. Mutations and changes in allele frequencies accumulate in the entire lineage as gene flow continues among its populations. In time, the species becomes so different from its ancestors that it may be classified as a new species, and the ancestral species is considered extinct.

EVOLUTIONARY TREES AND RATES OF CHANGE

Evolutionary trees summarize information about the continuity of relationships among groups. The sketch below is a simple way to start thinking about how tree diagrams are constructed. A *branch* in a tree represents a single line of descent from a common ancestor. Each *branch point* represents a time of genetic divergence:

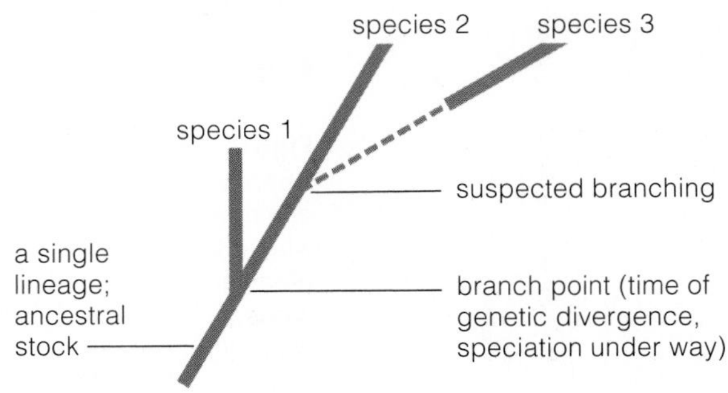

When plotted against time, a branch that ends before the present (the "treetop") signifies that the lineage has become extinct. A dashed line signifies that we know something about a lineage but not exactly where it fits in the tree.

The **gradual model of speciation** holds that species originate by small changes over long time spans. It fits

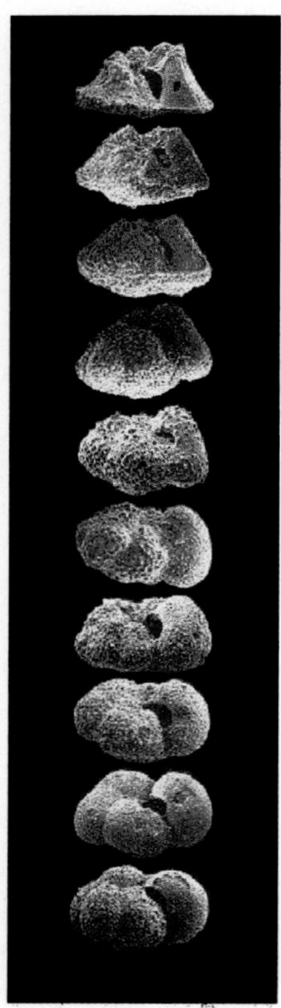

with many fossil sequences. For example, sedimentary rock layers often have vertical sequences of fossilized foraminiferan shells that show incremental changes in structure (Figure 13.22).

The **punctuation model of speciation** offers an explanation for different patterns of speciation. Most morphological changes are said to occur only during a brief period when populations start to diverge—within hundreds or thousands of years. Bottlenecks, founder effects, directional selection, or a combination of these three foster rapid speciation (Sections 12.5 and 12.7). Daughter species recover quickly from the adaptive wrenching, then do not change much over the next 2 to 6 million years.

Stability has indeed prevailed for all but 1 percent of the history of most lineages but the fossil record also has episodes of abrupt change. Both models help explain the speciation patterns because change has been gradual, abrupt, or both. Species have originated at different times, and have differed in how long they last. Some did not change much over millions of years; others were the start of adaptive radiations.

Figure 13.22 Fossilized foraminiferan shells from a vertical sequence of sedimentary rock layers. The first shell (*bottom*) is 64.5 million years old. The most recent (*top*) is 58 million years old. Microscopic analysis of shell patterns confirmed that their evolutionary order matches their geological sequence.

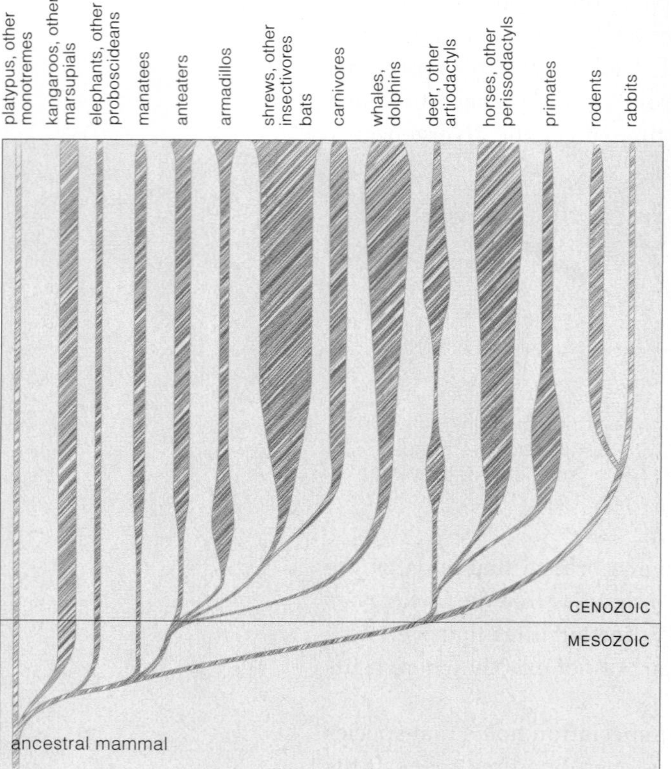

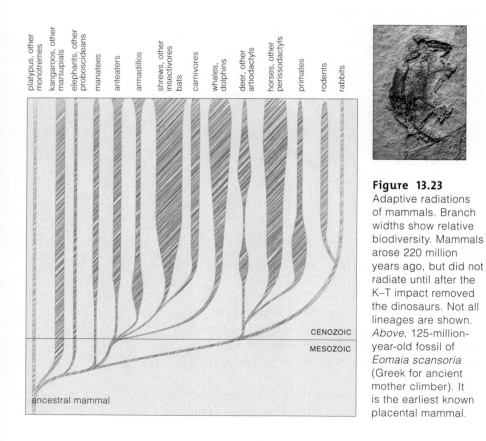

Figure 13.23 Adaptive radiations of mammals. Branch widths show relative biodiversity. Mammals arose 220 million years ago, but did not radiate until after the K–T impact removed the dinosaurs. Not all lineages are shown. *Above*, 125-million-year-old fossil of *Eomaia scansoria* (Greek for ancient mother climber). It is the earliest known placental mammal.

ADAPTIVE RADIATIONS

An **adaptive radiation** is a burst of divergences from a single lineage. It gives rise to many new species, such as the Hawaiian honeycreepers shown in Figure 13.18. The burst requires **adaptive zones**, or a set of different niches that become filled typically by a related group of species. Think of a *niche* as a particular way of life, such as "burrowing in ocean sediments" or "catching winged insects in the air at night." A lineage can enter a vacant adaptive zone or it competes well enough to displace the resident species.

First, a species must have physical access to a niche when it opens up. For example, mammals were once distributed in the tropical regions of the supercontinent Pangea. When Pangea broke up into land masses that drifted to different parts of the globe, its habitats and resources changed. This set the stage for independent adaptive radiations of mammals (Figure 13.23).

Second, a species may enter an adaptive zone by **key innovation**: A chance modification in some body structure or function gives it the opportunity to exploit the environment more efficiently or in a novel way.

Once a species enters an adaptive zone, genetic divergences give rise to other species that can fill a variety of niches within the zone. For example, when the forelimbs of certain vertebrates evolved into wings, novel niches opened up for the ancestors of modern birds and bats.

EXTINCTIONS—END OF THE LINE

An **extinction** is an irrevocable loss of a species. By current estimates, more than 99 percent of all species that ever lived are now extinct. In addition to ongoing, small-scale extinctions, the fossil record indicates that there were at least twenty or more **mass extinctions**. These are catastrophic losses of entire families or other major groups. For example, 95 percent of all species were abruptly lost 250 million years ago. Afterward, biodiversity slowly recovered as new species filled the vacant adaptive zones.

Lineages have changed. Species originated at different times and have differed in how long they have persisted.

An adaptive radiation is the rapid origin of many species from a single lineage. It happens when an adaptive zone, a new set of niches, opens up and the lineage has physical, evolutionary, and ecological access to it.

Repeated and often large extinctions happened in the past. After times of reduced biodiversity, new species formed and occupied new or vacated adaptive zones.

Bacteria	Protista	Plantae	Plantae	Animalia	Animalia	KINGDOM
Proteobacteria	Sarcomastigophora	Coniferophyta	Anthophyta	Arthropoda	Chordata	PHYLUM
Epsilonproteobacteria	Zoomastigophorea	Coniferopsida	Monocotyledonae	Insecta	Mammalia	CLASS
Campylobacterales	Trichomonadida	Coniferales	Asparagales	Diptera	Primates	ORDER
Helicobacteraceae	Trichomonadidae	Cupressaceae	Orchidaceae	Muscidae	Hominidae	FAMILY
Helicobacter	*Trichomonas*	*Juniperus*	*Vanilla*	*Musca*	*Homo*	GENUS
H. felis	*T. vaginalis*	*J. occidentalis*	*V. planifolia*	*M. domestica*	*H. sapiens*	SPECIES
none	none	western juniper	vanilla orchid	housefly	human	COMMON NAME

Figure 13.24 Taxonomic classification of six organisms. Each has been assigned to ever more inclusive categories (higher taxa)—in this case, from species to kingdom.

13.10 Organizing Information About Species

So far, you have been thinking about what species are and how they arise and vanish. Turn now to how taxonomists organize this information in ways that help us understand relationships between species.

NAMING, IDENTIFYING, AND CLASSIFYING SPECIES

Taxonomists identify, name, and classify species. They assign each a two-part name. Junipers, vanilla orchids, houseflies, humans—each has a scientific name that gives people all over the world a way to know they are talking about the same species (Figure 13.24). The name's first part is descriptive of similar species of the same kind of organism that we group together. As you saw in Chapter 1, a group of species is a **genus** (plural, genera). The second part of the name is the specific epithet. Together with the genus name, it designates one species alone.

All **classification systems** are organized systems of retrieving information about how species fit in the big picture of living things. Ever more inclusive groupings of species are **higher taxa** (singular, taxon). Family, order, class, phylum, and kingdom are examples. Now, most classification systems are phylogenetic, meaning that they reflect perceived evolutionary connections within and between higher taxa in addition to patterns of evolutionary change.

A **six-kingdom classification system** prevailed for some time. All prokaryotes were assigned to kingdoms Bacteria and Archaea, and all single-celled eukaryotes (and some multicelled forms) to kingdom Protista. Like the animals, nearly all plants and fungi are multicelled, and were each ranked as a separate kingdom. Figure 13.25*a* shows these groupings.

LINK TO SECTION 1.3

New fossil finds, and new insights from geology, morphological studies, and biochemical comparisons, caused many researchers to reconsider the six-kingdom system. Most of them decided to subsume the groups into a **three-domain system**, in which the three highest taxa are Bacteria, Archaea, and Eukarya (Figure 13.25*b*).

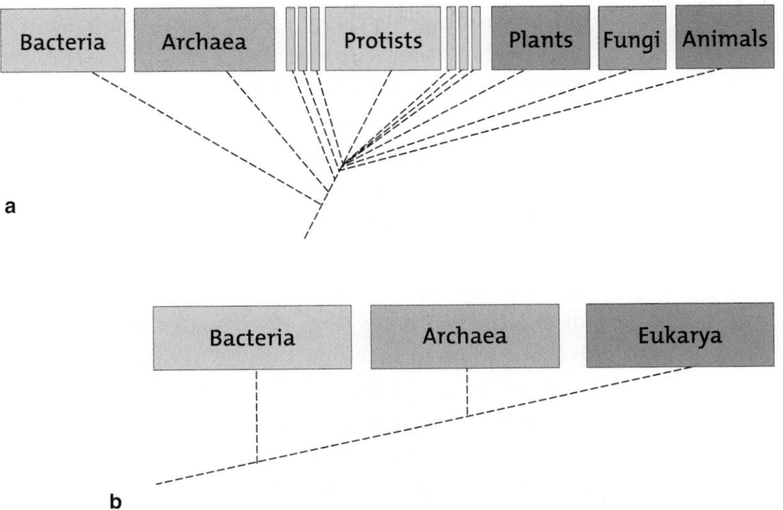

Figure 13.25 ***Animated!*** (**a**) Six-kingdom system of classification. The kingdom Protists is now being divided into more kingdoms. (**b**) The three-domain system of classification. Protists, plants, fungi, and animals share features that unite them in domain Eukarya.

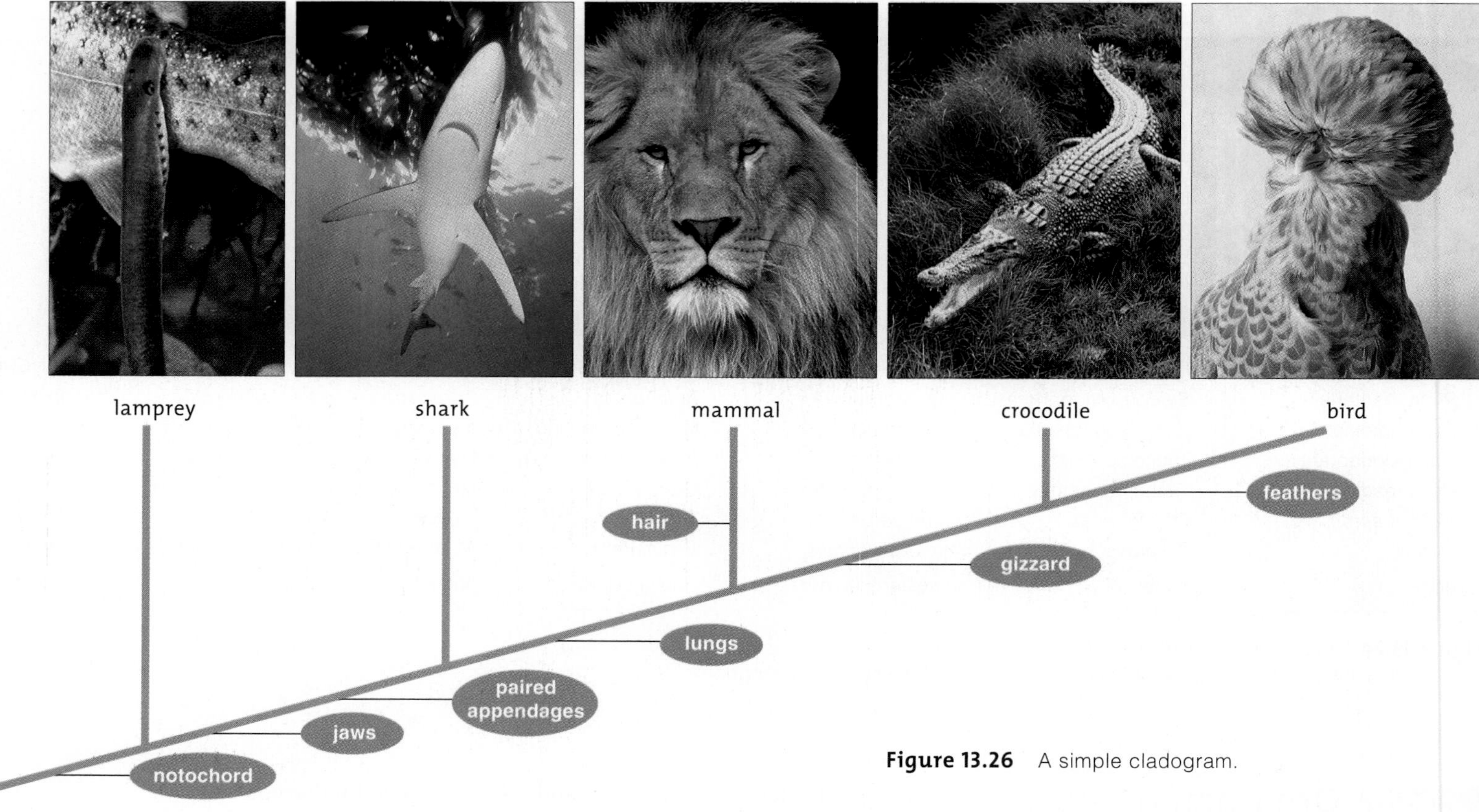

Figure 13.26 A simple cladogram.

WHAT'S IN A NAME? A CLADISTIC VIEW

Each set of species descended from just one ancestral species is called a **monophyletic group**, or **clade** (from *klados*, the Greek word for branch or twig). A cladistic classification system defines clades in terms of their history of divergences, or branch points in time. Only species that share traits derived from the last common ancestor are in the same clade. A **derived trait** is a novel feature that evolved only in one species, and is present only in its descendants. This emphasis means that descendants within a clade can differ—sometimes exuberantly so—in other traits.

Evolutionary tree diagrams called **cladograms** use the position of the branch point from the last shared ancestor to convey inferred evolutionary relationships (phylogenies) among taxa. A cladogram is an estimate of "who came from whom." It has no time bar with absolute dates, so it cannot convey differences in rates of evolution among taxa. Even so, a cladistic approach has already reinforced part of the fossil record, and it is making us reevaluate interpretations of the past.

As a very simple example, sharks, crocodiles, birds, and mammals have paired appendages. Sharks do not have lungs, but crocodiles, birds, and mammals do. All crocodiles and birds have a gizzard, but no mammal does. Birds alone have feathers, and mammals alone have fur. You can use these traits to put together a cladogram, as in Figure 13.26. From it, you can see that birds and crocodiles share a more recent ancestor than the one they share with mammals, so they are more closely related to each other than they are to mammals. Crocodiles are not ancestors of birds (both are modern clades), but birds, crocodiles, and mammals all share a more recent ancestor than they do with, say, sharks.

Cladograms reflect morphological *and* biochemical comparisons. Their simple branchings do not convey the thousands of traits that are analyzed to make the connections. Cladograms such as the one that is shown in Figure 13.27 are designed only to help us visualize monophyletic groups as sets within sets, but not as absolute rankings. As such they change as quickly as phylogenies become clarified. This is not a problem if we accept that the rankings are arbitrary.

Taxonomists identify, name, and classify sets of organisms into ever more inclusive categories, the higher taxa.

Classification systems organize information about species. Phylogenetic systems show our understanding of the evolutionary relationships among species.

Reconstructing the history of a given lineage is based on the fossil record, and morphological and biochemical comparisons.

Recent evidence, especially from comparative biochemistry, favors grouping organisms into a three-domain system of classification—the archaea, bacteria, and eukarya (protists, plants, fungi, and animals).

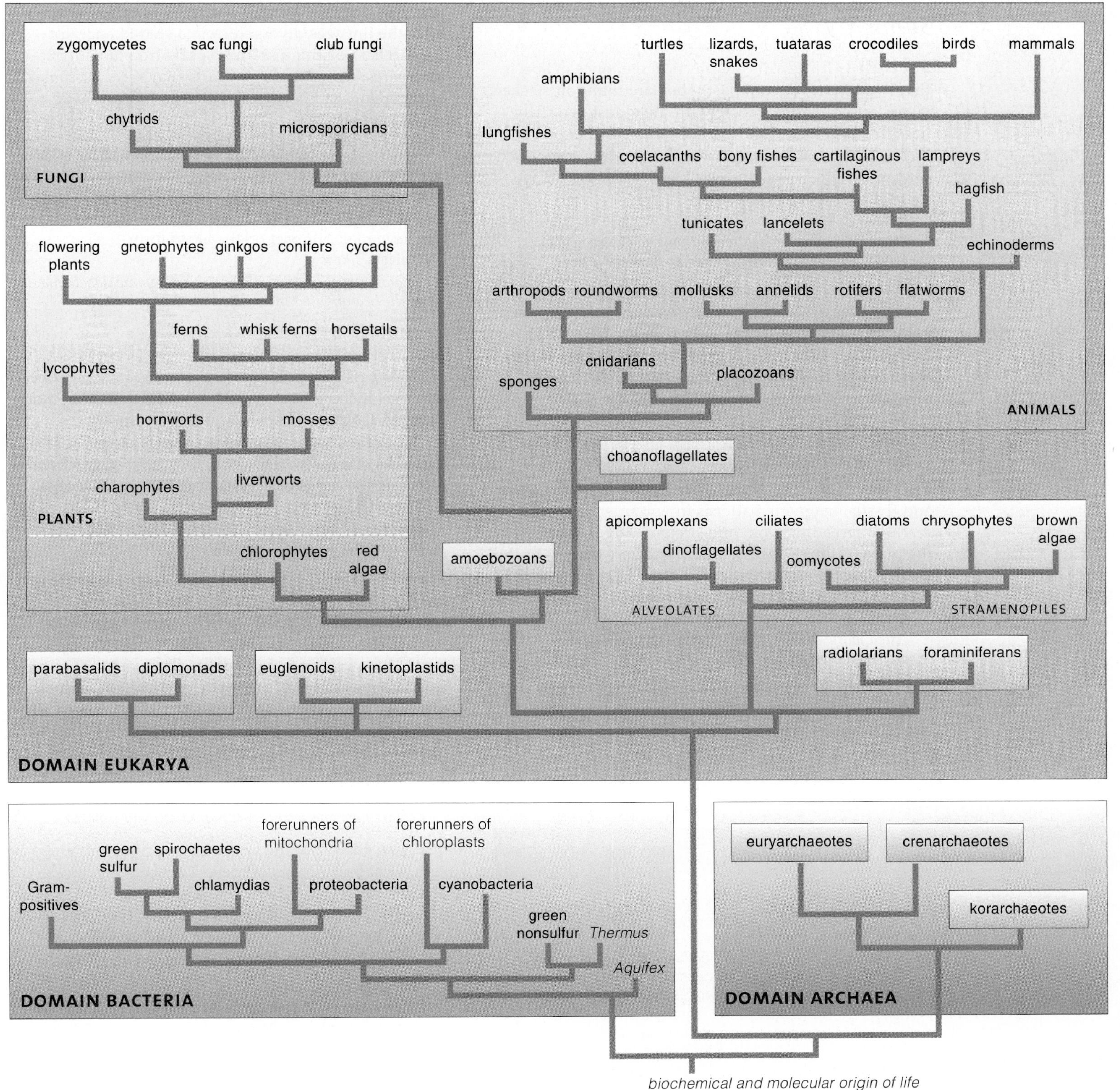

Figure 13.27 ***Animated!*** Evolutionary tree based on morphological and molecular comparisons of most major groups. Each set of organisms (taxon) has living representatives. Each branch point represents the last common ancestor of the set above it.

http://biology.brookscole.com/BTT2

Summary

Section 13.1 Fossils are physical evidence of species that lived in the distant past. They are found in stacked layers of sedimentary rock, with the oldest generally near the bottom and the most recent near the top.

Although the fossil record is incomplete and biased toward certain types of species, it does reveal many patterns in the history of life.

BiologyNow

Learn more about fossil formation and the geologic time scale with the animations on BiologyNow.

Section 13.2 Examining fossil layers provides a relative time scale that helps explain macroevolution: patterns, rates, and trends in evolution (Table 13.1). The geologic time scale uses abrupt transitions in the fossil record as boundaries. Radiometric dating has allowed us to assign absolute dates to the scale.

BiologyNow

Learn more about the half-life of a radioisotope's atoms with the animated interaction on BiologyNow.

Section 13.3 The global distribution of land masses and fossils, magnetic patterns in volcanic rocks, and seafloor spreading from mid-oceanic ridges support the plate tectonics theory. Over time, movements of the Earth's crustal plates shifted continents in ways that profoundly influenced life's evolution.

BiologyNow

Learn more about drifting continents with the interaction on BiologyNow.

Section 13.4 Comparative morphology reveals evidence of evolution. Homologous structures are one of the clues. These body parts recur in different lineages, but they became modified in different ways after the lineages diverged from a shared ancestor. Researchers distinguish these parts from analogous structures: dissimilar body parts that became similar in independent lineages and are not indicative of shared ancestry.

Section 13.5 Similarities in patterns and structures of embryonic development suggest common ancestry. Even minor genetic changes can alter the onset, rate, and completion time of developmental stages. They can have major impact on the adult form.

BiologyNow

Explore proportional changes in embryonic development with the animated interaction on BiologyNow.

Section 13.6 We are now clarifying evolutionary relationships through comparisons of DNA, RNA, and proteins between different species. Investigative methods include nucleic acid hybridization and, more recently, DNA sequencing and fingerprinting.

Mutations accumulate at predictable rates in DNA; like ticks of a molecular clock, they help researchers calculate the times of divergences between lineages.

BiologyNow

Learn more about amino acid comparisons with the interaction on BiologyNow.

Section 13.7 All populations of a species share a unique common ancestor and a gene pool, and they can interbreed and produce fertile offspring under natural conditions. Speciation may occur after gene flow ceases between populations. Then, reproductive isolation may develop gradually, as mutation, natural selection, and genetic drift operate independently in each population. In other cases, reproductive isolation happens within a few generations.

BiologyNow

Use the animation and interaction on BiologyNow to explore how species become reproductively isolated.

Section 13.8 By the allopatric speciation model, a physical barrier cuts off gene flow between two or more populations. By sympatric speciation, genetic divergence starts while populations are still in physical contact. By parapatric speciation, divergences start between populations that share a common border.

BiologyNow

Learn more about speciation on an archipelago, and the effects of sympatric speciation in wheat, with animations on BiologyNow.

Section 13.9 Speciation has differed in its timing, rate, and direction among lineages. Adaptive radiation occurs when a new habitat spurs rapid divergence from a single lineage.

Most species have by now become extinct. Mass extinctions and recoveries have occurred in the past.

Section 13.10 Each species has a unique, two-part scientific name. Taxonomy deals with identifying, naming, and classifying species. In classification

Table 13.1 Summary of Processes and Patterns of Evolution

Microevolutionary Processes		
Mutation	Original source of alleles	Stability or change in a species is the outcome of balances or imbalances among all of these processes, the effects of which are influenced by population size and by the prevailing environmental conditions.
Gene flow	Preserves species cohesion	
Genetic drift	Erodes species cohesion	
Natural selection	Preserves or erodes species cohesion, depending on environmental pressures	

Macroevolutionary Processes	
Genetic persistence	Basis of the unity of life. The biochemical and molecular basis of inheritance extends from the origin of first cells through all subsequent lines of descent.
Genetic divergence	Basis of life's diversity, as brought about by adaptive shifts, branchings, and radiations. Rates and times of change varied within and between lineages.
Genetic disconnect	Extinction. End of the line for a species. Mass extinctions are catastrophic events in which major groups are lost abruptly and simultaneously.

systems, sets of organisms (taxa) are organized into ever more inclusive categories.

A current three-domain classification system is based largely on phylogenetic evidence. It recognizes three domains: Bacteria, Archaea, and Eukarya. The Eukarya includes diverse lineages known informally as protists, as well as plants, fungi, and animals.

BiologyNow
Review biological classification systems with the animation on BiologyNow.

Figure 13.28 Rama the cama, a llama–camel hybrid, displays his short temper.

Self-Quiz

Answers in Appendix I

1. Life originated in the ______ eon.
   a. Archean   c. Phanerozoic
   b. Proterozoic   d. Cambrian

2. Which of these supercontinents formed first: Pangea or Gondwana?

3. Morphological convergences may lead to ______ .
   a. analogous structures   c. divergent structures
   b. homologous structures   d. both a and c

4. Reproductive isolating mechanisms ______ .
   a. stop interbreeding   c. reinforce genetic divergence
   b. stop gene flow   d. all of the above

5. Most species originate by a(n) ______ route.
   a. allopatric   c. parapatric
   b. sympatric   d. parametric

6. In evolutionary trees, each branch point represents a(n) ______ .
   a. single species   c. time of divergence
   b. extinction   d. adaptive radiation

7. *Pinus banksiana, Pinus strobus,* and *Pinus radiata* are ______ .
   a. three families of pine trees
   b. three different names for the same organism
   c. three species belonging to the same genus
   d. both a and c

8. Individuals of a monophyletic group ______ .
   a. are all descended from an ancestral species
   b. demonstrate morphological convergence
   c. have a derived trait that first evolved in their last shared ancestor
   d. both a and c

9. Match these terms suitably.

____ phylogeny	a. burst of divergences from one lineage into an adaptive zone
____ stratification	b. tree of branching lineages
____ fossils	c. similar body parts in different lineages with shared ancestor
____ homologous structures	d. insect wing and bird wing
____ cladogram	e. evolutionary relationships among species, from ancestors through descendants
____ analogous structures	f. layers of sedimentary rock
____ adaptive radiation	g. evidence of life in distant past

Additional questions are available on **BiologyNow™**

Critical Thinking

1. At the end of your backbone is a coccyx, a few small bones that are fused together. Could the human coccyx be a vestigial structure—all that is left of a tail of some distant vertebrate ancestors? Or is it the start of a newly evolving structure? Formulate a hypothesis, then design a way to test predictions based on the hypothesis.

2. Speculate on what might have been a key innovation in human evolution. Describe how that innovation might be the basis of an adaptive radiation in environments of the distant future.

3. Shannon thinks there are too many kingdoms and sees no reason to make another one for something as small as archaea. "Keep them with the other prokaryotes!" she says. Taxonomists would call her a "lumper." But Ann is a "splitter." She sees no reason to withhold kingdom status from archaea simply because they are part of a microscopic world that not many people know about. Which may be the most useful: more or fewer boundaries between groups? Explain your answer.

4. *Rama the cama,* a llama–camel hybrid, was born in 1997 (Figure 13.28). Camels and llamas have a shared ancestor but have been separated for 30 million years. Veterinarians collected semen from a male camel weighing close to 1,000 pounds, then used it to artificially inseminate a female llama one-sixth his weight. The idea was to breed an animal having a camel's strength and endurance and a llama's gentle disposition.

Instead of being large, strong, and sweet, Rama is smaller than expected and has a camel's short temper. Rama resembles both parents, with a camel's long tail and short ears but no hump, and llama-like hooves rather than camel footpads. Now old enough to mate, he is too short to get together with a female camel and too heavy to mount a female llama. He has his eye on Kamilah, a female cama born in early 2002, but will have to wait several years for her to mature. The question is, will any offspring from such a match be fertile?

What does Rama's story tell you about the genetic changes required for irreversible reproductive isolation in nature? Explain why a biologist might not view Rama as evidence that llamas and camels are the same species.

14 EARLY LIFE

IMPACTS, ISSUES

Looking for Life in All the Odd Places

In the 1960s, microbiologist Thomas Brock looked into Yellowstone National Park's hot springs and pools. He found a simple ecosystem of microscopically small cells, including *Thermus aquaticus*. This prokaryote lives in *really* hot water, on the order of 80°C (176°F)!

Brock's work had two unexpected results. First, it pointed the way to discovery of the Archaea, one of the three domains of life. Second, it led to a faster way to amplify DNA to useful quantities. *T. aquaticus* happens to make a heat-resistant enzyme. And that enzyme can catalyze the polymerase chain reaction—PCR—which became the premier tool for making multiple copies of a desired piece of DNA (Section 11.2).

So *bioprospecting* became the new game in town. Many companies started looking closely at thermal pools and other harsh environments for species that might yield more valuable products. What did they find? Forms of life that can withstand extraordinary levels of temperature, pH, salinity, and pressure.

Extreme thermophiles would find Yellowstone's hot springs too cool. They prefer life at hydrothermal vents on the seafloor, where water temperatures reach 110°C (230°F). Other species live in acidic springs, where pH approaches zero, and in highly alkaline soda lakes. In polar regions, some bacteria cling to life in salt ponds that never freeze and glacial ice that never melts.

It's not just prokaryotes that survive at extremes. Populations of snow algae tint mountain glaciers red. Another red alga, *Cyanidium caldarium*, grows in hot acidic springs. Free-living cells called diatoms inhabit extremely salty lakes, where the hypertonicity would make cells of most forms of life shrivel up and die. Some euglenoids thrive in acidic, metal-tainted lakes.

Life has taken hold in almost every place where there is a source of energy. Bacteria have even been filtered from the atmosphere 41 kilometers (25 miles) above the Earth. The nanobes in the filmstrip at right were found 3.8 kilometers (3 miles) below the Earth's surface in truly hot rocks—170°C (338°F).

Being one-tenth the size of most bacteria, nanobes can't be observed without electron microscopes. They look like the simplest fungi. Some argue that they are too small to be alive, to hold the metabolic machinery that runs life processes. But nanobes do have DNA and they seem to grow. Are they more like the proto-cells that preceded the origin of life? Maybe.

The rest of this unit presents a slice through time. It invites you to look at life in new ways and to trace the many lines of descent that led to the present range of biodiversity. The story is incomplete. Yet evidence from many avenues points to a principle that helps us organize information about its immense journey: *Life is a magnificent continuation of the physical and chemical evolution of the universe and the planet Earth.*

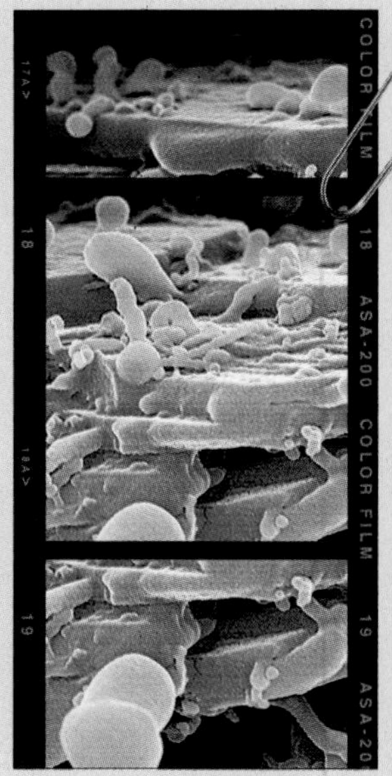

☑ ***How Would You Vote?*** *Private companies make millions of dollars selling an enzyme first isolated from cells at Yellowstone National Park. Should the government let companies bioprospect in national parks, as long as it shares in profits from any discoveries? See BiologyNow for details, then vote online.*

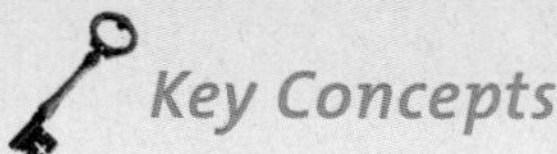

Key Concepts

ORIGIN OF LIFE
Experiments show that organic molecules that serve as the building blocks of life can form spontaneously, as can cell-like structures. There were the first steps toward the origin of life.

PROKARYOTES
The first cells were prokaryotic. Existing prokaryotes are the most widespread, numerous, and metabolically diverse organisms. We recognize two distinct prokaryotic lineages.

THE SIMPLEST EUKARYOTES
Protists, a diverse collection of producers and consumers, are now being sorted into smaller groups of relatives. Fungi are single-celled and multicelled consumers.

THE BAD BUNCH
Viruses are infectious particles that use a host cell's metabolic machinery to replicate. Like other disease-causing organisms, they coevolve with specific hosts. Viroids and prions are even simpler infectious agents.

Links to Earlier Concepts

Sections 1.2 and 1.3 introduced the characteristics of life and the major groups of organisms. Here we begin to fill in the details. This chapter describes how major kinds of organic molecules (2.5) could have originated and how cell membranes (3.3) and metabolic pathways (4.2) arose. It expands on the earlier introductions to prokaryotic and eukaryotic cell structure (3.4, 3.5), endosymbiosis (3.6), and classification (13.10).

Section 5.1 explained how evolution of the non-cyclic pathway of photosynthesis changed the Earth's atmosphere. Here you will meet the cells that brought about that change. You will also learn more about the metabolic diversity of life (5.4).

14.1 Origin of the First Living Cells

Life originated when Earth was a thin-crusted inferno, so we probably will never find evidence of the first cells. Yet we can arrive at a plausible explanation of how they emerged by considering three questions. What were the prevailing conditions? Do known principles of physics, chemistry, and evolution support the hypothesis that the first cells could have arisen through known chemical and molecular processes? And can we design experiments to test that hypothesis? Let's take a look.

CONDITIONS ON THE EARLY EARTH

Figure 14.1*a* shows part of one of the many vast clouds in the universe. It is mostly hydrogen gas, with water, iron, silicates, hydrogen cyanide, ammonia, methane, formaldehyde, and other small inorganic and organic substances. Between 4.6 billion and 4.5 billion years ago, the cloud that would become our solar system most likely had a similar composition. Clumps of minerals and ice at that cloud's edges grew more massive as they swept up asteroids and other objects. Over time, one of these clumps condensed into the early Earth.

Four billion years ago, hot gases blanketed the first patches of Earth's thin, fiery crust. Most likely, this first atmosphere was a mixture of gaseous hydrogen, nitrogen, carbon monoxide, and carbon dioxide. There was little free oxygen. How can we tell? When oxygen is present, some binds to iron in rocks and forms iron oxides. But geologists have found that such oxides did not appear until fairly recently in Earth's history.

The relatively low oxygen levels on the early Earth probably made the origin of life possible. Free oxygen is highly reactive. If it had been present, the organic compounds characteristic of life would not have been able to form and persist. Oxygen atoms would have attacked and destroyed compounds as they formed.

What about water? At first, all of the water that fell on the molten surface would have evaporated at once. After Earth's crust cooled and became solid, however, rainfall and runoff eroded mineral salts from rocks. Over millions of years, salty waters collected in crustal depressions and formed early seas. If liquid water had not accumulated, membranes could not have formed, because they take on their bilayer structure in water. No membrane, no cell, and no life.

Cells appeared less than 200 million years after the crust solidified, so carbohydrates, lipids, proteins, and nucleic acids must have formed by then. The synthesis of organic molecules always requires an energy input. What energy sources powered the assembly of the first organic molecules? Lightning, sunlight, or heat from hydrothermal vents may have fueled reactions.

Stanley Miller was the first to test the hypothesis that the simple organic molecules that now serve as the building blocks of life can form by chemical processes. He put water, methane, hydrogen, and ammonia in a reaction chamber (Figure 14.1*b*) and zapped the mix with sparks to simulate lightning. In less than a week, amino acids and other organic molecules had formed. Later, other researchers carried out experiments using mixes that were more like the early atmosphere. These tests also yielded organic compounds, including some that can act as building blocks of nucleic acids.

By another hypothesis, simple organic compounds formed in outer space. Scientists have detected amino acids in interstellar clouds and some meteorites. One meteorite that fell to Earth in Australia contained eight amino acids identical to those in living organisms.

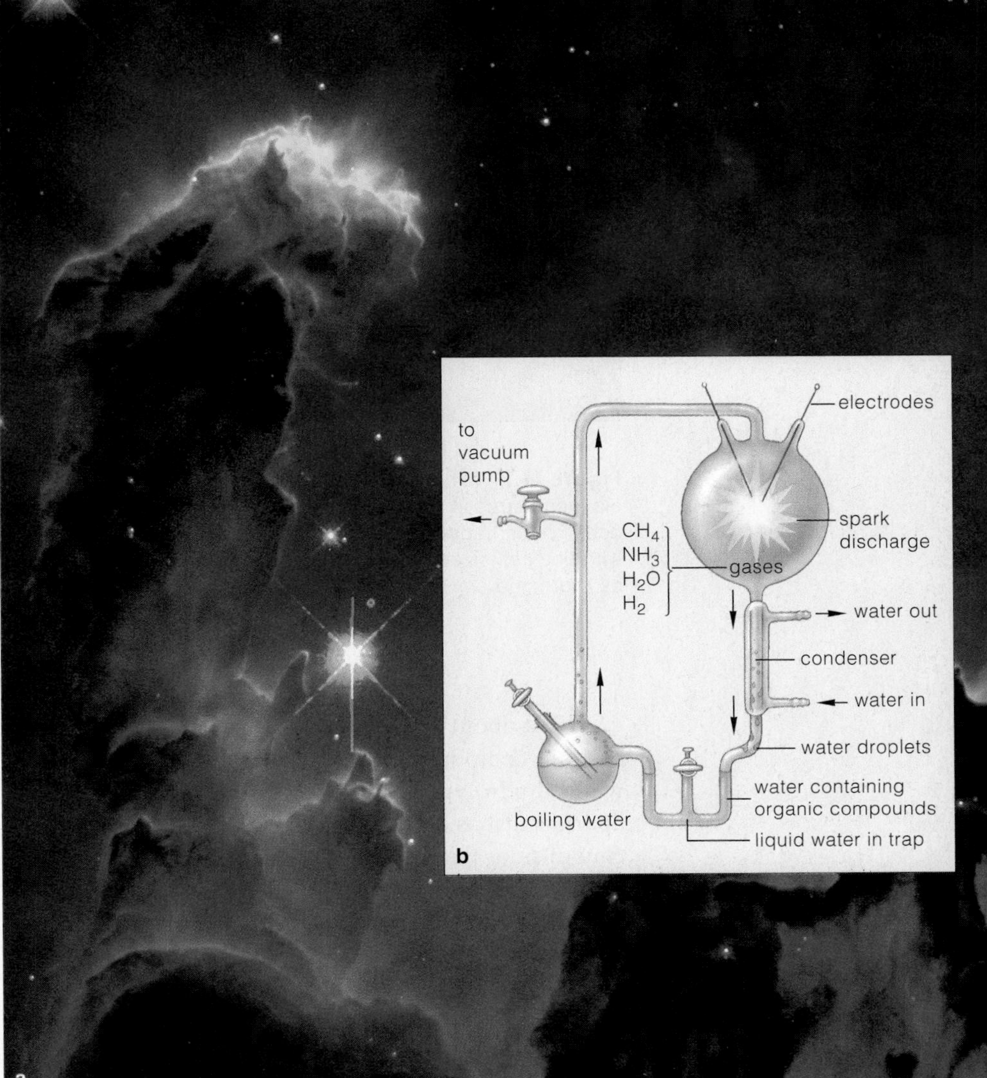

Figure 14.1 ***Animated!*** (**a**) Part of the Eagle nebula, a hotbed of star formation. Each of these pillars is wider than our solar system. New stars shine brightly on the tips of gaseous streamers.

(**b**) A diagram of the reaction chamber Stanley Miller used to test whether simple organic molecules, like those used to build living cells, can form spontaneously through chemical processes.

LINKS TO SECTIONS 1.2, 2.5, 3.3, 4.3

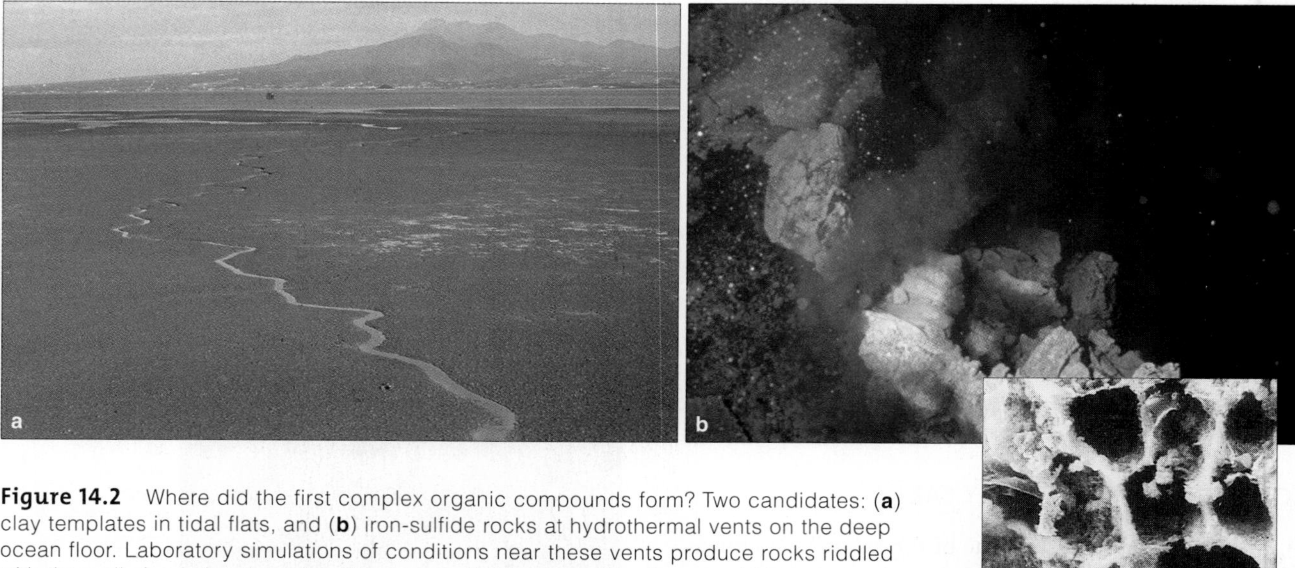

Figure 14.2 Where did the first complex organic compounds form? Two candidates: (**a**) clay templates in tidal flats, and (**b**) iron-sulfide rocks at hydrothermal vents on the deep ocean floor. Laboratory simulations of conditions near these vents produce rocks riddled with tiny cell-sized chambers (**c**). By one hypothesis, such chambers could have served as protected environments in which early membranes formed and reactions took place.

What about proteins, DNA, and the other complex organic compounds? Where did they form? In open water, hydrolysis reactions would have broken them apart as fast as they assembled. By one hypothesis, the clay of tidal flats bound and protected newly forming polymers (Figure 14.2*a*). Clays are rich in mineral ions that can attract amino acids or nucleotides. Once a few of these molecules bind to clay, others bond to them, forming chains that resemble proteins or nucleic acids.

By another hypothesis, biological molecules could have formed at deep-sea hydrothermal vents. (Figure 14.2*b*). The depths of ancient seas were oxygen-poor. Experiments show that amino acids will condense into proteinlike structures when heated in water.

ORIGIN OF AGENTS OF METABOLISM

Metabolism is a key defining characteristic of life. The word refers to all the reactions by which cells harness and use energy. In the first 600 million years or so of Earth history, enzymes, ATP, and other crucial organic compounds probably assembled spontaneously. If they did so in the same places, their close association might have promoted the start of metabolic pathways.

Imagine an ancient estuary, where seawater mixes with mineral-rich water being drained from the land. Beneath the sun's rays, organic molecules stick to clay. At first there are quantities of an amino acid; call it D. Molecules of D are added to proteins assembling on the clay—until D starts to run out. Imagine an enzyme-like protein has already been assembled. It makes D by acting on a plentiful, simpler substance C.

By chance, clumps of organic molecules include the enzyme-like protein. Now suppose C molecules start to run out. Molecular clumps that can produce C from simpler substances B and A will continue to persist and enlarge, while those that lack this ability will not. Suppose that B and A are carbon dioxide and water. The air and seas contain unlimited amounts of both.

In this way, a synthetic pathway could have arisen by chance:

$$A + B \longrightarrow C \longrightarrow D$$

This sort of "chemical evolution" may have been at work before cells appeared, increasing the frequency of certain enzymes and complex organic compounds.

ORIGIN OF SELF-REPLICATING SYSTEMS

Another defining characteristic of life is a capacity for reproduction, which now starts with protein-building instructions in DNA. As you learned in Section 10.4, using the DNA instructions to build proteins requires RNA, enzymes, and other molecules.

Did an **RNA world** in which RNA was a template for protein synthesis *precede* the origin of DNA? It's possible. As you know, RNA and DNA are a lot alike. Their sugars (ribose and deoxyribose) differ by only a hydroxyl group. Three of four bases are identical, and RNA's uracil differs only a bit from DNA's thymine.

Strands of RNA form spontaneously if nucleotides are mixed with phosphates and heated. Many existing enzymes get help from coenzymes, some of which are identical to RNA subunits.

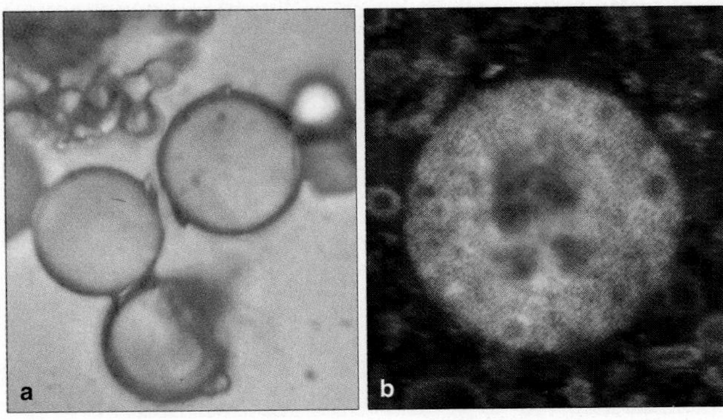

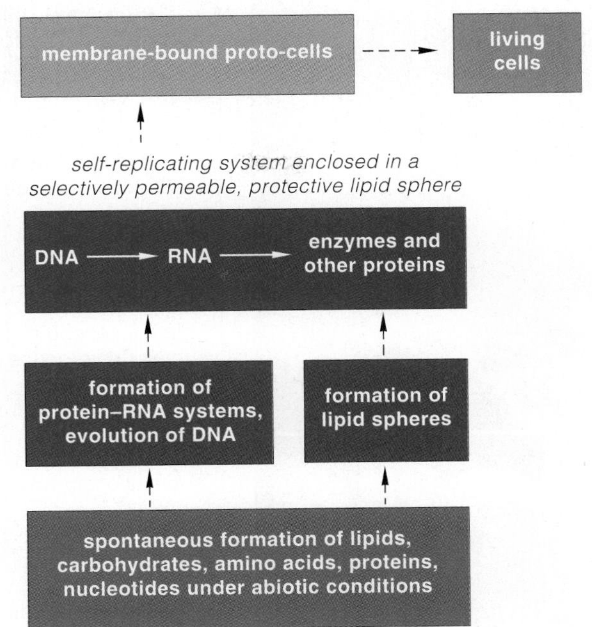

Figure 14.3 Laboratory-formed proto-cells. (**a**) Selectively permeable vesicles with a protein outer membrane were formed by heating amino acids, then wetting the resulting protein chains. (**b**) RNA-coated clay (*red*) surrounded by a membrane of fatty acids and alcohols (*green*). The mineral-rich clay catalyzes RNA polymerization and promotes the formation of membranous vesicles.

(**c**) Model for the steps that may have led to living cells.

Simple self-replicating systems of RNA, enzymes, and coenzymes have been synthesized in laboratories. So we know RNA can indeed serve as an information-storing template for making simple proteins. Also, the rRNA in ribosomes catalyzes peptide bond formation during protein synthesis (Figure 10.4). Ribosomes, you will recall, are highly conserved structures. The ones in eukaryotic cells are not that much different from those in prokaryotic cells. This suggests that the ability of rRNA to speed such reactions could have emerged early in Earth history.

If we assume that the first self-replicating systems were RNA-based, the question arises, why was there a switch to DNA? The structure of DNA may hold the answer. DNA molecules are more stable than RNAs and hold more information. As Section 9.2 describes, DNA is a double-stranded molecule that coils into helices. Helical coiling of a molecule maximizes the number of subunits that can be packed into a small space. Being single-stranded, RNA cannot form a helix.

The story of life's origin will always be incomplete because no human was there to witness it. But many details are filling in. For instance, researchers fed data on inorganic compounds and energy sources into an advanced supercomputer program. They asked the computer to put selected compounds through random chemical competition and natural selection—as might have happened untold billions of times in the distant past. They ran the program again and again. And they always got the same result: *Simple precursors evolved, the precursors spontaneously organized themselves as large, complex molecules, and they started interacting as systems.*

ORIGIN OF CELL MEMBRANES

The outermost membrane of every living cell is a lipid bilayer with diverse proteins, a plasma membrane. It controls which substances move into and out of a cell (Section 3.1). Without its control, cells can neither exist nor reproduce. Therefore, a step in life's origin had to be the formation of **proto-cells**, simple membrane sacs that could surround and protect information-storing templates and metabolic events.

In one experiment, amino acids formed protein-like chains. These were placed in heated water. After being cooled, the chains spontaneously assembled into tiny, stable, selectively permeable spheres (Figure 14.3*a*). In other experiments, fatty acids and alcohols formed into membranous vesicles around bits of clay (Figure 14.3*b*). These vesicles "grew," by adding more fatty acids, and the clay inside catalyzed the formation of RNA.

Also, researchers found out something intriguing about hydrothermal vents. Smokestacks at these vents are honeycombed with cell-sized chambers, as shown in Figure 14.2*c*. Mineral ions that projected into such chambers could have been templates on which RNA, proteins, DNA, and lipids could replicate themselves. If molecules became concentrated and protected from their surroundings inside such chambers, this could have favored the spontaneous formation of cells.

In short, there are gaps in the story of life's origin. But there also is experimental evidence that chemical evolution led to the organic molecules and structures characteristic of life. Figure 14.3*c* shows likely steps in the chemical evolution that preceded the first cells.

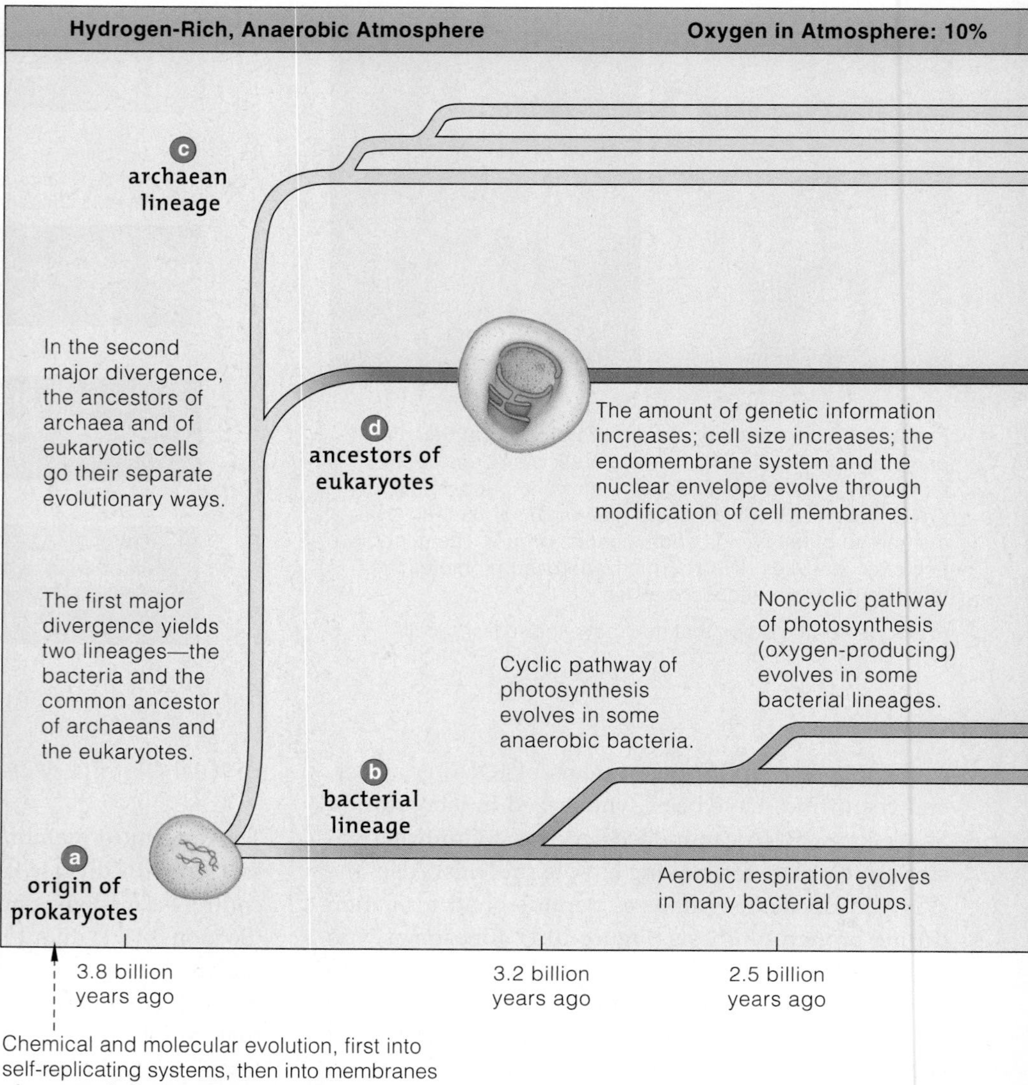

Hydrogen-Rich, Anaerobic Atmosphere
Oxygen in Atmosphere: 10%
c archaean lineage
In the second major divergence, the ancestors of archaea and of eukaryotic cells go their separate evolutionary ways.
d ancestors of eukaryotes
The amount of genetic information increases; cell size increases; the endomembrane system and the nuclear envelope evolve through modification of cell membranes.
The first major divergence yields two lineages—the bacteria and the common ancestor of archaeans and the eukaryotes.
Noncyclic pathway of photosynthesis (oxygen-producing) evolves in some bacterial lineages.
Cyclic pathway of photosynthesis evolves in some anaerobic bacteria.
b bacterial lineage
a origin of prokaryotes
Aerobic respiration evolves in many bacterial groups.
3.8 billion years ago
3.2 billion years ago
2.5 billion years ago
Chemical and molecular evolution, first into self-replicating systems, then into membranes of proto-cells before 3.8 billion years ago

LINKS TO SECTIONS 3.6, 5.1, 5.2, 6.1

THE FIRST CELLS AND BEYOND

The first living cells originated in the **Archean** eon, which lasted from 3.8 billion to 2.5 billion years ago. Fossils show they were like existing **prokaryotic cells**, with no nucleus. Maybe they were membrane-bound, self-replicating sacs of RNA, DNA, and other organic molecules. Free oxygen was absent, so they got energy by anaerobic pathways such as fermentation. Geologic processes had enriched the seas naturally with organic compounds. So "food" was available, there were no predators, and the cells were free from oxygen attacks.

Some populations diverged in two major directions shortly after the time of origin. One lineage gave rise to **bacteria**. The other gave rise to the shared ancestor of **archaea** and **eukaryotic cells** (Figure 14.4*a-d*).

By 3.2 billion years ago, light-trapping pigments, electron transfer chains, and other innovations led to the cyclic pathway of photosynthesis in some bacteria. These new cells tapped into sunlight, an unlimited energy source. Some formed flattened mats in which sediments collected. The mats accumulated, one atop the other. Minerals hardened and preserved the mats, which are now called **stromatolites** (Figure 14.4).

By 2.5 billion years ago, photosynthetic machinery had become altered in some bacterial species, and the noncyclic pathway of photosynthesis emerged. Free oxygen—a by-product of this pathway—started to accumulate. Here we pick up the threads of a story that we began in Chapters 3 and 5.

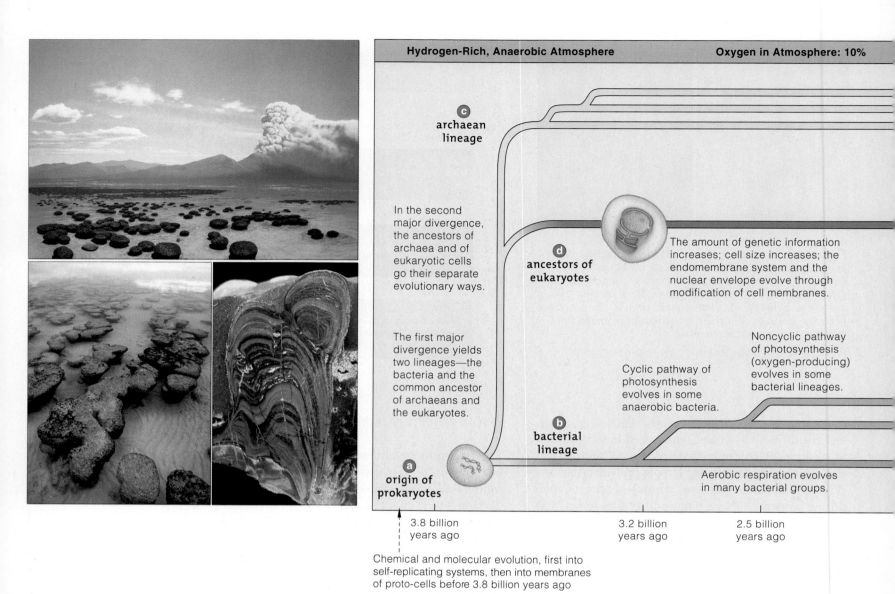

An atmosphere enriched with free oxygen had two irreversible effects. First, *it stopped the further chemical*

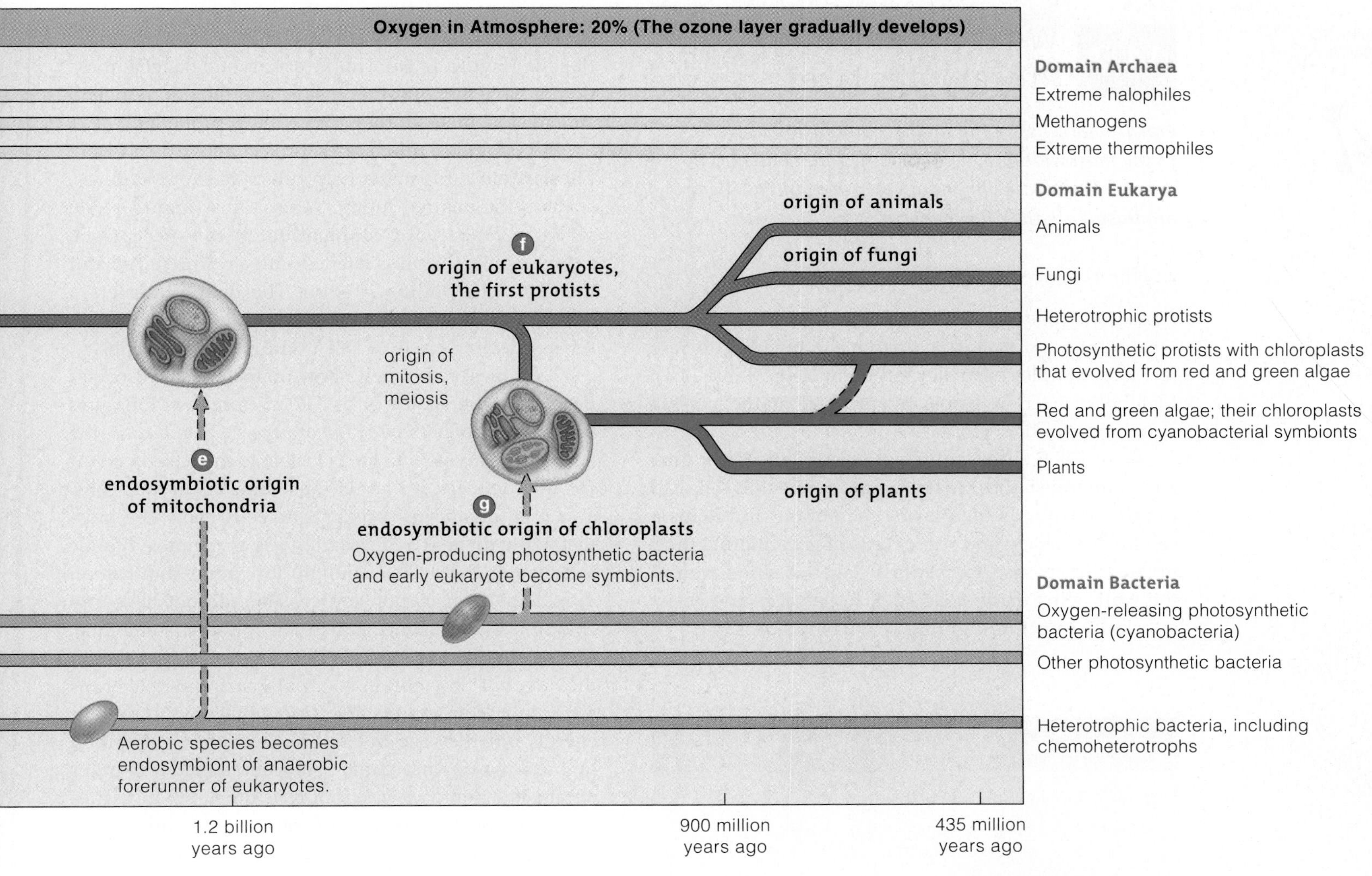

Figure 14.4 ***Animated!*** Milestones in the history of life. The photograph at *far left* shows stromatolites at low tide in Australia's Shark Bay. These mounds, 2,000 years old, are structurally similar to fossils more than 3 billion years old; the painting at the top shows how they may have looked then. The photo of the cut stromatolite shows layers of fossilized cyanobacterial mats.

origin of living cells. Except in a few anaerobic habitats, complex organic compounds could no longer assemble spontaneously without being oxidized. Second, *aerobic respiration became the dominant energy-releasing pathway.* It was a key innovation in the evolution of eukaryotes. As described in Section 3.6, some aerobic bacteria gave rise to mitochondria by endosymbiosis (Figure 14.4*e*). Photosynthetic bacteria that carried out the noncyclic pathway gave rise to the chloroplasts (Figure 14.4*g*).

By 900 million years ago, the major lineages that would give rise to protists, plants, fungi, and animals had diverged. All of these groups arose in the seas.

By 600 million years ago, tiny multicelled animals barely visible to the naked eye were scooting around the ocean floor. Some even had tissues and a digestive system, like most modern animals. Later, during the Cambrian, some of their descendants would start the first adaptive radiation of animals.

Experiments provide indirect evidence that the complex organic molecules characteristic of life could have formed under conditions that probably prevailed on the early Earth.

The first cells evolved by about 3.8 billion years ago. They were prokaryotic and made ATP by anaerobic pathways.

Early on, the lineage that led to the archaea and eukaryotes diverged from the lineage that led to modern bacteria.

Oxygen-releasing photosynthetic bacteria evolved. In time, the oxygen-enriched atmosphere put an end to the further spontaneous chemical origin of life. That atmosphere was a key selection pressure in the evolution of eukaryotic cells.

LINK TO SECTION 3.4

14.2 What Are Existing Prokaryotes Like?

Prokaryotic cells are the smallest organisms, as well as the most widespread, abundant, and metabolically diverse. They arose billions of years ago, and all living organisms are their descendants.

GENERAL CHARACTERISTICS

Prokaryotic cells arose before nucleated cells did. (*Pro–* means before; *karyon* is taken to mean nucleus.) A rare few have simple organelles but no nucleus (Table 14.1). Metabolic reactions typically proceed at the plasma membrane, in the cytoplasm, or at ribosomes.

Figure 14.5 shows a typical prokaryotic body plan and common shapes. A semirigid, permeable cell wall usually surrounds the plasma membrane and helps a cell hold its shape and resist rupturing as internal fluid pressure increases. A capsule or layer of slime around the wall helps many cells stick to surfaces and resist attacks by infection-fighting white blood cells.

Table 14.1 Characteristics of Prokaryotic Cells

1. No membrane-bound nucleus.
2. Generally a single chromosome (a circular DNA molecule) with no associated proteins; many species also contain plasmids.
3. Cell wall present in most species.
4. Reproduction mainly by prokaryotic fission.
5. Collectively, great metabolic diversity among species.

Many prokaryotes move by means of one or more flagella (Figure 14.6*a*). Unlike eukaryotic flagella, these do not have microtubules inside, and they do not bend side to side. Instead, they rotate like a propeller.

Pili (singular, pilus) often project above a cell wall. These protein filaments help cells adhere to surfaces, such as the vaginal lining. A sex pilus (Figure 14.6*b*) serves in **prokaryotic conjugation**. By this mechanism, a donor cell's sex pilus latches onto a recipient cell and then retracts, pulling it close. The donor transfers a plasmid to the recipient. A **plasmid**, recall, is a small, self-replicating circle of DNA with just a few genes.

The tiniest cells barely show up in light microscopes. *E. coli* is more typical; it is 1–2 micrometers long and 0.5–1.0 micrometers wide. *Thiomargarita namibiensis*, the largest prokaryote known, is visible to the naked eye. At 750 micrometers, it's larger than some eukaryotic cells.

Of all organisms, prokaryotic cells show the most metabolic diversity. *Photoautotrophs* are photosynthetic self-feeders that utilize sunlight for energy and carbon dioxide as their carbon source. They do not have any chloroplasts; pigments are in the plasma membrane. *Chemoautotrophs* are self-feeders that also use carbon dioxide, but they obtain energy by stripping electrons away from compounds. *Photoheterotrophs* capture light energy, but they are not self-feeders; they get carbon by breaking down organic compounds made by other organisms. Some *chemoheterotrophs* break down tissues of a living host to obtain necessary energy and carbon. Others survive by breaking down organic products, wastes, or remains.

Among the chemoheterotrophs we find **pathogens**. These infectious, disease-causing agents invade target species and multiply in or on them. Disease follows if a pathogen's metabolic activities damage host tissues.

coccus

bacillus

spirillum

a

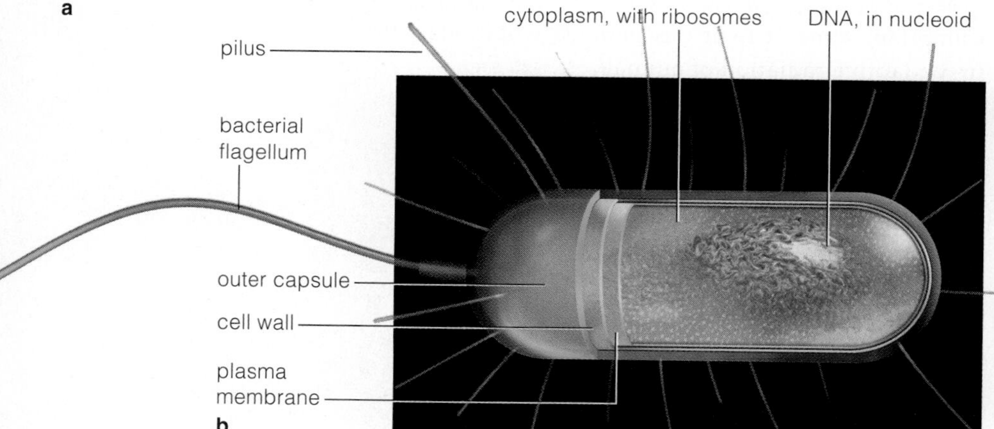

Figure 14.5 ***Animated!*** (**a**) The three most common shapes among the prokaryotic cells: spheres, rods, and spirals. (**b**) Generalized body plan of a prokaryotic cell.

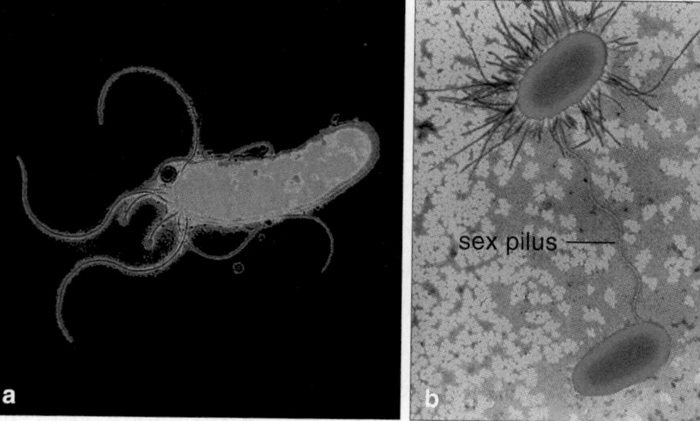

Figure 14.6 (**a**) *Helicobacter pylori*, a flagellated bacterial pathogen that is the most common cause of stomach ulcers. (**b**) A sex pilus connects these two *Escherichia coli* cells.

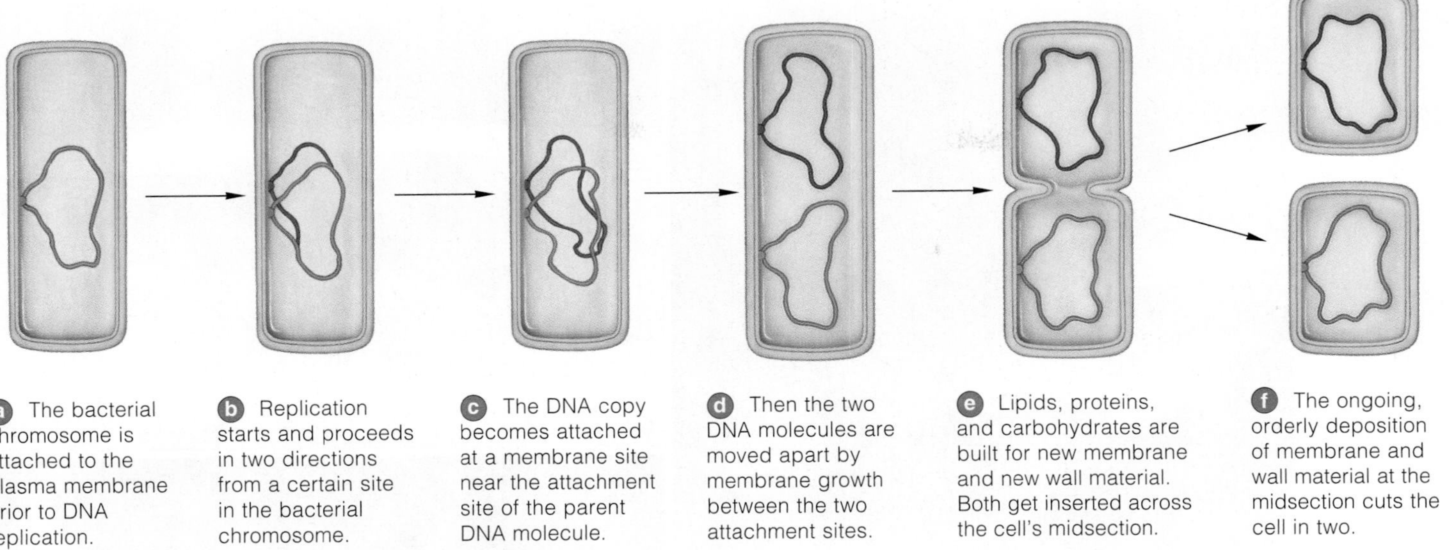

a The bacterial chromosome is attached to the plasma membrane prior to DNA replication.
b Replication starts and proceeds in two directions from a certain site in the bacterial chromosome.
c The DNA copy becomes attached at a membrane site near the attachment site of the parent DNA molecule.
d Then the two DNA molecules are moved apart by membrane growth between the two attachment sites.
e Lipids, proteins, and carbohydrates are built for new membrane and new wall material. Both get inserted across the cell's midsection.
f The ongoing, orderly deposition of membrane and wall material at the midsection cuts the cell in two.

Figure 14.7 ***Animated!*** Reproduction by way of prokaryotic fission. Only bacteria and archaea reproduce by this cell division mechanism.

When we describe bacterial pathogens in this and later chapters keep the following in mind: *Bacteria that cause human diseases are a tiny minority among the prokaryotes.* Many more prokaryotic species are harmless or even beneficial. Among other services, they put oxygen into the atmosphere, make essential nutrients available to plants, and break down wastes and remains of nearly all organisms. Without the prokaryotes, organic debris would pile up and the nutrient cycles that sustain all ecosystems would grind to a halt. Prokaryotes may be mostly microscopic, but their collective actions have enormous impacts on life on Earth.

GROWTH AND REPRODUCTION

The reproductive potential of prokaryotic cells can be staggering. Some types divide every twenty minutes. A cell divides in two, two become four, four become eight, and so forth. Before the cell divides, it nearly doubles in size. After division, each daughter cell has a single bacterial chromosome—one circular double-stranded molecule of DNA with few proteins.

In some species, a daughter cell buds from a parent cell. More commonly, a cell reproduces by prokaryotic fission (Figure 14.7). Briefly, a parent cell replicates its single chromosome, and this DNA replica attaches to the plasma membrane next to the parent molecule. As more membrane is added, the two DNA molecules are moved apart. Eventually, the membrane and cell wall extend into the cell's midsection and divide it into two genetically equivalent daughter cells.

PROKARYOTIC CLASSIFICATION

Once, reconstruction of the evolutionary history of the prokaryotes seemed an impossible task. Most ancient prokaryotic groups are either poorly represented in the fossil record or missing entirely.

Today, comparisons of gene sequences can clarify relationships. For example, studies of genes for rRNA opened the way for the division of all organisms into three domains: Archaea, Bacteria, and Eukarya:

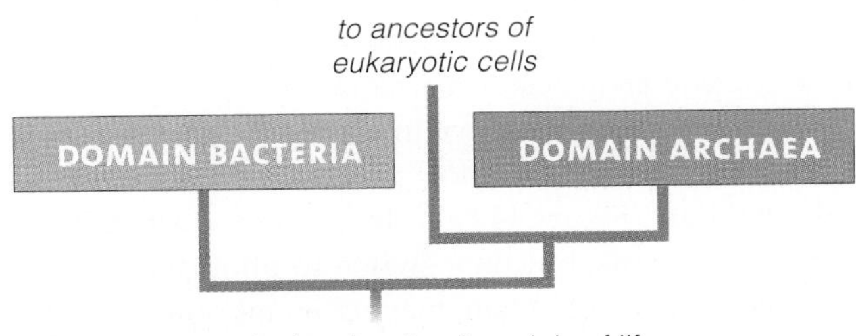

The two prokaryotic domains show some marked differences in genes, membranes, and metabolism. In some ways, archaeans are more similar to eukaryotes than to the bacteria. For example, only eukaryotes and archaeans wrap their DNA around histone proteins.

A recent discovery that most prokaryotic genomes contain hereditary material from more than one source complicates classification efforts a bit. Mixed genomes arise by **lateral gene transfer**, the transfer of genetic information between cells (often of different species) by conjugation or some other process.

LINKS TO SECTIONS 5.2, 11.2

Figure 14.8 Life in extreme environments. (**a**) In saltwater evaporation ponds in Utah's Great Salt Lake, extreme halophiles (certain archaea and red algae) tint the water pink. (**b**) Extreme thermophiles live in Yellowstone National Park's hot springs. (**c**) The parasitic *Nanoarchaeum equitans* (smaller *blue* spheres) survives only when attached to *Ignicoccus* (larger spheres). Both archaeans were isolated from 100°C water near a hydrothermal vent.

IMPACTS, ISSUES

Hot stuff! *Thermus aquaticus* (above) was discovered in outflow from a geyser. Some bacteria can survive at 95 °C, but in habitats where temperatures climb above 100 °C, the only life forms are archaeans.

EXAMPLES OF ARCHAEAN DIVERSITY

Many of the archaea withstand conditions as harsh as those on the early Earth, so they may resemble the first cells. Hence their group name; *archae–* means "beginning." Physiologically, many of the archaea are extreme halophiles (salt lovers), methanogens (cells that make methane), or extreme thermophiles (heat lovers). Species that live in more hospitable habitats tend to have an ancestor in one of these groups.

Extreme halophiles live in the Dead Sea, Great Salt Lake, seawater evaporation ponds, and other highly salty habitats (Figure 14.8*a*). Most can produce ATP by aerobic reactions, but they switch to photosynthesis if oxygen gets scarce. Their plasma membrane includes bacteriorhodopsin, a unique light-absorbing pigment.

Methanogens live in oxygen-free habitats, such as wetland mud, the gut of termites and mammals, and hydrothermal vents. They produce ATP by stripping electrons from hydrogen gas or from alcohols. Most use carbon dioxide as their carbon source and as the final electron acceptor for reactions that yield methane. All are anaerobes; free oxygen kills them. As a group, the methanogens release about 2 billion tons of methane per year. Their activity affects atmospheric levels of carbon dioxide and the global carbon cycle.

Sulfur-rich hot springs typify the homes of **extreme thermophiles** (Figure 14.8*b*). Like the methanogens, they are strict anaerobes. Unlike them, they use sulfur as the electron acceptor or donor for making ATP. All thrive at temperatures above 80°C. The first one to be discovered, *Solfolobus*, flourishes in acidic hot springs. Some bacteria, such as *Thermus aquaticus*, the source of the enzyme used in PCR (Section 11.2) also survive in hot spring environments.

Some archaean extreme thermophiles are the basis for food webs in the sediments around hydrothermal vents, where water temperatures exceed 110°C. These cells obtain electrons from hydrogen sulfide released from the vents. The existence of archaeans that can tap into this energy source suggests that life could have originated near such vents on the ancient seafloor.

Nanoarchaeum equitans, the smallest known cell, was discovered beside a hydrothermal vent off the coast of Iceland (Figure 14.8*c*). It also has the tiniest genome, with fewer than 500,000 base pairs in its chromosome.

EXAMPLES OF BACTERIAL DIVERSITY

By far, most of the 400 named genera of prokaryotes are bacteria. Among them, the cyanobacteria have the longest fossil record. As described in Section 14.1, they were already forming stromatolites more than three billion years ago. Cyanobacteria contain chlorophyll *a*, carry out photosynthesis by the noncyclic pathway, and release oxygen as a by-product. As you learned in

Figure 14.9 Cyanobacteria. (**a**) Cyanobacterial populations often form dense mats near the surface of nutrient-enriched ponds and lakes. (**b**) Some cells develop into resting spores and nitrogen-fixing heterocysts when conditions restrict growth.

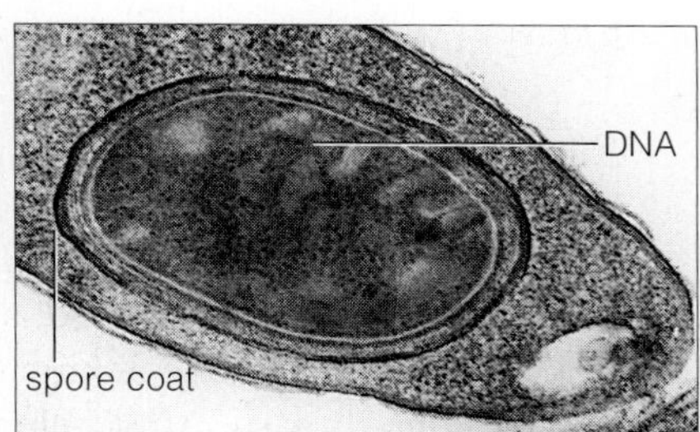

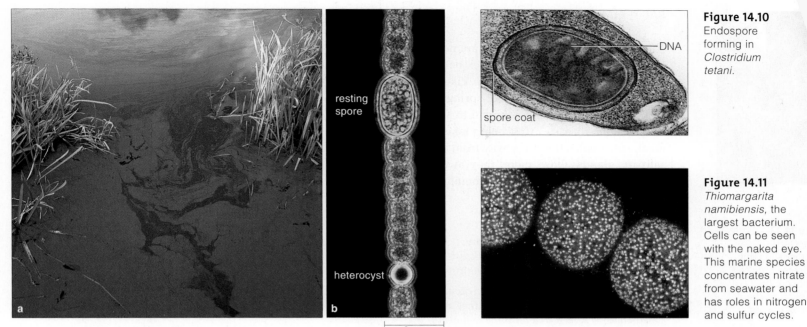

Figure 14.10 Endospore forming in *Clostridium tetani.*

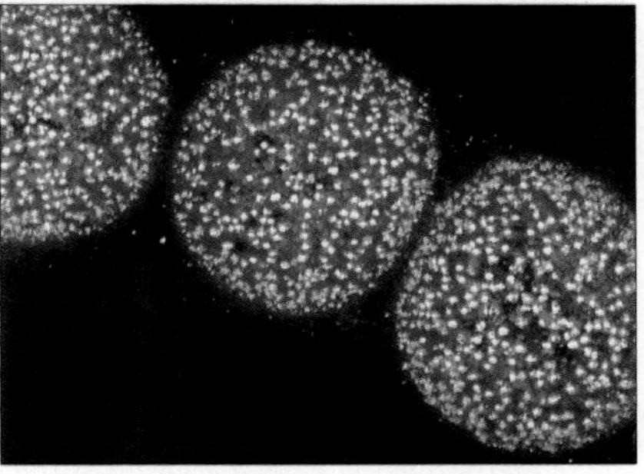

Figure 14.11 *Thiomargarita namibiensis*, the largest bacterium. Cells can be seen with the naked eye. This marine species concentrates nitrate from seawater and has roles in nitrogen and sulfur cycles.

Section 3.6, cyanobacteria gave rise to chloroplasts by endosymbiosis. We can trace almost all free oxygen in the atmosphere back to free-living cyanobacteria or to chloroplasts that are descended from them.

Modern cyanobacteria are common in ponds. They often remain together as mucus-sheathed chains that form slimy mats in nutrient-enriched habitats (Figure 14.9*a*,*b*). *Anabaena* and other nitrogen-fixing kinds can convert nitrogen gas (N_2) to ammonia, which is used in biosynthesis. Some cells become heterocysts, which make and share nitrogen compounds with other cells in the chain. In return, they receive carbohydrates.

Gram-positive bacteria are named for the stain that colors their thick, multilayered walls purple. Most are chemoheterotrophs. *Lactobacillus* species form lactate by fermentation reactions. Some are used to make yogurt, pickles, and other foods. *L. acidophilus* lowers the pH of skin and of vaginal and intestinal linings. This helps keep disease-causing bacteria from taking hold. Use of antibiotics can inhibit *L. acidophilus* and other beneficial bacteria. The result is often vaginitis or diarrhea.

Clostridium and *Bacillus* are Gram-positive bacteria that survive hostile conditions by forming endospores. This resistant structure forms inside the cell, around the chromosome and a bit of cytoplasm (Figure 14.10). It survives boiling, irradiation, and drying out. When conditions improve, the endospore will germinate. A bacterium emerges, and normal function resumes.

Endospores that germinate in the human body can be lethal. In 2001, *Bacillus anthracis* endospores sent in the mail caused fatal cases of *anthrax* in five people. *Clostridium botulinum* often taints fermented grain and food in improperly sterilized or sealed cans and jars. Toxins produced by the bacteria that emerge after the endopores are eaten cause *botulism*, a potentially lethal food poisoning. A related species, *C. tetani*, causes the disease *tetanus*, which is described in Section 21.4.

Proteobacteria—the most diverse bacterial group—include photosynthetic cells that don't release oxygen, free-living heterotrophs with roles in nutrient cycles, and many pathogens. For example, *Helicobacter pylori* can cause stomach ulcers (Figure 14.6*a*), *Vibrio cholerae* infection leads to cholera (Section 31.5), and *Neisseria gonorrhoea* is the agent of gonorrhea (Section 27.7).

The best studied prokaryote is *E. coli* (Figure 14.6*b*). Grown in laboratories around the world, its natural habitat is the mammalian gut. Here it makes vitamin K and helps keep harmful bacteria at bay. But mutant strains can cause a deadly foodborne illness.

Thiomargarita namibiensis is the largest prokaryote ever discovered (Figure 14.11). Nitrate in its vacuole acts as the acceptor for electrons it strips from sulfur for energy. *Rhizobium* lives in nodules on the roots of peas and some other plants and converts nitrogen gas to nitrates, which plants take up and use. *Nitrobacter* lives in soil and converts nitrites to nitrates.

BACTERIAL BEHAVIOR

Bacteria are small. Their insides are not elaborate. But some display remarkably complex behavior. Consider the spirochete *Borrelia burgdorferi*. Like all spirochetes, it resembles an overstretched spring (Figure 14.12). Ticks spread it from one mammal to another, where it causes Lyme disease. After being taken up in a blood meal, spirochetes find their way from a tick's gut to its salivary glands. Once there, they await the bite that will carry them to their next mammalian host.

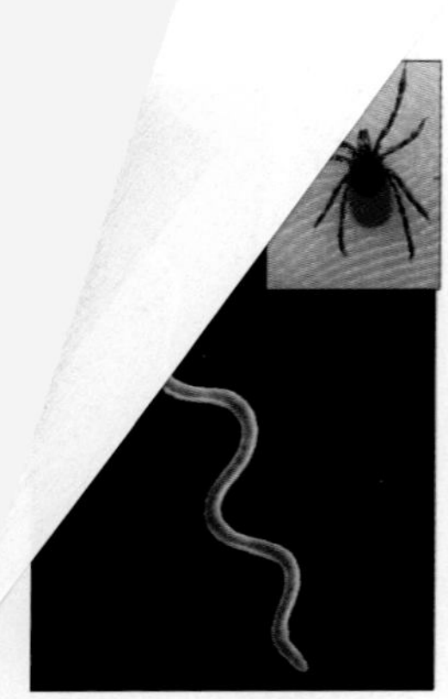

Figure 14.12 *Borrelia burgdorferi*, causes Lyme disease. This spirochete is spread from host to host by ticks, such as the deer tick, *Ixodes*.

Archaea and bacteria, the only prokaryote cells, make up two domains of life. Nearly all are microscopically small, and they alone reproduce by prokaryotic fission.

Prokaryotic cells inhabit almost all environments in and on Earth and its waters. Collectively, they are the most metabolically diverse and far-flung organisms.

14.3 The Curiously Classified Protists

Protists comprise a diverse group of eukaryote lineages. They include producers ranging from microscopic cells to seaweeds. Others are consumers: decomposers, parasites, or predators. The vast majority are single-celled species, but many lineages also have multicelled members.

Of all existing species, **protists** are the most like the first eukaryotic cells. Unlike prokaryotes, protist cells have a nucleus. Most also have mitochondria, ER, and Golgi bodies. They have more than one chromosome, each consisting of DNA with many proteins attached. They have a cytoskeleton that includes microtubules. Protist cells divide by way of mitosis, meiosis, or both.

Most protists are single-celled species. But there are colonial forms and lots of multicelled species. Many are photoautotrophs that have chloroplasts. Others are predators, parasites, and decomposers. A number of them can form spores.

Until recently, protists were defined largely by what they are *not*—not prokaryotes, not plants, not fungi, not animals. They were lumped together in a kingdom between prokaryotes and the "higher" forms of life. Thanks to improved comparative methods, protists are now being assigned to groups that reflect evolutionary relationships. Figure 14.13 shows these groupings.

FLAGELLATED PROTOZOANS AND EUGLENOIDS

Protozoans is a general term for heterotrophic protists that live as single cells. **Flagellated protozoans** move using one or more flagella.

Among the most ancient are the diplomonads and parabasalids. Both have multiple flagella at one end of the cell and a single flagellum at the other end. *Giardia lamblia* is a diplomonad. It has no mitochondria, Golgi bodies, or lysosomes. *G. lamblia* lives as an intestinal parasite inside humans, cattle, and some wild animals. The cells can survive outside the host body, in cysts. Among microbes, a **cyst** is a protective covering that formed from cell secretions. Drink water from a stream contaminated with *G. lamblia* cysts, and you could end up with *giardiasis*. Symptoms range from mild cramps to very severe diarrhea that can persist for weeks.

As an example of a parabasalid, consider *Trichomonas vaginalis*, a sexually transmitted parasite of humans (Figure 14.14). It causes *trichomoniasis*. Infection can permanently damage urinary and reproductive tracts.

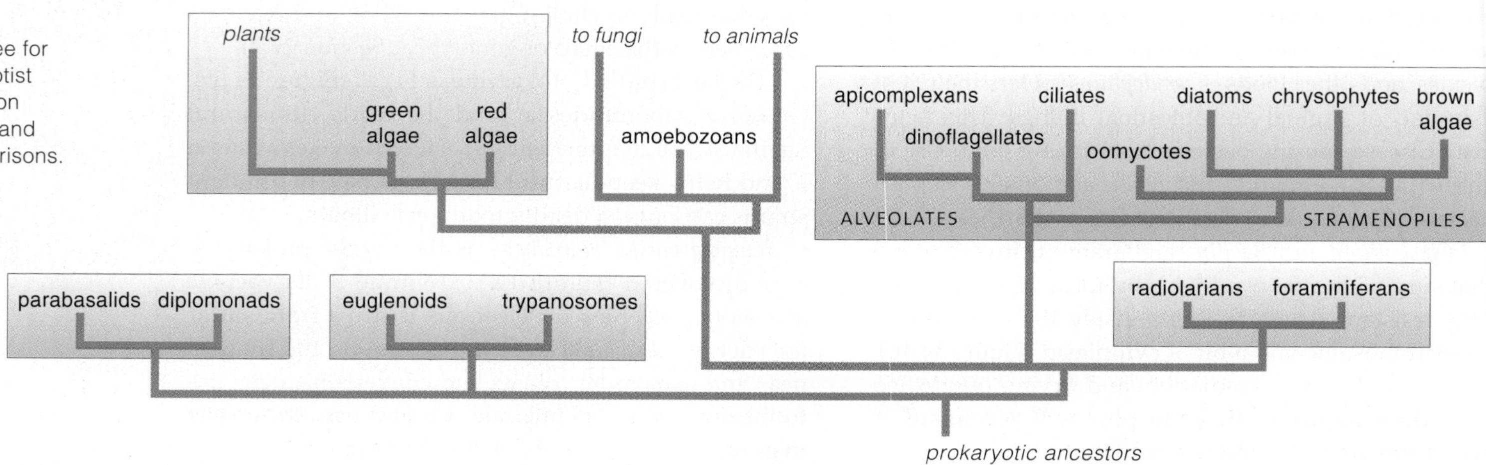

Figure 14.13 Evolutionary tree for major living protist groups based on morphological and genetic comparisons.

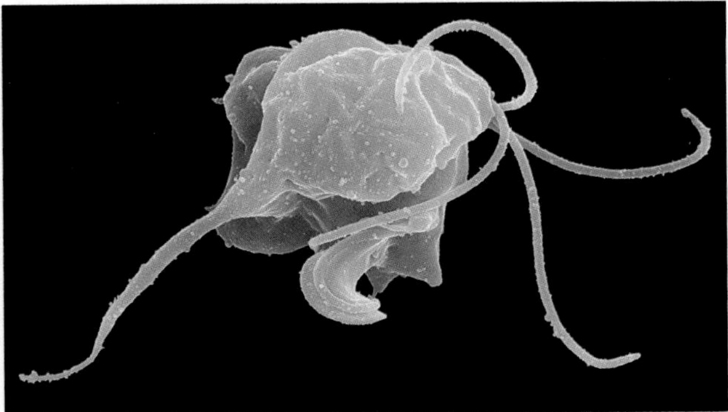

Figure 14.14 Scanning electron micrograph of *Trichomonas vaginalis*, which causes trichomoniasis, a common sexually transmitted disease.

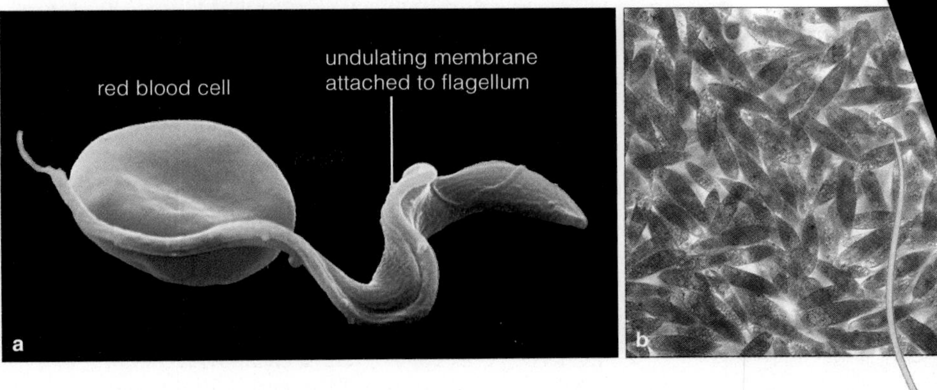

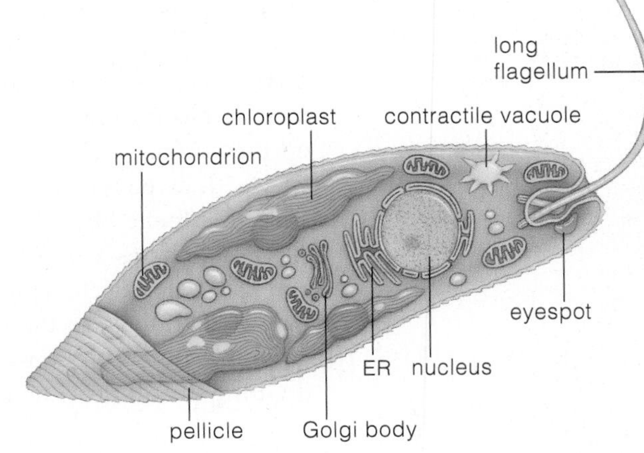

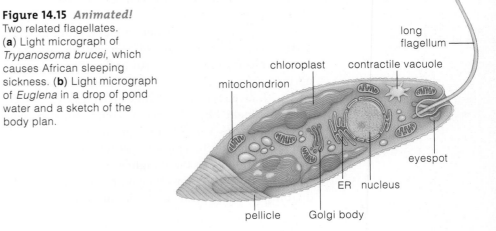

Figure 14.15 ***Animated!*** Two related flagellates. (**a**) Light micrograph of *Trypanosoma brucei*, which causes African sleeping sickness. (**b**) Light micrograph of *Euglena* in a drop of pond water and a sketch of the body plan.

Cells of *T. vaginalis* do not have mitochondria, but a similar organelle produces ATP by anaerobic means. Like mitochondria, this organelle is most likely the descendant of a symbiotic proteobacteria.

Trypanosomes are flagellated cells that live inside a diverse array of hosts, including plants, insects, and most vertebrates. Many infect humans. *Trypanosoma brucei* infection causes *African sleeping sickness*, which damages the central nervous system (Figure 14.15*a*). *T. cruzi* infection causes *Chagas disease*, which harms the liver, brain, and heart. Both diseases can be fatal.

Euglenoids are flagellated cells that live in ponds and lakes. They are related to trypanosomes, and the early ones were heterotrophs. Some still are, but many more have acquired chloroplasts by endosymbiosis. Like other freshwater protozoans, the euglenoids have a contractile vacuole that expels any excess water that diffuses inward (Figure 14.15*b*). You read about this organelle in Chapter 3.

As in many other protozoans, a **pellicle** holds the euglenoid body in and gives it shape. This is a flexible body covering constructed of parallel, spiraling strips of a translucent, protein-rich material. Pigments form an eyespot, which shields a light-sensitive receptor. This receptor helps photosynthetic euglenoids locate and stay in the places most suitable for photosynthesis.

RADIOLARIANS AND FORAMINIFERANS

Radiolarians and foraminiferans are related groups of mostly marine protists that secrete shells (Figure 14.16). Both groups use pseudopods to trap and engulf tiny prey, such as bacteria. Most foraminiferans live on the seafloor. The shell is hardened with calcium carbonate. Radiolarians have a silica-strengthened shell. Most are part of the marine **plankton**, microscopic organisms that drift or swim weakly in water.

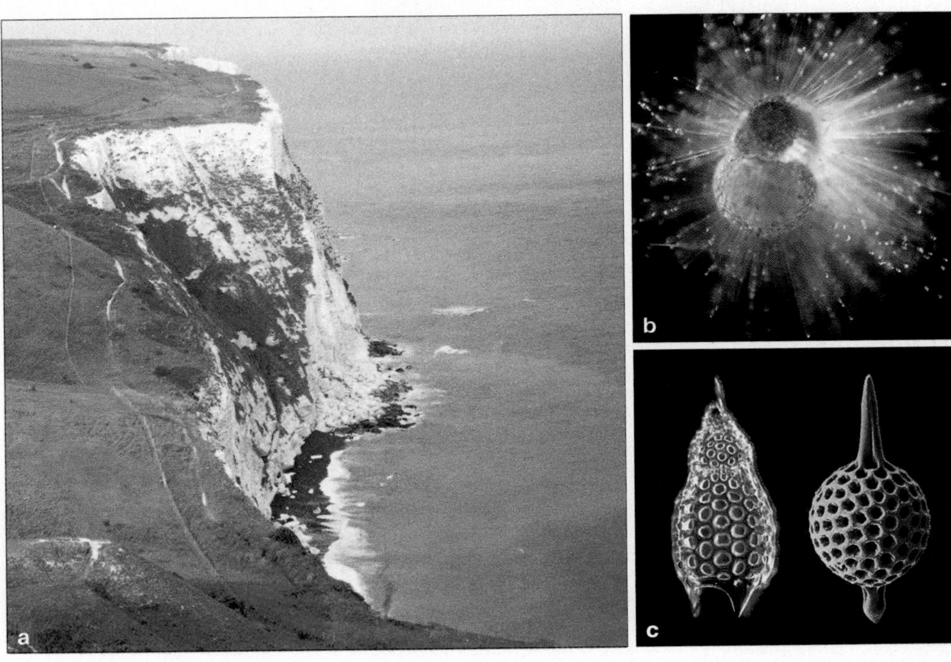

Figure 14.16 (**a**) England's white chalk cliffs at Dover are the calcium carbonate-rich remains of countless foraminiferans and other shelled protists. Remains were laid down over many millions of years, then uplifted. The cliffs now stand 300 meters high. (**b**) One living foraminiferan. The many tiny yellow spheres on its thin pseudopods are algal cells that live inside it. (**c**) The silica-rich shells of two radiolarians.

THE CILIATES

Molecular studies reveal that the ciliated protozoans, or **ciliates,** the dinoflagellates, and the apicomplexans are close relatives. Collectively, they are known as the *alveolates,* because all have alveoli—tiny membranous sacs—just beneath their outer membranes.

Ciliates live in fresh water, saltwater, and the soil. Most are free-living cells. Others attach to a surface or form colonies. About 30 percent live inside animals. A cell's cilia beat in synchrony to direct food toward an oral cavity and to move the cell. Ciliates eat bacteria, algae, and one another.

Paramecium is a freshwater species (Figure 14.17). A mature cell is about 200 micrometers long. Outer membranes form a pellicle. An oral cavity that opens on the surface leads to the gullet. Ingested food gets enclosed in enzyme-filled digestive vesicles. Excess water is expelled by one or more contractile vacuoles.

Most ciliates are able to reproduce both sexually and asexually. Each typically has two types of nuclei that it distributes to the daughter cells. Some ciliates engage in binary fission—they reproduce asexually by dividing in half near the midpoint of the cell.

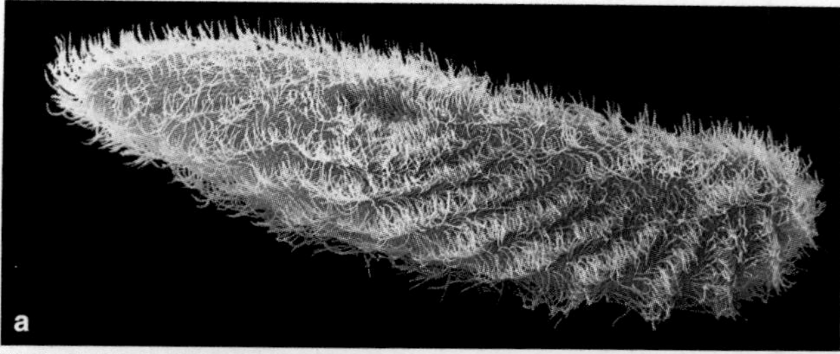

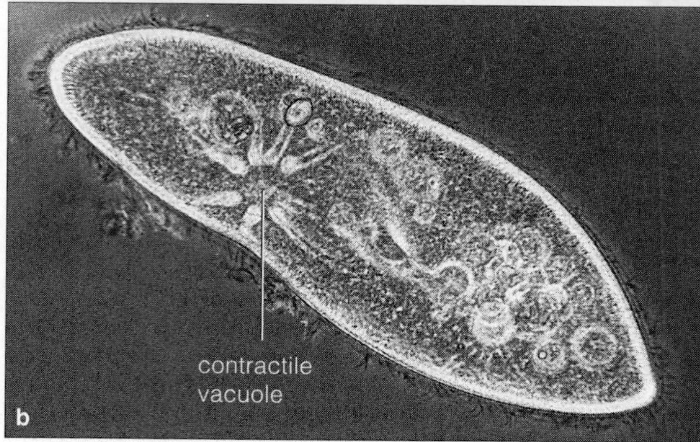

Figure 14.17 (**a**) Cilia at the surface of *Paramecium*, a single-celled predator. (**b**) A look at its organelles.

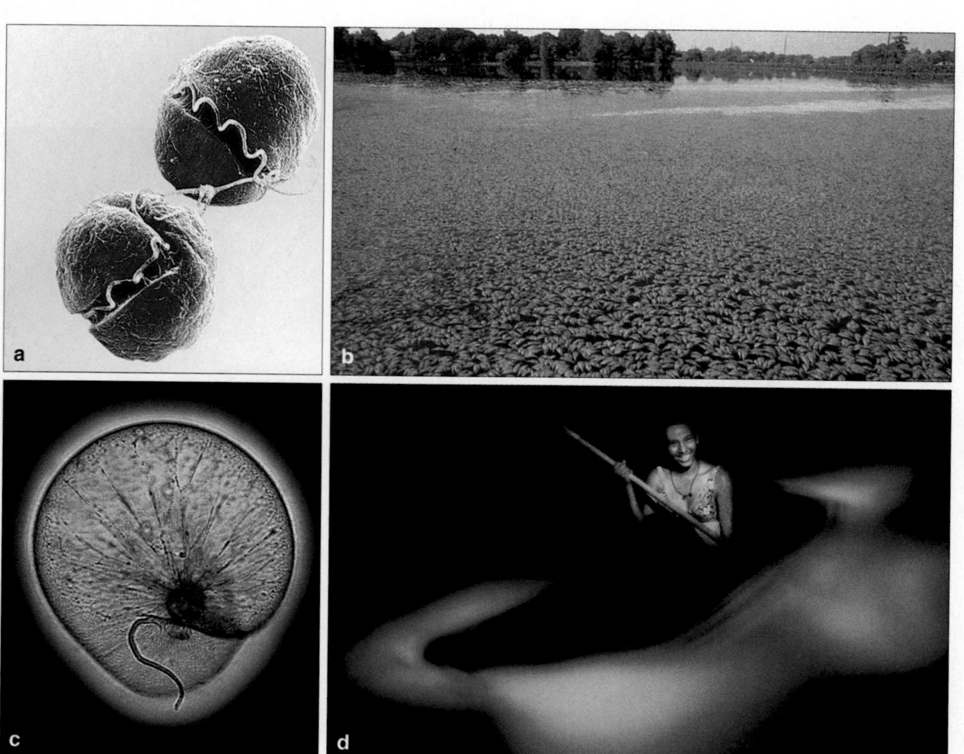

Figure 14.18 Effects of dinoflagellates. (**a**) *Karenia brevis* produces toxins that can cause fish kills (**b**) if populations reach high density during algal blooms. (**c**) *Noctiluca scintillans*, a bioluminescent dinoflagellate. (**d**) Near Vieques Island, Puerto Rico, each liter of water may hold 5,000 of these cells. Each flashes with blue light when agitated.

DINOFLAGELLATES

Dinoflagellates are single cells, usually with cellulose plates beneath their plasma membrane. Most species have a flagellum at the posterior end and another in a groove that runs around the cell. About half of the species are aquatic producers in freshwater or marine habitats. Their chloroplasts are descendants of red or green algae. A few photosynthetic species live inside corals and supply them with sugars. Most predatory dinoflagellates eat bacteria. Still others are parasites of marine or freshwater fish and invertebrates.

At high densities, certain dinoflagellates can have dramatic effects (Figure 14.18). Disturb bioluminescent species in tropical seas and the water will shimmer. Less pretty are the huge fish kills that result from algal blooms (also known as red tides). An **algal bloom** is a population explosion of a dinoflagellate or another aquatic protist. Algal blooms tend to occur in warm, shallow, nutrient-enriched waters. Increasingly, such nutrient enrichment is caused by pollutants.

APICOMPLEXANS AND MALARIA

Apicomplexans are parasitic alveolates equipped with a unique microtubular device that can attach to and penetrate a host cell. More than 4,000 apicomplexan species are known to parasitize animals ranging from insects to humans. Here we will focus on four species

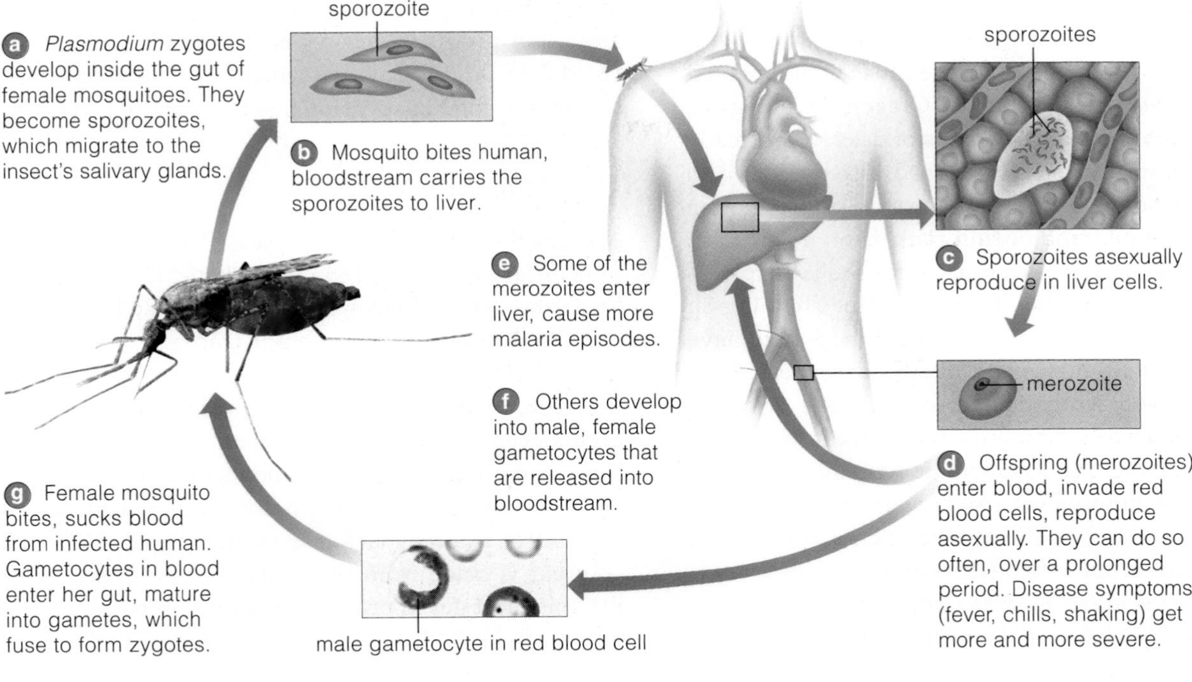

sporozoite
a Plasmodium zygotes develop inside the gut of female mosquitoes. They become sporozoites, which migrate to the insect's salivary glands.
b Mosquito bites human, bloodstream carries the sporozoites to liver.
sporozoites
c Sporozoites asexually reproduce in liver cells.
e Some of the merozoites enter liver, cause more malaria episodes.
merozoite
f Others develop into male, female gametocytes that are released into bloodstream.
d Offspring (merozoites) enter blood, invade red blood cells, reproduce asexually. They can do so often, over a prolonged period. Disease symptoms (fever, chills, shaking) get more and more severe.
g Female mosquito bites, sucks blood from infected human. Gametocytes in blood enter her gut, mature into gametes, which fuse to form zygotes.
male gametocyte in red blood cell

Figure 14.19 ***Animated!*** Life cycle of one of the *Plasmodium* species that cause malaria.

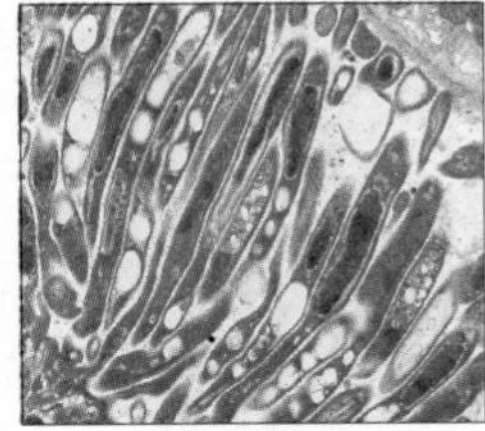

sporozoites

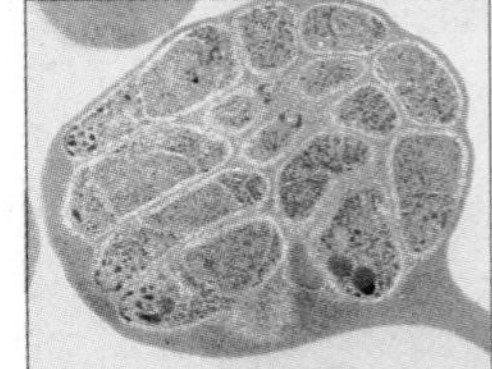

merozoites about to rupture a red blood cell

of *Plasmodium* that cause the disease *malaria*. These species have infected more than 300 million people in tropical parts of the world, especially Africa.

The bite of an infected female *Anopheles* mosquito puts the parasite in a host. Sporozoites, a motile stage, travel in the bloodstream to the liver. They reproduce asexually in liver cells until the cells burst and release another stage—the merozoites. Some merozoites will reinfect the liver. Others enter red blood cells, where they develop into immature gametes known as male and female gametocytes (Figure 14.19).

Plasmodium gametocytes cannot mature in humans because the parasites are sensitive to the temperature and oxygen levels inside their hosts' body. Humans are warm-bodied and their blood has little free oxygen; most of it is bound to hemoglobin. Gametocytes can mature only in a female mosquito, which has a lower body temperature and higher internal oxygen levels. In this insect's gut, zygotes form, divide, and give rise to sporozoites. This infective stage then migrates to the mosquito's salivary glands to await a new bite.

Malaria symptoms start when infected liver cells rupture and release merozoites, metabolic wastes, and cellular debris into the bloodstream. Shaking, chills, a burning fever, and drenching sweats follow. After one episode, symptoms usually subside for a few weeks or months. Infected individuals might feel healthy, but they should expect relapses. Jaundice, kidney failure, convulsions, and coma are outcomes of the disease.

Many *Plasmodium* strains are now resistant to older antimalarial drugs. Artemisinin, a synthetic version of a chemical in the wormwood plant (*Artemisia*) is a new drug. It lowers fever and reduces parasite numbers in the blood, so it can slow the course of infection.

Efforts to design a vaccine are ongoing. If things go smoothly and a vaccine seems effective, widespread clinical trials will probably get under way in 2006 or 2007. The sooner the better—every thirty seconds, one African child dies of malaria.

DIATOMS, BROWN ALGAE, AND RELATIVES

Colorless flagellates, water molds, and photosynthetic species—the diatoms, yellow-green, golden, and brown algae—are now grouped as stramenopiles. The shared trait of this diverse group is a shaggy-looking flagella.

Water molds, downy mildews, and white rusts are heterotrophs known as oomycotes. The name means egg fungus and refers to how the egg cells form. They are not actually fungi. Water molds help decompose organic debris and dead organisms in aquatic habitats. The downy mildews are important plant pathogens.

Another oomycote pathogen, *Phytophthora infestans*, causes *late blight*, a rotting of plant parts. During the

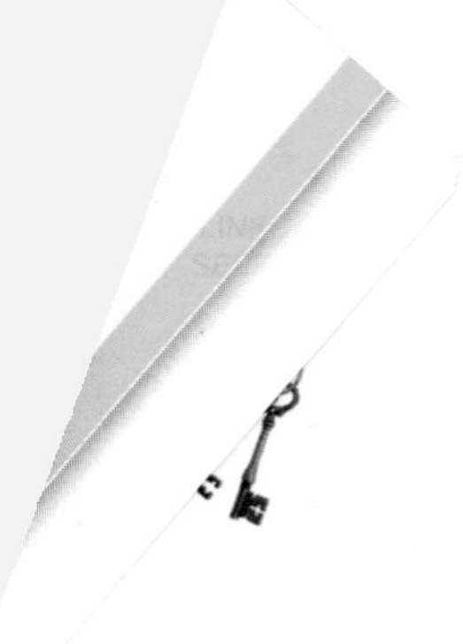

mid-1800s, it destroyed Irish potato crops, causing the famine that killed and displaced millions. *P. ramorum,* a related species, kills trees. It is the cause of an ongoing epidemic of *sudden oak death* in Oregon, Washington, and California. Tens of thousands of oaks have already died, and the pathogen has jumped to redwoods and other novel hosts. Signs of *P. ramorum* infection can include dead limbs and oozing cankers on tree trunks.

The diatoms have a two-part silica shell; the pieces fit together like a hat box (Figure 14.20). They live in freshwater and cool seas. The coccolithophores have calcium carbonate plates and live in warm seas. Both groups are photosynthetic and together release about as much oxygen as land plants. Both have survived for many millions of years and their remains form vast deposits. Coccolithophore shells contributed to those high cliffs at Dover, England. Silica-rich diatomaceous earth is quarried for use in filters and abrasives.

Most brown algae inhabit temperate or cool seas. Accessory pigments tint them olive, golden, or dark brown. In size, they range from microscopic filaments to giant kelps, such as *Macrocystis,* that reach thirty meters in height. Brown algae reproduce sexually and asexually, and a typical life cycle includes both gamete-producing and spore-producing stages.

The giant kelps are of great ecological importance. The sporophyte (spore-producing stage) of a giant kelp is the familiar seaweed (Figure 14.21*a*). It has stipes (stemlike parts), blades (leaflike parts), and holdfasts (structures that hold the alga in place). Hollow, gas-filled bladders impart buoyancy to stipes and blades, and keep them upright, near the sunlit surface waters. Diverse species feed on and take shelter in kelp beds.

Sargassum, Macrocystis, and other brown algae have commercial uses. Extracts are used to manufacture ice cream, pudding, jelly beans, toothpaste, cosmetics, and other products. Alginic acid from algal cell walls is used to produce algins, which serve as thickeners, emulsifiers, and suspension agents in many products.

Figure 14.20 Shells of two important aquatic producers. *Above*, silica shell of *Stephanodiscus niagarae*, a diatom that is common in temperate zone lakes. *Below*, the calcium carbonate shell from a coccolithophore.

RED ALGAE

Nearly all red algae live in the seas. They are abundant in tropical seas, as far 265 meters beneath the surface. Encrusting types help form coral reefs. Red algae are tinted red to black by phycobilins, accessory pigments that can absorb blue-green wavelengths that penetrate deep into water. Their chloroplasts are descended from cyanobacteria. In turn, some ancient single-celled types became the chloroplasts of dinoflagellates. Life cycles have sexual and asexual phases. Multicelled stages often form, but they lack true tissues (Figure 14.21*b*).

Agar is made from the walls of a few species. We use this inert gel to preserve moisture in baked goods and cosmetics, as a setting agent for jellies, and to package drugs in soft capsules. We use carrageenan extracts to stabilize paints, dairy products, and other emulsions. Some species of *Porphyra* are used in sushi as *nori*, a wrapping for rice and fish.

GREEN ALGAE

Of all protists, green algae are most plantlike in their structure and biochemistry, and they share a common ancestor with the land plants. Like land plants they are photosynthetic, with the same chlorophylls and starch grains inside their chloroplasts. Some have cell walls that include polysaccharides typical of plants.

Single-celled, sheetlike, tubular, filamentous, and colonial forms abound. Figures 14.22 and 14.23 show

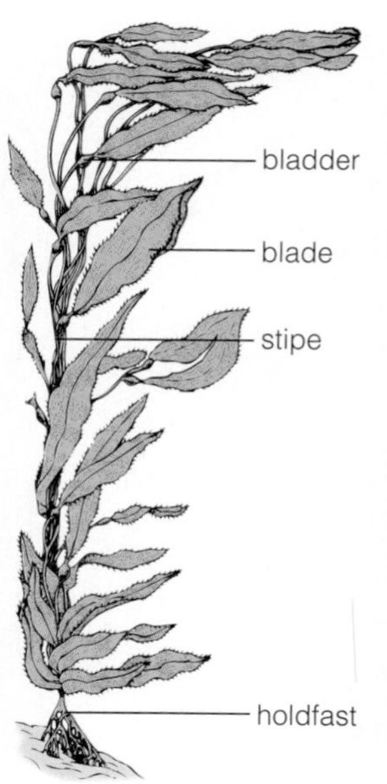

Figure 14.21 Examples of algae. (**a**) Body plan and photograph of the giant kelp, *Macrocystis*, one of the brown algae. (**b**) A red alga, *Antihamnion plumula*.

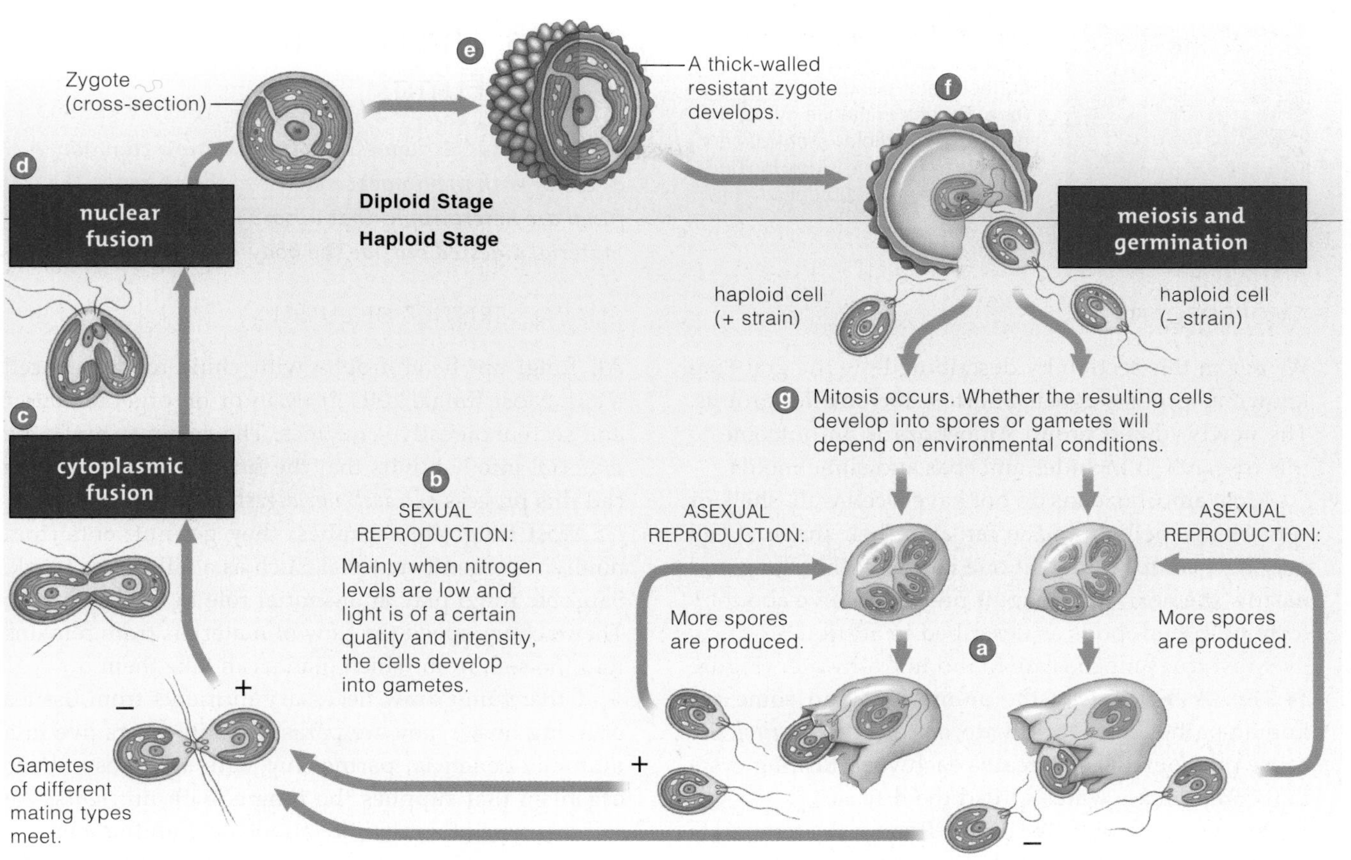
Zygote
(cross-section)
A thick-walled resistant zygote develops.
nuclear fusion
Diploid Stage
Haploid Stage
meiosis and germination
haploid cell (+ strain)
haploid cell (– strain)
cytoplasmic fusion
Mitosis occurs. Whether the resulting cells develop into spores or gametes will depend on environmental conditions.
SEXUAL REPRODUCTION:
Mainly when nitrogen levels are low and light is of a certain quality and intensity, the cells develop into gametes.
ASEXUAL REPRODUCTION:
ASEXUAL REPRODUCTION:
More spores are produced.
More spores are produced.
Gametes of different mating types meet.

Figure 14.22 ***Animated!*** Life cycle of *Chlamydomonas*, a single-celled freshwater green alga. When food is plentiful, haploid cells called spores multiply by mitosis (**a**). When nutrients get low, some haploid cells develop into gametes of two mating strains, (+ and –) (**b**). The cytoplasmic fusion of these gametes (**c**) is followed by fusion of their nuclei (**d**) to form a diploid zygote that develops into a thick-walled resting structure (**e**). Meiosis in this diploid cell (**f**) produces haploid cells that reproduce asexually (**g**).

just a few. Many cannot be seen without the aid of a microscope. Most live in freshwater. Green algae also grow in seawater, in soil, and on rocks, bark, and snow. Some are symbionts with fungi in lichens.

Green algae have diverse modes of reproduction. Consider *Chlamydomonas* (Figure 14.22), a single-celled alga about twenty-five micrometers wide. It sexually reproduces. It also forms sixteen or so daughter cells asexually, by mitotic cell divisions that take place within the parent cell wall. Daughter cells live in the parent for a while. Then they get away by secreting enzymes that digest what's left of the parent cell.

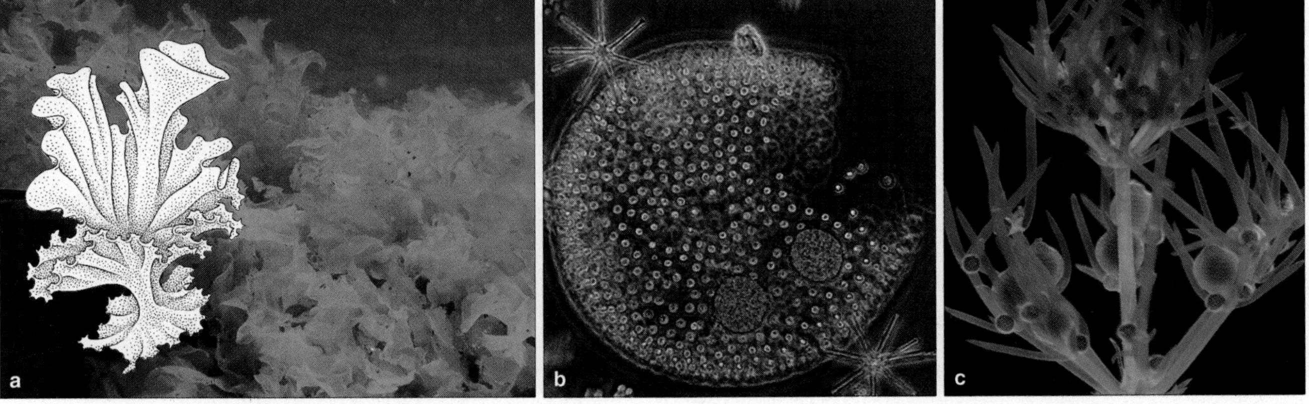

Figure 14.23 Green algae. (**a**) Sea lettuce (*Ulva*) grows in estuaries and attaches to kelps in the seas. (**b**) *Volvox*, a colony of cells that resemble free-living *Chlamydomonas* cells. This colony is rupturing. Released cells will found new colonies. (**a**) *Chara*, a charophyte. Members of this group are the closest relatives of plants. This one is commonly known as muskweed for its skunky odor.

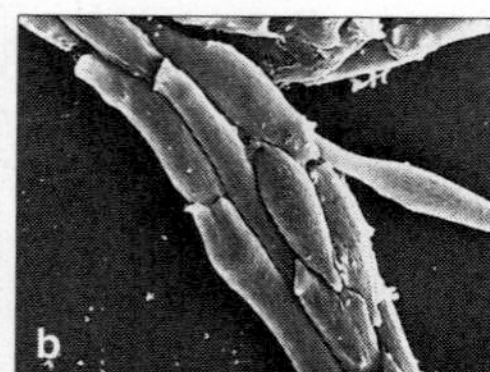
aggregating cells

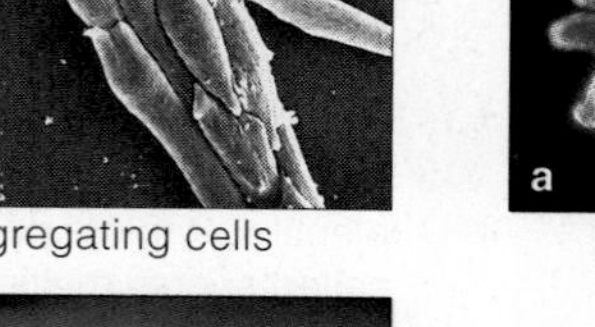

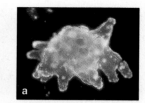

Figure 14.24 (**a**) *Amoeba proteus*, a true amoeba that lives in freshwater ponds and streams. (**b–e**) Stages in the life cycle of a cellular slime mold, *Dictyostelium discoideum*. Some slugs include as many as 100,000 cells.

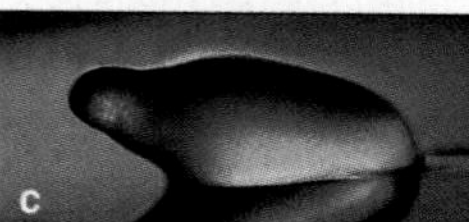
thousands of cells migrating as a "slug"

forming a fruiting body

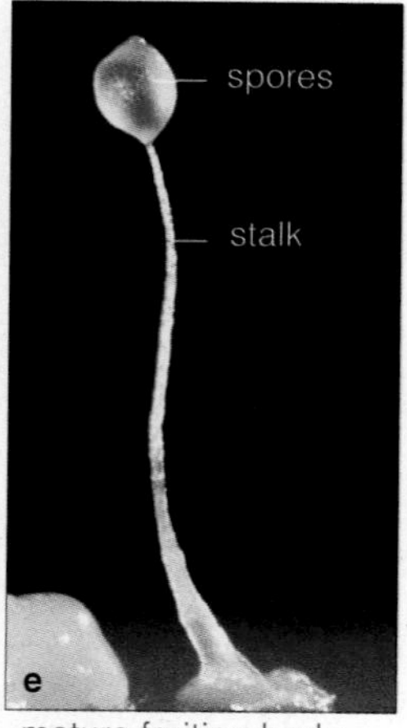

mature fruiting body

AMOEBOZOANS

We began this section by describing how the grab-bag known as protists is being sorted into smaller groups. The newly united group Amoeboza is an outcome of this research. It includes amoebas and slime molds.

Most amoebozoans do not have a cell wall, shell, or constrictive pellicle and so can alter their shape. A cell may be a compact blob at one moment, then long and narrow the next. Cells engulf prey and move about by forming pseudopods as described in Section 3.7.

Most true amoebas are found in freshwater (Figure 14.24*a*). A few live in the animal gut and some are human pathogens. Worldwide, *amoebic dysentery* is the cause of about 100,000 deaths each year. Resting cysts that contaminate water spread the disease.

Most slime molds live on the floor of the temperate zone forests. In plasmodial slime molds, an amoeba-like cell goes through multiple rounds of mitosis but does not divide its cytoplasm. A multinucleated mass called a plasmodium is formed by these divisions. It streams along the forest floor, engulfing bacteria. If conditions get bad, some cells become resting spores.

Cellular slime molds spend most of the life cycle as free-living amoeba-like cells that engulf and devour bacteria. When this food runs out, the cells secrete a chemical signal and come together to form a tiny sluglike mass that can migrate as a unit. The "slug" responds to light and to gradients in moisture. A fruiting body forms as cells differentiate into spores or part of the stalk that holds them. Air currents disperse the spores. Figure 14.24*b-e* shows stages in the life cycle of one cellular slime mold.

Diverse lineages of single-celled eukaryotic organisms were traditionally grouped in a single kingdom as protists. Spore-forming parasitic and predatory molds, and free-living single-celled decomposers, predators, grazers, and parasites, were placed in this kingdom. So were free-living, colonial, and multicelled photoautotrophs of aquatic habitats.

Molecular and biochemical comparisons are now clarifying relationships. The revised classification systems being constructed are far better reflections of the origin and evolution of these lineages.

14.4 The Fabulous Fungi

Molecular studies suggest the fungi share a common ancestor with the amoebozoans and the animals. The fungi are heterotrophs that absorb nutrients from material digested outside the body.

CHARACTERISTICS OF FUNGI

All fungi are heterotrophs with chitin-reinforced cell walls. Most fungal cells grow in or on organic matter and secrete digestive enzymes. The enzymes break the material into subunits that the fungus can absorb. We call this process *extracellular digestion and absorption.*

Most fungi are **saprobes**; they get nutrients from nonliving organic material, such as a fallen tree trunk. Saprobic fungi play an essential role in nutrient cyles. Their actions speed the flow of materials from remains into the soil, from which plants can take them up.

Other fungi draw necessary nutrients from tissues of living hosts; they are parasites. Still others live in a mutually beneficial partnership with a photosynthetic organism that supplies the fungus with nutrients. We call this type of win-win relationship a **mutualism**.

Most fungi reproduce both sexually and asexually. Both forms of reproduction may involve formation of resting spores. A fungal spore consists of a cell or cells surrounded by a protective wall.

In multicelled species, germination of a spore gives rise to a **mycelium** (plural, mycelia). This network of branched filaments has a high surface-to-volume ratio and grows over or into organic matter. A filament in a mycelium is a **hypha** (plural, hyphae). Cytoplasm of cells in a hypha often interconnects, so nutrients flow easily throughout a mycelium.

FUNGAL DIVERSITY

About 56,000 fungal species have been named, and there may be a million more. Fungi have been around

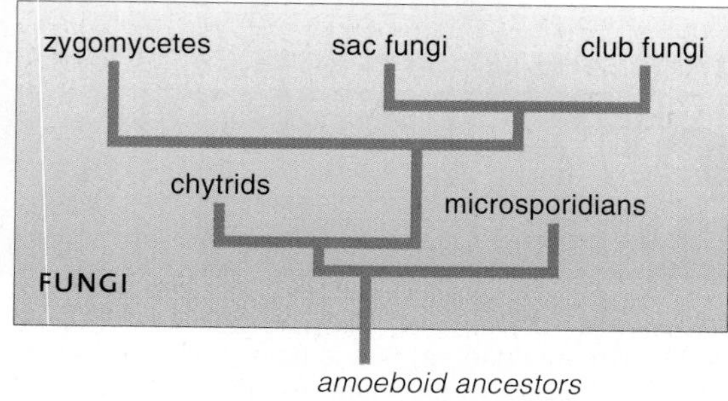

Figure 14.25 Fungal family tree.

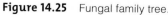

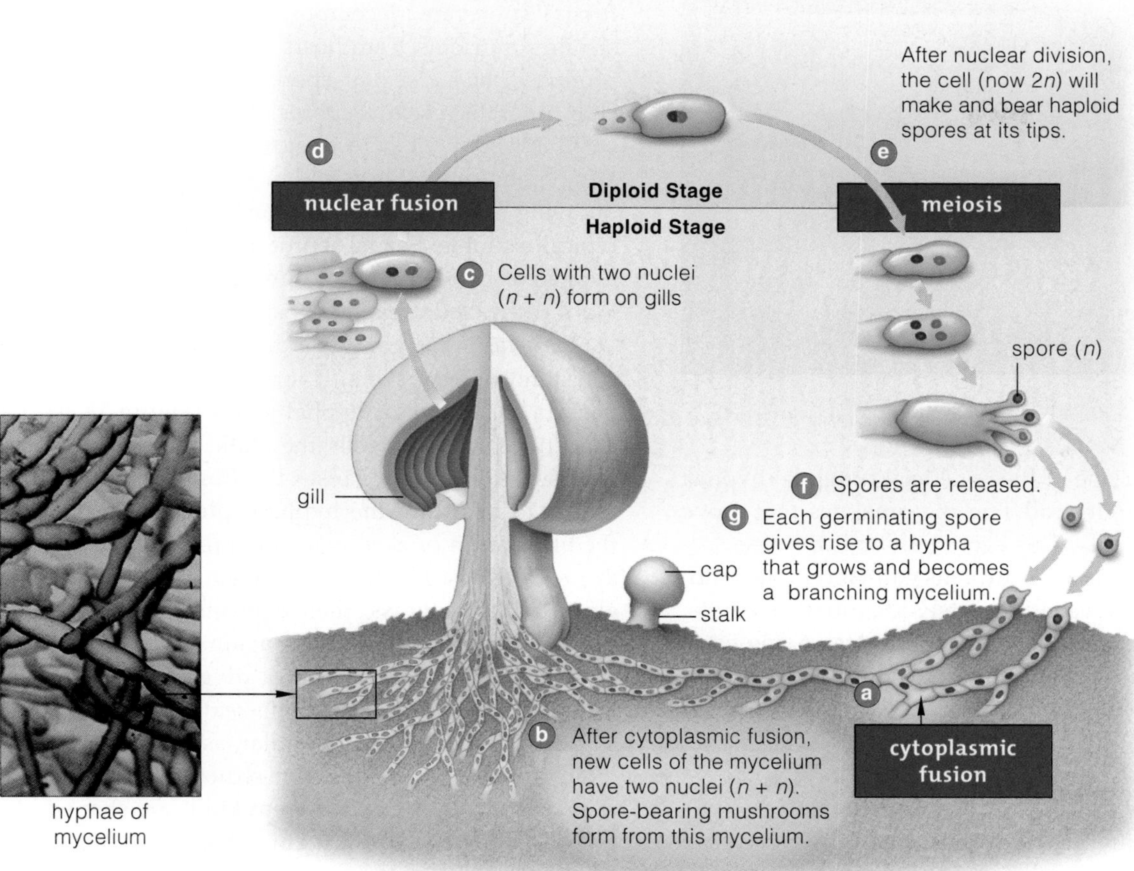

d
nuclear fusion
Diploid Stage
Haploid Stage
After nuclear division, the cell (now 2n) will make and bear haploid spores at its tips.
e
meiosis
c Cells with two nuclei (n + n) form on gills
spore (n)
gill
f Spores are released.
g Each germinating spore gives rise to a hypha that grows and becomes a branching mycelium.
cap
stalk
a
b After cytoplasmic fusion, new cells of the mycelium have two nuclei (n + n). Spore-bearing mushrooms form from this mycelium.
cytoplasmic fusion
hyphae of mycelium

Figure 14.26 ***Animated!*** Generalized life cycle of many club fungi. Hyphae of different mating strains often grow through the same patch of soil. (**a**) When hyphal cells of two compatible strains contact each other, their cytoplasm fuses but their nuclei do not. (**b**) Mitotic cell divisions yield a mycelium in which each cell contains two nuclei. When conditions are favorable, many hyphae of this mycelium start intertwining into the type of fruiting body called a mushroom. (**c**) Club-shaped structures form on the mushroom's gills. (**d**) Inside the terminal cell of each structure, two nuclei fuse and form a diploid zygote. (**e**) Four haploid spores form by meiosis. They migrate into the cytoplasmic extensions at the tip of the club-shaped structure. After spores are released (**f**), each may germinate and give rise to a new haploid mycelium (**g**).

and diversifying for a very long time. Some fossilized forms date to 900 million years ago. Microsporidians and chytrids are considered the most evolutionarily ancient fungal lineages (Figure 14.25). The spread of the parasitic chytrid *Batrachochytrium dendrobatidis* is believed to be contributing to a decline in amphibian populations (Section 29.5). Some microsporidians, such as *Enterocytozoon bieneusi*, can cause chronic diarrhea in AIDS patients.

Besides these two basal groups, three major fungal lineages became established by 300 million years ago. They are the **club fungi**, **sac fungi**, and **zygomycetes**. These groups are the focus of the rest of the section.

THE CLUB FUNGI

Most familiar mushrooms are reproductive parts of club fungi, and they usually appear only very briefly. Most of the fungus is a mycelium buried in soil or decaying wood. Each mushroom has a stalk and a cap with fine tissue sheets (gills) lining its underside. Spores form on gills, and air currents disperse them. When they alight and conditions favor germination, spores give rise to a haploid mycelium.

A mycelium may grow extensively, and then give rise to mushrooms when nutrients and moisture favor reproduction. Figure 14.26 shows how two different "mating" strains of hyphae start the process of sexual reproduction when they come together in the soil.

Mushrooms, shelf fungi, coral fungi, puffballs, and stinkhorns are among the thousands of species of club fungi. Figure 14.27 shows examples. Some saprobes are decomposers of litter on and in soil. Other species live as symbionts with young tree roots. Fungal rusts and smuts often ruin crops of wheat, corn, and other plants meant to feed the human population. Cultivation of *Agaricus bisporus*, the mushroom most commonly sold in grocery stores, is a multimillion-dollar business.

One honey mushroom, *Armillaria ostoyae*, is among the oldest and largest organisms alive. This individual has been spreading in an Oregon forest for 2,400 years. It now extends through 2,200 acres to an average depth of 3 feet. Imagine 1,665 football fields side by side and you get an idea of how big that is. *A. ostoyae*

a

b

Figure 14.27 Club fungi. (**a**) Purple coral fungus (*Clavaria*) and (**b**) sulfur shelf fungus (*Polyporus*).

Figure 14.28 (**a**) The fly agaric mushroom (*Amanita muscaria*), a hallucinogenic species. In India, Russia, and Central America, it was used in ancient rituals to induce trances. (**b**) *A. phalloides*, the death cap mushroom. Nausea and diarrhea occur 5 to 48 hours after eating it. Initial symptoms pass, but without treatment, the liver is severely damaged, kidneys fail, and coma and death can follow.

definitely is not one of the beneficial fungi. Its hyphae penetrate tree roots and clog the pipelines that move water and nutrients. This eventually kills the tree.

Before eating any mushrooms gathered in the wild, be sure they have been accurately identified as edible. Many can be deadly (Figure 14.28). As the saying goes, there are old mushroom hunters and bold mushroom hunters, but no old, bold mushroom hunters.

SPORES AND MORE SPORES

Fungal spores are usually miniscule and dry, and air currents disperse them. Each germinating spore may start a hypha, then a mycelium. Stalked reproductive structures may form on many hyphae, then produce asexual spores. After the spores germinate, each may be the start of *another* mycelium. In no time at all, the staggering numbers of descendants from that single fungal spore are decomposing organic materials or stealing nutrients from hosts. Each fungal group also produces *unique* sexual spores. As you just learned, club fungi form sexual spores in club-shaped cells.

Only about one percent of the named fungal species are zygomycetes. Most live in soils, dung, and other decaying matter. *R. stolonifer*, the black bread mold, is a typical example. Sexual reproduction occurs when the hyphae of two mating types meet and fuse. The result is a thick-walled diploid zygospore (Figure 14.29*a*).

Pilobolus is a zygomycete that grows on the dung of cows and other grazers. It forms its asexual spores in packets at the tips of hyphal stalks (Figure 14.29*b*). A fluid-filled region swells the stalk below the spores. Exposure to sunlight causes the fluid pressure in this region to rise until the hypha explodes. The force of the blast can propel spores onto a fresh patch of grass as much as two meters away. When a cow eats spores along with the grass, then deposits them in a fresh pile of dung, the cycle begins again.

The sac fungi are the most diverse fungal group. Multicelled species form their sexual spores in sac-shaped cells called asci (singular, ascus). The cells are often enclosed in reproductive structures composed of tightly woven hyphae. Different kinds are shaped like flasks, globes, and shallow cups (Figure 14.29*c,d*).

Among the sac fungi are the truffles. These highly prized culinary delicacies grow as symbionts with the roots of oak trees. Some sell for hundreds of dollars an ounce. Most food-spoiling molds are multicelled sac fungi. So are many of the single-celled fungi that we

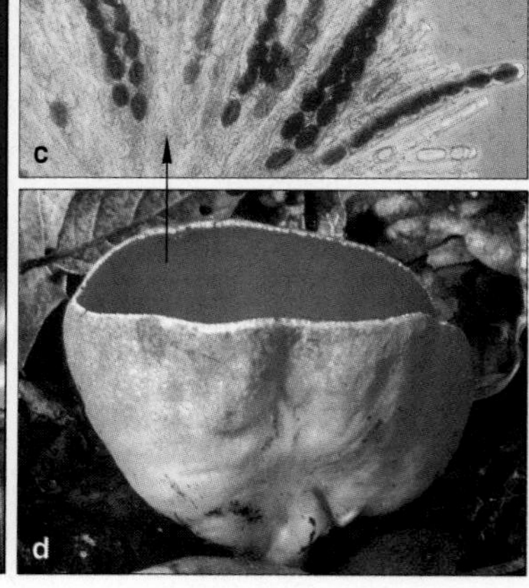

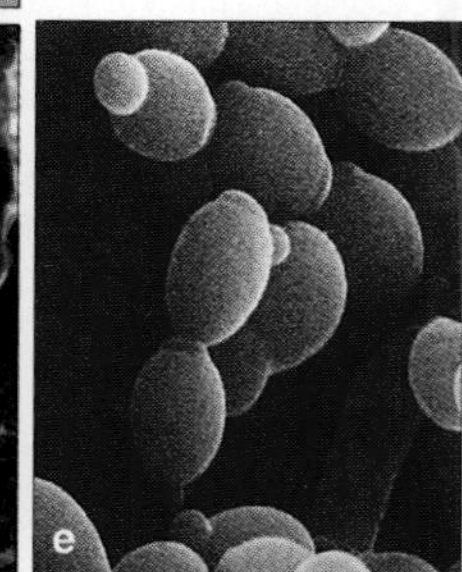

Figure 14.29 (**a**) *Rhizopus stolonifer*, the black bread mold, produces zygospores. (**b**) Hyphae of *Pilobolus*, another zygomycete. When the fluid-filled part of a hypha bursts, spores on top are propelled into the air. (**c,d**) *Sarcoscypha coccinia*, the scarlet cup fungus. On the cup's inner surface, spores form in saclike structures. (**e**) Another sac fungus: *Candida albicans* causes "yeast" infections of the vagina, mouth, and skin. When yeast cells fuse, they become a spore-producing sac.

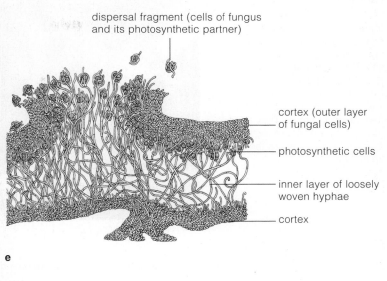

Figure 14.30 (**a**) One of the encrusting lichens, growing on a sunlit rock. (**b**) *Cladonia rangiferina*, an erect, branching lichen. (**c**) *Lobaria oregana*, a leaflike lichen that converts nitrogen gas to forms that trees in the Pacific Northwest's old-growth forests can use for growth. (**d**) *Usnea*, a pendant lichen commonly called old man's beard. (**e**) Body plan of a stratified lichen, cross-section.

call yeasts (others are club fungi). A packet of baker's yeast holds thousands of tiny spores of *Saccharomyces cerevisiae*. In a warm, moist spot, such as bread dough set out to rise, spores germinate. The cells that emerge will reproduce asexually by budding. Carbon dioxide released by their aerobic respiration causes the bread dough to rise. Fermentation by other yeast species is used to make alcoholic beverages. Still other yeasts are human pathogens (Figure 14.29*e*).

FUNGAL SYMBIONTS

Ever since plants invaded the land, fungi were there, many as symbionts. **Symbiosis**, recall, refers to species that interact closely during their life cycles. In cases of mutualism, their interaction benefits both the partners. Lichens and mycorrhizae generally are like this.

LICHENS Each **lichen** is a vegetative body in which a fungus intertwines with one or more photosynthetic organisms—usually green algae and cyanobacteria. It forms after the tip of a fungal hypha binds with a host cell. Both lose their wall, and their cytoplasm fuses or the hypha induces the host to cup around it. The two species multiply together into a multicelled form that usually has distinct layers. The growth pattern may be flattened, erect, leaflike, or pendulous (Figure 14.30).

Lichens typically colonize sites that are too hostile for most organisms, such as sunbaked or frozen rocks, fences, gravestones, and plants, even the tops of giant Douglas firs. Almost always, the fungus is the larger component. Cyanobacteria typically live in a separate structure inside or outside the main body. The fungus gets a fine long-term source of nutrients. Its nutrient withdrawals do hurt its food-producing partner a bit, but that partner may benefit by getting shelter in return.

Lichens absorb minerals. They produce antibiotics against bacteria that can decompose them and toxins that can kill invertebrate larvae that graze upon them. Coincidentally, their metabolic products enrich soil or help form new soil from bedrock. This is what happens when lichens colonize barren sites, including bedrock exposed by retreating glaciers. Conditions improve, so other species move in and replace the pioneers. This is probably what happened when plants invaded land.

Lichens that include cyanobacteria are important in the cycling of nitrogen. They capture nitrogen gas and convert it to a form that plants use (Figure 14.30*c*).

Lichens also provide early warning of deteriorating environmental conditions. How? They all absorb toxins but cannot get rid of them. From studies in New York City and in England, we know that when lichens die around human habitats, air pollution is getting bad.

MYCORRHIZAE Some fungi and young tree roots also are mutualists. Their interaction is a **mycorrhiza** (plural, mycorrhizae), which means "fungus-root." Underground parts of a fungus grow through the soil and functionally

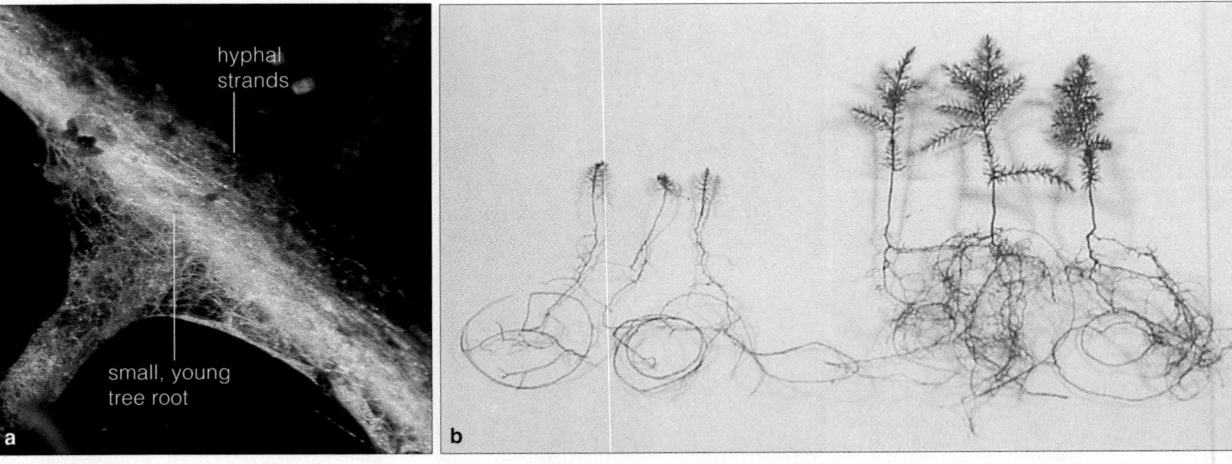

Figure 14.31 (**a**) Mycorrhiza on a hemlock root. (**b**) Effects of the presence or absence of mycorrhizae on plant growth in sterilized, phosphorus-poor soil. The juniper seedlings at left were controls; they grew without a mycorrhizal fungus. The seedlings at right are the experimental group. They were grown with a partner fungus. All of the seedlings are the same age.

increases the tree's absorptive surface area. Young roots start out with hairlike absorptive structures, but fungal hyphae are thinner and better at growing around soil particles. When phosphorus and other nutrients are in good supply, hyphae take them up from the soil. When these nutrients get scarce, the hyphae release some of them to their plant partner. The fungus benefits because the plant gives it some sugars. The loss is a trade-off; the plant does not grow as efficiently without the fungus.

Figure 14.31 shows an example of a mycorrhiza and its effects on plant growth. We'll take a closer look at this mutualistic interaction in Chapter 18.

AS FUNGI GO, SO GO THE FORESTS Since the early 1900s, collectors have recorded data on wild mushroom populations in many European forests. These records show that the number and kinds of fungi are declining. The mushroom gatherers are not to blame; inedible as well as edible species are disappearing. But the decline does correlate with rising levels of air pollution.

Under natural conditions, just about all trees have fungal symbionts on or in their roots. Compared to tree roots, fungal hyphae are smaller and better able to grow into small crevices where they can absorb inorganic minerals that they share with the tree.

As a tree ages, one mycorrhizal fungus gives way to another, in predictable patterns. When fungi die, trees lose this essential support system, and become more vulnerable to damage by severe frost, drought, and acid rain. Are North American forests at risk too? Conditions there are changing in similar ways.

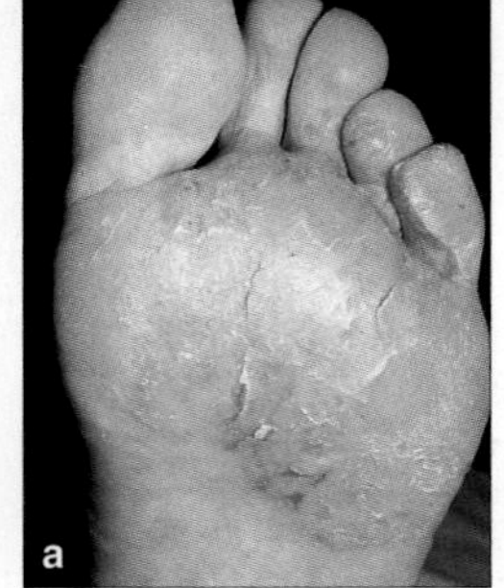
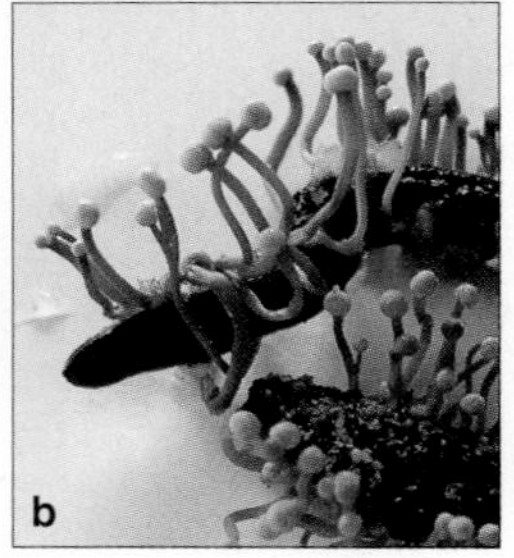

Figure 14.32 Love those fungi! (**a**) On one young individual, athlete's foot, caused by *Epidermophyton floccosum*. (**b**) Rye plant infected by the historically notable fungus *Claviceps purpurea*.

THE UNLOVED FEW

You know that you are a serious student of biology when you view organisms in terms of their roles in nature, rather than their effects on humans in general and you in particular. You salute the saprobic fungi as essential decomposers and praise parasitic ones that help keep populations in check. The true test is when you open the refrigerator to get your strawberries and a fungus has beaten you to them. Or when a fungus starts feeding on the warm, damp tissues between your toes (Figure 14.32*a*).

Which home gardeners wax poetic about black spot or powdery mildew on roses? Which farmers happily hand over millions of dollars a year to sac fungi that infect corn, wheat, peaches, and apples? Who rejoices that a certain sac fungus, *Cryphonectria parasitica*, killed off the chestnut trees in eastern North America?

Who willingly inhales airborne spores of *Ajellomyces capsulatus*? In soil this fungus forms a mycelium. In lung tissues it causes *histoplasmosis*, a potentially fatal respiratory disease.

And household molds! Allergies to mold promote sinus, ear, and lung infections, cause rashes, and can increase the frequency of asthma attacks.

Some fungi have even altered human history. One species, *Claviceps purpurea*, is a parasite of rye and other cereal grains (Figure 14.32*b*). Eating rye bread that was made from contaminated flour can lead to *ergotism*. Symptoms include vomiting, diarrhea, hallucinations, hysteria, and convulsions. Untreated, it can be fatal.

Ergotism outbreaks were common in Europe in the Middle Ages. They thwarted Peter the Great, a Russian czar who became obsessed with conquering ports along the Black Sea for his nearly landlocked empire. Soldiers laying siege to the ports ate mostly rye bread and fed rye to their horses. Soldiers went into convulsions and horses into "blind staggers."

Ever since their time of origin, fungal decomposers have been cyclers of nutrients that help sustain webs of life just about everywhere. It might help to keep this global view in mind when the few unloved ones make life miserable.

14.5 Viruses, Viroids, and Prions

The noncellular particles we call viruses are like stripped-down versions of life. There's no fossil record for viruses and biologists are still debating whether they arose before cellular life forms did or are derived from the genes of cellular organisms. Either way, molecular comparisons can be used to group them into related lineages and to study their ongoing evolution.

CHARACTERISTICS OF VIRUSES

In ancient Rome, *virus* meant "poison" or "venomous secretion." In the late 1800s, this rather nasty word was bestowed on newly discovered pathogens, smaller than bacteria. Many viruses deserve the name. They attack animals, plants, fungi, protists, bacteria—you name it, there are viruses that infect it.

A **virus** is a noncellular infectious agent with two characteristics. Each viral particle consists of a protein coat wrapped around genetic material and a few viral enzymes. It can't reproduce itself; a host cell is tricked into making copies of the virus.

Viral genetic material is DNA *or* RNA. The coat has one or more types of protein subunits organized into a rodlike or many-sided shape, as in Figure 14.33. The coat protects the genetic material during the journey to a new host cell. It also holds proteins that can bind to specific receptors on host cells. An envelope of mostly membrane remnants from the previously infected cell encloses some types of coats. It bristles with spikes of glycoproteins. Coats of complex viruses have sheaths, tail fibers, and other accessory structures attached.

The vertebrate immune system detects certain viral proteins. But genes for many of them mutate often, so a virus can slip past the immune fighters. For instance, people with respiratory problems might get new "flu shots" each year because spikes on the enveloped coat of influenza viruses keep changing.

Each kind of virus multiplies only in certain hosts. It cannot be studied easily except in living cells. This is why researchers study interactions between certain bacterial cells and **bacteriophages**, the class of viruses that infect them. Unlike the cells of humans and other complex, multicelled organisms, both can be cultured easily and fast. Both were used in early experiments to determine the function of DNA. They are still used as research tools and in genetic engineering.

Animal viruses contain double- or single-stranded DNA or RNA. They range in size from parvoviruses (a tiny 18 nanometers) to brick-shaped poxviruses (350 nanometers). Many kinds cause diseases, including the common cold, warts, herpes, influenza, and certain cancers. One virus, HIV, causes *AIDS* by infecting white blood cells. It weakens the immune system's ability to fight infections that normally are not life threatening.

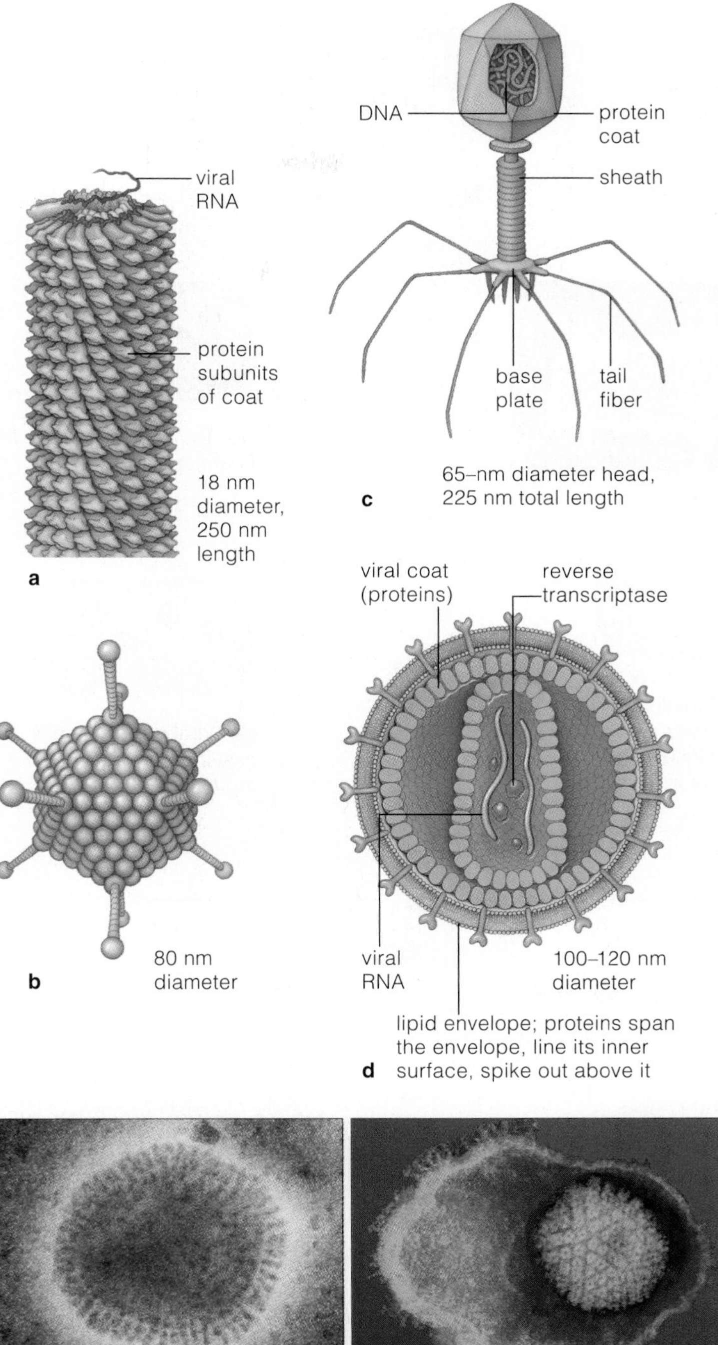

Figure 14.33 Body plans of viruses. (**a**) Protein subunits bound to nucleic acid form a spiral in rod-shaped viruses, including this tobacco mosaic virus. (**b**) Complex viruses, such as this bacteriophage, have structures attached to the coat. (**c**) Polyhedral (many-sided) viruses include this adenovirus. (**d**) A lipid envelope derived from membrane fragments of a lysed host cell surrounds HIV and other enveloped viruses. A capsid of regularly arrayed protein subunits encloses the viral RNA and a few enzymes.

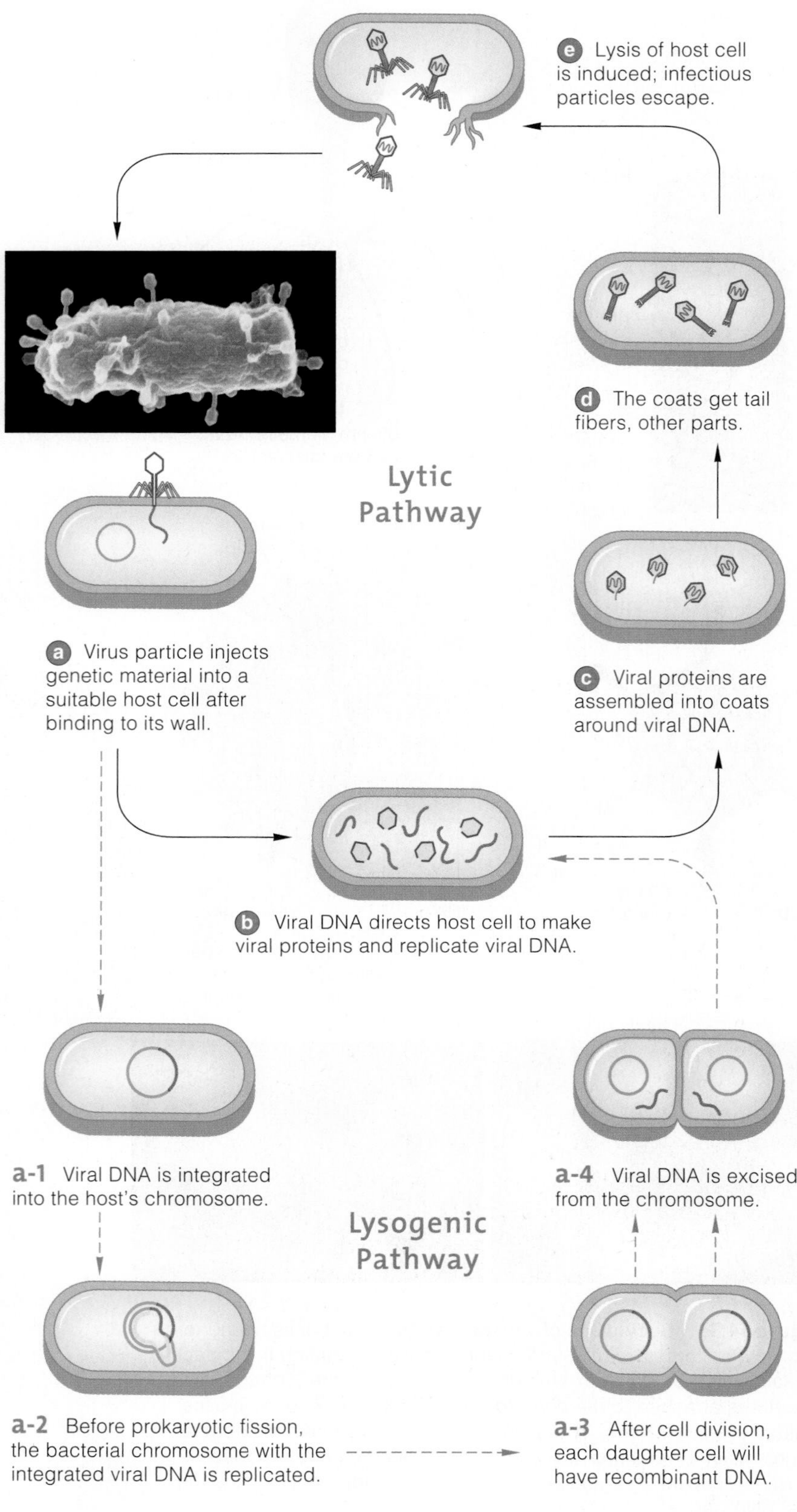

e Lysis of host cell is induced; infectious particles escape.
d The coats get tail fibers, other parts.
Lytic Pathway
a Virus particle injects genetic material into a suitable host cell after binding to its wall.
c Viral proteins are assembled into coats around viral DNA.
b Viral DNA directs host cell to make viral proteins and replicate viral DNA.
a-1 Viral DNA is integrated into the host's chromosome.
a-4 Viral DNA is excised from the chromosome.
Lysogenic Pathway
a-2 Before prokaryotic fission, the bacterial chromosome with the integrated viral DNA is replicated.
a-3 After cell division, each daughter cell will have recombinant DNA.

Figure 14.34 Generalized multiplication cycle for some bacteriophages. Virus particles may enter a lytic pathway. The lytic pathway may expand to include a lysogenic pathway for some types of viruses.

VIRAL MULTIPLICATION CYCLES

What happens during viral infections? Once host cells are infected, different viruses multiply in a variety of ways. Even so, nearly all multiplication cycles proceed through five basic steps, as outlined here:

1. *Attachment.* A virus attaches to a host cell through molecular groups that chemically recognize and lock on to specific molecular groups at the cell surface.

2. *Penetration.* Either the complete virus or its genetic material alone penetrates the cell's cytoplasm.

3. *Replication and protein synthesis.* In an act of molecular piracy, the viral DNA or RNA directs the host cell into producing many copies of the virus's nucleic acids and proteins, including enzymes.

4. *Assembly.* Viral nucleic acids and proteins become organized as new infectious particles.

5. *Release.* New virus particles are released from the cell by one mechanism or another.

Figure 14.34 shows two pathways that are common among bacteriophages. Steps 1 through 4 occur rapidly in a *lytic* pathway, and new particles are released when a host cell undergoes **lysis**. Lysis means that damage to a host cell's plasma membrane, wall, or both lets the cytoplasm dribble out. New virus particles dribble out also as the cell dies. The host itself built a viral enzyme that caused the damage and swift destruction.

In a *lysogenic* pathway, a latent period extends the cycle. The virus doesn't kill a host cell outright. A viral enzyme cuts the host chromosome, then integrates the viral genes into it. Before the cell divides, it replicates the recombinant molecule. Miniature time bombs thus await all of its descendants. Later, a molecular signal or some other stimulus may reactivate the cycle.

Latency is typical of many viruses, such as Type I *Herpes simplex*, which causes *cold sores* (fever blisters). This virus hides in sensory nerves. When stress or a sunburn reactivates it, the virus moves to epithelial cells in skin and causes painful eruptions. Like other enveloped viruses, it enters a host cell by a version of endocytosis, then leaves by budding from the surface.

RNA viruses have a different multiplication cycle. Viral RNA functions as a template for making mRNA or DNA. One of the enzymes that the retrovirus HIV carries into host cells catalyzes the formation of DNA by reverse transcription (Sections 11.1 and 23.8).

VIROIDS AND PRIONS

Stripped-down as viruses are, **viroids** are simpler still. They are circles of RNA with no protein coat and no protein-coding genes. Many viroids are pathogens that kill millions of dollars' worth of crops annually. So far, only one type has been found in humans. It helps to cause hepatitis D, a deadly liver disease.

A **prion** is an infectious protein. The best studied ones cause "mad cow disease," or bovine spongiform encephalopathy (BSE), and human variant Creutzfeldt-Jacob disease (vCJD). Normal versions of these prion proteins occur in the brain and spinal cord, where they are folded in a particular way. Infectious prions fold up wrong, and they induce normal versions to misfold too. The result is a degenerative nervous disease that is invariably fatal.

A virus consists of nucleic acid enclosed in a protein coat and sometimes an outer membranous envelope. It can replicate only in specific types of host cells. Viroids and prions are even simpler infectious particles.

14.6 Evolution and Infectious Diseases

Just by being human, you are a potential host for the kinds of diverse pathogens and parasites you have just read about. Before leaving the chapter, reflect on what this means from an evolutionary perspective.

Infection is the invasion of a cell or multicelled body by a pathogen (or by the pathogen's genetic material). **Disease** follows when the pathogen multiplies and the activities of its descendants disrupt body activities. Table 14.2 and Figure 14.35 give some examples.

It takes direct contact with an infected person or their secretions for a contagious disease to spread. The *sporadic* diseases, such as whooping cough, break out irregularly among very few people. *Endemic* diseases, such as tuberculosis, are more or less always present in a population but are confined to a small part of it.

During an *epidemic*, some disease quickly spreads through part of a population for a limited time, then subsides. *Pandemic* refers to many epidemics of the same disease breaking out in several countries during the same interval.

We discuss the ongoing AIDS pandemic in detail in Chapter 23. The 2003 outbreak of SARS (Severe Acute Respiratory Syndrome) was a brief pandemic caused by a coronavirus. Before quarantines halted its spread, 774 died. A new threat is H5N1, a bird influenza virus that has infected and killed people in Asia. Bird flu is not yet easily transmitted from person to person, but a mutation could change this, with deadly results.

Table 14.2 The Eight Deadliest Infectious Diseases

Disease	Main Agents	Estimated New Cases per Year	Estimated Deaths per Year
Acute respiratory infection*	Bacteria, viruses	1 billion	4.7 million
Diarrhea**	Bacteria, viruses, protozoans	1.8 billion	3.1 million
Tuberculosis	Bacteria	9 million	3.1 million
Malaria	Sporozoans	110 million	2.5–2.7 million
AIDS	Virus (HIV)	5.6 million	2.6 million
Measles	Viruses	200 million	1 million
Hepatitis B	Virus	200 million	1 million
Tetanus	Bacteria	1 million	500,000

* Includes pneumonia, influenza, and whooping cough.
** Includes amoebic dysentery, cryptosporidiosis, and gastroenteritis.

Now consider disease in terms of the pathogen's prospects for survival. A pathogen stays around only for as long as it has access to outside sources of energy and raw materials. To a pathogen, the human body is a jackpot. Inside it, a pathogen may multiply to huge population sizes. The point? Evolutionarily speaking, the ones that leave the most descendants win.

Two barriers prevent pathogens from taking over the world. First, any species with a history of being attacked by a specific pathogen has coevolved with it and has built-in defenses against it. The immune system of vertebrates is one example. Second, if some pathogen kills its host too quickly, it might vanish along with that individual. This is one reason most pathogens are not fatal. Infected individuals who live will spread the pathogen and thus contribute to its reproductive or replicative success.

An individual is most likely to die if it becomes host to overwhelming numbers of a pathogen, if it is a novel host that lacks coevolved defenses, or if a novel pathogen emerges and breaches current defenses.

Antibiotics are compounds synthesized by one organism that can kill another, and we use them as weapons against bacterial pathogens. Penicillin and other antibiotics have saved millions of human lives. However, these drugs act as agents of selection, and antibiotic-resistant forms of bacteria are increasing.

Lateral gene transfers among bacterial cells of the same or different species contribute to the problem. Many bacterial pathogens swap genes like teens swap music files. The genes that confer antibiotic resistance move about in plasmids or as a free piece of DNA.

In evolutionary terms, those pathogens that leave the most descendants win. They are at a selective advantage.

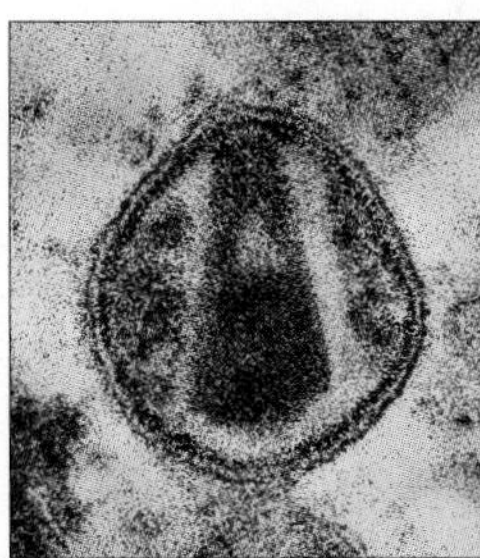

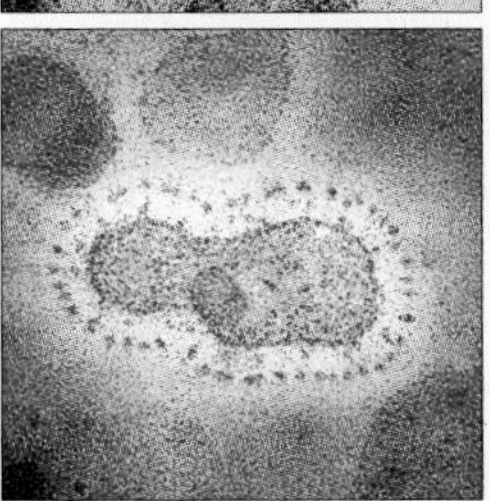

Figure 14.35 *Above*, HIV, the agent of AIDS. *Below*, two particles of the coronavirus that causes SARS.

Summary

Section 14.1 Experiments demonstrate that complex organic compounds could have self-assembled under conditions on early Earth. Chemical and molecular processes may have given rise to proto-cells. The first cells were prokaryotic. An early divergence gave rise to bacteria and to the shared ancestors of archaean and eukaryotic cells.

BiologyNow
Learn how organic compounds can spontaneously assemble with the animation on BiologyNow.

Section 14.2 Archaea and bacteria are the only prokaryotes. They are metabolically diverse. They alone can reproduce by prokaryotic fission. Many archaea live in places too hostile for most organisms. Bacteria are the most common prokaryotes. Many function as decomposers and photosynthesizers; some can cause disease. Table 14.3 compares features of prokaryotes and eukaryotes.

BiologyNow
Review the components of a typical prokaryotic cell with the animated interaction on BiologyNow.

Section 14.3 The "protist kingdom" is now being revamped to reflect new understanding of evolutionary relationships. Most protists are single-celled, but many lineages also have multicelled species. Nearly all types live in water or moist habitats, including host tissues.

Ancient lineages include euglenoids and related flagellated protozoans. The amoeboid protozoans use pseudopods to move about and capture prey. Ciliates, dinoflagellates, and apicomplexans are close relatives. Algal blooms can kill fish and other aquatic organisms; they are population explosions of dinoflagellates or other protists. Apicomplexans are parasites that spend part of their life cycle inside cells of their host.

Diatoms and coccolithophores are single-celled, shelled, aquatic producers. They are close relatives of the oomycotes—which include many plant pathogens—and the brown algae. Brown algae include the giant kelps, which are the largest protists. Green algae are the closest relatives of plants. Phycobilins mask chlorophylls in red algae. Slime molds spend part of their life cycle as a single cell and part in a larger cohesive group that can migrate and differentiate to form a spore-bearing body.

BiologyNow
View the body plan of a euglena and the life cyle of a green alga with the animation on BiologyNow.

Section 14.4 All fungi are heterotrophs. Most are decomposers. A few are parasites or pathogens. All engage in extracellular digestion. Sexual and asexual spores form during the life cycle of the three major lineages: zygomycetes, sac fungi, and club fungi. In multicelled species, germinating spores undergo cell divisions that produce a mycelium.

Lichens and mycorrhizae are cases of mutualism. In lichens, a fungal symbiont lives with a photosynthetic species, most often a green alga or cyanobacterium. In mycorrhizae, a fungus and young roots are symbionts.

BiologyNow
Observe the life cycle of a club fungus with the animation on BiologyNow.

Section 14.5 A virus is a noncellular infectious particle that can multiply only by pirating metabolic machinery of a host cell. It is DNA or RNA packaged in a protein coat. Some replication cycles of viruses include a latent phase, when viral genetic material is integrated into the host genome. Viroids are infectious RNA particles. Prions are infectious proteins.

BiologyNow
Use the animation on BiologyNow to explore viral structure and multiplication pathways.

Table 14.3 Comparison of Prokaryotes With Eukaryotes

	Prokaryotes	Eukaryotes
Organisms represented:	Archaea, bacteria	Protists, fungi, plants, and animals
Ancestry:	Two major lineages that evolved more than 3.5 billion years ago	Ancient prokaryotic ancestors gave rise to forerunners of eukaryotes, which evolved more than 1.2 billion years ago
Level of organization:	Single-celled	Protists, single-celled or multicelled. Nearly all others multicelled; division of labor among differentiated cells, tissues, and often organs
Typical cell size:	Small (1–10 micrometers)	Large (10–100 micrometers)
Cell wall:	Most with distinctive wall	Cellulose or chitin; none in animal cells
Membrane-enclosed organelles:	Rarely; no nucleus, no mitochondria	Typically profuse; nucleus present; most with mitochondria, many with chloroplasts
Modes of metabolism:	Both anaerobic and aerobic	Aerobic modes predominate
Genetic material:	One chromosome; plasmids in some	Chromosomes of DNA plus many associated proteins in a nucleus
Mode of cell division:	Prokaryotic fission, mostly; some reproduce by budding	Nuclear division (mitosis, meiosis, or both) associated with one of various modes of cytoplasmic division

Section 14.6 Evolution shapes the characteristics of pathogens, as it does for all organisms. Pathogens and their hosts have coevolved. Antibiotics select for antibiotic-resistant pathogens.

Self-Quiz

Answers in Appendix I

1. The ______ are major decomposers.
   a. fungi b. viruses c. ciliates d. dinoflagellates

2. All of the ______ are prokaryotic.
   a. archaea c. protists e. all are correct
   b. bacteria d. a and b

3. Chloroplasts are descended from ______.
   a. mitochondria c. archaeans
   b. cyanobacteria d. viroids

4. The ______ are parasitic eukaryotes that spend part of their life cycle inside specific cells of host organisms.
   a. viruses c. euglenoids e. both a and b
   b. apicomplexans d. slime molds f. all are correct

5. Chitin reinforces the cell walls of ______.
   a. coccolithophores c. foraminiferans
   b. diatoms d. fungi

6. Some of the ______ are human pathogens.
   a. slime molds c. flagellated protozoans
   b. archaea d. both a and c

7. When a spore of a multicelled fungus germinates, it gives rise to a ______.
   a. viroid c. mycelium
   b. fruiting body d. mycorrhiza

8. The genetic material of a ______ may be DNA or RNA.
   a. zygomycete b. ciliate c. dinoflagellate d. virus

9. Match these terms suitably.

____ green algae	a. protist population explosion
____ virus	b. silica-shelled producer
____ bacteria	c. most diverse prokaryotes
____ brown algae	d. noncellular, not alive
____ mycorrhiza	e. include the largest protists
____ lichen	f. fungus plus producer cells
____ algal bloom	g. closest relative of plants
____ mushroom	h. fossilized cyanobacteria
____ diatom	i. a "fungus-root"
____ stromatolite	j. spore-bearing fruiting body of a club fungus

Additional questions are available on **BiologyNow™**

Critical Thinking

1. The cellular slime mold *Dictyostelium discoideum* is now being studied by more than 660 biologists. Each year, hundreds of papers about its biology appear in scientific journals. It is one of the "model organisms" that help us learn about cell differentiation and other developmental processes. A single cell can be allowed to divide again and again to form a huge population. When such a population is starved, cells aggregate and form fruiting bodies, as shown in Figure 14.24. Although genetically identical, some cells will go on to become part of the stalk, while others become spores. How is fruiting body formation of genetically identical slime mold cells similar to the process by which a multicelled organism develops?

2. Runoff from heavily fertilized fields, animal waste, and raw sewage promotes algal blooms that can result in massive kills of fish and other aquatic species. If you find this environmental cost unacceptable, how would you stop the pollution? Bear in mind that we now depend on high-yield (and heavily fertilized) crops. How do you suggest we dispose of the waste from farms and cities?

3. The fungus *Fusarium oxysporum* is a plant pathogen. Some view it as a potential weapon in the war on drugs. Why? Strains of the fungus attack and kill only specific plants, such as the coca plants used to produce cocaine. Proposals to spray *F. oxysporum* to kill marijuana plants in Florida were abandoned after public outcries. What concerns, if any, would you have about the widespread spraying of natural mycoherbicides to kill off plants that are sources of illegal drugs?

4. Curtis Suttle studies microbial communities in ocean water. In one experiment, he selectively removed viruses from seawater and found that algae stopped growing. Further studies revealed that the algae required nutrients released by dying bacteria that lysed after viral infection. Make a list of some of the other ways in which viruses might help sustain natural ecosystems or be beneficial to human health.

5. One way to prevent bacterial food poisoning is by *food irradiation*, exposing food to high-energy rays that kill pathogens. The process also slows spoilage and can prolong shelf life. Some think this is a safe way to protect consumers. Others worry that irradiation could produce harmful chemicals. They point out that this treatment does not kill endospores that can cause botulism. In their view, a better approach would be to tighten and enforce food safety standards.

Irradiated meat, poultry, spices, and fruits are now in supermarkets. By law, they must be marked with a special symbol (Figure 14.36). Would noticing this symbol on a package influence your likelihood of purchasing that product? Explain your answer.

Figure 14.36 A radura, international symbol for irradiated food. It means that a product has been exposed to high-energy radiation to kill pathogens. Irradiation cannot make food itself radioactive.

15 PLANT EVOLUTION

IMPACTS, ISSUES

Beginnings, and Endings

Change is the way of life. About 300 million years ago, in the Carboniferous, swamp forests carpeted the wet, warm lowlands of continents. Tree-high ancestors of today's tiny club mosses and horsetails were dominant. Then things changed. As the global climate became cooler and drier, moisture-loving plants declined. Hardier plants—cycads, ginkgos, and conifers—rose to dominance. They were the gymnosperms, and they flourished for millions of years.

Things changed again. Other plants had evolved, and they held far more potential to radiate into diverse environments. The angiosperms—flowering plants—started to take over. They still dominate most regions. But in the far north, at high elevations, and in parts of the Southern Hemisphere, some conifers retained the competitive edge and held on as great forests.

Then small bands of humans learned to grow some flowering plants as crops, which became the foundation for population growth. Human populations did grow, spectacularly, and they required far more resources than crops. Conifers had the misfortune to become premier sources of lumber, paper, resins, and other goods. They became vulnerable to **deforestation**, the removal of all trees from large tracts of land.

In California, only 4 percent of the original coastal redwood forest is still undisturbed. In Maine, an area the size of Delaware has been logged over in just the past fifteen years. Many logs end up as paper or pulp. The United States also exports logs to lumber mills overseas. At the same time, it imports timber from tropical rain forests as far away as New Zealand.

Clearing the world's forests has cascading ecological effects. Nutrients wash away from exposed soils, and sediments clog streams. Herbicides applied to stop nontimber species from taking over interfere with ecosystem recovery. Species that depend on the forest for food and shelter become threatened or extinct.

Almost 750 species of plants in the United States alone are now on the endangered list. And with this bit of perspective on change, we turn to the beginning—and end of the line—of some ancient lineages.

How Would You Vote? *Demand for paper is a big factor in deforestation. But using recyled paper can add to the cost of a product. Would you be willing to pay more for papers, books, and magazines that are printed on recycled paper? See BiologyNow for details, then vote online.*

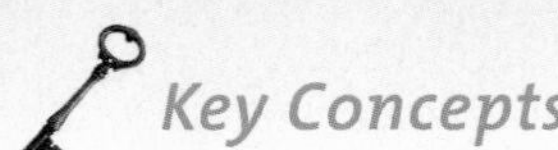

Key Concepts

EVOLUTIONARY TRENDS
Plants arose from a group of green algae. Their evolutionary story is one of increasing adaptation to life in drier environments.

BRYOPHYTES
All bryophytes are small plants, such as mosses. The gamete-producing stage is the longer lived and more conspicuous part of their life cycle.

THE RISE OF VASCULAR PLANTS.
The life cycle of vascular plants is dominated by a spore-producing body with specialized tissues that serve as pipelines for moving water and dissolved sugars.

THE RISE OF SEED PLANTS
Pollen grains and seeds were key innovations that allowed seed plants (gymnosperms and angiosperms) to radiate into diverse habitats. Only angiosperms produce flowers and fruits.

Links to Earlier Concepts

In Section 14.3, you learned about green algae, which include the closest relatives of the plants. You studied gamete formation in Section 7.7, and a brief look back may help you to understand how plant life cycles changed as lineages moved to increasingly drier habitats.

Much of what we know about plant evolution is based on the fossil record (13.1). This is a story of adaptive radiations and extinctions (13.9). A working knowledge of the overall history of life and the geologic time scale (13.2) may help you put the events we discuss in this chapter into a larger perspective.

15.1 Pioneers in a New World

We share the world with at least 295,000 kinds of plants. Within their tremendously diverse kingdom, we find recurring structures that correlate with present and past functions.

LINKS TO SECTIONS 7.7, 13.2, 13.3

About five hundred million years ago, the invasion of land was getting under way. Why then? As described in the previous chapter, the noncyclic photosynthetic pathway had evolved in cyanobacteria and the oxygen released as a by-product had changed the atmosphere. Far above Earth, the sun's energy had converted some oxygen into the ozone layer—a shield against deadly doses of ultraviolet radiation. Until that time, life had survived only beneath the surface of water and mud.

Green algae were evolving at the water's edge, and one group gave rise to plants. By 430 million years ago, *Cooksonia*, a simple branching plant a few centimeters high, was common (Figure 15.1*a*). It took another 160 million years for the much taller *Psilophyton* to evolve (Figure 15.1*b*,*c*). Then a great adaptive radiation began. It took only 60 million more years for plants to radiate from lowland swamps to mountain peaks and nearly everywhere in between. Changes in structure, function, and reproductive modes made this radiation possible.

ROOTS, STEMS, AND LEAVES

Underground absorptive structures evolved as plants colonized the land, and in some lineages they became systems of roots. A root system helped plants absorb water and dissolved mineral ions from soil. As they do today, young roots formed a close association with mycorrhizal fungi. Roots also helped anchor the plant.

Aboveground, shoot systems evolved. Stems and leaves intercepted sunlight energy and took in carbon dioxide from the air. A glue-like polymer called lignin appeared in the cell walls of some lineages. It allowed stems to grow taller and plants to stand more erect.

For many lineages, internal pipelines contributed to the evolution of roots and shoots. They evolved as components of vascular tissues—xylem and phloem. Xylem carries water and mineral ions throughout the plant. Phloem distributes sugars and other products of photosynthesis to all tissues.

Having enough water for metabolism was not a problem in most aquatic habitats. It became a greater challenge on the land. A waxy outer layer—a **cuticle**—evolved and helped plants to conserve water inside shoots on hot, dry days. Water and gases could enter or leave shoots only at **stomata**, gaps that spanned the cuticle. Evolution of stomata allowed control of water loss. A stomata-pierced cuticle helps to regulate water balance in nearly all existing land plants (Section 6.6).

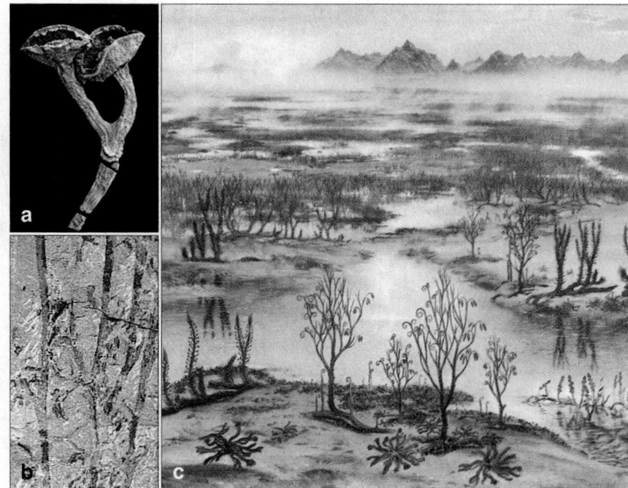

Figure 15.1 Early plants. (**a**) Scanning electron micrograph of a *Cooksonia* fossil shows a leafless branching structure. Spores formed at the branch tips. (**b**) A fossil of *Psilophyton*, which was common in swamps during the Devonian period (**c**).

FROM HAPLOID TO DIPLOID DOMINANCE

Life cycles were altered as plants evolved. We find multicelled haploid and diploid body forms in the life cycles of all plants, but the relative size of the two stages varies. The **gametophyte** is a haploid gamete-making body. When two gametes fuse at fertilization, a diploid zygote forms. It grows and develops into a **sporophyte**, in which haploid spores are produced by meiosis. A **spore** is a resting structure that withstands unfavorable environmental conditions. A plant spore gives rise to a new gametophyte (Figure 15.2).

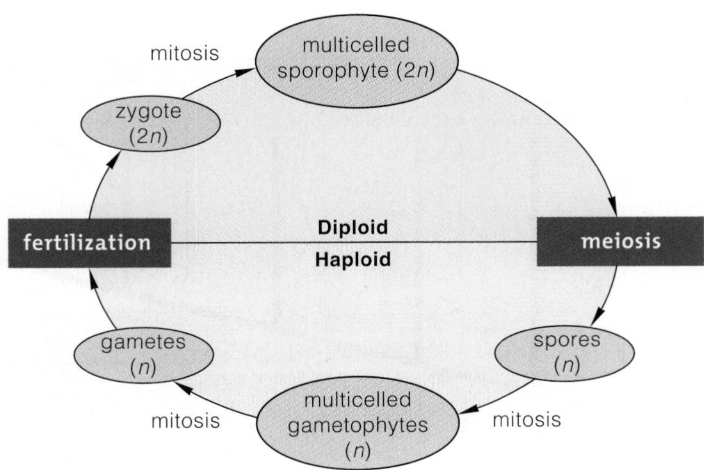

Figure 15.2 Generalized life cycle for all plants.

LINKS TO SECTIONS 7.7, 14.3

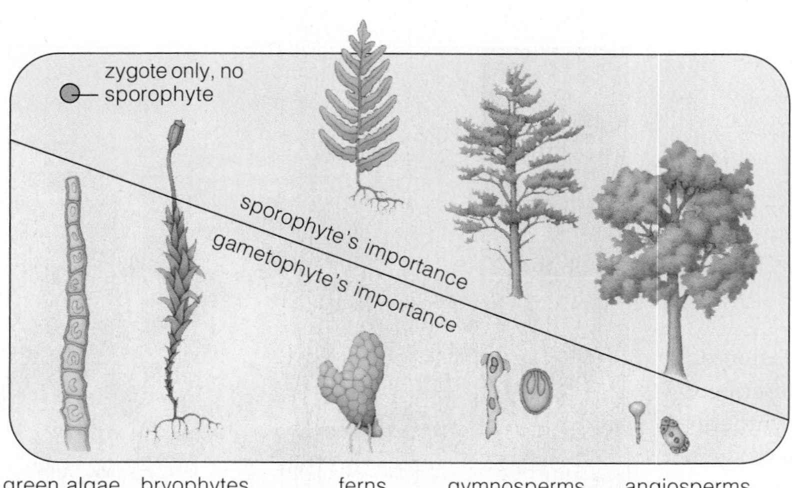

Figure 15.3 ***Animated!*** One major evolutionary trend in plant life cycles. Each plant's gametophyte (haploid) stage is on the *gold* background. Its sporophyte stage (diploid) is on the *green* background. For each group, the relative size of the green and gold areas indicates relative size and life span of the stages. Most algae, and all bryophytes, put more of their energy into gamete-producing structures. The other groups evolved in seasonally dry habitats on land. They put the most energy into complex structures that produce spores and also retain, nourish, and protect gametes through harsh times.

In the charophyte algae that gave rise to plants, the diploid phase of the life cycle is unicellular; a haploid, gamete-producing stage dominates the life cycle. As plants radiated onto land, and into higher and drier habitats, production and dispersal of spores took on greater importance. The fossil record reveals a trend toward sporophyte dominance.

In most plant lineages, the sporophyte became the larger, longer lived portion of the life cycle. Shrubs, trees, and other large forms having internal transport pipelines evolved. The spores became enclosed within hard capsules. Larger-bodied plants were better able to retain, nourish, and protect gametophytes, as well as the developing sporophytes. Figure 15.3 shows this evolutionary trend from haploid to diploid dominance.

SEEDS AND POLLEN GRAINS

Figure 15.4 is an evolutionary tree diagram for major groups of plants. Only 24,000 of the 295,000 existing species are *nonvascular* plants, or bryophytes. *Vascular* types have internal tissue systems that conduct water and solutes. They became the success stories on land. The lycophytes, horsetails, and ferns are among the *seedless* vascular plants. Cycads, ginkgos, conifers, and gnetophytes are gymnosperms, a group of *seed-bearing* vascular plants. The other group of seed plants is the angiosperms. They alone produce flowers.

Seed-bearing plants produce two types of spores. The larger ones are megaspores, which develop into egg-producing female gametophytes. Microspores are smaller. They develop into the **pollen grains**. Sperm-bearing male gametophytes form inside pollen grains.

Male gametes of nonvascular plants and seedless vascular plants must swim to eggs for fertilization to occur. Most of the seed-bearing plants have sperm that do not have flagella. Air currents, insects, birds, and so on deliver their pollen grains to eggs. Pollen grains are a major reason why the seed-bearing plants were able to radiate into higher and drier environments.

Seeds were also useful in diverse habitats. Female gametophytes of seed plants form within the tissues of

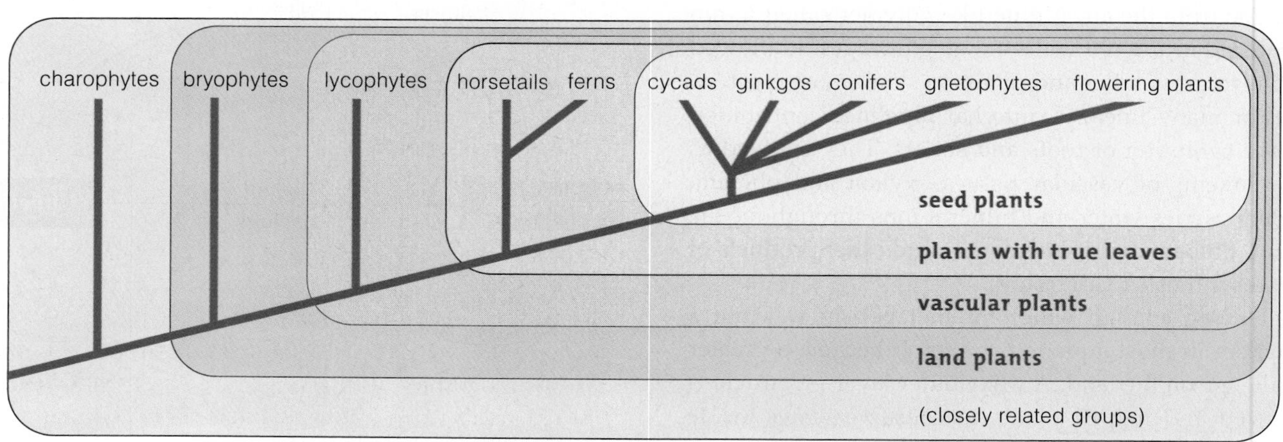

Figure 15.4 An evolutionary tree for plants, showing nested monophyletic groups.

sporophytes. Each **seed** is an embryo sporophyte, the tissues that nourish it, and an outer coat. Inside its coat, the embryo waits out seasons that do not favor its growth. When conditions become more favorable, the seed germinates and a new sporophyte develops.

The plant kingdom includes multicelled, photosynthetic species called bryophytes, seedless vascular plants, and seed-bearing vascular plants.

Most plant lineages became structurally adapted to life on land. They have root and shoot systems, a waxy cuticle, stomata, vascular tissues, and lignin-reinforced tissues.

Sporophytes with well-developed roots, stems, and leaves came to dominate the life cycles of most land plants. Parts of these complex sporophytes nourish and protect the new generation until conditions favor dispersal and growth.

The most recently evolved groups produce pollen grains and seeds. These were key innovations that allowed the seed plants to spread widely into diverse habitats.

Figure 15.5 Representative bryophytes. (**a**) Moss plants growing on damp rocks. (**b**) Peat moss (*Sphagnum*). A few sporophytes—brown, jacketed structures on white stalks—are attached to a pale gametophyte. (**c**,**d**) *Marchantia.* Like other liverworts, this one reproduces sexually as well as asexually. Unlike the other types, it has (**c**) female and (**d**) male reproductive parts on different plants.

15.2 The Bryophytes—No Vascular Tissues

The first plants were bryophytes. A gamete-forming body dominates their life cycle. There are no specialized tissues to distribute water and sugars, so plants remain small.

Modern **bryophytes** include 24,000 species of mosses, liverworts, and hornworts. None of these nonvascular plants is taller than twenty centimeters (eight inches). Despite their small size, they have leaflike, stemlike, and usually rootlike parts. Their **rhizoids** are threadlike cells or structures that absorb water and dissolved mineral ions. They also attach gametophytes to soil.

Bryophytes do not have specialized vascular tissues and most grow in moist habitats. Some can dry out, then revive after absorbing water. This is one way that some hardy species survive in deserts and windswept plateaus in Antarctica.

Mosses are the bryophytes that most people know about (Figure 15.5*a*). Like others of this group, they are highly sensitive to air pollution. Where air quality is poor, mosses are few. Most moss gametophytes have a low mounded or feathery form. Gametes develop at their shoot tips, in vessels enclosed in a tissue jacket. Rain causes these jackets to open, and sperm swim to the eggs through tiny water droplets that cling to the plants. The moss sporophyte consists of a stalk and a jacketed structure in which haploid spores develop.

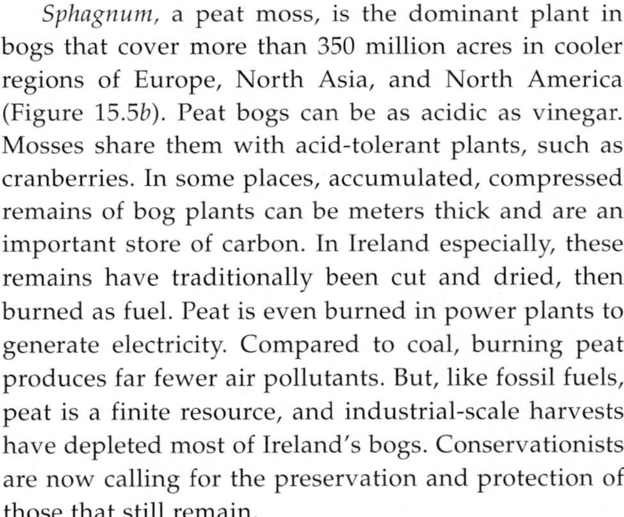

Sphagnum, a peat moss, is the dominant plant in bogs that cover more than 350 million acres in cooler regions of Europe, North Asia, and North America (Figure 15.5*b*). Peat bogs can be as acidic as vinegar. Mosses share them with acid-tolerant plants, such as cranberries. In some places, accumulated, compressed remains of bog plants can be meters thick and are an important store of carbon. In Ireland especially, these remains have traditionally been cut and dried, then burned as fuel. Peat is even burned in power plants to generate electricity. Compared to coal, burning peat produces far fewer air pollutants. But, like fossil fuels, peat is a finite resource, and industrial-scale harvests have depleted most of Ireland's bogs. Conservationists are now calling for the preservation and protection of those that still remain.

Acidic metabolic products of bryophytes hamper the growth of bacterial and fungal decomposers. Well-preserved human bodies about 2,000 to 3,000 years old have been found in peat bogs in Europe. These bogs may have been the sites for ritual human sacrifices. The high acidity of the bogs slowed bacterial growth and kept the victims' bodies from decomposing.

Compared to cotton, *Sphagnum* soaks up five times more water, which enters into dead cells in the plant's leaflike parts. Because of the antiseptic properties and absorbency of peat moss, it was used during World War I as a substitute for unavailable cotton bandages.

Early land plants may have resembled hornworts and liverworts (Figure 15.5*c*,*d*). Even today, these are

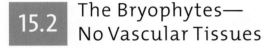

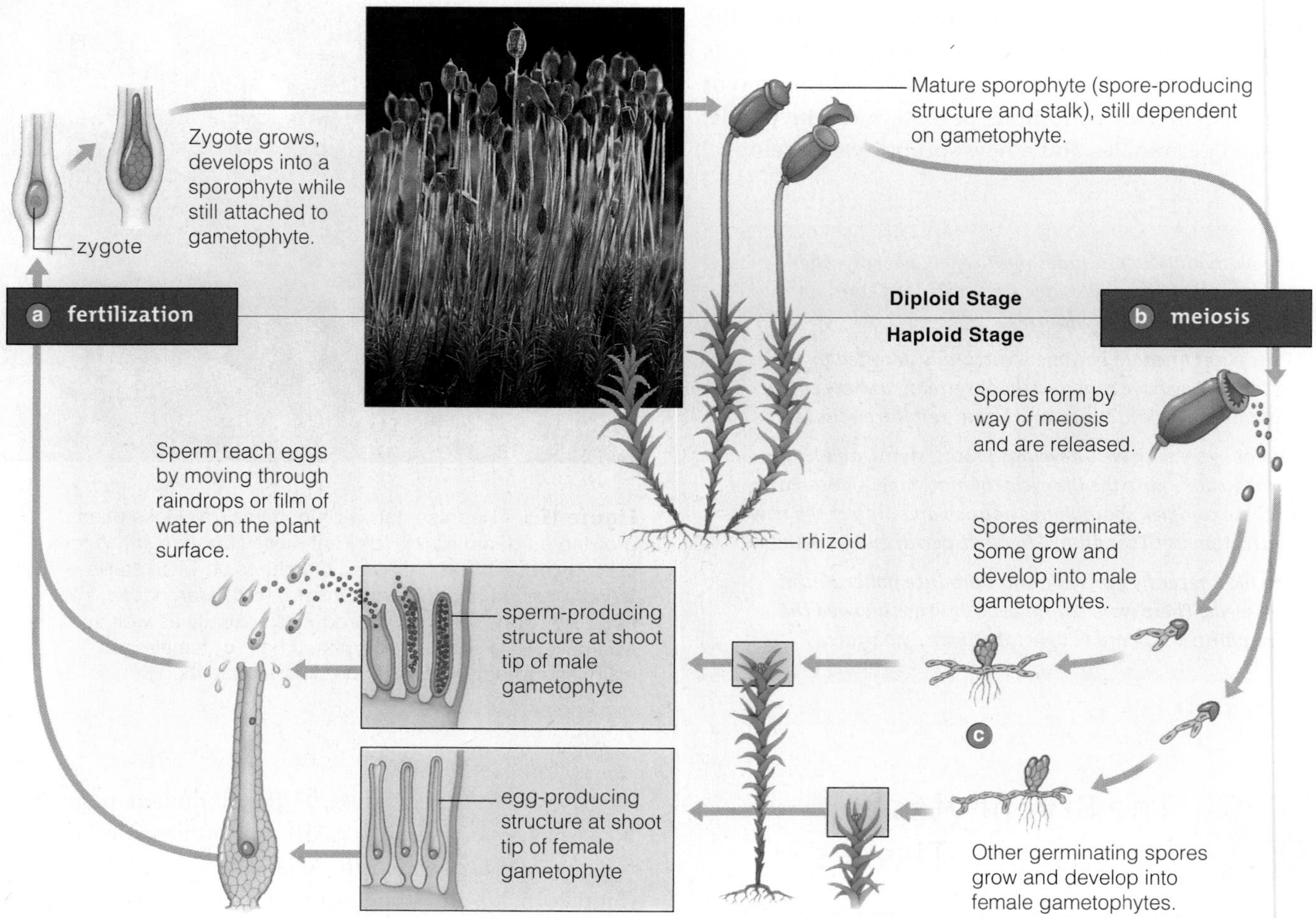

Figure 15.6 ***Animated!*** Life cycle of a moss (*Polytrichum*), one of the bryophytes. (**a**) Fertilization will occur after sperm swim to the eggs. It produces a zygote, which develops into a nonphotosynthetic sporophyte that will remain attached to and dependent on the gametophyte. (**b**) The sporophyte produces spores by meiosis. (**c**) These spores develop into new gametophytes, which will produce eggs or sperm.

zygote

Zygote grows, develops into a sporophyte while still attached to gametophyte.

Mature sporophyte (spore-producing structure and stalk), still dependent on gametophyte.

a fertilization

Diploid Stage

Haploid Stage

b meiosis

Spores form by way of meiosis and are released.

Sperm reach eggs by moving through raindrops or film of water on the plant surface.

rhizoid

Spores germinate. Some grow and develop into male gametophytes.

sperm-producing structure at shoot tip of male gametophyte

c

egg-producing structure at shoot tip of female gametophyte

Other germinating spores grow and develop into female gametophytes.

IMPACTS, ISSUES

The northern spotted owl and about 450 moss and 170 liverwort species live in the Pacific Northwest's old-growth forests. Logging of this region threatens both the owl and some bryophytes with extinction.

among the hardy pioneer species that colonize barren habitats. Gametophytes are most often small ribbon-like or "leafy" forms.

All bryophytes share three traits that evolved in early land plants. *First*, a waxy cuticle helps conserve water in aboveground parts. *Second*, a cellular jacket around the gamete-producing parts keeps in moisture. *Third*, the large gametophytes do not draw nutrients from sporophytes as they do in other plants. It is the other way around. In bryophytes, a sporophyte grows from and stays attached to the gametophyte. It draws nutrients from the gametophyte for part or all of the life cycle (Figure 15.6).

Bryophytes are nonvascular plants with flagellated sperm that require liquid water to reach and fertilize the eggs.

In these plants, a sporophyte develops within gametophyte tissues. It remains attached to the gametophyte and receives some nutritional support from it.

15.3 Seedless Vascular Plants

Seedless vascular plants differ from bryophytes in three respects. The sporophyte does not remain attached to a gametophyte, it has true vascular tissues, and it is the larger, longer lived phase of the life cycle.

About 350 million years ago, during the Carboniferous, mild climates prevailed. Swamp forests carpeted the warm, wet lowlands of continents. Consistently balmy temperatures allowed plants to grow through much of the year. Sporophytes that had lignin-braced tissues and a vascular system were at a selective advantage. These traits arose in the first seedless vascular plants. Lineages that survived to the present are **lycophytes**, **horsetails**, and **ferns**.

Lycophytes are known informally as club mosses. Some Carboniferous ones, such as *Lepidodendron*, were fifty-meter-tall giants (Figure 15.17). Some relatives of

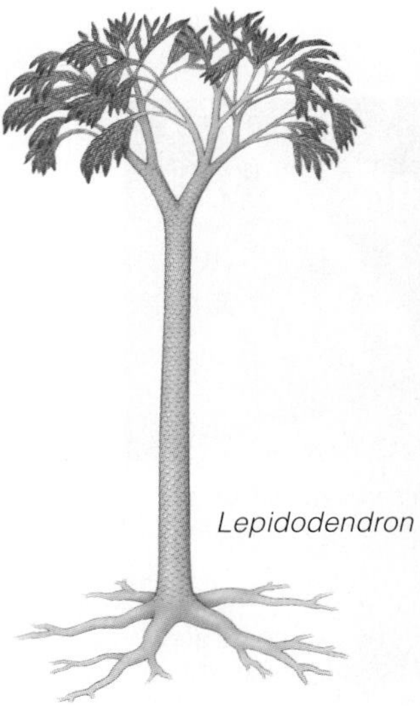

modern-day horsetails reached twenty meters high. Their aboveground stems grew from rapidly spreading **rhizomes,** or underground stems.

Most existing seedless vascular plants survive only in damp habitats. Gametophytes do not have vascular tissue and flagellated sperm require water to reach the egg. Species that live in drier places become dormant when rainfall is scarce and reproduce sexually during seasonal brief but heavy rains.

Figure 15.7 Reconstruction of a Carboniferous swamp forest. The sea level rose and fell fifty times during this geologic interval. Each time the sea receded, swamp forests flourished. When the sea moved back in, forest trees became submerged and buried in sediments, which protected them from decay.

Sediments slowly compressed the saturated, undecayed remains into peat. Each time more sediments accumulated, the increased heat and pressure made the peat more and more compact. In time, the compressed organic remains were transformed into great seams of coal. With its high percentage of carbon, coal is energy rich and one of our premier "fossil fuels."

It took a fantastic amount of photosynthesis, burial, and compaction to form each major coal seam. It has taken us only a few centuries to deplete much of the world's known coal deposits. Often you will hear about annual production rates for coal or some other fossil fuel. How much do we really produce in a year? None. We simply *extract* some portion. Fossil fuels are a nonrenewable source of energy.

CLUB MOSSES

About 1,100 far smaller lycophyte species made it to the present. Club mosses are the most familiar group. Their sporophytes have tiny leaves and a branching rhizome that gives rise to the vascularized roots and stems (Figure 15.8*a*).

On stem tips, tight clusters of modified leaves form spore-producing chambers. Such conelike reproductive structures made of modified leaves are found in many plant groups. Each is a **strobilus** (plural, strobili).

HORSETAILS

The tree-sized sphenophytes flourished in the ancient swamp forests. Twenty-five smaller species of a single genus (*Equisetum*) still survive today (Figure 15.8*b,c*). You may be familiar with them as "horsetails." They thrive along streams and roadsides, and in disturbed areas. Their sporophytes have rhizomes and hollow, jointed, silica-reinforced stems with a ringlike array of xylem and phloem. Some species produce spores at the tips of their photosynthetic stems. In others, spore-bearing stems with little chlorophyll appear early in the season, and photosynthetic ones come along later. Free-living gametophytes about the size of a pin head develop from the released spores.

Figure 15.8 (**a**) Sporophyte of a lycophyte (*Lycopodium*). (**b**) Mature vegetative stem of a horsetail (*Equisetum*). (**c**) Strobili at the tips of the nonphotosynthetic fertile stems produce and release spores. The young vegetative stem will later grow into the characteristic bushy form.

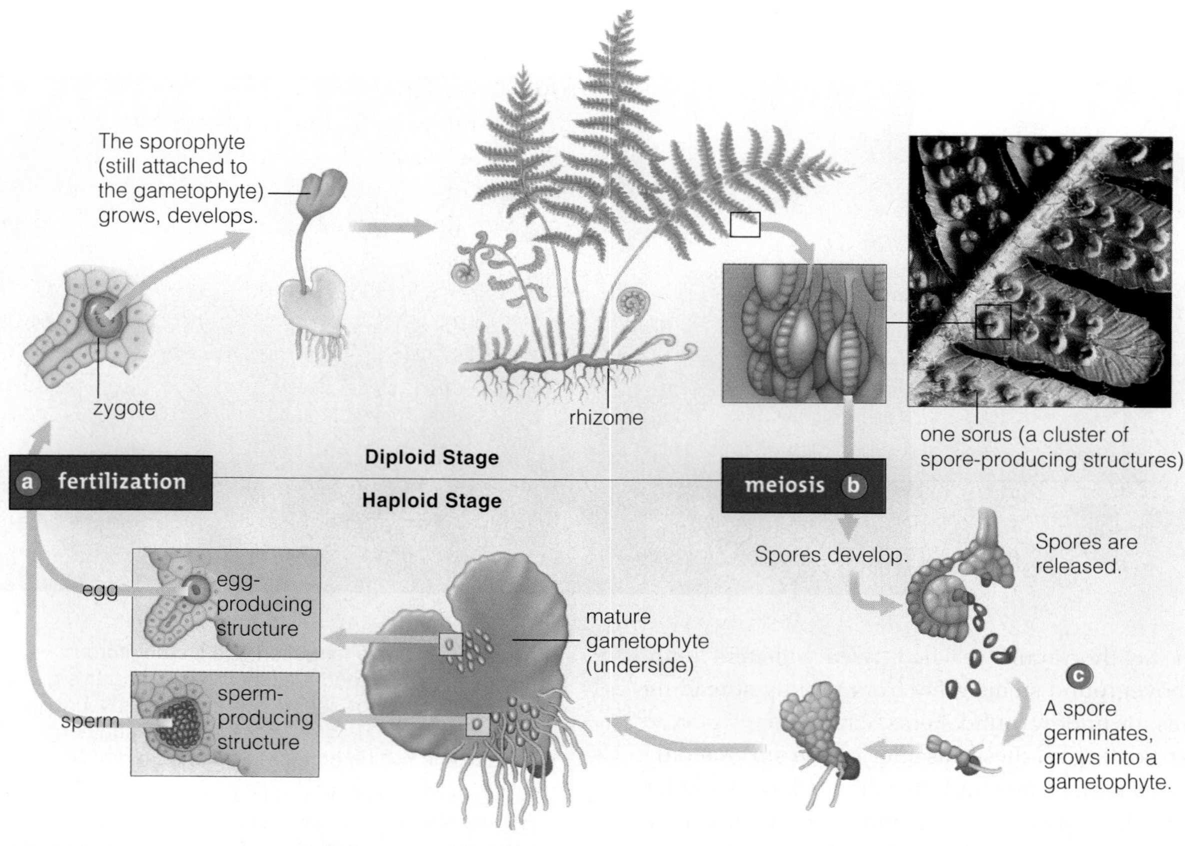

Figure 15.9 *Animated!* Life cycle of a chain fern (*Woodwardia*). (**a**) Fertilization occurs when sperm swim to eggs. It produces a diploid sporophyte, which grows into the familiar large, leafy photosynthetic form. (**b**) Meiosis in cells on the underside of fronds produces spores. (**c**) After their release, the spores germinate and grow into tiny heart-shaped gametophytes that produce eggs and sperm.

Figure 15.10 Forest of tree ferns (*Cyathea*) in Australia's Tarra-Bulga National Park.

Before the invention of modern abrasive cleansers, silica-rich stems of some *Equisetum* species were used to scrub pots and to polish metals. These species are commonly referred to as scouring rushes. Their high silica content may help them resist insect predators.

FERNS

With 12,000 or so species, the ferns are the largest and most diverse group of seedless vascular plants. All but about 380 species are native to the tropics, but we find them all over the world (Figures 15.9 and 15.10). There is an enormous variation in size. The fronds of certain floating ferns can be less than 1 centimeter wide. Some tree ferns grow 25 meters tall.

In most fern species, roots and leaves grow from vascularized rhizomes. Exceptions include the tropical tree ferns and species growing as epiphytes. *Epiphyte* refers to any aerial plant that grows attached to tree trunks or branches. The young fern leaves, or fronds, develop in a coiled pattern that looks somewhat like a fiddlehead, and are uncoiled by maturity. Fronds of many species, such as the chain ferns, are divided into leaflets (Figure 15.9).

Rust-colored patches appear on the lower surface of most fern fronds. Each patch, one tiny cluster of spore-forming chambers, is called a sorus (plural, sori). Most chamber walls are only one cell thick, and when they pop open, spores are catapulted through the air. After a spore germinates, it will develop into a heart-shaped gametophyte, only centimeters across.

Seedless vascular plants include lycophytes, horsetails, and ferns. Their sporophytes are adapted to conditions on land, yet they have not entirely escaped their aquatic ancestry. Films of water must be available at the time of sexual reproduction if their flagellated sperm are to reach eggs.

15.4 The Rise of Seed-Bearing Plants

Seed-bearing plants arose about 360 million years ago, as the Devonian gave way to the Carboniferous. In diversity, numbers, and distribution, they became the most successful groups of the plant kingdom.

LINKS TO SECTIONS 13.2, 13.3

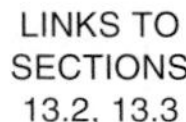

Seed-bearing plants made their entrance in the late Devonian. The first kinds were close relatives of ferns, but seeds, in some cases as big as walnuts, formed on their fernlike fronds. Those "seed ferns" flourished through the Carboniferous, and a few made it through a mass extinction that ended the Permian. Long before the last seed ferns went extinct, cycads, conifers, and other gymnosperms started adaptive radiations. They became the dominant plants of Mesozoic times. (What some folks call the Age of Dinosaurs, botanists call the Age of Cycads.) However, flowering plants originated during the late Jurassic or early Cretaceous. By 120 million years ago, the angiosperms were diversifying as gymnosperms were on the decline. They have now assumed supremacy in nearly all land habitats.

Think back on the factors that promoted the rise of land plants. Structural modifications, such as a water-conserving cuticle, were one reason they endured in seasonally dry, often cold habitats. Just as important were the extraordinarily adaptive ways in which two kinds of spores evolved in the seed-bearers.

Seed plants form two types of spores. The smaller **microspores** develop into pollen grains. Pollen grains are walled structures that enclose the immature male gametophytes. The male gametophyte is carried inside a pollen grain to female plant parts that contain the eggs. Pollen grains are tiny, and are easily spread by air currents or by animals. Regardless of how pollen grains are dispersed, **pollination** refers to the arrival of pollen on female reproductive parts of a seed plant. The evolution of pollen grains meant that sperm could travel to eggs without water. Keep in mind, most seed plants have sperm that cannot move by themselves.

The larger **megaspores** form on a sporophyte in the specialized structures called ovules. Each megaspore gives rise to a female gametophyte, which produces egg cells. Other gametophyte parts form a nutritious tissue and a seed coat. Sperm fertilize the egg and an embryo develops inside an ovule. *What we call a "seed" is a fully mature ovule.* The seed's outer coat protects an embryo sporophyte from harsh conditions. Inside each seed, there is a supply of nutrient-rich tissue that the embryo can draw upon to power its renewed growth once conditions favor germination.

A footnote to this story: Evolution of some seed plants has been shaped by interactions with humans. By half a million years ago, *Homo erectus* was stashing nuts and rose hips in caves, and roasting seeds. By 11,000 years ago, modern humans were domesticating seed plants as reliable sources of food. We now know 3,000 or so species as edible, and we plant staggering numbers of about 200 species to supply food (Figure 15.11). In this sense, modern agricultural practices are still contributing to the ongoing evolutionary success of at least some seed-bearing plants.

Seed-bearing plants are known for pollen grains that can be dispersed without water, ovules that mature into seeds, and tissue adaptations to dry conditions.

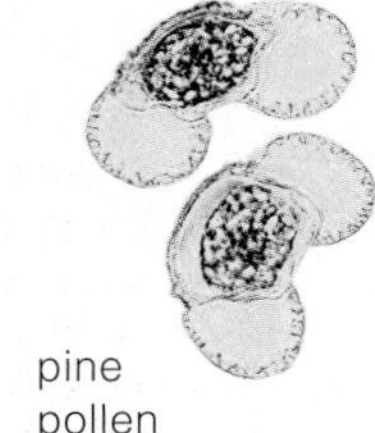

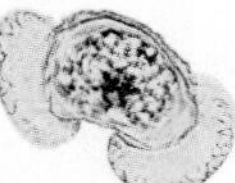

pine pollen grains

Figure 15.11 Edible treasures from flowering plants. (**a**) A sampling of fruits, which function in seed dispersal. (**b**) Mechanized harvesting of bread wheat, *Triticum*. (**c**) Indonesians picking shoots of tea plants. Only the terminal bud and two or three of the youngest leaves are picked for the finest teas. (**d**) From Hawaii, a field of sugarcane, *Saccharum officinarum*. Sap extracted from its stems is boiled to make sucrose crystals (table sugar).

15.5 Gymnosperms—Plants With "Naked" Seeds

The modern gymnosperms include diverse types from vines to shrubs to trees (Figure 15.12). Seeds form in ovules that are not enclosed within a flower or a fruit. Gymnos *means naked;* sperma *is taken to mean seed.*

The cycads and ginkgos were diverse in dinosaur times. Today, they are the only existing seed plants in which pollen releases ciliated sperm that can swim to eggs.

About 130 species of **cycads** made it to the present. Their pollen-bearing and seed-bearing cones form on separate plants (Figure 15.12*b*). Pollen is carried from male to female plants mainly by beetles. Most cycads live in tropical and subtropical regions, where people grind their seeds and trunks into a flour. Gardeners employ many cycads as ornamental plants, and some cycad species are disappearing from the wild because of over-collection of plants for this market.

Only a single species of ginkgo still survives, the maidenhair tree, *Ginkgo biloba* (Figure 15.12*c-f*). Like a few other gymnosperms, ginkgos are deciduous; they shed all their leaves in the fall. Several thousand years ago, ginkgos were widely planted around temples in China. Natural populations nearly became extinct; they may have been logged for firewood. Today, male trees are often planted along city streets. They have pretty fan-shaped leaves, are resistant to insects and disease, and tolerate air pollutants well. Few female trees are planted. Their fleshy seeds are the size of plums and give off a sharp, offensive smell as they decay.

Gnetophytes include tropical trees and leathery leafed vines as well as desert shrubs. They resemble flowering plants in some respects and are thought to be their closest living relatives. *Ephedra* (Figure 15.12*g*) forms photosynthetic stems. It is the source of the now-banned herbal supplement ephedrine. *Welwitschia* lives in hot African deserts. The sporophyte has a long taproot and a woody stem with two strap-shaped leaves that split lengthwise as a plant ages (Figure 15.12*h*).

Conifers are the most diverse gymnosperms. Most are woody trees or shrubs with needlelike or scalelike leaves. Most conifers shed some leaves all year long yet still stay leafy, or evergreen. A few are deciduous. Only the conifers have true **cones**. These reproductive structures—clusters of papery or woody scales—bear exposed ovules on their upper surface. The conifers include the tallest trees in the Northern Hemisphere (redwoods), as well as the most abundant (pines), and the oldest (bristlecone pines) (Figure 15.12*a*).

Figure 15.13 shows the life cycle of a conifer. The familiar tree is the sporophyte. Gametophytes form in cones (Figure 15.13*a,b*). In female cones, megaspores develop into female gametophytes that produce eggs. In male cones, microspores become pollen grains. Each spring, pollen grains are dispersed from the tree by the

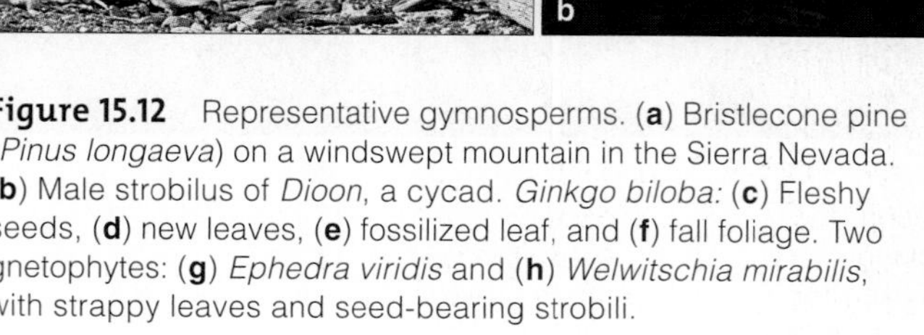

Figure 15.12 Representative gymnosperms. (**a**) Bristlecone pine (*Pinus longaeva*) on a windswept mountain in the Sierra Nevada. (**b**) Male strobilus of *Dioon*, a cycad. *Ginkgo biloba:* (**c**) Fleshy seeds, (**d**) new leaves, (**e**) fossilized leaf, and (**f**) fall foliage. Two gnetophytes: (**g**) *Ephedra viridis* and (**h**) *Welwitschia mirabilis*, with strappy leaves and seed-bearing strobili.

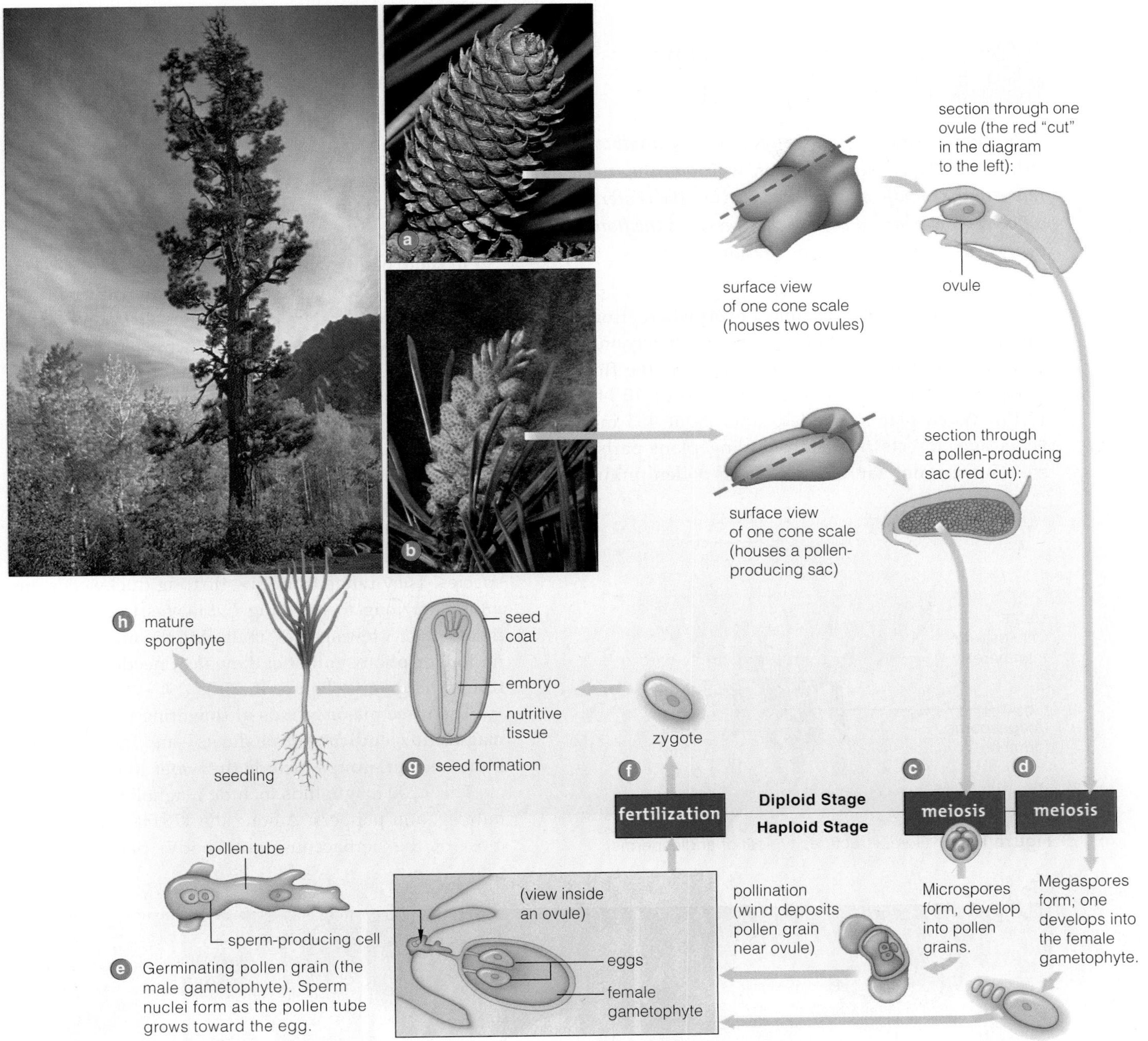

Figure 15.13 ***Animated!*** Life cycle of a conifer, the ponderosa pine.

wind. Pollination occurs when a pollen grain lands on an exposed ovule of a female cone scale. The pollen grain germinates and a tubular structure grows from it—the sperm-bearing male gametophyte. After a year, growth of the tube delivers nonmotile sperm to an egg, and fertilization takes place. The resulting zygote develops into an embryo sporophyte, which—along with female gametophyte tissues—becomes a seed.

After fertilization occurs, it takes yet another year for the seed to mature and be released from the parent pine. After seed release, the embryo sporophyte inside is nourished by tissues derived from the gametophyte. Other gametophyte tissues form a hardened coat that encloses and protects the seed.

Like their Permian ancestors, gymnosperms are adapted to seasonally dry climates. They bear seeds on the exposed surfaces of cones or other spore-producing structures.

LINK TO SECTION 13.9

15.6 Angiosperms—The Flowering Plants

Only angiosperms produce the specialized reproductive structures called flowers. Angeion, *which means vessel, refers to the female reproductive parts at the center of a flower. The enlarged base of the "vessel" is the floral ovary, where ovules and seeds develop.*

We find flowering plants almost everywhere, from icy tundra to deserts to oceanic islands. What accounts for their distribution and diversity? Consider the **flower**, a specialized reproductive shoot (Figures 15.14 and 15.15). When plants invaded land about 435 million years ago, insects that ate decaying plant parts and spores were not far behind. After pollen-producing plants evolved, insects made the connection between "plant parts with pollen" and "food." Plants lost some pollen to the insects but gained a reproductive edge: The insects delivered pollen to the plants' ovules. So plants evolved in ways that made them more enticing to insects. Insects became specialized in locating and gathering the pollen from particular plants, instead of searching at random for sources of food.

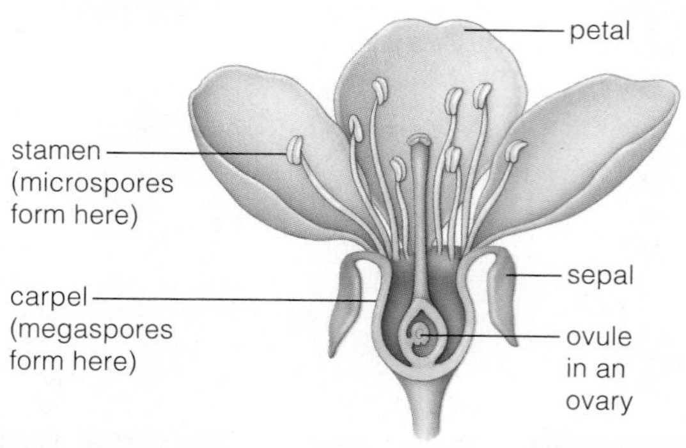

Figure 15.14 Floral structures typical of angiosperms.

In time, plants with flowers, fragrances, and nectar coevolved with diverse pollinators. A **pollinator** is any agent that transfers pollen to female floral parts of the same species. In addition to insects, pollinators include bats, birds, and other animals. **Coevolution** is a term for two or more species evolving jointly as a result of some close ecological interaction. Because of this interaction, a heritable change in one species can select for changes in the other, which evolves also.

Recruiting pollinators as assistants in reproduction contributed to the dominance of flowering plants for the past 100 million years. We know of at least 260,000 species. They range from tiny floating duckweeds one millimeter long to towering *Eucalyptus* trees over 100 meters tall. A few species, including the mistletoes, are no longer photosynthetic; they steal needed nutrients from other plants (Figure 15.15*d*).

The three major groups of flowering plants are the **magnoliids**, **eudicots**—true dicots—and the **monocots**. More ancient groups include the water lilies.

The 9,200 magnoliids include magnolias, avocados, nutmeg, and peppers. Among the 170,000 eudicots are most of the herbaceous (nonwoody) plants such as

Figure 15.15 (**a**) Examples of flowers. Their colors, patterns, and shapes attract pollinators. (**b**) Sacred lotus (*Nelumbo nucifera*), an aquatic species. The flower's radial pattern is typical of ancient lineages. (**c**) More recently evolved lineages, including the pansies (*Viola*), display a bilateral pattern, with symmetrical left and right parts. (**d**) Dwarf mistletoe (*Arceuthobium*). It draws nutrients from trees and stunts their growth.

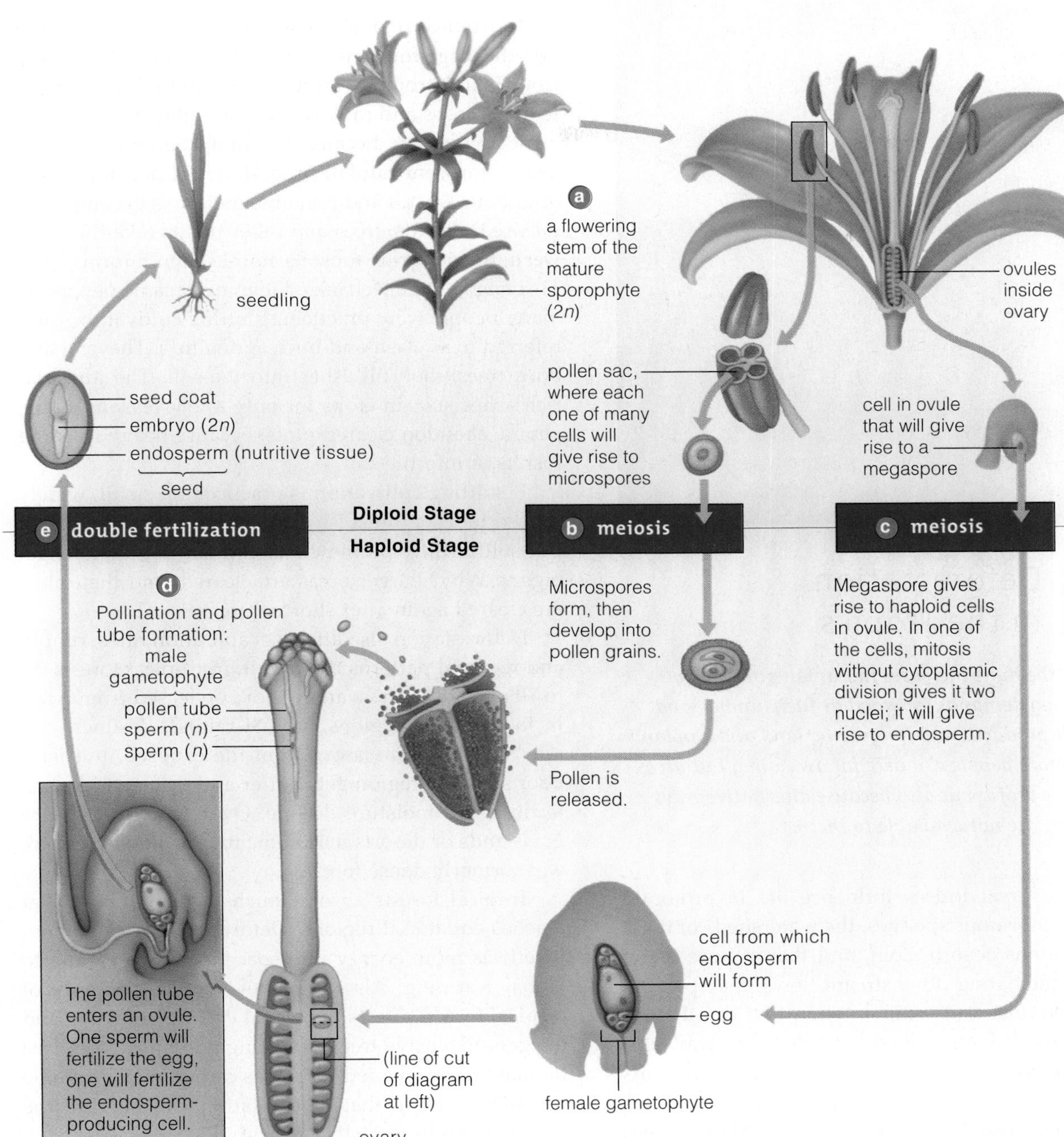

Figure 15.16 ***Animated!*** Example of a flowering plant life cycle. This lily (*Lilium*) is one of the monocots.

(**a**) The sporophyte is the dominant part of the life cycle. Gametophytes form inside flowers. Pollen grains develop from microspores in pollen sacs (**b**) and contain male gametophytes. Female gametophytes develop from megaspores inside an ovary (**c**). After pollination, a tube grows from the pollen grain into the ovule and delivers two sperm (**d**). (**e**) "Double" fertilization occurs in all flowering plant life cycles. One sperm fertilizes the egg. The other fertilizes a cell that gives rise to endosperm, a tissue that will nourish the embryo. Sections 19.1 and 19.2 detail the flowering plant life cycles, using a dicot as the example.

tomatoes, cabbages, roses, and daisies, most flowering shrubs and trees, and cacti. The 80,000 or so monocots include orchids, palms, lilies, and grasses, such as rye, wheat, corn, rice, sugarcane, and other valued plants.

The next unit takes a close look at the structure and function of flowering plants. As in the other vascular plants, the sporophyte dominates the life cycle. The flowering plants are unique in that they nourish new generations with a nutritious tissue called endosperm. It surrounds an embryo sporophyte in a seed. As the seed develops, ovary tissues mature into fruit, which functions in seed dispersal. Figure 15.16 illustrates a monocot life cycle. We will show a eudicot life cycle in Figure 19.4.

Angiosperms are the most successful plants in terms of diversity, numbers, and distribution. They alone produce flowers. Most species coevolved with animal pollinators.

Monocots and eudicots are the two most diverse lineages.

Figure 15.17 Tropical rain forest in Southeast Asia.

15.7 Deforestation in the Tropics

Paralleling the huge increases in human population are increasing demands for wood as fuel, lumber, and other forest products, and for grazing land and cropland. More and more people compete for dwindling resources for economic profit, but also because alternative ways of life simply are not available to them.

The world's great forests influence life in profound ways. Like enormous sponges, the watersheds of these forested regions absorb, hold, and then release water slowly. By mediating downstream flows, they prevent erosion, flooding, and sedimentation that can disrupt rivers, lakes, and reservoirs. Strip forests away, and the exposed soil is vulnerable to leaching of nutrients and to erosion, especially on steep slopes.

Again, **deforestation** means the removal of all trees from large tracts for logging, agriculture, and grazing. This chapter started with a look at what is going on in once-sweeping temperate forests of the United States, Canada, Europe, Siberia, and elsewhere. But tropical forests are in even more peril (Figure 15.15). For 10,000 years or more, these forests have been habitat for 50 to 90 percent of all land-dwelling species. In less than four decades, humans have destroyed more than half of them. Logging over millions of acres per year is like leveling about thirty-four city blocks every minute.

Deforestation is now greatest in Brazil, Indonesia, Colombia, and Mexico. If the clearing and destruction continue at present rates, then by 2010, only Brazil and Zaire will still have large tropical forests. By 2035, most of their forests also will be gone.

In tropical regions, clearing forests for agriculture sets the stage for long-term losses in productivity. The irony is that tropical forests are one of the worst places to grow crops and raise animals. In intact forests, litter can't accumulate because the high temperatures and heavy, frequent rainfall promote rapid decomposition of organic wastes and remains. As fast as decomposers release nutrients, trees and other plants take them up. Deep, nutrient-rich topsoils simply cannot form.

Long before mechanized logging became pervasive, many people were practicing shifting cultivation, once referred to as slash-and-burn agriculture. They cut and burn trees, then till ashes into the soil. The nutrient-rich ashes sustain crops for only a few seasons. Later, people abandon cleared plots, because heavy leaching results in infertile soil.

If shifting cultivation is practiced on small, widely spaced plots, a forest isn't necessarily damaged much. But soil fertility plummets as human populations get bigger. Why? Larger areas are cleared, and then plots are cleared again after shorter intervals.

Deforestation also affects evaporation rates, runoff, and regional patterns of rainfall. In tropical forests, 50 to 80 percent of the water vapor is released from trees. In logged-over regions, annual rainfall declines, and rain swiftly drains away from the exposed, nutrient-poor soil. The region gets hotter and drier, and its soil fertility and moisture decline. Over time, sparse, dry grasslands or deserts may come to prevail where there was formerly dense forest.

Tropical forests absorb much of the sunlight that reaches equatorial regions. Deforested land is shinier; it reflects more energy into space, possibly adding to global warming. Also, the photosynthetic activity of tropical forests affects the global cycling of carbon and oxygen. Extensive tree harvesting and burning of tree biomass release stored carbon as carbon dioxide. Some researchers think that deforestation will amplify the greenhouse effect for that reason.

Conservation biologists are attempting to reverse the trend. As three examples, a coalition of 500 groups is dedicated to preserving Brazil's remaining tropical forests. In India, women have already constructed and installed 300,000 inexpensive, smokeless wood stoves. In the past decade, the stoves saved more than 182,000 metric tons of trees by cutting demands for fuelwood. In Kenya, women have planted millions of trees as a source of wood and to counter erosion.

Once-vast forests helped sustain rapid increases in human population sizes. But lineages that began hundreds of millions of years ago are ending because of deforestation.

Summary

Section 15.1 Plants evolved from green algae about 500 million years ago. Nearly all existing plants are multicelled photoautotrophs. Table 15.1 summarizes the major groups.

Plants arose near the water's edge, then radiated into drier habitats. Key innovations included a waterproof cuticle with stomata, and internal pipelines of vascular tissue (xylem and phloem).

Complex diploid sporophytes evolved; they hold, nourish, and protect spores and gametophytes. Male gametes that can be dispersed without water (pollen grains) and seeds were key innovations that allowed the radiation of gymnosperms and flowering plants. Table 15.2 summarizes these trends.

BiologyNow
Compare the relative contributions of haploid and diploid stages to different plant life cycles with the interaction on BiologyNow.

Section 15.2 Mosses, liverworts, and hornworts are bryophytes. These nonvascular plants lack complex vascular tissues. Their flagellated sperm can reach the eggs only by swimming through films or droplets of water. The sporophyte begins development inside the gametophyte tissues. It remains attached to and often dependent upon the gametophyte even when mature.

BiologyNow
Observe the life cycle of a typical moss with the animation on BiologyNow.

Section 15.3 Lycophytes, horsetails, and ferns are among the seedless vascular plants. Their sporophytes have vascular tissues and are the larger, longer lived phase of the life cycle. Like bryophytes, they produce flagellated sperm and require free water for fertilization.

BiologyNow
Observe the life cycle of a typical fern with the animation on BiologyNow.

Section 15.4 Gymnosperms and flowering plants (angiosperms) are seed-bearing vascular plants. They produce two types of spores. Microspores give rise to pollen grains, the sperm-bearing male gametophytes. Even in the absence of water, pollen grains can deliver non-flagellated sperm to egg cells. Megaspores form in ovules, and consist of a female gametophyte (with egg) and tissues that will become the seed coat.

Section 15.5 The conifers, cycads, ginkgos, and gnetophytes are gymnosperms. They are adapted to dry climates and bear seeds on exposed surfaces of spore-bearing structures. In conifers, these spore-bearing structures are distinctive cones.

BiologyNow
Learn about the stages in the life cycle of a typical gymnosperm with the animation on BiologyNow.

Table 15.1 Comparison of Major Plant Groups

Bryophytes	24,000 species. Moist, humid habitats.
	Nonvascular land plants. Fertilization requires free water. Haploid dominance. Cuticle and stomata present in some.
Club mosses	1,100 species with simple leaves. Mostly wet or shady habitats.
Horsetails	25 species of single genus. Swamps and disturbed habitats.
Ferns	12,000 species. Wet, humid habitats in mostly tropical and temperate regions.
	Seedless vascular plants. Fertilization requires free water. Diploid dominance. Cuticle and stomata present.
Conifers	550 species, mostly evergreen, woody trees and shrubs with pollen- and seed-bearing cones. Widespread distribution.
Cycads	130 slow-growing tropical, subtropical species.
Ginkgo	1 species, a tree with fleshy-coated seeds.
Gnetophytes	70 species. Limited to some deserts, tropics.
	Gymnosperms—vascular plants with "naked seeds." Free water not required for fertilization. Diploid dominance. Cuticle and stomata present.
Flowering plants	
Monocots	80,000 species. Floral parts often arranged in threes or in multiples of three; one seed leaf; parallel leaf veins common; pollen with one furrow.
Eudicots (true dicots)	At least 170,000 species. Floral parts often arranged in fours, fives, or multiples of these; two seed leaves; net-veined leaves common; pollen with three or more pores and/or furrows.
Magnoliids and basal groups	9,200 species. Many spirally arranged floral parts or fewer, in threes; two seed leaves, net-veined leaves common; pollen with one furrow.
	Angiosperms—vascular plants with flowers, protected seeds. Free water not required for fertilization. Diploid dominance. Cuticle and stomata present.

Table 15.2 Evolutionary Trends Among Plants

Bryophytes	Ferns	Gymnosperms	Angiosperms
Nonvascular	Vascular		
Haploid dominance	Diploid dominance		
Spores of one type		Spores of two types	
Seedless		Seeds	
Flagellated sperm			Nonmotile sperm*

* Require pollination by wind, insects, animals, etc.

Section 15.6 Angiosperms are the dominant land plants. They alone produce flowers inside of ovaries. Most kinds coevolved with pollinators. Endosperm, a nutritious tissue, surrounds the embryo sporophyte inside seeds, which have a hard coat. Fruit encases most seeds and aids in dispersal.

BiologyNow
Explore the angiosperm life cycle with the animation on BiologyNow.

Section 15.7 Deforestation threatens ecosystems worldwide. Forests play a role in the global cycling of water, carbon dioxide, and oxygen. Their destruction may be amplifying the greenhouse effect.

Self-Quiz

Answers in Appendix I

1. Which of the following statements is *not* correct?
   a. Gymnosperms are the simplest vascular plants.
   b. Bryophytes are nonvascular plants.
   c. Lycophytes and angiosperms are vascular plants.
   d. Only angiosperms produce flowers.

2. Which does *not* apply to gymnosperms or angiosperms?
   a. vascular tissues
   b. diploid dominance
   c. single spore type
   d. all of the above

3. Bryophytes have independent ______ and attached, dependent ______ .
   a. sporophytes; gametophytes
   b. gametophytes; sporophytes
   c. rhizoids; zygotes
   d. rhizoids; stalked sporangia

4. Lycophytes, horsetails, and ferns are classified as ______ plants.
   a. multicelled aquatic
   b. nonvascular seed
   c. seedless vascular
   d. seed-bearing vascular

5. The ______ produce flagellated sperm.
   a. ferns
   b. conifers
   c. monocots
   d. a and c

6. The ______ produced in the male cones of a conifer develop into pollen grains.
   a. sori
   b. microspores
   c. megaspores
   d. a and c

7. A seed is ______ .
   a. a female gametophyte
   b. a mature ovule
   c. a mature pollen tube
   d. an immature spore

8. The majority of ______ rely on animal pollinators.
   a. lycophytes
   b. gymnosperms
   c. angiosperms
   d. bryophytes

9. Match the terms appropriately.

____ gymnosperm	a. gamete-producing body
____ sporophyte	b. help control water loss
____ lycophyte	c. "naked" seeds
____ bryophyte	d. spore-producing body
____ gametophyte	e. nonvascular land plant
____ stomata	f. seedless vascular plant
____ angiosperm	g. flowering plant

Additional questions are available on **BiologyNow™**

Figure 15.18 Kenyan Wangari Maathai, who in 2004 won a Nobel Peace Prize for her work in support of women's rights and sustainable development.

Critical Thinking

1. The 2004 Nobel Peace Prize was awarded to Wangari Maathai, who founded the Green Belt Movement in 1977 (Figure 15.18). Since then, group members—mostly poor rural women —have planted more than 25 million trees. Reforestation efforts help halt soil erosion, provide sources of fruits, food, and firewood, and empower women to work together for the long-term good of their communities. Find out if any nonprofit groups are working to protect forests in your area. Do you support their goals and methods?

2. Early botanists admired ferns but found their life cycle perplexing. In the 1700s, they learned to propagate them by sowing what appeared to be tiny dustlike "seeds"from the undersides of fronds. Despite many attempts, the scientists could not find the pollen source, which they assumed must stimulate the "seeds" to develop. Imagine you could write to one of these botanists. Compose a note that would clear up their confusion.

3. Today, the tallest bryophytes reach a maximum height of 20 centimeters or so. So far as we know from fossils, there were no giants among their ancestors. Lignin and vascular tissue first evolved in the lycophytes, and some extinct species stood 40 meters high. Among seed plants, *Sequoia* (a gymnosperm) and *Eucalyptus* (an angiosperm) can be more than 100 meters high. Explain why evolution of vascular tissues and lignin would have allowed such a dramatic increase in maximum plant height. How might an increase in maximum height give one plant species a competitive advantage over another?

4. In cycads and ginkgos, after a pollen grain lands on a receptive female cone, a pollen tube grows toward the ovary. Once the tube gets to the ovary, it releases flagellated sperm that swim a short distance through a fluid-filled egg chamber to reach the egg. In all other gymnosperms and all angiosperms, the sperm do not have flagella. The pollen tube delivers them directly to the egg. Speculate about how the transition from flagellated to nonflagellated sperm came about. How might it be selectively advantageous?

16 ANIMAL EVOLUTION

IMPACTS, ISSUES

Interpreting and Misinterpreting the Past

In Darwin's time, the presumed absence of transitional forms was an obstacle to acceptance of his new theory of evolution. Skeptics wondered, if new species evolve from older ones, then where are the transitional forms in the fossil record, the "missing links"? Where are the fossils that bridge major groups? Ironically, one such form was unearthed by workers at a limestone quarry in Germany just one year after Darwin's *On the Origin of Species* was published.

The fossil, about the size of a large crow, looked like a small meat-eating dinosaur. It had a long, bony tail, three long, clawed fingers on each forelimb, and a heavy jaw fringed with short, spiky teeth. Later, diggers found another fossil of the same type. Later still, someone noticed the feathers. If these were dinosaur fossils, what were they doing with feathers? The feathers looked like those of modern birds. The specimen type for this new fossil was named *Archaeopteryx* (ancient winged one).

Between 1860 and 1988, six *Archaeopteryx* specimens and a fossilized feather were found. Anti-evolutionists tried to dismiss them as forgeries. Someone, they said, had pressed modern bird bones and feathers against wet plaster. The dried, imprinted plaster casts merely looked like fossils. But microscopic examination confirmed the fossils are real. Further confirmation of their age came from the remains of obviously ancient jellyfishes, worms, and other organisms in the same limestone layers.

Today, radiometric dating shows *Archaeopteryx* lived 150 million years ago. Why are its remains preserved so well after so much time? *Archaeopteryx* lived in forests near a shallow lagoon. The lagoon was large, warm, and stagnant; coral reefs barred inputs of oxygenated water from the sea. So it was not a particularly inviting place for scavenging animals that might have feasted on—and obliterated the remains of—*Archaeopteryx* and other organisms that fell from the sky or drifted offshore.

Each storm surge drove fine sediments over the reefs. They gently buried the carcasses littered at the bottom of the lagoon. In time, the soft, muddy sediments were slowly compacted and hardened. They became a lasting limestone tomb that preserved more than 600 species, including *Archaeopteryx*.

No one was around to witness such transitions in the history of life. But fossils are real, just as the morphology, biochemistry, and molecular makeup of living organisms are real. Radiometric dating assigns fossils to their place in time. *Life's evolution is not "just a theory."* Remember, a scientific theory differs from speculation because its predictive power has been tested in nature many times, in many different ways that might disprove it. If a theory stands, there is a high probability it is not wrong.

In this chapter, you will learn that Darwin's theory of evolution by natural selection goes a long way toward explaining the patterns of animal evolution, including the evolution of our own lineage.

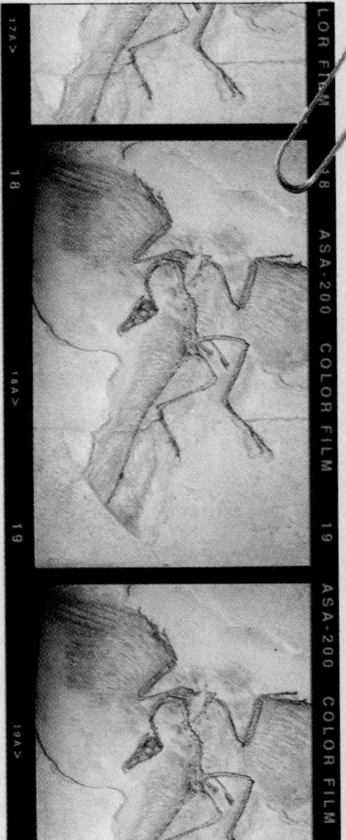

How Would You Vote? *Private fossil hunters have discovered important fossils, but some poach fossils from public lands for sale. Should sale of vertebrate fossils be regulated? See BiologyNow for details, then vote online.*

Key Concepts

ANIMAL CHARACTERISTICS
Animals are multicelled heterotrophs that move about during some part of the life cycle. Most have tissues and organs and undergo development.

INVERTEBRATES
The first animals were invertebrates and, in terms of both species and number of individuals, most still are. Animals arose in the seas, but many groups now have representatives on land.

VERTEBRATES
A backbone, jaws, paired fins, and lungs were key innovations in the evolution of fishes, amphibians, reptiles, birds, and mammals.

HUMAN EVOLUTION
Human traits emerged through modification of traits that evolved in earlier mammalian, primate, and then early hominid lineages.

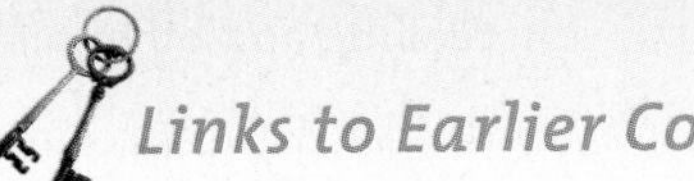

Links to Earlier Concepts

In this chapter you will investigate the diversity of animals, a group you met in Section 1.3. Much information about animal evolution comes from fossils (13.1, 13.2). You will learn how asteroid impacts (Chapter 13 Introduction) and changes in the distribution of continents (13.3) affected animal diversity, and how we use the evidence derived from comparative methods (13.4–13.6).

You may wish to review Sections 13.9 and 13.10. They introduced the patterns of speciation and extinction, and the main methods of organizing information about biological diversity.

LINKS TO SECTIONS 1.3, 13.4, 13.5, 13.10

16.1 Overview of the Animal Kingdom

We already know of more than 2 million animal species, and no simple definition can cover that much diversity.

We define **animals** by a list of their characteristics:

1. Animals are multicelled, and most (not all) species have tissues, organs, and organ systems.

2. All animals are aerobic and heterotrophic. They eat other organisms or withdraw nutrients from them.

3. Animals are sexual reproducers. Many species also reproduce asexually. Life cycles usually include stages of development. In embryos, cell divisions give rise to primary tissues—**ectoderm**, **endoderm**, and in most cases **mesoderm**—from which adult tissues develop.

4. Animals are motile; they actively move about during some or all of the life cycle.

All animal lineages are a monophyletic group; they all share a common ancestor, as shown in Figure 16.1. The ones traditionally known as mammals, birds, reptiles, amphibians, and fishes are **vertebrates**—animals that have a backbone. Yet all but 50,000 of the 2 million or so named species are backboneless **invertebrates**.

ANIMAL ORIGINS

How did animals arise? Most likely, they evolved from colonies of flagellated cells, much like the ones formed today by choanoflagellates. These single-celled protists resemble the body cells of sponges, the simplest living animals, and DNA sequence comparisons point to an evolutionary connection. In some ancient colonies, cells could have mutated in ways that led to specializations in feeding and other tasks that now define "animals."

However the animals arose, tiny, soft-bodied species shaped like disks, fronds, and blobs were living on or in seafloor sediments by 610 million years ago. We call them the **Ediacarans**. Some resembled existing animal groups and are probably their earliest representatives.

During the Cambrian (543 to 490 million years ago), animals underwent an enormous adaptive radiation. Before it was over, all major groups were represented by diverse forms in the seas. What caused this great explosion of diversity? Perhaps changes in the climate, sea level, and distribution of land masses triggered it. Or maybe mutations in genes that act in development kicked it off. Possibly predators and prey exerted new selective pressures on one another. Or, these and other factors may have collectively promoted speciations.

A LOOK AT BODY PLANS

How might we get a conceptual handle on animals as different as flatworms and dinosaurs, hummingbirds and humans? Comparing similarities and differences with respect to some basic features is one way to start.

BODY SYMMETRY AND CEPHALIZATION Animals can be described in part by their symmetry (Figure 16.2). With **radial symmetry**, body parts are arrayed around a central point, like the spokes of a wheel. Usually the mouth is at the center. Radial animals live in the water and respond to stimuli that arrive from all directions. In contrast, animals with **bilateral symmetry** have two halves that mirror each other. Their main body axis runs from *anterior* (front) to *posterior*. This axis divides a body into right and left halves, with a *ventral* surface (underside) and *dorsal* surface (the back). Most of the bilateral groups show **cephalization**—a concentration of nerve and sensory cells in the anterior end, the end that is the first to encounter stimuli.

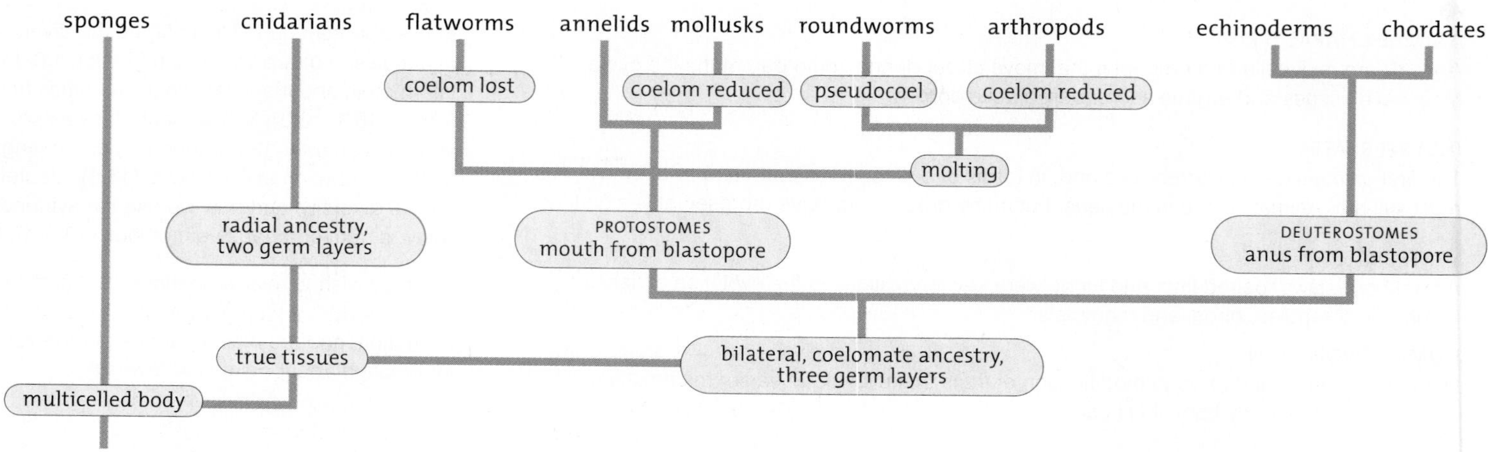

Figure 16.1 An evolutionary tree for animal groups covered in this text.

TYPE OF GUT The animal **gut** digests food and then absorbs the released nutrients. An *incomplete* digestive system, in which the gut is like a sac, evolved in early animals. Cnidarians, such as sea anemones and jellyfish, still have this type of gut. So do the flatworms.

A *complete* digestive system has a tubelike gut that starts with a mouth and ends at an anus. How this gut develops is one way that we define two major lineages of bilateral animals. In both lineages, the first opening that appears on the embryo's surface develops into one of the gut's openings. In the **protostomes**, which include flatworms, mollusks, annelids, roundworms, and arthropods, the initial embryonic opening (called the blastopore) becomes the mouth. In echinoderms, chordates, and other **deuterostomes**, this first opening becomes the anus.

A tubular gut has distinct advantages. The mouth can be modified with teeth or other structures instead of being a single, generic opening for taking in food and expelling residues. Also, regions of the tube can become specialized for food breakdown, absorption, storage, and elimination of undigestible wastes.

BODY CAVITIES A body cavity called a **coelom** lies between the body wall and the gut of most bilateral animals (Figure 16.3). Peritoneum, a tissue that arises from mesoderm, lines the coelom. This lining helps to hold the soft internal organs in place.

The coelom was a key innovation in the evolution of larger, active animals from smaller ancestral forms. Why? It cushions and protects internal organs, yet lets them move independently of the body wall. Some of the less complex invertebrates do not have any body cavity; tissues fill the region between the gut and the body wall. Others have a pseudocoel, a "false coelom." This is a body cavity that is not fully lined.

SEGMENTATION Many of the bilateral groups show some degree of **segmentation**, the division of a body into interconnecting units that are repeated one after the next along the main body axis. We clearly see body segments in annelids, such as earthworms. We also find clues to our own segmented origins in the segmented muscles that appear in human embryos.

> *Animals are multicelled, aerobically respiring heterotrophs. Nearly all animals have tissues, organs, and organ systems.*
>
> *Animals reproduce sexually and, in many cases, asexually. They go through a period of embryonic development, and most are motile during at least part of the life cycle.*
>
> *Animal bodies differ in their symmetry, cephalization, type of gut, type of body cavity, and degree of segmentation.*

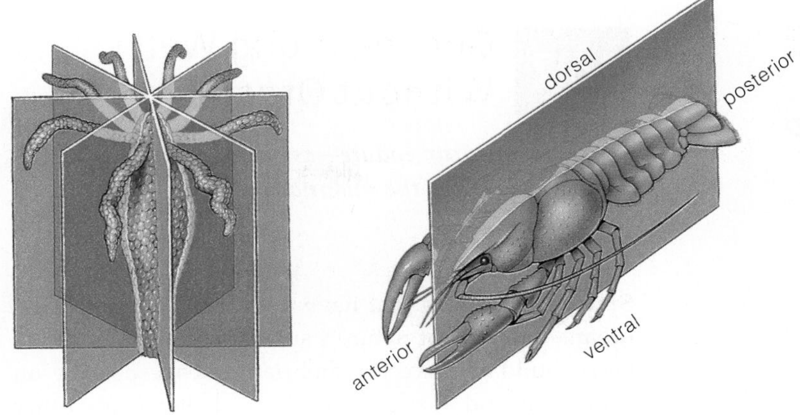

Figure 16.2 *Left*, radial body symmetry of *Hydra*, a cnidarian. *Right*, bilateral symmetry of a lobster, an arthropod. Like other bilateral animals, a lobster shows cephalization. Nervous and sensory organs—a brain, eyes, and antennae—are concentrated at the anterior or head end.

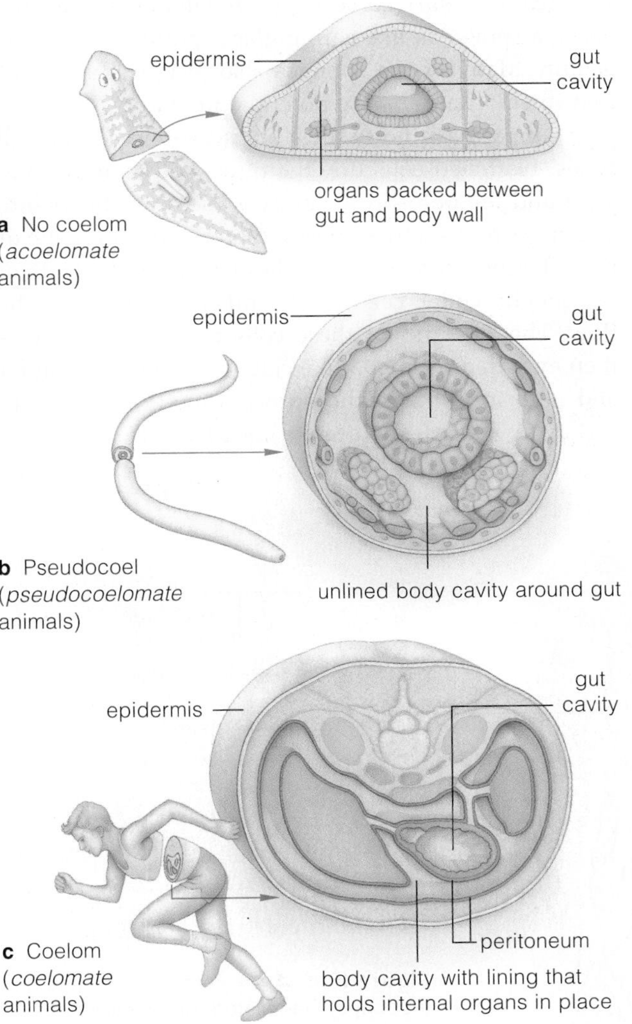

Figure 16.3 Types of body cavity. (**a**) A flatworm is acoelomate; its gut is surrounded by tissues. (**b**) Pseudocoelomates, such as roundworms, have an unlined cavity around the gut. (**c**) Like all vertebrates, humans have a lined body cavity, or coelom.

LINK TO SECTION 13.5

16.2 Getting Along Well Without Organs

Simple animals can endure—as demonstrated by the sponges and the cnidarians.

SPONGES

Sponges—animals that have no symmetry, tissues, or organs—are one of nature's success stories. They have been abundant since Precambrian times, especially on coral reefs and along coasts. They shelter many other animals, including worms and shrimps. Some are large enough to sit in; others are tiny. They can be flattened, lobed, tubelike, or vaselike (Figure 16.4).

Tough protein fibers and sharp, glassy spicules of calcium carbonate or silica often stiffen a sponge body. These protective components may be one reason why sponges have surived so long. Predators discover that biting a sponge is like eating glass splinters in fibrous gelatin. Many of the sponges also produce chemicals that smell or taste bad to potential predators.

Sponges have no tissues. There are three main cell types. Flattened cells line the outer body surface and surround the pores in the body wall. Water flows into a sponge through these pores, then out through one or more larger openings. Flagella of many thousands of phagocytic collar cells of the inner lining drive the flow. The "collars" of these cells trap food, which is then engulfed. Amoeba-like cells prowl between inner and outer cell layers; they process food, distribute it, cart away wastes, and secrete spicules.

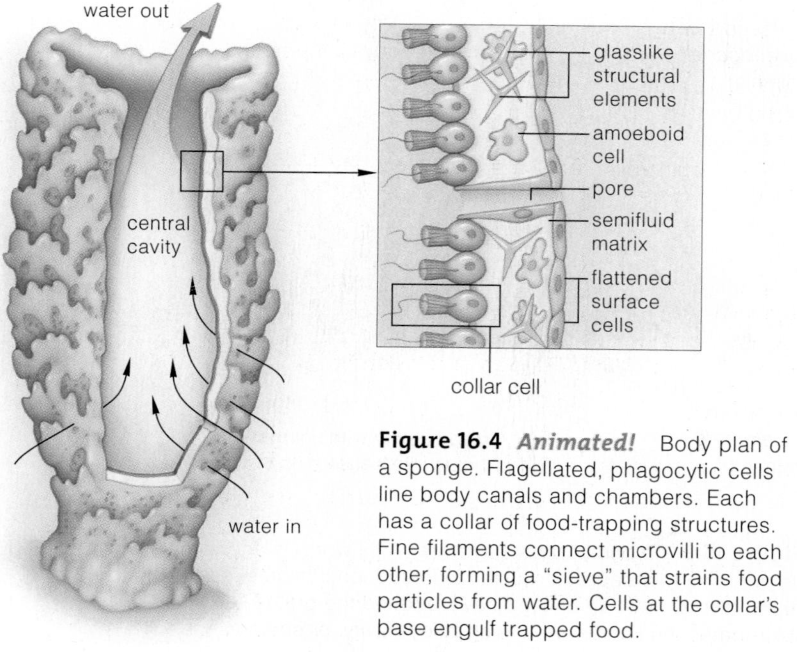

Figure 16.4 ***Animated!*** Body plan of a sponge. Flagellated, phagocytic cells line body canals and chambers. Each has a collar of food-trapping structures. Fine filaments connect microvilli to each other, forming a "sieve" that strains food particles from water. Cells at the collar's base engulf trapped food.

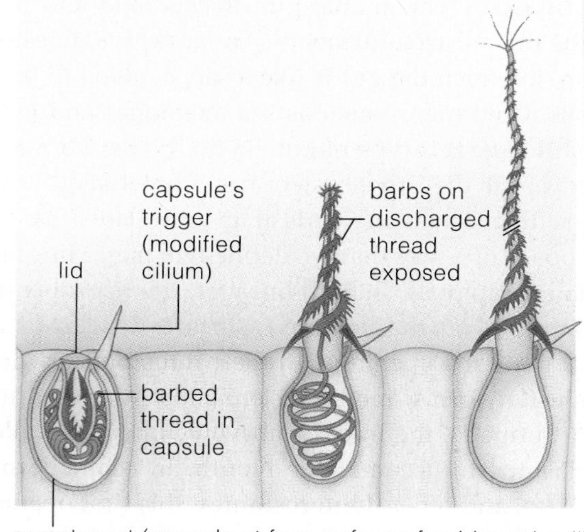

Figure 16.5 ***Animated!*** Action of a type of nematocyst. When prey brushes against the trigger, the capsule gets "leakier." As water diffuses in, internal pressure rises, and the thread is forced outward. Its barbed tip pierces the prey.

Most sponges reproduce sexually. They release their sperm into the water but retain fertilized eggs, which develop into a microscopic, swimming larva. A **larva** (plural, larvae) is a sexually immature stage that has a different body form than the adult. Many sponges also reproduce asexually by budding. The new individual develops on the parent body, then drops or is broken off and lives on its own.

CNIDARIANS

Cnidarians are radial, tentacled animals that include jellyfishes, sea anemones, corals, and hydras. Nearly all are marine. All have cells organized as tissues but no organs. Tentacles have distinctive capsules called nematocysts. If something brushes against a tentacle, the capsules snap open and substances pierce, poison, or entangle prey (Figure 16.5).

Cnidarian bodies take two main forms (Figure 16.6). The sea anemones, corals, and hydras are tube-shaped polyps. Usually one end of the polyp attaches to some surface. Tentacles surround a mouth at the other end. A jellyfish is a medusa, which appears like a saucer or bell. Medusae swim or drift through the water.

Like all animals except the sponges, cnidarians have layers of **epithelium**, a tissue that lines internal and external body surfaces. In cnidarians, epithelial tissue called *epidermis* covers the outer surface. *Gastrodermis* lines the saclike gut. Running through the epidermal layers are nerve cells that function as a *nerve net*. This simple nervous system interacts with contractile cells to alter body shape and move the body or its parts.

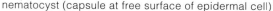

The "jelly" inside a jellyfish is mesoglea, a secreted gelatinous material found between the epidermis and the gastrodermis. Mesoglea makes a jellyfish buoyant and contractile cells exert force against it. As a jellyfish bell contracts, water streams out from beneath the bell, and the jellyfish is propelled forward. Any fluid-filled cavity or cell mass that contractile cells can act against is known as a *hydrostatic* skeleton.

Cnidarians reproduce both asexually and sexually. Sexual reproduction yields a planula, a bilateral ciliated larva. The larva moves about, and then settles down and develops into a polyp. In jellyfish, medusae develop on and are released by specialized polyps. The life cycle of some other cnidarians does not include a medusa stage.

Many cnidarians live in colonies. The reef-building coral polyps are cemented side by side in a structure built up from their own calcium carbonate secretions. Photosynthetic dinoflagellates in their tissues provide sugars that supplement feeding activities.

Sponges have no symmetry, tissues, or organs; they are at the cellular level of construction.

Cnidarians are the simplest living animals with cells that are structurally and functionally organized in tissue layers.

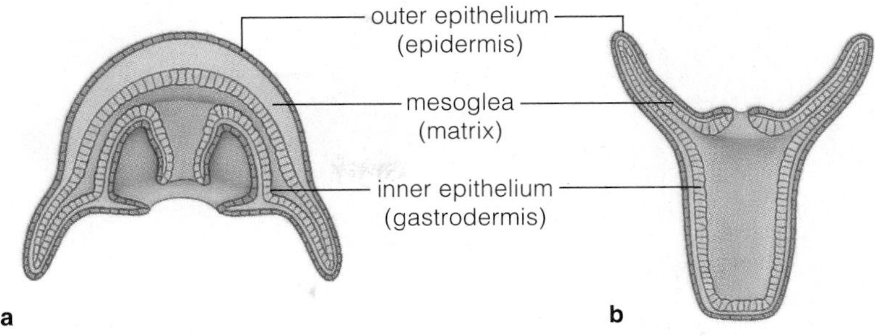

Figure 16.6 ***Animated!*** Cnidarian body plans: (**a**) medusa and (**b**) polyp. The *Obelia* life cycle includes both free-swimming medusae (**c**) and polyps. Sea anemones (**d**) and reef-building corals (**e**) live only as polyps.

16.3 Flatworms—Introducing Organ Systems

Beyond the cnidarians are animals ranging from simple worms to humans. Most are bilateral; all have organs.

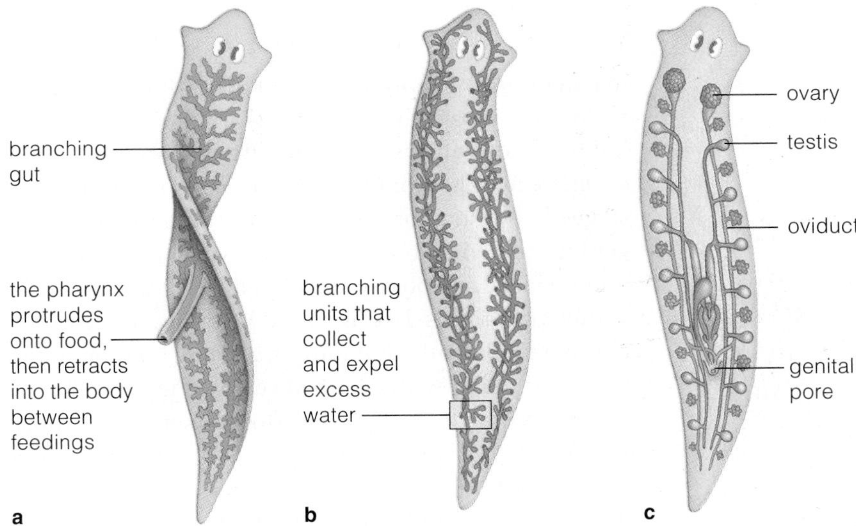

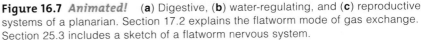

Figure 16.7 ***Animated!*** (**a**) Digestive, (**b**) water-regulating, and (**c**) reproductive systems of a planarian. Section 17.2 explains the flatworm mode of gas exchange. Section 25.3 includes a sketch of a flatworm nervous system.

With this section, we begin our survey of protostomes, one of the two lineages of animals that develop from a three-layered embryo. Embryonic tissues—exoderm, endoderm, and mesoderm—develop into organs. An **organ** is two or more tissues that interact in a common task, such as maintaining fluid balance. One or more organs interact to maintain conditions in the internal environment. Protostomes include flatworms, annelids, mollusks, roundworms, and arthropods.

Flatworms, the simplest animals to have any organ systems, include turbellarians, flukes, and tapeworms. These cephalized, bilateral animals all have a flattened body with longitudinal and circular muscles. As you saw in Figure 16.3*a*, there is no body cavity; the area between their gut and body wall is filled with tissues. Their digestive system is a saclike branching gut with a pharynx, a muscular tube that sucks in food or lets out waste (Figure 16.7). Most are hermaphrodites with female *and* male gonads and a penis. During mating, each gives some sperm to the other. The turbellarians called planarians also reproduce by transverse fission. The body splits in two crosswise and each piece grows the missing parts.

Planarians live in freshwater, but most of the other turbellarians are marine. All prey on or scavenge tiny animals. The flukes and tapeworms are parasites that

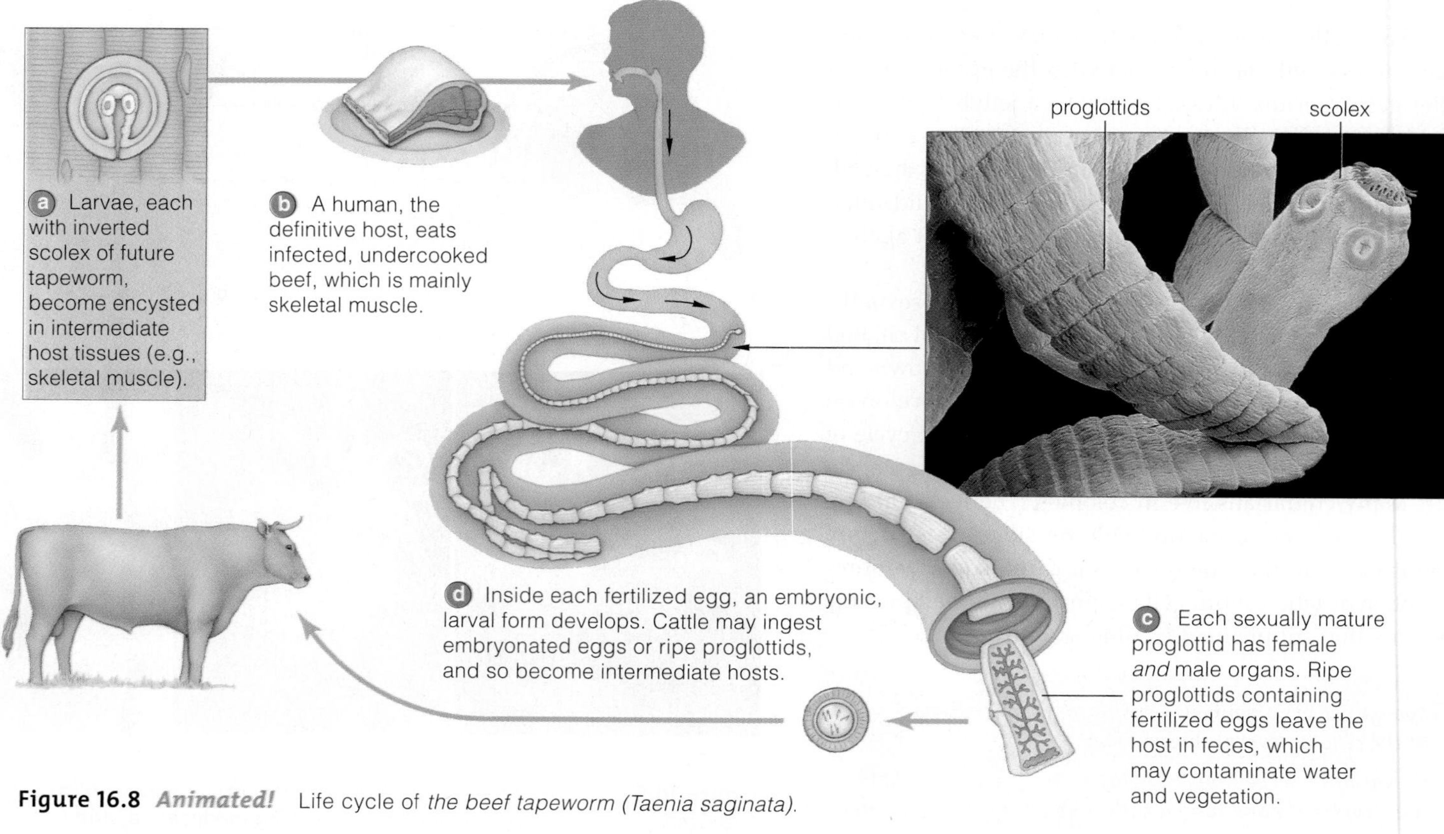

a Larvae, each with inverted scolex of future tapeworm, become encysted in intermediate host tissues (e.g., skeletal muscle).
b A human, the definitive host, eats infected, undercooked beef, which is mainly skeletal muscle.
proglottids
scolex
d Inside each fertilized egg, an embryonic, larval form develops. Cattle may ingest embryonated eggs or ripe proglottids, and so become intermediate hosts.
c Each sexually mature proglottid has female *and* male organs. Ripe proglottids containing fertilized eggs leave the host in feces, which may contaminate water and vegetation.

Figure 16.8 ***Animated!*** Life cycle of *the beef tapeworm (Taenia saginata).*

live in or on one to four different hosts during their life cycle. Like most parasites, tapeworms usually do not kill their host, at least not until after they reproduce. A *definitive* host harbors the mature parasites. Immature stages live in one or more *intermediate* hosts. Humans get infected when they ingest tapeworm larvae in raw, undercooked, or poorly preserved pork, beef, or fish. Adult tapeworms live in the host's gut, where they are constantly bathed in nutrients. They hold on to the intestinal lining with a barbed scolex (Figure 16.8).

In tropical regions, blood flukes commonly cause *schistosomiasis*. Freshwater larvae bore into the skin of humans wading in the water. Inside this host, flukes mature, mate, and lay eggs. Immune responses to the fluke eggs cause rash, fever, and muscle aches and can damage the liver, spleen, and other organs. Fertilized eggs pass in the feces and hatch into ciliated larvae that enter freshwater snails. Flukes multiply asexually in the snails, then develop into swimming forms that can attack humans and start the cycle once again.

Flatworms are simple bilateral, cephalized, acoelomate animals with organ systems. They include free-living turbellarians, as well as parasitic flukes and tapeworms.

16.4 Annelids—Segments Galore

Annelids have so many segments, it makes you wonder about controls over how the animal body plan develops.

Have you ever seen an earthworm wriggling out of its flooded burrow after a heavy rain (Figure 16.9)? It is a familiar example of 15,000 or so species of bilateral, segmented animals called **annelids**. Other members of this group are leeches and the less well-known, mostly marine polychaetes (Figure 16.10).

Except in the leeches, most annelid segments have two or more chitin-reinforced bristles, or setae, on each side. Bristles extended into sediments provide the traction for crawling and burrowing. Earthworms, with only a few bristles per segment, are oligochaetes. Polychaete worms typically have numerous bristles per segment. (*Oligo–* means few; *poly–* means many).

Earthworms have a complete gut and they eat their way through the soil. They draw in dirt through the muscular pharynx, digest any organic bits, and excrete wastes through an anus. Their activities help to aerate the soil. Like other annelids, they have a hydrostatic skeleton. Muscles work against their fluid-cushioned

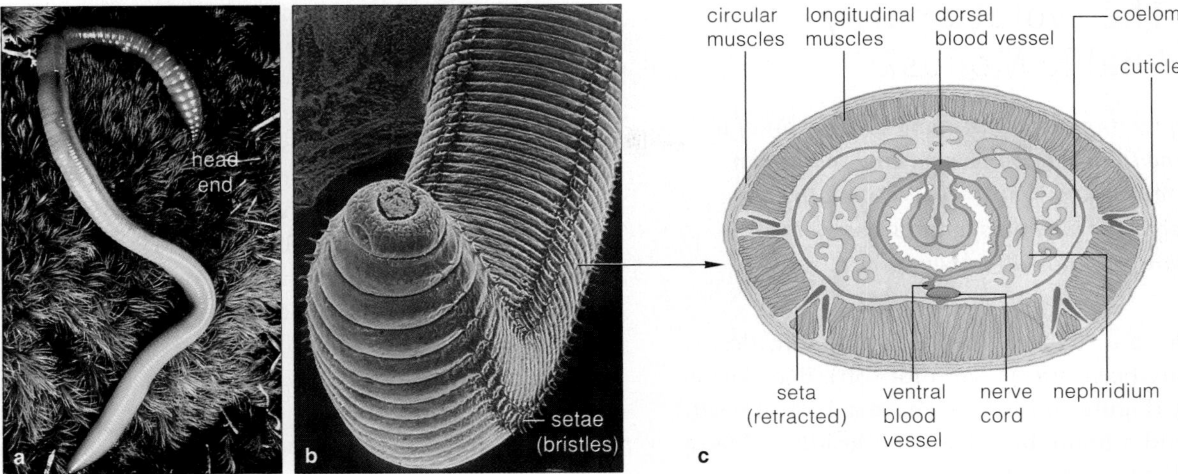

Figure 16.9 (**a**) A familiar annelid—one of the earthworms. (**b**) Externally, nearly all of the body segments of this bilateral worm look alike. Inside are specialized regions along the main body axis. (**c**) Cross-section through one segment. An outer cuticle surrounds two muscle layers, which surround internal organs. A nerve cord runs along the lower side of the body, connecting a cluster of nerves in each segment. Blood vessels running the length of the body supply small vessels that service each segment. Each segment also has an excretory organ or nephridium.

coelomic chambers. Inner and outer layers of muscles contract and relax in ways that shorten and fatten the segments, one after the other. The bristles alternately plunge into soil to gain traction, then retract, moving the worm forward.

In oligochaetes and polychaetes, heartlike pumps push blood through a continuous system of vessels in a closed circulatory system. The rudimentary **brain**, a cluster of nerve cell bodies, integrates sensory input and outputs calling for muscle responses. Also, a pair of **nerve cords** extend out from the brain and run the length of the body. They allow rapid communication among body regions and coordinated movements.

Annelids provide a glimpse into the evolutionary potential of a segmented body plan. Segmentation sets the stage for the evolution of regional specializations. Some segments can become modified and take on new functions, while others carry out their original tasks. In an earthworm, most segments look alike on the outside, but there are internal specializations. The leeches have specialized suckers at both ends. Polychaetes often have an elaborate head and many segments have fleshy-lobed parapods ("closely resembling feet").

Annelids are bilateral, coelomate, segmented worms that have complex organ systems. Some show the great degree of specialization possible with a segmented body plan.

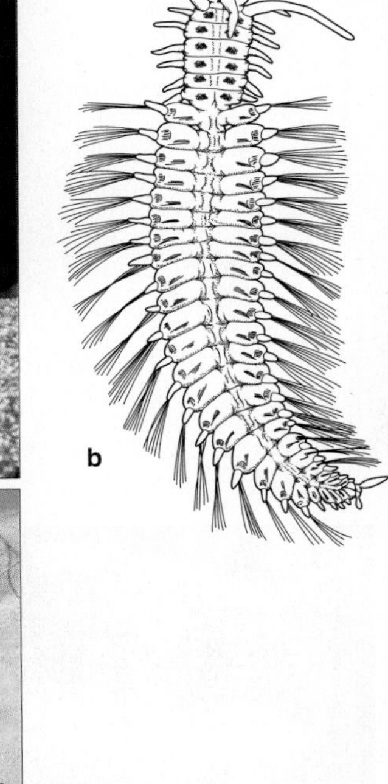

Figure 16.10 Annelid diversity. (**a**) A tube-dwelling polychaete with mucus-coated, featherlike structures at its head end. (**b**) As a group, the polychaetes have a dizzying variety of modifed segments. They are one of the most common animals along coasts. Many species are predators or scavengers; others dine on algae. Similarly, most species of leeches are scavengers or predators of small invertebrates, but *Hirundo medicinalis* has a taste for mammalian blood (**c**,**d**). It is used in medicine to draw off pooled blood and to slow clotting.

LINKS TO SECTIONS 13.2, 13.3

16.5 The Evolutionarily Pliable Mollusks

From our explorations in gardens and books, most of us would have no trouble recognizing a land snail when we see one. Yet few know much about its relatives in one of the largest of all animal groups, the mollusks. There are more than 110,000 known species.

Mollusks have a reduced coelom and a soft, fleshy, bilateral body (*molluscus* is Latin for soft). Body plans vary widely (Figure 16.11). Some have a distinct head, with eyes and a brain, but others are headless. Many have a shell or a reduced version of one. All have a mantle, which drapes like a skirt over the main body mass. This fleshy covering is unique to mollusks.

Mollusks range from tiny snails in treetops to huge predators of the open oceans. Most are gastropods, or "belly foots." The name refers to the underside of their body mass, which many use like a giant foot to glide across surfaces. Snails, slugs, and nudibranchs are the familiar members of this spectacularly diverse group.

Maybe it was their fleshy, soft bodies, so forgiving of chance evolutionary alterations, that gave mollusks the potential to radiate in so many ways. Let's explore this idea by focusing on a few representatives.

HIDING OUT, OR NOT

If you were small, soft of body, and *tasty*, a shell might discourage predators. Scallops, oysters, and mussels are bivalves. Members of this molluscan group have a hinged two-part shell that can snap shut when there is a threat. Scallops can also "swim" away by clapping the two shell halves together. Water forced out propels them backward, ideally far enough to get away from a predator that would like to make a meal of them.

If you were small, soft of body, and *toxic*, a shell would not be needed if there were a way to advertise your toxicity to predators. Some nudibranchs do just that. Gland cells of some species secrete sulfuric acid and other noxious substances. Other species graze on cnidarians, then incorporate nematocysts in their own tissues. They advertise the toxicity with vivid warning colors. Any predators that nibble at a colorful Spanish shawl nudibranch get stung by the stored nematocysts and quickly learn to avoid the species (Figure 16.11*c*).

ON THE CEPHALOPOD NEED FOR SPEED

About 500 million years ago cephalopods, with their buoyant, chambered shells, were dominant predators in Ordovician seas. Their living descendants include the fastest invertebrates (squids), the largest (the giant squid), and the smartest (octopuses) (Figure 16.11*d*).

In all but one existing descendant of thousands of ancestral species, the shell is reduced or entirely gone. Fossils show that disappearance of the shell coincided with an adaptive radiation of fishes that preyed upon cephalopods or competed with them for prey. In what might have been a long-term battle of speed and wits, most cephalopods lost their shell and became highly streamlined and—as invertebrates go—smart. Of all mollusks, cephalopods have the largest brain relative

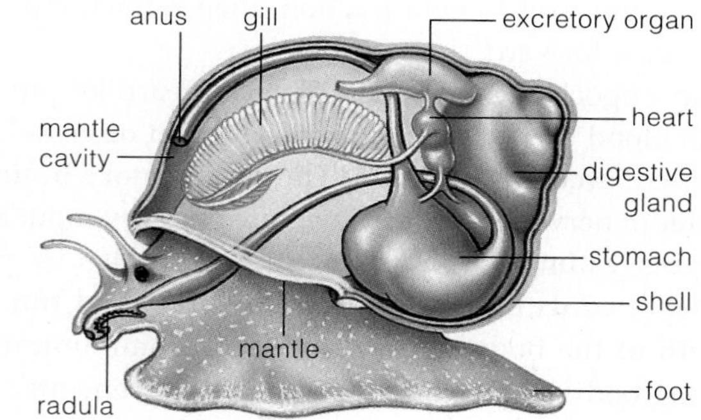

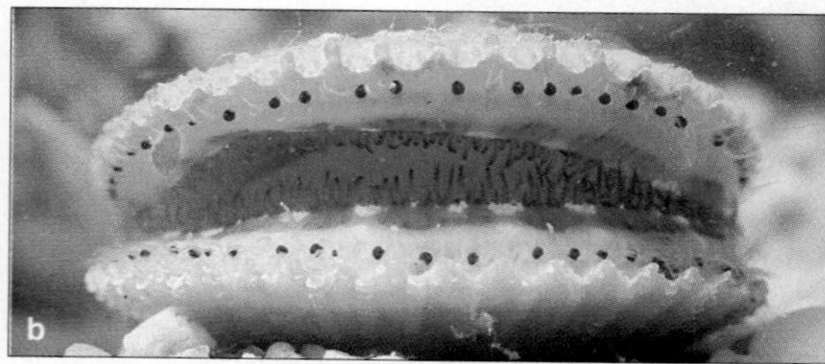

Figure 16.11 A medley of mollusks. (**a**) Body plan of a snail, one of the gastropods. (**b**) Scallop, a bivalve, with light sensitive "eyes" (blue dots) fringing its shell. (**c**) Spanish shawl nudibranchs (*Flabellina iodinea*) store nematocysts in their red respiratory organs. (**d**) An octopus. Like other cephalopods, it has a complex nervous system.

to their body size, and demonstrate the most complex and varied behavior.

For cephalopods, *jet propulsion* became the name of the game. Each forces a jet of water from the mantle cavity through the funnel-shaped siphon. Water is drawn into the cavity as mantle muscles relax. Then muscles contract while the mantle's free edge closes over the head. These contractions shoot a jet of water out through the siphon. By guiding siphon action, the brain controls the direction of escape or pursuit.

Cephalopods are the only mollusks with a closed circulatory system. Blood remains inside vessels, as it is pumped over gills and to and from body tissues by three hearts. This system delivers plenty of oxygen to the highly active tissues and supports the metabolic activities that fuel bursts of speed.

Mollusks are bilateral, soft-bodied, coelomate animals that differ tremendously in body size, body details, and life-styles. They include the largest, fastest, and smartest invertebrates.

16.6 Amazingly Abundant Roundworms

Roundworms are among the most abundant living animals. Sediments in shallow water often hold a million per square meter; a cupful of topsoil teems with them.

Roundworms, more formally known as nematodes, are more closely related to the insects than they are to other worms. At least 12,000 species are known. Most are microscopic or only millimeters long. The body is cylindrical and bilateral. There is a complete gut and a false coelom, which is filled by reproductive organs. Roundworms have a flexible cuticle, which is molted repeatedly as the animal grows. **Molting** is the name for a periodic casting off of an outer covering.

Most roundworms are free-living species that are decomposers. Many play vital roles in the cycling of nutrients through ecosystems. *Caenorhabditis elegans* is a favorite experimental organism among biologists. It has the same tissue types as more complex animals, but it is tiny and transparent Researchers can monitor the developmental fate of each of its 950 or so somatic cells. Also, this worm has a very small genome, and hermaphroditic forms can be self-bred to yield worms that are homozygous for specific desired traits.

Many soil-dwelling roundworms infect plant roots and are agricultural pests. Others parasitize animals including humans, dogs, and insects. More than one billion people are infected by *Ascaris lumbricoides*, an extremely large roundworm (Figure 16.12*a*). Infection is most common in Latin America and Asia.

Pigs and game animals can carry *Trichinella spiralis*. Adults of this parasite live attached to the intestinal lining. Their eggs develop into juveniles, which move through blood vessels to muscles, where they become encysted (Figure 16.12*b*). *Trichinosis*, the disease that results from infection, can be fatal. It is extremely rare in the United States, but common worldwide.

Repeated infections by *Wuchereria bancrofti* causes *elephantiasis*, a severe case of edema. Worms obstruct flow through lymph nodes, so fluid backs up in tissue spaces and enlarges lower body parts (Figure 16.12*c*). Mosquitoes are intermediate hosts for this parasite.

Roundworms are mostly tiny worms that have a false coelom. They shed their cuticle as they grow. Many are abundant in soil and some are plant pests. Others live as parasites of animals, including humans.

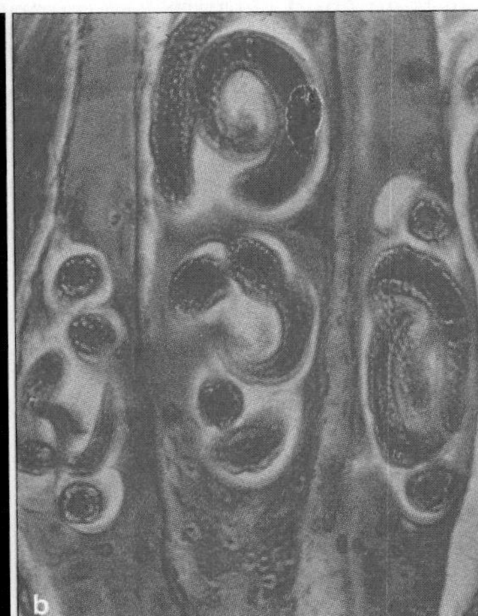

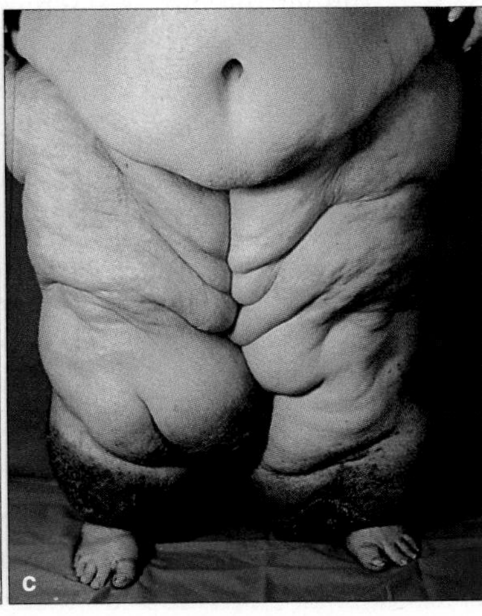

Figure 16.12 (**a**) Living roundworms (*Ascaris lumbricoides*). Worms can block ducts that empty into the intestinal tract, or the intestine itself. Larvae can migrate into and damage the lungs. (**b**) Juveniles of *Trichinella spiralis* in the muscle tissue of a host animal. (**c**) A case of elephantiasis. This type of severe edema results from an infection by the parasitic roundworm *Wuchereria bancrofti*.

LINKS TO SECTIONS 13.2, 13.9

16.7 Arthropods—The Most Successful Animals

Evolutionarily speaking, "success" means having the most offspring, species, and habitats; fending off threats and competition efficiently; and having the capacity to exploit the greatest amounts and kinds of food. Arthropods own these features.

Researchers have identified more than a million kinds of arthropods, mostly insects, and they are discovering new ones weekly. Of five lineages, trilobites are extinct. Chelicerates (spiders and mites), crustaceans, insects, and myriopods (centipedes and millipedes) survive.

KEYS TO SUCCESS

The success of arthropods—insects especially—can be traced to six key features: a hard exoskeleton, jointed appendages, fused and modified segments, specialized respiratory structures, efficient sensory systems, and often a division of labor in the life cycle.

In the next section, we will survey the diversity of arthropod groups. But first, let us consider the six key adaptations that contributed to the great success of the lineage as a whole.

A hardened exoskeleton. Arthropod cuticle acts as an external skeleton, called an **exoskeleton**. Exoskeletons might have evolved as defenses against predation. They took on added functions when some of the arthropods invaded the land. They support a body no longer made buoyant by water, and their waxes block water loss. Arthopods' muscles attach to and act against the exoskeleton.

An arthropod's cuticle would limit growth if it were not periodically molted. Hormones control molting. During certain stages of the life cycle, hormones induce a soft new cuticle to form under the old one, which is then shed (Figure 16.13). Body size is expanded by the uptake of air or water and by increased rapid cell divisions before the new cuticle hardens. Molting of the cuticle is a trait that the arthropods share with roundworms.

Figure 16.13 Molting by a red-orange centipede wriggling out of its old exoskeleton.

Jointed appendages. If cuticle was rigid everywhere, it would prevent all movement. Arthropod cuticle, however, thins in places where it spans *joints,* or regions where two body parts abut. Muscle contraction makes cuticle at joints bend, and moves body parts. The jointed exoskeleton was a key innovation that allowed the evolution of specialized appendages, including wings, antennae, and diverse types of legs. ("Arthropod" means jointed leg.)

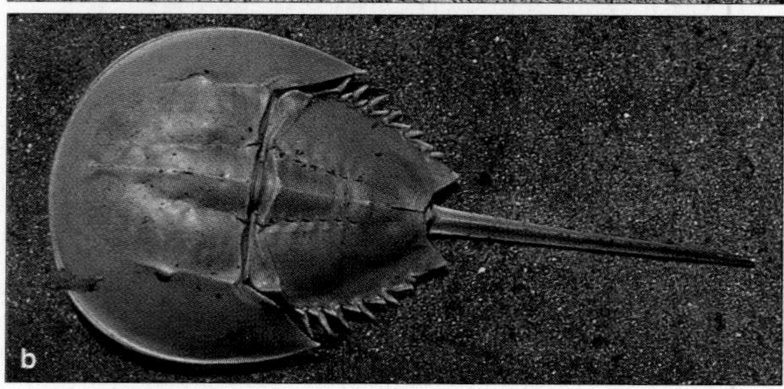

Figure 16.14 Two marine arthropods. (**a**) Fossil of a trilobite, a now extinct group, and (**b**) a horseshoe crab (*Limulus*). Horseshoe crabs are the closest living relatives of trilobites. They have survived basically unchanged for more than 250 million years, but are now threatened by human activities.

Specialized segments and fused-together segments. The first arthropods were segmented, and their segments were all more or less similar. In most existing lineages, some of segments became fused together or modified in specialized ways. For example, compare the highly similar body segments of a centipede with the winged and otherwise modified segments of a butterfly.

Respiratory structures. Many freshwater and marine arthropods use gills as respiratory organs. Among the insects and other arthropods on land, air-conducting tubes called tracheae evolved. The tubes start at pores on the body surface. Their extensive branching ends in fluid near tissues. Diffusion of oxygen into this fluid supports the high rates of aerobic respiration essential for flight and other energy-demanding activities.

Specialized sensory structures. For many arthropods, diverse sensory structures also contributed to success. For instance, some evolved into eyes that can sample a big part of the visual field and provide the brain with information from many directions.

Specialized developmental stages. Many arthropods divide the task of surviving and reproducing among different stages of development. The division of labor is most common among insects. For some species, the new individual is a *juvenile*—a miniaturized version of the adult that simply grows in size until it is sexually mature. Other species undergo **metamorphosis**. Their body changes—often in dramatic ways—between the

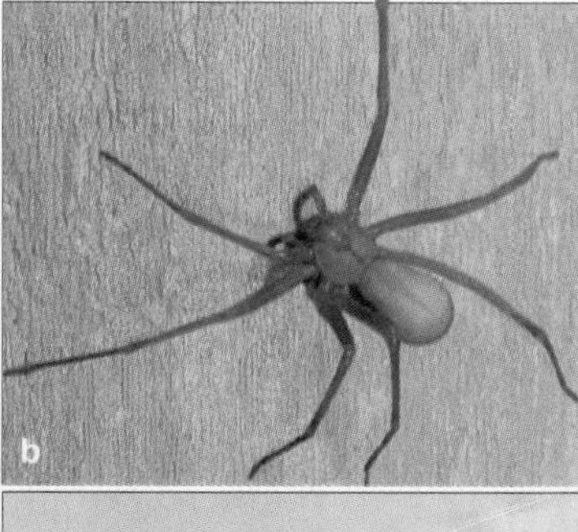
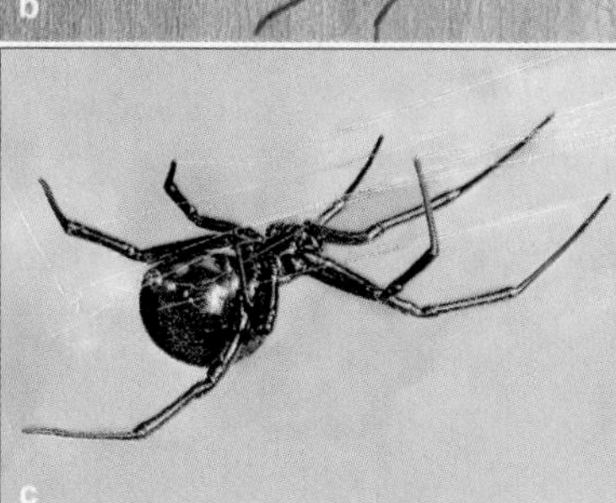
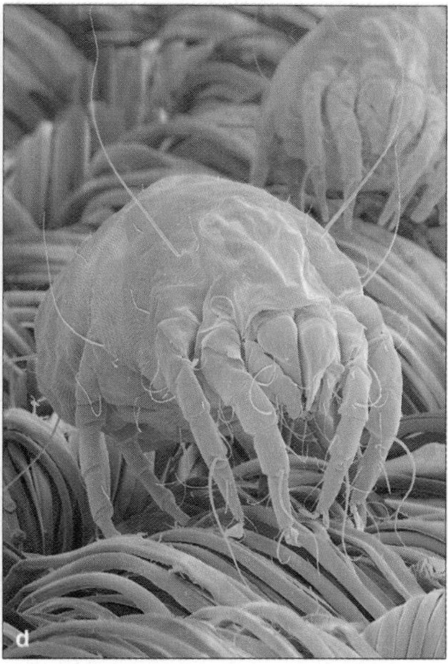

Figure 16.15 Arachnids. (**a**) Tarantula with one fearless female. Like all spiders, it helps keep insects in check. Its bite has not harmed humans. (**b**) Brown recluse, with a violin-shaped mark on its forebody. Its bite can be severe to fatal. (**c**) Female black widow, with a red hourglass-marking on the underside of a shiny black abdomen. The males are smaller and don't bite. (**d**) Dust mite. Dust mite feces and corpses cause allergies in susceptible people.

embryo and adult forms. Hormones help regulate this tissue reorganization and remodeling.

Eating to sustain rapid growth is the specialty of the immature stages; adults specialize in reproduction and in dispersal. (Think of how flightless caterpillars devour leaves, then turn into butterflies that fly away, mate, and deposit eggs.) The different developmental stages are adapted to different conditions, especially seasonal shifts in food sources, water supplies, and availability of potential mates.

SPIDERS AND THEIR RELATIVES

Like the extinct trilobites, the chelicerates originated in shallow seas during the early Paleozoic. The only surviving marine species are a few mites, sea spiders, and horseshoe crabs (Figures 16.14). Familiar kinds on land—spiders, scorpions, ticks, and chigger mites—are all arachnids (Figure 16.15). And we might say this: Never have so many been loved by so few.

Spiders and scorpions are predators that sting, bite, or subdue prey with venom. Spiders have appendages that spin silk threads for webs and egg cases. Most webs are netlike. The bola spider spins a thread with a ball of sticky material at the end, releases a scent that mimics a female moth's sex attractant, then uses a leg to swing the ball at male moths lured by the scent!

All spiders of the genus *Loxosceles* are poisonous to humans and other mammals, but we seldom see most of them. One, the brown recluse (*L. reclusa*), often hides in warm places in houses, especially near refrigerator motors and clothes dryers, and in closets. Its body has a distinctive violin-shaped marking (Figure 16.15*b*). Its bite ulcerates the skin and the wound is slow to heal.

About five people die each year in the United States as the result of black widow spider (*Latrodectus*) bites (Figure 16.15*c*). This spider's venom contains a toxin that interferes with nervous control of muscle tissue. A bite can paralyze muscles involved in breathing, so medical attention should be prompt.

Of all arachnids, mites are among the smallest, most diverse, and most widely distributed. Of an estimated 500,000 mites, the few parasites—ticks especially—are harmful; many transmit pathogens that cause serious diseases in humans. As you learned earlier, one tick is the vector for *Borrelia burgdorferi,* the bacterium that causes Lyme disease (Figure 14.12). Others transmit the agents of Rocky Mountain spotted fever, typhus, tularemia, encephalitis, and other diseases. Ticks do not fly or jump. They cling to grasses or shrubs, then crawl onto animals brushing past. Look for them after a walk in the wild. Finally, house dust mites are tiny but their remains and feces can cause allergic reactions and trigger asthma in some people (Figure 16.15*d*).

CRUSTACEANS

Shrimps, lobsters, crabs, barnacles, pillbugs, and other crustaceans got their name because they have a tough yet flexible "crust"—a cuticle hardened with calcium. The vast majority live in marine habitats, where they are so abundant they are sometimes called the insects of the seas. Humans harvest many edible types.

All crustaceans have two pairs of antennae that they use to feel and smell. There are many body segments, more than sixty in some groups. Usually, the cuticle extends back from the head as a shield above some or all segments. The head has jawlike appendages, and appendages used to handle food. "Decapods," such as the lobsters, shrimps, and crabs, have ten walking legs (five pairs). Figure 16.16 shows a typical crab life cycle.

Barnacles are the only arthropods that secrete an external "shell." This structure consists of plates that can close up tightly to protect them from predators, drying winds, and surf. Adults glue themselves by their antennae to rocks, piers, and even whales, then filter food from the seawater using their feet (Figure 16.17*b*). Given that a barnacle is glued in place at its head end, you might think sex could be challenging. But barnacles tend to cluster in groups and most are hermaphrodites. Each extends its penis—which can be several times the body length—out to its neighbors.

Copepods are less than two millimeters long. These crustaceans are the most numerous animals in aquatic habitats and are abundant on land (Figure 16.17*c*). A number parasitize different invertebrates and fishes. Most ingest phytoplankton. Copepods are themselves eaten by other invertebrates, fishes, and whales.

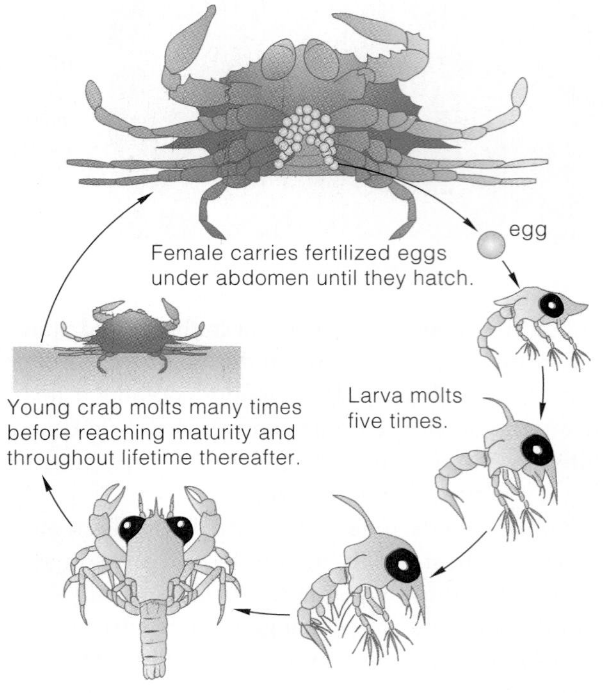
egg
Female carries fertilized eggs under abdomen until they hatch.
Larva molts five times.
Young crab molts many times before reaching maturity and throughout lifetime thereafter.

Figure 16.16 ***Animated!*** Life cycle of a crab.

A LOOK AT INSECT DIVERSITY

All insects share the six adaptations listed earlier. Their segmented body is divided into a head, an abdomen, and a thorax with three pairs of legs and usually two pairs of wings. They have paired mouthparts for biting, chewing, and other tasks, and one pair of antennae. The only winged invertebrates are in this group.

An insect body has a foregut, a midgut where food is digested, and a hindgut where water is reabsorbed. A system of small tubes disposes of waste materials. After proteins are digested, the residues diffuse from blood into the tubes. There, enzymes convert them to uric acid, which is eliminated with feces. This system lets land-dwelling insects get rid of potentially toxic wastes without losing precious water.

Figure 16.18 shows a few of the more than 800,000 species. If numbers and distribution are the measure, the most successful ones are small and reproduce often. By one estimate, if all the offspring of one female fly survived and reproduced for six more generations, she would have more than 5 trillion descendants!

a
b
c

Figure 16.17 (**a**) Lobster in the Caribbean Sea, off the coast of Belize. (**b**) Goose barnacles. Adults cement themselves to one spot. You might mistake them for mollusks, but when they open their hinged shell to filter-feed, you see jointed appendages—the hallmark feature of arthropods. (**c**) One of the copepods. Different kinds are free-living filter feeders, predators, and parasites.

The most successful insects are winged. They move among food sources that are too widely scattered to be exploited by other kinds of animals. The capacity for flight contributed enormously to their success on land.

Insect life cycles often include stages that permit exploitation of different resources at different times. In some groups, such as the springtails, eggs hatch into young that look like minature adults. In others, such as butterflies, a larva that is specialized for feeding turns into a pupa, which undergoes metamorphosis. This is resumption of growth and the transformation of the post-embryonic stages into the sexually mature, adult form. Tissue reorganization and a total remodeling of many body parts occur during this process.

The very features that contribute to insect success also make them our most aggressive competitors for resources. Insects destroy much of what we value, such as crops, stored food, wool, paper, and timber. Some transmit pathogens that make us sick.

Yet bees and other insects are necessary to pollinate flowering plants, including highly valued crop plants. And many "good" insects attack or parasitize the ones we would rather do without.

Figure 16.18 Examples of insects. (**a**) Newly hatched stinkbugs. (**b**) In the center of this group of honeybee workers, one honeybee is signaling the position of a food source as described in Section 32.2.

With more than 300,000 species of beetles, Coleoptera is the largest order of animals. Ladybird beetles (**c**) and a staghorn beetle from New Guinea (**d**) are two examples.

(**e**) A swallowtail butterfly and (**f**) a luna moth. (**g**) A damselfly. (**h**) Mediterranean fruit fly. Its larvae destroy citrus and other crops.

A duck louse (**i**) eats bits of feathers and skin. A flea (**j**) has legs adapted for jumping onto and off of its animal hosts. Fleas transmit diseases, including bubonic plague.

Arachnids, crustaceans, insects, and the millipedes and centipedes are living arthropods. As a group, they show enormously diverse modifications on a basic body plan.

All arthropods have an exoskeleton of hardened cuticle, jointed appendages, fused and modified body segments, respiratory structures, specialized sensory organs, and often a division of labor during the life cycle.

In terms of distribution, number of species, population sizes, competitive adaptations, and exploitation of diverse foods, insects are the most successful animals.

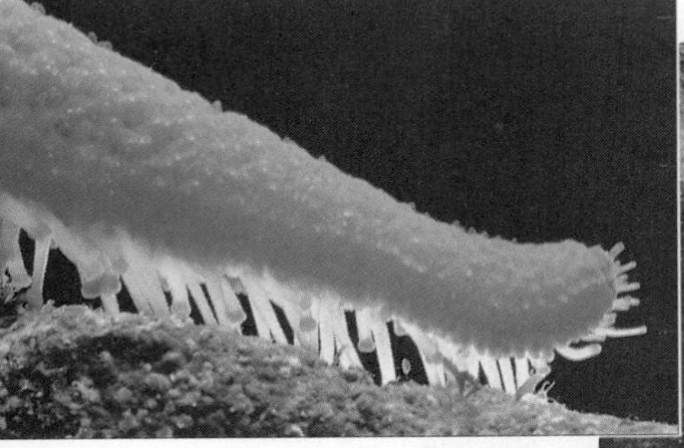

Figure 16.19 A few echinoderms. (**a**) Sea star, with a close-up of its little tube feet. (**b**) Sea urchins, which can move about on spines and a few tube feet. (**c**) Brittle star. Its slender arms (rays) make rapid, snakelike movements. (**d**) Sea cucumber, with rows of tube feet along its elongated body.

LINKS TO SECTIONS 13.4, 13.5, 13.10

16.8 The Puzzling Echinoderms

Reflect, for a moment, on Figure 16.1. Echinoderms and chordates show similar patterns of early development, which reflect their shared ancestry. Together, they are classified as deuterostomes. But unlike chordates, the adult echinoderms have radial features.

Echinoderm means spiny skinned. The body wall of these coelomate animals has spines, spicules, or plates of calcium carbonate. Included in this marine group are sea stars, brittle stars, sea urchins, sea cucumbers, and sea lilies (crinoids). Nearly all echinoderms have an internal skeleton of calcium carbonate and other substances. Adults have a generally radial body with some bilateral features, but the larval body is bilateral. Along with DNA data, this suggests that echinoderms arose from bilateral ancestors.

Adult echinoderms have no brain. A decentralized nervous system receives and responds to signals about touch, light, and chemicals. For instance, any arm of a sea star that touches the shell of a mussel can become the leader, directing the rest of the body to move in a direction suitable for capturing prey.

Echinoderms have tube feet. These are fluid-filled, muscular structures that have suckerlike adhesive disks (Figure 16.19). Sea stars use their feet for burrowing, walking, clinging to rocks, or gripping a clam or snail at feeding time. Eager sea stars swallow prey whole. Others push part of their stomach outside the mouth and start digesting prey before swallowing. Sea stars eliminate any coarse, undigested residues through the mouth. Their have an anus, but it is too small to expel chunks of shell.

Echinoderm sexes are usually separate. Eggs and sperm are released into the water. After fertilization, a bilateral free-swimming larva develops. The larva will grow and develop into the mostly radial adult form.

The radial echinoderms have bilateral larvae and probably had bilateral ancestors. Though brainless and spiny-skinned, they belong to the same lineage as the chordates.

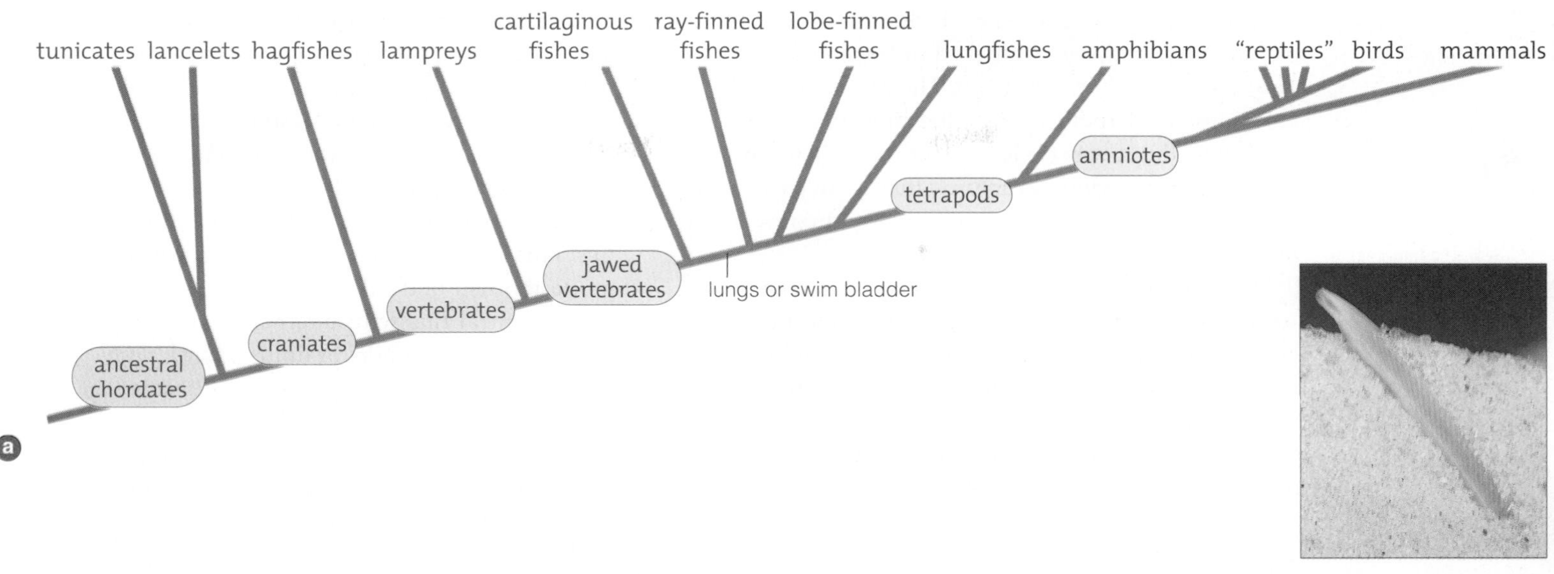

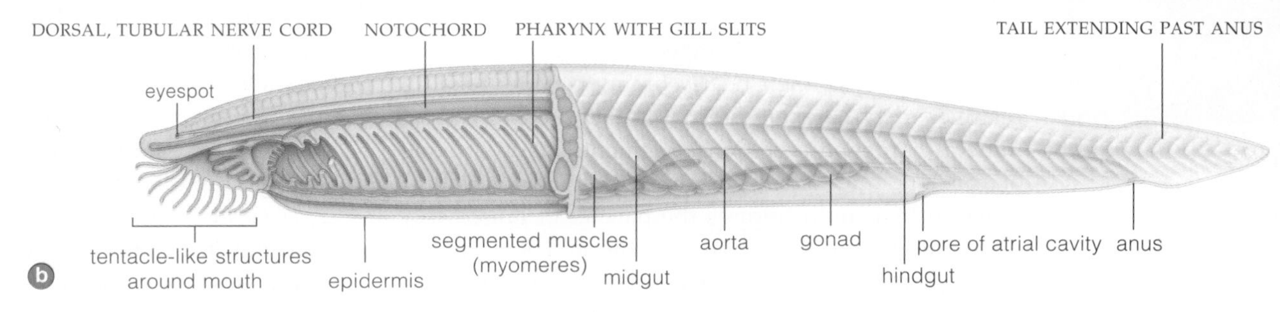

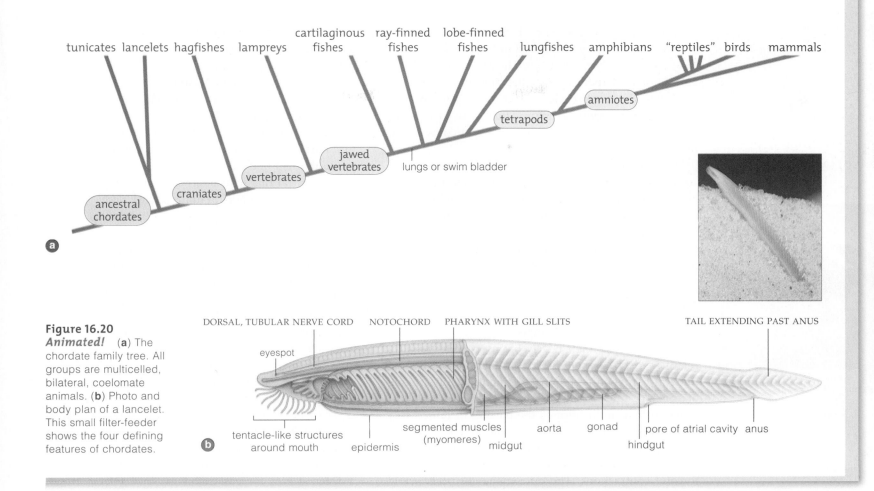

Figure 16.20 ***Animated!*** (**a**) The chordate family tree. All groups are multicelled, bilateral, coelomate animals. (**b**) Photo and body plan of a lancelet. This small filter-feeder shows the four defining features of chordates.

16.9 Evolutionary Trends Among Vertebrates

We start our look at vertebrates and their chordate origins with a reminder: Each kind of organism—including humans—is a mosaic of traits. Many traits are conserved from remote ancestors, and others are unique to its branch on the family tree.

Figure 16.20*a* is an evolutionary tree for the **chordates**. These coelomate, bilateral animals have four unique features that appear in their embryos and often persist into adulthood:

First, a notochord—a strong but flexible rod—runs lengthwise through the body and helps to support it. *Second*, a nerve cord parallels the notochord and gut; its anterior end develops into a brain. In contrast to the ventral nerve cords of many invertebrates, this one is dorsal. *Third*, embryos have gill slits in the wall of their pharynx, a type of tube that functions in feeding, respiration, or both. *Fourth*, a tail develops in embryos and extends past the anus.

The two groups of invertebrate chordates, tunicates and lancelets, filter food from currents of water that pass through gill slits in their pharynx. In lancelets, the characteristic chordate traits are visible in the adult, as shown in Figure 16.20*b*.

In the tunicates, the notochord, tail, and nerve cord occur only in a tadpole-shaped larval stage. The adults retain only the pharynx with gill slits (Figure 16.21). Water flows into an adult's saclike body through one siphon, is filtered through gill slits in the pharynx, then exits the body by way of the other siphon. Tunicates known as sea squirts are commonly found attached to piers and boats.

EARLY CRANIATES

Fishes, amphibians, reptiles, birds, and mammals are **craniates**. The distinguishing feature of this group is a cranium, a chamber of cartilage or bone that encloses the brain. The first craniates arose before 530 million years ago. They had fins, a notochord, and segmented muscles. These active swimmers resembled larvae of lampreys, one group of modern jawless fishes.

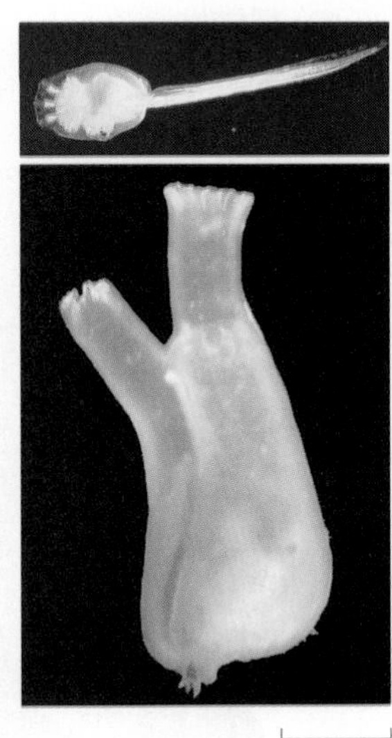

Figure 16.21 *Above*, A free-swimming tunicate larva. It will attach by its head to some surface and then take on the adult form (*below*). Like lancelets, both stages are filter feeders.

LINKS TO SECTIONS 13.4, 13.5, 13.10

The earliest jawless fishes, including ostracoderms, had armor-like plates composed of bone and dentin, a hard tissue still found in teeth. The armor plating may have protected them against the pincers of giant sea scorpions. But it was far less effective against **jaws**, the hinged, bony feeding structures that evolved in other lineages (Figure 16.22). When jawed craniates began their radiation, ostracoderms vanished.

Placoderms were among the earliest jawed fishes. Armor plates protected their head, but not much else. Before, chordate feeding strategies had been limited to filtering, sucking, and rasping food. Some placoderms evolved into huge predators. The changes sparked the evolution of offensive and defensive body structures that has continued to the present.

THE KEY INNOVATIONS

Vertebrates profoundly changed the course of animal evolution. Unlike their nearly brainless, filter-feeding relatives, they have an internal skeleton composed of mineral-hardened secretions. Segments made of bone or cartilage and called **vertebrae** (singular, vertebra) replaced their notochord as the partner of muscles. A vertebral column was more flexible and less clunky than external armor plates. The union of vertebrae and muscles promoted maneuverability, and allowed more forceful contractions. The outcome? *Agile, fast-moving fishes began their dominance of the seas.*

Jaws were part of a related trend. They had evolved from the first of a series of structural elements that had supported pharyngeal gill slits (Figure 16.22*b*). Their appearance started a race for better ways to overcome and escape them. For example, selection now favored better vision for detecting predators and bigger brains to guide escapes. *A trend toward complex sensory organs and nervous systems started among the ancient fishes and continued among vertebrates on land.*

Another trend began with the evolution of paired fins. **Fins** are appendages that help propel, stabilize, and guide the fish body through water. Some groups had fleshy ventral fins with internal skeletal supports that became the forerunners of limbs. *Paired, fleshy fins were a starting point for all legs, arms, and wings that evolved among amphibians, reptiles, birds, and mammals.*

A change in respiration started another trend. In lancelets, oxygen and carbon dioxide simply diffuse across the skin. In most early vertebrates, gills evolved. **Gills** are respiratory organs with moist, thin folds and adjacent blood vessels. They have a large surface area that exchanges gases between the environment and the body. Oxygen diffuses *from* water in a fish mouth into gills, as carbon dioxide diffuses out, *into* the water.

Among the larger, active fishes, gills became more efficient. But gills do not work out of water; their thin surfaces stick together without a flow of water to keep them moist. In the fishes that were ancestral to land vertebrates, two small outpouchings formed on the gut wall. The pouches evolved into **lungs**—internally moistened sacs for gas exchange. In a related trend, modifications to the heart and blood vessels enhanced the pumping of oxygen and carbon dioxide through their body. *The ancestors of land vertebrates relied more on paired lungs than on gills—and more efficient circulatory systems accompanied the evolution of lungs.*

Placoderms and many of the jawless fishes became extinct during the Carboniferous. Through adaptation, key innovations, and plain luck, some descendants of some jawless and jawed groups made it to the present.

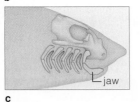

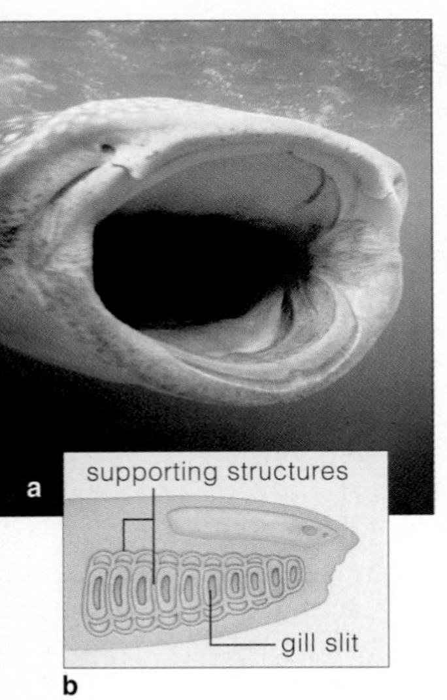

Figure 16.22 (**a**) A whale shark, with a mouth used in capturing zooplankton, squids, and sometimes fishes, including tuna. (**b**) In early jawless fishes, supporting elements reinforced a series of gill slits on both sides of the body. (**c**) In placoderms, the first elements in the series had become modified and functioned as jaws. (**d**) The human at left gives an idea of the size of one ancient jawed fish, the placoderm *Dunkleosteus*.

Chordate embryos alone have this combination of traits: a notochord, a tubular dorsal nerve cord, a pharynx with fine slits in its wall, and a tail extending past the anus. These traits are legacies from a shared ancestor, an early invertebrate chordate.

Tunicates and lancelets are living invertebrate chordates. They are aquatic filter feeders that strain food from water that flows through gill slits in the wall of their pharynx.

Craniates are chordates with a brain chamber of cartilage or bone. Vertebrates are craniates with a vertebral column (cartilaginous or bony segments).

Jaws, paired fins, and later, lungs, were key innovations that led to the adaptive radiation of vertebrates.

16.10 Major Groups of Jawed Fishes

Unless you study underwater life, you may not know that fishes are the world's dominant vertebrates. Their numbers exceed those of all other vertebrate groups combined. They are also the most diverse; they include about half of all vertebrates.

There are two major groups of living fishes, one with a skeleton of cartilage, the other with a skeleton of bone. Their form and behavior hint at the challenges of life in water. Water is 800 times denser than air and resists motion through it. In response to this constraint, many fish species became streamlined for pursuit or escape. The predatory sharks are an example (Figure 16.23*a*). A long, trim body reduces friction; their tail muscles provide the propulsive force for rapid movement. By contrast, a manta ray has a flattened body and glides through the water by flapping fins attached along the length of the body (Figure 16.23*b*).

Sharks and rays are the most well-known kinds of **cartilaginous fishes**. All have a skeleton of cartilage, and five to seven pairs of gill slits. Sharks shed and replace their teeth on an ongoing basis. These teeth are modified scales that are most often used to grab and tear chunks of flesh. Human surfers and swimmers can resemble typical prey, and rare attacks by a few species have given all of the sharks a bad reputation. Worldwide, shark attacks kill about 25 people a year.

Bony fishes have a skeleton made of bone, which is heavier than cartilage. A gas-filled swim bladder or lung helps these fish maintain buoyancy. By the early Devonian—about 400 million years ago—three main lineages had emerged: ray-finned fishes, lung fishes, and lobe-finned fishes.

In **ray-finned fishes** flexible fins are supported by rays that are derived from skin rather than cartilage or bones (Figure 16.24). With more than 21,000 species, these fishes make up nearly half of all the vertebrates. Ray-fined fishes exhibit a diverse variety of colors and body shapes that adapt them to a wide array of aquatic habitats. Elongated eels squeeze into crevices, flattened flounders lie on sandy bottoms, and box-shaped fishes can wedge themselves among corals.

Most of the fish we eat are ray-finned fish, such as salmon, sardines, sea bass, swordfish, snapper, trout, tuna, halibut, carp, and cod. Many species have been over-fished. Fish have been harvested faster than they can reproduce, so stocks have declined. For example, it is estimated that the population of Atlantic cod (*Gadus morhua*) off the coast of Nova Scotia has declined by 96 percent or more since the 1850s.

Figure 16.23 Cartilaginous fishes. (**a**) Galápagos shark and (**b**) a manta ray. Two projections on the ray's head unfurl and waft plankton to a mouth between them. Notice the pairs of gill slits.

Figure 16.24 Ray-finned bony fishes. (**a**) A perch showing its ray-supported fins. (**b**) Coral grouper. (**c**) Flying fish. Like bird wings, the modified pectoral fins help these fishes fly out of the water when they are being pursued by predators.

LINKS TO SECTIONS 13.4,13.5, 13.10

a

b

c

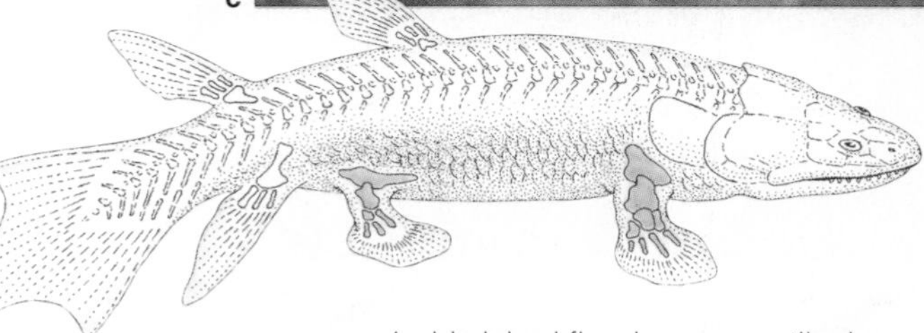

d inside lobed fins, bony or cartilaginous structures undergoing modification

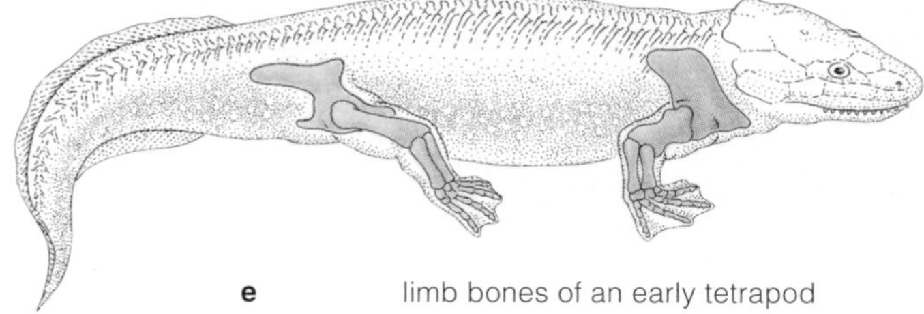

e limb bones of an early tetrapod

Coelacanths are the only living lobe-finned fishes. Their ventral fins include fleshy body extensions with bony elements inside them (Figure 16.25*a*). Lungfishes have similar fins (Figure 16.25*b*). As the name suggests, they have gills *and* one or two lunglike sacs. The fish inflates these sacs by gulping air. Like the air in our lungs, air in the sacs exchanges gases with the blood. Some lungfishes are totally dependent on this type of gas exchange; they drown if kept underwater!

Morphological and genetic evidence suggests that lungfishes are probably the sister group to **tetrapods**: amphibians, reptiles, birds, and mammals. Both have a similar arrangement of skull bones and the same type of circulation, with a separate circuit that carries blood to and from lungs. The "limbs" of lobe-finned fishes, lungfishes, and tetrapods are about the same in their size, position, and structure (Figure 16.25*d,e*).

Limb joints, digits, and other traits of early tetrapod fossils indicate that walking began in the water, rather than on land. By the late Devonian, some species were using their limbs for crawling, as well as swimming.

Ray-finned fishes are now the most diverse and abundant vertebrates. The lobe-finned fishes have fleshy ventral fins reinforced with skeletal parts. The lungfishes have simple lunglike sacs that supplement respiration by gills.

Walking probably originated in water during the Devonian, among the aquatic forerunners of land tetrapods.

16.11 Early Amphibious Tetrapods

Even a small genetic change could have transformed lobed fins into limbs. Remember those enhancers that control gene transcription? One of them influences how digits form on limb bones. Even a single mutation in one master gene can lead to a big change in morphology.

Amphibians are vertebrates with four legs (or a four-legged aquatic ancestor); they were the first tetrapods on land. Their body plan and reproductive mode are somewhere between fishes and reptiles.

Figure 16.25 (**a**) Living coelacanth (*Latimeria*), of the only lineage of existing lobe-finned fishes. (**b**) Australian lungfish. (**c**) Painting of Devonian tetrapods. *Acanthostega*, submerged, and *Ichthyostega*. Their skull, caudal tail, and fins were fishlike. Unlike fishes, they both had a short neck and four limbs with digits. Compare Figure 13.10. (**d**,**e**) Proposed evolution of skeletal elements inside fins into the limb bones of early amphibians.

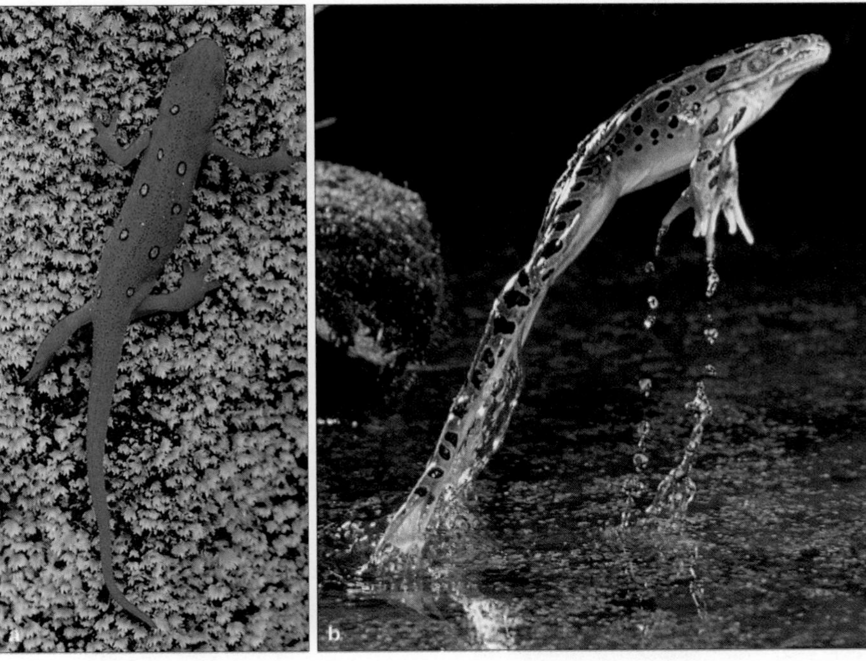

Figure 16.26 Some familiar amphibians. (**a**) Red-spotted salamander (*Notophthalamus*). (**b**) A frog jumping and displaying its tetrapod heritage. (**c**) The American toad.

The earliest tetrapods probably spent most or all of their time in water. But by the end of the Devonian, some types were semiaquatic and spending time on land. Life in this new habitat was both dangerous and promising. Temperatures shifted much more on land than in the water, air did not support the body as well as water did, and water was not always plentiful. But air is richer in oxygen. Amphibian lungs continued to be modified in ways that enhanced oxygen uptake. The heart, newly divided into three chambers, pumped more oxygen-rich blood to cells.

New sensory information also challenged the early amphibians. Swamp forests supported vast numbers of insects and other invertebrate prey. On land, vision, hearing, and balance became the most advantageous senses. Brain regions concerned with interpreting and responding to sensory input expanded.

All frogs, toads, and salamanders alive today are descended from those first amphibians (Figure 16.26). Most still live in water or in moist habitats. Even when species use gills or lungs, they also exchange oxygen and carbon dioxide across their thin skin. Respiratory surfaces must be kept moist to function, however, and skin dries easily in air.

Some existing amphibians spend their entire life in water. Others release eggs in water or produce aquatic larvae. The species adapted to dry habitats either lay eggs in moist places or, in a few cases, protect embryos during development inside the moist adult body.

Many amphibians are now in trouble (Figure 16.27) Six frog, four toad, and eleven salamander species are now listed as threatened or endangered in the United States and Puerto Rico. One, the California red-legged frog, is the largest frog native to the western United States. This was the species that inspired Mark Twain's famous short story "The Celebrated Jumping Frog of Calaveras County."

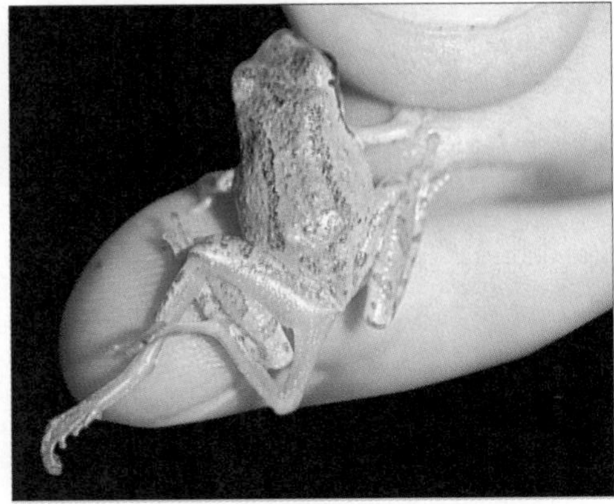

Figure 16.27 Frog deformity. A parasitic fluke (*Ribeiroia*) burrows into tadpole limb buds and alters cells in their embryos. The infected tadpoles grow extra legs or no legs at all. Fluke cysts have been found in weirdly legged frogs and salamanders across North America.

Researchers correlate these declines with shrinking habitats. Developers and farmers commonly fill in low-lying ground that once filled up seasonally and provided pools of standing water. Nearly all species of amphibians cannot breed unless they can deposit their eggs in water, in which the larvae will develop.

In body form and behavior, amphibians show resemblances to aquatic tetrapods and reptiles. Most species have not fully escaped dependency on aquatic or moist habitats to complete their life cycle.

LINKS TO SECTIONS 13.4, 13.9

Figure 16.28 (**a**) Jurassic scene, with *Archaeopteryx* gliding in the foreground. Left to right: Plant eaters *Stegosaurus* and *Apatosaurus*, the carnivore *Saurophaganax* ("king of reptile eaters"), and *Camptosaurus*, a horny-beaked plant eater. A climbing mammal looks out from the shrubbery. *Apatosaurus* was one of the largest land animals ever. This tiny-brained plant eater grew as long as 27 meters (90 feet) and weighed 33 to 38 tons. (**b**) Galápagos tortoise. (**c**) Spectacled caiman, a crocodilian. Its peglike upper and lower teeth do not match up. (**d**) Two hognosed snakes emerge from leathery amniote eggs. (**e**) *Sphaerodactylus ariasae*, the smallest lizard known.

16.12 The Rise of Amniotes

Amniotes were the first vertebrates to adapt to dry land habitats, through modifications in their organ systems and eggs. The eggs have four membranes that conserve water and support the embryo's development.

THE "REPTILES"

Late in the Carboniferous, some amphibians gave rise to **amniotes**, which were better adapted to life on dry land. The membranes in an amniote egg help conserve water and protect a developing embryo. Amniote skin is rich in the protein keratin, which helps waterproof it. A pair of well-developed kidneys help to conserve water, and fertilization is always internal.

Synapsids and sauropsids are two major groups of amniotes. Existing mammals and extinct mammal-like groups are synapsids. Sauropsids are "**reptiles**," and birds. We put "reptile" in quotes because this is not a formal taxonomic group. It is not monophyletic; the name refers to several lineages that all share the basic amniote features but not the derived traits that define birds or mammals.

The first **dinosaurs** descended from Triassic reptiles. They were turkey-sized and possibly warm-blooded. Many sprinted about on two legs. Adaptive zones opened up for dinosaurs about 213 million years ago, when huge fragments from an asteroid or comet fell to Earth. Craters in France, Canada, and North Dakota mark where they hit. Nearly all of the animals that survived the impacts were small, had high rates of metabolism, and tolerated big temperature changes.

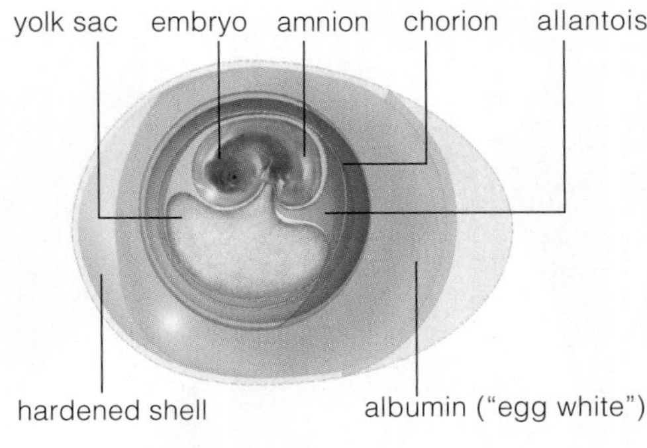

Figure 16.29 (**a**) Sketch of a typical amniote egg. (**b**) A Laysan albatross with a wingspan of more than two meters weighs less than 10 kilograms (22 pounds). Its bones are lightweight and attach to powerful muscles.

Descendants of the surviving dinosaurs became the ruling reptiles—which endured for 140 million years. Some reached monstrous proportions (Figure 16.28*a*). Many dinosaurs died in a mass extinction at the end of the Jurassic, others during a pulse of extinctions in the Cretaceous. Some lineages recovered, and new lineages evolved in the forests, swamps, and open habitats.

The last of the reptiles traditionally called dinosaurs vanished 65 million years ago, with a mass extinction. According to the **K–T asteroid impact theory**, collision with an asteroid as big as Mount Everest caused this extinction. A crater wider than Connecticut, in what is now Mexico's Yucatán peninsula, supports the theory. So does iridium in sediments from this time. Iridium is an element common on asteroids, but rare on Earth.

Many reptiles survived the asteriod impact. Figure 16.28*b–e* shows some of their descendants. The turtles and tortoises have a shell attached to their skeleton. If threatened, many can pull their head and limbs inside. All lay eggs that develop outside the body.

Lizards are the most diverse of the living reptiles. The largest, the monitor lizard known as the komodo dragon, can grow up to ten feet long. Most lizards lay eggs that develop outside the body. Some species give birth to live young; although eggs develop inside the mother, they are not nourished by her tissues.

Snakes arose from ancestral lizards. All the modern snakes are carnivores. Some retain bony remnants of ancestral hindlimbs. Like lizards, most lay eggs, but some give birth to live young.

Crocodilians are the most intelligent of the reptiles. Crocodiles, alligators, and caimans all live in or near water and are predators with powerful jaws, a long snout, and sharp teeth. The crocodilians are the only reptiles with a four-chambered heart, like mammals and birds. They lay eggs, and like many of the birds, they show parental behavior, such as guarding nests and assisting the hatchlings.

BIRDS

Birds are vertebrates with feathers. They originated when Mesozoic reptiles began an adaptive radiation. They diverged from a lineage of theropod dinosaurs, a type of small carnivore that ran about on two legs. Like existing reptiles, birds have a cloaca, an opening that serves in both excretion and reproduction. Birds make amniote eggs that are fertilized internally and develop outside the body (Figure 16.29*a*).

The evolution of bird flight involved more than feathers. It involved modification of the entire body, from internal bone structure to highly efficient modes of respiration and circulation. Elastic sacs that connect to lungs greatly enhance the uptake of oxygen. Like mammals, birds have a large, four-chambered heart. The heart pumps oxygen-rich blood to the lungs and to the rest of the body in separate circuits.

Each bird wing, a modified forelimb, consists of feathers and lightweight bones attached to powerful muscles. Air cavities in bones make them lightweight. The flight muscles attach to an enlarged breastbone (sternum) and to the upper limb bones attached to it. When the muscles contract, they produce a powerful downstroke (Figure 16.29*b*).

Modern birds differ in size, proportions, colors, and their capacity for flight. The tiny bee hummingbird weighs 1.6 grams (0.6 ounces). The ostrich, the largest living bird, can weigh 150 kilograms (330 pounds). It cannot fly, but it can sprint on powerful legs. Many of the birds show impressive capacities for learning and for social behavior. Bird song and other behaviors are topics in Chapter 32.

IMPACTS, ISSUES

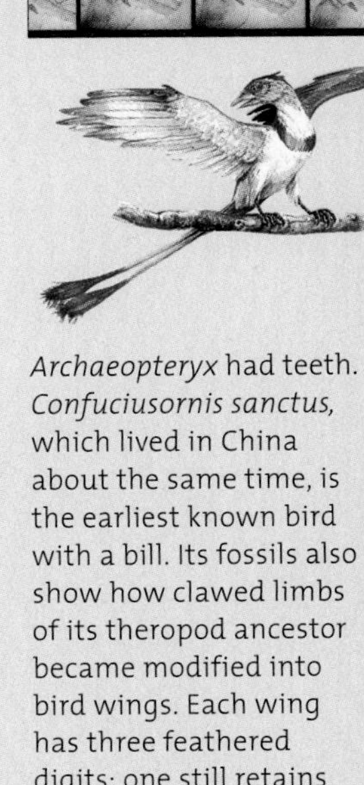

Archaeopteryx had teeth. *Confuciusornis sanctus*, which lived in China about the same time, is the earliest known bird with a bill. Its fossils also show how clawed limbs of its theropod ancestor became modified into bird wings. Each wing has three feathered digits; one still retains a large reptile-like claw.

Figure 16.30 Representatives of the three mammalian groups: (**a**) Humans are placental mammals. (**b**) An egg-laying mammal, the platypus. (**c**) Koala, a pouched mammal.

The smallest and the largest living placental mammals: (**d**) Kitti's hognosed bat, at 1.5 grams; and (**e**) blue whale, up to 200 tons.

MAMMALS

How do we define **mammals**? They alone have hair, although they differ in its distribution, amount, and type. Only mammalian females can secrete milk from mammary glands. Jaws and teeth are also distinctive. Mammals have a single lower jawbone and their four types of upper and lower teeth match up. The reptiles have a hinged two-part jaw, all teeth are one type only, and the teeth do not match up.

Where did mammals come from? Remember, the extinct mammal-like reptiles and all extinct and living species of mammals are *synapsids*. The defining trait of this group is an opening in a particular location on each side of the skull that functions in the attachment and alignment of jaw muscles. The trait arose in the Carboniferous. Later on, in the Permian, the synapsids underwent adaptive radiations and became the major land vertebrates. At the close of the Permian, many groups died out in a mass extinction.

Fortunately for humans, the group called *therapsids* endured. Over time, limbs became positioned upright under their trunk. Their skeletal arrangement made it easier to walk erect, and brain regions that function in balance and spatial positioning increased in size.

The earliest groups of modern mammals evolved early in the Jurassic. These were mouselike forms that coexisted with dinosaurs. By 130 million years ago, three lineages had evolved: **monotremes** (egg-laying mammals), **marsupials** (the pouched mammals), and **eutherians** (placental mammals). Figure 16.30 shows examples of each group.

Three species of monotremes survive and—with the exception of opossums—most pouched mammals are in Australia or New Zealand. In contrast, the placental mammals are widely distributed. What gave them the competitive edge? The placental mammals had higher metabolic rates, better body temperature control, and a more efficient method to nourish embryos. You will read more about this in later chapters. For now, note that the success of placental mammals was achieved at the expense of the other mammal lineages. When the placental mammals entered areas where monotremes or marsupials had evolved in their absence, placental mammals drove most nonplacental ones to extinction.

Amniotes, the first vertebrates to escape reproductive dependence on outside water, have water-conserving skin and kidneys, and amniote eggs. Synapsids (mammals and extinct mammal-like forms) are one major group. Sauropsids ("reptiles" and birds) are another.

Much of amniote history was a matter of luck, of species with particular traits being in the right or wrong places on a changing geologic stage at particular times.

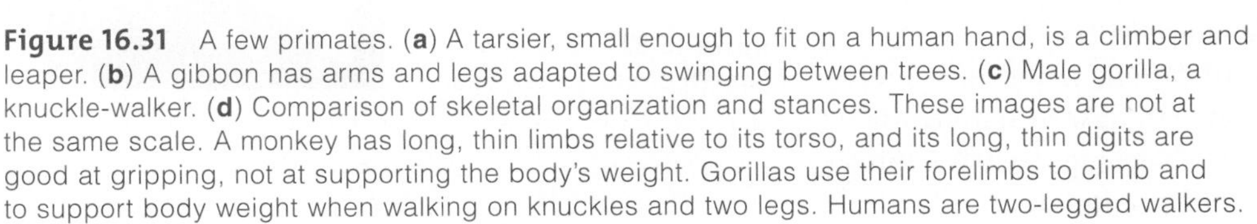

Figure 16.31 A few primates. (**a**) A tarsier, small enough to fit on a human hand, is a climber and leaper. (**b**) A gibbon has arms and legs adapted to swinging between trees. (**c**) Male gorilla, a knuckle-walker. (**d**) Comparison of skeletal organization and stances. These images are not at the same scale. A monkey has long, thin limbs relative to its torso, and its long, thin digits are good at gripping, not at supporting the body's weight. Gorillas use their forelimbs to climb and to support body weight when walking on knuckles and two legs. Humans are two-legged walkers.

16.13 From Early Primates to Humans

So far, you've traveled 570 million years through time, from tiny flattened animals to craniates, then on to the jawed fishes with a backbone. You met some early tetrapods, then amniotes—reptiles, birds, and mammals. Now you are about to access what we know about an evolutionary road that led to modern humans.

Tarsiers, prosimians, and anthropoids are the living **primates**. Prosimians include lemurs and lorises. The anthropoids are monkeys, apes, and humans. Apes are gibbons, siamangs, orangutans, gorillas, chimpanzees, and bonobos (Figure 16.31).

The oldest primate fossils are 55 million years old, but molecular evidence suggests the lineage arose 30 or so million years earlier. Molecular comparisons also reveal that apes and humans are close relatives. They are grouped together as hominoids. All extinct human-like species and modern humans are **hominids**.

Most primate species live in tropical or subtropical forests, woodlands, or grasslands with scattered trees. Most are tree dwellers. Yet no one feature sets them apart from other mammals, and each lineage has its defining traits. Five trends define the lineage that led to hominids, and then to humans.

First, there was less reliance on the sense of smell and more on daytime vision. *Second,* skeletal changes promoted **bipedalism**—upright walking—which freed hands for novel tasks. *Third,* bone and muscle changes led to diverse hand motions. *Fourth,* teeth became less specialized. *Fifth,* the evolution of the brain, behavior, and culture became interlinked. **Culture** is the sum of all behavioral patterns of a social group, passed on through generations by way of learning and symbolic behavior. In brief, *our uniquely human traits emerged by modification of traits that arose earlier, in ancestral species.*

Enhanced daytime vision. Early primates had an eye on each side of their head. Later species had forward-directed eyes, which were better at sampling shapes and movements in three dimensions. Other changes in vision helped individuals respond to variations in light intensity (dim to bright) and color.

Upright walking. How a primate walks depends on arm length and the shape and position of the shoulder blades, pelvic girdle, and backbone. With arm and leg bones about the same length, a monkey can climb and run on four legs. A gorilla walks on two legs and the knuckles of its long arms. Humans and bonobos stride on two legs. Humans have the shortest, most flexible backbone, which has an S-shaped curve. Bipedalism contributed to such structural differences.

Power grip and precision grip. So how did we get our amazing hands? Early mammals spread their toes

LINKS TO SECTIONS 13.3, 13.4

apart to support the body as they walked or ran on four legs. Among the ancient tree-dwelling primates, modifications to hand bones allowed them to curl their fingers around an object (*prehensile* movements) and to touch their thumb to fingertips (*opposable* movements). Later, power and precision grips evolved among other lineages of primates:

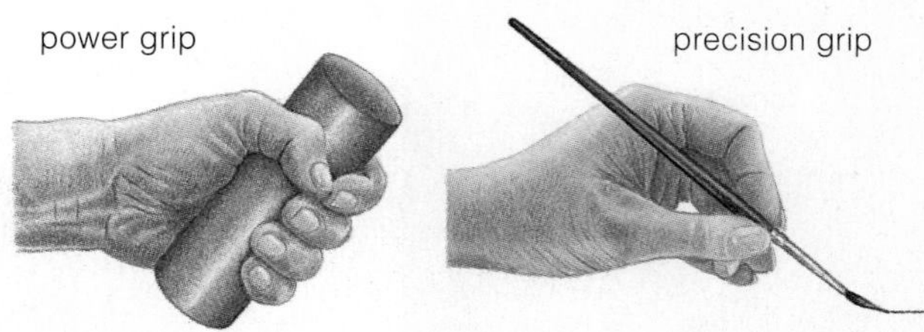

These hand positions gave human ancestors a capacity to make and to use tools. They were a basis for unique technologies and cultural development.

Teeth for all occasions. Modifications to the jaws and teeth accompanied a shift from eating insects to fruits and leaves, then to a mixed diet. Rectangular jaws and lengthy canines evolved in monkeys and apes. A bow-shaped jaw evolved in early hominid lineages. So did teeth that were smaller, and all about the same length.

Brains, behavior, and culture. Shifts in reproductive and social behavior accompanied the shift to life in the trees. In many lineages, parents invested greater effort in fewer offspring. Maternal care became more intense, and the young required longer periods of dependency before setting out on their own.

New forms of behavior promoted new connections in brain regions that deal with sensory inputs. A brain having more intricate wiring favored new behavior. As the brain coevolved with behavior, culture also evolved over time. It did so through enrichment in learning and symbolic behaviors, especially language. The capacity for language arose among hominids. It required more volume in the bony chamber holding the brain, as well as greater brain complexity.

ORIGINS AND EARLY DIVERGENCES

Fossils suggest that primates arose in tropical forests. Like the tiny rodents and tree shrews they resembled (Figure 16.32), they had voracious appetites and spent their nights foraging for eggs, insects, seeds, and buds beneath trees or scrambling about in lower branches. They had a long snout and an acute sense of smell.

During the Eocene (about 55 to 37 million years ago), some primates began to radiate into a new part of the habitat. They began to climb higher up in the trees. Compared with the earlier primates, these high climbers had a shorter snout, a larger brain, far better daytime vision, and much better ways to grasp and hold objects. *How did these traits evolve?*

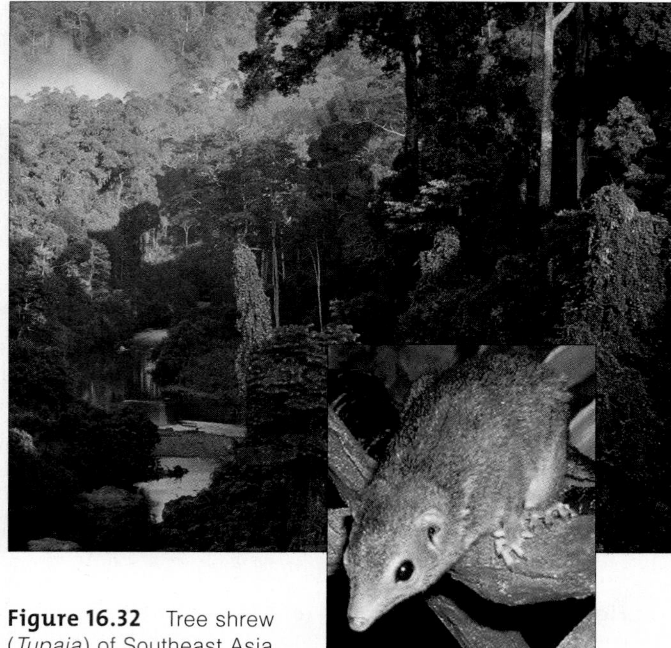

Figure 16.32 Tree shrew (*Tupaia*) of Southeast Asia.

Consider the trees. Trees offered food and safety from predators on the ground. They also were adaptive zones of uncompromising selection. Picture an Eocene morning with dappled leaves swaying in the breeze, colorful fruit, and predatory birds. An odor-sensitive, long snout would not have been of much use; breezes disperse odors. But a brain that could assess motion, depth, shape, and color was favored. So were skeletal changes that increased fitness. For instance, eye sockets facing forward instead of on the sides of the skull are better for depth perception.

By 36 million years ago, tree-dwelling anthropoids had evolved. They were on or close to the lineage that led to monkeys and apes. One had forward-directed eyes, a snoutless, flattened face, and an upper jaw that sported shovel-shaped front teeth. It probably used its hands to grab food. Some early anthropoids lived in swamps infested with predatory reptiles. Was that one reason why it became so imperative to think fast, grip strongly, and avoid adventures on the ground? Maybe.

Between 23 and 5 million years ago, in the Miocene, apelike forms—*the first hominoids*—evolved and spread through Africa, Asia, and Europe. Shifts in land masses and ocean circulation were causing a long-term change in climate. Africa became cooler and drier, with more seasonal change. Tropical forests, with their edible soft fruits, leaves, and abundant insects, gave way to open woodlands and later to grasslands. Food had become drier, harder, and more difficult to find.

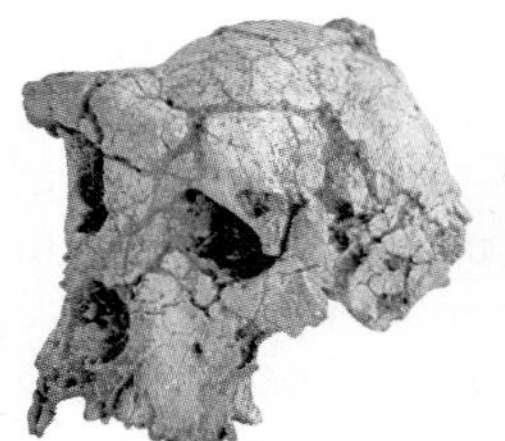

Sahelanthropus tchadensis
7–6 million years

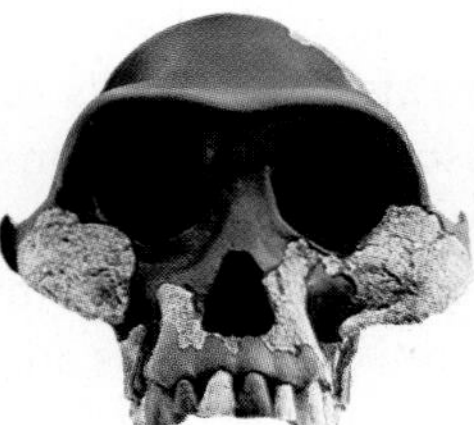

Australopithecus afarensis
3.6–2.9 million years

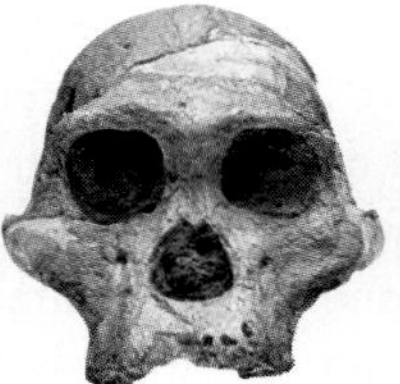

A. africanus
3.2–2.3 million years

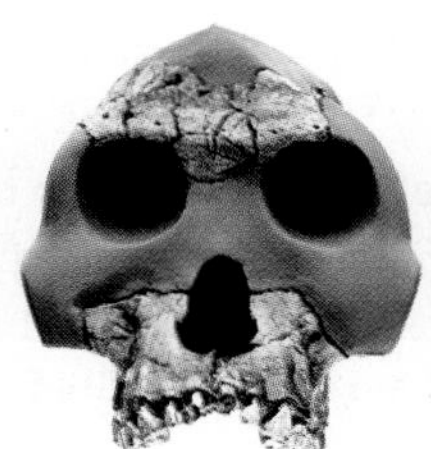

A. garhi
2.5 million years
(first tool user?)

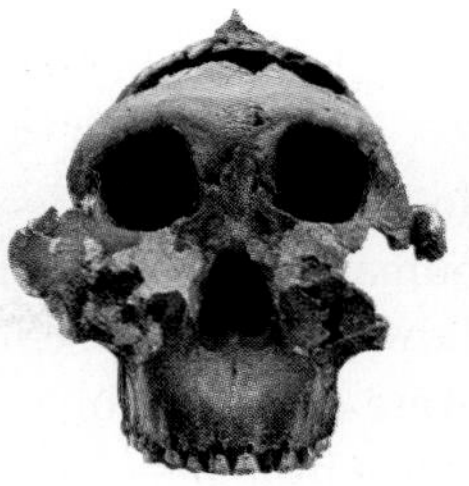

Paranthropus boisei
2.3–1.4 million years
(huge molars)

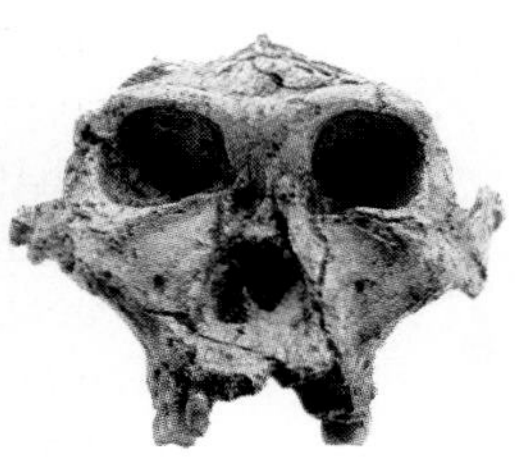

P. robustus
1.9–1.5 million years

a

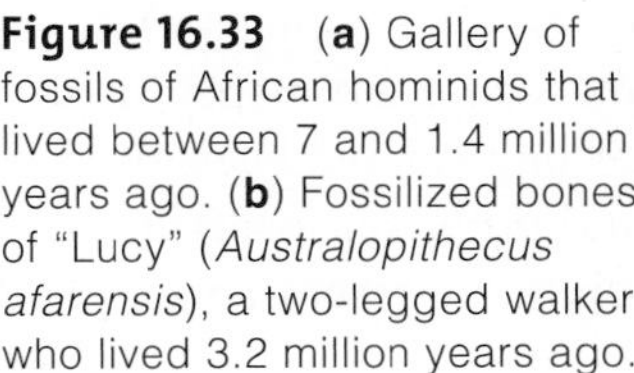

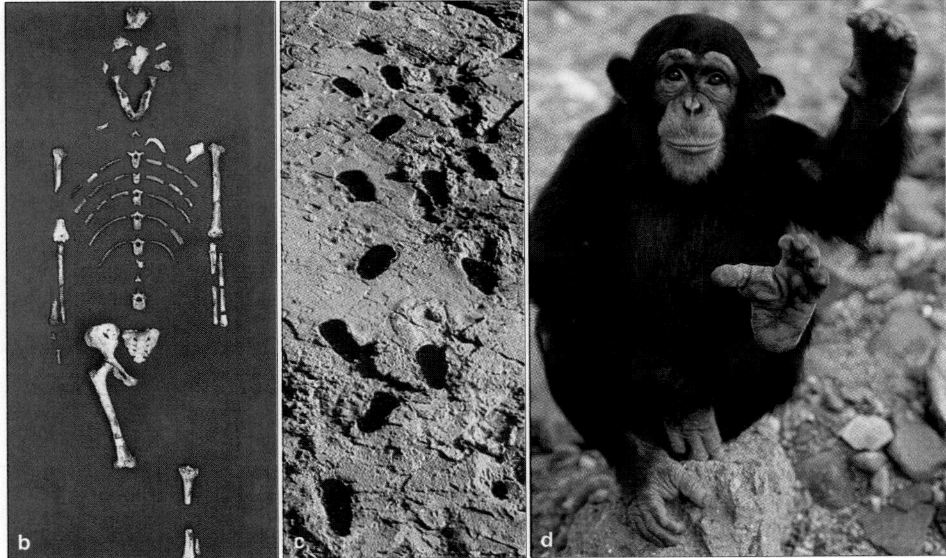

Figure 16.33 (**a**) Gallery of fossils of African hominids that lived between 7 and 1.4 million years ago. (**b**) Fossilized bones of "Lucy" (*Australopithecus afarensis*), a two-legged walker who lived 3.2 million years ago.

(**c**) At Laetoli in Tanzania, Mary Leakey found these footprints, made in soft, damp, volcanic ash 3.7 million years ago. The arch, big toe, and heel marks of these footprints are signs of bipedal hominids. Unlike apes, the early hominids didn't have a splayed-out big toe, as the chimpanzee in (**d**) obligingly demonstrates.

Hominoids that had evolved in the lush forests had two options: Move into new adaptive zones or die out. Most died out, but not the common ancestor of apes and, much later in time, modern humans.

THE FIRST HOMINIDS

When did the hominid lineage arise? *Sahelanthropus tchadensis* evolved in Central Africa as early as 6 or 7 million years ago (Figure 16.33*a*). Is it an ape or early hominid? Opinions vary. The braincase of the single fossil discovered thus far is no bigger than that of a chimpanzee. Yet, like the bigger brained hominids that evolved later in time, it has a shorter, flattened face, a pronounced brow ridge, and smaller canines.

The Miocene on through the Pliocene was a "bushy" time of evolution for hominids in central, eastern, and southern Africa. By this we mean that many diverse forms were evolving. *Ardipithecus ramidus* had evolved by 4.4 million years ago. It is the earliest of the fossil hominids now informally known as **australopiths** (for "southern apes"). *Australopithecus* and *Paranthropus* are others (Figure 16.33*a*,*b*).

We don't know exactly how australopiths are related to one another or to humans. But one or more species of *Australopithecus* are most likely among our ancestors.

Like apes, australopiths had a large face relative to their braincase, a relatively small brain, and jaws that protruded. But they differed in critical respects from earlier hominoids. For one thing, their thick-enameled molars could grind up harder food. For another, they walked upright. We know this from fossil pelvis and limb bones—and from footprints. About 3.7 million years ago, a group of *A. afarensis* walked across some newly fallen volcanic ash. A light rain transformed the ash to a stone-like hardness and preserved their tracks for millions of years (Figure 16.33*c*).

As bipedal hominids evolved in the late Miocene, they were still adapted to forest life. Their hands were adapted for strong gripping, not precision movement. Over time, descendants grew more fully upright and the dexterity of their hands increased. Some may have made tools. In an area of Ethiopian desert, researchers discovered antelope bones that appear to have been cut using stone tools. Fossils of *A. garhi* were found in the same sediments and date back 2.5 million years.

LINKS TO SECTIONS 13.3, 13.4, 13.7

EMERGENCE OF EARLY HUMANS

What do fossilized fragments of early hominids tell us about human origins? The record is still too sketchy to reveal how all the diverse forms were related, let alone which were our ancestors. Besides, exactly which traits characterize **humans**—members of the genus *Homo*?

What about brains? Our brain provides the basis of analytical and verbal skills, complex social behavior, and technological innovation. It sets us apart from apes that have a far smaller skull volume and brain size. Yet this feature alone can't tell us when certain hominids made the evolutionary leap to becoming human. Their brain size fell within the range of apes. They did make simple tools—but so do chimps and some birds. We have few clues to their social behavior.

We are left to speculate on physical traits—skeletal evidence of bipedalism, manual dexterity, larger brain volume, small face, and small, more thickly enameled teeth. These traits emerged late in the Miocene. They are features of what might be the first humans—*Homo habilis*. The name means "handy man."

Between 2.4 and 1.6 million years ago, australopiths and early forms of *Homo* lived together in woodlands bordering the savannas of eastern and southern Africa (Figure 16.34). Fossilized teeth suggest that they ate hard-shelled nuts, dry seeds, soft fruits and leaves, and insects—all seasonal foods. Most likely, *H. habilis* had to think ahead, to plan when to gather and store foods for cool, dry seasons ahead.

H. habilis shared its habitat with carnivores such as saber-tooth cats. A big cat's teeth could impale prey and tear flesh but could not break open bones to get at the marrow. Marrow and bits of meat still clinging to bones may have provided an important protein source to early hominids. *H. habilis* and other forms may have enriched their diet by scavenging carcasses, perhaps breaking open bones by striking them with stones.

We have many stone tools dating to the time of *H. habilis*, although we cannot be sure if they made them. Paleontologist Mary Leakey was the first to discover evidence of stone toolmaking. In Tanzania's Olduvai Gorge, she showed that the deepest sediments include crudely chipped stones that might have been used to smash food, dig up roots, and probe for insects in bark (Figure 16.35). The shallower, more recent layers yield more complex tools.

Production of even simple stone tools required a fair amount of brainpower and dexterity. Tool makers did not simply slam rocks together, but rather chose as raw materials rocks that were suited for sharpening, then shaped them with repeated angled blows from another rock.

EMERGENCE OF MODERN HUMANS

Between 1.9 and 1.6 million years ago, strikingly new forms evolved on the plains of Africa. Compared to *H. habilis*, they had bigger brains, stood taller, and were better adapted for walking upright for long distances. The most impressive fossil is the nearly complete skeleton of a boy, estimated to be nine to twelve years old, who died 1.6 million years ago. It is thought that had he lived, he would have stood nearly six feet tall.

Some researchers place all of these more modern-looking hominids into one species, *H. erectus*. Others distribute them into *H. erectus* and *H. ergaster* (Figure 16.36). All now agree that the new hominids were more creative toolmakers, and that social organization and communication skills must have been well developed.

H. erectus was the first hominid to leave Africa. By 1.8 million years ago, this species had expanded its range into Soviet Georgia. It eventually spread as far as Indonesia. *H. ergaster* stayed put in Africa and gave rise to *H. heidelbergensis*. Some members of this new species stayed in Africa; they are considered the most likely ancestors of *H. sapiens*. Other populations made their way into Europe, where they probably gave rise to Neandertals (*H. neanderthalensis*) (Figure 16.37).

H. habilis
2.4?–1.6 million years

a

b

Figure 16.34 (**a**) Fossil of an early human form. (**b**) Reconstruction of *Homo habilis* in an East African woodland. Two australopiths are in the distance.

Figure 16.35 Olduvai Gorge in Africa, and a sampling of stone tools. *Left to right:* crude chopper, more refined chopper, hand ax, and cleaver.

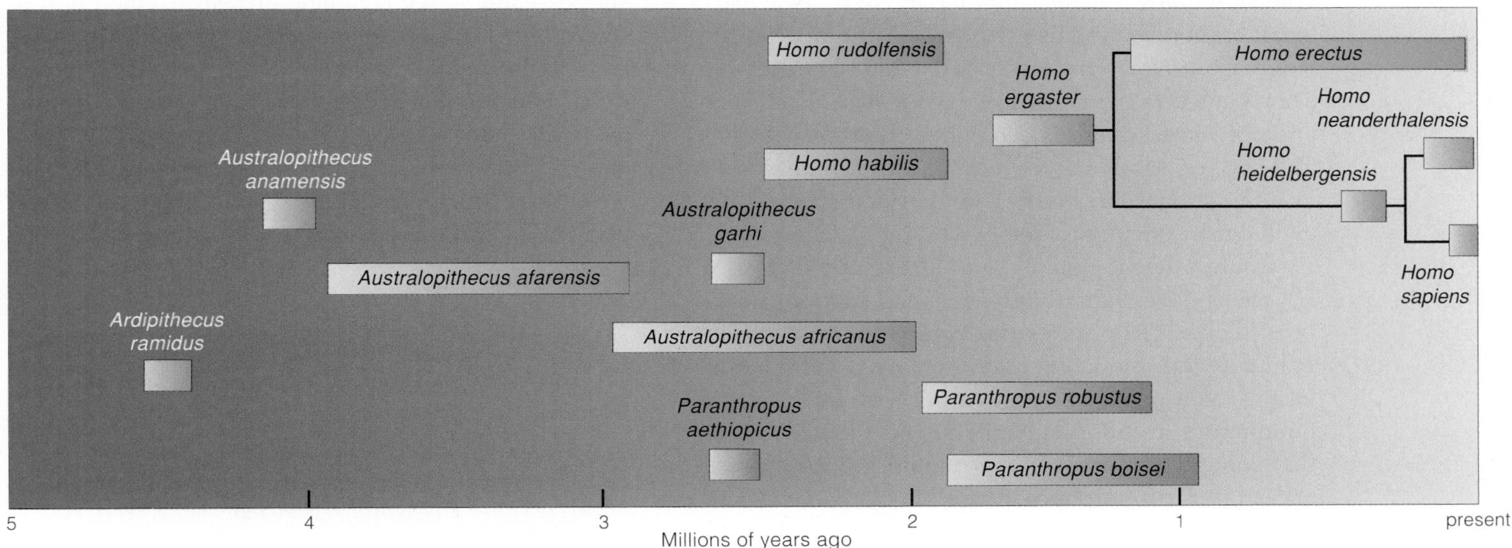

Figure 16.36 A timeline for the appearance and extinction of major hominid species (currently placed in four genera). The lines at the upper right depict one view of the evolutionary relationships among the most recent species. The number of hominid species, which fossils should be assigned to each species, and how the species are connected are all matters of heated debate among paleontologists.

At this writing, analysis of fossils from Ethiopia has pushed the earliest appearance of *Homo sapiens* back to 190,000 years ago. This new species had smaller teeth, facial bones, and jawbones. Many individuals also had a novel feature—a prominent chin. They had a higher, rounder skull and a bigger brain.

Like *H. sapiens*, the Neandertals had big brains, but they were heavier built, probably as an adaptation to life in far colder climates. They became extinct about 30,000 years ago, about the time that modern humans entered their range. There is no evidence that matings between the two species produced offspring or had any significant effect on the human gene pool. DNA from Neandertal fossils includes many sequences that are not represented in any present-day humans.

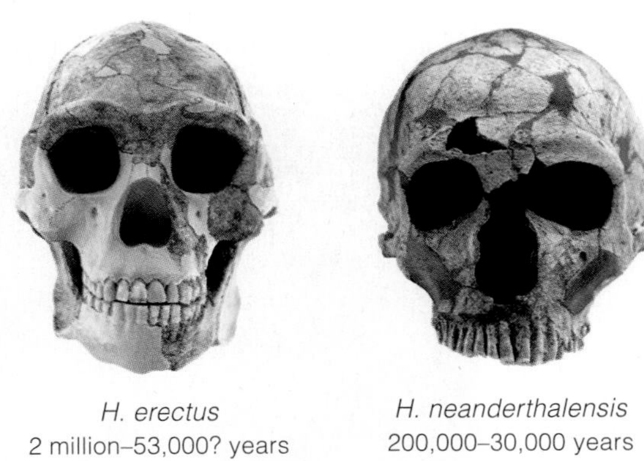

Figure 16.37 Fossils of two species of early humans. Evidence indicates that both species coexisted for a time with fully modern humans, *Homo sapiens.*

LINK TO SECTION 13.7

WHERE DID MODERN HUMANS ORIGINATE?

If we are interpreting the fossil record correctly, then Africa was the cradle for us all. At this writing, every human fossil older than 2 million years is from Africa. Between 2 million and 500,000 years ago, *H. erectus* populations left Africa in waves. They dispersed into the grasslands, forests, and mountains of Europe and then continued along into Asia.

Did the geographically separated populations of *H. erectus* evolve in regionally distinct ways in response to local selection pressures? Probably. In 2003, a team working on the Indonesian island of Flores found the 18,000-year-old remains of an adult female hominid who stood only a meter tall. Like *H. erectus*, she had a small brain (even relative to her size) and heavy brow ridges—but *H. erectus* disappeared 200,000 years ago. They named the fossil as a new species, *H. floresiensis*, and postulated that it is descended from *Homo erectus*. The species' small stature may be a trait that arose as an adaptation to life on the island. Island species often evolve to be smaller than their continental ancestors.

Could other regionally specialized populations of *H. erectus* have similarly given rise to subpopulations of *H. sapiens*? This is the idea behind the *multiregional* model. By this model, *H. sapiens* emerged in numerous locales, as a result of selective pressure on *H. erectus* populations in differing habitats. This model suggests that some differences among human groups arose in ancestral *H. erectus* populations.

The *African emergence* model has modern humans arising in sub-Saharan Africa between 200,000 and 100,000 years ago, then moving out of Africa (Figure 16.38). In each region where they settled, they replaced *H. erectus* populations that had preceded them. Then regional phenotypic differences were added on top of a basic *H. sapiens* body plan. The multiregional model emphasizes the differences among subpopulations of humans; this model emphasizes similarities.

In support of this model, the oldest *H. sapiens* fossils are from Africa. Also, in Zaire, finely wrought barbed-bone tools indicate that African populations were as skilled at producing tools as any *Homo* populations in Europe. Gene patterns from forty-three ethnic groups in Asia suggest that modern humans migrated from Central Asia, along India's coast, then into Southeast Asia and southern China. They later dispersed north and west across China and Russia into Siberia, where they crossed the Bering Strait into the Americas.

For the past 40,000 years, our cultural evolution has been outpacing macroevolution of the human species—and so we complete our story. One point to remember: Humans spread rapidly through the world by devising *cultural* means to deal with their diverse range of environments. Hunters and gatherers persist in certain parts of the world even as other groups have moved from "stone-age" technology to the age of high tech. Such differences highlight the great behavioral plasticity and depth of human adaptations.

Primates evolved from small, rodentlike mammals that moved into the trees 60 million years ago. By 7 million years ago, divergences had given rise to the first hominids.

These key adaptations evolved along the pathway from arboreal primates to modern humans: complex, forward-directed vision; bipedalism; refined hand movements; more generalized teeth; and an interlocked elaboration of brain regions, behavior, and culture.

Cultural evolution has outpaced the biological evolution of the only remaining human species, H. sapiens. *Today, humans everywhere rely on cultural innovation to adapt rapidly to a broad range of environmental challenges.*

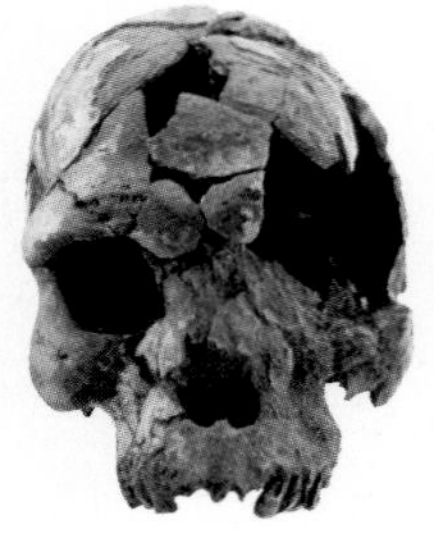

H. sapiens fossil from Ethiopia, 160,000 years old

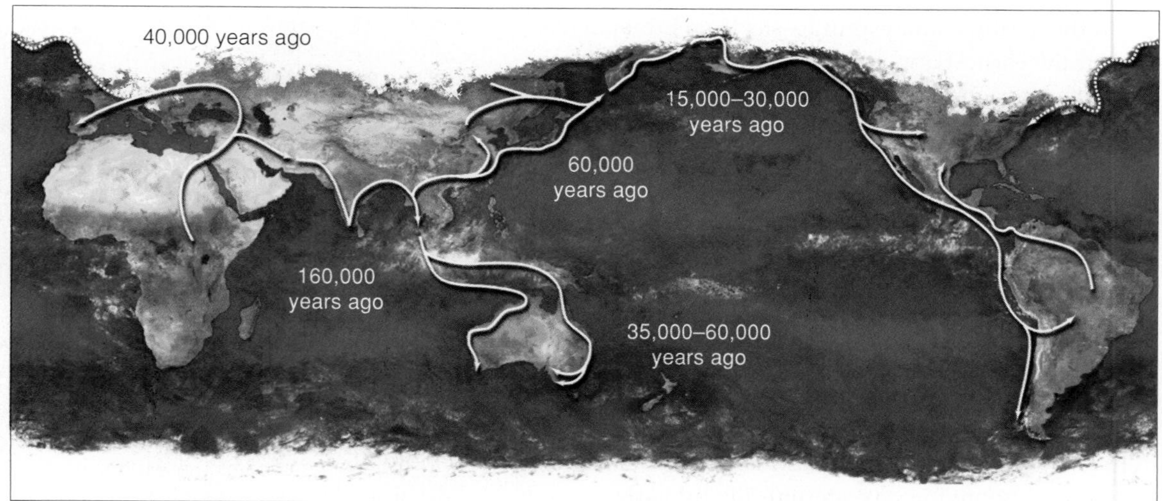

Figure 16.38 The estimated times when early *H. sapiens* reached different parts of the world, based on the radiometric dating of fossils. White lines show the presumed dispersal routes. Molecular data show that very few populations left Africa. Genetically, all living humans are extremely similar.

Summary

Section 16.1 Animals are multicelled heterotrophs that ingest or parasitize other organisms. Most have diploid body cells organized into tissues, organs, and organ systems. Animals reproduce sexually; many also reproduce asexually. They develop through embryonic stages. Most are motile during at least part of the life cycle. Table 16.1 summarizes the major groups.

Section 16.2 Sponges are simple filter feeders with no body symmetry or tissues. The radially symmetrical cnidarians, such as jellyfishes, sea anemones, corals, and hydras, are at the tissue level of organization. They are the only animals that make nematocysts: capsules with tubular, dischargeable threads that help capture prey.

Biology Now

See the water flow through a sponge and the action of a cndiaran nematocyst with animation on BiologyNow.

Sections 16.3, 16.4 Flatworms have the simplest organ systems. They have a saclike gut and no coelom. Free-living and parasitic turbellarians, tapeworms, and flukes are in this group. Annelids, highly segmented worms, have bristled segments. They are oligochaetes (such as earthworms), polychaetes, and leeches.

Biology Now

Learn about the organ systems of a flatworm and the life cycle of a tapeworm with animation on BiologyNow.

Section 16.5 Only mollusks form a mantle, which is an extension of the body mass that drapes back on itself like a skirt. Bivalves, gastropods, and cephalopods (that move by water-jet propulsion) are representatives.

Sections 16.6, 16.7 Roundworms and arthropods have a cuticle that they molt as they grow. Roundworms are free-living or parasitic, have a complete gut, and have a false coelom. Arthropods have a coelom and a jointed exoskeleton. The body has modified segments and specialized appendages. Living arthropods are the arachnids, crustaceans, centipedes, millipedes, and insects. Insects include the only winged invertebrates and are the most numerous and diverse animals.

Section 16.8 Echinoderms such as sea stars have an exoskeleton of spines, spicules, or plates of calcium carbonate. Adults are radial, but bilateral ancestry is evident in their larval stages and other features.

Section 16.9 Four features help define the chordates. The embryos develop a notochord, a dorsal hollow nerve cord (which becomes a brain and spinal cord), a pharynx with gill slits, and a tail extending past the anus. Some or all of these features persist in adults.

Lancelets and tunicates are invertebrate filter-feeding chordates. Craniates are chordates with a brain chamber of cartilage or bone. Most craniates are vertebrates; they have a vertebral column (backbone) of cartilage or bone. Tetrapods are vertebrates with four limbs or with four-limbed ancestors. Amniotes are the only vertebrates that make amniote eggs. Jaws, paired fins, and later, lungs, were key innovations that led to the adaptive radiation of vertebrates.

Biology Now

Explore the body plan and chordate features of a lancelet with animation on BiologyNow.

Section 16.10 Cartilaginous fishes and bony fishes are aquatic vertebrates. Bony fishes are the most diverse vertebrates, and one group gave rise to the tetrapods.

Section 16.11 Amphibians, the first tetrapods on land, share traits with aquatic tetrapods and reptiles. Existing groups include salamanders, frogs, and toads. Many species are now threatened or endangered.

Section 16.12 Amniotes were the first vertebrates that did not require external water for reproduction. They have water-conserving skin and kidneys, and produce amniote eggs. Mammals and their ancestors are one amniote lineage. "Reptiles" and birds (the only animals that have feathers) are another.

There are three lineages of mammals: egg-laying mammals, pouched mammals, and placental mammals. Today, placental mammals are the most diverse group and have displaced others in most habitats.

Table 16.1 Major Animal Groups

Group	Some Representatives	Existing Species
Poriferans	Sponges	8,000
Cnidarians	Hydras, jellyfishes	11,000
Flatworms	Turbellarians, flukes, tapeworms	15,000
Annelids	Leeches, earthworms, polychaetes	15,000
Mollusks	Snails, slugs, squids, octopuses	110,000
Arthropods	Crustaceans, spiders, insects	1,000,000+
Roundworms	Pinworms	20,000
Echinoderms	Sea stars, sea urchins, sea cucumbers	6,000
Chordates	Invertebrate chordates:	
	Tunicates, lancelets	2,100
	Vertebrate chordates:	
	Jawless fishes	84
	Jawed fishes	21,000
	Amphibians	4,900
	Reptiles	7,000
	Birds	8,600
	Mammals	4,500

Table 16.2 Primate Classification

Prosimians (lemurs, lorises)
Tarsiers
Anthropoids
- New World monkeys
- Old World monkeys
- Hominoids
  - Apes
  - Hominids

Section 16.13 The hominoid branch of the primate family tree evolved in Africa as the climate became cooler and drier. It includes apes and the hominids: humans and their most recent ancestors (Table 16.2).

Australopiths were early hominids, and they walked upright. The relationships among the diverse forms are sketchy. The first known members of the genus *Homo* lived in Africa between 2.4 million and 1.6 million years ago. *H. erectus* was the first hominid to migrate out of Africa. *H. sapiens* evolved by 190,000 years ago.

Self-Quiz

Answers in Appendix I

1. Only _______ have a notochord, a tubular dorsal nerve cord, a pharynx with gill slits, and a tail past the anus.
   a. earthworms c. chordates
   b. echinoderms d. both b and c

2. The _______ have a cuticle and molt as they grow.
   a. sponges d. arthropods
   b. cnidarians e. both c and d
   c. annelids

3. Reptiles moved fully onto land owing to _______ .
   a. tough skin d. amniote eggs
   b. internal fertilization e. a and c
   c. good kidneys f. all of the above

4. _______ preceded the emergence of modern humans.
   a. Craniates d. Amniotes
   b. Jawed vertebrates e. Hominoids
   c. Tetrapods f. all of the above

5. A coelom is a _______ .
   a. lined body cavity c. sensory organ
   b. resting stage d. bristle

6. The _______ are filter feeders.
   a. cephalopods c. lancelets
   b. flatworms d. sea stars

7. Earthworms are most closely related to_______ .
   a. leeches c. tapeworms
   b. flukes d. roundworms

8. All animals _______ .
   a. are motile for at least some stage in the life cycle
   b. consist of tissues arranged as organs
   c. can reproduce asexually as well as sexually
   d. both a and b

9. Match the organisms with the appropriate description.
   ____ sponges a. most diverse vertebrates
   ____ cnidarians b. no true tissues, no organs
   ____ flatworms c. jointed exoskeleton
   ____ roundworms d. mantle over body mass
   ____ annelids e. segmented worms
   ____ arthropods f. tube feet, spiny skin
   ____ mollusks g. nematocyst producers
   ____ echinoderms h. first with amniote eggs
   ____ amphibians i. hair or fur
   ____ fishes j. complete gut, false coelom
   ____ "reptiles" k. first terrestrial tetrapods
   ____ mammals l. saclike gut, no coelom

Additional questions are available on **Biology Now™**

Figure 16.39 Where did everybody go? One of the two species of tuatara (*Sphenodon*) that survived to the present.

Critical Thinking

1. In the summer of 2000, only 10 percent of the lobster population off Long Island Sound survived after a massive die-off. Many lobstermen in New York and Connecticut lost small businesses that their families had owned for generations. Some believe the die-off followed heavier sprays of pesticides, used to control mosquitoes that carry the West Nile virus. Explain how a chemical substance that targets mosquitoes might also harm or kill lobsters.

2. During the Triassic, hundreds of species of reptiles that resemble living tuataras roamed Pangea. Now only two species survive on a few small cold islands off the coast of New Zealand (Figure 16.39). Speculate on why these relic populations have survived there, despite disappearing from the rest of their once extensive range.

3. In 1798, a stuffed platypus specimen was delivered to the British Museum. At the time, many biologists were sure it was a fake, assembled by a clever taxidermist. Soft brown fur and beaverlike tail put the animal firmly in the mammalian camp. But a ducklike bill and webbed feet suggested an affinity with birds. Reports that the animal laid eggs only added to the confusion.

   We now know that platypuses burrow in riverbanks and forage for prey under water. Webbing on their feet can be retracted to reveal claws. The highly sensitive bill allows the animal to detect prey even with its eyes and ears tightly shut.

   To modern biologists, a platypus is clearly a mammal. Like other mammals, it has fur and the females produce milk. Young animals have more typical mammalian teeth that are replaced by hardened pads as the animal matures. Why do you think modern biologists can more easily accept the idea that a mammal can have some reptilelike traits, such as laying eggs? What do they know that gives them an advantage over scientists living in 1798?

4. Some paleontologists assign fossils to only four species in the genus *Homo*. Others split the genus into a many as seven species. Why is it more difficult to classify fossils than it is to classify existing species? What kinds of evidence would be useful in classifying new fossils?

17 PLANTS AND ANIMALS: COMMON CHALLENGES

IMPACTS, ISSUES

Too Hot To Handle

Heat can kill. In the summer of 2001 Korey Stringer, a football player for the Minnesota Vikings, collapsed after morning practice. His team had been working out in full uniform on a field where temperatures were in the high 90s. High humidity put the heat index above 100.

With an internal body temperature of 108.8°F and blood pressure too low to record, Stringer was rushed to the hospital. Doctors immersed him in an ice-water bath, then wrapped him in cold, wet towels. It was too late.

Stringer's blood-clotting mechanism shut down and he started to bleed internally. Then his kidneys faltered and he was placed on dialysis. He stopped breathing on his own and was put on a respirator. His heart gave up. Less than twenty-four hours after football practice had started, Stringer was pronounced dead. He was twenty-seven years old.

All organisms function best within a limited range of internal operating conditions. For us, "best" is when the body's internal temperature remains between about 97°F and 100°F. Past 104°F, blood is transporting a great deal of metabolically generated heat. Transport controls divert that heat away from the brain and other internal organs to the skin. A person's skin transfers heat to the outside environment—as long as it isn't too hot outside. Profuse sweating can dissipate more heat—but only if humidity is low enough for sweat to evaporate quickly.

As happened in Stringer's case, these normal cooling mechanisms fail above 105°F. The body's production of sweat falters, and its temperature starts to climb rapidly. The heart beats faster and often the individual becomes confused or faints. These are signs of *heat stroke*, which denatures the enzymes and other proteins that keep us alive. When heat stroke is not countered fast enough, brain damage or death is the expected outcome.

We use this sobering example as our passport to the world of anatomy and physiology. **Anatomy** is the study of body form—that is, its morphology. **Physiology** is the study of patterns and processes by which an individual survives and reproduces in the environment. It deals with how structures are put to use and how metabolism and behavior are adjusted when conditions change. It also deals with how, and to what extent, physiological processes can be controlled.

You can use a lot of this information to interpret what goes on in your own body. More broadly, you can use it to perceive commonalities among all animals and plants. Regardless of the species, *the structure of a given body part almost always has something to do with a present or past function*. Most aspects of form and function are long-standing adaptations that evolved as responses to environmental challenges.

That said, nothing in the evolutionary history of the human species suggests that the organ systems making up our body are finely tuned to handle intense football practice on hot, humid days.

How Would You Vote? *Should a coach or team doctor who does not stop team practice or a game when the heat index soars be held liable if a player dies from heat stroke? See BiologyNow for details, then vote online.*

Key Concepts

MULTICELLULAR ORGANIZATION

Each cell of a multicelled organism engages in activities that help assure its own survival. The structural and functional organization of all cells collectively sustains the whole body.

RECURRING CHALLENGES

Plants and animals both require mechanisms of internal transport, gas exchange, nutrition, ion-water balance with the external environment, and defense.

HOMEOSTASIS

Homeostasis is the maintenance of internal conditions within limits that allow the body to function normally. Both plants and animals have mechanisms of homeostasis, including systems of negative feedback control.

CELL-TO-CELL COMMUNICATION

Signaling mechanisms control and integrate body activities. Cells send messages that are received and interpreted by other cells, which alter their activities in response.

Links to Earlier Concepts

This chapter emphasizes the common processes that take place in cells and bodies of plants and animals. You learned about limitations on cell size in Section 3.2 and about the cellular effects of diffusion and osmosis in Sections 4.4 to 4.6. Keep in mind the trends in plant evolution (15.1) and animal evolution (16.1) that were introduced earlier. We will again touch on some examples of eukaryotic controls over gene expression (10.6) and how loss of controls can cause cancer (7.8).

LINKS TO SECTIONS 15.1, 16.1

17.1 Levels of Structural Organization

This chapter introduces some important concepts that will be applied in the next two units of the book. They can help deepen your understanding of how plants and animals function under stressful as well as favorable environmental conditions.

FROM CELLS TO MULTICELLED ORGANISMS

In most plants and animals, cells, tissues, organs, and organ systems split up the work of keeping the body alive. A separate cell lineage gives rise to body parts that will function in reproduction. Said another way, the plant or animal body shows a *division of labor*.

A **tissue** is a collection of particular kinds of cells and intercellular substances that interact in a task or tasks. Wood tissue and bone tissue support the body. All wood is composed of the same types of cells, just as the composition of all bone is similar.

An **organ** consists of two or more tissues, organized in specific proportions and patterns, that perform a task or tasks. A leaf that carries out photosynthesis and an eye that responds to light are examples of organs. An **organ system** is two or more organs that interact physically, chemically, or both. A plant's shoot system, with organs of photosynthesis and reproduction, is an example (Figure 17.1). Another is an animal's digestive system, which takes in food, breaks it down, absorbs breakdown products, and expels the residues.

GROWTH VERSUS DEVELOPMENT

A plant or animal body becomes structurally organized during growth and development. Generally, **growth** of a multicelled organism means its cells increase in number, size, and volume. **Development** is a succession of stages, during which specialized tissues, organs, and organ systems form. In other words, we measure growth in *quantitative* terms, and development in *qualitative* terms.

STRUCTURAL ORGANIZATION HAS A HISTORY

The structural organization of each tissue, organ, and organ system has an evolutionary history. Think back on how plants invaded the land and you have some idea of how their structure relates to function. Plants that left aquatic habitats found plenty of sunlight and carbon dioxide for photosynthesis, as well as oxygen for aerobic respiration. As they dispersed farther away from their aquatic cradle, however, they faced a new challenge: how to keep from drying out in air.

Consider that challenge when you see micrographs that reveal the internal structure of stems, and leaves (Figure 17.1). These pipelines carry water from soil to leaves. Stomata, small gaps across a leaf's epidermis, open and close in ways that conserve water (Section 5.3). Cells that have lignin-reinforced walls collectively support the upright growth of stems. Remember that

Figure 17.1 Morphology of a tomato plant (*Lycopersicon esculentum*). Different cell types make up vascular tissues that conduct either water and dissolved minerals or organic substances. Vascular tissues thread through other tissues making up most of the plant. A different tissue covers the surfaces of root and shoot systems.

Figure 17.2 Some of the structures that function in human respiration. Their cells carry out specialized tasks. Airways to a pair of organs called lungs are lined with ciliated cells that whisk away bacteria and other airborne particles that might cause infections. Inside the lungs are tubes (blood capillaries), a fluid connective tissue called blood, and thin air sacs (epithelial tissue), all with roles in gas exchange.

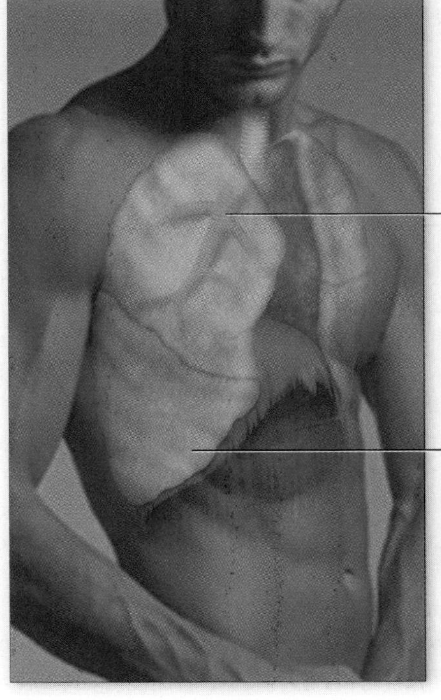

organs (lungs), part of an organ system (the respiratory tract) of a whole organism

ciliated cells and mucus-secreting cells that line respiratory airways

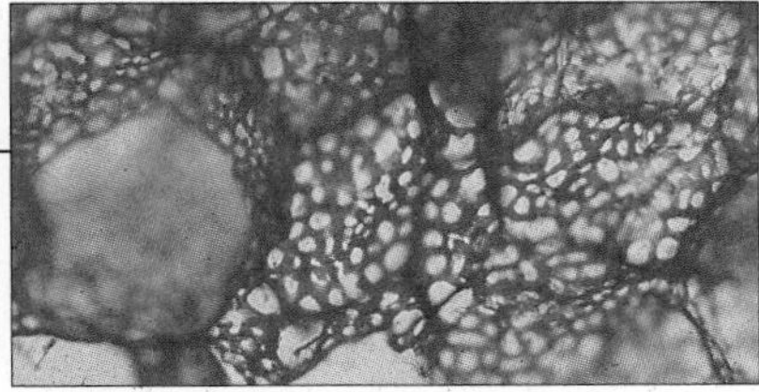

lung tissue (tiny air sacs) laced with blood capillaries—one-cell-thick tubular structures that hold blood, which is a fluid connective tissue

challenge also when you examine root systems. A soil region that has a greater concentration of water and dissolved mineral ions will stimulate root growth in its direction; hence the branching roots.

Similarly, respiratory systems of land animals are adaptations to life in air. Gases can move into and out of the body only by diffusing across a moist surface. This is not a problem for organisms that live in water, but moist surfaces dry out in air. Land animals have moist sacs for gas exchange *inside* their body (Figure 17.2). This brings us to the body's own environment.

THE BODY'S INTERNAL ENVIRONMENT

To stay alive, plant and animal cells must be bathed in a fluid that provides nutrients and carries away metabolic wastes. In this they are no different from the free-living single cells. But a plant or animal consists of thousands to many trillions of living cells. All must draw nutrients from the fluid that bathes them and dump wastes into it.

Fluid not inside body cells—extracellular fluid—is an **internal environment**. Changes in its composition and volume affect cell activities. The concentration and number of solutes in the internal environment must be maintained within the range that supports metabolism. *Survival of any plant or animal, simple or complex, requires a stable fluid environment for all of its living cells.* As you will see, this concept is central to understanding how plants and animals work.

HOW DO PARTS CONTRIBUTE TO THE WHOLE?

The next two units describe how a plant or an animal functions. The body provides its cells with a stable fluid environment. It acquires energy and essential nutrients and other raw materials, distributes them, and disposes of wastes. It has ways to protect itself against injury and attack. It also has the capacity to reproduce, and often it helps nourish and protect new individuals during their early growth and development.

The big picture is this: Each living plant or animal cell engages in metabolic activities that ensure its own survival. But the structural organization and collective activities of cells in tissues, organs, and organ systems sustain the whole body. They work to keep operating conditions in the internal environment within tolerable limits for individual cells—a state called **homeostasis**.

The structural organization of plants and animals emerges during growth and development.

Cells, tissues, and organs all require a favorable internal environment, and they collectively maintain it. All body fluids not contained in cells make up that environment.

Basic body functions include maintaining a stable internal environment, securing substances and distributing them to cells, disposing of wastes, protecting cells and tissues, reproducing, and often nurturing offspring.

LINKS TO SECTIONS 3.2, 4.4–4.6

17.2 Recurring Challenges to Survival

Plants and animals have such spectacularly diverse body plans that we sometimes forget how much they have in common.

How often do you think about the similarities between Tina Turner and a tulip (Figure 17.3)? Connections are there. They begin with what individual cells are doing inside both of those multicelled bodies. Nearly all are working in the best interest of the whole body. At the same time, the whole body is working in the best interest of individual cells.

Figure 17.3 What do these organisms have in common besides their good looks?

REQUIREMENTS FOR GAS EXCHANGE

Cells in Tina and the tulip would not last long if they could not continually exchange certain gases across their plasma membrane. Like most other heterotrophs, for instance, Tina must supply her aerobically respiring cells with oxygen and dispose of carbon dioxide end products. Like most other autotrophs, the plant must supply its photosynthetic cells with carbon dioxide. In addition, it must keep enough oxygen around for *its* aerobically respiring cells and dispose of the excess.

Certain aspects of the structure and function of all multicelled bodies are responses to this challenge: *How can gaseous molecules move to and from individual cells?*

Remember **diffusion**? This is a net movement of molecules or ions of a substance from a region where they are concentrated to an adjacent region where they are not as concentrated. Animals and plants have ways to keep gases diffusing in the directions most suitable for metabolism and cell survival. How? That question will lead you to the stomata at leaf surfaces (Chapter 18) and to the respiratory and circulatory systems of animals (Chapter 22).

REQUIREMENTS FOR INTERNAL TRANSPORT

For cells in a plant or animal to stay alive, they need a constant supply of oxygen and nutrients. If it takes too long for these substances to diffuse through the body, or from its surface, the body will shut down. This is one reason that cells and multicelled organisms have the sizes and shapes that they do.

Recall from Section 3.2 that with every increase in size, the ratio of a cell or body's surface to its volume decreases. That is the essence of the **surface-to-volume ratio**. If a cell or body were to get too big, it would not have enough surface area, relative to volume, for fast, efficient exchanges with the environment.

When a body or some body part is thin, as it is for the flatworm and lily pads in Figure 17.4, substances can easily diffuse between its individual cells and the environment. In massive bodies, however, individual cells far from an exchange point with the environment depend on systems of rapid internal transport.

In most plants and animals, substances move to and from cells through vascular tissues that act as internal pipelines. A vein in a plant leaf is a bundle of vascular tissue (Figure 17.5*a*). One tissue in the vein distributes soil water and mineral ions. Another distributes sugars.

In large animals, vascular tissues are usually blood vessels, which function as part of a circulatory system. Each time Tina takes a breath, she moves oxygen into her lungs and carbon dioxide out of them. Her blood vessels weave through lung tissues where gas exchange takes place. Blood vessels also supply her body tissues, where gases diffuse to and from the interstitial fluid and cells (Figure 17.5*b*).

As in plants, the animal vascular system transports nutrients and water. It also connects with other organ systems. Cells of the immune system use the blood vessels as their highway to move about the body. A urinary system adjusts the concentration and volume of the internal environment.

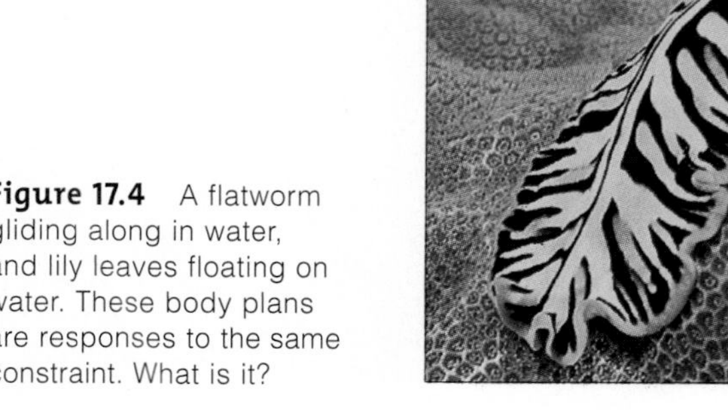

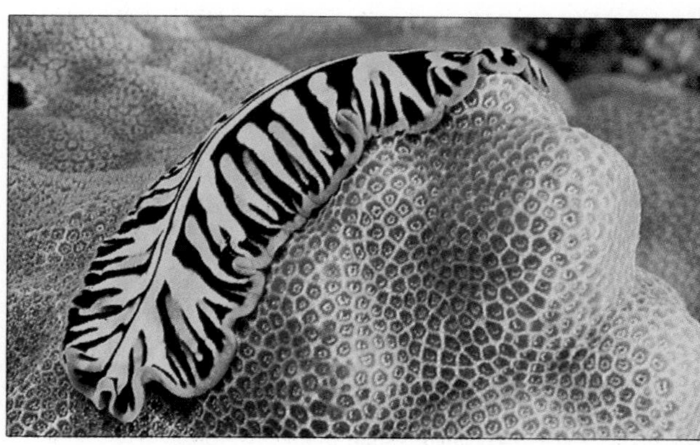

Figure 17.4 A flatworm gliding along in water, and lily leaves floating on water. These body plans are responses to the same constraint. What is it?

MAINTAINING A SOLUTE—WATER BALANCE

Plants and animals continually take in and lose water and solutes. They regularly produce metabolic wastes. Given these inputs and outputs, how do they keep the volume and composition of their internal environment within a tolerable range? Plants and animals deal with this challenge differently. Yet we can find similarities in their responses at the molecular level.

Substances tend to follow concentration gradients as they move into and out of the body, or from one of its compartments into another. At such interfaces, we see sheetlike tissues where cells passively and actively transport substances (Section 4.5). **Active transport**, recall, is a pumping of substances *against* gradients.

Active transport mechanisms in roots help control which solutes move into the plant. In leaves, they help control water loss and gas exchange by opening the stomata only at certain times. In animals, we find such mechanisms in kidneys and other organs. The active transport of ions across the membranes of cells in the kidney sets up osmotic gradients that cause water to move in ways that concentrate the urine. As you will see again and again in the next two units, *active and passive transport help maintain homeostasis in the internal environment by controlling the movements of substances into and out of all body cells.*

REQUIREMENTS FOR INTEGRATION AND CONTROL

We could go on about the structural and functional similarities between plants and animals. But we will end this discussion with a final point. In both groups, cells make signals that influence other cells and thus coordinate the body as a whole. These intercellular signaling mechanisms control how an animal or plant grows, develops, maintains, and even protects itself. The remaining sections in this chapter provide some examples of these signaling mechanisms.

ON VARIATIONS IN RESOURCES AND THREATS

Now move beyond the common challenges. Resources and dangers differ greatly among habitats. A **habitat** is the place where individuals of a species normally live.

What are the physical and chemical characteristics of the habitat? Is water plentiful, with suitable kinds and amounts of solutes? Is the habitat rich or poor in nutrients? Is it sunlit, shady, or dark? Is it warm or icy, windy or calm? How much do outside temperatures change from day to night? Are seasonal changes slight or highly pronounced?

Consider also the biotic (living) components of the habitat. What are the numbers and types of producers, predators, prey, pathogens, and parasites? Are there other species that compete for sunlight, nesting sites, or some other resource? These sorts of interactions can shape a species' anatomy and physiology.

Even in all that diversity, we often observe similar responses to similar challenges. Sharp cactus spines or sharp porcupine quills are effective deterrents to most animals that might eat the cactus or porcupine. Both arise from specialized epidermal cells. Vascular tissues and other body parts helped nurture those cells. And by contributing to building defensive structures at the body surface, epidermal cells help protect the body as a whole against an environmental threat.

Plant and animal cells function in ways that help ensure survival of the body as a whole. At the same time, tissues and organs that make up the body function in ways that help ensure survival of individual living cells.

The connection between each cell and the body as a whole is evident in the requirements for—and contributions to—gas exchange, nutrition, internal transport, stability in the internal environment, and defense.

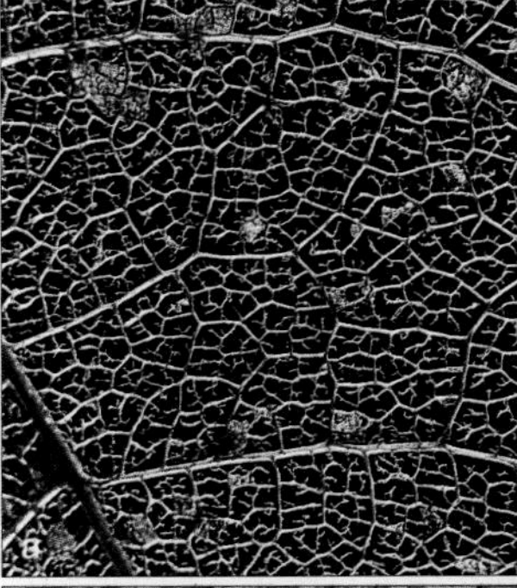

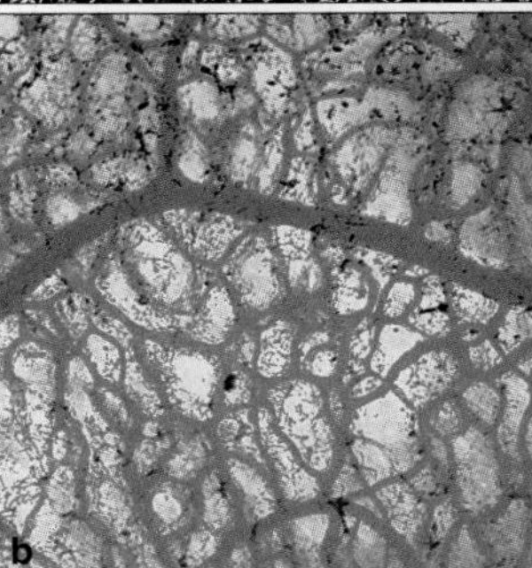

Figure 17.5 Transport lines to and from cells. (**a**) Veins visible in a decaying dicot leaf, and (**b**) some human veins and capillaries.

17.3 Homeostasis in Animals

In preparation for your trek through the next two units, take a moment to get a firm grasp on what homeostasis means to survival. Animal physiologists were the first to identify this state and describe how it is maintained, so we will start with animals.

Like other adult humans, your body has many trillions of living cells. Each cell must draw nutrients from and dump wastes into the same fifteen liters of fluid—less than sixteen quarts. Remember, fluid not inside cells is the extracellular fluid. Much is **interstitial fluid**; it fills spaces between cells and tissues. The rest is **plasma**, the fluid portion of blood. Interstitial fluid exchanges many substances with cells and with blood.

Homeostasis, again, is a state in which the body's internal environment is being maintained within the range that its cells can tolerate. In nearly all animals, three components interact and maintain homeostasis. They are sensory receptors, integrators, and effectors.

Sensory receptors are cells or cell parts that detect stimuli, which are specific forms of energy. The brain is one example of an **integrator**, a central command

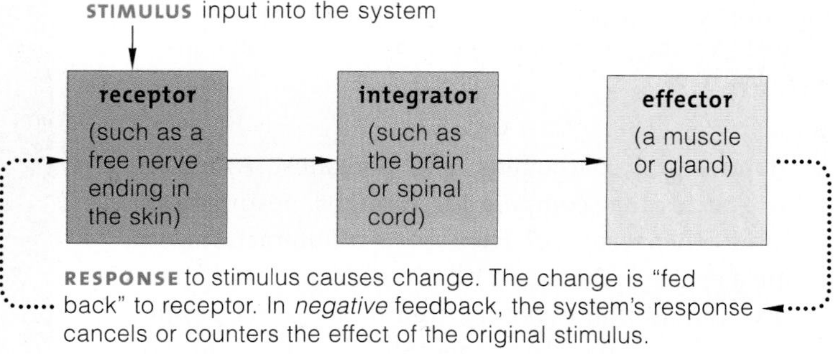

Figure 17.6 ***Animated!*** Three components of negative feedback systems in multicelled organisms.

post that pulls together information about stimuli and issues signals to the muscles, glands, or both. Muscles and glands are **effectors**. In reaction to signals from an integrator, they bring about a suitable response.

Consider how levels of carbon dioxide and oxygen are controlled in the blood. In this example, the brain is the integrator. It receives information from receptors that sense how much carbon dioxide is dissolved in blood. The brain compares this information against a set point for what the carbon dioxide level should be, and sends signals to effectors—the muscles involved in breathing. In response, breathing is altered.

Now think about what happens when a climber scales a mountain. As altitude increases, the density of the atmosphere declines, so the climber takes in fewer oxygen molecules in every breath. At altitudes above 3,300 meters, air is about half as dense as it is at sea level, and about half of climbers suffer *altitude sickness*. Receptors alert the brain, which calls on the body to hyperventilate (breathe deeper and more rapidly). If oxygen levels do not increase, headache, nausea, and vomiting will follow. All are warnings that cells are starved for oxygen.

NEGATIVE FEEDBACK

Feedback mechanisms are key homeostatic controls. In a **negative feedback mechanism**, an activity changes some condition, and when that condition shifts past a certain level, it triggers a response that reverses the change. Figure 17.6 shows one model for such control.

Think of a furnace with a thermostat. A thermostat senses the surrounding air's temperature relative to a preset point on a thermometer built into the furnace's control system. When the temperature falls below the preset point, it sends signals to a switching mechanism that turns on the furnace. When the air becomes heated enough to match the prescribed level, the thermostat signals the switch, which shuts off the furnace.

Similarly, feedback mechanisms work to keep the internal body temperature of humans and many other

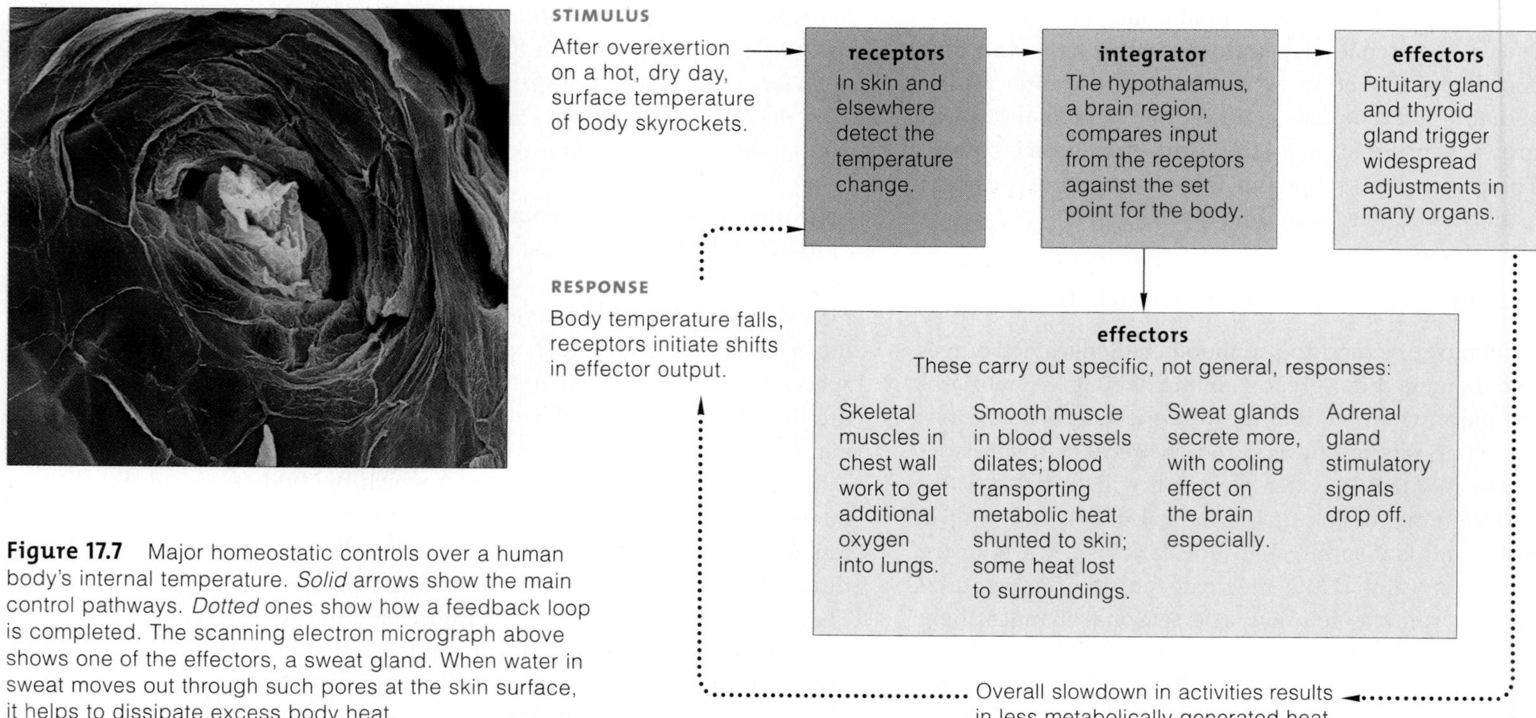

STIMULUS
After overexertion on a hot, dry day, surface temperature of body skyrockets.
receptors
In skin and elsewhere detect the temperature change.
STIMULUS input into the system
(such as a free nerve ending in the skin)
integrator
RESPONSE to stimulus causes change. The change is "fed back" to receptor. In negative feedback, the system's response cancels or counters the effect of the original stimulus.
The hypothalamus, a brain region, compares input from the receptors against the set point for the body.
effectors
Pituitary gland and thyroid gland trigger widespread adjustments in many organs.
RESPONSE
Body temperature falls, receptors initiate shifts in effector output.
effectors
These carry out specific, not general, responses:
Skeletal muscles in chest wall work to get additional oxygen into lungs.
Smooth muscle in blood vessels dilates; blood transporting metabolic heat shunted to skin; some heat lost to surroundings.
Sweat glands secrete more, with cooling effect on the brain especially.
Adrenal gland stimulatory signals drop off.
Overall slowdown in activities results in less metabolically generated heat.

Figure 17.7 Major homeostatic controls over a human body's internal temperature. *Solid* arrows show the main control pathways. *Dotted* ones show how a feedback loop is completed. The scanning electron micrograph above shows one of the effectors, a sweat gland. When water in sweat moves out through such pores at the skin surface, it helps to dissipate excess body heat.

mammals near 98.6°F (37°C) even during hot or cold weather. When a football player runs around on a hot summer day, the body becomes hot. Receptors trigger changes that should slow down the entire body *and* its cells (Figure 17.7). Under typical conditions, these and other control mechanisms avert overheating. They curb activities that naturally generate metabolic heat, and increase heat transfer to the surrounding air.

POSITIVE FEEDBACK

With a **positive feedback mechanism**, some change in conditions causes the body to respond in a manner that increases the intensity of the change. As a result, the system moves farther and farther from its initial state. Positive feedback often culminates in some event that returns the system to its starting state.

For example, the flow of sodium across the plasma membrane of a neuron allows the body to send and receive signals quickly. A stimulus causes the opening of a few sodium channels in the neuron membrane. As sodium ions flow in through the open channels, they increase the sodium levels inside the cell. This leads to the opening of additional channels and the flow of still more sodium inward, as you will learn in Chapter 25.

As another example, during labor, the fetus exerts pressure on the wall of the chamber enclosing it, the uterus. Pressure induces the production and secretion of a hormone, oxytocin, that makes muscle cells in the wall contract. Contractions exert pressure on the fetus, which puts more pressure on the wall, and so on until the fetus is expelled from the mother's body.

We have been introducing a pattern of detecting, evaluating, and responding to a flow of information about the internal and external environment. During this activity, organ systems work together. In time you will be asking these questions about their operation:

1. Which physical or chemical aspects of the internal environment are organ systems working to maintain?
2. How are organ systems kept informed of change?
3. By what means do they process the information?
4. What mechanisms are set in motion in response?

As you will read in Unit Five, organ systems of nearly all animals are under nervous and endocrine control.

Homeostatic controls maintain the body's internal environment within ranges that favor cell activities.

With negative feedback a change triggers a response that reverses the change. With positive feedback a response intensifies the change that stimulated it.

17.4 Does Homeostasis Occur in Plants?

Plants differ from animals in important respects. Even so, they, too, have controls over the internal environment.

Direct comparisons between plants and animals are not always possible. In young plants, new tissues arise only at the tips of actively growing roots and shoots. In animal embryos, tissues form all through the body. Plants do not have a centralized integrator, such as a brain. Yet plants have decentralized mechanisms that protect the internal environment and help ensure the survival for the body as a whole. It may be a bit of a leap, but we also can apply the concept of homeostasis to plants. Two examples will make the point.

RESISTING PATHOGENS

Like animals, plants are subject to infection by diverse pathogens—viruses, bacteria, fungi, and other agents that cause disease. In animals, white blood cells and blood proteins act against pathogens by mechanisms you will learn about in Chapter 23. But plants do not have any blood or blood cells. Instead, a wound or infection sets in motion a mechanism we call **systemic acquired resistance**. Signaling molecules are released at a damaged site. Here, the signals call for production of defensive compounds. Signals are also distributed to other undamaged tissues in more distant regions. Here, the signals induce cells to synthesize and release compounds that make tissues less vulnerable to future attacks for days to months. Synthetic forms of some of the compounds that act in systemic defenses are now being used to enhance the disease resistance of crops and ornamental plants.

Many woody plants can also protect their internal environment by building a fortress of thickened cell walls around wounds. They exude phenols and other toxic compounds. Many also secrete resins. The heavy flow of these gooey compounds saturates and protects bark and wood at an attack site. It also can seep into soil around roots. This type of plant response to attack is called **compartmentalization** (Figure 17.8).

Some toxins are so potent that they also kill cells of the tree itself. As compartments form around injured or infected or poisoned sites, the tree lays down new tissues over them. This works well when the tree acts

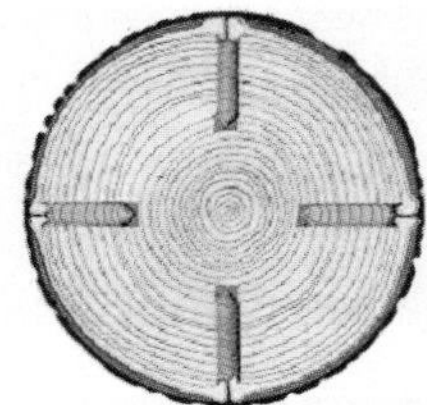

strong

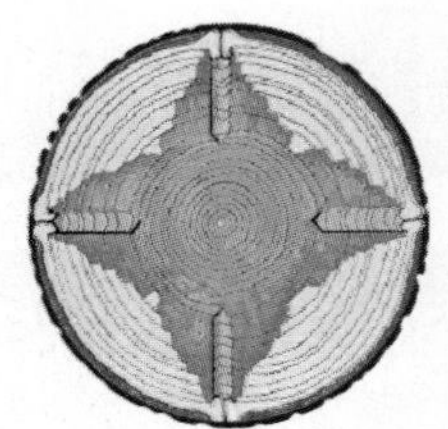

moderate

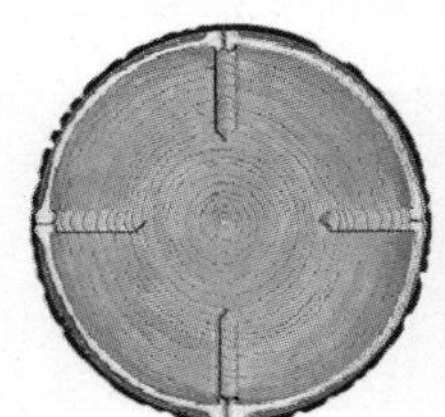

weak

Figure 17.8 ***Animated!*** Patterns of drilling into stems for an experiment to test compartmentalization responses. From top to bottom, the decay patterns (*green*) for three species that make strong, moderate, or weak compartmentalizing responses.

Figure 17.9 ***Animated!*** (**a**,**b**) Yellow bush lupine, *Lupinus arboreus*, in a sandy shore habitat. Introduced into northern California in the early 1900s, this exotic species is taking over the sand dunes. It outcompetes native species.

(**c**) Observational test of the rhythmic movements of the leaves of a young bean plant (*Phaseolus*). The investigator, Frank Salisbury, kept this plant in complete darkness for twenty-four hours. Its leaves continued to fold and unfold rhythmically even without the cues from changes in sunlight.

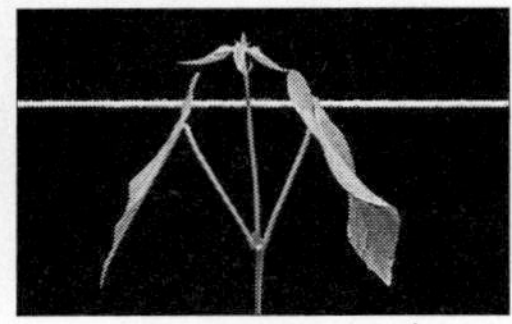

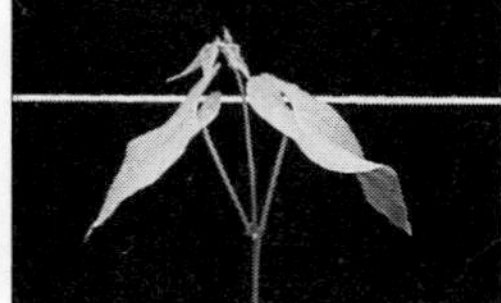

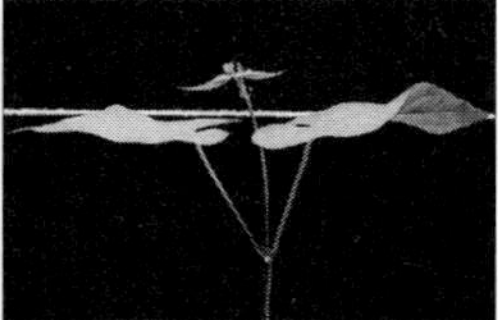

strongly or is not under massive attack. Drill holes in a tree species that makes a strong response and it walls off the wound rapidly. Drill holes in a tree species that makes a moderate response, and it decays lengthwise and somewhat laterally. Drill holes in a species that is a weak compartmentalizer, though, and spread of fungi and bacteria from the wound will cause massive decay. Even the strong compartmentalizers live only so long. When they form too many walls, they shut off the vital flow of water and solutes to living cells.

SAND, WIND, AND THE YELLOW BUSH LUPINE

Anyone who has tiptoed barefoot across sand near the coast on a hot, dry day can understand why very few plants grow in it. One of the exceptions is the yellow bush lupine, *Lupinus arboreus* (Figure 17.9*a*).

L. arboreus is a colonizer of soil exposed by fires or abandoned after being cleared for agriculture. It also has become common along windswept, sandy shores of the Pacific Northwest. Nitrogen-fixing bacteria live in association with the lupine's roots. This gives these plants a competitive edge in nitrogen-deficient soils.

A big environmental challenge near the beach is the scarcity of fresh water. Leaves of a yellow bush lupine are structurally adapted for water conservation. Each leaf has a surprisingly thin cuticle, but a dense array of fine epidermal hairs projects above it, especially on the lower leaf surface. Collectively, the hairs trap moisture escaping from stomata. The trapped, moist air slows evaporation and helps keep water inside the leaf.

These leaves also maintain homeostasis by folding up lengthwise. This helps them to resist the drying force of the wind. The folded leaf is better at stopping moisture loss from stomata than a broad unfolded leaf would be (Figure 17.9*b*).

Leaf folding by *L. arboreus* is a controlled response to changing conditions. When strong winds blow and the potential for water loss is great, these leaves fold tightly. The least-folded leaves are near the center of the plant or on the side most sheltered from the wind. Folding is a response to heat as well as to wind. When air temperature is highest during the day, leaves fold at an angle that helps reflect the sun's rays from their surface. The response minimizes heat absorption.

ABOUT RHYTHMIC LEAF FOLDING

For another example of leaf folding as a coordinated homeostatic response, look at Figure 17.9*c*. This plant holds its leaves out horizontally in the day and folds them closer to its stem at night. Even if you keep the plant in full darkness for a few days, it will continue to move its leaves into and out of the "sleep" position, at the appropriate times. This folding response might be a way to reduce heat losses during the night, when the air becomes cooler. It may help to maintain the plant's internal temperature within a tolerable range.

Rhythmic leaf movements are just one example of **circadian rhythm**, a biological activity that is repeated in cycles, each lasting for close to twenty-four hours. Circadian means "about a day." A pigment molecule called phytochrome, described in Chapter 19, may be part of homeostatic controls over leaf folding.

Homeostatic mechanisms are at work in plants, although they are not governed from central command posts as they are in most animals. Compartmentalization and rhythmic leaf movements are two examples of responses to specific environmental challenges.

Signal to
die docks
at receptor.
Signal leads
to activation
of enzymes
that break
down proteins
and others that
cut up DNA.

Figure 17.10 Artist's depiction of apoptosis. Body cells self-destruct when they finish their functions or if they become infected or cancerous and could threaten the body as a whole.

17.5 How Cells Receive and Respond to Signals

Signal reception, transduction, and response—this is a fancy way of saying cells chatter among themselves and make things happen.

Reflect on the earlier overview of how adjoining cells communicate, as by plasmodesmata in plants and gap junctions in animals (Section 3.8). Also think back on how free-living *Dictyostelium* cells issue signals to get together and make a spore-bearing structure. They do so in response to dwindling supplies of food, which is an environmental cue for change (Figure 14.28).

In all large multicelled organisms, one cell signals others in response to cues from the internal or external environment. Assorted local and long-distance signals lead to local and regional changes in metabolism, gene expression, growth, and development.

The molecular mechanisms by which cells "talk" to one another evolved early in the history of life. They involve three events: 1) activation of a specific receptor as by reversible binding of a signaling molecule, 2) the transduction of the signal into a molecular form that can operate in the cell, and 3) a functional response.

Most receptors are membrane proteins of the sort shown in Section 3.3. When activated, many change shape and start signal transduction. One enzyme might activate many molecules of a different enzyme, which then activates many molecules of another kind, and so on in cascading reactions that amplify the signal.

In the next two units, you will come across diverse cases of signal reception, transduction, and response. For now, a simple example will give you a sense of the kinds of events the signals set in motion.

The very first cell of a new multicelled individual holds the blueprints that will guide its descendants through growth, development, and reproduction, and often death. As part of that program, many cells heed calls to self-destruct when they finish their prescribed function. They also can execute themselves when they become infected, cancerous, or otherwise pose a threat to the body as a whole.

Apoptosis (app-uh-TOE-sis) is the formal name for programmed cell death. It starts up with signals that unleash enzymes of self-destruction, which each body cell synthesizes and stores in case they are needed.

Proteases are among the self-destruction enzymes. They can break apart structural proteins, including the protein subunits of the cytoskeleton and the histones that keep DNA structurally organized (Figure 17.10). Other enzymes called nucleases chop up the DNA.

A cell in the act of apoptosis shrinks away from its neighbors. Its surface bubbles outward and inward. No longer are its chromosomes extended through the nucleoplasm; they are bunched up near the nuclear envelope. The nucleus and then the cell break apart. Phagocytic cells that patrol and protect body tissues swiftly engulf the suicidal cells and remnants of them.

Cells were busily committing suicide when your own hands were developing. The human hand, recall, starts out as a paddlelike structure. If all goes well, apoptosis makes the paddle split up into individual

LINKS TO SECTIONS 7.8, 10.6

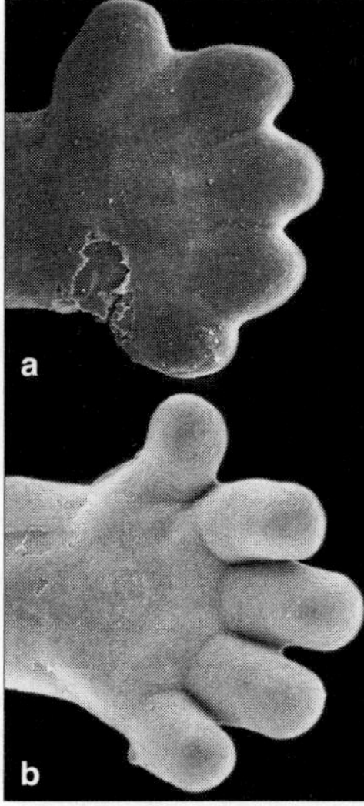

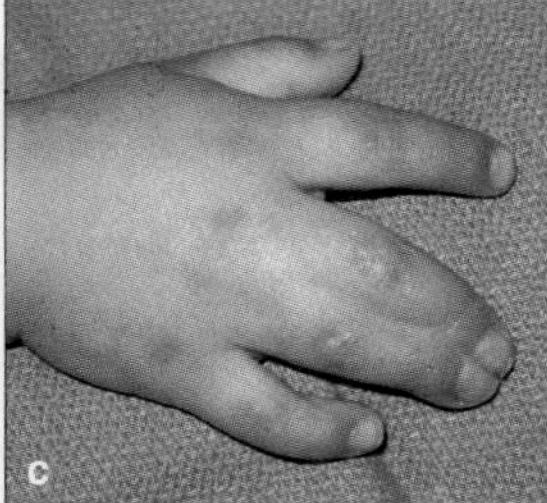

Figure 17.11 ***Animated!*** Human finger development. (**a**) Forty-eight days after fertilization, webs still hold embryonic fingers together. (**b**) Three days later, after apoptosis, no more webs. (**c**) Fingers stay stuck together if cells do not die on cue.

fingers over the course of just a few days. Figure 17.11 shows the outcome.

Often the timing of cell death can be predicted. For instance, keratinocytes are the most common cells in the skin; they usually survive for about three weeks. Dead ones are continually sloughed off and replaced at the skin surface. But these and other cells can be induced to die well ahead of schedule. All it takes is a sensitivity to signals that can call up the enzymes that carry out apoptosis.

Control genes suppress or trigger programmed cell death. One, *bcl-2,* helps keep normal body cells from dying before their time. Self-destruction fails in cancer cells, which are supposed to respond to signals to die, but do not. Maybe they lose the normal receptors that keep them in contact with signaling neighbors. Maybe they get abnormal signals about when to grow, divide, or stop dividing. In some kinds of cancer, for instance, the suppressor gene *bcl-2* has been tampered with or shut down. Apoptosis is no longer a control option.

What about plants? The controlled death of xylem cells creates pipelines that carry water throughout the plant. Also, controlled cell death helps plants to slow the spread of pathogens as part of acquired systemic resistance and compartmentalization.

Cells within a plant or animal body communicate with one another by secreting signaling molecules into extracellular fluid and selectively responding to signals from other cells.

Communication involves receiving signals, transducing them, and bringing about change in a target cell's activity.

Signal transduction may involve shape changes in receptors and other membrane proteins. It also involves activation of enzymes and other molecules in the cytoplasm.

http://biology.brookscole.com/BTT2

Summary

Section 17.1 Anatomy is the study of body form. Physiology is the study of how the body functions in the environment. The structure of most body parts correlates with current or past functions.

A plant or animal's structural organization emerges during growth and development. Each cell in the body performs metabolic functions that ensure its survival. Cells are organized in tissues, organs, and often organ systems. These structural units function in coordinated ways in the performance of specific tasks.

Section 17.2 Plants and animals have responded in similar ways to certain environmental challenges. All have the means for exchanging gases, transporting substances to and from cells, maintaining the body's water and solute concentrations within tolerable limits, and integrating and controlling body parts in ways that ensure survival of the whole organism. Both plant and animal cells have mechanisms for sensing and responding to environmental changes.

Section 17.3 Homeostasis is a state in which the body's internal environment is being maintained within a range that keeps cells alive. In animal cells, sensory receptors, integrators, and effectors interact to maintain homeostasis, often by feedback controls.

With negative feedback mechanisms, a change in some condition, such as body temperature, triggers a response that results in reversal of the change. With positive feedback mechanisms, a change leads to a response that intensifies the condition that caused it.

BiologyNow

Learn about negative feedback and the control of body temperature with animation on BiologyNow.

Section 17.4 Plants have decentralized mechanisms that help maintain the internal environment and ensure their survival. Plants respond to pathogens and wounds with protective responses, such as systemic acquired resistance and compartmentalization. Some respond to changes in environmental conditions by folding leaves. Rhythmic leaf folding is a circadian rhythm that may help a plant maintain its temperature at night.

BiologyNow

Observe compartmentalization and rhythmic leaf folding with animation on BiologyNow.

Section 17.5 Cell communication often involves signal reception and transduction, then a response by the target cell. Many signals are transduced through shape changes in membrane proteins that trigger reactions in the cell. Reactions may alter the activity of genes or enzymes that take part in cellular events. For example, after a signal unleashes protein-cleaving enzymes during apoptosis, a target cell self-destructs.

BiologyNow

See how apoptosis helps shape a human hand with video on BiologyNow.

Figure 17.12 (**a**) Biologist Consuelo De Moraes, who studies plant stress responses. (**b**) Caterpillar (*Heliothis virescens*) chewing a tobacco plant leaf. In response, the plant secretes chemicals. (**c**) The airborne chemicals attract a parasitic wasp, *Cardiochiles nigriceps*. (**d**) The wasp grabs the caterpillar, then inserts one egg inside it. The egg will grow and develop into a larva that will feed on the caterpillar's tissues and kill it.

Self-Quiz

Answers in Appendix I

1. An increase in the number, size, and volume of plant cells or animal cells is called ______.
   a. growth c. differentiation
   b. development d. all of the above

2. The internal environment consists of ______.
   a. all body fluids c. all body fluids outside cells
   b. all fluids in cells d. interstitial fluid

3. ______ influences the concentrations of water and solutes in the internal environment.
   a. Diffusion c. Passive transport
   b. Active transport d. All are correct

4. As basic functions, a plant or animal must ______.
   a. maintain a stable internal environment
   b. obtain and distribute water and solutes
   c. dispose of wastes
   d. defend the body against threats
   e. all of the above

5. Interstitial fluid is found ______.
   a. inside cells c. in spaces between cells
   b. in blood vessels d. all of the above

6. Cell communication typically involves signal ______.
   a. reception c. response
   b. transduction d. all of the above

7. A sweat gland is an example of ______.
   a. an integrator c. a receptor
   b. an effector d. none of the above

8. Match the terms with their most suitable description.

___ circadian rhythm	a. programmed cell death
___ homeostasis	b. 24-hour or so cyclic activity
___ integrator	c. stable internal environment
___ apoptosis	d. central command center
___ negative feedback	e. an activity changes some condition, then the change causes its own reversal
___ effector	f. detects some stimulus
___ receptor	g. carries out some response

Additional questions are available on **Biology Now™**

Critical Thinking

1. Consuelo De Moraes studies the interactions among plants, caterpillars, and parasitoid wasps. A *parasitoid* is a kind of parasite that lays its eggs in its host. After eggs hatch, they develop into larvae that feed on the host's tissues, killing it by devouring it from the inside out.

A caterpillar chewing on a tobacco plant leaf secretes saliva. Some chemicals in the saliva trigger the leaves to release molecules that diffuse through the air. A parasitoid wasp follows the concentration gradient to the leaves, grabs the caterpillar, and deposits an egg in it. Eggs will develop into caterpillar-munching larvae (Figure 17.12).

Plant responses are highly specific. Leaf cells release different chemicals in response to different caterpillar species. Each chemical attracts only the wasps that attack the kind of caterpillar that triggers the chemical's release.

This interaction benefits the wasps and the plants. But how could it evolve? Give a possible explanation for this plant–wasp interaction in terms of benefits to each species and natural selection theory.

2. The Arabian oryx (*Oryx leucoryx*), an endangered antelope, evolved in the harsh deserts of the Middle East. Most of the year there is no free water, and temperatures routinely reach 117°F (47°C). The most common tree in the region is the umbrella thorn tree (*Acacia tortilis*). List common challenges that the oryx and the acacia face. Also research the morphological, physiological, and behavioral responses of both organisms to the challenges.

3. High humidity increases the risk of heat stroke; it reduces the rate of evaporative cooling. So does wearing tight clothing that doesn't "breathe," like the uniform Korey Stringer wore during his last practice.

The elderly and the very young also are susceptible to heat stroke. Other factors that increase the risk are obesity, poor circulation, dehydration, and alcohol intake. Using Figure 17.7 as a reference, suggest how these risk factors might interfere with the homeostatic controls over the body's internal temperature.

18 PLANT FORM AND FUNCTION

IMPACTS, ISSUES

Drought Versus Civilization

The more we dig up records of past climates, the more we wonder about what is happening now. In any given year there are bad *droughts*, or significantly less rainfall than we expect in a given region. Drought is a normal climactic cycle, but it has been much worse than usual in Asia, Africa, Australia, and elsewhere—enough to cause mass starvation and cripple economies.

Humans built the whole of modern civilization on a vast agricultural base. Today, droughts that last two, five, or seven years challenge us. Imagine a drought lasting *200 years*. It happens. About 3,400 years ago, such a drought brought the Akkadian civilization in northern Mesopotamia to an end. Another drought that lasted over 150 years contributed to the collapse of the Mayan civilization in the ninth century.

More recently, Afghanistan was scorched by the worst drought in a hundred years. There, subsistence farmers make up 70 percent of the population. Seven years of drought wiped out their harvests, dried up their wells, and starved their livestock. In desperation, many rural families sold their land, their possessions, and their daughters. Despite imports, starvation has affected about 37 percent of the population.

Even brief episodes of drought reduce crop yield. Plants conserve water by closing stomata, which stops carbon dioxide from moving in. No carbon dioxide, no sugars. When stressed, a flowering plant also makes fewer flowers. The flowers that form do not open fully, so they may not get pollinated. Even if they do, seeds and fruits may drop from the plant before they ripen.

This unit focuses on the seed-bearing vascular plants, with emphasis on the flowering types that are intimately interconnected with human life. You will be taking a look at how these plants function and consider their growth, development, and reproduction. Their structure and physiology help them survive short-term water loss—but not long-term drought (*left*).

How Would You Vote? *In dry regions of the U.S., large-scale farms and big cities compete for water. Should cities restrict urban growth? Should farming be restricted to areas with sufficient rainfall to sustain agriculture? See BiologyNow for details, then vote online.*

Key Concepts

PLANT TISSUES
Flowering plants are all made of the same simple and complex tissues, but these are arranged somewhat differently among the major groups.

GROWTH PATTERNS
Mitotic divisions of cells at meristems make roots and shoots lengthen and thicken. Different patterns of growth among the descendants of meristem cells lead to the structural organization of plant parts.

HOW PLANT PARTS FUNCTION
Plants have root structures, cellular mechanisms, and often symbionts that help them take up water and dissolved nutrients, which then move to photosynthetic cells throughout the plant. Plants also have mechanisms that distribute the products of photosynthesis to all cells that are growing or storing food.

Links to Earlier Concepts

Chapter 15 introduced plant structures and correlated them with their present and past functions. In this chapter, you will be integrating your knowledge of cell structure (Section 3.5), photosynthesis (5.2, 5.3), and homeostasis in plants (17.4, 17.5) to understand adaptations that sustain photosynthesis—the fine structure of leaves and responses to environmental conditions. You already had glimpses of these adaptations in ecological (5.4) and evolutionary (15.1) contexts.

You will also draw on your knowledge of ions and hydrogen bonding (2.2), water's cohesive properties (2.3), membrane transport mechanisms (4.4, 4.5), and osmosis (4.6) to understand mechanisms of water and solute transport in plants.

18.1 Overview of the Plant Body

Earlier, we surveyed some of the 295,000 or so species of plants. Even that sprint through plant diversity is enough to tell you that no one species can be viewed as typical of the group. Nevertheless, when we hear the word "plant," we usually think of flowering plants, and they will be the main focus here.

With 260,000 species, flowering plants dominate the plant kingdom. Its major groups are the magnoliids, eudicots (true dicots), and monocots (Section 15.6). We focus here on eudicots and monocots.

Many flowering plants have a body plan similar to the one in Figure 18.1. Above ground are **shoots**, which consist of plant stems together with their leaves. Stems provide a structural framework for plant growth, and pipelines inside conduct water and nutrients. Flowers are reproductive structures.

In most species, **roots** are plant structures that grow downward and outward through soil. An underground root system absorbs water and dissolved minerals, and it anchors the aboveground parts of many plants. Root cells store and release photosynthetically–derived food for their own use and often for the rest of the plant.

THREE PLANT TISSUE SYSTEMS

Plant stems, branches, leaves, and roots have the same three kinds of tissue systems (Figure 18.1). The *ground tissue* system makes up the bulk of the plant, and has many essential functions, including food storage. The *vascular tissue* system threads through ground tissue of all plant parts and delivers water and solutes to cells. The *dermal tissue* system protects all exposed surfaces.

Parenchyma, collenchyma, and sclerenchyma are the simplest tissues in these systems; each has only one cell type. Xylem, phloem, and dermal tissues are complex, with organized arrays of two or more types of cells. How do these plant tissues arise? All of them originate at **meristems**, which are localized regions of actively dividing cells.

Apical meristems are in shoot and root tips, where all young plant parts begin lengthening. Populations of cells that form here are the immature forerunners of dermal tissues, primary vascular tissues, and ground tissue. Apical meristem activity and the lengthening of shoots and roots together are a plant's *primary growth.*

Stems and roots of many plants also become thicker over time. Such thickening is *secondary growth.* Increases in girth begin with cell divisions in **lateral meristems**, cylindrical arrays of cells that form in stems and roots.

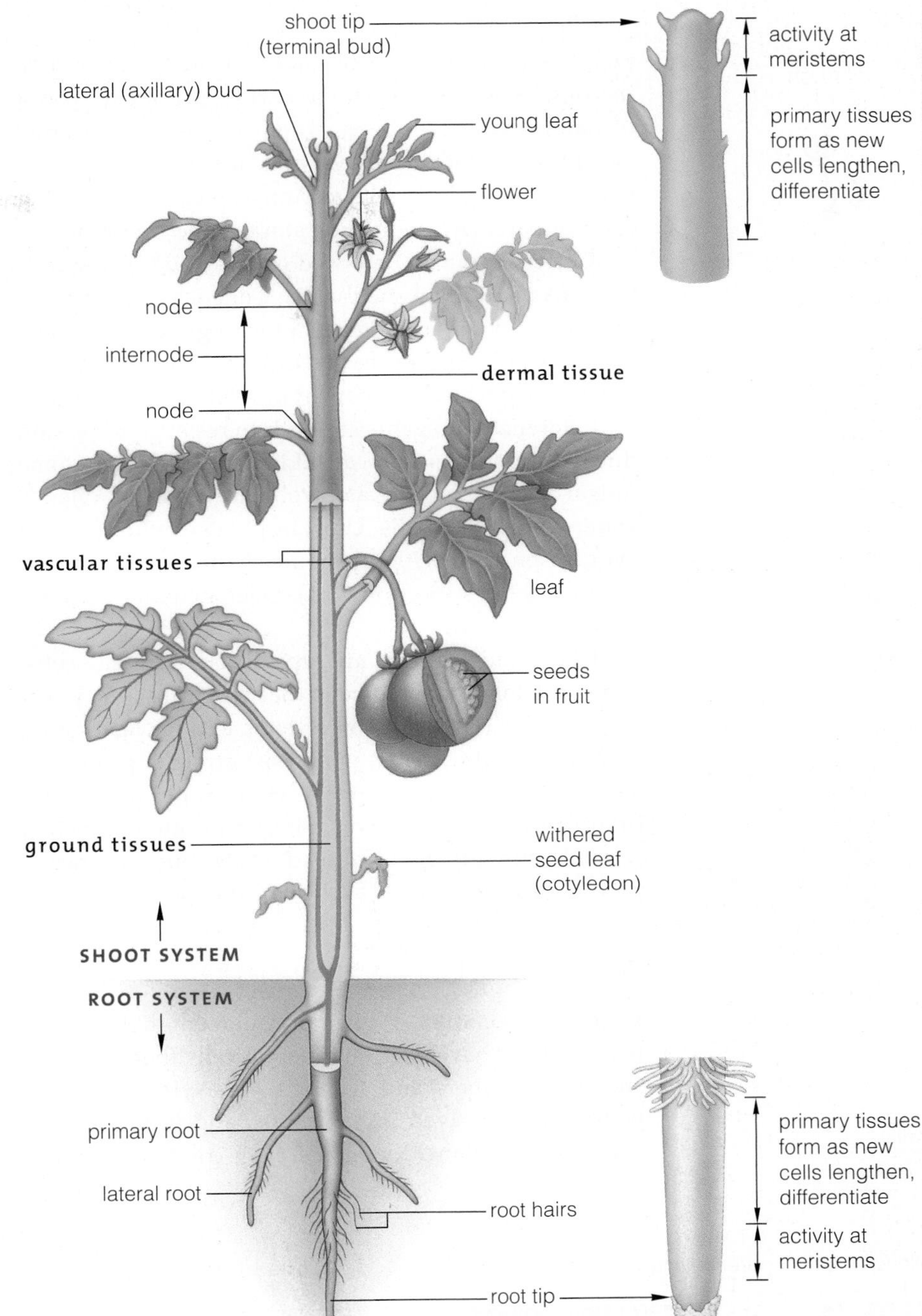

Figure 18.1 ***Animated!*** Body plan for the commercially grown tomato plant (*Lycopersicon esculentum*). Its vascular tissues (*purple*) conduct water, dissolved minerals, and organic substances. They thread through ground tissues that make up most of the plant body. Epidermis, a type of dermal tissue, covers root and shoot surfaces.

SIMPLE TISSUES

Parenchyma tissue makes up most of the soft, new primary growth of roots, stems, leaves, and flowers. Parenchyma cells tend to be thin-walled, pliable, and

LINK TO SECTION 3.5

many-sided. Parenchyma is alive at maturity, and so its cells retain the capacity to divide. Mitotic divisions of parenchyma cells repair plant wounds. Mesophyll is a photosynthetic parenchyma in leaves. Air spaces between mesophyll cells permit gas exchange. Other roles of parenchyma include storage and secretion.

Collenchyma is composed of elongated, living cells with unevenly thickened walls. Collenchyma forms as patches or rings in many lengthening stems and leaf stalks. A polysaccharide (pectin) imparts flexibility to the primary walls of collenchyma cells (Figure 18.2*a*).

Sclerenchyma gains strength to resist compression due to lignin in its cell walls. Lignin also thwarts some fungal attacks, and it waterproofs the walls of xylem's water-conducting cells. Upright plants could not have evolved on land without the mechanical support that lignin provides (Section 15.1). Sclerenchyma cells are dead at maturity.

Fibers and sclereids are typical sclerenchyma cells. *Fibers* are long, tapered cells that structurally support the vascular tissues of some stems and leaves (Figure 18.2*b*). They flex, twist, and resist stretching. We use some kinds of fibers to make cloth, rope, paper, and other commercial products. Short and often branched *sclereids* strengthen hard seed coats, such as peach pits. They make pears gritty (Figure 18.2*c*).

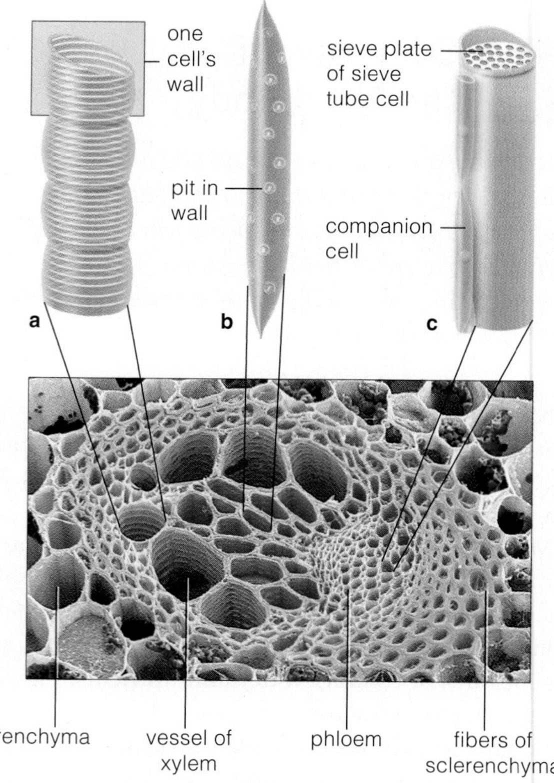

Figure 18.3 Simple and complex tissues in a stem. In xylem, (**a**) part of a column of vessel members, and (**b**) a tracheid. (**c**) One of the living cells that interconnect as sieve tubes in phloem. Companion cells closely associate with sieve tubes.

COMPLEX TISSUES

VASCULAR TISSUES Xylem and phloem are vascular tissues. They have long conducting cells that are often sheathed in fibers and parenchyma. **Xylem** conducts water and dissolved mineral ions, and it structurally supports the plant. Xylem's sturdy conducting tubes are interconnected walls of two kinds of cells, *vessel members* and *tracheids* (Figure 18.3*a*,*b*). Both types of xylem cells have perforated secondary walls, and both types are dead at maturity. Lignin deposited in various patterns stiffens their walls. Many of the perforations

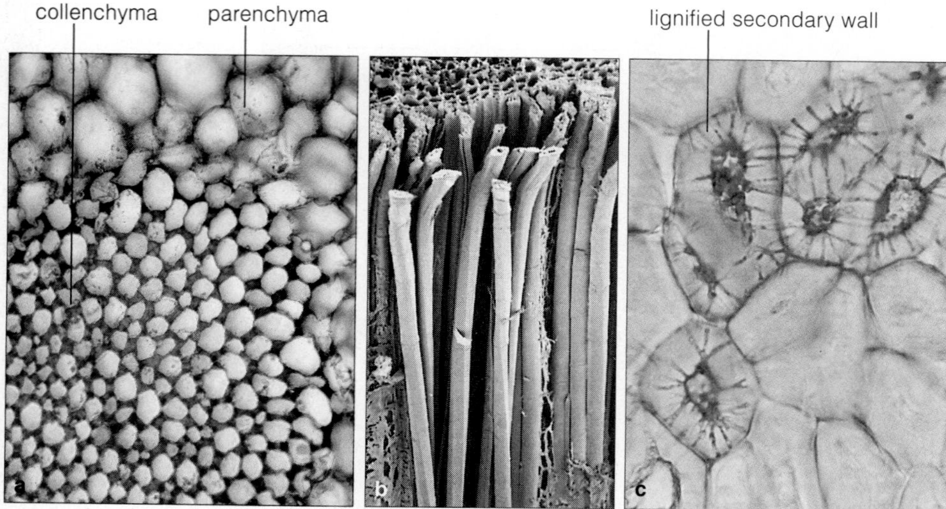

Figure 18.2 Some simple tissues. (**a**) Collenchyma from a supporting strand in a celery stem. Sclerenchyma: (**b**) strong flax stem fibers, and (**c**) stone cells, a type of sclereid in pears.

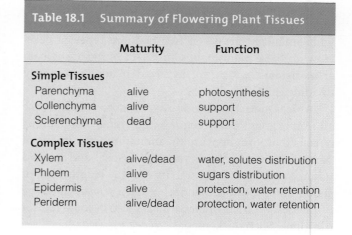

Table 18.1 Summary of Flowering Plant Tissues

	Maturity	Function
Simple Tissues		
Parenchyma	alive	photosynthesis
Collenchyma	alive	support
Sclerenchyma	dead	support
Complex Tissues		
Xylem	alive/dead	water, solutes distribution
Phloem	alive	sugars distribution
Epidermis	alive	protection, water retention
Periderm	alive/dead	protection, water retention

in adjoining walls match up, so water moves laterally between tubes as well as upward through them.

Phloem conducts sugars and other organic solutes. Its main cells, *sieve-tube members*, are alive at maturity. They connect end to end, at *sieve plates*, to form tubes that are used for transporting sugars to cells in all the growing or storage regions of the plant (Figure 18.3*c*). Special parenchyma cells called *companion cells* load the products of photosynthesis into sieve tubes.

DERMAL TISSUES The first dermal tissue to form is the **epidermis**, which is typically a single layer of cells. In mature roots and stems of woody plants, a thicker dermal tissue called periderm replaces the epidermis. Epidermal cells secrete waxes and cutin onto their wall that faces to the outside of the plant. These secretions form a **cuticle**, a covering that helps the plant conserve water and deflect attacks by some pathogens.

The epidermis of leaves and young stems contains specialized cells. For example, a small gap, or **stoma** (plural, stomata), forms across the epidermis when paired *guard cells* swell and move apart. Most plants have many stomata. Each is a point where diffusion of water vapor, oxygen, and carbon dioxide gases across the epidermis is controlled. Another tissue, periderm, replaces the epidermis of woody stems and roots.

Table 18.1 summarizes simple and complex plant tissues and their functions in angiosperms.

EUDICOTS AND MONOCOTS—SAME TISSUES, DIFFERENT FEATURES

Roses, buttercups, and beans are among the eudicots. Lilies, ryegrass, corn, and wheat are typical monocots. Both eudicots and monocots are composed of the same tissues, organized in different ways (Figure 18.4). For example, seeds of eudicots have two cotyledons, and seeds of monocots have one. **Cotyledons** are leaflike structures that store or absorb food for plant embryos. They wither after the seed germinates and new leaves begin to make their own food.

Vascular plants typically have stems for upright growth and internal transport, leaves that function in photosynthesis, specialized reproductive shoots, and other structures. They have roots that absorb and conduct water and nutrients.

Plant tissues originate at meristems, localized regions of dividing cells in shoot and root tips. Ground tissues make up most of a plant. Vascular tissues distribute water and solutes. Dermal tissues cover and protect all plant surfaces.

Eudicots and monocots have the same cells and tissues, organized in distinctive ways.

a Eudicots

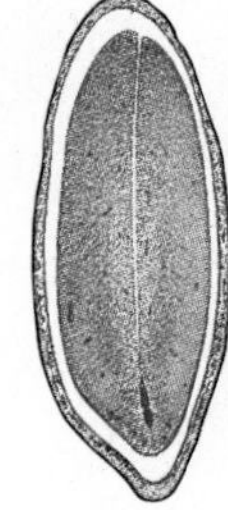

Inside seeds, two cotyledons (seed leaves of embryo)

Usually four or five floral parts (or multiples of four or five)

Leaf veins usually in a netlike array

Three pores and/or furrows in the pollen grain surface

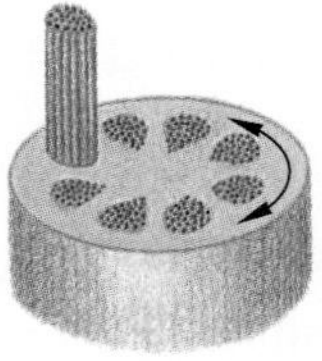

Vascular bundles organized as a ring in ground tissue

b Monocots

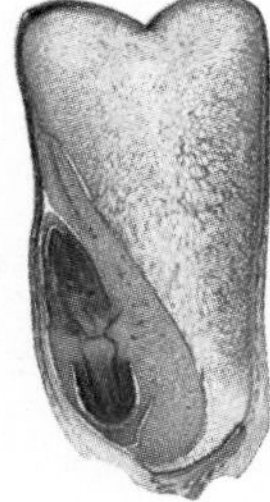

Inside seeds, one cotyledon (seed leaf of embryo)

Usually three floral parts (or multiples of threes)

Leaf veins usually running parallel with one another

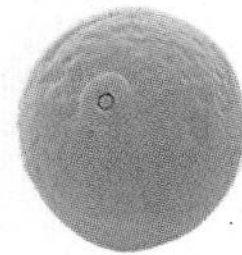

One pore or furrow in the pollen grain surface

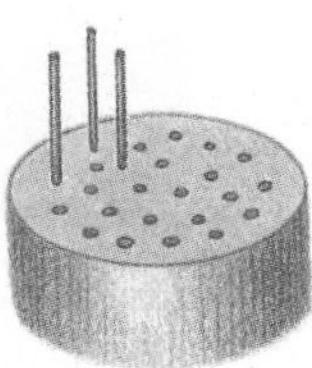

Vascular bundles distributed throughout ground tissue

Figure 18.4 ***Animated!*** Some of the differences among eudicots and monocots. Compare Table 15.1.

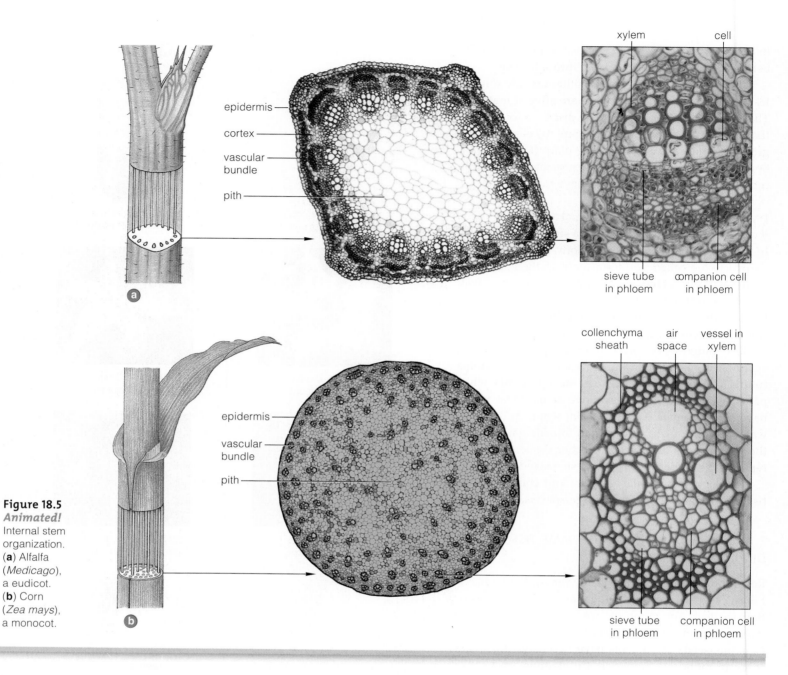

epidermis
cortex
vascular bundle
pith
xylem
cell
sieve tube in phloem
companion cell in phloem
a

epidermis
vascular bundle
pith
collenchyma sheath
air space
vessel in xylem
sieve tube in phloem
companion cell in phloem
b

Figure 18.5 ***Animated!*** Internal stem organization. (**a**) Alfalfa (*Medicago*), a eudicot. (**b**) Corn (*Zea mays*), a monocot.

LINKS TO SECTIONS 5.2, 5.3, 15.1

18.2 Primary Structure of Shoots

Ground, vascular, and dermal tissues of monocot and eudicot stems become organized in characteristic patterns during growth.

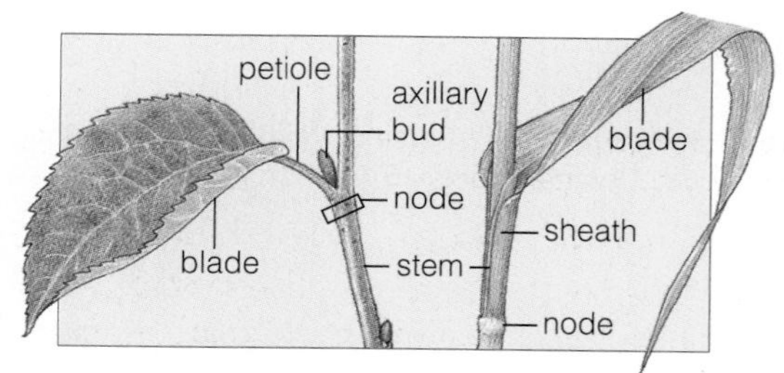

INSIDE THE STEM

During primary growth, primary xylem and phloem develop within the same cylindrical sheath of cells. They form the conducting lines of **vascular bundles**, multistranded cords that thread lengthwise through all ground tissue systems. In most eudicot stems and roots, long vascular bundles form a ring that divides the ground tissue into a cortex and pith (Figure 18.5*a*). A stem *cortex* is the parenchyma between the vascular bundles and the epidermis. The *pith* is the parenchyma inside the ring of vascular bundles. The stems of most monocots have a different arrangement, in which their long vascular bundles are distributed throughout the ground tissue (Figure 18.5*b*).

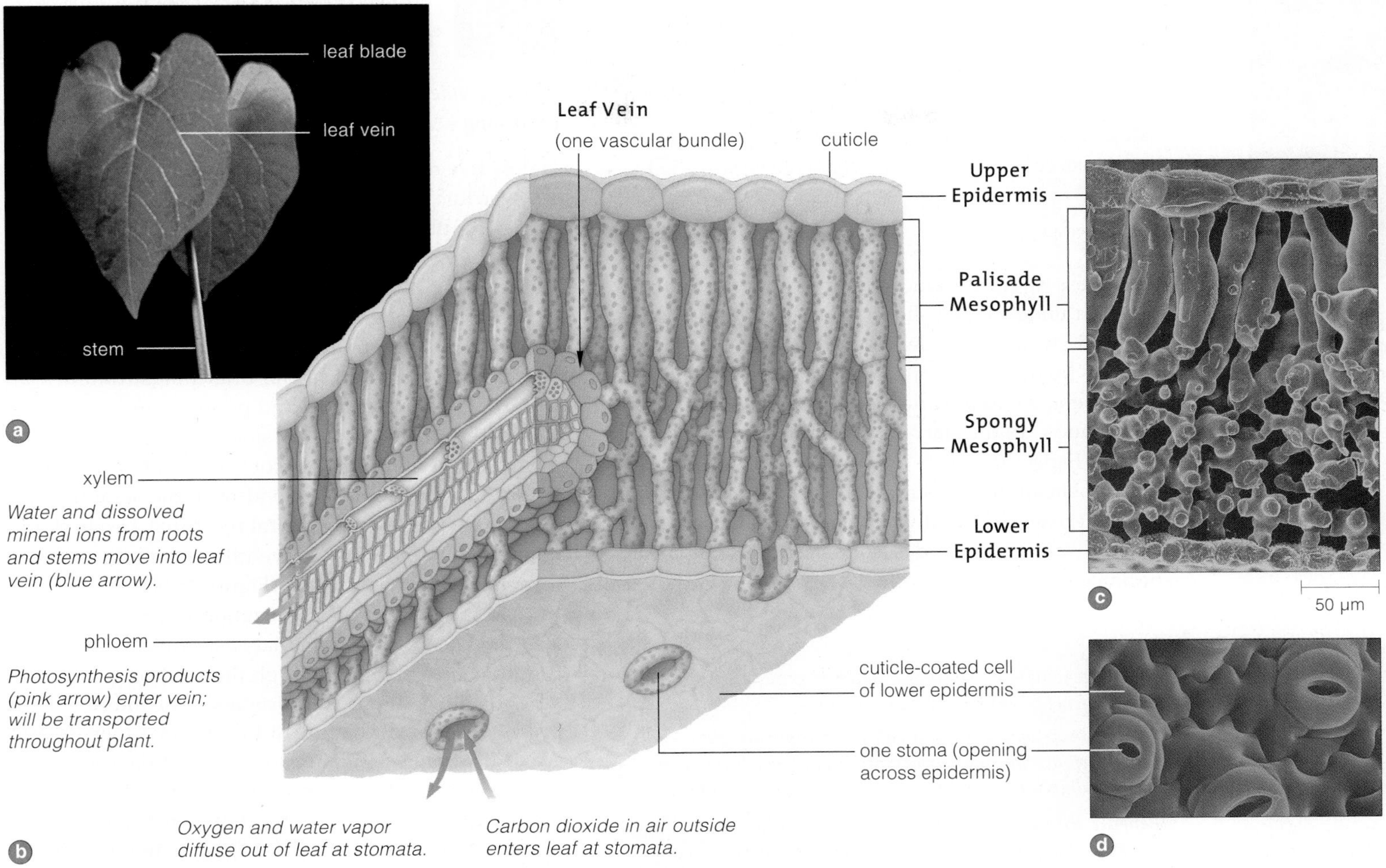

leaf blade

leaf vein

stem

a

Leaf Vein
(one vascular bundle)

cuticle

Upper Epidermis

Palisade Mesophyll

Spongy Mesophyll

Lower Epidermis

xylem

Water and dissolved mineral ions from roots and stems move into leaf vein (blue arrow).

phloem

Photosynthesis products (pink arrow) enter vein; will be transported throughout plant.

c

50 µm

cuticle-coated cell of lower epidermis

one stoma (opening across epidermis)

Oxygen and water vapor diffuse out of leaf at stomata.

Carbon dioxide in air outside enters leaf at stomata.

b

d

Figure 18.6 ***Animated!*** Leaf organization for *Phaseolus*, a bean plant. (**a**) Leaves. (**b**,**c**) Leaf fine structure. (**d**) Open stomata.

LEAF STRUCTURE

A **leaf** is a sugar factory that has many photosynthetic cells. Most leaves have a flat blade (Figure 18.6*a*) with a stalk, or petiole, attached to the stem. The orientation and shape of a leaf help it intercept sunlight, conserve water, and exchange gases. Most leaves also are thin, with a high surface-to-volume ratio, and many reorient themselves during the day according to the angle of the sun. In hot, arid places like deserts, the leaves of many plants remain parallel with the sun's rays in order to reduce heat absorption. Thick or needle-like leaves can conserve water.

Inside the leaf, cells of mesophyll, a photosynthetic parenchyma, are exposed to air spaces (Figure 18.6*b*,*c*). Carbon dioxide reaches them by diffusing into the leaf through stomata (Figure 18.6*d*); oxygen diffuses out the same way. Leaves of eudicots have two layers of mesophyll tissue. Under the upper epidermis are the columnar parenchyma cells of the *palisade* mesophyll. These cells contain more chloroplasts than cells of the *spongy* mesophyll layer below. In monocot leaves such as grass blades, mesophyll is not organized in layers. Monocot leaves grow vertically and can intercept light from all directions.

Leaf **veins** are vascular bundles in leaves. In them, continuous strands of xylem speedily move water and dissolved nutrients to mesophyll cells, and continuous strands of phloem carry their photosynthetic products away. Most eudicot veins branch into several minor

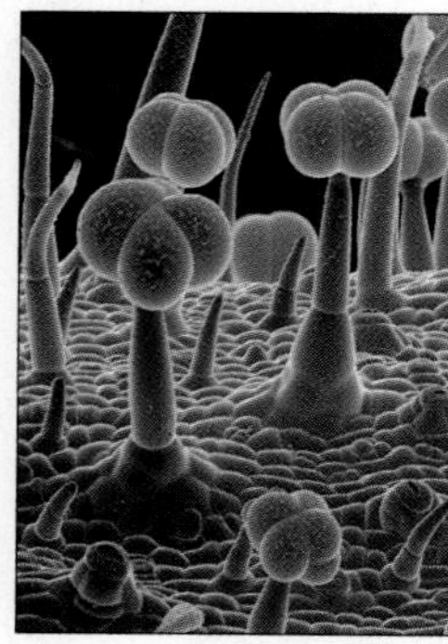

Figure 18.7 Tomato gland cells that secrete insect repellents onto the epidermal surface.

veins embedded in mesophyll. In monocots, all veins are similar in length and run parallel with the leaf's long axis.

Epidermis covers every leaf surface exposed to the air. It may be smooth, sticky, or slimy, and have hairs, scales, spikes, hooks, and other surface specializations that conserve water or protect the leaf (Figure 18.7).

HOW DO STEMS AND LEAVES FORM?

Buds are a shoot's main zone of primary growth. A bud is an undeveloped shoot of meristem tissue, often protected by modified leaves called bud scales. The growing tips of shoots are *terminal buds. Lateral buds* form where leaves are attached to a lengthening stem. Lateral buds are the start of side branches, leaves, and flowers (Figure 18.8*a*).

Just beneath the surface of a bud, the cells of shoot apical meristem tissue divide continuously during the growing season. Their descendants differentiate into specialized tissues (Figure 18.8*b*).

The arrangement of vascular bundles in ground tissues differs inside monocot and eudicot stems.

Leaves are structurally adapted to intercept sunlight and distribute water and nutrients. Components include mesophyll (photosynthetic cells), veins, and epidermis.

Beneath the surface of a bud, cells of apical meristem tissue divide and differentiate into all of the specialized cells and tissues of the plant body.

18.3 Primary Structure of Roots

Roots provide plants with a large surface area for absorbing water and dissolved mineral ions.

Unless tree roots start to buckle a sidewalk or choke off a sewer line, most of us do not pay much attention to flowering plant root systems. While we are walking above them, roots are busily mining the soil for water and minerals. Tree roots can grow 100 meters deep.

Root primary growth results in one of two kinds of root systems. The **taproot system** of eudicots consists of a primary root and lateral branchings from it. Oak trees, dandelions, carrots, and poppies are examples of plants that have taproot systems (Figure 18.9*a*). The primary root of most monocots, as in grasses, is short-lived. *Adventitious roots* extend from the stem instead. These give rise to many lateral roots that are similar in diameter and in length. Adventitious and lateral roots form a **fibrous root system** (Figure 18.9*b*).

A root's structural organization is arranged inside a seed. When a seed germinates, a primary root is the first structure to poke through the seed coat. In nearly all eudicot seedlings, the primary root thickens as it grows downward. Figure 18.10*a* shows a meristem in a root tip. Some cellular descendants of the root apical meristem give rise to a root cap, a dome-shaped mass of cells that protects a young root as it is lengthening through the soil. Other cells differentiate into dermal, ground, and vascular tissues. As in stems, the ground tissue of roots becomes divided into cortex and pith.

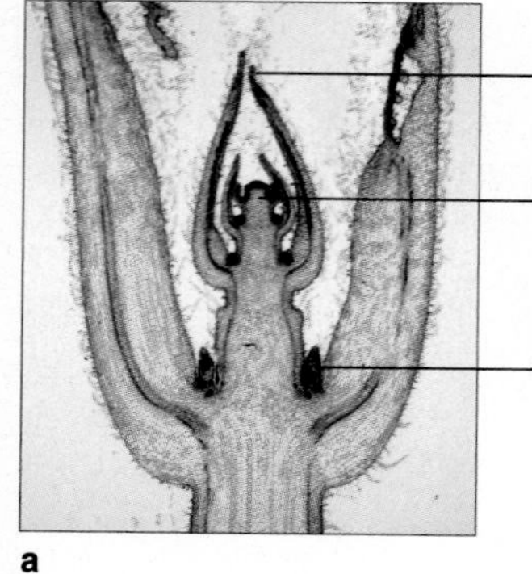

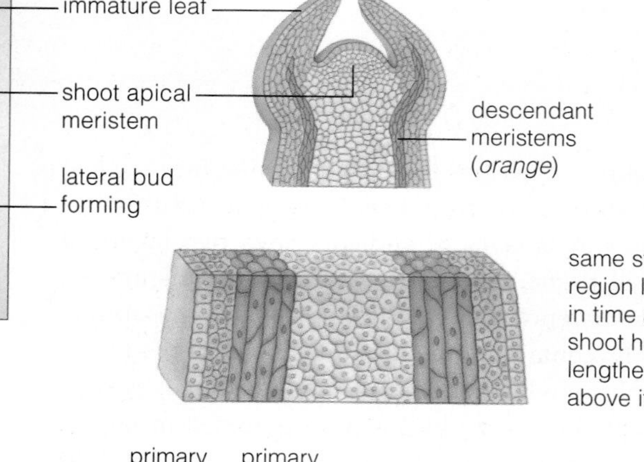

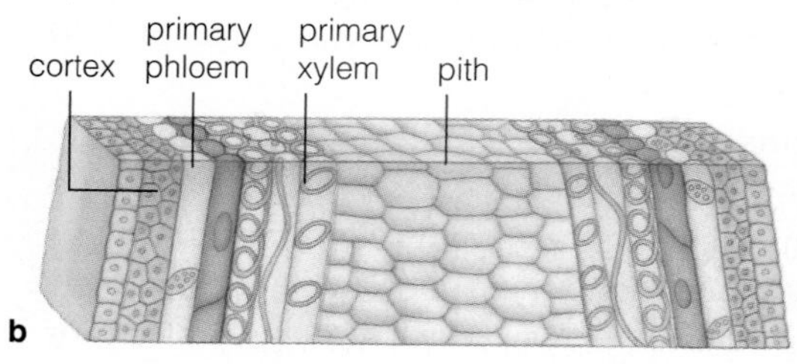

Figure 18.8 Eudicot stem structure. (**a**) This cut through the center of a stem shows the apical meristem, lateral buds, and immature leaves. (**b**) Successive stages of the stem's primary growth, starting at the shoot apical meristem.

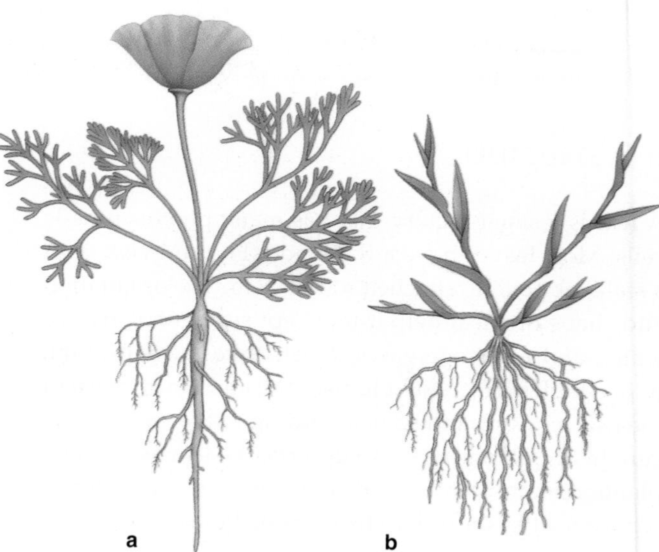

Figure 18.9 (**a**) Taproot system of a eudicot, the California poppy. (**b**) Fibrous root system of a grass plant, a monocot.

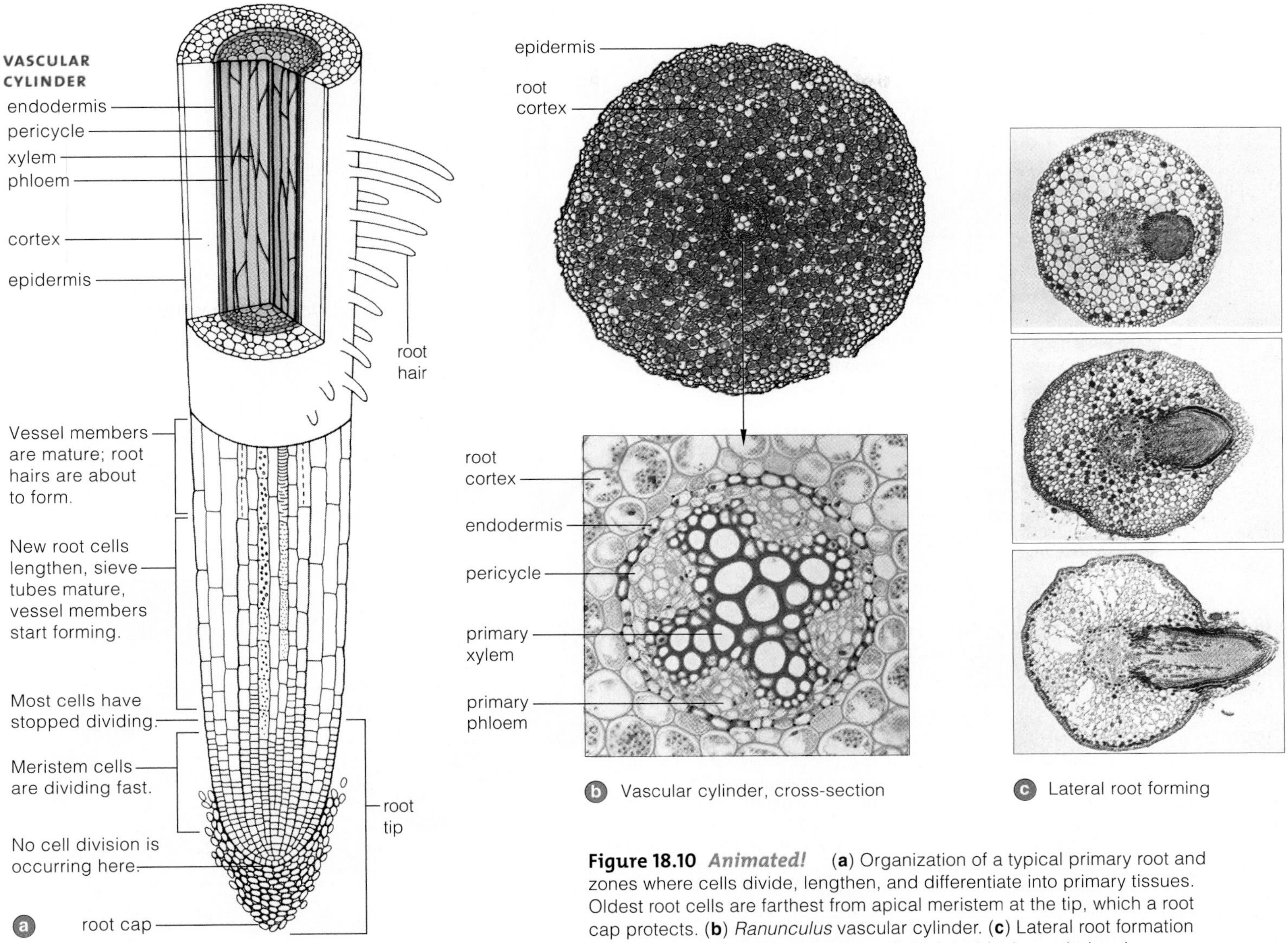

VASCULAR CYLINDER
endodermis
pericycle
xylem
phloem
cortex
epidermis
root hair
Vessel members are mature; root hairs are about to form.
New root cells lengthen, sieve tubes mature, vessel members start forming.
Most cells have stopped dividing.
Meristem cells are dividing fast.
No cell division is occurring here.
root tip
root cap

epidermis
root cortex
root cortex
endodermis
pericycle
primary xylem
primary phloem

(a)

(b) Vascular cylinder, cross-section

(c) Lateral root forming

Figure 18.10 ***Animated!*** (**a**) Organization of a typical primary root and zones where cells divide, lengthen, and differentiate into primary tissues. Oldest root cells are farthest from apical meristem at the tip, which a root cap protects. (**b**) *Ranunculus* vascular cylinder. (**c**) Lateral root formation from the pericycle, a cylinder one cell thick inside the endodermis.

Root apical meristem also gives rise to the ground tissue system and the **vascular cylinder**. The vascular cylinder is composed of primary xylem and phloem, and a layer of cells called the pericycle. Cells of the pericycle divide repeatedly to form lateral roots that erupt through epidermis (Figure 18.10*b,c*).

Air spaces between the cells of the ground tissue system allow air to diffuse throughout a root. Like all other living cells in the plant, root cells make ATP by way of aerobic respiration, and they use oxygen in air for the reactions.

Root epidermis is the plant's absorptive interface with the soil. Some epidermal cells send out extensions called **root hairs**. Collectively, root hairs enormously increase the surface area available for taking up water and dissolved nutrients. When water enters a root, it moves from cell to cell until it reaches the **endodermis**, a layer of cells around the vascular cylinder. Wherever endodermal cells adjoin, their walls are waterproofed, which forces incoming water to pass through the cells' cytoplasm. This arrangement allows endodermal cells to control the movement of water and the substances dissolved in it into the vascular cylinder. We return to this topic in Section 18.6.

Roots provide a plant with a tremendous surface area for absorbing water and solutes. Inside each is a vascular cylinder, with long strands of primary xylem and phloem.

cork cambium

vascular cambium

thickening

a Two lateral meristems in older stems and roots of woody plants produce secondary growth, or increases in diameter. *Vascular cambium* gives rise to secondary vascular tissues. *Cork cambium* gives rise to periderm.

VASCULAR CAMBIUM

stem surface

primary xylem

primary phloem

VASCULAR CAMBIUM

secondary xylem

secondary phloem

b Pattern of activity at vascular cambium. Reading left to right, ongoing cell divisions enlarge the inner core of secondary xylem and displace vascular cambium toward the stem or root surface.

outer surface of stem or root

division

division

One of the cells of vascular cambium at the start of secondary growth.

One of the two daughter cells differentiates into a xylem cell (coded *blue*), and the other remains meristematic.

One of the two daughter cells differentiates into a phloem cell (coded *pink*), and the other remains meristematic.

The same pattern of cell division and differentiation into xylem and phloem cells continues through the growing season.

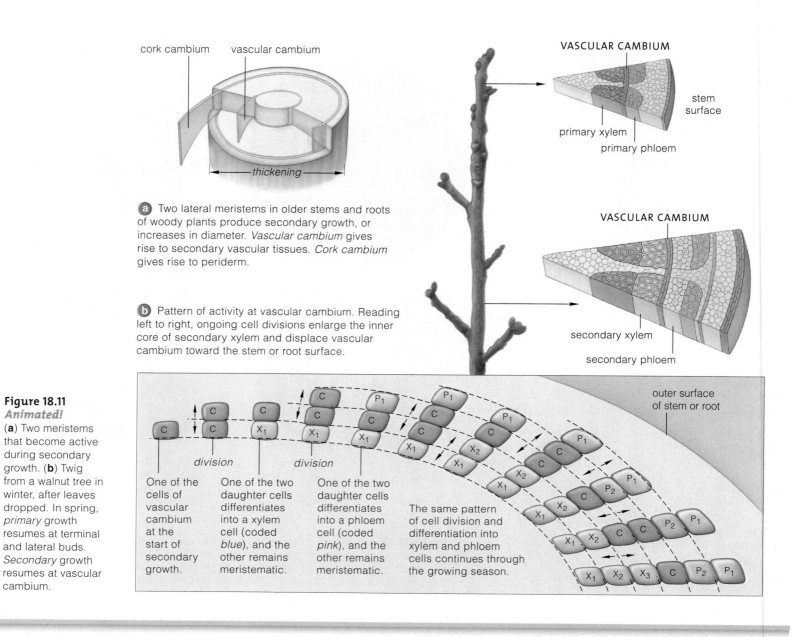

Figure 18.11 ***Animated!*** (**a**) Two meristems that become active during secondary growth. (**b**) Twig from a walnut tree in winter, after leaves dropped. In spring, *primary* growth resumes at terminal and lateral buds. *Secondary* growth resumes at vascular cambium.

LINK TO SECTION 10.6

18.4 Secondary Growth—The Woody Plants

Thickening of roots and stems—secondary growth—starts with two lateral meristems. One gives rise to secondary xylem and phloem, and the other to a sturdier surface covering that replaces epidermis.

Flowering plant life cycles extend from germination to seed formation, then death. Annuals finish the cycle in one growing season and usually are "nonwoody," or herbaceous. Biennials such as carrots form roots, stems, and leaves in a single season; in the next season, they form flowers and seeds, and then they die. Perennials continue growth and seed formation year after year.

Many perennial plants, including trees, go through a dormant period during the cold or dry seasons. Each spring, primary growth resumes at buds on twigs and at roots. Inside buds, new tissues are formed by cell divisions at apical meristems. The woody plants add secondary growth in two or more growing seasons. These include all gymnosperms, some monocots, and many eudicots. Early in life, stems and roots of woody plants are like those of nonwoody plants. Differences occur as their lateral meristems—**vascular cambium** and **cork cambium**—become active. They give rise to secondary vascular tissues (Figure 18.11).

Secondary growth adds girth to a perennial plant from the inside. Secondary xylem, or **wood**, forms on the *inner* surface of the vascular cambium. Secondary phloem forms on the *outer* surface. Some meristematic

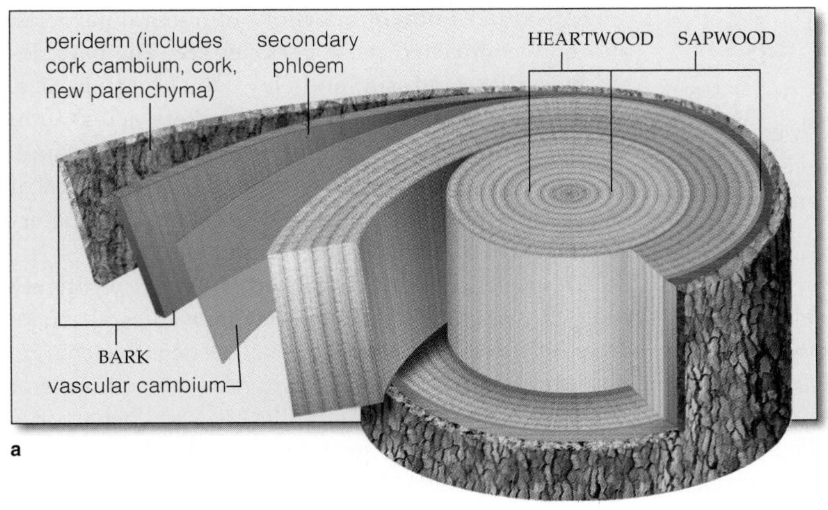

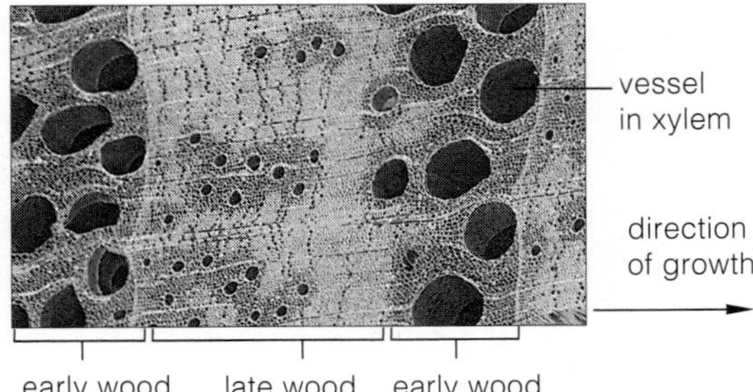

Figure 18.12 ***Animated!*** (**a**) One type of woody stem. (**b**) Early and late wood cut from red oak. Late wood is evidence that a tree did not waste energy making large-diameter xylem cells for water uptake during a dry summer or drought.

cells of vascular cambium give rise to vascular tissue that extends lengthwise through the stem. Other cells of the vascular cambium give rise to horizontal rays of parenchyma. The rays transport water "sideways" in the stem, in a pattern like spokes of a bicycle wheel. Water and nutrients move through woody stems and roots in these secondary vascular tissues.

As seasons pass and the plant ages, its inner core of xylem expands outward, and the meristematic cells are displaced toward root and stem surfaces. Some cortex and epidermis split away. Then a new covering, *periderm*, forms from cork cambium. The periderm and secondary phloem compose **bark**, which is essentially all of the living cells and dead tissue on the outside of the vascular cambium (Figure 18.12*a*).

Cell divisions at cork cambium produce **cork**. This tissue has densely packed rows of dead cells, the walls of which are thickened with a fatty substance called suberin. Cork protects, insulates, and waterproofs the stem or root surface. Cork also forms over wounded tissues. When leaves are about to drop from the plant, cork forms at the place where petioles attach to stems.

Wood's appearance and function change as a stem or root ages. Metabolic wastes such as resins, tannins, gums, and oils, clog and fill the oldest xylem so much that it no longer is able to transport water and solutes. These substances often darken and strengthen wood, which becomes more aromatic and prized by furniture builders as **heartwood**.

Sapwood forms through secondary growth. It is the functioning xylem between heartwood and vascular cambium (Figure 18.12*a*). In trees of temperate zones, dissolved sugars that had been stored in roots during winter travel through sapwood's secondary xylem to flower buds in spring. This sugar-rich fluid is called **sap**. Each spring, New Englanders collect and boil sap from maple trees to make maple syrup.

Vascular cambium becomes inactive during cold winters or drought. *Early* wood, with large-diameter, thin-walled cells, starts forming with the first rains of a growing season. *Late* wood forms in dry summers and has smaller-diameter, thick-walled xylem cells (Figure 18.12*b*). Early and late wood reflect light differently, so a cut tree trunk shows alternating bands. The bands are called growth rings, or "tree rings" (Figure 18.13).

Trees growing where seasonal change is predictable usually add one growth ring per year. Trees living in deserts can add more than one ring of early wood per season, in response to unusual extra rainfall. Seasonal change in the tropics is almost nonexistent, so growth rings are not a feature of tropical trees.

Oak, hickory, and other eudicot trees with vessels, tracheids, and fibers in their xylem are **hardwoods**. Pines, redwoods, and other conifers are **softwoods**; their xylem has tracheids and rays of parenchyma, but no vessels or fibers, so it is less dense than hardwood.

In woody plants, secondary vascular tissues form at a ring of vascular cambium inside older stems and roots. Wood is an accumulation of secondary xylem.

Periderm and secondary phloem form bark.

Wood may be classified by its location and functions (as in heartwood versus sapwood) and by the type of plant (many eudicot trees are hardwood, and conifers are softwood).

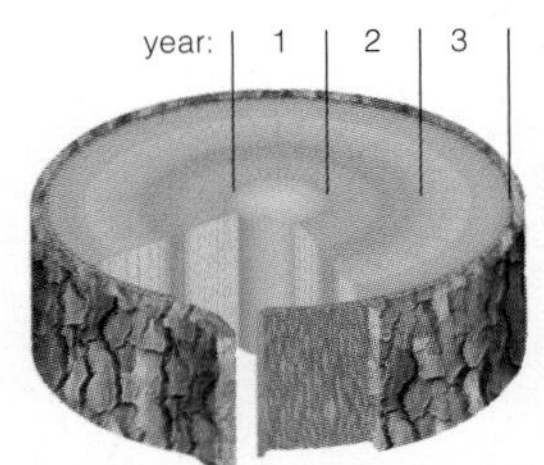

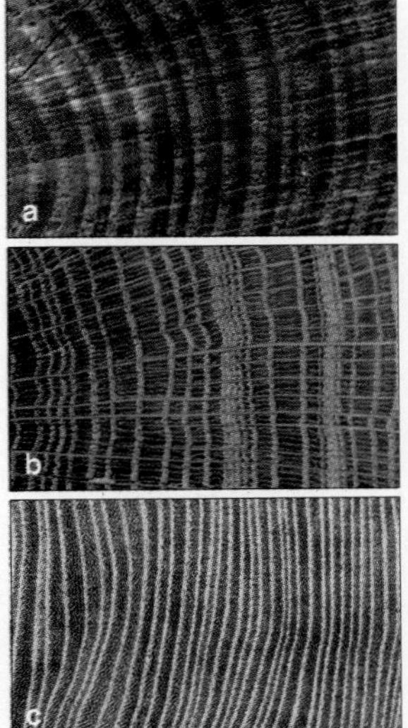

Figure 18.13 Tree rings of different woods, (**a**) pine, (**b**) oak, and (**c**) elm. The series of rings on the elm tree was made between 1911 and 1950. Count all of the rings, and you will find the age of the tree. The differences in tree ring widths are clues to weather conditions in seasons past.

LINKS TO SECTIONS 2.2, 2.4

18.5 Plant Nutrients and Availability in Soil

We have mentioned nutrients in passing. Exactly what are they? A ***nutrient*** *is any element that is essential to an organism's life, because no other element can indirectly or directly fulfill its metabolic role.*

Plant growth depends on sixteen essential elements, as listed in Table 18.2. Soil usually holds dissolved ions, including calcium and potassium. Nine elements are *macro*nutrients. These are required in amounts above 0.5 percent of the plant's dry weight (its weight after all the water has been removed from the plant body). Seven other elements are *micro*nutrients; they make up traces—usually a few parts per million—of the plant's dry weight. Every one of these nutrients is essential for normal plant growth.

PROPERTIES OF SOIL

Soil consists of mineral particles mixed with variable amounts of decomposing organic material, or **humus**. The minerals form by the weathering of hard rocks. Humus forms from dead organisms and organic litter: fallen leaves, feces, and so on. Water and air occupy spaces between the particles and organic bits.

Soils differ in the proportions of mineral particles, and how compacted those particles are. The particles are primarily sand, silt, and clay, which vary in size. Grains of sand are 0.05 to 2 millimeters across; you can see individual ones by sifting beach sand through your fingers. You cannot see individual particles of silt because they are only 0.002 to 0.05 millimeters across. Particles of clay are even smaller.

Clay consists of thin, stacked layers of negatively charged crystals. Sheets of water molecules alternate between layers of clay. Because of its negative charge, clay can temporarily bind positively charged mineral ions dissolved in soil water. This chemical behavior is the reason that plants grow in soil. Clay latches onto dissolved nutrients that would otherwise drain past roots too quickly to be absorbed.

Although they do not bind mineral ions as well as clay, sand and silt are necessary for growing plants as well. Without enough intervening sand and silt, tiny clay particles can pack so tightly that they exclude air. Without enough air space in the soil, root cells cannot secure oxygen to sustain aerobic respiration.

How suitable is a given soil for plant growth? The answer depends partly upon the proportions of sand, silt, and clay. Soils with the best oxygen and water penetration for growing plants are **loams**, which have roughly equal proportions of the three particle types. Decomposing organic matter, or **humus**, also affects

Table 18.2 Plant Nutrients and Symptoms of Deficiencies

MACRONUTRIENT	FUNCTIONS	SYMPTOMS OF DEFICIENCY
Carbon, Hydrogen, Oxygen	Raw materials for second stage of photosynthesis	None; all are abundantly available (water, carbon dioxide are sources)
Nitrogen	Component of proteins, coenzymes, chlorophyll, nucleic acids	Stunted growth; light-green older leaves; older leaves yellow and die (these are symptoms that define a condition called chlorosis)
Potassium	Activation of enzymes; contributes to water–solute balances that influence osmosis*	Reduced growth; curled, mottled, or spotted older leaves; burned leaf edges; weakened plant
Calcium	Component in control of many cell functions; cementing cell walls	Terminal buds wither, die; deformed leaves; poor root growth
Magnesium	Chlorophyll component; activation of enzymes	Chlorosis; drooped leaves
Phosphorus	Component of nucleic acids, ATP, phospholipids	Purplish veins; stunted growth; fewer seeds, fruits
Sulfur	Component of most proteins, two vitamins	Light-green or yellowed leaves; reduced growth

MICRONUTRIENT	FUNCTIONS	SYMPTOMS OF DEFICIENCY
Chlorine	Role in root and shoot growth and photolysis	Wilting; chlorosis; some leaves die
Iron	Roles in chlorophyll synthesis and in electron transport	Chlorosis; yellow and green striping in leaves of grass species
Boron	Roles in germination, flowering, fruiting, cell division, nitrogen metabolism	Terminal buds, lateral branches die; leaves thicken, curl, become brittle
Manganese	Chlorophyll synthesis; coenzyme action	Dark veins, but leaves whiten and fall off
Zinc	Role in forming auxin, chloroplasts, starch; enzyme component	Chlorosis; mottled or bronzed leaves; root abnormalities
Copper	Component of several enzymes	Chlorosis; dead spots in leaves; stunted growth
Molybdenum	Part of enzyme used in nitrogen metabolism	Pale green, rolled or cupped leaves

* All mineral elements contribute to water–solute balances.

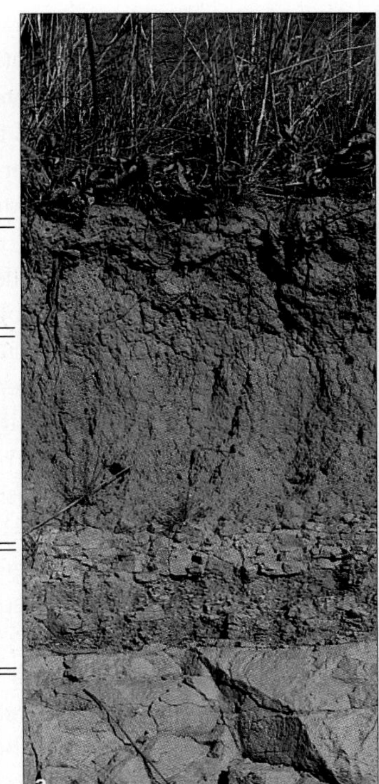
O HORIZON
Fallen leaves and other organic material littering the surface of mineral soil
A HORIZON
Topsoil, with decomposed organic material; variably deep (only a few centimeters in deserts, elsewhere extending as far as thirty centimeters below the soil surface)
B HORIZON
Compared with A horizon, larger soil particles, not much organic material, more minerals; extends thirty to sixty centimeters below soil surface
C HORIZON
No organic material, but partially weathered fragments and grains of rock from which soil forms; extends to underlying bedrock
BEDROCK
a

b

Figure 18.14 (**a**) Soil horizons that developed in one habitat in Africa. (**b**) Erosion in Chiapas, Mexico. Forests that once sponged up water from soil were cut down. Out-of-control runoff from rains cut deeper and wider gullies and carried away topsoil.

plant growth because it releases nutrients that plants absorb. Humus contains a lot of negatively charged organic acids, so positively-charged mineral ions stick to it. Humus also soaks up water, swelling as it does. Its alternate swelling and shrinking aerates the soil, meaning that it creates beneficial air spaces.

Most plants grow vigorously in soils that contain between 10 and 20 percent humus. Soil with less than 10 percent humus may not have enough nutrients to support plant growth. Soil with more than 90 percent humus tends to stay heavily saturated with water—so much so that air, and the oxygen that comes with it, is excluded. Swamps and bogs are like this. Very few kinds of plants grow in them.

Soils develop slowly, over thousands of years, and they are in different stages of development in different places. They generally form in layers, or horizons, that are distinct in color and composition (Figure 18.14*a*). Identifying the layers and how deep they are helps us compare soils in different areas. For instance, the A horizon is **topsoil**. It is the one most essential for plant growth, because it contains the bulk of decomposing organic material, and therefore has the most nutrients available for plants to absorb.

LEACHING AND EROSION

Water absorbs salts, mineral ions, and other nutrients as it filters through soil. **Leaching** is such downward percolation of water, and small quantities of dissolved substances, through soil. Leaching is fastest in sandy soils, which do not bind nutrients as well as clay soils. During heavy rains and the resulting runoff, leaching occurs more in forests than in grasslands. Why? Grass plants grow faster than trees and thus take up water more quickly, so there is less runoff.

Soil erosion is a loss of soil under the force of wind and water (Figure 18.14*b*). Strong winds, fast-moving water, and poor vegetation cover can cause the greatest losses. For example, erosion from croplands puts about 25 billion metric tons of topsoil into the Mississippi River, then the Gulf of Mexico each year. The nutrient losses not only affect plant growth, they also affect all organisms that depend on plants for survival.

Nutrients are essential elements. Each has a nutritional role that no other element can perform. Plants require nine macronutrients and seven micronutrients.

The mineral component of soil includes particles ranging from large-grained sand to silt and fine-grained clay. Clay reversibly binds water molecules and dissolved mineral ions, thereby making them accessible to plant roots.

Soil also contains humus, which is a reservoir of organic material, rich in organic acids, in different stages of decay. Most plants grow best in soils having equal proportions of sand, silt, and clay, as well as 10 to 20 percent humus.

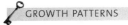

IMPACTS, ISSUES

Droughts and high winds are facts of life in North America's Great Plains. In the 1850s, native prairies were plowed under for farms. Since then, a third of the once-deep topsoil and half of the nutrients have disappeared. The greatest loss came in the 1930s. People witnessing huge clouds of topsoil blown from the exposed earth named the region the "Dust Bowl." Will the current big drought and ongoing erosion invite more of the same?

LINKS TO SECTIONS 4.4, 4.5, 14.4

18.6 How Do Roots Absorb Water and Mineral Ions?

Mining soil for mineral ions and water molecules takes energy. Wherever soil texture and composition change, new roots must form to infiltrate the different regions. Patches of soil with more favorable concentrations of water and mineral ions stimulate greater root growth.

SPECIALIZED ABSORPTIVE STRUCTURES

Plants require a lot of water and dissolved nutrients. For instance, a mature corn plant absorbs as much as three liters of water per day. Root hairs, mycorrhizae, and root nodules help with the uptake.

As a plant adds primary growth, its root system may develop billions of tiny root hairs (Figure 18.15). Collectively, these thin extensions of root epidermal cells enormously increase the surface area available for absorption of water and nutrients.

You already know something about "fungus-roots," or mycorrhizae (singular, mycorrhiza). As introduced in Section 14.4, each is a symbiotic interaction between a young root and fungus. Fungal filaments—hyphae—form a velvety cover around roots or penetrate the root cells. Collectively, the hyphae have a large surface area that absorbs mineral ions from a larger volume of soil than roots alone can (Figure 18.16). The fungus absorbs some sugars and nitrogen-rich compounds from root cells. The root cells get some scarce minerals that the fungus is better able to absorb.

Certain bacteria in soil are mutualists with clover, peas, and other legumes, which, like other plants, do not grow when nitrogen is scarce. Nitrogen ($N{\equiv}N$, or N_2) is plentiful in air, but it is very stable. Plants do not have enzymes that can break apart this gas. The bacteria do. By using ATP and a special enzyme, they can convert nitrogen gas to ammonia (NH_3). Metabolic conversion of gaseous nitrogen to ammonia is called nitrogen fixation. Other soil bacteria convert ammonia to nitrate (NO_3), a form of nitrogen that plants readily absorb. You will read more about mutualism and other two-way benefits between species in Chapter 30.

Nitrogen-fixing bacteria infect roots and thus become symbionts in localized swellings called **root nodules**

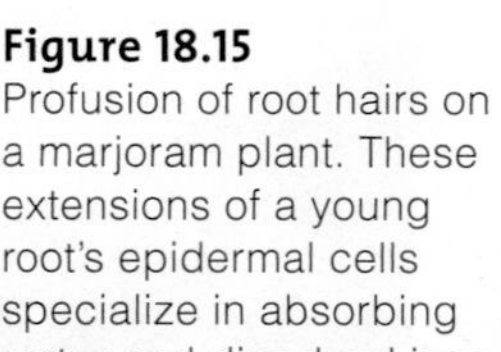

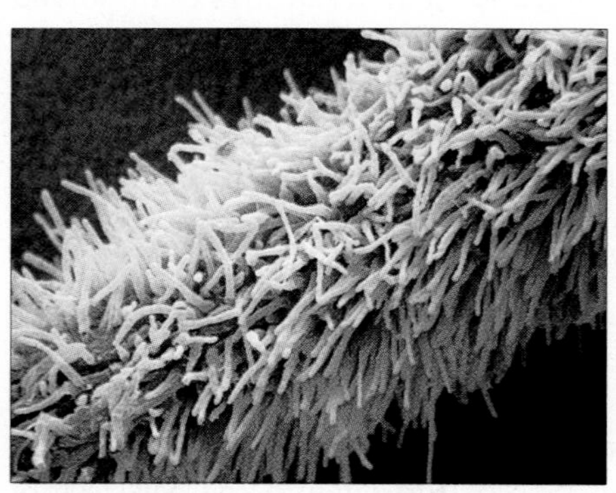

Figure 18.15 Profusion of root hairs on a marjoram plant. These extensions of a young root's epidermal cells specialize in absorbing water and dissolved ions.

Figure 18.16 Lodgepole pine (*Pinus contorta*) seedling. Notice the extent of mycorrhizae compared to the shoot system, which is only four centimeters tall.

Figure 18.17 (**a**) Root nodules of a soybean plant. (**b**) Effect of root nodules on plant growth. Soybean plants growing in nitrogen-poor soil show the effect of root nodules on growth. Only the plants in the rows at *right* were inoculated with nitrogen-fixing bacteria and formed nodules.

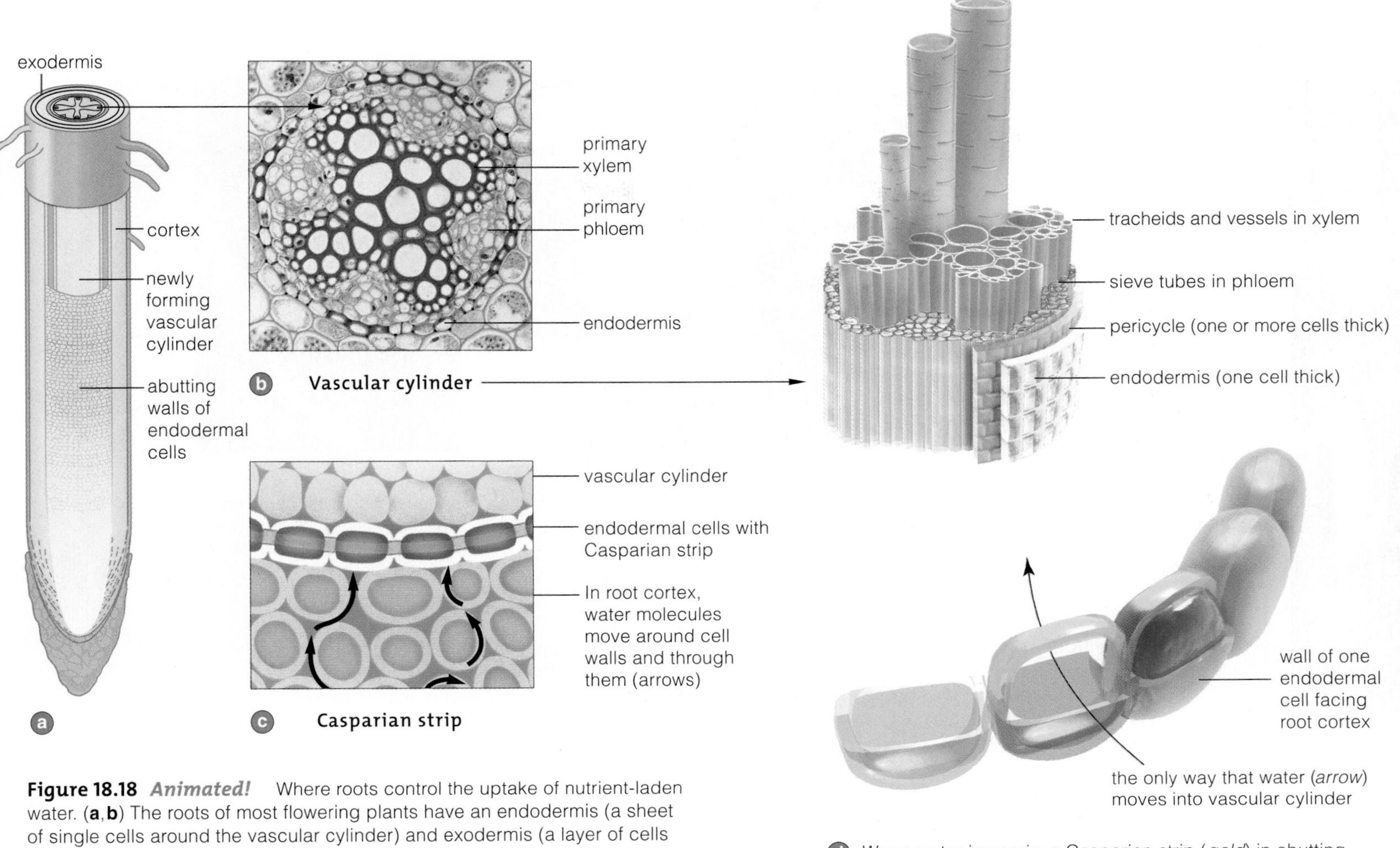

Figure 18.18 ***Animated!*** Where roots control the uptake of nutrient-laden water. (**a**, **b**) The roots of most flowering plants have an endodermis (a sheet of single cells around the vascular cylinder) and exodermis (a layer of cells beneath the epidermis).

(**c**) Abutting walls of cells of both layers have a waxy Casparian strip that keeps water from slipping past cells. Water must move through cytoplasm of endodermal cells. (**d**) Transport proteins in the plasma membrane of endodermal cells control water and nutrient uptake.

(Figure 18.17). The bacteria pilfer some photosynthetic products. The plants absorb some of the nitrogen that the bacteria assimilated from the air.

HOW ROOTS CONTROL WATER UPTAKE

Section 18.3 introduced the structure of a typical root. Now let's take a look at its function. Water molecules in soil cross the root epidermis and continue on to the vascular cylinder. A cylindrical sheet of endodermis encases the vascular cylinder. This sheet, a single layer of endodermal cells, is all that intervenes between root cortex and the xylem and phloem pipelines to the rest of the plant (Figure 18.18).

Wherever endodermal cells abut, a band of waxy, waterproof deposits called a **Casparian strip** prevents water from seeping around endodermal cells into the vascular cylinder. Water and dissolved nutrients thus enter the vascular cylinder only through the plasma membranes and cytoplasm of endodermal cells. Like all other living cells, endodermal cells have transport proteins embedded in their plasma membrane. These transport proteins are the points at which a plant can adjust the quantity and types of solutes that it absorbs from the soil water.

Roots of many plants also have an **exodermis**, a cell layer just beneath their surface. Walls of exodermal cells commonly have a Casparian strip that functions like the one embedded in endodermal cell walls.

Root hairs, root nodules, and mycorrhizae greatly enhance the uptake of water and dissolved nutrients.

A waxy Casparian strip prevents water and nutrients from penetrating the vascular cylinder except via transport proteins in the plasma membranes of endodermal cells.

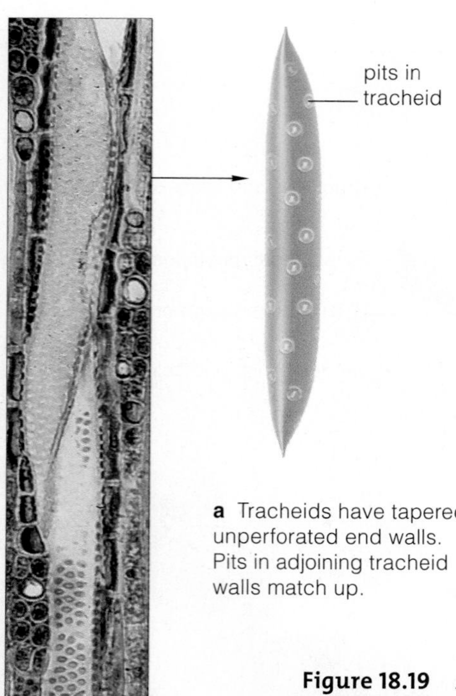

a Tracheids have tapered, unperforated end walls. Pits in adjoining tracheid walls match up.

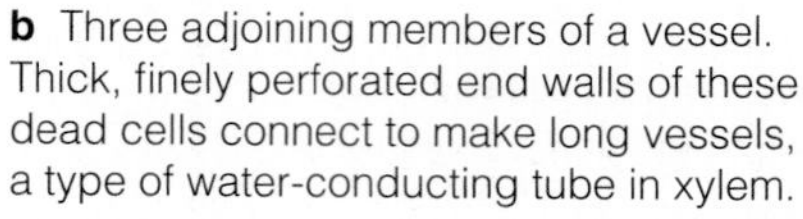

b Three adjoining members of a vessel. Thick, finely perforated end walls of these dead cells connect to make long vessels, a type of water-conducting tube in xylem.

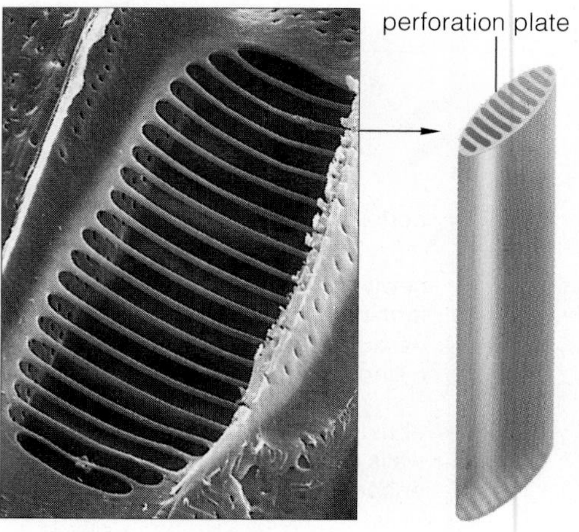

c Perforation plate at the end wall of a vessel member. The perforated ends allow water to flow unimpeded.

Figure 18.19 A few types of tracheids and vessel members from xylem. Interconnected, perforated walls of cells that died at maturity form these water-conducting tubes. The pectin-coated perforations may help control water distribution to specific regions. When hydrated, the pectins swell and close off water flow. During droughts, they shrink, and water moves freely through open perforations to leaves.

LINKS TO
SECTIONS
2.3, 4.6

18.7 Water Transport Through Plants

By now, you have a sense of how the distribution of water and dissolved mineral ions to all living cells is central to plant growth and functioning. Turn now to a model that explains how water moves throughout the plant.

TRANSPIRATION DEFINED

Plants secure only a fraction of the water they absorb for growth and metabolism. Most water is lost, mainly through the many stomata in leaves. The evaporation of water molecules from leaves, stems, and other plant parts is a process called **transpiration**. Far from being a waste, the process is essential for water uptake.

COHESION–TENSION THEORY

Start with a basic question: How does water actually get all the way up *to* leaves in the first place? What gets individual water molecules to the uppermost leaves of plants, including redwoods and other trees that may be more than 100 meters tall? Inside vascular tissues, water moves through xylem. Section 18.1 introduced you to the **tracheids** and **vessel members,** the water-conducting cells of xylem. Take a closer look at them in Figure 18.19. These cells are dead at maturity; only their lignin-impregnated walls are left. This means that xylem's conducting cells cannot be expending energy to pull water "uphill."

Some time ago, the botanist Henry Dixon devised a theory that explains how water is transported within plants. By his **cohesion–tension theory**, water inside xylem is pulled upward by air's drying power, which creates a continuous negative pressure called tension. The tension extends all the way from leaves to roots. Figure 18.20 illustrates Dixon's theory. As you review this figure, consider the following points.

First, air's drying power causes transpiration: the evaporation of water from all parts of the plant that are exposed to the air—but most notably at stomata. Transpiration puts water molecules that are inside the waterproof conducting tubes of xylem into a state of tension. The tension extends from veins inside leaves, down through the stems, and on into the young roots where water is being absorbed.

Second, continuous columns of water show *cohesion*, which means they resist breaking into droplets while they are pulled upward under tension. Remember the cohesive property of water (Section 2.3)? The collective strength of many hydrogen bonds imparts cohesion to the columns of water in xylem.

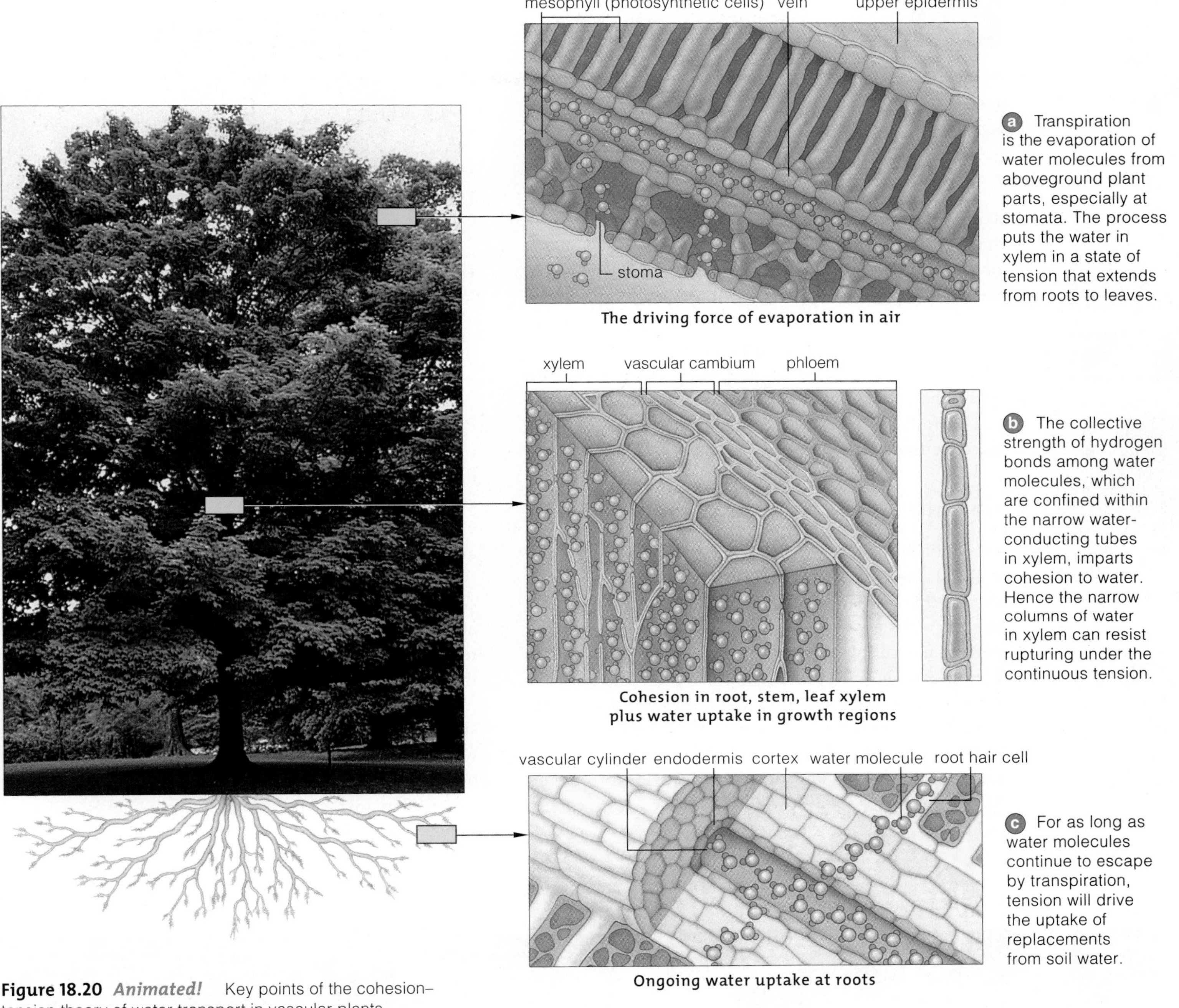

mesophyll (photosynthetic cells)
vein
upper epidermis
stoma
The driving force of evaporation in air
a Transpiration is the evaporation of water molecules from aboveground plant parts, especially at stomata. The process puts the water in xylem in a state of tension that extends from roots to leaves.
xylem
vascular cambium
phloem
Cohesion in root, stem, leaf xylem plus water uptake in growth regions
b The collective strength of hydrogen bonds among water molecules, which are confined within the narrow water-conducting tubes in xylem, imparts cohesion to water. Hence the narrow columns of water in xylem can resist rupturing under the continuous tension.
vascular cylinder
endodermis
cortex
water molecule
root hair cell
Ongoing water uptake at roots
c For as long as water molecules continue to escape by transpiration, tension will drive the uptake of replacements from soil water.

Figure 18.20 ***Animated!*** Key points of the cohesion–tension theory of water transport in vascular plants.

Third, for as long as individual molecules of water escape from a plant, the continuous tension inside the xylem permits more molecules to be pulled up from the roots, and therefore to replace them.

Hydrogen bonds are strong enough to hold water molecules together inside the water-conducting tubes of xylem. However, the bonds are not strong enough to stop the water molecules from breaking away from one another during transpiration and then escaping from leaves, through stomata.

Transpiration is a process of evaporation from plant parts.

By a cohesion–tension theory, transpiration puts water in xylem into a state of tension from leaf veins down to roots where water is being absorbed.

Tension pulls columns of water upward through the plant.

The collective strength of hydrogen bonds between water molecules keeps the columns of water from breaking into droplets as they rise through xylem tubes.

LINKS TO SECTIONS 5.3, 17.4, 17.5

18.8 How Do Stems and Leaves Conserve Water?

At least 90 percent of the water transported from roots to a leaf evaporates right out. Only about 2 percent gets used in photosynthesis, membrane functions, and other activities, but that amount must be maintained.

Think of a young plant cell, with a soft primary wall. As it grows, water diffuses in, fluid pressure increases on the wall and makes it expand, and the cell grows in volume. Expansion ends when there is enough *turgor*, which is the internal fluid pressure against the wall, to counterbalance the water uptake. When soil dries out or gets too salty, however, the balance can end, with bad effects on the plant (Figure 18.21). Yet plants are not entirely at the mercy of their surroundings. They have a cuticle, and they have stomata.

IMPACTS, ISSUES

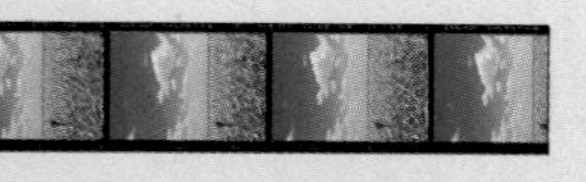

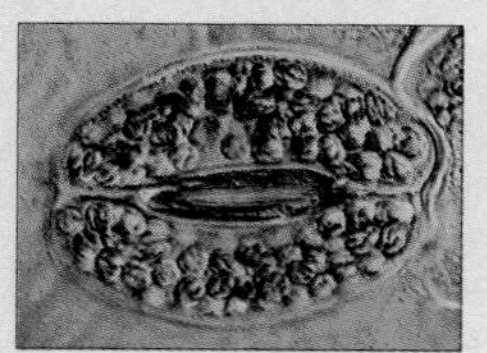

When calcium levels in guard cells oscillate at specific frequencies, stomata close and stay shut. Researchers at San Diego State University would like to pinpoint the genetic basis of this response as a way to design crop plants that can respond faster to drought conditions.

THE WATER-CONSERVING CUTICLE

Even mildly water-stressed plants would wilt and die without their cuticle (Figure 18.22). Epidermal cells secrete this translucent, waterproof layer, which coats the cell walls exposed to air. Cuticle is made of waxes, pectin, and cellulose fibers embedded in cutin, which is an insoluble lipid polymer.

A cuticle does not stop light rays from reaching photosynthetic cells. It does restrict water loss. It also restricts inward diffusion of the carbon dioxide used for photosynthesis, and outward diffusion of oxygen that is formed as its by-product.

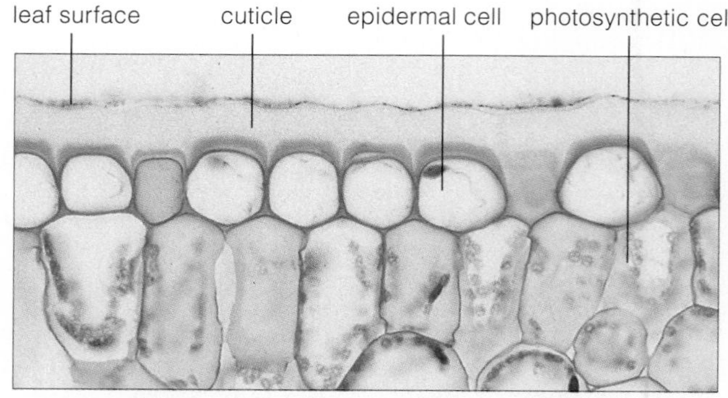

Figure 18.22 A typical plant cuticle, with epidermal cells and photosynthetic cells just beneath it.

CONTROLLED WATER LOSS AT STOMATA

Photosynthesis uses carbon dioxide and releases free oxygen. Both gases can diffuse across cuticle-covered epidermis at the stomata. Water is lost as well through stomata, but roots can replace the losses when there is enough water in the soil. Stomata of most plants close at night; CAM plants are exceptions (Section 5.3).

Water is conserved when stomata are closed, and carbon dioxide collects in the leaves as cells engage in aerobic respiration. Like you, plants use this energy-releasing pathway to make ATP, which drives most of their metabolic reactions.

A pair of specialized parenchyma cells define each stoma (Figure 18.23). When these two *guard cells* swell with water, they bend and a gap—the stoma—forms between them. When the guard cells lose water, they collapse against each other, so the gap closes.

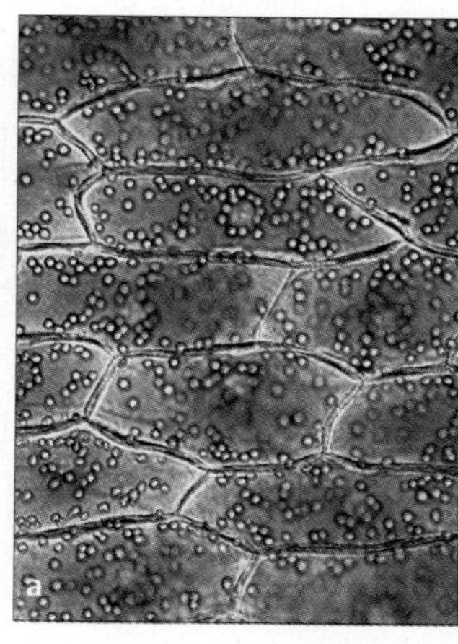

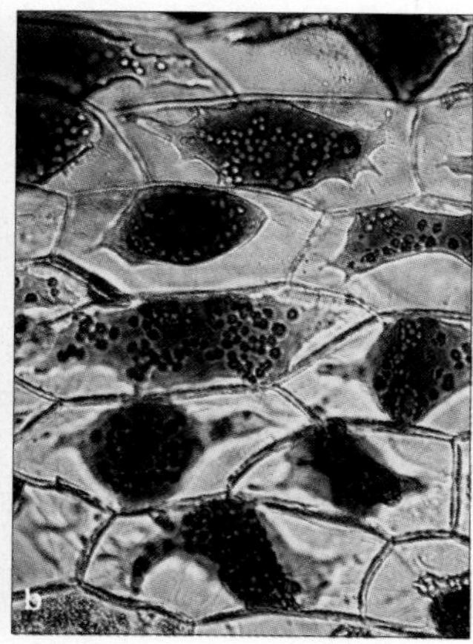

Figure 18.21 Wilting. (**a**) Cells from an iris petal, plump with water. (**b**) Cells from a wilted iris petal. The cytoplasm and central vacuole shrank, and the plasma membrane moved away from the wall.

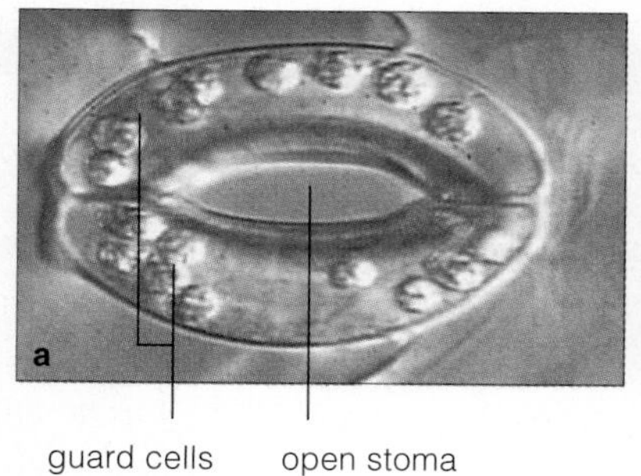

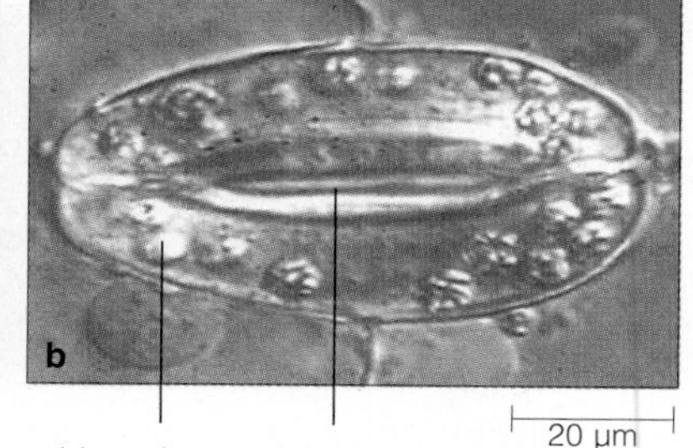

Figure 18.23 Stomata in action. Whether a stoma is open or closed depends on the shape of two guard cells, one on each side of this small gap across a cuticle-covered leaf epidermis. (**a**) This stoma is open. Water entered collapsed guard cells, which swelled and moved apart, forming the stoma. (**b**) This stoma is closed. Water left the two guard cells, which then collapsed against each other and closed the stoma. Chloroplasts are visible as round objects. Guard cells are the only epidermal cells that have chloroplasts.

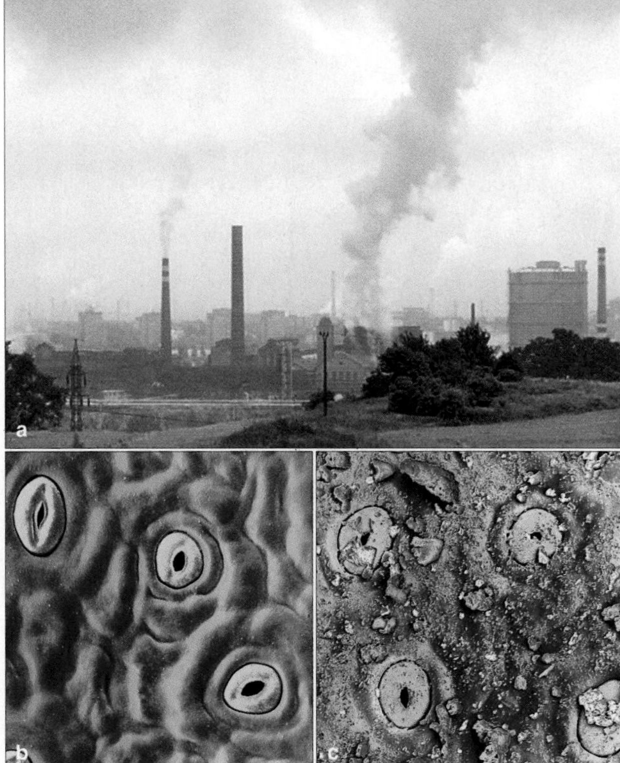

Figure 18.24 (**a**) Smog in Central Europe. (**b**) Stomata at a holly leaf surface. (**c**) A holly plant growing in industrialized regions becomes covered with gritty airborne pollutants that clog stomata and keep sunlight from reaching photosynthetic cells in the leaf.

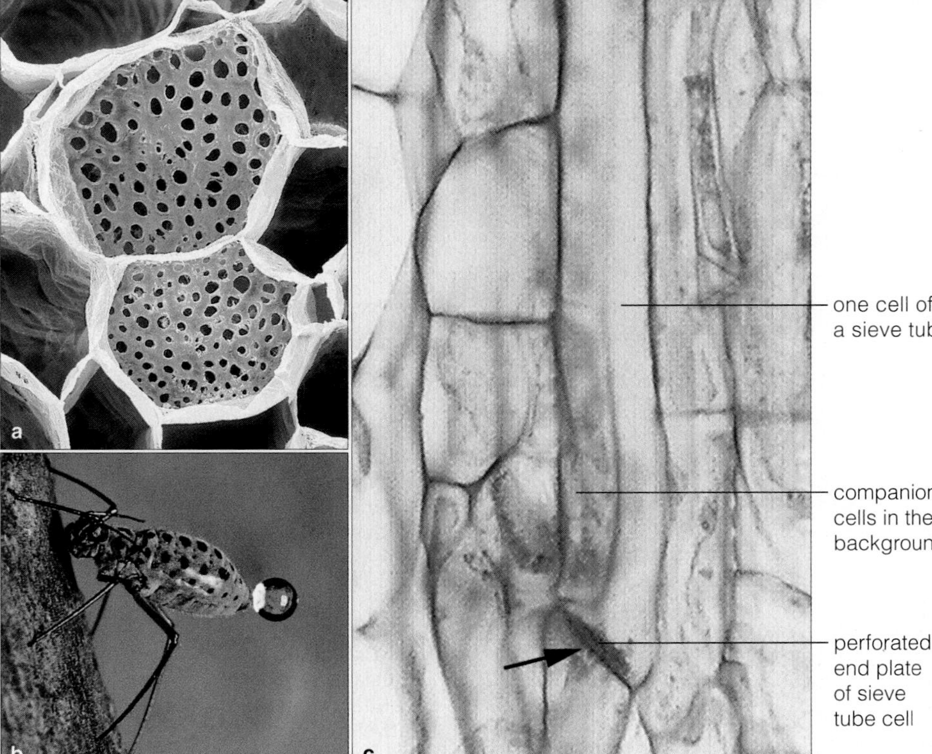

Figure 18.25 (**a**) Close-up of sieve plates on the ends of sieve tube members. (**b**) Honeydew exuding from an aphid after this insect's mouthparts penetrated a sieve tube. High pressure in phloem forced this droplet of sugary fluid out through the terminal opening of the aphid gut. (**c**) Part of a sieve tube inside phloem. Arrow points to perforated ends of sieve tube members.

Environmental cues, such as water availability, the level of carbon dioxide inside the leaf, and light, affect whether stomata remain open or closed. For example, photosynthesis starts after the sun comes up. Carbon dioxide levels fall in all photosynthetic cells, including guard cells. This decrease provokes stomata to open.

CAM plants, including most cacti, conserve water differently. They open stomata at night, when they fix carbon dioxide by the metabolic C4 pathway described in Section 5.3. The next day, when their stomata close, CAM plants use carbon dioxide in photosynthesis.

Plant survival depends on the function of stomata. Think about it when you are out and about on smog-shrouded days (Figure 18.24).

Water-dependent events in plants are severely disrupted when water loss exceeds uptake at roots for extended periods. Wilting is one observable outcome.

Transpiration and gas exchange occur mainly at stomata. These numerous small openings span the waxy cuticle, which covers all plant epidermal surfaces exposed to air.

Plants open and close stomata at different times to control water loss, carbon dioxide uptake, and oxygen disposal, all of which affect rates of photosynthesis and plant growth.

18.9 How Organic Compounds Move Through Plants

LINKS TO SECTIONS 2.6, 5.1

Xylem distributes water and minerals through the plant. The vascular tissue called phloem distributes organic products of photosynthesis.

Like xylem, phloem has many conducting tubes, fibers, and strands of parenchyma cells. Unlike xylem, it has **sieve tubes** through which organic compounds rapidly flow. These long tubes are formed by living cells that are positioned side by side and end to end. Where the end walls abut, they form porous sieve plates (Figure 18.25*a,c*). **Companion cells** load organic compounds into neighboring sieve tubes.

Some carbohydrate products of photosynthesis are used inside the leaf cells that make them. The rest get transported to roots, stems, buds, flowers, and fruits. In plants, carbohydrates are stored as starch, which is too big and insoluble to be transported. When energy reserves are required, the stored starch is converted to sucrose, which is easily transported. Experiments with aphids (Figure 18.25*b*) showed that sucrose is the main carbohydrate transported in phloem.

Translocation is the formal name for the process that moves sucrose and all other dissolved compounds

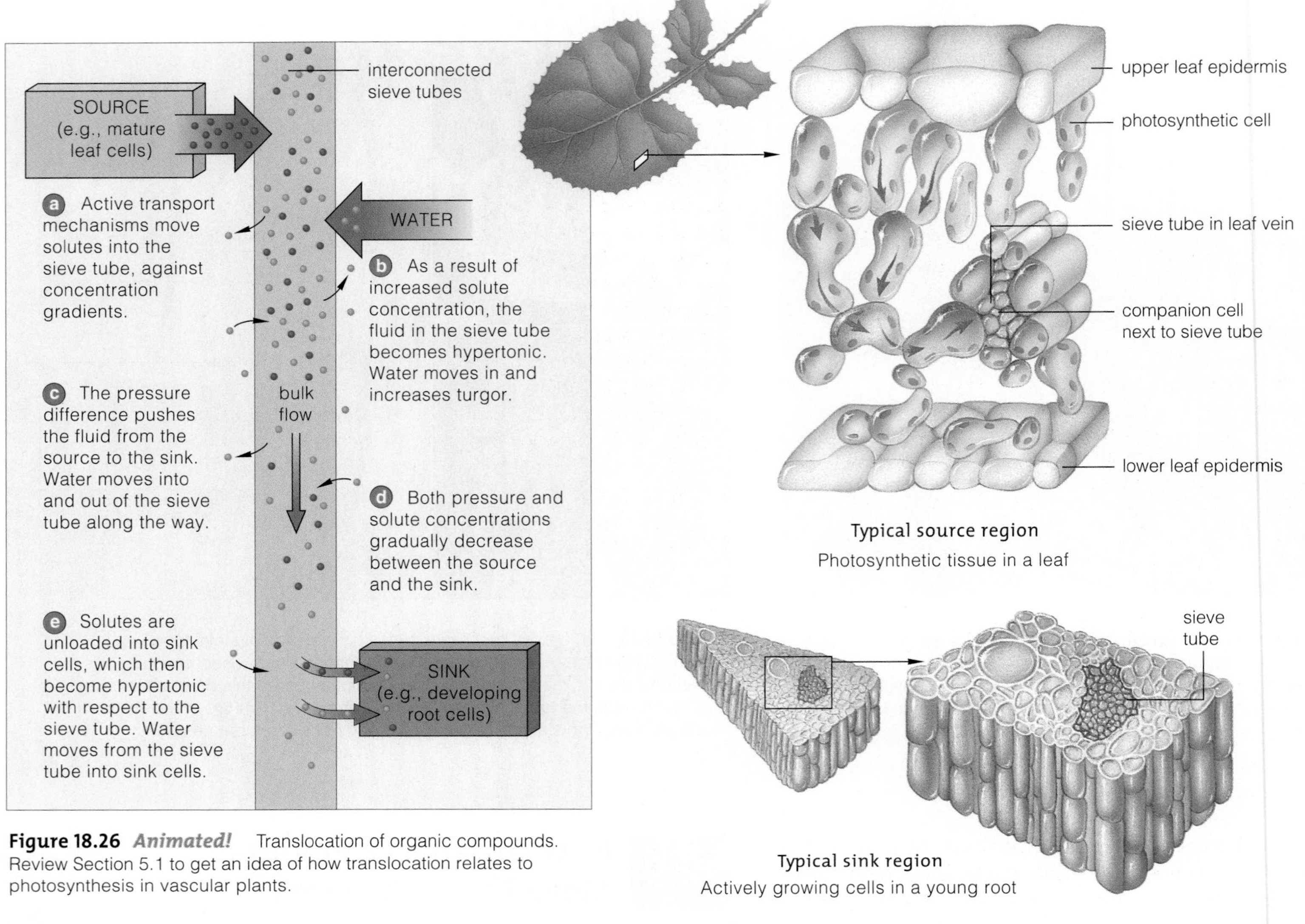

SOURCE (e.g., mature leaf cells)
interconnected sieve tubes
a Active transport mechanisms move solutes into the sieve tube, against concentration gradients.
WATER
b As a result of increased solute concentration, the fluid in the sieve tube becomes hypertonic. Water moves in and increases turgor.
c The pressure difference pushes the fluid from the source to the sink. Water moves into and out of the sieve tube along the way.
bulk flow
d Both pressure and solute concentrations gradually decrease between the source and the sink.
e Solutes are unloaded into sink cells, which then become hypertonic with respect to the sieve tube. Water moves from the sieve tube into sink cells.
SINK (e.g., developing root cells)
upper leaf epidermis
photosynthetic cell
sieve tube in leaf vein
companion cell next to sieve tube
lower leaf epidermis
Typical source region
Photosynthetic tissue in a leaf
sieve tube
Typical sink region
Actively growing cells in a young root

Figure 18.26 ***Animated!*** Translocation of organic compounds. Review Section 5.1 to get an idea of how translocation relates to photosynthesis in vascular plants.

through the phloem of a vascular plant. High pressure drives translocation. Pressure inside the conducting tubes of phloem can be as much as five times higher than the pressure inside automobile tires.

Phloem translocates photosynthetic products along declining pressure and solute concentration gradients. The *source* of the flow is any region of the plant where companion cells are loading organic compounds into the sieve tubes. Common sources are mesophylls—the photosynthetic tissues in leaves. The flow ends at a *sink*, which is any plant region where the products are being unloaded, used, or stored. For example, while flowers and fruits are growing during development, they are sink regions.

Why do organic compounds flow from a source to a sink? According to the **pressure flow theory**, internal pressure builds up at the source end of the sieve tube system and *pushes* the solute-rich solution on toward any sink, where solutes are being removed. You can use Figure 18.26 to track what happens after sucrose moves from the photosynthetic cells into small veins in leaves. By energy-requiring reactions, companion cells in the veins load sugars into sieve tube members. As the solute concentration increases inside the tubes, water also moves in, by osmosis. Turgor (internal fluid pressure) builds up against the sieve tube walls. As the pressure increases, it pushes sucrose-laden fluid out of the leaf, into the stem, and on to the sink.

Plants transport carbohydrates as starch but distribute them mainly in the form of sucrose.

Translocation is the movement of dissolved compounds to different parts inside a plant. It is driven by concentration and pressure gradients in the sieve tube system of phloem.

Concentration and pressure gradients form in phloem between sources of organic compounds and places the compounds are being used (sinks).

Summary

Section 18.1 The aboveground shoot systems of most plants include stems, leaves, and the case of angiosperms, reproductive parts called flowers. The vascular tissue system distributes nutrients and water through the plant. The dermal system protects plant surfaces. The ground tissue system makes up the bulk of the plant body.

Simple plant tissues have only one type of cell. Complex plant tissues have two or more. Monocots and eudicots (true dicots) are the primary angiosperm groups. They have the same tissues organized in somewhat different ways.

All tissues originate at meristems, regions of cells that are undifferentiated and retain a capacity for division. Primary growth at apical meristems lengthens shoots and roots. Secondary growth thickens older stems and roots. Wood is secondary xylem.

BiologyNow

Explore the tomato body plan and compare monocot and eudicot tissues with the animation on BiologyNow.

Section 18.2 Stems support upright growth and conduct water and solutes through vascular bundles. Most eudicot stems have a ring of bundles dividing the ground tissue into cortex and pith. Monocot stems often have vascular bundles distributed through ground tissue.

Leaves have mesophyll (photosynthetic parenchyma) and veins (vascular bundles) between their upper and lower epidermis. Water vapor, oxygen, and carbon dioxide cross the cuticle-covered epidermis at stomata.

BiologyNow

Look inside plant stems and the structure of a leaf with the animations on BiologyNow.

Section 18.3 Roots absorb water and dissolved mineral ions, which become distributed to aboveground parts. Eudicots typically have a taproot system, and many monocots have a fibrous root system.

BiologyNow

Learn about root structure and function with the animation on BiologyNow.

Section 18.4 Woody plants add secondary growth by cell divisions at lateral meristems. Wood is classified by its location and function (as in heartwood or sapwood).

BiologyNow

Learn about the structure of wood with the animation on BiologyNow.

Section 18.5 Plant nutrition requires water, mineral ions, and carbon dioxide. Mineral-laden water and the products of photosynthesis are distributed throughout the plant. Nutrients are essential elements; no other element can perform their metabolic functions.

Plants obtain nine macronutrients (such as carbon, oxygen, hydrogen, nitrogen, and phosphorus) and seven micronutrients (such as iron) from air, water, and soil. The properties of a given soil greatly affect the accessibility of water, oxygen, and nutrients to plants.

Section 18.6 Root hairs tremendously increase the surface area for absorption of water and nutrients. For many land plants, mycorrhizae and bacterial symbionts in root nodules assist in nutrient uptake.

Roots exert some control over absorption. A waxy Casparian strip seals abutting walls of endodermal (and exodermal) cells that form one-cell-thick cylinders inside roots. Water and dissolved mineral ions reach the vascular cylinder for distribution through the plant by moving through the cytoplasm of endodermal cells. Their inward diffusion is controlled to a large extent at active transport proteins of these cells.

BiologyNow

See how vascular plant roots control nutrient uptake with the animation on BiologyNow.

Section 18.7 Plants distribute nutrient-laden water through tracheids and vessel members of the vascular tissue called xylem. Both cell types are dead at maturity, but their walls connect as narrow pipelines.

Transpiration is the evaporation of water from plant parts, mainly at stomata, into air. A cohesion–tension theory explains it as a force that pulls water upward through xylem by causing continuous negative pressure (tension) from leaves to roots. Water molecules escape from leaves, but more are pulled into the leaf under tension. Collectively, hydrogen bonds among water molecules resist rupturing; they impart cohesion, so water is pulled upward as continuous fluid columns.

BiologyNow

Learn about water transport in vascular plants with the animation on BiologyNow.

Section 18.8 A cuticle is a waxy, waterproof cover on all plant surfaces exposed to the surroundings. It helps the plant conserve water on hot, dry days.

Gas exchange and transpiration occur at stomata, openings across the cuticle-covered epidermis of leaves and many stems. In response to environmental cues, water moves into and out of paired guard cells, which plump up or collapse against each other. This opens or closes the gap (the stoma) between them.

When stomata are open, they permit gas exchange. Plants conserve water when stomata are closed.

Section 18.9 Sugars and other organic compounds are distributed throughout the plant, driven by pressure and concentration gradients. Sugars from photosynthetic cells (sources) are loaded into sieve tubes of phloem by companion cells. They are unloaded at actively growing regions or at storage regions (the sinks).

BiologyNow

Observe how vascular plants distribute organic compounds with the animation on BiologyNow.

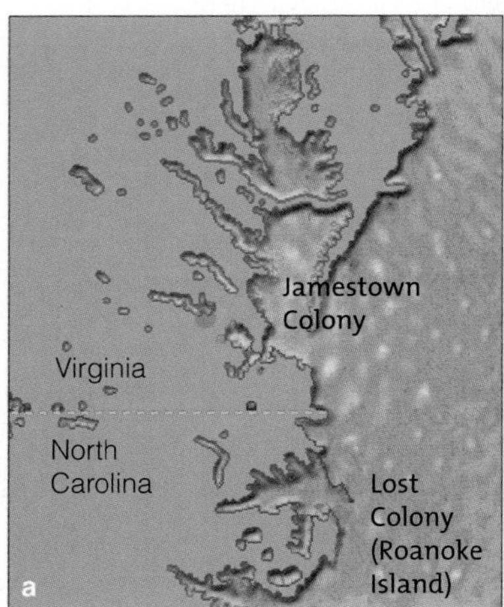

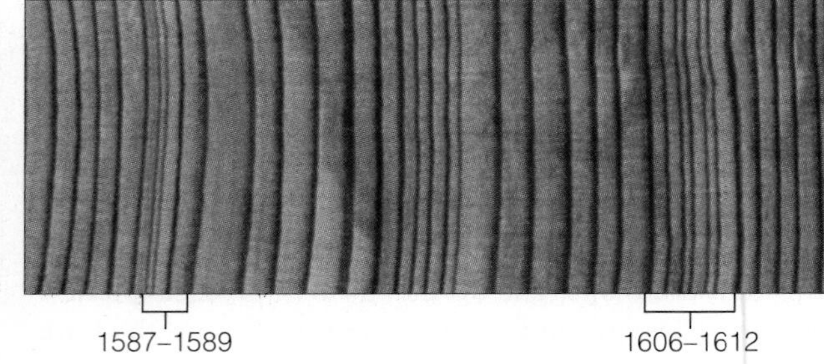

Figure 18.27 (**a**) Location of two early American colonies. (**b**) Bald cypress swamp. (**c**) Bald cypress growth layers, cross-section. This tree was living when English colonists first settled in North America. Narrow annual rings mark severe drought years, which stunted tree growth.

Self-Quiz

Answers in Appendix I

1. Roots and shoots lengthen through activity at _______ .
   a. apical meristems c. vascular cambium
   b. lateral meristems d. cork cambium

2. In many plant species, older roots and stems thicken by activity at _______ .
   a. apical meristems c. vascular cambium
   b. cork cambium d. both b and c

3. A _______ strip in endodermal cell walls forces water and solutes to move through root cells, not around them.
   a. cutin c. Casparian
   b. lignin d. cellulose

4. _______ conducts mainly water and ions and _______ conducts mainly sugars.
   a. Phloem; xylem b. Xylem; phloem

5. The nutrition of some plants depends on a root–fungus association known as a _______ .
   a. root nodule c. root hair
   b. mycorrhiza d. root hypha

6. In phloem, organic compounds flow through _______ .
   a. collenchyma cells c. vessels
   b. sieve tubes d. tracheids

7. Water evaporation from plant parts is called _______ .
   a. translocation c. transpiration
   b. expiration d. tension

8. Water transport from roots to leaves occurs by _______ .
   a. pressure flow
   b. differences in source and sink solute concentrations
   c. the pumping force of xylem vessels
   d. cohesion–tension among water molecules

9. Match the plant parts with the suitable description.
   ____ apical meristem a. mass of secondary xylem
   ____ lateral meristem b. source of primary growth
   ____ xylem c. distribution of sugars
   ____ phloem d. source of secondary growth
   ____ vascular cylinder e. distribution of water
   ____ wood f. central column in roots

Additional questions are available on **Biology Now™**

Critical Thinking

1. Home gardeners, like farmers, must be sure that their plants have access to nitrogen from either nitrogen-fixing bacteria or fertilizer. Insufficient nitrogen stunts plant growth; leaves yellow and die. Which biological molecules incorporate nitrogen? How would nitrogen deficiency affect biosynthesis and cause these symptoms?

2. You just returned home from a three-day vacation. Your severely wilted plants tell you they weren't watered before you left. Being aware of the cohesion–tension theory of water transport, explain what happened to them.

3. When moving a plant from one location to another, it helps to include some native soil around the roots. Explain why, in terms of mycorrhizae and root hairs.

4. In 1587, about 150 English settlers arrived at Roanoke Island off the coast of North Carolina. When ships arrived in 1589 to resupply the colony, they found the island had been abandoned. Searches up and down the coast failed to turn up the missing colonists. About twenty years later, the English established a colony at Jamestown, Virginia. Although this colony survived, the initial years were hard. In the summer of 1610 alone, more than 40 percent of the colonists died, many of starvation.

Scientists examined wood cores from bald cypress (*Taxodium distichum*) that had been growing at the time the Roanoke and Jamestown colonies were founded. Growth rings revealed the colonists were in the wrong place at the wrong time (Figure 18.27). They arrived at Roanoke just in time for the worst droughts in eight hundred years. Nearly a decade of severe drought struck Jamestown.

We know the corn crop of the Jamestown colony failed totally. Drought-related crop failures probably occurred at Roanoke. The settlers also had trouble finding fresh water. Jamestown was established at the head of an estuary; when river levels dropped, the water became brackish.

If secrets locked in trees intrigue you, look into a field of study called dendroclimatology. For example, find out what growth layers reveal about fluctuations in climate where you live. See if you can correlate it with human events at the time the changes were being recorded.

19 PLANT REPRODUCTION AND DEVELOPMENT

IMPACTS, ISSUES

Imperiled Sexual Partners

Imagine a world without chocolate. No chocolate bars, hot cocoa, brownies, or chocolate cake. Too far-fetched? Chocolate comes from seeds of *Theobroma cacao* trees. Although this tree is native to tropical rain forests, it does not have to grow in them. It grows nicely in places where rain forests were cleared to make way for sunny plantations. There is just one problem. Plantation trees do not produce many seeds.

As plantation owners found out, the cacao tree's only pollinators are midges, tiny flies the size of a pinhead. By carrying pollen from flower to flower, midges play a significant role in the sexual reproduction of *T. cacao*. Midges live and breed only in the deep, damp shade of tropical rain forests. No forests, no midges. No midges, no *T. cacao* seeds. That is a problem for the commercial production of chocolate.

Like *T. cacao*, three-fourths of the world's crop plants complete their life cycles with the help of pollinators—mostly insects such as bees, moths, butterflies, beetles, and flies. Honeybees (*Apis mellifera*, in the filmstrip at *right*) pollinate most of the crops in the United States. Some crop plants are best pollinated by different bee species. For example, the blueberry bee (*Habropoda laboriosa*) of the southeastern United States specializes in pollinating native blueberry plants.

The reproductive success of most flowering plants depends on the animal pollinators that coevolved with them. That same dependence puts many plant species at risk as their pollinators are clobbered by pesticide use and habitat loss. About 570 pollinator species in the United States alone are endangered, and another 140 are threatened. Pollinator shortages are contributing to the current decline in seed production in all flowering plant species, the impact of which ripples throughout communities. Flowers are an outcome of coevolution. With pollinators on the decline, how many flowering plants will disappear with them?

How Would You Vote? *Besides targeting pests, microencapsulated pesticides also wipe out populations of important pollinators. Should their use be restricted? See BiologyNow for details, then vote online.*

Key Concepts

FLOWERING PLANT LIFE CYCLES

Flowering plant life cycles alternate between spore-producing bodies and gamete-producing bodies. Although most flowering plants reproduce primarily by sexual reproduction, many have asexual phases. Pollen and eggs, then seeds and fruits develop on specialized floral shoots. Most flowering plant species complete their life cycle with the help of coevolved pollinators.

DEVELOPMENTAL CUES

Flowering plant life cycles are orchestrated by genes, hormones and other signaling molecules, and environmental cues. Interactions among hormones control plant growth, development, and reproduction.

PERSPECTIVE

Current research, an extension of thousands of years of breeding practices, is generating new varieties of crop plants that have global benefits.

Links to Earlier Concepts

A review of sexual and asexual reproduction (Section 7.5) will orient you for this chapter, which returns to the reproductive mode of flowering plants (15.6). You may wish to reflect on the evolutionary history of flowering plants and their pollinators (15.1, 15.4). You will be using your knowledge of plant tissues and the meristems that give rise to them (18.1), as well as re-encountering gene control (10.6) and cloning techniques (11.6).

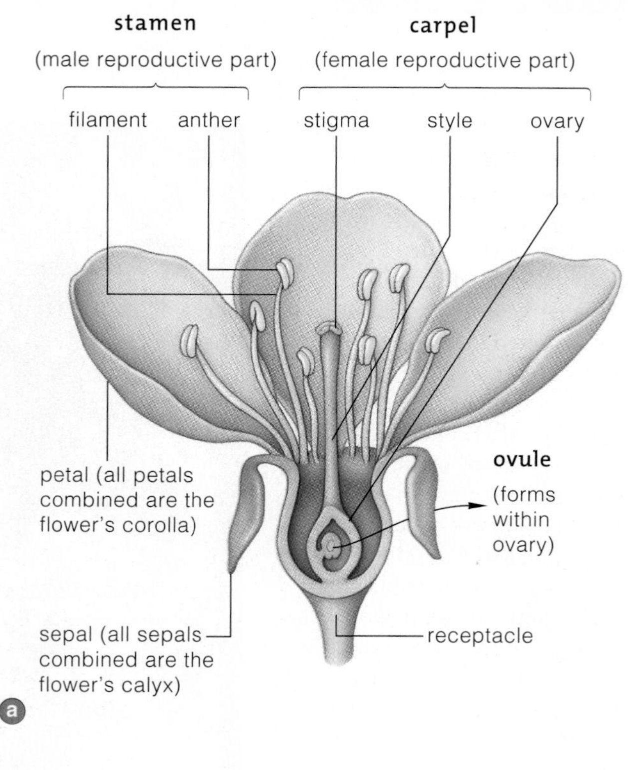

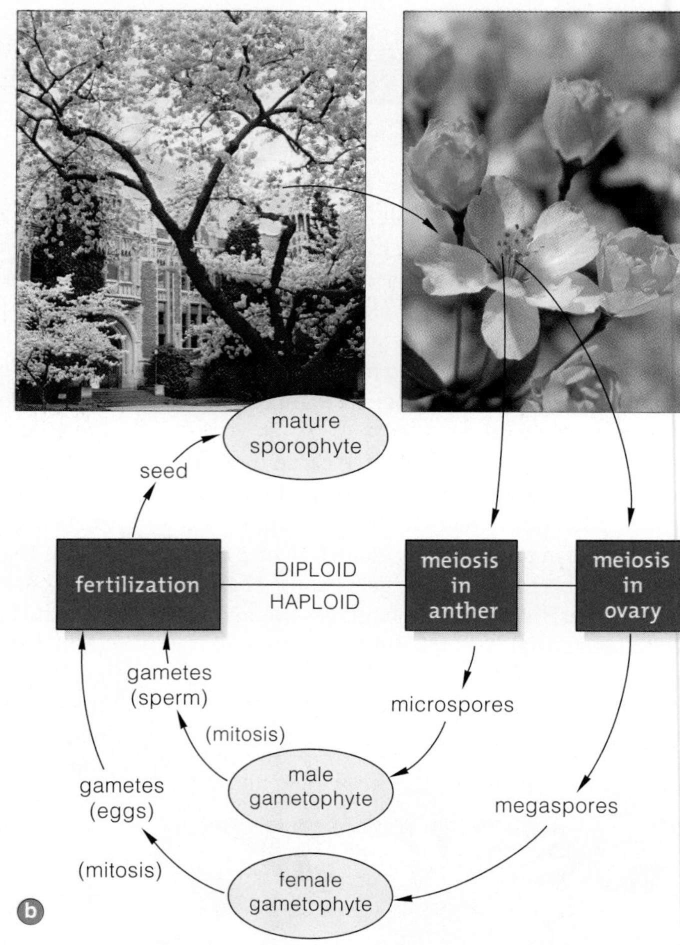

Figure 19.1 ***Animated!*** Typical flowering plant life cycle. A cherry tree is the example.

LINKS TO SECTIONS 7.5, 15.1, 15.6

19.1 Sexual Reproduction in Flowering Plants

Like us, flowering plants engage in sex. They produce and protect gametes. Like human females, they nourish and protect developing embryos. By issuing various sensory invitations to pollinators, flowers improve a plant's odds of reproductive success.

REGARDING THE FLOWERS

We would like to call the flower in Figure 19.1 typical, but it is only one of more than a quarter of a million kinds. All **flowers** are specialized reproductive shoots of a diploid **sporophyte**, a plant body that grows by mitotic divisions of a fertilized egg. Sporophytes make haploid reproductive bodies called spores. Two kinds of spores develop into male and female **gametophytes**. These haploid structures give rise to male or female gametes during the life cycle (Sections 15.1 and 15.6).

Flowers are whorls of modified leaves attached to the top of the floral shoot: the receptacle. In a common pattern, the leaves differentiate as an outermost whorl of sepals, a ring of leaflike petals (nonfertile parts) and, innermost, the stamens and carpels (fertile parts).

Stamens are male reproductive parts of a flower. They usually consist of an anther on top of a thin stalk, or filament. Inside a typical **anther** are two to six pairs of pouches called pollen sacs. Inside the sacs, meiosis of diploid cells gives rise to haploid, walled spores. These spores differentiate into pollen grains, structures that contain the male gametophytes (Figure 19.2).

A flower's female reproductive parts, or **carpels**, are located at its center. The upper portion of a carpel, or of a fused group of carpels, is known as the stigma. The stigma's sticky or hairy surface tissue gently traps pollen grains long enough for them to resume growth after a period of dormancy, or **germinate**. The carpel's lower, swollen portion is an **ovary**. Inside the ovary are one or more **ovules**, structures in which a haploid, egg-producing female gametophyte forms.

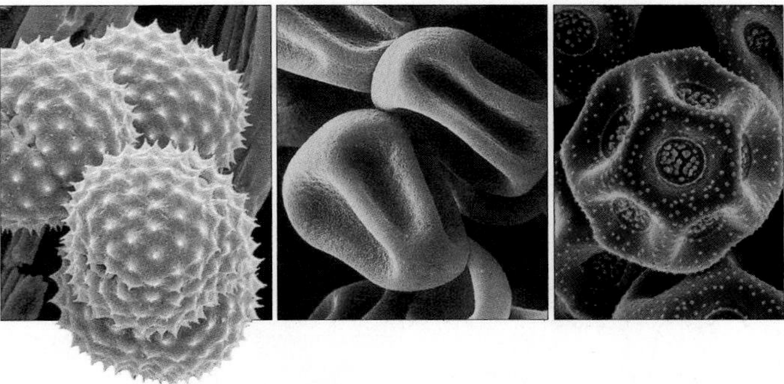

Figure 19.2 From left to right, ragweed, grass, and chickweed pollen. The pollen grains of many families or species of flowering plants are distinctive in size, wall sculpturing, and number of wall pores.

REGARDING THE POLLINATORS

Flowering plants release pollen at specific times of year. You are aware of this reproductive event if you are one of millions of people who experience hay fever, which is an allergic reaction to proteins on the wall of pollen grains. These proteins did not evolve to make people suffer. They help female floral parts identify pollen of the same species.

Pollinators are air currents, animals, or any other agent that transfers pollen grains from male to female reproductive parts of flowering plants. Section 15.6 introduced coevolution of flowering plants and their animal pollinators. The structure, coloration, patterns, fragrances, and sugar-rich nectar of flowers coevolved with specific kinds of pollinators that transfer pollen most efficiently.

Consider how stamens and carpels are positioned where pollinating agents can brush against them. For instance, in hummingbird-pollinated flowers, stamens and carpels are located above nectar-filled floral tubes Consider as well that birds have acute daytime vision and a poor sense of smell. Bird-pollinated plants will typically have flowers with vivid red components and do not waste energy making fragrances.

Flowers with nectar cups deep enough for beetles to drown in do not attract beetles. Beetles are attracted to flowers that smell like pungent dung or decaying litter and rotting fruit on a damp forest floor. This is the type of place where beetles first evolved.

Fragrant, colorful flowers with sturdy petals attract day-flying butterflies and moths (Figure 19.3*a*). Night-flying nectar-sipping bats and moths pollinate flowers with intense, sweet odors and white or pale petals that show up in the dark (Figure 19.3*b*).

Bees are attracted mainly to flowers that have sweet odors and yellow, blue, or purple parts. In such flowers, pigment molecules that absorb or reflect ultraviolet light have been organized in distinctive patterns. Unlike us, bees see the patterns, which they recognize as guides to the nectar. Unlike beetles, bees have mouthparts that extend down into long and skinny floral cups, and thus reach the nectar.

IMPACTS, ISSUES

Wind-pollinated plants release huge numbers of lightweight pollen grains to counter the hit-or-miss nature of reaching female floral parts. Pollen drifting through air is the cause of most allergic reactions called "hay fever." Insect-pollinated plants make fewer pollen grains. These are often sticky or barbed and tend to adhere to a pollinator's surface parts, such as the insect hairs shown here.

Figure 19.3 Pollinators. (**a**) Bright, fragrant flowers with sturdy petals attract day-flying moths and butterflies. Here, burnet moths enjoy a wild bergamot flower. (**b**) Large, white flowers tip the long, spiny arms of the giant saguaro. Insects and birds visit them by day, and bats by night. The cactus offers nectar. The animals transport pollen grains, which stick to their body, from one plant to another.

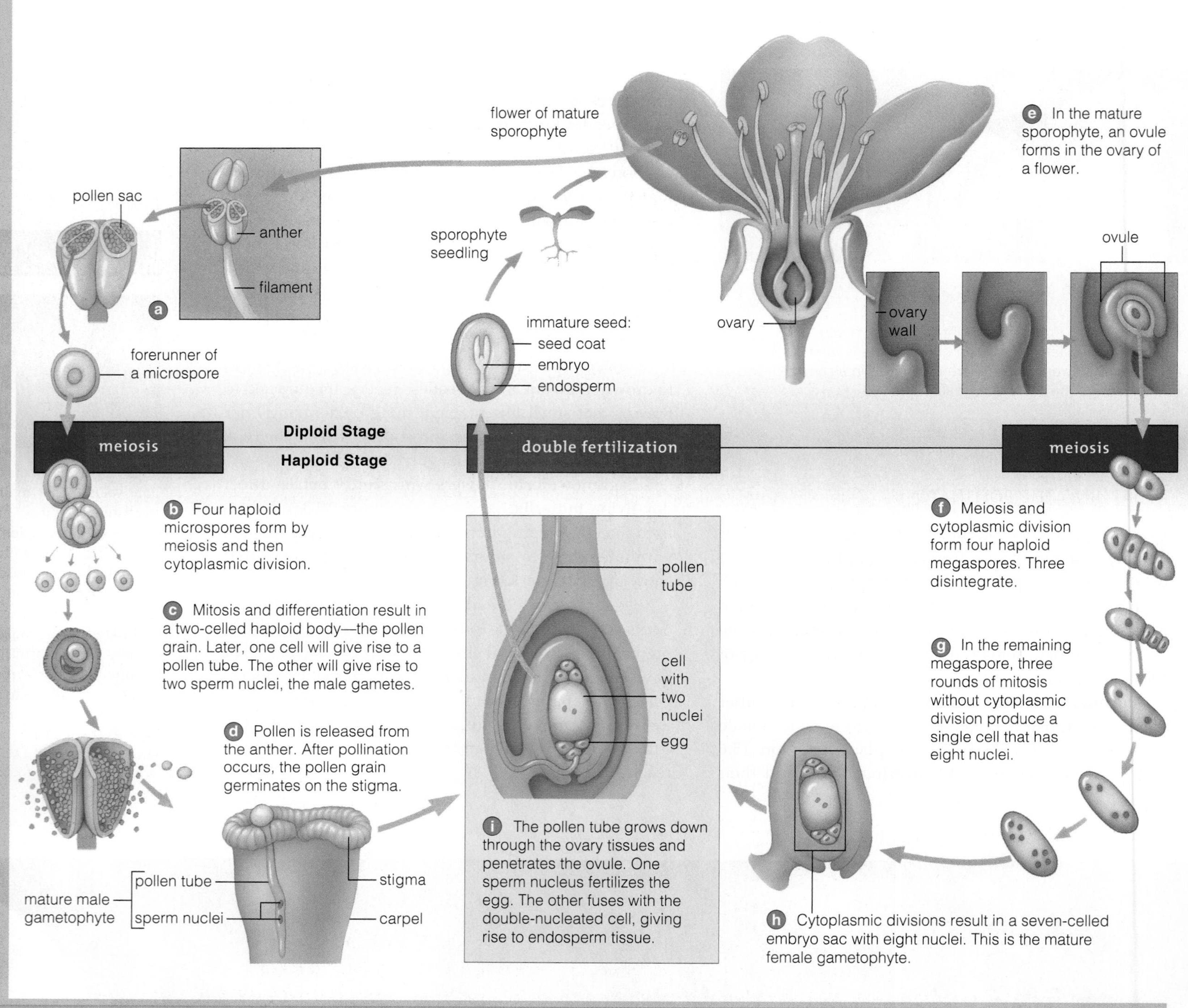

flower of mature sporophyte
e In the mature sporophyte, an ovule forms in the ovary of a flower.
pollen sac
anther
filament
a
sporophyte seedling
ovule
immature seed:
seed coat
embryo
endosperm
ovary
ovary wall
forerunner of a microspore
meiosis
Diploid Stage
Haploid Stage
double fertilization
meiosis
b Four haploid microspores form by meiosis and then cytoplasmic division.
pollen tube
f Meiosis and cytoplasmic division form four haploid megaspores. Three disintegrate.
c Mitosis and differentiation result in a two-celled haploid body—the pollen grain. Later, one cell will give rise to a pollen tube. The other will give rise to two sperm nuclei, the male gametes.
cell with two nuclei
egg
g In the remaining megaspore, three rounds of mitosis without cytoplasmic division produce a single cell that has eight nuclei.
d Pollen is released from the anther. After pollination occurs, the pollen grain germinates on the stigma.
i The pollen tube grows down through the ovary tissues and penetrates the ovule. One sperm nucleus fertilizes the egg. The other fuses with the double-nucleated cell, giving rise to endosperm tissue.
pollen tube
stigma
mature male gametophyte
sperm nuclei
carpel
h Cytoplasmic divisions result in a seven-celled embryo sac with eight nuclei. This is the mature female gametophyte.

Figure 19.4 ***Animated!*** Closer look at the life cycle of cherry (*Prunus*).

FROM SPORES TO ZYGOTES

How is pollen produced? In a growing anther, masses of diploid, spore-producing cells form by mitotic cell divisions. Walls develop around each mass, forming a pollen sac (Figure 19.4*a*). Every spore-producing cell inside the sacs undergoes meiosis to form four haploid **microspores** (Figure 19.4*b*). Each microspore has a cell wall that displays species-specific molecules. Mitosis and differentiation of the microspores produce pollen grains containing male gametophytes (Figure 19.4*c*). After a period of dormancy, the pollen sacs split open, and the pollen grains are released from the anther to be taken up by pollinators (Figure 19.4*d*).

In the meantime, one or more ovules form inside a cell mass on the inner wall of a flower's ovary (Figure 19.4*e*). Protective layers grow around the mass. One cell inside each ovule undergoes meiosis to form four haploid **megaspores** (Figure 19.4*f*). Three megaspores typically disintegrate. Mitosis and nuclear division of the remaining megaspore give rise to a cell with eight haploid nuclei (Figure 19.4*g*). This multinucleated cell divides to form a seven-celled embryo sac that is the mature female gametophyte (Figure 19.4*h*). One of the

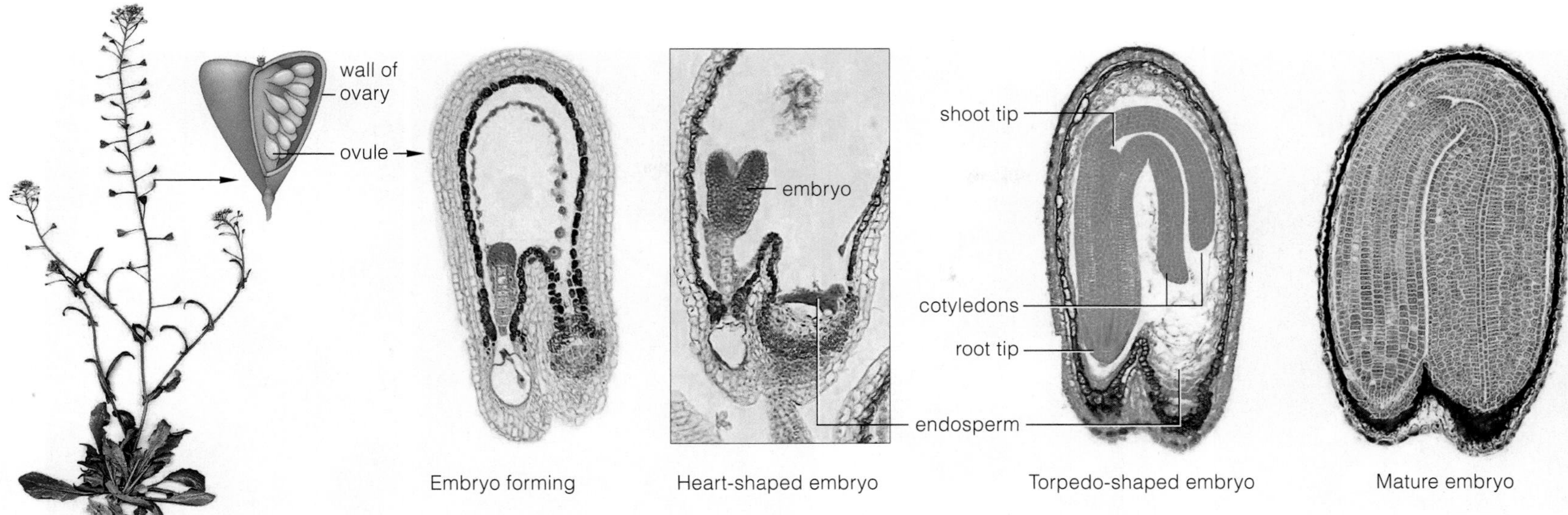

Figure 19.5 ***Animated!*** Mature sporophyte and developing seeds of shepherd's purse (*Capsella*), a eudicot.

cells inside the embryo sac has two nuclei. This cell will help form triploid **endosperm**, a nutritive tissue. Another cell is the female gamete, the egg.

Meanwhile, back at the anthers, pollen grains have been released. **Pollination** occurs when one lands on a receptive stigma. The pollen grain then germinates and starts to grow. A pollen tube, a tubular outgrowth of the pollen grain, develops from the gametophyte. Two sperm nuclei, or male gametes, form inside the pollen tube. Taken together, a pollen tube and its contents of male gametes constitute the mature male gametophyte (Figure 19.4*d*).

The pollen tube grows down through the carpel and ovary toward the ovule, carrying with it the two sperm nuclei. Plant hormones in the ovule guide the tube's growth to the embryo sac. The gametes are then released into the sac (Figure 19.4*i*).

Double fertilization occurs in flowering plants. One of the sperm nuclei from the pollen tube fuses with (fertilizes) the egg to form a diploid zygote. The other fuses with the double-nucleated cell in the embryo sac. This cell gives rise to the $3n$ endosperm tissue that will help nourish the embryo later, as it grows. After fertilization, the ovule matures into a seed.

Sexual reproduction is the dominant reproductive mode of flowering plant life cycles. Such cycles alternate between the formation of sporophytes (spore-producing bodies) and gametophytes (gamete-producing bodies).

Male gametophytes form in stamens and are contained in pollen grains. Female gametophytes with egg cells form as part of ovules, which develop in the ovaries of carpels.

After pollination and double fertilization, an embryo and nutritive tissue grow in the ovule, which matures as a seed.

19.2 From Zygotes to Seeds Packaged in Fruit

LINK TO SECTION 18.1

Angiosperms are the only plants that form seeds inside of ovaries. They are also the only plants that can produce fruit—the mature ovary wall with or without additional layers of tissue.

How do seeds and fruits develop? Consider a *Capsella* embryo. Its **cotyledons**, or seed leaves, develop from tiny lobes of meristematic tissue (Figure 19.5). Eudicot embryos have two cotyledons; monocot embryos have one. As is the case for many other eudicots, a *Capsella* embryo absorbs nutrients from endosperm and stores them in its two cotyledons. By contrast, the embryos of corn, wheat, and most other monocots do not tap nutritive tissue until the seed germinates.

After fertilization, nutrients from the parent plant are moved into the maturing ovule to nourish the new embryo sporophyte. The food reserves accumulate in endosperm or in cotyledons. Eventually, the ovule wall pulls away from the ovary wall to become a protective seed coat. The embryo, its food reserves, and the seed coat now constitute the mature ovule, a self-contained package called a **seed**.

As seeds grow, other floral parts form fruits. *Simple* fruits are dry or fleshy, derived from one ovary, like pea pods, acorns, and *Capsella*. *Aggregate* fruits form from separate ovaries of a single flower. Raspberries are aggregate fruits. *Multiple* fruits form from many ovaries of separate flowers. The pineapple, a multiple fruit, forms as the ovaries of many flowers fuse into a fleshy, enlarging mass that becomes surrounded by a waxy, hardened rind. *Accessory* fruits form from the

Figure 19.6 (**a**) Simple, dry fruits of maple (*Acer*). (**b**) Cedar waxwing eating mountain ash berries, a simple fleshy fruit. (**c**) Cacao fruits, or pods. Each of these simple, fleshy fruits has up to forty seeds, the cacao "beans" piled up here. Seeds are processed into cocoa butter and essences of chocolate. At least a thousand compounds in chocolate exert compelling (some say addictive) effects on the human brain. The average American buys 8 to 10 pounds of chocolate annually.

(**d**) Cocklebur, a prickly accessory fruit that sticks to fur of animals brushing against it. (**e**) Strawberry (*Fragaria*). All of its fleshy tissues develop from a receptacle. At maturity, the receptacle has many small, dry, fruit-enclosed seeds on its surface.

IMPACTS, ISSUES

California's leading export crop, almonds, depends heavily on honeybees. Two kinds of trees are planted in an orchard—one to supply pollen, the other to bear fruit. A million or so honeybees in hives are trucked in to carry pollen between trees. If current declines in the honeybee populations continue, the California almond farmers will be hard hit by economic losses.

ovary and other floral structures. Apples are accessory fruits: the core is derived from the ovary of a flower, and the rest of the fruit is a fleshy receptacle. Other examples of fruits are shown in Figure 19.6.

All fruits have a common function—*seed dispersal*. They are adapted to specific dispersing agents, such as air or water currents, or passing animals. For example, the fruits of maple trees (*Acer*) are dispersed by wind. Part of the maple fruit extends out, like a pair of thin, lightweight wings. It breaks in half, drops from a tree, and spins sideways as it interacts with air currents. The spinning motion moves the seed far enough away from the parent plant that seedlings do not have to compete with it for water, mineral ions, and sunlight.

Many mammals, birds, insects, and other animals disperse seeds. Some fruits have hooks, spines, barbs, hairs, sticky substances, or other specializations that can adhere to fur, feathers, or hair. The fruits of bur clover, cocklebur, and bedstraw are examples.

Animals also disperse seeds by ingesting them. For instance, as strawberry seeds pass through an animal gut, digestive enzymes break down only a part of the hard seed coats. The seeds are expelled from the body later, in feces, often far from the parent plant. Then, the thinned regions of the seed coats are easier for the embryos to break through when the seeds germinate.

Some water-dispersed fruits have heavy wax coats, and others have sacs of air that help them float. With its thick, water-repelling tissue, a coconut palm fruit is adapted to bob along in the open ocean. It can drift for hundreds of kilometers before saltwater, which could kill the embryo, penetrates the fruit.

Humans are grand agents of dispersal. In the past, explorers carried many more seeds all over the world than they do now. Cacao (Figure 19.6*c*), strawberries, oranges, corn, pineapple, pistachios, and many other desirable plants were coaxed to grow in new places, in tremendous numbers. Humans have greatly improved the reproductive success of all of these plants, which, in evolutionary terms, is what life is all about.

A mature ovule, which encases an embryo sporophyte and food reserves inside a protective coat, is a seed.

A mature ovary, with or without additional floral parts that have become incorporated into it, is a fruit.

Different kinds of fruits develop from one or more ovaries, and many incorporate other tissues in addition to those of the ovary wall.

Seeds and fruits are structurally adapted for dispersal by air currents, water currents, and many kinds of animals.

19.3 Asexual Reproduction of Flowering Plants

Sexual reproduction dominates most flowering plant life cycles, but many species also can reproduce asexually. Because they grow by way of mitosis, the offspring are clones, genetically identical to the parent.

ASEXUAL REPRODUCTION IN NATURE

Many flowering plant species reproduce asexually by modes of vegetative growth. In essence, new roots and shoots grow from the extensions or fragments of parent plants. For example, runners (horizontal, aboveground stems) develop on strawberry plants, then new roots and shoots develop at every other node. Each new plant is a genetic replica, or clone, of its parent.

Forests of quaking aspen (*Populus tremuloides*) are a different example of vegetative reproduction. Figure 19.7 shows shoot systems of an individual plant. That parent plant's root system keeps on giving rise to new shoots. One aspen forest in Colorado stretches across hundreds of acres and includes 47,000 shoots. Barring rare mutations, trees at the north end of this clone are genetically identical to those at the south end. Water travels from roots near a lake to shoot systems in drier soil. Dissolved ions travel in the opposite direction.

No one knows how old the aspen clones are. As long as conditions in the environment favor growth, such clones are as close as any organism gets to being immortal. One very old clone is a creosote bush (*Larrea divaricata*) that has been growing in the Mojave Desert for about 11,700 years.

INDUCED PROPAGATION

LINKS TO SECTIONS 7.1, 11.6

Most of our familiar houseplants, woody ornamentals, and orchard trees are clones, propagated from cuttings or fragments of shoots. For example, most of the navel oranges we eat are produced by descendants of one particular California tree.

Twigs or buds may be grafted onto, or joined to, a plant of a closely related species. In 1862, phylloxera was accidentally introduced into France via imported American grapevines. European grapevines had little resistance to this plant louse, a tiny insect that attacks and kills the root systems of vines. By 1900, phylloxera had destroyed two thirds of the vineyards in Europe, devastating the wine-making industry there. Today, French vintners routinely graft their prized grapevines onto the roots of phylloxera-resistant American vines.

Researchers now use shoot tips, leaf or stem pieces, and other parts of individual plants for **tissue culture propagation**, in which an entire plant is cloned from a single cell. This technique is useful when a desirable mutation arises. Tissue culture can yield thousands to millions of identical plants from a single specimen. The technique is being employed in efforts to improve food crops, including corn, wheat, rice, and soybeans. It also is used to propagate rare or hybrid ornamental plants such as orchids and lilies.

Flowering plants reproduce asexually by modes of vegetative growth, which yield genetically identical offspring. Such propagation in nature has resulted in the world's largest individual plants, and the oldest.

Figure 19.7 Quaking aspen (*Populus tremuloides*), a plant that reproduces asexually. Runners gave rise to this impressive stand of clones.

LINKS TO SECTIONS 10.6, 18.1

19.4 Patterns of Early Growth and Development

An embryo sporophyte inside its seed coat lies idle after its dispersal from the parent plant, until conditions favor germination. Then, in response to environmental cues, the embryo's growth and development resume. The sporophyte matures, and the life cycle turns again.

HOW DO SEEDS GERMINATE?

Figure 19.8 shows an embryo in a grain of corn, a type of dry fruit. Such an embryo can remain dormant for a long time, even for thousands of years. When suitable environmental conditions occur, the embryo resumes growth; it germinates.

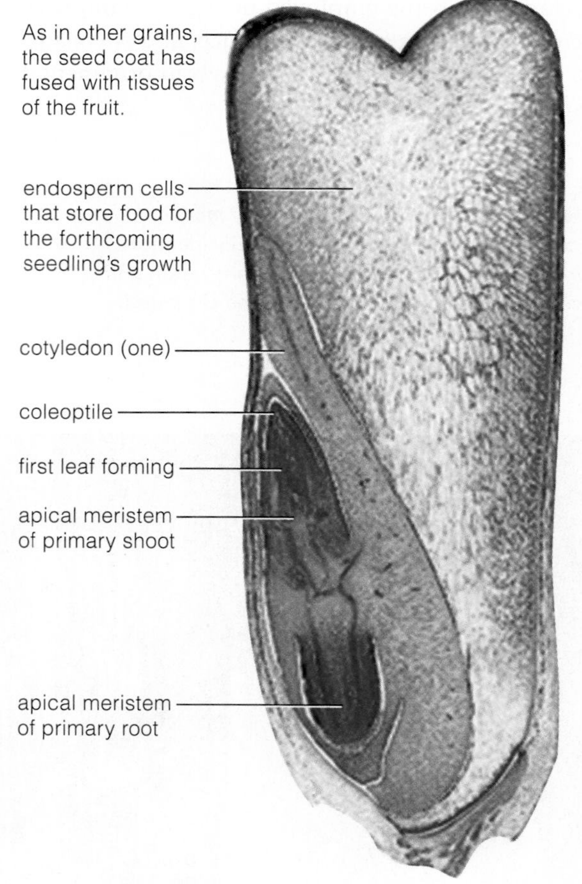

Figure 19.8 Embryo sporophyte and its food reserves inside a grain of corn (*Zea mays*).

Germination occurs in response to seasonal factors such as the length of day, temperature, and oxygen and moisture levels in soil. Mature seeds, for instance, do not hold enough water to support cell expansion or metabolism. Where water is ample only on a seasonal basis, germination coincides with rain. At that time, water seeps into seeds. After a seed swells with water, its coat ruptures.

Once the seed coat splits, more oxygen reaches the embryo, and aerobic respiration moves into high gear. The embryo's meristematic cells now divide rapidly. Germination is over when the primary root breaks out of the seed coat.

GENETIC PROGRAMS, ENVIRONMENTAL CUES

Have you ever noticed the buds along a young stem? Each holds a community of undifferentiated cells, a meristem (Section 18.1). Some meristematic cells never differentiate; they keep dividing. Others are the basis for development. Their descendants give rise to plant tissues according to a pattern established while the embryo sporophyte was still inside its seed coat.

That pattern has a genetic basis. All cells in a plant are descended from the same cell—the zygote—so all of them have the same genes. But, guided by various signals, different cells in a growing plant begin to use different subsets of those genes. As they do so, they become specialized for certain functions. This process, remember, is called differentiation (Section 10.6). The plant body develops as specialized cells form tissues (Figures 19.9 and 19.10).

You might be surprised that development in plants depends on extensive coordination among cells, just as animal development does. Plant cells communicate with other plant cells often some distance away in the body. As in animals, hormones are the primary signals for cell–cell communication.

For instance, the cells of apical meristem make a growth-stimulating hormone that diffuses downward into new tissues below. This hormone stimulates cells in the tissues to divide and elongate in a direction that lengthens the shoot.

Environmental cues—especially water availability, gravity, seasonal shifts in the hours of darkness versus daylight, and temperature—also guide plant growth and development.

The body plan of flowering plants starts taking form while the embryo sporophyte is still part of a seed.

Interactions among genes, hormones, and the environment govern how each individual plant develops.

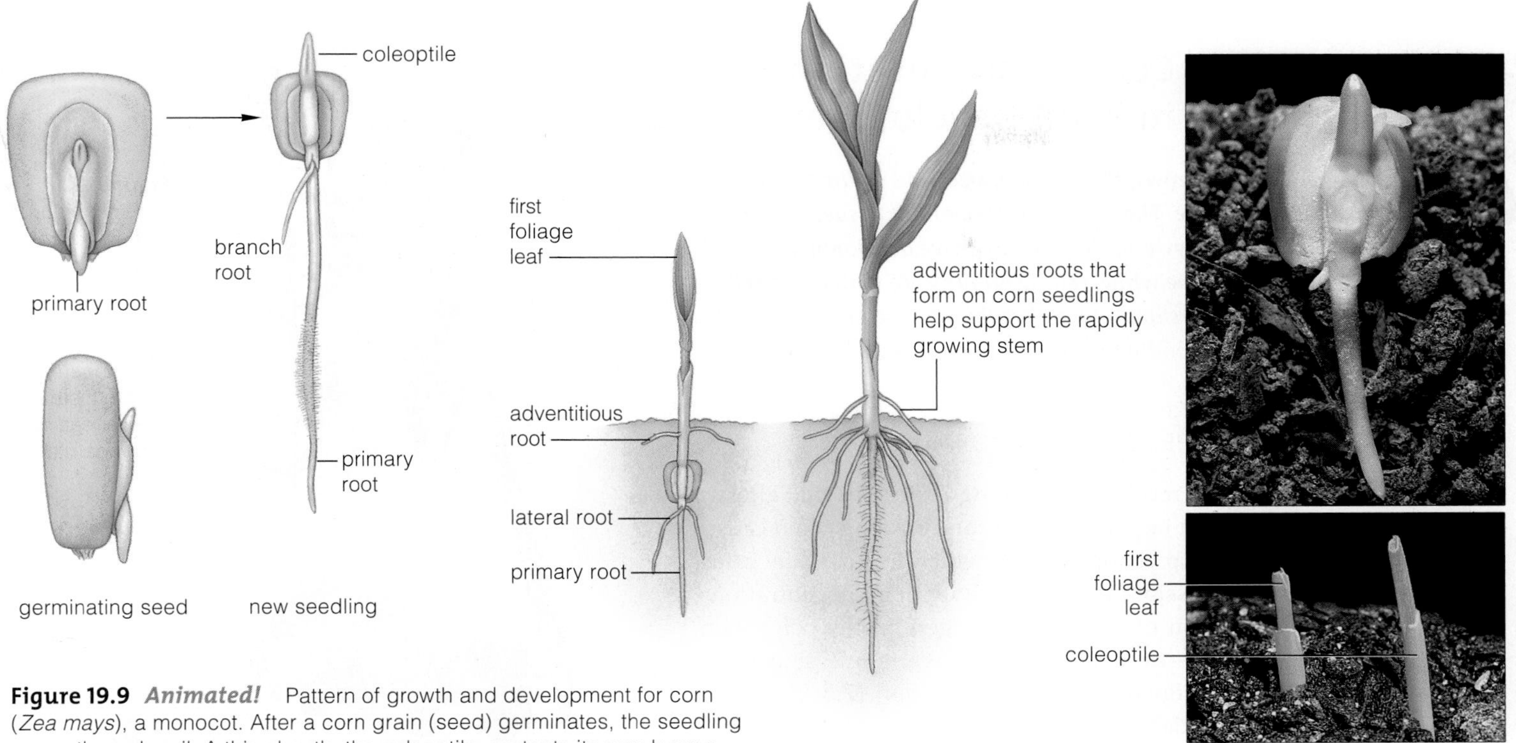

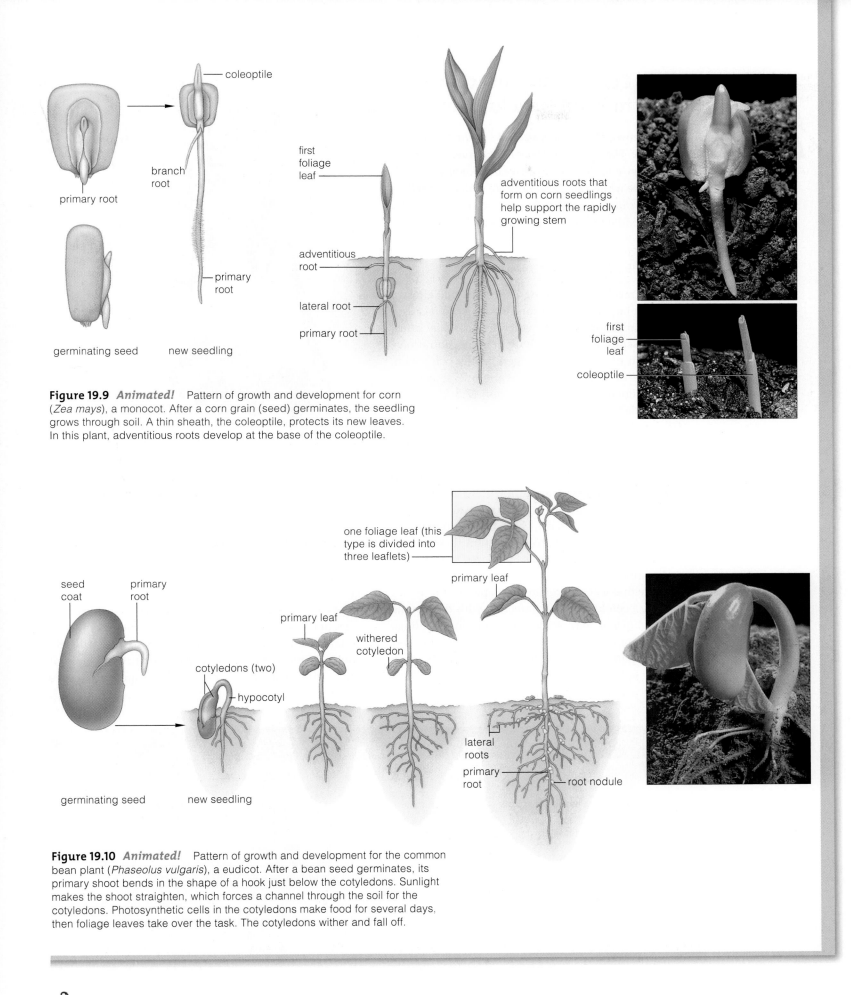

Figure 19.9 ***Animated!*** Pattern of growth and development for corn (*Zea mays*), a monocot. After a corn grain (seed) germinates, the seedling grows through soil. A thin sheath, the coleoptile, protects its new leaves. In this plant, adventitious roots develop at the base of the coleoptile.

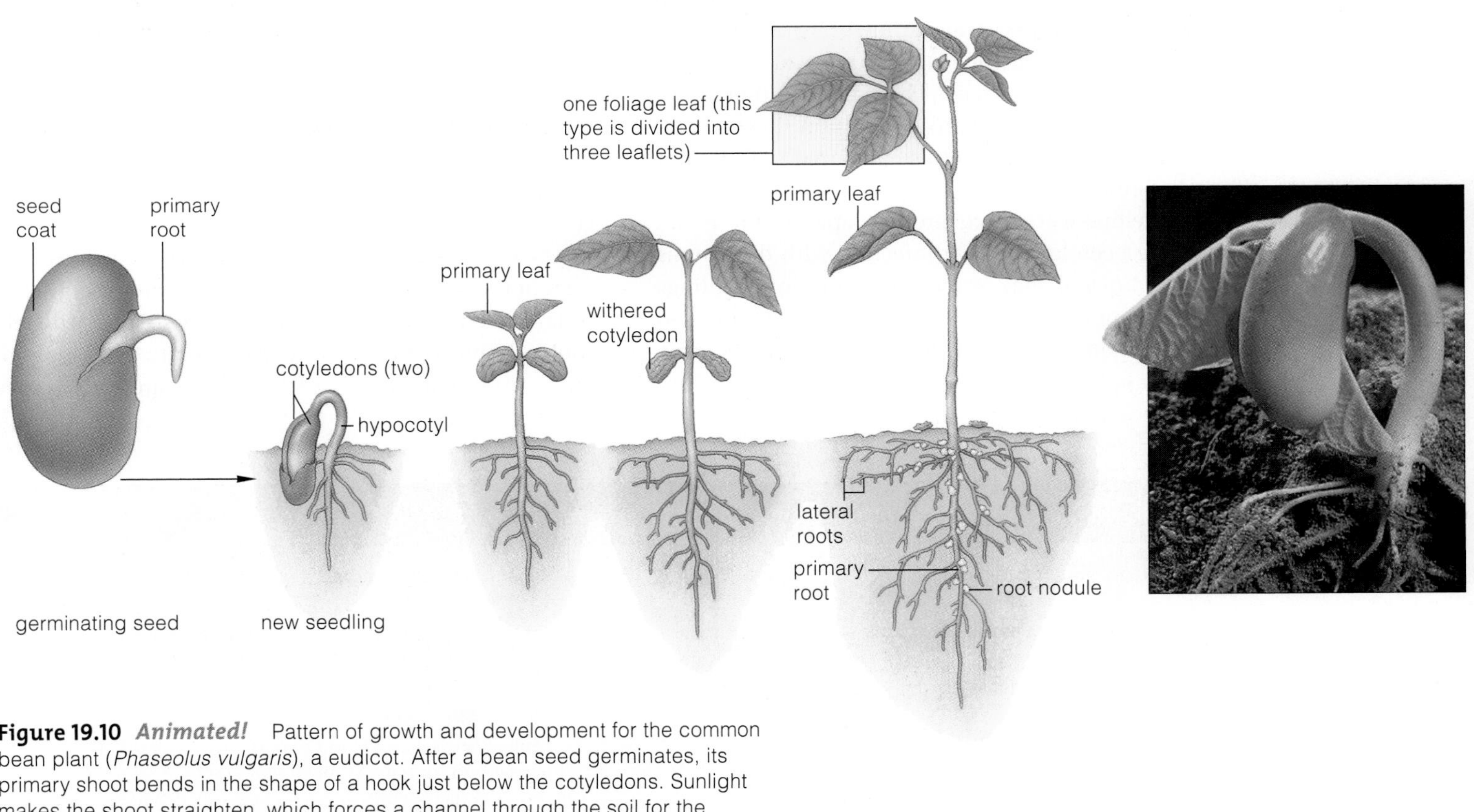

Figure 19.10 ***Animated!*** Pattern of growth and development for the common bean plant (*Phaseolus vulgaris*), a eudicot. After a bean seed germinates, its primary shoot bends in the shape of a hook just below the cotyledons. Sunlight makes the shoot straighten, which forces a channel through the soil for the cotyledons. Photosynthetic cells in the cotyledons make food for several days, then foliage leaves take over the task. The cotyledons wither and fall off.

LINKS TO SECTIONS 10.1, 17.5

19.5 Cell Communication in Plant Development

As a plant grows, the number, size, and volume of its cells increase. The cells form specialized tissues, organs, and organ systems during successive developmental stages. All the while, cells communicate with other cells by secreting signaling molecules into extracellular fluid and by responding to signals themselves.

Figure 19.11 Demonstration of the effect of gibberellins. The three tall cabbage plants were treated with gibberellins. Next to them are two untreated cabbage plants.

MAJOR TYPES OF PLANT HORMONES

Hormones, recall, are signaling molecules that alter an activity of a target cell. Hormone binding often causes a receptor to change shape, which in turn may affect gene expression, enzyme activity, ion concentrations, or activation of second messengers in the cytoplasm. The process by which a cell converts one signal into another is signal transduction (Section 17.5). There are five major classes of plant hormones (Table 19.1).

Gibberellins are growth hormones that are found in flowering plants, ferns, mosses, gymnosperms, and some fungi. By stimulating division and elongation of cells, gibberellins make stems lengthen. They also help seeds germinate and stimulate flowering of biennials and other plants. Remember the dwarf pea plants that Mendel used in his studies of heredity (Section 10.1)? Their short stems result from a mutation that reduces the rate of gibberellin synthesis compared with that of normal plants.

Gibberellins were discovered in Japan in the 1930s. It had long been known that infection with the fungus *Gibberella fujikuroi* makes the stems of rice seedlings lengthen so much that the plants eventually topple and die. Researchers discovered they could cause the same effect by applying extracts of the fungus to healthy plants (Figure 19.11). They identified the compound in the extracts that triggered stem elongation, and named it after the fungus.

Auxins form in the apical meristems of shoots and coleoptiles. These hormones cause both structures to elongate by cell divisions (Figure 19.12*a–e*). Auxins also induce cell division and differentiation in the vascular cambium, and stimulate fruit formation.

Auxins also have inhibitory effects. For example, secretion of auxin by a shoot tip prevents the growth of lateral buds along the lengthening stem. Gardeners routinely pinch off shoot tips. This ends the supply of auxin in a main stem, so lateral buds are free to give

Table 19.1 Major Classes of Plant Hormones and Their Effects

Hormone	Source	Stimulatory or Inhibitory Effects
Gibberellins	Young tissues of shoots, seeds, roots	Make stems lengthen greatly through cell division and elongation; help seeds germinate; help induce flowering in some plants
Auxins	Apical meristems of shoots, coleoptiles, seeds	Lengthen shoots, coleoptiles; induce vascular cambium formation, fruit formation; block lateral bud formation; role in abscission
Cytokinins	Mainly the root tip	Stimulate cell division in roots and shoots, leaf expansion; promote lateral bud formation, inhibit leaf aging
Ethylene (a gas)	Tissues undergoing ripening, aging, stress	Stimulates fruit ripening, senescence of leaves, flowers; leaf and fruit abscission
Abscisic acid (ABA)	Mature leaves in response to water stress; may be produced in seeds	Stimulates stomatal closure; induces transport of photosynthetic products from leaves to seeds; stimulates embryo formation in seeds; may induce and maintain dormancy in some species

oat seeds

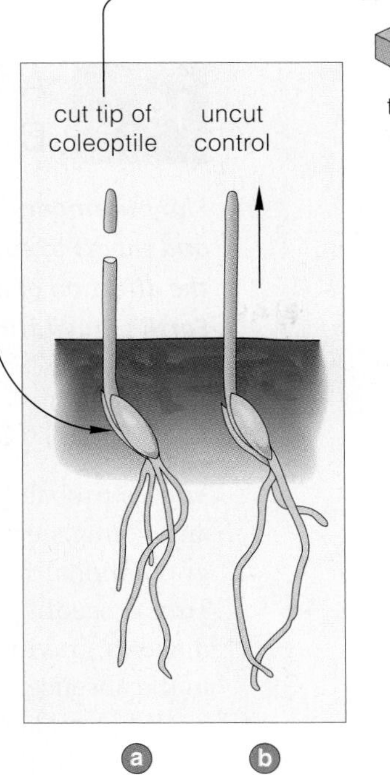

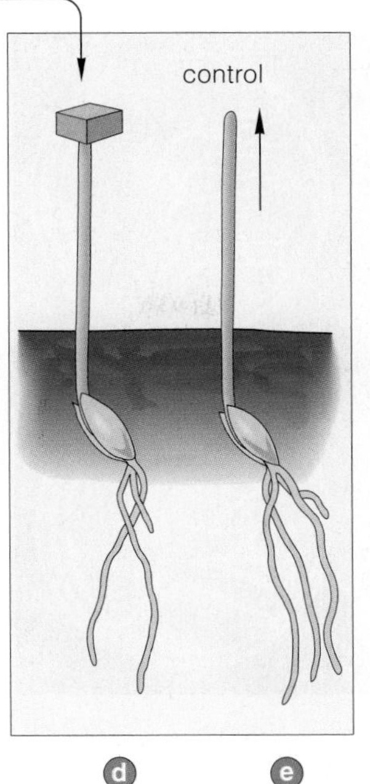

Figure 19.12 ***Animated!*** Auxin diffusing down from a coleoptile tip makes a seedling's cells lengthen.

(**a**) Cut off its tip and a seedling will not lengthen as much as (**b**) an uncut control seedling. (**c**) A block of agar positioned under the cut tip for several hours will absorb auxin from the tip. (**d**) If the agar is placed on top of a different de-tipped coleoptile, cells below it will lengthen about as fast as (**e**) an uncut seedling.

(**f**) A plant cutting dipped into a rooting powder with auxin is induced to form new roots.

rise to branches. Auxins stop **abscission** (dropping of leaves, flowers, and fruits). They also inhibit growth of main roots, yet enhance the growth of roots from stems of plant cuttings (Figure 19.12*f*).

IAA (indoleacetic acid) is the most common natural auxin. Applying it in an orchard in springtime results in fewer but larger fruits. It also is used to keep mature fruit from dropping from trees so that all the fruit can be picked at the same time.

Synthetic auxins are used as herbicides. One, 2,4-D, is widely used to kill eudicots. It causes uncontrolled, disorganized growth that leads to plant death. During the Vietnam conflict, 2,4-D was mixed with equal parts of another related compound, 2,4,5-T, to make *Agent Orange*. About 19 million gallons of this herbicide were sprayed on forested war zones, effectively removing enemy-concealing foliage. Later, dioxin, a carcinogen and trace contaminant of 2,4,5-T, was linked to liver cancer, miscarriages, abnormal births, and leukemias. 2,4,5-T is now banned in the United States.

Cytokinins stimulate rapid cell division in root and shoot meristems and in maturing fruits. They oppose auxin's effects by making lateral buds grow. They also keep leaves from aging, and so are used to prolong the shelf life of cut flowers and other horticultural prizes.

Ethylene, the only gaseous plant hormone, exerts control over the growth of most tissues, fruit ripening, leaf dropping, and other aging responses. Today, food distributors use ethylene to ripen tomatoes and other green fruit after they ship them to grocery stores (fruit that is picked green does not bruise or deteriorate as fast as ripe fruit). Ethylene also is used to brighten the rinds of citrus fruits to be displayed in markets.

Abscisic acid (ABA) was misnamed; it is a growth inhibitor that has very little to do with abscission. ABA makes stomata close in water-stressed plants. It induces and maintains dormancy in certain buds and seeds. ABA also stimulates the transport of photosynthetic products to seeds, and it is involved in maturation of embryos. Growers apply ABA to nursery plants before they are shipped in order to induce dormancy, because plants in a dormant state bruise less easily when they are handled.

Plant hormones are signaling molecules secreted by plant cells that alter the activity of other plant cells.

The main classes of plant hormones are gibberellins, auxins, cytokinins, ethylene, and abscisic acid.

Interactions among hormones and other kinds of signaling molecules govern the normal growth, development, day-to-day functioning, and reproduction of plants.

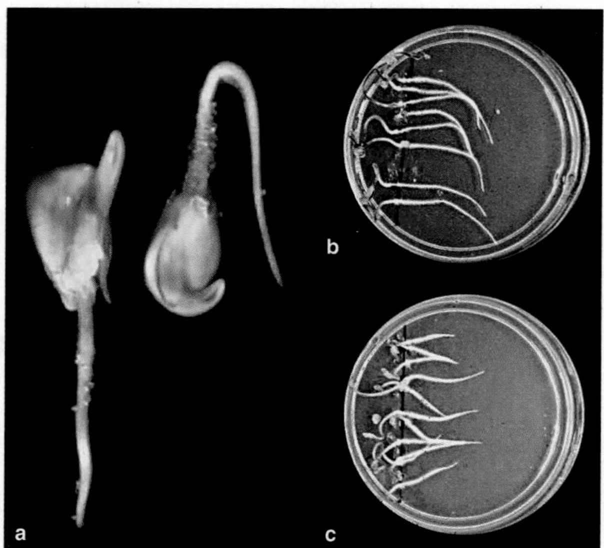

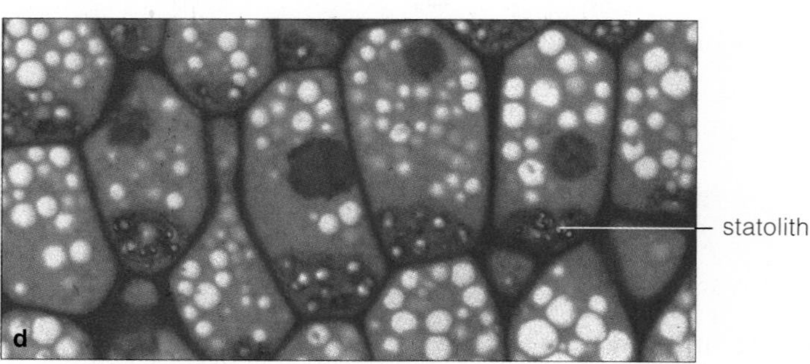

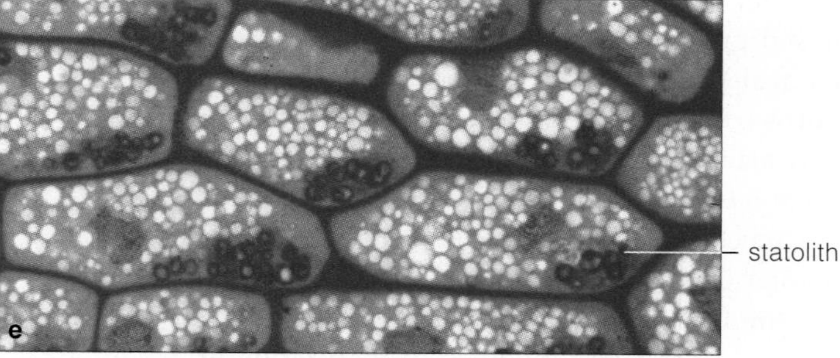

Figure 19.13 ***Animated!*** Gravitropism in corn. (**a**) A corn primary root and shoot growing in a normal orientation, and one turned upside-down. (**b**) In primary roots turned on their side, auxin moves to the down-facing side, where it stops cells from lengthening. Cells of the up-facing side continue to lengthen and the roots bend down. (**c**) Seedlings exposed to auxin transport inhibitors do not bend in response to gravity.

The redistribution of auxin is a response to redistribution of statoliths in gravity-sensing cells. (**d**) Starch-packed statoliths settle to the bottom of cells in a corn root cap. (**e**) Ten minutes after the root is turned sideways, the statoliths have settled to the new "bottom" of the cells.

19.6 Adjusting Rates and Directions of Growth

Plant hormones mediate "tropic" responses of roots and shoots to cues from the environment. They guide the direction of growth toward or away from light, Earth's gravitational force, or some other stimulus.

RESPONSES TO GRAVITY

You've probably noticed that roots tend to grow down and young stems to curve up. Both do so in response to gravitational force; they are examples of **gravitropism**. Turn a seedling on its side in a dark room and it still makes a gravitropic response. Its shoot curves up even in the absence of light (Figure 19.13*a*).

What makes the plant do this? In a stem oriented horizontally, the cells on the lower side elongate faster than the cells on the upper side. The different rates of elongation on the top and bottom of the stem cause the shoot to bend up. Gravitropism in plants involves auxin and proteins that transport it (Figure 19.13*b,c*).

How can plant cells "know" whether they are on the bottom or the top? Gravity-sensing mechanisms of many organisms involve specialized cells containing organelles called *statoliths*. Plant statoliths are stuffed with clusters of heavy starch grains. In response to gravity, statoliths in specialized cells settle downward until they rest in the lowest region of the cytoplasm (Figure 19.13*d,e*). If the plant is reoriented, the shifting statoliths cause redistribution of auxin.

RESPONSES TO LIGHT

When light streams in from one direction only, a stem or leaf adjusts its rate and direction of growth so that it grows toward the light source. This response is a form of **phototropism**. Phototropism orients photosynthetic parts of a plant to maximize light interception.

Phototropism in plants occurs in response to blue light. Nonphotosynthetic pigments called *flavoproteins* absorb blue light, and transduce its energy into signals that cause auxin to be redistributed to the shaded side of a shoot or coleoptile. As auxin moves into cells that are less exposed to light, it makes them elongate faster than the cells on the illuminated side. The difference in growth rates of cells on either side bends the plant toward the light (Figure 19.14*a,b*).

You can observe a phototropic response by putting seedlings of sun-loving plants in a dark room next to a window through which rays of sunlight are streaming. They will begin curving in the direction of the light (Figure 19.14*c*).

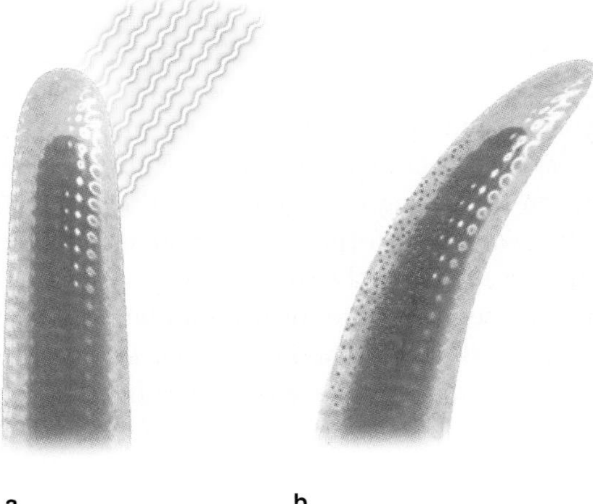

Rays of sunlight strike one side of a coleoptile.
Auxin (*red*) moves to cells on the shaded side. The cells on the shaded side elongate more quickly than those on the lit side, and the coleoptile bends.
a
b

Figure 19.14 ***Animated!*** Phototropism. (**a**,**b**) Hormone-mediated differences in cell elongation rates along its length will induce a coleoptile to bend toward light. (**c**) Flowering shamrock (*Oxalis*) seedlings responding to directional light.

RESPONSES TO CONTACT

Hairs and other epidermal cell surface specializations may be part of sensory mechanisms in some plants that can detect contact. In response, redistribution of auxin and ethylene induces the plants to adjust the direction of growth. In vines and tendrils, cells on the contact side stop elongating and the cells on the other side keep growing. Their unequal rates of growth make the shoot curl around the object, and often more than once (Figure 19.15). Roots have a similar mechanism, but tend to grow away from contact rather than toward it.

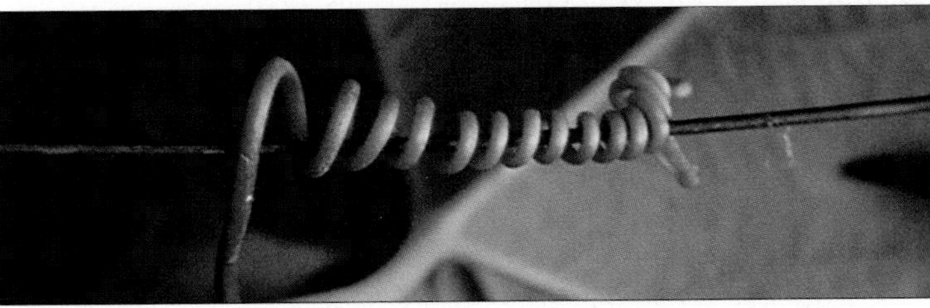

Figure 19.15 *Passiflora* tendril twisting in response to contact.

RESPONSES TO MECHANICAL STRESS

Mechanical stress, as inflicted by prevailing winds and grazing animals, inhibits stem lengthening. Trees high up in windswept mountains are stubbier than sheltered trees of the same species at lower elevations. Similarly, plants grown outdoors commonly have shorter stems than the same kinds of plants grown in a greenhouse. Briefly shake a plant every day and you will inhibit its overall growth (Figure 19.16).

Figure 19.16 Effect of mechanical stress on tomato plants. (**a**) This plant grew in a greenhouse. (**b**) Each day for twenty-eight days, this plant was mechanically shaken for thirty seconds. (**c**) This one had two shakings each day.

Plants adjust the direction and rate of growth in response to environmental stimuli that include gravity, light, contact, and mechanical stress.

LINK TO SECTION 17.4

19.7 Meanwhile, Back at the Flower . . .

Hormones activate the genes for flower formation. But what activates the hormone-secreting cells?

HOW DO PLANTS KNOW WHEN TO FLOWER?

Like all other eukaryotes, plants have various internal mechanisms that govern recurring changes in their biochemistry. In flowering plants, these mechanisms help bring about seasonal adjustments in the patterns of growth, development, and reproduction—which, in the case of flowering plants, means forming flowers.

Figure 19.18 Red and far-red light in the visible light spectrum.

You have probably noticed that different species of plants flower at different times of the year. Summer-blooming irises and spinach are *long-day* plants. The cockleburs and chrysanthemums are *short-day* plants; they flower in the spring or autumn. Sunflowers and other *day-neutral* plants form flowers whenever they mature, without regard to the season.

Actually, those names are misleading, because the cue that triggers flower formation is *night length,* not daylength. "Short-day" plants will flower only when nights are *longer* than a critical value, and "long-day" plants flower when nights are *shorter* than a critical value (Figure 19.17).

Photoperiodism is a biological response to change in the length of daylight relative to darkness. A blue-green pigment, phytochrome, helps a plant detect the length of a day. It is one of several photoreceptors that maintains circadian rhythm in plants (Section 17.4).

Phytochrome is activated by red light; far-red light inactivates it (Figure 19.18). The relative amounts of red light and far-red light in the sunshine that reaches a given environment vary with the time of day and through the seasons. In winter, when there are fewer hours of daylight in the twenty-four hour cycle, far-red light predominates.

The connection between photoperiodism and the timing of flowering remained elusive for more than 50 years. Now we know that several transcription factors link a plant's circadian rhythms to flowering.

Circadian rhythms change with the seasons as the length of daylight changes. In response to a change in circadian rhythms, leaf cells transcribe more or less of a flowering gene. The product of that gene activates a transcription factor in shoots, which in turn activates master genes controlling formation of floral structures. The master gene products cause cellular descendants of an uncommitted bud of meristem tissue to develop into a flower.

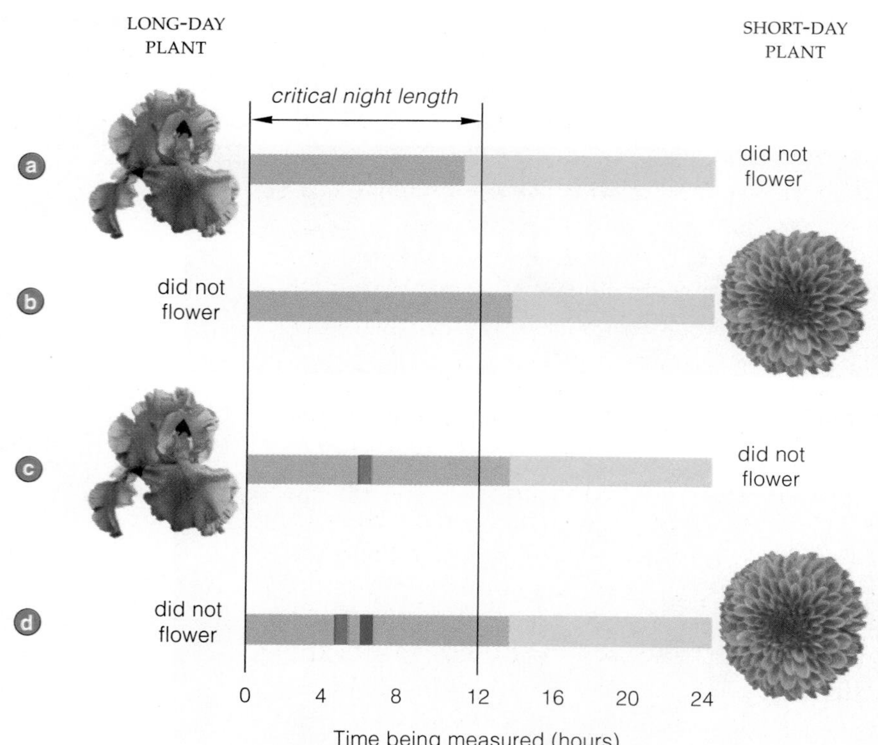

Figure 19.17 ***Animated!*** Experiments showing that short-day plants flower by measuring night length. Each horizontal bar represents 24 hours. *Yellow* signifies daylight; *blue-gray* signifies night. (**a**) Long-day plants flower when the night is *shorter* than a length critical for flowering. (**b**) Short-day plants flower when the night is *longer* than a critical value.

(**c**) When an intense red flash interrupts a long night, both plants respond as if it were a short night. (**d**) A short pulse of far-red light after the red flash cancels the disruptive effect of the red flash.

Like all other organisms, flowering plants have internal mechanisms that regulate recurring cycles of growth, development, and reproduction, in response to cues from the environment.

The main environmental cue for flowering is the length of night—the hours of darkness, which vary seasonally.

Figure 19.19 Leaves of a horse chestnut tree (*Aesculus hippocastanum*) changing color in autumn. The leaf scar at *right* is an abscission zone. Before the leaf detached from the tree, a tissue formed in a horseshoe-shaped area (hence the tree's name). The tissue helps prevent water loss and infection after plant parts drop away.

Potted plant grown inside a greenhouse did not flower.
Branch exposed to cold outside air flowered.

Figure 19.20 Localized effect of cold temperature on dormant buds of a lilac (*Syringa*). For this experiment, which started during the cold winter months, a lilac plant was kept inside a greenhouse, where the temperature was warmer than the outside air. But one of its branches protruded outside the greenhouse. Only the buds on that branch flowered in spring.

19.8 Life Cycles End, and Turn Again

In midsummer, days almost imperceptibly begin to grow shorter, and many plants stop growing. They do so even when temperatures are mild, the sky is bright, and water is plentiful. They prepare for ***dormancy****, a period of arrested growth that will not end until spring.*

As leaves and fruits grow in early summer, their cells make auxin, which moves into stems. The auxin helps maintain growth. By midsummer, daylight hours are decreasing. Deciduous plants begin to divert nutrients away from leaves, stems, and roots to flowers, fruits, and seeds. In late summer, auxin production declines in leaves and fruits, and nutrients become routed to storage sites in twigs, stems, and roots.

Eventually, flowers, leaves, fruits, and stems drop. The process by which plant parts are shed is called abscission (Figure 19.19). It occurs at zones laid out as the plant developed—the bases of leaves, fruit stalks, and twigs. Abscission occurs in response to shorter daylight hours, injury, high temperatures, or water or nutrient deficiency.

Auxin-deprived leaves, fruits, and other structures release ethylene that diffuses into nearby abscission zones. The ethylene is a signal for cells in the zone to make enzymes that digest their own walls, and also the middle lamella (Section 3.8). The cells enlarge as their walls soften. Bulging cells separate from one another as the cementing layer between them dissolves. Tissue in the zone weakens, and the structure above it drops. **Senescence** refers to the phase from full maturity until the eventual death of plant parts or the whole plant. It is part of the plant life cycle.

LINK TO SECTION 3.8

For many species, growth stops in autumn as a plant enters **dormancy**; metabolic activities idle. Long nights, cold temperatures, and dry, nitrogen-poor soil are strong cues for dormancy. A dormant plant will not resume growth until certain environmental cues occur, usually at the start of spring.

Dormancy-breaking mechanisms operate between fall and spring. For dormancy to end, a few species require exposure of the dormant plant to many hours of cold temperature, but more typical cues include the return of milder temperatures and plentiful water and nutrients. With the return of favorable conditions, life cycles turn once more. Seeds germinate, buds resume growth, and new leaves form.

Flowering is, in part, often a response to seasonal changes in temperatures. For instance, unless the buds of some biennials and perennials are exposed to cold air in the winter, flowers will not form on their stems in spring. Low-temperature stimulation of flowering is called **vernalization** (from *vernalis*, which means "to make springlike"). Figure 19.20 shows experimental evidence of this effect.

Multiple cues from the environment influence processes of plant growth, senescence, and dormancy. Changes in daylength, nutrient and moisture availability, and temperature are such cues.

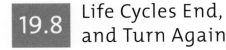

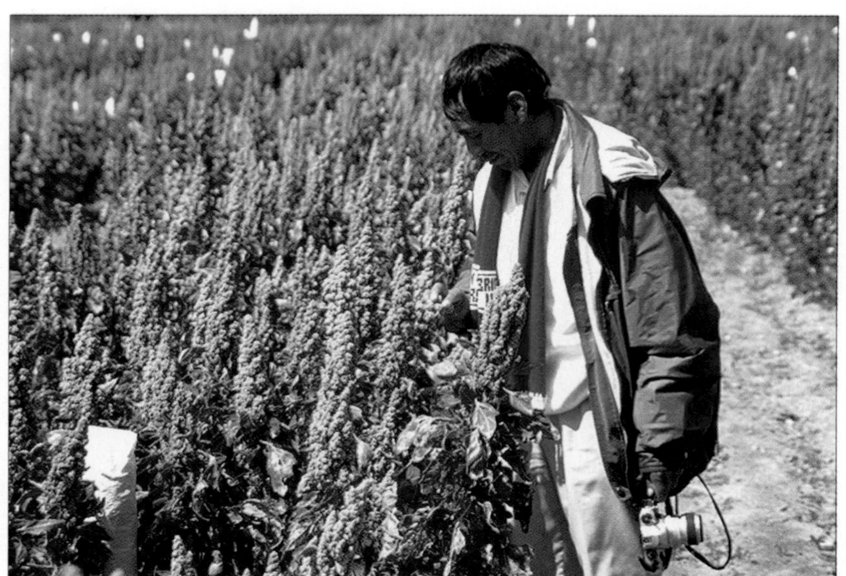

Figure 19.21 Alejandro Bonifacio in a field of hybrid quinoa plants.

19.9 Regarding the World's Most Nutritious Plant

In this unit, you explored the nature of plant structure, function, and behavior. Consider, finally, an example of how individuals put knowledge of plants to use in ways that have tremendous impact on human lives.

Alejandro Bonifacio grew up in poverty-stricken rural Bolivia. As a child he spoke Aymaran, a language that predates the Incas. He learned Quechua, the language of the Incas, and also learned Spanish before entering college. He earned a bachelor's degree in agriculture and became a plant breeder for the Bolivian Institute of Agricultural Technology.

His research interest is *Chenopodium quinoa*, a plant that originated in the Andes. Quinoa is a leafy eudicot, a distant relative of spinach and beets. Its seeds are not a cereal grain, but they are nutritional treasures. For many thousands of years, quinoa seeds have been a staple of Latin American diets.

Quinoa seeds are 16 percent protein, on average. By comparison, wheat seeds are about 12 percent protein and rice seeds, about 8 percent. However, the proteins in both wheat and rice are deficient in the amino acid lysine. The array of amino acids in quinoa is as good as that in milk. The seeds also contain more iron than most cereal grains do, and are a good source of calcium, phosphorus, and many B vitamins. In addition, quinoa plants are highly resistant to drought, frost, and saline soils. It is the only food crop that can grow in the arid salt deserts that prevail in much of Bolivia.

Far to the north, even before Alejandro received his college scholarship, Dr. Daniel Fairbanks had accepted a position as a botanist at Brigham Young University (BYU). He, too, became interested in quinoa because of its potential to feed millions in Bolivia and Peru, the most impoverished countries of Latin America. There, the majority of families eke out a living as subsistence farmers, and *kwashiorkor* is common.

Kwashiorkor is a form of malnutrition that results from protein-deficient diets. Fatigue, drowsiness, and irritability are its early symptoms. Continued protein deficiency adversely affects body growth. Edema sets in, and muscle mass decreases. The immune system weakens. The abdomen of a kwashiorkor victim often protrudes. Epidermal conditions, such as vitiligo and thinning hair, are common. Severe cases can result in mental and physical problems that lead to coma and death. Kwashiorkor most commonly affects people in impoverished countries, but a large percentage of the elderly in other countries may be affected.

In 1991, Bonifacio and Fairbanks met at a conference on Andean crops and became friends. Later, the World Bank awarded Bonifacio a fellowship to study in the United States, and Fairbanks became his advisor at BYU. Bonifacio earned his Ph.D. and learned a fourth language—English.

Bonifacio is now the principal investigator of an international quinoa research program funded by the McKnight Foundation. The program takes a systemic approach to improving quinoa production for farmers locked in poverty. About twenty scientists, including Fairbanks, now collaborate in research encompassing conservation, improvement, and utilization of genetic diversity, the economic impact of new quinoa strains, agricultural technologies, nutrition, health, and local community programs.

Thousands of Bolivian families are now producing more food, thanks to the new quinoa strains (Figure 19.21). Children who would otherwise have died from kwashiorkor are now attending school.

Plants enrich our world in uncountable ways. Researchers around the world are taking part in wide-ranging efforts to develop new varieties that can improve human life.

Summary

Section 19.1 Sporophytes, usually with roots, stems, and leaves, dominate flowering plant life cycles. Flowers, which are specialized reproductive shoots, have nonfertile parts (sepals and petals), fertile male reproductive parts (stamens), and female reproductive parts (carpels). Two kinds of haploid spores that form in different floral structures give rise to male and female gametophytes.

The anthers of stamens contain pollen sacs. Inside the sacs, haploid microspores form by way of meiosis. A male gametophyte develops inside the microspore that differentiates into a pollen grain.

Carpels contain ovaries. One or more cell masses that form on an ovary wall become ovules. One haploid cell in each mass, a megaspore, gives rise to the female gametophyte. One gametophyte cell becomes an egg.

Pollinators are agents that transfer pollen from male to female reproductive parts of flowering plants of the same species. Pollination is the arrival of pollen grains on a receptive stigma (a carpel's sticky top). A germinating pollen grain forms a pollen tube that carries with it two male gametes as it grows through tissues of the ovary, toward the egg. With double fertilization, one male gamete fertilizes the egg, forming a zygote. The other male gamete fuses with the endosperm mother cell.

BiologyNow
Take a closer look at the life cycle of a eudicot, and investigate the plant life cycle and the structure of a flower, with the animations on BiologyNow.

Section 19.2 A seed is a mature ovule. It consists of the embryo sporophyte and its food reserves inside a protective seed coat. Eudicot embryos have two seed leaves, or cotyledons. Monocot embryos have one. The ovary wall and sometimes other tissues form fruits around seeds. Fruits function in seed dispersal.

BiologyNow
See how an embryo sporophyte develops inside a eudicot seed with the animation on BiologyNow.

Section 19.3 Many flowering plants can reproduce asexually by sending out runners or by other types of vegetative growth. Tissue culture propagation is a laboratory technique that can be used to propagate thousands to millions of plants from vegetative cells.

Section 19.4 Embryos remain dormant in seeds until conditions favor germination, or resumption of growth. While the embryo sporophyte is still encased in its seed coat, the body plan of the future plant has already been laid out. Genes, signaling molecules, and environmental cues control and orchestrate subsequent growth and development.

BiologyNow
Compare the growth and development of a monocot and a eudicot with the animation on BiologyNow.

Section 19.5 Like signaling molecules in other eukaryotes, hormones secreted by one plant cell alter the activity of a different cell. At specific times in the plant life cycle, hormones promote or arrest growth by stimulating or inhibiting cell division, elongation, differentiation, and reproduction.

The main plant hormones are gibberellins, auxins, cytokinins, ethylene, and abscissic acid.

BiologyNow
Observe the effect of auxin on plant growth with the animation on BiologyNow.

Section 19.6 In addition to controlling patterns of growth and development, hormones also trigger adjustments in the direction and rate of growth in response to environmental stimuli.

In gravitropism, roots grow downward and stems grow upright in response to Earth's gravity. Statoliths (clusters of particles in cells) are gravity-sensing mechanisms in some plant cells.

In phototropism, stems and leaves adjust rates and directions of growth toward or away from light. A flavoprotein is the receptor for blue wavelengths of light that trigger the strongest response.

Plants adjust their direction of growth in response to contact with solid objects. They also respond to mechanical stress.

BiologyNow
Investigate plant tropisms with the animation on BiologyNow.

Section 19.7 Environmental cues trigger internal mechanisms that govern recurring cycles of growth, development, and reproduction in plants. The main cue for flowering is the number of hours of darkness, which varies seasonally. Phytochrome, a pigment, has a role in photoperiodic responses. Phytochrome is activated by red light and inactivated by far-red light.

Short-day plants flower mainly in spring or fall, when there are more hours of darkness relative to daylight in the circadian cycle. Long-day plants flower in summer when there are fewer hours of darkness relative to daylight hours. Day-neutral plants flower whenever they are mature enough to do so.

Photoperiodism factors in leaves are linked to expression of genes for flowering in shoot tips.

BiologyNow
Learn how plants respond to daylength (night length) with the animation on BiologyNow.

Section 19.8 Dormancy is a period of arrested growth that does not end until the arrival of specific dormancy-breaking environmental cues. Dormancy typically involves the dropping of flowers, leaves, fruits, and other plant parts, by a hormone-mediated process called abscission.

Senescence is the part of the life cycle of a plant that occurs between maturity and death of the plant or plant parts.

Section 19.9 New varieties of food-bearing crops continue to be developed. Plants having enhanced nutritional value benefit people around the world.

Figure 19.22 Field of sunflowers (*Helianthus*) that are busily demonstrating solar tracking.

Self-Quiz

Answers in Appendix I

1. The ______, which bears flowers, roots, stems, and leaves, dominates the life cycle of flowering plants.
   a. sporophyte c. megaspore
   b. gametophyte d. microspore

2. Seeds are mature ______; fruits are mature ______.
   a. ovaries; ovules c. ovules; ovaries, mostly
   b. ovules; stamens d. stamens; ovaries, mostly

3. After meiosis within pollen sacs, haploid ______ form.
   a. megaspores c. stamens
   b. microspores d. carpels

4. Cotyledons develop as part of ______.
   a. endosperm c. embryo sporophytes
   b. stamens d. ovaries

5. A whole plant grows from a tissue or structure that became separated from a parent plant. This is ______.
   a. double fertilization
   b. photoperiodism
   c. vegetative propagation
   d. imbibition

6. Plant hormones ______.
   a. interact with one another
   b. are active in plant embryos within seeds
   c. are active in adult plants
   d. all of the above

7. ______ stops lateral buds from branching out.
   a. Gibberellin c. Cytokinin
   b. Auxin d. Ethylene

8. In Michigan, a long-day plant will flower ______.
   a. in July
   b. in April
   c. all year round
   d. whenever temperatures are warm enough

9. Match the terms with the most suitable description.

   ____ phototropism
   ____ photoperiodism
   ____ double fertilization
   ____ ovule
   ____ carpel
   ____ abscission
   ____ pollinator
   ____ senescence

   a. flowers, leaves, fruits drop from plant in autumn
   b. formation of zygote and first cell of endosperm
   c. agent of pollen transfer
   d. starts out as cell mass in ovary; may become a seed
   e. flower's female reproductive structure
   f. light makes stem or leaf adjust direction of growth
   g. response to change in length of daylight relative to darkness
   h. programmed aging of plants or plant parts

Additional questions are available on **Biology Now™**

Critical Thinking

1. Photosynthesis sustains plant growth, and inputs of sunlight energy sustain photosynthesis. Why, then, do seedlings that germinated in a fully darkened room grow taller than different seedlings of the same species that germinated in the sun?

2. *Solar tracking* refers to the observation that many plants maintain the flat blades of their leaves at right angles to the sun throughout the day. Figure 19.22 shows an example. This tropic response maximizes the plant's harvesting of the sun's rays. Name one type of molecule that might be involved in the response.

3. Scientists isolated a mutated plant that produces excess amounts of auxin. Predict what some of the phenotypic traits of this plant might be.

4. All but one species of large-billed birds native to New Zealand's tropical forests are now extinct. Numbers of the surviving species, the kereru, are declining rapidly due to the habitat loss, poaching, predation, and interspecies competition that wiped out the other native birds. The kereru remains the sole dispersing agent for several native trees that produce big seeds and fruits. One tree, the Puriri (*Vitex lucens*), is New Zealand's most valued hardwood. Explain, in terms of natural selection, why we might expect to see no new Puriri trees in New Zealand.

5. Cattle typically are given somatotropin, an animal hormone that makes them grow bigger (the added weight means greater profits). There is a major concern that such hormones may have unforeseen side effects on beef-eating humans. Do you think plant hormones applied to crop plants can affect humans? Why or why not?

6. Before cherries, apples, peaches, and many other fruits ripen and the seeds inside them mature, their flesh is bitter or sour. Only later does it become tasty to animals that help disperse its seeds. Develop a hypothesis about how this feature can help the plant's reproductive success.

20 ANIMAL TISSUES AND ORGAN SYSTEMS

IMPACTS, ISSUES

It's All About Potential

Each year in the United States, about 10,000 people injure their spinal cords. In 1995 actor Christopher Reeve, best known for his role as Superman, was one of them. A bad fall from a horse left him paralyzed and unable to breathe on his own. He became an advocate for the disabled and was a vocal champion of human stem cell research until his death in 2004.

Stem cells are undifferentiated cells; they have not yet traveled down a developmental road that would specialize them for one particular function. Unlike other cells, they and their daughter cells retain the potential to develop into many different cell types depending on their surroundings and other factors.

The first cells that form when a fertilized animal egg begins its divisions are the most versatile. These *embryonic stem cells* are the forerunners of all body cells. They can be cultured in the laboratory and —in theory—their descendants could be coaxed to develop into cells of any type. Many scientists think that lab-grown human embryonic stem cells could one day provide replacement cells for people who suffer from spinal cord injuries, diabetes, heart attacks, Parkinson's disease, and other ailments.

Some people object to such research because the stem cells are derived from aborted embryos, human cell clusters formed by cloning methods, or cell clusters that were created for *in vitro* fertilization but never used. For this and other reasons, researchers continue to investigate the potential of adult stem cells.

Adult stem cells persist in bone marrow, fat, the nasal linings, and other tissues. They are more set in their ways than embryonic cells, but it is possible that with the right culture methods, they could be stimulated to produce a number of cell types.

Stem cells can start you thinking more closely about this unit's topics: how the animal body and its parts are put together (anatomy) and the ways that they function (physiology). This chapter introduces you to the types of animal tissues and organ systems. Whether you look at a flatworm or butterfly, a bird or human, keep in mind that survival and reproduction depend on the ability of the animal body to perform four essential tasks:

1. Maintain internal conditions within ranges that help keep cells, tissues, and organs functioning.
2. Acquire water, nutrients, and other materials; distribute them; and dispose of wastes.
3. Protect against injury or attack.
4. Reproduce and, often, help nourish and protect offspring through early growth and development.

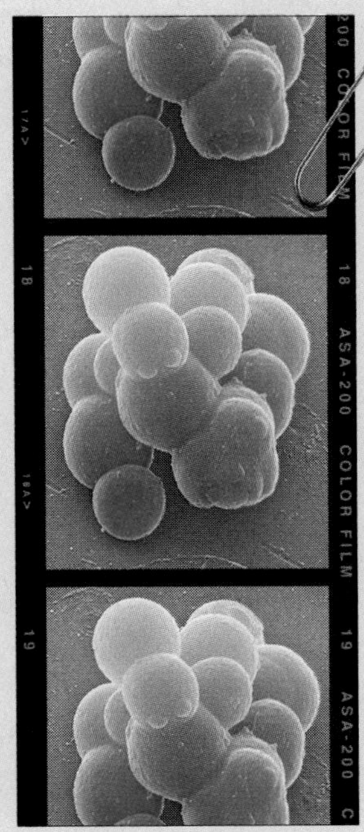

How Would You Vote? *Should scientists be allowed to start new embryonic stem cell lines from cell clusters created for* in vitro *fertilization but not used? See BiologyNow for details, then vote online.*

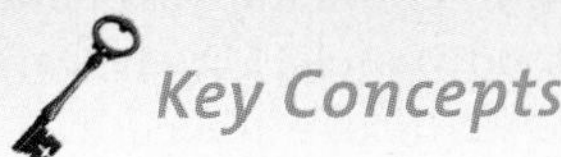

Key Concepts

TYPES OF TISSUES
In nearly all animals, four tissue types are organized to form organs that interact in maintaining homeostasis. Sheetlike epithelial tissues cover body surfaces and line inner cavities and tubes. Connective tissues bind, support, and insulate other tissues. Muscle tissues contract and move the body or its parts. Nervous tissue detects stimuli, integrates information, and calls for responses from other tissues.

ORGAN SYSTEMS
All vertebrates have the same assortment of organ systems. Each organ system has specialized functions, but all interact and contribute to survival of the body as a whole. Three embryonic tissues are the source of all adult organs.

Links to Earlier Concepts

Sections 1.1 and 17.1 introduced you to levels of organization in living organisms and the concept of homeostasis. Section 16.1 described trends in the evolution of animal bodies. We expand on all of these themes in this chapter.

You will see that many differences among tissues arise from differences in what proteins they produce (2.8). Tissues vary in their capacity for secretion (3.5), their number of mitochondria (3.7), whether they have cilia (3.7), and whether transport proteins in their membrane make them "excitable" (4.5).

You learned about the importance of temperature regulation in Section 17.3. In this chapter, we will consider other functions of human skin in light of what we know about human evolution (16.13).

LINKS TO SECTIONS 1.1, 2.8, 17.1

20.1 Organization and Control in Animal Bodies

Animal bodies show a hierarchical organization. Cells make up tissues, which compose organs, which are components of organ systems. All these parts interact to keep the internal environment just right for life.

Recall that a **tissue** is a community of cells interacting in the performance of one or more tasks, such as the task of contraction by muscle tissue. An **organ** consists of two or more tissues organized in proportions and patterns that allow it to carry out specific activities. For example, a vertebrate heart includes four types of tissues. They are arranged in predictable proportions and configurations, and interact to carry out the task of rhythmic contraction. In an **organ system**, two or more organs interact physically, chemically, or both in a task. For example, in the circulatory system, many interconnected vessels transport blood throughout the body, driven by the force of a beating heart.

As an animal goes about its activities, it gains and loses solutes and water, and is exposed to temperature variations. Through all this, organs and organ systems interact to keep solute concentrations and temperature of the internal environment within the range that cells can tolerate. In other words, the interactions among organs ensure **homeostasis**, the state in which internal conditions are maintained within tolerable limits.

Animal cells, tissues, organs, and organ systems work to maintain individual cells and the body as a whole.

free surface of epithelium

a

b

simple epithelium

basement membrane

connective tissue

Figure 20.1 Epithelium. This animal tissue has a free surface exposed to a body fluid or to the outside environment. The sketch below the athlete shows this arrangement for simple epithelium, which has only a single layer of flattened cells. (**a**) Stratified epithelium consists of multiple layers of cells, which are most flattened near the tissue's free surface. (**b**) Other epithelial cells, such as these from the lining of the small intestine, are shaped like columns. Still other epithelial cells are cube-shaped.

20.2 Four Basic Types of Tissues

We know of more than 2 million spectacularly diverse kinds of animals. Nearly all of them are composed of four basic types of tissues. These are epithelial tissue, connective tissue, muscle tissue, and nervous tissue.

EPITHELIAL TISSUES

Most of what you see when you look in the mirror—your skin, hair, and nails—is either **epithelial tissue** or the structures derived from it. Epithelium also lines tubes and cavities, such as blood vessels, airways, and the digestive, urinary, and reproductive tracts. It is a sheetlike tissue with one surface that faces the outside environment or some internal body fluid. The opposite surface rests upon a protein-rich noncellular basement membrane, which attaches it to the underlying tissues. Figure 20.1 illustrates this arrangement.

Cells in epithelium are cube-shaped, flattened, or column-shaped. *Simple* epithelium is just one cell layer thick; it lines body cavities, ducts, and tubes. *Stratified* epithelium has two or more layers of cells and usually functions in protection. Some kinds of epithelium are ciliated, pigmented, or have *microvilli*, tiny projections that increase the surface area of the plasma membrane.

Most animals have glandular epithelium, a sheetlike tissue that includes secretory cells. Secretory products of these cells are put to use outside the cells. As one example, glandular epithelium in the stomach's lining secretes acid that aids in digestion.

Most animals also secrete material through **glands**, secretory organs derived from epithelium. There are two kinds of glands. The **exocrine glands** secrete their products to the surface of an epithelium through ducts. The salivary glands, sweat glands, and milk-producing mammary glands are examples. The **endocrine glands** do not have ducts. They release their products, called hormones, into some body fluid. Most commonly, the molecules enter the bloodstream, which carries them throughout the body. As you will learn in Chapter 26, hormones are signaling molecules that influence the activity of all organs and organ systems.

Cells in epithelial tissue are usually linked by cell junctions. The junctions are structural and functional links between cells. There are three major types: tight junctions, adhering junctions, and gap junctions. Cells in most tissues form one or more of these types.

Tight junctions prevent fluids from leaking across a tissue because these branching protein filaments seal the lateral walls of cells together. Cells joined by tight junctions often act as a selective barrier. The junctions prevent material from moving *between* these cells, but transport proteins at the cells' plasma membrane allow specific ions or molecules to move *through* the cells.

Tight junctions protect your organs from the acidic fluid in your stomach. If this acid leaked across the stomach's epithelial lining, it would denature proteins in surrounding tissues. This is what happens to people who have a *peptic ulcer*. A bacterial infection damages the epithelial lining of the stomach and then acid leaks out into the adjacent tissues.

Adhering junctions are clumps or bands of protein filaments that act like spot welds to hold cells together (Figure 20.2*b*). Adhering junctions abound in tissues, such as skin, that are frequently pulled and stretched.

Ions and certain small molecules move though **gap junctions**, which are protein complexes that connect the cytoplasm of adjacent cells (Figure 20.2*c*). During development, signal flow through gap junctions helps determine the fate of embryonic cells. In cardiac and smooth muscle, an abundance of gap junctions allows the coordinated contraction of millions of cells.

Epithelia are sheetlike tissues that cover body surfaces and line cavities, ducts, and tubes. Epithelia have one free surface facing a body fluid or the outside environment.

Glands are secretory organs derived from epithelium.

Most cells are connected to others in the same tissue by one or more types of cell junctions.

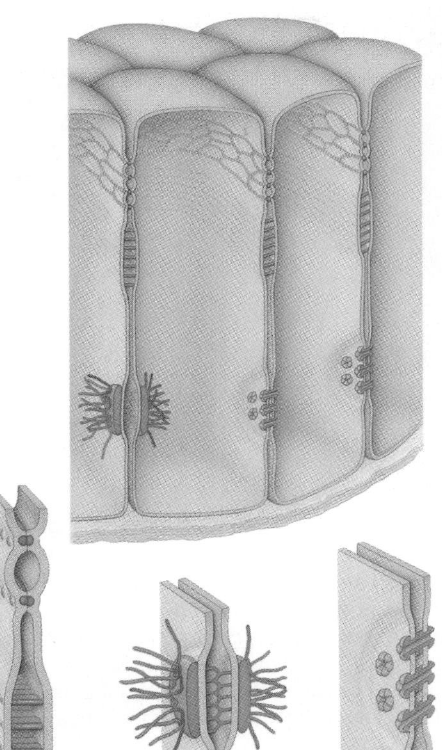

Figure 20.2 ***Animated!*** Cell junctions in animal tissues. (**a**) In tight junctions, proteins run parallel with the tissue's free surface and prevent leaks between adjoining cells. (**b**) Adhering junctions are spot welds or collars between adjoining cells. Intermediate filaments of the cytoskeleton anchor them to the plasma membrane. (**c**) Gap junctions are cylindrical protein arrays between plasma membranes of two adjacent cells. These open channels connect the cytoplasm of cells, so signals flow rapidly between them.

CONNECTIVE TISSUES

Connective tissues are the most abundant and widely distributed vertebrate tissues. They range from the soft connective tissues to the specialized types called bone tissue, cartilage, adipose tissue, and blood. In all types except blood, cells called fibroblasts make and secrete fibers of collagen, the body's most abundant protein. Fibroblasts also secrete polysaccharides that form the tissue's "ground substance." This noncellular material surrounds and supports the tissue's cells.

SOFT CONNECTIVE TISSUES **Loose connective tissue** is found beneath the skin and in areas between organs. It often attaches an epithelium to the wall of an organ. Its relatively few fibers and fibroblasts are dispersed in a semifluid ground substance. White blood cells that patrol this tissue serve as one of the body's first lines of defense against pathogens (Section 23.2).

Dense, irregular connective tissue has fibroblasts and many collagen fibers oriented every which way. It occurs in the lower layer of the skin and in capsules around organs that do not stretch, such as the kidney. In **dense, regular connective tissue**, fibers lie parallel to one another in an orderly array, with fibroblasts in

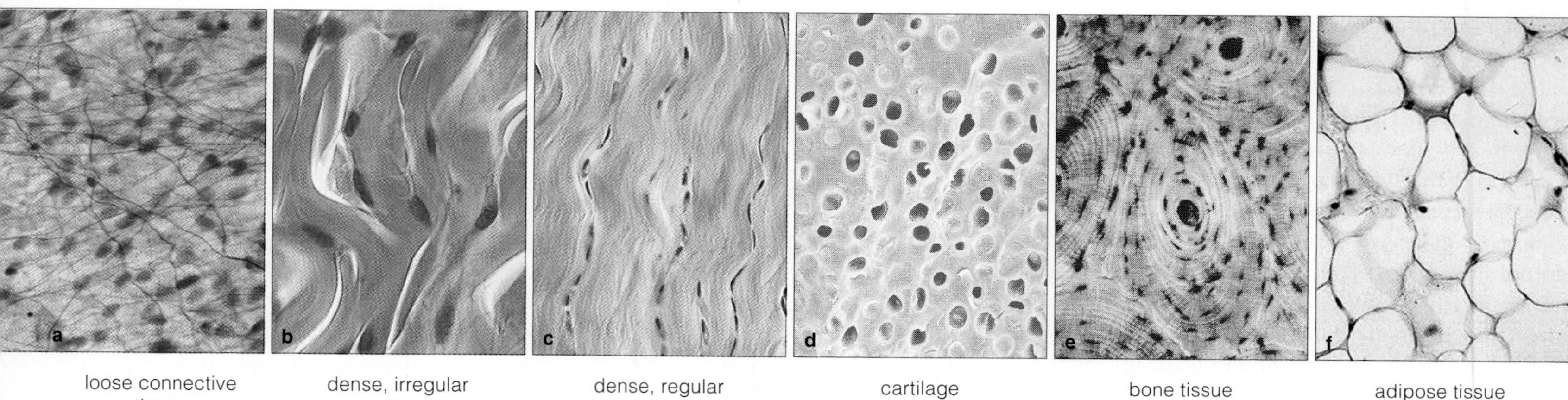

loose connective tissue

dense, irregular connective tissue

dense, regular connective tissue

cartilage

bone tissue

adipose tissue

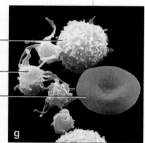

white blood cell

platelet

red blood cell

cells and platelets of blood

g

Figure 20.3 Examples of soft connective tissues with three patterns of fibers, fibroblasts, and other cells in the ground substance: (**a**) loose, (**b**) dense and irregular, and (**c**) dense and regular. Examples of specialized connective tissues: (**d**) cartilage and (**e**) bone tissue, both with living cells imprisoned in an extensive matrix of their own secretions. (**f**) Densely packed fat cells in adipose tissue, and (**g**) the cells and cell fragments (platelets) of blood, a fluid connective tissue.

between. The parallel arrays of fibers make the tissue resistant to tearing. It is found in the straplike tendons that connect skeletal muscle to bone and in ligaments, which attach one bone to another. This tissue also forms tough sheets that enclose muscle. Figure 20.3*a–c* shows examples of soft connective tissues.

SPECIALIZED CONNECTIVE TISSUES In all vertebrate embryos, **cartilage** forms as fibroblasts of a specialized type secrete an intercellular material that accumulates and encloses them (Figure 20.3*d*). Cartilage is rubbery, pliable, and able to resist compression. It maintains the shape of the outer ear, the nose, and other body parts. It cushions the joints between limb bones, between the bony segments of the vertebral column, and elsewhere. In adults, cartilage cells seldom divide, so injuries that damage cartilage are very slow to heal.

Bone tissue is mineral hardened; its collagen fibers and ground substance are strengthened with calcium salts (Figure 20.3*e*). This is the major tissue of organs called bones. Bones of the vertebrate skeleton enclose, support, and protect soft internal tissues and organs. Limb bones, such as your long leg bones, have weight-bearing functions. They interact with skeletal muscles attached to them to move the body. Certain bones also have inner regions where blood cells form.

A few fat droplets form in the cytoplasm of many cells while they are converting excess carbohydrates and lipids to fats. By contrast, fat droplets nearly fill cells in **adipose tissue** (Figure 20.3*f*). This tissue is the body's main reservoir of energy. Fats move rapidly to and from the cells by way of blood vessels. Adipose tissue also insulates and cushions certain body parts. In some animals, one type of adipose tissue also helps warm the body by its metabolic heat production.

Blood is generally classified as a connective tissue because it is derived from cells in bone. Blood serves transport functions. Its fluid portion, called plasma, is mainly water with dissolved proteins, ions, and other substances. The blood also contains great numbers of red blood cells, white blood cells, and platelets (Figure 20.3*g*). Red blood cells function in gas transport and exchange. They deliver oxygen to metabolically active tissues, then carry carbon dioxide away from them. The various types of white blood cells help defend the body. The platelets act in blood clotting.

Diverse types of connective tissues bind together, support, strengthen, protect, and insulate other body tissues.

Soft connective tissues consist of protein fibers as well as a variety of cells arranged in a ground substance. They differ in the proportions and arrangements of these components.

Cartilage, bone, blood, and adipose tissue are specialized connective tissues. Cartilage and bone are both structural materials. Adipose tissue is a reservoir of stored energy. Blood is a fluid that serves transport and immune functions.

IMPACTS, ISSUES

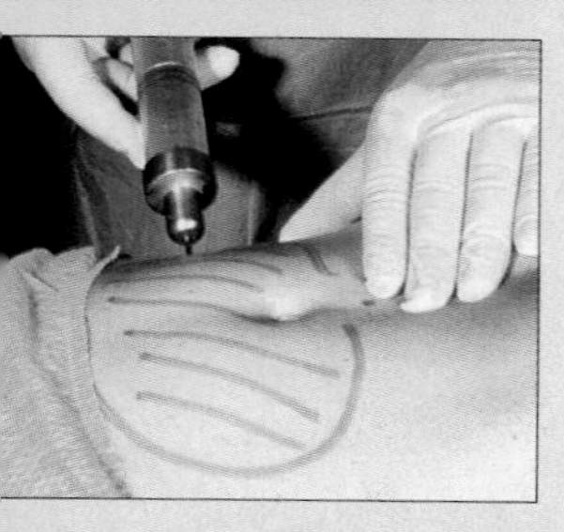

Researchers can turn fat into muscle—at least in the lab. They begin with some stem cells in the fat suctioned from patients (liposuction). They also turn fat cells into bone, cartilage, and nerve cells. Fat, cartilage, and bone are connective tissues, but nerve cells are a different tissue type. This suggests that adult stem cells may be more flexible and useful than previously thought.

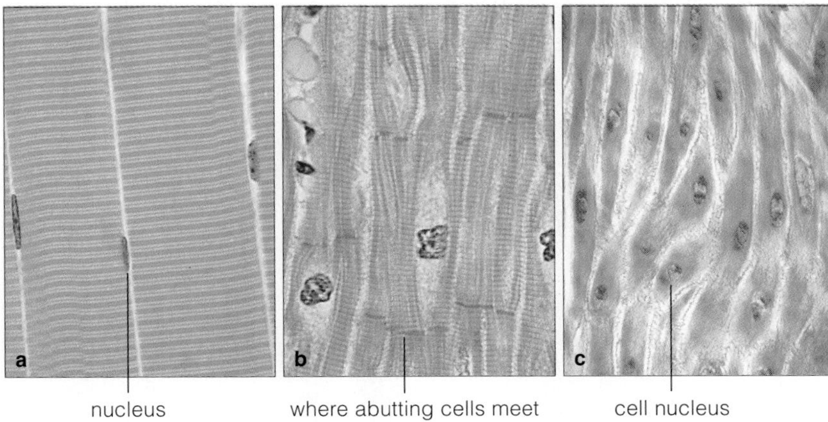

Figure 20.4 Three types of muscle tissue. (**a**) Part of two of the long muscle fibers bundled up in skeletal muscle tissue. Row after row of contractile units give them a striped appearance. (**b**) Adjacent striped cells of cardiac muscle tissue. The dark horizontal bands where cell ends meet are rich in adhering junctions. (**c**) Smooth muscle tissue, with no stripes.

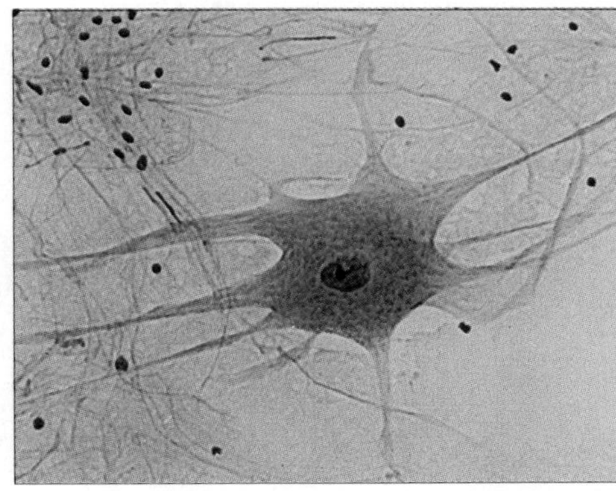

Figure 20.5 Motor neuron. This type of cell relays signals from the brain or spinal cord to muscles and glands. Different neurons interact to detect and process information about internal and external conditions, and initiate responses.

MUSCLE TISSUES

In an all vertebrates, we find three classes of muscle. These are skeletal muscle, cardiac muscle, and smooth muscle. Each type has unique properties that reflect its specific functions.

Skeletal muscle tissue attaches to skeletal elements and moves the body or its parts. It has parallel arrays of muscle fibers (Figure 20.4*a*). Muscle fibers form by fusion of muscle cells during development, so a fiber has multiple nuclei. Inside each muscle fiber are long strands, each divided crosswise into many contractile units. Skeletal muscle looks striated (striped) because of these repeating units.

Skeletal muscle tissue makes up about 40 percent of the weight of an average human. We can make our skeletal muscles contract at will, so they are sometimes referred to as "voluntary" muscles.

Only the heart wall contains **cardiac muscle tissue** (Figure 20.4*b*). Its cells contract as a unit in response to signals that flow through gap junctions between their abutting plasma membranes. Each branching cardiac cell has a single nucleus. It has far more mitochondria than other types of muscle cells, because the heart's constant contractions require a steady supply of ATP.

Cardiac muscle tissue also is striated, although less conspicuously so. Cardiac muscle and smooth muscle are both said to be "involuntary" muscle; we usually cannot make them contract just by thinking about it.

We find layers of **smooth muscle tissue** in the wall of many soft internal organs, including the stomach, bladder, and uterus. Cells are unbranched and tapered at both ends, and have a single centrally positioned nucleus. Contractile units are not arranged in orderly repeating fashion, so smooth muscle does not appear striated (Figure 20.4*c*). Contractions of smooth muscle propel material through the gut, shrink the diameter of blood vessels, and close sphincters.

Muscle tissue, which can contract (shorten) in response to stimulation, helps move the body and specific body parts.

Skeletal muscle attaches to and moves bones; it is the only type of muscle that is under our voluntary control. Smooth muscle tissue occurs in soft internal organs. Cardiac muscle makes up the contractile walls of the heart.

NERVOUS TISSUE

Of the four types of animal tissues, **nervous tissue** exerts the most control over how the body responds to changing conditions. Its cells make up communication lines of nervous systems. Figure 20.5 shows a motor neuron, one kind of excitable cell in nervous tissue. In this context, "excitable" means that a cell has specific transport proteins in its plasma membrane that allow electrical disturbances to move along this membrane. In neurons, these electrical disturbances alter the cell's activity in ways that can transmit information to other neurons, or to muscles or glands.

Chapter 25 will give you a sense of how more than a hundred billion neurons form communication lines through your body. You will learn how different kinds detect and assess changes in the internal and external environments, and then call up responses.

IMPACTS, ISSUES

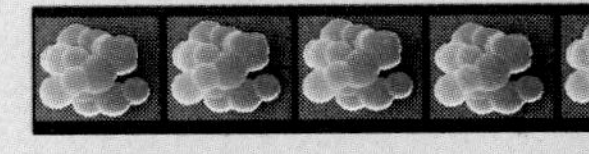

Muscular dystrophy, an X-linked genetic disorder, is caused by mutation in a muscle protein—dystrophin. Johnny Huard is using stem cells to treat mice with this disorder. He injects muscle stem cells from normal newborn mice into muscles of affected mice. Early results are promising. The transplanted cells take hold, integrate into the muscle, and make normal dystrophin.

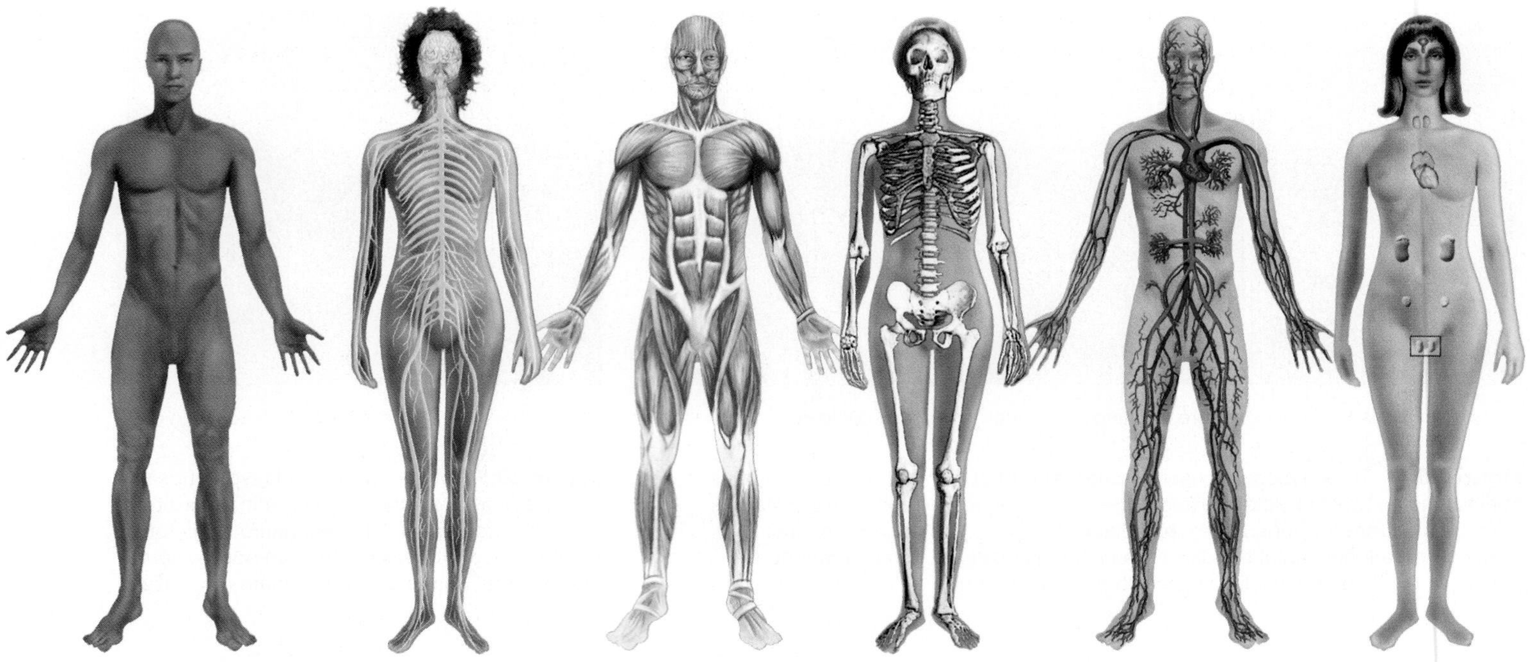

Integumentary System
Protects body from injury, dehydration, and some pathogens; controls its temperature; excretes certain wastes; receives some external stimuli.

Nervous System
Detects external and internal stimuli; controls and coordinates the responses to stimuli; integrates all organ system activities.

Muscular System
Moves body and its internal parts; maintains posture; generates heat by increases in metabolic activity.

Skeletal System
Supports and protects body parts; provides muscle attachment sites; produces red blood cells; stores calcium, phosphorus.

Circulatory System
Rapidly transports many materials to and from cells; helps stabilize internal pH and temperature.

Endocrine System
Hormonally controls body functioning; works with nervous system to integrate short-term and long-term activities.

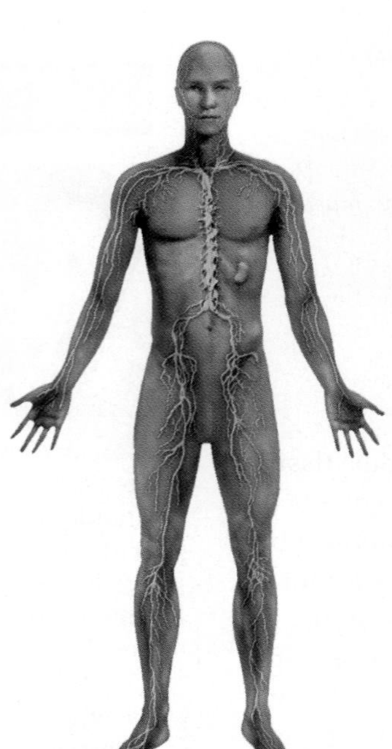

Lymphatic System
Collects and returns some tissue fluid to the bloodstream; defends the body against infection and tissue damage.

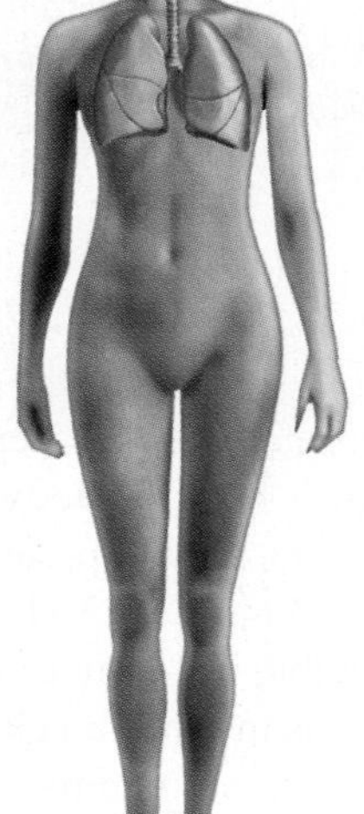

Respiratory System
Rapidly delivers oxygen to the tissue fluid that bathes all living cells; removes carbon dioxide wastes of cells; helps regulate pH.

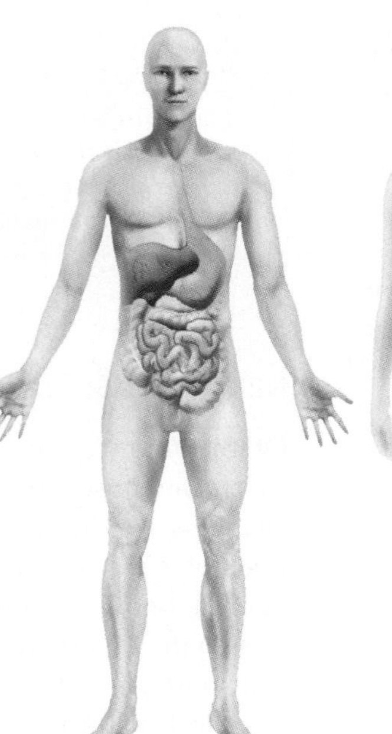

Digestive System
Ingests water and food; breaks food down into component molecules and absorbs them into internal environment; eliminates food residues.

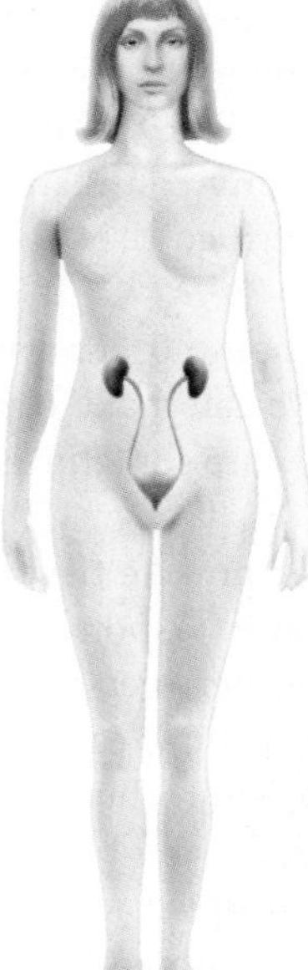

Urinary System
Maintains the volume and composition of internal environment; excretes excess fluid and wastes filtered from the blood.

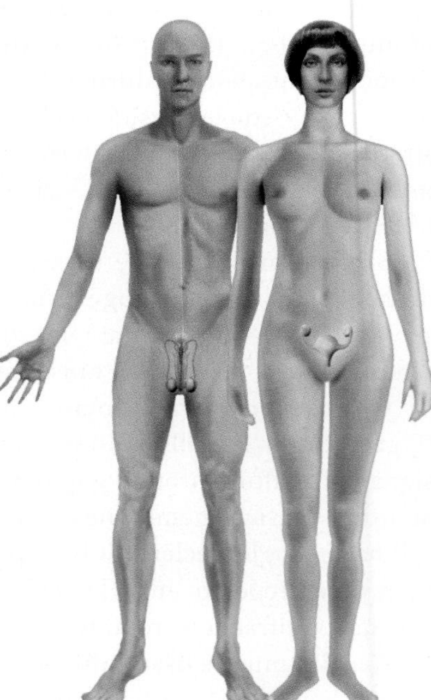

Reproductive System
Female: Produces eggs; after fertilization, affords a protected, nutritive environment for the development of new individuals. *Male:* Produces and transfers sperm to the female. Hormones of both systems also influence other organ systems.

a

Neurons are not the only cells in nervous tissue. More than half of the volume of your nervous system consists of diverse neuroglial cells. These cells support and protect the neurons.

Neurons are the basic units of communication in nervous tissue. They detect specific stimuli, integrate information, and issue or relay commands for response.

20.3 Organ Systems Made From Tissues

Nearly all animals have tissues arranged in organs, and most have organ systems. Each organ system has specialized functions, such as the transport and exchange of substances.

OVERVIEW OF MAJOR ORGAN SYSTEMS

All vertebrates have the same types of organ systems (Figure 20.6*a*). The systems interact to sustain all their body's cells. You might think this is an overstatement. For example, how could muscles and bones be helping each small, individual cell stay alive? Yet interactions between the skeletal system and muscular system let us move, say, toward sources of nutrients and water. Parts of these organ systems help keep blood moving to cells, as when muscles in legs contract and help to squeeze blood in nearby veins back toward the heart. The circulatory system transports essential substances, such as oxygen and nutrients, to cells and moves any metabolic products and wastes away from them. With each breath, the respiratory system delivers oxygen to the circulatory system and gets rid of carbon dioxide waste. Muscular and skeletal systems work in concert with the respiratory system in breathing—and so on.

LINK TO SECTION 16.1

Figure 20.6*b* and *c* is a guide to some terms used in describing the positions of an animal's organs. Figure 20.6*d* shows the location of the body cavities in which most vertebrate organs are located.

TISSUE AND ORGAN FORMATION

Where do tissues and organ systems come from? Their development begins with sperm and eggs, which arise from *germ* cells, or immature reproductive cells. (All other body cells are *somatic*, after *soma*, the Greek word for body.) The zygote formed by fertilization undergoes repeated mitotic divisions to yield a hollow ball of cells. In most animals, these cells become arranged as three primary tissues—ectoderm, mesoderm, and endoderm. These in turn give rise to all adult tissues.

Ectoderm gives rise to the outer layer of skin and tissues of the nervous system. **Mesoderm** gives rise to the body's muscles, bones, and most of the circulatory, reproductive, and urinary systems. **Endoderm** gives rise to the lining of the digestive tract and to organs derived from it. We will return to the fates of these three embryonic tissues in Chapter 27.

In general, all vertebrates have the same organ systems.

Vertebrate tissues, organs, and organ systems arise from three primary tissues that form in the early embryo. These tissues are ectoderm, mesoderm, and endoderm.

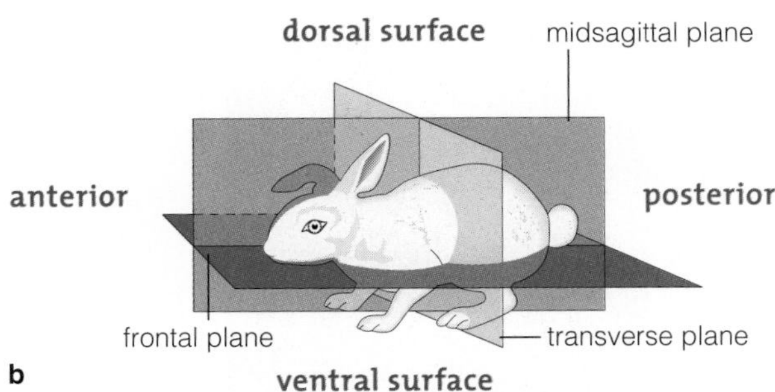

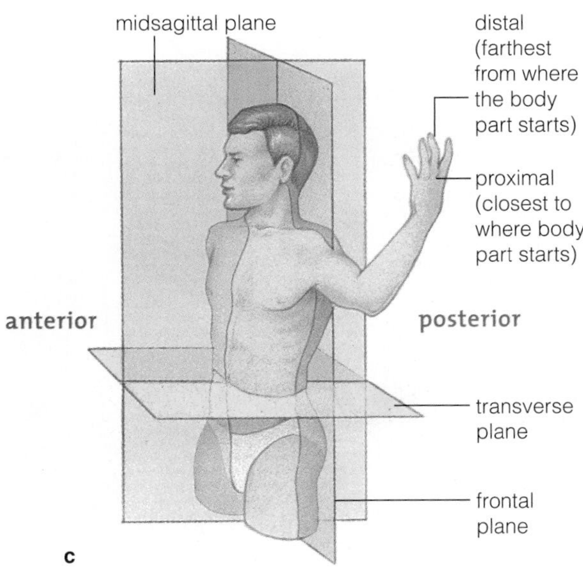

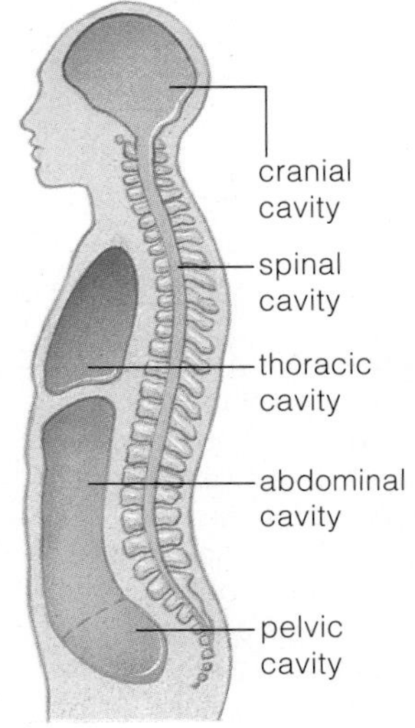

Figure 20.6 ***Animated!*** (**a**) Major vertebrate organ systems. (**b**) Most vertebrates move with the main body axis parallel with the Earth's surface. For them, *dorsal* refers to their back or upper surface, and *ventral* to the opposite, lower surface. (**c**) Humans walk upright, with their main body axis perpendicular to the ground. *Anterior* refers to the front of an upright walker; it corresponds to ventral. *Posterior* refers to the back; it corresponds to dorsal. (**d**) Major body cavities inside which many of the organs are located.

LINKS TO SECTIONS 16.12, 17.3

20.4 Skin—Example of an Organ System

A waterproof, keratin-toughened, yet highly sensitive, skin is our body's main interface with the environment. It is an organ with a remarkable variety of functions.

IMPACTS, ISSUES

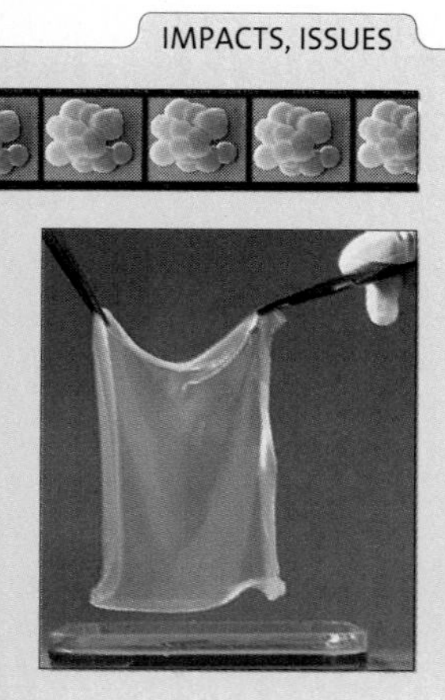

Researchers dream of one day using stem cells to grow organs such as kidneys and hearts. But lab-grown epidermis is already a reality. *Artificial skin* is being used to treat burns and to cover and help heal some chronic wounds.

One of the organ systems listed in Figure 20.6*a* is the integumentary system, which in vertebrates is called skin. **Skin** consists of an outer epidermis (one or more epithelial sheets) and an underlying dermis of dense connective tissue. In your case, two of its key features evolved when vertebrates moved onto dry land. The outer epidermal layers of skin became keratinized, and glands with ducts became recessed below its surface. Keratinocytes are cells that make the water-resistant protein keratin. Melanocytes are cells that make and give up melanin to keratinocytes. **Melanin**, recall, is a brownish-black pigment, one of the body's barriers to harmful ultraviolet (UV) radiation from the sun.

Skin is your body's largest organ. Skin can stretch, conserve body water, and repair small cuts or burns. It helps you get vitamin D and can help cool you on hot days by dissipating metabolic heat. Sensory receptors in skin help your brain monitor events in the outside world. Some cells in skin even help defend the body.

The **epidermis** is a stratified epithelium with many cell junctions and no extracellular matrix (Figure 20.7). Ongoing mitotic cell divisions rapidly push cells from the deeper epidermal layers to the sheet's free surface. Abrasion and the pressure exerted by the growing cell mass flatten and kill older cells before they reach the surface. The oldest cells are rubbed off or flake away.

The **dermis**, mainly a dense connective tissue, has stretch-resisting elastin and supportive collagen fibers. Blood vessels, lymph vessels, and sensory receptors thread through it. The hypodermis, a continuous layer beneath skin, anchors the dermis to structures below it. It consists of loose connective tissue and adipose tissue that has insulative and cushioning roles.

Skin can be very dark to light, depending on how melanocytes are distributed and how active they are. In pale skin, little melanin forms. Such skin often looks pink because hemoglobin's color shows through thin-walled blood vessels and epidermis. A yellow-orange pigment, carotene, also contributes to skin color.

Human skin has many exocrine glands, including mammary glands and about 2.5 million sweat glands that dissipate body heat. Oil gland secretions lubricate and soften hair and skin, and kill many bacteria.

An average human scalp has about 100,000 hairs. We find living, dividing cells in the follicle at a hair's base. As these cells divide, they push cells above them up, lengthening the hair. The portion of a hair that you can see—the portion that sticks out beyond the skin—is the keratin-rich remains of now dead cells.

THE VITAMIN CONNECTION When you sit out in the sun, exposure to UV light stimulates epidermal cells to make cholecalciferol, which later can be converted to vitamin D. But UV exposure damages DNA, which increases risk of skin cancer. It also causes breakdown of folate, one of the B vitamins. An embryo's nervous system cannot develop normally without folate.

By one hypothesis, variations in skin color among human groups are adaptations to regional differences in exposure to sunlight. Humans first evolved beneath

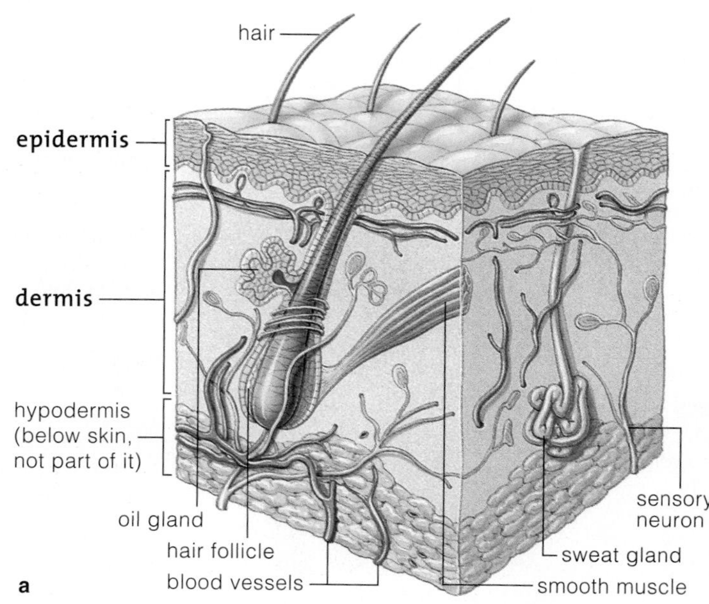

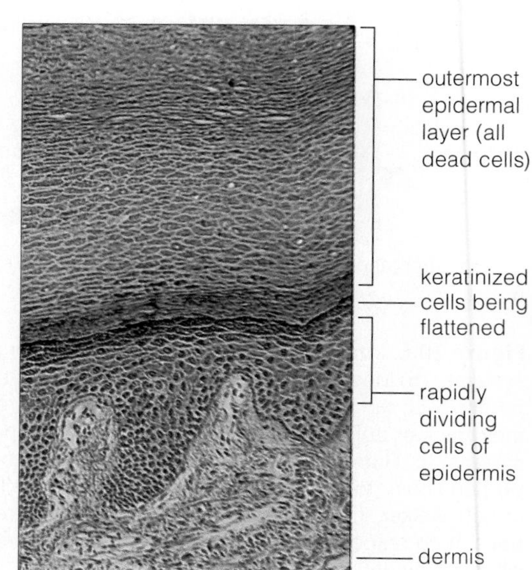

Figure 20.7 ***Animated!*** (**a**) Skin structure typical of body parts with hair, oil glands, and sweat glands. (**b**) A section through human skin. Thickened skin is typical of the palms of the hands and soles of the feet, which are both subject to abrasion.

the intense sunlight of African savannas. Melanin-rich skin reduced cancer risks and kept folate levels high, and people could still synthesize plenty of vitamin D. Later on, some human populations migrated to colder regions, where bodies were often covered or indoors. In this habitat, the UV protection afforded by melanin darkened skin became less important and paler skin that facilitated vitamin D production came to prevail.

This hypothesis has been tested by comparing skin color among people native to more than fifty regions that vary in annual UV levels. The data supported the hypothesis. The lower the annual levels of UV light in a region, the lighter the native people's skin color.

ON SUNTANS AND SHOE-LEATHER SKIN Melanocytes make melanin when exposed to UV radiation. Melanin production accelerates slowly, then peaks about ten days after the first exposure. Before then, unprotected skin can become mildly to severely burned.

An increase in melanin helps shield the skin's inner cell layers from harmful UV radiation, but some UV still gets through. UV damage makes skin less elastic; it gives longtime sun-lovers a shoe leather look. It also alters DNA and can lead to skin cancers.

Skin changes naturally as we age. Epidermal cells divide less often, so the skin thins and recovers more slowly from injury. Glandular secretions slow, so the skin is drier. Collagen and elastin fibers become fewer and wrinkles deepen. As melanocytes dwindle, hair grays and the skin burns more easily. Smoking can accelerate these changes; it restricts blood delivery to the skin, starving it of vital oxygen and nutrients.

THE FRONT LINE OF DEFENSE Some cells in skin help protect us against pathogens. The **Langerhans cells** are white blood cells that prowl the epidermis and act as sentries. If a Langerhans cell encounters viruses or bacteria, it engulfs them and releases molecular alarm signals that mobilize the immune system.

UV radiation damages Langerhans cells. This may be why *cold sores* often follow a sunburn. These small, painful blisters announce a recurring *Herpes simplex* infection. Nearly all of us harbor this virus. It hides in neurons inside facial tissues. Sunburns and other kinds of stress factors reactivate virus particles, which escape from neuron endings and infect epithelial cells causing painful blisters.

Skin is the body's protective, pliable covering. Its keratinized, melanin-shielded cell layers help the body conserve water, minimize damage by ultraviolet radiation, produce vitamin D, and maintain body temperature.

Summary

Section 20.1 Tissues, organs, and organ systems interact to maintain the stable internal environment required for individual cell survival. At homeostasis, conditions in the internal environment are being maintained at levels suitable for cell activities.

Section 20.2 A tissue is an aggregation of cells and intercellular substances that perform a common task. Most animals are constructed of four types: epithelial, connective, muscle, and nervous tissue.

Epithelial tissues cover external body surfaces and line internal cavities and tubes. Epithelium has one free surface exposed to some body fluid or the environment. Gland cells and glands are derived from epithelium.

Animal tissues have a variety of cell-to-cell junctions. Tight junctions prevent substances from leaking across the tissue. Adhering junctions cement neighboring cells together. Gap junctions are open channels that connect the cytoplasm of abutting cells. They permit the rapid transfer of ions and small molecules between cells.

Connective tissues bind, support, strengthen, protect, and insulate other tissues. Soft connective tissues have different proportions and arrangements of protein fibers, fibroblasts, and other cells in a ground substance. The specialized connective tissues include cartilage, bone tissue, adipose tissue, and blood.

Muscle tissues contract (shorten) and then relax, or return to their resting position. They help move the body or parts of it. The three types of muscle tissue are skeletal muscle, smooth muscle, and cardiac muscle.

Neurons are communication lines of nervous tissue. Different kinds detect, integrate, and assess internal and external conditions, and deliver commands to glands and muscles that carry out responses. Diverse, abundant neuroglial cells protect and support the neurons.

BiologyNow
Use the interaction on BiologyNow to investigate the different types of tissues.

Section 20.3 An organ is a structural unit of different tissues combined in definite proportions and patterns that allow them to perform a common task. An organ system has two or more organs interacting chemically, physically, or both in tasks that keep individual cells as well as the whole body functioning. All tissues and organs are derived from just three embryonic tissues: ectoderm, mesoderm, and endoderm.

BiologyNow
Survey the organ systems of the vertebrate body with animation on BiologyNow.

Section 20.4 The organ system called skin functions in protection, temperature control, detection of shifts in the environment, vitamin D production, and defense.

BiologyNow
Explore the structure and function of skin with the animation on BiologyNow.

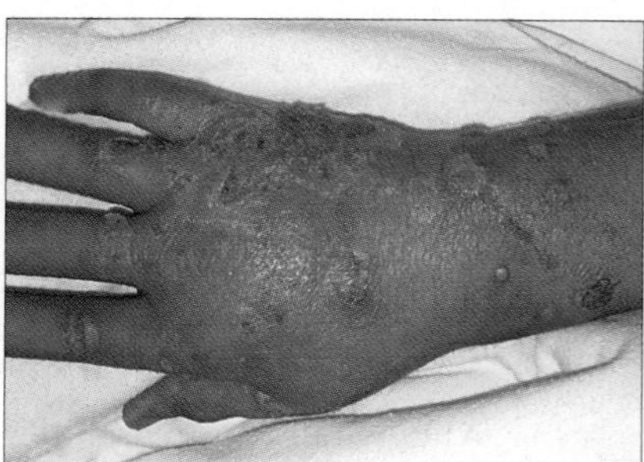

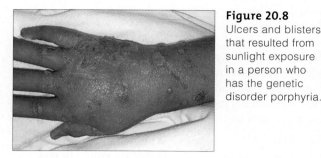

Figure 20.8 Ulcers and blisters that resulted from sunlight exposure in a person who has the genetic disorder porphyria.

Figure 20.9 Mucus-secreting and ciliated cells from an airway.

Self-Quiz

Answers in Appendix I

1. _______ tissues are sheetlike with one free surface.
   a. Epithelial c. Nervous
   b. Connective d. Muscle

2. _______ function in cell-to-cell communication.
   a. Tight junctions c. Gap junctions
   b. Adhering junctions d. all of the above

3. In most animals, glands are located in _______ tissue.
   a. epithelial c. muscle
   b. connective d. nervous

4. Most _______ have fibroblasts that secrete collagen.
   a. epithelial tissues c. muscle tissues
   b. connective tissues d. nervous tissues

5. _______ tissues are the body's most abundant and widely distributed tissue.
   a. Epithelial c. Nervous
   b. Connective d. Muscle

6. _______ is mostly plasma.
   a. Irregular connective tissue c. Cartilage
   b. Blood d. Bone

7. Your body converts excess carbohydrates and proteins to fats, which accumulate in _______.
   a. Langerhans cells c. adipose tissue cells
   b. neurons d. melanocytes

8. Cells of _______ can shorten (contract).
   a. epithelial tissue c. muscle tissue
   b. connective tissue d. nervous tissue

9. Only _______ muscle tissue has a striated appearance.
   a. skeletal c. cardiac
   b. smooth d. a and c are correct

10. _______ detects and integrates information about changes and controls responses to those changes.
   a. Epithelial tissue c. Muscle tissue
   b. Connective tissue d. Nervous tissue

11. Match the terms with the most suitable description.

____ exocrine gland	a. outermost skin layer
____ endocrine gland	b. secretes through duct
____ epidermis	c. innermost primary tissue
____ endoderm	d. contracts, not striated
____ smooth muscle	e. cements cells together
____ blood	f. fluid connective tissue
____ adhering junction	g. ductless secretion

Additional questions are available on **Biology Now™**

Critical Thinking

1. *Porphyria* is the name for a collection of rare genetic disorders. Affected people lack one of the enzymes in the metabolic pathway that forms heme, the iron-containing group in hemoglobin. As a result, intermediates of heme synthesis (porphyrins) accumulate in the body. In some forms of the disorder, porphyrins pile up in the bones, skin, and teeth. When porphyrins in the skin are exposed to sunlight, they absorb energy and release energized electrons. As the electrons careen around the cell, they can break bonds and cause free radicals to form. The most noticeable result is formation of lesions and scars on the skin (Figure 20.8). In the most extreme cases, gums and lips can recede, causing canine teeth to look more fanglike. Affected individuals must avoid sunlight, and compounds in garlic can exacerbate their symptoms.

   By one hypothesis, people suffering from the most extreme forms of porphyria may have been the source for vampire stories that date back to even before the Middle Ages. Is this plausible? What kind of additional information would support this historical hypothesis?

2. Adipose tissue and blood are often said to be atypical connective tissues. Compared to other types of connective tissues, which features are typical and which are not?

3. The most common ovarian tumors in young women are ovarian *teratomas*. The name comes from the Greek work *teraton*, which means monster. What makes these tumors "monstrous" is the presence of well-differentiated tissues, most commonly bones, teeth, fat, and hair. Early physicians suggested that teratomas—which they found mainly in young women—arose as a result of nightmares, witchcraft, or sex with the devil. In fact, teratomas are derived from germ cells. Other kinds of tumors arise from somatic cells and include cells derived from one primary tissue. Only teratomas have cells derived from more than one primary tissue. Explain why this allows teratomas to produce a greater variety of cell types.

4. Opponents of using animals to test the effects of chemicals used in cosmetics and other products argue that alternative methods are available. Human cells that are being grown in culture are used for some studies of this type. Given what you have learned in this chapter about how tissues, organs, and organ systems interact, discuss advantages and disadvantages of tissue-specific tests compared to tests using whole living animals.

5. Figure 20.9 is a micrograph of cells that line an airway. What type of tissue is this? How can you tell?

21 HOW ANIMALS MOVE

IMPACTS, ISSUES

Pumping Up Muscles

In 1998, Mark McGwire broke Major League Baseball's home-run record. His admission that he used a dietary supplement containing androstenedione helped boost the product's sales and focused attention on the role of unregulated supplements in sports.

Natural androstenedione is an intermediate that forms when the body is synthesizing testosterone. This sex hormone has well-documented tissue-building (anabolic) effects. According to promoters, andro also raises testosterone levels, which in turn boosts the rate of protein synthesis in muscles.

Does andro work? Probably not. In controlled studies, men who took the supplement did not increase muscle mass or strength more than those who took a placebo. At most, andro raised testosterone levels for a few hours.

Andro also is an intermediate in the synthesis of estrogen, a sex hormone with feminizing effects. Its known side effects on males include shriveling testicles, the development of female-like breasts, and loss of hair. Both sexes are at risk for liver damage, acne, and a decline in the blood level of "good" (HDL) cholesterol. Women can develop masculinized patterns of hair growth and speech, and their menstrual cycle can be disrupted. Reports of these negative health effects caused the Food and Drug Administration to place a ban on nonprescription sale of androstenedione beginning in January 2005.

Supplement makers also market creatine phosphate as a performance enhancer. The body makes some of this compound and gets more from food. When a muscle is relaxed, ATP donates a phosphate to creatine. When the muscle needs ATP to contract, this reaction is reversed. Creatine phosphate donates a phosphate group to ADP, forming ATP. Studies show that creatine supplements do improve performance in brief, high-intensity exercise, such as sprinting. But they put a strain on kidneys, and short term side effects include weight gain, high blood pressure, muscle cramps, and stomach upsets. The long-term effects of creatine supplementation are not known.

In this chapter you will learn how the skeletal and muscular systems evolved and function. This can help you to evaluate how far systems should be pushed in pursuit of enhanced performance.

How Would You Vote? *Manufacturers of dietary supplements are not required to prove that their claims for these products are accurate. Should dietary supplements be subject to more stringent testing for effectiveness and safety? See BiologyNow for details, then vote online.*

Key Concepts

WHAT SKELETONS DO

Animals move when the force of muscle contraction is applied to a skeleton. The skeleton may be a fluid enclosed within a cavity, hard external parts, or internal bones. Bones of the vertebrate skeleton function in movement, mineral storage, and protection and support of soft organs. Some bones are the sites of blood cell formation.

HOW MUSCLES WORK

Vertebrate skeletal muscle is the functional partner of bones. Muscles are made up of parallel arrays of muscle fibers. Each fiber is divided along its length into many contractile units called sarcomeres. A sarcomere contracts (shortens) as a result of ATP-driven sliding interactions between protein filaments.

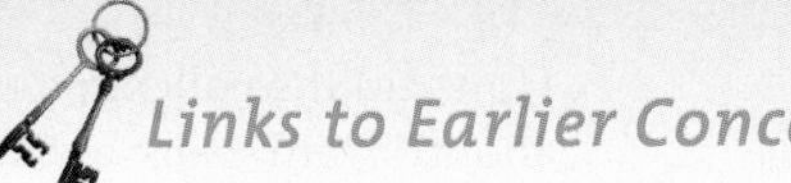

Links to Earlier Concepts

Sections 16.10–16.13 explained how the vertebrate skeleton became modified with the move to land and as hominids began to walk upright. We will fill in some details of that story here.

You already know about the microfilaments and motor proteins involved in muscle contraction (3.7). Energy to fuel their movements comes from aerobic respiration and lactate fermentation (6.1–6.4).

A review of the mechanism of X-linked inheritance will help you understand how one important muscle disorder is inherited (8.7).

LINKS TO SECTIONS 16.10–16.13

21.1 So What Is a Skeleton?

Nearly all animals move by contracting and relaxing muscle fibers. But muscles cannot move the body on their own. Force of contraction must be applied against something else. A skeleton fulfills this requirement.

Figure 21.1 Invertebrate skeletal systems. *Above*, Sea anemones have an hydrostatic skeleton. The body changes shape when muscles redistribute fluid in an internal cavity. *Below*, An insect, such as a fly, has a hard cuticle that acts as an exoskeleton. Fly wings move up and down when muscles pull on the cuticle.

TYPES OF SKELETONS

There are three types of skeletons. With a **hydrostatic skeleton**, muscles work against a fluid, which becomes redistributed within a confined space in the body. The fluid resists compression, like a waterbed. Many soft-bodied invertebrates, such as the earthworms and sea anemones, utilize a hydrostatic skeleton (Figure 21.1). Some of the contractile cells in a sea anemone's body wall run along the body's long axis; others encircle the gut. In resting position, radial cells are relaxed and the longitudinal ones are contracted. When the radial cells contract and longitudinal ones relax, the body extends upward to its feeding position.

With an **exoskeleton**, a cuticle, shell, or some other rigid, *external* body part receives the applied force of a muscle contraction. An arthropod, recall, has a hinged exoskeleton (Figure 21.1). Muscles attached to parts of an exoskeleton move or bend its parts. As an example, a fly's wings move up and down as muscles attached to its thorax contract and relax. The changes in thorax shape cause the attached wings to flap up and down as fast as 200 times a second.

All vertebrates have an **endoskeleton**, with *internal* body parts accepting the applied force of contraction.

EVOLUTION OF VERTEBRATE SKELETONS

How did the collection of bones in your skeleton come about? Think back to Chapter 16. From a notochord to vertebral column, from the gill supports to jaws, from structural elements in lobed fins to limbs—these key evolutionary trends emerged in vertebrates that made the transition to life on land.

Deprived of water's buoyancy, skeletons evolved to meet the new challenges. For example, the pelvic and pectoral girdles of fishes function as a stable base for moving fins, which propel, guide, and stabilize the body in water. Among early four-legged vertebrates, the two girdles transferred body weight to the limbs. Limbs that became positioned closer to the body mass helped hold it upright and move it forward. A cage of hardened bones, attached to the backbone, protected the heart, lungs, and other soft internal organs from collapsing under the body's weight.

Figure 21.2 shows the kind of bony endoskeletons that evolved among land vertebrates. Comparing this to Figure 21.3, can you identify the homologous features in your own skeleton? Notice the pectoral girdle (at the shoulders), pelvic girdle (at the hips), and the paired arms, hands, legs, and feet. This part of your skeleton is the legacy of early tetrapods. So are the flat shoulder blades and slender collarbone of the pectoral girdles. It is easy for a "fish out of water" to dislocate a shoulder blade or to fracture a collarbone—the one broken most often—after falling hard on an outstretched arm.

The rest of a human skeleton—jaws and other skull bones, twelve pairs of ribs, a breastbone, and twenty-six vertebrae—are the legacies of ancient vertebrates. **Vertebrae** (singular, vertebra), those bony segments of a backbone, extend from the skull to the pelvic girdle. **Intervertebral disks** made of cartilage lie between the bones and act as shock absorbers and flex points.

The vertebral column's bony parts are attachment sites for paired muscles and form a protective canal for the spinal cord. These bones also transmit the weight of a human's upright torso to the lower limbs.

For reasons still under debate, ancestral members of our lineage began walking upright more than 4 million years ago. The transition to an upright posture put a notably S-shaped curve into the backbone. Ever since, the bones and disks of the vertebral column have been stacked up against gravity. Sometimes a disk can slip or rupture—*herniate*—harming spinal nerves and causing chronic numbness or pain. No longer aquatic, no longer four-legged, we must live with costs as well as benefits of getting around in the world on two legs.

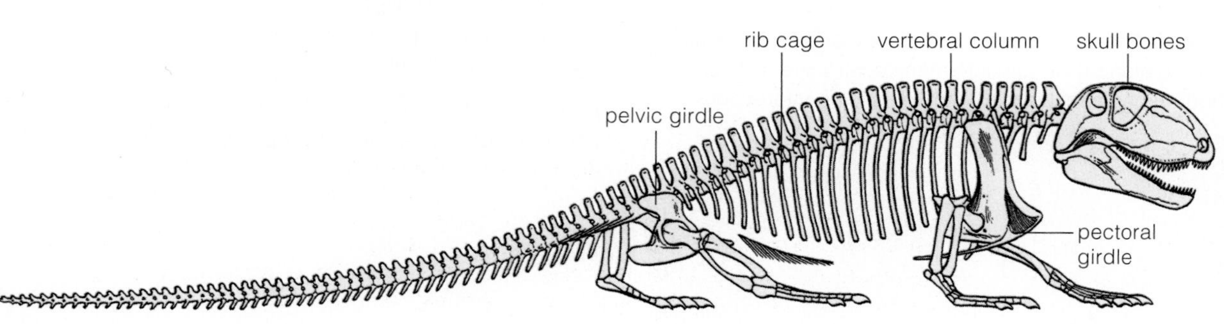

Figure 21.2 Endoskeleton of a typical early reptile. All vertebrates have an internal skeleton of cartilage and, in most cases, bone. Compare these components with those of the human skeleton in Figure 21.3.

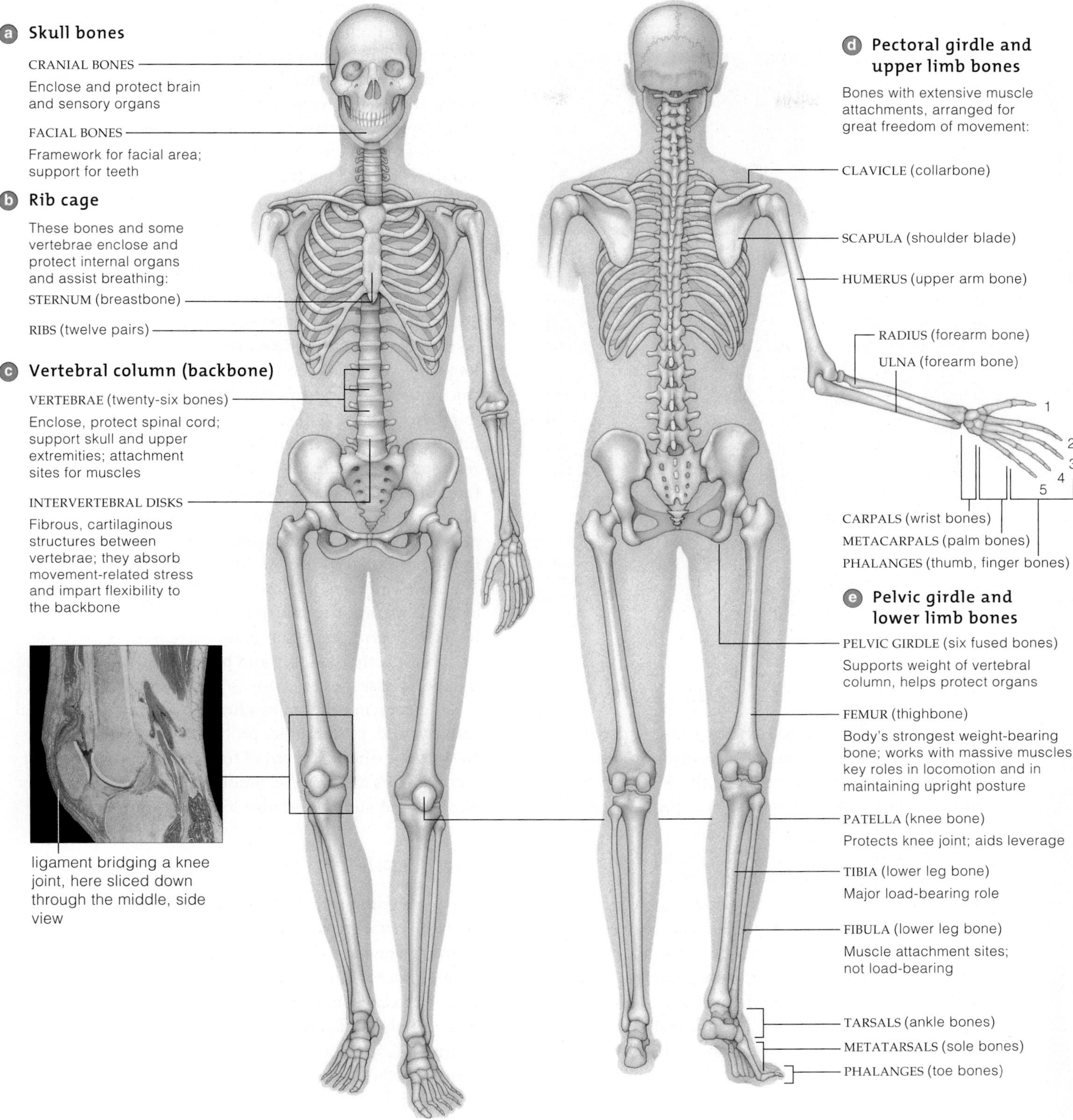

Animal skeletons have structural elements or body fluids against which the force of contraction may be applied. The skeleton of land vertebrates is a collection of earlier adaptations to life in water and later modifications to a life deprived of water's buoyancy.

Figure 21.3 *Animated!* Bones of the human skeletal system. Major bones of the *axial* portion are listed at left. Those of the *appendicular* portion are listed at right. Overall, there are 206 bones. The smallest ones are inside the middle ear. The largest is the femur, which attaches the hip to the lower leg at the knee. Ligaments and other kinds of connective tissue structures bridge skeletal joints, the areas of contact or near-contact between bones.

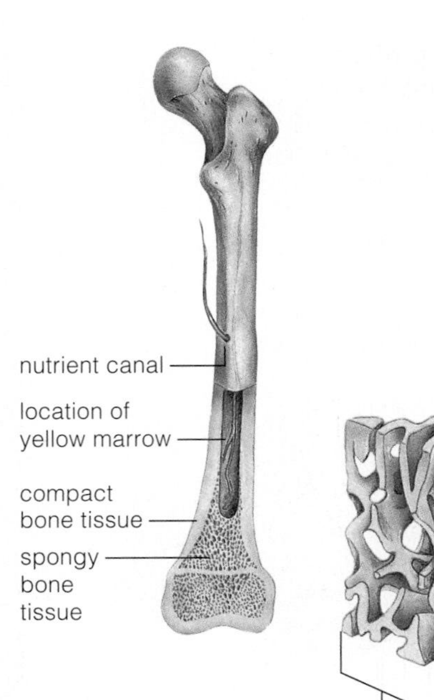

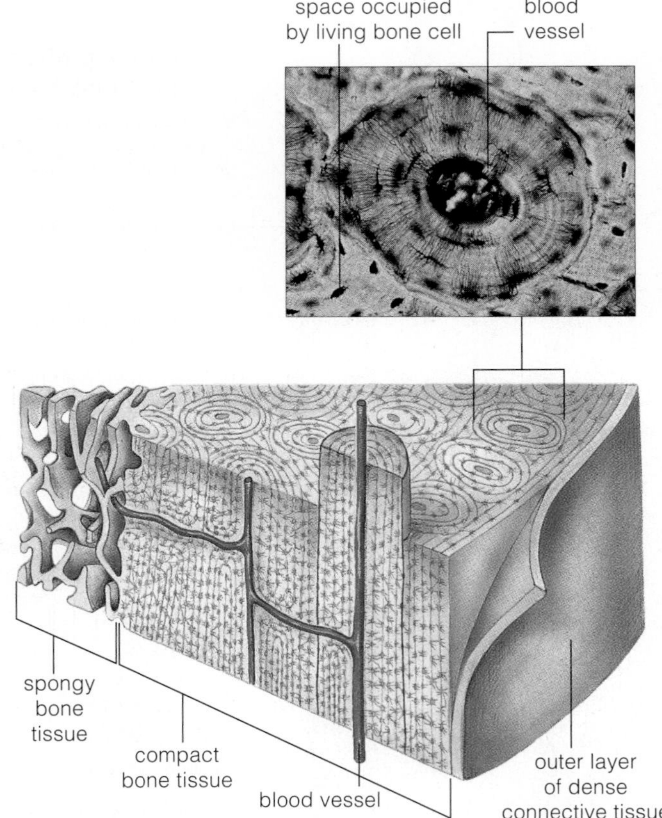

Figure 21.4 ***Animated!*** Structure of a human thighbone (femur).

BONE STRUCTURE AND FUNCTION

Bones are organs that have many roles. Some attach to and support skeletal muscles; they help to maintain or change the orientation of the body or its parts. Some form chambers that surround and protect vulnerable organs. Some are sites for formation of blood cells. All serve as reservoirs for calcium and phosphorus ions.

Bone is specialized connective tissue. Its cells lie in a calcium-hardened, collagen-rich organic matrix. The three kinds of bone cells are osteoblasts, osteocytes, and osteoclasts. The *osteoblasts* are bone-forming cells; they secrete the material that forms the matrix. When these cells are imprisoned by their own secretions, they are transformed into *osteocytes*. This is the main cell type in adult bones. *Osteoclasts* specialize in the breakdown of bone. They secrete substances that dissolve the matrix.

There are two types of bone. *Compact* bone makes up the outer, weight-bearing part of a femur (Figure 21.4). It is made of many thin, cylindrical, dense layers surrounding canals for nerves and blood vessels that service living bone cells. The shaft and bone ends also have *spongy* bone tissue that is lightweight but strong.

Red marrow fills spaces in the spongy bone of long bones and some others, such as the breastbone. It is the main site for blood cell formation. The main cavity of adult long bones is filled by **yellow marrow**, which is mostly fat cells. Yellow marrow can convert to blood cell-making red marrow in times of severe blood loss.

The bone mass in a healthy, young adult is usually fairly stable. In a continual balancing act, osteoblasts build up the matrix—and osteoclasts secrete enzymes that break it down. The ongoing mineral deposits and removals help to maintain the blood concentrations of calcium and phosphorus, fine tune bone strength, and support essential metabolic activities.

The level of calcium in the blood is one of the most tightly controlled aspects of metabolism. Calcium ions play a role in nerve cell function, muscle contraction, and other important processes. Bones and teeth store just about all of the body's calcium. Negative feedback loops between the blood and endocrine glands control calcium release and uptake. If the calcium level rises, the thyroid gland will secrete calcitonin. This hormone inhibits the action of osteoclasts, thus slowing calcium release into the blood. When the calcium level falls, the parathyroid glands secrete parathyroid hormone. This stimulates osteoclasts to break down matrix and also decreases calcium losses in urine. We will return to the hormonal control of bone density in Chapter 26.

Until a person is about twenty-four years old, the osteoblasts produce matrix faster than the osteoclasts break it down, and bone mass increases. As we age, osteoblast activity falters and bone density declines. In the worst cases, *osteoporosis* is the result (Figure 21.5). A deficiency in calcium or vitamin D, limited physical activity, and parathyroid problems increase the risk. So does the diminishment of female sex hormones that accompanies menopause. Smoking, excessive intake of alcohol, and steroid use also hinder bone deposition.

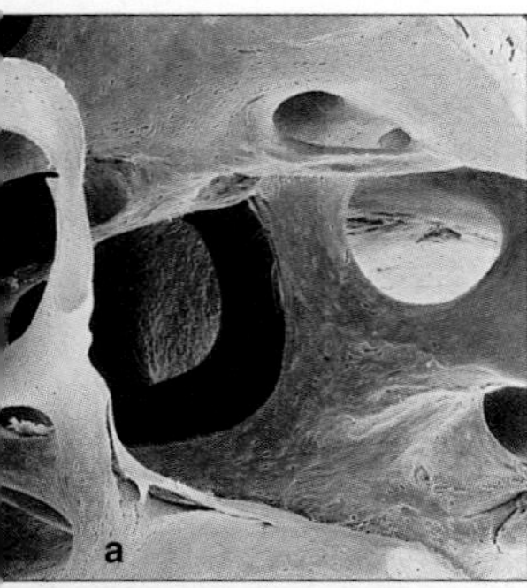

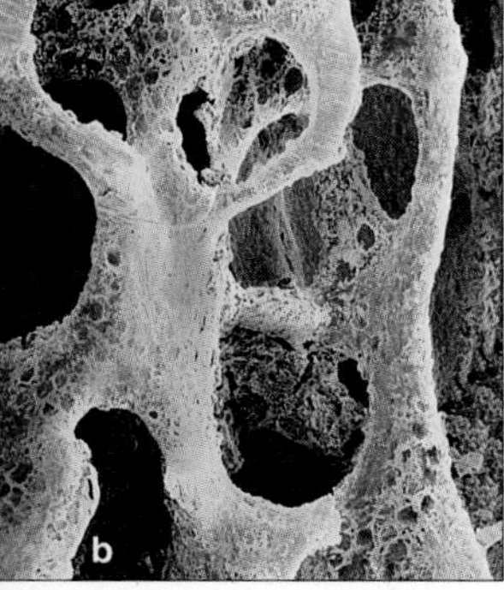

Figure 21.5 (**a**) Normal bone tissue. (**b**) Bone affected by osteoporosis.

WHERE BONES MEET—SKELETAL JOINTS

Connective tissue bridges **joints**, the areas of contact or near-contact between bones. *Cartilaginous* joints, such as those between vertebrae, hold bones together and cushion them, but allow only a tiny bit of movement. *Synovial* joints, such as the knee joint, allow bones to move freely. Stability of these joints is reinforced by **ligaments**, straps of dense connective tissue that hold bones together and prevent unwanted movement.

Knee joints are especially vulnerable to stress. They let you swing, bend, and rotate the bones below them. Overstretch or tear muscles or tendons and you suffer a *strain*. Tear ligaments and you have a *sprain*. Move the wrong way and you may dislocate bones. A severe blow to the knee can sever ligaments. For a successful recovery, they must be surgically repaired within ten days. Why? Phagocytic white blood cells patrol the joint's lubricating fluid and clean it up after everyday wear and tear. Unless the damaged ligaments are fixed immediately, these cells treat them as debris and turn them to mush.

Joint inflammation and degenerative disorders are collectively called arthritis. In *osteoarthritis*, cartilage at freely movable joints wears off. Joints in the fingers, knees, hips, and vertebral column are affected most. In *rheumatoid arthritis*, the membranes in joints become inflamed and thicken, cartilage degenerates, and bone deposits accumulate. This is an autoimmune response. It has a heritable basis, but bacterial or viral infection might trigger it. Rheumatoid arthritis can develop at any age, but symptoms usually start before age fifty.

Bones are collagen-rich, mineralized organs that function in movement, support, protection, storage of calcium and other minerals, and blood cell formation.

Ongoing calcium deposits and withdrawals are a balancing act that maintains bone mass and the concentration of calcium in blood.

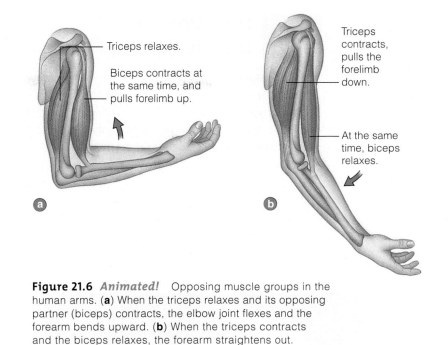

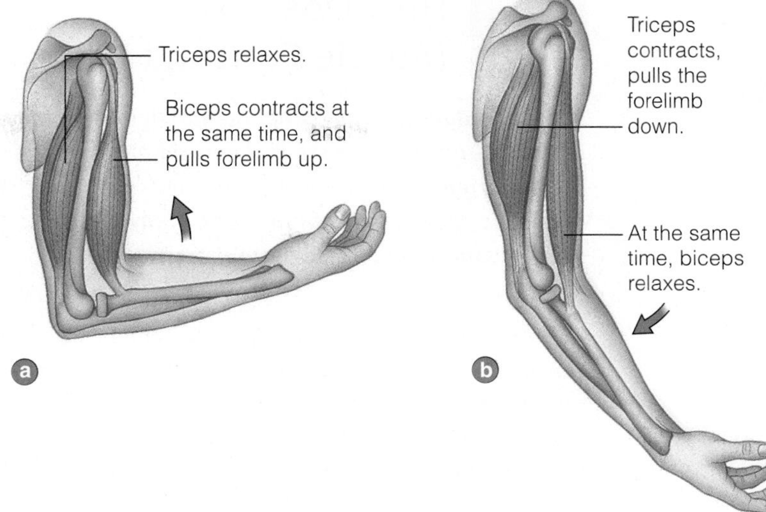

Triceps relaxes.
Biceps contracts at the same time, and pulls forelimb up.
Triceps contracts, pulls the forelimb down.
At the same time, biceps relaxes.
a
b

Figure 21.6 ***Animated!*** Opposing muscle groups in the human arms. (**a**) When the triceps relaxes and its opposing partner (biceps) contracts, the elbow joint flexes and the forearm bends upward. (**b**) When the triceps contracts and the biceps relaxes, the forearm straightens out.

21.2 How Do Bones and Muscles Interact?

Fibers in muscle contract (shorten) in response to suitable stimulation from the nervous system. Then they lengthen as a response to gravity and other loads. When you run, dance, breathe, squint, or scribble notes, many muscles are working to move your body or change the positions of some of its parts.

Skeletal muscles are the functional partners of bones. Each skeletal muscle is made up of muscle fibers, which were formed in the embryo as many undifferentiated, immature muscle cells fused together. These fibers are bundled up inside dense connective tissue that extends beyond them. The extension is a cordlike or straplike **tendon** that attaches the bundle to bone. Many of the attachment sites are like a car's gearshift. *They act like a lever system, in which a rigid rod is attached to a fixed point and moves about it.* Muscles connect to bones (rigid rods) near a joint (fixed point). When they contract, they transmit force to bones and make them move.

Skeletal muscles also interact with one another. Some work in pairs or groups in ways that promote a movement. Others work in opposition—the action of one opposes or reverses another muscle's action. Look at Figure 21.6. Extend your right arm forward, then place your left hand over the biceps in your upper right arm and slowly bend the elbow. Feel the biceps contract? Even when a biceps contracts only a bit, it causes a large motion of the bone connected to it. This is the case for most leverlike arrangements.

Bear in mind, only *skeletal* muscle is the functional partner of bone. As mentioned earlier, *smooth* muscle is mainly a component of the stomach, bladder, and other internal organs (Section 20.2). *Cardiac* muscle is present only in the heart wall. We look more closely at smooth muscle and cardiac muscle in later chapters.

The human body has close to 700 skeletal muscles, some near the surface, others deep inside the body wall. The trunk has muscles of the thorax, backbone, abdominal wall, and pelvic cavity. Other muscles are attached to upper and lower limb bones. Collectively, skeletal muscles account for about 40 percent of body weight in a young man of average fitness.

Later chapters explain how various skeletal muscles interact with and assist other organ systems. For now, let's look at mechanisms that bring about contraction.

Cordlike or straplike tendons of dense connective tissue attach skeletal muscles to bones. Only skeletal muscles transmit contractile force to bones.

Many skeletal muscles work as a group to promote a movement; many others work to oppose or reverse the action of others.

LINK TO SECTION 3.7

21.3 How Does Skeletal Muscle Contract?

Bones of a dancer or any other human in motion move in some direction when skeletal muscles attached to them shorten. A muscle shortens when its cells shorten. And when each muscle cell shortens, many individual units of contraction inside it are shortening.

In both skeletal and cardiac muscles, the basic units of contraction are called **sarcomeres**. Organization of the sarcomere starts with muscle fibers that run the length of the muscle (Figure 21.7*a*). Each of the fibers contains many **myofibrils**, threadlike, cross-banded structures that also parallel the muscle's long axis. When muscle tissue is stained, these bands alternate in a dark and light pattern and give the muscle cells a striped, or striated, appearance (Figure 21.7*b*). Extending across each light band is a dense array of proteins—a Z band. Two Z bands at either end of the sarcomere anchor its parts (Figure 21.7*c*).

Each sarcomere contains thin and thick filaments. Two coiled strands of **actin**, a globular protein, make up each thin filament (Figure 21.7*d*). **Myosin**, a motor protein that has a club-shaped head, makes up thick filaments (Figure 21.7*e*).

Both the thick and thin filaments run parallel with the myofibril. Thick ones have their myosin heads just a few nanometers away from the thin ones. Anchored

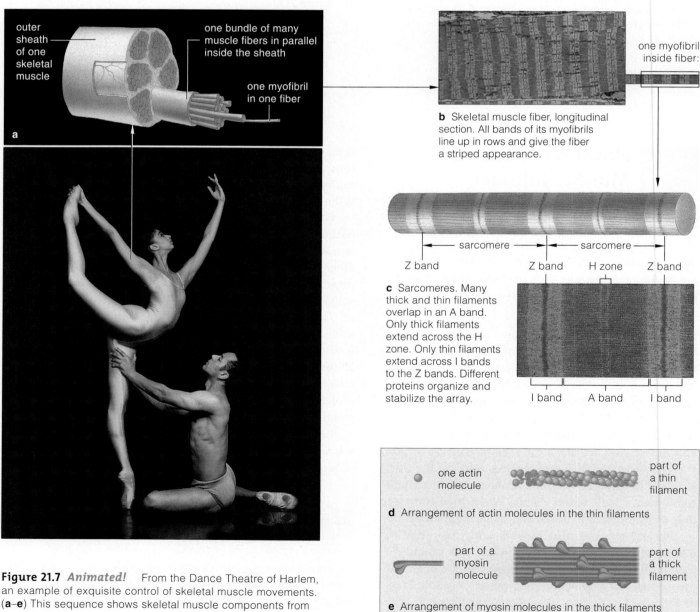

outer sheath of one skeletal muscle

one bundle of many muscle fibers in parallel inside the sheath

one myofibril in one fiber

a

one myofibril inside fiber:

b Skeletal muscle fiber, longitudinal section. All bands of its myofibrils line up in rows and give the fiber a striped appearance.

sarcomere | sarcomere

Z band | Z band | H zone | Z band

I band | A band | I band

c Sarcomeres. Many thick and thin filaments overlap in an A band. Only thick filaments extend across the H zone. Only thin filaments extend across I bands to the Z bands. Different proteins organize and stabilize the array.

one actin molecule

part of a thin filament

d Arrangement of actin molecules in the thin filaments

part of a myosin molecule

part of a thick filament

e Arrangement of myosin molecules in the thick filaments

Figure 21.7 ***Animated!*** From the Dance Theatre of Harlem, an example of exquisite control of skeletal muscle movements. (**a–e**) This sequence shows skeletal muscle components from a biceps down to molecules with contractile properties.

to each Z band is a set of actin filaments that extend toward the center of the sarcomere. Elastic filaments anchored to the Z band keep myosin positioned at the sarcomere's center. These filaments also help a relaxed muscle resist stretching.

Muscle bundles, muscle fibers, myofibrils, and the thick, thin, and elastic filaments are all in the same parallel orientation. What's the point of this? *It focuses the force of contraction on a bone in a particular direction.*

SLIDING-FILAMENT MODEL FOR CONTRACTION

How do sarcomeres shorten and contract a muscle? By the **sliding-filament model** for contraction, all of the myosin filaments stay in place. They use short, ATP-driven power strokes to slide the actin filaments over them, toward the sarcomere's center. As the amount of overlap between actin and myosin filaments increases, the sarcomere shortens (Figure 21.8).

Part of a myosin head has ATPase activity, so it can bind and split molecules of ATP. When ATP binds to a myosin head, it is broken into ADP and phosphate—this primes the myosin for contraction (Figure 21.8*b*).

All skeletal muscles contract in response to signals from motor neurons. These signals trigger the release of calcium ions from a specialized type of endoplasmic reticulum called the **sarcoplasmic reticulum**.

The increase in intracellular calcium exposes sites on actin filaments that myosin heads can bind to. The heads form cross-bridges to actin (Figure 21.8*c*). ADP and phosphate bound to a myosin head are released as the head tilts toward the sarcomere center, pulling the bound actin with it (Figure 21.8*d*).

Binding of a new ATP frees the myosin head from actin and it resumes its original position (Figure 21.8*e*). It then attaches to another binding site on actin, tilts in another stroke, and so on. Contraction of a sarcomere requires hundreds of myosin heads to perform a series of strokes all down the length of the actin filaments.

We will return to the subject of how motor neurons call for muscle contraction later, in Chapter 25.

GETTING ENERGY FOR CONTRACTION

All cells depend on ATP, but only in muscles does the demand skyrocket. When a muscle cell at rest is called on to contract, its phosphate donations from ATP must increase twenty to a hundred fold. The cell has only a small supply of ATP at the start of contractile activity. This is when the creatine phosphate you learned about in the opening essay comes into play. An enzyme can strip a phosphate group from creative phosphate and transfer it to ADP. The muscle cell has about five times as much creatine phosphate as ATP, so the reaction is

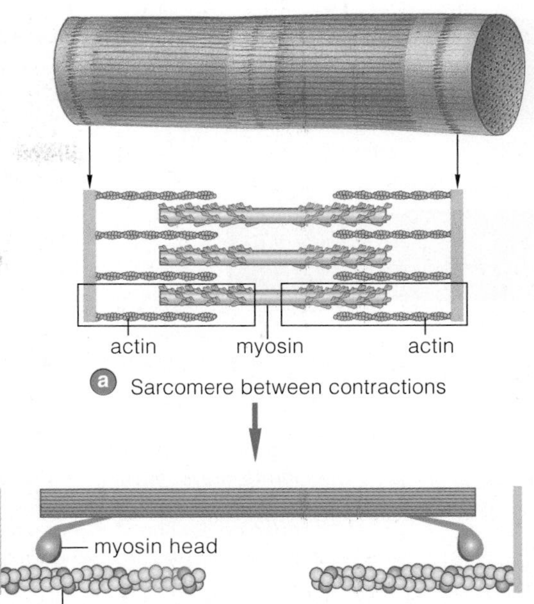

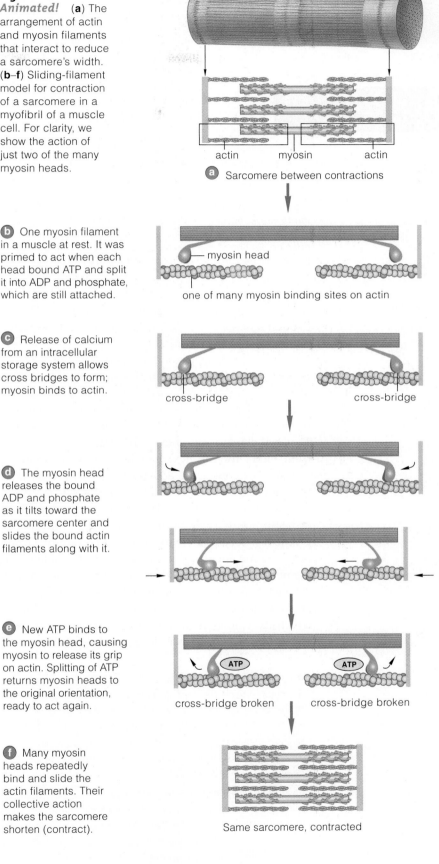

Figure 21.8 ***Animated!*** (**a**) The arrangement of actin and myosin filaments that interact to reduce a sarcomere's width. (**b–f**) Sliding-filament model for contraction of a sarcomere in a myofibril of a muscle cell. For clarity, we show the action of just two of the many myosin heads.

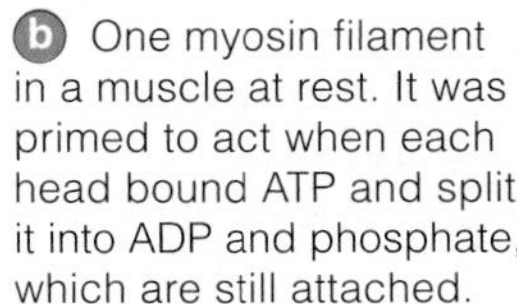

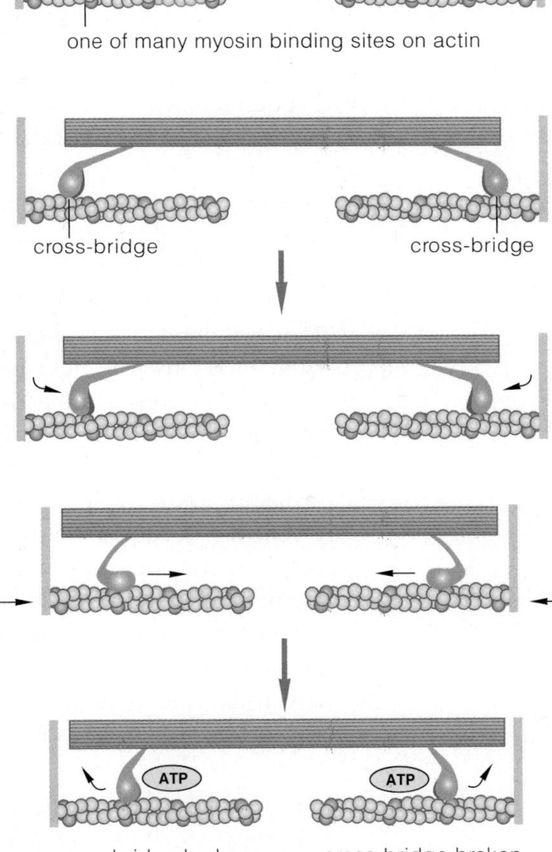

(**b**) One myosin filament in a muscle at rest. It was primed to act when each head bound ATP and split it into ADP and phosphate, which are still attached.

(**c**) Release of calcium from an intracellular storage system allows cross bridges to form; myosin binds to actin.

(**d**) The myosin head releases the bound ADP and phosphate as it tilts toward the sarcomere center and slides the bound actin filaments along with it.

(**e**) New ATP binds to the myosin head, causing myosin to release its grip on actin. Splitting of ATP returns myosin heads to the original orientation, ready to act again.

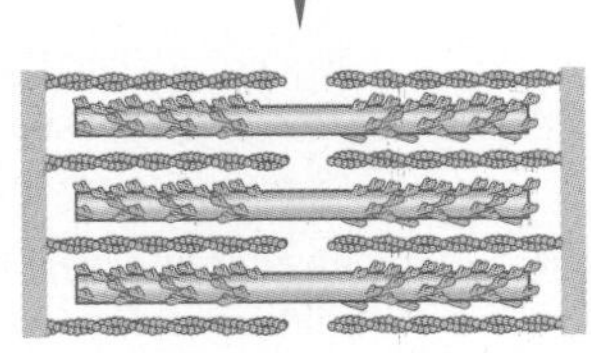

(**f**) Many myosin heads repeatedly bind and slide the actin filaments. Their collective action makes the sarcomere shorten (contract).

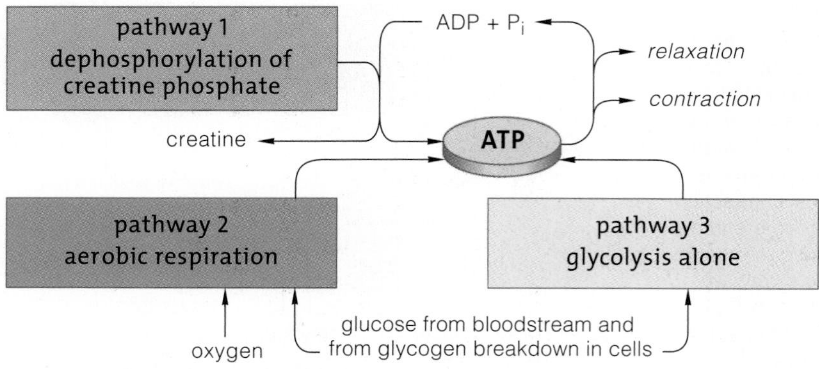

Figure 21.9 ***Animated!*** Metabolic pathways by which ATP forms inside muscles in response to the demands of physical exercise. Taking creatine supplements, as described in the opening section, can increase the amount of stored creatine phosphate available to power muscle contractions.

LINKS TO SECTIONS 16.1–16.5

good for the first few muscle contractions. That is enough to keep things going until slower ATP-forming pathways kick in (Figure 21.9).

During prolonged, moderate exercise, the oxygen-requiring reactions of aerobic respiration typically can supply most of the ATP used in contraction. Suppose a muscle cell exhausts its store of glycogen for glucose (the starting substrate) in the first five to ten minutes. That cell depends on glucose and fatty acid deliveries from blood for the next half hour or so of ongoing activity. For contractile activity longer than this, fatty acids are the main fuel source, as Section 6.5 describes.

Not all fuel is burned in aerobic respiration. Even in resting muscle, some of the pyruvate produced by glycolysis is converted to lactate. The rate of lactate fermentation increases with exercise. Remember that the net ATP yield from this anaerobic process is small. But lactate is a transportable fuel. It can be moved out of skeletal muscle fibers and picked up by the blood, which distributes it to other muscle fibers or tissues. There, lactate is used as fuel or stored as glycogen.

A skeletal muscle shortens through combined decreases in the length of its numerous sarcomeres. Sarcomeres are the basic units of skeletal and cardiac muscle contraction.

The parallel orientation of a skeletal muscle's components directs the force of contraction toward a bone that must be pulled in some direction.

By energy-driven interactions between myosin and actin filaments, the many sarcomeres of a muscle cell shorten and collectively account for its contraction.

During exercise, ATP availability in muscle cells affects whether contraction will proceed, and for how long.

21.4 Properties of Whole Muscles

Understanding how cross-bridges form in sarcomeres provides insight into how muscles can be strengthened. It also can explain causes of muscle fatigue, weakness, paralysis, and degeneration.

MUSCLE TENSION

A motor neuron has many endings that send signals to different fibers inside each muscle. A motor neuron and all muscle fibers it controls constitute one **motor unit**. Brief stimulation from one motor neuron causes all fibers in the motor unit that it controls to contract for a few milliseconds. This brief contraction and then relaxation is a **muscle twitch** (Figure 21.10).

When a new stimulus is applied before a response ends, the muscle twitches once again. When a motor unit is stimulated repeatedly during a short interval, the twitches all run together. The result is a sustained contraction called **tetanus**. The contraction produces three or four times the force of a single twitch. Figure 21.10*c* shows a recording of tetanic contraction.

Muscle tension is the mechanical force exerted by a muscle on an object. It is affected by the number of fibers recruited into action. Opposing this force is a load, either an object's weight or gravity's pull on the muscle. Only when muscle tension exceeds opposing forces does the stimulated muscle shorten. *Isotonically*

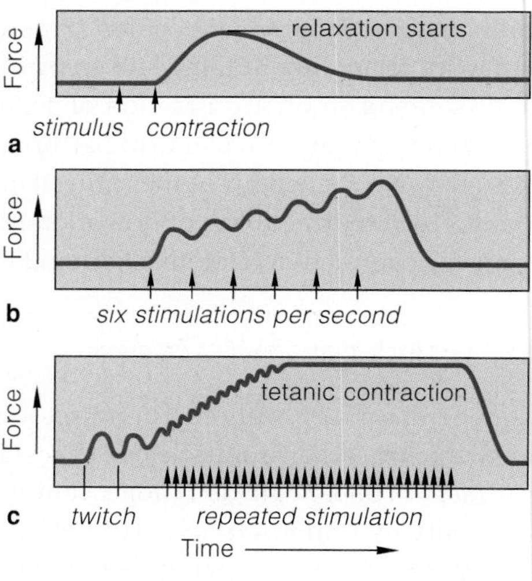

Figure 21.10 ***Animated!*** Recordings of twitches in muscles that were artificially stimulated at different frequencies. (**a**) A single twitch. (**b**) Summation of twitches after six stimulations per second. (**c**) Tetanic contraction occurs as a response to rapidly repeated stimulation.

contracting muscles shorten and move a load (Figure 21.11*a*). *Isometrically* contracting muscles do develop tension but cannot shorten, as when you attempt to lift something that is too heavy (Figure 21.11*b*).

A BAD CASE OF TETANIC CONTRACTION

An anaerobic, endospore-forming bacterium known as *Clostridium botulinum* can cause one dangerous type of food poisoning. Improperly canned food may include *C. botulinum* endospores. After spores germinate, the bacteria release an odorless, tasteless toxin known as botulinum toxin. If a person ingests toxin-tainted food, this poison enters motor neurons and stops them from releasing acetylcholine (ACh). This molecule normally signals skeletal muscle to contract. Without ACh, no contraction can occur, and muscles become flaccid and paralyzed. These symptoms of *botulism* usually appear within 36 hours after tainted food was eaten. Affected people must be treated with an antitoxin to keep the skeletal muscles that act in breathing from failing.

The related bacterium, *C. tetani*, lives in the gut of horses, cattle, and other grazing animals, and also in many people. Its endospores often abound in soils, especially manure-rich ones. They can withstand heat, boiling, and disinfectants. If endospores contaminate a deep puncture wound or a cut, they may germinate in the tissues. Although the bacteria cannot spread from a wound, they secrete a toxin that gets into the blood, and then into peripheral nerves. It travels along these nerves to the spinal cord and brain.

In the spinal cord, the toxin acts on nerve cells that help control motor neurons. It blocks release of two signaling molecules, GABA and glycine, so the motor neurons are freed up from normal inhibitory control. That is when symptoms of the disease *tetanus* begin.

After four to ten days, the overstimulated muscles stiffen and go into spasm—they contract continually. Prolonged, spastic paralysis follows. Fists and the jaw clench. (Lockjaw is another name for this disease.) The body becomes rigid, and the back freezes into an arch (Figure 21.12). If muscles involved in respiration are affected, death nearly always follows.

Today, vaccines have all but eradicated tetanus in the United States. Worldwide, the annual death toll is more than 200,000. Most deaths are in newborns who became infected during an unsanitary delivery.

WHAT IS MUSCLE FATIGUE?

When ongoing, strong stimulation keeps a muscle in a state of tetanic contraction, muscle fatigue follows. **Muscle fatigue** is a decrease in a muscle's capacity to generate force, a decline in tension.

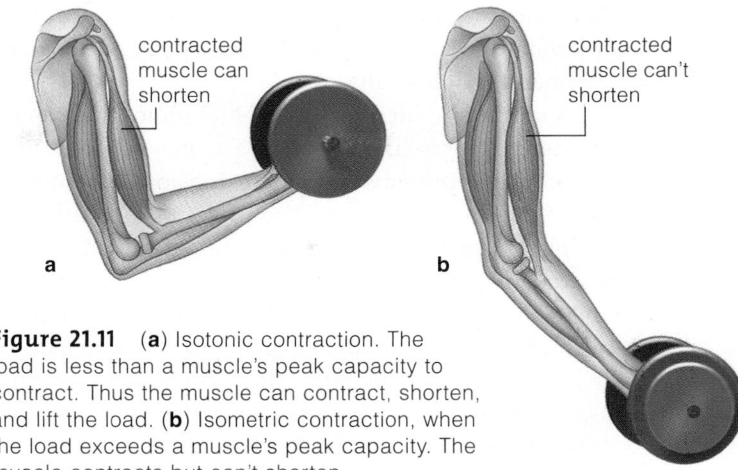

Figure 21.11 (**a**) Isotonic contraction. The load is less than a muscle's peak capacity to contract. Thus the muscle can contract, shorten, and lift the load. (**b**) Isometric contraction, when the load exceeds a muscle's peak capacity. The muscle contracts but can't shorten.

Figure 21.12 Painting of a young victim of a contaminated battle wound dying of tetanus in a military hospital.

After a few minutes of rest, a fatigued muscle will contract again in response to stimulation. The extent of recovery depends largely upon how long and how often the muscle was stimulated previously. Muscles trained by a pattern of brief, intense exercise fatigue and recover fast. This happens during weight lifting. Muscles used in prolonged, moderate exercise fatigue slowly and take longer to recover, often up to a day. Glycogen depletion is one factor that causes this long-term muscle fatigue.

A *muscle cramp* is a sudden, involuntary, painful contraction that resists relaxation. Any skeletal muscle can cramp, but muscles of the calves and thighs are affected most often. Motion aggravates cramping, but gentle stretching, massage, and heat may relieve it. To prevent cramps, stretch before strenuous activity and avoid overexertion and dehydration.

WHAT ARE MUSCULAR DYSTROPHIES?

Muscular dystrophies are a class of genetic disorders in which muscles progressively weaken and degenerate.

LINK TO SECTION 8.7

Duchenne muscular dystrophy usually strikes during childhood. Myotonic muscular dystrophy is the most common kind in adults.

A single mutant gene on the X chromosome causes Duchenne muscular dystrophy. The gene encodes one of a group of proteins found in extensions of a muscle fiber's plasma membrane. The protein group attaches the plasma membrane to actin filaments at Z bands of sarcomeres. The mutant form makes this attachment unstable, so muscles gradually deteriorate. Muscular dystrophy occurs mostly in males, at a frequency of about 1 in 3,500. Generally, affected individuals must use a wheelchair for mobility by their teens and live only into their twenties. There is no cure.

MUSCLES, EXERCISE, AND AGING

A muscle's properties depend on a person's age and how often, how long, and how intensively the muscle is made to contract. Regular exercise does not make more muscle fibers. Existing cells increase in size and metabolic activity, and they become slower to fatigue. Think of *aerobic exercise*—in which large muscles are caused to contract repeatedly during a long interval. Such exercise increases the number of mitochondria in both fast and slow muscle fibers, and increases the blood capillaries that service them. Such physiological changes will improve endurance. By contrast, *strength training* (intense, short-duration exercise such as the lifting of weights) stimulates fast-acting muscle fibers to produce more myofibrils and enzymes of glycolysis. Strong, bulging muscles develop, but these muscles do not have much endurance. They fatigue rapidly.

As people age, their muscles generally begin to shrink. The number of muscle fibers declines and the remaining fibers increase their diameter more slowly in response to exercise. Injuries to muscle take longer to heal. Even so, exercise can be helpful at any age. Strength training can slow the loss of muscle tissue. Aerobic exercise can improve circulation. Also, as one beneficial side effect, aerobic exercise by the middle-aged and elderly can lift major depression as well as many drugs can. Aerobic exercise may help improve memory and capacity to plan and organize complex tasks. In short, exercise is good for more than just muscles. It is good for the brain.

IMPACTS, ISSUES

Several large clinical trials are under way to test the hypothesis that creatine supplements counter some of the deteriorating muscle function in people affected by muscular dystrophy. In one trial, children with Duchenne muscular dystrophy are being given creatine or a placebo for several months. They will be periodically tested for muscle strength. The results will be compared to determine whether creatine increased muscle strength in the experimental groups.

Collectively, cross-bridges that form in sarcomeres exert a mechanical force, called muscle tension, in response to an external force acting on the muscle.

Properties of muscles vary with age and activity. Exercise can be beneficial at any age.

Summary

Section 21.1 Three categories of skeletal systems are common in animals—hydrostatic skeletons, exoskeletons, and endoskeletons. Different kinds apply contractile force against body fluids or structural elements, such as bones.

Humans, like other land vertebrates, have a bony endoskeleton with skull bones, a pelvic girdle, a pectoral girdle, and paired limbs. Skeletal modifications in early ancestors of humans allowed upright walking.

Bones are collagen-rich, mineralized organs. They function in mineral storage, movement, and protection and support of soft organs.

Blood cells form in bones that contain red marrow. Ongoing mineral deposits and removals help maintain blood levels of calcium and phosphorus, and also adjust bone strength. The osteoblasts deposit minerals in bone, osteoclasts make enzymes that break bone down, and hormones guide the release and uptake.

Ligaments or cartilage attach bone to bone at joints. Ligaments let the bones move freely; cartilage does not.

BiologyNow
Use the interaction on BiologyNow to investigate the bones of the human skeleton.

Section 21.2 Tendons attach skeletal muscles to bones. When skeletal muscles contract, they transmit force that makes the bones move. Many muscles work as opposing pairs. The internal organization of skeletal muscle and cardiac muscle promotes strong, directional contraction. Smooth muscle contractions move soft internal organs.

BiologyNow
Watch opposing muscle groups interact with the animation on BiologyNow.

Section 21.3 In response to signals from the nervous system, muscles contract, or shorten. In each skeletal muscle fiber (and cardiac muscle cell), many threadlike structures called myofibrils are divided into sarcomeres, the basic unit of contraction. A sarcomere has parallel arrays of actin and myosin filaments. ATP-driven sliding interactions between actin and myosin filaments shorten the sarcomeres of the fibers in a muscle. This collective shortening results in the muscle's contraction.

Dephosphorylation of creatine phosphate, aerobic respiration, and glycolysis yield ATP for contraction.

BiologyNow
Explore the structure of a muscle, see what happens when it contracts, and learn what fuels that contraction with the animation on BiologyNow.

Section 21.4 Muscle tension is a mechanical force caused by cross-bridge formation. A muscle shortens only when muscle tension exceeds an opposing load.

A motor neuron and all muscle fibers that form junctions with its endings are a motor unit. All fibers of a motor unit contract at the same time. Repeated stimulation of a motor unit results in tetanus, which is a sustained contraction.

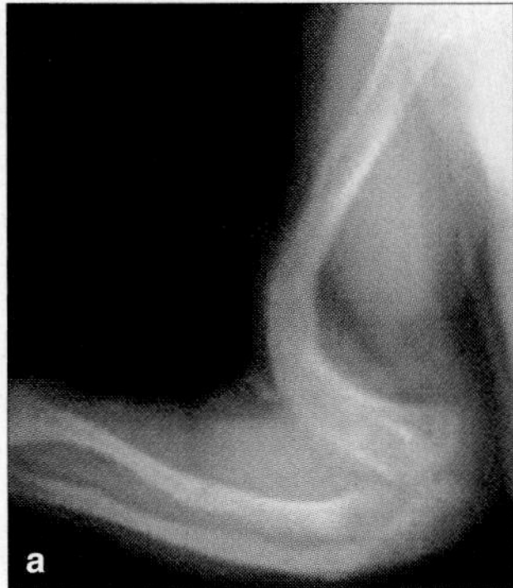

Figure 21.13 (**a**) X-ray of an arm bone severely deformed by osteogenesis imperfecta (OI), a genetic bone disorder. (**b**) Tiffany is affected by OI. She was born with multiple fractures in her arms and legs. By age six, she had undergone surgery to correct more than 200 bone fractures and to implant steel rods in her legs. Every three months she receives infusions of an experimental drug that doctors hope will strengthen her bones.

Self-Quiz

Answers in Appendix I

1. Only vertebrates have ______.
   a. an endoskeleton   c. a hydrostatic skeleton
   b. an ectoskeleton   d. a skeleton

2. Bone functions in ______.
   a. mineral storage   c. blood cell formation
   b. movement   d. all of the above

3. The ______ is the basic unit of contraction.
   a. osteoblast   c. muscle fiber
   b. sarcomere   d. myosin filament

4. Bones move when ______ muscles contract.
   a. cardiac   c. smooth
   b. skeletal   d. all of the above

5. A part of the ______ molecule can bind and break down ATP to release energy.
   a. actin   c. both
   b. myosin   d. neither

6. A sarcomere shortens when ______.
   a. thick filaments shorten
   b. thin filaments shorten
   c. both thick and thin filaments shorten
   d. none of the above

7. Release of ______ from intracellular storage allows actin and myosin to interact.
   a. sodium ions   c. calcium ions
   b. potassium ions   d. calcitonin

8. ATP for muscle contraction can be formed by______.
   a. aerobic respiration
   b. glycolysis
   c. creatine phosphate breakdown
   d. all of the above

9. Match the words with their defining feature.

____ osteoblast	a. actin's partner
____ muscle twitch	b. all in the hands
____ muscle tension	c. blood cell production
____ joint	d. decline in tension
____ myosin	e. bone forming cell
____ marrow	f. motor unit response
____ metacarpals	g. force exerted by cross-bridges
____ myofibrils	h. area of contact between bones
____ muscle fatigue	i. muscle fiber's threadlike parts

Additional questions are available on **Biology Now™**

Critical Thinking

1. The genetic disorder *osteogenesis imperfecta* (OI) arises from a mutant form of collagen. As bones develop, this protein forms a scaffold for deposition of mineralized bone tissue. The scaffold becomes formed improperly in children with OI. They end up with fragile, brittle, easily broken bones (Figure 21.13).

OI is rare; most doctors never come across it. Parents are often accused of child abuse when they take an OI child with multiple bone breaks to an emergency room. Prepare a brief handout about OI that could be used to alert the staff in emergency rooms to the condition. You may wish to use the resources on the Osteogenesis Imperfecta Foundation web site at www.oif.org as a starting point.

2. Lydia is training for a marathon. While training, she plans to take creatine supplements because she's heard that they give athletes extra energy. She asks you whether you think this is a good idea. What is your response?

3. Compared to most people, Lydia and other long-distance runners have more cells in their skeletal muscles, and the cells have more mitochondria. By comparison, sprinters have skeletal muscle fibers that have more of the enzymes necessary for glycolysis, but fewer mitochondria. Think about the differences between these two forms of exercise. Explain why the two kinds of runners have such pronounced differences in their muscles.

4. Botulinum toxin can kill you if you eat it, but some people pay hundreds of dollars to have it injected into their face. These *Botox injections* prevent particular facial muscles from contracting and pulling on the skin in ways that cause wrinkling. Most commonly, Botox is injected into muscles between the eyebrows. In about 3 percent of patients the eyelids droop for weeks as a side effect. Can you explain why?

5. When the nervous system signals a motor unit to contract, all muscle fibers in that motor unit act together. The nervous system cannot call for contraction of just a portion of a motor unit's fibers. The number of muscle fibers controlled by a single motor neuron varies. Among motor units that control eye muscles, one motor neuron signals only 5 or so muscle fibers. In the biceps of the arm, there are about 700 fibers per motor unit. What would be the advantage of having only a few fibers per motor unit? What would be the advantage of having many?

22 CIRCULATION AND RESPIRATION

IMPACTS, ISSUES

Up In Smoke

Each day 3,000 or so teenagers join the ranks of habitual smokers in the United States. Most are not even fifteen years old. The first time they light up, they cough and choke on irritants in the smoke. They typically become dizzy and nauseated, and develop headaches.

Sound like fun? Hardly. So why do they ignore the signals of the threat to the body and work so hard to be a smoker? Mainly to fit in. To many adolescents, the misguided perception of social benefits overwhelms the seemingly remote threat to health.

Yet fulfillment of that threat starts right away. Cilia that line the airways to the lungs normally sweep out airborne pathogens and pollutants. Smoke from just a single cigarette immobilizes them for hours. Smoke also kills the white blood cells that patrol and defend the respiratory system's lining. In this way, gunk starts to clump up in the airways of young smokers. Pathogens become entrenched in the gunk, as biofilms. The smoker is now more susceptible to colds, asthma, and bronchitis.

Each smoke-filled inhalation delivers nicotine that quickly reaches the circulatory system and brain. This highly addictive stimulant constricts blood vessels, which makes the heart beat harder and raises blood pressure. Smoking also raises the blood level of "bad" cholesterol (LDL) and lowers that of "good" cholesterol (HDL). It makes blood stickier and more sluggish, and it promotes the formation of clots. Clogged arteries, heart attacks, and strokes are among the physiological costs of a social pressure that leads to addiction.

Which smokers have not heard that they are inviting painful, deadly lung cancers? Maybe they do not know that carcinogens in cigarette smoke can induce cancers in organs all through their body. For instance, we now know that females who start smoking when they are teenagers are about 70 percent more likely to develop breast cancer than those who never took up the habit.

The damaging effects of cigarette smoke are not confined to smokers. Their families, coworkers, and friends often get an unfiltered dose of carcinogens in tobacco smoke. Urine samples from nonsmokers who live with smokers show that their body contains high levels of carcinogens. Each year in the United States, secondhand smoke causes an estimated 3,000 deaths from lung cancer. Children exposed to secondhand smoke are more prone to develop respiratory ailments such as asthma, as well as middle ear infections.

This chapter focuses on respiration and circulation, and how they contribute to homeostasis in the internal environment. If you or someone you know has joined the culture of smoking, you might use the chapter as a preview of what it does to internal operating conditions. For a more graphic preview, find out what goes on every day with smokers in emergency rooms and intensive care units. No glamour there. It is not cool, and it is not pretty.

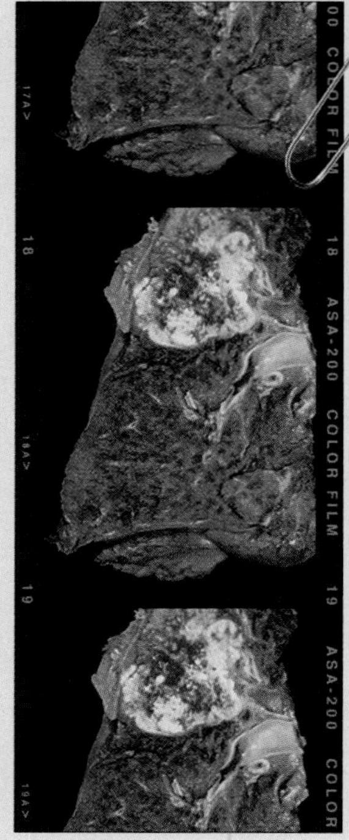

How Would You Vote? *Tobacco is a worldwide threat to health and a profitable product for American companies. As tobacco use in the United States declines, should this country support international efforts to reduce tobacco use around the globe? See BiologyNow for details, then vote online.*

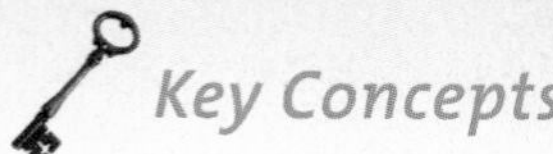

Key Concepts

BLOOD CIRCULATION

In large animals, diffusion alone is not enough to move oxygen, nutrients, and other solutes to cells and carry away wastes and secretions. Rapid transport of the substances takes place along vascular highways under the driving force of a muscular pump.

GAS EXCHANGE

All animals engage in aerobic respiration, which requires rapid oxygen uptake and carbon dioxide removal. In vertebrates, a respiratory system and circulatory system interact in the transport and exchange of these gases.

EFFECTS OF SMOKING

Inhaling tobacco smoke into the lungs can lead to cardiovascular disorders; bronchitis, emphysema, and other respiratory disorders; and cancers.

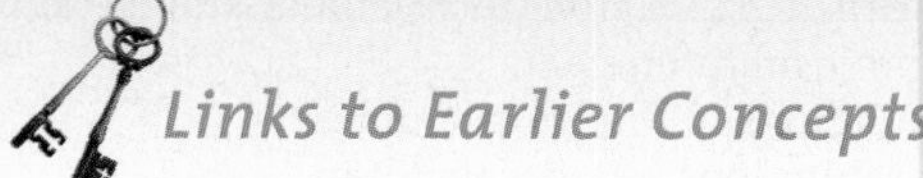

Links to Earlier Concepts

This chapter describes how the respiratory and circulatory systems maintain conditions in the internal environment (Sections 17.1,17.3). You may want to review the concepts of diffusion and osmosis (4.4, 4.6) and look back at information about cholesterol and hemoglobin (2.6, 2.9).

Keep in mind the trends in vertebrate evolution discussed in Section 16.9. Remember too the properties of blood, cardiac muscle, and other tissues, and of cell junctions (20.2) as well as interactions among organ systems (20.3).

Finally, an understanding of cancer (7.8) will help you put the effects of cigarette smoking in context.

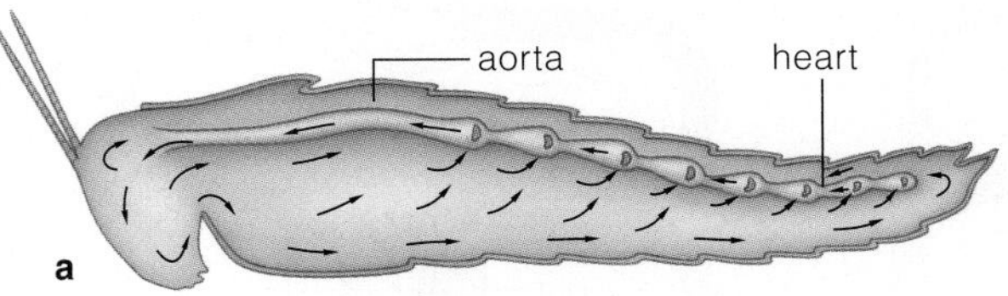

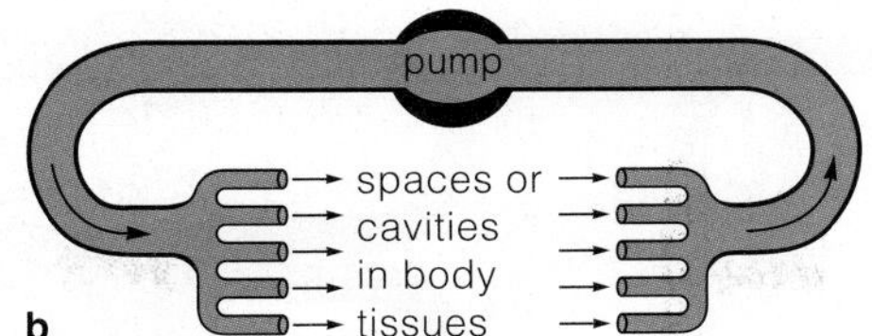

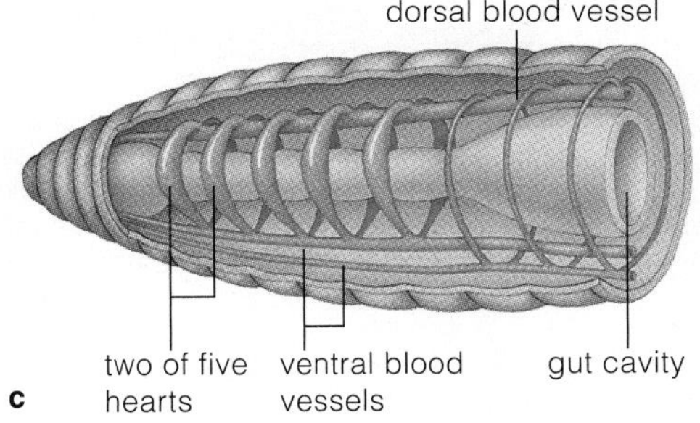

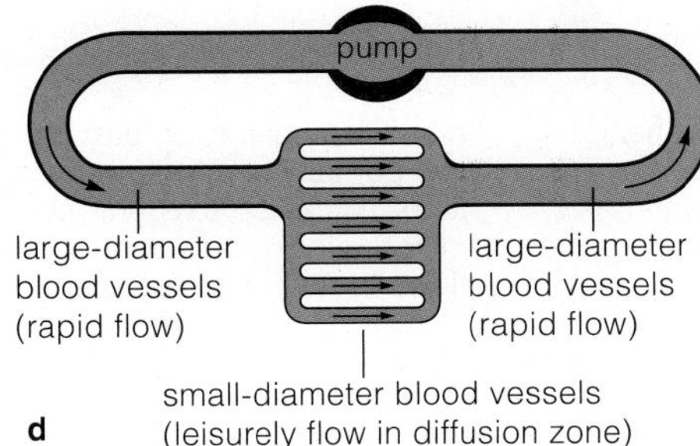

Figure 22.1 ***Animated!*** Comparison of open and closed circulatory systems. (**a**,**b**) In a grasshopper's open system, hearts pump blood through a vessel (an aorta). From there, blood moves into tissue spaces. It mingles with fluid bathing cells, then reenters hearts through openings in its wall. (**c**,**d**) The closed system of an earthworm keeps blood in pairs of muscular hearts near the head end and inside many blood vessels.

22.1 The Nature of Blood Circulation

Imagine that an earthquake has closed off highways around your neighborhood. Grocery trucks cannot enter and garbage trucks cannot leave, so food supplies dwindle and wastes pile up. Cells would face a similar predicament if your circulatory system was disrupted.

Your body's highways are part of a circulatory system, which swiftly moves substances to and from cells. The system helps maintain a favorable neighborhood, so to speak, and this is vital work. The circulatory system works together with other organ systems to maintain the volume, solute concentration, and temperature of **interstitial fluid**, which fills the tissue spaces between cells. Exchanges between interstitial fluid and **blood**, a circulating connective tissue, keep conditions tolerable for enzymes and other molecules that carry out cell activities. Together, interstitial fluid and blood are the body's *internal* environment.

A circulatory system consists of vessels and one or more hearts. A **heart** is a muscular pump that drives blood through the blood vessels. Most of the mollusks and all arthropods have an *open* circulatory system. Blood is pumped out of blood vessels and into a body cavity. It directly contacts the tissues before reentering a heart or hearts (Figure 22.1*a*,*b*). All vertebrates and many invertebrates have a *closed* circulatory system (Figure 22.1*c*,*d*). Blood stays inside a heart and blood vessels. Gas exchange between the blood and tissues takes place across the walls of small blood vessels.

The rate of blood flow varies with the diameter of the vessels that contain it. The total volume of blood is continuously pumped out of the heart, through the vessels, and back to the heart. The rate of flow is faster through large-diameter vessels. It slows as the blood spreads out into many smaller vessels. The greatest slowdown is at the small-diameter **capillaries**. Here, the blood exchanges substances with interstitial fluid.

LINKS TO SECTIONS 16.8, 17.1, 17.3

THE CIRCULATORY AND RESPIRATORY SYSTEMS OF VERTEBRATES EVOLVED TOGETHER

All vertebrates have a closed circulatory system, but fishes, amphibians, birds, and mammals differ in the details of their pumps and plumbing. The differences evolved over hundreds of millions of years as some of the vertebrates moved onto land.

Gases are exchanged at moist surfaces, where they dissolve in a fluid, then diffuse across cell membranes. In early vertebrates, gills functioned as the respiratory surface. Later, internally moistened sacs called lungs evolved, and circulatory systems became modified to move blood quickly to and from these lungs.

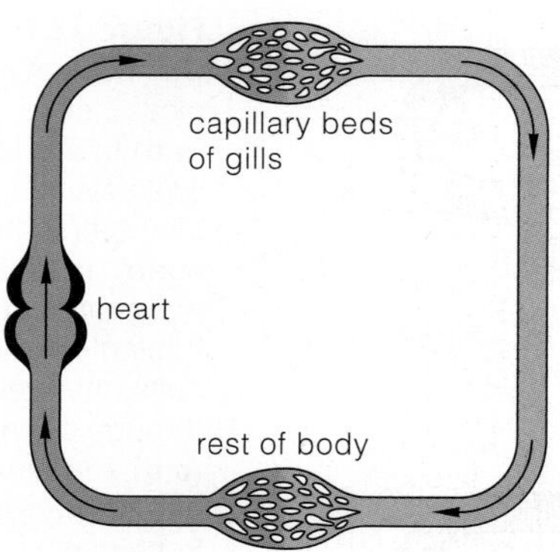

a In fishes, a two-chambered heart (atrium, ventricle) pumps blood in one circuit. Blood picks up oxygen in gills, delivers it to rest of body. Oxygen-poor blood flows back to heart.

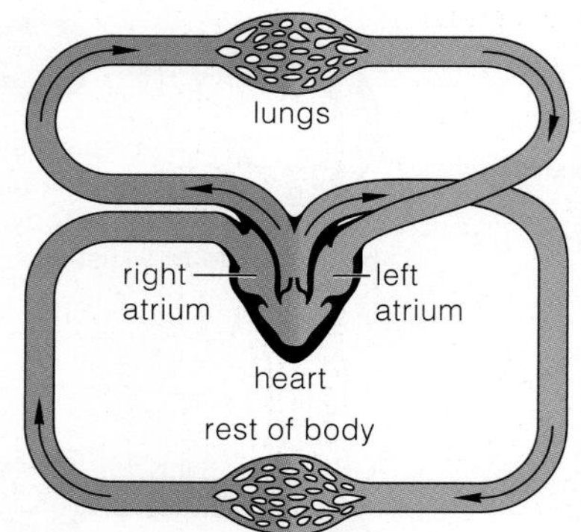

b In amphibians, a heart pumps blood through two partially separate circuits. Blood flows to lungs, picks up oxygen, returns to heart. But it mixes with oxygen-poor blood still in the heart, flows to rest of body, returns to heart.

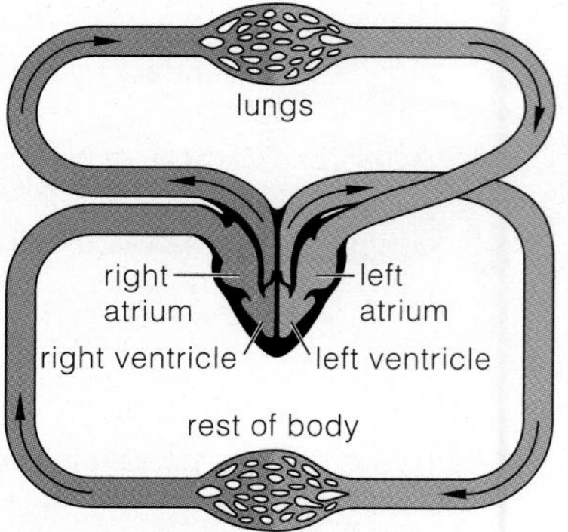

c In birds and mammals, the heart is fully partitioned into two halves. Blood circulates in two circuits: from the heart's right half to lungs and back, then from the heart's left half to oxygen-requiring tissues and back.

Figure 22.2 ***Animated!*** The closed circulatory systems of fishes, amphibians, birds, and mammals. *Red* indicates oxygen-rich blood. *Blue* indicates blood is oxygen-poor.

LINKS TO SECTIONS 16.9, 20.2

In fishes, blood still flows in a single circuit (Figure 22.2*a*). The two-chambered heart drives blood through two capillary beds. Blood is oxygenated at capillaries in gills. It then flows through capillary beds in the body before returning to the heart. As blood travels through gill capillaries, pressure drops and flow slows. For this reason, the movement of blood through the organs and back to the heart is relatively leisurely.

In amphibians, the heart became partitioned into three chambers. Two chambers called atria empty into a single ventricle. Oxygenated and oxygen-poor blood flow more quickly in these *two* circuits, but they mix together in the ventricle (Figure 22.2*b*).

Even more efficient hearts, with two fully divided, side-by-side pumps and circuits of vessels, evolved in the birds and mammals (Figure 22.2*c*). The heart's *right* half pumps oxygen-poor blood into the lungs. There, blood picks up oxygen and gives up carbon dioxide. The oxygenated blood then flows into the heart's left half. This route is known as the **pulmonary circuit**.

In the **systemic circuit**, the heart's *left* half pumps the oxygenated blood to all tissues and organs where oxygen is used and carbon dioxide forms. After giving up oxygen and taking on carbon dioxide, blood flows into the heart's right half, then on into the lungs.

A double circuit is a fast, efficient mode of blood delivery. It supports the high levels of activity typical of vertebrates whose ancestors evolved on land.

LINKS WITH THE LYMPHATIC SYSTEM

Pressure generated by the heart's pumping action not only moves blood through vessels, it also forces small amounts of water, ions, and certain soluble proteins into interstitial fluid. The excess fluid and reclaimable solutes move back into capillaries or they enter lymph vessels that return them to the blood. The **lymphatic system** consists of these vessels, lymph nodes, and other structures, as Chapter 23 describes.

Circulatory systems move substances to and from living cells in a multicelled body. A closed system confines blood inside one or more pumps (hearts) and blood vessels.

In vertebrates, blood flows fastest in large-diameter vessels between the heart and capillary beds. In capillaries, flow slows enough for efficient exchanges with interstitial fluid.

Vertebrate closed systems coevolved with the respiratory and lymphatic systems. In fishes, blood flows in one circuit. In birds and mammals, it flows in two, through a heart divided into two side-by-side pumps.

The double circuit keeps oxygen-rich blood separate, and supports the high levels of activity of most land vertebrates.

22.2 Characteristics of Human Blood

Blood, a connective tissue, transports oxygen, nutrients, and other solutes to cells. It carries away metabolic wastes and secretions such as hormones. Blood helps stabilize internal pH and transports cells that defend and repair tissues. In birds and mammals, it helps to maintain the core body temperature by moving excess heat to the skin, where it can be dissipated.

How much blood is in your body? That depends on body size and the concentrations of water and solutes. If you are average sized, you hold four or five quarts. Blood is a viscous fluid, thicker than water and slower flowing. Its components are plasma, red blood cells, white blood cells, and platelets. All blood cells and the platelets arise from stem cells in red bone marrow. As described in Chapter 20, stem cells stay unspecialized and retain a capacity for mitotic cell division. Some of their daughter cells will remain stem cells, but others go on to differentiate into specialized types.

Plasma is mainly water and constitutes about 50 to 60 percent of the total blood volume. It is the transport medium for blood cells and platelets. Vital nutrients, wastes, signaling molecules, and hundreds of plasma proteins are dissolved in it. Some of these proteins transport lipids and fat-soluble vitamins. Others help blood to clot or act against pathogens. Sugars, amino acids, vitamins, and mineral ions are dissolved in the plasma. So are oxygen, carbon dioxide, and nitrogen.

Disk-shaped erythrocytes, or **red blood cells**, are the most numerous cells in blood (Figure 22.3). Their red color comes from hemoglobin, an iron-containing pigment that can reversibly bind oxygen. Red blood cells carry oxygen from lungs to aerobically respiring tissues and help to move carbon dioxide away from these tissues. Oxygenated blood is bright red. Oxygen-poor blood is darker red but it appears blue through blood vessel walls near the body surface.

As red blood cells mature, they lose their nucleus and other organelles. The mature cell contains mostly hemoglobin. There are also a few enzymes. Such cells circulate for about four months. Ongoing replacements keep their number fairly stable. About 3 million new red blood cells enter your bloodstream every second.

A **cell count** is a measure of the quantity of cells of a specified type in 1 cubic millimeter of blood. For example, the average man has about 5.4 million red blood cells per microliter. A woman has about 4.8 million.

Leukocytes, or **white blood cells**, have a variety of roles in housekeeping and defense. Some types patrol tissues, where they target or engulf damaged or dead cells and anything chemically recognized as "nonself." Others are massed in organs of the lymphatic systems, such as the lymph nodes and spleen. They defend the body against viruses, bacteria, and other pathogens.

White blood cells differ in size, nucleus shape, and other characteristics. A few examples will give you a sense of what different types do. The neutrophils and basophils engulf invaders and damaged cells and take part in the first line responses to tissue damage. The macrophages and dendritic cells summon up immune responses to specific threats. Lymphocytes named B and T cells carry out immune responses. The natural killer cells induce self-destruction of cancer cells and cells infected by viruses.

Some stem cells give rise to megakaryocytes. These "giant" cells shed cytoplasmic fragments wrapped in plasma membrane. The fragments are platelets. Each platelet lasts only five to nine days, but hundreds of thousands circulate in the blood. If an injury occurs, platelets release substances that initiate blood clotting.

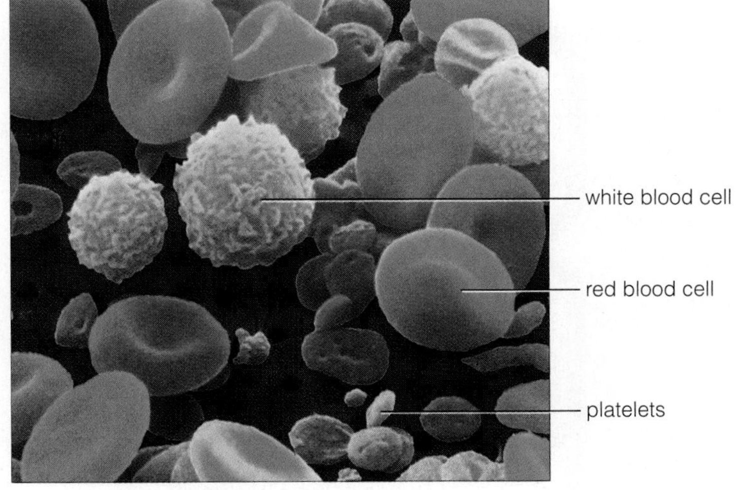

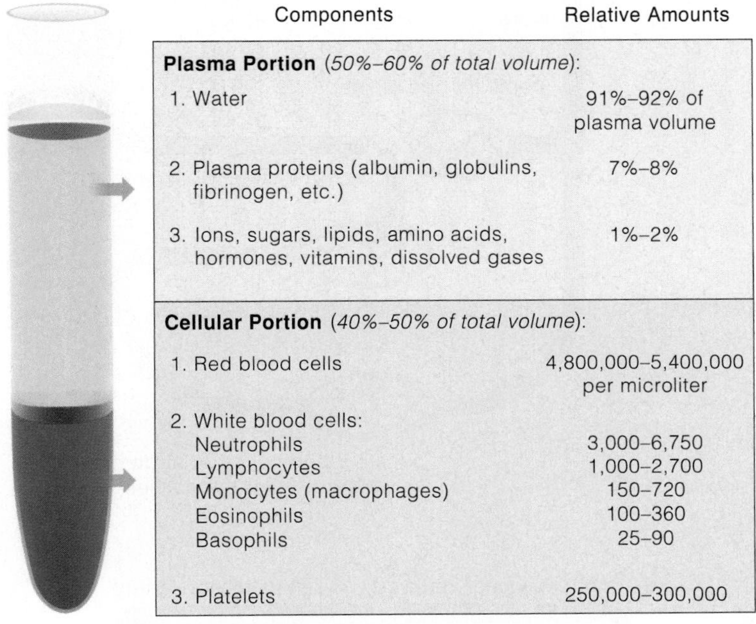

Components	Relative Amounts
Plasma Portion (*50%–60% of total volume*):	
1. Water	91%–92% of plasma volume
2. Plasma proteins (albumin, globulins, fibrinogen, etc.)	7%–8%
3. Ions, sugars, lipids, amino acids, hormones, vitamins, dissolved gases	1%–2%
Cellular Portion (*40%–50% of total volume*):	
1. Red blood cells	4,800,000–5,400,000 per microliter
2. White blood cells:	
Neutrophils	3,000–6,750
Lymphocytes	1,000–2,700
Monocytes (macrophages)	150–720
Eosinophils	100–360
Basophils	25–90
3. Platelets	250,000–300,000

Figure 22.3 Typical components of human blood. The scanning electron micrograph above shows a few of the cellular components.

The sketch of a test tube shows what happens when you keep a blood sample from clotting. The sample separates into the straw-colored plasma, which floats on a reddish-colored cellular portion.

Vertebrate blood has key roles in transporting substances, defending against pathogens, and maintaining the volume, composition, and temperature of the internal environment.

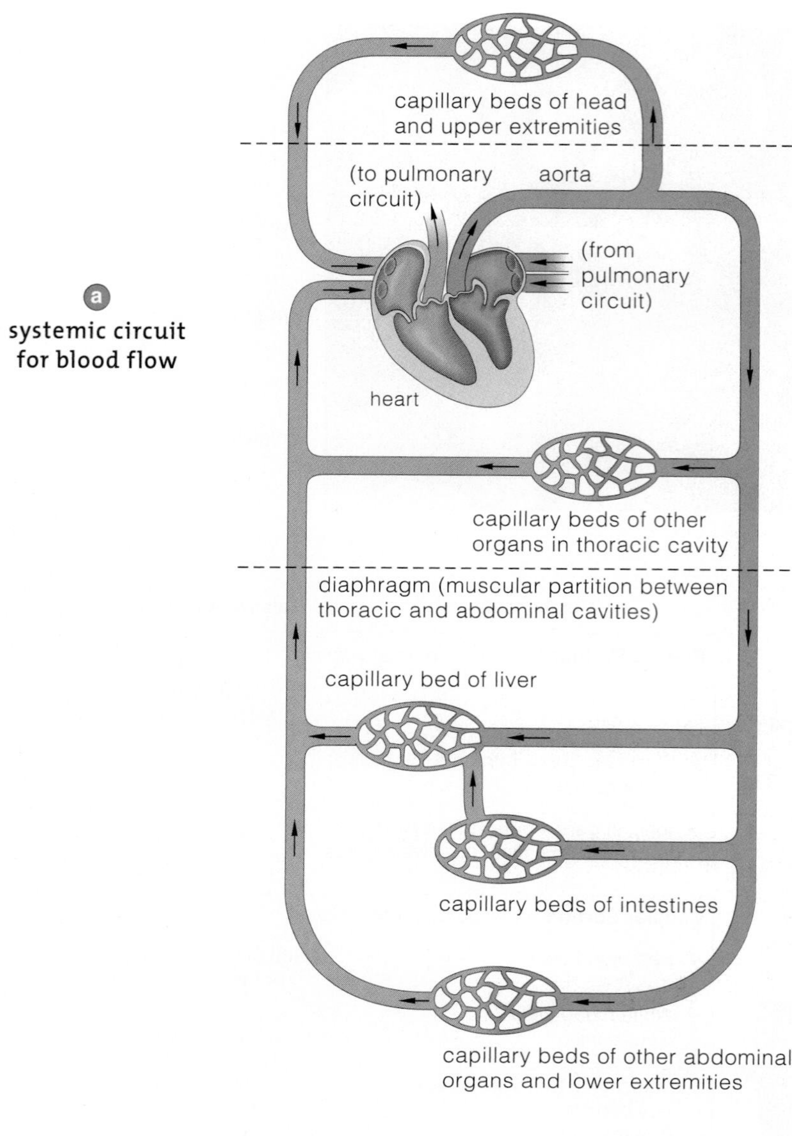

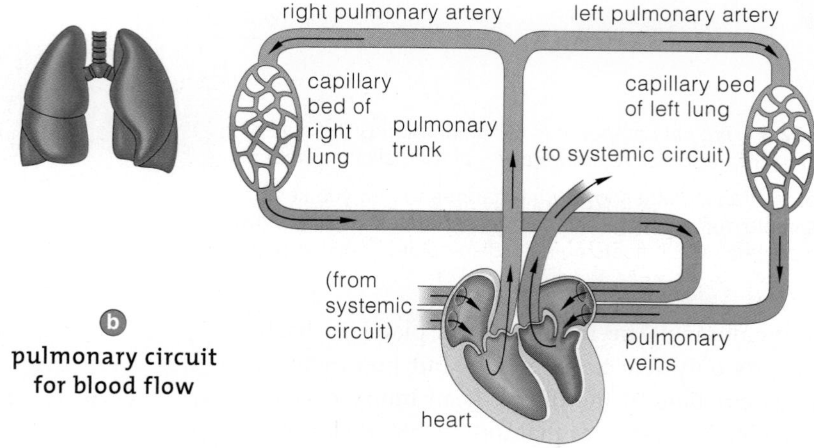

Figure 22.4 ***Animated!*** (**a**,**b**) Systemic and pulmonary circuits for blood flow through the human cardiovascular system. Blood vessels carrying oxygenated blood are color-coded *red*. Those carrying oxygen-poor blood are colored *blue*.

22.3 Human Cardiovascular System

"Cardiovascular" comes from the Greek kardia *(heart) and Latin* vasculum *(vessel). In a human cardiovascular system, blood flows from a muscular heart into large arteries, then to small, muscular arterioles, which branch into smaller capillaries. It continues into small venules, then large-diameter veins return it to the heart.*

ONE BIG CIRCUIT AND A SPECIAL CIRCUIT TO THE LUNGS

A partition divides a human heart into a double pump that drives blood through two cardiovascular circuits (Figure 22.4). Each circuit includes arteries, arterioles, capillaries, venules, and veins. Figure 22.5 shows the largest of these blood vessels.

The shorter loop, the pulmonary circuit, oxygenates blood. It leads from the heart's right half to capillary beds in both lungs, then returns to the heart's left half. The systemic circuit is a longer loop. The heart's left half pumps oxygenated blood into the main artery, the **aorta**. That blood gives up oxygen in all tissues, then oxygen-poor blood returns to the heart's right half.

Most blood making a circuit through the systemic circuit flows through only one capillary bed. But some flows through two. Blood picks up glucose and other absorbed substances as it flows through capillaries in the intestine, then gives some of them up as it flows through a second capillary bed in the liver. This organ stores nutrients and neutralizes toxins (Chapter 4).

The distribution of blood flow among body regions can be adjusted to meet changing needs. For example, the volume of blood that flows through the muscles in your arms varies, depending on your activities. Do a few push-ups and flow to arm muscles increases. Flow adjustments also help regulate the body's temperature. When you get hot, blood flow to capillaries in the skin increases and metabolic heat radiates away from the body's surface. When you get too cold, the blood flow to capillaries near the skin decreases. Only the brain is exempt from such flow adjustments; its high oxygen needs can only be met by a constant blood supply.

THE HEART IS A LONELY PUMPER

A human heart is a durable pump that beats 2.5 billion times in a seventy-year life span. A double-layered sac of tough connective tissue (pericardium) protects and anchors the heart. Fluid between the layers of the sac is lubrication that allows perpetual wringing motions. The inner layer of the sac is the heart wall (Figure

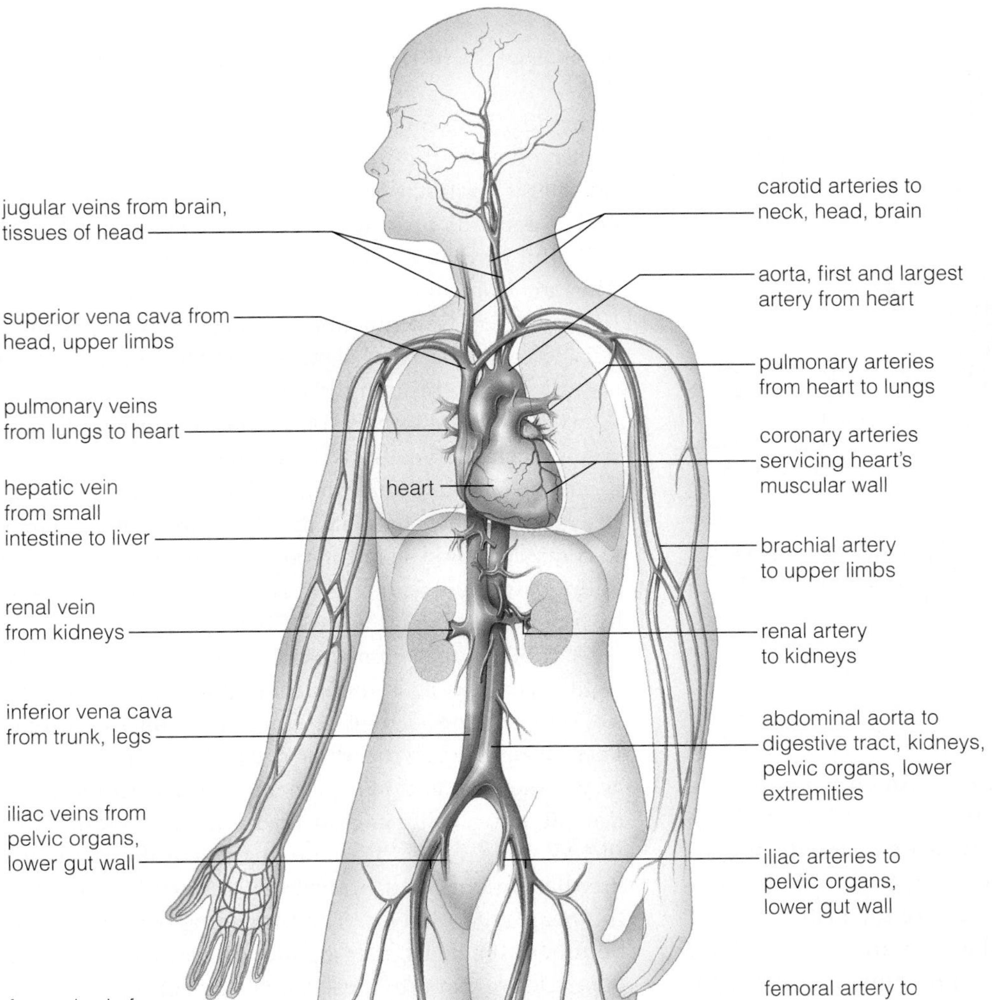

Figure 22.5 ***Animated!*** The major blood vessels of the human cardiovascular system. Not all are shown. For example, each arm has a set of the arteries and veins listed, and so does each leg. Humans, recall, have bilateral ancestry.

LINK TO SECTION 20.2

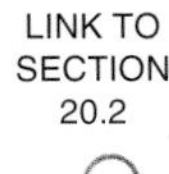

IMPACTS, ISSUES

Nicotine affects how the heart beats. It increases heart rate and interferes with the nervous system's ability to adjust the rate in response to exercise. The more someone smokes, the greater the impact. Compromised neural control of the heart rate increases the risk of a heart attack.

22.6). It is mostly myocardium—cardiac muscle cells attached to elastin and collagen fibers. These densely crisscrossed fibers serve as a "skeleton" against which contractile force can be applied. Endothelium, a type of epithelium, lines the heart's inner wall, as well as blood vessels. Coronary arteries that branch off from the aorta deliver oxygen to capillaries that service the highly demanding, always active cardiac muscle.

Each side of the heart has two chambers: an atrium (plural, atria) that receives blood, and a ventricle that expels it. One-way heart valves prevent backflow from one chamber to another and help keep blood moving forward. Fluid pressure alternately forces these valves open and shut. An atrioventricular (AV) valve controls blood flow out of each atrium and into a ventricle. A semilunar valve controls flow from each ventricle into an artery carrying blood away from the heart.

The heart's four chambers go through a sequence of contraction and relaxation called a **cardiac cycle**. First, the relaxed atria fill. Fluid pressure forces AV

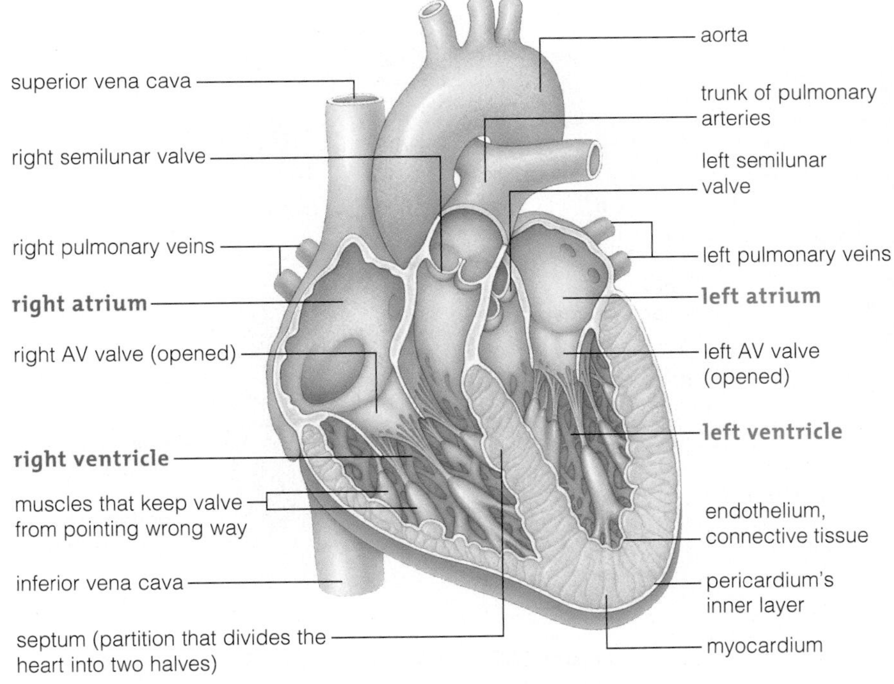

Figure 22.6 ***Animated!*** Cutaway view of a human heart.

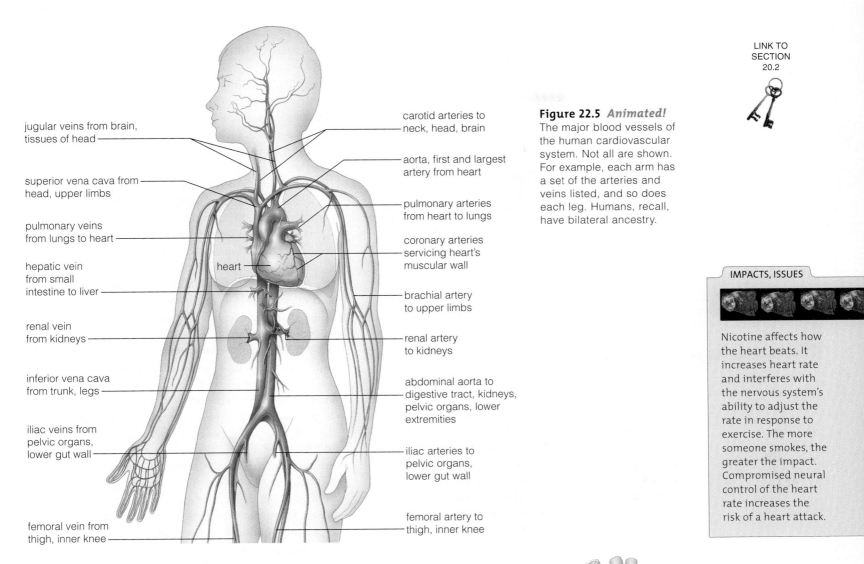

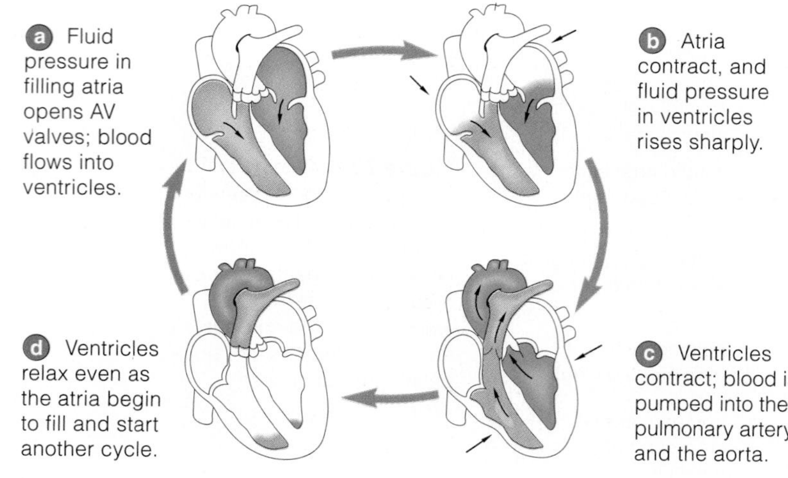

a Fluid pressure in filling atria opens AV valves; blood flows into ventricles.
b Atria contract, and fluid pressure in ventricles rises sharply.
c Ventricles contract; blood is pumped into the pulmonary artery and the aorta.
d Ventricles relax even as the atria begin to fill and start another cycle.

Figure 22.7 ***Animated!*** Cardiac cycle. Blood and heart movements make a "lub-dup" sound at the chest wall. At each "lub," AV valves close as ventricles contract. At each "dup," semilunar valves close as ventricles relax.

LINK TO SECTION 20.2

valves to open and blood flows through them into the ventricles, which fill as the atria contract (Figure 22.7). As ventricles contract, fluid pressure inside them rises above that in the arteries they empty into. Semilunar valves open, and blood flows out of the heart.

As the ventricles relax, the atria are already filling to begin a new cycle. Atrial contraction only helps fill the ventricles. *Contraction of the ventricles is the driving force for blood circulation.*

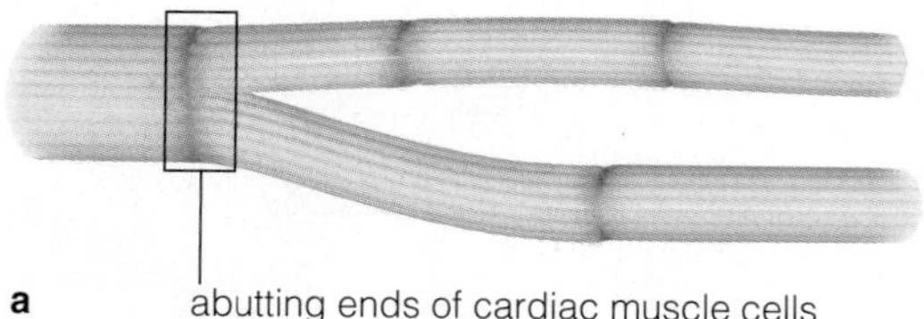

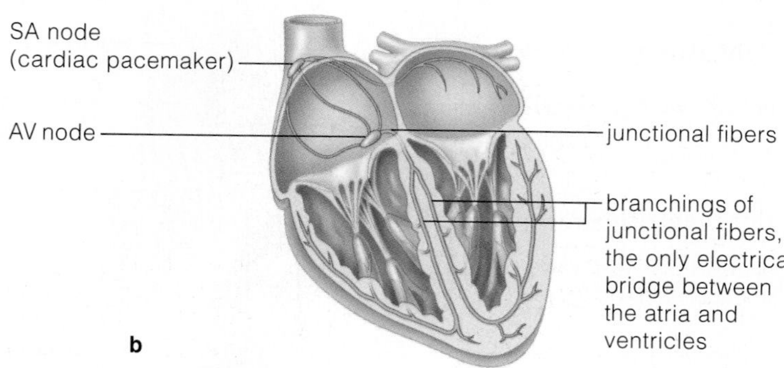

Figure 22.8 (**a**) Adhering junctions connect cardiac muscle cells end to end. Gap junctions located nearby enhance the rapid spread of excitation from cell to cell. (**b**) The cardiac conduction system.

HOW DOES CARDIAC MUSCLE CONTRACT?

Heart (cardiac) muscle, remember, has orderly arrays of sarcomeres, the units of contraction. Like skeletal muscle, it contracts by a sliding-filament mechanism, and it depends on abundant mitochondria to provide ATP. But cardiac muscle cells are structurally unique. They are branching, short, and connected at their ends (Figure 22.8*a*). Gap junctions connect the cytoplasm of adjacent cells and allow signals calling for contraction to spread swiftly through the tissue.

Where do the signals come from? In cardiac muscle, some specialized cells do not contract. They are part of a **cardiac conduction system**, which initiates and then relays signals that stimulate the other cardiac muscle cells to contract. This system consists of a sinoatrial (SA) node and an atrioventricular (AV) node, functionally linked by junctional fibers (Figure 22.8*b*).

The SA node, a clump of noncontracting cells in the right atrium wall, is the **cardiac pacemaker**. Its cells fire off excitatory signals seventy or so times a minute. Each firing triggers a cardiac cycle. The signal spreads across the atria and makes them contract. At the same time, it activates junctional fibers, which conduct it to the AV node. From this node, the signal flows along a bundle of fibers that branch in the partition between the heart's two halves. The branchings extend to the bottom of the heart, then up along ventricle walls. The ventricles respond to these signals by contracting with a twisting motion that ejects blood into the aorta and the pulmonary arteries.

The nervous system can only adjust the rate and strength of contractions set by the natural pacemaker. If a heart is removed from the body, it will continue to beat for a short time, as a result of SA node activity.

The human cardiovascular system has two separated circuits, pulmonary and systemic, for blood flow.

The pulmonary circuit is a short loop from the heart's right half, through both lungs, to the heart's left half. Oxygen-poor blood flowing into the circuit rapidly picks up oxygen and gives up carbon dioxide at capillary beds in the lungs.

The longer systemic circuit starts at the heart's left half and aorta. It delivers oxygen and accepts carbon dioxide at capillary beds of all metabolically active regions. Then its veins deliver oxygen-poor blood to the heart's right half.

The heart has four chambers (each half has one atrium and one ventricle). Contraction of the ventricles is the driving force for blood circulation away from the heart.

The heart beats spontaneously at a rate governed by its own pacemaker. The nervous system only adjusts the rate.

22.4 Structure and Function of Blood Vessels

Arteries are large-diameter transporters of oxygenated blood through the body. Arterioles are sites of control over the flow volume through organs. Blood capillaries and to some extent venules are diffusion zones. Veins are blood volume reservoirs and transport oxygen-poor blood back to the heart.

Figure 22.9 compares the structure of human arteries, arterioles, capillaries, and veins. Fluid pressure drops as blood flows through these vessels. It is highest in contracting ventricles, still high at the start of arteries, and lowest in the veins.

Blood pressure is fluid pressure imparted to blood by the ventricular contractions. In the pulmonary and systemic circuits, the difference in pressure between two points determines the flow rate. The beating heart establishes high pressure at the start of a circuit. Then, as blood rubs against enclosing vessel walls, friction impedes its flow. The resistance to blood flow depends mainly on how *wide* the blood vessels are. Simply put, the resistance rises as the vessels narrow down. Just a twofold decrease in blood vessel radius increases the resistance to flow sixteenfold.

RAPID TRANSPORT IN ARTERIES

Compare the structure of the blood vessels in Figure 22.9. With their large diameter and low resistance to flow, arteries are efficient transporters of oxygenated blood. They also are pressure reservoirs that smooth out pulsations in pressure that are generated by each cardiac cycle. An artery's thick, muscular, elastic wall bulges from the large volume of blood that ventricular contraction forces into them. The wall recoils, and this forces oxygen-rich blood forward through the circuit as the heart is relaxing.

ADJUSTING RESISTANCE AT ARTERIOLES

When you are resting, the distribution of blood flow is close to that in Figure 22.10. If you are running, more of the total volume is going to leg muscles to supply their increased oxygen needs. The flow is adjusted by widening the arterioles to leg muscles, and narrowing others, such as those that supply the digestive tract.

The nervous and endocrine systems issue signaling molecules that act on cells of smooth muscle ringing the arteriole walls (Figure 22.9*b*). Specific signals make these cells relax, causing **vasodilation**—enlargement, or dilation, of the blood vessel diameter. Others make

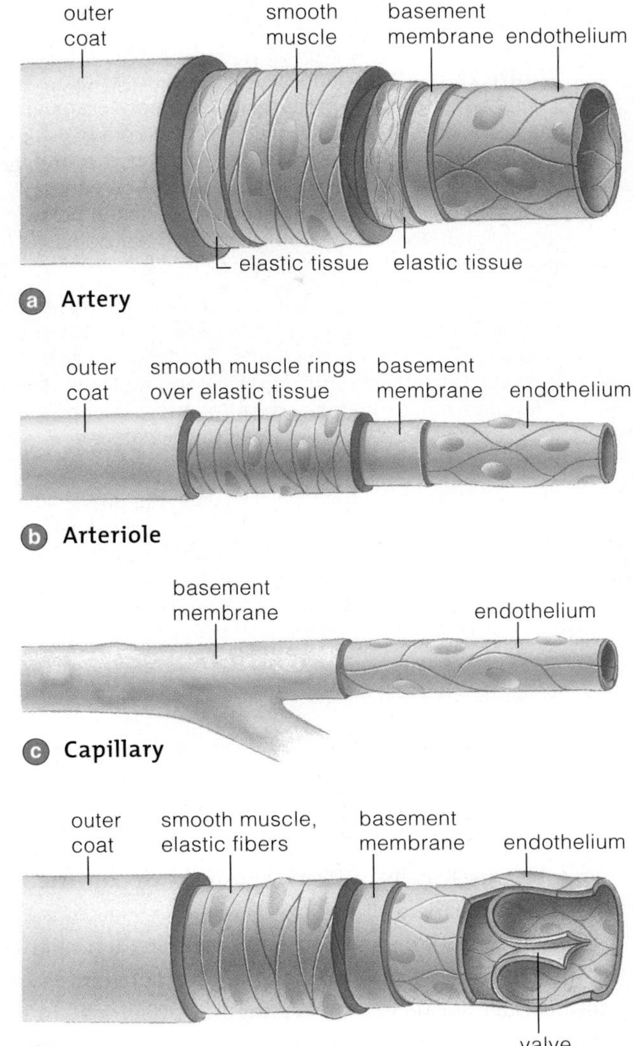

Figure 22.9 Structure of blood vessels. These are not drawn to the same scale. The basement membrane, a noncellular layer, is rich in proteins and polysaccharides. Small venules are like capillaries in structure and function.

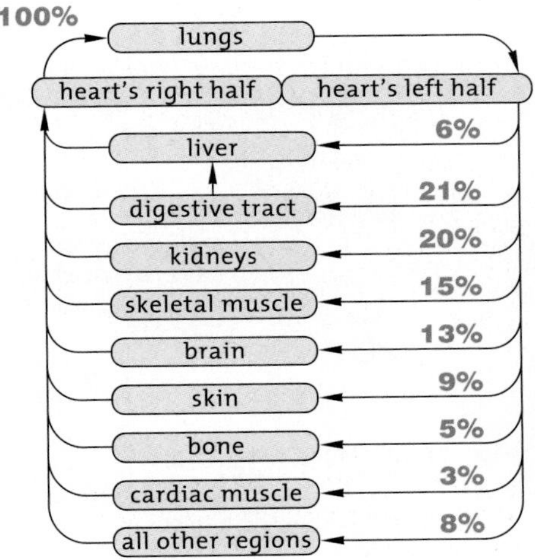

Figure 22.10 Distribution of the heart's output in a person at rest. How much of the output flows through a given region is adjusted by selectively vasodilating and vasoconstricting at many arterioles all along the systemic circuit.

Figure 22.11 Measuring blood pressure. A hollow cuff attached to a pressure gauge is wrapped around the upper arm. The cuff is inflated with air to a pressure above the highest pressure of the cardiac cycle, when ventricles contract. Above this pressure, you cannot hear sounds through a stethoscope positioned below the cuff and above the brachial artery, because no blood is flowing through the vessel.

Air in the cuff is slowly released, so some blood flows into the artery. The turbulent flow causes soft tapping sounds. When the tapping starts, the gauge's value is about 120 mm mercury (Hg) in young adults at rest. This value means that the measured pressure would force mercury to move upward 120 millimeters in a narrow glass column.

More air is released from the cuff. Just after the sounds grow dull and muffled, blood is flowing continuously, so the turbulence and tapping end. Silence signifies the end of each cardiac cycle, before the heart pumps out blood. A typical reading is 80 mm Hg.

Blood pressure is usually written as systolic pressure over diastolic pressure. For the example above, it would be 120/80 mmHg.

LINKS TO SECTIONS 4.4, 4,6

smooth muscle cells contract, causing **vasoconstriction**. The word means a shrinking of blood vessel diameter.

Arteriole diameter also is adjusted when changes in metabolism shift concentrations of substances in a tissue. Such local chemical changes are "selfish" in that they invite or divert blood flow to meet a tissue's own metabolic needs. As you run, skeletal muscle cells use up oxygen, and the levels of carbon dioxide, hydrogen and potassium ions, and other solutes rise. The shift makes arterioles in the tissue dilate. More blood flows through active muscles to deliver more raw materials and carry away cell products and metabolic wastes. When the muscles relax and demand less oxygen, the oxygen level rises and makes the arterioles constrict.

MEASURING BLOOD PRESSURE

As noted above, blood pressure is highest in arteries, declines as blood moves through the systemic circuit, and is lowest in the veins. Typically, blood pressure is measured in the brachial artery of an upper arm while a person is at rest (Figure 22.11). During each cardiac cycle, *systolic* pressure is the highest pressure that the contracting ventricles exert against wall of the artery. *Diastolic* pressure is the lowest arterial blood pressure of a cardiac cycle, reached as ventricles are relaxing.

FROM CAPILLARIES BACK TO THE HEART

Every capillary bed is a diffusion zone for exchanges between blood and interstitial fluid. Living cells in any body tissue will die quickly if they are deprived of the exchanges. Brain cells start to die within five minutes.

Between 10 billion and 40 billion capillaries service the human body. Collectively, they offer a tremendous surface area for gas exchanges. At least one is located nearby every living cell. Proximity to the cells is vital; shorter distances mean faster diffusion rates. It would take years for oxygen to diffuse on its own from lungs all the way down your legs. By then your toes, and the rest of you, would long be dead.

Red blood cells are eight micrometers across. This means that they have to squeeze single file through capillaries, which are about three to seven micrometers across. This tight fit puts the oxygen-transporting red blood cells and the solutes in plasma in direct contact

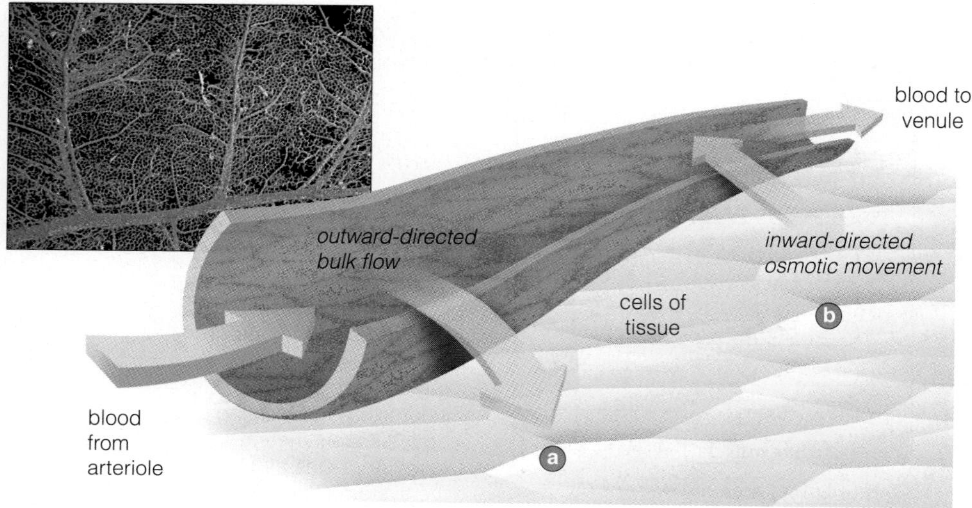

Figure 22.12 Bulk flow in a capillary bed. Fluid crosses a capillary wall by ultrafiltration and reabsorption. (**a**) At a capillary's arteriole end, ultrafiltration occurs. The difference between blood pressure and interstitial fluid pressure forces plasma fluid out through clefts between endothelial cells of the capillary wall. Most plasma proteins remain behind in the vessel, increasing the osmolarity of the plasma.

(**b**) Reabsorption is an osmotic movement of some interstitial fluid *into* a capillary. It occurs because the plasma, with its dissolved proteins, has a greater solute concentration than the interstitial fluid. Reabsorption near the end of a capillary bed tends to balance ultrafiltration at the start. Normally there is only a small *net* movement of fluid from the blood to the interstitial fluid. The lymphatic system returns this escaped fluid to the blood.

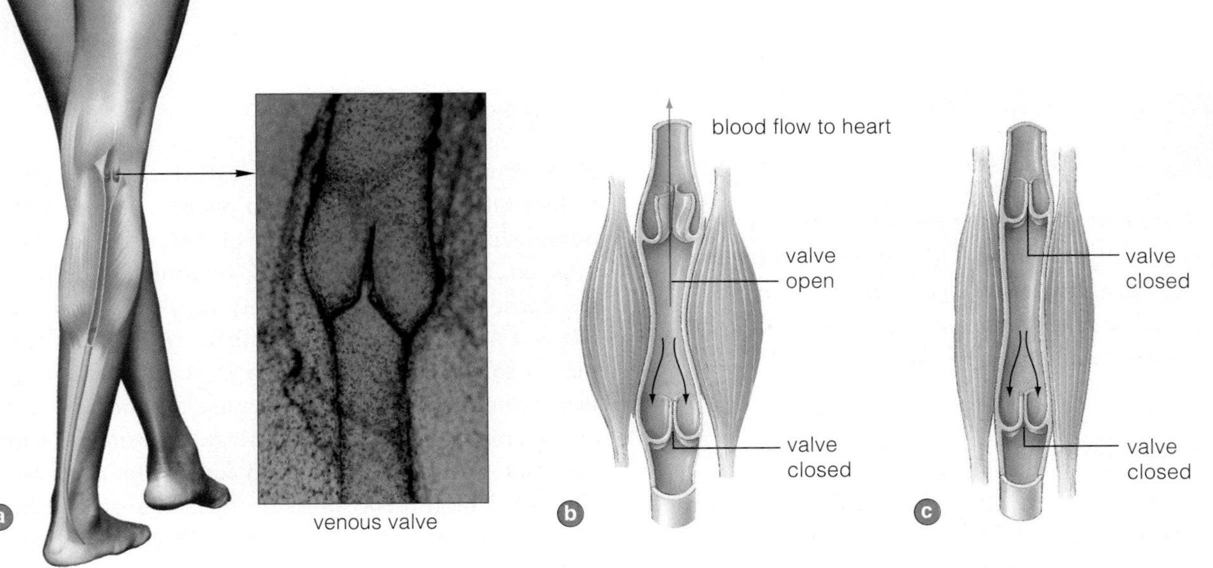

Figure 22.13 Vein function. (**a**) Valves in medium-sized veins prevent backflow of blood. Bulging of adjacent contracting skeletal muscles helps raise fluid pressure in a vein. (**b**) Skeletal muscles next to the vein contract, help blood flow forward. (**c**) Skeletal muscles relax and valves in vein shut. Backflow is prevented.

with the exchange area—the capillary wall—or only a short distance away from it.

Each capillary is a cylindrical sheet of endothelial cells organized as a single layer (Figure 22.9*c*). So how do oxygen and carbon dioxide get from the blood in a capillary, across these cells, and into interstitial fluid? Like many other small lipid-soluble molecules, they can diffuse through the lipid bilayer of an endothelial cell's plasma membrane, move through its cytoplasm, then diffuse out on the other side.

Also, movement of fluid into and out of a capillary bed allows direct exchanges between the plasma and the interstitial fluid. In most tissues, the capillaries are leaky—the cells in their walls are separated by narrow clefts. At the arterial end of a capillary bed, pressure exerted by the beating heart forces fluid out through the clefts in the walls and into surrounding interstitial fluid. This process is called *ultrafiltration*. Fluid that is forced out is high in oxygen, ions, and essential nutrients, such as glucose. Most proteins are too big to exit through the clefts and remain in the plasma.

As the blood flows through a capillary bed, blood pressure declines. Near the venous end of the capillary bed, osmotic pressure is greater than blood pressure. Here, water moves by osmosis from the interstitial fluid into the protein-rich plasma. We call this *reabsorption*. Effects of the two processes are shown in Figure 22.12. There is a small *net* outward flow from a capillary bed, which the lymphatic system returns to blood.

High blood pressure forces additional fluid out of the capillaries. If the fluid pools in interstitial spaces, it causes swelling known as *edema*. Exercise sometimes results in a temporary edema, because many arterioles dilate. Edema also can result from an obstructed vein or from heart failure. Extreme edema is one symptom of *elephantiasis*; a roundworm infection interferes with the return of lymph back to the blood (Section 16.6).

VENOUS PRESSURE

Blood moving out of capillary beds enters venules, or "little veins," which merge into large-diameter veins. Functionally, venules are a bit like capillaries—some solutes can diffuse across them—but they are larger in diameter. Several capillaries feed into each venule.

Veins are large-diameter, low-resistance transport tubes to the heart (Figure 22.9*d*). Their valves prevent backflow. When the force of gravity pulls the blood in veins backward between heart contractions, the valves close up tight. The vein wall can bulge greatly under pressure, more so than an arterial wall. Thus veins are reservoirs for variable volumes of blood. Collectively, the veins of an adult human can hold up to 50 to 60 percent of the total blood volume.

The vein wall contains some smooth muscle. When blood must be circulated faster, as during exercise, the muscle contracts. The wall stiffens and the vein bulges less. This increases the pressure in veins and drives more blood back to the heart. Contractions of skeletal muscles also faciliate flow. As these muscles contract, they bulge and press against nearby veins. This helps to push blood back toward the heart (Figure 22.13).

Arteries are major transporters of oxygenated blood and also are pressure reservoirs. The total resistance to blood flow, and thus the distribution of blood among regions of the vascular system, is adjusted mainly at arterioles.

Capillary beds are diffusion zones for exchanges between blood and interstitial fluid. Blood pressure forces fluid out from a capillary at one end, then osmotic pressure causes most of the fluid to return to the blood at the other end.

Veins are highly distensible blood volume reservoirs, and they help adjust flow volume back to the heart.

IMPACTS, ISSUES

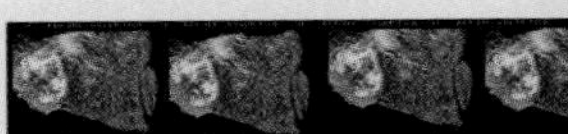

Nicotine causes the smooth muscle in blood vessel walls to constrict, which raises blood pressure. Over the long term, high blood pressure can damage the kidneys and increase the risk of strokes. In males, high blood pressure doubles the risk of impotence.

LINK TO SECTION 2.6

22.5 Cardiovascular Disorders

The most common disorders of the cardiovascular system are hypertension, or sustained high blood pressure, and atherosclerosis, a progressive thickening of the arterial wall that narrows the lumen. Both cause heart attacks and the type of brain damage called strokes.

GOOD CLOT, BAD CLOT

When small vessels are cut or torn, **hemostasis** is the process that stops blood loss from them and constructs a framework for repairs. Immediately after the injury, smooth muscle in the damaged vessel wall contracts in an automatic response called a spasm. This can slow or even temporarily pause the blood flow. Next, platelets clump together and temporarily fill the breach. They release substances that prolong the spasm and attract more platelets. In the final phase, blood coagulates—proteins in the plasma convert blood to a gel and help to form a clot. Rod-shaped proteins called fibrinogens respond to the collagen fibers exposed by the damage. The rods stick together as long threads that adhere to the exposed collagen. They form a net that traps blood cells and platelets (Figure 22.14). The entire mass is a blood clot. It retracts and forms a compact patch that seals the breach in the blood vessel wall.

Clot formation is vital for repairing blood vessels. But clots cause problems when they completely block blood flow through a vessel, as you will see shortly.

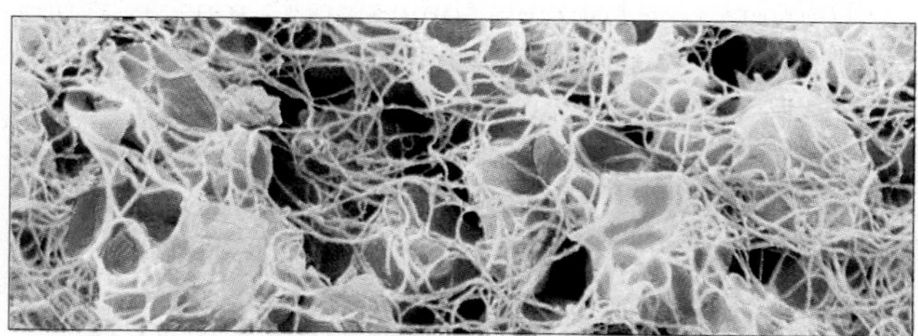

Figure 22.14 Fibrous protein net that helps blood clots form.

THE SILENT KILLER

Hypertension, or high blood pressure, is known as the "silent killer" because affected people often reveal no obvious symptoms. A quarter of American adults are affected, but most do not seek treatment. Commonly, the cause is unknown. Heredity may be a factor, and African Americans are especially at risk. Diet can also play a role. Eating foods high in sodium encourages water retention, which in turn raises the blood volume and increases fluid pressure. Regular exercise, a low-salt diet, and medications can lower blood pressure. Untreated high blood pressure raises the risk of heart disease, stroke, and kidney failure. Overweight people with high blood pressure are most vulnerable.

ATHEROSCLEROSIS

In *arteriosclerosis*, arteries thicken and lose elasticity. With *atherosclerosis*, this condition worsens as lipids accumulate in the arterial wall and narrow the inside diameter. You probably already know that cholesterol plays a role in this "hardening of the arteries." Your body uses some cholesterol to make cell membranes, myelin sheaths, bile salts, and steroid hormones. The liver makes enough cholesterol to meet normal needs, but more cholesterol is absorbed from food in the gut. People differ in how the excess is handled.

In the blood, most cholesterol is bound to proteins as *low-density lipoproteins*, or LDLs. Most cells take up this form of cholesterol. A smaller amount is bound as *high-density lipoproteins*, or HDLs, which the liver takes up. HDL is metabolized in the liver, and the residues are eliminated in bile secreted into the gut. As LDL levels rise, so does risk of atherosclerosis. Lipids build up on an artery's wall, smooth muscle proliferates, the site becomes inflamed, and fibrous connective tissue is laid down. The *atherosclerotic plaque* that results makes the arterial wall bulge inward, narrowing the vessel's diameter (Figure 22.15).

Hardened plaque can rupture the artery wall. If the resulting blood clot stays in one place, it is a *thrombus*.

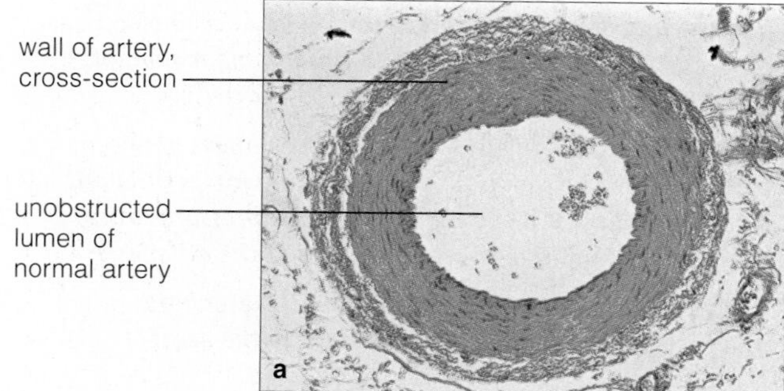

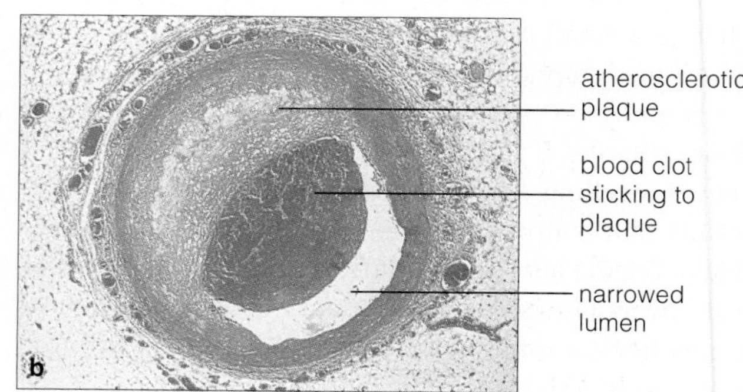

Figure 22.15 Sections from (**a**) a normal artery and (**b**) an artery with a lumen narrowed by an atherosclerotic plaque. A blood clot almost clogs this one.

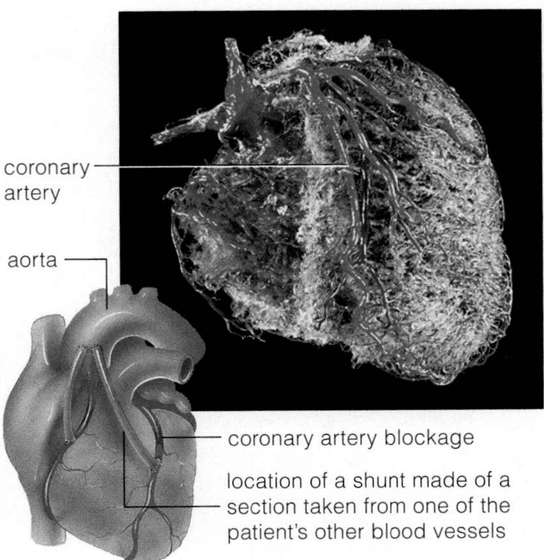

Figure 22.16 Coronary arteries and other blood vessels of the heart. Resins were injected into them, then the rest of the cardiac tissues were dissolved, leaving this accurate, three-dimensional corrosion cast. The sketch shows two coronary bypasses (artificially colored *green*), which extend from the aorta and past two clogged parts of the coronary arteries.

If it is dislodged and travels in blood, it is an *embolus*. An embolus can obstruct a small blood vessel. If this happens in the brain, it can cause a *stroke*.

Coronary arteries can be sites of atherosclerosis. If they narrow by 25 percent, they invite *angina pectoris* (chest pain) or a heart attack. During an attack, some of the cardiac muscle cells die from lack of blood. The first attack is fatal about 30 percent of the time. Those who survive often have a badly weakened heart.

With *coronary bypass surgery*, a bit of blood vessel from elsewhere in the body is stitched to the aorta and to the coronary artery below a clogged region (Figure 22.16). With *laser angioplasty*, laser beams directed at the plaques vaporize them. With *balloon angioplasty*, a balloon is inflated inside the blocked artery and serves to flatten the plaques.

RISK FACTORS

Cardiovascular disorders are the leading cause of death in the United States. Each year, about 40 million people have cardiovascular disorders, and about one million die. There are many risk factors. As Section 22.9 explains, smoking is the most important one. Other factors include genetic predisposition, a high blood level of cholesterol, high blood pressure, obesity (Chapter 24), and diabetes mellitus, a condition described in Section 26.4. Increased age also is a risk factor; the older you get, the greater your risk. Finally, gender initially is a factor; until age fifty, males are at greater risk.

22.6 The Nature of Respiration

The high energy demands of animal cells are most often fueled by ATP produced by aerobic respiration. This pathway requires oxygen and produces carbon dioxide. How do animals supply oxygen to their cells and dispose of the carbon dioxide wastes?

THE BASIS OF GAS EXCHANGE

It takes a large amount of energy from ATP to support the movements of a multicelled animal body. Aerobic respiration is the only metabolic pathway that makes enough ATP to meet these needs. Recall from Chapter 7 that this pathway needs oxygen and releases carbon dioxide. In most animals, circulatory and respiratory systems work together to deliver oxygen to body cells and to get rid of the cells' carbon dioxide wastes.

Aerobic respiration is a metabolic pathway. Do not confuse it with **respiration**, a process that provides an ongoing flow of oxygen from outside the body to a moist boundary with the internal environment, and a flow of carbon dioxide the other way—out of the body.

As Section 4.4 explains, the direction that any solute diffuses depends upon its concentration gradient. This concept also applies to gases. Each gas contributes its own *partial pressure* to the total pressure of a mixture of gases. A gas diffuses in the direction set by its own partial pressure gradient. For example, oxygen diffuses from air into any region where its partial pressure is lower. A steeper gradient causes faster diffusion.

Gases enter and leave an animal body by crossing a **respiratory surface**, usually a thin layer of epithelium. The surface must be moist; gases cannot diffuse across it unless they are dissolved in fluid. In some animals, all exchanges occur across the body surface (Figure 22.17). But most animals have structures that function to increase the diffusion rate of gases. As you will see, many factors influence how many molecules of gas cross a respiratory surface during any given interval.

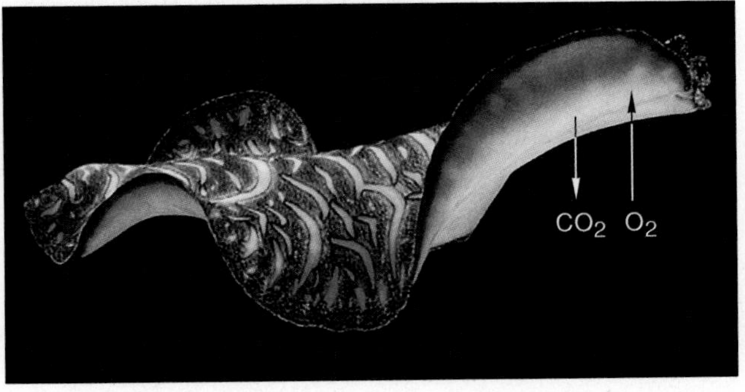

Figure 22.17 In this marine flatworm, the respiratory surface is the body's outermost layer. Oxygen diffuses into the animal, and carbon dioxide diffuses out across this thin, moist epithelium.

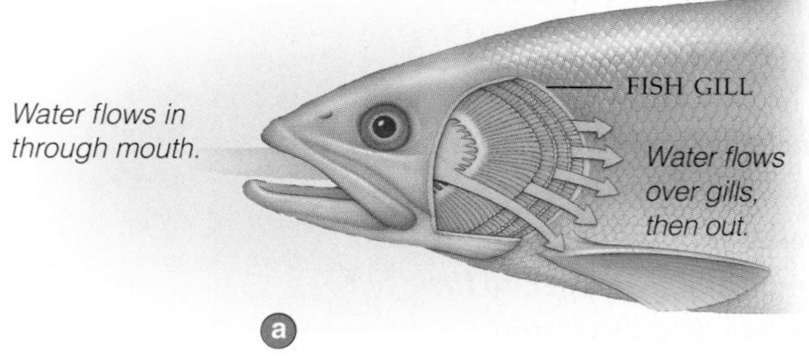

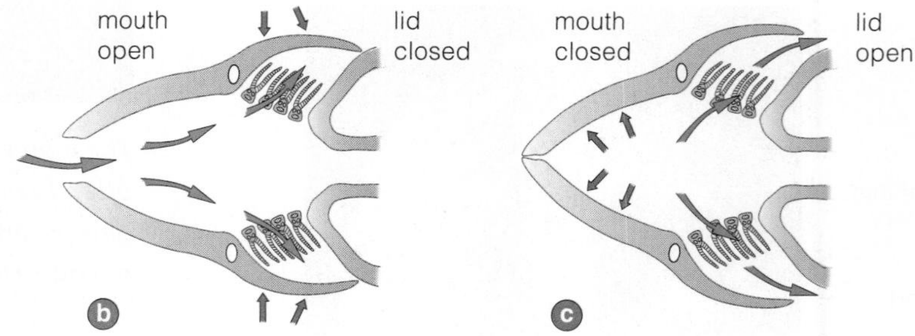

Figure 22.18 ***Animated!*** (**a**) One of a pair of fish gills. A bony lid, removed for this sketch, protects it.

(**b**) Gill ventilation. Opening the mouth and closing the lid pulls flowing water across the gill. (**c**) Closing the mouth and opening the lid forces water outside.

(**d**) Gas exchange. Filaments in gills have a vascularized respiratory surface. The filament organization directs the incoming water past the gas exchange surfaces.

(**e**) A blood vessel carries oxygen-poor blood from the body to each filament. Another vessel carries oxygenated blood into the body. Blood flows from one into the other, counter to the flow of water over the gill's respiratory surfaces. The countercurrent flow favors the movement of oxygen (down its partial pressure gradient) from water into the blood.

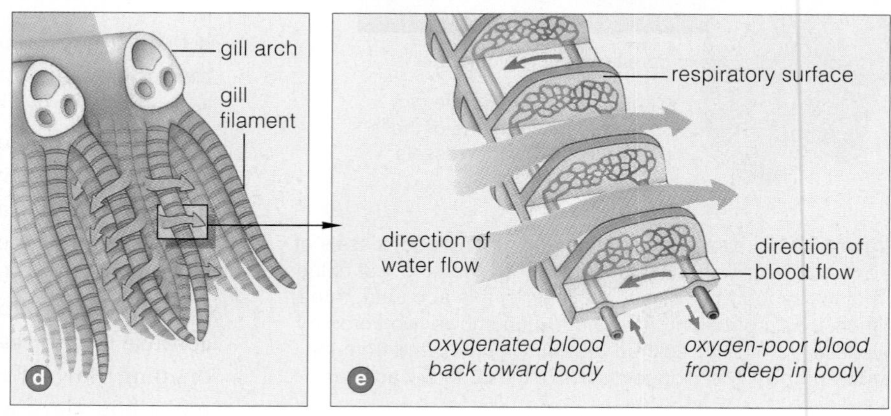

LINKS TO SECTIONS 2.9, 3.2, 4.4, 17.2

WHICH FACTORS INFLUENCE GAS EXCHANGE?

SURFACE-TO-VOLUME RATIO The surface-to-volume ratio (Section 3.2) influences gas exchange. Unless they have respiratory organs, animals are tiny, tubelike, or flat; gases diffuse directly across their body surface.

Think back on that flatworm in Figure 17.4. Lack of a circulatory system constrains its body form and size. Recall that as a body increases in size, its surface area increases less quickly than its volume. If the worm got larger, the diffusion distance between the body surface and its internal cells would be too great. Gas exchange would be too slow, and the worm would die.

VENTILATION Large-bodied, active animals require faster gas exchange than diffusion alone offers. Diverse adaptations enhance the exchange rate. Many aquatic invertebrates, including mollusks, have cilia that drive water across gills. A **gill** is a thin, moist, blood vessel-rich, respiratory organ. Crustaceans, such as lobsters, have gills too. In insects, air is moved into and out of a system of branched tracheal tubes. Gas exchange takes place at a moist surface at the tubes' tips. In mammals, muscles move air into and out of paired lungs.

RESPIRATORY PIGMENTS Oxygen binds reversibly to **hemoglobin** and to other respiratory pigments. These molecules increase gas exchange rates by keeping the pressure gradients across a respiratory surface steep. For example, at the respiratory surface of human lungs, hemoglobin in blood binds oxygen and carries it away. This lowers the oxygen level at this surface. When this blood arrives at oxygen-poor tissues, oxygen diffuses away from hemoglobin. The blood, now oxygen-poor, flows back to the lungs. **Myoglobin**, another pigment, also binds oxygen. Abundant in cardiac and skeletal muscle cells, it contributes to their red color. It serves as an oxygen store. When the oxygen level in a muscle drops, myoglobin releases some and helps meet needs.

GILLS OF FISHES AND AMPHIBIANS

A few kinds of fish larvae and a few amphibians have *external* gills projecting into water. Adult fishes have a pair of *internal* gills: rows of slits or pockets at the back of the mouth that extend to the body's surface (Figure 22.18). Whatever its form, each gill has a wall of moist, thin, blood vessel-rich epithelium that functions as a respiratory surface.

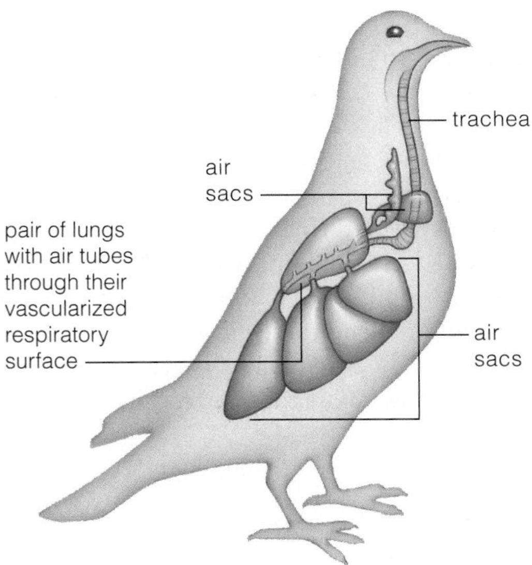

Figure 22.19 ***Animated!*** A bird's respiratory system. Two small, inelastic lungs have a number of attached air sacs. A bird inhales and draws air into the sacs through tubes, open at both ends, that thread through the lung tissue. This tissue is the respiratory surface.

The bird exhales, forcing air out of the sacs, through the tubes, and out of the trachea. Air is moved continually through the lungs and and across respiratory surfaces. This system supports the high metabolic activity that birds require for flight.

In fishes, water flows into the mouth and pharynx, then over arrays of gill filaments. Blood moves into and out of each filament. First, water flows past one blood vessel that moves blood from the gill back to the rest of the body. The blood inside has less oxygen than the water does, so oxygen diffuses inward. Next, the water flows over a vessel moving blood into the gill. Although the water already gave up some oxygen, it still contains more than the oxygen-poor blood in this vessel. So a second helping of oxygen diffuses in.

An exchange of substances between two fluids that are moving in opposite directions is a **countercurrent exchange**. In this case, it allows a fish to extract more oxygen from water than it could using a one-way flow mechanism. It increases the efficiency of gas exchange.

EVOLUTION OF PAIRED LUNGS

Some fishes, nearly all amphibians, and all mammals and birds have **lungs**, internal respiratory surfaces in the shape of a cavity or a sac. More than 450 million years ago, lungs started to form as outpouchings of the gut wall of some fishes. In oxygen-poor habitats, lungs afforded a larger surface area for gas exchange. This proved advantageous for the first tetrapods that moved onto land, because air holds far more oxygen than water does. Besides, gills would not have worked on land. Gill filaments stick together when water does not flow through them and keep them moist.

Amphibians never fully completed their transition to land. Salamanders especially still use the skin as a respiratory surface for integumentary exchange. Some do not even have lungs; all gases are exchanged across the body surface and the lining of the mouth. Frogs and toads rely more on small lungs for oxygen uptake, but most of the carbon dioxide wastes diffuse outward across their skin. Unlike us, frogs do not pull air into lungs. Instead, they raise the floor of the mouth and throat, pushing air inside this cavity into their lungs.

In reptiles, birds, and mammals, the skin is waterproof and has no role in gas exchange. Lungs are the sole respiratory organs (Figures 22.19 and 22.20). The contraction of skeletal muscles causes air to be sucked into the lungs. Compared to reptile lungs, the lungs of birds and mammals have more internal branches and a more extensive respiratory surface. Dissolved oxygen diffuses across this surface, then enters capillaries that carry it throughout the body. In body tissues, where oxygen levels are low, this gas diffuses into interstitial fluid, then into cells. Carbon dioxide moves quickly in the other direction and is expelled from the lungs.

This mode of gas exchange is notably embellished only in birds (Figure 22.19). Birds have a system of air sacs connected to their lungs. Gases are not exchanged in these sacs. Instead, the sacs increase the rate of gas exchange by causing air to move *through* lungs, rather than into and out of them. Birds have a constant flow of fresh air across their respiratory surface. In contrast, mammals have to mix inhaled fresh air with oxygen-depleted air already present in their lungs.

Respiration is a process by which animals move oxygen into the internal environment, and move carbon dioxide wastes from aerobic respiration out of it.

Oxygen and carbon dioxide enter and leave by diffusing across a moist respiratory surface. They tend to move down their respective pressure gradients.

Gas exchange requires steep partial pressure gradients between the outside and inside of the animal body. The greater the area of the respiratory surface and the larger the partial pressure gradient, the faster diffusion will be.

A countercurrent flow mechanism in fish gills compensates for low oxygen levels in aquatic habitats. Internal air sacs—lungs—are more efficient in dry habitats on land.

Amphibians use integumentary exchange and force air into and out of small lungs. Ventilation of paired lungs is the mode of respiration in reptiles, birds, and mammals.

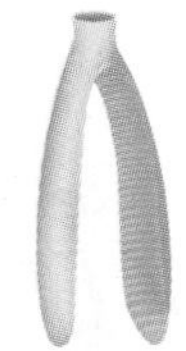

Amphibian
Salamander; like fishes, early amphibians

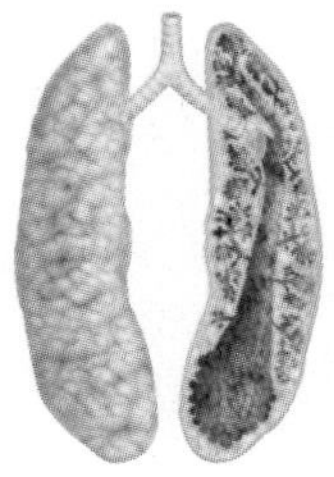

Reptile
Lizard; adapted to dry habitats

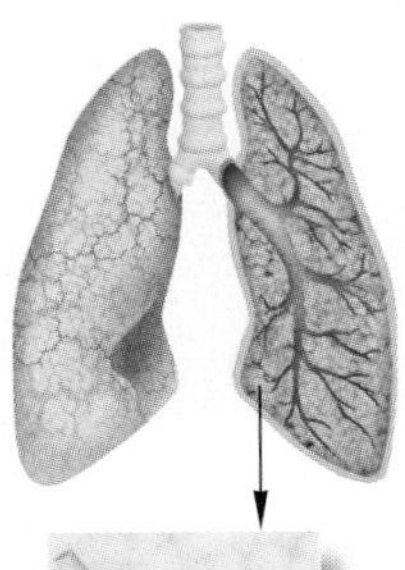

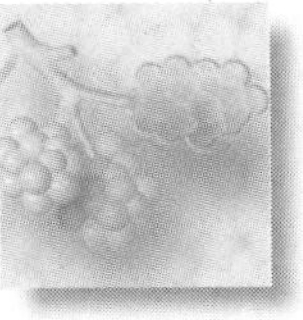

Mammal
Human; adapted to dry habitats

Figure 22.20 Lung structure for three kinds of vertebrates, suggestive of an evolutionary trend from sacs for simple gas exchange to larger, more complex respiratory surfaces.

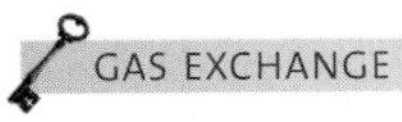

LINK TO SECTION 20.3

22.7 Human Respiratory System

It will take at least 300 million breaths to get you to age seventy-five. You might go without food for a few hours or days. However, stop breathing for even five minutes and normal brain function is over.

THE SYSTEM'S MANY FUNCTIONS

Figure 22.21 shows the parts of the human respiratory system and lists their functions. Its two paired lungs combined have about 600 million outpouchings, or air sacs called **alveoli** (singular, alveolus). Controls adjust breathing rates so that the inflow and outflow of air at alveoli match metabolic demands for gas exchange.

The respiratory system functions in more than gas exchange. It is the basis of vocalizations, such as speech. It helps blood in veins return to the heart and helps the body dispose of excess heat and water. Controls over breathing also contribute to maintaining the acid–base balance in the internal environment. Some parts of the system intercept many potentially dangerous airborne agents and substances before they can reach the lungs.

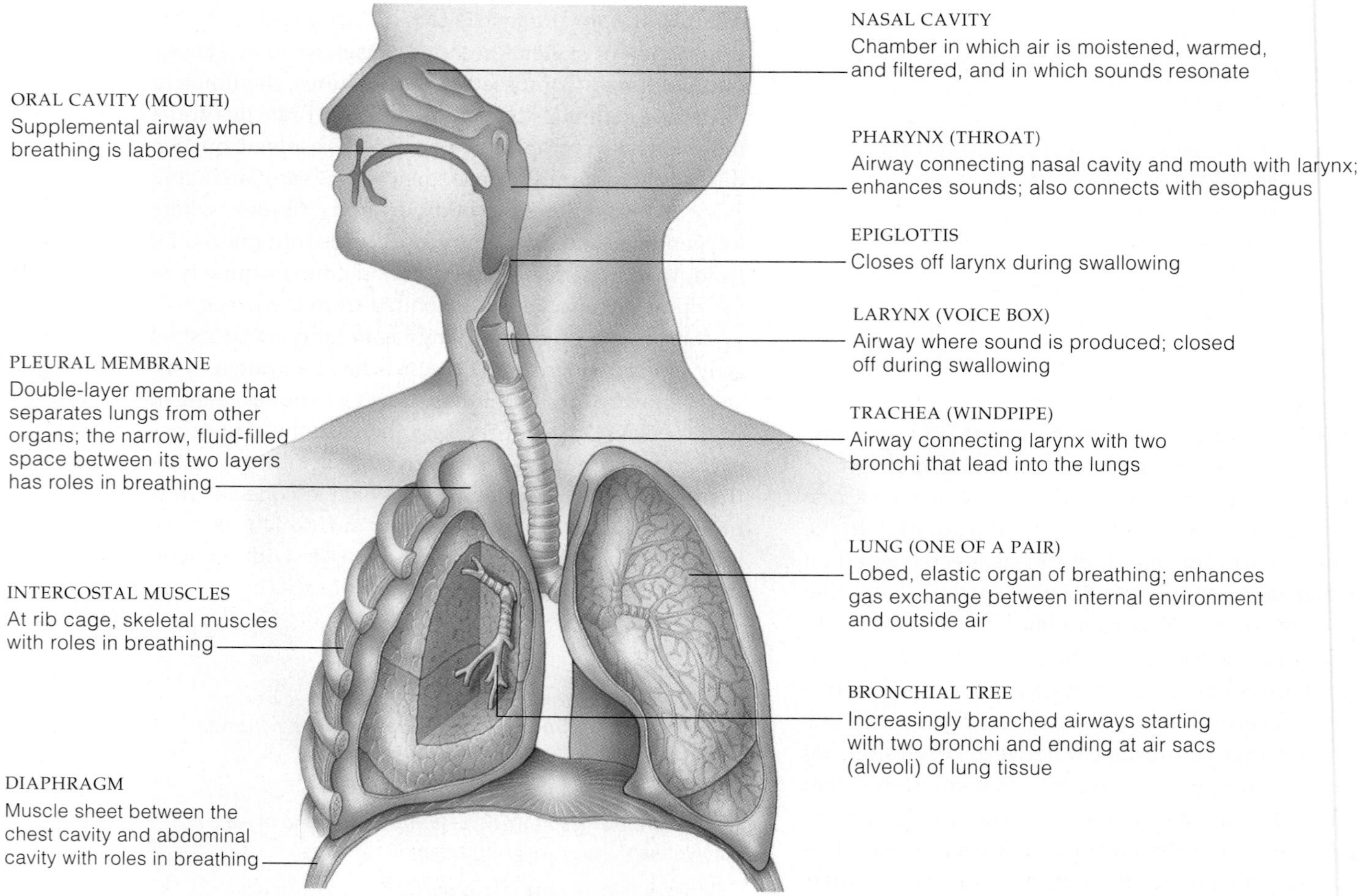

NASAL CAVITY
Chamber in which air is moistened, warmed, and filtered, and in which sounds resonate

ORAL CAVITY (MOUTH)
Supplemental airway when breathing is labored

PHARYNX (THROAT)
Airway connecting nasal cavity and mouth with larynx; enhances sounds; also connects with esophagus

EPIGLOTTIS
Closes off larynx during swallowing

LARYNX (VOICE BOX)
Airway where sound is produced; closed off during swallowing

PLEURAL MEMBRANE
Double-layer membrane that separates lungs from other organs; the narrow, fluid-filled space between its two layers has roles in breathing

TRACHEA (WINDPIPE)
Airway connecting larynx with two bronchi that lead into the lungs

LUNG (ONE OF A PAIR)
Lobed, elastic organ of breathing; enhances gas exchange between internal environment and outside air

INTERCOSTAL MUSCLES
At rib cage, skeletal muscles with roles in breathing

BRONCHIAL TREE
Increasingly branched airways starting with two bronchi and ending at air sacs (alveoli) of lung tissue

DIAPHRAGM
Muscle sheet between the chest cavity and abdominal cavity with roles in breathing

a Human respiratory system

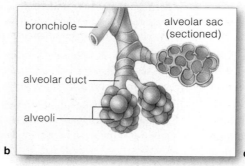

b
bronchiole
alveolar sac (sectioned)
alveolar duct
alveoli

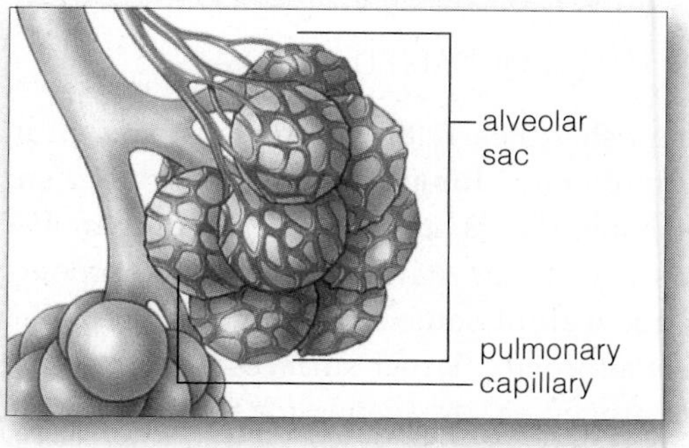

c
alveolar sac
pulmonary capillary

Figure 22.21 ***Animated!*** (**a**) Components of the human respiratory system. Muscles, including the diaphragm, and parts of the axial skeleton have secondary roles in respiration. (**b**,**c**) Location of alveoli relative to bronchioles and to lung (pulmonary) capillaries.

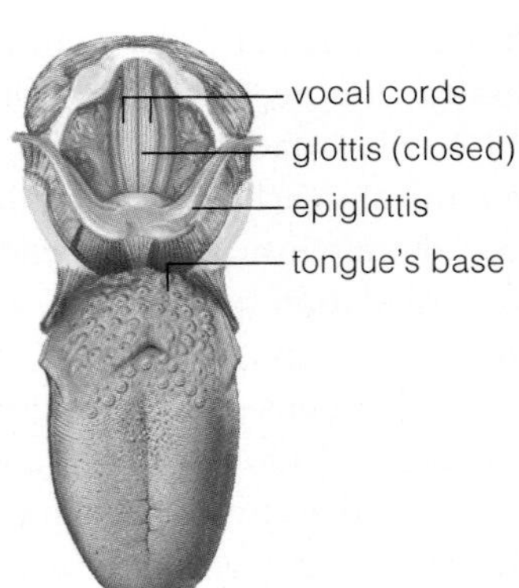

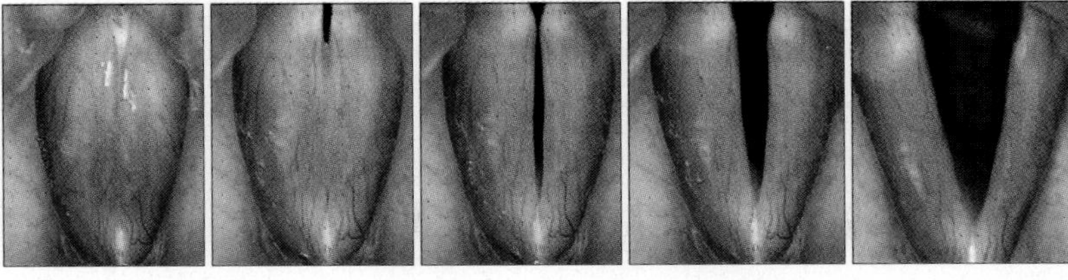

Figure 22.22 Human vocal cords, where the sounds used in speech originate. Upon commands from the nervous system, skeletal muscle action changes the glottis (the gap between them). The glottis is shut tight when you swallow, open a bit when you speak, and fully open when you breathe silently.

The respiratory system ends at alveoli, and there it exchanges gases with the pulmonary capillaries. (The Latin *pulmo* means lung.) This is where the circulatory system takes over the task of gas exchange for the rest of the body.

FROM AIRWAYS INTO THE LUNGS

Take a deep breath. Now look at Figure 22.21 to get an idea of where the air will travel in your respiratory system. Unless you are out of breath and panting, air has entered two nasal cavities, not your mouth. In the nose, mucus warms and moistens the air, hairs and ciliated epithelium filter dust and particles from it, and sensory receptors detect odors. Air flows into the **pharynx**, or throat. This is the entrance to the **larynx**, a big airway with two paired folds of mucus-covered membrane projecting into it. The lower pair are vocal cords (Figure 22.22). When you breathe, air is forced in and out through the glottis, the gap between the cords. Air flow can make the vocal cords vibrate, producing the sounds that are the basis for speech.

While the larynx muscles contract and relax, elastic ligaments in the vocal cords tighten or slacken to alter how much the folds are stretched. The nervous system coordinates the narrowing and widening of the glottis. When its signals cause increased tension in the larynx muscles, for example, the gap between the vocal cords shrinks and you make high-pitched sounds. The lips, teeth, tongue, and the soft roof above the tongue are enlisted to modify different patterns of sounds.

Infected or irritated vocal cords swell up, which interferes with their capacity to vibrate. If the swelling causes hoarseness, this condition is called *laryngitis*.

At the entrance to the larynx is an epiglottis. When this tissue flap points up, air moves into the **trachea**, or windpipe. This air pathway actually crosses a food pathway leading from the esophagus and pharynx to the stomach. Normally, the epiglottis tilts down and blocks the trachea's entrance when you swallow.

The trachea branches into two airways to the lung tissue. Each airway is a **bronchus** (plural, bronchi). Its epithelial lining has many cilia and mucus-secreting cells that serve as a barrier to infection. Bacteria and airborne particles stick to the mucus, and cilia sweep the mucus to the mouth, where it can be expelled.

Human lungs are elastic, cone-shaped organs of gas exchange, located inside the rib cage on both sides of the heart and above the diaphragm. The **diaphragm** is a broad, muscular partition between the thoracic and abdominal cavities. A thin, saclike pleural membrane lines the outer surface of each lung and also the inner surface of the thoracic cavity's wall. The thoracic wall and lungs press the sac's two surfaces together.

A film of lubricating fluid cuts friction between the pleural membrane surfaces. In *pleurisy*, a respiratory ailment, the membrane becomes inflamed and swollen. Its two surfaces rub each other, so breathing is painful.

Inside each lung, air moves through finer and finer branchings. These branched, epithelium-lined airways are **bronchioles**. At their tips, alveolar ducts lead into alveolar sacs. These are clusters of cup-shaped alveoli. Gases diffuse easily across the moistened epithelium of the alveoli. Collectively, these alveolar sacs provide an enormous surface area for the exchange of gases with the blood. If your 600 million or so alveoli were spread out in a single layer, they would cover the surface of a racquetball court!

Oxygen uptake and carbon dioxide removal are the major functions of the human respiratory system. Inside its pair of lungs, the circulatory system takes over the remaining tasks of respiration.

The respiratory system also has roles in moving venous blood to the heart, vocalizing, adjusting the body's acid–base balance, defending against harmful airborne agents or substances, removing or modifying many bloodborne substances, and detecting odors.

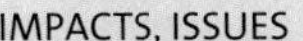
IMPACTS, ISSUES

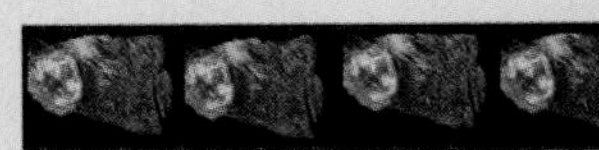

Lung cancer is the top cause of cancer deaths in the United States for both men and women. More than forty carcinogens are found in tobacco smoke, and smokers account for over 80 percent of lung cancer cases. Women smokers are more susceptible than men. On average, they develop cancers earlier and with lower exposure to tobacco. Once diagnosed with lung cancer, fewer than 15 percent are alive five years later.

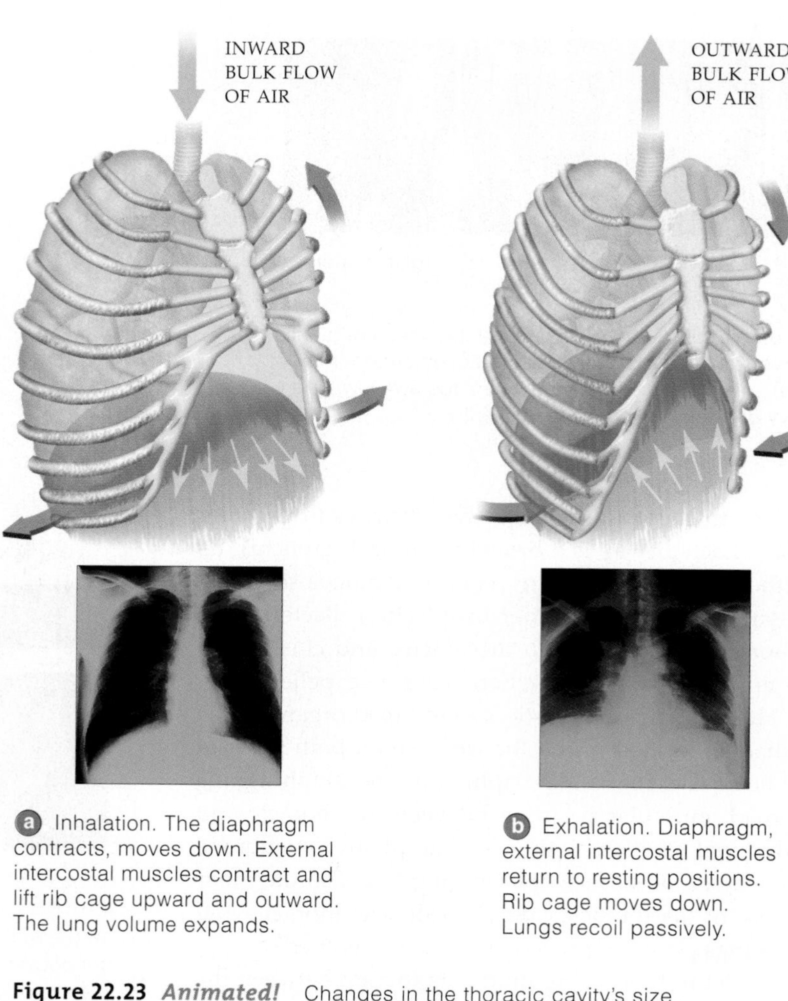

ⓐ Inhalation. The diaphragm contracts, moves down. External intercostal muscles contract and lift rib cage upward and outward. The lung volume expands.

ⓑ Exhalation. Diaphragm, external intercostal muscles return to resting positions. Rib cage moves down. Lungs recoil passively.

Figure 22.23 ***Animated!*** Changes in the thoracic cavity's size during inhalation (**a**) and exhalation (**b**). The x-ray images show how the maximum inhalation possible changes the thoracic cavity volume.

LINK TO SECTION 4.4

22.8 Moving Air and Transporting Gases

Breathing enhances the flow of gases between the atmosphere and the alveolar sacs. A partial pressure gradient drives oxygen from these sacs into interstitial fluid, then into blood, then into tissues throughout the body. A partial pressure gradient drives carbon dioxide in the opposite direction.

THE RESPIRATORY CYCLE

Each **respiratory cycle** is one breath in (inhalation) and one breath out (exhalation). *Inhalation is an active, energy-requiring action.* Think of what happens as you breathe quietly. Your diaphragm and muscles between the rib cage contract and increase your thoracic cavity volume (Figure 22.23). Breathe hard, and the volume increases more. Why? Contracting neck muscles raise the sternum and the first two ribs attached to them.

The changes in volume during each cycle shift the pressure gradients between the air inside and outside the respiratory tract. Before inhalation, pressure in the alveoli is equal to atmospheric pressure.

Figure 22.23*a* shows what happens as you start to inhale. The diaphragm flattens and moves down, and the rib cage is lifted upward and outward. As lungs expand along with the thoracic cavity, pressure in the alveoli falls below atmospheric pressure. Air follows the pressure gradient and flows down into the airways.

The second part of the respiratory cycle, exhalation, is passive when you breathe quietly. Muscles that brought about inhalation relax and the lungs passively recoil, with no further energy outlays. The decrease in lung volume compresses the air in alveolar sacs. Now the pressure inside is greater than atmospheric pressure, so air moves out of the lungs (Figure 22.23*b*).

Exhalation becomes active and energy-demanding only when you are vigorously exercising and so must expel more air. The abdominal wall muscles contract during active exhalation. Abdominal pressure increases and exerts an upward-directed force on the diaphragm, which is thereby pushed upward. Also, the internal intercostal muscles contract. When they do, they pull the thoracic wall inward and downward, and their action flattens the chest wall. In all such ways, the thoracic cavity's dimensions shrink. The lung volume shrinks as well when elastic tissue of the lungs passively recoils.

EXCHANGES AT THE RESPIRATORY MEMBRANE

An alveolus is a cupped sheet of epithelial cells with a basement membrane. A pulmonary capillary consists of endothelial cells, and its basement membrane fuses with that of the alveolus. The alveolar and capillary endothelia, together with their basement membranes, form the respiratory membrane (Figure 22.24). Gases diffuse quickly across this thin membrane.

OXYGEN AND CARBON DIOXIDE TRANSPORT

Some oxygen can dissolve in plasma, but not enough to meet the body's ongoing needs. In all vertebrates and many invertebrates, hemoglobin increases blood's oxygen-carrying capacity. It fills the red blood cells of all vertebrates. Recall once again that a hemoglobin molecule is made up of four polypeptide chains and four **heme groups**, each with one iron atom that can bind reversibly with oxygen (Section 2.9). *Of all the oxygen inhaled into the human body, 98.5 percent is bound to heme groups of hemoglobin.*

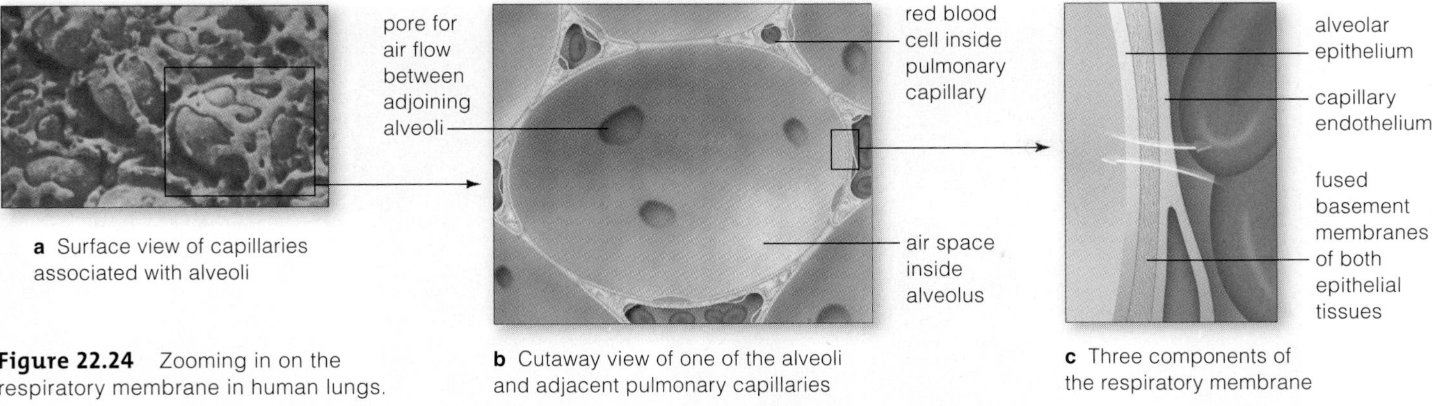

Figure 22.24 Zooming in on the respiratory membrane in human lungs.

The inhaled air that flows into alveoli has a high concentration of oxygen. The opposite is true of blood in pulmonary capillaries. Thus, in lungs, oxygen tends to diffuse across the respiratory membrane and into the plasma. It diffuses into red blood cells, where it binds reversibly with hemoglobin. In this way, *oxyhemoglobin*, or HbO_2, forms.

Oxygen is bound only weakly in HbO_2 molecules. These molecules release oxygen where partial pressure of this gas is lower than in the lungs. They give it up faster in tissue regions where blood is warmer, the pH is lower, and carbon dioxide's partial pressure is high. Such conditions are found in contracting muscles and other metabolically active tissues.

Carbon dioxide diffuses into the capillaries from any tissue where its concentration is higher than it is in the blood. From there, three mechanisms transport it to lungs. About 10 percent stays dissolved in the blood. Another 30 percent becomes bound to hemoglobin to form *carbaminohemoglobin* ($HbCO_2$). However, most of it—60 percent—is transported as bicarbonate (HCO_3^-). Carbon dioxide and water react to form carbonic acid, which separates immediately into bicarbonate and H^+. A few molecules of bicarbonate form spontaneously, but the vast majority form inside red blood cells where carbonic anhydrase speeds the reaction. Action of this enzyme keeps carbon dioxide levels in the blood low. This maintains the gradient that promotes diffusion of carbon dioxide from interstitial fluid into the blood.

Most bicarbonate that forms inside red blood cells diffuses into the blood plasma. It flows to the alveoli, where the partial pressure of carbon dioxide is lower than it is in lung capillaries. Here, reactions reverse—bicarbonate is broken into carbon dioxide and water. Both leave the alveolar sacs in exhalations.

At this point, reflect on Figure 22.25. It summarizes the partial pressure gradients for oxygen and carbon dioxide through the human respiratory system.

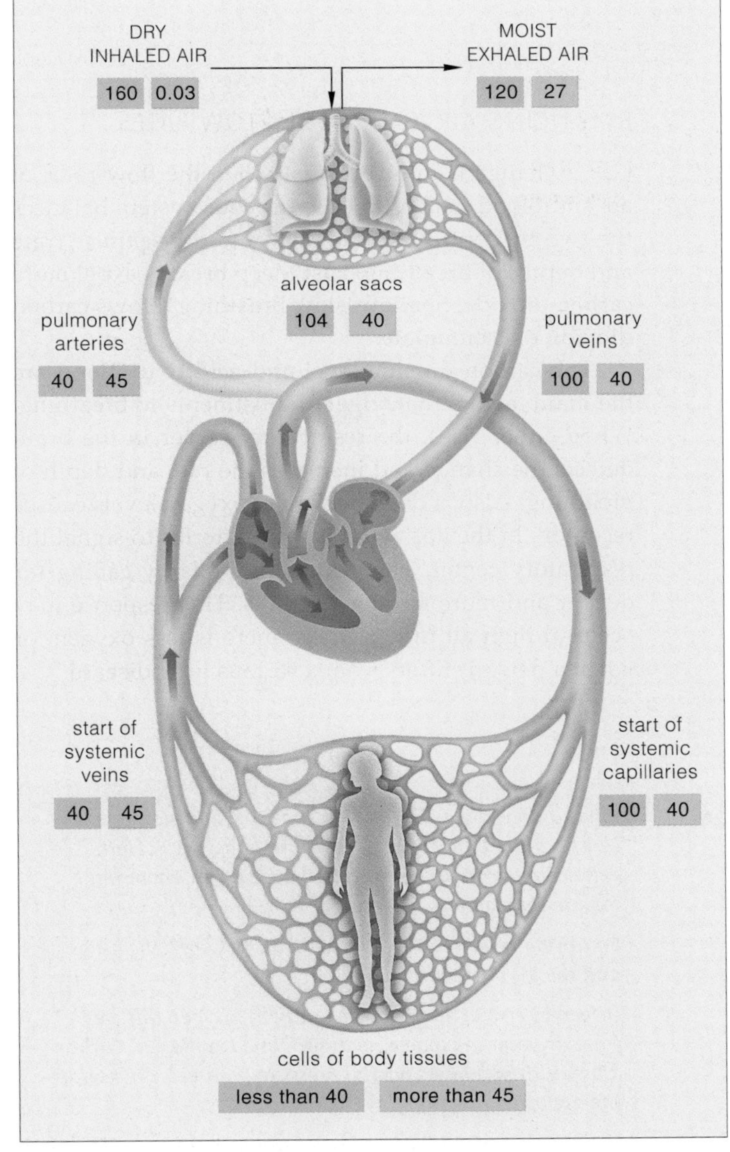

Figure 22.25 ***Animated!*** Partial pressures (in mm Hg) for oxygen (*pink* boxes) and carbon dioxide (*blue* boxes) in the respiratory tract.

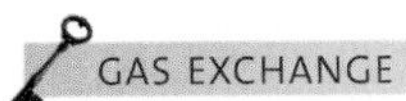

LINK TO SECTION 7.8

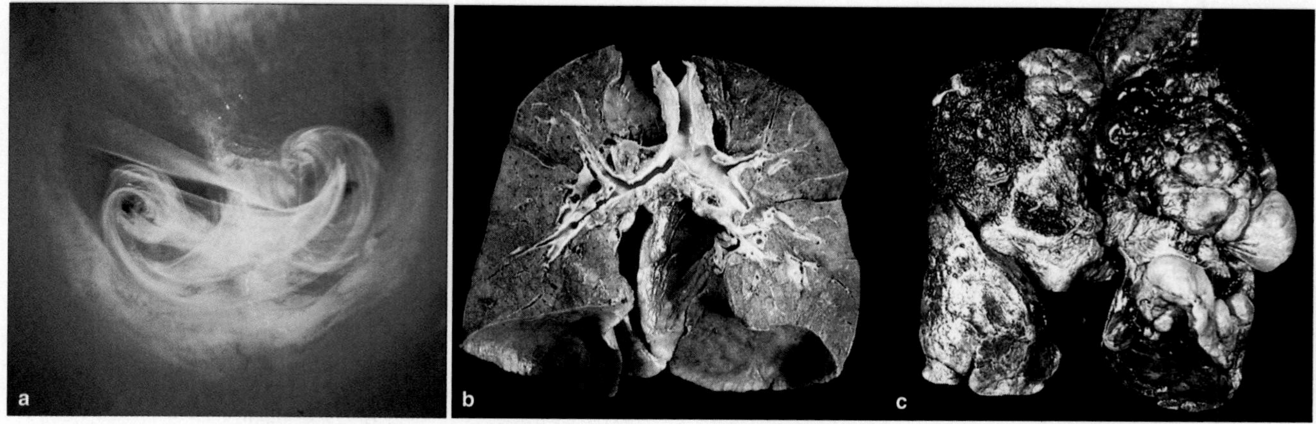

Figure 22.26 (**a**) A toxic cloud of smoke swirls toward a smoker's bronchi. (**b**) Normal human lungs. (**c**) Lung tissues from a person affected by emphysema.

BALANCING AIR AND BLOOD FLOW RATES

Gas exchange is most efficient when the flow rates of air and blood match up. The nervous system balances the two by controlling the *rhythm* and *magnitude* (rate and depth) of breathing. Fast, deep breaths expel more carbon dioxide. Shallow, slow breathing allows carbon dioxide to accumulate.

The carbon dioxide level and acidity of blood are the main factors that trigger adjustments to breathing. When either rises, the respiratory center in the brain detects the change and increases the rate and depth of breathing. Also, a deep decline in oxygen levels causes receptors in the aorta and carotid arteries to signal the respiratory center. The center responds by calling for deeper and more frequent breaths. This response may occur at high altitudes, where there is less oxygen, or when a person suffers from a serious lung disease.

Inhalation is always an active, energy-requiring process. During quiet breathing, exhalation is passive. Forceful exhalation is an active process that requires abdominal muscle contraction.

Breathing reverses pressure gradients between the lungs and the air outside the body.

Driven by its partial pressure gradient, oxygen diffuses from alveolar air spaces, and into lung capillaries. Carbon dioxide, driven by its partial pressure gradient, diffuses in the opposite direction.

Hemoglobin in red blood cells enormously enhances the oxygen-carrying capacity of blood. Most carbon dioxide is transported in blood in the form of bicarbonate.

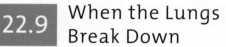

22.9 When the Lungs Break Down

In large cities, in certain workplaces, and even in the cloud around a cigarette smoker, airborne particles and gases are present in abnormally high concentrations. They put extra workloads on the respiratory system.

BRONCHITIS AND EMPHYSEMA

Ciliated and mucus-secreting epithelial cells line the walls of your bronchioles. This is one of the defenses that protect you from respiratory infections. Cigarette smoke and other airborne pollutants harm this lining (Figure 22.26). Smoking may contribute to *bronchitis*—epithelial cells lining the airways become irritated and they secrete too much mucus, which accumulates. Tiny particles get stuck in the mucus and bacteria begin to grow in it. Coughing brings up some of this gunk, but if irritation persists, so does the coughing.

Initial attacks of bronchitis are usually treated with antibiotics, which help to keep bacteria in check. But if irritation caused by smoking continues, bronchioles are chronically inflamed. Bacteria, chemical agents, or both degrade the bronchiole walls, ciliated cells die off, and mucus-secreting cells multiply. Scar tissue forms and narrows or obstructs the airways.

In chronic bronchitis, thick mucus clogs the airways. Tissue-destroying bacterial enzymes go to work on the thin, stretchable walls of the alveoli. As walls crumble, inelastic fibrous tissue forms around them. Alveoli are enlarged, and fewer exchange gases. In time, the lungs become distended and inelastic. It becomes difficult to run, to walk, and even to exhale. These problems are

EFFECTS OF SMOKING

Risks Associated With Smoking	Reduction in Risks by Quitting
SHORTENED LIFE EXPECTANCY: Nonsmokers live 8.3 years longer on average than those who smoke two packs daily from their midtwenties on.	Cumulative risk reduction; after 10 to 15 years, life expectancy of ex-smokers approaches that of nonsmokers.
CHRONIC BRONCHITIS, EMPHYSEMA: Smokers have 4–25 times more risk of dying from these diseases than do nonsmokers.	Greater chance of improving lung function and slowing down rate of deterioration.
CANCER OF LUNGS: Cigarette smoking is the major cause.	After 10 to 15 years, risk approaches that of nonsmokers.
CANCER OF MOUTH: 3–10 times greater risk among smokers.	After 10 to 15 years, risk is reduced to that of nonsmokers.
CANCER OF LARYNX: 2.9–17.7 times more frequent among smokers.	After 10 years, risk is reduced to that of nonsmokers.
CANCER OF ESOPHAGUS: 2–9 times greater risk of dying from this.	Risk proportional to amount smoked; quitting should reduce it.
CANCER OF PANCREAS: 2–5 times greater risk of dying from this.	Risk proportional to amount smoked; quitting should reduce it.
CANCER OF BLADDER: 7–10 times greater risk for smokers.	Risk decreases gradually over 7 years to that of nonsmokers.
CORONARY HEART DISEASE: Cigarette smoking is a major contributing factor.	Risk drops sharply after a year; after 10 years, risk reduced to that of nonsmokers.
EFFECTS ON OFFSPRING: Women who smoke during pregnancy have more stillbirths, and weight of liveborns averages less (which makes babies more vulnerable to disease, death).	When smoking stops before fourth month of pregnancy, risk of stillbirth and lower birthweight eliminated.
IMPAIRED IMMUNE SYSTEM FUNCTION: Increase in allergic responses, destruction of defensive cells (macrophages) in respiratory tract.	Avoidable by not smoking.
BONE HEALING: Evidence suggests that surgically cut or broken bones require up to 30 percent longer to heal in smokers, possibly because smoking depletes the body of vitamin C and reduces the amount of oxygen reaching body tissues. Reduced vitamin C and reduced oxygen interfere with production of collagen fibers, a key component of bone. Research in this area is continuing.	Avoidable by not smoking.

a

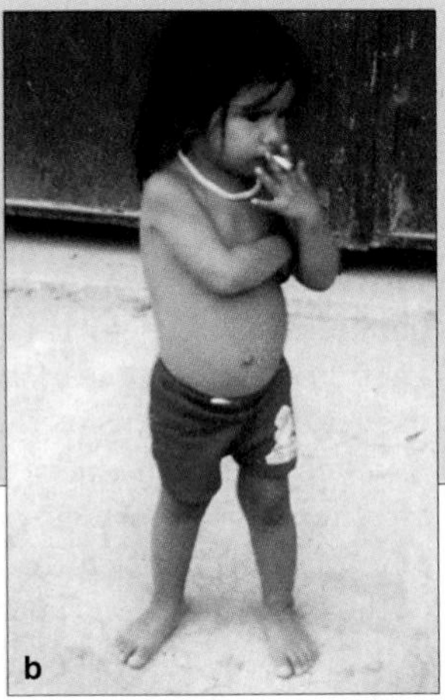

Figure 22.27 (**a**) From the American Cancer Society, a list of the major risks incurred by smoking and benefits of quitting. (**b**) A child in Mexico City already proficient at smoking cigarettes, a behavior that ultimately will endanger her capacity to breathe.

among the symptoms of *emphysema*, a chronic ailment that affects over a million people in the United States.

A few people are genetically predisposed to develop emphysema. They do not have a workable gene for the enzyme antitrypsin, which can inhibit bacterial attack on alveoli. Poor diet and persistent or recurring colds and other respiratory infections also invite emphysema later in life. However, *smoking is the major cause of the disease*. Emphysema may develop slowly, over twenty or thirty years. Once this damage has occurred, lung tissues cannot be repaired (Figure 22.26*c*).

SMOKING'S IMPACT

Tobacco use kills 4 million people annually around the world. The number will probably rise to 10 million by 2030. It is estimated that each year in the United States, the direct medical costs of treating tobacco-induced respiratory disorders drain 22 billion dollars from the economy. As G.H. Brundtland, the former director of the World Health Organization, once put it, tobacco remains the only legal consumer product that kills half of its regular users. If you are a smoker, you may wish to consider the information in Figure 22.27 carefully.

Increasingly, laws are being passed to prohibit any smoking in public buildings, workplaces, restaurants, airports, and similar enclosed spaces. Such regulations currently provide smokefree areas to a little more than a third of the population of the United States.

Worldwide, tobacco companies continue to expand their sales. Developing countries hold about 85 percent of the more than 1 billion current smokers. Women and children are increasingly among them. Mark Palmer, a former U.S. ambassador, argues that exporting tobacco is the single worst thing we do to the rest of the world.

Smoking marijuana (*Cannabis*) can also damage the respiratory system. Marijuana cigarettes are unfiltered and contain more tar than tobacco cigarettes. Smoke is inhaled more deeply and "joints" are smoked down to their very ends, where tar is the most concentrated. In addition to psychological effects, use of marijuana can cause or worsen asthma, bronchitis, and emphysema.

Summary

Section 22.1 A circulatory system moves substances to and from the interstitial fluid that bathes cells. All vertebrates have a closed circulatory system, in which a heart pumps blood through blood vessels. Fluid leaking from the blood vessels enters the lymphatic system, which filters it, and then returns fluid to the circulation.

BiologyNow

Use the animation on BiologyNow to compare open and closed circulatory systems and the kinds of closed systems of the different vertebrate groups.

Section 22.2 Blood consists of plasma, blood cells, and platelets. Plasma is water with dissolved ions and molecules. Red blood cells are packed with hemoglobin and function in the transport of oxygen and, to a lesser extent, carbon dioxide. Many kinds of white blood cells function in housekeeping and defense. Platelets release substances that initiate blood clotting. All blood cells and platelets arise from stem cells in bone marrow.

Section 22.3 The human heart pumps blood through two separate circuits. The pulmonary circuit carries oxygen-poor blood from the heart to the lungs, then returns oxygen-rich blood to the heart. The systemic circuit carries oxygen-rich blood from the heart to the tissues, then returns oxygen-poor blood to the heart.

Each half of the heart has two chambers: an upper atrium and a lower ventricle. In one cardiac cycle, action potentials from the cardiac pacemaker trigger contraction of the atria, then the ventricles. Ventricular contraction drives blood flow away from the heart.

BiologyNow

Investigate structure and function of the human cardiovascular system with the animation and interactions on BiologyNow.

Section 22.4 Blood pressure is highest in contracting ventricles, is slightly lower in the arteries, and declines as blood flows through each circuit. It is affected by changes in the heart's output and regulated mostly by adjustments in the diameter of the arterioles.

Capillary beds are diffusion zones for exchanges between blood and interstitial fluid. Veins are a blood volume reservoir and move blood back to the heart.

Section 22.5 Hypertension and atherosclerosis can lead to heart attacks or strokes. A healthy life-style can decrease the risk of cardiovascular diseases.

Section 22.6 Aerobic respiration uses oxygen and produces carbon dioxide. These gases enter and leave an animal body by crossing a respiratory surface. In small, flattened animals, this may be the body surface. Most animals exchange gases across the respiratory membranes of respiratory organs, such as gills or lungs.

BiologyNow

Compare the respiratory systems of fish and birds with the animation on BiologyNow.

Section 22.7 In human lungs, millions of alveoli make up the respiratory membrane. Air flows through nasal cavities, the pharynx, through the larynx, trachea, bronchi, and bronchioles, which end at alveolar sacs.

BiologyNow

Learn about components of the human respiratory system with the animation on BiologyNow.

Section 22.8 Each respiratory cycle consists of one inhalation and one exhalation. Inhalation requires energy. Muscle contractions expand the chest cavity, lung pressure decreases below atmospheric pressure, and air flows in. The events are reversed during exhalation, which is usually passive.

Oxygen follows its partial pressure gradient from alveolar air spaces into the pulmonary capillaries, then into red blood cells, where it binds with hemoglobin. In capillary beds in the tissues, hemoglobin releases oxygen, which diffuses across interstitial fluid into cells. Carbon dioxide diffuses from cells into the interstitial fluid, then into the blood. Here, it reacts with water to form bicarbonate. The reactions are reversed in the lungs; carbon dioxide diffuses from the blood into air in alveoli, then is expelled.

A respiratory center in the brain controls the rate and depth of breathing.

BiologyNow

See how pressure changes during the respiratory cycle and investigate the effects of partial pressure gradients with animation and interaction on BiologyNow.

Section 22.9 Bronchitis and emphysema are respiratory disorders caused or worsened by smoking. Worldwide, smoking is a leading cause of deaths.

Self-Quiz

Answers in Appendix I

1. All vertebrates have _______ .
   a. an open circulatory system
   b. a closed circulatory system
   c. a four-chambered heart
   d. both b and c

2. Blood flows from the left atrium to the _______ .
   a. aorta
   b. left ventricle
   c. left atrium
   d. pulmonary arteries

3. Blood pressure is highest in the _______ and lowest in the _______ .
   a. arteries, veins
   b. arterioles, venules
   c. veins, arteries
   d. capillaries, arterioles

4. Contraction of _______ drives the flow of blood away from the heart.
   a. the atria
   b. arterioles
   c. the ventricles
   d. skeletal muscle

5. At rest, the largest volume of blood is in the _______ .
   a. arteries
   b. capillaries
   c. veins
   d. arterioles

6. In the blood, most oxygen is transported _______ .
   a. in red blood cells
   b. in white blood cells
   c. bound to hemoglobin
   d. both a and c

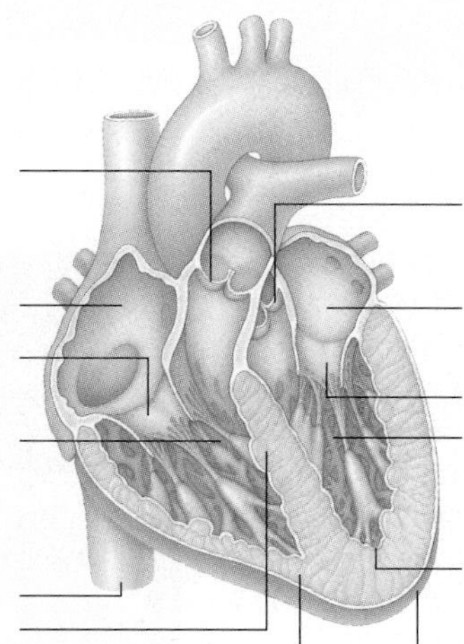

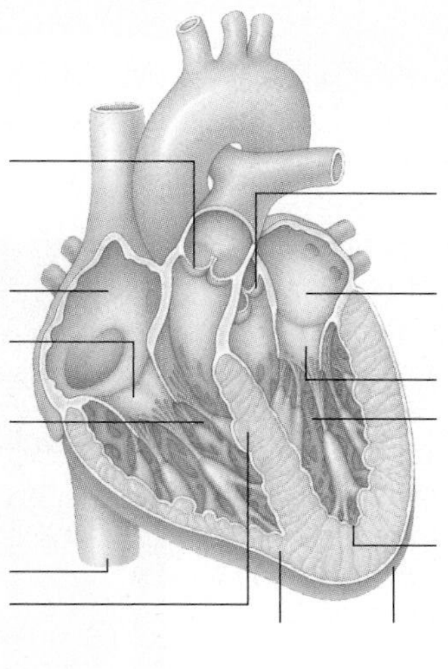

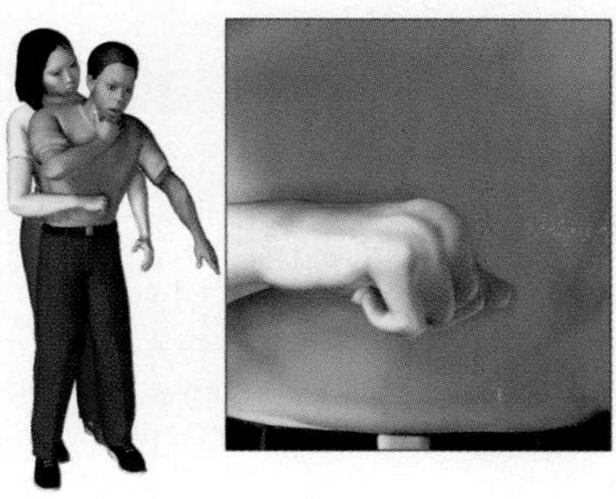

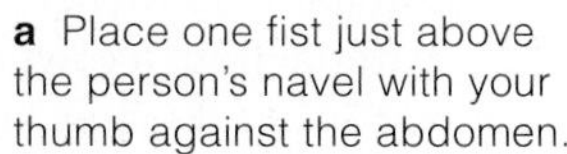

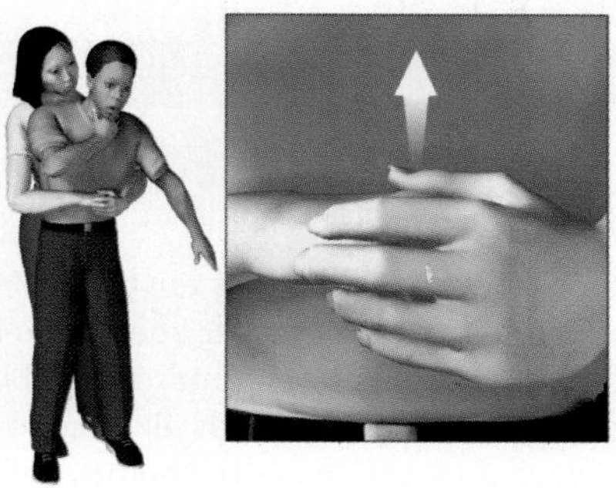

a Place one fist just above the person's navel with your thumb against the abdomen.

b Cover the fist with your other hand. Thrust both up and in with sufficient force to lift the person off his or her feet.

Figure 22.28 The Heimlich maneuver

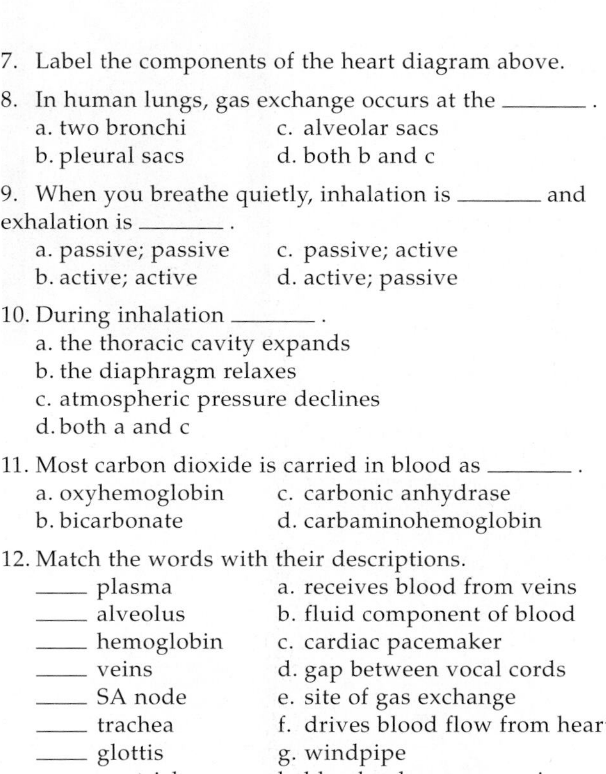

7. Label the components of the heart diagram above.

8. In human lungs, gas exchange occurs at the ______ .
   a. two bronchi c. alveolar sacs
   b. pleural sacs d. both b and c

9. When you breathe quietly, inhalation is ______ and exhalation is ______ .
   a. passive; passive c. passive; active
   b. active; active d. active; passive

10. During inhalation ______ .
   a. the thoracic cavity expands
   b. the diaphragm relaxes
   c. atmospheric pressure declines
   d. both a and c

11. Most carbon dioxide is carried in blood as ______ .
   a. oxyhemoglobin c. carbonic anhydrase
   b. bicarbonate d. carbaminohemoglobin

12. Match the words with their descriptions.

____ plasma	a. receives blood from veins
____ alveolus	b. fluid component of blood
____ hemoglobin	c. cardiac pacemaker
____ veins	d. gap between vocal cords
____ SA node	e. site of gas exchange
____ trachea	f. drives blood flow from heart
____ glottis	g. windpipe
____ ventricle	h. blood volume reservoir
____ atrium	i. respiratory pigment

Additional questions are available on **Biology Now™**

Critical Thinking

1. When you swallow food, contracting muscles force the epiglottis down, to its closed position. This prevents food from going down the trachea and blocking air flow. Yet each year, several thousand people choke to death after food enters the trachea and blocks air flow for as little as four or five minutes.

The *Heimlich maneuver*—an emergency procedure only—can often dislodge food stuck in the trachea. The rescuer stands behind the victim, makes a fist with one hand, then positions the fist, thumb-side in, against the abdomen of the person at risk (Figure 22.28). The fist must be slightly above the navel and well below the rib cage. The fist is pressed into the abdomen with a sudden upward thrust. The thrusts can be repeated if necessary.

When correctly performed, the Heimlich maneuver forcibly elevates the diaphragm, causing a sharp decrease in the thoracic cavity volume and an abrupt rise in alveolar pressure. Air forced up into the trachea by the pressure may be enough to dislodge the obstruction. Once the obstacle is dislodged, the victim must see a doctor at once because an inexperienced rescuer can inadvertently cause internal injuries or crack a rib.

Reflect on our current social climate, in which lawsuits abound. Would you perform the Heimlich maneuver to save a friend's life? A stranger's life? Why or why not?

2. The deaths of a few airline travelers have led to worry about *economy-class syndrome*. The theory is that sitting motionless for long periods on flights allows blood to pool and clots to form in the legs. Low oxygen levels in airline cabins may increase clotting. If a clot becomes large enough to block blood flow or breaks free and is carried to the lungs or the brain, the outcome can be deadly.

There might be a time lag between clot formation and health problems, so a connection to air travel might easily have been overlooked. Studies are now under way to determine whether economy-class travel represents a significant risk. Given what you know about blood flow in the veins, explain why periodically getting up and moving around in the plane's cabin during a long flight may lower the risk of clot formation.

3. Carbon monoxide is a colorless, odorless gas in exhaust fumes from burning fossil fuel, wood, and tobacco. Carbon monoxide can bind to hemoglobin's oxygen-binding sites. It does so far more strongly, so it does not dissociate from hemoglobin as easily as oxygen does. Exposure to even a small amount of this gas can make up to 50 percent of the oxygen-binding sites in hemoglobin unavailable for oxygen transport. What effect would carbon monoxide poisoning have on body tissues?

23 IMMUNITY

IMPACTS, ISSUES

The Face of AIDS

Chedo Gowero (*below*) was ten years old when both of her parents died from AIDS. She had to leave school to support herself, her grandmother, and her seven-year-old brother. She now spends all of her days gathering firewood and working in the homes and fields of her neighbors. To keep her younger brother in school, she had no choice but to assume adult responsibilities.

Chedo and her brother are among the 12 million African children orphaned because of AIDS. They are lucky. Many others have been forced into prostitution because they had no other way to support themselves.

By the end of 2004, AIDS had claimed the lives of an estimated 20 million people. At least 40 million are now infected with HIV, the virus that causes it (shown in the filmstrip). Even after twenty years of top-notch research all over the world, we still do not have an effective vaccine against AIDS.

Why not? After all, researchers have developed vaccines against twenty or so dangerous diseases. The very first vaccine was conjured up long ago as a defense against the smallpox epidemics that swept again and again through the world's cities. Some outbreaks were so severe that only half of the stricken survived. The survivors had permanent scars on their face, neck, arms, and shoulders, but seldom contracted the disease again. They were said to be protected, or "immune," from smallpox.

No one had a clue to what caused smallpox or any other disease, but the idea of acquiring immunity was immensely appealing. For centuries, people gambled with immunity by poking all manner of materials into their skin—pus, scabs, and the like. Many of them lost that gamble. Then, in 1796, Edward Jenner injected material from a cowpox sore into the arm of a healthy boy. Six weeks later, he injected the boy with *smallpox* scabs. Both were lucky; the boy did not get smallpox. As Jenner had predicted, the infectious agent that causes cowpox can provoke immunity to smallpox.

Jenner named his procedure **vaccination**, after the Latin word for cowpox (*vaccinia*). Though controversial, the use of Jenner's vaccine spread quickly throughout Europe, and then to the rest of the world. Worldwide use of the vaccine had eradicated smallpox by 1970.

Since the beginning of the twentieth century, major advances in microscopy, biochemistry, and molecular biology have increased our understanding of the body's defenses. As AIDS reminds us, however, we still have a lot to learn. And with this sobering thought, we invite you into the world of immunity.

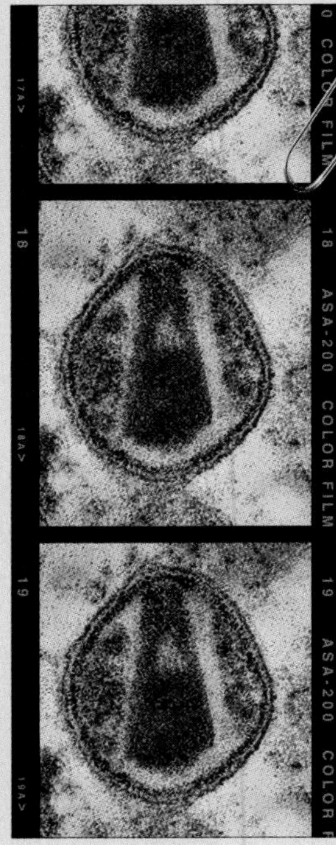

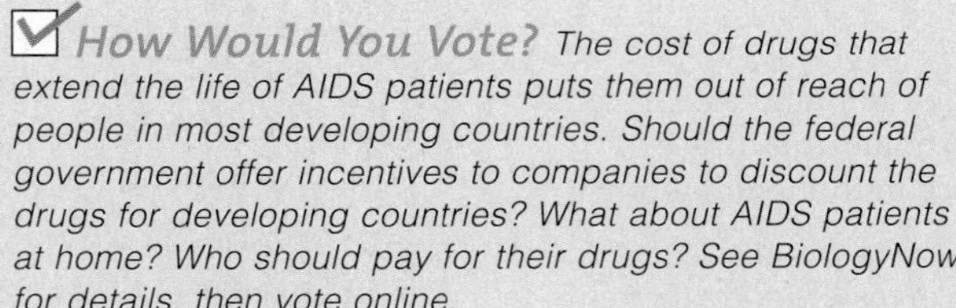

How Would You Vote? *The cost of drugs that extend the life of AIDS patients puts them out of reach of people in most developing countries. Should the federal government offer incentives to companies to discount the drugs for developing countries? What about AIDS patients at home? Who should pay for their drugs? See BiologyNow for details, then vote online.*

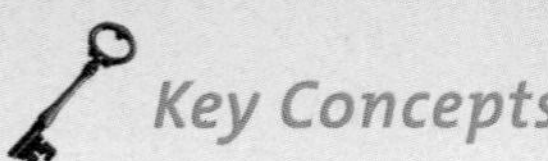

Key Concepts

THE BODY'S DEFENSES
Physical, chemical, and cellular responses counter a tremendous variety of pathogens and cancer cells.

MECHANISMS OF IMMUNITY
Skin, mucous membranes, chemical secretions, and normally harmless microbes colonizing body surfaces are all physical barriers that keep pathogens outside the body.

Innate immunity targets threats in general. Phagocytic cells, plasma proteins, inflammation, and fever quickly rid the body of many invaders.

If a threat persists, adaptive immunity begins against specific invaders. In an antibody-mediated immune response, B cells make and secrete antibodies. In a cell-mediated immune response, cytotoxic T cells directly destroy ailing body cells.

IMMUNITY GONE WRONG
Allergies, autoimmune disorders, and immune deficiencies are the outcomes of faulty or failed immune mechanisms.

Links to Earlier Concepts

In this chapter you will be integrating what you have learned about disease-causing microorganisms and their hosts. You will be applying your knowledge of prokaryotic cells (Section 14.2) and viruses (14.5) as you learn about their interactions with eukaryotic cells. You will be using what you know about protein structure (2.8), the endomembrane system (3.5), endocytosis and phagocytosis (4.7), and RNA processing (10.2) to understand eukaryotic cellular defenses. Gene expression control (10.6) and cell signaling (17.5) gave you the background to learn about immune signaling mechanisms. You will see how the body's systems, including skin (20.4), the circulatory system (22.1), white blood cells (22.2), and the respiratory system (22.7), work to fight infection. Finally, you may wish to review the section on the coevolution of pathogens and hosts (14.6).

23.1 Integrated Responses to Threats

Reflect on the evolution of the circulatory and lymphatic systems of vertebrates that colonized land. As body fluid circulation became more efficient, so did body defenses against pathogens. Cellular defenders and molecular weapons now intercept diverse pathogens tumbling along in blood or attacking internal tissues.

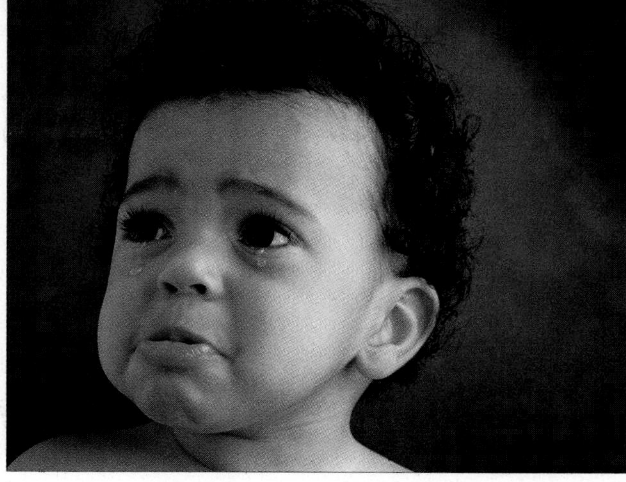

Figure 23.1 Yes, you have to read this chapter. But immunity need not be scary. Your body has three very effective lines of defense that keep it safe from pathogens. Even tears contain an antimicrobial protein.

EVOLUTION OF THE BODY'S DEFENSES

LINKS TO SECTIONS 4.7, 14.2, 14.6, 22.1

You continuously cross paths with a staggering array of viruses, bacteria, fungi, parasitic worms, and other agents of disease. They are in the air that you breathe, the food you eat, and everything you touch. You need not lose sleep over this, however (Figure 23.1). Having coevolved with most of them, you and all other animal species have very effective defenses.

Immunity, the body's ability to resist and combat infections, started millions of years ago as multicelled organisms evolved from free-living cells. Mutations in membrane proteins introduced new patterns that were unique to all cells of a given type. Mutations also gave rise to mechanisms that could recognize these proteins as belonging to *self*, meaning part of one's own body. Self-recognition gave cells the capacity to stay together to form tissues. Just as importantly, by default it gave them the capacity to recognize *nonself*.

Discriminating between self and nonself was a key innovation. It foreshadowed the detection of **antigen**: any molecule that the body recognizes as nonself and that provokes an immune response. Most antigens are proteins, lipids, and oligosaccharides. They typically are present on viruses, bacteria and other foreign cells, tumor cells, toxins, and allergens.

Antigen-detecting mechanisms evolved two billion years ago, before the divergences that led to diverse multicelled eukaryotes. Genes evolved that encoded a fixed set of receptors for molecular patterns that occur mainly on pathogens. Today, all multicelled organisms have these receptors.

By 700 million years ago, pattern receptors of early animals had become functionally linked with chemical responses. Anything that attached to pattern receptors on an animal cell triggered it to release complement. In humans, **complement** is a set of about thirty proteins circulating in the blood. Complement can kill microbes directly or tag them for engulfment by phagocytosis (Section 4.7). **Phagocytes** do the engulfing.

Microbial pattern receptors and complement offered **innate immunity**—fast, preset responses to a fixed set of nonself cues. Invertebrates still survive without any other kinds of immune defenses. Actually, until about 450 million years ago, there was no alternative. Novel pathogens went unrecognized, and as such they often proved fatal. Adapting to them was not possible in the individual's lifetime.

Then, in the class of vertebrates known as jawed fishes, signaling molecules called **cytokines** evolved. So did white blood cells called **lymphocytes**. *Cytokines and lymphocytes together tailor immune defenses to a vast array of specific pathogens that an individual may encounter during its lifetime.* This capacity is **adaptive immunity**. Unlike innate immunity, adaptive immunity is specific, selective, remembered, and regulated. All of the cells, body tissues, and proteins that take part in adaptive immunity evolved as an integrated system. Table 23.1 compares innate and adaptive responses.

THREE LINES OF DEFENSE

It is important to remember that adaptive immunity evolved within the context of innate immune systems. Researchers once thought immune cells and chemical defenses took part either in one response or the other, but now it is clear that the systems are integrated. We describe the integrated systems in terms of three lines of defense, detailed in the next few sections.

First, the intact skin and linings of body tubes and cavities are effective physical barriers that exclude most

Table 23.1 Comparison of Adaptive and Innate Immunity

	Innate Immunity	Adaptive Immunity
Response time:	Immediate	Slower
How antigen is detected:	Fixed gene sequences for unchanging set of about 1,000 receptors	Random recombinations of gene sequences give rise to tremendous receptor diversity
Triggers:	Damage to tissues; patterns on microbes	Pathogens, toxins, altered body cells
Memory:	None	Long-term

basophil

mast cell

T lymphocyte (T cell)

B lymphocyte (B cell)

neutrophil

eosinophil

dendritic cell

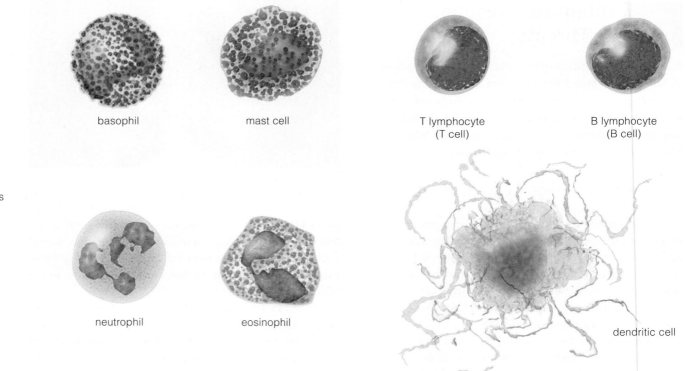

Figure 23.2 Gallery of cellular defenders: sketches of cells that carry out the immune responses. Nearly all function in both innate and adaptive immunity. Staining them reveals structural details. The cytoplasmic granules of basophils, mast cells, and eosinophils contain enzymes, toxins, and signaling molecules.

LINKS TO SECTIONS 4.7, 22.2

pathogens. Resident populations of microorganisms on skin and mucous membrane surfaces afford additional protection by displacing or competing with occasional dangerous intruders.

Innate immunity, the second line of defense, starts immediately after antigen has been detected internally. Specific invaders are not targeted; instead, pathogen-associated molecular patterns trigger some generalized responses, including engulfment by phagocytic white blood cells, activation of complement, fever, and acute inflammation. These mechanisms can eliminate many invaders before infection becomes established.

Table 23.2 Chemical Weapons of Immunity and Their Effects

Complement	Directly kills cells; enhances lymphocyte responses
Cytokines	Cell–cell and cell–tissue communication:
Interleukins	Cause inflammation and fever, cause T cells and B cells to undergo division and differentiation, stimulate bone marrow stem cells, attract phagocytes, activate NK cells
Interferons	Impart resistance to virus infection, activate NK cells
Tumor Necrosis Factor	Causes inflammation; kills tumor cells, causes T cells to accumulate in lymph nodes during infection
Other Chemicals	Various antimicrobial and defensive effects
Enzymes and peptides; clotting factors; protease inhibitors; toxins; hormones	

Activation of innate immunity sets in motion the third line of defense, adaptive immunity. Lymphocytes divide to form huge populations of cells that target specific threats. Many of them act immediately. Others persist long after an infection ends. If the same antigen returns, these memory cells will implement a faster, secondary response.

INTRODUCING THE DEFENDERS

White blood cells participate in both the adaptive and innate immune responses. All arise from stem cells in bone marrow. Many kinds circulate in the blood and lymph; others populate the lymph nodes, spleen, liver, kidneys, lungs, brain, and tissues in other organs. A number of them are phagocytic, and all are secretory (Figure 23.2). All white blood cells release cytokines and other signaling molecules that function in cell-to-cell communication. Cytokines, including interleukins, interferons, and tumor necrosis factor, control every aspect of vertebrate immunity. Some white blood cells also secrete various enzymes and toxins that directly kill microbes (Table 23.2).

Like miniature SWAT teams, white blood cells react to anything foreign in the body. **Neutrophils**, the most abundant of the circulating white blood cells, are fast acting. **Macrophages** are the big eaters of the bunch; each can engulf up to a hundred bacterial cells. They are mature forms of phagocytic monocytes that patrol

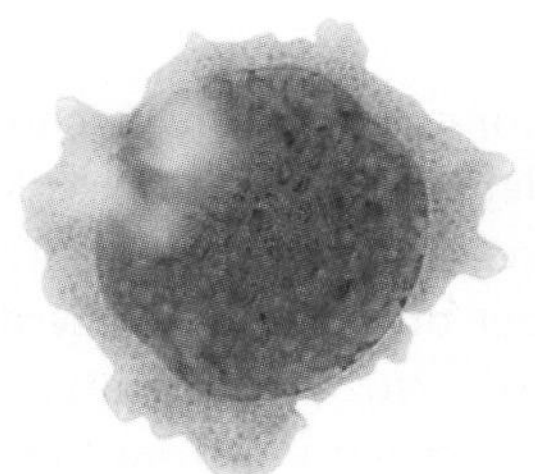

Natural Killer (NK) cell

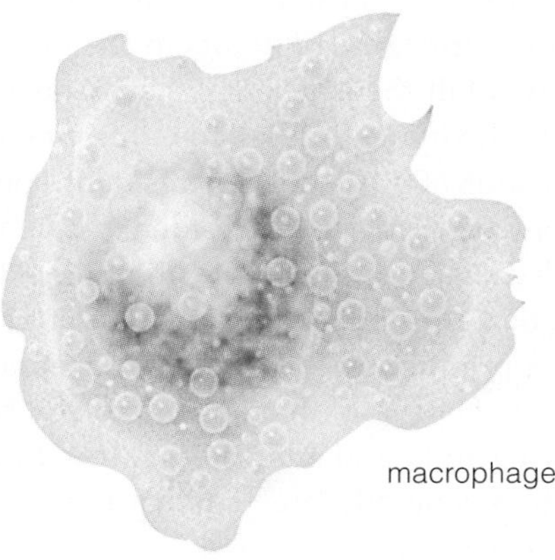

macrophage

blood. The main function of phagocytic **dendritic cells** is to alert the adaptive immune system to the presence of antigen in interstitial fluid of skin and body linings.

The granules inside some white blood cells contain enzymes and toxic proteins which, when released into local tissues, kill extracellular pathogens. **Basophils** that circulate in blood and **mast cells** in tissues release enzymes and histamines in response to antigen. Often associated with nerve tissues, mast cells may link the nervous and immune systems. **Eosinophils** specialize in parasite control, targeting organisms that are too big for phagocytosis—protozoa, fungi, and worms.

B and **T lymphocytes**—which we call B and T cells—are central to adaptive immunity. In these cells alone, the information in antigen-receptor genes gets shuffled and spliced into spectacularly diverse combinations. **Natural killer cells** (NK cells) are lymphocytes that are most active in innate responses, but they also have a pivotal role in adaptive responses. They destroy body cells that have been altered by viral infection or cancer, and so are unrecognized by B or T cells.

Body surfaces are a physical barrier to trespassers. Innate immune responses target nonspecific threats, and adaptive immunity provides lasting protection from specific invaders.

White blood cells and signaling molecules carry out both innate and adaptive immune responses.

23.2 Surface Barriers

LINKS TO SECTIONS 20.4, 22.4

A pathogen can only cause infection if it enters the body's internal environment. It must first penetrate the skin and other protective barriers on body surfaces (Table 23.3). These barriers are your body's first line of defense.

INTACT SKIN

Your skin is teeming with about 200 different kinds of microbial life forms. If you showered today, there are probably thousands of them per square inch of your external surfaces; if you did not, there may be billions. Your outer surfaces are in continuous contact with the external environment, so you cross paths with plenty of microorganisms that can (and do) live very well on your skin. They particularly favor warm, moist spots, such as between your toes.

Your skin provides these microbes with a relatively stable environment and a constant supply of nutrients. In return, their large populations deter more insidious types from colonizing your surfaces.

Microorganisms cannot get past healthy, intact skin. Vertebrate skin has tough surface layers made of dead, keratin-stuffed epithelial cells packed tightly together (Section 20.4). Microorganisms flourish on top of this waterproof, oily environment without harming you, unless your skin is damaged or weakened. Repeatedly enclose your toes in warm, damp shoes, for instance, and you are inviting some of the fungi to penetrate the sodden tissues and cause athlete's foot.

LININGS OF TUBES AND CAVITIES

The internal tissues of healthy people are typically free of microorganisms. Body cavities are another matter. Different populations of microbes inhabit the cavities and tubes opening on the body's surface, including the eyes, nose, mouth, and urinary and genital openings. Their presence provides some benefit, but only on the

Table 23.3 Examples of Surface Barriers

Physical Barriers	Intact skin and epithelia of body tubes and cavities, such as the gut and eye sockets
Mechanical Barriers	Sticky mucus; broomlike action of cilia; flushing action of tears, saliva, urination, diarrhea
Chemical Barriers	Protective secretions (sebum and other waxy coatings); low pH of gastric juices, urinary and vaginal tracts; chemicals such as lysozyme; established populations of resident microbes

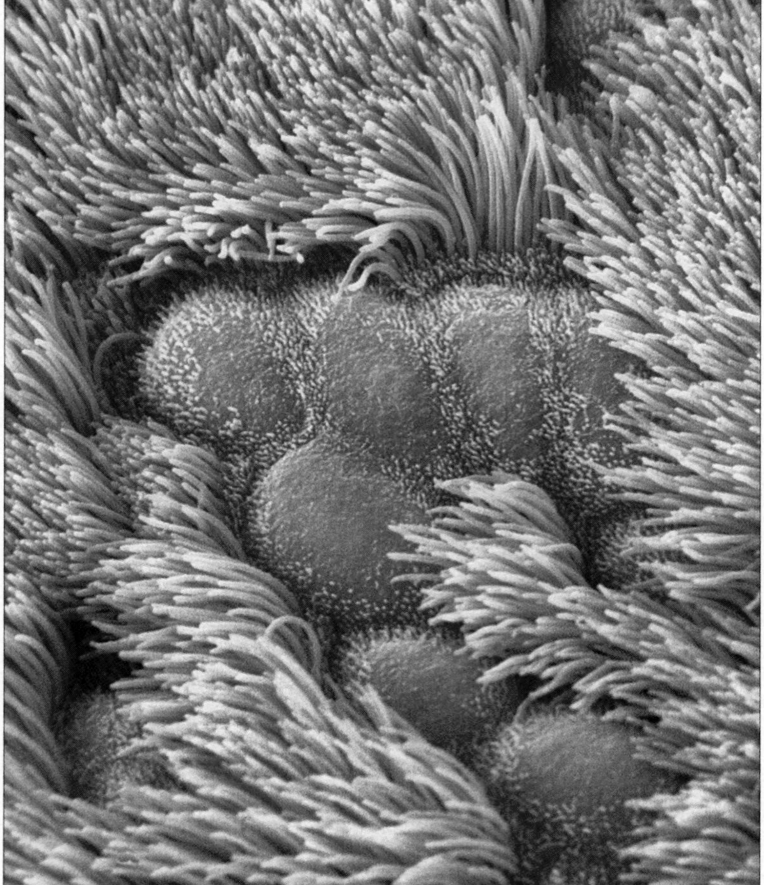

Figure 23.3 Part of the free surface of an airway to the lungs. Goblet cells (*gold*) are especially abundant in the glandular epithelia that line airways. Bacteria and particles get stuck in the mucus secreted by goblet cells, then ciliated cells (*pink*) sweep them toward the mouth for disposal.

outsides of membranes: Some can be quite dangerous if they invade internal tissues. Chemical and mechanical barriers keep them on the right side of the membranes.

A coating of thick, sticky mucus on the free surface of many epithelial linings is one such barrier. Mucus contains an enzyme called **lysozyme**, which cleaves polysaccharides in bacterial cell walls and so unravels their structure. The tears that rinse your eyes are rich with lysozyme. They help keep microbial populations around your eyes in check.

When you breathe in, air pressure forces airborne bacteria against the mucus-coated lining of airways to the lungs (Figure 23.3). Coughing flings many bacteria back outside. Lysozyme in the mucus ensures that the rest will not survive long enough to breach the walls of the sinuses and the lower respiratory tract. Many ciliated cells are components of the lining. Like tiny brooms, the cilia beat in synchrony at the lining's free surface. Their action sweeps debris and bacteria-laden mucus to the throat for disposal.

Your mouth is a particularly inviting habitat for microorganisms, offering nutrients, warmth, moisture, and epithelial surfaces for colonization. Accordingly, it harbors gigantic populations of various streptococci, lactobacilli, staphylococci, and other types of bacteria, including a great number of anaerobes. Microbes that colonize the mouth resist lysozyme in saliva. Most of the ones that enter the stomach die from the low pH. Thanks to the secretions from glandular cells of the stomach lining, gastric fluid is a potent mixture, rich in hydrochloric acid and protein-digesting enzymes. If any microbes can withstand that brew and survive to reach the small intestine, the bile salts in the intestinal lumen usually will kill them. The few hardy ones that manage to get to the lumen of the large intestine must compete with about 500 established resident species. Even then, if they do displace the residents, a flushing mechanism called diarrhea typically will remove them before they can damage and breach the gut lining.

Lactic acid, a product of *Lactobacillus* fermentation, keeps the vaginal pH beyond the range of tolerance for most species of bacteria and fungi. Also, the flushing action of urination generally prevents pathogens from colonizing the urinary tract.

AN UNEASY BALANCE

On the outside of your body, your normal microbial inhabitants can be either neutral or beneficial to your health, but *keeping* them on the outside is critical to it. Some of these microbes are not particularly harmless. If body surfaces fail, the resulting invasion may result in illnesses—both minor and severe.

Colonization of internal tissues by common bacteria has been proven to cause ulcers, meningitis, whooping cough, colitis, bacterial endocarditis, pneumonia, lung and brain abscesses, sepsis, and duodenal, gastric, and colon cancers.

The bacterial agent of tetanus, *Clostridium tetani*, passes through our digestive systems so often that we classify it as indigenous. The bacterium responsible for diphtheria, *Corynebacterium diphtheriae*, was considered to be a normal skin inhabitant before widespread use of the diphtheria vaccine obliterated the disease. The most common inhabitant of human external surfaces, *Staphylococcus epidermidis*, is also currently the leading cause of hospital-acquired bacterial disease.

Our intact, healthy skin and mucous membranes usually keep these microorganisms at bay. But changes in physiology that accompany maturation, aging, and illnesses may alter our surface defense mechanisms.

For example, *Propionibacterium acnes*, a rod-shaped bacterium, is a pervasive skin occupant (Figure 23.4*a*). It feeds on sebum, a greasy mixture of glycerides, fats, and waxes that lubricates the skin and hair. Sebaceous glands secrete sebum at hair follicles. During puberty,

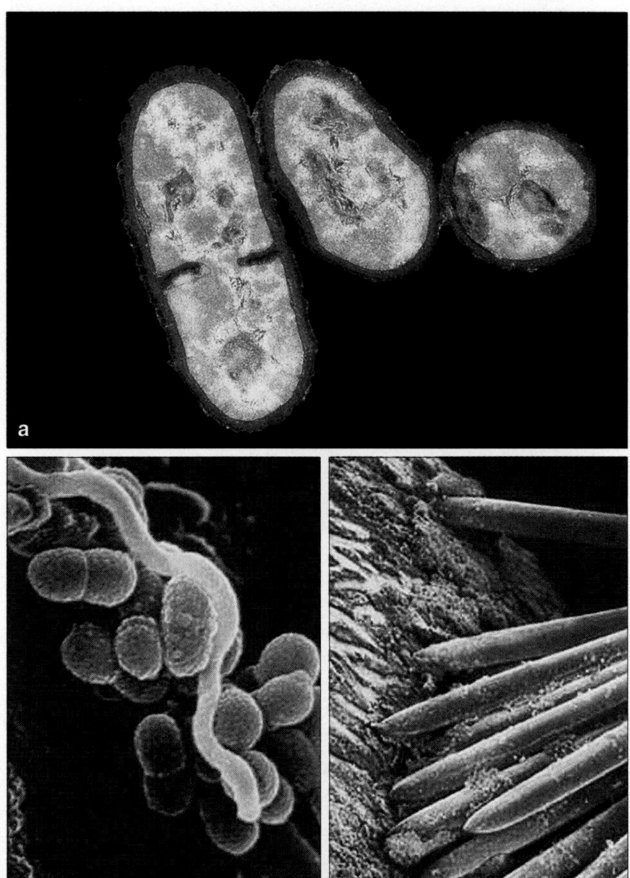

Figure 23.4 Common inhabitants of external human surfaces. (**a**) Overgrowth of this common skin bacterium, *Propionibacterium acnes*, causes acne. (**b**) Plaque is caused by *Streptococcus mutans*, which is actually a group of related bacteria. (**c**) Micrograph of toothbrush bristles scrubbing plaque on a tooth surface.

the levels of sex hormones in the bloodstream increase, causing sebaceous glands to make more sebum than before. Excess sebum, along with dead skin cells that flake away at the skin surface, obstruct the openings of hair follicles. *P. acnes* can survive on the surface of the skin, but far prefer anaerobic habitats. They multiply to tremendous numbers inside the closed hair follicles. Secretions of the flourishing *P. acnes* populations leak into internal tissues, initiating inflammation around the follicles. The resulting pustules are called *acne.*

Arguably the most costly negative effect of normal flora is dental **plaque**, a thick mixture of bacteria, their extracellular products, and saliva glycoproteins that sticks tenaciously to teeth. Normal bacterial residents of the mouth, *Streptococcus mutans* and *S. sanguis*, are responsible for it (Figure 23.4*b,c*). Other bacteria that live in the plaque deposits are fermenters. They break down bits of carbohydrate that stick to teeth and then secrete organic acids, which etch tooth enamel away.

In young, healthy people, tight junctions between the gum epithelium and teeth form a barrier that keeps bacteria out of the internal environment. As we age, connective tissue beneath gum epithelium thins, and so the barrier becomes vulnerable. Deep pockets form between the teeth and gums, and a very nasty gang of anaerobic bacteria accumulates in these pockets. Their noxious secretions, including destructive enzymes and acids, cause inflammation of surrounding gum tissues. This is *periodontitis*, and it forms an oral wound that is a portal for pathogenic invasion of internal tissues.

As you can see, body surface barriers are in uneasy balance with actual and potential colonizers. That is why it is important to remember that antibiotics often kill more than their intended targets. They can tip the balance by killing off resident bacterial populations, which allows pathogenic opportunists to move in.

Surface barriers protect the internal environment of all vertebrates. The barriers include intact skin, lysozyme and other secretions, and linings of internal tubes and cavities. The linings have mucus-secreting cells and often ciliated cells.

Also, populations of normally harmless bacteria prevent pathogens from colonizing body surfaces.

23.3 The Innate Immune Response

Phagocytosis, inflammation, and fever are the body's fixed, off-the-shelf set of mechanisms that act at once to counter threats in general and prevent infection. Every kind of vertebrate starts out life with these mechanisms.

LINKS TO SECTIONS 3.8, 4.3, 17.3

Macrophages in interstitial fluid are usually the first to enter the scene of an invasion. They engulf and digest essentially anything other than undamaged body cells (Figure 23.5). If their pattern receptors bind an antigen, they secrete cytokines that, among other effects, attract dendritic cells, neutrophils, and more macrophages.

Complement is also an important aspect of innate immunity. Vertebrates make about 30 different types of complement proteins on an ongoing basis, mainly in the liver. Circulating throughout blood and interstitial fluid, the different complement proteins bind various molecular patterns that are characteristic of pathogens. A complement molecule becomes "activated" when it binds to one of these antigens. Activated complement activates more complement molecules, which activate more complement, and so on. Such cascading reactions produce tremendous local concentrations of activated complement molecules very quickly.

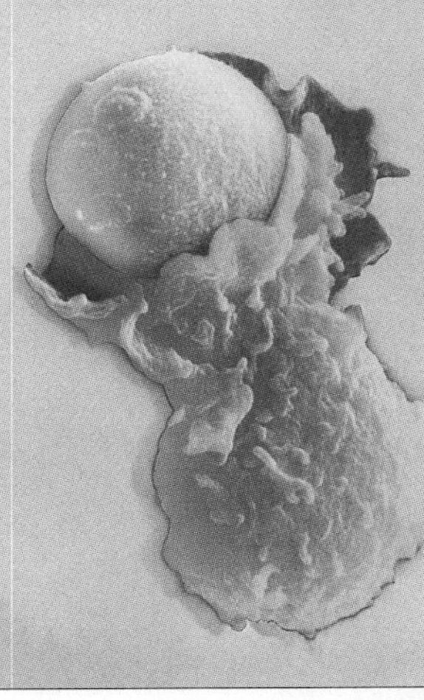

Figure 23.5 Macrophage caught in the act of engulfing a pathogenic yeast cell.

The activated complement proteins have big effects. They are chemotactic, meaning they attract phagocytic cells. Like snuffling bloodhounds, phagocytes follow the complement gradients to their most concentrated source, in the threatened tissue. Complement proteins also attach themselves to invaders. Phagocytic white blood cells have complement receptors, so a microbe covered with complement is recognized and engulfed more quickly than one that has no complement coat. Some complement molecules form membrane attack complexes. These structures insert themselves into the cell wall or plasma membrane of a bacterium and so promote its disintegration (Figure 23.6).

Activated complement and cytokines secreted by macrophages both trigger **acute inflammation**, a fast, local, nonspecific response to tissue invasion (Figure 23.7). Outward symptoms of inflammation are redness, swelling, warmth, and pain. A series of internal events causes these symptoms.

First, mast cells in connective tissue respond either to complement cascades or directly to an antigen by secreting histamines and cytokines into the interstitial fluid. Histamines make arterioles in the tissue dilate. As a result, more blood flows through them, carrying a supply of phagocytes following the cytokine trail. As blood engorges the arterioles, the tissue reddens and warms with bloodborne metabolic heat.

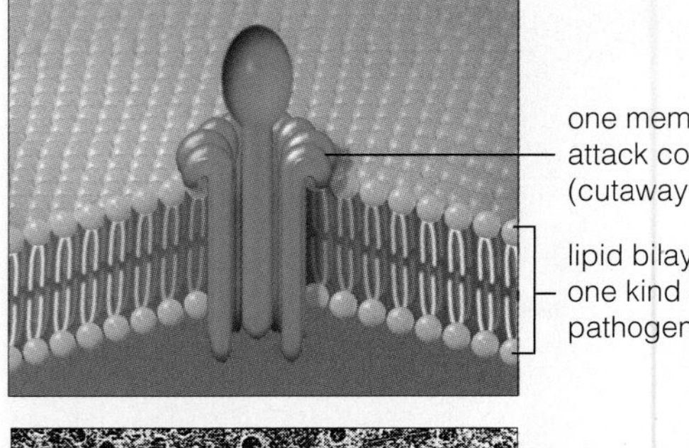

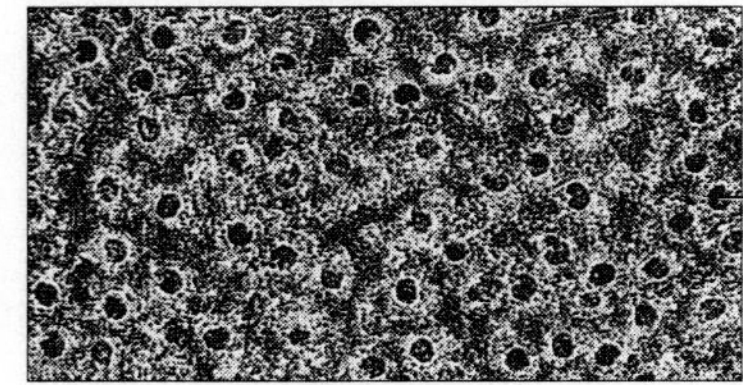

Figure 23.6 Activated complement assembles into membrane attack complexes that insert themselves into the lipid bilayer of pathogens. The resulting pores make the cell disintegrate.

Histamines also make small blood vessels "leaky." They induce endothelial cells composing blood vessel walls to shrink and pull apart, so plasma proteins and phagocytes escape the vessels (Figure 23.8). Osmotic pressure in interstitial fluid rises, and fluid balance across vessel walls shifts, swelling the tissue. Pressure on free nerve endings gives rise to sensations of pain.

In addition to complement proteins, other plasma proteins leaking into interstitial fluid include clotting

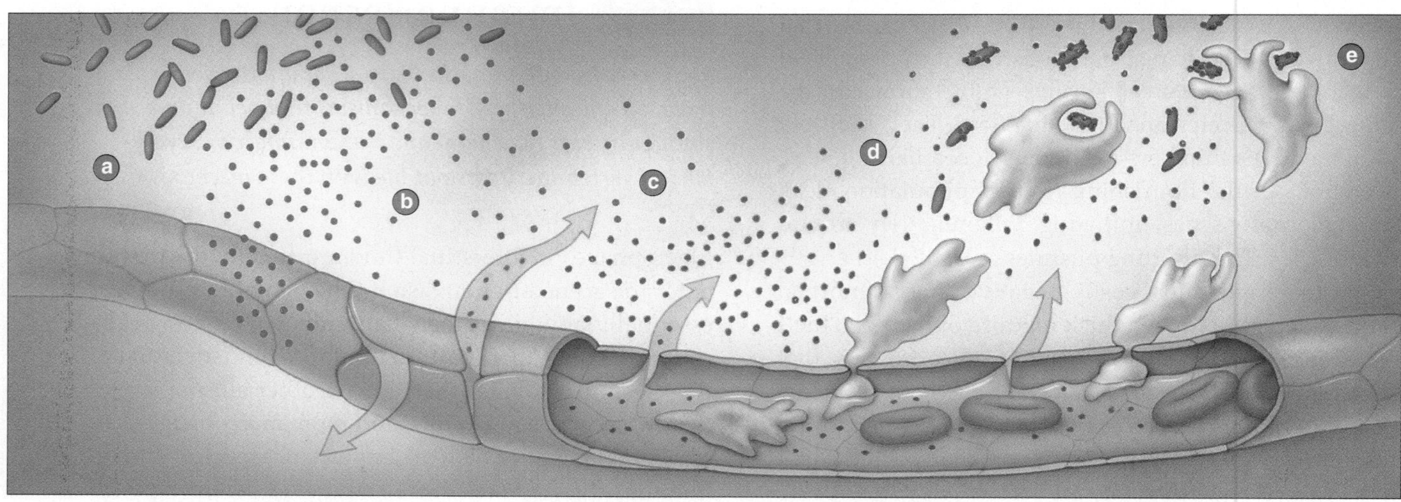

a Bacteria invade a tissue and directly kill cells or release metabolic products that damage tissue.

b Mast cells in tissue release histamines, which then trigger arteriole vasodilation (hence redness and warmth) as well as increased capillary permeability.

c Fluid and plasma proteins leak out of capillaries; localized edema (tissue swelling) and pain result.

d Complement proteins attack bacteria. Clotting factors wall off inflamed area.

e Neutrophils and macrophages engulf invaders and debris. Some macrophage secretions kill targets, attract more lymphocytes, and call for fever.

Figure 23.7 ***Animated!*** Inflammation in response to bacterial infection. White blood cells and plasma proteins enter an infected tissue.

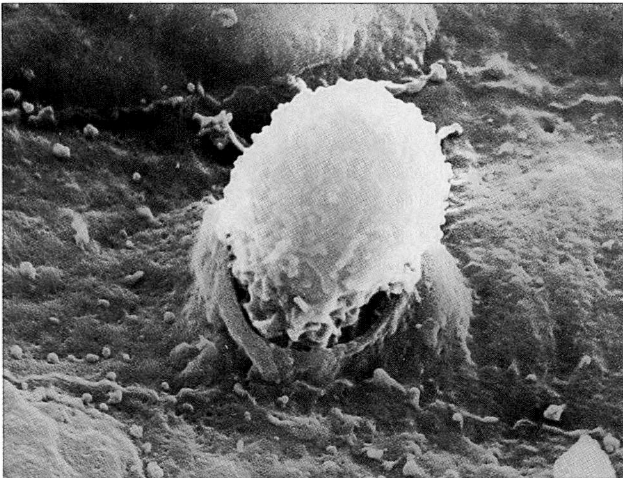

Figure 23.8 Phagocyte squeezing through a blood vessel wall. Such cells kill invaders, remove cellular debris, and alert the immune system to threats.

factors that are activated by macrophage secretions. Fibrin threads form, stick to exposed collagen fibers in damaged tissue, and trap blood cells and platelets to form a clot. Clots wall off inflamed areas and delay the spread of invaders into surrounding tissues.

Fever is an internal body temperature that exceeds the normal 37°C (98.6°F). Some cytokines secreted by macrophages stimulate the brain to make and release prostaglandins. These signaling molecules then raise the set point on the hypothalamic thermostat, which controls internal temperature (Section 17.3).

Fevers are not usually harmful. Quite the contrary, in humans, a fever of 39°C (102°F) enhances immunity by increasing the rate of enzyme activity (Section 4.3), thus speeding metabolism. The rate of production and the activity of phagocytes increases, and tissue repair accelerates. Many microorganisms grow more slowly at the elevated temperature, thus giving immune cells a head start in the proliferation race. Some bacteria also secrete heat-shock proteins that are recognized by lymphocytes, further triggering the immune response.

Phagocytosis, inflammation, and fever rid the body of most pathogens before they become established. If pathogens do persist, adaptive immunity takes over. Dendritic cells and macrophages continue working in the adaptive response, which is discussed in the next section. Activated complement proteins also mediate immune cell interactions and help control maturation of immune cells during adaptive immune responses.

Innate immunity is the body's immediate, nonspecific response to internal threats. Complement, phagocytes, acute inflammation, and fever work together to eliminate most invaders from the body before infection is established.

23.4 Tailoring Responses to Specific Antigens

Sometimes surface barriers, fever, and inflammation are not enough to end a threat, and an infection becomes established. Long-term, specific defenses now begin.

FEATURES OF ADAPTIVE IMMUNITY

Given life's diversity, the number of different antigens is essentially infinite, and as such no living system can recognize them all. But vertebrate adaptive immunity comes close. Lymphocytes, phagocytes, and signaling chemicals interact to effect four defining characteristics of vertebrate adaptive immunity. They are self/nonself recognition, specificity, diversity, and memory.

Self versus nonself recognition starts with the specific molecular configurations that impart a unique identity to all the cells of each organism. Plasma membranes of vertebrate cells are studded with **MHC markers**, self-recognition proteins named for the group of genes that encode them (the *M*ajor *H*istocompatibility *C*omplex). **T cell receptors**, or TCRs, are antigen receptors on the surface of T cells. Part of every TCR recognizes MHCs and other such cell-surface markers as self, and part of it also recognizes antigen as nonself.

Specificity means each new B cell or T cell makes receptors for one—and only one—kind of antigen.

Diversity refers to the collection of antigen receptors on all B and T cells in the body. There are potentially billions of different antigen receptors, so an individual has the potential to counter billions of different threats.

Memory refers to the immune system's capacity to "remember" an antigen that it encountered before. The first time an antigen appears in the body, lymphocyte armies form after a few days. If that antigen shows up later, a faster, heightened response occurs. That is why you do not get as sick the second time a disease agent infects you.

FIRST STEP—THE ANTIGEN ALERT

Recognition of a specific antigen is only the first step in adaptive immunity. It triggers repeated mitotic cell divisions that lead to significant populations of B and T cells, all primed to recognize the same antigen.

T cell receptors do not recognize antigen unless it is presented by an antigen-presenting cell. Macrophages, B cells, and dendritic cells do the presenting. First, by phagocytosis, these cells engulf something bearing an antigen. Vesicles containing the antigenic particle form inside the cell, then fuse with lysosomes. Lysosomal enzymes digest the antigen-bearing particle into bits.

IMPACTS, ISSUES

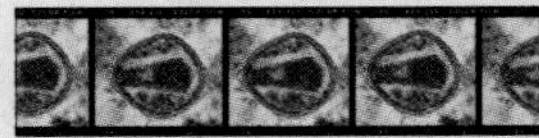

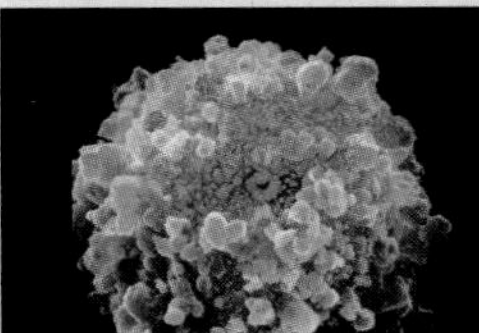

HIV (*red*) has infected this T cell. It replicates mainly in dendritic cells, macrophages, and helper T cells. The number of helper T cells declines in infected people. When helper T cells decline below a certain point, the immune system is compromised and symptoms of AIDS start to emerge.

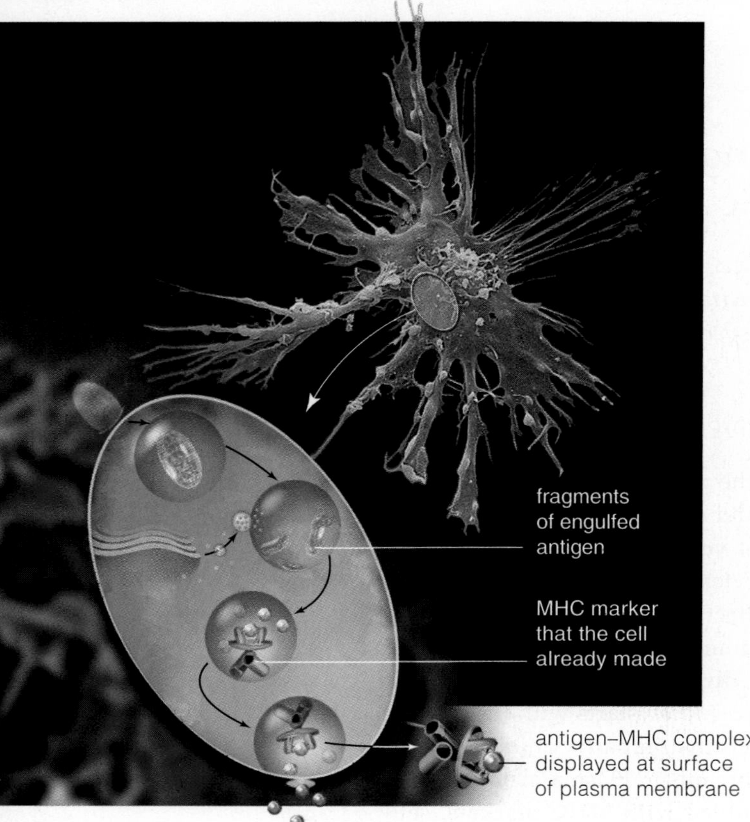

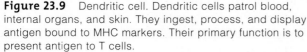

Figure 23.9 Dendritic cell. Dendritic cells patrol blood, internal organs, and skin. They ingest, process, and display antigen bound to MHC markers. Their primary function is to present antigen to T cells.

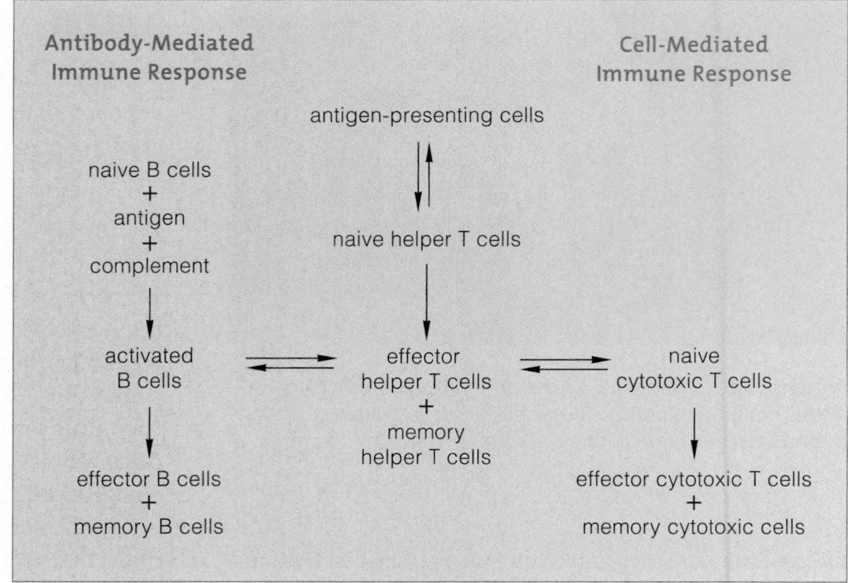

Figure 23.10 Preview of key interactions between the antibody-mediated and cell-mediated responses—the two arms of adaptive immunity. A "naive" cell simply is one that has not made contact with its specific antigen.

LINKS TO
SECTIONS
3.5, 4.7, 10.6

Some of the antigen bits bind with MHC markers inside lysosomes. The antigen–MHC complexes move to the plasma membrane, and there they are displayed at the cell's surface (Figure 23.9). *A cell's MHC markers paired with antigen fragments are a call to arms.*

Odds are that at least one T cell will have a receptor that binds this particular antigen–MHC complex. The T cell then secretes cytokine signals that act on any other B or T cell sensitive to the same antigen. These divide until huge populations of B and T cell clones form. Most of them become *effector* cells: differentiated lymphocytes that act immediately against the antigen. *Memory* cells are long-lived B and T cells that develop during the first exposure to an antigen. They are set aside for any future encounters with the same antigen.

TWO ARMS OF ADAPTIVE IMMUNITY

Like a boxer's one-two punch, adaptive immunity has two separate arms. These two arms work together to eliminate diverse threats (Figure 23.10).

Not all threats present themselves in the same way. For example, bacteria, fungi, or toxins circulating in the blood or interstitial fluid are intercepted quickly by phagocytes and B cells. These cells interact during an **antibody-mediated immune response**. As you will see, B cells execute most of this response.

However, some pathogens evade B cells. They can hide in body cells, drain the life out of them, and often reproduce inside of them. These are called *intracellular* pathogens, and they are vulnerable to immune system defenses only briefly, between slipping out of one cell and infecting others.

From previous chapters, you already know about some viruses, bacteria, fungi, and apicomplexans that are among the many intracellular pathogens. You know about tumor cells and other genetically altered body cells that pose different kinds of threats. All are the targets of the **cell-mediated immune response**, which does not involve antibodies. Instead, antigens on the surface of ailing body cells are detected and killed by phagocytes and cytotoxic T cells.

SO WHERE IS ANTIGEN INTERCEPTED?

After engulfing an antigenic particle in a local tissue, a dendritic cell or macrophage migrates to a lymph node (Figure 23.11). There, it alerts T cells that an invader has entered the body. About 25 billion lymphocytes pass through each lymph node per day.

Also, free antigen in interstitial fluid enters lymph vessels, which deliver it to lymph nodes. Inside the nodes, it trickles past arrays of B cells, macrophages, and dendritic cells—which bind, process, and present

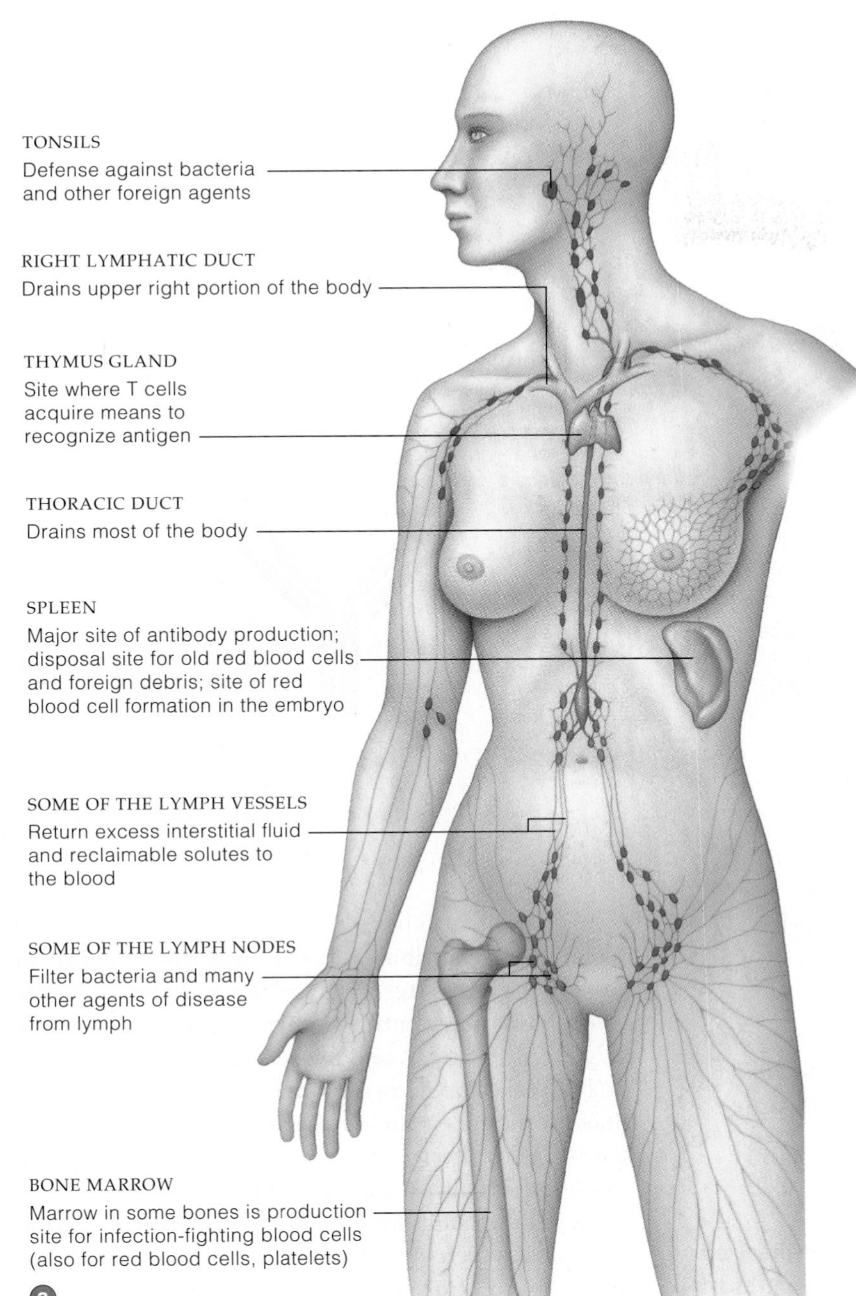
TONSILS
Defense against bacteria and other foreign agents
RIGHT LYMPHATIC DUCT
Drains upper right portion of the body
THYMUS GLAND
Site where T cells acquire means to recognize antigen
THORACIC DUCT
Drains most of the body
SPLEEN
Major site of antibody production; disposal site for old red blood cells and foreign debris; site of red blood cell formation in the embryo
SOME OF THE LYMPH VESSELS
Return excess interstitial fluid and reclaimable solutes to the blood
SOME OF THE LYMPH NODES
Filter bacteria and many other agents of disease from lymph
BONE MARROW
Marrow in some bones is production site for infection-fighting blood cells (also for red blood cells, platelets)
a

Figure 23.11 ***Animated!*** Big battlegrounds in adaptive immunity. Lymph nodes strategically positioned along the lymph vascular highways host macrophages, dendritic cells, B cells, and T cells.

(**a**) The human lymphatic system, consisting of vascular highways and lymphoid organs. Lymph capillaries thread through every capillary bed. They merge into larger lymph vessels. These deliver foreign agents and cellular debris from tissues to disposal centers, the lymph nodes.

Immature T cells differentiate and acquire unique antigen receptors in the thymus gland. The thymus also produces hormones that affect these cells.

The spleen filters pathogens and used-up blood cells from blood. Part of it is an extensive reservoir of red blood cells and even produces them in human embryos. Another part has fixed masses of macrophages and dendritic cells that engulf any passing foreign matter and pathogens. Antigen that becomes displayed on their surfaces is detected by circulating populations of B cells and T cells.

(**b**) A lymph node. Lymph nodes occur at intervals along lymph vessels. Before flowing into blood, lymph is filtered through at least one node. As in the spleen, macrophages and dendritic cells that have taken up residence in the node present antigen to billions of T cells and B cells that filter by every day. Lymph nodes become swollen with T cells during an infection.

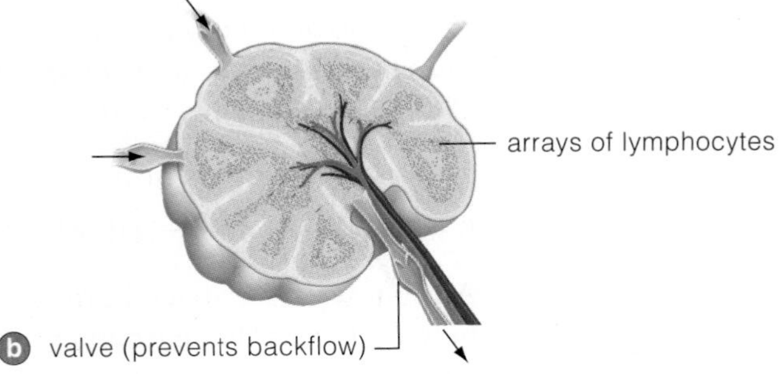
arrays of lymphocytes
b
valve (prevents backflow)

it to passing T cells. Because lymph flows on into the bloodstream, there is a chance that free antigen may circulate to body tissues. However, lymph nodes trap most of it. Antigen that does manage to slip through is filtered out as blood flows through the spleen.

During infection, antigen-presenting T cells become trapped in lymph nodes, hence the swollen "lumps" you may have noticed from time to time under your jaw and elsewhere when you are sick.

Immune responses subside once antigen is cleared away. The tide of battle turns as effector cells and their secretions kill most antigen-bearing agents. With less antigen present, fewer immune fighters are recruited. Complement proteins assist in the cleanup by helping to clear immune complexes from blood and deposit them into the liver and spleen for disposal.

The four characteristics of adaptive immunity are self–nonself recognition, specificity, diversity, and memory.

Antibody-mediated responses and cell-mediated responses are the two interacting arms of adaptive immunity. The first targets antigen detected in blood or interstitial fluid. The second directly kills infected or altered body cells.

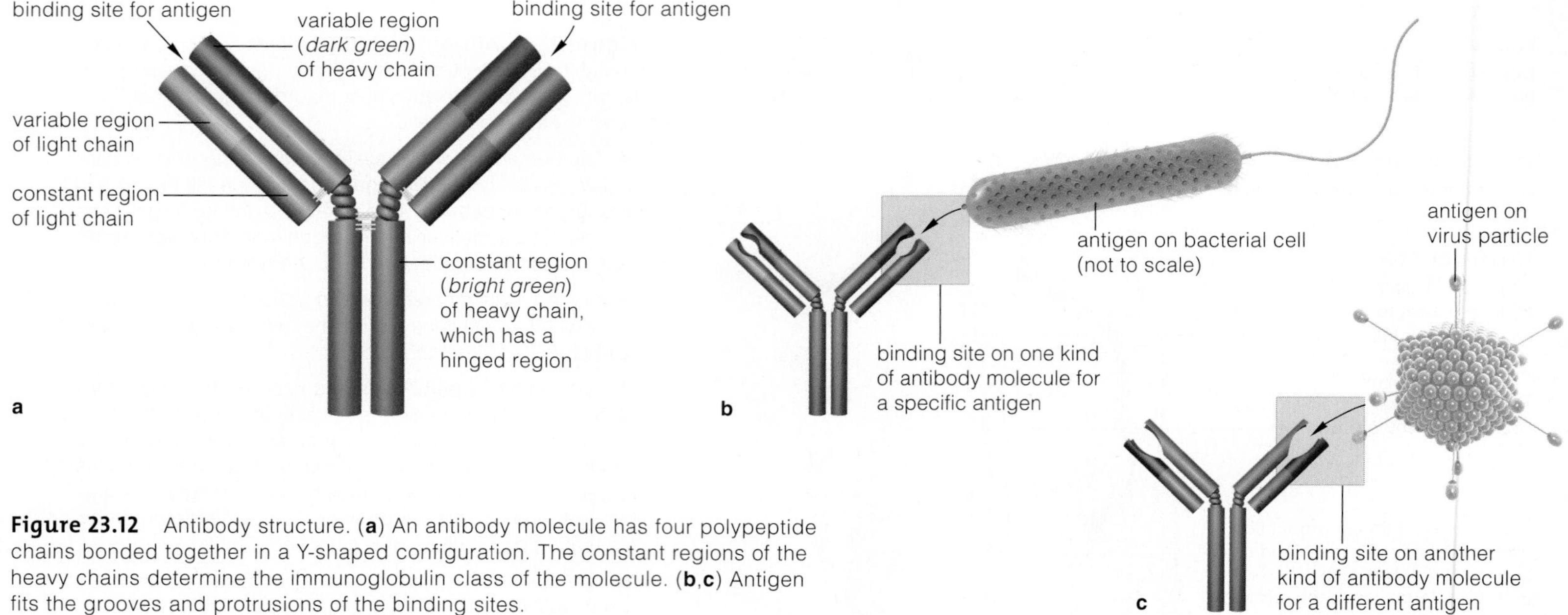

Figure 23.12 Antibody structure. (**a**) An antibody molecule has four polypeptide chains bonded together in a Y-shaped configuration. The constant regions of the heavy chains determine the immunoglobulin class of the molecule. (**b**,**c**) Antigen fits the grooves and protrusions of the binding sites.

LINKS TO SECTIONS 2.8, 10.2

23.5 Antibodies and Other Antigen Receptors

Your body continually produces B and T lymphocytes that collectively bear receptors capable of recognizing billions of different antigens. How does receptor diversity arise?

B CELL WEAPONS—THE ANTIBODIES

Antibodies are Y-shaped proteins that are synthesized only by B cells that encounter and bind antigen. Most antibodies circulate in blood and enter interstitial fluid during inflammation. Each specifically binds to one antigen—the one that prompted the B cell to make the antibody in the first place. Antibodies help phagocytic cells bind to antigen-bearing particles or cells, activate complement, and can also prevent pathogen binding to host cells.

Think back on Section 2.8, which explains the levels of protein structure. An antibody molecule consists of four polypeptides chains—two identical "light" chains and two identical "heavy" chains. Each has a unique *variable* region, a domain that binds to antigen (Figure 23.14). The variable regions match up and fold into an array of bumps and grooves with a unique distribution of charge. The variable region of an antibody will only bind antigen that has a complementary set of grooves, bumps, and distribution of charge.

Each chain also has a *constant* region, an unvarying domain that contributes to the structural identity, or class, of the antibody. The five classes of antibodies are called **immunoglobulins**, or **Igs**. We abbreviate these as *IgG*, *IgA*, *IgE*, *IgM*, and *IgD*.

IgG makes up 80 percent of all immunoglobulins in blood. It induces complement cascades and neutralizes toxins. IgG protects a fetus with its mother's acquired immunities. It also is secreted into early milk.

IgA is the main immunoglobulin in exocrine gland secretions, which include tears, saliva, and milk. It is a chemical protector in the mucus that coats the linings of the respiratory, digestive, and reproductive tracts. Invading bacteria and viruses cannot bind to cells if *IgA* binds to them first.

IgE induces inflammation after invasion by parasitic worms or other pathogens. The constant regions of its heavy chains become anchored at the surface of mast cells, basophils, monocytes, or dendritic cells. Antigen binding to its variable region causes the anchoring cell to release histamines and cytokines. This property of *IgE* contributes to allergic reactions and HIV infection.

IgM is the first immunoglobulin to be made in an immune response and the first made by newborns. The surface of each new B cell is covered with hundreds of thousands of *IgM* and *IgD* antibodies, each of which recognizes the same antigen. We call these antibodies *B cell receptors*, the surface immunoglobulins that are the antigen receptors for that cell.

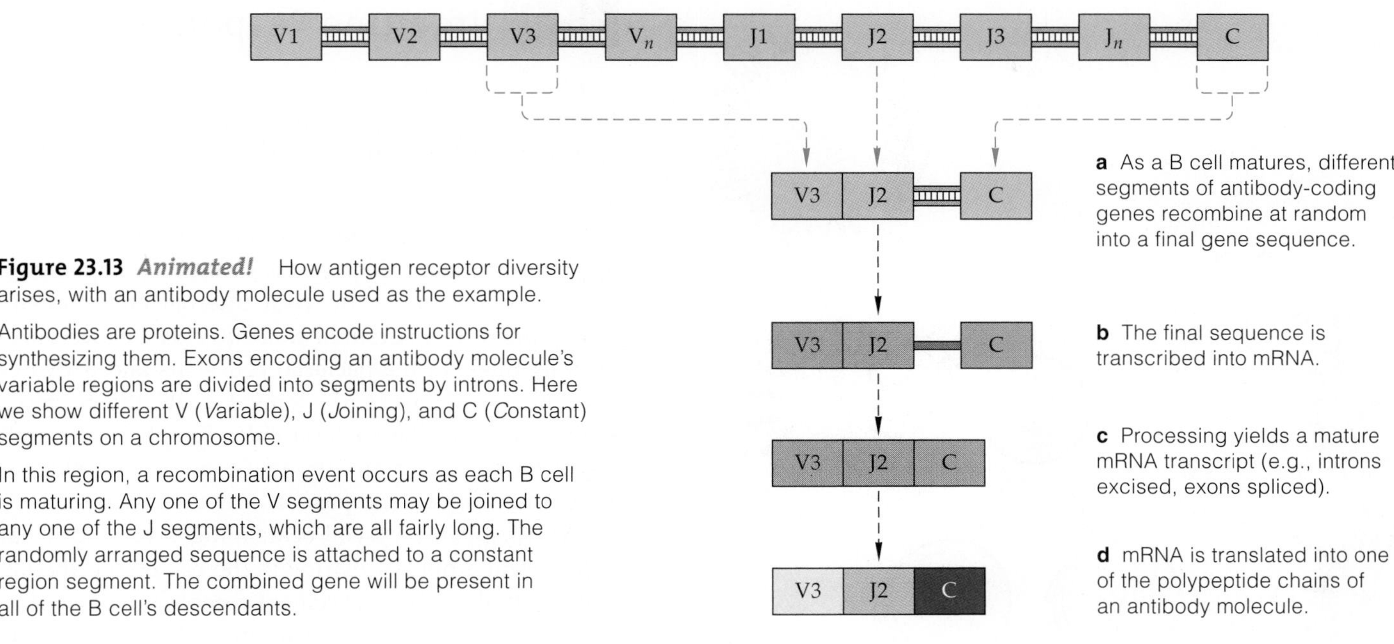

V1
V2
V3
V_n
J1
J2
J3
J_n
C
a As a B cell matures, different segments of antibody-coding genes recombine at random into a final gene sequence.
b The final sequence is transcribed into mRNA.
c Processing yields a mature mRNA transcript (e.g., introns excised, exons spliced).
d mRNA is translated into one of the polypeptide chains of an antibody molecule.

Figure 23.13 ***Animated!*** How antigen receptor diversity arises, with an antibody molecule used as the example.

Antibodies are proteins. Genes encode instructions for synthesizing them. Exons encoding an antibody molecule's variable regions are divided into segments by introns. Here we show different V (*V*ariable), J (*J*oining), and C (*C*onstant) segments on a chromosome.

In this region, a recombination event occurs as each B cell is maturing. Any one of the V segments may be joined to any one of the J segments, which are all fairly long. The randomly arranged sequence is attached to a constant region segment. The combined gene will be present in all of the B cell's descendants.

THE MAKING OF ANTIGEN RECEPTORS

A gene encoding an antigen receptor does not start out as one continuous sequence. Instead, antigen receptor gene segments (exons) are divided by introns. During differentiation of each B or T cell, an antigen receptor gene is assembled from segments selected at random (Figure 23.13). The combined sequence is transcribed into mRNA. *These random assortments are the source of B and T cell receptor diversity.*

Before each new B cell leaves the bone marrow, it is already synthesizing its unique antigen receptors. The constant region of each one is positioned in the lipid bilayer of the B cell's plasma membrane, and the two variable arms project above it. In time the cell bristles with more than 100,000 identical antigen receptors. It is a "naive" B cell: It has not yet met its antigen.

What about T cells? They, too, form inside the bone marrow, but they do not mature until they take a tour through the thymus gland. There, T cells start making receptors for MHC markers. They also begin to make unique antigen receptors—TCRs—by gene splicing events similar to the ones that occur in B cells. These recombinations are random, however, so many of the TCRs are defective: they bind MHC markers instead of antigen, or they do not recognize MHC markers at all.

So how does an individual have a functional set of T cells that do not rampage through the body attacking body cells? Thymus cells make small peptides that are assembled from a variety of the body's own proteins. Attached to MHC markers, they function as built-in quality controls that weed out the "bad" TCRs.

Any passing T cell that binds too tightly to one of the complexes has TCRs that recognize a self peptide. The ones that do not bind at all cannot recognize MHC markers. Both types of cells will die. Thus, by the time a naive T cell leaves the thymus to begin its journey through the circulatory system, its surface bristles with functional TCRs. Under most circumstances, none of the T cells will perceive the body's own cells as foreign. As you will see, though, occasional T cells do not follow this pattern. These rogue T cells recognize self and will attack body cells. Immune disorders are the result.

Antibodies are antigen-receptor proteins made by B cells.

Each individual has the potential to make B and T cell receptors that recognize billions of different antigens.

By random recombinations of segments from receptor-encoding regions of DNA, each B and T cell gets a gene sequence that codes for one unique antigen receptor.

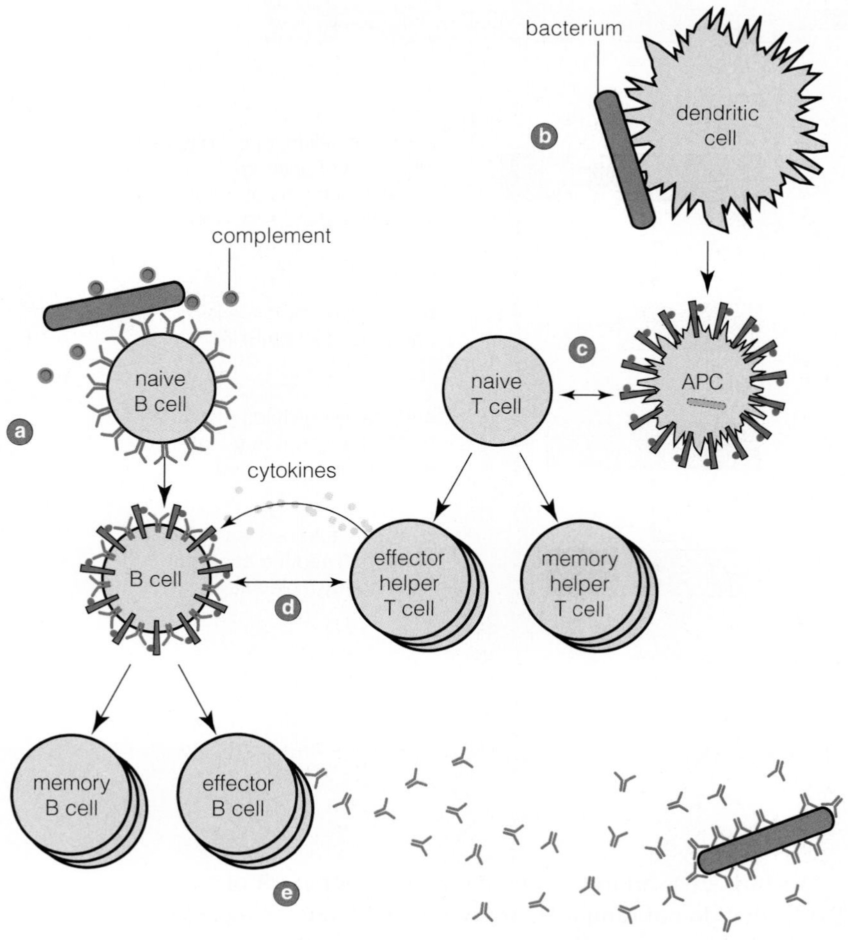

Figure 23.14 ***Animated!*** Example of an antibody-mediated immune response.

(**a**) A naive B cell binds its specific antigen on the surface of a bacterium. After also binding complement, the B cell engulfs the bacterium. Fragments of the bacterium bind MHC markers, and the complexes become displayed at the B cell's surface.

(**b**) A dendritic cell engulfs the same kind of bacterium that the B cell encountered. Digested fragments of the bacterium bind to MHC markers, and the complexes become displayed at the dendritic cell's surface. It is now an antigen-presenting cell (APC). The dendritic cell migrates to a lymph node.

(**c**) Binding to this antigen-presenting dendritic cell induces naive helper T cells to proliferate and differentiate into effector helper T cells and memory helper T cells.

(**d**) TCRs of an effector helper T cell bind to the antigen-MHC complexes on the B cell. Binding makes the helper T cell secrete cytokines. These signals induce the B cell to divide, giving rise to a huge population of clones with identical antigen receptors. The cells differentiate into effector B cells and memory B cells.

(**e**) Effector B cells begin making and secreting huge numbers of *IgA, IgG,* or *IgE*, all of which recognize the same antigen as the original B cell receptor. The new antibodies circulate throughout the body, and bind to any remaining bacteria.

23.6 Antibody-Mediated Immune Response

Think of B cells as assassins. Each has a genetic assignment to liquidate one particular target—an extracellular pathogen or toxin. Antibodies are their molecular bullets.

Suppose that you accidentally nick your finger. Being opportunists, some *Staphyloccocus aureus* cells hanging out on your skin invade your internal environment. Complement in interstitial fluid quickly latches on to carbohydrates in the bacterial cell walls and activates cascading reactions. Complement coats the bacteria. Within an hour, the bacteria enter lymph vessels and are captured by a lymph node in one of your elbows. There they are paraded past an army of naive B cells.

As it happens, a B cell bears antigen receptors that tightly bind a polysaccharide in the bacterial cell wall. The B cell also has complement receptors that bind to complement coating the bacterium. These two events provoke receptor-mediated endocytosis, and the B cell engulfs the bacterium. The B cell is no longer naive; it is now activated (Figure 23.14*a*).

Meanwhile, undaunted by complement and tissue inflammation, more *S. aureus* cells have been secreting metabolic products into interstitial fluid around your cut. The secretions attract phagocytes. One of them, a dendritic cell, engulfs a few bacteria, then migrates to that same lymph node in your elbow. It digests the bacterial cells and displays antigen fragments bound to MHC markers on its surface (Figure 23.14*b*).

Each hour, about 500 different naive T cells travel through the lymph node, inspecting resident dendritic cells. In this case, one of those T cells has TCRs that tightly bind the *S. aureus* antigen–MHC complexes on the dendritic cell. For the next twenty-four hours, the T cell and the dendritic cell interact. Transcription factors become activated in the T cell; then it disengages and returns to the circulatory system (Figure 23.14*c*).

By a theory of *clonal selection*, the *S. aureus* antigen "chose" that T cell because it bears a receptor able to bind with it. Interaction with the antigen-presenting cell stimulated mitosis of the T cell. Many descendants now form a gigantic population of genetically identical cells. This population differentiates into *helper T cells*, all with identical TCRs specific for *S. aureus* antigen.

Let's go back to that busy B cell in the lymph node. By now, digested bits of *S. aureus* are bound to MHC markers and displayed at the B cell surface. One of the new helper T cells recognizes these *S. aureus* antigen–MHC complexes on the B cell. Like long-lost friends, the two cells bind and exchange costimulatory signals. The helper T cell begins to secrete several interleukins, which the B cell recognizes as a signal to divide and differentiate (Figure 23.14*d*).

When the cells disengage, the B cell divides again and again to form a huge population of clones, all with identical receptors (Figure 23.15*a*). The cells differentiate into effector and memory B cells (Figures 23.14*e* and 23.15*b,c*). The effector cells function immediately in the *primary* immune response against this initial exposure to antigen. They stop making membrane-bound B cell receptors. Instead, they start secreting *IgG*, *IgA*, or *IgE*. Each secreted antibody molecule recognizes the same antigen as the original B cell receptor.

Huge numbers of antibodies specific for *S. aureus* now circulate through the body. In droves, they bind to all of the bacteria that remain in the bloodstream and interstitial fluid (Figure 23.14*e*). A coating of antibodies physically prevents a bacterium from binding to body cells. Antibody covered bacteria are recognized easily by phagocytic cells, and so are engulfed quickly. Also, antibodies bound to bacteria activate complement, and membrane attack complexes form.

In an antibody-mediated immune response, a naive B cell becomes activated after it binds to antigen, to complement, and to an activated helper T cell.

The activated B cell divides, forming a huge number of clones, all with the same antigen receptor. These cells differentiate into effector and memory cell populations.

Effector B cells secrete antibodies that bind any antigen remaining in the body, tagging it for destruction.

23.7 The Cell-Mediated Immune Response

If B cells are assassins, cytotoxic T cells are warriors that specialize in cell-to-cell combat. Their targets are altered body cells that evade antibody-mediated responses.

Remember, the main targets of an antibody-mediated response are extracellular pathogens and toxins freely circulating in blood and interstitial fluid. Viruses and certain bacteria, fungi, and protists can slip into body cells and remain hidden from an antibody-mediated response. Cell-mediated defenses against these more insidious threats begin in interstitial fluid, during an acute inflammatory response. The plasma membrane of infected body cells usually displays antigen: viral or bacterial peptides from an infectious agent, or body proteins altered during a cancerous transformation.

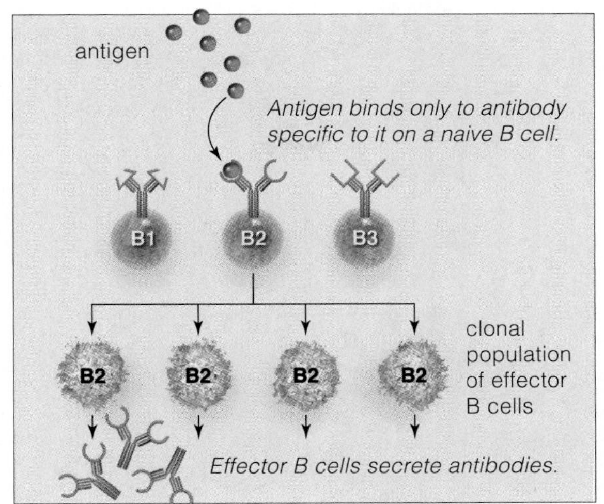

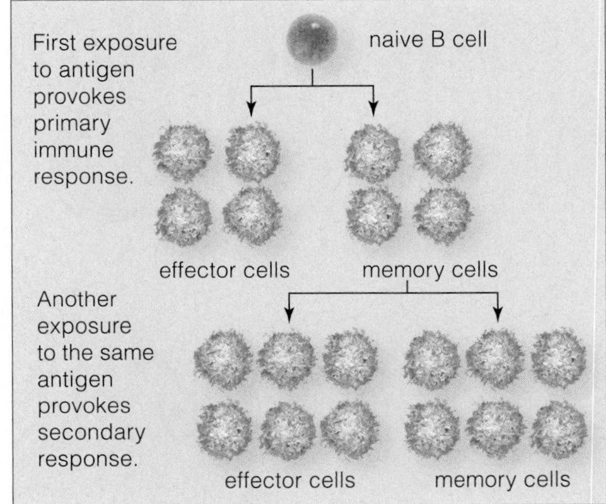

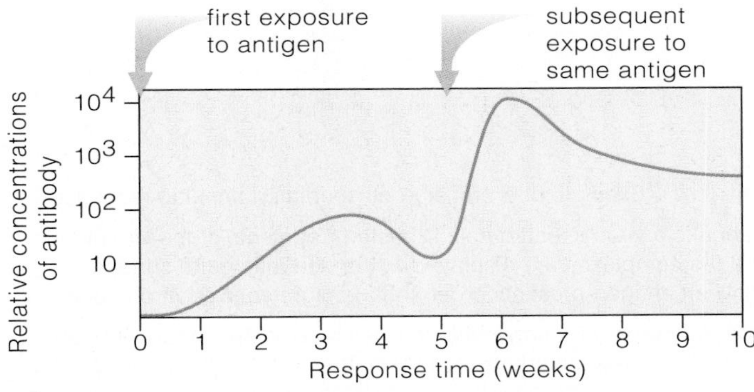

Figure 23.15 ***Animated!*** (**a**) Clonal selection of a B cell. Only B cells with receptors that bind antigen divide and differentiate. (**b**) The first exposure to antigen generates a primary immune response in which effector cells fight infection. Memory cells also form in a primary response but are set aside. When antigen is encountered again, they mount a secondary response.

(**c**) A secondary immune response is faster and stronger than the primary response that preceded it.

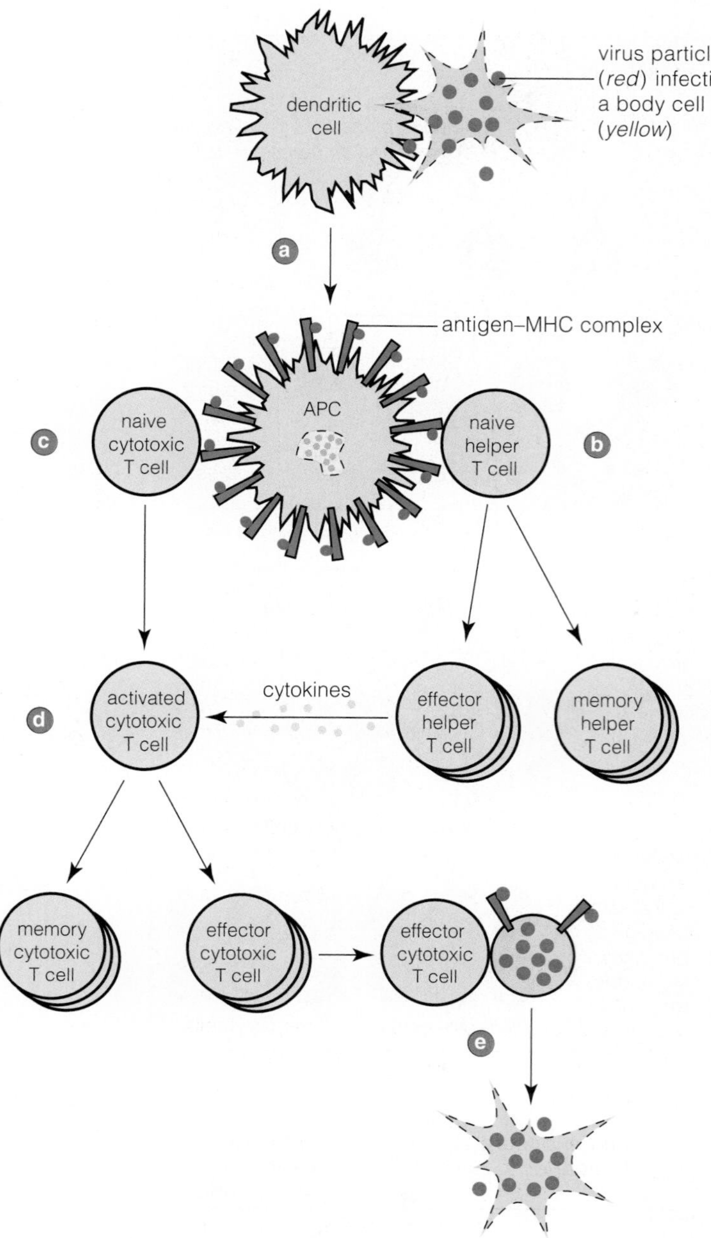

Figure 23.16 ***Animated!*** Example of a primary cell–mediated immune response.

(**a**) A dendritic cell engulfs a virus-infected cell. Digested fragments of the virus bind to MHC markers, and the complexes are displayed at the dendritic cell's surface. The dendritic cell, now an antigen-presenting cell (APC), migrates to a lymph node.

(**b**) In the lymph node, receptors of a naive helper T cell bind to the antigen–MHC complexes on the APC. The interaction activates the helper T cell, which then divides to form a large clonal population. The daughter cells differentiate into effector and memory helper T cells.

(**c**) In the lymph node, TCR receptors on a naive cytotoxic T cell bind to the antigen–MHC complexes on the APC. The interaction activates the cytotoxic T cell.

(**d**) Cytokines secreted by effector helper T cells stimulate the activated cytotoxic T cell to divide as it returns to the circulatory system. Its clonal descendants differentiate into large populations of effector and memory cytotoxic T cells.

(**e**) One of the new cytotoxic T cells encounters antigen–MHC complexes on the surface of a cell infected with the same virus. The cytotoxic T cell touch-kills the infected cell.

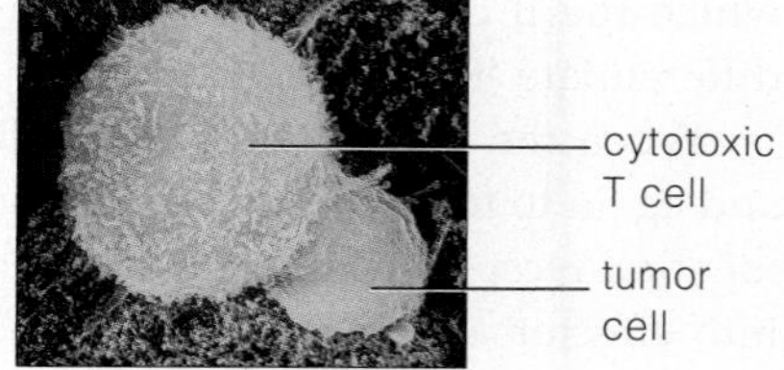

Figure 23.17 Cytotoxic T cell caught in the act of touch-killing a tumor cell.

Dendritic cells recognize, engulf, and digest these antigens as part of the diseased cells or their remains (Figure 23.16*a*). The dendritic cells migrate to lymph nodes, where they present antigen–MHC complexes to two populations of T cells: naive helper T cells and naive cytotoxic T cells (Figure 23.16*b*,*c*). Both types of T cell become activated if their receptors bind antigen–MHC complexes on the surface of a dendritic cell.

The descendants of naive helper T cells differentiate into effector helper T cells, which secrete interleukins and other cytokines. These signaling molecules induce naive, activated cytotoxic T cells to divide and mature into cytotoxic T cells (Figure 23.16*d*).

Cytotoxic T cells circulate throughout the body in blood and interstitial fluid. They bind any body cell that displays the original antigen together with MHC markers, and then inject it with perforin and proteases. These toxins poke holes in the cell and induce it to die by apoptosis (Figure 23.16*e*). Cytotoxic T cells are the body's primary weapons against infected body cells and tumors (Figure 23.17). They are also implicated in the rejection of tissue and organ transplants.

Cytokines secreted by certain helper T cells enhance macrophage function by stimulating them to secrete more histamines, cytokines, and toxins, thus enabling them to kill tumor cells and larger parasites. Helper T cell cytokines also stimulate cell divisions of NK cells. These "natural killers" attack cells that are tagged for destruction by antibodies. NK cells also detect stress markers on infected and cancerous body cells. A cell that has normal MHC markers is not killed. But if an infection or malignant transformation alters its MHC markers, the cell will be destroyed. Both NK cells and macrophages are crucial in fighting these cells because, unlike cytotoxic T cells, they can kill body cells that do not display MHC markers.

In a primary cell-mediated response, antigen-presenting dendritic cells activate helper T cells.

Helper T cells secrete cytokines that induce formation of cytotoxic T cells, proliferation of NK cells, and enhance the activity of macrophages. These three cell types destroy infected or altered body cells.

23.8 Defenses Enhanced or Compromised

Sometimes immune responses are not strong enough or are misdirected or compromised. Here are examples of what we can and cannot do about it.

IMMUNIZATION

Immunization refers to processes designed to induce immunity. In *active* immunization, a preparation that contains an antigen, a **vaccine**, is administered orally or injected into the body. The first injection provokes a primary immune response, just as an infection would. A booster, or second immunization, elicits a secondary immune response in which memory cells form.

Many vaccines are weakened or killed pathogens. Others are harmless viruses that have some genes of pathogens inserted into their DNA. After vaccination, body cells produce the antigen encoded by those genes, and immunity is established.

Passive immunization helps those who are already battling hepatitis B, tetanus, rabies, and a few other diseases. It requires an injection of antibody purified from the blood of someone who already fought the disease. These antibodies fight the infection but do not activate the adaptive immune system, so memory cells do not form. Protection ends when the body disposes of injected antibody molecules.

ALLERGIES

In millions of people, exposure to harmless proteins triggers an immune response. Any substance that is ordinarily harmless yet provokes such responses is an **allergen**. Sensitivity to allergens is called an **allergy**. Drugs, foods, pollen, dust mites, fungal spores, and venom from bees, wasps, and some other insects are among the most common allergens (Figure 23.18).

In some people, exposure to an allergen stimulates the immune system to make *IgE*, which binds to mast cells. When an allergen attaches to this bound *IgE*, the mast cells secrete histamines and cytokines, initiating an inflammatory response. Histamines released in the respiratory tract cause copious amounts of mucus to be secreted. Stuffed sinuses, sneezing, and a drippy nose are symptoms of *asthma* and *hay fever*.

Anaphylactic shock is a life-threatening response to an allergen. For instance, someone allergic to wasp or bee venom can die within minutes of a sting. Airways constrict, making it hard to breathe. As fluids leak out of permeable capillaries, blood pressure plummets. The circulatory shock may quickly lead to death.

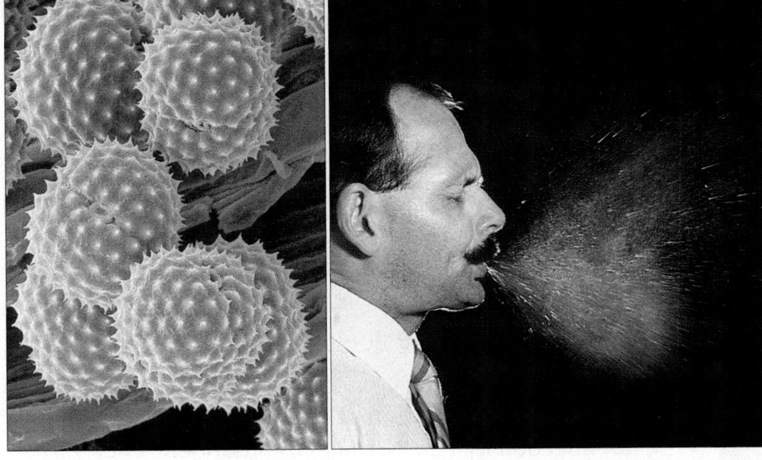

Figure 23.18 An effect of ragweed pollen, an allergen, on a sensitive person.

AUTOIMMUNE DISORDERS

Sometimes lymphocytes and antibody molecules fail to discriminate between self and nonself. When that happens, they mount an **autoimmune response**, or a misdirected attack against one's own tissues.

In some cases, auto-antibodies can bind to hormone receptors. *Graves' disease* is like this. Auto-antibodies bind to the stimulatory receptors on the thyroid gland, causing it to produce excess thyroid hormone. This quickens the body's overall metabolic rate. Feedback loops that normally balance hormone production do not include antibodies, so antibody binding continues, excess thyroid hormone is produced, and metabolism spins out of control.

A common neurological disorder, *multiple sclerosis*, develops after autoreactive T cells attack myelin, the material that insulates nerve cells. Symptoms include numbness, paralysis, and blindness. At least eight genes heighten susceptibility to multiple sclerosis, but activation of white blood cells against a viral infection may trigger the disorder.

DEFICIENT IMMUNE RESPONSES

Loss of immune function can have a lethal outcome. Altered genes or abnormal developmental steps result in *primary* immune deficiencies that are present from birth. *Secondary* immune deficiencies result from the loss of immune function after exposure to an agent such as a virus. Immune deficiencies make individuals more vulnerable to infections by infectious agents that are otherwise harmless to those in good health. *AIDS*, or acquired immunodeficiency syndrome, is the most common secondary immune deficiency.

Immunization programs are designed to boost immunity to specific diseases.

Some heritable disorders, developmental abnormalities, and attacks by viruses and other outside agents result in misdirected, compromised, or nonexistent immunity.

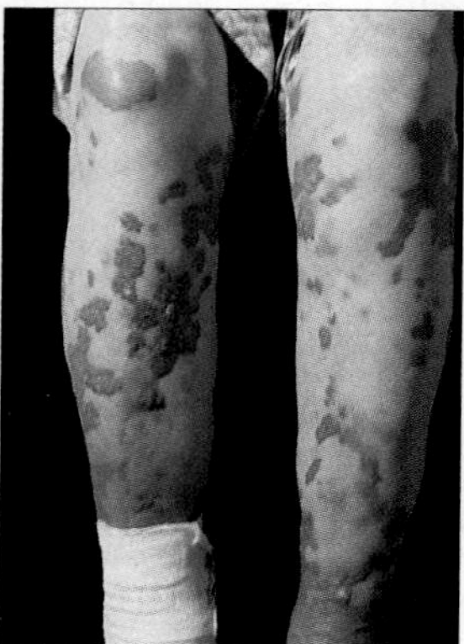

Figure 23.19 The lesions that are a sign of Kaposi's sarcoma.

23.9 AIDS: Immunity Lost

Worldwide, HIV infection rates continue to skyrocket. An effective vaccine still eludes us. For now, the best protection is avoiding unsafe behavior.

AIDS is a constellation of disorders that develop after an infection by HIV (*H*uman *I*mmunodeficiency *V*irus). This virus cripples the immune system, making the body highly susceptible to infections and rare forms of cancer. Worldwide, HIV has infected an estimated 34 million to 46 million individuals.

There is no way to rid the body of known forms of the virus, HIV–I and HIV–II. *There is no cure for those already infected.* At first, an infected person appears to be in good health, maybe fighting "a bout of the flu." But symptoms emerge that foreshadow AIDS: fever, many enlarged lymph nodes, fatigue, chronic weight loss, and drenching night sweats. Then opportunistic infections strike. They include yeast infections of the mouth, esophagus, vagina, and elsewhere, and a form of pneumonia caused by *Pneumocystis carinii*. Painless colored lesions often erupt, especially on the legs and feet (Figure 23.19). The lesions are evidence of *Kaposi's sarcoma*, a cancer common among AIDS patients.

HIV INFECTION—A TITANIC STRUGGLE BEGINS

HIV, a retrovirus, has a lipid envelope, a bit of plasma membrane acquired when the virus budded from an infected cell (Section 14.5). Proteins spike out from the envelope, span it, and line its inner surface. Just beneath the envelope, more viral proteins enclose two strands of RNA and copies of reverse transcriptase. Figure 23.20 shows this structural arrangement.

HIV primarily infects macrophages, dendritic cells, and helper T cells. When it first enters the body, the virus is engulfed by dendritic cells in skin or in the linings of the respiratory and gastrointestinal tracts. The cells migrate to lymph nodes, where they present HIV antigen to naive T cells. Any that bind divide and differentiate into helper T cells, which interact with naive B cells and naive cytotoxic T cells. The results are virus-neutralizing *IgG* antibodies and cytotoxic T cells that can destroy HIV-infected cells.

It is a typical adaptive response, and it clears the body of most of the HIV. However, during this first response, HIV infects a few T helper cells in a few lymph nodes. For years or even decades, the immune system functions normally. The antiviral antibody and cytotoxic T cells keep HIV levels relatively low. But HIV multiplying in the lymph nodes continually shed envelope proteins into the bloodstream. In response,

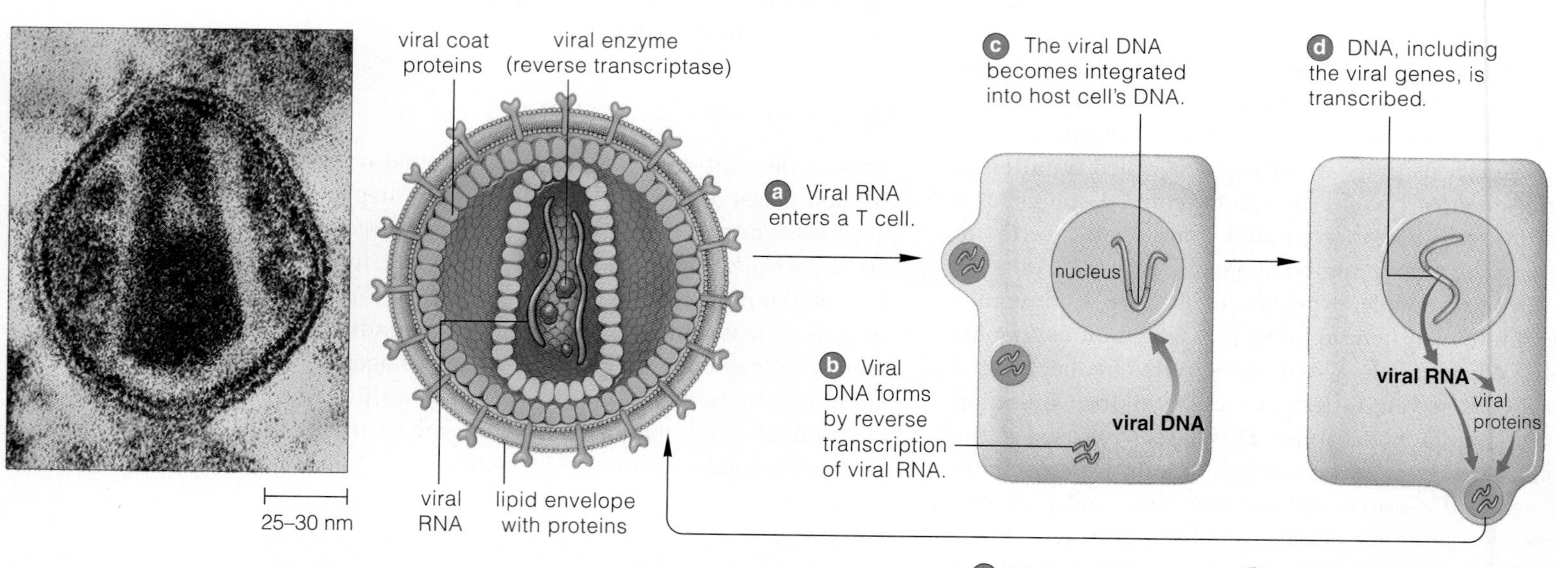
viral coat proteins
viral enzyme (reverse transcriptase)
viral RNA
lipid envelope with proteins
25–30 nm
a Viral RNA enters a T cell.
b Viral DNA forms by reverse transcription of viral RNA.
c The viral DNA becomes integrated into host cell's DNA.
nucleus
viral DNA
d DNA, including the viral genes, is transcribed.
viral RNA
viral proteins
e Some transcripts are new viral RNA, others are translated into proteins. Both self-assemble as new virus particles.
f Virus particles that bud from the infected cell may attack a new one.

Figure 23.20 ***Animated!*** Replication cycle of HIV, the retrovirus that causes AIDS. The micrograph at *left* shows one viral particle.

helper T cells in uninfected lymph nodes secrete an interleukin called IL-4. This cytokine causes B cells to stop producing virus-neutralizing *IgG,* and to start producing HIV-specific *IgE* instead.

IgE becomes anchored to mast cells, basophils, monocytes, and dendritic cells. When it also binds to antigen, the *IgE* prompts its anchoring cell to release histamines and cytokines. Ordinarily, this response protects mucus membranes from invasions of worms or arthropods, because it induces lymphocytes and plasma proteins to enter and defend infested tissues.

In an HIV infection, however, *IgE* binding to viral envelope proteins causes the anchoring cell to release its contents. One of the chemicals it releases is IL-4, which signals B cells to make *IgE* instead of *IgG*. Less *IgG* becomes available to fight HIV as more *IgE* binds viral proteins—which stimulates more B cells to make *IgE,* and so forth. This vicious cycle makes adaptive immunity mechanisms less and less effective.

The number of virus particles rises. Up to 2 billion helper T cells become infected. Up to 1 billion virus particles are built every day. Every two days, half of the virus particles are destroyed and half of the helper T cells lost in battle are replaced. HIV reservoirs and masses of infected T cells collect in the lymph nodes.

Eventually the battle tilts as the body makes fewer replacement helper T cells. It may take a decade or more, but the erosion of helper T cell counts destroys the body's capacity for adaptive immunity.

Some viruses make far more particles in a day, but the immune system usually wins. Other kinds may persist for a lifetime, but the immune system keeps them in check. HIV demolishes the immune system. Secondary infections and tumors kill the patient.

Table 23.4 Global Cases of HIV and AIDS*

Region	AIDS Cases	New HIV Cases
Sub-Saharan Africa	25,400,000	3,100,000
South/Southeast Asia	7,100,000	890,000
Latin America	1,700,000	240,000
Central Asia/East Europe	1,400,000	210,000
North America	1,000,000	44,000
East Asia	1,100,000	290,000
Western/Central Europe	610,000	21,000
Middle East/North Africa	540,000	92,000
Caribbean Islands	440,000	53,000
Australia/Pacific Islands	35,000	5,000

*Global estimates as of December 2004, *www.unaids.org*

HOW IS HIV TRANSMITTED?

LINKS TO SECTIONS 14.5, 14.6

Most often, HIV is transmitted by having unprotected sex with an infected partner. HIV virus in semen and vaginal secretions enters a partner through epithelial linings of the penis, vagina, rectum, and (rarely) the mouth. The likelihood of transmission is increased by damaged epithelial linings, as from rough sex, anal intercourse, or other sexually transmitted diseases.

Infected mothers can transmit HIV to their children. HIV also travels in tiny amounts of infected blood in syringes that are shared by intravenous drug abusers and by patients in hospitals of poor countries. HIV is not transmitted by casual contact.

WHAT ABOUT DRUGS AND VACCINES?

Current drugs cannot cure HIV, but some can slow its progression by interfering with processes unique to viral replication. For example, nucleotide phosphate analogs like AZT interrupt viral replication when they replace normal nucleotides in HIV cDNA. The cost of current treatments can be as much as 15,000 dollars per year. That price puts treatments beyond the reach of most of the infected people in developing countries, where the AIDS pandemic is now raging (Table 23.4).

At present, the best option for halting the spread of HIV appears to be the development of a vaccine. This may not be on the immediate horizon. Traditional vaccine strategies do not work with HIV because the virus replicates in the immune system; the very same mechanism that fights the virus is also susceptible to infection. Also, HIV mutates very quickly. Antibodies exert a selective pressure on the virus. Given the very fast pace of viral replication, HIV genes evolve faster than the immune system can make antibodies to their mutated products, vaccine or no vaccine.

Even so, persistent researchers are using several strategies to develop an HIV vaccine. An immediate, strong immune response to a primary HIV exposure might clear the virus before it has a chance to infect helper T cells. The best option for a vaccine might be an engineered virus with enough HIV components to trigger the desired immune response, but not so many components that vaccination would pose a threat to an individual's health.

HIV infects T cells, which are central to all immune responses. Billions of these cells engage in battle during each round of infection, but the battle eventually tilts in favor of the virus. At present there is no cure.

IMPACTS, ISSUES

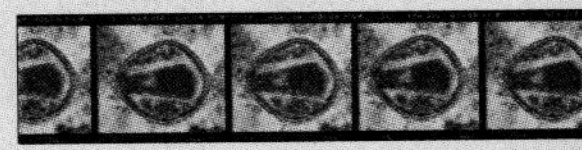

Scientists used to think that once a person reached adulthood, the thymus became inactive and adults could never replace the memory helper T cells killed by HIV. But in some AIDS patients, a decline in T cells may reawaken the thymus. That is the good news. The bad news is that HIV stays dormant in the thymus, so any new T cells that form may already carry dormant virus particles. When antigen activates these cells, the virus also may be reactivated.

Summary

Section 23.1 Vertebrates fend off pathogens with physical and chemical barriers at body surfaces. They also are protected by innate responses to tissue damage and invasion. Adaptive responses of white blood cells counter specific threats.

Antigen is any molecule or particle recognized as foreign that elicits an immune response.

Section 23.2 Skin and linings of the body's tubes and cavities are physical barriers to infection. Chemical barriers include glandular secretions such as mucus as well as lysozyme in saliva, tears, and gastric fluid. Microbes that colonize these external surfaces usually have neutral or beneficial effects. Some cause disease when they breach the surface barriers.

Section 23.3 Innate immunity uses microbial pattern receptors and complement in fast, off-the-shelf responses to a fixed set of nonself cues.

The complement system helps coordinate defenses in both the innate and adaptive immune responses. Complement proteins kill intruders directly, or tag them for destruction by phagocytes.

Phagocytic cells function in innate immunity. Macrophages and dendritic cells engulf and digest anything other than healthy tissue. The complement system helps coordinate defenses in both innate and adaptive immune responses.

Inflammation is a fast innate response to tissue damage, toxins, or invasion. It starts when mast cells release histamines, which increase blood flow to the tissue and make capillaries leaky to phagocytes and plasma proteins.

Fever, another innate response, fights infection by increasing the metabolic rate while slowing the growth of microorganisms.

BiologyNow

View the animation of inflammation and complement action on BiologyNow.

Section 23.4 Four main characteristics of adaptive immunity are self/nonself recognition, sensitivity to specific threats, potential to intercept a tremendous number of diverse pathogens, and memory. B and T lymphocytes (B cells and T cells) are central to it.

An adaptive immune response involves recognition of antigen and repeated cell divisions that form large populations of short- and long-lived lymphocytes, all of which recognize the same antigen.

Macrophages, dendritic cells, and B cells function as antigen-presenting cells. They engulf and digest antigen, then display its fragments in association with MHC markers at their surface.

Section 23.5 Antibodies are Y-shaped proteins made only by B cells. They bind specific antigens, tagging them for destruction or neutralizing them.

B and T cells form in bone marrow. Each receives a unique antigen-receptor sequence from random gene recombination events during its formation. Billions of specific antigen receptors are possible.

T cells mature in the thymus gland. Normally, only those T cells that properly distinguish between self and nonself survive.

BiologyNow

Learn how antibody diversity is generated with the animation on BiologyNow.

Section 23.6 B cells carry out antibody-mediated responses, assisted by T cells and cytokines. First, antigen presented by an antigen-presenting cell prompts a naive T cell to proliferate and differentiate into huge populations of effector helper T cells and memory T cells, all of which recognize the same antigen.

Next, a naive B cell that recognizes and binds both antigen and complement ingests the antigen. Digested bits of antigen become displayed on the surface of the B cell along with MHC markers.

Binding to an effector helper T cell induces this antigen-presenting B cell to divide and differentiate into huge clonal populations of antibody-secreting and memory B cells.

After a primary response to antigen, memory cells persist. A later encounter with the same antigen will provoke a faster, more intense secondary response.

BiologyNow

Observe an antibody-mediated response with the animation on BiologyNow.

Section 23.7 T cells make cell-mediated responses to fight intracellular pathogens and altered body cells. Helper T cells secrete cytokines that provoke naive T cells to divide and differentiate into populations of cytotoxic T cells and memory T cells.

Cytotoxic T cells touch-kill their targets by injecting them with perforin and enzymes that cause the cell to die by apoptosis.

Cytokines secreted by helper T cells also cause NK cells to divide, and enhance macrophage activity.

BiologyNow

Observe a cell-mediated response with the animation on BiologyNow.

Section 23.8 Immunization is designed to provoke immunity to specific diseases. Allergies are immune responses to normally harmless substances.

In autoimmunity, the body's own cells become misidentified as foreign. Immune deficiency is a weakened or nonexistent capacity to mount an immune response.

Section 23.9 HIV, a retrovirus, causes AIDS. The virus destroys the immune system mainly by infecting helper T cells. At present, AIDS cannot be cured.

BiologyNow

See how HIV invades and replicates inside a cell with animation on BiologyNow.

Self-Quiz

Answers in Appendix I

1. _______ are the first line of defense against threats.
   a. Skin, mucous membranes
   b. Tears, saliva, gastric fluid
   c. Urine flow
   d. Resident bacteria
   e. a through c
   f. all of the above

2. Complement proteins _______.
   a. form pore complexes
   b. enhance resident bacteria
   c. promote inflammation
   d. neutralize toxins
   e. a and c
   f. a, c, and d

3. _______ trigger immune responses.
   a. Interleukins
   b. Lysozymes
   c. Immunoglobulins
   d. Antigens
   e. Histamines
   f. all of the above

4. Recognition of specific threats is based on _______.
   a. antigen receptor diversity
   b. recombinations of receptor genes
   c. liver proliferation
   d. both a and b
   e. all of the above

5. Antibody-mediated responses work against _______.
   a. intracellular pathogens
   b. extracellular pathogens
   c. extracellular toxins
   d. both a and c
   e. both b and c
   f. all of the above

6. Immunoglobulin _______ promotes antimicrobial activity in mucus-coated surfaces of some organ systems.
   a. IgA b. IgE c. IgG d. IgM e. IgD

7. _______ are targets of cytotoxic T cells.
   a. Extracellular virus particles in blood
   b. Virus-infected body cells or tumor cells
   c. Parasitic flukes in the liver
   d. Bacterial cells in pus
   e. Pollen grains in nasal mucus

8. Match the immunity concepts.
   ___ inflammation — a. neutrophil
   ___ antibody secretion — b. effector B cell
   ___ fast-acting phagocyte — c. nonspecific response
   ___ immunological memory — d. immune response against own body
   ___ autoimmunity — e. secondary responses

Additional questions are available on **Biology Now™**

Critical Thinking

1. As described in the chapter opener, Edward Jenner was lucky. He performed a potentially harmful experiment on a boy who managed to survive it. Why would it be ill-advised for a would-be Jenner to do the same thing today?

2. New research links some of our resident microbes with seemingly unrelated major illnesses. Periodontitis itself is not life-threatening, for instance, but it is a fairly good predictor for heart attacks. Why do you think that is so?

3. Pigs are considered the most likely animal candidates as future organ donors for humans. As a first step, some researchers are developing transgenic pigs—animals that received and are expressing foreign genes (Figure 23.21). Which genes might be inserted, or deleted, that would stop the human immune system from attacking a pig-to-human transplant? Explain your reasoning.

Figure 23.21 Two piglets developed from the same cell lineage at the University of Missouri. The yellow hooves and snout of the piglet at left are evidence that it is transgenic.

4. As Section 23.5 explains, generation of tremendously diverse B cell and T cell antigen receptors is the basis of pathogen-specific immune responses. The genetic roots of this adaptive immunity are deep in the chordate family tree. Remember the lancelets (Section 16.9)? Researchers have found that these invertebrate chordates have genes that resemble the V-region genes of vertebrates. However, the ones in lancelets cannot undergo rearrangements, as they do in vertebrates. Lancelets can make innate immune responses, but not adaptive immune responses. Why?

5. Elena developed *chicken pox* when she was in first grade. Later in life, when her children developed chicken pox, she remained healthy even though she was exposed to countless virus particles daily. Explain why.

6. Before each flu season, you get a flu shot, or influenza vaccination. This year, you come down with "the flu" anyway. What do you suppose happened? There are at least three explanations.

7. Remember, cancer arises when normal body cells lose control over the cell cycle. Researchers have been trying for decades without much success to formulate vaccines targeting various types of cancer.

New cancer treatments based on *cell-therapy* are promising, however. The premise of cell therapy is to replace defective or deficient body cells with functional ones. One new cell therapy method developed at Johnson & Johnson Pharmaceutical Research & Development is to culture some of a cancer patient's own naive T cells with tumor-specific antigens. The resulting "super" cytotoxic T cells are very effective at killing cancer cells when they are infused back into the patient.

What do you think is a reason that traditional vaccines have not been as universally effective against cancer as the vaccine developed for smallpox? Why do you think cell therapy might be more effective?

8. Monoclonal antibodies are made by immunizing a mouse with a particular antigen, then removing its spleen. Individual B cells producing mouse antibodies specific for the antigen are isolated from the spleen and then fused with myeloma cells that grow *in vitro*. The resulting *hybridoma* cells can grow and divide indefinitely; they also produce and secrete antibodies that target the antigen.

Monoclonal antibodies may be used for passive immunization (Section 23.8), but they tend to fight their target only for a short while before the body eliminates them. Explain why.

5. *Elimination.* Expulsion of undigested, unabsorbed wastes from the end of the gut.

digest. In all birds, waste products of digestion collect in a *cloaca* before being expelled from the body.

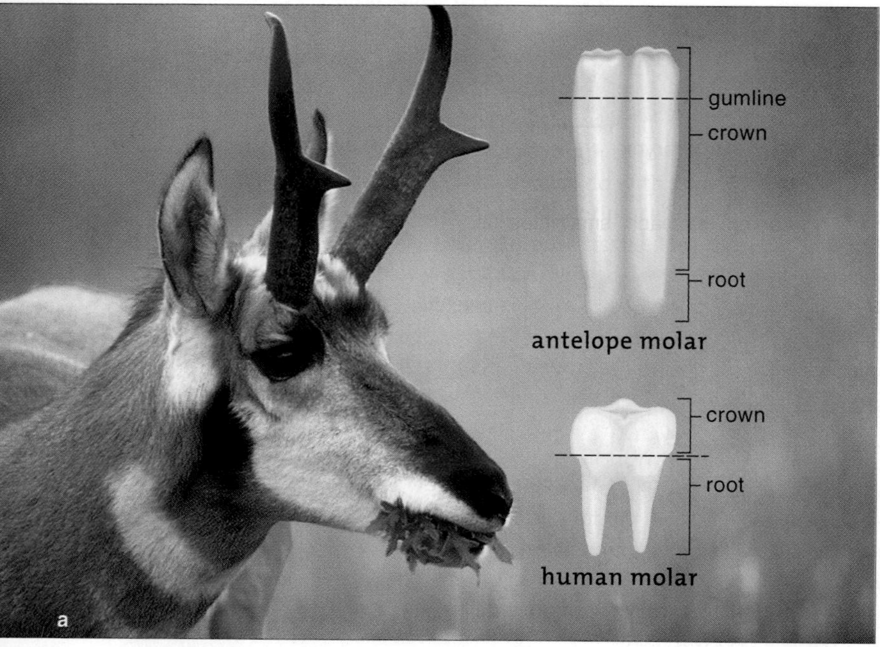

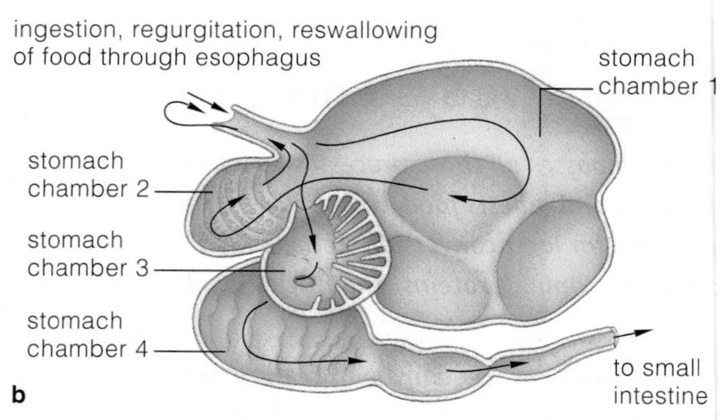

Figure 24.2 ***Animated!*** (**a**) Lower molar of a pronghorn antelope compared with a human lower molar. (**b**) Chambers of the antelope stomach. In the first two chambers, food is mixed with fluid, kneaded, and exposed to fermentation action of bacterial, protozoan, and fungal symbionts. Some symbionts degrade cellulose; others make organic compounds, fatty acids, and vitamins. The host uses some of these substances. Kneaded food is regurgitated into the mouth, rechewed, and swallowed. It enters the third chamber and is digested again before entering the last stomach chamber.

LINK TO SECTION 20.4

Pronghorn antelopes are grazers; they are adapted to feed on grass and low shrubs. A pronghorn's cheek teeth, its molars, have a flattened crown that serves as a grinding platform. Compared to the crown part of a human molar, an antelope's is relatively taller (Figure 24.2*a*). Why? You do not brush your mouth against the ground as you eat. Antelopes do. Abrasive soil gets in their mouths along with tough plant material, so the crown wears down quickly. Because of this, natural selection has favored a taller crown.

Antelopes are *ruminants*—hoofed mammals having multiple stomach chambers in which cellulose slowly undergoes digestion (Figure 24.2*b*.) These chambers steadily accept food during extended feeding periods and slowly liberate nutrients when the animal rests.

Carnivores tend to have shorter intestines than do herbivores. Meat is easier to digest than plant parts, which often include cellulose and lignin. In many meat eaters, the stomach expands to accept a huge amount of food in a single sitting. For example, a male lion that weighs 250 kg is able to devour as much as 40 kg of meat at a single time.

Digestive systems function in movements that break up, mix, and directionally propel food; secretion of substances with roles in digestion; nutrient absorption; and expulsion of undigested residues.

Incomplete digestive systems are a saclike cavity with only one opening. Complete digestive systems are a tube with two openings and regional specializations in between.

24.2 Human Digestive System

Humans have a complete digestive system, with a mouth at one end, and an anus at the other. About 10 meters of gut connect them. Accessory glands and organs secrete substances into the gut, as do cells of the gut lining.

INTO THE MOUTH, DOWN THE TUBE

Figure 24.3 lists major organs of the human digestive system and their functions. This is a tubular system, and a mucus-coated epithelium lines all free surfaces facing the gut lumen (the area in the tube's middle).

The digestive system begins with the mouth, which in adults has thirty-two teeth (Figure 24.4). The body's hardest tissue is a calcium-rich *enamel* that forms each tooth's thin outer layer. Beneath this is *dentin*, a thicker, bonelike layer. Nerves and blood vessels run through loose connective tissue in the *pulp* at a tooth's center.

Incisors and canine teeth are used to bite off bits of food. Premolars and molars grind and crush these bits. Food is positioned for chewing by a muscular tongue attached to the floor of the mouth. Taste buds dot the tongue's surface. They enclose sensory receptors that respond to particular molecules or ions in the food.

Paired salivary glands lie in back of and beneath the tongue. They secrete saliva in response to nervous signals. Saliva consists of a combination of water, ions, and proteins. It moistens food and begins the process

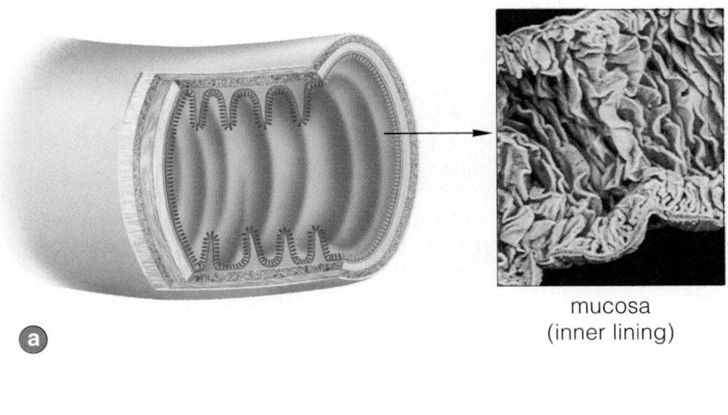

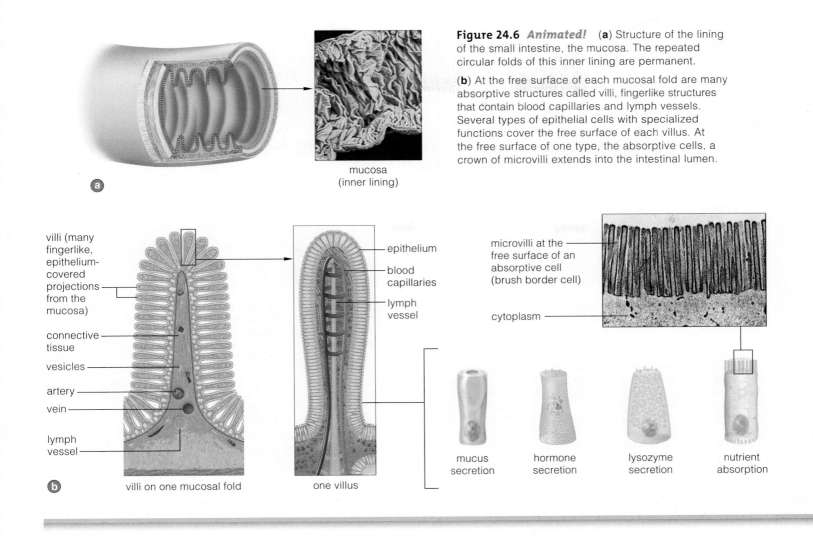

Figure 24.6 *Animated!* (**a**) Structure of the lining of the small intestine, the mucosa. The repeated circular folds of this inner lining are permanent.

(**b**) At the free surface of each mucosal fold are many absorptive structures called villi, fingerlike structures that contain blood capillaries and lymph vessels. Several types of epithelial cells with specialized functions cover the free surface of each villus. At the free surface of one type, the absorptive cells, a crown of microvilli extends into the intestinal lumen.

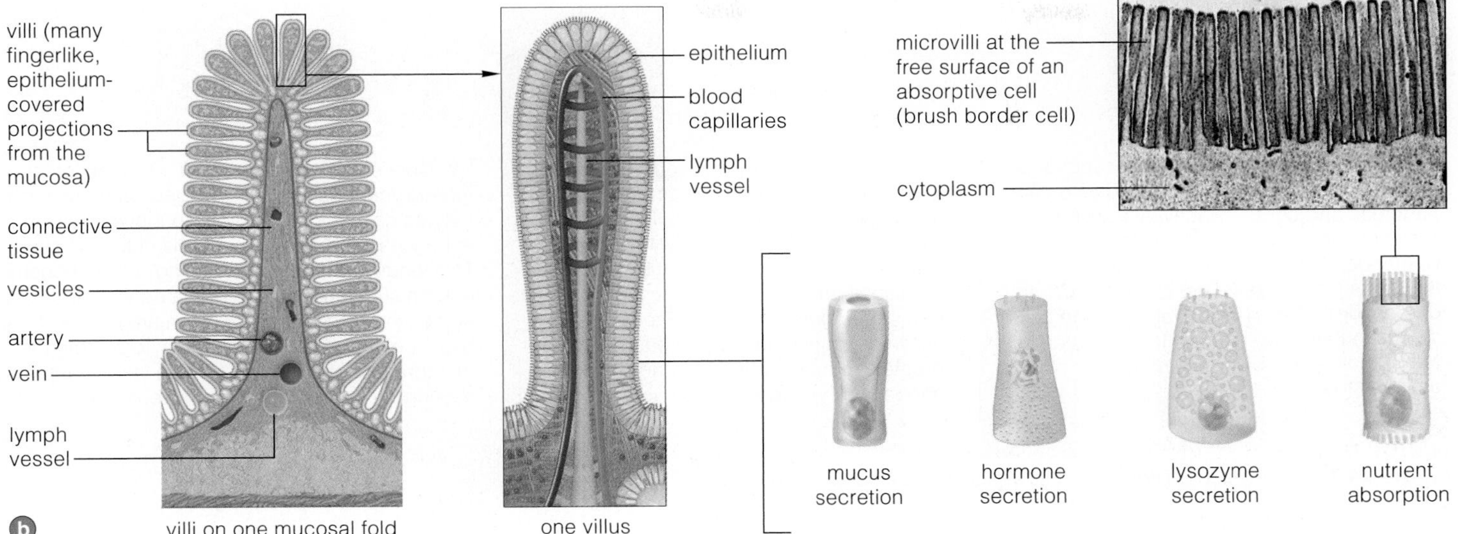

The sides and tip of each villus are covered mostly with cells that have elongated absorptive projections called **microvilli** (singular, microvillus). Each cell has 1,700 or so microvilli, which make its outer edge look somewhat like a brush. Hence these cells are known as "brush border" cells. Collectively, the many folds and projections of the small intestinal lining increase its surface area hundreds of times over.

In addition to brush border cells, the lining of the small intestine has secretory cells that produce mucus or hormones. Other lining cells act against bacterial pathogens. They engulf bacteria and secrete lysozyme, an enzyme that destroys bacterial cell walls.

HOW NUTRIENTS ARE ABSORBED

Absorption is the passage of nutrients, water, salts, and vitamins into the internal environment. The large absorptive surface of the intestinal wall facilitates the process. So does the action of rings of smooth muscle in this wall. Their repeated contraction and relaxation sloshes chyme back and forth, mixing it with enzymes and forcing it against the absorptive lining.

Different substances are absorbed in different ways. Water from food and drinks moves across the intestinal lining by osmosis. Amino acids and monosaccharides cross through transport proteins. **Bile** acts in digestion of fats and helps the monoglycerides and fatty acids to diffuse across the lipid bilayer of cell membranes.

Bile, produced by the liver, is a mixture of salts, bile pigments, the phospholipid lecithin, and cholesterol. Production of bile is the body's main way of getting rid of excess cholesterol. One iron-rich bile pigment, called bilirubin, is derived from the hemoglobin of dead red blood cells. In between meals, the main bile duct from the liver to the small intestine is closed off. Bile backs up inside of the gallbladder, which stores it. Eating a fatty meal stimulates the gall bladder to release stored bile into the small intestine.

Bile aids fat digestion by allowing emulsification, the dispersion of fat particles in a solution. Most fats in food are triglycerides. They are insoluble in water

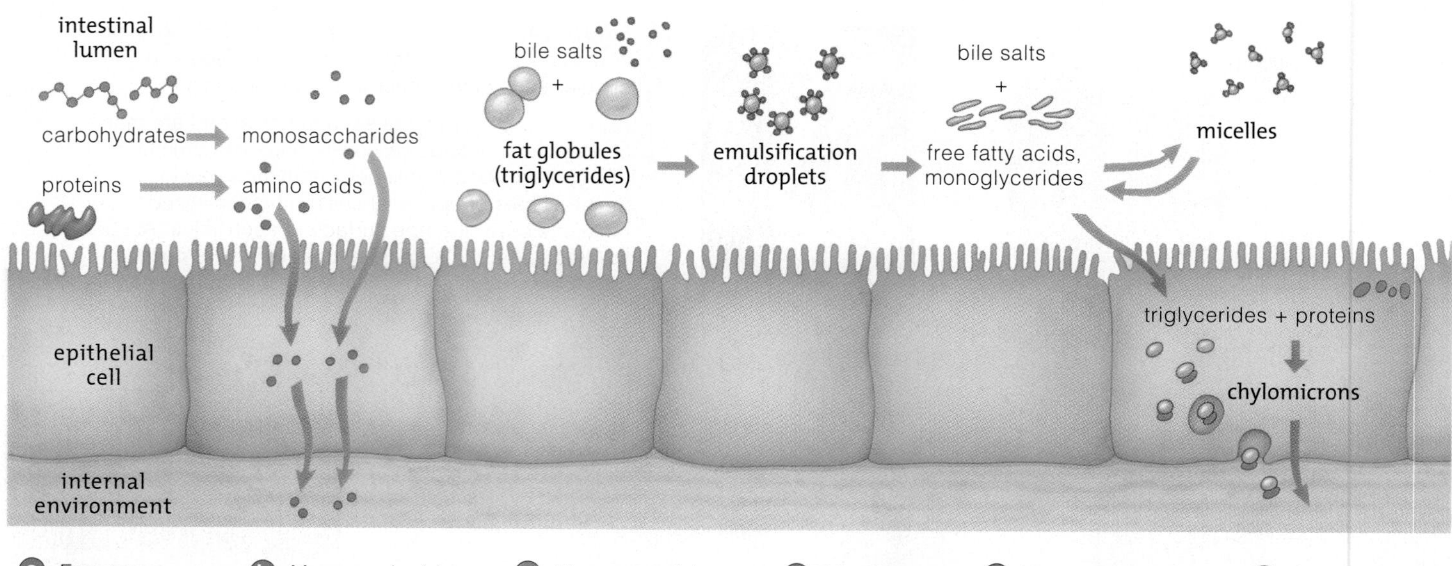

intestinal lumen
bile salts
+
bile salts
+
carbohydrates
monosaccharides
proteins
amino acids
fat globules (triglycerides)
emulsification droplets
free fatty acids, monoglycerides
micelles
triglycerides + proteins
epithelial cell
chylomicrons
internal environment
a Enzymes secreted by the pancreas and cells of the epithelial lining complete the digestion of carbohydrates to monosaccharides, and proteins to amino acids.
b Monosaccharides and amino acids are actively transported across the plasma membrane of the epithelial cells, then out of the same cells and into the internal environment.
c Movement of the intestinal wall breaks up fat globules into small droplets. Bile salts prevent the globules from re-forming. Pancreatic enzymes digest the droplets to fatty acids and monoglycerides.
d Micelles form as bile salts combine with digestion products and phospholipids. Products can readily slip into and out of the micelles.
e Concentration of monoglycerides and fatty acids in micelles enhances gradients. This leads to diffusion of both substances across the lipid bilayer of the plasma membrane of cells making up the lining.
f Products of fat digestion reassemble into triglycerides in cells of the intestinal lining. These become coated with proteins, then are expelled (by exocytosis from the cells) into the internal environment.

Figure 24.7 ***Animated!*** Digestion and absorption in the small intestine.

LINKS TO SECTIONS 2.5–2.7, 4.5, 4.6

and cluster as large globules in chyme. As the muscle layers in the intestinal wall move and agitate chyme, fat globules break into smaller droplets that become coated with bile salts. Bile salts carry negative charges, so these droplets repel each other and stay separated as the "emulsion." Compared to bigger globules, tiny droplets present a greater surface area to fat-digesting enzymes. These enzymes cleave triglycerides into free fatty acids and monoglycerides (Figure 24.7).

In **micelle formation**, bile salts, phospholipids, and products of fat digestion combine into tiny spherical droplets, or micelles. When combined this way, these molecules can reach high concentrations right next to the intestinal lining. When they do, they diffuse out of micelles and into the lining's epithelial cells. The fatty acids and monoglycerides recombine inside the cells, as triglycerides, which bind to proteins. The resulting small particles, called chylomicrons, leave the cells by exocytosis and enter the internal environment.

Glucose and amino acids move from the interstitial fluid into blood vessels. Triglycerides enter into lymph vessels, which eventually drain into blood vessels.

CONTROLS OVER DIGESTION

As you know, homeostatic controls counter shifts in the internal environment. But controls over digestion act *before* digested nutrients flood in. The nervous and endocrine systems regulate digestive functions.

The digestive system is affected by nervous signals from the brain. The sight, smell, or even the thought of a meal can stimulate salivation and increase secretion of gastrin, acid, and enzymes by the stomach lining. Emotion affects digestion. Anxiety and fear cause the diversion of blood away from the gut, slow secretion of acid and enzymes, and decrease the rate of smooth muscle contractions that mix and move gut contents.

A local nervous system interacts with the brain in the regulation of digestion. This is a network of nerves that extends through the walls of the digestive tract. When food enters the gut, the stretching of these walls activates sensory receptors. Sensory signals are sent to smooth muscles, glands, and to the brain. In response, muscles increase their rate of contraction, and gland cells secrete enzymes into the lumen or hormones into the blood. A big meal activates more receptors, so the contractions become more forceful.

Hormones influence appetite and digestion. Recall that **leptin** suppresses appetite and **ghrelin** enhances it. The hormone **gastrin** is produced by some stomach cells, and promotes secretion of HCl and enzymes by others. When HCl moves into the small intestine, cells in that lining release the hormone **secretin**, which tells the pancreas to secrete bicarbonate. Another intestinal hormone, **cholecystokinin** (CCK), calls for gallbladder contractions and secretion of pancreatic enzymes. It also suppress appetite.

Mechanical breakdown of food begins in the mouth. The saliva contains enzymes that start carbohydrate digestion.

Gastric fluid contains acid and digestive enzymes that start the process of protein digestion in the stomach.

Digestion is completed in the small intestine with the help of enzymes made here and in the pancreas. Bile secreted by the gallbladder aids in fat digestion and absorption.

With its richly folded intestinal mucosa, millions of villi, and hundreds of millions of microvilli, the small intestine has a vast surface area for absorbing nutrients.

THE LARGE INTESTINE

Not everything that enters the small intestine can be or should be absorbed. Undigestible material, such as cellulose fibers, is called "bulk." It is moved from the small intestine into the **large intestine** (colon) where, together with cells shed from the gut lining and some other wastes, it becomes concentrated as feces.

Most of the water that enters the gut is absorbed in the small intestine, but a final concentration of waste takes place in the colon. Cells in the lining of the colon actively transport sodium ions out of the lumen across the cell and then into the interstitial fluid, from which they enter the blood. Wastes become concentrated as water follows the sodium ions by osmosis. Other ions and vitamins made by bacteria that live in the colon are also absorbed from the colon into the blood.

Following a meal, nervous signals stimulate large regions of the large intestine to contract at once. This moves material already in the lumen farther along and makes room for the incoming batch.

The final length of the large intestine is the **rectum**. The rectum stores feces until enough accumulates to stretch rectal walls and trigger a defecation reflex. The brain controls the timing of defecation by sending out signals that cause contraction or relaxation of skeletal muscle in the anal sphincter at the end of the gut.

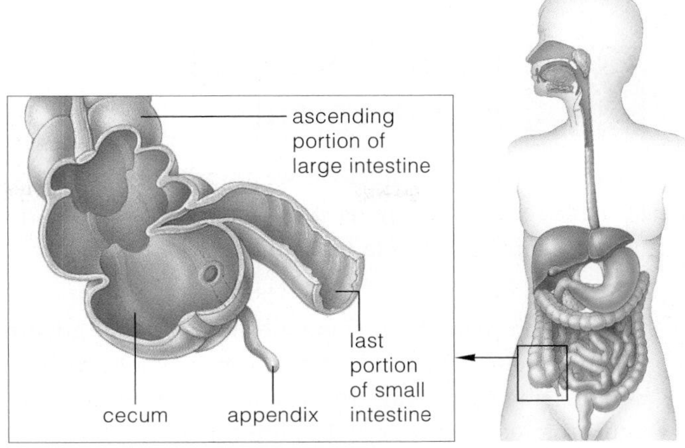

Figure 24.8 Cecum and appendix of the start of the large intestine (colon).

Extending from the first part of the colon, or cecum, is a wormlike projection called the **appendix** (Figure 24.8). It has no known digestive function, but is full of white blood cells that participate in body defenses. In the United States, about 7 percent of people will have an inflamed appendix, or *appendicitis*, at some time in their life. Usually this occurs when bits of waste clog the opening to the appendix, obstructing blood flow and allowing fluids to build up inside it. Tissue begins to die and infection may set in. Unless the inflamed appendix is removed by surgery, it can rupture. If it does, bacteria enter the abdominal cavity and cause infections that may be life-threatening.

A diet high in fiber can help prevent appendicitis. It also helps speed movement of material through the colon. A low-fiber diet can contribute to *constipation*, a decrease in the frequency of bowel movements.

Some people are genetically predisposed to develop *colon polyps*, small growths that project from the wall of the colon. Polyps are benign growths, but some can become cancerous. If colon cancer is detected early, it is highly curable. Even so, it kills about 50,000 people in the United States every year. Blood in the stool and changes in bowel habits can be symptoms and should be discussed with your physician. Anyone over the age of 50 should have a *colonoscopy*, a procedure in which the colon is examined using a camera at the end of a long flexible tube. A newer version of this procedure, *virtual colonoscopy*, uses X-rays to provide images that are analyzed to detect polyps or cancer.

The large intestine, or colon, absorbs water and mineral ions from the gut, and compacts undigested residues.

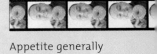

IMPACTS, ISSUES

Appetite generally declines with age. Among some of the elderly, the decline is so great it endangers health. The hormone CCK plays a role. When released by the small intestine, it stimulates pancreatic secretions, the release of bile, and appetite suppression by way of brain signals. CCK output rises with age. A new drug that blocks CCK may help restore appetite and prevent malnutrition in seniors. Also under study are anti-obesity drugs that stimulate CCK secretion.

LINKS TO SECTIONS 2.5–2.7, 6.5

24.3 Human Nutritional Requirements

What happens to the nutrients we absorb? They are burned as fuel in energy-releasing pathways and used as building blocks to build and replace tissues.

Earlier, you read about some mechanisms that govern organic metabolism for the body as a whole. Section 6.5 described pathways by which carbohydrates, fats, and proteins are converted to intermediates that enter the ATP-producing pathway of aerobic respiration.

Cells also cycle molecules through other metabolic pathways that use them as energy sources and building blocks. This molecular turnover is continuous, and it is massive. Figure 24.9 summarizes the major routes by which organic compounds obtained from food can be shuffled and reshuffled in the body as a whole.

After a meal, digestion and absorption replenishes the body's supplies of water, minerals, and organic compounds. We will look at vitamins and minerals in the next section. Here our focus is the main classes of nutrients: carbohydrates, fats, and proteins.

CARBOHYDRATES

Most complex carbohydrates are readily broken down into glucose, the body's main energy source. Starch is abundant in fleshy fruits, cereal grains, and legumes, such as peas and beans. Foods that are rich in complex carbohydrates are typically high in fiber, which helps to prevent constipation and colon problems. As the epidemiologist Denis Burkitt put it, where people pass small volumes of feces, you have large hospitals.

Refined sugars, such as table sugar and corn syrup, are sometimes referred to as "empty calories" because they supply no vitamins, minerals, or fiber. They add to caloric intake, but meet no other nutritional needs.

When glucose absorption from the gut exceeds the body's immediate requirements, some is used to make glycogen in liver and muscle cells. The rest is used to synthesize fats that become stored in adipose tissue.

While compounds are being absorbed and stored, most cells use glucose for energy. Between meals, the brain takes up two-thirds of the circulating glucose. Other body cells use fat and glycogen stores. Adipose tissue cells dismantle fats to glycerol and fatty acids, which enter blood. Liver cells break down glycogen to release glucose, which also enters blood. Most cells take up fatty acids and glucose for ATP production.

GOOD FAT, BAD FAT

You cannot stay alive without lipids. Cell membranes incorporate the phospholipid lecithin and the sterol cholesterol. Fat stored in adipose tissue acts as energy reserves, cushions many internal organs, and provides

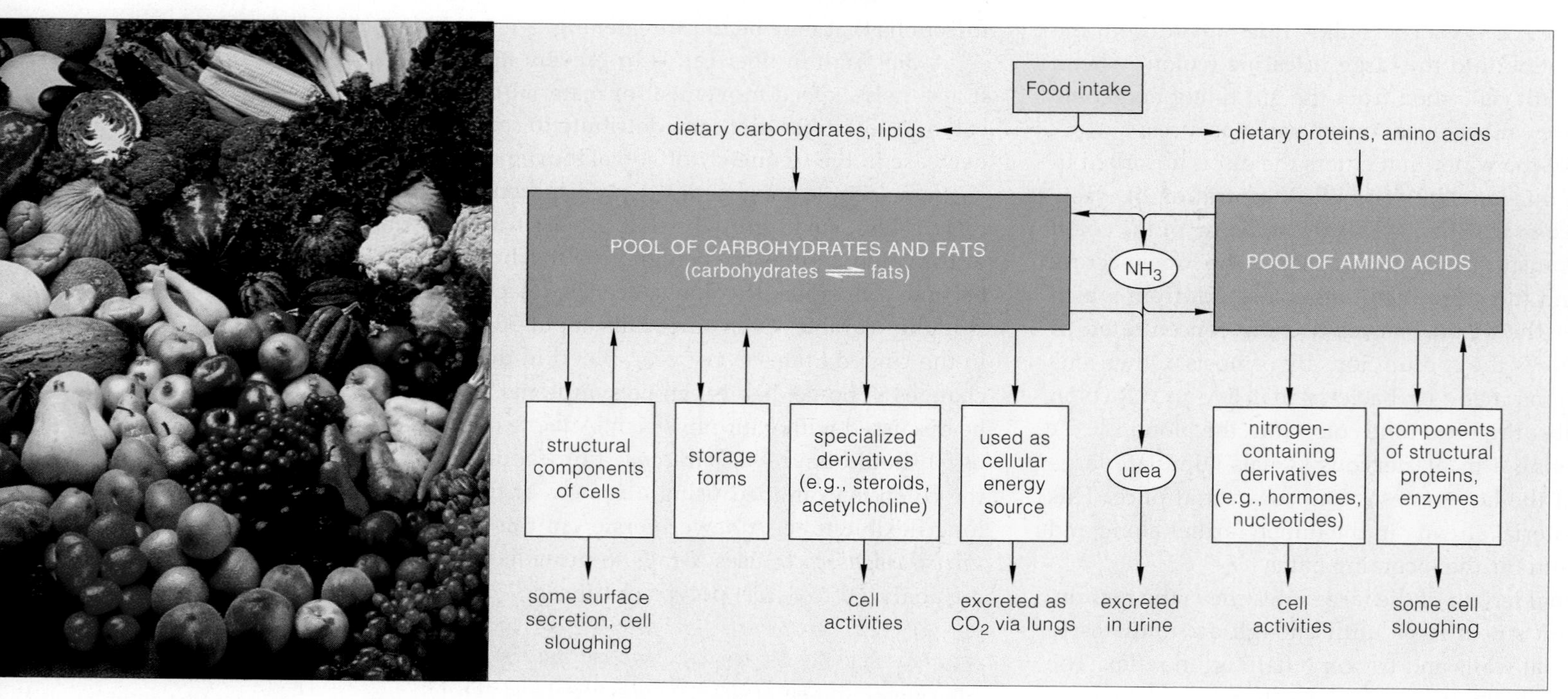

Figure 24.9 Major pathways of organic metabolism. Cells continually synthesize and tear down carbohydrates, fats, and proteins. Urea forms mainly in the liver as a result of protein breakdown.

insulation beneath the skin. Your body makes most fats you need from proteins and carbohydrates. Those it cannot make, such as linoleic acid, are called **essential fatty acids**. Vegetable oils are one good source of these. Dietary fats also help you absorb fat-soluble vitamins.

Butter, cheese, and fatty meats are rich in saturated fats and cholesterol. Overconsumption of these foods increases the risk for heart disease and stroke, as well as for cancers. The trans fatty acids, or "trans fats," are also bad for the heart. All food labels are now required to show the amounts of trans fats, saturated fats, and cholesterol per serving.

BODY-BUILDING PROTEINS

Your body also cannot function without proteins. It requires amino acid components of dietary proteins for its own protein-building programs. Of the twenty common types, eight are **essential amino acids**. Your cells cannot synthesize them; you must get them from food. They are methionine (or cysteine, its metabolic equivalent), isoleucine, leucine, lysine, phenylalanine (or tyrosine), threonine, tryptophan, and valine.

Most proteins in animal products are *complete*; their amino acid ratios can satisfy human nutritional needs. Nearly all plant proteins are *incomplete*—they lack one or more of the essential amino acids. To meet all your amino acid requirements with plant-based foods alone requires combining them so that amino acids missing from one component are provided by the others. As an example, rice and beans provide all necessary amino acids, but rice alone or beans alone do not.

DIETARY RECOMMENDATIONS

In January of 2005, the United States Food and Drug Administration issued a set of nutritional guidelines to replace its earlier "food pyramid." The guidelines are based on nutritional research and are designed to educate consumers about how a healthy diet can help reduce the risk for chronic diseases, such as diabetes, hypertension, heart disease, and certain cancers. The recommended number of servings of the various food groups are shown in Table 24.1. In comparison with the diet of a typical American, the guidelines call for a lowered intake of refined grains, saturated and trans fats, and sugars. The guidelines also suggest eating more whole grains, legumes, dark green and orange vegetables, fruits, and milk products. The full report can be downloaded from www.health.gov.

The new USDA guidelines call for about 55 percent of daily caloric intake to come from carbohydrates. Yet "low carb" diets are now enormously popular. About 10 million Americans report that they are following such a program. These diets call for a reduced intake of carbohydrates and a greater intake of proteins and fats. There is no question that such diets can promote rapid weight loss. However, the long-term effects of these low-carb diets are still being debated, in some cases heatedly, by health professionals.

Why? Digestion of proteins will produce toxic ammonia. Enzymes in the liver convert the ammonia to urea, which is then filtered out by kidneys and excreted in the urine. Also, the breakdown of fats yields acidic ketones that must be filtered out of the blood and then excreted. The National Kidney Foundation cautions that ketone acidity and all the extra filtering that is required to get rid of excess urea can strain kidneys and increase risk for kidney stones. Anyone who has impaired kidney function should avoid a high-protein diet.

IMPACTS, ISSUES

The use of the sugar fructose in soft drinks and other foods may be contributing to rising obesity levels. Eating or drinking a product sweetened with glucose stimulates the release of leptin from fat cells and also decreases the stomach's ghrelin output. The combined result is a decrease in feelings of hunger. In contrast, fructose delivers just as many calories as glucose, but does not affect these hormones, so a person continues to feel hungry. This encourages overconsumption and, possibly, weight gain.

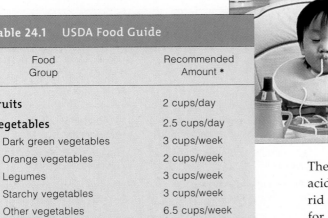

Table 24.1 USDA Food Guide

Food Group	Recommended Amount *
Fruits	2 cups/day
Vegetables	2.5 cups/day
Dark green vegetables	3 cups/week
Orange vegetables	2 cups/week
Legumes	3 cups/week
Starchy vegetables	3 cups/week
Other vegetables	6.5 cups/week
Grains	6 ounces/day
Whole grains	3 ounces/day
Other grains	3 ounces/day
Meat, Beans, Egg	5.5 ounces/day
Milk Products	3 cups/day
Oils	24 grams/day

* Based on a 2,000 calorie/day intake, recommended for sedentary females ages 10 to 30. Recommended caloric intake and serving sizes are more for males and highly active females and less for older females.

The body requires inputs of carbohydrates, fats, and proteins for normal growth and maintenance.

A healthy diet can help to prevent chronic diseases. Most people should lower their intake of sugars, saturated fats and trans fats. At the same time, they should increase their intake of whole grains, fruits, and vegetables.

LINKS TO SECTIONS 2.9, 4.1

VITAMINS AND MINERALS

Vitamins are organic substances that are essential for growth and survival; nothing else can carry out their metabolic functions, which Table 24.2 lists. Most plants synthesize vitamins on their own. Over evolutionary time, different animals lost the ability to make certain vitamins and so must get them from food.

At the minimum, human cells require the thirteen vitamins listed in Table 24.2. Each vitamin has specific metabolic roles. Many reactions use several types, and the absence of one affects the functions of others.

Some vitamins, including vitamin C and vitamin E, act as antioxidants. They can prevent free radicals from reacting with and damaging DNA and other essential molecules. A free radical, recall, is an atom or group of atoms with an unpaired electron. It can steal electrons from other molecules (a reaction called oxidation), and thus disrupt their function.

Minerals are *inorganic* substances that are essential for growth, development, and metabolism (Table 24.3). For example, all cells require iron for use in electron transfer chains. Red blood cells need extra iron; they use it in hemoglobin synthesis. Also, neurons cannot send signals unless there is just the right amount of sodium and potassium.

People who are in good health get all the vitamins and minerals they need from a balanced diet of whole

Table 24.2 Major Vitamins: Sources, Functions, and Effects of Deficiencies or Excesses*

Vitamin	Common Sources	Main Functions	Effects of Chronic Deficiency	Effects of Extreme Excess
Fat-soluble vitamins				
A	Beta-carotene, yellow fruits, yellow or green leafy vegetables, fortified milk, egg yolk, fish, liver	Used in synthesis of visual pigments, bone, teeth; maintains the skin	Dry, scaly skin; lowered resistance to infections; night blindness; permanent blindness	Malformed fetuses; hair loss; changes in skin; liver and bone damage; bone pain
D	Fish liver oils, egg yolk fortified milk; also made in skin	Promotes bone growth, mineralization; enhances calcium absorption	Bone deformities (rickets) in children, bone softening in adults	Retarded growth; kidney damage; calcium deposits in soft tissues
E	Whole grains, dark green vegetables, vegetable oils	Counters effects of free radicals; helps maintain cell membranes; blocks breakdown of vitamins A, C in gut	Lysis of red blood cells; nerve damage	Muscle weakness, fatigue, nausea, headaches
K	Enterobacteria; also in green leafy vegetables, cabbage	Blood clotting; ATP formation via electron transfer chains	Abnormal blood clotting; severe bleeding (hemorrhaging)	Anemia; liver damage, jaundice
Water-soluble vitamins				
B_1 (thiamin)	Legumes, whole grains, green leafy vegetables, lean meats, eggs	Connective tissue formation; folate utilization; coenzyme action	Water retention in tissues; tingling sensations; heart disturbances; poor coordination	None reported from food; possible shock reaction from repeated injections
B_2 (riboflavin)	Whole grains, poultry, fish, egg white, milk	Coenzyme action	Skin lesions	None reported
Niacin	Green leafy vegetables, potatoes, peanuts, poultry, fish, pork, beef	Coenzyme action	Contributes to pellagra (damage to skin, gut, nervous system, etc.)	Skin flushing; possible liver damage
B_6	Spinach, tomatoes, potatoes, meats	Coenzyme in amino acid metabolism	Skin, muscle, and nerve damage; anemia	Impaired coordination; numbness in feet
Pantothenic acid	In many foods (meats, yeast, egg yolk especially)	Coenzyme in glucose metabolism, fatty acid and steroid synthesis	Fatigue, tingling in hands, headaches, nausea	None reported; may cause diarrhea occasionally
Folate (folic acid)	Dark green vegetables, whole grains, yeast, lean meats; enterobacteria	Coenzyme in nucleic acid and amino acid metabolism	A type of anemia; inflamed tongue; diarrhea; impaired growth; mental disorders	Masks vitamin B_{12} deficiency
B_{12}	Poultry, fish, red meat, dairy foods (not butter)	Coenzyme in nucleic acid metabolism	A type of anemia; impaired nerve function	None reported
Biotin	Legumes, egg yolk; colon bacteria produce some	Coenzyme in fat, glycogen formation and in amino acid metabolism	Scaly skin (dermatitis), sore tongue, depression, anemia	None reported
C (ascorbic acid)	Fruits and vegetables, especially citrus, berries, cantaloupe, green pepper, cabbage, broccoli	Collagen synthesis; used in carbohydrate metabolism; structural role in bone, teeth, cartilage; may counter free radicals	Scurvy, poor wound healing, impaired immunity	Diarrhea, other digestive upsets; may skew results of some diagnostic tests

* Guidelines for appropriate daily intakes have not been set by the Food and Drug Administration.

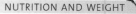

foods. Vitamin and mineral supplements are necessary only for strict vegetarians, the elderly, and people who are chronically ill or taking medication that interferes with how their body processes nutrients. For example, vitamin K supplements can promote calcium retention and lessen osteoporosis in elderly women.

However, metabolism varies in its details from one individual to the next, so no one should take massive amounts of any vitamin or mineral supplement except under medical supervision. Also, the excessive intake of many vitamins and minerals can be harmful (Tables 24.2 and 24.3). Any excess of a water-soluble vitamin is eliminated in urine. However, vitamin A and other fat-soluble vitamins accumulate in fatty tissues, where they can disrupt normal metabolic activities. Similarly, sodium is present in plant and animal foods, and it is a component of table salt. It contributes to the body's salt–water balance, muscle activity, and nerve signals. Even so, a prolonged, excessive intake of this mineral may contribute to high blood pressure.

24.4 Weighty Questions, Tantalizing Answers

The percentage of Americans who are overweight or obese is increasing steadily. To reverse this trend and stem the rising tide of obesity-related disease will require changes in eating habits and activity levels.

When you consider your ideal weight, you probably think about how you would like to look. As a culture, we are obsessed with being thin, but we are also fond of fast and fatty foods. Billions of dollars in sales of diet books, plans, and products reflect our fat phobia. Steadily increasing waistlines demonstrate the failure of most current approaches to weight control or loss.

According to the government's latest reports, about two thirds of Americans are overweight; nearly a third are obese. Looks aside, this is a pressing public health issue. Each year, weight-related cases of hypertension,

Table 24.3 Major Minerals: Sources, Functions, and Effects of Deficiencies or Excesses*

Mineral	Common Sources	Main Functions	Effects of Chronic Deficiency	Effects of Extreme Excess
Calcium	Dairy products, dark green vegetables, dried legumes	Bone, tooth formation; blood clotting; neural and muscle action	Stunted growth; possibly diminished bone mass	Impaired absorption of other minerals; kidney stones in susceptible people
Chloride	Table salt (usually too much in diet)	HCl formation in stomach; contributes to body's acid–base balance; neural action	Muscle cramps; impaired growth; poor appetite	Contributes to high blood pressure in susceptible people
Copper	Nuts, legumes, seafood, drinking water	Used in synthesis of melanin, hemoglobin; some transfer chain components	Anemia, changes in bone and blood vessels	Nausea, liver damage
Fluorine	Fluoridated water, tea, seafood	Bone, tooth maintenance	Tooth decay	Digestive upsets; mottled teeth; in chronic cases, deformed skeleton
Iodine	Marine fish, shellfish, iodized salt, dairy products	Thyroid hormone formation	Enlarged thyroid (goiter), with metabolic disorders	Toxic goiter
Iron	Whole grains, green leafy vegetables, legumes, nuts, eggs, lean meat, molasses, dried fruit, shellfish	Formation of hemoglobin and cytochrome (transfer chain component)	Iron-deficiency anemia, impaired immune function	Liver damage, shock, heart failure
Magnesium	Whole grains, legumes, nuts, dairy products	Coenzyme role in ATP–ADP cycle; roles in muscle, nerve function	Weak, sore muscles; impaired neural function	Impaired neural function
Phosphorus	Whole grains, poultry, red meat	Component of bone, teeth, nucleic acids, ATP, phospholipids	Muscular weakness; loss of minerals from bone	Impaired absorption of minerals into bone
Potassium	Most foods, diet provides ample amounts	Muscle and neural function; roles in protein synthesis and body's acid–base balance	Muscular weakness	Muscular weakness, paralysis, heart failure
Sodium	Table salt; diet provides ample to excessive amounts	Key role in body's salt–water balance; roles in muscle and neural function	Muscle cramps	High blood pressure in susceptible people
Sulfur	Proteins in diet	Component of body proteins	None reported	None likely
Zinc	Whole grains, legumes, nuts, meats, seafood	Component of digestive enzymes; roles in normal growth, wound healing, sperm formation, and taste and smell	Impaired growth, scaly skin, compromised immune function	Nausea, vomiting, diarrhea; impaired immune function and anemia

* The guidelines for appropriate daily intakes have not been set by the Food and Drug Administration.

LINKS TO SECTIONS 4.5, 4.6

Figure 24.10 How to estimate "ideal" weights for adults. Values shown are consistent with a long-term Harvard study into the link between excess weight and risk of cardiovascular disorders. The "ideal" varies. It is influenced by specific factors such as having a small, medium, or large skeletal frame.

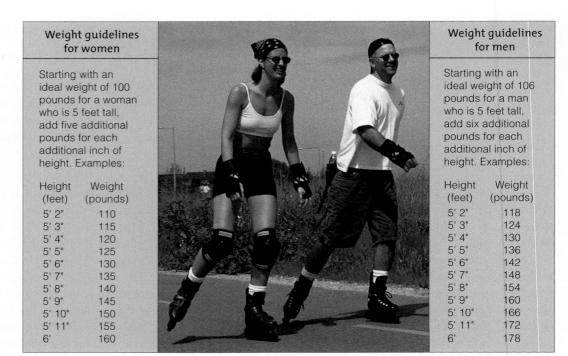

Weight guidelines for women

Starting with an ideal weight of 100 pounds for a woman who is 5 feet tall, add five additional pounds for each additional inch of height. Examples:

Height (feet)	Weight (pounds)
5' 2"	110
5' 3"	115
5' 4"	120
5' 5"	125
5' 6"	130
5' 7"	135
5' 8"	140
5' 9"	145
5' 10"	150
5' 11"	155
6'	160

Weight guidelines for men

Starting with an ideal weight of 106 pounds for a man who is 5 feet tall, add six additional pounds for each additional inch of height. Examples:

Height (feet)	Weight (pounds)
5' 2"	118
5' 3"	124
5' 4"	130
5' 5"	136
5' 6"	142
5' 7"	148
5' 8"	154
5' 9"	160
5' 10"	166
5' 11"	172
6'	178

type 2 diabetes, heart disease, gallstones, gout, breast and colon cancers, and osteoarthritis put an estimated 100-billion-dollar drain on this nation's economy. The future does not look much rosier. Since the 1970s, the percentage of overweight children and adolescents has doubled. One recent study concluded that if current trends continue, weight-related deaths could decrease the life expectancy in the United States by as much as 5 years over the next decade.

So what is a healthy weight? Figure 24.10 shows one set of standards, based on long-term studies of the effect of weight on cardiovascular health. Another approach is to consider the *body mass index* (BMI), as determined by this formula:

$$\text{BMI} = \frac{\text{weight (pounds)} \times 700}{\text{height (inches)}^2}$$

A BMI of 25 to 29.9 is defined as overweight and a BMI of 30 or more as obese. Having a BMI higher than 25 carries an increased risk of premature death due to weight-related illness. Fat distribution can affect risk, with an expanded waistline carrying more health risks than fat in other body regions. Problems are especially likely among males who have a waist circumference of more than 40 inches (120 cm) and females with a waist size that exceeds 35 inches (88 cm).

It is difficult to lower weight-related risks with diet alone. Eat less, and the body slows its metabolic rate to conserve energy. To function normally over the longer term while maintaining an optimal weight, you must balance energy intake with energy used for metabolic activities. For most people, this means controlling their portions of low-calorie, nutritious foods *and* exercising regularly. Bear in mind that the energy stored in food is measured as kilocalories or Calories (with a big C). A **kilocalorie** is 1,000 calories, or units of heat energy.

To figure out how many kilocalories you should take in daily to maintain a preferred weight, multiply that weight (in pounds) by 10 if you are not active physically, 15 if moderately active, and by 20 if highly active. Next, subtract one of the following amounts from the multiplication value:

Age:	*Subtract:*
25–34	0
35–44	100
45–54	200
55–64	300
Over 65	400

If you want to weigh 120 pounds and are highly active, 120 × 20 = 2,400 kilocalories. Such calculations provide only a rough estimate of caloric needs. Other factors, including height, must also be considered. An active person who stands 5 feet, 2 inches tall does not require as many calories as an active 6-footer of the same weight. Age is a factor too. Resting metabolic rate drops as a person gets older, so less food is required.

Excessive weight gain comes at a high health-related cost. To lose weight requires taking in fewer calories than are burned by metabolism and physical activity.

IMPACTS, ISSUES

The latest candidate for appetite-affecting hormones is $PYY_{3\text{-}36}$. Glandular cells of the stomach and small intestine release it after a meal. It acts in the brain to suppress appetite. Volunteers who were given an intravenous dose of $PYY_{3\text{-}36}$ before a buffet meal ate less than a control group that got a saline dose. A nasal spray that delivers the hormone to suppress appetite is now being tested in large clinical trials.

24.5 Urinary System of Mammals

All animals must control solute concentrations in their internal environment. In land vertebrates, including mammals, shifts in fluid–solute balance are managed primarily by the actions of the urinary system.

SHIFTS IN WATER AND SOLUTES

Over the course of any day, as a mammal eats and drinks, different solid foods and fluids enter its gut. Afterward, the absorbed water, nutrients, and other substances move into the blood, then interstitial fluid, then individual cells. Blood and interstitial fluid make up the *extracellular* fluid. Keeping the composition and the volume of this fluid within limits that body cells can tolerate is an important component of maintaining homeostasis. To keep the extracellular fluid within the tolerance of living cells, any gains or losses of water or solutes must be balanced. During any interval, the body—and the extracellular fluid inside it—must gain the same quantities of water and solutes that it loses.

WATER GAINS AND LOSSES Like all other mammals, humans absorb water from the gut. Water also forms as a by-product of many metabolic reactions. In land mammals, the volume of water that enters the gut in the first place depends largely on a thirst mechanism. If these mammals lose too much water, they drink.

Urinary excretion eliminates water, together with wastes and excess solutes, in urine. Water evaporates from the lungs, and most mammals also lose water in sweat. A healthy mammal loses little water from the gut; most is absorbed, rather than eliminated in feces.

SOLUTE GAINS AND LOSSES Absorption from the gut, secretions from cells, breathing, and metabolism all put solutes into the extracellular fluid. Breathing increases the level of oxygen. Aerobic respiration by body cells adds carbon dioxide.

Solutes are lost by urinary excretion, respiration, and sweating. Urine includes various wastes formed by the breakdown of organic compounds. For instance, as noted earlier, ammonia forms when amino groups are split from amino acids during protein breakdown. **Urea** forms in the liver when two ammonia molecules join with carbon dioxide. Also in urine are uric acid (from nucleic acid and amino acid degradation) and breakdown products from hemoglobin, food additives, and often drugs. Mineral ions are lost in sweat, and all mammals exhale carbon dioxide, the most abundant metabolic waste, from their lungs.

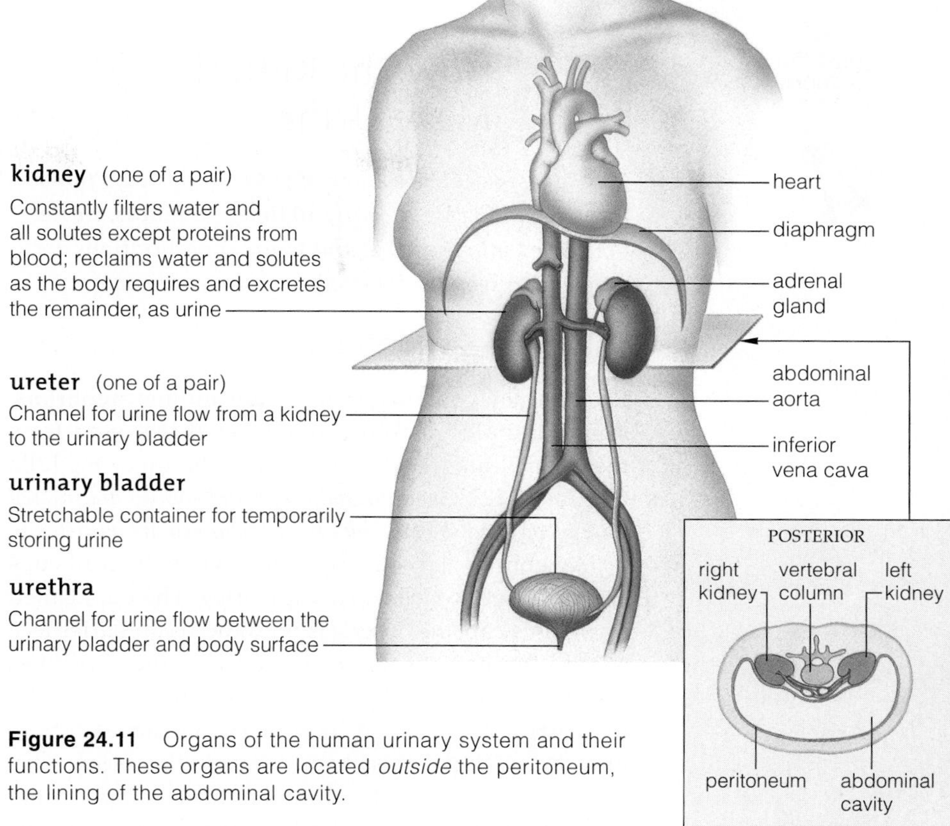
kidney (one of a pair)
Constantly filters water and all solutes except proteins from blood; reclaims water and solutes as the body requires and excretes the remainder, as urine
ureter (one of a pair)
Channel for urine flow from a kidney to the urinary bladder
urinary bladder
Stretchable container for temporarily storing urine
urethra
Channel for urine flow between the urinary bladder and body surface
heart
diaphragm
adrenal gland
abdominal aorta
inferior vena cava
POSTERIOR
right kidney
vertebral column
left kidney
peritoneum
abdominal cavity
ANTERIOR

Figure 24.11 Organs of the human urinary system and their functions. These organs are located *outside* the peritoneum, the lining of the abdominal cavity.

COMPONENTS OF THE HUMAN URINARY SYSTEM

The human **urinary system** consists of two kidneys, two ureters, a bladder, and a urethra (Figure 24.11). The **kidneys** are a pair of bean-shaped organs, each about as large as an adult's fist. Each kidney is enclosed in a protective outer capsule of connective tissue.

The kidneys filter water, mineral ions, nitrogen-rich wastes, and other substances from blood. Then they adjust the composition of the filtrate and return nearly all water and non-waste solutes to blood. **Urine** is the fluid that is left. It consists of urea and other wastes, water, and any excess solutes.

Urine flows from each kidney into a ureter, a tube leading to the urinary bladder. Urine is briefly stored in this muscular sac before flowing into the urethra, a tube that opens at the body surface.

An urge to urinate arises when stretch receptors in the bladder are stimulated and—in a reflex response—smooth muscle in the bladder wall contracts. After age 2 or 3 years, urination is voluntary. It occurs when a sphincter between the bladder and urethra is relaxed voluntarily so that bladder contraction can force urine into the urethra and out of the body.

A urinary system regulates the composition and volume of extracellular fluid. Kidneys filter water and solutes from blood, then return most. Any excess leaves the body as urine.

LINKS TO SECTIONS 4.5, 4.6, 22.4

24.6 How The Kidneys Make Urine

Every day, the entire volume of your bloodstream flows through your kidneys nearly 30 times. As it does, wastes are moved into the urine and solute concentrations are adjusted to keep the internal environment in balance.

NEPHRONS—FUNCTIONAL UNITS OF KIDNEYS

Each kidney contains more than a million **nephrons**, slender tubules packed inside lobes that extend from the outer kidney cortex down into the inner medulla (Figure 24.12). *At nephrons, water and solutes are filtered from blood, and the amounts to be reclaimed are adjusted.*

A nephron starts in the cortex, where its wall cups around a set of *glomerular* capillaries. The cup, called Bowman's capsule, interacts with these capillaries as a blood-filtering unit. Fluid filtered into the cup flows into the proximal tubule, which extends deeper into the cortex. A U-shaped loop of Henle carries filtrate into the medulla, then back toward the cortex. This loop connects to the distal tubule, which empties into a collecting duct. All collecting ducts empty into the renal pelvis, a central cavity. The urine that collects in this cavity is funneled into one of the ureters.

A network of *peritubular* capillaries weaves around each nephron's tubules. During urine formation, most water and solutes that leave nephron tubules and enter the interstitial fluid are absorbed by these capillaries and returned to the general circulation.

FILTRATION, REABSORPTION, AND SECRETION

Blood pressure generated by heart contractions is the force that drives **glomerular filtration**, the first step in urine formation (Figure 24.13*a*). By this process, water and solutes are forced out through spaces between cells in the walls of glomerular capillaries. These capillaries are structurally specialized for filtering; they are much "leakier" than most capillaries. Even so, most proteins are too large to leak out; they remain in the blood that exits the glomerular capillary bed.

The solute-rich filtrate forced out of the glomerular capillaries is collected within the interior of Bowman's capsule. This filtrate enters the proximal tubule, where **tubular reabsorption** begins. By this process, solutes and water move from filtrate in a kidney tubule into

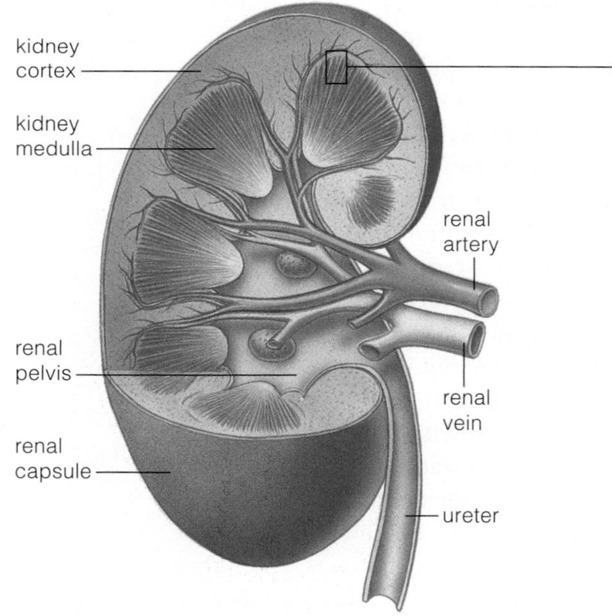

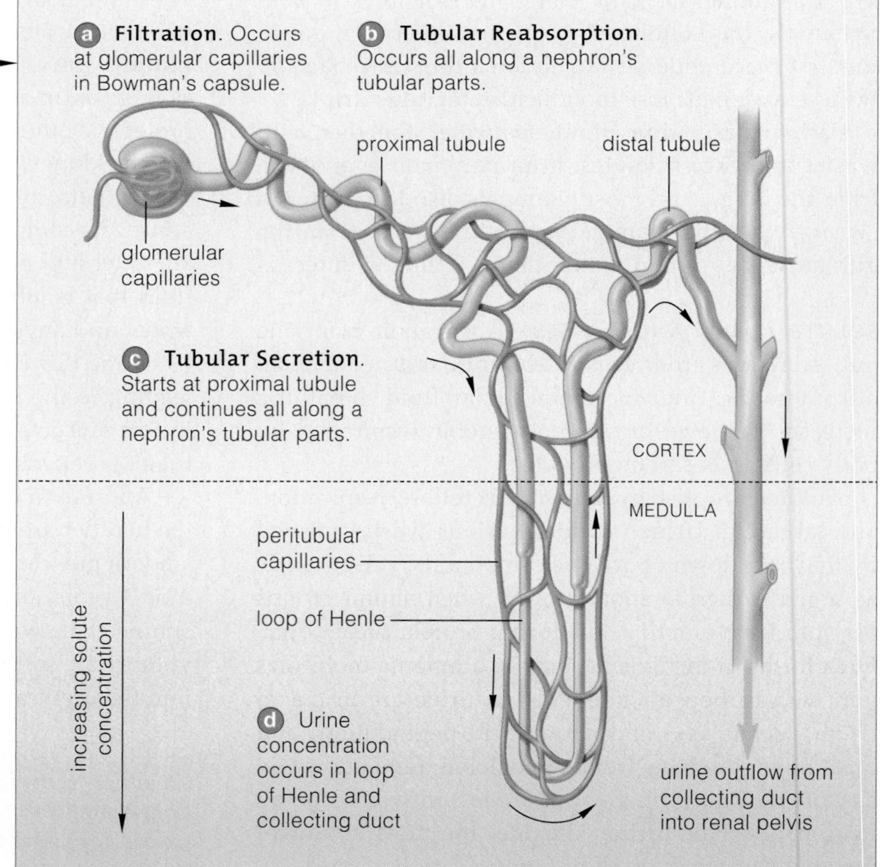

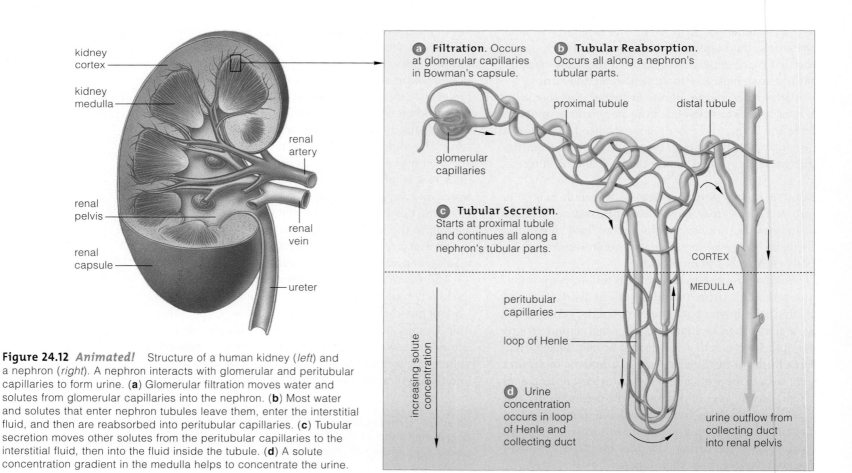

Figure 24.12 ***Animated!*** Structure of a human kidney (*left*) and a nephron (*right*). A nephron interacts with glomerular and peritubular capillaries to form urine. (**a**) Glomerular filtration moves water and solutes from glomerular capillaries into the nephron. (**b**) Most water and solutes that enter nephron tubules leave them, enter the interstitial fluid, and then are reabsorbed into peritubular capillaries. (**c**) Tubular secretion moves other solutes from the peritubular capillaries to the interstitial fluid, then into the fluid inside the tubule. (**d**) A solute concentration gradient in the medulla helps to concentrate the urine.

the interstitial fluid, then into the adjacent peritubular capillaries (Figure 24.13*b*).

How do substances get out of the filtrate and into the interstitial fluid? Water moves across tubule cells in response to osmotic gradients created after ions and molecules, such as Na^+, Cl^-, and glucose, are helped across by transport proteins.

About two thirds of the water and solutes that left glomerular capillaries is returned to the blood as the filtrate flows through proximal tubules. Reabsorption continues as filtrate flows through the nephron's other tubular regions. About 99 percent of the water will be returned to the blood, along with all the glucose and amino acids. Only half the urea is reabsorbed; this is a waste product that the body needs to get rid of.

Tubular secretion begins in the proximal tubule, as H^+ and K^+ are moved from blood in peritubular capillaries into accumulating filtrate (Figure 24.13*c*). Secretion continues as these ions and other wastes are added to the filtrate in other tubule regions, including the distal tubules and collection duct.

Tubular secretion is essential to regulating the pH of the internal environment. Recall that enzymes can function only within a limited pH range. An excess of H^+ can disrupt many essential metabolic processes. Recall also that buffer systems can help to neutralize excess H^+ (Section 2.3). However, the effects of buffers are temporary; they cannot eliminate H^+. If too much H^+ accumulates, buffer systems can be overwhelmed, and pH falls rapidly to life-threatening levels. Kidneys help maintain acid–base balance by regulating how much bicarbonate is reabsorbed into the blood and how much H^+ is secreted and excreted into the urine.

ADJUSTING URINE CONCENTRATION

You probably already know that the concentration of urine can vary. Sip soda all day and your urine will be dilute. Spend 8 hours or so without drinking, as when you sleep, and your urine becomes concentrated. But even when it is most dilute, urine has far more solutes than the plasma. How does the body concentrate urine and how does it adjust this concentration?

For water to leave a kidney tubule by osmosis, the interstitial fluid around a tubule must be hypertonic relative to the filtrate. In other words, the interstitial fluid has to be saltier than the filtrate. This is what happens in the kidney medulla. Fluid here shows a gradient in solute concentration; it is saltiest near the turn in the loop of Henle (Figures 24.12*d* and 24.13*d*).

Urine gets more and more concentrated as it moves down through the loop. Water moves out of filtrate by osmosis *before* the turn and filtrate in the tubule gets saltier until it matches interstitial fluid. The loop *after* the turn does not let water out, but transport proteins pump out salt. This makes the interstitial fluid even saltier—so it attracts more water out of the filtrate that is just entering the loop.

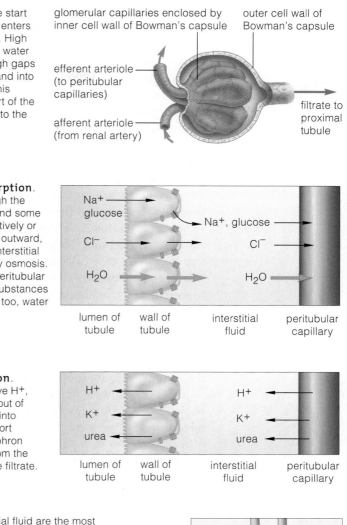

a Filtration. At the start of the nephron, blood enters glomerular capillaries. High blood pressure forces water and solutes out through gaps in the capillary walls and into Bowman's capsule. This capsule is the first part of the nephron and drains into the proximal tubule.

glomerular capillaries enclosed by inner cell wall of Bowman's capsule

outer cell wall of Bowman's capsule

efferent arteriole (to peritubular capillaries)

afferent arteriole (from renal artery)

filtrate to proximal tubule

b Tubular Reabsorption. As filtrate flows through the kidney tubules, ions and some nutrients are either actively or passively transported outward, into the surrounding interstitial fluid. Water follows, by osmosis. Cells making up the peritubular capillaries transport substances back into blood. Here too, water follows by osmosis.

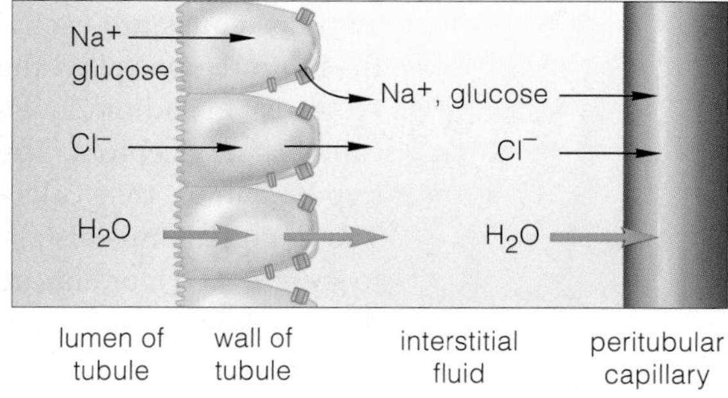

c Tubular Secretion. Transport proteins move H^+, K^+, urea, and wastes out of peritubular capillaries into interstitial fluid. Transport proteins in tubular nephron regions move them from the interstitial fluid into the filtrate.

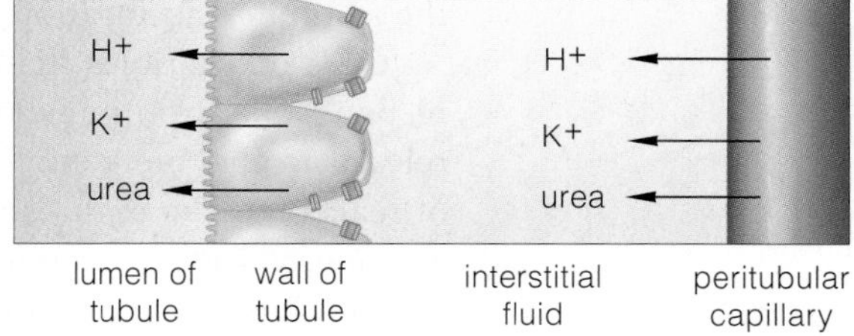

d Solutes in interstitial fluid are the most concentrated deep in the kidney medulla, The gradient draws more and more water from the *descending* limb of the loop of Henle, by osmosis. But Na^+, Cl^-, urea, and other solutes cannot cross the descending limb wall, so the filtrate becomes more concentrated as it approaches the loop of Henle's hairpin turn.

Transporters in the thick-walled *ascending* limb actively pump Na^+ out of the filtrate. Cl^- follows passively. Interstitial fluid gets saltier and attracts *more* water out of the descending limb and the collecting duct.

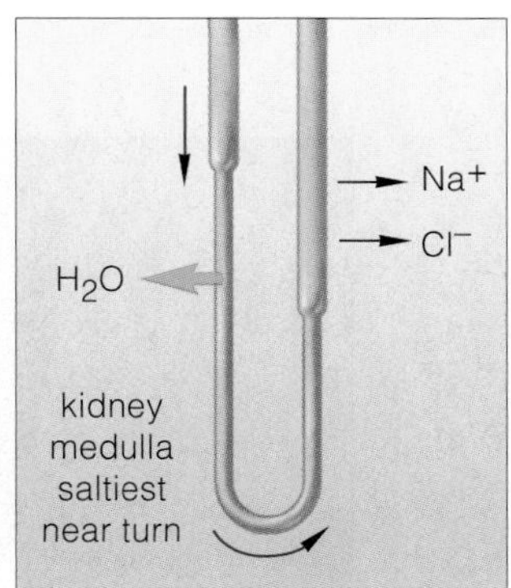

Figure 24.13 ***Animated!*** How urine is formed. (**a**) Filtration puts water, ions, and solutes into kidney tubules. (**b**) Reabsorption returns most of these to the blood. (**c**) Secretion moves other solutes from the blood to the filtrate. (**d**) The solute concentration gradient around the loop of Henle helps concentrate urine.

LINKS TO SECTIONS 4.5, 4.6, 22.6

The different permeabilities of the ascending and descending loops of Henle and their close proximity allow them to exchange solutes, thus concentrating the urine. We call this type of system, in which two fluids flowing in opposite directions exchange substances, a *countercurrent exchange system*. Recall that a similar sort of system operates in the gills of fishes. It maximizes the amount of oxygen that diffuses from water into a fish's bloodstream (Section 22.6).

Adjusting how permeable the kidney tubules are to water and sodium makes the urine more or less dilute. Two hormones are central to these adjustments, as is the brain region called the hypothalamus.

A rise in sodium levels stimulates sensory neurons called osmoreceptors. They signal a *thirst center* in the hypothalamus that calls for water-seeking behavior. The hypothalamus also stimulates the pituitary gland to secrete **ADH**, or antidiuretic hormone. ADH acts on distal tubules and collecting ducts, making them more permeable to water. The body reabsorbs more water, so less departs in urine (Figure 24.14). ADH secretion slows when the volume of extracellular fluid rises and the sodium concentration declines.

Cells in arterioles that carry blood into the kidney respond to a drop in extracellular fluid volume. They release an enzyme, renin, which sets in motion a chain of reactions. The end result is that the adrenal glands above each kidney secrete **aldosterone**. This hormone acts on cells of the distal tubules and increases sodium reabsorption. The result? Less sodium is excreted, and as water follows sodium back into the blood, the urine becomes more concentrated.

Hormonal disorders affect kidney function. *Diabetes insipidus* results when the pituitary gland secretes too little ADH or ADH receptors do not respond properly. Affected people produce huge volumes of dilute urine and suffer from extraordinary thirst. In contrast, too much aldosterone, or *hyperaldosteronism*, leads to fluid retention and high blood pressure.

Filtration, reabsorption, and secretion in the kidneys help maintain the composition and volume of extracellular fluid.

Glomerular filtration forces water and small solutes into the kidney tubules. The driving force for filtration is the pressure generated by the beating heart.

The proximal tubule reabsorbs most of the filtered water and solutes. Ions and molecules that are transported out of the tubule enter peritubular capillaries. Water follows by osmosis.

Certain excess solutes diffuse out of peritubular capillaries and into interstitial fluid. Then, by tubular secretion, they move into the nephron for subsequent excretion. Secretion into the urine is the only way the body can rid itself of excess H^+ to maintain acid–base balance.

ADH action counters decreases in the extracellular fluid volume; it promotes water conservation. Aldosterone counters decreases in the extracellular fluid volume; it promotes sodium conservation.

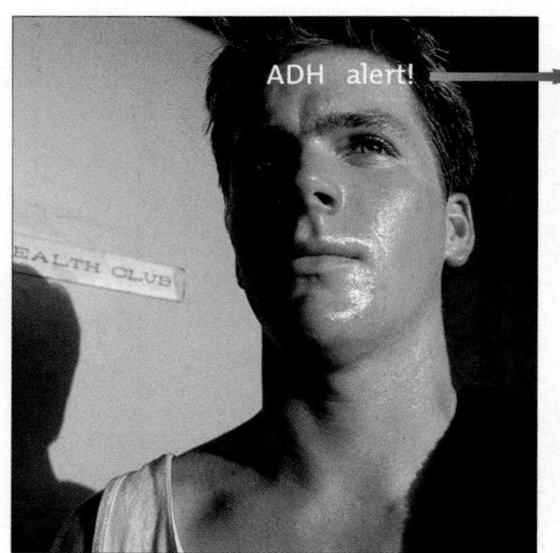

Stimulus

a Water loss lowers the blood volume. Sensory receptors in the hypothalamus detect a big deviation from the set point.

b The hypothalamus stimulates the pituitary gland to step up its secretion of ADH.

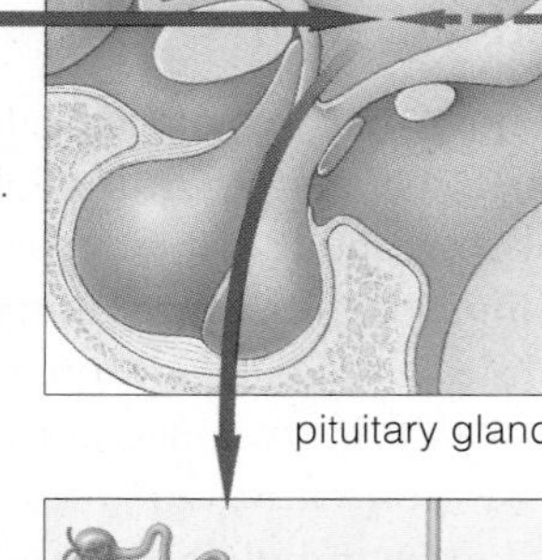

c ADH circulates in blood, reaches nephrons in the kidneys. By acting on cells of distal tubules and collecting ducts, it makes the tube walls more permeable to water.

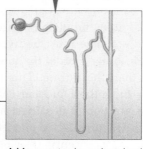

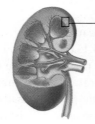

d More water is reabsorbed by peritubular capillaries around the nephrons, so less water is lost in urine.

e The blood volume rises.

Response

f Sensory receptors in hypothalamus detect the increase in blood volume. Signals calling for ADH secretion slow down.

Figure 24.14 Feedback control of ADH secretion, one of the negative feedback loops from kidneys to the brain that helps adjust the volume of extracellular fluid. Nephrons in the kidneys reabsorb more water when we do not take in enough water or lose too much, as by profuse sweating.

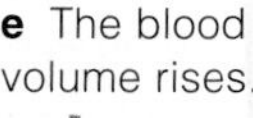

24.7 When Kidneys Break Down

At this point, you probably have figured out that good health depends on nephron function. Whether by illness or accident, when the nephrons of both kidneys become damaged and no longer perform their regulatory and excretory functions, we call this renal failure. Chronic renal failure is irreversible.

In the United States, the single largest cause of kidney problems is diabetes. High blood pressure accounts for many additional cases. These chronic diseases damage blood vessels, including the capillaries in the kidney.

Certain genetic disorders make people more likely to have infections or conditions that damage kidneys. Chronic exposure to lead, arsenic, pesticides, or other toxins is also a risk factor. So is repeated use of certain medications, including high doses of aspirin.

As noted earlier, high protein diets produce extra urea and other wastes. This forces the kidneys to work overtime. A high protein diet also increases the risk for *kidney stones*. These are hard deposits that form when uric acid, calcium, and other wastes settle out of the urine and collect in the renal pelvis. Most stones leave the body in urine, but some become lodged in a ureter or the urethra. Such a stone can cause great pain. If it blocks urine flow, it can open the way to an infection and permanent kidney damage.

Kidney health is usually described in terms of the rate of filtration through the glomerular capillaries. If filtration rate drops by half—regardless of whether the cause is low blood flow to the kidney, damaged kidney vessels, or damaged tubules—a person is said to be in *renal failure*. This can be fatal. Wastes build up in the blood and interstitial fluid. The pH rises. Changes in the concentrations of other ions, such as Na^+ and K^+, interfere with normal metabolic function.

About 13 million people in the United States have experienced renal failure. A *kidney dialysis machine* may restore the proper solute balances. Like the kidneys, it selectively adjusts the solutes in blood. Dialysis refers to exchanges of solutes across any artificial membrane between two different solutions.

In *hemodialysis*, the machine is connected to a vein or an artery. Blood is pumped through semipermeable tubes submerged in a warm solution of salts, glucose, and other substances. As blood is flowing through the tubes, wastes diffuse out and solute concentrations are returned to normal levels. The blood is then returned to the patient's body.

In *peritoneal dialysis*, fluid of suitable composition is pumped into the abdominal cavity, and then drained out. In this case, the tissue lining the abdominal cavity functions as the membrane for dialysis (Figure 24.15).

Figure 24.15 What kidney trouble? Karole Hurtley has attitude as well as energy. At age 13 she was the national champion for her age group and karate belt level. She also lives with renal failure.

Karole receives regular checkups from a home dialysis nurse for Denver's Children's Hospital. Her dad jokes that Karole takes so many pills that she rattles when she walks and is pricked "a gazillion times a year" for blood tests.

Eight or nine times a night, while Karole sleeps, a peritoneal dialysis machine beside her bed pumps a special solution into her torso and filters out waste products that her kidneys no longer can remove. Inspirational attitude? You bet.

Kidney dialysis can keep a person alive through an episode of temporary kidney failure. But if the kidney damage is permanent, dialysis must be continued for life—or until a transplant provides a kidney.

Each year in the United States, about 12,000 people receive kidney transplants. More than 40,000 remain on a waiting list because there is a shortage of donors. Most kidneys come from people who agreed to donate organs after death. But an increasing number are from living donors, most often a relative, spouse, or friend. A transplant from a living donor stands a better chance of success than one from a deceased donor. A person needs only one kidney to be healthy, so the risks to a donor are mainly related to the surgery. Advantages of kidneys from living donors, too few donated organs, and the high costs of dialysis have even led some to suggest that people should be allowed to sell a kidney.

http://biology.brookscole.com/BTT2

Summary

Section 24.1 A digestive system breaks down food to molecules small enough to be absorbed into the internal environment, and it eliminates unabsorbed residues. It works with other organ systems in ways that promote homeostasis (Figure 24.16).

Most animals have a complete digestive system. This is basically a tube with two openings (mouth and anus) and specialized areas between them.

BiologyNow

Use the animation on BiologyNow to compare animal digestive systems.

Section 24.2 Table 24.4 lists the components and accessory organs of the human digestive system, along with their functions.

Starch digestion starts in the mouth. Chewing mixes food with amylase-containing saliva. Protein digestion starts in the stomach, a muscular sac with a glandular lining that secretes gastric fluid.

The pancreas secretes digestive enzymes. Bile, which assists in fat digestion, is made in the liver and stored in the gallbladder. Ducts leading from the pancreas and gallbladder empty into the small intestine, where the digestion of all nutrients is completed.

Most nutrients are absorbed in the small intestine. Glucose and amino acids are actively transported across the intestinal lining, then enter the blood.

Table 24.4 Summary of the Digestive System

Mouth (oral cavity)	Start of digestive system, where food is chewed, moistened; polysaccharide digestion begins
Pharynx	Entrance to tubular parts of digestive and respiratory systems
Esophagus	Muscular tube, moistened by saliva, that moves food from pharynx to stomach
Stomach	Sac where food mixes with gastric fluid and protein digestion begins; stretches to store food taken in faster than can be processed; gastric fluid destroys many microbes
Small intestine	Receives secretions from liver, gallbladder, and pancreas; completes digestion; absorbs most nutrients; delivers unabsorbed waste to colon
Colon (large intestine)	Concentrates and stores undigested matter (by absorbing mineral ions and water)
Rectum	Stores feces until distension triggers explusion
Anus	Terminal opening of digestive system
Accessory Organs	
Salivary glands	Glands that secrete saliva, a fluid with polysaccharide-digesting enzymes, buffers, and mucus (which moistens and lubricates ingested food)
Pancreas	Secretes enzymes that digest all major food molecules and buffers against HCl secretions from cells of stomach lining
Liver	Secretes bile; roles in carbohydrate, fat, and protein metabolism
Gallbladder	Stores and concentrates bile from the liver

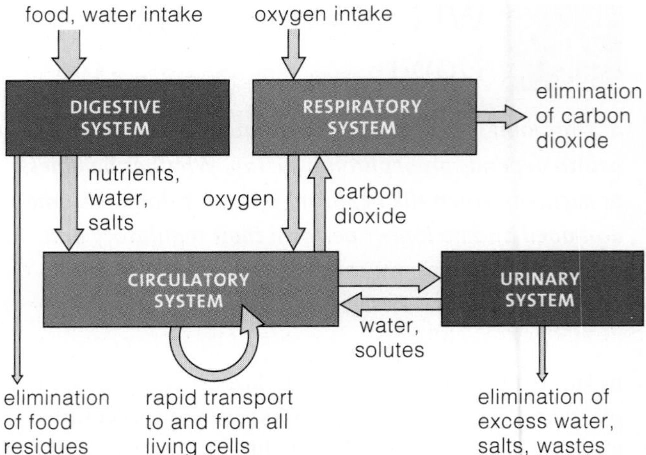

Figure 24.16 Summary of links between the digestive, respiratory, circulatory, and urinary systems. These organ systems and others work to supply cells with raw materials and eliminate wastes. They contribute to homeostasis—stability in the internal environment.

Monoglycerides and fatty acids diffuse into cells of the lining. They recombine to form triglycerides, which are secreted into interstitial fluid. From there, they enter lymph vessels that deliver them to blood.

The large intestine absorbs minerals and water, and concentrates undigested residues for expulsion.

Local controls as well as the nervous and endocrine systems respond to the volume and composition of food in the gut. They cause changes in muscle activity and in secretion rates for hormones and enzymes.

BiologyNow

Investigate the structure and function of the human digestive system with the animation on BiologyNow.

Section 24.3 A healthy diet supplies the body with the nutrients it needs and helps to reduce the risk for chronic diseases. The vitamins and minerals necessary for normal metabolism usually can be obtained from a balanced diet.

Section 24.4 To maintain body weight, energy intake must balance energy output. Obesity raises the risk of health problems and shortens life expectancy.

Section 24.5 The human urinary system interacts with other organ systems, mainly those shown in Figure 24.16, to balance the intake and output of solutes and water. The urinary system consists of a pair of kidneys, a pair of ureters, a urinary bladder, and a urethra.

BiologyNow

Study the components of the human urinary system with the animation on BiologyNow.

Section 24.6 Kidneys contain an enormous number of nephrons, or blood-filtering units. A nephron has a cup-shaped entrance around one set of blood capillaries and tubular regions that associate with another set of blood capillaries. Urine forms in nephrons by three processes: filtration, reabsorption, and secretion.

Filtration. Blood pressure drives all water and all solutes except proteins out of the capillaries at the start of a nephron. These enter the nephron's tubular parts.

Reabsorption. Water and solutes to be conserved move out of the tubular parts, then into the capillaries that thread around the nephron. A small volume of water and solutes remains in the nephron.

Two hormones, ADH (antidiuretic hormone) and aldosterone adjust urine concentration. ADH promotes water reabsorption across the duct walls. This returns more water to the blood and concentrates the urine. Aldosterone also concentrates the urine. It promotes reabsorption of sodium, and water follows by osmosis.

Secretion. H^+, K^+, and some other substances move from capillaries into the nephron, for excretion.

BiologyNow

Explore the structure of a nephron and see how urine is formed with the animation on BiologyNow.

Section 24.7 Good kidney function is essential to life. Dialysis or a transplant are the only options for those who have permanently damaged kidneys.

Self-Quiz

Answers in Appendix I

1. A bird's ______ helps it to grind up hard foods.
   a. crop c. cloaca
   b. gizzard d. gall bladder

2. Digestion is completed and most nutrients are absorbed in the ______ .
   a. mouth c. small intestine
   b. stomach d. colon

3. Bile has roles in ______ digestion and absorption.
   a. carbohydrate c. protein
   b. fat d. amino acid

4. Glucose and amino acids are absorbed ______ .
   a. into lymph vessels c. with the help of bile salts
   b. by diffusion d. through transport proteins

5. Match each organ with a digestive function.
   ___ gallbladder a. makes bile
   ___ colon b. concentrates undigested residues
   ___ liver c. secretes most digestive enzymes
   ___ small intestine d. absorbs most nutrients
   ___ stomach e. secretes gastric fluid
   ___ pancreas f. stores, secretes bile

6. Urea forms as a breakdown product of ______ .
   a. nucleic acids d. proteins
   b. simple sugars e. complex carbohydrates
   c. saturated fats f. a through d

7. Glomerular filtration moves fluid into ______ .
   a. collecting ducts c. glomerular capillaries
   b. proximal tubules d. Bowman's capsule

8. Kidneys return water and small solutes to the blood by the process of ______ .
   a. filtration c. tubular secretion
   b. tubular reabsorption d. both a and b

9. ______ promotes sodium reabsorption into blood.
   a. ADH d. High extracellular fluid volume
   b. Aldosterone e. both a and b
   c. Secretin

10. Match each structure with a function.
   ___ ureter a. start of nephron
   ___ distal tubule b. delivers urine to body surface
   ___ urethra c. carries urine from kidney to bladder
   ___ Bowman's capsule d. secretes ADH
   ___ pituitary gland e. target of ADH

Additional questions are available on **BiologyNow™**

Critical Thinking

1. *Anorexia nervosa* is an eating disorder in which people, most often young women, starve themselves. The name means "nervous loss of appetite," but affected people are usually obsessed with food and continually hungry.

The causes of anorexia nervosa are complicated, but genes certainly play a role. Australian researchers looked at a gene for a protein that transports norepinephrine, one of the brain's signaling molecules, across cell membranes. Individuals carrying one mutant allele were twice as likely to develop the disorder. This discovery may lead to new treatments. Even now, many recovered anorexics enjoy normal lives (Figure 24.17).

Like people who undergo stomach stapling to lose weight, anorexics also are vulnerable to bone loss and brittle bones. Speculate on the underlying causes of the problems in these individuals.

Figure 24.17 One success story: In 2000, after recovering from anorexia, Dutch cyclist Leontien Zijlaard won three Olympic gold medals. Four years earlier, she had been too malnourished and weak to compete.

2. Fat cells continually produce leptin. In healthy people, the kidneys filter leptin from blood, and enzymes in the nephrons break it down. Suppose an individual has a suppressed kidney function. What would be the expected effect on leptin levels? On appetite?

3. Holiday meals often are larger and have a higher fat content than everyday meals. After stuffing themselves at an early dinner on Thanksgiving Day, Richard and other members of his family feel uncomfortably full for the rest of the afternoon. Based on what you have learned about the controls over digestion, propose an explanation for their discomfort.

4. A glassful of whole milk contains lactose, proteins, butterfat, vitamins, and minerals. Explain what will happen to each component in your digestive tract.

5. Label the components of the diagram below:

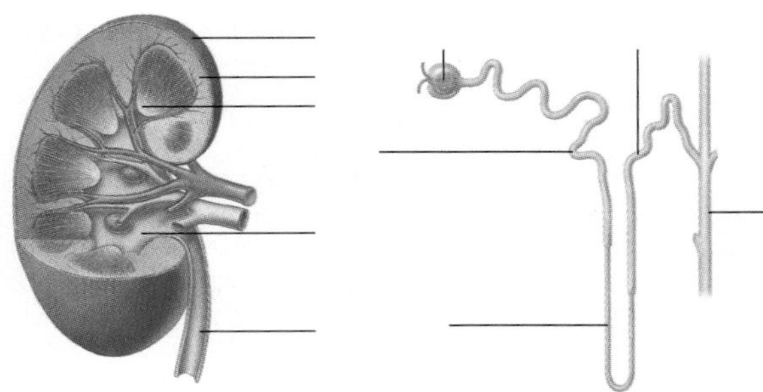

25 NEURAL CONTROL AND THE SENSES

IMPACTS, ISSUES

In Pursuit of Ecstasy

Ecstasy is a drug that makes you feel socially accepted, relieves anxiety, and sharpens the senses while giving you a mild high. It also can leave you dying in a hospital bed, foaming at the mouth and bleeding from every orifice as your temperature skyrockets. It can send your family and friends spiraling into horror and disbelief as they watch you stop breathing. Lorna Spinks ended life that way. She was nineteen years old.

Her anguished parents released the photograph at right, taken just minutes after her death. They wanted others to know what Lorna did not: Ecstasy can kill.

Ecstasy's main active chemical ingredient, MDMA, is related to methamphetamine, or "speed." MDMA causes neurons in the brain to release extra serotonin, a signal that contributes to feelings of pleasure. It also disrupts processes that clear this signal away, so users feel good. At the same time, users are energized by MDMA's effect on norepinephrine, a signal that primes the body for a fight or to run away.

Lorna Spinks overdosed on MDMA. Waves of panic and seizures set in. Her heart pounded, her blood pressure soared, and her temperature increased so much that it shut down organ systems one after the other.

An MDMA overdose does not often end in death, but other problems are common. For example, neurons cannot rebound quickly when the brain's serotonin stores are depleted. We know that repeated MDMA doses given to laboratory animals alter the number and structure of their serotonin-secreting neurons.

We also know that below-normal levels of serotonin in humans can contribute to attention deficit disorders, depression, and memory problems. As studies of Ecstasy users reveal, the degree of memory loss increases with the frequency of drug use. Giving up Ecstasy use entirely allows the memory to recover, but the return to normal function can take many months.

Another concern is that MDMA also alters neurons that secrete dopamine. Parkinson's disease arises when people do not have enough of this signal because the neurons that should release it have been damaged. Will long-term Ecstasy users one day have tremors and other Parkinson's-like symptoms? It is too early to know.

The human nervous system is a legacy of millions of years of evolution in cellular signaling pathways, first in invertebrates, then through vertebrate lineages that led to humans. Among living species, we alone have a sense of history and of destiny. And we alone can choose to act in ways that we know will nuture or harm the nervous system that gave us our remarkable capacities.

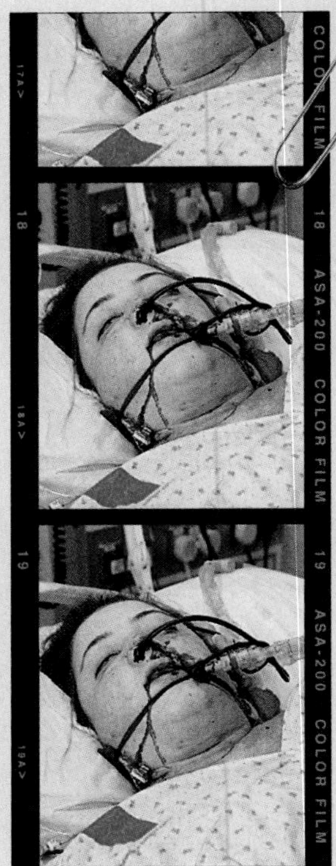

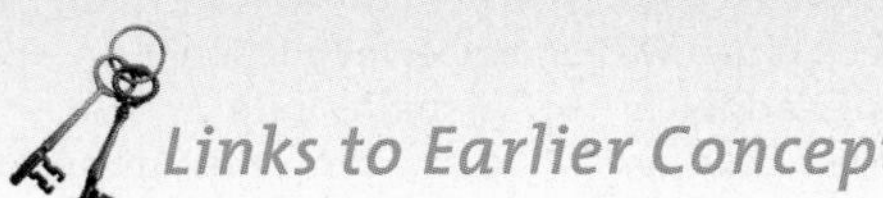

How Would You Vote? *Would you support laws that allow nonviolent drug offenders to enter into drug rehabilitation programs as an alternative to jail? See BiologyNow for details, then vote online.*

Key Concepts

HOW NEURONS WORK

Neurons are cells that are specialized for intercellular communication. Messages travel along a neuron's plasma membrane in the form of an electrical disturbance. Chemical signals carry messages between neurons or from a neuron to another cell.

NERVOUS SYSTEMS

Nearly all animals have a nervous system. Different components of the system detect and process information about conditions inside and outside the body, then command muscles and glands to make suitable responses.

SENSORY SYSTEMS

Sensory receptors detect forms of energy and send signals along nerve pathways to integrating centers. There, processing of input from many receptors gives rise to hearing, vision, taste, smell, and other sensations.

Links to Earlier Concepts

Understanding how neurons send signals begins with knowledge of diffusion and concentration gradients (Section 4.4), and how substances cross cell membranes (4.5). Positive feedback (17.3) and signaling mechanisms (17.5) come into play, as do concepts of energy (4.1). This chapter expands on the functions of sensory receptors (1.2) and the mechanism of color blindness (8.7). It revisits body plans of a few invertebrates (16.2, 16.3), as well as trends in animal evolution (16.1,16.9).

25.1 Neurons—The Great Communicators

In Chapter 16 you learned about how nervous systems evolved as increasingly precise ways of sensing and responding to conditions inside and outside the body. Here we turn to the intracellular communication lines that make up these nervous systems.

LINKS TO SECTIONS 1.2, 4.1, 17.3

NEURONS AND THEIR FUNCTIONAL ZONES

Sensing and responding to sounds and sights, hunger and passion, fear and rage—no matter what an animal does, it all begins with the cells of the nervous system. Even before you were born, cells were differentiating and organizing themselves in newly forming nervous tissue. During that time, they started to chatter among themselves. All through your life, in times of danger or excitement, reflection or sleep, their chattering has not stopped, and it will continue for as long as you do.

The cells are **neurons**, the communication units of nervous systems (Figures 25.1 and 25.2). They detect and integrate information about external and internal conditions, then alert the various muscles and glands that carry out responses. Vertebrate nervous systems have three classes of neurons. **Sensory neurons** detect and relay information about stimuli to the spinal cord and brain. A *stimulus* is a form of energy, such as light, detected by a specific receptor. **Interneurons**, mostly inside the spinal cord and brain, receive and process this information and send messages to other neurons. **Motor neurons** transmit signals from interneurons to the **effectors**—muscles and glands—that carry out the appropriate responses.

Each neuron has cytoplasmic extensions that differ in number as well as length. The cell body and slender extensions called **dendrites** are typical *input* zones for information. Nearby is a *trigger* zone. From this area of plasma membrane, information travels along a slender and often long extension called an **axon**—the neuron's *conducting* zone. The axon's endings are *output* zones, where a message may be sent to other cells.

All neurons are metabolically assisted, protected, insulated, and held in place by the many diverse cells collectively known as **neuroglia**. This cellular support staff makes up about half the volume of your nervous system. In your brain, neuroglial cells outnumber the neurons by about ten to one.

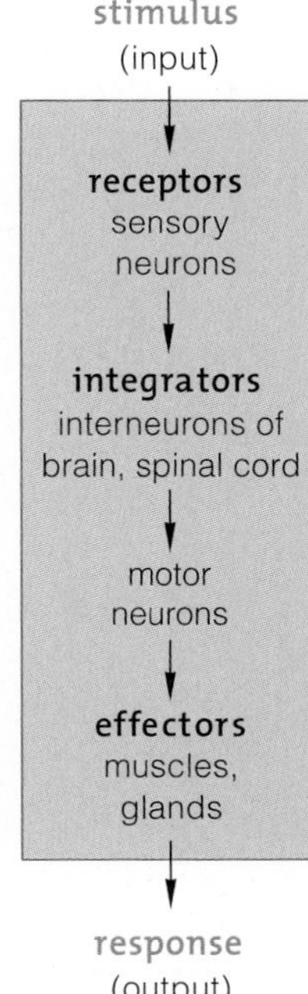

Figure 25.2 How the three classes of neurons interact in vertebrate nervous systems.

RESTING MEMBRANE POTENTIAL

When a neuron is "at rest," or not being stimulated, mechanisms are maintaining a slight electric gradient across the cell's plasma membrane. The cytoplasmic fluid inside the neuron is just a bit more negatively charged than the interstitial fluid outside. As in a car's battery, these separated charges are a form of potential energy, which we can measure in millivolts (mV). (A millivolt is one thousandth of a volt.) The potential energy of the gradient across the plasma membrane of a resting neuron is the **resting membrane potential**. For many neurons, it is −70 mV. As you will learn shortly, electrical potential across the membrane can be reversed if the neuron becomes excited by a signal of the appropriate type and size.

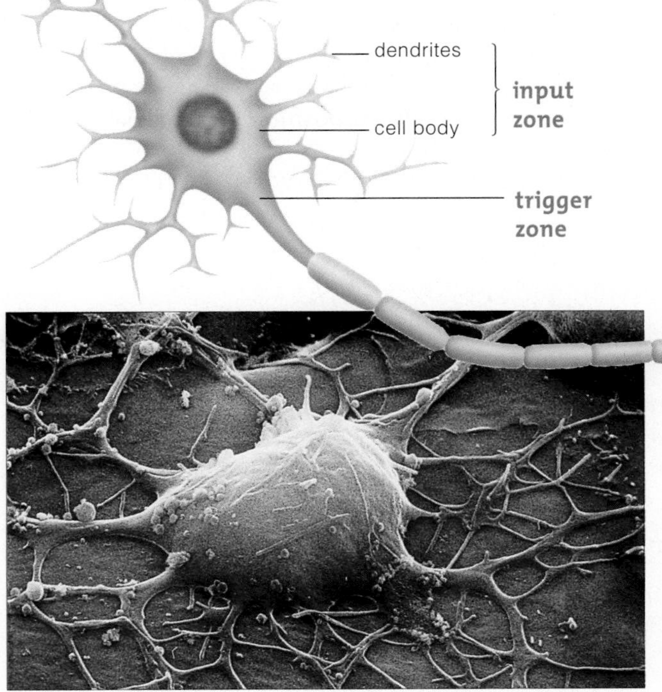

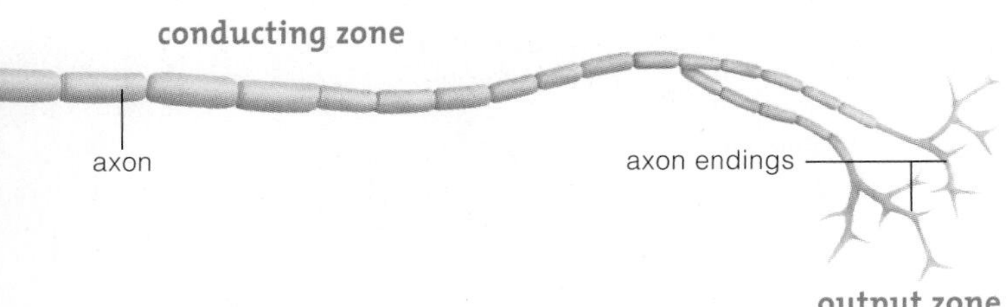

Figure 25.1 ***Animated!*** Functional zones of a neuron. The sketch and scanning electron micrograph show a motor neuron. This type of neuron carries messages away from the brain and spinal cord to effectors, such as muscles or glands. Information flows in one direction, from dendrites and the cell body to the trigger zone, then along the axon to its endings.

LINKS TO SECTIONS 4.4, 4.5, 17.3

Recall from Section 4.4 that movement of any ion across a membrane is dependent on the concentration gradient for that ion. There are more sodium ions in the fluid just outside the neuron than there are inside it; the reverse is true for potassium ions. We can show these ion concentration gradients this way (with larger text representing higher concentrations):

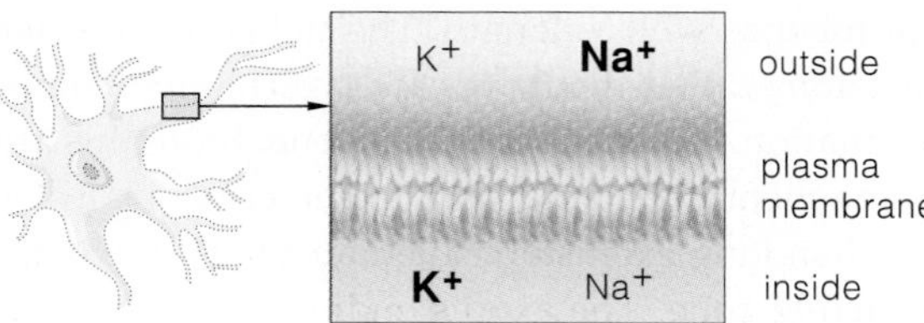
K^+
Na^+
outside
plasma membrane
K^+
Na^+
inside

Mechanisms built into the membrane create and maintain the ion gradients. Recall that ions cannot just diffuse across a membrane's lipid bilayer. They cross only with the help of the transport proteins that span the plasma membrane. Some passive transporters in a neuron membrane are like open channels. Ions flow through them continuously. Others have gates that are closed when the neuron is at rest. Neuron membranes have gated channels for K^+ and Na^+, as well as some always open K^+ channels (Figure 25.3).

The neuron membrane also has sodium-potassium pumps. These active transport proteins pump Na^+out of the neuron at the same time they move K^+ in.

THE ACTION POTENTIAL

All living cells maintain electrical and concentration gradients across their plasma membrane, but only in *excitable* cells will the electric gradient briefly reverse in response to adequate stimulation. Such a reversal is called an **action potential**.

Tap your wrist gently. Under your skin, pressure slightly deforms the membrane of receptor endings—input zones of sensory neurons—and a few ions slip across it. As a result, the voltage difference across the membrane shifts just a bit. In this case, light pressure has produced a local signal.

Signals at an input zone vary in size and they do not spread far from the point of stimulation. It takes certain gated ion channels to spread a signal farther, and input zones of neurons do not have them.

When a stimulus *is* intense or long-lasting, signals spread into the trigger zone. This patch of membrane is richly endowed with gated channels for sodium ions (Figure 25.4*a*). If the potential across the membrane changes enough to cause these gates to open, we say the cell has reached *threshold potential*.

Once a neuron reaches threshold potential, an action potential always follows. Opened gates allow Na^+ to rush down electrical and concentration gradients into the neuron (Figure 25.4*b*). This causes more gates to open, so more Na^+ enters. The ever increasing, inward flow of sodium ions is a type of **positive feedback**. The event intensifies as a result of its own occurrence:

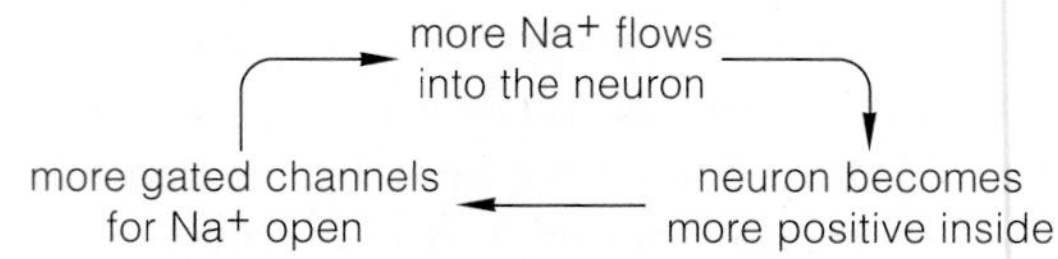
more Na^+ flows into the neuron
neuron becomes more positive inside
more gated channels for Na^+ open

Inward flow of Na^+ reverses the electrical gradient across the neuron membrane—the cytoplasmic side of the membrane becomes more positively charged than interstitial fluid outside. But this reversal lasts only a millisecond. Why? The charge reversal causes gated K^+ channels to open and also returns gates on the Na^+ channels to their closed position (Figure 25.4*c*). As K^+

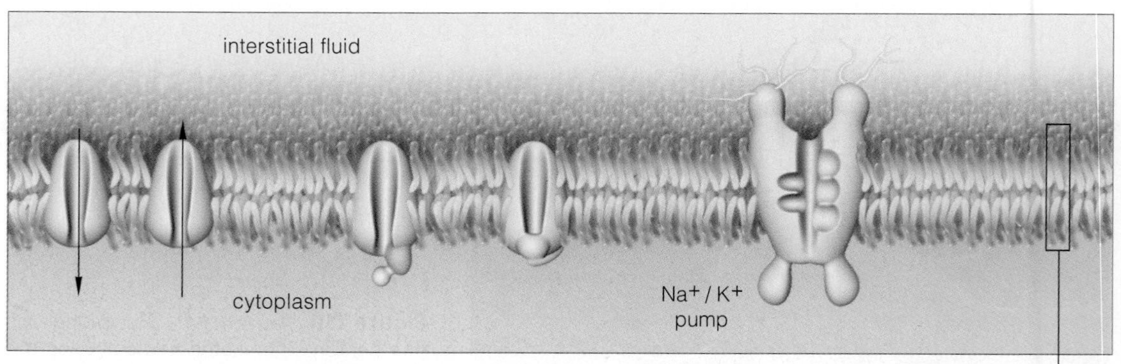
interstitial fluid
cytoplasm
Na^+ / K^+ pump
Passive transporters with open channels let ions continually leak across the membrane.
Other passive transporters have voltage-sensitive gated channels that open and shut. They assist diffusion of Na^+ and K^+ across the membrane, down their concentration gradients.
Active transporters pump Na^+ and K^+ across the membrane, against their concentration gradients. They counter ion leaks and restore resting membrane conditions.
lipid bilayer of neural membrane

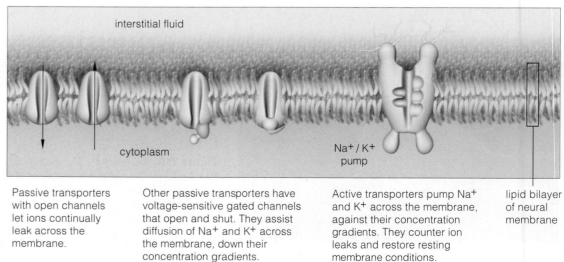

Figure 25.3 ***Animated!*** How ions move across the plasma membrane of a neuron. Selective ion channels and pumps span the membrane. They set up and maintain ion concentration gradients across the membrane.

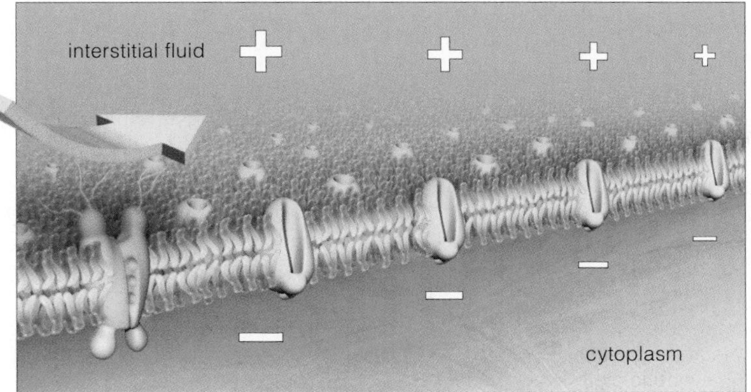

ⓐ Membrane at rest; inside of neuron negative with respect to the outside. An electrical disturbance (*yellow* arrow) spreads from an input zone to an adjacent trigger region of the membrane, which has a great number of gated sodium channels.

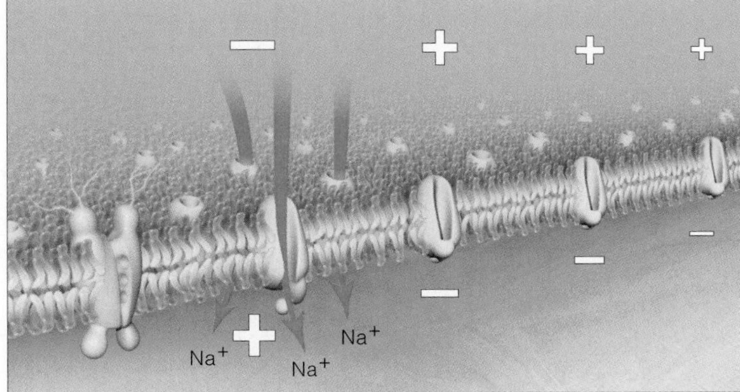

ⓑ A strong disturbance initiates an action potential. Sodium gates open. The sodium inflow decreases the negativity inside the neuron. The change causes more gates to open, and so on until threshold is reached and the voltage difference across the membrane reverses.

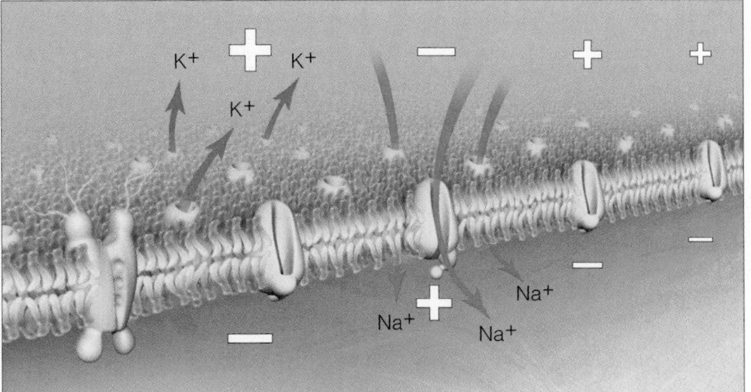

ⓒ With the reversal, sodium gates shut and potassium gates open (*red* arrows). Potassium follows its gradient out of the neuron. Voltage is restored. The disturbance triggers an action potential at the adjacent site, and so on, away from the point of stimulation.

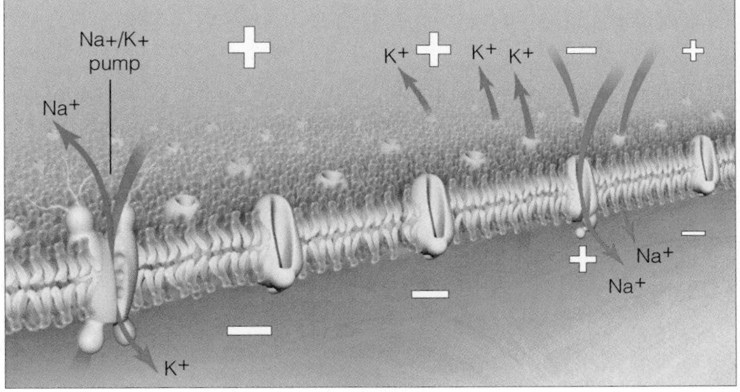

ⓓ Following each action potential, the inside of the plasma membrane becomes negative once again. However, the sodium and potassium concentration gradients are not yet fully restored. Active transport at sodium–potassium pumps restores them.

Figure 25.4 ***Animated!*** Propagation of an action potential along the axon of a motor neuron.

diffuses out of the axon, the neuron's cytoplasm once again becomes negative relative to the interstitial fluid. Action of sodium-potassium pumps helps to return the neuron to its resting potential (Figure 25.4*d*).

An action potential is "all-or-nothing." Whenever a patch of axon membrane reaches threshold level, an action potential occurs. In every action potential, gates open the same way and the electrical gradient across the membrane changes by exactly the same amount. This means action potentials are all the same size.

A signal is transmitted along the length of an axon with no decrease in magnitude as an action potential is triggered in one membrane region after another. Na^+ inflow at one patch of neuron membrane pushes the potential in an adjacent region to threshold level. Gated channels open in the neighboring region and an action potential happens there. This in turn pushes the next membrane patch to threshold, and so on.

Action potentials are one-way signals; they move *away* from the trigger zone and *toward* axon endings. This is because for a brief interval after Na^+ inflow, voltage-gated Na^+ channels cannot be reopened. This prevents an action potential from traveling backward.

Neurons are specialized for communication. The inside of a resting neuron has a slight negative charge compared to the outside. During an action potential, ion flow across the plasma membrane briefly reverses this voltage difference.

Each action potential affects the voltage difference at the next patch of membrane, and the next. That is how an action potential moves along an axon.

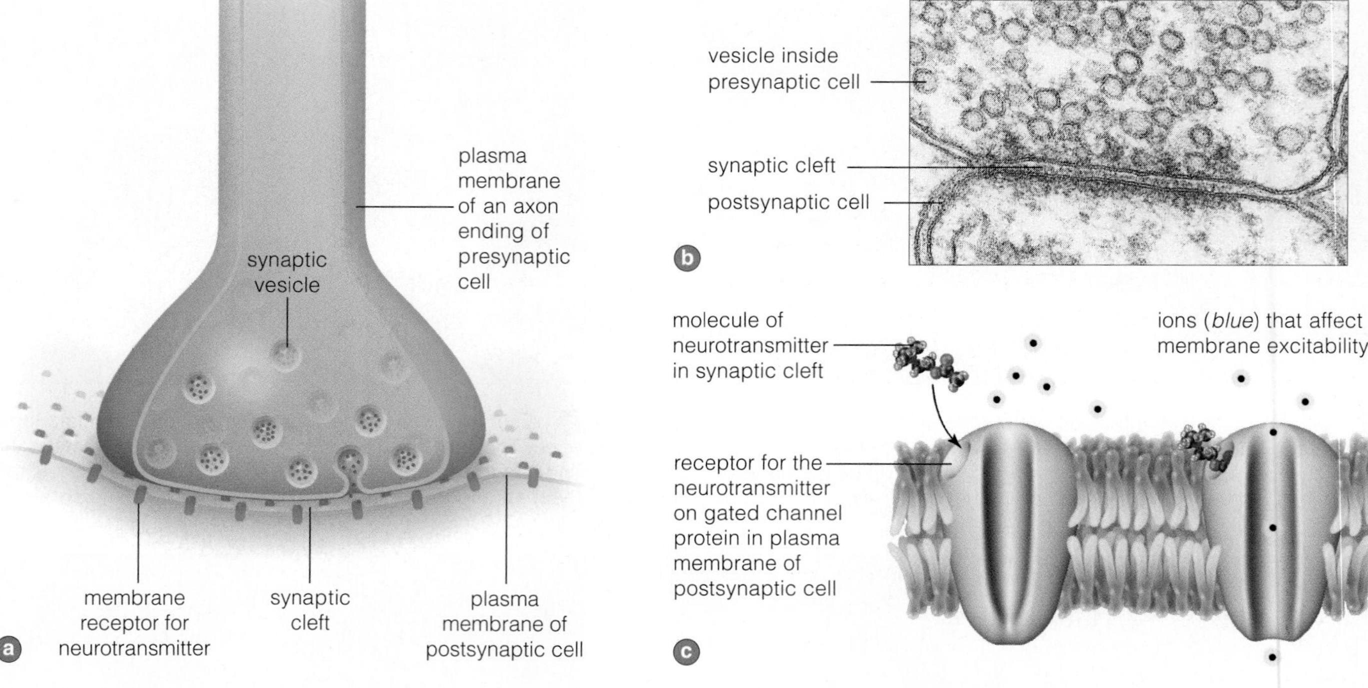

Figure 25.5 ***Animated!*** Example of a chemical synapse. (**a**,**b**) Only a thin cleft separates a presynaptic cell from a postsynaptic cell. Arrival of an action potential at an axon ending triggers exocytosis of molecules of neurotransmitter from the presynaptic cell. (**c**) The molecules diffuse across the cleft. They bind to receptors that are part of gated proteins of the postsynaptic cell membrane. The gates open and ions flow in, causing a localized change in membrane potential.

LINKS TO
SECTIONS
1.2, 4.5, 17.5

25.2 How Messages Flow From Cell to Cell

Messages travel along an axon as the electrical signals we call action potentials. But action potentials cannot jump from cell to cell. A neuron communicates with another neuron or an effector cell through the use of chemical signals called neurotransmitters.

CHEMICAL SYNAPSES

A **chemical synapse** is a region where axon endings of a neuron pass signals to another neuron or to a muscle or gland (Figure 25.5*a*,*b*). The signal-sending neuron is the *pre*synaptic cell. A tiny synaptic cleft separates it from the *post*synaptic (signal-receiving) cell.

Just inside the axon endings of the presynaptic cell are many vesicles that contain **neurotransmitter**. Only axon endings of neurons make and release this type of signaling molecule. The arrival of an action potential causes synaptic vesicles to move through the cytoplasm, to the plasma membrane. As these vesicles fuse with the neuron membrane, neurotransmitter is released into the synaptic cleft.

Neurotransmitter molecules diffuse across the cleft and bind to receptors on the postsynaptic cell's plasma membrane. Some receptors serve as ion channels. The binding of a neurotransmitter opens a tunnel through their interior, and allows ions to diffuse into or out of the postsynaptic cell (Figure 25.5*c*). Different receptors influence membrane permeability in other ways.

Some neurotransmitters excite a postsynaptic cell. They push it closer to threshold potential by allowing ions in. For example, a neurotransmitter that increases the permeability of a postsynaptic cell to sodium ions would have this effect. Other neurotransmitters have an inhibitory effect, as by increasing potassium influx.

Different kinds of cells may respond differently to the same neurotransmitter. It all depends on the kind of receptors they have. Consider what goes on at a neuromuscular junction, a synapse between a motor neuron and skeletal muscle (Figure 25.6). The neuron releases acetylcholine, or ACh, which binds to specific receptors on muscle fibers. In this case, ACh has an excitatory effect; it calls for contraction of the muscle

(Section 21.3). Cardiac muscle cells also have receptors for ACh, but their receptors are a different type. As a result, ACh makes cardiac cells less likely to contract.

SYNAPTIC INTEGRATION

Neurons connect with one another in a complex web and each receives signals from many others. At any moment, countless excitatory and inhibitory signals are washing over input zones of a postsynaptic cell. Some of these signals drive the cell's membrane closer to threshold potential. Others drive it farther away or keep it near the resting potential.

With **synaptic integration**, all of the signals that reach the input zone of a neuron at the same instant are summed together. By this process, signals arriving at a neuron are dampened or reinforced. Excitatory signals may sum together to push a cell to threshold, as when signals that increase sodium permeability arrive from many presynaptic cells. Excitatory signals can be canceled out by inhibitory ones. Arrival of a signal that increases permeability to potassium ions can offset one that increases sodium permeability. In this way, a neuron's response to stimuli is finely tuned.

A SAMPLING OF SIGNALS

ACh is one of many neurotransmitters. Others include norepinephrine, epinephrine (also called adrenaline), dopamine, serotonin, and GABA. Norepinephrine and epinephrine act to arouse the body to respond to a stressful situation, as by fleeing or fighting.

Dopamine affects learning, fine motor control, and feelings of pleasure. The death of dopamine-secreting neurons in part of the brain causes *Parkinson's disease*. Minor hand tremors are commonly the earliest sign. Balance is affected and walking may become difficult.

Serotonin is a neurotransmitter that affects mood and memory. Like norepinephrine and dopamine, its effects are disrupted by Ecstasy use. Low serotonin levels can result in depression. Prozac (fluoxetine) and similar antidepressants elevate serotonin levels.

GABA (gamma amino butyric acid) is an inhibitor of neurotransmitter release by other neurons. Section 21.4 described how the clenched muscles symptomatic of *tetanus* arise from inhibition of GABA's function.

Neurons also release molecules that modify effects of certain neurotransmitters. For example, substance P allows pain perception. Enkephalins and endorphins suppress substance P. They are secreted in response to exertion or injury. Endorphins are also released when you laugh, get a massage, cuddle, or have an orgasm. In addition to suppressing pain, release of endorphins improves mood and may enhance immune function.

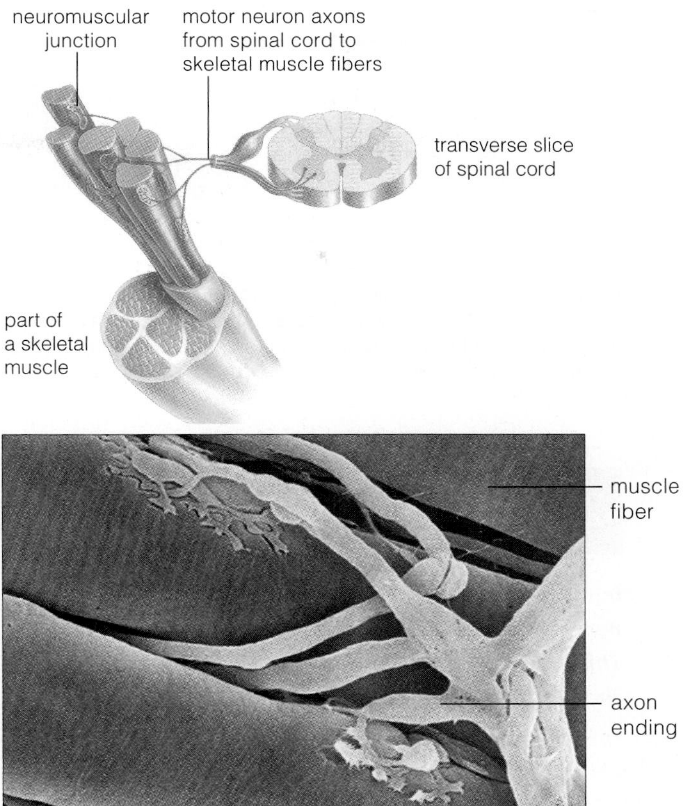

Figure 25.6 A neuromuscular junction, a synapse where a motor neuron relays signals to a skeletal muscle fiber.

After neurotransmitter molecules have acted, they must be quickly removed from synaptic clefts so that new signals can be sent. Some of the molecules simply diffuse away. Membrane transport proteins actively pump other molecules back into the presynaptic cell or into neighboring neuroglial cells. Enzymes secreted into the cleft break down others. For example, one enzyme—acetylcholinesterase—breaks down ACh.

Neurotransmitters are signaling molecules that cross a synaptic cleft between two neurons or between a neuron and a muscle cell or gland cell.

The type and amount of neurotransmitter influence whether a signal will have excitatory or inhibitory effects on the postsynaptic cell. So do the type and amount of receptors and gated ion channels on the postsynaptic cell membrane.

Synaptic integration is the summation of all excitatory and inhibitory signals that arrive at a postsynaptic cell's input zone. Summation allows messages to be reinforced or downplayed. It occurs through the competing effects of neurotransmitters on membrane permeability to ions.

IMPACTS, ISSUES

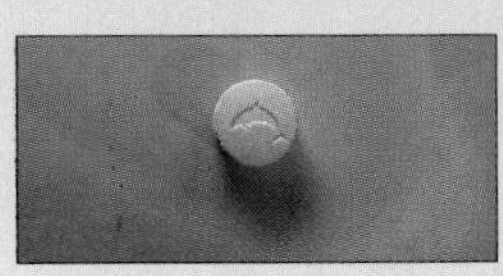

The MDMA in Ecstasy acts mainly at axon endings that release serotonin. It stimulates the release of *all* stored serotonin, which floods the chemical synapses and saturates receptors on postsynaptic cells. Ecstasy also slows the reuptake of serotonin by the same axon endings. Repeatedly using this drug might cause the axon endings to wither and die. Endings may grow back damaged, or in the wrong places.

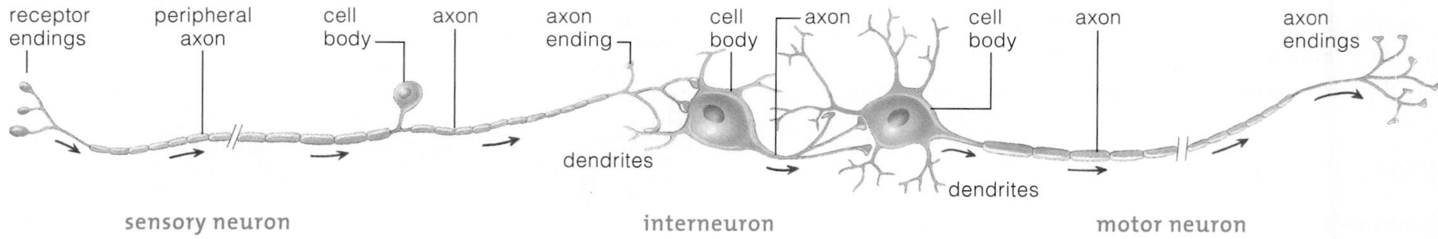

Figure 25.7 General direction of information flow in nervous systems. Sensory neurons relay signals *into* the spinal cord and brain, where they form chemical synapses with interneurons. Interneurons *within* the spinal cord and brain integrate the signals. Many synapse with motor neurons, which carry signals *away* from the spinal cord and brain.

LINK TO SECTION 17.5

25.3 The Paths of Information Flow

In most organisms, neurons are bundled together in nerves and organized into circuits and reflex pathways. This structural organization lays the groundwork for the functional specialization seen in most nervous systems. Different neurons interact in different tasks.

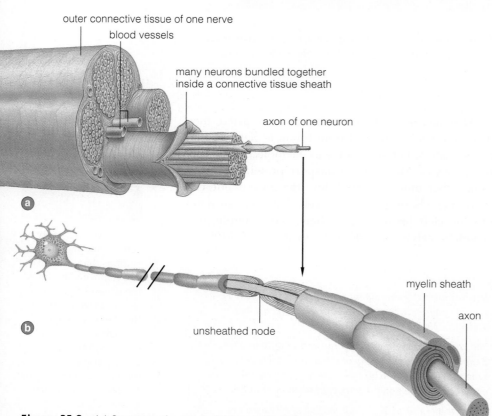

outer connective tissue of one nerve
blood vessels
many neurons bundled together inside a connective tissue sheath
axon of one neuron
a
b
unsheathed node
myelin sheath
axon

Figure 25.8 (**a**) Structure of a nerve. (**b**) In myelinated axons, ions flow across the neuron membrane only at unsheathed nodes, which have many gated sodium channels. When the disturbance caused by an action potential reaches a node, sodium gates open and start a new action potential. The disturbance spreads swiftly to the next node, where it triggers a new action potential, and so on down the neuron.

BLOCKS AND CABLES OF NEURONS

As in other complex animals, information generally flows through your body from sensory receptors, to interneurons, and then to motor neurons (Figure 25.7). However, many signals loop about in amazing ways.

Consider the billions of interneurons in your brain. In *diverging* circuits, dendrites and axons of neurons fan out from one brain region and communicate with other parts of the brain. In a *converging* circuit, signals from numerous neurons flow to just a few. In some circuits, neurons synapse back on themselves. Some of these *reverberating* circuits cause eye muscles to twitch rhythmically while you are asleep.

Information also flows rapidly through **nerves**, the long-distance cables between body regions. In nerves, the axons of sensory neurons, axons of motor neurons, or both are bundled inside connective tissues. Figure 25.8*a* is an example of this structural organization.

Certain types of neuroglial cells wrap like jelly rolls around most axons. They enclose the axon in a myelin sheath that acts like insulation around an electric wire. Myelin enhances conduction of action potentials along an axon (Figure 25.8*b*). With myelin in place, messages move as fast as 120 meters per second. In contrast, the conduction speed in axons that do not have a myelin wrapping seldom exceeds 2 meters per second.

In *multiple sclerosis* (MS), certain white blood cells wrongly identify a type of protein in myelin sheaths as foreign and destroy it. This autoimmune response can damage the axons in the brain and spinal cord. The resulting disruption in information flow can lead to muscle weakness, severe fatigue, numbness, tremors dizziness, and visual problems. An estimated 500,000 people in the United States are afflicted. Genes have a role in susceptibility to MS, but viral infection or other environmental factors may set it in motion. The incidence of the disease rises with the distance from the equator. There is no cure, but new treatments can prevent or slow the otherwise progressive disability.

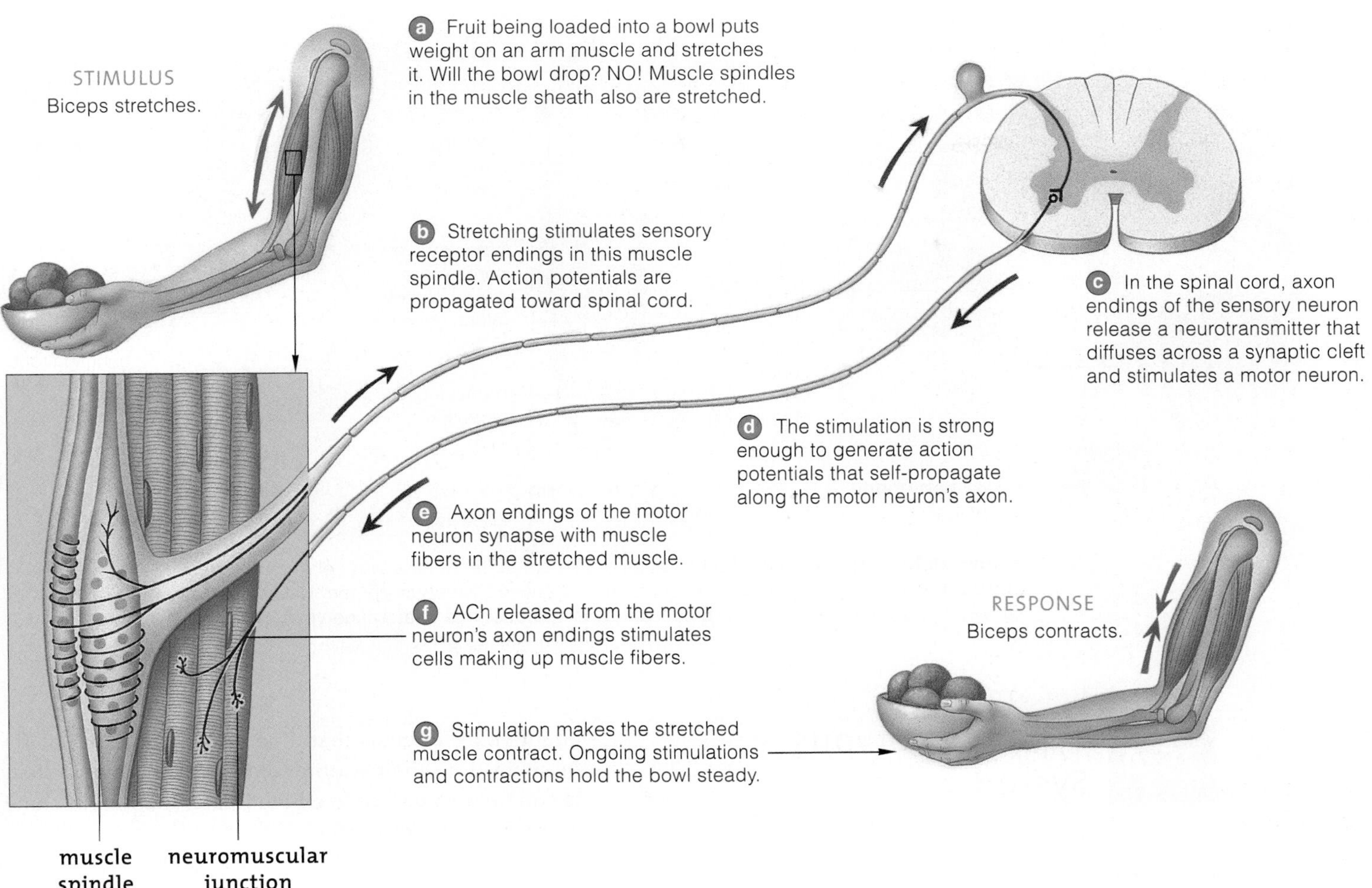

STIMULUS
Biceps stretches.
a Fruit being loaded into a bowl puts weight on an arm muscle and stretches it. Will the bowl drop? NO! Muscle spindles in the muscle sheath also are stretched.
b Stretching stimulates sensory receptor endings in this muscle spindle. Action potentials are propagated toward spinal cord.
c In the spinal cord, axon endings of the sensory neuron release a neurotransmitter that diffuses across a synaptic cleft and stimulates a motor neuron.
d The stimulation is strong enough to generate action potentials that self-propagate along the motor neuron's axon.
e Axon endings of the motor neuron synapse with muscle fibers in the stretched muscle.
f ACh released from the motor neuron's axon endings stimulates cells making up muscle fibers.
g Stimulation makes the stretched muscle contract. Ongoing stimulations and contractions hold the bowl steady.
RESPONSE
Biceps contracts.
muscle spindle
neuromuscular junction

Figure 25.9 ***Animated!*** Stretch reflex. In skeletal muscle, stretch-sensitive receptors of sensory neurons (one is shown here) are located in muscle spindles. Stretching generates action potentials, which reach sensory axon endings in the spinal cord. These synapse with a motor neuron that carries signals to contract, from the spinal cord back to the stretched muscle.

REFLEX ARCS

Reflexes are the most ancient paths of information flow. A **reflex** is a movement or some other response to a stimulus that happens automatically, no thought required. In the simplest reflex arcs, sensory neurons synapse directly on motor neurons. In more complex reflexes, sensory neurons interact with one or more interneurons that can activate or suppress all motor neurons required for a coordinated response.

The *stretch reflex* is a simple reflex arc that causes a muscle to contract after gravity or some other load stretches it. Suppose you hold a bowl while someone puts peaches into it. The load makes your hand drop a bit, stretching the bicep muscle (Figure 25.9). This involuntary stretching stimulates **muscle spindles** in the bicep muscle. These are sensory organs composed of muscle and nervous tissue. The receptor endings of muscle spindles are input zones of sensory neurons.

In the spinal cord, axons of these sensory neurons synapse with motor neurons, the axons of which lead back to the stretched bicep. Arrival of action potentials at the axon endings of the motor neurons triggers ACh release, which in turn causes contraction of the bicep. This steadies the arm against the added gravitational force exerted by each additional peach.

The vertebrate brain has blocks of information-processing interneurons. Nerves are long-distance communication lines consisting of long fibers of sensory neurons, motor neurons, or both. Nerves extend throughout the body.

Reflex arcs, in which sensory neurons synapse directly on motor neurons, are the simplest paths of information flow.

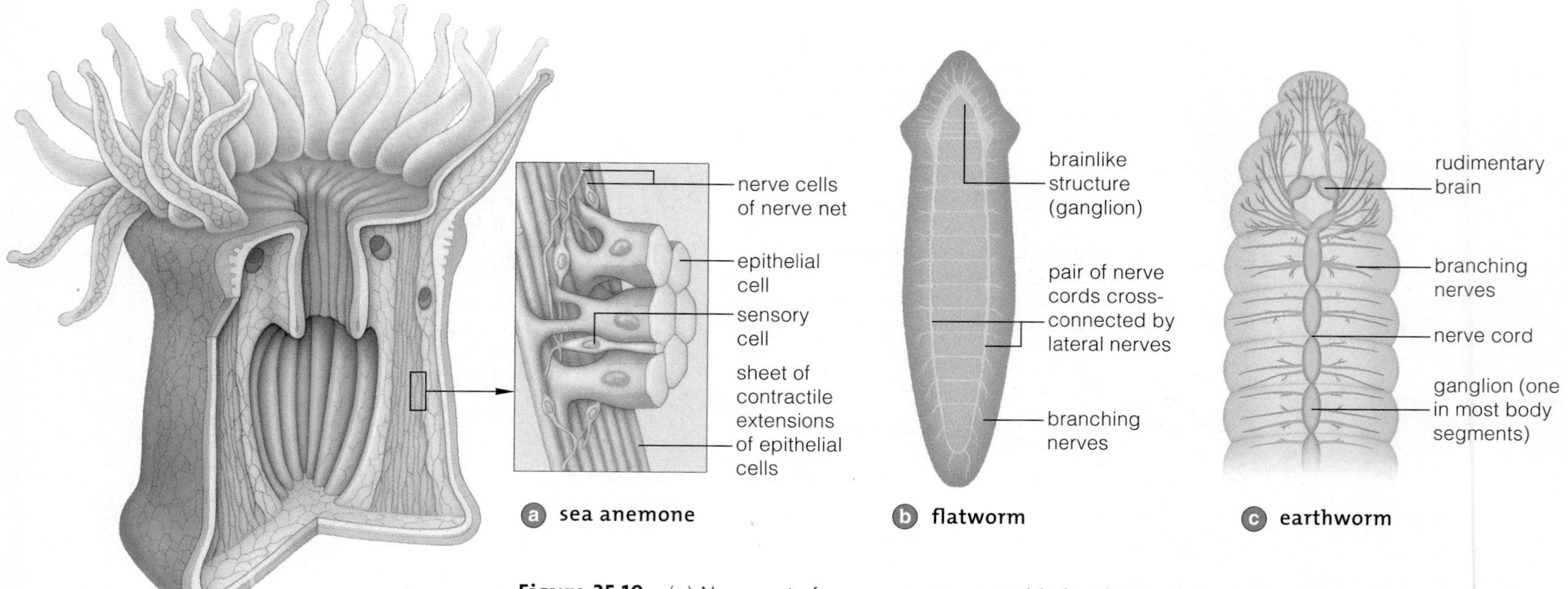

Figure 25.10 (**a**) Nerve net of a sea anemone, a cnidarian. Interacting nerve cells, sensory cells, and contractile cells are organized in epithelium between the outer epidermis and the body wall's jellylike midlayer, the mesoglea. (**b**,**c**) Two invertebrates with a bilateral nervous system.

LINKS TO SECTIONS 16.1–16.3, 16.9

25.4 Types of Nervous Systems

All animals with tissues have a nervous system, in which neurons interact with one another to detect stimuli, process information, and call for responses.

REGARDING THE NERVE NET

To appreciate the diversity of nervous systems, start by recalling that animals first evolved in the seas. We still find animals with the simplest nervous systems in aquatic habitats. They are the cnidarians, such as sea anemones. These invertebrates show *radial* symmetry. Similar body parts, such as tentacles, are positioned regularly around the central body axis, something like spokes of a bike wheel (Section 16.1).

Radial animals have a **nerve net**, a loose mesh of nerve cells intimately associated with epithelial tissue (Figure 25.10*a*). Sensory cells and contractile cells form reflex pathways in which sensory stimulation triggers simple, stereotyped motions. For example, one reflex pathway controls feeding behavior in sea anemones. It extends from sensory receptors in tentacles to nerve cells that control contractile cells that surround the opening to the saclike gut.

A nerve net extends through the body but does not focus information. It does little more than bring about weak contractions of the body wall or cause tentacles to bend slowly. Its owners will never dazzle you with speed or precision acrobatics. Even so, the nerve net is adaptive for animals that dine on any food that drifts or swims into their waiting tentacles. A radial system is equally responsive to edible tidbits that might come along from any direction.

ON THE IMPORTANCE OF HAVING A HEAD

Today, the flatworms are the simplest animals with a bilateral nervous system. Two nerve cords connected by branching nerves form a ladderlike system (Figure 25.10*b*). In some species, nerve cords expand to form ganglia in their head end. A **ganglion** (plural, ganglia) is a group of nerve cell bodies that function as a local integrating center. In flatworms, ganglia in the head end integrate signals from paired sensory organs, such as eyespots. Other ganglia elsewhere in the body exert local control over different nerves.

Cephalization—concentration of sensory cells in a head end—evolved together with a bilateral nervous system in nearly all animal groups. We see evidence of bilateral heritage in the rudimentary brains and the paired ganglia of most invertebrates (Figure 25.10*c*). We see it also in the paired sensory structures, brain centers, nerves, and skeletal muscles of all vertebrates.

EVOLUTION OF THE SPINAL CORD AND BRAIN

Hundreds of millions of years ago, the earliest fishlike vertebrates were evolving (Section 16.8). A column of bony segments was taking over the functions of their notochord, a long rod of stiffened tissue that worked with segmented muscles to bring about movement. Above the notochord, the hollow, tubular nerve cord was evolving into the spinal cord and brain. These were key innovations that opened the way for diverse new body plans and behaviors.

At first, simple reflex pathways prevailed. Sensory neurons synapsed directly with motor neurons, which directly signaled muscles to contract. However, among fast-moving vertebrates, predators or prey that could better assess and respond to food and danger had the competitive edge. The senses of smell, hearing, and balance became keener on land. Bones and muscles evolved in ways that allowed specialized movements. The brain thickened with tissues that could integrate sensory input and orchestrate complex responses.

The oldest vertebrate brain regions still govern the coordination of respiration and other basic functions. Many more interneurons now synapse on the sensory and motor neurons of ancient pathways and on one another. In complex vertebrates, interneurons receive, store, retrieve, and evaluate information. They give us our capacity to reason, remember, and learn.

The vertebrate nervous system has a central and a peripheral region (Figure 25.11). Most interneurons are inside the *central* nervous system—the spinal cord and brain. The *peripheral* nervous system includes nerves that extend through the rest of the body. All sensory fibers are *afferent*; they convey signals into the central nervous system. The motor fibers are *efferent*; they carry signals away from the central regions.

Figure 25.12 illustrates the location of your brain, spinal cord, and some peripheral nerves. Each of these nerves has both sensory and motor components. As an example, a sciatic nerve extends from the spinal cord down each leg. This nerve is composed of the bundled extensions of hundreds of neurons. Some axons in this bundle carry signals *to* skeletal muscles that move the leg, or *to* smooth muscle that adjusts the flow through the leg's arterioles. Others convey messages *away* from sensory receptors in the leg's muscles, joints, and skin.

Simple radial animals have a nerve net. This diffuse mesh of nerve cells extends through the body and interacts with sensory and contractile cells in simple reflex pathways.

Cephalized, bilateral animals have a nervous system with a brain or ganglia at the head end, paired nerves, and paired sensory structures.

The central nervous system of vertebrates consists of the brain and spinal cord. The peripheral nervous system consists of nerves that thread through the rest of the body and carry signals into and out of the central region.

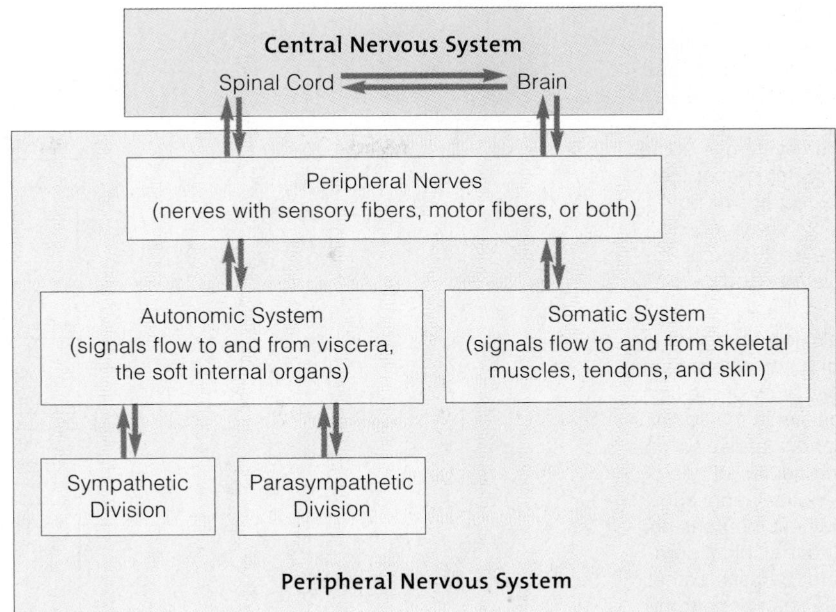

Figure 25.11 Functional divisions of vertebrate nervous systems. The spinal cord and brain are its central portion. The peripheral nervous system consists of spinal nerves, cranial nerves, and their branchings, which extend through the rest of the body. They carry signals to and from the spinal cord and brain.

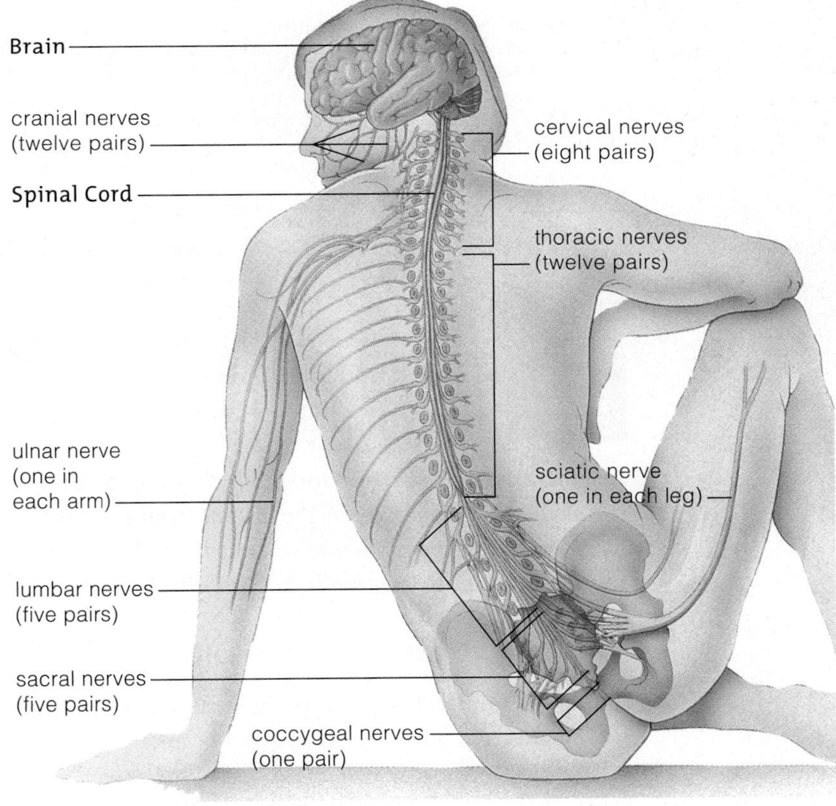

Figure 25.12 Some of the major nerves of the human nervous system.

Figure 25.13 ***Animated!*** The major sympathetic nerves and parasympathetic nerves connecting the central nervous system to some organs. Remember that there are *pairs* of both kinds of nerves, one to service each side of the bilateral body.

Signals flow to organs by way of a two-neuron path. The cell body of the first neuron lies in the spinal cord and synapses on the second neuron at one of the ganglia. Ganglia for sympathetic neurons are close to the spinal cord. Those of parasympathetic ganglia lie in or near the organs they affect. The second neuron synapses with the organ.

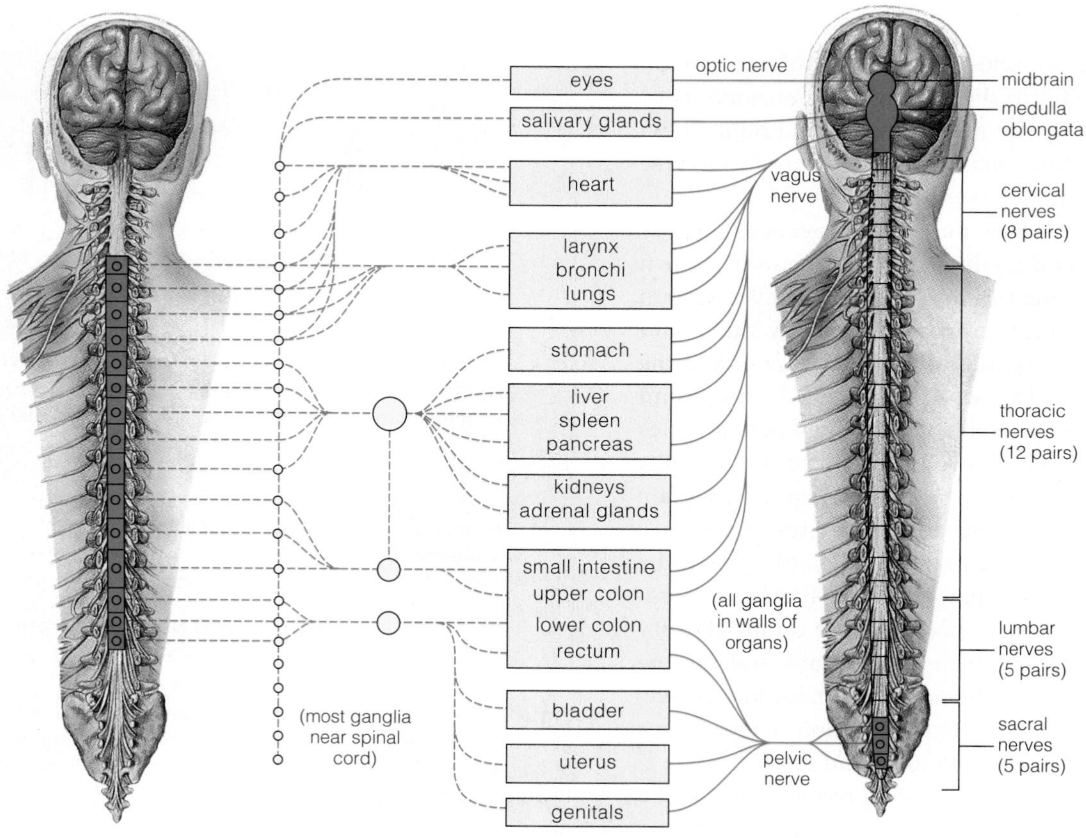

sympathetic outflow from the spinal cord

Examples of Responses

- Heart rate increases
- Pupils of eyes dilate (widen, let in more light)
- Glandular secretions decrease in airways to lungs
- Salivary gland secretions thicken
- Stomach and intestinal movements slow down
- Sphincters (rings of muscle) contract

Examples of Responses

- Heart rate decreases
- Pupils of eyes constrict (keep more light out)
- Glandular secretions increase in airways to lungs
- Salivary gland secretions become dilute
- Stomach and intestinal movements increase
- Sphincters (rings of muscle) relax

parasympathetic outflow from the spinal cord and brain

LINK TO SECTION 17.5

25.5 The Peripheral Nervous System

Peripheral nerves branch throughout the body. Some axons carry messages about the internal and external environment to the brain and spinal cord. Others carry commands to muscles, glands, and internal organs.

SOMATIC AND AUTONOMIC DIVISIONS

Inside the human body, the peripheral nervous system includes thirty-one pairs of *spinal* nerves that connect with the spinal cord. This system also includes twelve pairs of *cranial* nerves that connect with the brain.

Cranial and spinal nerves are further classified by function. Those that carry signals about movement of the head, trunk, or limbs are **somatic nerves**. Sensory axons of these nerves deliver messages from receptors in skin, skeletal muscles, and the joints to the central nervous system. Their motor axons deliver commands from the brain and spinal cord to skeletal muscles.

By contrast, the spinal and cranial nerves that are connected to smooth muscle, cardiac (heart) muscle, and glands are **autonomic nerves**. They carry signals to and from internal organs (Figure 25.13).

SUBDIVISIONS OF THE AUTONOMIC SYSTEM

The neurons in the autonomic system are of two types: sympathetic and parasympathetic. Most organs receive signals from both types, and the signals usually work antagonistically—signals from one type oppose those from the other. **Sympathetic neurons** are most active in times of sharpened awareness, stress, excitement, or danger. They release norepinephrine at synapses with organs. In contrast, **parasympathetic neurons** are most active during more tranquil times. Release of ACh by these neurons fosters housekeeping processes, such as urine production and digestion.

What happens when something startles or scares you? Parasympathetic input slows down, giving way to sympathetic signals that raise your heart rate and blood pressure. They also make you sweat more, and breathe faster. The signals stimulate adrenal glands, which secrete epinephrine. This helps put you into a state of intense arousal—primed to fight or to make a quick getaway. Hence the term *fight–flight response*.

An organ is often receiving opposing sympathetic and parasympathetic signals. Consider smooth muscle in the gut wall. At the same instant that sympathetic neurons are releasing norepinephrine at synapses with these smooth muscle cells, parasympathetic neurons are releasing ACh at other synapses with the same cells. One signal tells the gut to slow its contraction, while the other calls for increased activity. Synaptic integration determines the muscle's actual response.

Nerves of the peripheral nervous system connect the brain and spinal cord with the rest of the body.

The somatic nerves of the peripheral nervous system govern skeletal muscle movements. Its autonomic nerves control smooth muscle, cardiac muscle, and glands. Most organs receive both sympathetic and parasympathetic signals, which have opposing effects.

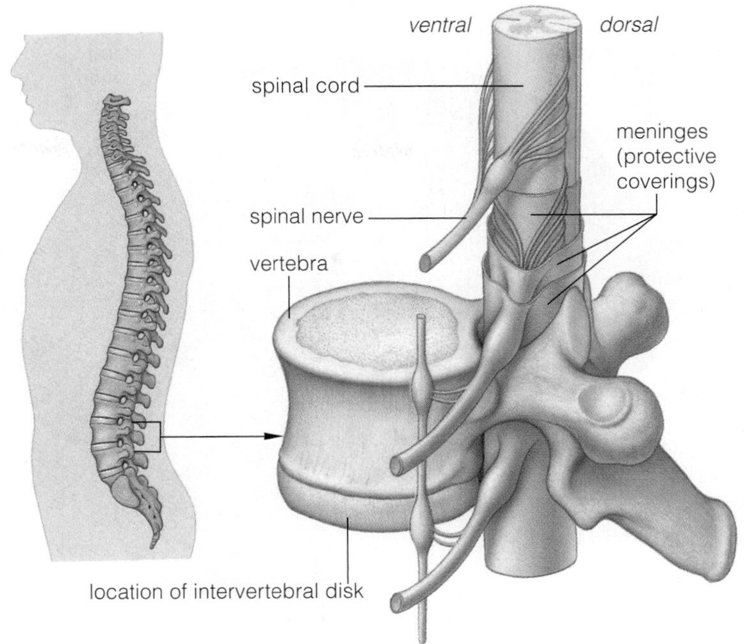

Figure 25.14 Organization of the spinal cord and its location relative to the vertebral column. The cord is a cylinder of tissue that runs through the column. Three layers of membranous covering called meninges also enclose and help protect the cord. Spinal nerves carry signals to and away from the cord.

25.6 The Central Nervous System

The brain and spinal cord make up the body's central nervous system. Most of the body's interneurons are in these organs. They receive and integrate information, and issue the necessary commands for responses.

THE SPINAL CORD

Inside the brain and spinal cord, communication lines are called tracts, not nerves. Tracts that make up the *white* matter are the myelin-sheathed axons of sensory or motor neurons. *Gray* matter consists of cell bodies, dendrites, and unmyelinated axons of interneurons, along with supporting neuroglia.

The **spinal cord** serves as an expressway for signals traveling between the peripheral nervous system and the brain. Some reflex connections occur in the spinal cord too. Recall that stretch reflex (Figure 25.8)? The knee-jerk reflex is another spinal reflex. A tap below the knee sends signals to the spinal cord. Signals sent back to leg muscles cause the leg to jerk in response.

The spinal cord is enclosed within and protected by bones of the vertebral column (Figure 25.14). The cord also is protected by three meninges—coverings that extend around the brain as well. In *meningitis*, a viral or bacterial infection inflames these coverings. Fever, severe headaches, a stiff neck, and nausea are among the symptoms that follow.

An injury that disrupts the flow of signals through the spinal cord can cause sensation loss and paralysis. The extent of symptoms depends on where the cord is damaged. Spinal nerves that carry information to and from the lower body emerge from the spinal cord lower than nerves that connect to the upper body. An injury to the lumbar part of the spinal cord may result in paralysis of the legs or *paraplegia*. Injury to higher spinal regions can paralyze all limbs (*tetraplegia*), as well as the muscles of breathing. Christopher Reeve's injury, as described in Chapter 20, was like this. A fall from a horse shattered vertebrae near the top of his vertebral column and harmed the spinal cord there.

About 250,000 Americans are currently paralyzed by spinal cord injuries. What keeps their axons from growing back after the injury? A component of myelin prevents axon renewal and regrowth. Researchers are looking for a way to block its effect.

IMPACTS, ISSUES

An MDMA overdose throws the entire sympathetic system into high gear. The heart races, body temperature rises, and panic sets in. If you suspect someone has overdosed, get medical help fast. And tell the doctor what you think happened. Immediate, informed action may save a life.

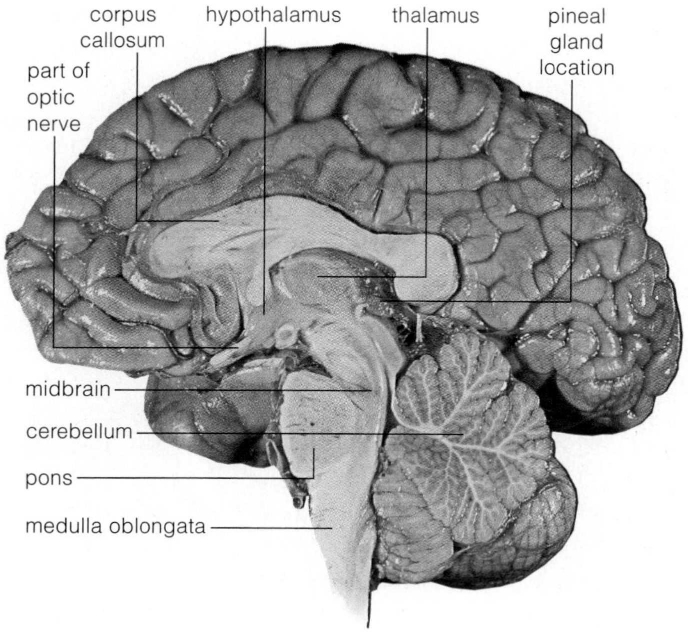

Figure 25.15 Right half of a human brain.

Table 25.1 Major Components of the Central Nervous System

FOREBRAIN	Cerebrum	Localizes and processes sensory inputs; initiates, controls, and coordinates skeletal muscle action. Governs memory, emotions,abstract thought in the most complex vertebrates
	Olfactory lobe	Relays sensory input from nose to olfactory centers in the cerebrum
	Thalamus	Has relay stations for conducting sensory signals to and from cerebral cortex; has role in memory
	Hypothalamus	With the pituitary gland, a homeostatic control center that adjusts fluid volume, composition, and temperature of the internal environment. Governs behaviors affecting organ functions (e.g., thirst, hunger, and sex) and emotion
	Limbic system	Governs emotions; has roles in memory
	Pituitary gland	Interacts with the hypothalamus in endocrine control of metabolism, growth, and development
	Pineal gland	Helps control some circadian rhythms; also has role in mammalian reproductive physiology
MIDBRAIN	Roof of midbrain (tectum)	In fishes and amphibians, its centers coordinate sensory input (as from optic lobes) and motor response. In mammals, it relays sensory input to forebrain
HINDBRAIN	Pons	A "bridge" between the cerebrum and cerebellum, and between the spinal cord and the forebrain. Works with the medulla oblongata in control of breathing
	Cerebellum	Coordinates limb movement and posture
	Medulla oblongata	Relays signals between spinal cord and pons; reflex centers help control breathing and other functions
SPINAL CORD		Makes reflex connections for limb movement. Relays signals between the brain and peripheral nervous system

REGIONS OF THE BRAIN

The hollow, tubular nerve cord that forms in every chordate embryo becomes a brain and spinal cord in vertebrates. Genes controlling segmentation subdivide this tube into the forebrain, midbrain, and hindbrain regions (Table 25.1). The most ancient tissue, the **brain stem**, persists in all brain regions and connects with the spinal cord. Figure 25.15 shows a human brain.

The hindbrain's **medulla oblongata** houses reflex centers for respiration, circulation, and other essential tasks. It integrates motor responses and some complex reflexes, such as coughing. The **cerebellum** helps to control motor skills and posture. Axons from its two halves reach the **pons** (Latin for bridge). The pons is like a traffic officer; it controls signal flow between the cerebellum and integrating centers in the forebrain.

In fishes and amphibians, the midbrain is heavily involved in receiving and integrating visual input, as well as initiating motor responses. As land vertebrates evolved, much of the responsibility for assessing and responding to visual stimuli shifted to the forebrain.

Vertebrates first evolved in water, where chemical odors diffusing from predators, prey, and mates were vital cues. They relied heavily on olfactory lobes and paired outgrowths from the brain stem that integrated olfactory input and responses to it. Especially among land vertebrates, the outgrowths expanded into two halves of the **cerebrum**: the two cerebral hemispheres.

The **thalamus** became a forebrain center for sorting out sensory input and relaying it to the cerebrum. The **hypothalamus** ("beneath the thalamus") evolved into the main center for homeostatic control of the internal environment. It is assisted in this role by the pituitary gland which is attached to it. A light-sensitive pineal (pine-cone shaped) gland is located in the center of the brain. We will discuss the endocrine aspects of the brain in detail in Chapter 26.

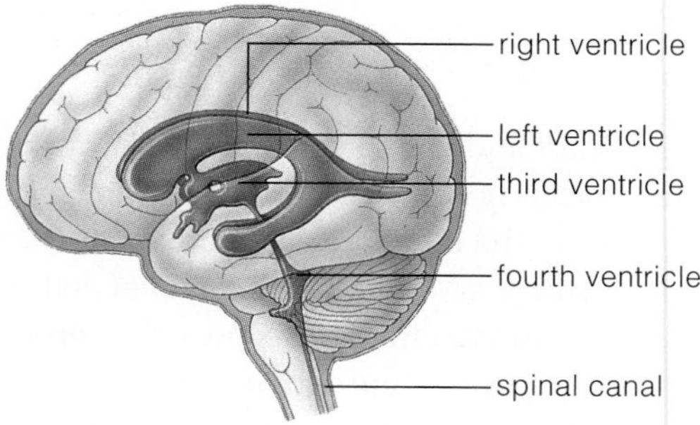

Figure 25.16 Cerebrospinal fluid (*blue*). This extracellular fluid is produced in the four interconnected ventricles (cavities) in the brain and in the spinal cord's central canal.

THE BLOOD—BRAIN BARRIER

The hollow neural tube of chordate embryos persists in adults as a system of cavities and canals filled with cerebrospinal fluid (Figure 25.16). This fluid cushions the brain and spinal cord against jarring movement.

A **blood–brain barrier** protects the brain and spinal cord from harmful substances. It exerts some control over which solutes enter the cerebrospinal fluid. Tight junctions link cells in the walls of the capillaries that supply the brain. Membrane transport proteins in the capillary walls allow glucose and other nutrients, as well as some ions, to move across the cells. The barrier prevents many toxins and wastes from getting in, but does not stop the diffusion of fat-soluble molecules, such as oxygen, carbon dioxide, and alcohol.

FOCUS ON THE CEREBRAL CORTEX

A fissure divides the human cerebrum into left and right hemispheres. Each half has an outer layer of gray matter, the cerebral cortex. The left half deals mainly with analytical skills, speaking, and mathematics. In most people, it dominates the right half, which has more to do with visual–spatial relationships and with music. Each half responds to sensory input mainly from the opposite side of the body. For example, any signals about the right arm go to the left hemisphere. Signals flow back and forth along a connecting band of nerve tracts, the corpus callosum, and coordinate activities of the two halves. Each hemisphere has four lobes (frontal, occipital, temporal, and parietal lobes) that receive and process signals (Figure 25.17).

The cerebral cortex governs conscious behavior. *Motor* areas in the layer affect voluntary motor activity. The body is mapped out in the primary motor cortex of each hemisphere's frontal lobe. This area controls the movements of skeletal muscles.

The frontal lobe also includes the premotor cortex and Broca's area. The premotor cortex governs learned motor skill patterns, such as riding a bicycle. Broca's area is the basis of our capacity to understand and to produce grammatically complex sentences. In most people, it is located in the left cerebral hemisphere.

Sensory areas allow us to perceive sensations. In the parietal lobe, neurons of the primary somatosensory cortex are laid out in regions that correspond to regions of the human body. The neurons receive sensory input from the skin and joints. Another sensory area in this lobe receives signals about taste. In the occipital lobe, the primary visual cortex integrates signals from both eyes. The perception of sounds and odors arises in the primary sensory areas in each temporal lobe.

Association areas integrate information that brings about conscious actions. These areas occur throughout the cortex, but not in the primary motor and sensory areas. Each integrates numerous inputs. For instance, the visual association area that surrounds the primary visual cortex can help us to recognize something we observe by comparing it with our visual memories. The most complex area, the prefrontal cortex, is the basis of complex learning, intellect, and personality. Without it, we would be incapable of abstract thought, judgment, planning, and concern for others.

CONNECTIONS WITH THE LIMBIC SYSTEM

Encircling the upper brain stem is a **limbic system** that controls emotions and has roles in memory. It includes the hypothalamus, amygdala, hippocampus, cingulate gyrus, and parts of the thalamus (Figure 25.18). The hypothalamus is a clearinghouse for our emotions and visceral activity. The amygdala is vital for emotional stability and interpreting social situations. It receives and integrates sensory input. If these stimuli suggest that there is a threat, the amygdala sends signals to

MDMA produces an upbeat mood and increased feeling of empathy by its actions in the limbic system, especially the amygdala and hippocampus. It promotes serotonin release in these regions. Emotional aftereffects —mainly depression and anxiety—result from depleted serotonin in the same areas. The hippocampus is also involved in memory, and memory problems are typical side effects.

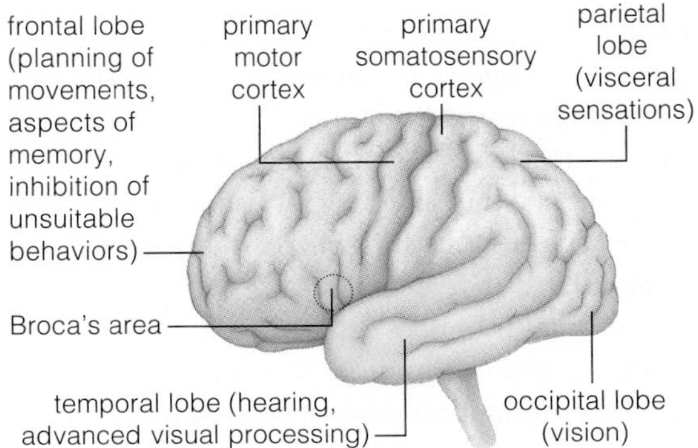

Figure 25.17 Primary receiving and integrating centers of the human cerebral cortex. Association areas coordinate and process sensory input from diverse receptors.

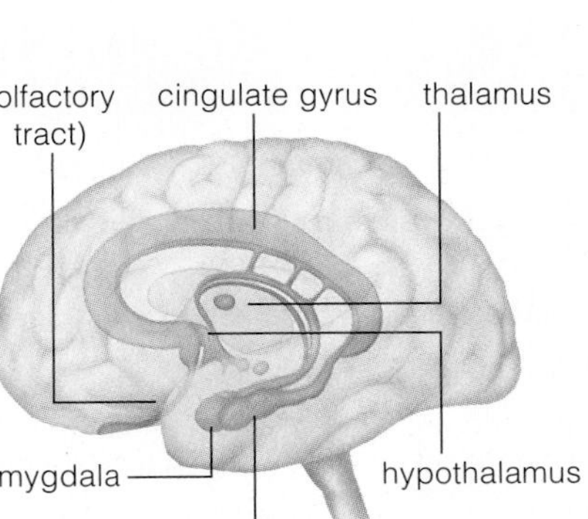

Figure 25.18 Limbic system components that govern your emotional responses to stimuli. When you are frightened, the amygdala calls for increased sympathetic output.

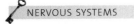

the hypothalamus calling for sympathetic output. This readies the body for a fight or flight response. The amygdala also interacts with the hippocampus to put information into the memory. The cingulate gyrus is a ridge of tissue that is active in goal-directed behavior and helps us to maintain attention on specific tasks.

The limbic system is evolutionarily related to the olfactory lobes and still deals with the sense of smell. That is one reason why smells often summon up the memory of a well-loved place or person.

By its connections with the prefrontal cortex and other brain centers, the limbic system correlates organ activities with self-gratifying behavior, such as eating and sex. That is why the limbic system is called our emotional–visceral brain. It can set your heart on fire with passion or fire up your appetite. Centers in the cerebral cortex override or dampen many these and other "gut reactions."

The vertebrate brain develops from a hollow neural tube, which persists in adults as a system of cavities and canals filled with cerebrospinal fluid. The fluid cushions nervous tissue from sudden, jarring movements.

Nervous tissue is subdivided into the hindbrain, forebrain, and midbrain. The brain stem, the most ancient tissue in all three segments, offers reflex control of functions basic to survival. The forebrain holds the highest integrative centers.

The cerebral cortex, each hemisphere's outermost layer of gray matter, contains motor, sensory, and association areas. Interactions among these areas govern conscious behavior.

The cerebral cortex also interacts with the limbic system, which governs emotions and contributes to memory.

Table 25.2 Warning Signs of Drug Addiction*

1. Tolerance—it takes increasing amounts of the drug to produce the same effect.
2. Habituation—it takes continued drug use over time to maintain the self-perception of functioning normally.
3. Inability to stop or curtail use of the drug, even if there is persistent desire to do so.
4. Concealment—not wanting others to know of the drug use.
5. Extreme or dangerous actions to get and use a drug, as by stealing, asking more than one doctor for prescriptions, or jeopardizing employment by using drugs at work.
6. Deterioration of professional and personal relationships.
7. Anger and defensive behavior when someone suggests there may be a problem.
8. Drug use preferred over previous customary activities.

* Three or more of these signs may be cause for concern.

25.7 Drugging the Brain

Psychoactive drugs are substances that affect the action of neurotransmitters. Some induce a neurotransmitter's release. Others prevent its breakdown or reuptake. Still others block receptors on the postsynaptic cell where the neurotransmitter would normally bind.

Pyschoactive drugs, both legal and illegal, are taken to alleviate pain, relieve stress, induce relaxation, or just for pleasure. Many are habit-forming. Even if the body functions well without them, the user continues to take them for real or imagined effects. The body often develops tolerance; it takes larger or more frequent doses to get the same effect. Habituation and tolerance can lead to **drug addiction**, a chemical dependence in which some drug assumes a essential biochemical role in the body. Table 25.2 lists warning signs of addiction; three or more may be cause for concern.

All major addictive drugs stimulate the release of dopamine, the neurotransmitter directly involved in the sense of pleasure. For example, most people who smoke think this is just a habit. Actually, they smoke because of the dopamine spike when nicotine hits the brain's neurons. *All* addicts specifically crave the spike and the vast majority cannot stop the craving.

Addicts abruptly deprived of their drugs undergo biochemical upheaval, which causes physical pain and mental anguish. But continued addiction is harmful too. In addition to their effects on the brain, drugs that are inhaled can harm the airways and lungs. Sharing needles can spread AIDS, hepatitis B, and hepatitis C, while doing damage to blood vessels. Detoxifying and eliminating any type of drug, legal or illegal, puts a strain on the liver and kidneys.

EFFECTS OF SOME PSYCHOACTIVE DRUGS

STIMULANTS Stimulants make you more alert, then slow you down. *Caffeine* in coffee, tea, chocolate, and many soft drinks is one. Low doses act at the cerebral cortex and increase alertness. Larger doses act at the medulla oblongata; they induce clumsiness and mental incoherence. The *nicotine* in tobacco mimics ACh and directly stimulates diverse sensory receptors. Section 22.10 lists the health impact of nicotine addiction.

Between 1.2 and 3.7 million Americans are *cocaine* abusers. This stimulant causes pleasure by blocking the reabsorption of norepinephrine, dopamine, and other neurotransmitters, so postsynaptic cells are continually stimulated. Blood pressure, and sexual appetite, rise. After the drug wears off, the body's neurotransmitter

25.9 Somatic Sensations

Somatic sensations start with receptors in body surface tissues, skeletal muscles, and the wall of internal organs. These receptors are most highly developed in birds and mammals. Amphibians and fishes have only a few.

Somatic sensations arise when signals reach sensory areas of the cerebral cortex, the cerebrum's outermost layer of gray matter (Section 25.5). In all these areas, interneurons are organized like maps for individual parts of the body surface, just as they are for the motor cortex. Map regions with the largest areas correspond to parts with the greatest sensory acuity such as the fingers, thumbs, and lips (Figure 25.21).

You and other mammals detect touch, pressure, cold, warmth, and pain near the body surface. Regions with the most sensory receptors, such as the fingertips, are most the sensitive. Less sensitive regions, such as the back of the hand, have far fewer receptors.

Free nerve endings are thinly myelinated or naked branched endings of sensory neurons in the skin and internal tissues. They serve as mechanoreceptors, pain receptors, or thermoreceptors. For example, some give rise to a sense of prickling pain, as when a finger gets jabbed with a pin. Others contribute to the itching or warming sensations that some chemicals elicit.

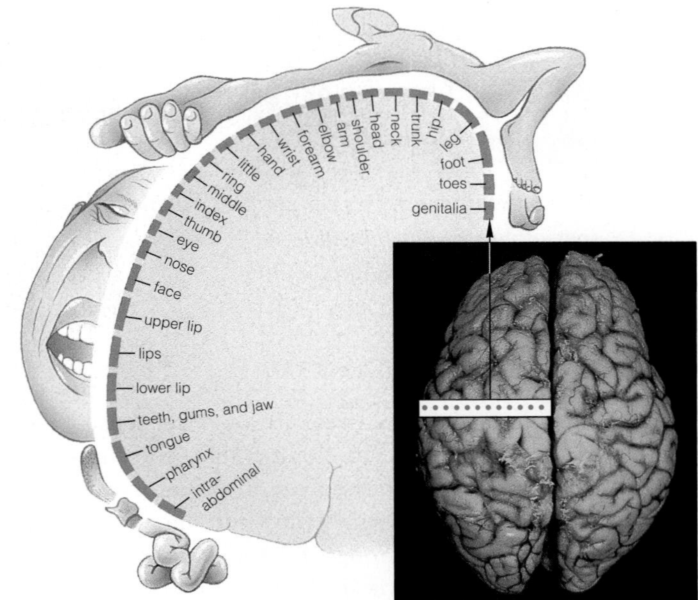

Figure 25.21 Differences in representation of different body parts in the primary somatosensory cortex. This strip of cerebral cortex is about an inch wide and extends from the top of the head to above the ear.

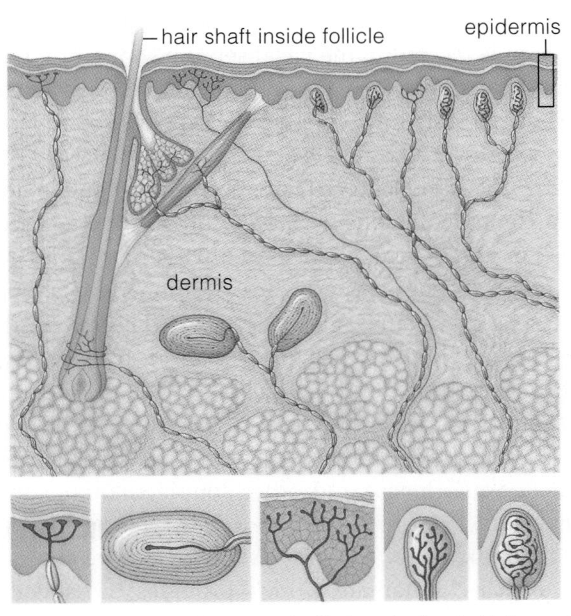

Figure 25.22 A sampling of receptors in human skin.

Figure 25.22 shows sensory receptors that are near the body surface. Meissner's corpuscle adapts very slowly to vibrations of low frequency. It is abundant in the fingertips, lips, eyelids, nipples, and genitals. The bulb of Krause detects cold temperature. Ruffini endings are excited by steady touch, pressure, and high temperatures. Pacinian corpuscles in the skin help you to sense fine textures. Others occur near joints and in some internal organs. Their onionlike layers alternate with fluid-filled spaces. This structure is responsive to pressure changes resulting from touch and vibrations.

Mechanoreceptors detect limb motions and how the body is positioned in space. They are in the skin and in or near skeletal muscle, joints, and ligaments. Stretch receptors of muscle spindles, which lie parallel to skeletal muscle fibers, are among them (Figure 25.9). The response of such receptors depends on how much and how fast a muscle is stretched.

Pain is a perception of injury to some body region. Sensations of *somatic* pain start with signals from pain receptors in skin, skeletal muscles, tendons, and joints. Sensations of *visceral* pain are associated with internal organs. Excess chemical stimulation, muscle spasms, muscle fatigue, gross distension of the gut wall, and poor blood flow to organs are common triggers.

Sensory receptors near the body surface and in internal organs detect touch, pressure, temperature, pain, motion, and positional changes of body parts.

LINK TO SECTION 17.5

25.10 The Special Senses

In the remainder of this chapter, we sample the special senses, which include the senses of smell, taste, vision, hearing, and balance.

SENSES OF SMELL AND TASTE

Smell (olfaction) and taste are chemical senses. Stimuli are detected by chemoreceptors that send signals when a specific chemical dissolved in the surrounding fluid binds to them. The pathways for signals about smell and taste travel through the thalamus, then on to the cerebral cortex. It is in the cortex that perceptions of these stimuli take on shape and undergo fine-tuning. Sensory information about smell and taste also travels to the limbic system, which integrates it with stored memories and emotional state (Section 25.6).

Olfactory receptors have action potentials when they are exposed to water-soluble or easily vaporized chemicals. The lining of your own nose has about 5 million of these receptors. That of a bloodhound nose has more than 200 million. The receptor axons lead into one of two olfactory bulbs. In these small brain structures, axons synapse with groups of cells that sort out components of a scent. From there, information flows along the olfactory tract to the cerebrum, where it is further processed (Figure 25.23).

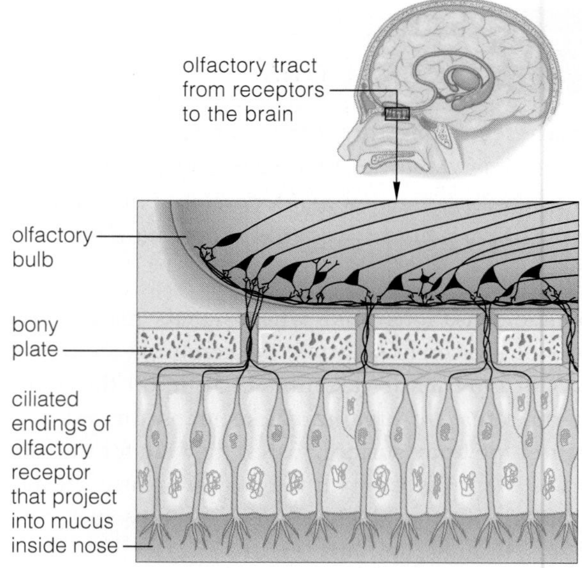

Figure 25.23 Pathway from sensory endings of olfactory receptors in the human nose to the cerebral cortex and the limbic system. Receptor axons pass through holes in a bony plate between the lining of the nasal cavities and the brain.

Many animals use olfactory cues to navigate, locate food, and communicate socially, as with pheromones. **Pheromones** are signaling molecules that are secreted by one individual that influence the social behavior of other individuals of its species. For example, olfactory receptors on antennae of a male silk moth allow him to locate a pheromone-secreting female that is more than a kilometer upwind.

In reptiles and most mammals (but not primates), a cluster of sensory cells forms a *vomeronasal organ* that detects pheromones. In humans, a reduced version of this organ is located in the nasal septum separating the two nostrils. Whether it is still functional, and what role if any it plays in human behavior, is a matter of continued study and debate.

Different animals have **taste receptors** on antennae, legs, or tentacles, or inside the mouth. On the surface of your mouth, throat, and especially the upper part of the tongue, chemoreceptors are positioned in about 10,000 sensory organs called taste buds (Figure 25.24). Fluids in the mouth reach the receptors by entering a pore in each taste bud.

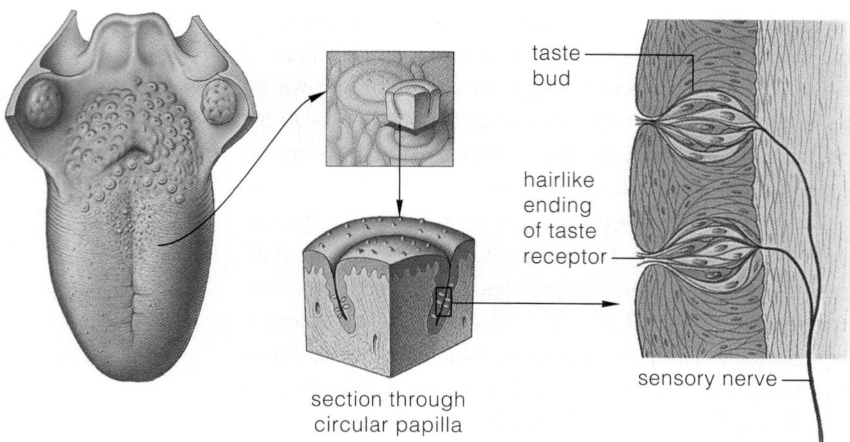

Figure 25.24 Taste receptors in the human tongue. The structures called circular papillae enclose epithelial tissue that contains taste buds. A human tongue has approximately 10,000 of these sensory organs, each of which has as many as 150 chemoreceptors.

You perceive many different tastes, but they all are some combination of five primary sensations: sweet (as elicited by glucose and other simple sugars), sour (acids), salty (NaCl or other salts), bitter (plant toxins, including alkaloids), and *umami* (elicited by amino acids such as glutamate, which provides the savory taste typical of aged cheese and meat). With its strong savory taste and no odor or texture of its own, MSG (monosodium glutamate) is now a commonly used flavor enhancer.

Wall of eyeball	(three layers)
Outermost layer	*Sclera.* Protects eyeball
	Cornea. Focuses light
Middle layer	*Choroid.* Blood vessels nutritionally support wall cells; pigments prevent light scattering
	Ciliary body. Its muscles control lens shape; its fine fibers hold lens upright
	Iris. Controls size of the pupil
	Pupil. Serves as entrance for light
Innermost layer	*Retina.* Absorbs, transduces light energy
	Fovea. Increases visual acuity
	Start of optic nerve. Carries signals to brain

Interior of eyeball	
Lens	Focuses light on photoreceptors
Aqueous humor	Transmits light, maintains pressure
Vitreous body	Transmits light, supports lens and eyeball

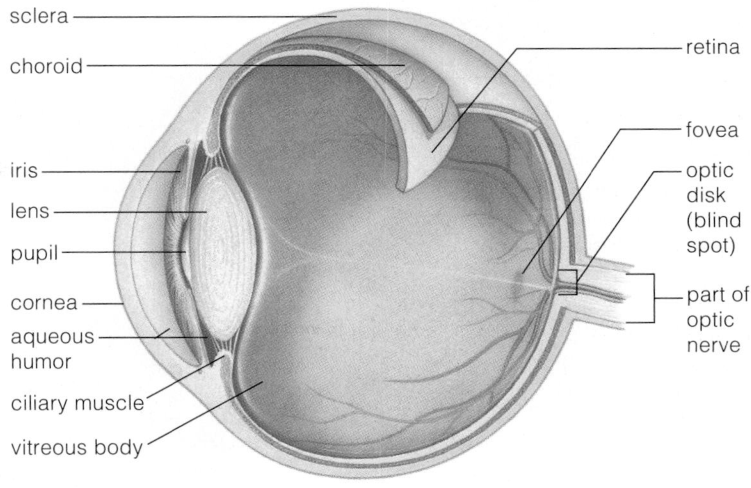

Figure 25.25 ***Animated!*** Structure of the human eye, cutaway view.

THE SENSE OF VISION

At the minimum, the sense of **vision** requires eyes and brain centers that receive and interpret patterns of visual stimulation. The brain pieces together an image from information about the shape, brightness, position, and movement of visual stimuli.

Eyes are sensory organs that include a tissue with a dense array of photoreceptors. All photoreceptors have this in common: *They incorporate pigment molecules that can absorb photon energy, which is converted to excitation energy in sensory neurons.*

Remember planarians (Section 16.3)? They and many other invertebrates have eye spots. These are clusters of photoreceptors that allow the animal to detect changes in light intensity in the visual field. A **visual field** is the area of space that the animal sees. Forming images of the visual field is easier when the eye has a **lens**, a transparent body that bends light rays from any point in the visual field so they fall on the photoreceptors.

Except in cephalopods and some of the arthropods, invertebrate vision is fuzzy. Detailed images require a large collection of photoreceptors. The planarian's eye spot holds only about 200 photoreceptors, compared with 70,000 per square millimeter in an octopus eye.

VERTEBRATE EYES In vertebrates, vision starts at the **retina**, a tissue densely packed with photoreceptors. Vision involves image formation in brain centers that accept and interpret stimulation from photoreceptors in different parts of each eye. Consider Figure 25.25, which shows the structure of a human eye. As in most vertebrates, a human eyeball has a three-layered wall. Its outer layer is the sclera and cornea. The sclera, the densely fibrous "white" part, protects most of the eyeball. A transparent dome-shaped cornea composed of protein fibers covers the front of the eye. The cornea's curvature helps to direct light into the eye's interior.

The middle layer has a choroid, ciliary body, iris, and pupil. The choroid is a darkly pigmented tissue, which absorbs light wavelengths that photoreceptors missed and stops them from scattering in the eye.

Suspended just behind the cornea is the doughnut-shaped, pigmented iris. Light enters the eye through the pupil, a dark "hole" in the iris's center. In bright sunlight, circular muscles in the iris contract, and the pupil's diameter shrinks. In dim light, the iris's radial muscles contract, and the pupils enlarge, allowing more light into the eye. Sympathetic stimulation also widens the pupils, as during a fight–flight response. Amphetamines and similar drugs have the same effect because they increase sympathetic signaling.

Once past the pupil, light rays pass through a lens kept moist by a clear fluid called the aqueous humor. A jellylike vitreous body fills the area behind the lens. At the very back of this chamber is the retina.

Because the cornea and lens have a curved surface, light rays coming from a particular point in the visual field strike them at different angles, so the paths that they are following are altered. The newly angled paths make light rays converge at the back of the retina, in a pattern that is upside-down and reversed left to right

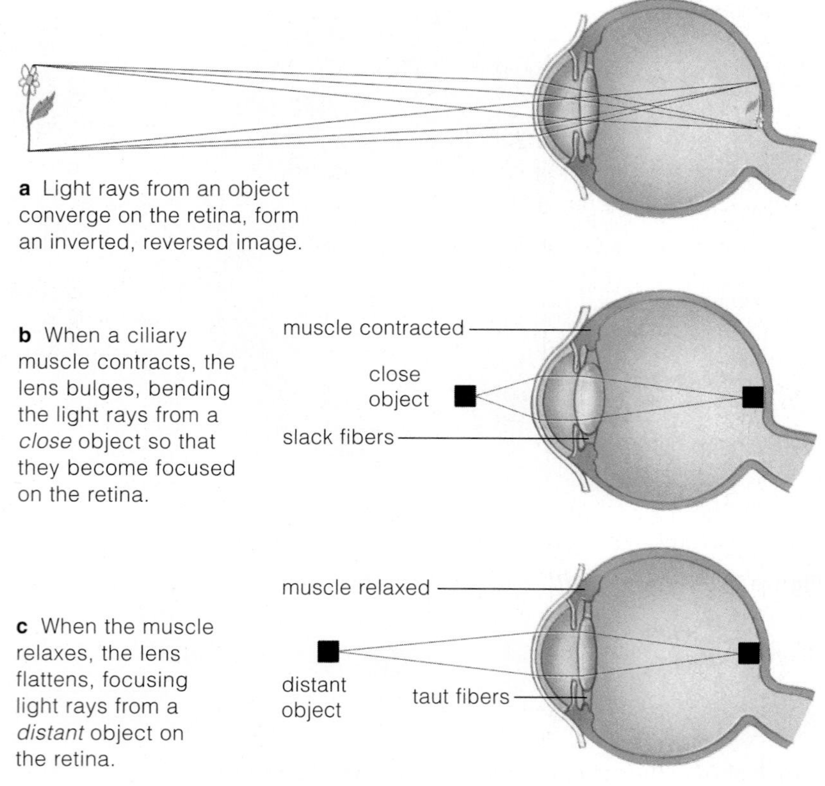

a Light rays from an object converge on the retina, form an inverted, reversed image.
b When a ciliary muscle contracts, the lens bulges, bending the light rays from a close object so that they become focused on the retina.
muscle contracted
close object
slack fibers
c When the muscle relaxes, the lens flattens, focusing light rays from a distant object on the retina.
muscle relaxed
distant object
taut fibers

Figure 25.26 ***Animated!*** (**a**) Pattern of retinal stimulation in a human eye. A curved, transparent cornea in front of the pupil changes the trajectories of light rays as they enter into the eye. (**b**,**c**) Two focusing mechanisms use a ciliary muscle that encircles the lens and attaches to it.

LINKS TO SECTIONS 1.2, 8.7

relative to the original source of the light rays (Figure 25.26*a*). This might seem confusing, but brain centers know how to interpret this turned about stimulation.

The closer an object is to the eye, the more rays of light reflected from that object will be diverging when they enter the eye. By changing the shape of the lens, light rays can be bent so that they all fall on the retina, regardless of how far they have traveled.

Lens adjustments are called **visual accommodation**. A ciliary muscle encircles the lens and attaches to it. When you focus on something close up, the ciliary muscle contracts, causing the lens to bulge outward. Rays of light that are diverging from the object are focused onto the retina (Figure 25.26*b*). When an object is farther away, light rays do not have to be bent as much to be focused. The ciliary muscles can relax a bit, which allows the lens to flatten (Figure 25.26*c*).

Sometimes the eyeball is not shaped quite right and the lens cannot be adjusted enough to make the focal point fall correctly on the retina. In *nearsightedness*, the eyeball is too long or the ciliary muscle that adjusts the lens contracts too strongly. Light from distant objects gets focused too soon—in front of the retina instead of right on it (Figure 25.27*a*).

In *farsightedness*, the eyeball is not long enough or the lens is "lazy." The focal point for light from close objects is behind the retina (Figure 25.27*b*). The latter often happens after age forty; the lens usually loses some of its natural flexibility as we grow older.

FOCUS ON THE RETINA What happens to light at the retina? The retina has a basement layer, a pigmented epithelium on top of the choroid. Resting on this layer are densely packed arrays of **rod cells** and **cone cells**, two classes of photoreceptors (Figure 25.28). Rod cells detect very dim light. At night or in dark places, they respond to changes in light intensity across the visual field—the start of coarse perception of motion.

A rod cell's outer segment is folded into several hundred membrane disks, each containing a hundred million molecules of the visual pigment rhodopsin. Membrane stacking and the very high pigment density increase the likelihood of intercepting packets of light energy—photons. Action potentials triggered by the absorption of even a few photons make us conscious of dimly lit objects, as in darkened rooms or late at night.

The sense of color and of acute daytime vision starts with photon absorption by *cone* cells. There are three types, each having a different pigment. One cone cell pigment absorbs mainly the red wavelengths, another absorbs mainly blue, and a third absorbs green.

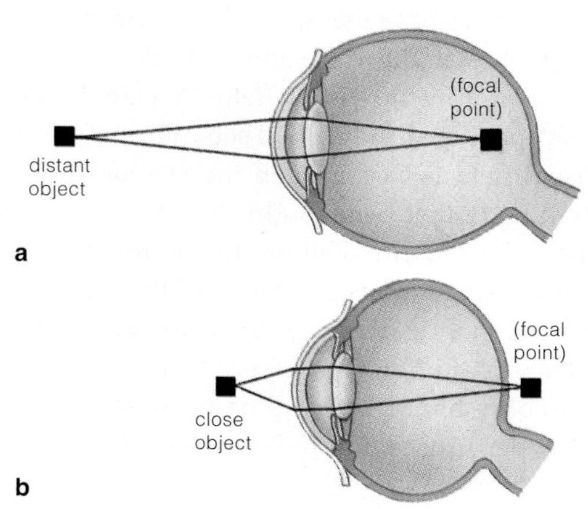

Figure 25.27 Causes of focusing problems. In nearsightedness (**a**) images from distant objects converge in front of the retina. In farsightedness (**b**) images from close objects have not yet converged when they reach the retina.

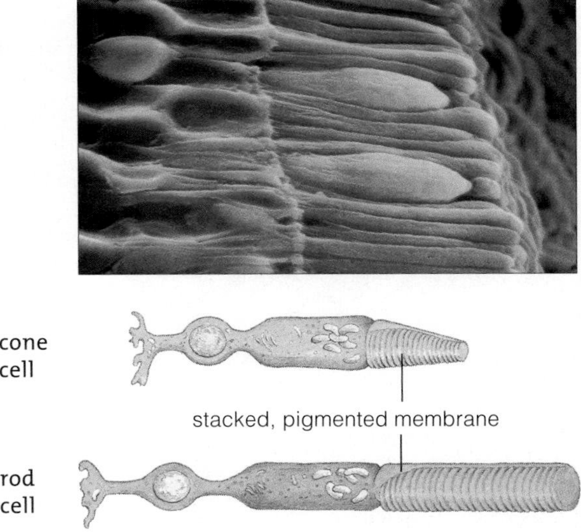

Figure 25.28 Scanning electron micrograph and sketches of rod cell and cone cell structure.

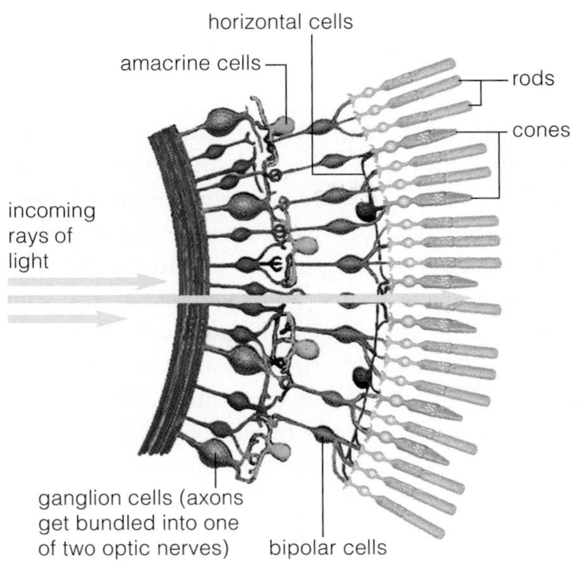

Figure 25.29 ***Animated!*** Organization of the photoreceptors and related nervous cells in the retina.

Normal human color vision requires all three kinds of cones. Individuals who lack one or more cone types are *color-blind*. We discussed the genetics of *red–green color blindness* in Section 8.7. It is an X-linked, recessive abnormality, and so appears most often in males. Some or all cone cells that detect light of red or green wavelengths are missing. Most affected people have trouble telling red from green in dim light, and some cannot distinguish between the two even in bright light.

Signal integration and processing start in the retina. Several layers of cells lie above rods and cones (Figure 25.29). The cells accept information and process signals from the photoreceptors. Input from about 125 million rods and cones synapses on bipolar cells. These signals are dampened or enhanced by the amacrine cells and horizontal cells, then the signals flow to ganglion cells, the axons of which start an optic nerve (Figure 25.30).

Bundled axons of ganglion cells leave the retina as the beginning of the optic nerve. The part of the retina where the optic nerve exits lacks photoreceptors and so cannot respond to light; it is a "blind spot." We all have a blind spot in each eye, but usually do not notice it. Information about the region that is missed by one eye is provided to the brain by the other eye.

The region of retina called the macula is richest in photoreceptors. Within the macula, the fovea, a funnel-shaped depression, has the greatest density of cones and contributes most to visual acuity (Figure 25.30).

Macular degeneration is one common cause of vision loss. In the United States, about 13 million people have age-related macular degeneration (AMD). Destruction of photoreceptor cells in the macula and fovea clouds the center of the visual field more than the periphery. The causes of AMD are not fully understood, although certain genes heighten the risk. So do smoking, obesity, and high blood pressure. AMD cannot be cured, but the vision loss can be postponed with eye drops, vitamins, and laser therapy.

Vision can also be impaired by a *cataract*, a gradual clouding of the lens. A cataract alters the amount of light that reaches the retina and where light is focused. A defective lens can be replaced with an artificial one.

Glaucoma happens when too much aqueous humor accumulates and pressure inside the eye increases. This damages blood vessels and ganglion cells, and it can interfere with peripheral vision and visual processing. Glaucoma can be treated with medication and surgery.

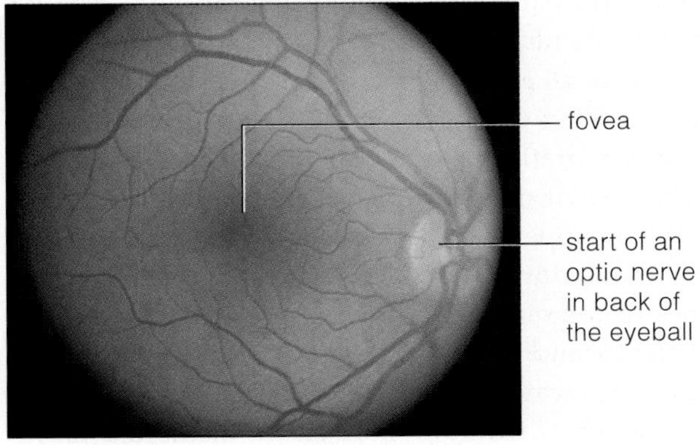

Figure 25.30 A view of the retina at the back of the eye. The fovea contains only cones. A photoreceptor-rich area, the macula, surrounds it. The optic nerve, made up of the axons of ganglion cells, exits the eye at the blind spot.

LINKS TO SECTIONS 1.2, 4.1

THE SENSE OF HEARING

PROPERTIES OF SOUND Many arthropods and most vertebrates can perceive sounds; they have a sense of **hearing**. Sounds are waves of compressed air, or one form of mechanical energy. Clap your hands and you compress air and create pressure variations that can be depicted as wave forms. The *amplitude* of any sound corresponds to its loudness, or intensity. We measure amplitude in decibels. The human ear can detect a one-decibel difference between sounds. The *frequency*, or pitch, of a sound is the number of wave cycles per second. Each cycle extends from the peak (or trough) of one wave to the corresponding peak (or trough) on the next wave in line (Figure 25.31).

Unlike a tuning fork's pure tone, most sounds are combinations of waves of different frequencies. Their timbre, or sound quality, is variable. You perceive one person's voice as nasal and another as deep because of differences in timbre. Variations in timbre are the main cue you use to recognize individuals by their voices.

THE VERTEBRATE EAR Any sound is transmitted much better in water than in air, so the vertebrate move onto land was accompanied by some remodeling. Parts of the ear were expanded into structures that trap and amplify sounds traveling through air.

Figure 25.32*a* shows the three parts of your own ears: the outer, middle, and inner regions. Your *outer* ear is adapted for gathering sounds from the air, as it is for most mammals. Its pinna, a skin-covered, sound-collecting, folded flap of cartilage, projects outward from the side of the head. The outer ear also includes a deep auditory canal leading to the middle ear.

Each *middle* ear amplifies and transmits air waves to the inner ear. It has an eardrum, which started out as a shallow depression on each side of the head in reptiles. The eardrum is a thin membrane, one that can rapidly vibrate in response to air waves. Behind the eardrum of all existing crocodiles, birds, and mammals is an air-filled cavity and a set of small bones, which transmit vibrations to the inner ear. These bones, the hammer, anvil, and stirrup, are shown in Figure 25.32*b*.

What happens to sounds collected at the outer ear? They move down the auditory canal to the eardrum. Bones of the middle ear pick up the vibrations. They interact to amplify a stimulus by transmitting the force of pressure waves to a smaller surface called the oval window. That "window" is an elastic membrane just in front of the cochlea.

The pea-sized cochlea, along with the vestibular apparatus described earlier, are key components of the *inner* ear. Figure 25.32*c* depicts an "uncoiled" cochlea, the ear region that sorts out the pressure waves. Pressure waves make the oval window bow in and out, which produces fluid pressure waves inside two ducts (scala vestibuli and scala tympani). These waves reach still another membrane known as the round window. The round window compensates for differences in pressure by passively bowing inward and outward.

The cochlear duct, the inner ear's third duct, sorts out the pressure waves. Its *basilar* membrane is narrow and stiff, then it broadens and becomes more flexible deeper in the coil. The structural differences along its length mean that the membrane can vibrate differently in response to incoming sounds of different frequencies. Higher-pitched sounds set up vibrations early in the coil. But sounds of lower frequency cause vibrations deeper in the coil.

Which sensory receptors detect these vibrations? Attached to the inner basilar membrane is the organ of Corti, with its arrays of vibration-sensitive acoustical receptors. These mechanoreceptors are hair cells. They have specialized cilia that are embedded in a *tectorial* membrane hanging above them (Figure 25.32*d*,*e*). Their

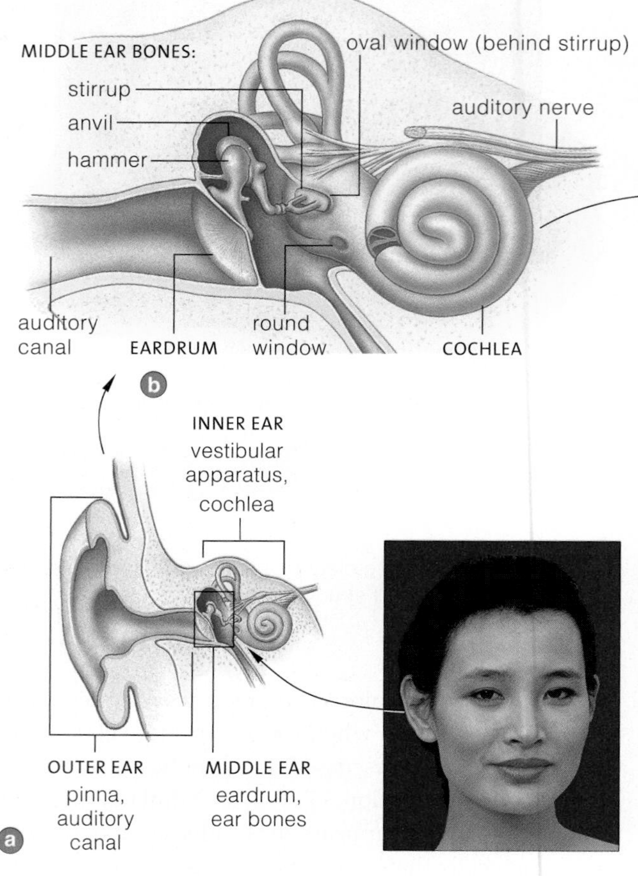

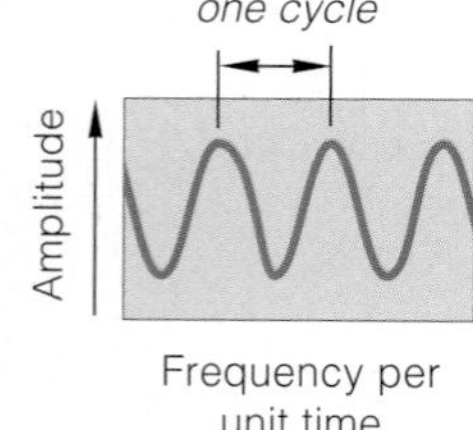

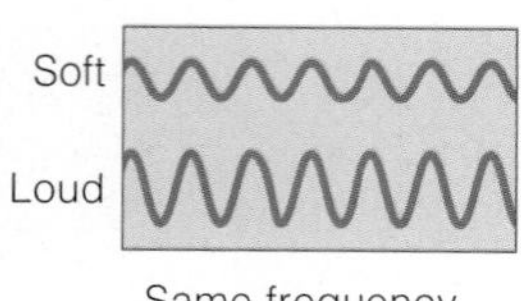

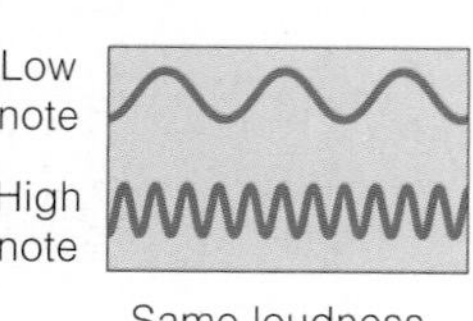

Figure 25.31 ***Animated!*** Properties of sound.

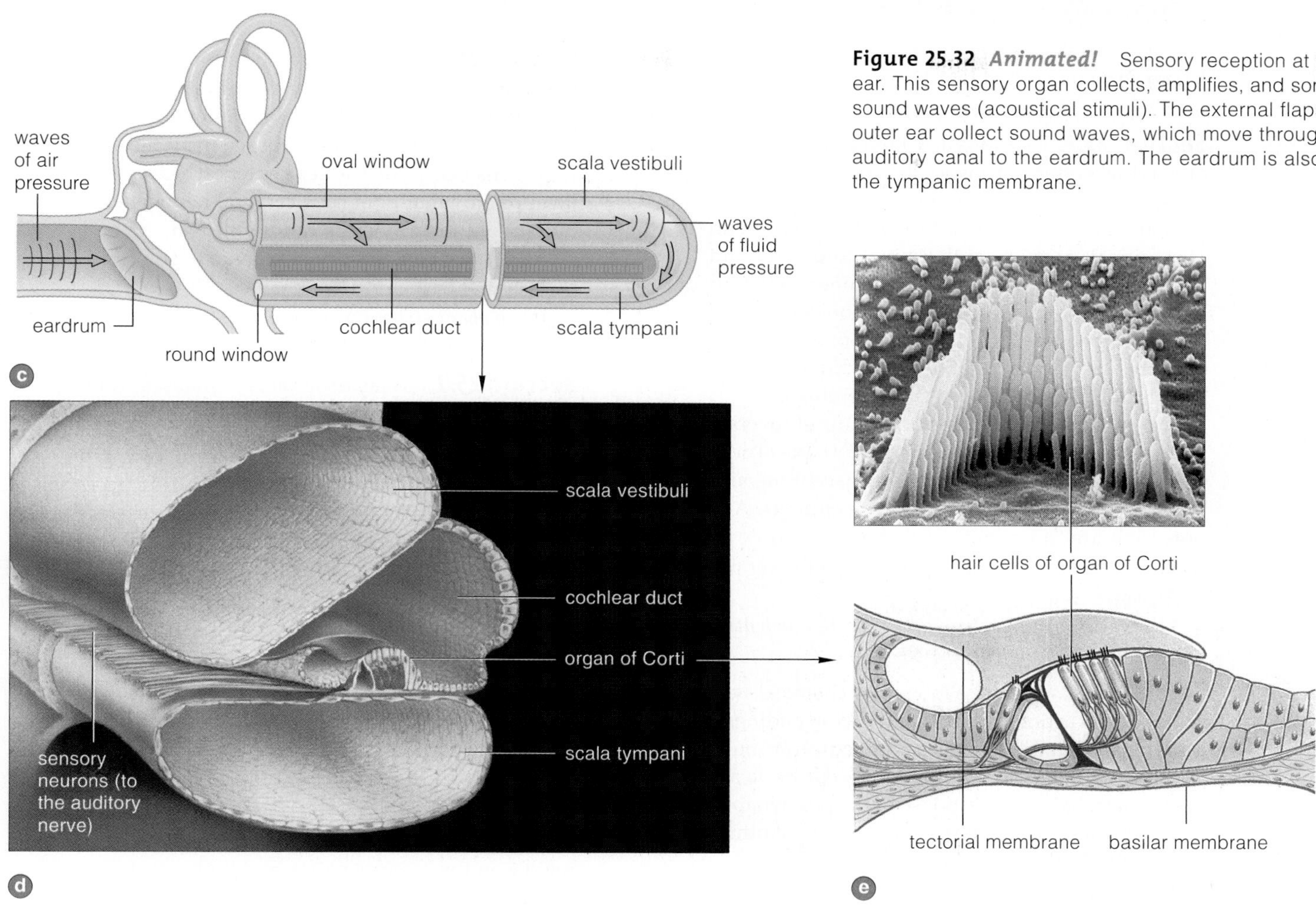

Figure 25.32 ***Animated!*** Sensory reception at the human ear. This sensory organ collects, amplifies, and sorts out sound waves (acoustical stimuli). The external flaps of the outer ear collect sound waves, which move through an auditory canal to the eardrum. The eardrum is also called the tympanic membrane.

cilia bend when pressure waves displace the basilar membrane. This bending transduces the mechanical energy of pressure waves into action potentials, which then travel along an auditory nerve into the brain.

A word of caution. Exposure to very loud sounds damages the inner ear's delicate hair cells and may lead to temporary or permanent loss of hearing. The louder the sound, the less time it takes to cause deafness.

THE SENSE OF BALANCE

All animals assess and respond to displacements from equilibrium, when the body is balanced in relation to gravity, acceleration, and other forces that may affect its positions and its movement. In humans, the organs of equilibrium are the vestibular apparatus and three semicircular canals, which form a continuous fluid-filled system. When the head rotates, the fluid shifts, stimulating some mechanoreceptors. Other receptors respond to starts and stops. The brain compares the signals from organs of equilibrium in both ears. At the same time, it receives input from receptors in the skin, joints, and tendons. All this information allows your brain to control eye muscles and keep the visual field in focus, even as you shift position or your head rotates. Integration also helps maintain awareness of the body's position and motion in space.

The senses of smell, taste, balance, hearing, and vision are special senses in that their sensory pathways start at receptors that are not widely distributed, but rather are located in specific tissues and organs.

All of these senses have receptors that respond to specific stimuli. In various ways, the stimuli disturb the voltage difference across the plasma membrane of the sensory receptor cells.

Strong disturbances generate action potentials that travel on sensory nerves to specific parts of the cerebral cortex for final processing and integration.

Summary

Section 25.1 Neurons are the communication units of nervous systems. Sensory neurons relay signals into the spinal cord and brain. Interneurons within the spinal cord and brain integrate information. Motor neurons carry signals away from the spinal cord and brain.

Signals reaching neuron input zones (cell body or dendrites) disturb the ion distribution across their plasma membrane. Strong stimulation may drive the disturbance to the threshold of an action potential that travels down to the neuron's axon endings.

An action potential is an abrupt, brief reversal in the voltage across a neuron plasma membrane. The reversal triggers an action potential at an adjacent membrane patch, then the next, and on to axon endings. Action potentials are "all or nothing;" they occur only if a cell reaches threshold level and are always the same size.

BiologyNow

Learn about neuron structure and action potentials with the animation on BiologyNow.

Section 25.2 Neurons interact at chemical synapses. Arrival of an action potential at the axon endings of a presynaptic cell triggers release of neurotransmitter. These signaling molecules bind to receptors on a postsynaptic cell. The response of the postsynaptic cell to any given signal depends in part on what other signals are arriving at the same time.

BiologyNow

See how signals are transmitted at a chemical synapse with the animation on BiologyNow.

Section 25.3 Most axons are enclosed in insulating myelin, which speeds signal transmission. A nerve is a bundle of axons. Reflexes are stereotyped movements in response to stimuli. In the simplest reflex arcs, sensory neurons synapse directly on motor neurons.

BiologyNow

Observe how signals flow through a nervous system with the animation on BiologyNow.

Section 25.4 Radial animals have simple nerve nets. Most animals have a bilateral nervous system with a cluster of neurons or a brain at their head end.

The vertebrate nervous system is functionally divided into a central nervous system (brain and spinal cord) and peripheral nervous system (nerves that connect the spinal cord and brain with the rest of the body).

Section 25.5 The somatic nerves of the peripheral system connect to skeletal muscles and to receptors in joints and the skin. The autonomic nerves (sympathetic and parasympathetic) connect to internal organs and often work in opposition. In a fight–flight situation, sympathetic signals predominate. In a relaxed state, parasympathetic signals predominate.

BiologyNow

Explore the divisions of the peripheral nervous system with the animation on BiologyNow.

Section 25.6 The embryonic neural tube gives rise to the spinal cord and to the three brain regions: hindbrain, forebrain, and midbrain. The brain stem is the most evolutionarily ancient neural tissue in all three regions. It governs many basic reflex centers. The thin surface layer of gray matter of each cerebral hemisphere is the most evolutionarily recent addition. This cerebral cortex governs most conscious behavior.

BiologyNow

Learn about structure and function of the brain and spinal cord with the animation on BiologyNow.

Section 25.7 Many drugs act at synapses. They stimulate or block release or uptake of neurotransmitter.

Section 25.8 Sensory receptors convert stimulus energy, such as light or pressure, into action potentials. The brain evaluates information from sensory receptors based on which nerve delivers them, signal frequency, and the number of axons firing in a given interval.

Section 25.9 Receptors for somatic sensations, such as touch, pressure, and warmth, are not localized in a special organ or tissue. They signal sensory areas of the cerebral cortex, where interneurons are organized like maps for individual parts of the body surface.

Section 25.10 Receptors for the special senses—taste, smell, hearing, balance, and vision—are located in special sensory organs. The senses of taste and smell involve pathways from chemoreceptors to processing regions in the cerebral cortex and limbic system.

Human vision requires focusing of light onto a dense array of photoreceptors (rods and cones) of the retina. Other cells of the retina coarsely integrate and process signals, which travel along optic nerves to the brain for final processing and interpretation.

In hearing, the outer ear collects sound waves, the middle ear amplifies them, and the inner ear sorts them and converts them to action potentials.

Organs of equilibrium in the inner ear detect movement and changes in the body's position.

BiologyNow

Investigate the special senses with the animation on BiologyNow.

Self-Quiz

Answers in Appendix I

1. Neurons send signals at ______ .
   a. axon endings
   b. the cell body
   c. dendrites
   d. myelin sheaths

2. ______ occur mainly in the brain and spinal cord.
   a. Sensory neurons
   b. Motor neurons
   c. Interneurons
   d. b and c

3. An action potential begins when ______ .
   a. a neuron receives adequate stimulation
   b. gated sodium channels open
   c. sodium–potassium pumps begin working
   d. both a and b

4. Neurotransmitters are released by ______ .
   a. dendrites c. myelin sheaths
   b. axon endings d. both a and b

5. When you sit quietly on the couch reading, output from the ______ system prevails.
   a. sympathetic c. both
   b. parasympathetic d. neither

6. Skeletal muscles are controlled by ______ nerves.
   a. sympathetic c. somatic nerves
   b. parasympathetic d. both a and b

7. In some areas of the ______ neurons are arranged like maps of the parts of the body.
   a. cerebral cortex c. brain stem
   b. limbic system d. cerebellum

8. The blood–brain barrier controls what enters ______ .
   a. blood c. peripheral nerves
   b. cerebrospinal fluid d. both a and b

9. Which is a somatic sensation (not a special sense)?
   a. hearing d. taste
   b. smell e. both a and c
   c. touch f. all of the above

10. ______ is a reduced response to an ongoing stimulus.
   a. Propagation c. Sensory adaptation
   b. Perception d. Synaptic integration

11. Chemoreceptors play a role in the sense of ______ .
   a. hearing b. smell c. vision d. pain

12. In a vertebrate eye, photoreceptors are in the ______ .
   a. retina c. lens and choroid
   b. sclera and cornea d. both a and b

13. Color blindness arises when ______ are missing or defective.
   a. hair cells c. cone cells
   b. rod cells d. neuroglial cells

14. Match each structure with its function.

____ rod cell	a. sorts out pressure waves
____ cochlea	b. connects to spinal cord
____ limbic system	c. protects brain and spinal cord from some toxins
____ brain stem	d. speeds signal transmission
____ cerebral cortex	e. contains chemoreceptors
____ taste bud	f. roles in emotion, memory
____ myelin	g. detects light
____ blood–brain barrier	h. governs higher thought

Additional questions are available on **Biology Now™**

Critical Thinking

1. In 2002, Charlene Singh was twenty-two years old, a citizen of the United Kingdom, and attending college in Florida. In that year she was diagnosed with *variant Creutzfeldt-Jakob disease* (vCJD) (Figure 25.33). The disease is related to bovine spongiform encephalopathy (BSE), commonly known as *mad cow disease*.

Charlene's nightmare began with memory problems. Within months, she was having trouble controlling her movements, including walking. Her mother brought her to England, where she was diagnosed as having vCJD. Charlene returned to Florida and became bedridden and unable to feed herself or communicate. She died in the summer of 2004.

The first case of vCJD was reported in 1995. By 2005, the disease had killed 150 people, with the vast majority in the United Kingdom. Most scientists now believe that vCJD is caused by a misshapen version of a prion, a type of infectious protein (Section 14.5). Build-up of the prion proteins kills brain cells.

Charlene grew up in the United Kingdom. Abnormal prions probably entered her body after she ate BSE-infected beef. Most countries, including the United States, now ban the use of discarded animal parts in animal feed, a practice that contributed to the BSE epidemic. Yet a few prion-infected cows have turned up in the United States. Some people argue that vCJD is a good reason to avoid eating beef altogether. Do you agree? What dietary and life-style precautions do you take to protect the health of your brain?

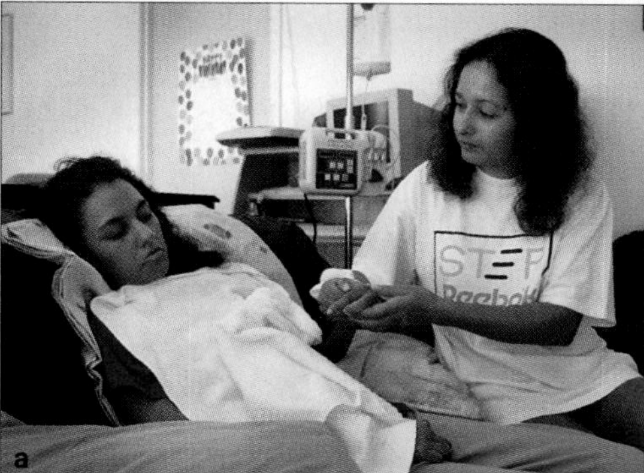
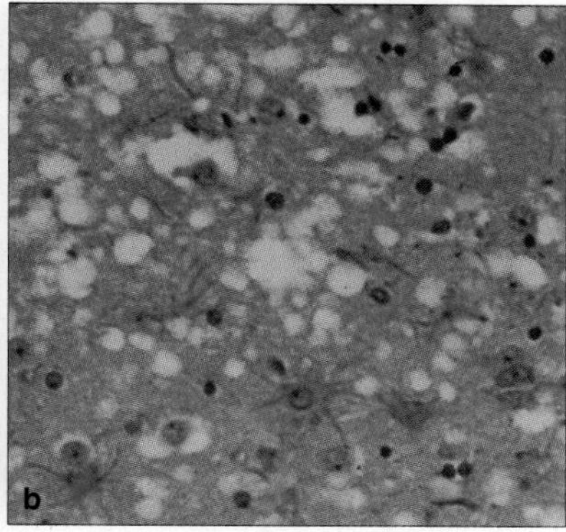

Figure 25.33 (**a**) Charlene Singh, thus far the only American resident diagnosed with vCJD, being cared for by her mother. The disease has since claimed Charlene's life. (**b**) Section through a brain damaged by bovine spongiform encephalopathy (BSE). The light "holes" are areas where tissue has been destroyed.

2. The blood–brain barrier in human newborns, especially premature ones, is not yet fully developed. Explain why this makes careful monitoring of diet and environmental chemical exposure particular important.

3. The axons of some human motor neurons extend from the base of the spinal cord to the big toe, a distance of more than a meter. In a giraffe, the longest axons are several meters long. What are some of the functional challenges involved in the development and maintenance of such lengthy cellular extensions?

4. In humans, photoreceptors are most concentrated at the very back of the eyeball. In birds of prey, such as owls and hawks, the greatest density of photoreceptors is in a region closer to the eyeball's roof. When these birds are on the ground, they cannot see objects above them unless they turn their head almost upside down (Figure 25.34). How is this retinal organization adaptive?

5. Frog of the genus *Dendrobates* secrete a poison that some hunters in South America use on arrow tips. The toxin is similar in structure to the neurotransmitter ACh, which is released by motor neurons at synapses with skeletal muscles. What effect do you think an injection of this toxin has on an animal that is being hunted?

Figure 25.34 Here's looking at you!

26 ENDOCRINE CONTROLS

IMPACTS, ISSUES

Hormones in the Balance

Atrazine has been widely used as an herbicide for more than forty years. Each year, people in the United States apply about 76 million pounds of it, mostly to fields of corn and other monocot crops. They also apply it to lawns at homes, parks, and golf courses. Where does it go from there? Into soil and water. Atrazine molecules break down in less than a year, but until they do, they turn up in groundwater, wells, ponds, and even in rain.

Is this a bad thing? Endocrinologist Tyrone Hayes, shown at right, thinks so. He suspects that atrazine is one of the synthetic organic compounds that mimics, blocks, or boosts effects of natural hormones. Such compounds are called *endocrine disruptors*.

Hayes began studying atrazine's effects on African clawed frogs (*Xenopus laevis*) and leopard frogs (*Rana pipiens*) in his laboratory. He exposed male tadpoles to atrazine, and found that some of them developed as hermaphrodites, with male *and* female reproductive organs. Results showed that atrazine levels as low as 0.1 parts per billion could cause males to develop ovaries.

Does atrazine have similar effects in the wild? To answer this question, Hayes collected frogs from ponds and ditches all across the Midwest. In every atrazine-contaminated pond he sampled, some male frogs had abnormal sex organs. In the pond with the highest atrazine concentration, 92 percent of male frogs had feminized sex organs.

Since Hayes' initial studies, other scientists have also reported that atrazine causes or contributes to frog deformities. The Environmental Protection Agency found the data intriguing. Among other things, this agency oversees application of chemicals in agriculture. It has called for further study of atrazine's potentially harmful effects on amphibians. It also is encouraging farmers to do more to minimize the runoff of atrazine-laden water from their fields.

The endocrine disruptors may be adversely affecting humans as well. Studies of males in Western countries found that sperm counts declined by about 40 percent between 1938 and 1990. Men in agricultural regions, where the chemical contamination of drinking water is highest, may have the lowest sperm counts. University of Missouri researchers discovered that semen samples from men in Columbia, Missouri, contained only about half as many healthy sperm as samples from men in New York City and Los Angeles.

This chapter focuses on hormones—their sources, targets, and interactions. All vertebrates have similar hormone-secreting glands and systems. If the details start to seem remote, remember endocrine disruptors. What you learn here can help you evaluate research that may influence your health and life-style.

How Would You Vote?

Farmers currently depend on agricultural chemicals to maximize crop yields. Some of these chemicals may disrupt hormone function in frogs and other untargeted species. Should chemicals that are under suspicion remain in use while researchers investigate them? See BiologyNow for details, then vote online.

Key Concepts

HOW HORMONES WORK
Hormones are signaling molecules secreted by endocrine glands. These ductless glands release hormones into the bloodstream, which distributes them throughout the body. Only cells that have receptors for a particular hormone will respond to it.

A MASTER CONTROL CENTER
The hypothalamus and pituitary glands lie deep within the brain and are structurally and functionally linked. They secrete hormones that encourage or suppress the activities of endocrine glands throughout the body.

OTHER HORMONE SOURCES
Many endocrine glands secrete hormones in response to signals from the hypothalamus and pituitary gland. Other endocrine glands secrete hormones in response to local changes in the internal environment.

Links to Earlier Concepts

You already learned how hormones affect appetite and urine formation in Chapter 24. Here, you will be drawing on knowledge of the plasma membrane (Section 3.3) to see how different hormones interact with target cells. You will use your knowledge of gene controls (10.6), feedback mechanisms (17.3), and cell-to-cell signaling (17.5). This chapter will build on your understanding of glucose metabolism (6.5), neurotransmitters (25.2), and the human brain (25.6). Earlier, you read about glandular epithelium (20.2). All hormone-secreting cells are embedded in this kind of tissue, which may form a gland or a lining of an organ.

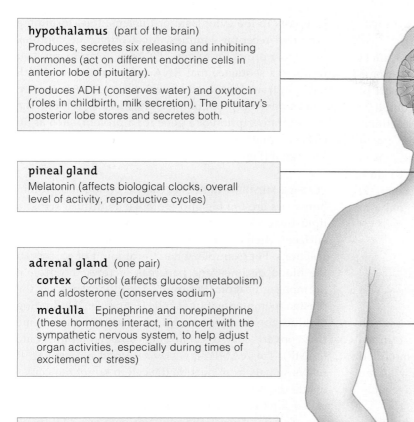

hypothalamus (part of the brain)
Produces, secretes six releasing and inhibiting hormones (act on different endocrine cells in anterior lobe of pituitary).
Produces ADH (conserves water) and oxytocin (roles in childbirth, milk secretion). The pituitary's posterior lobe stores and secretes both.

pineal gland
Melatonin (affects biological clocks, overall level of activity, reproductive cycles)

adrenal gland (one pair)
cortex Cortisol (affects glucose metabolism) and aldosterone (conserves sodium)
medulla Epinephrine and norepinephrine (these hormones interact, in concert with the sympathetic nervous system, to help adjust organ activities, especially during times of excitement or stress)

ovaries (one pair of female gonads)
Estrogens and progesterone (maintain primary sex organs, influence secondary sexual traits)

testes (one pair of male gonads)
Testosterone (develops and maintains primary sex organs, influences secondary sexual traits)

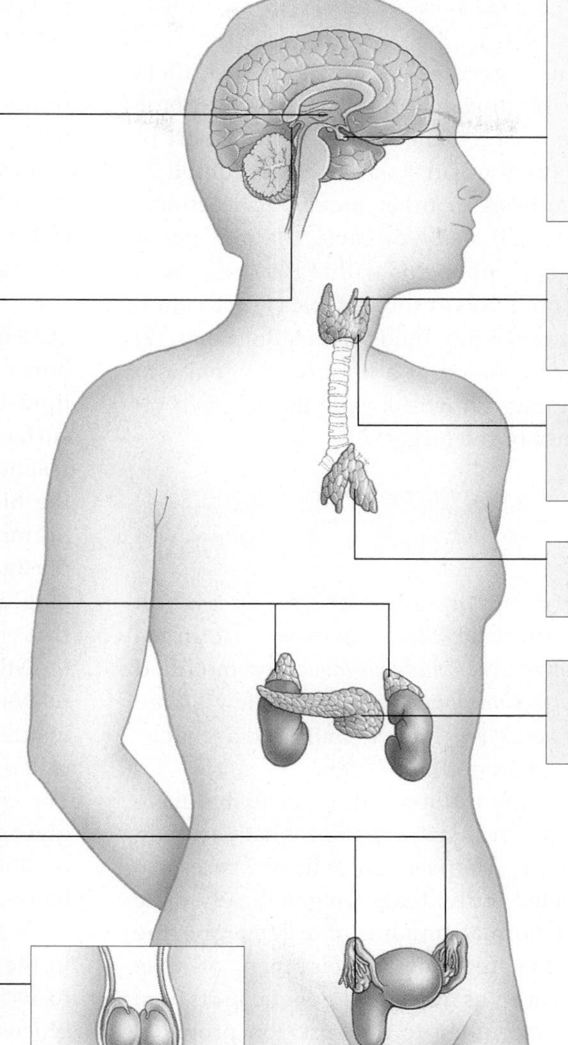

pituitary gland, anterior lobe
Four hormones (ACTH, TSH, FSH, LH) stimulate other glands. Two (prolactin, somatotropin) stimulate overall growth and development.
pituitary gland, posterior lobe
Stores, secretes two hypothalamic hormones: ADH and oxytocin

thyroid gland
Thyroid hormone (thyroxine and triiodothyronine); roles in development and metabolic control

parathyroid glands (four)
Parathyroid hormone (increases blood level of calcium)

thymus gland
Thymosins (roles in white blood cell functioning)

pancreatic islets
Insulin (lowers blood level of glucose), glucagon (raises blood level of glucose)

Figure 26.1 Overview of the major components of the human endocrine system and primary effects of their main hormonal secretions. The system also includes endocrine cells in many organs, such as the heart, kidneys, stomach, liver, small intestine, and skin. Chapter 24 described the effects of leptin and other hormones that affect body weight.

26.1 Hormones and Other Signaling Molecules

Hormones and other signaling molecules coordinate growth, development, and reproductive cycles of nearly all animals. Signaling molecules bind to receptors on target cells and bring about changes in their activities.

LINKS TO SECTIONS
3.3, 17.5, 25.2

CATEGORIES OF SIGNALING MOLECULES

Within any animal body, cells communicate with one another by means of intercellular signaling molecules. The main kinds are neurotransmitters, local signaling molecules, and hormones. Each is made by one type of cell and affects the behavior of other cells in the same body. Those cells that can be affected are "target cells." They have receptors that bind the signaling molecule.

Neurotransmitters, released from axon endings of neurons, act swiftly on target cells by diffusing across a tiny space between them. Many types of cells secrete **local signaling molecules** that modify the conditions in nearby tissues. Various *prostaglandins* relax smooth muscle in blood vessel walls, affect the release of acid by cells in the stomach, and modulate inflammation. *Nitric oxide* (*NO*), another local signaling molecule, is a gas. It acts on blood vessels and affects blood pressure and blood distribution, as when it allows more blood to flow into a male's penis to produce an erection.

In animals, **hormones** are signaling molecules that are secreted into the bloodstream by endocrine glands, endocrine cells, and hypothalamic neurons. The blood carries hormones to target cells inside the body, often some distance from their source. In humans and other vertebrates, the many sources of hormones interact as an **endocrine system** (Figure 26.1).

LINKS TO SECTIONS 3.3, 10.6, 24.6

HOW HORMONES EXERT THEIR EFFECTS

Like all other signaling molecules, the hormones affect cell activities by binding to protein receptors. Some hormones induce a target cell to alter its transport of glucose, sodium, or other substances. Others inhibit or speed protein synthesis or other metabolic reactions. Still others modify cell shape or cause the movement of proteins or other components within the cytoplasm.

Hormone action involves three steps: (1) activation of a receptor as it reversibly binds with a hormone, (2) transduction, or conversion of a hormonal signal into a molecular form that can work inside the cell, and (3) functional response of the target cell:

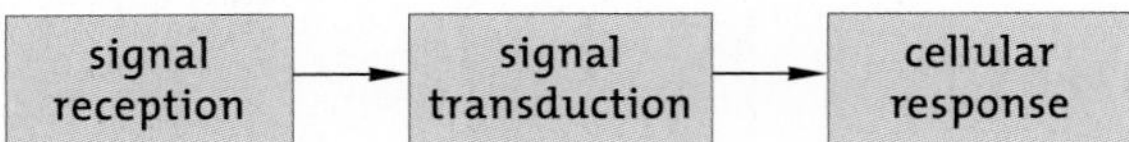

Hormones vary in their molecular structure, which affects how they function. *Steroid hormones* are lipids derived from cholesterol. *Amine hormones* are modified amino acids. *Peptide hormones* consist of a few amino acids. *Protein hormones* are longer amino acid chains. Table 26.1 lists some examples of each.

Hormones can only influence those cells that have functional receptors for them. Consider what occurs in *androgen insensitivity syndrome,* an X-linked recessive disorder. All affected individuals are genetically male (XY). They secrete normal amounts of testosterone, the steroid hormone that regulates development of male sexual traits. However, as a result of genetic mutation, their testosterone receptors do not function properly. As a result, the testes do not move from the abdomen into the scrotum. Genitals and secondary sexual traits will look slightly to fully female. Different mutations alter the receptor structure in different ways, and thus lead to a more or less feminized appearance.

INTRACELLULAR RECEPTORS Testosterone and other steroid hormones are lipids that easily diffuse across the lipid bilayer of a plasma membrane. They bind to receptors in the cytoplasm or nucleus, thus forming a hormone–receptor complex. Most often, the complex binds to a promoter region of a gene. A promoter, you recall, is a sequence that RNA polymerase binds with to begin transcription. By interacting with a promoter, a hormone–receptor complex enhances or inhibits the rate of transcription of a gene for a protein, such as an enzyme, that carries out the target cell's response to the signal (Figure 26.2*a*).

PLASMA MEMBRANE RECEPTORS Peptide and protein hormones are too big and too polar to diffuse across a lipid bilayer. They bind to receptors at a target cell's surface. Often this binding starts a cascade of enzyme reactions. For example, when the amount of glucose in the blood declines, the pancreas secretes the protein hormone glucagon. Receptors for glucagon span the plasma membrane of liver cells (Figure 26.2*b*). Binding of glucagon to these receptors activates an enzyme that converts ATP to cyclic adenosine monophosphate (cAMP). This cAMP serves as a **second messenger**. It relays signals from outside the cell to target molecules inside the cell. In liver cells, cAMP activates enzymes called kinases, which in turn activate other enzymes. The cascade of reactions activates enzymes that break glycogen into its glucose subunits. As glucose released by liver cells enters the interstitial fluid, and then the blood, concentrations return to normal.

When the blood level of glucose rises, certain cells in the pancreas secrete insulin. The binding of insulin to receptors on muscle cells causes vesicles to deliver glucose transporters that are stored in the cytoplasm to the muscle plasma membrane. The transporters are inserted into the membrane, glucose uptake by muscle cells increases, and glucose level in the blood declines.

Some steroid hormones with intracellular receptors also produce effects by binding to receptors at the cell membrane. The drug tamoxifen targets receptors for estrogen at the surface of breast cancer cells. It blocks these receptors and slows the growth of the cancer.

VARIATIONS IN HORMONE ACTION

No tissue in the vertebrate body is beyond the reach of hormones. Just about all cells have receptors for many different hormones, which may have reinforcing or opposing effects on the target. Each skeletal muscle cell, for instance, has receptors for glucagon, insulin, cortisol, epinephrine, estrogen, testosterone, thyroid hormone, and growth hormone, among others. What is happening in any muscle cell at any given moment depends in part on the blood concentrations of these many hormones, on any interactions among them, and on the effects of other signaling molecules in the body.

Table 26.1 Categories of Hormones and a Few Examples

Steroid hormones	Estrogens, progesterone, testosterone, aldosterone, cortisol
Amines	Melatonin, epinephrine, norepinephrine, thyroid hormone (thyroxine, triiodothyronine)
Peptides	Oxytocin, antidiuretic hormone, calcitonin, parathyroid hormone
Proteins	Growth hormone (somatotropin), insulin, prolactin, follicle-stimulating hormone, luteinizing hormone

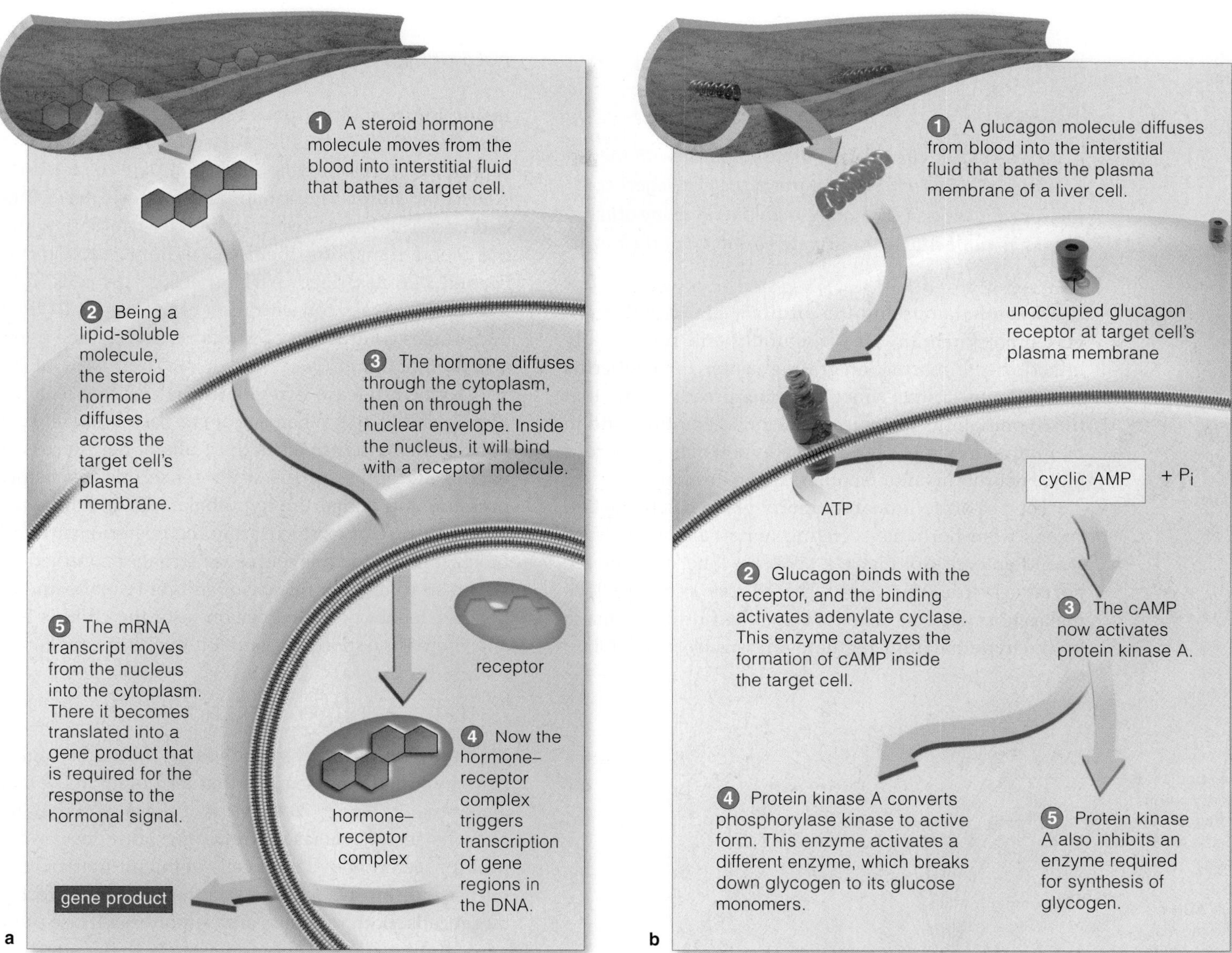

1 A steroid hormone molecule moves from the blood into interstitial fluid that bathes a target cell.
2 Being a lipid-soluble molecule, the steroid hormone diffuses across the target cell's plasma membrane.
3 The hormone diffuses through the cytoplasm, then on through the nuclear envelope. Inside the nucleus, it will bind with a receptor molecule.
receptor
hormone–receptor complex
4 Now the hormone–receptor complex triggers transcription of gene regions in the DNA.
5 The mRNA transcript moves from the nucleus into the cytoplasm. There it becomes translated into a gene product that is required for the response to the hormonal signal.
gene product
a
1 A glucagon molecule diffuses from blood into the interstitial fluid that bathes the plasma membrane of a liver cell.
unoccupied glucagon receptor at target cell's plasma membrane
ATP
cyclic AMP
+ Pi
2 Glucagon binds with the receptor, and the binding activates adenylate cyclase. This enzyme catalyzes the formation of cAMP inside the target cell.
3 The cAMP now activates protein kinase A.
4 Protein kinase A converts phosphorylase kinase to active form. This enzyme activates a different enzyme, which breaks down glycogen to its glucose monomers.
5 Protein kinase A also inhibits an enzyme required for synthesis of glycogen.
b

Figure 26.2 ***Animated!*** (**a**) Example of *steroid* hormone action inside a target cell. (**b**) Example of *peptide* hormone action. Cyclic AMP, a second messenger, relays the signal from a plasma membrane receptor into the cell.

Also, any given hormone may bind to receptors in many different tissues. The receptors can vary in their structure and binding may call up different responses. For example, ADH (antidiuretic hormone) secreted by the posterior lobe of the pituitary acts on kidney cells and causes the body to conserve water. But ADH also binds to other receptors in the wall of blood vessels. Here it causes the blood vessels to constrict. In many mammals (but not in humans), this is a major control over blood pressure. ADH also affects social behavior of some mammals by its effects on still other target cells in the brain. The diversity in response to ADH is a function of variation among ADH receptors and in what happens after the receptors bind this hormone.

Components of the endocrine system produce and secrete hormones, which the blood carries to target cells.

Hormone molecules reversibly bind with a target cell's receptors. Different receptors for different kinds of hormones are located at the surface of the plasma membrane, in the cytoplasm, or inside the nucleus.

Formation of a hormone–receptor complex leads to the conversion of a signal into a molecular form that brings about a response. Responses include adjusting specific metabolic reactions and altering transport mechanisms.

Most cells have receptors for many different hormones and most hormones have different kinds of receptors on different types of cells.

LINKS TO SECTIONS 17.3, 24.6, 25.6

26.2 The Hypothalamus and Pituitary Gland

In all vertebrates, the hypothalamus interacts with the pituitary gland as a major integrating center. Together, they coordinate and control the activities of many other organs, a number of which secrete hormones of their own.

The **hypothalamus** and the **pituitary gland** within the vertebrate forebrain are structurally and functionally linked as the master integrating center for endocrine and nervous system functions. You already know that the hypothalmus has roles in memory, emotion, and in maintaining homeostasis. Some hypothalamic neurons make neurotransmitters; others make hormones.

The pea-sized human pituitary gland has two lobes. Axons from hormone-secreting hypothalamic neurons extend into the pituitary's *posterior* lobe and release hormones from this lobe. The *anterior* pituitary lobe makes hormones of its own, but other hormones made by the hypothalamus regulate releases from this lobe.

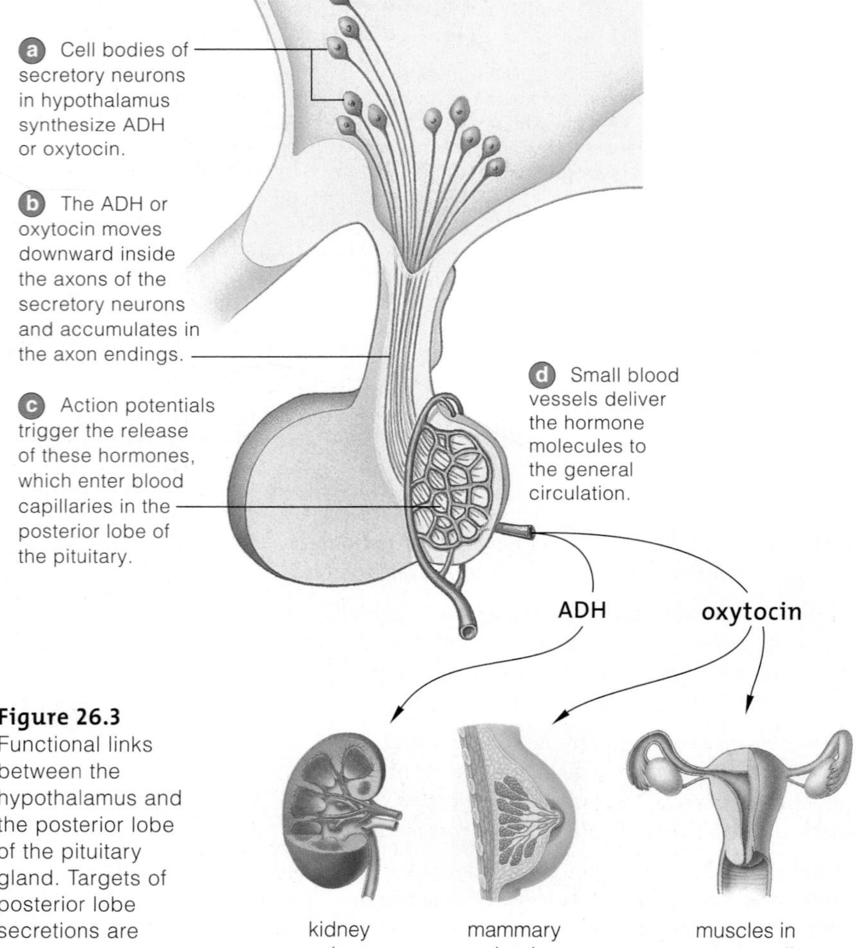
a Cell bodies of secretory neurons in hypothalamus synthesize ADH or oxytocin.
b The ADH or oxytocin moves downward inside the axons of the secretory neurons and accumulates in the axon endings.
c Action potentials trigger the release of these hormones, which enter blood capillaries in the posterior lobe of the pituitary.
d Small blood vessels deliver the hormone molecules to the general circulation.
ADH
oxytocin
kidney nephrons
mammary glands
muscles in uterus wall

Figure 26.3 Functional links between the hypothalamus and the posterior lobe of the pituitary gland. Targets of posterior lobe secretions are shown at right.

POSTERIOR PITUITARY SECRETIONS

Inside the pituitary's posterior lobe, axon endings of hypothalamic neurons lie adjacent to a capillary bed (Figure 26.3). Action potentials stimulate the release of ADH (antidiuretic hormone) and oxytocin from these endings. The hormones enter nearby capillaries, then the blood transports them throughout the body. As Section 24.6 described, ADH affects target cells in the kidneys; it causes concentration of the urine. Oxytocin encourages contraction of smooth muscle in the uterus and in milk ducts.

Oxytocin release is regulated by positive feedback during childbirth. When a woman goes into labor, the body of the fetus stretches the walls of her uterus and vagina. Receptors in the walls sense this mechanical pressure and signal the hypothalamus, which in turn causes release of oxytocin from the posterior pituitary. Increasing oxytocin levels cause stronger contractions, which stimulate the release of still more oxytocin. The positive feedback cycle goes on until the child is born and there is no more pressure on the birth canal.

ANTERIOR PITUITARY SECRETIONS

The anterior pituitary responds to hormonal signals from hypothalamic neurons that terminate in the stalk just above it (Figure 26.4). Most of these hypothalamic neurons secrete **releasers** (releasing hormones) which prompt secretion by target cells of the pituitary. Others release **inhibitors**, which have the opposite effect on target cells. Both releasors and inhibitors diffuse out of axon endings and into blood vessels that carry them into the anterior pituitary. Here, the hormones diffuse out into the pituitary tissue and regulate the secretion of six types of anterior lobe hormones. These anterior lobe hormones are:

Adrenocorticotropin	ACTH
Thyroid-stimulating hormone	TSH
Follicle-stimulating hormone	FSH
Luteinizing hormone	LH
Prolactin	PRL
Growth hormone (somatotropin)	GH (or STH)

ACTH acts on cells of the adrenal glands and TSH on the thyroid gland. FSH and LH act on the primary sex organs (ovaries and testes). They have roles in gamete formation and other aspects of reproduction in both sexes. Prolactin acts on diverse cells, but is best known for its role in stimulating and maintaining production of milk. GH, also called somatotropin, has targets in most tissues. Most notably, it affects the growth of bone and soft tissues.

ABNORMAL PITUITARY OUTPUTS

Pituitary disorders disrupt growth and development. Overproduction of growth hormone during childhood can lead to *pituitary gigantism*. Affected adults have normal proportions, but are unusually large (Figure 26.5*a*). Too little GH during childhood causes *pituitary dwarfism*. In this case, any affected adults have normal body form but are unusually small.

Excess secretion of GH during adulthood leads to *acromegaly*. GH regulates growth of connective tissues, but long bones cannot increase in length after puberty, so affected people do not get taller. Their hands and feet become enlarged and swollen. Facial features are distorted by overgrowth of cartilage and bone (Figure 26.5*b*,*c*). The skin thickens, and the liver, heart, kidney, and other internal organs may enlarge.

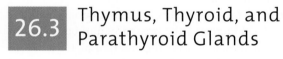

The posterior lobe of the pituitary gland secretes ADH and oxytocin, two hormones produced by the hypothalamus.

The anterior lobe of the pituitary produces and secretes ACTH, TSH, FSH, LH, PRL, and GH. Secretion is regulated by hypothalamic hormones.

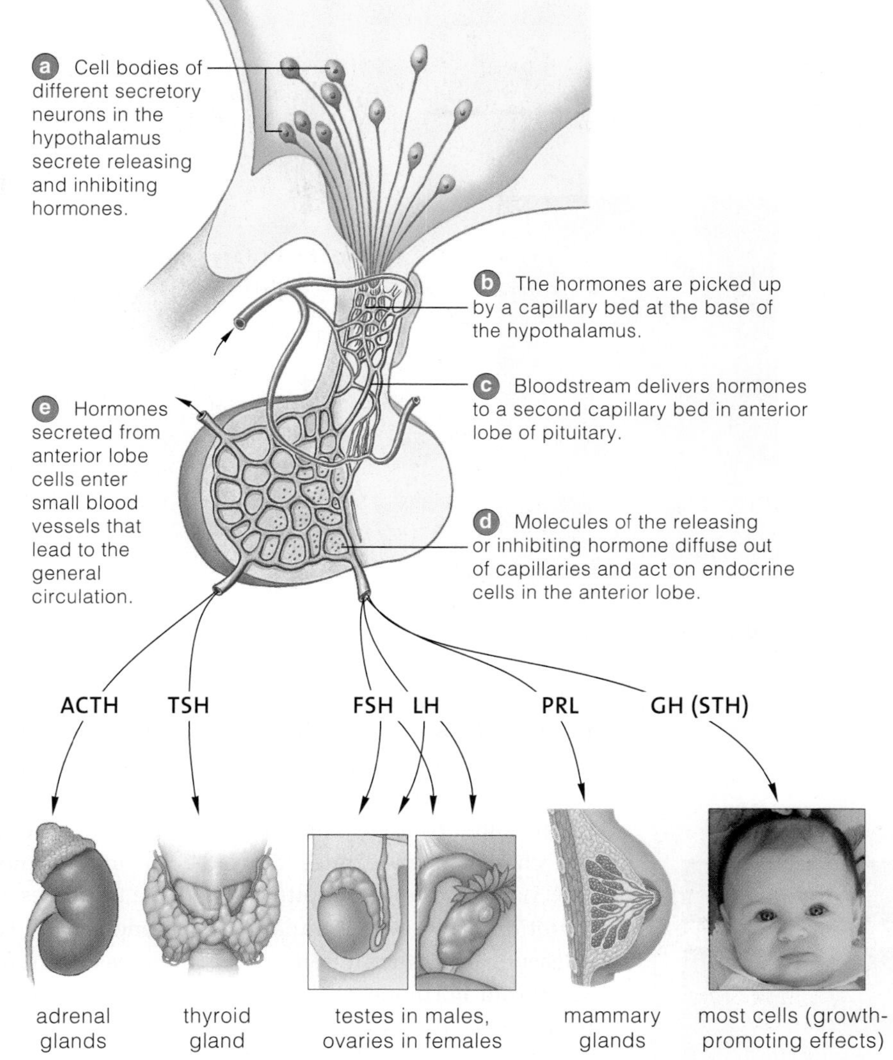

a Cell bodies of different secretory neurons in the hypothalamus secrete releasing and inhibiting hormones.
b The hormones are picked up by a capillary bed at the base of the hypothalamus.
c Bloodstream delivers hormones to a second capillary bed in anterior lobe of pituitary.
d Molecules of the releasing or inhibiting hormone diffuse out of capillaries and act on endocrine cells in the anterior lobe.
e Hormones secreted from anterior lobe cells enter small blood vessels that lead to the general circulation.
ACTH
TSH
FSH
LH
PRL
GH (STH)
adrenal glands
thyroid gland
testes in males, ovaries in females
mammary glands
most cells (growth-promoting effects)

Figure 26.4 Links between the anterior pituitary gland and the hypothalamus.

26.3 Thymus, Thyroid, and Parathyroid Glands

Thymus gland secretions help T cells mature. Hormones produced by the thyroid gland have roles in development and metabolism. A feedback loop to the pituitary and hypothalamus controls their secretion. Parathyroid glands help to regulate calcium levels.

The **thymus gland** lies beneath the upper part of the breastbone or sternum (Figure 26.1). The hormones it secretes help infection-fighting T cells mature (Section 23.5). The thymus gland reaches its maximum volume during late adolescence, then shrinks. However even in adults, it helps to regulate immune function.

FEEDBACK CONTROL OF THYROID SECRETION

The **thyroid gland** looks a bit like a butterfly resting at the base of the neck, in front of the trachea. It secretes two amines (thyroxine and triiodothyronine) that are referred to collectively as "thyroid hormone." Thyroid hormone is essential to normal development. In birds and mammals, it also regulates metabolic rates.

Thyroid hormone levels are controlled by negative feedback loops to the pituitary and hypothalamus. In

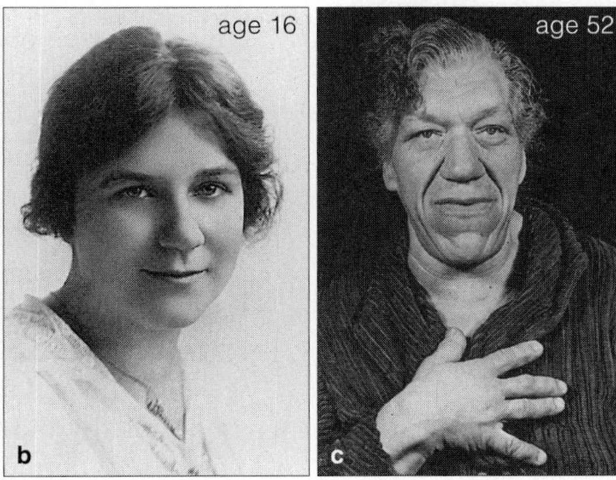

Figure 26.5 Abnormal pituitary outputs. (**a**) Mother with her son, who is affected by pituitary gigantism. At age 12, he already stands six feet five inches tall. (**b**,**c**) Symptoms of agromegaly usually appear in middle age. Excess growth hormone secretion thickens the fingers, enlarges ears, lips, and nose, and makes the brow and chin protrude.

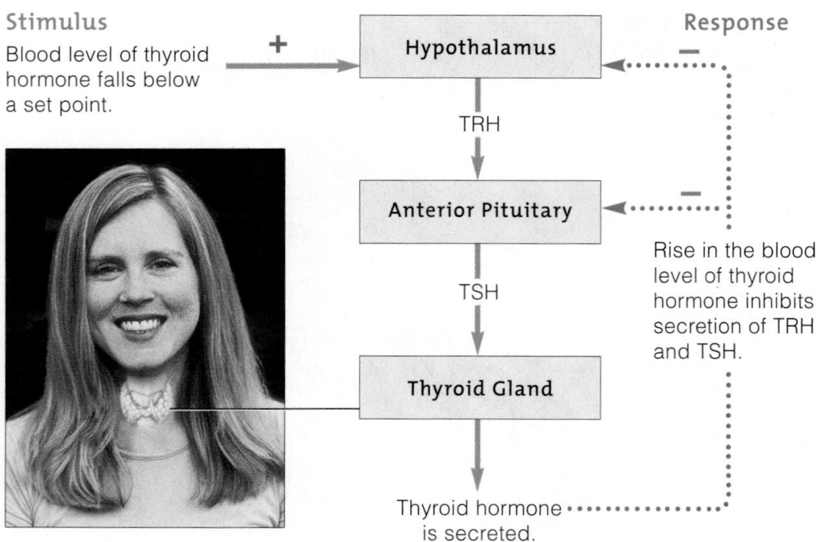

Figure 26.6 Feedback loop that governs thyroid hormone secretion.

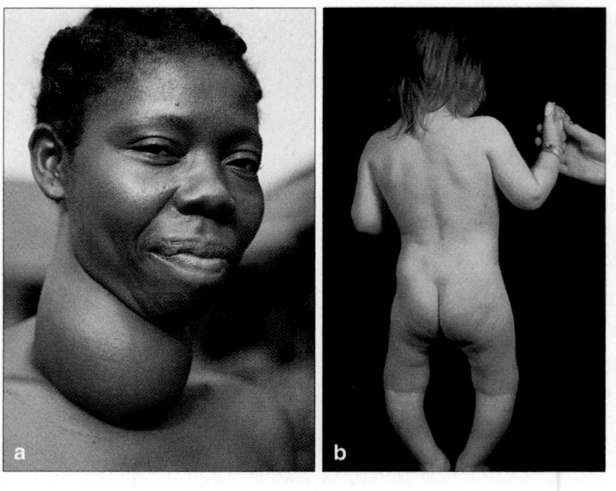

Figure 26.7 (**a**) A case of goiter caused by a diet low in the micronutrient iodine. (**b**) A child affected by rickets shows the characteristic bowed legs.

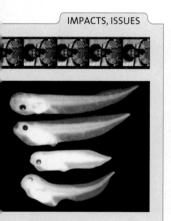

negative feedback, any increase in a hormone's level above a set point inhibits secretion of that hormone. Figure 26.6 shows what happens when the blood level of thyroid hormone falls. The hypothalamus senses the change and secretes thyroid-releasing hormone (TRH). In the anterior pituitary, TRH stimulates the secretion of thyroid-stimulating hormone (TSH). This releaser causes the thyroid gland to increase secretion of thyroid hormone, which raises concentration of the hormone in the bloodstream back to the set point. In response, the hypothalamus and pituitary slow their secretion of TRH and TSH.

The importance of feedback control of hormones is brought into sharp focus by cases where controls fail. Thyroid hormone incorporates the mineral iodine, which must be obtained from the diet. Low levels of iodine cause a thyroid enlargement known as *goiter* (Figure 26.7*a*). Goiter occurs because without iodine, the thyroid hormone levels in blood fall below a set point. This decline causes the pituitary to secrete TSH, which stimulates the thyroid. But without iodine, the thyroid gland cannot produce the thyroid hormone that could turn off TSH secretion. Over time, the lack of thyroid hormone causes the anterior pituitary to secrete more and more TSH, which results in excessive growth of thyroid tissue.

Thyroid hormone deficiency, or *hypothyroidism,* can disrupt development, slow growth, and delay sexual maturation. Worldwide, maternal iodine deficiency during pregnancy remains the most common cause of preventable mental impairment in newborns. Eating iodized salt is an inexpensive way to prevent dietary hypothyroidism, but such salt is not readily available in some regions of less-developed countries.

IMPACTS, ISSUES

Pollutants can disrupt thyroid function and interfere with frog metamorphosis. The uppermost *Xenopus laevis* tadpole in this photograph was raised in water from a lake with few deformed frogs. Tadpoles below it were raised in water taken from three lakes with increasingly higher concentrations of dissolved chemicals. Later tests showed that giving thyroid hormone to developing tadpoles lessens or eliminates some of the pollutant-associated deformities.

PARATHYROID GLANDS AND CALCIUM

Four **parathyroid glands** are located on the thyroid's posterior surface. They release parathyroid hormone (PTH) in response to a decline in the level of calcium in blood. Calcium ions, remember, are essential to many physiological and metabolic processes.

The main targets of PTH are cells in bones and the kidneys. PTH increases the breakdown of bone, which releases calcium ions into the interstitial fluid. From there they enter the bloodstream. In the kidneys, PTH increases calcium reabsorption, so less is lost in urine. PTH also causes activation of vitamin D. This vitamin helps the the intestine take up calcium from food.

Children who do not get enough vitamin D absorb too little calcium to build healthy new bone. Their low blood calcium level also causes oversecretion of PTH, which results in the breakdown of existing bone. The resulting disorder is *rickets*. Bowed legs and pelvic deformities are common symptoms (Figure 26.7*b*).

In many vertebrates, calcitonin, a peptide hormone secreted by the thyroid gland, opposes the effects of PTH. In humans, the thyroid gland has little role in calcium regulation. The gland can be removed with no apparent effect on blood calcium levels.

The thymus gland influences immune function.

Secretion of thyroid hormone is regulated by feedback loops to the hypothalamus and anterior pituitary gland. Thyroid hormone acts in development and affects metabolic rates. It includes iodine, which must be obtained in the diet.

Parathyroid glands secrete PTH, the main regulator of calcium levels in blood.

26.4 Adrenal Glands and Stress Responses

Adrenal gland secretions are governed by negative feedback loops. Signals from the nervous system can cause some secretions to rise in times of stress.

Humans have a pair of adrenal glands, one on top of each kidney (Figure 26.8). Cells of the **adrenal cortex**, the gland's outer layer, secrete the steroid hormones cortisol and aldosterone. As described in Section 24.5, aldosterone acts on the kidneys to regulate sodium and water reabsorption. Negative feedback loops with the anterior pituitary and hypothalamus maintain blood cortisol levels. Figure 26.8 shows what happens when the level of cortisol in blood decreases below the set point. This decline triggers secretion of corticotropin releasing hormone (CRH) by the hypothalamus. CRH causes cells in the anterior pituitary to secrete ACTH (adrenocorticotropin). ACTH in turn causes cells of the adrenal cortex to secrete cortisol.

Secretion of CRH and ACTH continues until the blood level of cortisol rises above the set point. Then, the hypothalamus and pituitary slow their secretions, and cortisol secretion declines in response.

Cortisol helps maintain the blood level of glucose. It causes stored glycogen breakdown by the liver, and breakdown of fatty acids and proteins in other tissues. Released fatty acids and amino acids that enter the blood serve as alternative energy sources for the body (Section 6.5). In this way, cortisol maintains a steady supply of glucose to fuel brain activity.

With injury, illness, or anxiety, the nervous system overrides the feedback loop described above and blood levels of cortisol soar. This helps ensure that the brain gets adequate glucose at times when circumstances make normal feeding activity less likely. Cortisol also helps to keep inflammatory responses in check.

Stress, excitement, or danger also triggers hormone secretion by the adrenal medulla, the inner part of an adrenal gland. Think back to the fight–flight response (Section 25.5). Sympathetic signals flow to the adrenal medulla and elicit the secretion of norepinephrine and epinephrine. These molecules enter the blood and are carried throughout the body. Their effects on target organs are like that of direct sympathetic stimulation.

In the short term, a fear or stress response can be highly adaptive, as when it allows an animal to flee from and escape a predator. *Long-term* stress causes problems. High cortisol levels disrupt the production and release of other hormones, suppress the immune system, and interfere with memory. We observe the impact of elevated cortisol levels in people affected by

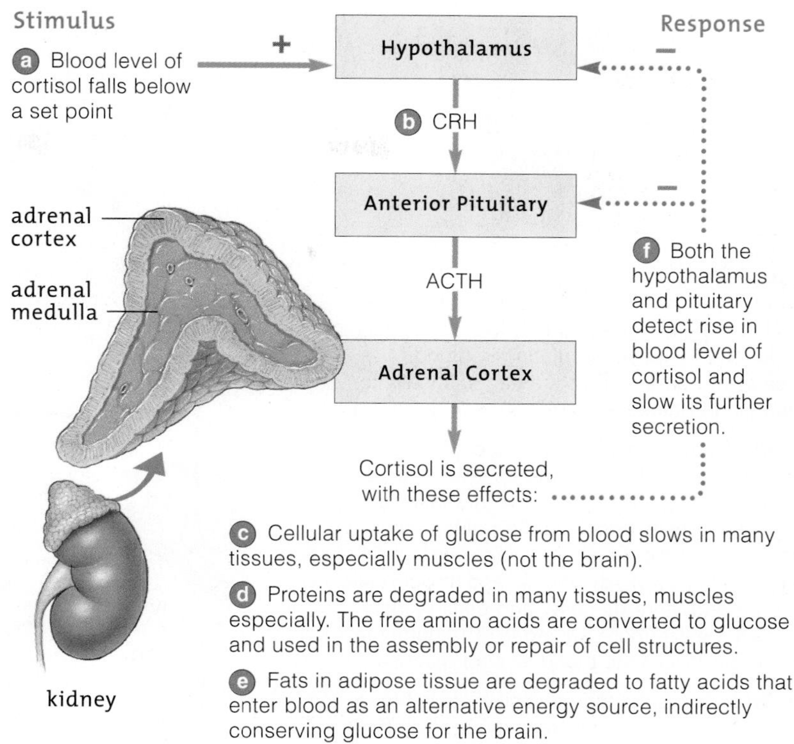

Stimulus
a Blood level of cortisol falls below a set point
+
Hypothalamus
b CRH
Anterior Pituitary
ACTH
Adrenal Cortex
Response
–
–
f Both the hypothalamus and pituitary detect rise in blood level of cortisol and slow its further secretion.
adrenal cortex
adrenal medulla
kidney
Cortisol is secreted, with these effects:
c Cellular uptake of glucose from blood slows in many tissues, especially muscles (not the brain).
d Proteins are degraded in many tissues, muscles especially. The free amino acids are converted to glucose and used in the assembly or repair of cell structures.
e Fats in adipose tissue are degraded to fatty acids that enter blood as an alternative energy source, indirectly conserving glucose for the brain.

Figure 26.8 ***Animated!*** Structure of the human adrenal gland. One gland rests on top of each kidney. The diagram shows a negative feedback loop that governs cortisol secretion.

Cushing syndrome. Symptoms include a puffy, rounded "moon face" and a fatty trunk. Blood sugar and blood pressure become abnormally high. White blood cell counts are low and infections are common. Women's menstrual cycles are erratic or nonexistent. Men are often impotent. Use of cortisone, a drug prescribed to relieve pain, inflammation, and other ailments, has a similar effect. The body converts cortisone to cortisol.

Persistent social stress that keeps cortisol levels high may harm human health. In social primates, such as baboons, low status is correlated with high cortisol levels and poor health. As a group, low income people have more health problems than those who are well off—even after factoring out obvious causes, such as variations in diet and access to health care. Could higher than normal cortisol levels be an important link between poverty and poor health? Maybe.

LINKS TO SECTIONS 6.5, 17.3, 25.6

The adrenal cortex secretes cortisol. Negative feedback loops to the hypothalamus and pituitary govern secretion, but nervous system stimulation can override these controls.

Secretion by the adrenal medulla is triggered by nervous signals and prepares the body to fight or run away.

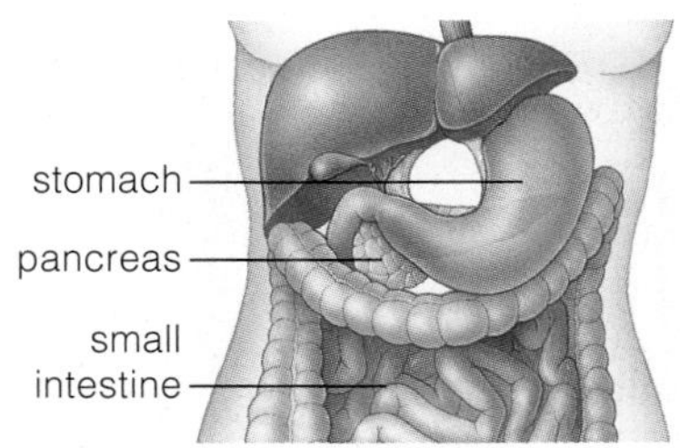

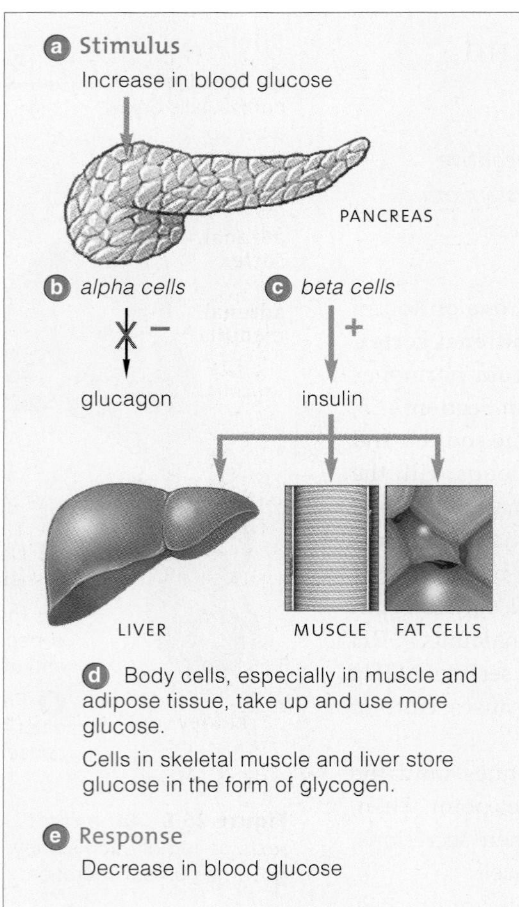

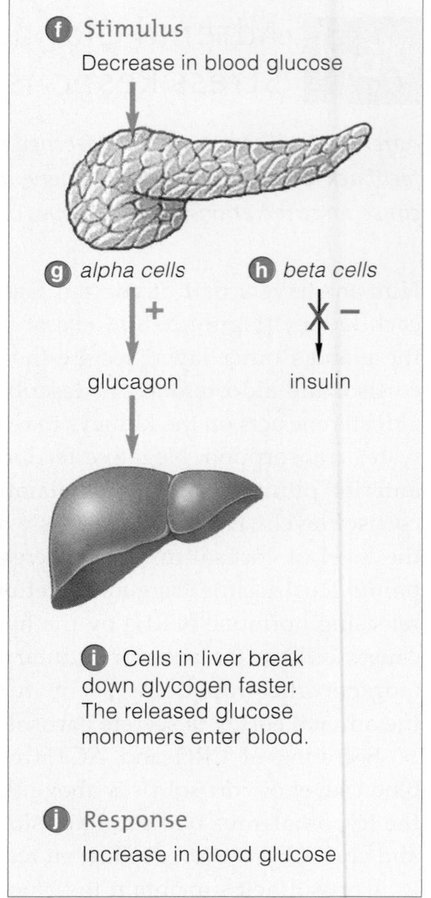

Figure 26.9 ***Animated!*** *Above*, location of the pancreas. *Right*, how cells that secrete insulin and glucagon respond to a change in the level of glucose circulating in blood. These two hormones work antagonistically to maintain the glucose level in its normal range.

(**a**) *After* a meal, glucose enters blood faster than cells can take it up and its level in blood increases. In the pancreas, the increase (**b**) stops alpha cells from secreting glucagon and (**c**) stimulates beta cells to secrete insulin. In response to insulin, (**d**) adipose and muscle cells take up and store glucose, and cells in the liver synthesize more glycogen. The outcome? (**e**) Insulin *lowers* the blood level of glucose.

(**f**) *Between* meals, the glucose level in blood declines. The decrease (**g**) stimulates alpha cells to secrete glucagon and (**h**) slows the insulin secretion by beta cells. (**i**) In the liver, glucagon causes cells to convert glycogen back to glucose, which enters the blood. The outcome? (**j**) Glucagon *raises* the blood level of glucose.

LINKS TO SECTIONS 6.5, 17.3

26.5 The Pancreas and Glucose Homeostasis

The actions of many hormones are antagonistic—their effects oppose one another. Two pancreatic hormones, insulin and glucagon, act in this way. Together, they maintain a healthy glucose concentration in the blood.

The pancreas sits in the abdominal cavity, just behind the stomach. This organ has exocrine and endocrine functions. Exocrine cells, which make up the bulk of the organ, secrete digestive enzymes into the adjacent small intestine. The endocrine cells are grouped in clusters called pancreatic islets. Each islet has three types of hormone-secreting cells.

Alpha cells secrete the peptide hormone glucagon, which raises the concentration of glucose in the blood. Cells throughout the body take up glucose from the blood. When the glucose blood level falls below a set point, alpha cells secrete glucagon, and the liver is the main target. Section 26.1 explained how the binding of glucagon activates enzymes that break down glycogen into its glucose subunits.

Beta cells, the most abundant cells in the pancreatic islets, secrete insulin. This protein hormone causes the glucose concentration in the blood to decline. After a meal, high blood glucose stimulates the beta cells to release insulin. The hormone's targets are mainly liver, adipose tissue, and skeletal muscle cells. It stimulates muscle and fat cells to take up glucose. In all target cells, insulin activates enzymes with roles in protein and fat synthesis, and it inhibits enzymes that catalyze protein and fat breakdown.

Delta cells secrete somatostatin. This hormone helps control digestion and nutrient absorption. It can also inhibit the secretion of insulin and glucagon.

In summary, food intake and intracellular demands have the potential to cause the amount of glucose in the blood to move outside its ideal range. Insulin and glucagon keep blood glucose near optimal levels by their opposing effects, as shown in Figure 26.9.

Table 26.2	Some Complications of Diabetes
Eyes	Changes in lens shape and vision; damage to blood vessels in retina; blindness
Skin	Increased susceptibility to bacterial and fungal infections; patches of discoloration; thickening of skin on the back of hands
Digestive system	Gum disease; delayed stomach emptying that causes heartburn, nausea, vomiting
Kidneys	Increased risk of kidney disease and failure
Heart and blood vessels	Increased risk of heart attack, stroke, high blood pressure, and atherosclerosis
Hands and feet	Impaired sensations of pain; formation of calluses, foot ulcers; possible amputation of a foot or leg because of tissue death caused by poor circulation

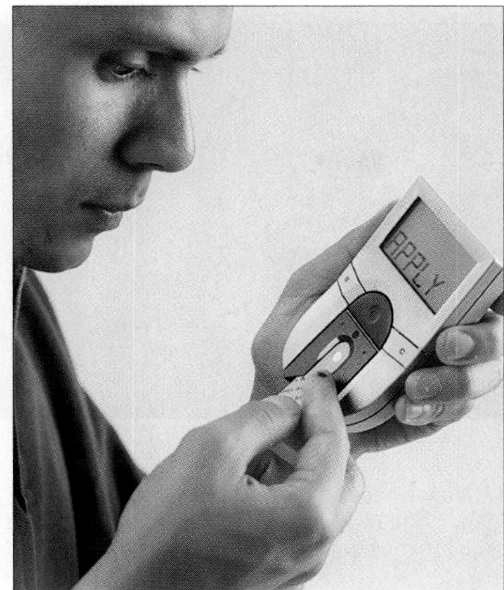

Figure 26.10 A diabetic man prepares to check his blood glucose level by placing a drop of blood in a glucometer. Compared to Caucasians, Hispanics and African Americans are about 1.5 times as likely to be diabetic. Asians and Native Americans are at even greater risk.

DIABETES MELLITUS

Glucose homeostasis is disrupted in *diabetes mellitus*, a metabolic disorder. Cells take up too little glucose, so blood sugar stays high. Tissues throughout the body may be damaged as a result (Table 26.2).

Type 1 diabetes occurs when an immune reaction, genetic defect, tumor, or toxin destroys beta cells. Loss of beta cells means that less or no insulin is produced. Type 1 diabetes accounts for only 5 to 10 percent of all reported cases. Symptoms most commonly appear in childhood and adolescence, which is why this type is known as juvenile-onset diabetes. Affected people monitor the amount of sugar in their blood and inject insulin to keep it near normal levels (Figure 26.10).

The more common form of diabetes mellitus is Type 2 diabetes, in which target cells fail to respond to insulin. As a result, blood sugar levels remain high. Symptoms typically appear in middle age, as insulin production declines. Genetics is a factor, but obesity increases the risk. Diet, exercise, and oral medications can control most cases of Type 2 diabetes. Some Type 2 diabetics also eventually require insulin injections.

Worldwide, rates of Type 2 diabetes are soaring. By one estimate, more than 150 million people are now affected. Preventing diabetes and its complications is among the most pressing public heath priorities.

Glucagon acts to raise blood sugar, insulin to lower it. Both of these hormones are produced by cells in pancreatic islets. Diabetes mellitus occurs when too little insulin is produced or target cells do not respond normally to insulin.

26.6 Hormones and Reproductive Behavior

Hormones influence growth, development, reproduction, and behavior. Environmental factors, such as daylength, can affect the onset and rates of hormone secretion in ways that regulate these aspects of the life cycle.

THE GONADS

The gonads are primary reproductive organs: testes (singular, testis) in males and ovaries in females. The testes and ovaries synthesize the same sex hormones—estrogens, progesterone, and testosterone—but relative amounts vary. Ovaries produce mostly estrogen and progesterone; the testes produce mainly testosterone. All are steroid hormones.

Synthesis of sex hormones steps up during **puberty**, the period of sexual maturation. In females, estrogen secretion results in development of female secondary sexual traits, such as enlarged breasts and wider hips. Estrogens and progesterone also regulate maturation of eggs and prepare the body for pregnancy.

The small amount of testosterone synthesized by a woman's ovaries influences *libido*—a desire for sex. Removal of the ovaries, as for some types of cancer treatment, decreases female libido.

In males, testosterone production rises at puberty. The hormone causes development of male secondary sexual traits and regulates production of sperm. The small amount of estrogen produced by the testes may play an essential role in sperm development.

We will return to the subject of sex hormones and their role in human reproduction in Chapter 27.

IMPACTS, ISSUES

Dioxin is a by-product of a variety of industrial processes. It is pervasive in the environment. The first evidence of its adverse effects came after the Vietnam War. Many servicemen had handled Agent Orange. This defoliant was used to clear dense tropical vegetation that kept enemy forces hidden from aerial surveillance. As it turned out, dioxin was a contaminant of Agent Orange. Those who had handled it developed diabetes twice as often as their fellow servicemen did. Given such evidence, how might you avoid dioxin? One way is to limit animal fats in the diet; dioxin accumulates in fatty tissues.

LINK TO SECTION 25.6

Figure 26.11 Male white-throated sparrow, belting out a song that began, indirectly, with an environmentally induced decrease in melatonin secretion from his pineal gland.

THE PINEAL GLAND

The **pineal gland** lies deep within the vertebrate brain (Figures 25.15 and 26.2). It secretes the amine hormone melatonin, which serves as part of an internal timing mechanism, or **biological clock**. Light sets this clock by slowing melatonin production. The amount of daylight varies seasonally and affects melatonin synthesis.

Variations in melatonin levels influence seasonal behavior in many species. For example, in some male songbirds long winter nights lead to an increase in melatonin production. The rise in melatonin indirectly prevents singing and other courtship behaviors by slowing testosterone secretion. As long as the levels of this male sex hormone stay low, the male bird will not sing. In the spring, there are fewer hours of darkness and the melatonin levels in the blood decline. With the inhibitory effects of melatonin lifted, the bird's testes secrete additional testosterone. The result is the chorus of birdsong that we enjoy in the spring (Figure 26.11).

Melatonin may also affect sex hormone production in humans. A drop in the blood level of melatonin is correlated with the onset of puberty and may trigger it. Also, certain endocrine disorders that increase the production of melatonin delay puberty. Others that lower melatonin production accelerate it.

Melatonin regulates human sleep–wake cycles, in part by affecting body temperature. Just after sunrise, melatonin secretion decreases, body temperature rises, and we awaken. At night, melatonin secretion causes body temperature to decline and we become sleepy. Because cycles of melatonin secretion that affect sleep–wake rhythms are set by exposure to light, travelers who fly across many time zones are advised to spend some time in the sun to reset their internal clock and minimize the effects of jet lag.

IMPACTS, ISSUES

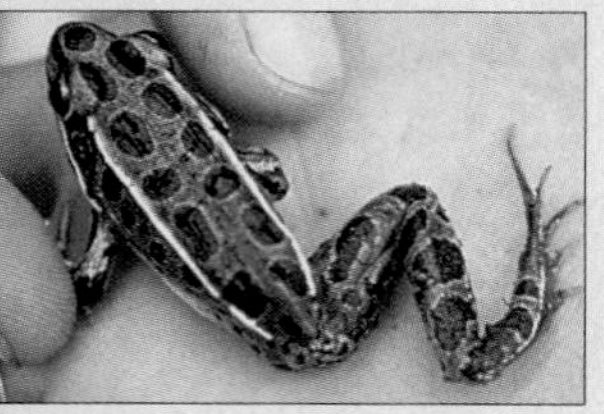

Deformed frogs are turning up all over the world. Many factors are involved, including increased levels of UV radiation, parasites, and water pollutants. A recent study found that malformed frogs have lowered levels of sex hormones and the releasing hormone GnRH. Another study shows that frogs infected with parasitic trematodes are more likely to be deformed if they live near runoff from agricultural fields. Both studies point to endocrine disruptors.

Seasonal affective disorder (SAD) affects some people when days get shorter. They are tired and depressed, and crave carbohydrates. These "winter blues" might develop because a biological clock is not functioning properly. Most people synthesize the same amount of melatonin year round, regardless of the daylength. However, people with SAD tend to secrete more in the winter. Some studies have found that carefully timed use of artificial light can lower melatonin levels in the winter and help alleviate SAD symptoms.

Conversely, staring at a bright computer screen late into the night or routinely working night shifts in an artificially lit setting can disrupt rhythms of melatonin secretion. The long-term health effects, if any, of this hormonal disruption are unknown. However, given what we know about melatonin's influence on mood and on other hormones, researchers urge caution.

Gonads produce hormones that promote development of sexual characteristics and have roles in sexual reproduction. Ovaries produce mostly progesterone and estrogens. Testes produce mainly testosterone.

Melatonin secreted by the pineal gland may influence the onset of puberty. It regulates biological clock activity, and production is inhibited by a well-lit environment.

Summary

Section 26.1 All vertebrates have an organ system of endocrine glands and cells that secrete hormones. In most cases, hormones travel through the bloodstream to nonadjacent target cells. A cell is a target of a hormone if it has receptors that can bind that hormone. Table 26.3 summarizes the hormones covered in this chapter.

Steroid hormones are lipid soluble; they can enter a target cell and interact directly with DNA. Peptide and protein hormones bind to membrane receptors at the cell surface. Binding commonly leads to the formation of a second messenger molecule. The second messenger causes a series of enzyme activations in the cytoplasm.

BiologyNow

Compare the mechanisms of steroid and protein hormones by viewing the animation on BiologyNow.

Section 26.2 The hypothalamus, a forebrain region, is structurally and functionally linked with the pituitary gland as a major center for homeostatic control.

The posterior pituitary releases antidiuretic hormone (ADH) and oxytocin made by cells in the hypothalamus. Oxytocin causes contraction of smooth muscle in milk ducts and the uterus. ADH acts on kidney tubules and promotes concentration of the urine.

The anterior pituitary makes and secretes its own hormones: Adrenocorticotropin (ACTH) tells the adrenal cortex to secrete cortisol, thyroid-stimulating hormone

(TSH) calls for thyroid hormone secretion, and follicle-stimulating hormone (FSH) and luteinizing hormone (LH) stimulate hormone production by male and female gonads and have roles in gamete formation. Prolactin (PRL) stimulates milk production by the mammary glands. Growth hormone (GH) has growth-promoting effects on cells throughout the body.

BiologyNow
Use the animation on BiologyNow to study how the hypothalamus and pituitary interact.

Section 26.3 The thymus functions in immunity.

A negative feedback loop to the anterior pituitary gland and hypothalamus governs the thyroid hormone. Iodine is required for normal thyroid gland function.

The parathyroid glands are the main regulators of calcium levels in the blood. They release parathyroid hormone (PTH) in response to low calcium levels. PTH acts on bone cells and kidney cells in ways that raise calcium levels in the blood.

Section 26.4 Secretion of aldosterone by the adrenal cortex helps regulate urine concentration. Cortisol secretion by the adrenal cortex is governed by a negative feedback loop to the anterior pituitary gland and hypothalamus. In times of stress, the central nervous system can override the feedback controls, allowing cortisol levels to rise.

The adrenal medulla secretes epinephrine and norepinephrine; they have the same effects on organs as sympathetic nervous stimulation.

BiologyNow
Watch the animation on BiologyNow to see how cortisol levels are maintained by negative feedback.

Table 26.3 Examples of Major Human Endocrine Glands and Hormones

Source	Hormone	Primary Actions
Pituitary gland		
Posterior lobe	Antidiuretic hormone (ADH)	Induces water conservation in the kidneys, concentrates the urine
	Oxytocin (OCT)	Induces contractions in milk ducts and the smooth muscle of uterus
Anterior lobe	Follicle-stimulating hormone (FSH)	Stimulates sex hormone secretion, egg maturation; sperm formation
	Luteinizing hormone (LH)	Stimulates sex hormone secretion, release of egg and sperm
	Thyroid-stimulating hormone (TSH)	Stimulates release of thyroid gland hormones
	Prolactin (PRL)	Stimulates and sustains milk production
	Growth hormone (GH) or somatotropin	Promotes growth in young; induces protein synthesis, cell division; has roles in glucose, protein metabolism in adults
Thymus gland	Thymosin, thympoetin	Regulate white blood cell (T cell) maturation
Thyroid gland	Thyroid hormone	Regulates metabolism; has roles in growth, development
Parathyroid glands	Parathyroid hormone	Elevates calcium concentration in blood
Adrenal glands		
Adrenal cortex	Aldosterone	Increases sodium retention, concentrates the urine, raises blood pressure
	Cortisol	Increases blood glucose level, alters metabolic activity throughout body
Adrenal medulla	Norepinephrine, epinephrine	Increase heart rate, blood pressure, and blood glucose
Pancreatic islets	Insulin	Lowers blood glucose concentration
	Glucagon	Increases blood glucose concentration
	Somatostatin	Inhibits digestion
Gonads	Testosterone	In males, promotes sperm formation and development of male secondary sexual characteristics; in females, has role in libido
	Estrogens	In females, promotes egg maturation and development of female secondary sexual characteristics; in males, may have a role in sperm maturation
	Progesterone	In females, prepares the uterus for pregnancy; no significant effect in males
Pineal gland	Melatonin	Influences daily biorhythms, possibly onset of puberty

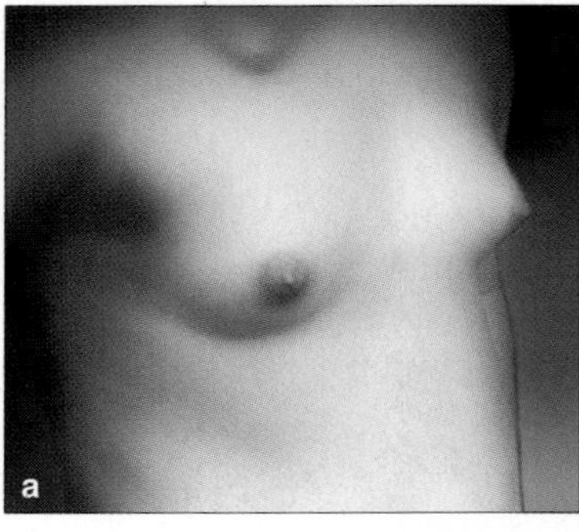

Figure 26.12 Two examples of hormone effects in humans. (**a**) At *twenty-three months*, this Puerto Rican girl already shows breast development. Industrial chemicals that mimic estrogen may be a factor. (**b**) Researcher Hiralal Maheshwari, with two men who have a heritable form of dwarfism.

Section 26.5 Insulin and glucagon are pancreatic hormones secreted in response to shifts in the blood level of glucose. Insulin stimulates glucose uptake by muscle and liver cells and thus lowers the blood glucose level. Glucagon stimulates the release of glucose, which increases blood levels. Diabetes mellitus is a metabolic disorder in which blood glucose is chronically high.

BiologyNow
Use the animation on BiologyNow to see how the actions of insulin and glucagon regulate blood sugar.

Section 26.6 Gonads (ovaries and testes) secrete estrogens, progesterone, and testosterone. These sex hormones are steroids that act in reproduction and in development of secondary sexual traits.

Melatonin secretion by the vertebrate pineal gland is part of a biological clock, a type of internal timing mechanism. In humans, it affects the daily sleep/wake cycle and perhaps the onset of puberty. Exposure to light suppresses melatonin production.

Self-Quiz

Answers in Appendix I

1. _______ are signaling molecules released from one type of cell that can alter target cell activities.
   a. Hormones　　c. Local signaling molecules
   b. Neurotransmitters　　d. all of the above

2. ADH and oxytocin are hormones produced in the hypothalamus but released from the _______ .
   a. anterior lobe of pituitary
   b. posterior lobe of pituitary
   c. pancreas
   d. pineal gland

3. Overproduction of _______ causes acromegaly and pituitary gigantism.
   a. growth hormone (GH)　　c. insulin
   b. thymosin　　d. melatonin

4. The _______ regulate(s) calcium levels in the blood.
   a. hypothalamus　　c. pineal gland
   b. pancreas　　d. parathyroid glands

5. _______ lowers blood sugar levels; _______ raises it.
   a. Glucagon; insulin　　c. Melatonin; insulin
   b. Insulin; glucagon　　d. Cortisol; glucagon

6. A rise in hormone concentration in the blood slows production of that hormone in a _______ feedback loop.
   a. positive　　c. long-term
   b. negative　　d. b and c

7. Match the hormone source listed at left with the most suitable description at right.
   _____ adrenal cortex　　a. affected by daylength
   _____ thyroid gland　　b. major control center
   _____ thymus gland　　c. increase blood calcium
   _____ parathyroid glands　　d. allows T cell maturation
   _____ pancreatic islets　　e. stress increases secretions
   _____ pineal gland　　f. output rises at puberty
   _____ hypothalamus　　g. hormones require iodine
   _____ testes　　h. insulin, glucagon

Additional questions are available on **BiologyNow™**

Critical Thinking

1. In the late 1990s pediatricians reported that the age at which girls in the United States begin to develop breasts has been declining. Evidence that chemicals play a role comes from Puerto Rico. These islands have the highest reported incidence of premature breast development in the world (Figure 26.12*a*). One study found that 68 percent of the girls who showed premature breast development also had high blood levels of phthalates—chemicals used in the manufacture of some plastics and pesticides. None of the girls in the control group had high phthalate levels.

Design an experiment for testing the hypothesis that exposure to phthalates may be contributing to early breast development among girls in the mainland United States.

2. In a remote village in Pakistan, men average about 130 centimeters (a little over 4 feet) tall, and the women are 115 centimeters (3 feet, 6 inches or so) tall. Northwestern University researchers visited the people, and realized that all were part of an extended family. The abnormality, a type of *dwarfism*, is one case of autosomal recessive inheritance (Figure 26.12*b*).

Biochemical and genetic analyses show that affected individuals have low levels of somatotrophin, yet their gene coding for this hormone is normal. The abnormality starts with a mutant gene that specifies a receptor for one of the hypothalamic releasing hormones. That receptor occurs on cells in the anterior lobe of the pituitary gland. Explain how the mutation causes the abnormality.

3. Maya is affected by Type 1 insulin-dependent diabetes. One day, she miscalculates and injects herself with too much insulin. She begins to feel confused and sluggish. She calls for medical assistance and injects herself with glucagon her doctor prescribed for such an emergency. An ambulance arrives and she is given dextrose (a sugar) intravenously. What caused her symptoms? How did the glucagon injection help?

27 REPRODUCTION AND DEVELOPMENT

IMPACTS, ISSUES

Mind-Boggling Births

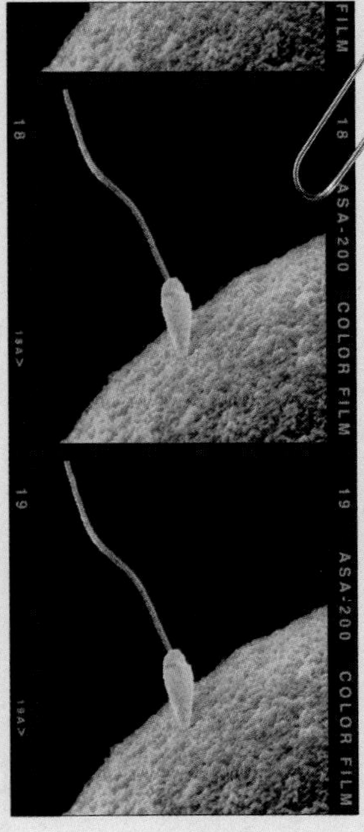

In December of 1998, Nkem Chukwu of Texas gave birth to octuplets ahead of schedule. The six girls and two boys survived their premature birth—the first octuplets to do so—but their *combined* weight was just over 4.5 kilograms (10 pounds). The smallest weighed 520 grams (less than a pound) and died of heart and lung failure six days afterward. The other newborns had to remain in the hospital for three months but are now healthy.

Chukwu had undergone hormone injections, which caused the maturation and release of many eggs at the same time. Doctors offered her the option of reducing the number of embryos but she chose to carry them all.

Multiple births are becoming increasingly common. In the past two decades, the number rose by almost 60 percent. The incidence of higher order multiple births (triplets or more) has quadrupled.

Delayed childbearing is one of the causes. A woman's fertility peaks during her mid-twenties. By the time she is thirty-nine years old, her chance of conceiving naturally has declined by about 50 percent. Even so, the number of first-time mothers who are more than forty years old has doubled in the past decade. Many probably would not have given birth if they had not chosen to use fertility drugs and other forms of reproductive interventions.

The rise in higher order multiple births worries some doctors. Carrying more than a single embryo increases the likelihood of miscarriage, premature delivery, and a surgical delivery by cesarean section. The oldest of the octuplets shown here was born naturally; all the others were delivered surgically. Compared to single births, the weights of multiple newborns are lower and mortality rates are higher. Their parents face additional physical, emotional, and financial burdens.

With this chapter, we examine how the offspring of sexually reproducing parents develop. We'll start with the principles that govern the reproduction and development of all animals, including humans. We will consider how their gametes and zygotes form, and how all of the specialized cells and structures of the adult arise. It is a developmental journey we all have made. Our ability to understand that journey and to manipulate it has profound implications for many generations yet unborn.

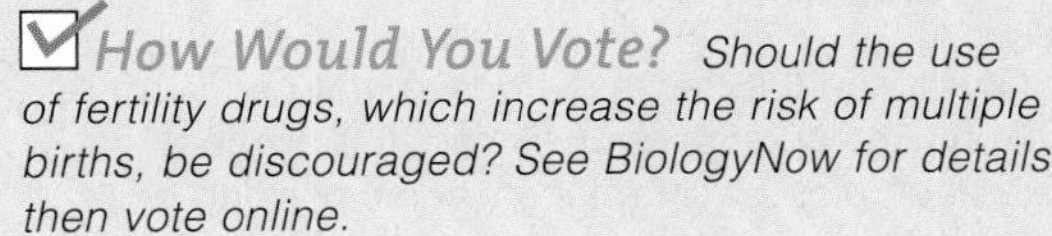

How Would You Vote? *Should the use of fertility drugs, which increase the risk of multiple births, be discouraged? See BiologyNow for details, then vote online.*

Key Concepts

BASIC PRINCIPLES

Nearly all animals engage in sexual reproduction, which produces genetically variable offspring. Patterns of development vary in their details, but the underlying processes are similar. All animals share certain master genes that map out body form.

HUMAN REPRODUCTIVE SYSTEMS

Male and female reproductive systems arise from the same embryonic structures. Each system consists of a pair of gamete-producing structures with accessory ducts and glands. Hormones govern gamete formation and other aspects of reproduction.

HUMAN DEVELOPMENT

Human development proceeds from a fertilized egg, to a ball of cells, to a bilateral embryo that shows the typical chordate features, to a recognizably human fetus. During prenatal development, the individual is nourished by substances that diffuse from the mother's bloodstream. Organs continue to develop after birth.

Links to Earlier Concepts

Section 7.5 introduced the advantages of sexual reproduction, and Section 7.7 described gamete formation. This chapter delves deeper into these topics. Mitosis and cytoplasmic division (7.3, 7.4) come into play, as do control of gene expression (10.6), apoptosis (17.5), and formation of tissues and organs (20.3). You will then recognize typical chordate (16.9) and amniote (16.12) features in human embryos, and revisit sex determination (8.4) and hormonal controls (26.2, 26.6).

The effects of sexually transmitted bacteria (14.2), protozoans (14.3), and viruses (14.5), including HIV (23.9), are also described in detail.

LINK TO SECTION 7.5

27.1 Methods of Reproduction

Sexual reproduction occurs in the life cycle of nearly all animals, even those that also reproduce asexually. When the environmental challenges vary from one generation to the next, benefits of producing variable offspring outweigh the associated costs.

SEXUAL VERSUS ASEXUAL REPRODUCTION

In earlier chapters, you considered the cellular basis of **sexual reproduction**. With this type of reproduction, meiosis and gamete formation typically occur in two prospective parents. At fertilization, a gamete from one parent fuses with a gamete from the other to form the zygote, or the first cell of the new individual. You also thought about **asexual reproduction**, whereby a single parent organism produces identical offspring by one of a variety of mechanisms. We now turn to just a few structural, behavioral, and ecological aspects of these two modes of animal reproduction.

Invertebrates often reproduce asexually. In sponges and cnidarians, new individuals can form by budding. Offspring form on the parent body, then detach and live on their own. Sponges also reproduce asexually by fragmentation. A piece breaks off and then grows into a new individual. Flatworms and echinoderms undergo transverse fission. The body splits and each part regenerates the missing regions (Figure 27.1).

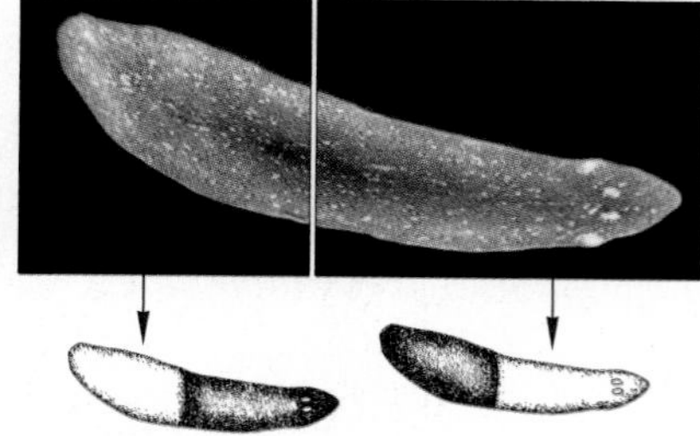

Figure 27.1 One of the flatworms (*Dugesia*) that can reproduce asexually by spontaneous fission. If it divides into two pieces, each piece will replace what is missing and will be genetically identical to the original flatworm.

In these cases of asexual reproduction, all offspring are genetically identical to the parent, unless there is a mutation. All have more or less the same phenotype. Phenotypic uniformity may increase offspring survival when each individual's gene-encoded traits are highly adapted to a relatively consistent set of environmental conditions. Under these circumstances, novel genetic combinations introduced by recombination are more likely to do harm than good.

Many animals live where opportunities, resources, and danger are quite variable. For the most part, they reproduce sexually, meaning female and male parents bestow different mixes of alleles on offspring. As you read earlier, in Section 7.7, the resulting variation in traits improves the odds that some offspring, at least, should survive and reproduce, especially if conditions change in the environment.

Many species switch between sexual and asexual reproduction. The plant-sucking insects called aphids are one example. In the summer, female aphids give birth to many daughters that arise by *parthenogenesis*, the development of offspring from an unfertilized egg. An aphid's daughters live on the same plant as their mother, or on an adjacent one, so they all experience more or less similar conditions. In autumn, males are produced and sexual reproduction yields genetically variable females. The females spend the winter in a resting state. In the spring, they seek out new plants on which to raise the new generation of genetically identical daughters.

Figure 27.2 Reproductive costs. (**a**) Colorful feathers of a male monal pheasant attract females. They also make males conspicuous to predators and to humans, who have now hunted the species to near extinction for its feathers. (**b**) A male sable antelope builds horns to fight off rivals.

COSTS OF SEXUAL REPRODUCTION

Separation into sexes has costs. For example, animals have to locate, recognize, and attract members of the opposite sex. Many expend energy in the production of sex attractants or receptors for these signals. Others invest in visual signals, such as bright feathers (Figure 27.2*a*), or in structures that allow the production of courtship calls or songs.

Fending off the competition for mates can also be costly. Many males invest in horns, spurs, or massive bodies to fight off rivals (Figure 27.2*b*).

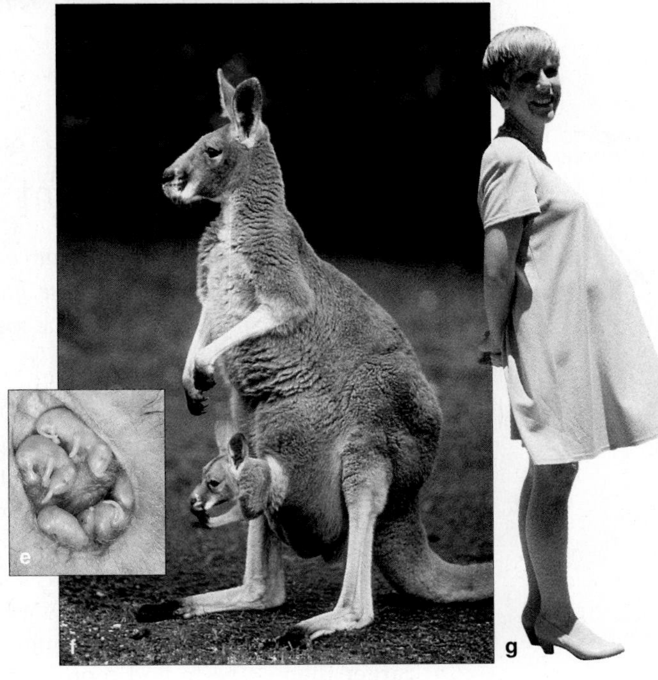

Figure 27.3 Adding to the examples in Chapter 16, another look at where some invertebrate and vertebrate embryos develop, how they are nourished, and how (if at all) parents protect them.

Snails (**a**) and spiders (**b**) release eggs from which the young later hatch. Snails are hermaphrodites and abandon their eggs. Spider eggs develop in a silk egg sac that the female anchors or carts around with her. Females often die soon after they make the sac. Some guard the sac, then carry spiderlings about for a few days and feed them.

(**c**) Ruby-throated hummingbirds and all other birds lay fertilized eggs with large yolk reserves. The eggs develop and hatch outside the mother. Unlike snails, one or both parent birds expend energy feeding and caring for the young.

(**d**) Embryos of most sharks, most lizards, and some snakes develop in their mother, receive nourishment continuously from yolk reserves, and are born live. Shown here, live birth of a lemon shark.

Embryos of marsupial and placental mammals draw nutrients from their mother's tissues as they develop. (**e**) In kangaroos and other marsupials, embryos are born at a relatively early stage and complete development in a pouch on the mother's ventral surface. Juvenile stages (joeys) continue to be nourished from mammary glands in the mother's pouch (**f**). (**g**) A human female, a placental mammal, retains a fertilized egg in her body. Her tissues nourish the developing individual until birth.

Assuring the survival of at least some offspring is also energetically costly. Many invertebrates, fishes, and amphibians release motile sperm and nonmotile eggs into the environment. If each individual were to release only *one* sperm or *one* egg at mating time, the chances of fertilization would not be very good. These animals invest energy in making many gametes, often thousands of them.

As another example, nearly all animals on land use internal fertilization, or the union of sperm and egg *within* the female body. They invest metabolic energy to construct elaborate reproductive organs, such as a penis and a uterus. A penis deposits sperm inside the female, and a uterus is a chamber in which an embryo develops inside the females of many mammals.

Finally, parents—especially mothers—often invest in nourishing their young (Figure 27.3). Most animal eggs include **yolk**, which is rich in lipids and proteins. Sea urchins eggs have very little yolk because they hatch in less than a day and larvae are immediately able to feed themselves. In contrast, mother birds lay notably yolky eggs. Nutrients in the yolk sustain a bird embryo through an extended period of development inside an eggshell that forms after fertilization.

Mammalian eggs have very little yolk. Instead, the individual develops inside the mother's body, where it is sustained by a direct exchange of substances with maternal tissues (Figure 27.3*e*–*g*). Your own mother put enormous demands on her body to protect and nourish you during your prenatal development.

Most animals reproduce sexually, at least part of the time. Sexual reproduction entails many costs, including those associated with attracting and keeping a mate, producing gametes, and nourishing the young as they develop.

A selective advantage—variation in traits among offspring —offsets the biological costs that are associated with the separation into two sexes.

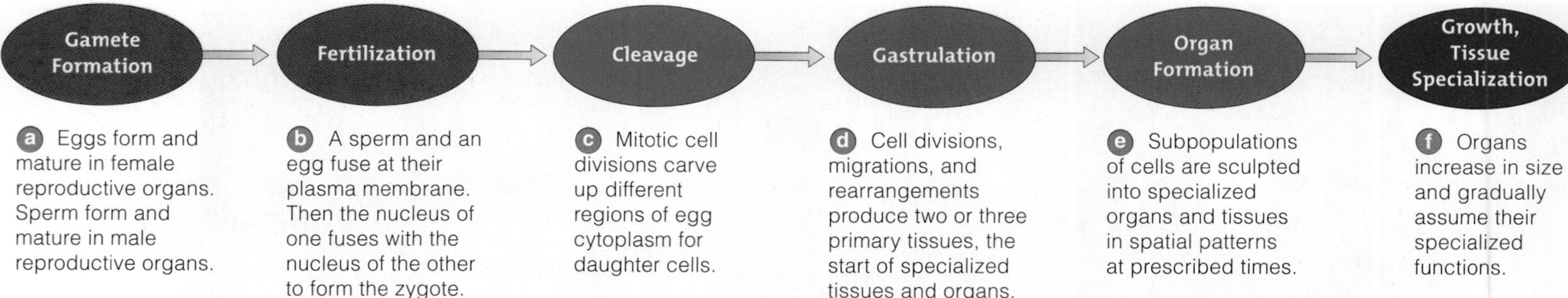

Figure 27.4 Overview of the stages of animal reproduction and development.

LINKS TO SECTIONS 7.2, 7.3, 7.7, 9.5, 20.3

27.2 Processes of Animal Development

Animal development proceeds from gamete formation and fertilization through cleavage, gastrulation, organ formation, then growth and tissue specialization.

Figure 27.4 is an overview of the six stages of animal reproduction and development. Figure 27.5 provides examples of these stages for one vertebrate, a frog.

During *gamete formation*, eggs or sperm develop in reproductive tissues or organs. Each sperm consists of the paternal DNA and a bit of cellular equipment that helps it reach and penetrate an egg. There is relatively little cytoplasm. An **oocyte**, or immature egg, is much larger than a sperm and more complex internally. As an oocyte matures, enzymes, mRNA transcripts, and other materials are made and stored in its cytoplasm.

Fertilization is the second stage. It starts as a sperm penetrates an egg. It ends when sperm and egg nuclei fuse, forming a zygote. A zygote's cytoplasm is almost entirely derived from the egg.

During *cleavage,* mitotic cell divisions increase the number of cells without adding to a zygote's original volume. These repeated cell divisions yield a **blastula**, a hollow ball composed of many smaller cells called blastomeres. Together, the blastomeres enclose a fluid-filled cavity or blastocoel (Figure 27.5*f*).

Materials are not randomly distributed in the egg cytoplasm, and during cleavage different blastomeres inherit different mRNAs. Which "maternal messages" are inherited helps determine the fate of a cell lineage. For example, cells of one lineage alone might receive maternal mRNA for a regulatory protein that turns on the embryo's gene for a particular hormone.

As cleavage draws to a close, cell division slows. The blastula undergoes *gastrulation*. This fourth stage is a time of structural reorganization. Cells become arranged into a **gastrula**, an embryonic form with two or three primary tissue layers (germ layers). Cellular descendants of these primary tissues give rise to all tissues and organs of the adult (Figure 27.5*g–j*).

Ectoderm, the outermost primary tissue, emerges first. It is the source of the nervous system and outer portions of the body coverings. Inner **endoderm** is the source of the gut's inner lining and organs derived from it. In most animal embryos, **mesoderm** forms in between the outer and inner primary tissue layers. It gives rise to muscles, to most of the skeleton, to the circulatory, reproductive, and excretory systems, and to connective tissues of the gut and integument.

After the primary tissue layers appear, their cells become distinct subpopulations. This marks the onset of *organ formation*. All cells of an embryo have the same number and kinds of genes, having descended from the same zygote. They all turn on genes that are essential for survival, such as those for histones and glucose-metabolizing enzymes. But from gastrulation onward, different cell lineages also engage in **selective gene expression**: They express some groups of genes and not others. This is the start of **cell differentiation**. By this process, cell lineages become specialized with regard to their composition, structure, and function.

Growth and tissue specialization is the sixth stage of animal development. Over time, tissues and organs mature in size, shape, proportion, and function. This final stage extends into adulthood.

HOW TISSUES AND ORGANS FORM

Tissues and organs develop in a programmed, orderly sequence by the process known as **morphogenesis**. During morphogenesis, the cells of various lineages divide, grow, migrate, and alter their size and shape. Tissues lengthen or widen and fold over. Cells die at specific times and in specific locations. Let's consider some examples of these processes.

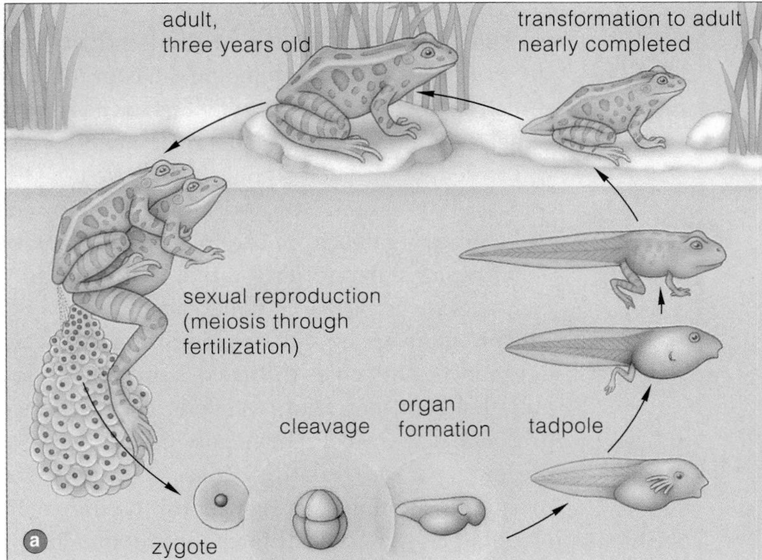

Figure 27.5 ***Animated!*** Reproduction and development of the leopard frog, *Rana pipiens*.

(**a**) We zoom in on the life cycle as a female releases her eggs into the surrounding water, and a male releases sperm over the eggs. A zygote forms at fertilization.

(**b–j**) Appearance of a frog embryo over time. (**b**) About one hour after fertilization, a surface feature—the gray crescent—appears opposite the point where the sperm entered. This region will become the "dorsal lip," the site where cells begin to migrate inward during gastrulation.

(**c–f**) Cleavage produces a blastula, a ball of cells that contains a fluid-filled cavity (a blastocoel).

(**g–i**) During gastrulation, cells actively migrate and rearrange themselves into three primary tissue layers. A primitive gut cavity in which internal organs will be suspended forms.

(**j**) As cell differentiation proceeds, a notochord, neural tube, and other organs form, moving the frog embryo on its way to becoming the larval stage called a tadpole.

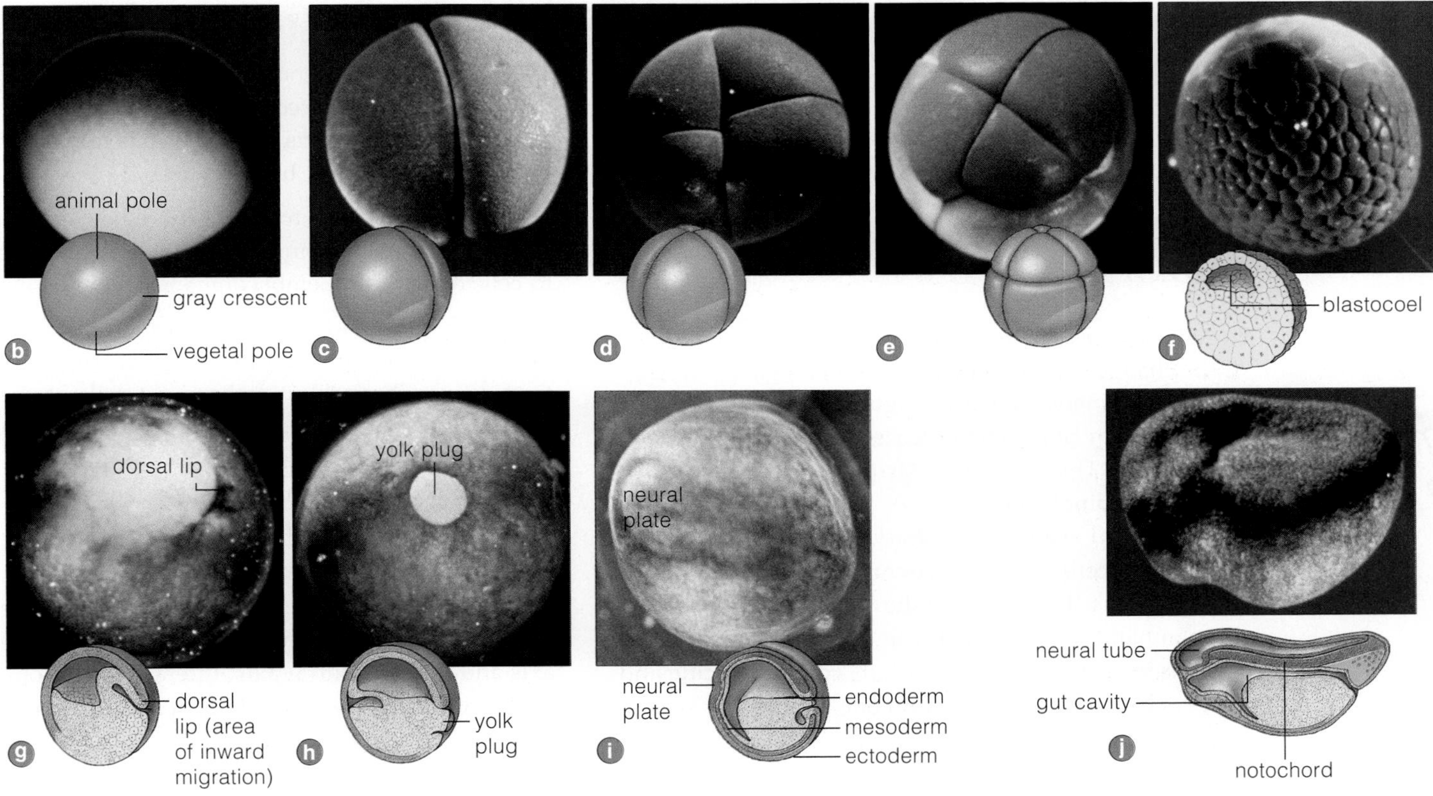

Some cells migrate along prescribed routes. When they reach their destination, these cells make connections with others already present. For example, embryonic neurons move into place during development of the nervous system. The direction and speed of neuron migration is determined by chemical gradients and the expression of adhesion proteins. A neuron may find its way to the proper position in the brain by responding to the "stickiness" of glial cells. Adhesion proteins on the neuron's plasma membrane stick to other proteins on the glial cells' surface. The neuron stops where these proteins are the most abundant.

Figure 27.6 *Animated!* How a neural tube forms. (**a**) As gastrulation ends, the ectoderm is a uniform sheet of cells. (**b**) Elongation of two bands of cells creates a slight indentation—the neural groove. (**c**) Microfilaments constrict in cells lining the groove, making the cells wedge-shaped. (**d**) As a result of the cellular shape changes, the sheet folds and its edges come together. (**e**) This forms the neural tube.

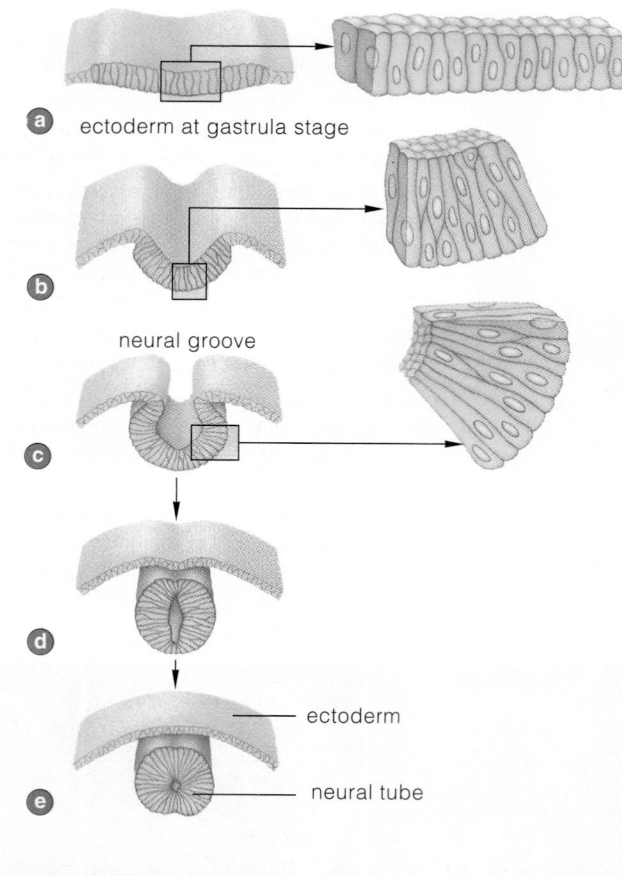

Many signals are short range, involving cell-to-cell contacts. For instance, signals that activate or inhibit genes for adhesion proteins and recognition proteins directly affect how cells interact in tissues and organs. Other signals cause cytoskeletal elements in the target cell to lengthen. Such signals can cause a cell to stick to neighbors, break free and move to a new place, or become segregated from the cells in adjoining tissue.

Other signals are long range. They act on control elements in the DNA of nonadjacent embryonic cells. **Morphogens** are among the long-range signals. These degradable molecules diffuse out of a signaling center, and their concentration weakens with distance. This gradient helps a cell chemically assess its position in the embryo and thus influences its fate. Differences in a signaling molecule's concentration induce lineages of cells positioned at different locations along certain gradients to read different parts of the same genome.

LINKS TO SECTIONS 10.6, 17.5

Sheets of cells expand, and some fold inward or outward. Cells change shape as microtubules elongate and rings of microfilaments constrict. Figure 27.6 illustrates how alterations in cell shape give rise to the neural tube of vertebrates. This embryonic structure is the start of the brain and spinal cord.

As a final example, *cell deaths help sculpt body parts.* Embryonic cells undergo programmed self-destruction by apoptosis. Figure 17.11 showed how recognizable human hands are sculpted from paddle-like structures as cells in webs between the digits commit suicide and fingers become separated from one another.

EMBRYONIC INDUCTION

With **embryonic induction**, the developmental fate of some cell lineage is affected by its exposure to gene products from cells in different tissues. For example, an embryonic cell may produce and secrete signaling molecules that cause changes in the developmental fate of adjacent cells. Or signals that diffuse through a cell's neighborhood may cause it to change in some way. Through these changes, specialized tissues and organs emerge from clumps of embryonic cells. Their emergence is called **pattern formation**.

A THEORY OF PATTERN FORMATION

The same kinds of genes are mapmakers for all major groups of animals. According to one theory of pattern formation, the products of these genes interact in ways that map out the basic body plan:

1. Tissues and organs form in ordered, spatial patterns as a result of cytoplasmic localization, short-range cell-to-cell contacts, and embryonic inductions.

2. Morphogens and other inducer molecules that form in certain tissues are long-range signals that weaken as they diffuse through an embryo. So cells at the start, middle, and end of the gradients get different chemical information, and they activate classes of **master genes** in sequence. The products of these genes interact with control elements to map out the body plan.

3. Products of **homeotic genes** and other master genes form when blocks of genes become turned on or off in different cells along the embryo's anterior–posterior axis and dorsal–ventral axis. Interactions among other gene products fill in the details of specific body parts.

In most animals, sexual reproduction involves six stages: gamete formation, fertilization, cleavage, gastrulation, organ formation, and growth and tissue specialization.

Selective gene expression is the basis of cell differentiation. It results in cell lineages with differing forms and functions.

Morphogenesis is a program of orderly changes in body size, shape, and proportion. It produces specialized tissues and organs at prescribed times in prescribed locations.

Master genes that turn on other genes and thus govern where body parts form are similar in all animals.

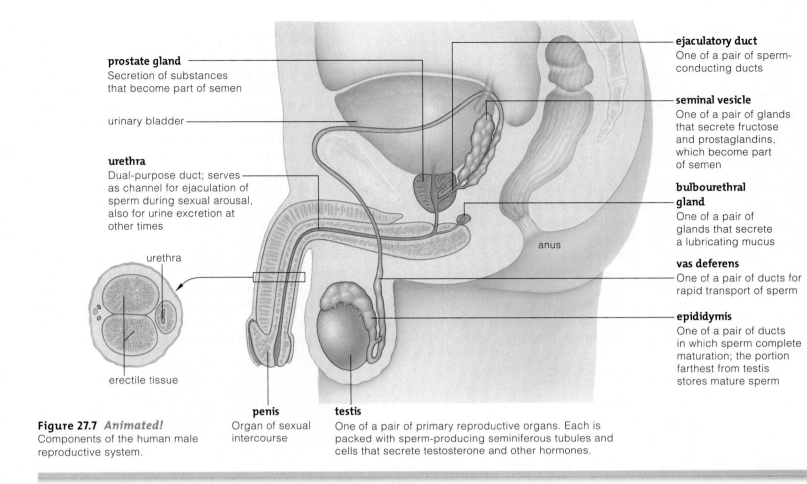

prostate gland
Secretion of substances that become part of semen

urinary bladder

urethra
Dual-purpose duct; serves as channel for ejaculation of sperm during sexual arousal, also for urine excretion at other times

urethra

erectile tissue

ejaculatory duct
One of a pair of sperm-conducting ducts

seminal vesicle
One of a pair of glands that secrete fructose and prostaglandins, which become part of semen

bulbourethral gland
One of a pair of glands that secrete a lubricating mucus

anus

vas deferens
One of a pair of ducts for rapid transport of sperm

epididymis
One of a pair of ducts in which sperm complete maturation; the portion farthest from testis stores mature sperm

penis
Organ of sexual intercourse

testis
One of a pair of primary reproductive organs. Each is packed with sperm-producing seminiferous tubules and cells that secrete testosterone and other hormones.

Figure 27.7 ***Animated!*** Components of the human male reproductive system.

27.3 Reproductive System of Human Males

So far, you have considered some principles of animal reproduction and development. The remainder of this chapter applies these principles to humans, starting with the male reproductive system.

Human males have two **testes** (singular, testis). These primary reproductive organs produce sperm and male sex hormones (Figures 27.7 and 27.8). Testes start to form on the wall of an XY embryo's abdominal cavity, as Figure 8.18 shows. Before birth, they descend into a scrotum, an outpouching of skin suspended below the pelvic region. Sperm production starts during puberty, when secondary sexual traits also emerge. Most boys enter puberty between ages twelve and sixteen. Signs include enlarging testes, growth spurts, hair sprouting on the face and pubic area, and a deepening voice.

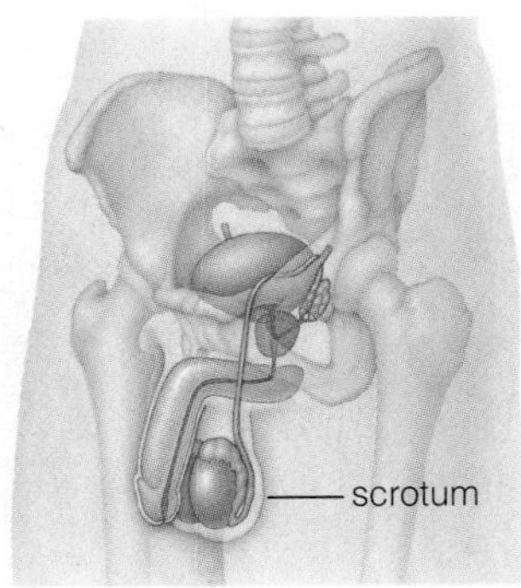

Figure 27.8 Location of a human male's reproductive system relative to the pelvic girdle and urinary bladder. The scrotum, an outpouching of skin, holds the testes. Its smooth muscles can contract and make the pouch move closer to the main body mass; muscle relaxation lowers it. Adjustments help maintain a temperature suitable for sperm formation.

A SPERM'S JOURNEY

Sperm travel from the testes through a series of ducts that lead to the urethra (Figure 27.7). Immature sperm enter the epididymides (singular, epididymis), a pair of long, coiling ducts. The last portion of each epididymis stores the mature sperm. During a male's reproductive years, about 100 million sperm mature every day. The unused ones die, break down, and are reabsorbed.

During ejaculation, contraction of smooth muscle in the walls of reproductive ducts propels sperm from the epididymides, through the vasa deferentia (singular, vas deferens), ejaculatory ducts, and urethra. Sperm exit the body from the urethra, which extends through the penis to its tip. A sphincter between the urethra and bladder closes to prevent urine flow during this time.

LINKS TO SECTIONS 8.4, 26.2, 26.6

On the way to the urethra, sperm are mixed with glandular secretions to form semen. Seminal vesicles add sugar that sperm use for energy. Prostate gland secretions contain signaling molecules and help buffer the acidic conditions in the female reproductive tract. Paired bulbourethral glands add mucus-rich fluid.

CANCERS OF THE PROSTATE AND TESTES

Prostate cancer is now the second most common cancer among men, surpassed only by lung cancers. *Testicular cancer* is much less common. Both cancers are painless in early stages and they may spread undetected into lymph nodes of the abdomen, chest, neck, then lungs.

Doctors detect prostate cancer by blood tests for increases in prostate-specific antigen (PSA) and with physical examinations. Adult males should examine their testes monthly—after a warm shower or bath—to check for enlargement, hardening, or new lumps.

HOW SPERM FORM

A testis holds 125 or so meters of seminiferous tubules (Figure 27.9*a*). Inside a tubule, diploid spermatogonia (singular, spermatogonium) will undergo mitosis and produce *primary* spermatocytes. These are diploid cells. Meiosis I in these cells forms *secondary* spermatocytes, then meiosis II yields spermatids. A haploid spermatid develops into a sperm—a mature male gamete (Figure 27.9*b*). This entire process, from the spermatogonium to the mature sperm, takes about nine weeks. Sertoli cells in the lining of the seminiferous tubules support and nourish the maturing sperm.

Each mature sperm is a flagellated cell. Its head is packed with DNA (Figure 27.9*c*). Extending over most of the tiny head is a cap. Enzymes in the cap function at fertilization. Mitochondria in the midpiece behind the head power the tail's whiplike motions.

The steroid hormone testosterone is necessary for sperm production. Leydig cells in the testes secrete testosterone when prompted by luteinizing hormone (LH). Follicle-stimulating hormone (FSH) starts sperm production at puberty. LH and FSH are secreted by the pituitary, and their secretion is in turn controlled by the hypothalamus (Section 26.2).

Human males have a pair of testes that produce sperm and sex hormones. A system of ducts carry sperm from the testes, where they form by meiosis, to the urethra, from which they exit the body. Substances secreted into the ducts by several accessory glands mix with the sperm to form semen.

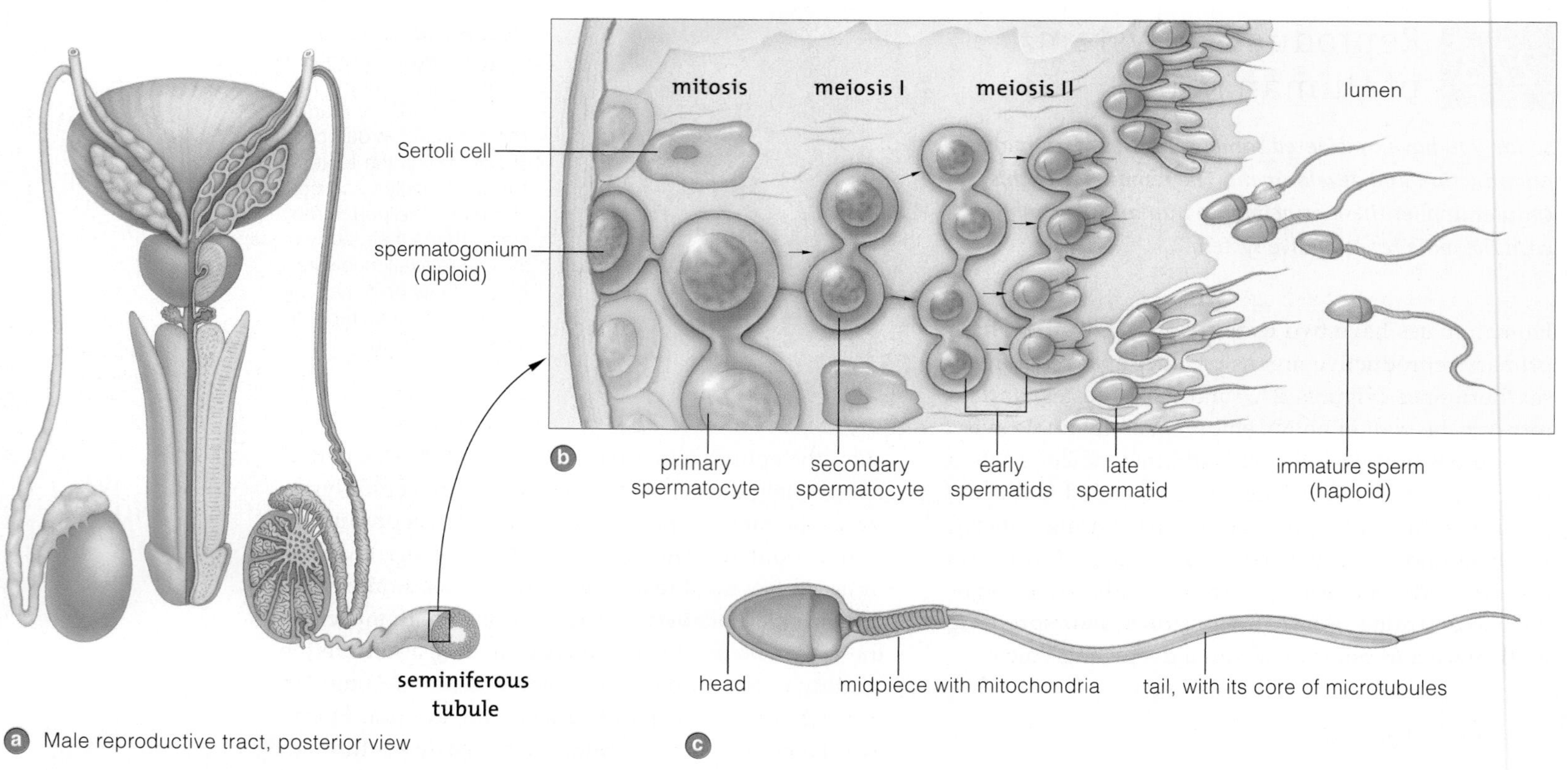

Figure 27.9 Where and how sperm form (**a**,**b**), and a mature sperm (**c**).

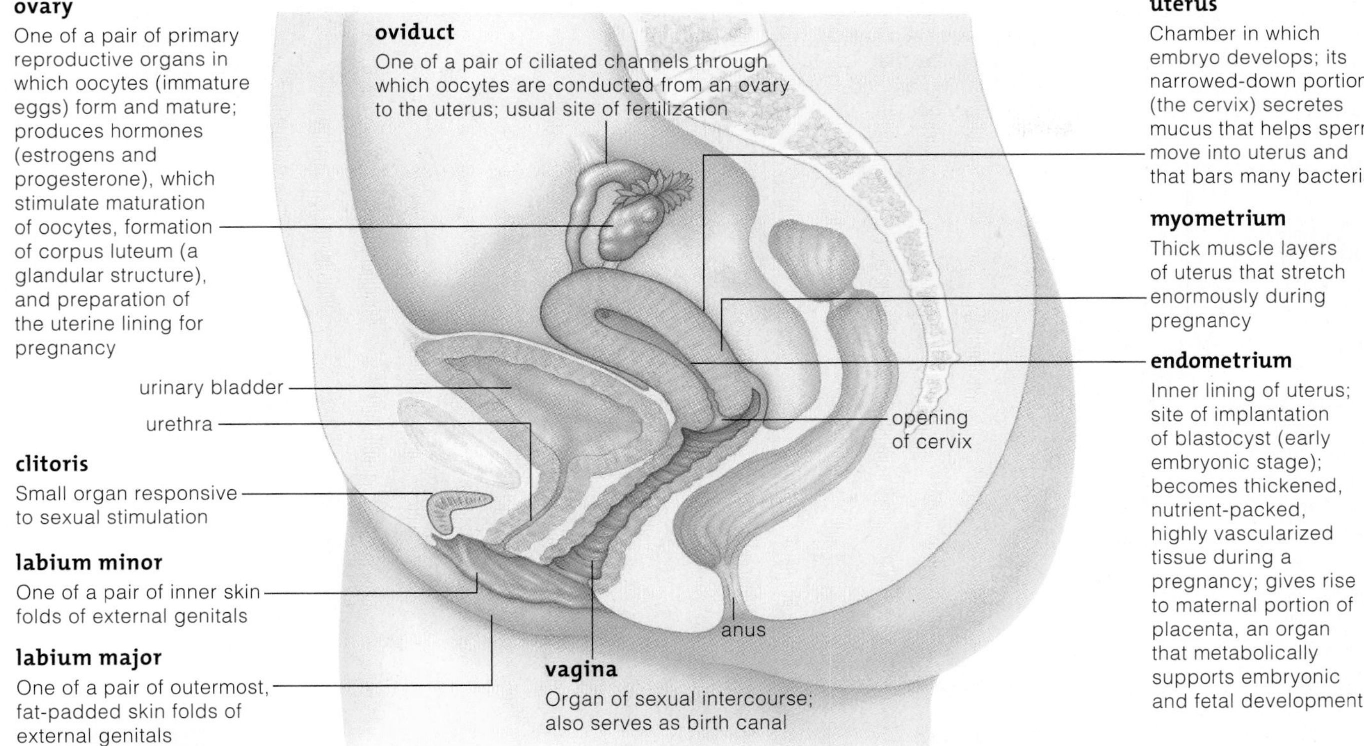
ovary
One of a pair of primary reproductive organs in which oocytes (immature eggs) form and mature; produces hormones (estrogens and progesterone), which stimulate maturation of oocytes, formation of corpus luteum (a glandular structure), and preparation of the uterine lining for pregnancy
oviduct
One of a pair of ciliated channels through which oocytes are conducted from an ovary to the uterus; usual site of fertilization
uterus
Chamber in which embryo develops; its narrowed-down portion (the cervix) secretes mucus that helps sperm move into uterus and that bars many bacteria
myometrium
Thick muscle layers of uterus that stretch enormously during pregnancy
endometrium
Inner lining of uterus; site of implantation of blastocyst (early embryonic stage); becomes thickened, nutrient-packed, highly vascularized tissue during a pregnancy; gives rise to maternal portion of placenta, an organ that metabolically supports embryonic and fetal development
urinary bladder
urethra
opening of cervix
clitoris
Small organ responsive to sexual stimulation
labium minor
One of a pair of inner skin folds of external genitals
labium major
One of a pair of outermost, fat-padded skin folds of external genitals
anus
vagina
Organ of sexual intercourse; also serves as birth canal

Figure 27.10 ***Animated!*** Components of the human female reproductive system.

27.4 Reproductive System of Human Females

We turn next to the reproductive system of the human female. Figures 27.10 and 27.11 show its components.

A human female's *primary* reproductive organs, a pair of ovaries, produce oocytes and secrete sex hormones. An oocyte released from an ovary moves into one of a pair of oviducts, also known as Fallopian tubes. Both oviducts open into the **uterus**, a hollow pear-shaped organ. Fertilization usually occurs in an oviduct, then embryos develop in the uterus.

Most of the uterine wall is a thick layer of smooth muscle called the myometrium. The uterine lining, or **endometrium**, consists of connective tissues, glands, and blood vessels. The cervix is the narrowed-down portion of the uterus that connects to the vagina. The vagina is a muscular tube that extends to the body's surface. It receives sperm during sexual intercourse and functions as part of the birth canal.

At the body's surface are genital organs of sexual stimulation. Outermost are the labia majora, a pair of skin folds padded with adipose tissue. They enclose the labia minora, a pair of smaller folds having many blood vessels but no adipose tissue. These folds partly enclose the clitoris. This female sex organ is derived from the same embryonic tissue as the male's penis. Like the penis, the clitoris has a great many sensory receptors and is sensitive to sexual stimulation. The urethra opens at the body surface midway between the vagina and the clitoris.

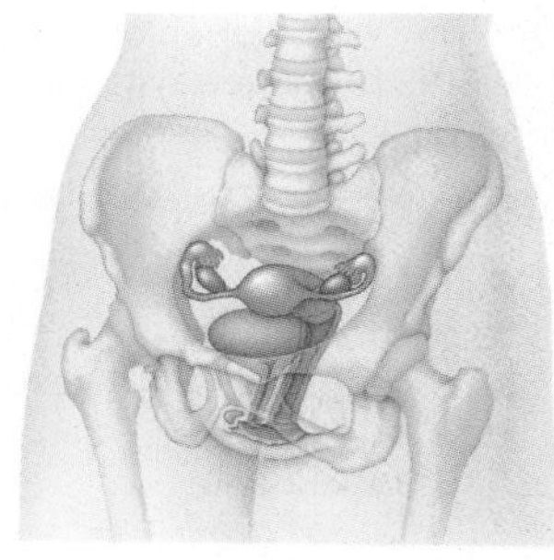

Figure 27.11 Location of the female reproductive system relative to the pelvic girdle and urinary bladder.

OVERVIEW OF THE MENSTRUAL CYCLE

When a female reaches puberty, usually around age ten to sixteen, she begins to follow a **menstrual cycle**. She becomes fertile for a few days at a time on a cyclic basis. Every twenty-eight days or so, an oocyte matures and is released from an ovary, and the uterus is prepared for pregnancy. If fertilization does not occur, the lining of the uterus breaks down. Discarded uterine tissue and a small amount of blood flow from the uterus, through the cervix, and out of the vagina. The recurring flow is called menstruation. It marks the first day of a new cycle.

Menstrual cycles continue until a woman is in her late forties or early fifties. Then cycles falter as production of the sex hormones **progesterone** and **estrogen** decline. This is the onset of *menopause*, the twilight of a female's reproductive capacity.

Figure 27.12 ***Animated!*** Cyclic events in a human ovary, cross-section. A follicle stays in the same place inside an ovary all through one menstrual cycle. It does not "move around" as in this diagram, which shows the *sequence* of events.

In the cycle's first phase, a follicle grows and matures. At ovulation, the second phase, the mature follicle ruptures and releases a secondary oocyte. In the third phase, a corpus luteum forms from remnants of the follicle.

The micrograph shows a secondary oocyte being released from an ovary. It will enter an oviduct, the passageway to the uterus.

ovary

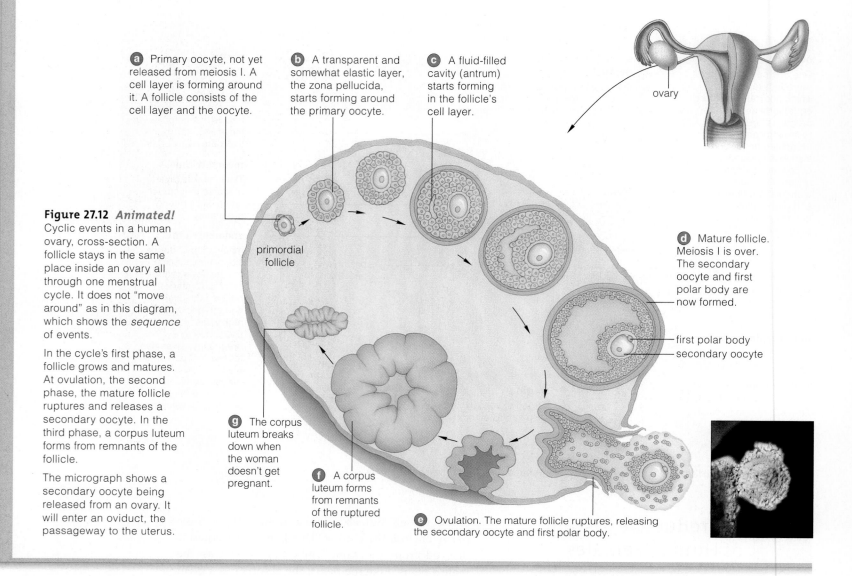

(a) Primary oocyte, not yet released from meiosis I. A cell layer is forming around it. A follicle consists of the cell layer and the oocyte.

(b) A transparent and somewhat elastic layer, the zona pellucida, starts forming around the primary oocyte.

(c) A fluid-filled cavity (antrum) starts forming in the follicle's cell layer.

primordial follicle

(d) Mature follicle. Meiosis I is over. The secondary oocyte and first polar body are now formed.

first polar body

secondary oocyte

(e) Ovulation. The mature follicle ruptures, releasing the secondary oocyte and first polar body.

(f) A corpus luteum forms from remnants of the ruptured follicle.

(g) The corpus luteum breaks down when the woman doesn't get pregnant.

LINKS TO SECTIONS 8.4, 26.2, 26.6

The menstrual cycle unfolds in three phases. In the *follicular* phase, menstruation occurs, the uterine lining regenerates, and an oocyte matures. The next phase, **ovulation**, is limited to the release of an oocyte from an ovary. In the third, or *luteal* phase, hormones make the endometrium thicken for a potential pregnancy.

All phases are regulated by feedback loops from ovaries to the hypothalamus and the anterior pituitary gland. Gonadotropin-releasing hormone (GnRH) from the hypothalamus induces the secretion of FSH and LH by the anterior pituitary. These hormones promote changes in the ovaries. They also stimulate the ovaries to secrete estrogens and progesterone that cause cyclic changes in the uterus.

CYCLIC CHANGES IN THE OVARY

Take a moment to review Figure 7.13, the overview of meiosis in an oocyte. Each normal baby girl has about 2 million primary oocytes in her ovaries. By the time she is seven years old, about 300,000 remain; her body reabsorbed the rest. Her primary oocytes have already started meiosis, but nuclear division has stopped short. It will resume, a few oocytes at a time, beginning with her first menstrual cycle. Usually, only a single oocyte matures during each cycle. Fraternal twins arise when two oocytes mature and both become fertilized.

Through her reproductive years, a woman releases about 400 to 500 oocytes. Eggs released late in life are at greater risk of changes in chromosome number or structure when meiosis resumes in them. One possible outcome is Down syndrome (Section 8.9).

Figure 27.12 depicts a follicle near the surface of an ovary. It consists of a primary oocyte and a layer of cells that surround and nourish it. As the menstrual cycle starts, the hypothalamus is secreting just enough GnRH to make cells in the anterior lobe of the pituitary step up their secretion of FSH and LH. The increases cause the oocyte to grow and more cell layers to form around it. Glycoproteins start to accumulate between the oocyte and the follicle cells. They form the zona pellucida, a noncellular, transparent layer (27.12*b*).

At the same time, FSH and LH are stimulating the developing follicle cells to secrete estrogen. The blood levels of this hormone begin to rise (Figure 27.13*a–c*).

Eight to ten hours before its release from an ovary, the oocyte completes meiosis I. Its cytoplasm divides in two. One division product is the **secondary oocyte**, a haploid cell with still-duplicated chromosomes and most of the cytoplasm. The other is the first of three **polar bodies** (Figure 27.12*d*).

Halfway through a cycle, the pituitary detects the rise in estrogens. It responds with a brief outpouring of LH (Figure 27.13*a*). The surge causes changes that make the follicle swell. It also makes enzymes digest the swollen wall. Rupture of the wall releases fluid and the secondary oocyte (Figures 27.12*e* and 27.13*b*). Thus, the midcycle surge of LH triggers ovulation—the release of a secondary oocyte from the ovary.

CYCLIC CHANGES IN THE UTERUS

The estrogens released early in the cycle help promote pregnancy in another way. They stimulate growth of the endometrium and its glands. Before the midcycle LH surge, follicle cells were secreting progesterone and estrogens (Figure 27.13*c*). Blood vessels were growing in the thick endometrium. At ovulation, the estrogens began acting on tissues of the cervix and the vagina. They stimulated the cervix to release a thin mucus, an ideal medium for sperm to swim through.

The midcycle LH surge also stimulates formation of a glandular **corpus luteum** from some cells of the now ruptured follicle (Figure 27.12*f*). Progesterone and some estrogen secreted by the corpus luteum promote endometrial thickening in preparation for arrival of a **blastocyst**. This is the specific type of blastula that is formed by cleavage of a fertilized mammalian egg.

Meanwhile, the hypothalamus has been preventing other follicles from maturing by slowing down FSH secretion. If a blastocyst does not implant itself in the uterus, the corpus luteum will persist for about twelve days. During the waning days of the cycle, the corpus luteum secretes prostaglandins. These local signaling molecules cause its self-destruction by apoptosis.

Once the corpus luteum has disappeared, levels of progesterone and estrogen drop and the endometrium starts to break down. With less oxygen and nutrients, its blood vessels constrict and the tissues die. Blood escapes as weakened capillary walls begin to rupture. Blood and sloughed tissues flow from the uterus out of the vagina for three to six days. Then a new cycle begins, with rising estrogen levels causing renewed growth of the endometrium (Figure 27.13*d*).

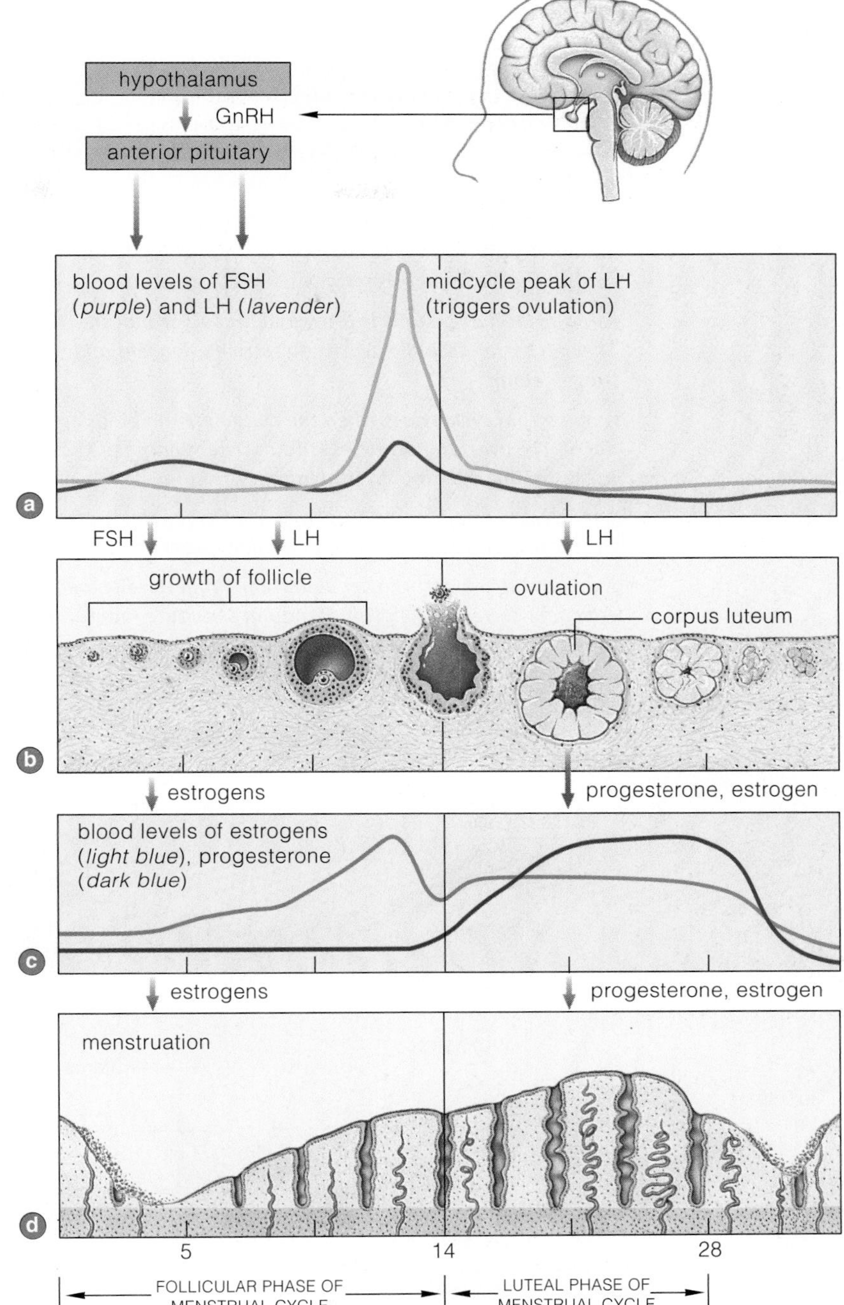

Figure 27.13 ***Animated!*** Changes that occur in the ovary and the uterus, correlated with changing hormone levels. Based on a twenty-eight day cycle in which day 1 is the onset of menstruation. Green arrows indicate which hormones dominate the cycle's first phase (the time when the follicle matures), then the second phase (the time when the corpus luteum forms).

(**a**,**b**) FSH and LH are secreted by the anterior pituitary in response to GnRH from the hypothalamus. FSH and LH act on the ovaries, stimulating a follicle to grow and an oocyte to mature. A midcycle surge of LH triggers ovulation and formation of a corpus luteum. A decline in FSH after ovulation prevents any additional follicles from maturing.

(**c**,**d**) Early in the cycle, secretion of estrogen by a maturing follicle stimulates repair and rebuilding of the lining of the uterus. After ovulation, the corpus luteum secretes some estrogen and more progesterone. The progesterone prepares the uterus for pregnancy.

Ovaries are the primary reproductive organs in human females. They produce oocytes (immature eggs) and sex hormones. The uterus is the chamber in which embryos develop.

Hormones control cyclic changes in the ovaries and the endometrium. The hypothalamus and pituitary gland control the hormonal secretions.

FSH and LH make a follicle grow and an oocyte mature. They also stimulate the ovaries to secrete estrogens and progesterone.

A menstrual cycle starts when the endometrium breaks down. The ovarian hormones stimulate rebuilding of the uterus lining, priming it for pregnancy.

A midcycle surge of LH triggers ovulation, the release of the secondary oocyte and one polar body from the ovary.

A corpus luteum forms after ovulation. Progesterone and estrogens secreted from this glandular structure maintain the uterus lining unless pregnancy does not occur.

27.5 How Pregnancy Happens

Sperm released into the vagina make their way toward the ovaries. Most commonly, fertilization occurs when sperm and oocyte meet up in the oviduct.

SEXUAL INTERCOURSE

Sexual intercourse, or *coitus*, can deliver semen into a female's vagina. For the male, intercourse requires an *erection*, whereby a normally limp penis stiffens and increases in size. It also involves *ejaculation*, expulsion of semen into the urethra and out from the penis.

As Figure 27.7 shows, cylinders of spongy tissue extend through the length of the penis. A great many friction-activated sensory receptors are located on its mushroom-shaped tip, the glans penis. In males who are *not* sexually aroused, the penis stays limp because the arterioles leading into the cylinders are narrowed.

As a person becomes sexually aroused, their heart rate and blood pressure rise. Blood flow to the genitals increases. In males, blood flows into the penis faster than it flows out, so the organ enlarges and stiffens. A similar mechanism enlarges a female's clitoris. At the same time, her secretion of cervical mucus increases, lubricating the vagina in preparation for intercourse.

During intercourse, mechanical stimulation of the male's penis and the female's clitoris and vaginal wall may lead to *orgasm*. Neurotransmitters flood the brain causing sensations of release and pleasure. In females, muscles of the vaginal wall contract rhythmically. In males, similar reflexive contractions force sperm and the contents of seminal vesicles and the prostate gland into the urethra. The substances mix and form semen, which is ejaculated in the vagina.

You may have heard that a female cannot become pregnant unless she reaches orgasm. Don't believe it!

FERTILIZATION

On average, an ejaculation can put 150 million to 350 million sperm into the vagina. Within thirty minutes some have made their way into the female's oviducts.

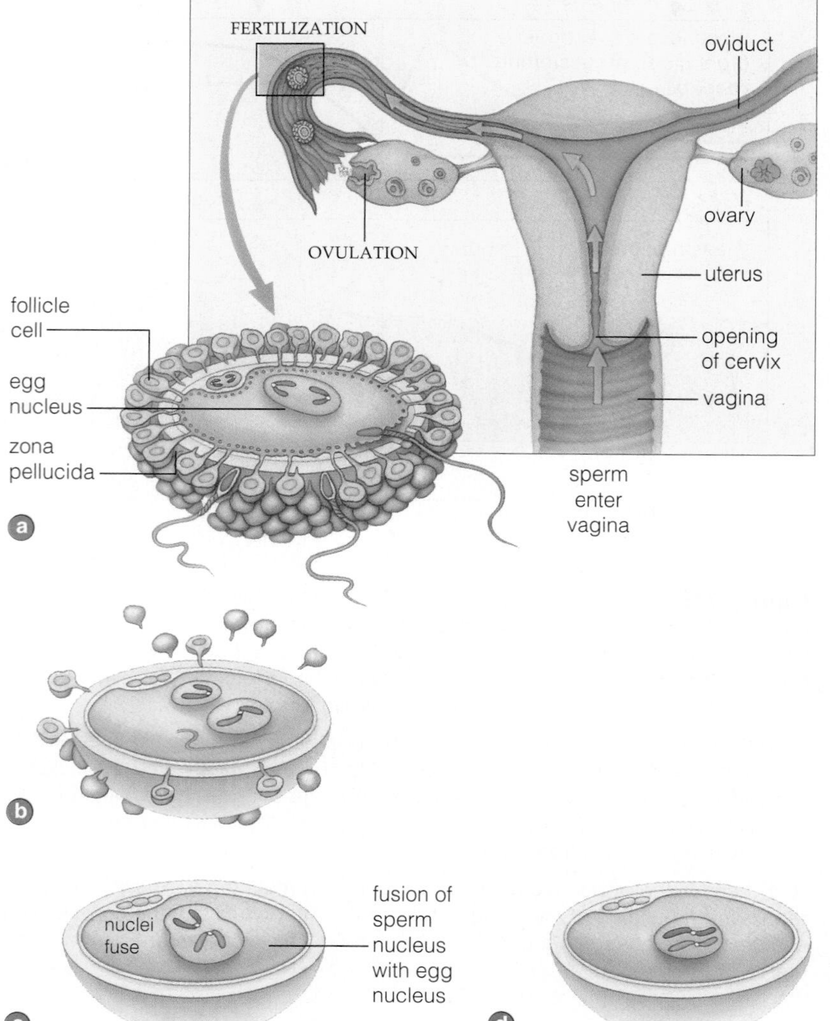

Figure 27.14 ***Animated!*** Fertilization. (**a**) Sperm surround a secondary oocyte and release enzymes that digest the zona pellucida. (**b**) When a sperm penetrates the oocyte, granules in the egg cortex move to the surface and the zona pellucida is shed. Penetration stimulates meiosis II of the oocyte's nucleus. (**c**) The sperm tail degenerates; its nucleus enlarges and fuses with the oocyte nucleus. (**d**) At nuclear fusion, a zygote forms and fertilization is completed.

Just a few hundred will survive the long journey to the upper region of the oviduct, where eggs usually are fertilized (Figure 27.14).

If sperm and oocyte meet, receptors in the plasma membrane of the sperm's head bind glycoproteins in the zona pellucida enclosing the oocyte. This activates these receptors, which triggers the release of digestive enzymes from the *acrosome*, the compartment that caps the front of a sperm's head. The activity of acrosomal enzymes clears a passage through the zona pellucida.

Usually only a single sperm enters the oocyte. The sperm's entry causes granules in the oocyte cortex (the region just beneath the plasma membrane) to move to the cell's surface (Figure 27.14*b*). The granules contain enzymes that digest the zona pellucida. With this outer coat gone, other sperm have nothing to attach to and are unable to enter the oocyte.

Once inside the oocyte, a sperm breaks down until only its nucleus and centrioles remain. Sperm entry prompts the secondary oocyte and the first polar body to complete meiosis II. The result is three polar bodies and a single mature egg, or **ovum** (plural, ova). The egg nucleus and sperm nucleus fuse. The combining of these haploid nuclei restores the diploid number for a brand new zygote.

PREVENTING PREGNANCY

The variety of methods used to prevent pregnancy are termed contraception. As illustrated by Figure 27.15, some of these methods are more effective than others. They also differ in their mode of action.

Abstinence, or no sex at all, is the most reliable way to avoid pregnancy, but it takes great self-discipline to override the neural and hormonal sense of urgency associated with sex. Versions of the *rhythm method* are all forms of abstinence. With these methods, a woman avoids sex at times when she could become pregnant. She monitors her menstrual periods, cervical mucus secretion, body temperature, hormone production, or a combination of these factors to determine when she is ovulating. Even with careful monitoring, mistakes occur. Sperm can live in the vagina for up to a week, so a few deposited just before ovulation may survive to fertilize an egg.

Withdrawal, or removing the penis from the vagina before ejaculation, requires a lot of willpower and can often fail, because fluid released from the penis before ejaculation contains some sperm.

Douching (chemically rinsing the vagina right after intercourse) is ineffective. Sperm move out of reach of a douche ninety seconds after ejaculation.

Surgical methods are designed for those who wish to end their fertility permanently. Men may opt for a *vasectomy*. A small incision in the scrotum allows the surgeon to cut or block each vas deferens. Reversal of this procedure is possible, but only about 60 percent of those who undergo reversal can father a child.

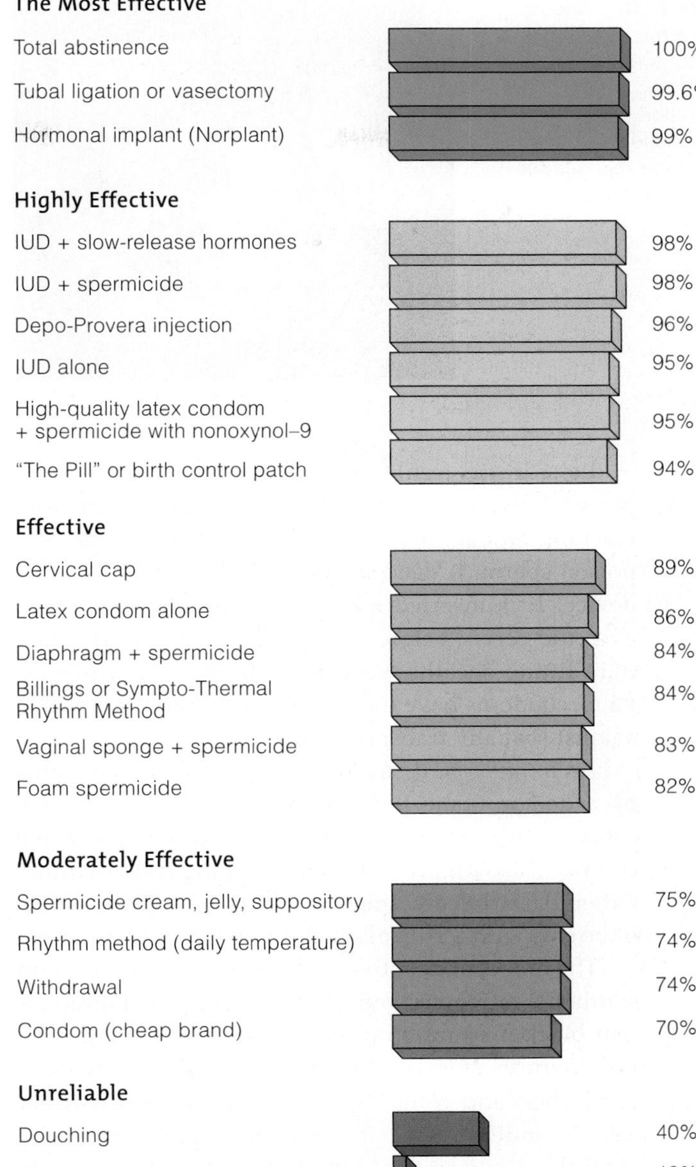

Figure 27.15 Comparison of the effectiveness of some common contraceptive methods. These percentages indicate the number of unplanned pregnancies per 100 couples who rely on that method of birth control for a year. For example, oral contraceptives have an effectiveness of 94%. That means that on average, 6 of every 100 women using this method will become pregnant within the year.

A *tubal ligation* is a female version of vasectomy. It too is designed to be permanent. Oviducts are blocked or cut to prevent eggs and sperm from meeting. This procedure is more common than vasectomy. Surgical reversal is about 70 percent successful.

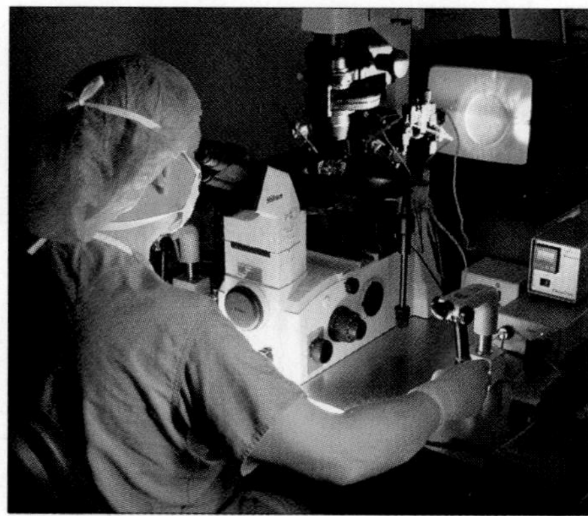

Figure 27.16 Doctor using a micromanipulator to inject a human sperm into an egg during *in vitro* fertilization (IVF). The monitor shows what he sees in the microscope.

Less drastic fertility control methods are based on physical and chemical barriers that stop sperm from reaching an egg. *Spermicidal foam* and *spermicidal jelly* poison sperm. A *diaphragm* is a flexible, dome-shaped device. It is inserted into the vagina and positioned over the cervix before intercourse. *Condoms* are thin, tight-fitting sheaths worn over the penis during sex. Latex condoms have the advantage of being protective against sexually transmitted diseases.

An *intrauterine device,* or IUD, is placed in the uterus by a doctor. Some IUDs release progesterone, which causes mucus secretions to thicken so they block the cervix. It also prevents thickening of the endometrium. Other IUDs release small amounts of copper, which interferes with fertilization and prevents implantation.

The oral contraceptives, or *birth control pills,* contain synthetic estrogens and progesterone-like hormones that block maturation of oocytes and ovulation. "The Pill" reduces menstrual cramps but can cause nausea, headaches, and weight gain in some women. With at least 50 million users, it is the most common fertility control method in the United States. It is very highly effective. Its use lowers the risk of ovarian cancer but may increase risk of breast, cervical, and liver cancers.

A *birth control patch* delivers hormones through the skin and acts in the same way as oral contraceptives. Injections or implants of synthetic progesterone also prevent ovulation. A *Depo-Provera* injection works for three months. *Norplant* rods inserted under the skin last for five years. Both methods are highly effective.

Morning-after pills are high doses of estrogen and progesterone. These pills are not meant for regular use and can cause nausea, vomiting, and headaches. The pills work best when taken as soon as possible, but are of some use as late as five days after intercourse.

IMPACTS, ISSUES

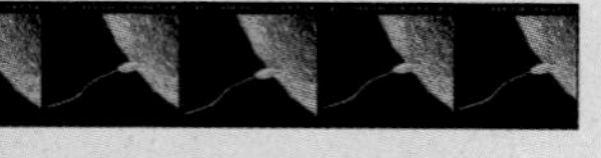

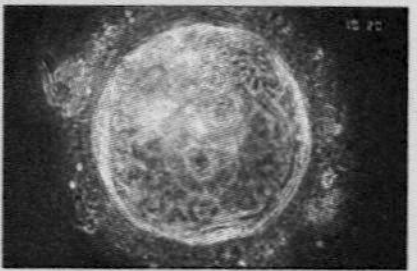

By traditional in vitro fertilization, several early cell clusters may be inserted into the uterus to improve the odds that at least one will survive. More than one often does. In one study, 40 percent of IVF patients gave birth to twins. A more recent technique, blastocyst transfer, allows the clusters to develop longer before transfer into a uterus. Survival chances are better, so fewer have to be implanted. This lowers the risk of multiple births.

SEEKING OR ENDING PREGNANCY

Some couples have trouble conceiving naturally and turn to *in vitro fertilization (IVF)*. The term refers to conception outside the body. "In vitro" literally means "in glass," as in a glass petri dish. Hormone injections cause maturation of oocytes. Before an oocyte can be released, it is withdrawn and a sperm is injected into it (Figure 27.16). If fertilization succeeds, cleavage gets underway. A few days later, the resulting tiny ball of cells is transferred into a woman's uterus to develop.

In vitro fertilization attempts are costly and those carried out for reasons of infertility usually fail. Is this a cost society should bear? Court battles have been waged over this issue. Another concern is the fate of cell clusters that are created for IVF, but never used.

About ten percent of women who learn that they are pregnant suffer a *spontaneous abortion*, or miscarriage. The risk is higher for women older than age 35. Many more pregnancies end without ever being detected. It is estimated that about 50 percent of fertilized eggs are spontaneously aborted, most often because they have genetic defects.

An *induced abortion* is the deliberate dislodging and removal of an embryo or a fetus from the uterus. In the United States, about half of all unplanned pregnancies end in induced abortion. From the clinical standpoint, abortion usually is a brief, safe, and relatively painless procedure, especially in the first trimester.

The drug mifepristone (RU 486) and a prostaglandin can be used to induce abortion during the first nine weeks of pregnancy. The two chemicals interfere with progesterone receptors in the uterus. The uterine lining cannot be maintained, and neither can pregnancy.

The intense physiological events that accompany coitus have one function: to put sperm on a collision course with an egg. Fertilization is over when a sperm nucleus and egg nucleus fuse, forming a diploid zygote.

Various methods can be used to prevent ovulation, interfere with fertilization, or prevent blastocyst implantation.

In vitro fertilization unites sperm and an egg outside the body. Abortion terminates a pregnancy.

27.6 Sexually Transmitted Diseases

The sex act often provides an opportunity for pathogens to travel from one individual to another.

CONSEQUENCES OF INFECTION

Each year, pathogens that cause **sexually transmitted diseases**, or STDs, infect about 15 million Americans (Table 27.1). Two-thirds of those infected are under

Table 27.1 Estimated New STD Cases Per Year

STD	U.S. Cases	Infectious Agent
HPV infection	5,500,000	Virus
Trichomoniasis	5,000,000	Protist
Chlamydia	3,000,000	Bacteria
Genital herpes	1,000,000	Virus
Gonorrhea	650,000	Bacteria
Syphilis	70,000	Bacteria
AIDS	40,000	Virus

age twenty-five; one-quarter are teenagers. More than 65 million Americans have a viral STD that cannot be cured. Treating STDs and their complications costs a staggering 8.4 billion dollars in an average year.

The social consequences are enormous. Women are more easily infected than men, and they develop more complications. Example: Pelvic inflammatory disease (PID), a secondary outcome of bacterial STDs, scars the female reproductive tract and can cause infertility, chronic pain, and tubal pregnancies (Figure 27.17*a*). Infected women commonly pass chlamydia on to their newborn (Figure 27.17*b*). Besides causing skin sores, type II *Herpes* virus kills half of infected fetuses and causes neural defects in one-fourth of the rest.

MAJOR AGENTS OF STDS

HPV Infection by human papillomaviruses (HPV) is the most widespread and fastest spreading STD in the United States. Of about 100 HPV strains, a few cause *genital warts* (Figure 27.18*c*). In women, the bumplike growths form on the vagina, cervix, and the external genitals. In men, they form on the penis and scrotum. A few strains of HPV cause cervical cancer. Cervical changes can be detected by a pap smear, in which cells from the cervix are examined using a microscope.

TRICHOMONIASIS Infection by *Trichomonas vaginalis,* a flagellated protozoan, causes a yellowish discharge, soreness, and itching of the vagina. Men are usually symptom free. Untreated infections can damage the urinary tract and result in infertility. A single dose of an antiprotozoal drug can quickly cure the infection. Both sexual partners must receive treatment.

CHLAMYDIA *Chlamydial infection* is primarily a young person's disease. Forty percent of those infected are still in their teens. A bacterium, *Chlamydia trachomatis,* causes the disease, and antibiotics quickly kill it. Most infected women have no symptoms and are not even diagnosed. Half of the infected men suffer discharges from the penis and painful urination. Untreated men risk inflammation of the epididymides and infertility.

GENITAL HERPES About 45 million Americans have *genital herpes*, caused by type II *Herpes simplex* virus. Infection requires contact with active *Herpes* viruses or with sores that contain them. Once a person has been infected, they harbor the virus for life. It is reactivated sporadically, which causes painful sores at or near the originally infected tissues. Menstruation, emotional stress, or other infections can trigger flare-ups.

GONORRHEA The bacterium *Neisseria gonorrhoeae* is the cause of *gonorrhea*. Less than one week after a male is infected, yellow pus oozes from the penis. Urination becomes frequent and painful. An infected female may have minimal symptoms at first, but the pathogen can infect oviducts, causing cramps, scarring, and sterility.

Oral contraceptives promote infection by altering vaginal pH. Populations of resident bacteria decline—and so *N. gonorrhoeae* is free to move in.

Even though antibiotic treatment quickly cures this disease, it still is rampant. Why? Many women tend to discount the slight symptoms of the early stages.

SYPHILIS *Syphilis* is caused by *Treponema pallidum*, one of the spirochete bacteria. Painless chancres (skin ulcers) appear in the primary stage. They heal, but the spirochetes multiply inside the spinal cord, brain, eyes, bones, joints, and mucous membranes. In a secondary infectious stage more chancres form (Figure 27.17*d*). Continued infection can damage the liver, bones, and other internal organs. Chronic immune reactions to the bacteria may severely damage the brain and spinal cord and lead to blindness and paralysis.

AIDS Infection by HIV (or human immunodeficiency virus) leads to *AIDS*. At first there may be no outward symptoms. Five to ten years later, chronic disorders develop. They are the symptoms of AIDS (*A*cquired *I*mmune *D*eficiency *S*yndrome). The immune system weakens (Section 23.9). Eventually infections overrun and kill the immune-compromised person.

HIV spreads by anal, vaginal, and oral intercourse and intravenous drug use. Virus particles in blood, semen, urine, or vaginal secretions usually enter their new host through cuts on the penis, vagina, or rectum.

Free or low-cost, confidential testing for exposure to HIV is available through public health facilities. It takes a few weeks to six months or more after first exposure before detectable amounts of antibodies can form in response to the infection. Anyone who tests positive for HIV can spread the virus.

LINKS TO SECTIONS
14.2, 14.3, 14.5, 23.9

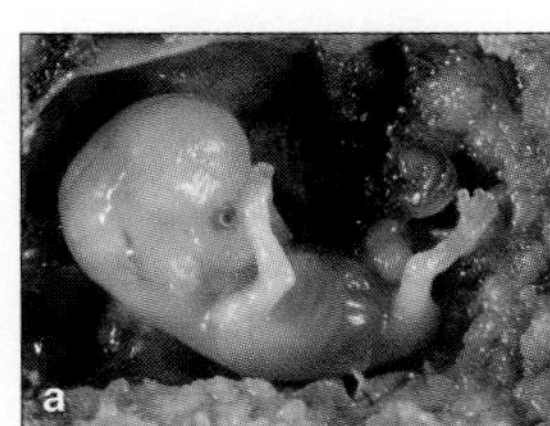

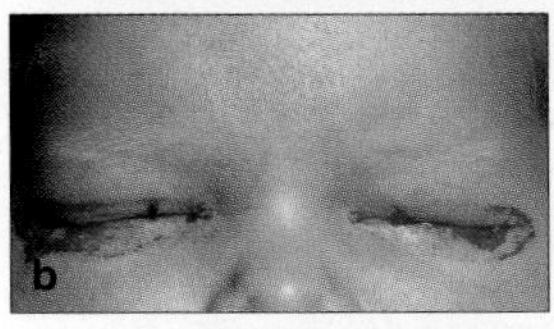

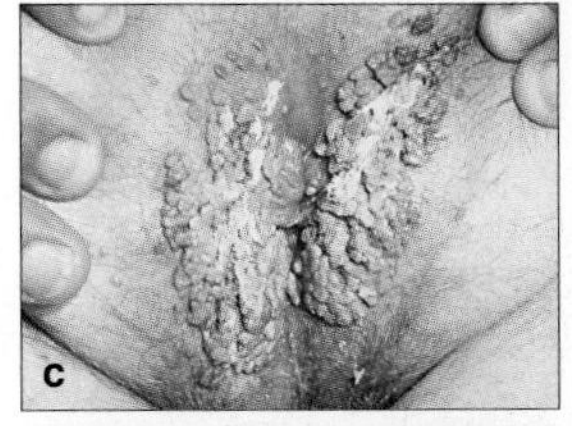

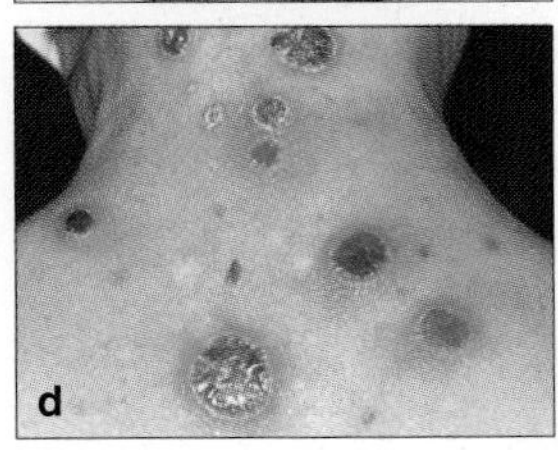

Figure 27.17 A few potential consequences of unsafe sex. (**a**) STDs that scar the oviduct can cause a tubal pregnancy. An embryo implants itself in an oviduct, instead of the uterus. Untreated tubal pregnancies can rupture an oviduct and cause bleeding, infection, and death. (**b**) One sign of a chlamydial infection transferred from a mother to her infant. (**c**) Genital warts. (**d**) The chancres typical of secondary syphilis.

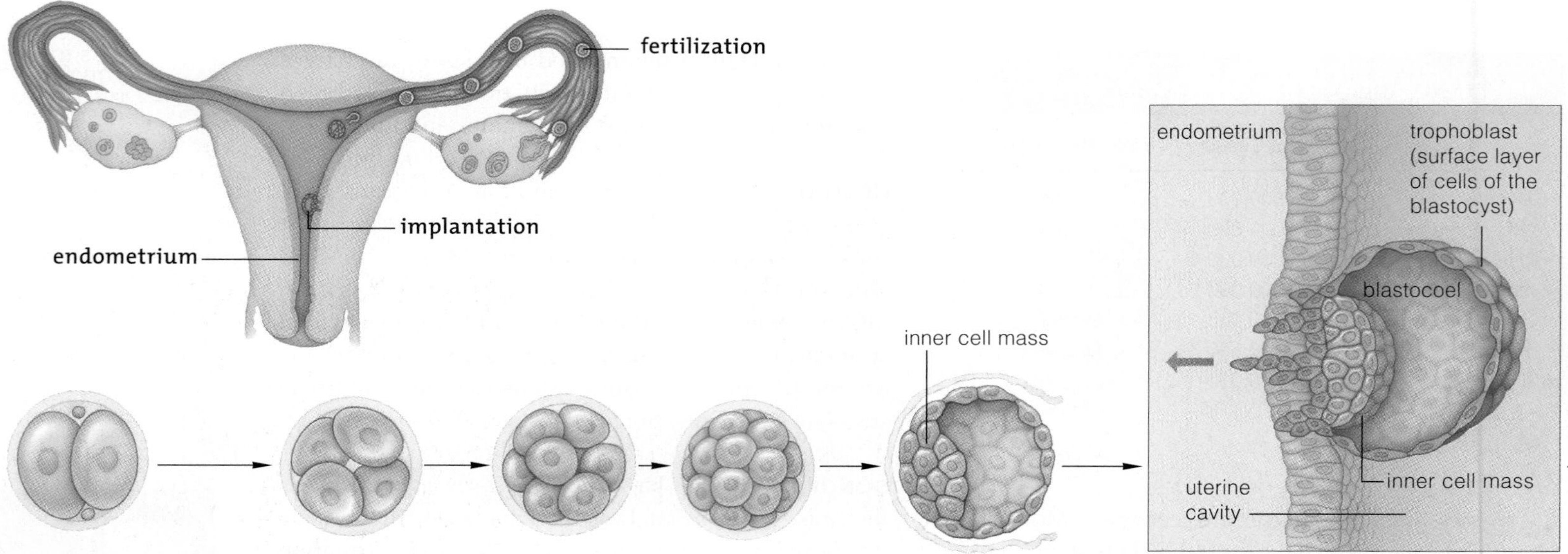

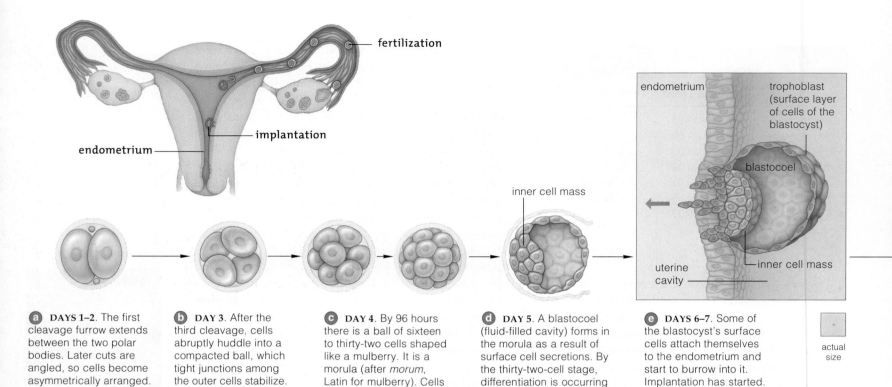

a **DAYS 1–2**. The first cleavage furrow extends between the two polar bodies. Later cuts are angled, so cells become asymmetrically arranged. Until the eight-cell stage forms, they are loosely organized, with space between them.

b **DAY 3**. After the third cleavage, cells abruptly huddle into a compacted ball, which tight junctions among the outer cells stabilize. Gap junctions formed along the interior cells enhance intercellular communication.

c **DAY 4**. By 96 hours there is a ball of sixteen to thirty-two cells shaped like a mulberry. It is a morula (after *morum*, Latin for mulberry). Cells of the surface layer will function in implantation and will give rise to a membrane, the chorion.

d **DAY 5**. A blastocoel (fluid-filled cavity) forms in the morula as a result of surface cell secretions. By the thirty-two-cell stage, differentiation is occurring in an inner cell mass that will give rise to the embryo proper. This embryonic stage is the blastocyst.

e **DAYS 6–7**. Some of the blastocyst's surface cells attach themselves to the endometrium and start to burrow into it. Implantation has started.

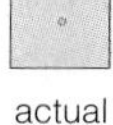

actual size

LINKS TO SECTIONS 7.2, 7.3, 16.12

Health-care workers advocate safe sex, but there is confusion about what this means. The best way to be safe is to use a high-quality latex condom *together with* a spermicide that has nonoxynol–9. This combination helps prevent HIV transmission but even so there is still a slight risk of infection. No one should engage in open-mouthed kissing with a person who has tested positive for HIV. Caressing is not risky *if* there are no lesions or cuts where body fluids with HIV can enter the body. Other sexually transmitted diseases increase susceptibility to HIV infection by causing sores that make it easier for a virus to get into body tissues.

In the United States, infection rates have begun to climb again, possibly because of a misperception that AIDS is no longer a threat. The new drug therapies do prolong life, but only at great cost and with many side effects. AIDS is still incurable and deadly. In 2004, it killed 15,000 people in the United States.

STDs are surprisingly common in the general population. Engaging in unsafe sex exposes you to viral, protistan, and bacterial pathogens that can endanger your health, your fertility, and even your survival.

27.7 Human Prenatal Development

During the nine months from fertilization to birth, cell divisions and developmental processes lay out a basic vertebrate body plan, then give it uniquely human characteristics. Nutrients and energy that fuel development are provided by the mother's diet.

Pregnancy lasts an average of thirty-eight weeks from the time of fertilization. It takes about one week for a blastocyst to form. Weeks three through eight are the *embryonic* period, during which all of the major organ systems form. When this interval ends, a developing individual appears distinctly human, and is called a **fetus**. In the *fetal* period, from the start of the ninth week until birth, organs enlarge and take on special functions. We often refer to the first three months of pregnancy as the first trimester. The second trimester extends from the start of the fourth month to the end of the sixth, and the third trimester from the seventh month until birth. Starting with Figure 27.18, the next series of illustrations shows typical features that arise during each stage of development.

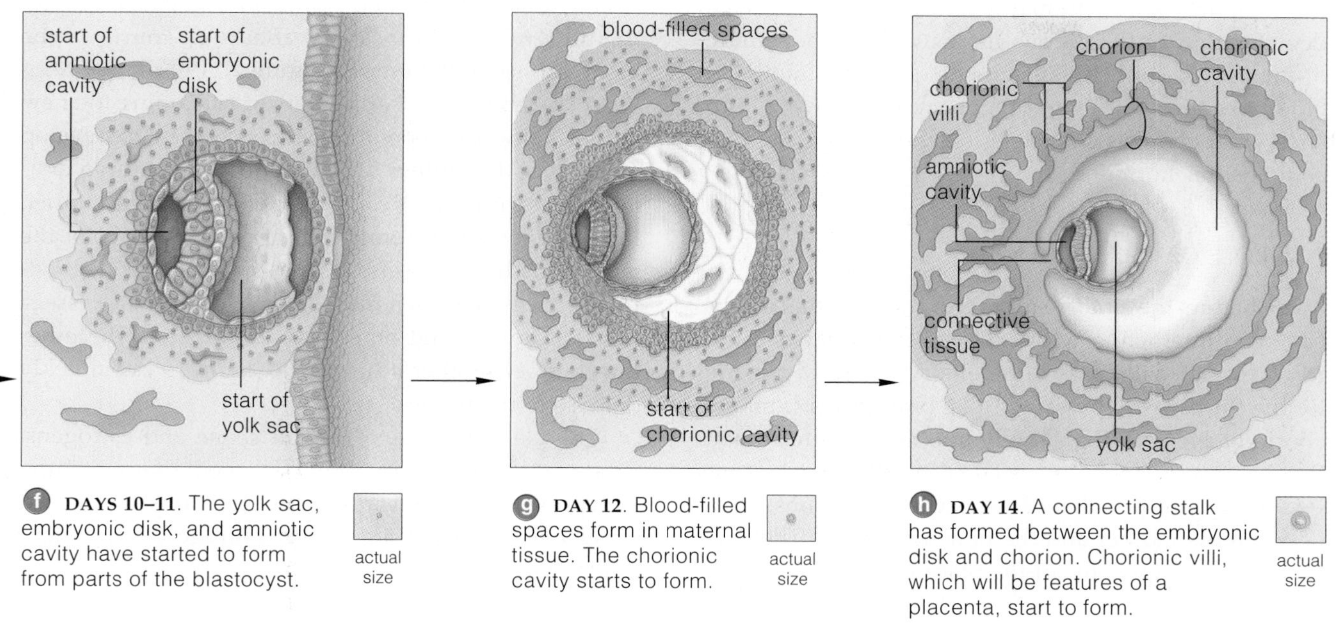

f DAYS 10–11. The yolk sac, embryonic disk, and amniotic cavity have started to form from parts of the blastocyst.

g DAY 12. Blood-filled spaces form in maternal tissue. The chorionic cavity starts to form.

h DAY 14. A connecting stalk has formed between the embryonic disk and chorion. Chorionic villi, which will be features of a placenta, start to form.

Figure 27.18 *Animated!* From fertilization through implantation. A blastocyst forms, and its inner cell mass will give rise to an embryo. Three extraembryonic membranes (the amnion, chorion, and yolk sac) start forming. A fourth membrane (allantois) forms after the blastocyst is implanted.

CLEAVAGE AND IMPLANTATION

Three to four days after fertilization, cleavage already has started in the zygote. Genes are being expressed; the early divisions require their products. At the eight-cell stage, the cells huddle into a ball. A blastocyst forms by the fifth day. It consists of a trophoblast (a surface layer of cells), a blastocoel filled with their secretions, and an inner cell mass (Figure 27.18*d*). One or two days after that, **implantation** is underway. The blastocyst adheres to the uterine lining and invades the mother's tissues, forming the connections that will metabolically support the pregnancy. By this time, the inner cell mass has become two flattened layers of cells in the shape of a disk. This embryonic disk will give rise to the embryo proper.

EXTRAEMBRYONIC MEMBRANES

Like the reptiles and birds, humans are amniotes and human development involves formation of the same four extraembryonic membranes described in Section 16.12—the amnion, allantois, yolk sac, and chorion.

As a human blastocyst begins to implant, a fluid-filled amniotic cavity appears between the embryonic disk and the blastocyst surface (Figure 27.18*f*). Many cells migrate around the wall of the cavity and form the **amnion**, a membrane that will enclose the embryo. Fluid in the amniotic cavity acts as a buoyant cradle in which an embryo grows and moves freely, protected from temperature changes and mechanical impacts.

As the amnion is forming, other cells move around the inner wall of the blastocyst, forming a lining that becomes a **yolk sac**. In animals that produce shelled eggs, this sac holds nutritive yolk. A human yolk sac does not hold yolk and does not support the embryo nutritionally. Some cells that form in a human yolk sac give rise to the embryo's first blood cells. Others give rise to the germ cells.

After a human blastocyst has implanted, an out-pouching of the yolk sac will become the **allantois**. Part of the allantois functions as the embryo's urinary bladder and part will be a component of the placenta.

The outermost membrane, the **chorion**, contacts the endometrium (Figure 27.18*g*,*h*). Chorionic villi, tiny fingerlike projections, extend from the chorion into the endometrium. They anchor a blastocyst and form part of the placenta.

Implantation of a blastocyst in the uterus prevents menstruation because some cells of the chorion secrete

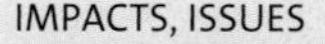

Identical twins form when an embryo splits, most often between days 3 and 8. Worldwide, the birth rate is about 4 per 1,000. It is similar among racial groups. Identical twins are descended from a single fertilized egg, and so they are genetically identical. Comparisons of twins are often used to sort out the extent to which a given trait is genetically determined.

LINKS TO
SECTIONS
16.9, 16.12

human chorionic gonadotropin (HCG). This is the first hormone that a new human produces and it prevents degeneration of the corpus luteum. By the beginning of the third week, HCG can be detected in a mother's blood or urine. At-home pregnancy tests have a "dip stick" that changes color if it is exposed to urine that contains this hormone.

THE ROLE OF THE PLACENTA

Even before the embryonic period starts, the uterus has been interacting with extraembryonic membranes in ways that will sustain the embryo's rapid growth. By the third week, tiny fingerlike projections from the chorion have grown profusely into the maternal blood that has pooled in endometrial spaces. The projections —chorionic villi—enhance the exchange of substances between the mother and the new individual. They are functional components of the placenta.

The **placenta** is a blood-engorged organ composed of endometrial tissue and extraembryonic membranes. At full term, it will make up about one-fourth of the inner surface of the uterus (Figure 27.19).

A placenta is the body's way of sustaining the new individual while allowing its blood vessels to develop apart from the mother's. The blood of an embryo does not come into direct contact with that of its mother. Instead, oxygen and nutrients must diffuse out of the mother's blood vessels, across the placenta's blood-filled spaces, and then into the embryo's blood vessels. Carbon dioxide and other wastes diffuse in the other direction. A mother's lungs and kidneys can quickly dispose of the wastes.

The placenta secretes progesterone and estrogens. Progesterone maintains the thickened uterine lining and prevents smooth muscle contractions that could expel the embryo. Estrogens encourage development of the mammary glands.

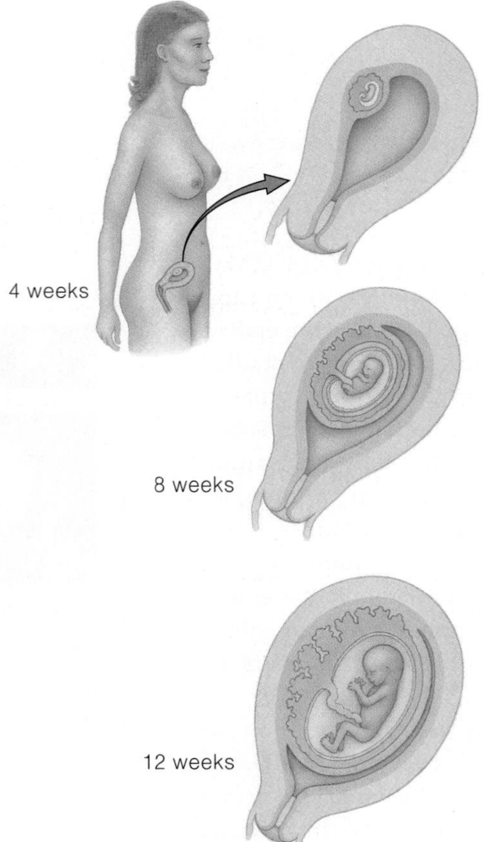

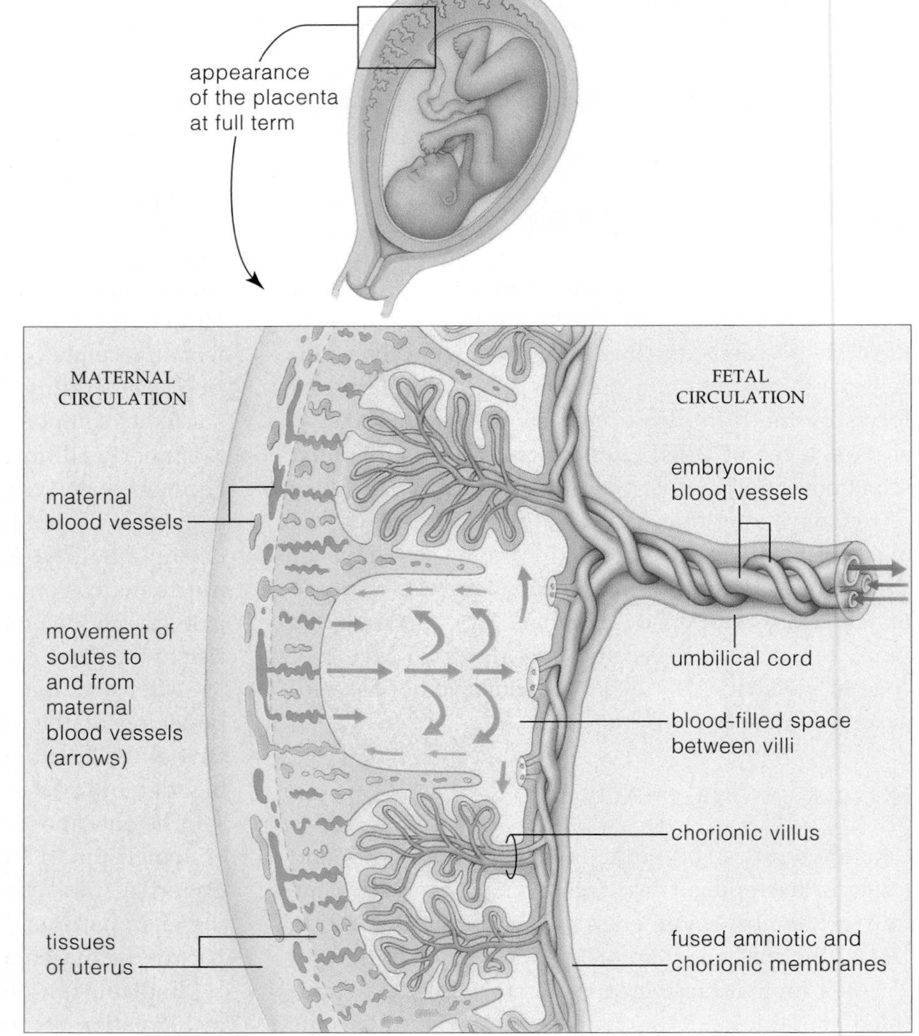

Figure 27.19 Relationship between fetal and maternal blood circulation in a full-term placenta. Blood vessels extend from the fetus, through the umbilical cord, and into chorionic villi. Maternal blood spurts into spaces between villi, but the two bloodstreams are not allowed to intermingle. Oxygen, carbon dioxide, and other small solutes diffuse across the placental membrane surface.

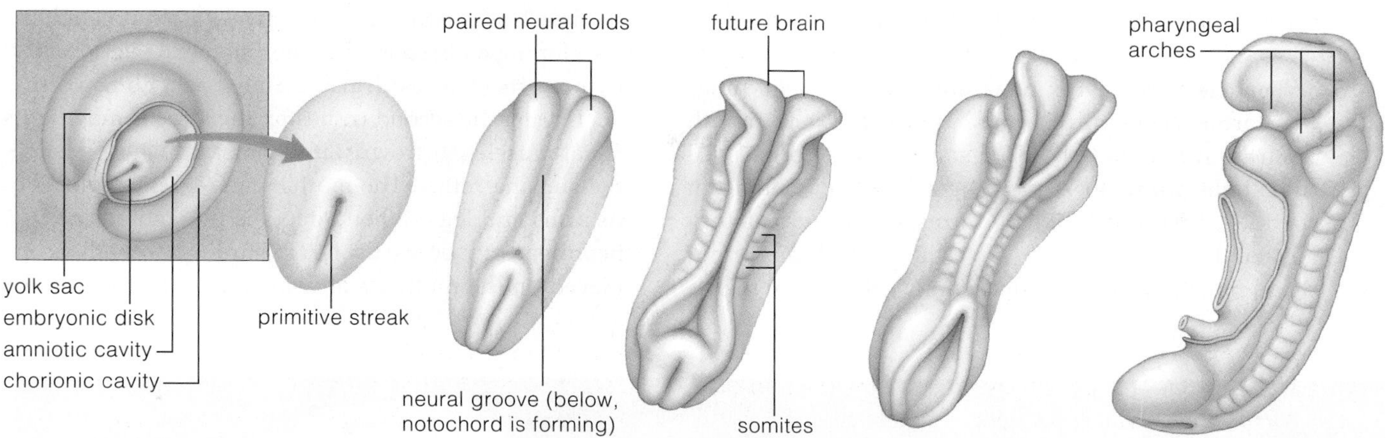

a **DAY 15.** A faint band appears around a depression along the axis of the embryonic disk. This is the primitive streak, and it marks the onset of gastrulation in vertebrate embryos.

b **DAYS 18–23.** Organs start to form through cell divisions, cell migrations, tissue folding, and other events of morphogenesis. Neural folds will merge to form the neural tube. Somites (bumps of mesoderm) appear near the embryo's dorsal surface. They will give rise to most of the skeleton's axial portion, skeletal muscles, and much of the dermis.

c **DAYS 24–25.** By now, some embryonic cells have given rise to pharyngeal arches. These will contribute to the formation of the face, neck, mouth, nasal cavities, larynx, and pharynx.

Figure 27.20 Hallmarks of the embryonic period of humans and other vertebrates. A primitive streak and then a notochord form. Neural folds, somites, and pharyngeal arches form later. (**a**,**b**) Dorsal views of the embryo's back. (**c**) Side view.

EMERGENCE OF THE VERTEBRATE BODY PLAN

By the time a woman misses her first menstrual period after fertilization, cleavage is over and gastrulation is under way. Gastrulation is the developmental stage when cell divisions, migrations, and rearrangements give rise to the primary tissue layers.

By now, the embryonic disk is surrounded by the amnion and chorion except where a stalk joins it to the chorion wall. The yolk sac lining has formed from one of the two layers by cell divisions and migrations.

The other layer now starts to generate the embryo proper. A depression appears on the embryonic disk, and its sides thicken by cell divisions and migrations. This "primitive streak" lengthens and thickens further the next day. It marks the onset of gastrulation (Figure 27.20*a*). It also defines the embryo's anterior–posterior axis and, in time, its bilateral symmetry.

Endoderm and mesoderm form from cells that are migrating inward at this axis. Now pattern formation begins, as outlined earlier in Section 27.2. Embryonic inductions and interactions among classes of master genes map out the basic body plan of all vertebrates. Specialized tissues and organs start forming in orderly steps in prescribed parts of the embryo.

For example, by the eighteenth day, the embryonic disk has two neural folds that will merge to form the neural tube (Figure 27.20*b*). The tube is the forerunner of the brain and spinal cord. Neural tube defects can arise if the tube does not form properly. *Spina bifida*, in which the spinal cord is not properly covered, is the most common of these defects.

Some mesoderm folds into a tube that becomes the notochord. In vertebrates, the notochord is a structural framework only; bony segments of a vertebral column will form around it. Toward the end of the third week, multiple paired segments, or **somites**, form from some of the mesoderm (Figure 27.20*b*). These are embryonic sources of most bones, of skeletal muscles of the head and trunk, and of the dermis overlying these regions.

Pharyngeal arches appear at an embryo's head end (Figure 27.20*c*). They look like the arches that give rise to fish gills, but they will develop into bone, cartilage, and muscles of the face, ears, larynx, and pharynx.

Cleavage produces a blastocyst, which implants itself in the uterus. Projections from its surface invade maternal tissue and connections form a placenta.

The placenta permits exchanges between the mother and embryo without intermingling their bloodstreams.

During the third week after fertilization—when a pregnant female has missed her first menstrual period—the basic vertebrate body plan emerges in the new individual.

A primitive streak, neural tube, somites, and pharyngeal arches form during the embryonic period of all vertebrates.

IMPACTS, ISSUES

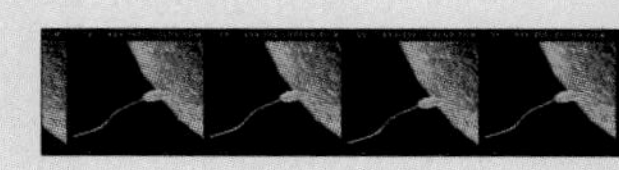

Conjoined twins form when an embryo splits after day 12. The split is not complete and the twins remain joined, most commonly at the chest or abdomen. In 2002, a team at UCLA's Mattel Hospital was able to successfully separate these two Guatemalan girls who had been joined at the head.

LINKS TO SECTIONS 16.9, 16.12

EMERGENCE OF DISTINCTLY HUMAN FEATURES

When the fourth week ends, the embryo is 500 times its starting size. The placenta has sustained the growth spurt, but now the pace slows as details of organs fill in. Limbs form; toes and fingers are sculpted from paddles. The umbilical cord forms, and the circulatory system becomes intricate. Growth of the all-important head now is faster than that of any other body region (Figure 27.21). After the eighth week is completed, the developing embryo is no longer merely "a vertebrate." By now, its features clearly define it as a human fetus.

During the second trimester, the fetus is moving its facial muscles. It frowns; it squints. It busily practices the sucking reflex. The mother can feel movements of its arms and legs. When a fetus is five months old, its heartbeat can be easily heard through a stethoscope placed on the mother's abdomen. Soft, fuzzy hair, the

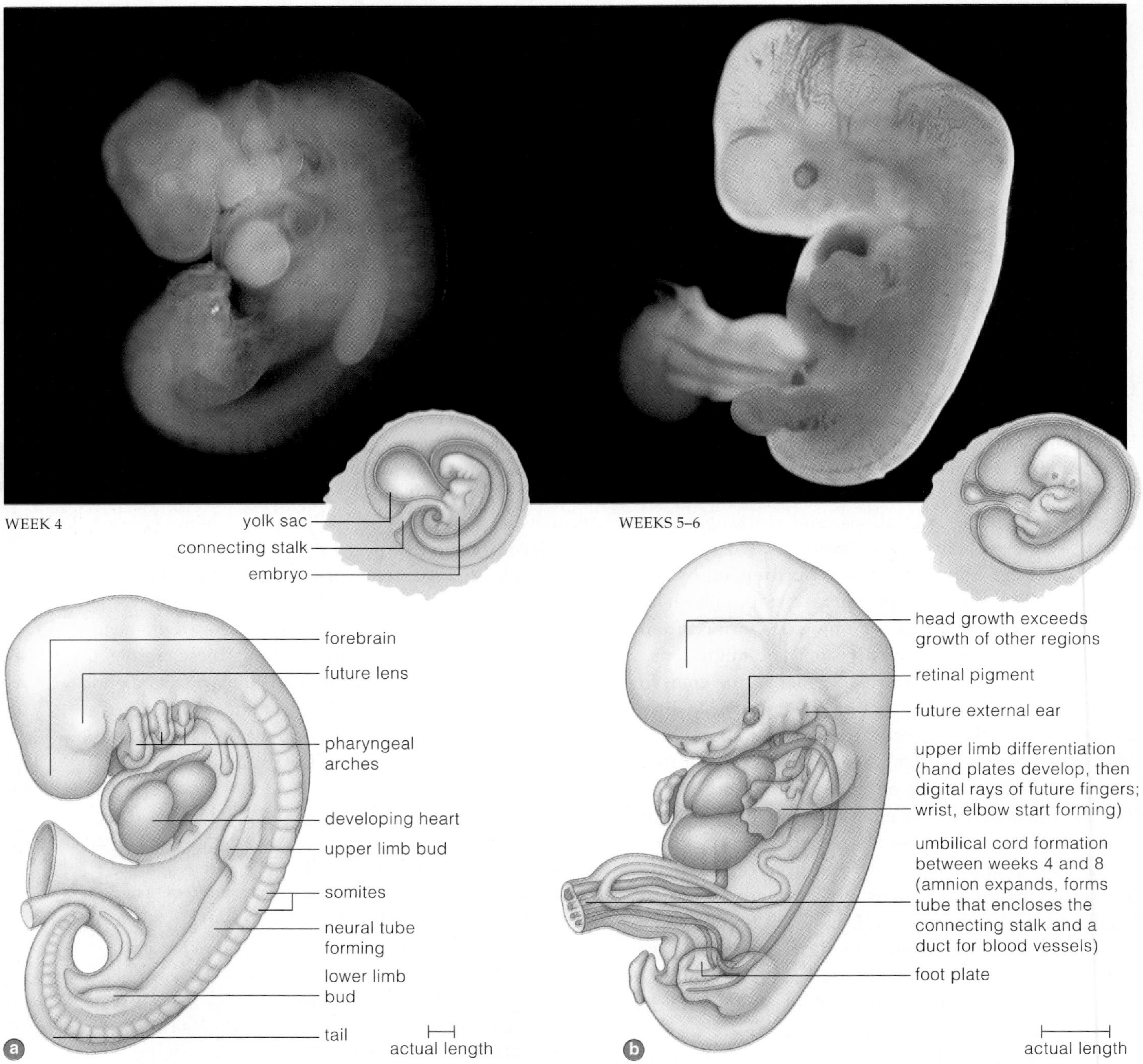

WEEK 4
yolk sac
connecting stalk
embryo
forebrain
future lens
pharyngeal arches
developing heart
upper limb bud
somites
neural tube forming
lower limb bud
tail
actual length
a

WEEKS 5–6
head growth exceeds growth of other regions
retinal pigment
future external ear
upper limb differentiation (hand plates develop, then digital rays of future fingers; wrist, elbow start forming)
umbilical cord formation between weeks 4 and 8 (amnion expands, forms tube that encloses the connecting stalk and a duct for blood vessels)
foot plate
actual length
b

Figure 27.21 Human development in the embryonic (**a–c**) and fetal (**d**) stages. All dates are given as weeks after fertilization.

lanugo, covers the fetus. The fetal skin is wrinkled, reddish, and protected from abrasion by a thickened, cheeselike coating. Eyelids and eyelashes form in the sixth month, and the eyes open in the seventh month.

The optimal birthing time, on average, is 38 weeks after the estimated time of fertilization. A fetus born prematurely (before 22 weeks) cannot survive. Births before 28 weeks are risky, mainly because the lungs have not developed enough. The risks start to drop after this. By thirty-six weeks, the liveborn survival rate is 95 percent. However, a fetus born between 36 and 38 weeks still has some trouble breathing and maintaining a normal core temperature.

In the fetal period, primary tissues that formed in the early embryo become sculpted into distinctly human features.

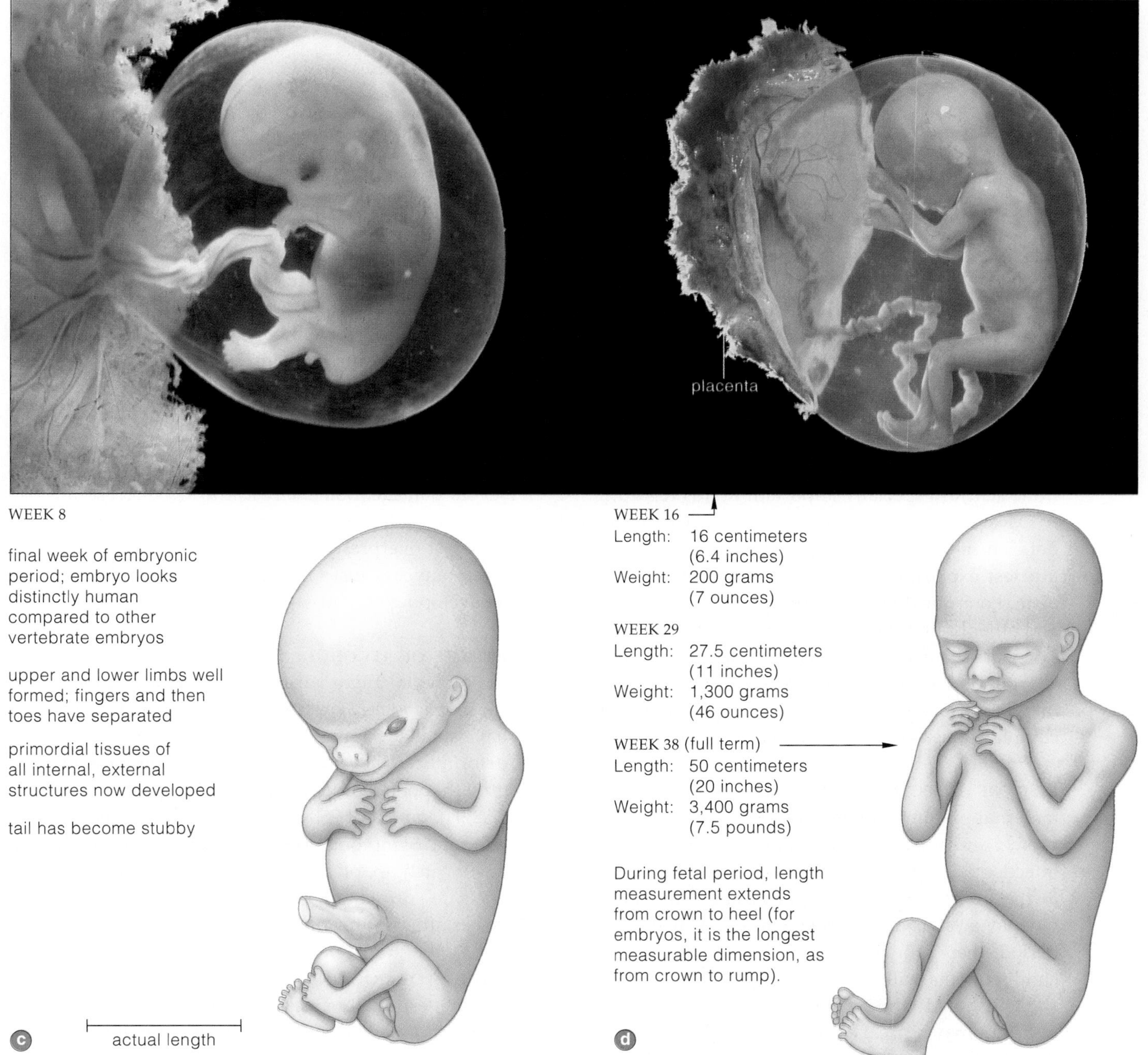

placenta
WEEK 8
final week of embryonic period; embryo looks distinctly human compared to other vertebrate embryos
upper and lower limbs well formed; fingers and then toes have separated
primordial tissues of all internal, external structures now developed
tail has become stubby
c
actual length
WEEK 16
Length: 16 centimeters (6.4 inches)
Weight: 200 grams (7 ounces)
WEEK 29
Length: 27.5 centimeters (11 inches)
Weight: 1,300 grams (46 ounces)
WEEK 38 (full term)
Length: 50 centimeters (20 inches)
Weight: 3,400 grams (7.5 pounds)
During fetal period, length measurement extends from crown to heel (for embryos, it is the longest measurable dimension, as from crown to rump).
d

LINKS TO SECTIONS 24.3, 25.6

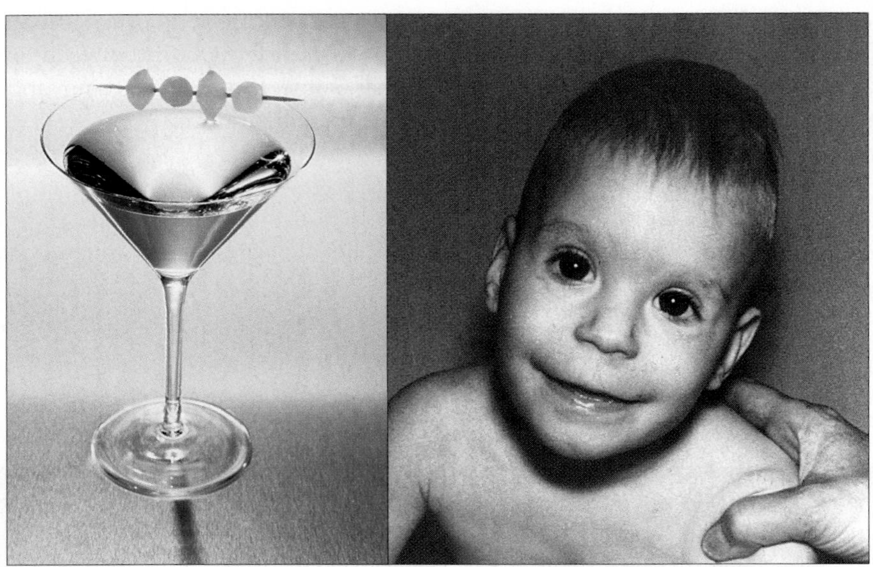

Figure 27.22 An infant with fetal alcohol syndrome—FAS. The obvious symptoms are low and prominently positioned ears, improperly formed cheekbones, and an abnormally wide, smooth upper lip. Growth-related complications and abnormalities of the nervous system can be expected.

MOTHER AS PROVIDER, PROTECTOR, POTENTIAL THREAT

NUTRITIONAL CONSIDERATIONS An embryo will get all of the proteins, carbohydrates, and lipids required for normal growth and development if a mother-to-be sticks to a well-balanced diet, as Section 24.3 outlines. However, supplements of some minerals and vitamins can help ensure that supplies are adequate for normal development. Increasing intake of B-complex vitamins before conception and in early pregnancy reduces the risk of severe neural tube defects in the embryo. Folate (folic acid) is especially important in this regard.

Most fetal organs are highly vulnerable to nutritional deficiencies. As one example, the brain undergoes its greatest expansion in the weeks just before and after birth. Poor nutrition at this time may affect intelligence and other brain functions later in life.

A pregnant woman must eat enough to gain 20 to 25 pounds, on average. If she does not, her newborn may be seriously underweight, at risk of postdelivery complications and, in time, impaired brain function.

RISK OF INFECTIONS Antibodies produced by the maternal immune system cross the placenta. They can protect an embryo or fetus from most serious bacterial infections. Some viral diseases are especially dangerous in the first six weeks after fertilization. If a pregnant woman gets *rubella* (German measles) in this critical period, there is a 50 percent chance that some of her child's organs will not form properly. For instance, if she is infected while embryonic ears are forming, her newborn may be deaf. If she is infected from the fourth month of pregnancy onward, the disease will have no notable effect. A woman can avoid this risk entirely by getting vaccinated against the virus before pregnancy.

EFFECTS OF PRESCRIPTION DRUGS Pregnant women should not take any drugs except under close medical supervision. To underscore this point, the tranquilizer *thalidomide* was routinely prescribed in Europe. Infants of women who used it during the first trimester had severely deformed arms and legs, or none at all. This drug has been withdrawn from the market. But other tranquilizers, sedatives, and barbiturates are still being prescribed, and they may cause similar although less severe damage. Some *anti-acne drugs* increase the risk of facial and cranial deformities. Tetracycline, one of the most overprescribed antibiotics, yellows the teeth. Streptomycin causes hearing problems and may harm a developing nervous system.

EFFECTS OF ALCOHOL Alcohol can pass freely across the placenta. Heavy drinking during a pregnancy can cause *fetal alcohol syndrome* (FAS). Affected individuals have smaller than normal heads and brains, and facial deformities. They suffer from mental impairment, slow growth, poor coordination, and heart problems (Figure 27.22). One out of 750 newborns in the United States is affected. Symptoms are not reversible; FAS children never catch up, physically or mentally. There might be no "safe" drinking level, and drinking as early as two weeks after fertilization may harm the fetus.

EFFECTS OF COCAINE A pregnant woman who uses cocaine, especially crack, disrupts the nervous system of her future child, who will be abnormally small and often unusually irritable early in life.

EFFECTS OF TOBACCO SMOKING Toxic compounds in tobacco smoke impair fetal growth and development. These compounds accumulate even inside the fetuses of pregnant nonsmokers who are simply exposed to secondhand smoke at home or in the workplace. One seven-year study of infants born during the same week in Great Britain revealed that even with income and other such variables factored out, children of mothers who smoked while pregnant were smaller, had more heart defects, and were more likely to have died. At age seven, the children of mothers who had smoked lagged behind those of nonsmokers in "reading age."

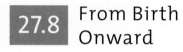

A pregnant woman shares drugs and toxins, as well as nutrients, with her developing child.

27.8 From Birth Onward

A woman gives birth nine months or so after fertilization. Yet human development does not end with birth, but rather continues for many years.

THE PROCESS OF BIRTH

Properties of the cervix are altered as a fetus nears full term. Up to this point, the cervix has remained firm and has helped keep the fetus from slipping out of the uterus prematurely. In the final weeks of pregnancy, the connective tissue of the cervix weakens. Collagen fibers become less tightly linked and the cervix gets thinner, softer, and more flexible. This prepares it to stretch enough to allow the passage of the fetus.

The birth process is known as **labor**, or parturition. Typically, the amnion ruptures just before birth, and the amniotic fluid drains out of the vagina. The cervix dilates and the fetus moves through it into the vagina, and then into the outside world (Figure 27.23).

The hormone **oxytocin** stimulates smooth muscle contractions during labor. As the fetus nears full term, it "drops," or shifts down; its head usually touches the cervix. Receptors in the cervix sense this mechanical pressure and signal the hypothalamus, which in turn causes oxytocin secretion by the posterior pituitary.

The binding of oxytocin to smooth muscles of the uterus increases the strength of contractions, which causes more mechanical pressure. This leads to more oxytocin secretion, and so on. The result is a positive feedback cycle—stretching of the cervix triggers more oxytocin secretion, which causes more stretching. This continues until the fetus has been expelled and there is no longer any mechanical pressure on the cervix. Synthetic oxytocin is sometimes given intravenously in order to induce or increase contractions.

Strong contractions help detach the placenta from the uterus and expel it, as the afterbirth. Contractions also help to stop the bleeding where the placenta was attached to the wall of the uterus. They cause blood vessels at this ruptured attachment site to constrict. The umbilical cord is cut and tied off. A few days after shriveling up, the cord's stump has become the navel.

Corticotropin-releasing hormone (CRH) affects the timing of labor, and it may contribute to *postpartum depression*. The hypothalamus makes this hormone in all humans, but during pregancy the placenta makes it as well. This additional source can increase the CRH levels in the blood by threefold.

CRH stimulates cortisol secretion by the adrenal cortex. Cortisol secretion, you will recall, is increased when the body is under stress. It may help a mother to cope with the strains of pregnancy and labor.

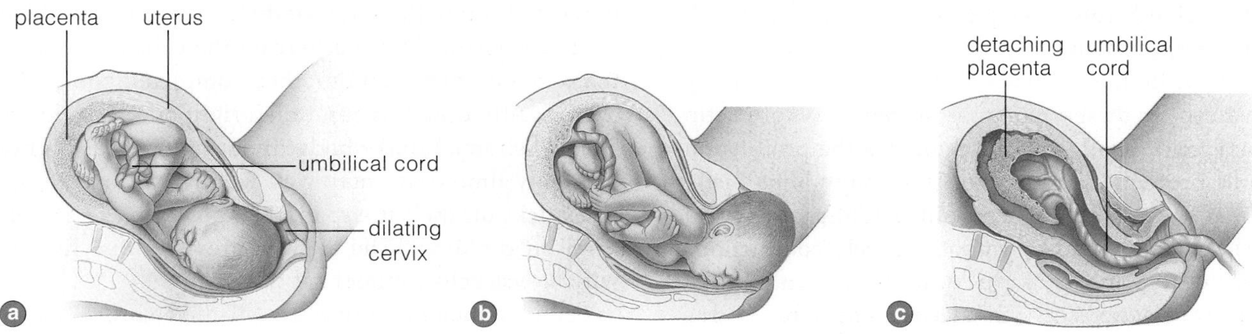

Figure 27.23 Labor and birth. (**a**) Most common position of a human fetus at the onset of labor. (**b**) Muscle contractions propel the fetus out of the uterus and through the vagina. (**c**) The afterbirth consists of the placenta, tissue fluid, and blood.

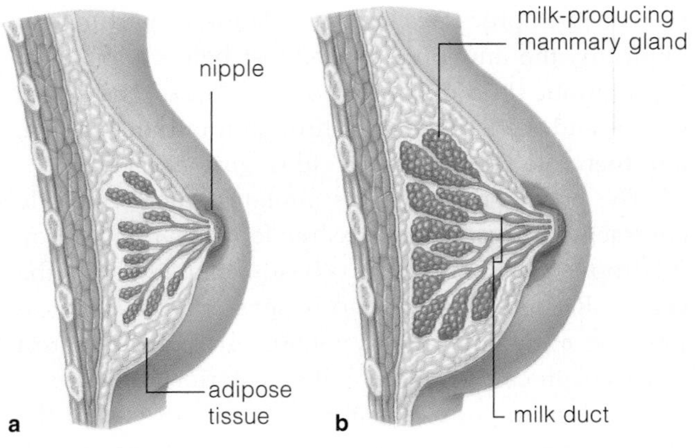

Figure 27.24 (**a**) Breast of a woman who is not pregnant. (**b**) Breast of a lactating woman.

During pregnancy, a high blood level of cortisol can suppress CRH production by the hypothalamus. After birth, with the CRH-secreting placenta suddenly gone, the level of cortisol plummets. Cortisol affects mood and its sudden decrease can trigger a brief postpartum depression. Depression ends after the hypothalamus resumes its role as the main source of CRH.

NOURISHING THE NEWBORN

Once the lifeline to the mother is severed, a newborn enters the extended time of dependency and learning that is typical of all primates. Early survival requires an ongoing supply of milk or a nutritional equivalent. **Lactation**, or milk production, occurs in mammary glands inside a mother's breasts (Figure 27.24). Before pregnancy, breast tissue is largely adipose tissue and a system of undeveloped ducts. Their size depends on how much fat they hold, not on their milk-producing ability. During pregnancy, estrogens and progesterone stimulate the development of a glandular system for milk production. **Prolactin** secreted by the anterior pituitary gland stimulates the growth of cells that will produce milk proteins.

After birth, the mechanical stimulus of suckling by a newborn increases maternal secretion of prolactin and oxytocin. The prolactin stimulates the production of milk proteins. For the first few days after birth, mammary glands produce a fluid rich in proteins and lactose. Oxytocin causes smooth muscle contractions that force fluid into milk ducts. It also causes smooth muscle contractions in the uterus, which helps this organ shrink to its pre-pregnancy size.

In addition to its nutrients, human breast milk has immunoglobulins that enhance resistance to infection. Other components of breast milk stimulate the growth of symbiotic bacteria in an infant's gut. Alcohol, drugs, mercury, and other toxins in a mother's body can also be secreted in milk. HIV and some other viruses are carried by milk as well.

POSTEMBRYONIC DEVELOPMENT

As is the case for many species, humans change in size and proportion until reaching sexual maturity. Figure 27.25 shows a few of the proportional changes that occur as the life cycle unfolds. Table 27.2 defines the prenatal ("before birth") stages and the postnatal ("after birth") stages of a human life.

Postnatal growth is most rapid between the years thirteen and nineteen. Sex hormone secretions increase and bring about the development of secondary sexual traits as well as sexual maturity. Not until adulthood are the bones fully mature.

Body tissues are usually well maintained into early adulthood. As more years pass, they slowly deteriorate through processes collectively described as *aging*. The main hypotheses for why we age are our next topic.

WHY DO WE AGE AND DIE?

As the years pass, all animals undergo aging; tissues become harder to maintain and repair. Each species has a maximum life span—150 years for tortoises, 20 years for dogs, 12 weeks for butterflies, 35 days for fruit flies, and so on. The oldest human for whom we can verify age lived for 122 years.

According to one hypothesis, aging occurs because cells can only divide a finite number of times. If so, an animal body might be analogous to a clock shop, with each type of cell, tissue, and organ ticking away at its own genetically set pace. Paul Moorhead and Leonard Hayflick tested this hypothesis. They cultured human embryonic cells—which generally divided about fifty times before dying out.

Hayflick also took cultured cells that were part of the way through the series of divisions and froze them for a few years. After he thawed the cells and placed them in a culture medium, they completed an *in vitro* cycle of fifty doublings and died right on schedule.

No cell in a human body divides more than eighty or ninety times. You may well wonder: If an internal clock ticks off their life span, then how can *cancer* cells keep on dividing? The answer provides insight into why *normal* cells cannot beat the clock.

Cells, remember, duplicate all their chromosomes before dividing. Capping each chromosome's ends are **telomeres** made of DNA and proteins. Telomeres keep the chromosome ends from unraveling. They act like glue or plastic protectors on the ends of shoelaces. A small piece of each telomere is lost with each nuclear division. When only a nub is left, cells can no longer divide and they die.

Cancer cells and germ cells are exceptions to the rule of limited cell divisions; both make telomerase, an enzyme that makes telomeres lengthen. When cells being grown in culture are given telomerase, they go on dividing well beyond the normal life span.

The premature death of some cloned animals, such as the sheep Dolly, also supports idea that telomeres set a cell divisions clock. Dolly began her life with telomeres from another sheep; they had already been frayed during that animal's lifetime.

By another hypothesis, the accumulation of damage at the molecular and cellular levels causes aging. Both environmental assaults and spontaneous mistakes by DNA repair mechanisms may cause the damage.

For example, the rogue molecular fragments called free radicals attack all biological molecules, including DNA. This includes the DNA of mitochondria, the power plants of eukaryotic cells. Structural changes in DNA compromise the synthesis of enzymes and other proteins necessary for metabolism. Free radicals are implicated in many age-related problems, including cataracts, atherosclerosis, and Alzheimer's disease.

DNA replication and repair problems also have been implicated in aging. *Werner's syndrome*, a genetic aging disorder, occurs when there is a mutation in the gene for Werner's protein. Among its functions, this protein serves as a helicase; it unwinds DNA. Affected people develop normally until puberty, when they fail to show a growth spurt.

The mutation associated with Werner's syndrome allows gene mutations to accumulate at high rates. Sooner or later, damage interferes with cell division. The result of this damage usually becomes apparent in the patient's thirties. The hair grays, the skin thins, cataracts form, and bone mass declines. Most affected people die in their late forties.

It is likely that both hypotheses of aging have some merit. Aging probably occurs as an outcome of many interconnected processes in which genes, hormones, environmental assaults, and a decline in DNA repair mechanisms come into play.

The human life cycle flows naturally from the time of birth, growth, and development, to production of the individual's own offspring, and on through aging to the time of death.

A biological clock ticks off the life span of all body cells, and DNA becomes increasingly damaged over time. These factors and others result in aging.

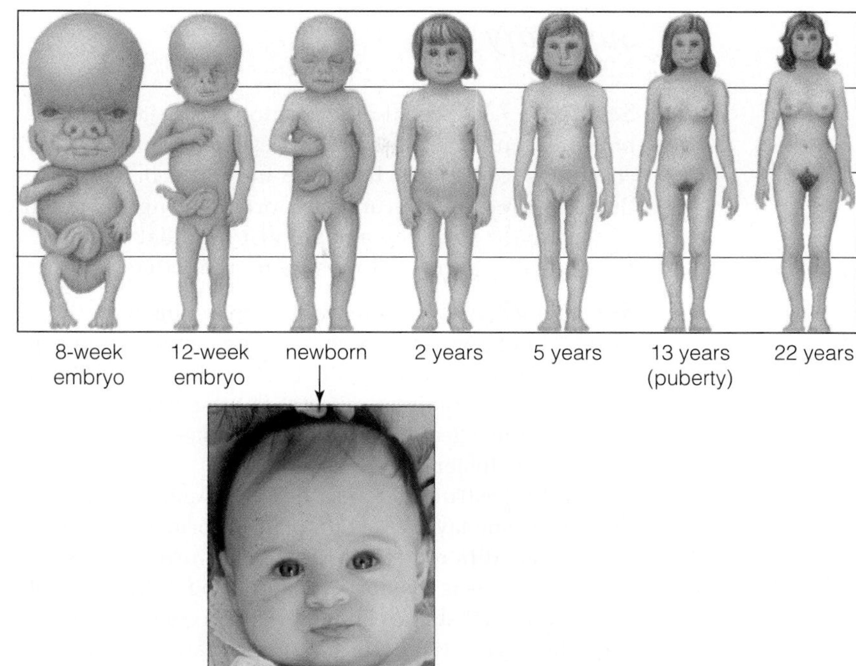

Figure 27.25 Observable, proportional changes in the human body during prenatal and postnatal growth. Changes in overall physical appearance are slow but noticeable until the teenage years. For example, compared to an embryo, the legs of teenagers are longer and the trunk shorter, so the head is proportionally smaller. Correlate these drawings with the stages in Table 27.2.

Table 27.2 Stages of Human Development

Prenatal period	
Zygote	Single cell resulting from fusion of sperm nucleus and egg nucleus at fertilization.
Morula	Solid ball of cells produced by cleavages.
Blastocyst	Ball of cells with surface layer, fluid-filled cavity, and inner cell mass (the mammalian blastula).
Embryo	All developmental stages from two weeks after fertilization until end of eighth week.
Fetus	All developmental stages from ninth week to birth (about thirty-eight weeks after fertilization).
Postnatal period	
Newborn	Individual during the first two weeks after birth.
Infant	Individual from two weeks to about fifteen months after birth.
Child	Individual from infancy to about ten or twelve years.
Pubescent	Individual at puberty; secondary sexual traits develop; girls between 10 and 15 years, boys between 12 and 16 years.
Adolescent	Individual from puberty until about 3 or 4 years later; physical, mental, emotional maturation.
Adult	Early adulthood (between 18 and 25 years); bone formation and growth finished. Changes proceed very slowly after this.
Old age	Aging processes result in expected tissue deterioration.

Summary

Section 27.1 Sexual reproduction has costs in terms of energy outlays. It requires specialized reproductive structures, control mechanisms, and forms of behavior that assist fertilization and support offspring. A benefit is phenotypic variation among offspring that may aid in survival and thus help assure reproductive success.

Section 27.2 Sexual reproduction begins with the formation of gametes, sperm and eggs, that combine at fertilization to form a zygote. During cleavage, mitosis increases the number of cells but not the original volume of the zygote. Cleavage yields a blastula—a hollow, fluid-filled cluster of cells.

During gastrulation, cells become organized into primary tissue layers. Organs start to form as cells in these layers differentiate (sets of genes turn on or off). Morphogenesis is the development of body form by cell migrations, cell shape changes, and apoptosis. In the final stage, growth and tissue specialization, organs enlarge and develop specialized properties.

Similar master genes control development in all animals. Diffusion of master gene products through a body sets up gradients that act in pattern formation.

BiologyNow
Watch a frog develop and see how the neural tube is formed with the animation on BiologyNow.

Section 27.3 Testes are primary male reproductive organs (Table 27.3). They produce sperm and the sex hormone, testosterone. Pituitary hormones (LH, FSH) and testosterone affect sperm production. Accessory glands produce the other components of semen. These are added as the sperm pass through a series of ducts that convey them out of the body.

BiologyNow
Investigate the structure and function of the male reproductive system with the animation BiologyNow.

Section 27.4 Ovaries are the female's primary reproductive organs (Table 27.3) They produce oocytes and sex hormones: estrogens and progesterone.

In the follicular phase of the menstrual cycle, FSH stimulates maturation of an ovarian follicle. A follicle secretes estrogen, which stimulates thickening of the lining of the uterus. A midcycle surge of LH triggers ovulation, release of a secondary oocyte from an ovary.

In the menstrual cycle's luteal phase, progesterone and estrogen secreted by the corpus luteum causes thickening of the endometrium. When fertilization does not occur, the corpus luteum degenerates, the uterine lining is shed, and the menstrual cycle starts again

BiologyNow
Use the animation and interaction on BiologyNow to learn about female reproductive structures and the cyclic changes in the ovary and uterus.

Section 27.5 Fertilization occurs when a secondary oocyte meets up with a sperm, usually in the oviduct. The oocyte completes meiosis II, thus becoming the mature egg, or ovum. It fuses with the sperm nucleus.

Pregnancy can be avoided by behavior, or through diverse methods that prevent ovulation, fertilization, or implantation. In vitro fertilization brings sperm and eggs together for fertilization outside the body. An abortion ends a pregnancy.

BiologyNow
See what happens during fertilization with the animation on BiologyNow.

Section 27.6 Sexually transmitted diseases (STDs) are caused by protozoan, bacterial, and viral pathogens and are spread by unsafe sex. Viral STDs are incurable. Bacterial and protozoal STDs can be cured. If untreated, they can cause sterility and harm health.

Section 27.7 After fertilization, a blastocyst forms by cleavage and burrows into the wall of the uterus. Membranes form around the blastocyst. Some combine with the maternal tissue to form a placenta that passes oxygen and nutrients between mother and embryo. The placenta can also deliver harmful toxins.

By the end of the eighth week, when the embryo becomes a fetus, it appears distinctly human.

BiologyNow
Observe fertilization and implantation and learn how the placenta forms and functions with the animation on BiologyNow.

Section 27.8 Hormones prepare a woman's body for pregnancy, for labor, and for nursing. During labor, uterine contractions expel the fetus and afterbirth. Milk supplies nutrients and contains antibodies that help a newborn resist infection. Development and growth continue until adulthood.

Table 27.3 Organs of the Human Reproductive Tracts

Male Reproductive Tract	
Testes	Sperm production; sex hormone production
Epididymides	Sperm maturation and storage
Vasa deferentia	Convey sperm from epididymides to ejaculatory ducts
Ejaculatory ducts	Convey sperm from vas deferens to the urethra
Urethra	Conveys sperm through penis and out of the body
Penis	Organ of sexual intercourse
Female Reproductive Tract	
Ovaries	Egg production, maturation; sex hormone production
Oviducts	Convey oocyte from ovary to uterus
Uterus	Chamber in which new individual develops
Cervix	Entrance into uterus; mucus secretion
Vagina	Organ of sexual intercourse; birth canal

Self-Quiz

Answers in Appendix I

1. Sexual reproduction among animals is ______.
   a. biologically costly c. evolutionarily beneficial
   b. diverse in its details d. all of the above

2. A cell formed during cleavage is a ______.
   a. zygote c. blastomere
   b. morula d. gastrula

3. ______ produces three primary tissue layers.
   a. Gametogenesis c. Gastrulation
   b. Implantation d. Pattern formation

4. Homeotic genes control ______.
   a. the rate of cleavage c. apoptosis
   b. overall body plan d. cell migrations

5. Meiotic divisions of germ cells in the ______ give rise to sperm.
   a. seminiferous tubule c. penis
   b. prostate gland d. epididymis

6. During a menstrual cycle, a midcycle surge of ______ triggers ovulation.
   a. estrogen b. progesterone c. LH d. FSH

7. During the luteal phase of the menstrual cycle, the corpus luteum secretes ______.
   a. LH c. progesterone
   b. FSH d. prolactin

8. A ______ implants in the lining of the uterus.
   a. zygote c. blastocyst
   b. gastrula d. fetus

9. Which of the following shows human developmental stages in the correct order?
   a. zygote, blastocyst, embryo, fetus
   b. zygote, embryo, blastocyst, fetus
   c. zygote, embryo, fetus, blastocyst
   d. blastocyst, zygote, embryo, fetus

10. Which of the following STDs are caused by bacteria?
   a. chlamydia d. trichomoniasis
   b. gonorrhea e. a and b
   c. genital warts f. all of the above

11. The embryonic neural tube will become the ______.
   a. digestive tract c. coelom lining
   b. skeletal muscles d. brain and spinal cord

12. Secretion of oxytocin stimulates ______.
   a. contraction of uterine smooth muscle
   b. flow of milk from mammary glands
   c. ovulation e. a and b
   d. implantation f. all of the above

13. Match each human reproductive structure with the most suitable description.

____ testis	a. maternal *and* fetal tissues
____ cervix	b. stores mature sperm
____ placenta	c. produces testosterone
____ vagina	d. produces estrogen and progesterone
____ ovary	e. usual site of fertilization
____ oviduct	f. lining of uterus
____ epididymis	g. birth canal
____ endometrium	h. entrance to uterus

Additional questions are available on **Biology Now™**

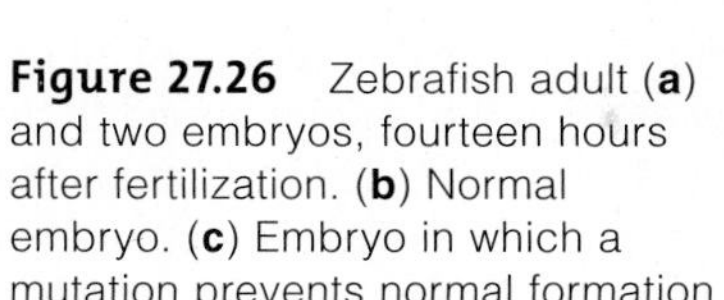

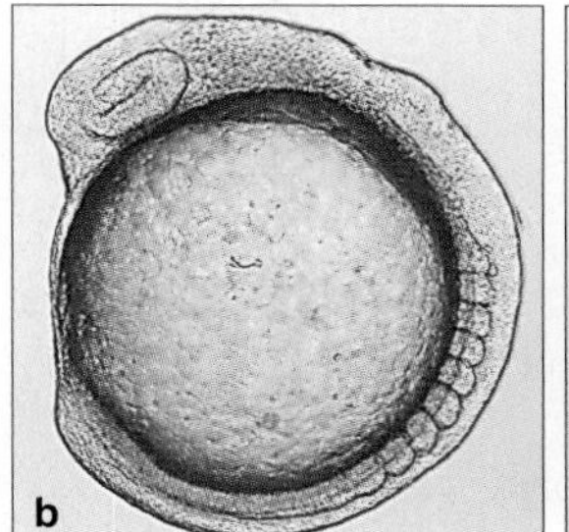

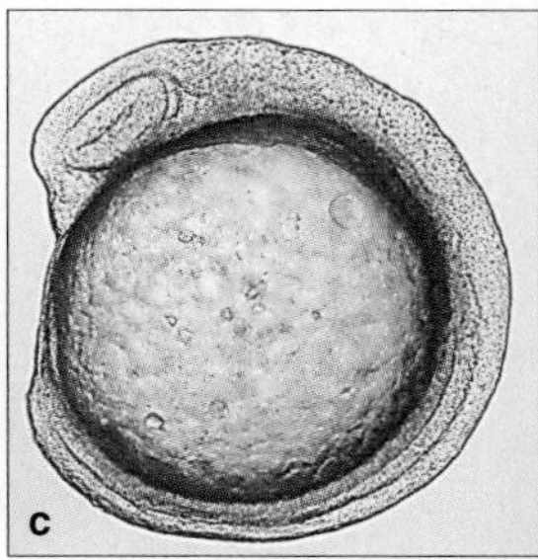

Figure 27.26 Zebrafish adult (**a**) and two embryos, fourteen hours after fertilization. (**b**) Normal embryo. (**c**) Embryo in which a mutation prevents normal formation of somites.

Critical Thinking

1. The zebrafish (*Danio rerio*) is special to developmental biologists. This small freshwater fish is easily maintained in tanks. A female produces hundreds of eggs, which can develop and hatch in three days. The transparent embryos let researchers directly observe developmental events. Cells can be injected with dye to see how they change position, or they can be killed or injected with genes and the effects observed. Zebrafish have even traveled in space aboard a space shuttle for experiments that assessed the effects of zero gravity on their development.

   One lethal mutation in zebrafish interferes with the formation of somites (Figure 27.26). Explain why an inability to form somites would be bad for an embryo.

2. Fraternal twins, which are nonidentical genetically, arise when two oocytes mature and are released and fertilized at the same time. Such twins run in families. The incidence varies among ethnic groups and is highest among blacks and lowest in Asians. The variation may be an outcome of differences in gene products that affect the FSH level in blood. Explain how a high FSH level would increase the likelihood of fraternal twins.

3. It is common knowledge that a woman's fertility declines with age. What about men? The record holder for oldest father is currently Les Cooley. He was almost ninety-four years old when his son was born. But new evidence suggests that delaying fatherhood can be risky. Researchers at UC Berkeley found that the number of motile sperm men produce decreases by about 0.7 percent per year after age thirty-five. Other studies suggest that increased paternal age is associated with higher incidences of birth defects and schizophrenia in offspring. Unlike women, who have all the oocytes they will ever have at birth, men continually produce new gametes. What do you think explains the changes in a man's sperm?

4. By UNICEF estimates, each year 110,000 people are born with abnormalities as a result of rubella infections. Major symptoms of *congenital rubella syndrome,* or CRS, are deafness, blindness, mental impairment, and heart problems. A child of a nonvaccinated woman infected in the first trimester of pregnancy is at risk, but infection during the final trimester usually has no ill effect on development. Explain how this is possible.

5. Every litter of nine-banded armadillos consists of four genetically identical siblings. Explain how this happens and why the young are not identical to their parents.

28 POPULATION ECOLOGY

IMPACTS, ISSUES

The Human Touch

In 1722, on Easter morning, a European explorer landed on a small volcanic island and found a few hundred skittish, hungry Polynesians living in caves. He saw dry grasses and scorched shrubs, but no trees. He noticed about 200 massive stone statues near the coast and 700 unfinished, abandoned ones in inland quarries. Some weighed fifty tons. What did they represent?

Two years later James Cook visited and saw only four canoes on the whole island. Nearly all of the statues had been tipped over, often onto face-shattering spikes.

Later, researchers solved the mystery of the statues. Easter Island, as it came to be called, is only 165 square kilometers (64 square miles) in size. Voyagers from the Marquesas discovered this eastern outpost of Polynesia around A.D. 350. The place was a paradise. Its fertile soil supported dense forests and lush grasses. New arrivals built canoes from long, straight palms strengthened with rope made of fibers from hauhau trees. They used wood as fuel to cook fish; they cleared forests to plant crops. They also had many children.

By 1400, possibly as many as 15,000 people were living on the island. Crop yields declined; erosion and harvesting had depleted the soil's nutrients. In time, fish vanished from nearshore waters around the island. All the native birds had been eaten and people were raising rats as food.

Survival was at stake; those in power appealed to the gods. They directed the people to carve divine images of unprecedented size, and to use a system of greased logs to move them over miles of rough terrain to the coast. By about 1550, no one ventured offshore to fish. No one could build canoes because people had cut down all the palms and used all the hauhau trees for firewood. Easter Islanders turned to their last source of protein. They started to hunt and eat one another.

Central authority crumbled and gang wars raged. Those on the rampage burned the remaining grasses to destroy hideouts. The dwindling population retreated to caves and launched raids against perceived enemies. Winners ate losers and tipped over the statues. There was nowhere to go. What could they have been thinking when they chopped down the last palm?

From archeological and historical records, we know Easter Island sustained a society that flourished amid abundant resources, and then abruptly fell apart as those resources were exhausted. What happened on this small bit of land in the vastness of the Pacific is a chilling reminder that all resources are finite and no population can soar forever.

With this example in mind, we turn to the principles that govern the growth of all populations. Later in the chapter, we'll consider how they apply to the 6.4 billion humans now inhabiting Earth, our own island in space.

How Would You Vote? *With few predators, deer populations are soaring so high that they are harming many forests. Deer hunting is the least expensive way to keep deer populations in check. Do you support encouraging hunting in areas where the presence of too many deer is degrading the habitat and threatening other species? See BiologyNow for details, then vote online.*

Key Concepts

DESCRIBING POPULATIONS
Beyond its shared phenotypic traits, every population has a characteristic size, density, and age structure. It also shows patterns of distribution and growth.

LIFE HISTORY PATTERNS
Longevity, fertility, and the age when reproduction begins are examples of life history traits. Like other heritable traits, they are subject to natural selection.

HUMAN POPULATIONS
Humans have sidestepped certain limits on population growth that affect other species. Our population size has soared as a result.

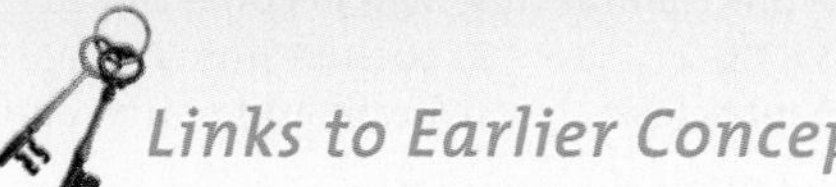

Links to Earlier Concepts

As described in Section 1.1, a population is one level of organization in nature. Populations grow when individuals reproduce sexually or asexually (7.1, 14.2, 27.1). Environmental challenges (17.2) can limit population growth, and so can infectious disease (14.6, 23.8). Think back on how humans evolved (16.13), as you read about how they sidestepped many controls on population growth, and keep contraception methods in mind as we discuss controlling human populations (27.6).

Recall too how natural selection can influence the frequency of traits that characterize a population (12.3, 12.5). This will help you understand how life history traits can be altered.

28.1 Characteristics of Populations

*Certain principles govern the growth and sustainability of populations. They are the bedrock of **ecology**, the systematic study of how organisms interact with one another and their physical and chemical environment. The interactions start within and between populations, and extend through communities, ecosystems, and the biosphere. These are topics of this last unit of the book.*

From earlier discussions, you know that a population consists of all the members of a species living in some defined area at a particular time. The gene pool of a population, along with environmental factors, is the basis for morphological, physiological, and behavioral traits. When studying a population, ecologists collect information about the genes, modes of reproduction, and behavior of its component individuals. They also consider **demographics**. These are vital statistics that describe the population, such as its size, age structure, density, and distribution.

OVERVIEW OF THE DEMOGRAPHICS

Population size refers to the number of individuals that make up a population's gene pool. A population's **age structure** is the number of individuals in each age category. For instance, members may be grouped into *pre-reproductive*, *reproductive*, and *post-reproductive* ages. Pre-reproductive individuals will have the capacity to produce offspring later, when they mature. Together with the individuals that are already able to reproduce, they make up the population's **reproductive base**.

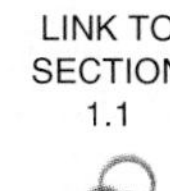
LINK TO SECTION 1.1

Population density is the number of individuals in some specified area or volume of a habitat, such as the number of frogs per acre of rainforest. A *habitat*, recall, is the type of place where a species normally lives. We characterize a habitat by physical features, chemical features, and other species sharing the same habitat.

Crude density is a measured number of individuals in some specified area. You'll see how ecologists make counts that can help them track changes in population density over time. The counts do not reveal how much of a habitat is utilized as living space. Even areas that seem uniform, such as a long, sandy beach, are more like tapestries of light, moisture, temperature, mineral composition, and other variables. Also, environmental conditions may change from day to night or with the seasons, so part of a habitat might be more suitable than other parts some or all of the time.

Different species that share the same area typically compete for energy, nutrients, living space, and other resources. As later chapters explain, species interact in ways that benefit or harm one another. For now, it's enough to know that species interactions influence a population's density and dispersion through a habitat.

Population distribution describes how individuals are dispersed in a specified area. There are three types of distribution patterns: clumped, nearly uniform, or random (Figure 28.1). In nature, individuals of most species tend to clump together, rather than spread out randomly or uniformly.

Why is a clumped distribution so common? First, each species is adapted to a limited set of ecological conditions that usually occur in patches. For example, animals often cluster in an area that provides access to food, water, or nesting sites. Plants survive only where soil, water, and sunlight are adequate. Second, many

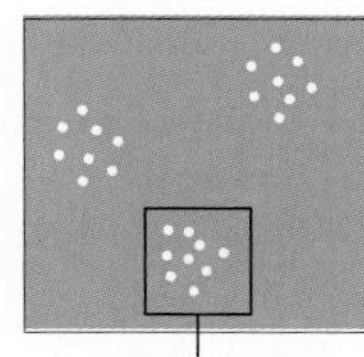
clumped

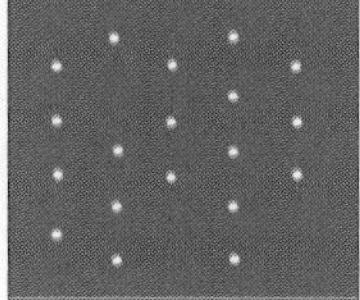
nearly uniform

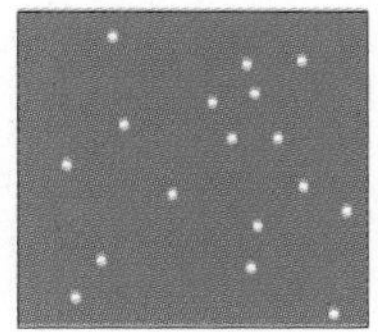
random

Figure 28.1 Three patterns of population distribution: clumped, as in squirrelfish schools; more or less uniform, as a penguin breeding colony; and random, as when wolf spiders prowl on a forest floor.

LINKS TO SECTIONS 7.1, 27.1

animal species form social groups. These groups can promote survival and reproduction, as by providing more opportunities for mating and mutual defense against predators. Third, the offspring of many species do not disperse far from their parents. As the saying goes, the nut doesn't fall very far from the tree. Many larvae and immature forms of animals also have little mobility. For example, sponges are usually clumped because the larvae settle down close to their parents. Asexual reproduction also results in clumping.

When individuals are more evenly spaced than we would expect them to be by chance, they show a near uniform dispersal pattern. This pattern often arises as a result of territoriality or stiff competition for some essential resource. Many nesting colonies of seabirds show this distribution because each pair actively keeps other individuals from settling too close to them.

We observe random dispersion only when habitat conditions are nearly uniform, resource availability is fairly constant, and members of a population neither attract nor avoid one another. Wolf spiders, which are solitary hunters on forest floors, are an example. Each generation may be randomly spaced.

The regional scale we use to consider distribution can affect what we observe. For example, although sea birds will display a nearly uniform distribution within their colony, their colony sites are usually clumped. In addition, seabirds that show a clumped distribution at mating time are randomly distributed during other times of the year.

ELUSIVE HEADS TO COUNT

From a practical standpoint, measuring a population's size or density, or determining how it is distributed, can can be challenging. Making an absolute count of all individuals within the population's range is nearly always impossible. Many species are difficult to locate, with members distributed over a wide area. With few exceptions, biologists study population characteristics by making their counts in some portion or portions of a population's range, and then using that data to make estimates about the population as a whole.

For example, you could get a map of some region and divide it into small plots, or quadrats. **Quadrats** are sampling areas of the same size and shape. You could then count and collect data about individuals in these plots and use that to make estimates about the general population. Ecologists often use this method when they are studying plants and other species that do not move around much.

Suppose you want to study a population of animals that roam from place to place. How can you determine if the individuals you count in a given plot are the same ones you counted earlier in a different plot?

The population density of mobile animals is very often estimated by **capture–mark–recapture methods**. The idea is to capture individuals and mark them in some way. Deer get collars, squirrels get dye, salmon get fin tags, birds get leg rings, butterflies get wing markers, and so on (Figure 28.2). Marked animals are released at time 1. Some time later, at time 2, the traps are reset. The proportion of marked individuals in the second group of captives is taken to be representative of the proportion marked in the whole population:

$$\frac{\text{Marked individuals in sampling at time 2}}{\text{Total captured in sampling 2}} = \frac{\text{Marked individuals in sampling at time 1}}{\text{Total population size}}$$

Ideally, the marked and unmarked individuals of the population are captured at random, none of the marked animals die during the course of the study, and none migrate out of the study area. In the real world, recapturing marked individuals might *not* be random. Squirrels marked after being attracted to bait in boxes might now avoid the boxes, or seek them out. In addition, markings may affect an animal's health or make it more conspicuous to predators.

The timing of such studies can affect the results. Few places yield abundant resources all year long, so many animal populations move between habitats as seasons change. In such cases, the capturing, marking, and then recapturing of individuals might be carried out more than once a year, for several years.

Figure 28.2 Animals marked for population studies. (**a**) Costa Rican owl butterfly (*Caligo*) and (**b**) Florida Key deer.

Each population has characteristic demographics: size, density, distribution pattern, and age structure.

Environmental conditions and species interactions shape these characteristics, which may change over time.

		Net Monthly Increase:	New Population Size:
G = r ×	3,920 =	1,568 =	5,488
r ×	5,488 =	2,195 =	7,683
r ×	7,683 =	3,073 =	10,756
r ×	10,756 =	4,302 =	15,058
r ×	15,058 =	6,023 =	21,081
r ×	21,081 =	8,432 =	29,513
r ×	29,513 =	11,805 =	41,318
r ×	41,318 =	16,527 =	57,845
r ×	57,845 =	23,138 =	80,983
r ×	80,983 =	32,393 =	113,376
r ×	113,376 =	45,350 =	158,726
r ×	158,726 =	63,490 =	222,216
r ×	222,216 =	88,887 =	311,103
r ×	311,103 =	124,441 =	435,544
r ×	435,544 =	174,218 =	609,762
r ×	609,762 =	243,905 =	853,667
r ×	853,677 =	341,467 =	1,195,134

a

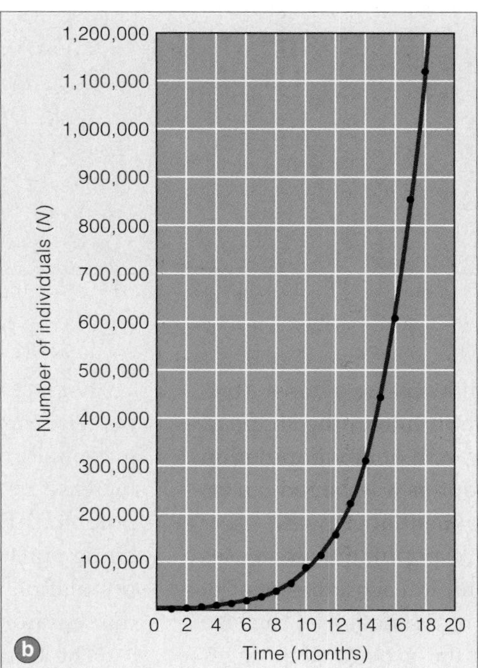

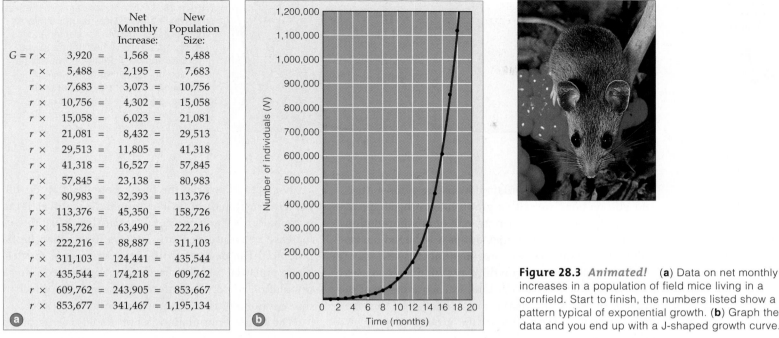

Figure 28.3 ***Animated!*** (**a**) Data on net monthly increases in a population of field mice living in a cornfield. Start to finish, the numbers listed show a pattern typical of exponential growth. (**b**) Graph the data and you end up with a J-shaped growth curve.

28.2 Population Size and Exponential Growth

When resources are abundant, birth rates often exceed death rates and a population grows faster and faster. Each generation includes more breeding individuals than the prior one did, so each produces more offspring.

FROM ZERO TO EXPONENTIAL GROWTH

We measure changes in population size in terms of birth rates, death rates, and how many individuals are entering and leaving during a specified interval. The size increases by births and immigration. **Immigration** is a permanent addition of individuals that previously belonged to another population. Population size falls by deaths and emigration. **Emigration** is a permanent loss of individuals to another population. **Migration** is a round trip between regions. Many species migrate daily or seasonally, but because individuals return to the starting point, we need not consider this transient effect in our initial study of population size.

For our purposes, assume immigration is balancing emigration over time, so we may ignore the effects of both on population size. Doing so allows us to define **zero population growth** as an interval during which the number of births and the number of deaths are balanced. The population's size is stabilized during an interval, with no overall increase or decrease.

Births, deaths, and other variables that might affect population size can be measured as **per capita** rates, or rates per individual. *Capita* means heads, as in head counts. Imagine 2,000 mice in a cornfield. Twenty or so days after their eggs get fertilized, the female mice produce a litter, nurse offspring for a month or so, and get pregnant again. If they give birth to 1,000 mice every month, the birth rate will be 1,000/2,000 = 0.5 per mouse per month. If 200 of the 2,000 mice die in that interval, the death rate would be 200/2,000 = 0.1 per mouse per month.

If we assume the birth rate and death rate remain constant, we can combine both into one variable: the net reproduction per individual per unit time, or ***r***. In this example, r is $0.5 - 0.1 = 0.4$ per mouse per month. In equation form, $G = rN$, which simply means:

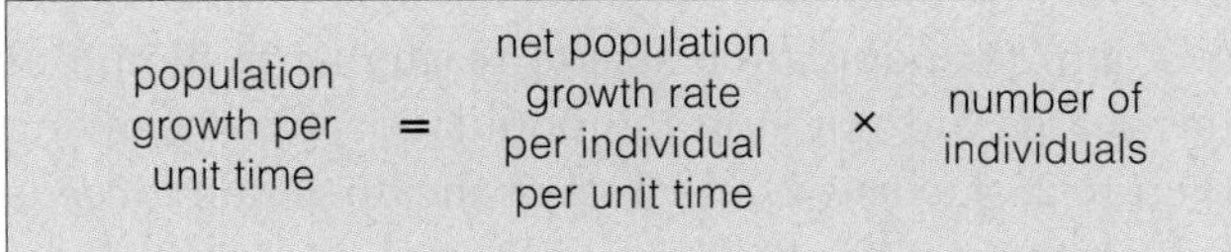

population growth per unit time
=
net population growth rate per individual per unit time
×
number of individuals

As the next month begins, 2,800 mice are scurrying about. With that net increase of 800 fertile mice, the reproductive base is larger. Assume r doesn't change. A net increase of $0.4 \times 2{,}800 = 1{,}120$ puts the population at 3,920 (Figure 28.3*a*). Over time, r stays constant, and a growth pattern emerges. In less than two years from the first count, the number of mice in the cornfield has increased from 2,000 to more than a million!

LINKS TO SECTIONS 14.2, 16.13

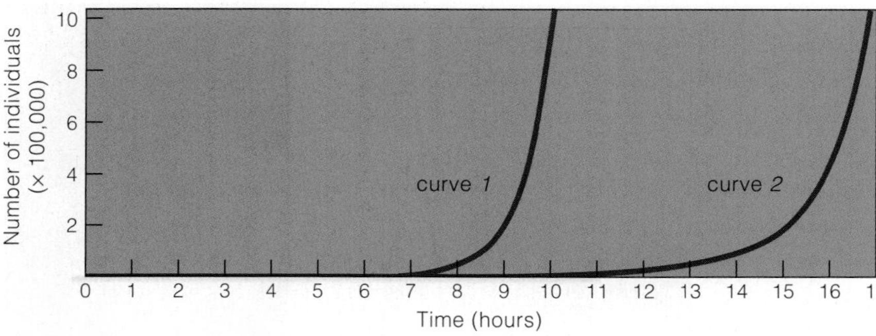

Figure 28.4 Effect of deaths on the rate of increase in two bacterial populations. Plot population growth for bacterial cells that reproduce every half hour and you get growth curve *1*. Plot the growth of a population of cells that divide every half hour, with 25 percent dying between divisions, and you get growth curve *2*. Deaths slow the rate of increase but can't directly prevent exponential growth.

Plot the monthly population increases against time, and you will end up with a graph line in the shape of a J, as in Figure 28.3*b*. When growth of any population over increments of time plots out as a J-shaped curve, you know you are tracking exponential growth.

With **exponential growth**, a population increases by a *fixed* percentage of the total in each interval. The number of births and deaths are a function of the size of a population. With exponential growth, the bigger the reproductive base of a population, the more new individuals will be added during a specified interval. The rate of increase for the population gets higher and higher, even though the per capita rate of increase does not change.

Consider a lone bacterium in a culture flask with all the nutrients it requires for growth. Thirty minutes later, the cell divides in two. Thirty minutes pass, and each of these two cells divides, and so on every thirty minutes. If no cells die between divisions, population size will double in each interval—from 1 to 2, then 4, 8, 16, 32, and so on. The length of time it takes for a population to double in size is its **doubling time**.

The larger the population gets, the more cells there are to divide. After 9–1/2 hours (nineteen doublings), it has more than 500,000 bacterial cells. After 10 hours (twenty doublings), there are more than 1 million cells. Plot the size of this population against time, and you will get Figure 28.4, curve *1*.

To see the effect of deaths on the growth rates, start over. Assume 25 percent of the first cell's descendants die every half hour. It now takes about seventeen hours (not ten) for population size to reach one million. But the deaths did not stop the population from growing (Figure 28.4, curve *2*). As long as the birth rate exceeds the death rate, a population will grow exponentially.

IMPACTS, ISSUES

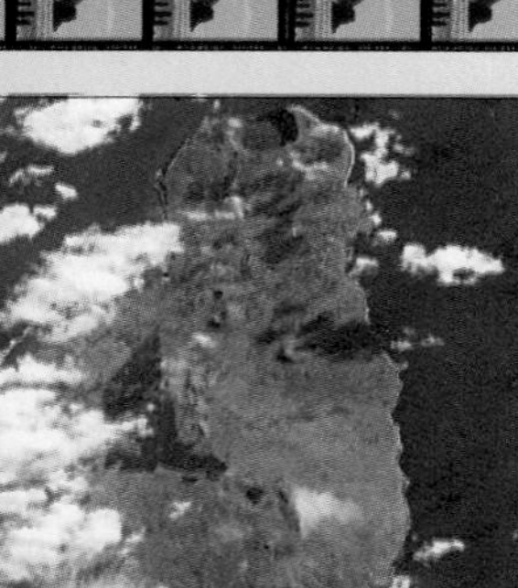

The carrying capacity of an environment can vary over time. Drought may have accelerated the unraveling of the agriculturally based society on Easter Island, shown in the satellite image above. Long-term drought may have been a factor in the crop failures, deforestation, and increased social unrest. This volcanic island has few streams. Most of the rainfall seeps into underground basins. Even today, islanders face severe water shortages, often for extended times.

WHAT IS THE BIOTIC POTENTIAL?

Every species has a **biotic potential**, a maximum rate of increase per individual under ideal conditions. This is the theoretical rate at which the population would increase if resources were unlimited and there were no predators, parasites, or pathogens around.

For many bacteria, the biotic potential can be 100 percent every half hour. For humans and other large mammals, it's 2 to 5 percent per year. The *actual* rate of increase in nature is nearly always lower than biotic potential. The degree of difference between a species' biotic potential and the observed rate of increase for a population in some environment shows the effect of that environment on that population.

The human population is not displaying its biotic potential. A woman is biologically able to bear twenty or more children over the course of her lifetime, but few do. Even so, our population size has been growing exponentially ever since the mid-eighteenth century, for reasons we will discuss later in this chapter.

During a specified interval, population size is generally an outcome of births, deaths, immigration, and emigration.

With exponential growth, population size increases by a fixed percentage of the whole in each interval, so its reproductive base gets larger and larger over time. A plot of population size against time produces a J-shaped curve.

As long as the per capita birth rate remains above the per capita death rate, a population will grow exponentially.

28.3 Limits on the Growth of Populations

In nature, limits on resource availability usually prevent population size from expanding above a certain point. The quantity of resources varies among environments, and over time within an environment.

WHAT ARE THE LIMITING FACTORS?

Nearly all the time, environmental circumstances keep any population from fulfilling its biotic potential. That is why sea stars, the females of which could produce 2,500,000 eggs each year, do not fill up the oceans with

Figure 28.5 Amazing response to a scarcity of nesting sites, a limiting factor for weaver bird populations in dry habitats in Africa.

(**a**) African weavers construct densely woven, cup-shaped nests that are only wide and deep enough for a hen and her nestlings. Many nests hang from the same spindly limbs.

(**b**) This nest in Namibia is like an apartment house in a place where few trees are available. Between 100 and 300 pairs of sparrow weavers occupy their own flask-shaped nests, each with its own tubular entrance.

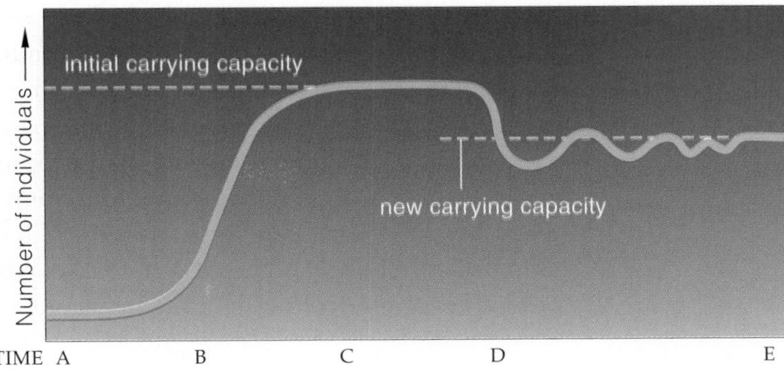

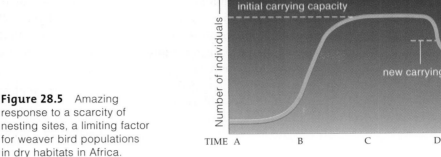

Figure 28.6 ***Animated!*** Idealized S-shaped curve characteristic of logistic growth. Growth slows after a phase of rapid increase (time B to C). The curve flattens out as the carrying capacity is reached (time C to D). S-shaped growth curves can show variations, as when changes in the environment lower the carrying capacity (time D to E).

sea stars. That also is why populations of humans will never fill up the planet.

To get a sense of what some of the constraints may be, start again with a bacterial cell in a culture flask, where you can control the variables. First you enrich the culture medium with glucose and other nutrients necessary for bacterial growth. Then you sit back and let bacterial cells reproduce for many generations.

At first the growth pattern seems to be exponential. Then growth slows, and population size is relatively stable. After the stable period, the size plummets until all the bacterial cells are dead. *What happened?* As the population grew ever larger, it used up more and more nutrients. Nutrient scarcity became an environmental signal for cells to stop dividing. When the supply was exhausted, the cells starved to death.

Any essential resource that is in short supply is a **limiting factor** on population growth. Food, mineral ions, refuge from predators, living space, and even a place to build a nest are examples (Figure 28.5). The number of limiting factors can be extensive, and their effects can vary. Even so, a scarcity of one factor alone often puts the brakes on population growth.

What if you had kept on freshening the supply of nutrients for the bacterial population? After growing exponentially, that population would have collapsed anyway. Like all other organisms, the bacteria produce metabolic wastes. Confined populations of bacteria drastically alter their own living conditions. By their own metabolic activities, they pollute experimentally designed habitats and, in so doing, put a stop to any further exponential growth.

CARRYING CAPACITY AND LOGISTIC GROWTH

Now visualize a small population, with its individuals dispersed throughout the habitat. As its size increases, more and more individuals must share nutrients, living quarters, and other resources. As the share available to each diminishes, fewer individuals may be born, and more may die by starvation or nutrient deficiencies. The population's growth rate will slow until births are balanced (or outnumbered) by deaths. Ultimately, the *sustainable* supply of resources determines how large the population can become. **Carrying capacity** is the maximum number of individuals of a species that a given environment can sustain indefinitely.

The pattern of **logistic growth** shows how carrying capacity can affect population size. By this pattern, a small population starts growing slowly in size, then it grows rapidly, and finally its size levels off once the carrying capacity is reached. Figure 28.6 reveals how the pattern plots out as an S-shaped curve. Here is a way to represent this pattern as an equation:

population growth per unit time	=	maximum net population growth rate per individual per unit time	×	number of individuals	×	proportion of resources not yet used

LINK TO SECTION 14.6

An S-shaped curve is only an approximation of what actually happens in nature. A population that is growing really fast may overshoot carrying capacity. If this happens, the death rate skyrockets, and the birth rate plummets. Population size may drop even below carrying capacity. The human population on Easter Island went through such a soar-and-crash experience. Figure 28.7 shows another example, the explosion and die-off of a reindeer population on an island.

DENSITY-DEPENDENT CONTROLS

The logistic growth equation just described deals with **density-dependent controls**. These controls are factors that come into play when population density increases. They adversely affect survival and reproduction.

When a small, rapidly growing population reaches its carrying capacity, dwindling resources work as a control to stabilize or decrease numbers. Besides the environmental factors, interactions among members of the population may keep the number of individuals below the maximum sustainable level.

Crowding often leads to interactions that increase death rates. Predators, parasites, and pathogens may exert greater effects when prey or host populations are at a high density. They usually reduce the numbers of prey or hosts. By thinning a population, they remove the condition that invited their controlling effect, and the population size may increase again.

Bubonic plague and *pneumonic plague*, two terrible diseases, are examples of density-dependent controls. Both are caused by a bacterium, *Yersinia pestis*. A large reservoir of *Y. pestis* persists inside rabbits, rats, and certain other small mammals. Fleas transmit the pest by biting new hosts. Bacterial cells reproduce rapidly in the flea gut. In time, their numbers and metabolic activities disrupt digestion and cause gnawing hunger sensations. They provoke the flea into feeding more often on more of its hosts, and so the disease spreads.

Bubonic plague devastated Europe in the fourteenth century. It swept through cities where many people were crowded together, sanitary conditions were poor, and rats were abundant. That one epidemic claimed 25 million lives. Plagues are still threats.

DENSITY-INDEPENDENT FACTORS

Sometimes events cause more deaths or fewer births regardless of a population's density. For instance, each year, millions of monarch butterflies fly down from Canada to Mexico's forested mountains, where they spend the winter (Figure 28.8). But logging cleared out many forest trees, which normally buffer temperature. In 2002 a sudden freeze, combined with deforestation, killed millions of butterflies. The events were **density-independent factors**, meaning that their effects came into play independently of population density.

Similarly, heavy applications of pesticides in your backyard may also kill most insects, mice, cats, birds, and other animals. They will do this regardless of how sparse or dense the populations are.

Resources in short supply put limits on population growth. Together, all of the limiting factors acting on a population dictate how many individuals can be sustained.

Carrying capacity is the maximum number of individuals of a population that can be sustained indefinitely by the resources in a given environment. The number may rise or fall with changes in resource availability.

The size of a low-density population may increase slowly, go through a rapid growth phase, then level off once the carrying capacity for the population is reached. This is a logistic growth pattern.

Density-dependent controls and density-independent factors can bring about decreases in population size.

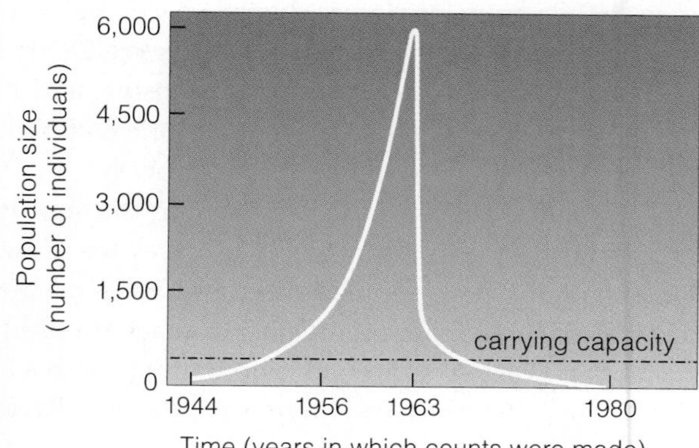

Figure 28.7 Carrying capacity of a reindeer herd on an island. In 1944, twenty-nine reindeer were introduced to St. Matthew Island in the Bering Sea. With no predators, numbers grew into the thousands. Then, during the winter of 1963–64, a scarcity of food coupled with a heavy snowfall decimated the herd. In 1966, the island supported only forty-two reindeer. The curve reflects how the population overshot carrying capacity, and then crashed.

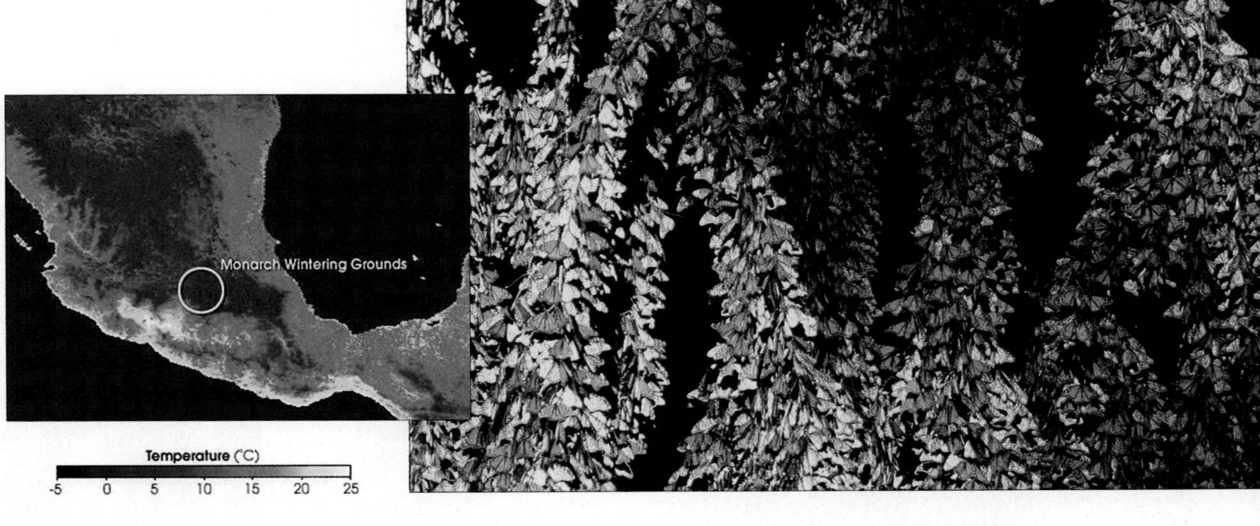

Figure 28.8 Monarch butterflies at their wintering ground in central Mexico. Each year these migratory insects travel hundreds of kilometers south to places that normally have been cool and humid in winter. If they were to stay in their northern breeding grounds, monarchs would risk being killed by more severe weather conditions.

Even so, deforestation and winter freezes in Mexico are now killing millions of them annually. The satellite image records the winter freeze in January 2002.

28.4 Life History Patterns

Researchers have identified age-specific adaptations that affect the survival, fertility, and reproduction of individuals of each species.

So far, we have looked at populations as if all of their members are identical during any given interval. For most species, however, the individuals of a population are at many different stages in a life cycle. Each stage may interact in a different way with other organisms and with the environment. Different stages may make use of different resources, as when larvae eat leaves and butterflies sip nectar. They may also be more or less vulnerable to predation or other threats.

In short, each species has a **life history pattern**, or a set of adaptations that influence longevity, fertility, and age at first reproduction. Each pattern reflects the individual's schedule of reproduction. In this section and the next, we look at a few of the environmental variables that underlie these age-specific patterns.

LIFE TABLES

Although each species has a characteristic life span, few of its individuals get to the maximum age possible. Death looms larger at some ages than others. Another variable: Individuals of a species tend to reproduce or emigrate during a characteristic age interval.

Age-specific patterns in populations are of interest to life insurance companies, as well as ecologists. Such investigators typically track a **cohort,** a collection of same-aged individuals, from the cohort's birth until all are dead. They also track the number of offspring born to all individuals during each interval. Life tables present this data as an age-specific death schedule. Often the death data are converted into much cheerier "survivorship" schedules: the number of individuals reaching some specified age (x). Table 28.1 is a typical example. It lists data for the human population of the United States in 2001.

PATTERNS OF SURVIVAL AND REPRODUCTION

Evolutionarily speaking, we measure the reproductive success of individuals in terms of the number of their surviving offspring. But that number will vary among species, which differ in (1) how much energy and time are allocated to producing gametes, securing mates, and parenting, and in (2) the size of offspring. It seems

Table 28.1 Life Table for the Human Population of the United States in 2001

Age Interval	Number at Start of Interval	Number Dying During Age Interval	Life Expectancy at Start of Interval	Reported Live Births
0–1	100,000	684	77.2	
1–5	99,316	132	76.7	
5–10	99,184	76	72.8	
10–15	99,108	96	67.9	7,315
15–20	99,012	330	62.9	525,493
20–25	98,682	468	58.1	1,022,106
25–30	98,214	471	53.4	1,060,391
30–35	97,743	554	48.6	951,219
35–40	97,189	801	43.9	453,927
40–45	96,388	1,154	39.2	95,788
45–50	95,234	1,682	34.7	5,244
50–55	93,552	2,373	30.3	263
55–60	91,179	3,474	26.0	
60–65	87,705	5,186	21.9	
65–70	82,519	7,397	18.1	
70–75	75,122	10,018	14.6	
75–80	65,104	13,284	11.5	
80–85	51,820	15,877	8.8	
85–90	35,943	16,147	6.5	
90–95	19,796	11,906	4.8	
95–100	7,890	5,845	3.6	
100+	2,045	2,045	2.7	

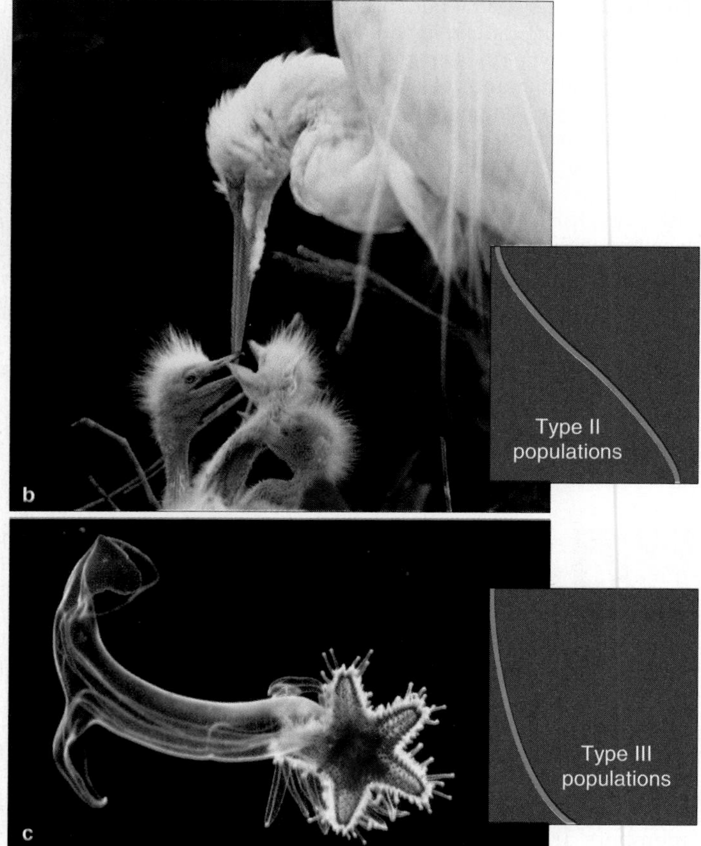

Figure 28.9 Three generalized survivorship curves. (**a**) Elephants and petunias are Type I populations. They both have high survivorship until some age (seventy years for an elephant, one for a petunia), then high mortality. (**b**) Snowy egrets are Type II populations. They have a fairly constant death rate. (**c**) Sea star larvae represent Type III populations, which show low survivorship early in life.

LINKS TO SECTIONS 12.3, 12.5

that trade-offs have been made in response to selection pressures, such as prevailing conditions in the habitat and the type of species interactions.

A **survivorship curve** is the graphical equivalent of a life table. It shows the percentage of cohort members that remain alive at each age interval. Three general types of survivorship curves are possible. Each type of curve is associated with a particular way of allocating resources to reproduction.

Type I curves reflect high survivorship until fairly late in life, then a large increase in deaths. Such curves typify annual plants as well as large mammals that bear only one or a few large offspring at a time, then engage in extended parental care (Figure 28.9*a*). As an example, a female elephant gives birth to four or five calves and devotes several years to parenting each. Type I curves also are typical of human populations with widespread access to modern health care services.

Type II curves reflect a more or less constant death rate throughout the life span. These curves are typical of organisms that are just as likely to be killed or to die of disease at any age, such as many lizards, small mammals, and large birds (Figure 28.9*b*).

Type III curves signify a death rate that is highest early in life. We see this for species that produce many small offspring and do little, if any, parenting. Most trees show this type of curve. So do nearly all marine invertebrates (Figure 28.9*c*). As one example, a sea star releases enormous numbers of eggs. The tiny larvae must feed, grow, and finish developing on their own without any support, protection, or guidance from parents. Most are devoured in the first days of life. The mortality rate drops as the sea stars increase in size and take on their adult spiny-skinned form.

At one time, it was thought that selection would tend to favor *either* early production of many small offspring *or* late production of a few large offspring. We now realize that the two patterns are extremes at opposite ends of a range of possible life histories. Both types of life history patterns, as well as intermediate ones, are sometimes evident in different populations of the same species, as the next example makes clear.

EVOLUTION OF LIFE HISTORY TRAITS

Any trait that has a genetic basis and varies among the members of a population has the potential to evolve in response to natural selection. Life history traits, such as age of first reproduction and number of offspring per reproductive event, are no exception.

In one long-term study, biologists David Reznick and John Endler documented the evolutionary effects of predation on the life history traits of one species of small fish, the guppy, *Poecilia reticulata*. Their study began with fieldwork in the mountains of Trinidad, an island in the southern Caribbean Sea. Here, guppies live in shallow freshwater streams (Figure 28.10).

Waterfalls in the streams act as barriers; they keep guppies from moving from one part of the stream to another, effectively isolating them genetically. Waterfalls also keep predators from moving from one part of a stream to another. As a result, different guppy populations are preyed upon by different predators.

The two kinds of guppy predators in this habitat, killifishes and pike-cichlids, differ in their body size. A killifish is a relatively small fish. It preys on the small, immature guppies but ignores big adults. Pike-cichlids are large fish. They tend to pursue big, mature guppies, and ignore small ones. Many of the streams have one of these predators, but not the other.

Reznick and Endler discovered that the guppies in streams with pike-cichlids grow faster and their body size is smaller at maturity, compared to the guppies in killifish-containing streams. Also, guppies in streams with pike-cichlids reproduce earlier, have more offspring at a time, and breed more often (Figure 28.11).

To determine whether these life history differences have a genetic basis, guppies from the two habitats were reared for several generations in aquariums. The results support the hypothesis that the differences are inherited. Descendants of guppies from streams where there were pike-cichlids showed different life history traits than descendants of guppies from streams with killifishes, even when both groups were raised under identical conditions.

Based on their field and laboratory observations, Reznick and Endler hypothesized that predation is a selective agent that shapes guppy life history patterns.

Figure 28.10 Evolutionary biologist David Reznick contemplating interactions among guppies (*Poecilia reticulata*) and their predators in a freshwater stream in Trinidad.

They set up a field experiment by introducing some guppies to a site upstream from one waterfall. Before the experiment, this waterfall had been an effective barrier to dispersal; it had prevented guppies and all predators except killifish from moving upstream. The guppies that were introduced to the experimental site were from a population that had evolved with pike-cichlids downstream from the waterfall. That part of the stream was designated the control site.

Eleven years later, researchers revisited the stream. They found that the guppy population at the upstream experimental site had evolved. Exposure to a novel predator had caused shifts in growth rate, age at first reproduction, and other life history traits. Guppies at the control site showed no such changes.

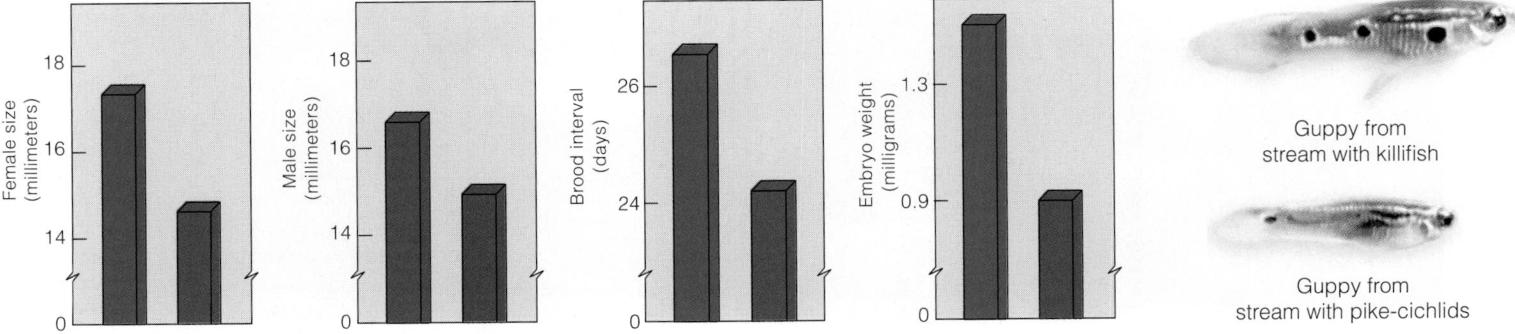

Figure 28.11 Variation in traits among guppies subject to different predation pressures. *Red* bars represent fish that evolved with killifish. *Green* bars are fish that evolved with pike-cichlids. Killifishes eat small guppies and pike-cichlids eat larger ones. The photos show males from two populations.

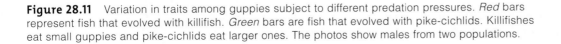

LINKS TO SECTIONS 16.13, 23.8, 27.6

The evolution of life history traits in response to predation pressure is not merely of theoretical interest. It can have enormous commercial importance. In the same way that predation caused evolution of guppies, fishing by humans may have caused rapid evolution in the North Atlantic codfish (*Gadus morhua*). From the mid-1980s to the early 1990s, as fishing pressure on cod increased, the fish matured at increasingly smaller ages and sizes.

In 1992, the Canadian government banned fishing for cod in some areas, but this and later closures came too late to prevent the population from plummeting. The population still has not recovered. Looking back, it seems that life history changes were an early sign that the cod population was in trouble. Had biologists recognized this, they might have been able to save the fishery and protect the livelihood of more than 35,000 fishers and associated workers. Ongoing monitoring of life history data for other economically important fishes may help prevent similar disasters in the future.

Tracking a group of individuals from birth until death provides data about age-specific survival and other life history characteristics. Like other traits, life history patterns can vary both between species and among populations of one species and can be altered by natural selection.

28.5 Human Population Growth

The world's human population has surpassed 6.4 billion. Now consider this: The annual additions to that base population will result in a larger absolute increase each year, well into the foreseeable future.

Figure 28.12 shows human population growth over time. Rapid population growth continues even though more than 1 billion people are already confronted by limits to growth. They are malnourished or starving. They are without clean water, adequate shelter, access to modern health care, or sewage treatment.

In 2004, the average rate of increase for the human population was 1.2 percent per year. As long as birth rate exceeds the death rate, these additions will result in a *larger* absolute increase each year.

WHY HAVE HUMAN POPULATIONS SOARED?

For most of its existence, the human population grew slowly. Then, during the past two centuries, growth rates skyrocketed. There are three main reasons. First, humans steadily developed their capacity to expand into many new habitats and into new climate zones.

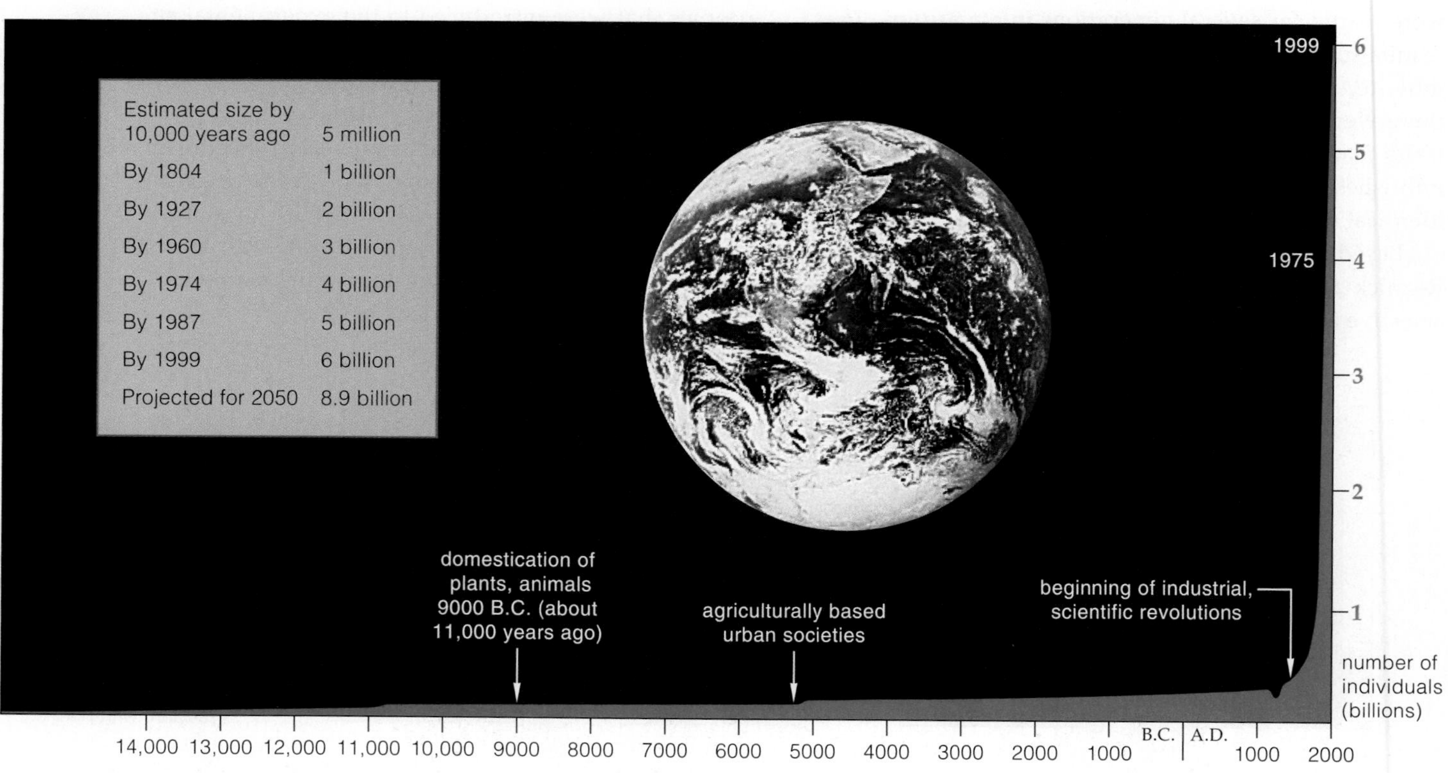

Figure 28.12 Growth curve (*red*) for the world human population. The *blue* box lists how long it took for the human population to increase from 5 million to 6 *billion*.

Second, humans developed new ways to increase the carrying capacity in existing habitats. Finally, human populations sidestepped limiting factors.

Early humans mostly gathered plant foods but also scavenged for meat. Small bands of hunter–gatherers started to disperse away from Africa about 2 million years ago. By 40,000 years ago, their descendants were established in much of the world (Section 16.12).

Most species can't expand into such a broad range of habitats. The early humans drew on a capacity for learning and memory to figure out how to make fires, shelters, and tools, and to plan out community hunts. Experience spread quickly from group to group by language. *The human population expanded into diverse environments very rapidly, as compared with the long-term geographic dispersal of other kinds of organisms.*

What about the second factor? Beginning 11,000 or so years ago, some human groups began to stay in one place and to domesticate various plants and animals. A pivotal factor was the domestication of wild grasses that were ancestral to modern wheat and rice.

Later in time, food supplies increased again by the use of fertilizers, herbicides, and pesticides. Methods of transportation improved; so did food distribution. *Thus, the management of food supplies through agriculture increased the carrying capacity for the human population.*

What about the third factor—the sidestepping of limiting factors? Think about what happened when medical practices and sanitary conditions improved. Until about 300 years ago, poor hygiene, malnutrition, and contagious diseases maintained high death rates. Diseases, a density-dependent control, swept through vermin-infested cities. Then came plumbing and water treatment. Vaccines and antibiotics were developed as weapons against pathogens. Death rates fell sharply, and rapid population growth was under way.

In the mid-eighteenth century, people discovered how to harness energy stored in fossil fuels, starting with coal. Industrialized societies began to form.

By controlling many disease agents and tapping into concentrated forms of energy (fossil fuels), humans have managed to sidestep major factors that previously limited their population growth.

It took 2.5 million years for the human population size to reach 1 billion, then 123 years to reach 2 billion, then another 33 to reach 3 billion, then 14 more to get up to 4 billion, and just 13 more to get to 5 billion. It took only 12 more years to arrive at 6 billion!

We expect a decline in growth as birth rates drop or as death rates rise. Alternatively, our population growth might continue if technological breakthroughs are able to further increase carrying capacity. *However, human population growth cannot be sustained indefinitely.* Why? Density-dependent controls will kick in.

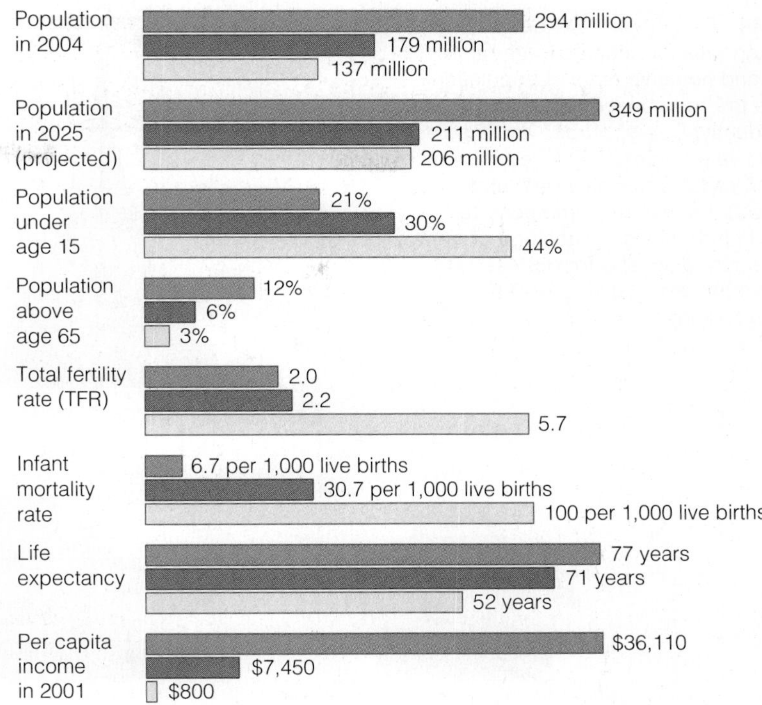

Figure 28.13 Key demographic indicators for three countries, mainly in 2003. The United States (*brown*) is highly developed, Brazil (*rust*) is moderately developed, and Nigeria (*gold*) is less developed.

FERTILITY RATES AND FAMILY PLANNING

Currently, most countries realize that their population growth, resource depletion, pollution, and quality of life are interconnected, as Figure 28.13 indicates. Many are attempting to lower long-term birth rates by providing information on methods of fertility control and access to these methods.

These attempts are having an impact; birth rates are declining worldwide. What about death rates? Overall, they are falling too, mainly because improved nutrition and health care are lowering infant mortality rates (the number of infants per 1,000 who die within the first year). However, the HIV/AIDS pandemic is causing death rates to soar in some African countries.

We still expect the world population to reach 8.9 billion by 2050. Think about the resources that will be required. We will have to boost food production and find more sources of energy and fresh water to meet even basic needs. These large-scale manipulations of resources will no doubt intensify pollution.

India, China, Pakistan, Nigeria, Bangladesh, and then Indonesia are expected to show the most growth, in that order. China (with 1.3 billion people) and India (with 1.1 billion) dwarf all other countries. Together, they make up 38 percent of the world population. The United States is next in line, with 295 million.

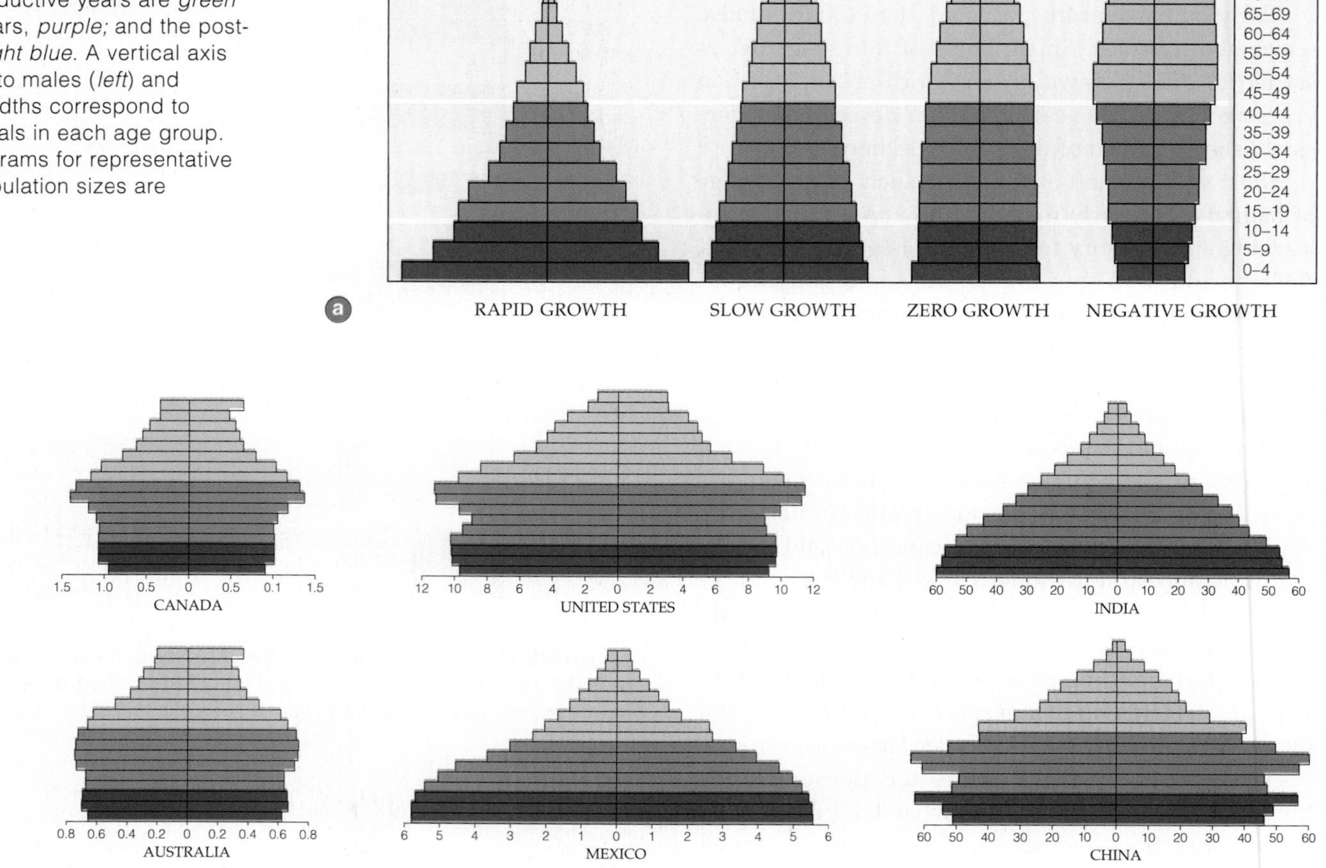

Figure 28.14 ***Animated!*** (**a**) General age structure diagrams for countries with rapid, slow, zero, and negative rates of population growth. The pre-reproductive years are *green* bars; reproductive years, *purple;* and the post-reproductive years, *light blue.* A vertical axis divides each graph into males (*left*) and females (*right*). Bar widths correspond to proportions of individuals in each age group. (**b**) Age structure diagrams for representative countries in 1997. Population sizes are measured in millions.

The **total fertility rate** (TFR) is the average number of children born to the women of a population during their reproductive years. TFR estimates are based on current age-specific rates. In 1950, the worldwide TFR averaged 6.5. By 2003, it had declined to 2.8 but was still above replacement levels, the number of children a couple must bear to replace themselves. At present, this replacement level is 2.1 for developed countries and as high as 2.5 in some developing countries. (It is higher in developing countries because more female children die before reaching reproductive age.)

The least developed countries continue to have the highest TFRs. In the African country of Niger, women average 7.5 births in their lifetime. About 12 percent of children born here die before their first birthday.

Comparing the age structure diagrams for different populations is revealing. In Figure 28.14, consider the size of the age groups that will be reproducing during the next fifteen years. The broader the base of an age structure diagram, the faster the population it depicts can be expected to increase in size. The United States population has a relatively narrow base. It is growing more slowly than countries with younger populations.

Even if every couple decides to bear no more than two children, world population growth will not slow for sixty years, because 1.9 billion are about to enter the reproductive age bracket. *More than one-third of the world population is in the broad pre-reproductive base.*

ECONOMICS AND POPULATION GROWTH

We can correlate changes in population growth with changes that often unfold in four stages of economic development. These four stages are at the heart of the **demographic transition model**. By this model, living conditions are harshest during the *preindustrial* stage, before technological and medical advances become widespread. Birth and death rates are both high, so the growth rate is low. Next, in the *transitional* stage, industrialization begins. Food production and health care improve. Death rates drop. But birth rates stay high in agricultural societies. Why? Large families are

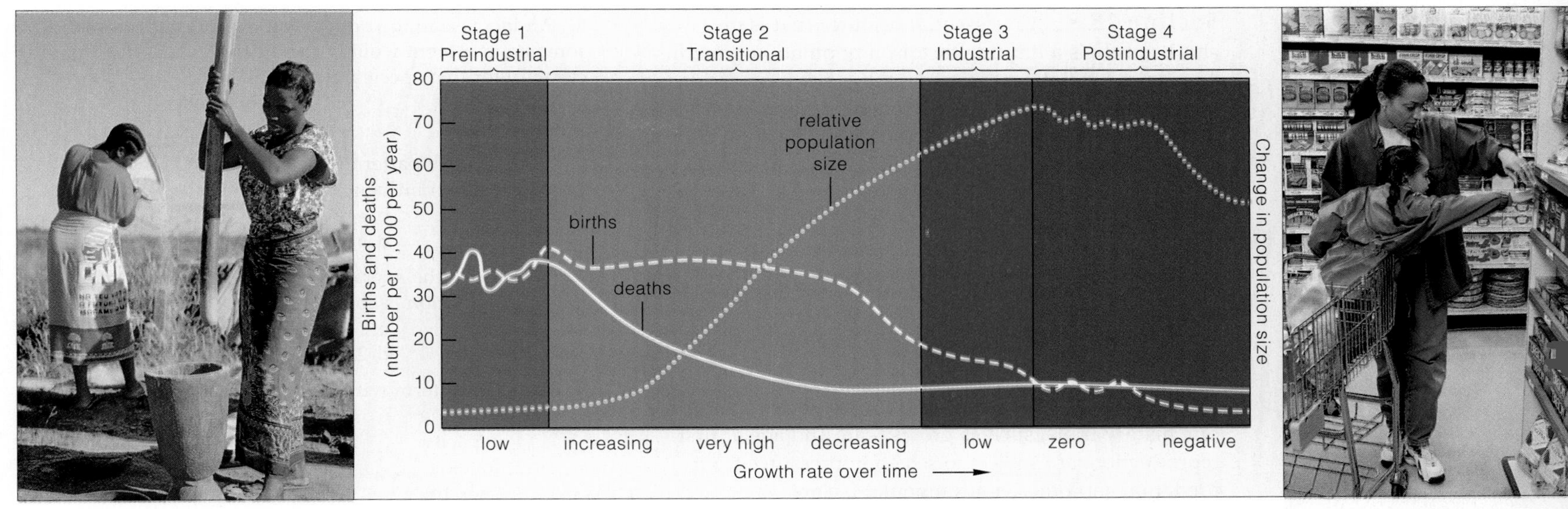

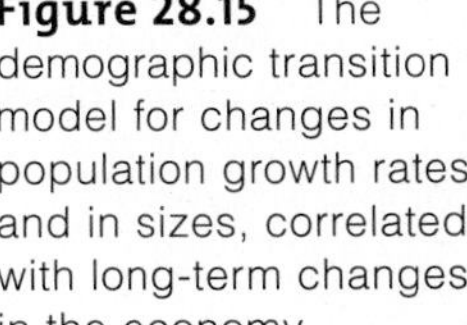

Figure 28.15 The demographic transition model for changes in population growth rates and in sizes, correlated with long-term changes in the economy.

a supply of inexpensive labor able to help in the fields. The population continues to increase rapidly for some time. As living conditions improve, birth rates decline, and growth starts to level off (Figure 28.15).

During the *industrial* stage, when industrialization is in full swing, population growth slows dramatically. People move from the country to cities, and urban couples tend to want smaller families.

In the *postindustrial* stage, population growth rates become negative. The birth rate falls below the death rate, and population size slowly decreases.

The United States, Canada, Australia, and most of western Europe, Japan, and countries of the former Soviet Union are in the industrial stage. Their growth rate is slowly falling. Mexico and other less developed countries are in the transitional stage. If population growth continues to outpace economic growth, then death rates will likely rise. Thus many countries might get stuck in the transitional stage or go back to the harsh conditions of the preceding stage.

ENVIRONMENTAL IMPACTS OF POPULATIONS

The United States represents about 4.6 percent of the world population. Yet the people of the United States produce 21 percent of the world's goods and services. They also consume about 25 percent of all processed minerals and a big portion of nonrenewable sources of energy. They generate about 25 percent of the world's pollution and trash.

By contrast, India is home to a little over 15 percent of the world population. Yet it produces only about 1 percent of all goods and services. It uses 3 percent of minerals and nonrenewable energy resources, and it generates only about 3 percent of the total pollution and waste products.

Extrapolating from these numbers, G. Tyler Miller estimated that it would take 12.9 billion individuals living in India to have the same environmental impact as 284 million people in the United States. Think about this estimate when you reflect on possible solutions to the biodiversity crisis and environmental challenges sketched out in later chapters.

> *Through expansion into new habitats, cultural intervention, and technological innovation, the human population has temporarily skirted environmental resistance to growth. Its size has been skyrocketing ever since.*
>
> *Birth rates fall as nations become more industrialized, but their per capita consumption of resources increases.*

Summary

Section 28.1 A population is a group of individuals of the same species occupying a given area. We typically measure its size, density, distribution, and age structure as well as characteristic ranges of heritable traits.

BiologyNow
Learn how to measure population size with the animation on BiologyNow.

Section 28.2 The growth rate for a population during a specified interval is determined by rates of birth, death, immigration, and emigration. In cases of exponential growth, the population increases by a fixed percentage of the whole with each successive interval. This trend plots out as a J-shaped curve. As long as a population's per capita birth rate is above its per capita death rate, it will show exponential growth.

BiologyNow
Observe the effects of exponential population growth with the animation on BiologyNow.

Figure 28.16 Saguaros growing very slowly in Arizona's Sonoran desert.

Section 28.3 Any essential resource that is in short supply is a limiting factor on population growth. The logistic growth equation describes how population growth is affected by density-dependent controls, such as disease or competition for resources. The population slowly increases in size, goes through a rapid growth phase, then levels off in size once carrying capacity is reached. Carrying capacity is the maximum number of individuals that can be sustained indefinitely by the resources available in their environment.

BiologyNow
Learn about logistic population growth with the animation on BiologyNow.

Section 28.4 Each species has a life history pattern. Life history traits, such as growth rate and age at first reproduction, are subject to selection pressures from predation and other environmental factors.

Section 28.5 The human population has surpassed 6.4 billion. Its rapid growth resulted from expansion into diverse habitats and agricultural and medical technologies that raised carrying capacity.

Worldwide, total fertility rate is declining. However, the broad pre-reproductive base of the human population means that population size will continue to increase for at least sixty years, even if all individuals have no more than a replacement number of children.

The demographic transition model is one way to correlate industrial and economic development with changes in population growth rates. Growth rates tend to decline as nations become more developed. But the per capita resource consumption of developed nations is many times higher.

BiologyNow
Compare the age structures of populations in different countries with the interaction on BiologyNow.

Self-Quiz

Answers in Appendix I

1. The rate at which a population grows or declines depends on the rate of ______ .
   a. births c. immigration e. a and b
   b. deaths d. emigration f. all of the above

2. Populations grow exponentially when ______ .
   a. population size expands by ever increasing increments through successive time intervals
   b. size of low-density population increases slowly, then quickly, then levels off once carrying capacity is reached
   c. a and b are characteristics of exponential growth

3. For a given species, the maximum rate of increase per individual under ideal conditions is the ______ .
   a. biotic potential c. environmental resistance
   b. carrying capacity d. density control

4. Resource competition, disease, and predation are ______ controls on population growth rates.
   a. density-independent c. age-specific
   b. population-sustaining d. density-dependent

5. An increase in the population of a prey species in some environment would most likely ______ the carrying capacity for that species' predators.
   a. increase c. not affect
   b. decrease d. stabilize

6. A life history pattern for a population is a set of adaptations that influence the individual's ______ .
   a. longevity c. age when it starts reproducing
   b. fertility d. all of the above

7. At present, the average natural rate of increase for the human population is about ______ percent per year.
   a. 0 c. 1.9 e. 3.8
   b. 1.2 d. 2.7 f. 4.0

8. Match each term with its most suitable description.
   ____ carrying capacity a. maximum rate of increase per individual under ideal conditions
   ____ logistic growth b. group of individuals born during the same period of time
   ____ exponential growth c. population growth plots out as an S-shaped curve
   ____ biotic potential d. largest number of individuals sustainable by the resources in a given environment
   ____ limiting factor e. population growth plots out as a J-shaped curve
   ____ cohort f. essential resource that restricts population growth when scarce

Additional questions are available on **BiologyNow™**

Critical Thinking

1. The Chittenango snail (*Novisuccinea chittenangoensis*) is New York State's most endangered animal. The dime-sized terrestrial snail lives only in the spray zone adjacent to a single waterfall. Water pollution and other disturbances threaten it. In 2002, a detailed census around the waterfall turned up 105 individuals. Each was marked and released. What type of information will marked-snail monitoring provide about this threatened species? How might this help save the snail from extinction?

2. Each summer, the giant saguaro cacti in deserts of the American Southwest produce tens of thousands of tiny black seeds apiece. Most die, but a few land in a sheltered spot and sprout the following spring. The saguaro is a CAM plant (Section 6.6) and growth is slow (Figure 28.16). After 15 years, the saguaro may be only knee high, and it will not flower until it is about age 30. The cactus may survive to age 200. Saguaros share their desert habitat with annuals, such as poppies, that sprout, produce seeds, and die in just a few weeks. How would you describe these two life history strategies? How could such different life histories both be adaptive in this environment?

3. A third of the world population is younger than fifteen. Describe the effect of this age distribution on the human population's future growth rate. If you think it will have a severe impact, what recommendations would you make to encourage these individuals to limit their family size? What are some of the factors that might prevent them from following the recommendations? How would you feel about following these recommendations yourself?

29 COMMUNITY STRUCTURE AND BIODIVERSITY

IMPACTS, ISSUES

Fire Ants in the Pants

Two species of Argentine ants, *Solenopsis richteri* and *S. invicta*, entered the United States during the 1930s, most likely as cargo ship stowaways. They infiltrated communities throughout the Northern Hemisphere, starting with the southeastern states. *S. invicta*, the red imported fire ant shown at the right, has recently colonized Southern California and New Mexico.

The aptly named fire ants inflict hot, painful stings. Disturb one, and it will bite down with its mandibles and inject venom with the stinger on its abdomen. A pus-filled bump on the victim's skin follows the pain. At one time or another, about half the people who live in areas where fire ants are common have been stung. The venom has killed more than eighty of them.

Imported fire ants do more than bite humans. They attack anything that disturbs them, including livestock, pets, and wildlife. They outcompete native ant species and other animals that prey on insects. Their dispersal may be contributing to the decline of bobwhite quail, horned lizards, and other endemic species.

Invicta means invincible in Latin. So far, this insect pest is living up to its species name. Broad-spectrum pesticides have done little to slow its march into new habitats. They actually may be helping the invader by killing off native competitors.

Ecologists are enlisting biological controls. Two fly species attack *S. invicta* in its native habitat. Both are parasitoids, a rather gruesome kind of parasite. A fly pierces an adult ant's protective cuticle and lays an egg in its soft tissues. The larva that hatches lives in and feeds on the ant's body. After it gets big enough, the fly larva moves into the ant's head and secretes an enzyme that makes the head fall off, as shown in the photo at the left. The larva metamorphoses into an adult inside the head.

In 1997, ecologists released one parasitoid fly species in Florida. In 2001, they released a second parasitoid species in the southern states. It is too soon to know if the flies work. As researchers wait for results, they are exploring other options. One is to use imported microbes that will infect *S. invicta* but not native ants. Another is to introduce another ant species that invades *S. invicta* colonies and kills their egg-laying queens.

As you will see in this chapter, every species is part of a complex web of relationships, some beneficial, some neutral, and some harmful. When species enter novel habitats far from home, they often shred this web, with unexpected consequences.

How Would You Vote? *Currently, only a small fraction of the crates being imported to the United States are inspected for exotic species. Should the inspections increase? See BiologyNow for details, then vote online.*

Key Concepts

SPECIES INTERACTIONS
A community includes all the populations of all species within some area. The structure of a community is determined by physical factors and by the interactions among species within it. Through their interactions, species exert selective pressure on one another.

STABILITY AND CHANGE
The kind and number of species in a community often change over time in response to species interactions, physical changes in the environment, and disturbances. Introduction of a new species into a community can have a dramatic effect on those already present.

ANALYZING BIODIVERSITY
Different communities include different numbers of species, and ecologists have identified some factors that shape global patterns of species diversity. Ongoing declines in species numbers in many communities are largely the result of human activities.

Links to Earlier Concepts

Section 1.1 introduced the concept of a biological community. Sections 13.8 and 13.9 described how species originate, change over time, and become extinct. Here you will learn how species interactions and human activities affect these processes, as well as their effect on population growth (28.3). You may wish to review some earlier examples of challenges faced by all species (17.2) and mutually beneficial interactions (3.6, 14.4, 15.6), as well as the evolution of infectious disease (14.6). Geologic forces also come into play (13.3), as does human population growth (28.5).

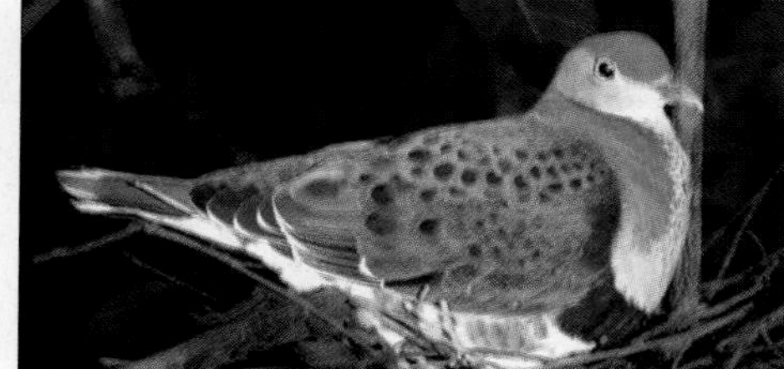

Figure 29.1 A sampling of the twelve or more fruit-eating pigeon species of New Guinea's tropical rain forests. Left to right, the turkey-sized Victoria crowned pigeon, the smaller superb crowned fruit pigeon, and the pied imperial pigeon.

Why are there so many? Wouldn't competition for fruit leave one the winner? Actually, the forest trees differ in the size of fruit and fruit-bearing branches. Big pigeons perch on fat branches and eat big fruit. Smaller pigeons, with smaller bills, cannot peck open big fruit. They eat small fruit hanging from branches too slender to hold big pigeons. They all coexist by *partitioning* the fruit supply.

Trees benefit, too. Tough-coated seeds in the fruit resist digestion when traveling through a pigeon gut. Pigeons fly about and dispense seed-containing droppings, often some distance from parent trees. Seeds dropped close to home cannot compete much with mature trees, which already have extensive roots and leafy crowns and have the edge in securing water, mineral ions, and sunlight.

29.1 Which Factors Shape Community Structure?

The type of place where each organism normally lives is its ***habitat****. Habitats have physical and chemical features, such as temperature, and an array of species. Directly or indirectly, populations of all species in each habitat associate with one another as a* ***community****.*

LINKS TO SECTIONS 1.1, 3.6, 14.4, 15.6

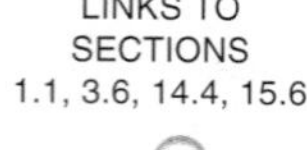

Every community has a characteristic structure, which arises as an outcome of five factors. *First*, climate and topography interact to dictate temperatures, rainfall, type of soil, and other habitat conditions. *Second*, the kinds and quantities of food and other resources that become available through the year affect which species live there. *Third*, adaptive traits of each species help it survive and exploit specific resources in the habitat, as in Figure 29.1. *Fourth*, species interact by competition, predation, and mutually beneficial actions. *Fifth*, the overall pattern of population sizes affects community structure. This pattern reflects the history of changes in population size, the arrivals and disappearances of species, and physical disturbances to the habitat.

THE NICHE

All organisms of a community share the same habitat, the same "address." But each kind has a "profession" that makes it different from the rest. It is distinct in the sum of its activities and relationships as it goes about getting and using resources. The **niche** of a species is a description of the way that it utilizes its habitat.

The *fundamental* niche of any species describes all the ways that this species could utilize resources if it did not have to compete with any other species. In the real world, competition influences the dimensions of a more limited, *realized* niche. Unlike the fundamental niche, a realized niche can change as competition and resource availability vary.

CATEGORIES OF SPECIES INTERACTIONS

Dozens to hundreds of species interact even in simple communities. Some of the interactions are indirect. As one example, when Canadian lynx prey on snowshoe hares, they indirectly help the plants that these hares eat. The plants indirectly help the Canadian lynx by fattening hares and providing cover for ambushes.

We recognize five categories of direct interactions. **Commensalism** directly helps one species but does not affect the other much, if at all. For instance, some birds use tree branches for roosting sites. The trees get nothing but are not harmed. With **mutualism**, benefits flow both ways between the interacting species. Do not think of this as cozy cooperation. Benefits arise from two-way exploitation. With **interspecific competition**, disadvantages flow both ways between species. Finally, **predation** and **parasitism** are interactions that directly benefit one species (the predator or the parasite) and directly hurt the other (the prey or the host).

Commensalism, mutualism, and parasitism are all types of **symbiosis** ("living together"). Two species are symbiotic if they live in close proximity (often in or on one another) for an extended period or for a portion of their life cycle.

Close interactions between species, whether they are beneficial or detrimental, can lead to **coevolution**. Changes in one species exert selective pressure on the other species. Pollinator preferences select for changes in flower form, and these in turn select for discerning pollinators. Predation selects for swifter prey, then the swifter prey favor speedier predators, and so on.

Community structure arises from a habitat's physical and chemical features, resource availability over time, adaptive traits of its species, how its species interact, and the history of the habitat and its occupants.

A niche is the sum of all activities and relationships in which individuals of a species engage as they secure and use the resources necessary to survive and reproduce.

Direct and indirect interactions among species help shape community structure. Close interactions between species can lead to coevolution—evolution in close concert.

Figure 29.2 Mutualism on a rocky slope of Colorado's high desert. Only one yucca moth species pollinates each *Yucca* plant species. Each moth matures when yucca flowers blossom. The female moth collects yucca pollen and rolls it into a ball. She flies to another flower, pierces its ovary, and lays her eggs inside. As the moth crawls out, she pushes the ball of pollen onto the flower's pollen-receiving platform. Pollen tubes grow through the ovary tissues and deliver sperm to the plant's eggs. Seeds develop after fertilization. Meanwhile, the moth eggs develop into larvae that eat a few seeds, then gnaw their way out of the ovary. Seeds that larvae do not eat give rise to new yucca plants.

29.2 Mutually Beneficial Interactions

Chapters 15 and 19 described how flowering plants and pollinators take part in mutually beneficial interactions. In this and other mutualisms, two species form a close partnership that benefits both. In extreme cases, each partner requires the other to survive and reproduce.

Some mutualisms are *obligatory*; members of a species can complete their life cycle only with the assistance of a member of a different species. Yucca plants and yucca moths are one example (Figure 29.2).

A lichen is a case of obligatory mutualism between certain fungi and a photosynthetic alga or bacteria. So is a mycorrhiza, in which a fungus's hyphae penetrate root cells or form a dense, velvety mat around them. As earlier chapters explain, the fungus is efficient at absorbing mineral ions from soil, and the plant grows better when it gets extra ions from the fungus. The fungus in turn gets some of the sugars that the plant makes by way of photosynthesis. The fungus depends on the plant for reproductive success; it will not make spores if the plant stops photosynthesizing.

An anemone fish that needs a sea anemone to live and reproduce is another example (Figure 29.3). The anemone's nematocyst-laden tentacles protect the fish and its eggs. The anemone can survive and reproduce on its own, but having a fish partner lessens predation by other fishes on the anemone's tentacles.

Think back on how endosymbiosis gave rise to cell organelles (Section 3.8). Certain bacteria were able to live inside other host cells. These symbionts got food from the host and reproduced on their own. In time, hosts came to rely on ATP produced by the symbionts, which were now called mitochondria and chloroplasts. Without these prokaryotic mutualists, no eukaryotic cells would be around today.

Figure 29.3 Mutualism near the Solomon Islands. The sea anemone (*Heteractis magnifica*) with a mutualistic pink anemone fish (*Amphiprion perideraion*). The anemone fish chases away butterfly fishes that try to eat bits of the sea anemone's tentacles. In return, the anemone fish gets protection and shelter for itself and for its eggs, which many reef dwellers consider tasty.

In cases of mutualism, each of the participating species reaps benefits from the interaction.

LINKS TO SECTIONS 12.1, 28.3

29.3 Competitive Interactions

Every habitat has only a limited supply of energy, nutrients, living space, and other essential resources. Often, many species must compete for access to them.

Intraspecific competition occurs between individuals of the same species. As Malthus and Darwin understood, these interactions can be fierce and they are important driving forces for natural selection (Section 12.1). In this chapter, our focus is on *interspecific* competition, the competition between members of different species. Usually this is not as intense. Why? *The requirements of two species may be similar, but they never can be as close as they are for individuals of the same species.*

With *interference* competition, the members of one species actively prevent members of another species from using a resource. For example, some chipmunks chase away other chipmunk species, thus preventing them from sharing seeds and shelter (Figure 29.4).

Plants can interfere with their competitors as well. Some plants, such as sagebrush, exude chemicals from their leaves that taint the surrounding soil and help to prevent potential competitors from taking root.

In *exploitative* competition, competing species have equal access to a resource and do not try to keep one another away. This is also called *scramble* competition. For example, blue jays, deer, and squirrels eat acorns in oak forests. The animals do not fight over acorns, but they do compete. By eating or storing acorns, each of these species reduces the number of acorns that is available for the others.

COMPETITIVE EXCLUSION

When species compete for some resource, each gets less of that resource than it would if it lived alone. In this way, competition has a negative impact on both competitors. The more alike two species are in their needs, the more intensely they will compete. If two species have identical needs, the superior competitor will drive the less competitive species to extinction in that environment. We call this **competitive exclusion**.

The concept of competitive exclusion grew out of experiments carried out by G. Gause in the 1930s. Gause grew two species of *Paramecium* separately and then in the same culture (Figure 29.5). Both of these protists required the same food—bacteria—and they competed intensely for it. In a few weeks, one species outgrew the other, driving it to extinction.

Gause also studied two other *Paramecium* species that did not overlap much in requirements. When he grew them together, one species tended to feed upon bacteria suspended in culture tube liquid. The other ate yeast cells at the tube's bottom. Population growth rates slowed for both of the species—but their overlap in resource use was not enough for one species to fully exclude the other. They continued to coexist.

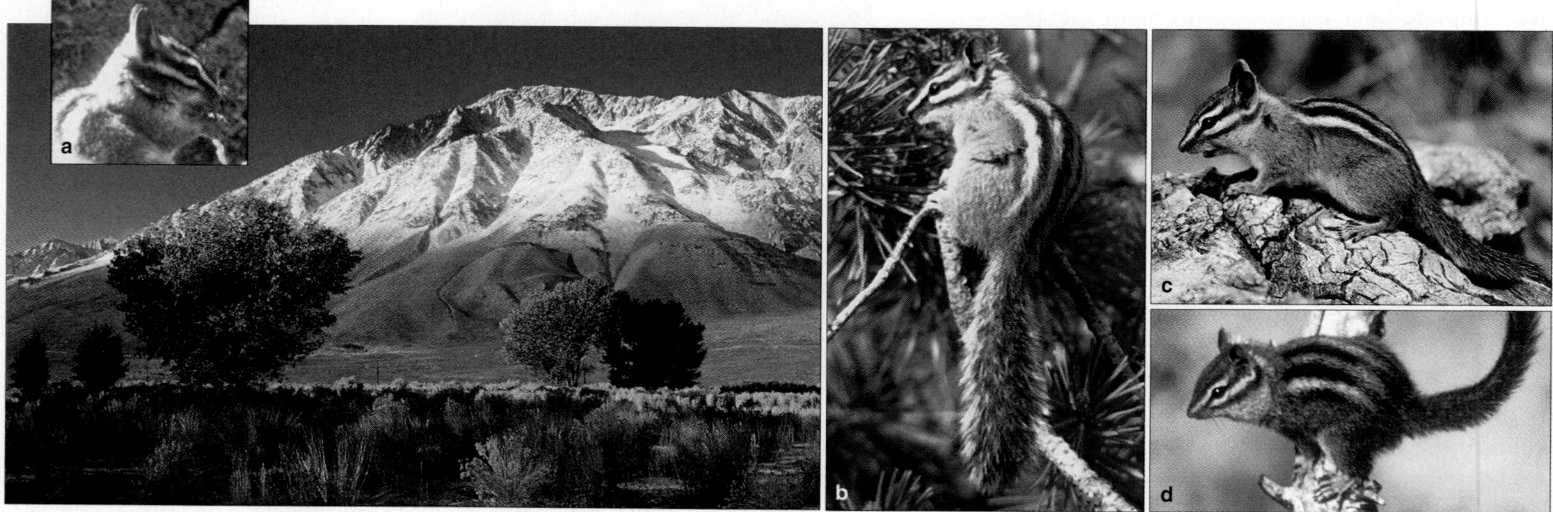

Figure 29.4 Example of interspecific competition in nature. On the slopes of the Sierra Nevada, competition helps keep nine species of chipmunks (*Tamias*) in different habitats. The alpine chipmunk (**a**) lives in the alpine zone, the highest elevation. Below it are the lodgepole pine, piñon pine, and then sagebrush habitat zones. Lodgepole pine chipmunks (**b**), least chipmunks (**c**), and other species live in the forest zones. Merriam's chipmunk (**d**) lives at the base of the mountains, in sagebrush. Its traits would allow it to move up into the pines, but the aggressive behavior of forest-dwelling species keep it from moving there. Food preferences keep the pine forest chipmunks out of the sagebrush habitat.

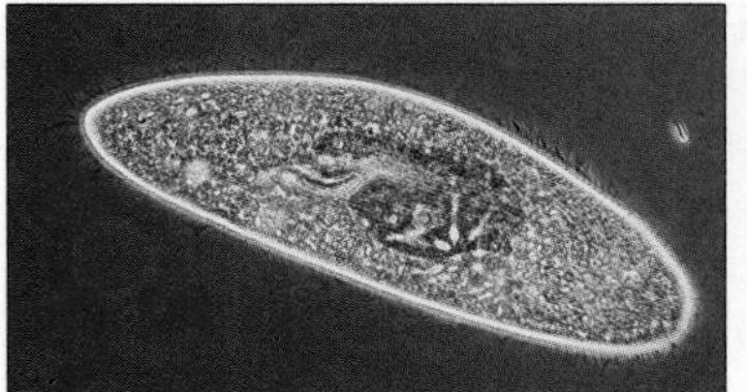

Paramecium caudatum

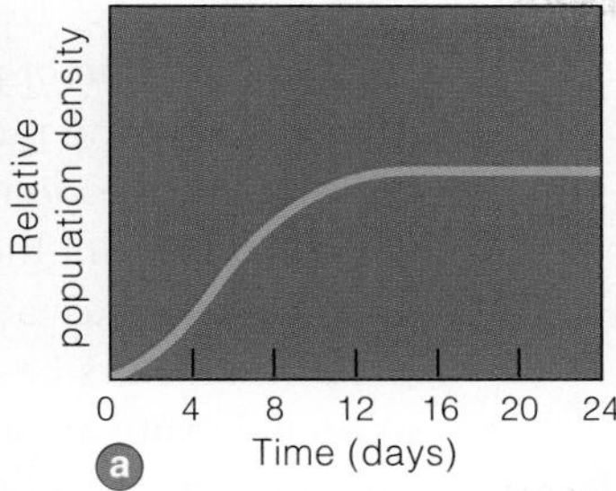

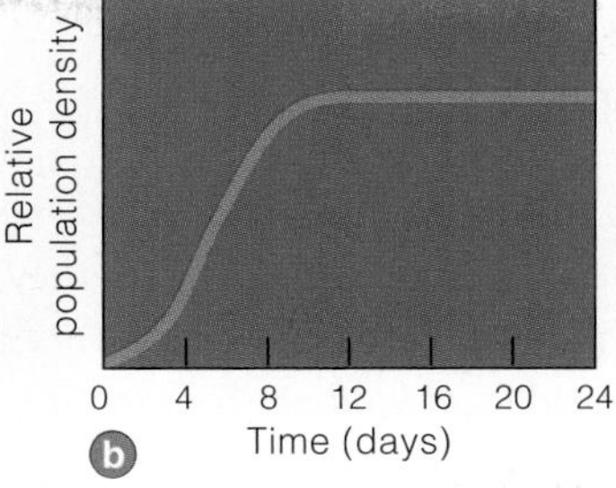

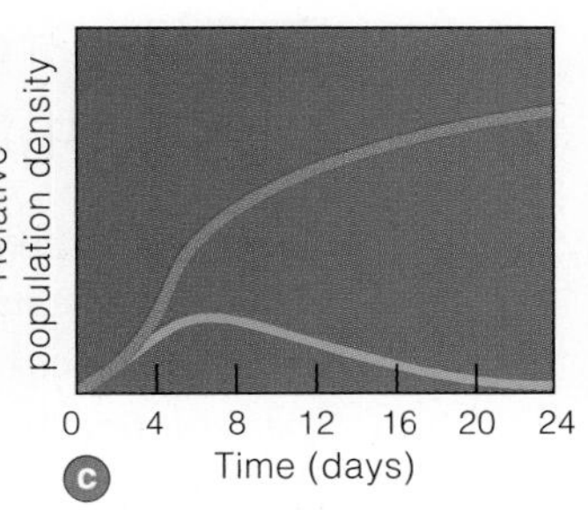

P. aurelia

Figure 29.5 ***Animated!*** Results of competitive exclusion between two protist species that compete for the same food. (**a**) *Paramecium caudatum* and (**b**) *P. aurelia* were grown in separate culture tubes and established stable populations. The S-shaped graph curves indicate logistic growth and stability. (**c**) Then the two species were grown together. *P. aurelia* (*brown* curve) drove *P. caudatum* toward extinction (*green* curve in **c**). This experiment and others suggest that two species cannot coexist indefinitely in the same habitat *when they require identical resources.* If their requirements do not overlap much, one might influence the population growth rate of the other, but they may still coexist.

COEXISTING COMPETITORS

Think back on those twelve fruit-eating pigeon species in the same New Guinea forest. They all use the same resource: fruit. Yet they overlap only slightly in their use of it, because each prefers fruits of a certain size. They are a fine example of **resource partitioning**—the *subdividing* of some category of similar resources in a way that allows competing species to coexist.

A similar competitive situation arises among three species of annual plants in a plowed, abandoned field. Like other plants, all three require sunlight, water, and dissolved mineral ions. Yet each species is adapted to exploiting a different part of the habitat (Figure 29.6). The bristly foxtail grasses have a shallow, fibrous root system, and they are drought tolerant. Indian mallow plants have a taproot system and grow in deeper soil. Smartweed's taproot system branches in topsoil and soil below the roots of the other species.

Field studies allow ecologists to determine if two species are competing in nature. A typical experiment involves removing one possible competitor from some portion of the environment. For example, N. Hairston studied slimy salamanders (*Plethodon glutinosus*), and Jordan's salamander (*P. jordani*), which overlap in their ranges (Figure 29.7). He removed *P. glutinosus* from some plots and *P. jordani* from others. Control plots were unaltered. Five years later, salamander density in the control plots had not changed. But in the plots that lacked *P. jordani*, population density of *P. glutinosus* had soared. Similarly, *P. jordani* numbers had increased in the *P. glutinosus*-free plots. Hairston concluded that under natural circumstances, the species compete and that competition reduces the population size of both.

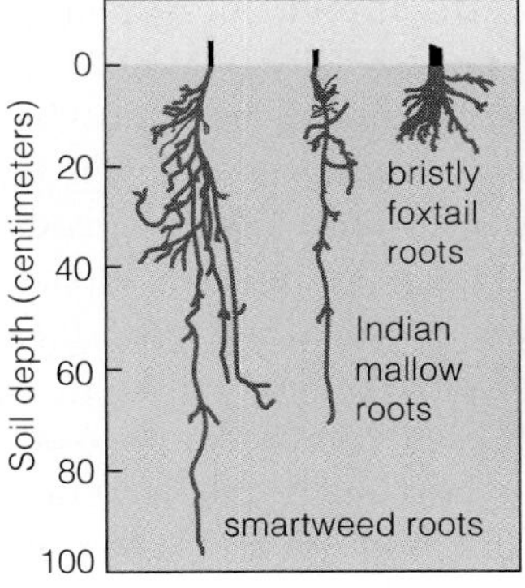

smartweed Indian mallow bristly foxtail

Figure 29.6 Resource partitioning among three annual plant species in an abandoned field. The plants differ in how they are adapted to secure soil water and mineral ions. The roots of each species tap into a different region of the soil.

Figure 29.7 Two competing species of salamander that coexist in the Great Smokey and Balsamic Mountains. (**a**) *Plethodon glutinosus* and (**b**) *P. jordani*.

Competition has negative effects on both participants. In some situations, one species actively prevents another from using a resource. In others, both species have equal access to the resource and each scrambles to get a bigger share.

Species that have dissimilar resource requirements are more likely to coexist than those having similar needs.

Two competing species also may coexist by partitioning a resource—accessing it at different times or locations.

LINK TO SECTION 28.2

29.4 Predator–Prey Interactions

Directly or indirectly, interactions among coexisting populations organize the community to which they belong. Let's look at patterns of species abundances, species interactions, and local biodiversity, starting with the nature of interactions between predators and prey.

Predators are consumers that feed on living organisms —their **prey**—and typically cause their prey's death. Diverse carnivores and omnivores fall in this category. Herbivores are similar to predators, but they often do not cause death of a plant. Even so, many herbivores and the plants they eat coevolve in the same way that prey and predators do.

MODELS FOR PREDATOR–PREY INTERACTIONS

There are three general patterns of predator responses to increases or decreases in prey density (Figure 29.8). In the type I response, a predator removes a constant proportion of prey over time, regardless of levels of prey abundance. The number of prey killed depends only on prey density. This type of response is typical of passive predators, such as spiders. The more flies there are, the more get tangled in spider webs.

In a type II response, the ability of predators to eat and digest prey determines how many they capture. As prey density increases, the proportion caught rises steeply at first, then slows as predators are exposed to more prey than they can deal with at one time. Figure 29.8*b* gives an example. A wolf that has just killed a caribou will not hunt another until it has eaten and digested the first one.

In a type III response, predators capture only a small proportion of prey when the prey density is low. They capture the most at intermediate prey densities, and then the response levels off. This response occurs among predators that switch to other prey when low abundance makes a prey species difficult to locate.

Other factors besides individual predator response to prey density are at work. For example, predator and prey reproductive rates influence the interaction. So do hiding places for prey, the presence of other prey or predators, and carrying capacities (Section 28.3).

THE CANADIAN LYNX AND SNOWSHOE HARE

Often predator and prey populations will oscillate. For example, there is a ten-year oscillation in populations of one predator, the Canadian lynx, and the snowshoe hare that is its main prey (Figure 29.9). To identify the causes behind this cyclic pattern, the ecologist Charles Krebs and coworkers recorded hare densities for ten years in the Yukon River Valley in Alaska. They set up 1-square-kilometer control plots. They also set up a number of experimental plots. Some plots had electric fences that kept out predatory mammals. Other plots had extra food added or were fertilized to increase plant growth. The team captured and released more than a thousand snowshoe hares, lynx, squirrels, and coyotes after putting radio collars on each one.

In predator-free plots, hare density doubled. In plots with extra food, it tripled. In plots with extra food *and* fewer predators, hare density increased elevenfold.

Nevertheless, the experimental manipulations only delayed the cyclic declines in population density. They did not stop them. Why? The fence could not keep out owls and other raptors. Only 9 percent of the collared hares starved to death; predators devoured 83 percent

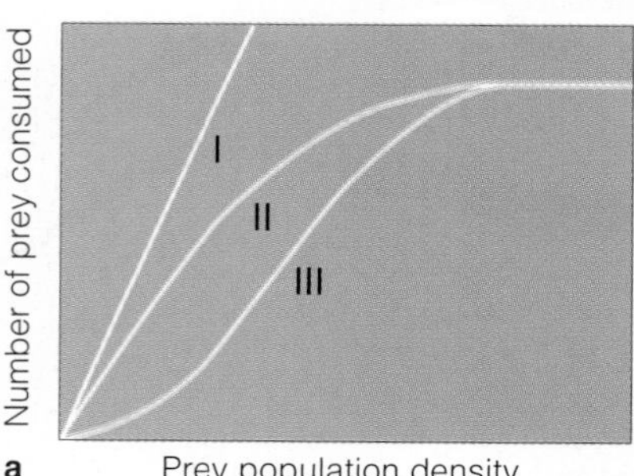

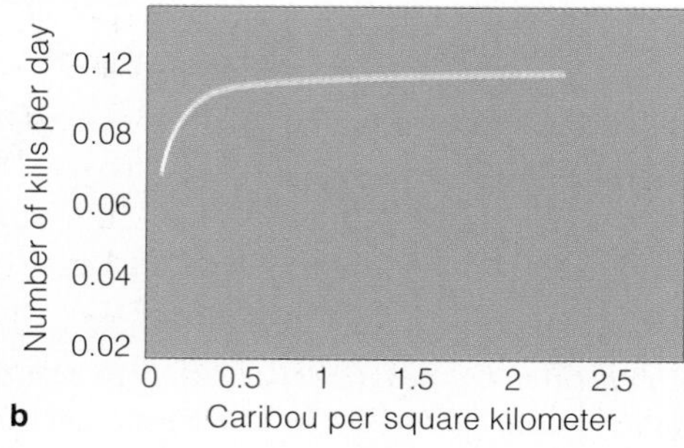

Figure 29.8 ***Animated!*** (**a**) Three alternative models for responses of predators to prey density. Type I: Prey consumption rises linearly as prey density rises. Type II: Prey consumption is high at first, then levels off as predator bellies stay full. Type III: When prey density is low, it takes longer to hunt prey, so the predator response is low. (**b**) A type II response in nature. For one winter month in Alaska, B. W. Dale and coworkers observed four wolf packs (*Canis lupus*) feeding on caribou (*Rangifer taranuds*). This interaction fits the type II model for the response of predators to prey density.

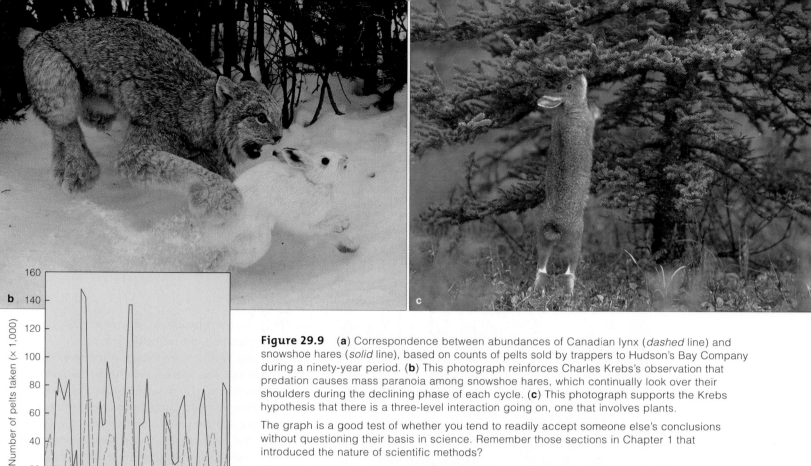

Figure 29.9 (**a**) Correspondence between abundances of Canadian lynx (*dashed* line) and snowshoe hares (*solid* line), based on counts of pelts sold by trappers to Hudson's Bay Company during a ninety-year period. (**b**) This photograph reinforces Charles Krebs's observation that predation causes mass paranoia among snowshoe hares, which continually look over their shoulders during the declining phase of each cycle. (**c**) This photograph supports the Krebs hypothesis that there is a three-level interaction going on, one that involves plants.

The graph is a good test of whether you tend to readily accept someone else's conclusions without questioning their basis in science. Remember those sections in Chapter 1 that introduced the nature of scientific methods?

What other factors may have had impact on the cycle? Did the weather vary, with more severe winters imposing greater demand for hares (to keep lynx warmer) and higher death rates? Did the lynx compete with other predators, such as owls? Did the predators turn to alternative prey during low points of the hare cycle? When fur prices rose in Europe, did the trapping increase? When the pelt supply outstripped the demand, did trapping decline?

of the rest. Krebs concluded that a simple predator–prey or plant–herbivore model could not explain the results of his experiments in the Yukon River Valley. Rather, in the Canadian lynx and snowshoe hare cycle, other variables are at work in a *three-level* interaction. Additional experiments may shed light on the causes of these cycles.

AN EVOLUTIONARY ARMS RACE

Predator and prey species often coevolve. Suppose a gene mutation that provides a novel means of defense arises in a prey species. Over generations, directional selection causes this gene to spread through the prey population. Suppose also that some members of the predator species have a gene that makes them better at thwarting the new defense. These predators, and their descendants, eat more often, hence are more likely to survive and reproduce. The predators exert selection pressure that favors better defenses by prey, which exert selection pressure on predators, and so it goes over generations. Let us now consider a few outcomes of such evolutionary arms races.

PREY DEFENSES You have already learned about some defensive adaptations. Many species have hard parts that enclose them or otherwise make them difficult to eat. Think back on the prickly elements in the bodies of sponges, the shells of mollusks, and the exoskeletons that protect arthropods and echinoderms.

Other prey species contain chemicals that taste bad or sicken predators. Some produce the toxins through their own metabolic processes. Others use chemical or physical weapons that they obtained from their prey, as when sea slugs store nematocysts from cnidarians (Figure 16.11*c*).

The leaves, stems, and seeds of many plants contain bitter, hard-to-digest, or poisonous substances that help to deter herbivores. Ricin-producing castor bean plants (Chapter 10) are a good example. Ricin sickens or kills animals that eat the seeds. Caffeine in coffee beans and nicotine in tobacco leaves are both defensive compounds for which humans have developed a taste.

Well-defended prey often show **warning coloration**, conspicuous patterns and colors that predators learn to recognize as *AVOID ME!* signals. An inexperienced bird that eats an orange-and-black patterned monarch

LINKS TO SECTIONS 1.6, 17.2

a A dangerous model

b One of its edible mimics

c Another edible mimic

d And another edible mimic

Figure 29.10 An example of mimicry. Edible insect species often resemble toxic or unpalatable species that are not at all closely related. (**a**) A yellowjacket can deliver a painful sting. It may be the model for nonstinging wasps (**b**) beetles (**c**), and flies (**d**) that have a similar appearance.

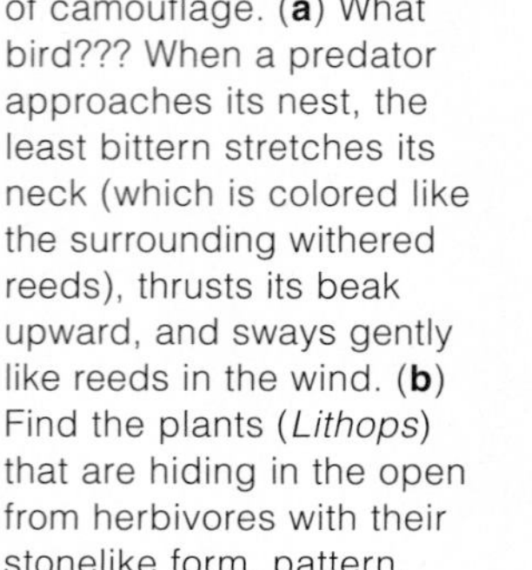

Figure 29.11 Prey demonstrating the fine art of camouflage. (**a**) What bird??? When a predator approaches its nest, the least bittern stretches its neck (which is colored like the surrounding withered reeds), thrusts its beak upward, and sways gently like reeds in the wind. (**b**) Find the plants (*Lithops*) that are hiding in the open from herbivores with their stonelike form, pattern, and coloring.

butterfly learns quickly to associate this patterning with the vomiting of foul-tasting toxins.

Mimicry is an evolutionary convergence in body form, in which one species benefits by its resemblance to another. Section 1.6 described how the resemblance between two unrelated unpalatable butterflies protects both species against predation. In other cases, prey that are tasty and harmless evolve to resemble unpalatable or well-defended species (Figure 29.10). Predators may ignore the mimic after contacting the model's repelling taste, toxic secretion, or painful bite or sting.

Some prey engage in **camouflage**, or hiding in the open. Such species have a form, patterning, color—and often behavior—that allows them to blend into their surroundings and avoid detection. Figure 29.11 shows two examples. One, *Lithops,* is a desert plant that looks like a rock. It flowers only during a brief rainy season, when other plants grow profusely and there is plenty of drinking water, so herbivores are less interested in its juicy water-filled tissues.

When a prey animal is cornered or under attack, survival may turn on a last-chance trick. Some animals startle the predator, as by hissing, puffing up, showing teeth, or flashing big eye-shaped spots. The tail may detach from the body and wiggle a bit as a distraction. Opossums pretend to be dead. So do hognose snakes, which also secrete smelly fluid. Skunks squirt a foul-smelling, irritating repellent.

PREDATOR RESPONSES TO PREY In a coevolutionary arms race, predators counter prey defenses with their own adaptations. Stealth, camouflage, and ingenious ways of avoiding the repellents are countermeasures. Consider edible beetles that direct sprays of noxious chemicals at assailants (Figure 29.12*a*). Grasshopper mice plunge the beetle's sprayer end into the ground, then munch on the unprotected head (Figure 29.12*b*).

Many prey can outrun predators when they get a head start. But consider the cheetah, the fastest land

Figure 29.12 Predator responses to prey defenses. (**a**) Some beetles spray noxious chemicals at attackers, which deters them some of the time. (**b**) At other times, grasshopper mice plunge the chemical-spraying tail end of their beetle prey into the ground and feast on the head end. (**c**) Find the scorpionfish, a venomous predator with camouflaging fleshy flaps, multiple colors, and profuse spines. (**d**) Where do the pink flowers end and the pink praying mantis begin?

animal. One cheetah was clocked at 114 kilometers per hour (70 mph). A cheetah's body is highly adapted for sprinting. Compared to the other large cats, it legs are longer relative to its body size. Nonretractable claws act like cleats that increase traction. Coevolution with fast prey contributed to the cheetah's need for speed. Its preferred prey, the Thomson's gazelle, is among the fastest land animals and can maintain a speed of 80 kilometers per hour.

Camouflage helps some predators catch their prey. Think how snow white polar bears stalk seals, striped tigers crouch in tall, golden grasses, and the mottled scorpionfish sits on the seafloor (Figure 29.12*c*). Many predatory insects are remarkably difficult to recognize (Figure 29.12*d*). Camouflaged predators may select for enhanced sensory systems in prey. The visual systems of many prey can detect even the slightest motion that could give away a predator.

Different kinds of predators show different behavioral responses to changes in prey density.

Carrying capacities for predators and prey, the response of predators to changes in prey density, the availability of prey refuges, and the presence of alternative prey will all influence the stability of predator and prey populations.

Predators and prey exert selective pressure on one another. The result is an evolutionary arms race.

29.5 Parasites and Parasitoids

Parasites spend all or part of their life cycle in or on other living organisms, from which they draw nutrients. The host is harmed, but usually is not killed outright. Parasites help keep host populations in check, and we sometimes call them into service in our battle against pests.

Parasites have wide-reaching ecological impacts. By draining nutrients from hosts, they alter the amount of energy and nutrients that a host population demands from a habitat. Also, weakened hosts are usually more vulnerable to predation and less attractive to potential mates. Some parasite infections cause sterility. Others shift the ratio of host males to females. In such ways, parasitic infections lower birth rates, raise death rates, and affect intraspecific and interspecific competition.

Sometimes the gradual drain of nutrients during a parasitic infection indirectly leads to death. The host becomes so weak that it cannot fight off secondary infections. Nevertheless, in evolutionary terms, killing a host too quickly is bad for a parasite's reproductive success. A parasitic infection must last long enough to give the parasite time to produce some offspring. The longer it survives in the host, the more offspring. We therefore expect natural selection to favor individual parasites that have less-than-fatal effects on hosts.

Only two types of interactions typically kill a host. First, a parasite may attack a novel host, one with no coevolved defenses against it. Second, the host may be supporting too many individual parasites at the same time. The collective assault may overwhelm the body.

The previous chapters introduced diverse parasites. Some spend their entire life cycle in or on a single host. Others have different hosts during different stages of their life cycle. Insects and other arthropods are often **vectors** that convey a parasite from host to host.

LINK TO SECTION 14.6

All viruses and some of the bacteria, protists, and fungi are parasites. Figure 29.13 describes a parasitic fungus that infects amphibians of tropical rain forests. Tapeworms and flukes are all parasites. So are some roundworms, insects and crustaceans, and all ticks.

Figure 29.13 Examples of parasites. (**a**) Harlequin frog (*Atelopus varius*) from Central America. (**b**) Section through skin of a harlequin frog affected by *chytridiomycosis*, an infectious disease caused by a fungus. *Arrows* point to flask-shaped cells of the fungal parasite. Each cell holds spores that will be released to the skin surface. They may be dispersed to other host frogs. (**c**,**d**) Dodder (*Cuscuta*) is also known as strangleweed or devil's hair. This parasitic flowering plant lacks chlorophylls. It winds around a host plant during growth. Modified roots penetrate the host's vascular tissues and draw water and nutrients from them.

Figure 29.14 A brown-headed cowbird nestling (*Molothrus ater*) with its foster parent. A female cowbird laid an egg in this smaller bird's nest. The host species does not recognize a cowbird nestling as an intruder and feeds it as if were its own offspring.

Figure 29.15 Biological control—a commercially raised parasitoid wasp about to deposit an egg in an aphid. This wasp cuts aphid numbers by crippling the egg laying capacity of its host even before the larva hatches inside and kills the aphid.

Even a few plants are parasitic. Nonphotosynthetic types, such as dodders, obtain sugars and nutrients from other plants (Figure 29.13*c*,*d*). The mistletoes and similar types carry out photosynthesis but tap into the nutrients and water in a host plant's tissues.

The larvae of **parasitoids** develop inside another species and devour it from the inside as they grow. The adults are free-living. Unlike parasites, parasitoids usually kill their hosts. Most parasitoids are insects. The ant-killing flies described earlier are one example.

Social parasites are animals that take advantage of the social behavior of a host species to carry out their life cycle. For example, cuckoos and cowbirds lay their eggs in the nests of other birds. In this way, they steal parental care from these hosts (Figure 29.14).

BIOLOGICAL CONTROL AGENTS

Parasites and parasitoids are commercially raised and released in target areas as *biological controls*. They are promoted as a workable alternative to pesticides. The chapter introduction and Figure 29.15 give examples.

Effective biological controls display five attributes. The agents are adapted to a specific host species and to the host's habitat; they are good at locating hosts; their population growth rate is high compared to the host's; their offspring are good at dispersing; and they make a type III functional response (Section 29.4), with little lag time after shifts occur in the host population size.

Use of biological control agents is not without risks. Releasing more than one species of biological control agent in an area may trigger competition among them and lessen their effectiveness. Introduced control agents sometimes parasitize nontargeted species in addition to, or instead of, species they were expected to attack.

For example, in Hawaii, introduction of parasitoids to control a non-native stink bug caused the extinction of the koa bug, the state's largest native bug. No koa bugs have been sighted since 1978. Apparently, the exotic parasitoids found the koa bugs a more tempting target than the pests they were imported to attack. An estimated 13 percent of the now established biological control agents imported into Hawaii harm native or beneficial species as well as their intended targets.

Parasites live on and draw nutrition from a living host. They usually weaken but do not kill the host.

Parasitoids are free-living as adults, but larvae live in and feed on another organism, which they usually kill.

Social parasites steal parental care from another species.

Parasites and parasitoids can be used to control pests.

29.6 Changes in Community Structure Over Time

You probably already realize that the array of species in a community changes over time. Where there are now meadow grasses, there may one day be trees. Or, after a fire, grasses may thrive where trees now stand.

ECOLOGICAL SUCCESSION

The transformation, over time, of a meadow to a forest is an example of **ecological succession**. By this process, one array of species is replaced by another, which is in turn replaced by another, and so on.

Primary succession occurs in habitats that lack soil and have few or no existing species. A newly formed volcanic island or a rocky area exposed by the retreat of a glacier will undergo primary succession (Figure 29.16). Pioneer species are the first to take hold. They include lichens, mosses, and some weedy annuals. The early arrivals are easily dispersed. They have a short life cycle and are able to tolerate intense sunlight, big swings in temperature, and nutrient-poor soil.

By their actions, pioneer species can help to create or improve the soil. In doing so, they may set the stage for their own replacement. Many are mutualists with nitrogen-fixing bacteria, so they can grow in nitrogen-poor habitats. Seeds of later species find shelter inside mats of the pioneers, which do not grow high enough to shade out the new seedlings.

Over time, organic wastes and remains accumulate. They add volume and nutrients to soil, thus creating conditions that make it easier for additional species to take hold. Later successional species often crowd out the early ones, whose spores and seeds then travel as fugitives on wind and water—destined, perhaps, for another new but temporary habitat.

If an existing community is altered by a natural or man-made disturbance, **secondary succession** occurs. We observe this kind of succession after a fire destroys a forest or a farm is abandoned.

INTERMEDIATE DISTURBANCE HYPOTHESIS

Early studies of primary succession led ecologists to the view that a community goes through predictable stages, determined by a habitat's typography, climate, and soil, and the interactions among species. By this view, succession ends with the formation of a stable *climax* community that does not change over time. In any particular habitat, succession always produces the same climax community, and if something destroys this community, the process of secondary succession will eventually restore it.

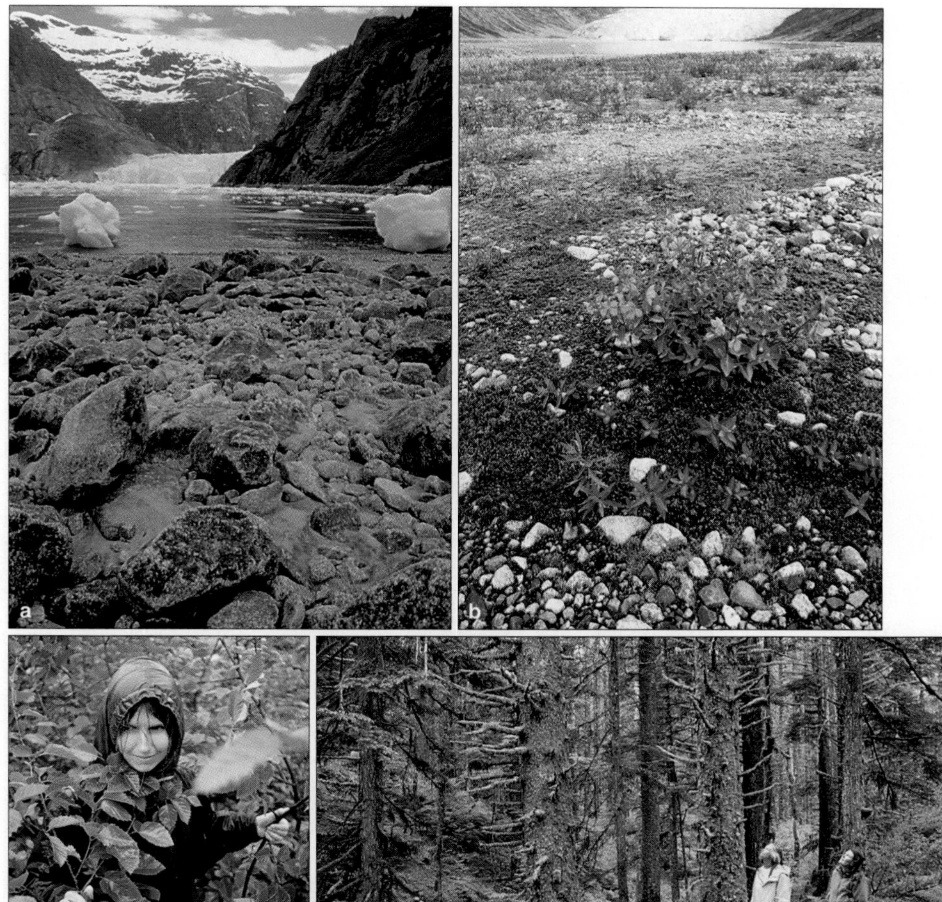

Figure 29.16 One of three pathways of primary succession observed in Alaska's Glacier Bay region. (**a**) A glacier is retreating from the sea. Meltwater leaches nitrogen and other minerals from the newly exposed soil. (**b**) The first invaders of nutrient-poor soil are lichens, horsetails, mosses, and fireweed and mountain avens. Some pioneer species are mutualists with nitrogen-fixing microbes. They grow and spread fast over the glacial till. (**c**) Alders, too, are symbionts with nitrogen-fixing microbes. Within fifty years, they have formed dense, mature thickets in which cottonwood, hemlock, and a few evergreen spruce grow quickly. (**d**) After eighty years, western hemlock and spruce trees crowd out the mature alders. (**e**) After a century, dense forests of spruce dominate. In the other successional pathways, different species of trees dominate at each stage.

Figure 29.17 A natural laboratory for the study of succession. (**a**) Eruption of Mount Saint Helens in 1980. Nothing remained of the rich climax community at the base of this Cascade volcano. (**b**) In less than a decade, many pioneer species were established. (**c**) Twelve years after the eruption, seedlings of Douglas fir were taking hold in soils enriched with volcanic ash.

In recent times, ideas about community structure have changed. Experiments have shown that communities are not reconstituted after disturbances, and random factors—such the order in which species arrive—can determine community structure. Ecologists now regard communities as dynamic entities that are rarely, if ever, in a stable state. In this "non-equilibrium" view, fires, storms, freezes, droughts, and other disturbances play an important role in shaping community structure.

The **intermediate disturbance hypothesis** holds that the number of species in a community is influenced by how frequent and severe disturbances are. The highest species richness occurs in habitats that are subject to moderate disturbances with an intermediate frequency. With this type of pattern, there are many opportunities for new species to enter the habitat after disturbances but not enough time between disturbances for many species to become competitively excluded from it.

SPECIES INTERACTIONS IN SUCCESSION

Facilitation, inhibition, and tolerance are interspecific interactions that influence succession. With *facilitation*, the presence of one species alters habitat conditions in a way that allows another to take root. For example, after Teresa Turner removed attached algae from plots in the intertidal zone in Oregon, surf grass could not move in. Algae has to be present to provide anchoring sites for the grasses.

With *inhibition*, a species alters the habitat in a way that discourages another species from growing there. As described earlier, sagebrush secretes chemicals into the soil that deter germination and growth of other plants. Wayne Sousa showed that the presence of one species of algae prevents another species of algae from growing on rocks along Southern California shores.

With *tolerance*, the presence of one species has no effect on the arrival of another species. For example, a fern that can tolerate a broad range of light levels will sprout regardless of whether or not trees are blocking its access to sunlight.

In 1980, an eruption of Mount Saint Helens leveled about 600 square kilometers of forest (Figure 29.17). This gave ecologists the chance to monitor succession firsthand. Researchers observed and recorded natural patterns of colonization. They also manipulated plots inside the blast zone. William Morris and David Wood showed that facilitation and inhibition affected the pattern of plant succession. By adding seeds of certain plant species to some of the plots and keeping other plots barren, they showed that some early colonizers helped different plant species become established and some kept different plant species out.

By the process of ecological succession, one array of species replaces another in a sequential fashion. Primary succession occurs in habitats more or less devoid of life. Secondary succession occurs in areas where an existing community has been severely disturbed.

Traditionally, communities were expected to eventually reach a predictable stable state—the climax community. Ecologists now emphasize the role of random events and disturbances in determining the numbers and kinds of species in communities that are constantly changing.

The presence of one species in a habitat may aid, prevent, or have no effect on the success of later arriving species.

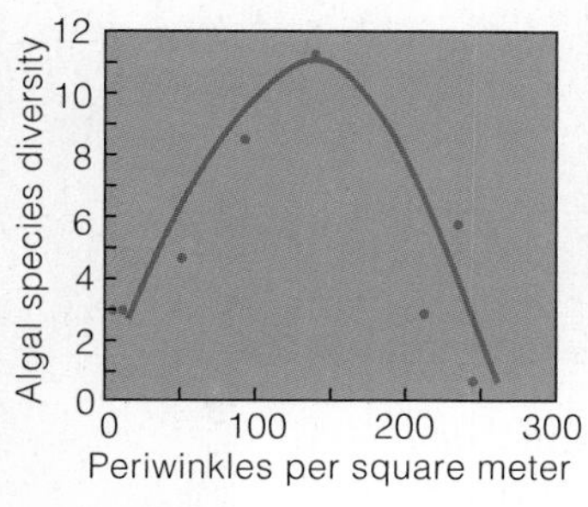

d Algal diversity in tidepools

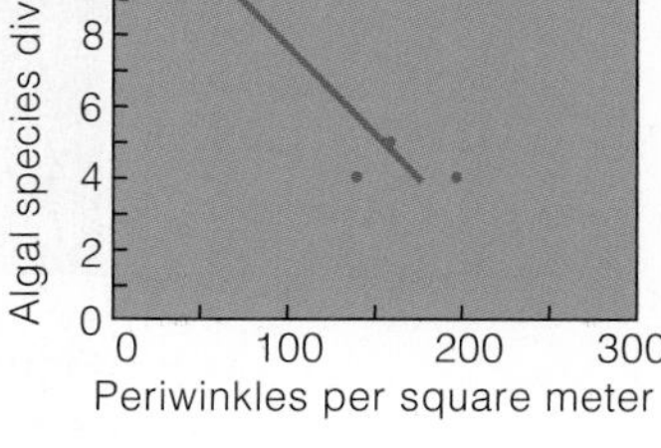

e Algal diversity on rocks exposed at low tide

Figure 29.18 Effect of competition and predation on the community structure of an intertidal zone. (**a**) Grazing periwinkles (*Littorina littorea*) affect the number of algal species in different ways in different marine habitats. (**b**) *Chondrus* and (**c**) *Enteromorpha*, two kinds of algae in their natural habitat. (**d**) Periwinkles graze on *Enteromorpha*, the dominant alga in tidepools; less competitive algae that might otherwise be overwhelmed have an advantage. (**e**) Algal diversity is lower on exposed rock surfaces nearby. There, periwinkles do not graze on *Chondrus* and other kinds of red algae, the dominant species.

29.7 Forces Contributing to Community Instability

The importance of species interaction in community structure is illustrated by removing keystone species or by adding exotic invaders.

Community stability is an outcome of forces that have come into uneasy balance. Resources are sustained as long as populations do not reach carrying capacity. Predator and prey populations rise and fall together. Competitors continually engage in tugs-of-war over resources. In the short term, disturbances might slow the growth of some populations. Also, any long-term changes in climate typically have destabilizing effects.

If instability becomes great enough, the community might change in ways that will persist even when the disturbance ends or is reversed. If some of its species happen to be rare or do not compete as well as others, they might be driven to extinction.

THE ROLE OF KEYSTONE SPECIES

The uneasy balancing of forces in a community comes into focus when we observe the effects of a **keystone species**, a species that has a disproportionately large effect on a community relative to its actual abundance. Robert Paine was the first to describe the importance of a keystone species. He was studying interactions on the rocky shores of California's coast. Species in this rocky intertidal zone survive by clinging to rocks, and access to spaces to hold on to is a limiting factor. Paine set up control plots with the sea star *Pisaster ochraceus* and its prey—chitons, limpets, barnacles, and mussels. He removed all sea stars from his experimental plots.

With the sea stars gone, the mussels (*Mytilus*) took over Paine's experimental plots, and they crowded out seven other invertebrate species. The mussels are the preferred prey of the sea star, and the best competitors in its absence. Normally, predation by sea stars keeps the diversity of prey species high, because it restricts competitive exclusion by mussels. Remove all the sea stars, and the community shrinks from fifteen species down to eight.

The impact of a keystone species can vary between habitats that differ in their species arrays. Periwinkles (*Littorina littorea*) are alga-eating snails of intertidal zones. Jane Lubchenco showed that their removal can increase *or* decrease the diversity of algal species in different habitats (Figure 29.18).

In tidepools, the periwinkles prefer to eat the alga *Enteromorpha*, which can outgrow other algal species. By keeping *Enteromorpha* in check, periwinkles help less competitive algae survive. On rocks in the lower intertidal zone, *Chondrus* and other tough, red algae that periwinkles avoid are dominant. In this habitat, periwinkles prefer to graze on competitively weaker algal species. Thus, periwinkles *promote* algal diversity in tidepools but *reduce* it on exposed rock surfaces.

Figure 29.19 (**a**) Kudzu (*Pueraria lobata*) taking over part of Lyman, South Carolina. It has spread from East Texas to Florida and Pennsylvania. Ruth Duncan of Alabama, who makes 200 kudzu vine baskets a year, just can't keep up. (**b**) *Caulerpa taxifolia* suffocating yet another richly diverse marine ecosystem.

IMPACTS, ISSUES

The Texas horned lizard, or "horny toad," is a fierce-looking predator and the state's official reptile. The mainstay of its diet is harvester ants. However, these native ants cannot compete with imported fire ants, which are displacing them. That's bad news because the horned lizards do not eat the invaders. Imported fire ants also compete directly with the lizards for other insect prey.

HOW SPECIES INTRODUCTIONS TIP THE BALANCE

Sometimes, one member of an established community moves out from its home range and successfully takes up residence elsewhere. This directional movement, called **geographic dispersal**, happens in three ways.

First, over a number of generations, a population might expand its home range by slowly moving into outlying regions that prove hospitable. Second, some individuals might be rapidly transported across great distances. This *jump* dispersal often takes individuals across regions they could not cross on their own, as when tree snakes traveled from New Guinea to Guam inside a ship's cargo hold. Third, with imperceptible slowness, a population might be moved away from its home range over geologic time, as by continental drift.

Successful dispersal and colonization of a vacant adaptive zone are often rapid. Consider one of Amy Schoener's experiments in the Bahamas. She set out plastic sponges on the barren sand at the bottom of Bimini Lagoon. How fast did aquatic species take up residence in the artificial hotels? Schoener recorded occupancy by 220 species within thirty days.

When you hear someone bubbling enthusiastically about an exotic species, you can safely bet the speaker is not an ecologist. An **exotic species** is a resident of an established community that has dispersed from its home range and become established elsewhere. Unlike most imports, which never do take hold outside their home range, an exotic species permanently insinuates itself into a new community. Here, with no coevolved parasites, pathogens, or predators to keep the species in check, its numbers may soar.

More than 4,500 exotic species have established themselves in the United States after jump dispersal. These are just the ones we know about. We put some of the new arrivals, including soybeans, rice, wheat, corn, and potatoes, to use as food crops.

These and many other additions proved harmless or have had beneficial effects. However, other imports had adverse effects. Consider the following examples of how an imported plant, marine alga, and mammal wreaked havoc on community structure.

THE PLANTS THAT ATE GEORGIA In 1876, a vine called kudzu (*Pueraria lobata*) from Japan was introduced to the United States. In temperate regions of Asia, it is a well-behaved legume with a well-developed root system. It *seemed* like a good idea to use it for forage and to control erosion. But kudzu grew faster in the Southeast, where herbivores, pathogens, and less competitive plants posed no serious threat to it.

With nothing to discourage it, kudzu shoots grew sixty meters a year. Its vines now blanket streambanks, trees, telephone poles, houses, and almost everything else in their path (Figure 29.19*a*). Attempts to dig up or burn it are futile. Grazing goats and herbicides help. But goats eat other plants too. Herbicides contaminate water supplies. Kudzu may spread as far north as the Great Lakes by the year 2040.

On the bright side, Asians use a starch extracted from kudzu in drinks, herbal medicines, and candy. A kudzu processing plant in Alabama may export this starch to Asia, where demand currently exceeds the supply. Also, kudzu may help save forests; it can be used as an alternative source for paper.

THE ALGA TRIUMPHANT They just looked so perfect in saltwater aquariums, those bright green, long, and feathery branches of the green alga *Caulerpa taxifolia*. The Stuttgart Aquarium in Germany even developed a hybrid strain that was sterile, and generously shared it with other marine institutions. Was it from Monaco's Oceanographic Museum that the strain escaped into the wild? Some say yes, Monaco says no.

The aquarium strain of *C. taxifolia* cannot reproduce sexually. It grows quickly, a few centimeters a day, and reproduces asexually by runners. Boat propellers and fishing nets help to disperse it. Since 1984, the alga has colonized more than 10,000 acres of coastal seafloor in the Mediterranean Sea (Figure 29.19*b*). It thrives along sandy or rocky shores and in mud, and survives ten days out of water. Unlike its tropical parents, this strain thrives in cool water, and tolerates pollutants. Its toxin repels herbivores. It displaces endemic algae, invertebrates, and fishes. It has been found in coastal waters of Southern Australia, where it has forced the closure of some waters to fishing and boating.

C. taxifolia has also invaded the coastal waters of the United States, but quick action may have beaten it back. In 2000, some scuba divers discovered the alga growing in areas off the coast of Southern California. Someone might have drained water from an aquarium into a storm drain or into the lagoon itself. In response to the discovery, a coalition of government and private groups sprang into action. They tarped over the area to shut out sunlight, pumped chlorine bleach into the mud to poison the alga, and used welders to boil it. The aggressive measures seem to have been effective. No sightings of the exotic invader have been reported in these waters since September of 2002.

Importing *C. taxifolia* or any closely related species of *Caulerpa* into the United States is now illegal. To protect native aquatic communities, aquarium water should never be dumped into storm drains or natural waterways. It should be discarded into a sink or toilet so that wastewater treatment can kill any algal spores.

THE RABBITS THAT ATE AUSTRALIA During the 1800s, British settlers in Australia just could not bond with koalas and kangaroos, and so they imported familiar animals from home. In 1859, in what would be the start of a major disaster, a northern Australian landowner imported and then released two dozen wild European rabbits (*Oryctolagus cuniculus*). Good food and sport hunting, that was the idea. An ideal rabbit habitat with no natural predators—that was the reality.

Six years later, the landowner had killed 20,000 rabbits and was besieged by 20,000 more. The rabbits displaced livestock and native wildlife. Now 200 to 300 million are hippity-hopping through the southern half of the country. Their collective feeding is turning grasslands and shrublands into eroded deserts.

Figure 29.20 Rabbit-proof fence? Hardly. This is part of a fence that was built to keep rabbits from spreading and destroying more of Australia's vegetation. The effort was a failure.

Rabbits have been shot and their warrens plowed under, fumigated, and dynamited. When a fence 2,000 miles long was built to keep rabbits out of western Australia, rabbits made it from one side to the other before workers could finish the job (Figure 29.20).

In 1951, the Australian government introduced a myxoma virus that infects South American rabbits and causes *myxomatosis*. Mosquitoes and fleas transmit the virus from host to host. Having no coevolved defenses against the import, European rabbits died in droves. But natural selection has since favored a rise in rabbit populations resistant to the imported virus.

In 1991, on an uninhabited island in Spencer Gulf, Australian researchers began an experimental study of a calicivirus. The virus kills rabbits by affecting their heart and lungs. In 1995, the virus made an unforeseen jump to the mainland, perhaps carried by insects. In 1996, it was deliberately released into other areas. By 2001, rabbit populations were 80 to 85 percent below their peak values. Grasses and shrubs have begun to rebound and native herbivores have begun increasing their numbers.

The jury is still out on the long-term impact of the viral releases. Some Australians fear it might infect animals other than rabbits—so far this seems unlikely.

The presence or absence of a single species can have a profound effect on the structure of a community. Competition from introduced species is driving many native species toward extinction.

LINKS TO SECTIONS 13.3, 28.1

29.8 Patterns of Species Diversity

You already learned that the frequency of disturbances affects the number of different species in a community. Ecologists have also documented other factors that cause species richness to vary from one region to the next.

In large part, the number of species in any region is an outcome of the evolutionary history of each member species and its resource requirements, physiology, and dispersal capacity. Rates of birth, death, immigration, and emigration also shape species numbers. In turn, those rates are shaped by conditions in the local and regional environments, and by species interactions.

The species richness of a region is also affected by geographic and climactic variables, such as sunlight intensity, rainfall, and temperature along gradients in latitude, elevation, and—for aquatic habitats—depth. Within broad regions, microenvironments give rise to local variations in the overall patterns.

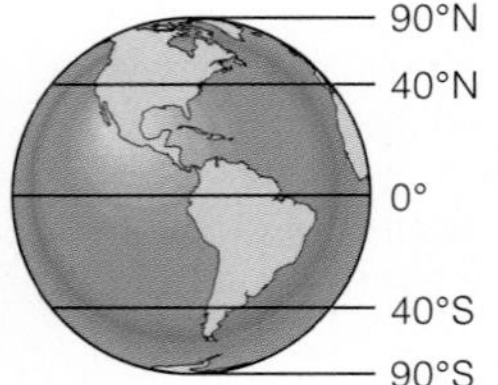

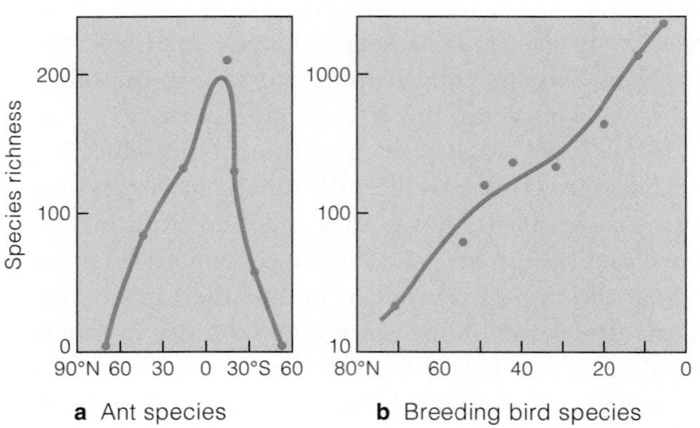

Figure 29.21 Two patterns of species diversity corresponding to latitude, as revealed by the number of ant species (**a**) and birds (**b**) in Central and North America.

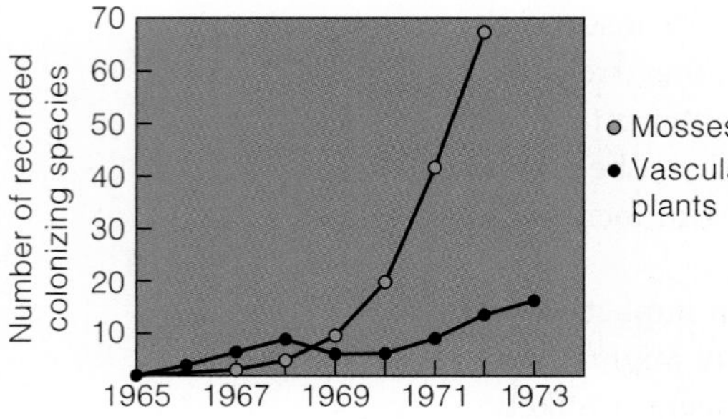

Figure 29.22 Surtsey, a volcanic island, at the time of its formation. The chart gives the number of colonizing species between 1965 and 1973. Newly formed, isolated islands are natural laboratories for ecologists.

MAINLAND AND MARINE PATTERNS

The most striking pattern of species richness relates to distance from the equator. For most groups of plants and animals, *the number of coexisting species on land and in the seas is greatest in the tropics, and it systematically declines from the equator to the poles.* Figure 29.21 shows examples of this pattern. Here we focus on the factors that help bring about the pattern and also maintain it.

First, for reasons explained in Chapter 31, tropical latitudes intercept more intense sunlight and receive more rainfall, and their growing season is longer. As one outcome, resource availability tends to be greater and more reliable in the tropics than elsewhere. This favors a degree of specialized interrelationships not possible where species are active for shorter periods.

Second, tropical communities have been evolving for a longer time than temperate ones, some of which did not begin forming until the end of the last ice age.

Third, species diversity might be self-reinforcing. Diversity of tree species in tropical forests is much greater than in comparable forests at higher latitudes. When more plant species compete and coexist, more kinds of herbivores also compete and coexist, partly because no single kind of herbivore can overcome all the chemical defenses of all the different plants. More predatory and parasitic species evolve in response to the larger array of prey and hosts. The same type of effect may promote diversity on tropical reefs.

ISLAND PATTERNS

Volcanic islands often serve as natural laboratories for biological studies. As an example, in 1965, a volcanic eruption produced a new island, called Surtsey, just southwest of Iceland. Within just six months, bacteria, fungi, seeds, flies, and some seabirds were established on it. One vascular plant appeared after two years, and mosses two years after that (Figure 29.22). The number of plant species rose after soils had become enriched. Seabirds began to nest on the island in 1970. Today, five species nest there and migratory birds pass through by the thousands. Some carry seeds in their feathers or bellies and so introduce new plant species.

As is the case for other islands, the number of new species on Surtsey did not increase indefinitely. Why not? Studies of community patterns on islands around the world provide us with two models.

First, any islands distant from a source of potential colonists receive few colonizing species, and the few that do arrive are adapted for long-distance dispersal (Figure 29.23). This is known as the **distance effect**. Second, larger islands tend to support more species than smaller islands that are the same distances from

Figure 29.23 *Animated!* (**a**) Two island patterns. Distance effect: Species richness on islands of a specified size declines with increasing distance from a source of colonizing species. *Green* circles signify islands less than 300 kilometers from the colonizing source. *Orange* triangles signify islands more than 300 kilometers from the source areas. Area effect: Among islands the same distance from a source of colonizing species, the larger ones support more species. (**b**) Wandering albatross, one travel agent for jump dispersals. Seabirds that island-hop long distances often have seeds stuck to their feathers. Seeds that successfully germinate in a new island community may give rise to a population of new immigrants.

a colonizing source. Figure 29.23*a* has examples of this **area effect**. Larger islands tend to offer more habitats with more variation. Most of them have more complex topography and higher elevations. These are the kinds of variations that favor species diversity. Also, being bigger targets, big islands intercept more colonists.

Most importantly, extinctions suppress biodiversity on small islands. There, populations are small and far more vulnerable to severe storms, volcanic eruptions, disease, and random shifts in birth and death rates.

Again, on any island, the number of species reflects a balance between immigration rates for new species and extinction rates for established ones. Small islands that are distant from a source of colonists have low immigration rates and high extinction rates. The result is that, over the long term, they support fewer species than do larger islands that lie closer to the mainland.

The number of species in a region—species richness—is affected by gradients in latitude, elevation, and (for the aquatic species) depth.

Species richness in a given area also is an outcome of the evolutionary history of each species, its requirements for resources, its physiology, its capacity for dispersal, and its rates of birth, death, immigration, and emigration.

Generally, communities in the tropics have the most species and those at the poles have the least. The species diversity on an island also depends on its size and distance from a source of potential colonists.

29.9 Conservation Biology

Based on many lines of evidence accumulated over the past few centuries, an estimated 99 percent of all species that have ever lived are extinct. Even so, the full range of species diversity is greater now than it has been at any time in the past.

Reflect on the story of life's evolution, as told in Unit Three. So immense were five of the mass extinctions that we use them as the major boundaries for geologic time. After each event, species diversity plummeted on land and in the seas. In the aftermath, other species embarked on adaptive radiations. Recovery to the same level of species richness was exceedingly slow, on the order of 20 million to 100 million years.

The five major mass extinctions and long recoveries overshadow other, less spectacular extinction events. For example, the K–T asteroid impact was the final blow for some lineages. But other factors had been at work. Many dinosaur lineages became extinct during the 10 million years that *preceded* the impact. Geologic changes may have been a factor. Pangea was breaking up even before the K–T impact. Eventually, the deep, coldwater currents from the south started moving into warm equatorial seas. The change in ocean circulation altered air circulation and climate. A wide variety of groups perished during that global reorganization.

LINKS TO SECTIONS 13.8–13.10

ABOUT THE NEWLY ENDANGERED SPECIES

We recognize three levels of **biodiversity**—diversity encompassed in diverse genes, diverse species, and diverse habitats. All three components of biodiversity are now in decline. Human activities are causing this biodiversity crisis.

No giant asteroids have hit the Earth for 65 million years. Yet a sixth major extinction event is under way. Although mammals survived the asteroid impact that did in the dinosaurs, many have not been as lucky in their encounters with humans. Shortly after the first humans migrated into the Americas, three quarters of their large mammal species, known as the *megafauna*, disappeared. Climate changes may have played a role, but hunting was certainly a contributing factor.

The International Union for the Conservation of Nature (IUCN) defines an **endangered species** as one that has lost 80 percent of its population, has suffered a drastic reduction in its range, and is likely to become extinct in the next 20 to 50 years. By this definition, about 10 percent of mammals, 5 percent of birds, and 20 percent of amphibians are endangered. Nearly all are *endemic* species, that is, they evolved in one region and are found nowhere else.

THE MAJOR THREATS

Naturalist Edward O. Wilson defines **habitat loss** as physical reduction in suitable places to live, as well as the loss of suitable habitat through chemical pollution. Habitat loss is putting major pressure on more than 90 percent of the species that now face extinction.

Species-rich tropical forests are being decimated (Chapter 15). As other examples, 98 percent of tallgrass prairies, 50 percent of wetlands, and 85 to 95 percent of old-growth forests of the United States alone are gone. Even arid lands are not exempt from human intrusion. Figure 29.24 is a case in point.

Habitats often become fragmented, or chopped into patches, each with a separate periphery. Species at a patch's periphery are more vulnerable to temperature changes, winds, fires, predators, and pathogens. Also, fragmentation often creates patches that are not large enough to support the population sizes required for successful breeding. They may not have enough food or other resources to sustain a species.

Think back on the factors shaping species diversity on islands. Even though islands make up a fraction of the Earth's surface, about half the plants and animals driven to extinction since 1600 were island natives that just had nowhere else to go. R. MacArthur and E. O. Wilson used an island biogeography model to estimate the likelihood of extinctions in habitat islands—natural areas that are surrounded by a "sea" of habitat that has been degraded by human activities. The model is also used to estimate how large an area must be set aside as protected reserves to ensure survival of a species.

Some species help warn us of changes in habitats that could lead to local extinctions. Birds are one of the key **indicator species**. Different types live in all major land regions and climate zones, they react quickly to changes, and they are relatively easy to track.

Species introductions are another major threat to endemic species. Remember, each habitat has only so many resources, and its occupants must compete for a share. Introduced species often grow to large numbers and overrun endemic competitors. They have been a key factor in almost 70 percent of the cases where endemic species are being driven to extinction.

Also, in a sad commentary on human nature, the more rare a wild animal becomes, the more its value soars in the black market. Figure 29.25 gives you an idea of the money involved in illegal wildlife trading. As a final point, biodiversity also is threatened by overharvesting of species that have commercial value. Whales are a prime example. Humans kill whales for food, lubricating oil, fuel, cosmetics, fertilizer, even corset stays. Substitutes exist for all whale products. Yet fishermen of several nations ignore a moratorium on slaughtering large whales, and they routinely cross boundaries of sanctuaries set aside for recoveries.

Figure 29.24 Humans as competitor species—example of dominance in the competition for living space in a habitat.

EMERGENCE OF CONSERVATION BIOLOGY

Awareness of the impending extinction crisis gave rise to **conservation biology**. We define this field of pure and applied research as (1) a systematic survey of the full range of biological diversity, (2) efforts to decipher biodiversity's evolutionary and ecological origins, and (3) attempts to identify methods that might maintain and use biodiversity to benefit the human population. Its goal is to conserve and use, in sustainable ways, as much biodiversity as possible.

THE ROLE OF SYSTEMATICS Reflect on Figure 29.26. Our current taxonomic knowledge is fairly complete for many groups, including flowering plants, conifers, and fishes, birds, mammals, and other vertebrates. We tend to know more about large organisms, especially the land dwellers, and the showy ones, such as birds and butterflies. Yet a staggering number of arthropods, fungi, protists, and prokaryotes are probably still out there, waiting to be identified.

For example, biologists recently surveyed a single hot spring in Yellowstone National Park. They turned up more archaean species than were already known for the rest of Earth—which gives you an idea of how much work is to be done.

When the inventorying of biodiversity is weighed against society's demands for, say, social services and protection from bioterrorism, guess which issue takes a back seat? Realistically, we just do not have enough time, money, or workers to complete a global survey of biodiversity. Possibly 99 percent of existing species have not even been described, and we do not know where most of them live. That is why the initial work has focused on identifying **hot spots**, or habitats of a large number of species that are found nowhere else and are in greatest danger of extinction.

At the first survey level, workers target a limited area, such as an isolated valley. A complete survey is out of the question, so they inventory flowering plants, birds, mammals, fishes, butterflies, and certain other indicator species within the habitat. At the next level, broader areas that are major or multiple hot spots are systematically explored. The widely separated forests of Mexico are an example of a key hot spot. Research stations have been set up at different latitudes and at different elevations across the region.

At the highest survey level, hot spot inventories are combined with existing knowledge of biodiversity in ecoregions. An **ecoregion** is a broad land or ocean region defined by climate, geography, and producer species. Figure 29.27 on the following page is a work in progress. It identifies 238 regions that are essential for global biodiversity: 142 land, 53 freshwater, and 43 marine. Initially, 25 are selected for rapid conservation efforts, which will be refined over the coming decades.

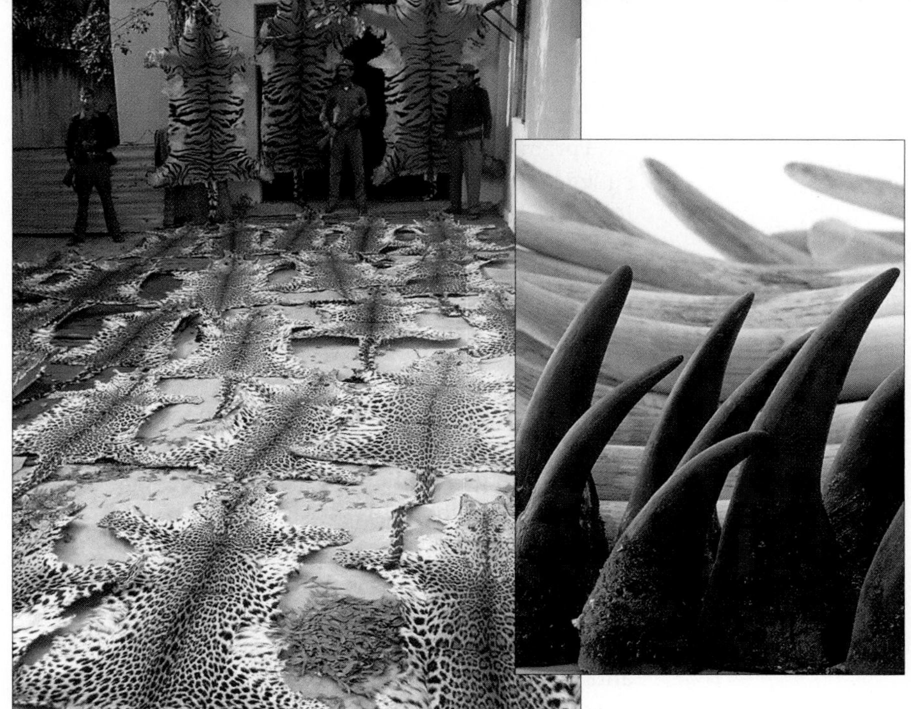

Figure 29.25 Humans as predator species—confiscated tiger skins, leopard skins, rhinoceros horns, and elephant tusks. An illegal wildlife trade threatens the survival of more than 600 species, yet all but 10 percent of the trade escapes detection. A few of the black market prices: Live, mature saguaro cactus ($5,000–15,000), bighorn sheep, head only ($10,000–60,000), polar bear ($6,000), grizzly ($5,000), bald eagle ($2,500), peregrine falcon ($10,000), live chimpanzee ($50,000), live mountain gorilla ($150,000), Bengal tiger hide ($100,000), and rhinoceros horn ($28,600 per kilogram).

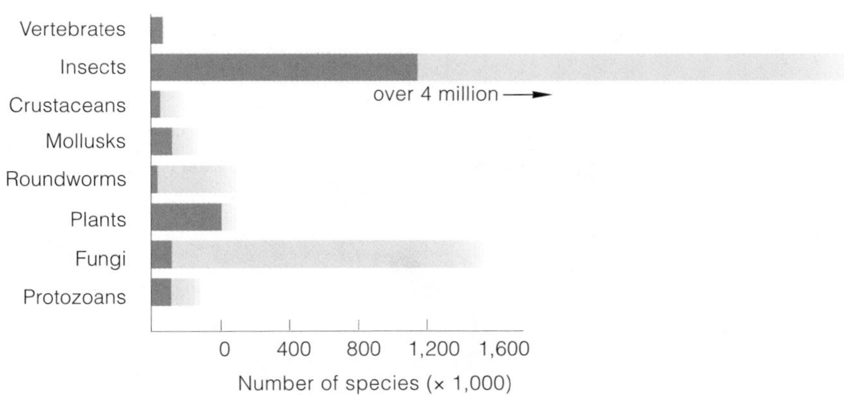

Figure 29.26 Current species diversity for a few major taxa. *Red* is the number of named species; *gold* is the estimated number of species not yet discovered.

BIOECONOMIC ANALYSIS All nations have three forms of wealth—material, cultural, and biological. Until recently, many undervalued their biological wealth—*their biodiversity*, which is a source of food, medicine, and other goods. No one knows how much biological wealth they have already lost.

LINK TO SECTION 28.5

For instance, one particularly observant Mexican college student located *Zea diploperennis*, a wild maize long believed to be extinct. It had disappeared from most of its range, but a relic population clung to life in a 900-acre region of mountainous terrain near Jalisco. Unlike domesticated corn, *Z. diploperennis* is perennial and resistant to most viruses. Gene transfers from the wild species into crop plants may boost production of corn for hungry people in Mexico and elsewhere.

In recognition of the potential value of this species, the Mexican government has set aside its mountainous habitat as a biological reserve, the first ever created to protect a wild relative of an important crop plant.

In the next fifty years, human population size may reach 8.9 billion, with most growth in the developing countries. Each person requires raw materials, energy, and living space. How many species will be crowded out? The countries with the greatest monetary wealth also have the least biological wealth and use most of the world's natural resources. Countries with the least monetary wealth and the most biological wealth have the fastest-growing populations.

People who are locked in poverty commonly must choose between saving themselves or an endangered species. Deprived of education and resources that the people in developed countries take for granted, they clear tropical forests. They hunt animals and dig up plants in "protected" parks and reserves. They try to raise crops and herds on marginal land.

Knowing that asking economically pressed people to stop is simplistic at best, conservation biologists are working to identify ways in which people might earn a living instead from their biological wealth—from the biodiversity in threatened habitats.

The systematic expansion of species inventories is a baseline for economic analysis, for assigning future value to ecoregions. Governments—and individuals—can protect a habitat and still withdraw resources in sustainable fashion. Threatened species can be studied for what they may offer. For example, one might be a source of new medicines and other chemical products. Most developing countries do not have laboratories for chemical analysis. However, large pharmaceutical companies do. The National Institute of Biodiversity of Costa Rica collects and identifies species that look promising and sends in chemical samples extracted from them to one company. If natural chemicals end up being marketed, Costa Rica will share the royalties, which are earmarked for conservation programs.

SUSTAINABLE DEVELOPMENT Identifying the possible uses of a country's biological wealth and putting them into practice is not easy. Yet there are successes, such as those in the next two examples.

Strip logging. Chapter 15 hinted at the destruction in once-great tropical rain forests, the richest homes of biodiversity on land. For now, simply think about one approach to minimizing the losses.

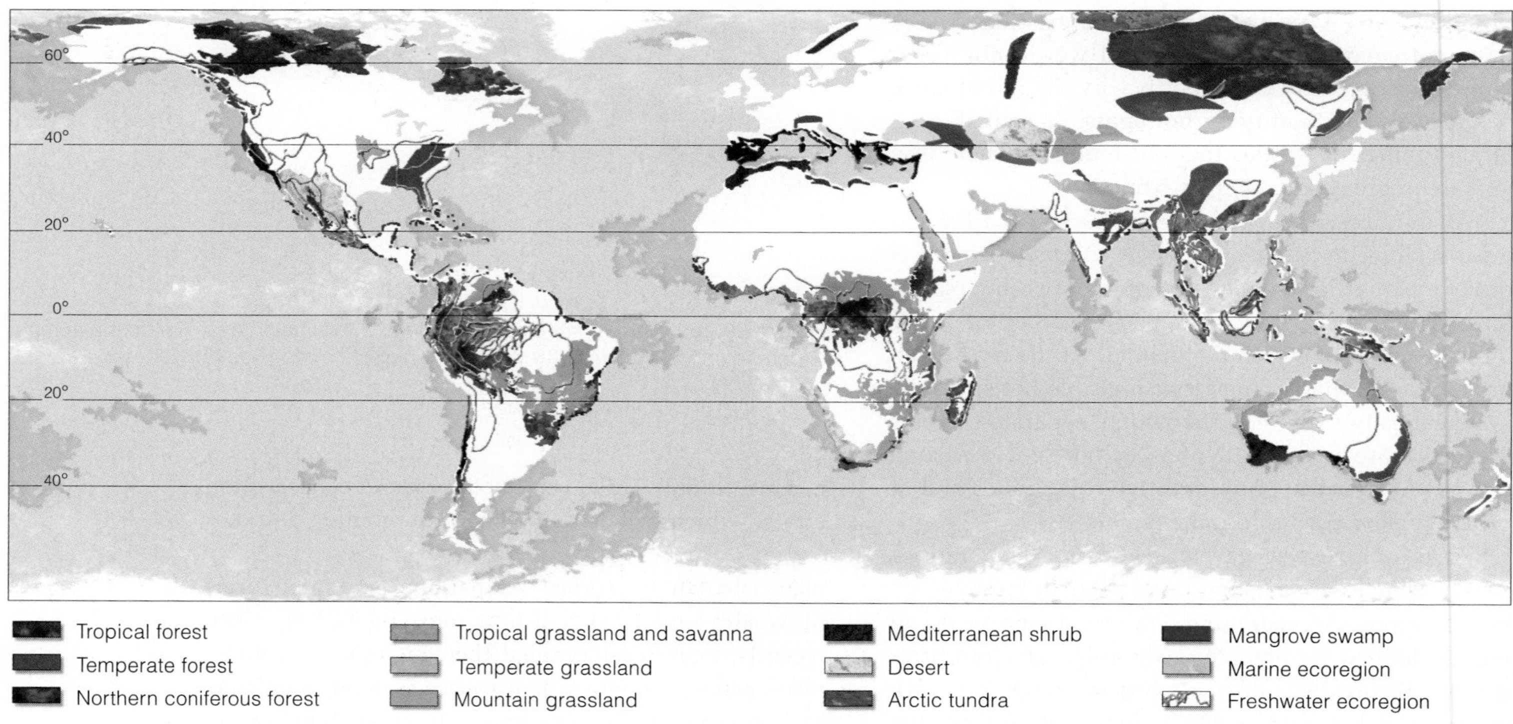

Figure 29.27 Map of the most vulnerable regions of land and seas, compiled by the World Wildlife Fund.

Besides providing wood for local economies, trees of the tropical rain forests yield diverse, exotic woods prized by developed countries. What if they could be logged in a profitable, sustainable way that preserved biodiversity? Gary Hartshorn was the first to propose **strip logging** in portions of forests that are sloped and have many streams (Figure 29.28). The procedure is to clear a narrow corridor that parallels contours of the land, then cut a narrow road at the top to haul logs. In a few years, saplings grow in this cleared corridor. A second corridor is cleared above the road. Nutrients that leach from exposed soil here trickle into the first corridor and are taken up by the saplings and promote their growth. Later, a third corridor is cut above the second one—and so on in a profitable cycle of logging, which the habitat sustains over time.

Ranching in a riparian zone. Reflect on the map in Figure 29.27, and you see that the biodiversity crisis is not confined to developing countries. Now take a closer look at a riparian zone in the western United States. A **riparian zone** is a relatively narrow corridor of vegetation along a stream or a river, as in Figure 29.29. Its plants afford a major line of defense against flood damage by sponging up water during spring runoffs and summer storms. Shade cast by the canopy of taller shrubs and trees in the zone helps conserve precious water during droughts.

A riparian zone offers wildlife food, shelter, and shade, particularly in arid and semiarid regions. For instance, in the western United States, 67 to 75 percent of endemic species spend all or part of their life cycle in riparian zones. They include 136 songbird species, some of which nest only in riparian vegetation.

Many cattle are raised in the American West. In comparison with wild species, they have higher water needs and tend to congregate at rivers and streams. They trample and feed until grasses and herbaceous shrubs are gone. It takes only a few head of cattle to ruin a riparian zone. The damage can be reversed if cattle are excluded. Environmentally aware ranchers provide food and water away from streams and rotate grazing areas, so each area has time to recover.

The current range of global biodiversity is an outcome of an overall pattern of extinctions and slow recoveries.

For the past forty years, human activities have raised rates of extinction through habitat losses, species introductions, overharvesting, and illegal wildlife trading.

Conservation biology entails a systematic survey of the full range of biodiversity, analysis of its evolutionary and ecological origins, and identification of methods to maintain and use biodiversity for the benefit of humans.

Figure 29.28 Strip logging. This practice may sustain biodiversity even as it enhances the regeneration of economically valued forest trees.

Figure 29.29 Riparian zone restoration. This example comes from Arizona's San Pedro River, shown before and after restoration efforts.

Summary

Section 29.1 Each species occupies a certain habitat characterized by physical and chemical features and by the array of other species living in it. All populations of all species occupying a habitat are a community. Each species in the community has its own niche, or way of living.

Species in a community are linked by interactions of the types summarized in Table 29.1. Commensalism, mutualism, and parasitism are types of symbiosis: a relationship in which organisms live together for some part or all of the life cycle.

Section 29.2 Mutualism is a species interaction that benefits both participants. Some mutualists cannot complete their life cycle without their partner species.

Section 29.3 Two species that require the same limited resource tend to compete, as by exploiting it as fast or efficiently as possible or interfering with the other's use of it. Interspecific competition harms both participants. By the competitive exclusion concept, when two (or more) species require identical resources, they cannot coexist indefinitely. Species are more likely to coexist when they differ in their use of resources. Resource partitioning allows similar species to coexist by using different portions of shared resources.

BiologyNow

Learn about competitive exclusion with the animation on BiologyNow.

Section 29.4 Predators are free-living and typically kill their prey. Predator and prey populations often show cyclic oscillations in their density. The carrying capacity for prey, likely refuges for prey, predator feeding efficiency, and often alternative prey sources influence the cycles.

Predators and prey exert selection pressure on each other. Evolved prey defenses include threat displays, chemical weapons, camouflage, and mimicry. Predators overcome defenses of prey by adaptive behavior, such as stealth and by camouflage.

BiologyNow

Learn how predator and prey populations interact with the animation on BiologyNow.

Section 29.5 Parasites live in or on other living hosts and withdraw nutrients from host tissues for all or part of their life cycle. Hosts may or may not die as a result. Social parasites leave eggs and offspring in the care of a host species. Parasitoids are free-living as adults, but the larvae develop inside and feed on a host species, which they eventually kill. Parasites and parasitoids are used as biological control agents.

Section 29.6 Ecological succession is the sequential replacement of arrays of species within a community. Primary succession occurs in new habitats. Secondary succession occurs in disturbed ones. The first species in a community are pioneer species. Their presence may help, hinder, or have no effect on other colonists.

The view that all communities eventually reach a stable climax state has been replaced by new models that emphasize continual change and acknowledge the importance of disturbances. Communities with an intermediate rate of disturbance tend to include more species than those disturbed more or less often.

Section 29.7 Community structure reflects an uneasy balance of forces that operate over time. Major forces are competition and predation, as by keystone species. Introduction of species that in one way or another expand their geographic range can change community structure in permanent ways.

Section 29.8 Global patterns of species diversity tend to follow environmental gradients. The species richness in a region is influenced by the evolutionary history of each species and its resource requirements, physiology, capacity for dispersal, and demographics. It also is influenced by species interactions.

BiologyNow

Learn about the area effect and distance effect with the interaction on BiologyNow.

Section 29.9 The current range of biodiversity is an outcome of an overall pattern of mass extinctions and slow recoveries. Human activities are increasing rates of extinction. Conservation biologists seek ways to conserve and use biodiversity in a sustainable fashion.

Table 29.1 Types of Species Interactions

Type of Interaction	Direct Effect on Species 1	Direct Effect on Species 2
Commensalism	Benefits	None
Mutualism	Benefits	Benefits
Interspecific competition	Harmed	Harmed
Predation	Benefits	Harmed
Parasitism	Benefits	Harmed

Self-Quiz

Answers in Appendix I

1. A realized niche ______ .
   a. is smaller than a species' fundamental niche
   b. is affected by the presence of other species
   c. can shift over time
   d. all of the above

2. Which of the following interspecific interactions often lowers the population size of both participants?
   a. mutualism
   b. commensalism
   c. parasitism
   d. predation
   e. interspecific competition

Figure 29.30 Two phasmids: (**a**) A stick insect from South Africa. (**b**) A leaf insect from Java. (**c**) Many phasmid eggs resemble seeds.

Figure 29.31 One of the new realities: Casket in the ground or a concrete foundation for an artificial reef?

3. Which of the following is not a type of symbiosis?
   a. predation c. parasitism
   b. mutualism d. commensalism
4. With ______ both species benefit.
   a. predation c. mutualism
   b. interspecific competition d. commensalism
5. If one of two competing species is removed from a portion of a shared habitat, the population size of the species that remains will ______ over time.
   a. increase c. remain the same
   b. decrease
6. A predator population and prey population ______.
   a. always coexist at relatively stable levels
   b. may undergo cyclic or irregular changes in density
   c. cannot coexist indefinitely in the same habitat
   d. both b and c
7. With ______, one species modifies a habitat in a way that makes it easier for some other species to move in.
   a. tolerance c. inhibition
   b. facilitation d. a and c
8. The species diversity of an area is affected by ______.
   a. climate and topography d. both a and b
   b. possibilities for dispersal e. a through c
   c. evolutionary history
9. Following mass extinctions, recovery to the same level of species diversity has taken many ______ of years.
   a. hundreds b. millions c. billions
10. Match the terms with the most suitable descriptions.

____	geographic dispersal	a. early colonist of barren or disturbed places
____	area effect	b. dominates community
____	pioneer species	c. individuals leave home range, and get established elsewhere
____	symbiotic species	d. island size influences number of species it can support
____	keystone species	e. live together and interact for some or all of life cycle
____	endemic species	f. area near river or stream
____	riparian zone	g. originated in one region and found nowhere else
____	niche	h. activities, interactions, and needs of one particular species

Additional questions are available on **Biology Now**™

Critical Thinking

1. With antibiotic resistance rising, researchers are looking for ways to reduce use of antibiotics. Some cattle that once would have been fed antibiotic-laced food are now being given feed that contains cultures of bacteria. These *probiotic feeds* are designed to establish or bolster populations of helpful bacteria in the animal's gut. The idea is that if a large population of beneficial bacteria is in place, then the harmful bacteria cannot become established or thrive. Which ecological principle is guiding this research?

2. Insects called phasmids are extraordinary masters of disguise. Most resemble sticks or leaves. Even their eggs may be camouflaged (Figure 29.30). The phasmids are all herbivores. Most spend their days motionless, moving about and feeding only at night. If disturbed, a phasmid will fall to the ground and remain motionless.

   Speculate on the selective pressures that might have shaped phasmid morphology and behavior. Suggest an experiment that could help determine whether the cryptic appearance and behavior of one species have adaptive value.

3. Does focusing on species-rich hot spots cause us to undervalue regions where species diversity is lower? For example, broad bands of coniferous forests in the Northern Hemisphere have low species diversity, but photosynthesis here affects global carbon cycles. How can we ensure that regions that provide such biological services are protected?

4. Mentally transport yourself to a tropical rain forest of South America. Imagine you do not have a job. There are no jobs in sight, not even in the overcrowded cities some distance away. You have no money or contacts to get you anywhere else. And yet you are the sole supporter of a large family. Today a stranger approaches you. He tells you he will pay good money if you can capture alive a certain brilliantly feathered parrot in the forest. You know the parrot is rare; it is seldom seen. However, you have an idea of where it lives. What will you do?

5. Here's an alternative to formal burial on land. Ashes of cremated people are mixed with pH-neutral concrete and used to build *artificial reefs*. Each perforated concrete ball weighs 400 to 4,000 pounds and is designed to last 500 years. So far, 100,000 of these reefs are helping marine life in 1,500 locations (Figure 29.31). Proponents say that the burial costs less, does not waste land, and is ecologically friendly. Your thoughts?

30 ECOSYSTEMS

IMPACTS, ISSUES

Bye-Bye Bayou

Each Labor Day, the coastal Louisiana town of Morgan City celebrates the region's economic mainstays with the Louisiana Shrimp and Petroleum Festival. The state is the nation's third-largest petroleum producer and the leader in shrimp production. But the petroleum industry's success may be contributing to the decline of the state's fisheries.

The global air temperature is rising, and fossil fuel burning is a major culprit. Because warm air heats the surface waters of the ocean, and water expands when it is heated, the sea level is rising. Warmer air also is melting ancient glaciers and vast ice caps, and this meltwater further increases the ocean's volume.

If current trends persist, then low elevations along the United States coastline—including *14,720,000 acres* next to the Gulf of Mexico and Atlantic Ocean—will be one to three feet under water within fifty years!

With more than 40 percent of the country's coastal marshes, Louisiana has the most to lose. Its wetlands are already sinking, because dams and levees along its rivers interfere with the deposition of sediments that could replace those washed out to sea. The anticipated rise in sea level will obliterate about 70 percent of these wetlands.

Is an ecological and economic disaster unfolding? Consider: More than 3 billion dollars' worth of shellfish and fish are harvested from Louisiana's wetlands each year. About 40 percent of North America's migratory ducks—more than 5 million birds—overwinter here. Louisiana's wetlands also buffer low inland elevations from storm surges and hurricanes.

There's more. The warmer water will promote algal blooms that cause massive fish kills. It will favor the growth of bacterial pathogens, and more people will get sick after swimming in contaminated water or after eating contaminated shellfish.

Inland, heat waves and wildfires will become more intense, and deaths from heat stroke will climb. Warmer temperatures will allow disease-causing mosquitoes to extend their ranges farther inland.

Global warming is expected to increase evaporation and alter rainfall patterns, leading to droughts in some places and floods in others. Hurricanes are expected to increase in intensity. The damage from hurricanes in 2005 may be a preview of things to come.

This chapter will start you thinking about a major concept: The atmosphere, ocean, and land interact in ways that influence the living conditions of organisms throughout the biosphere. Driven by energy streaming in from the sun, these interactions give rise to globe-spanning cycles upon which life depends. The chapter also will start you thinking about a related concept of equal importance: We have become major players in these global interactions even though we still do not fully comprehend how they work.

How Would You Vote? *Emissions from motor vehicles contain greenhouse gases. Should the government raise minimum fuel economy standards for cars and SUVs? See BiologyNow for details, then vote online.*

Key Concepts

ECOSYSTEM STRUCTURE

Organisms and their physical environment make up an ecosystem. Energy flows in one way through organisms in an ecosystem. Producers capture energy from the sun and use inorganic compounds from the environment to build tissues. Consumers get energy and nutrients by eating other organisms. Nutrients are cycled among organisms.

GLOBAL CYCLES

Earth's waters are cycled on a vast scale through environmental reservoirs and the living components of ecosystems. So are nutrients, such as carbon, nitrogen, and phosphorus. Human activities are disrupting these global cycles.

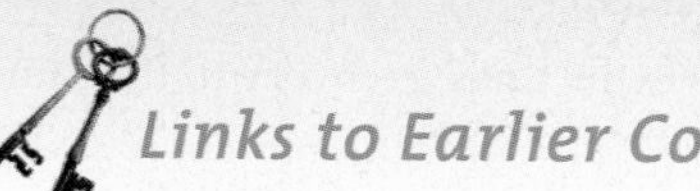

Links to Earlier Concepts

Sections 1.1 and 1.2 introduced you to the idea of an ecosystem. Sections 18.5 and 24.3 described the mineral needs of plants and animals. Here you will learn how these minerals cycle between living organisms and their environment. Carbon fixation by plants (5.3) and protists (14.3), and the many ecological roles of prokaryotes (14.2) are relevant too. Laws governing energy (4.1) come into play as energy is transferred among organisms (29.4). We consider again the effects of pesticides (19.9) and the runaway growth of human populations (28.5).

30.1 The Nature of Ecosystems

In the preceding chapter, we focused on community structure, how species interact in a habitat. This chapter continues with those interactions, but at a higher level of organization. Each community, with its physical environment, works as an ecosystem.

OVERVIEW OF THE PARTICIPANTS

An **ecosystem** is an array of organisms together with their physical environment. They interact through a one-way flow of energy and the cycling of vital raw materials. It is an open system; it is unable to sustain itself without continual inputs of energy.

Diverse ecosystems abound on Earth's surface. In climate, landforms, soil, vegetation, animal life, and other features, deserts differ from hardwood forests, which differ from tundra and the prairies. Reefs differ from the open ocean, which differs from streams and lakes. Despite their many differences, all ecosystems are alike in certain aspects of structure and function.

Energy input, from sun

Producers
Autotrophs (plants and other self-feeding organisms)

Nutrient Cycling

Consumers
Heterotrophs (animals, most fungi, many protists, many bacteria)

Energy output (mainly metabolic heat)

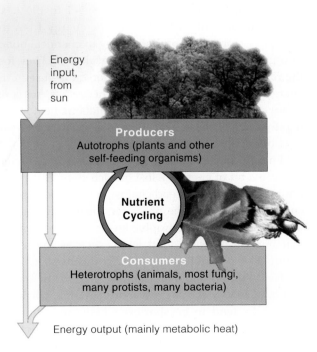

Figure 30.1 Model for ecosystems. Energy flows in one direction: into an ecosystem, through living organisms, and out from it. Nutrients cycle among the producers and consumers. In nearly all ecosystems, energy flow starts with photoautotrophs that capture energy from the sun.

Ecologists investigate these similarities and can use them to make predictions about how ecosystems will be affected by human-caused and natural changes.

With a few exceptions, ecosystems run on sunlight energy captured by photoautotrophs—mostly plants and phytoplankton. The photoautotrophs are **primary producers** (Figure 30.1). They convert sunlight energy to chemical bond energy, using it to synthesize organic compounds from inorganic materials.

All other organisms in the system are **consumers**. These heterotrophs feed on the tissues, products, and remains of other organisms. We describe consumers by their diets. *Herbivores* eat plants. *Carnivores* eat flesh. *Parasites* live in or on a host and feed on its tissues. **Detritivores**, such as crabs, eat detritus—particles of decomposing organic matter. Finally, the wastes and remains of all organisms are degraded to inorganic building blocks by the bacterial, protist, and fungal **decomposers**. Decomposers absorb most breakdown products, but some spill out into the environment.

Some consumers do not fall into simple categories. Many are *omnivores*, which dine on animals, plants, fungi, protists, and even bacteria. A red fox is like this (Figure 30.2). If it comes across a dead bird, the fox will scavenge it. A *scavenger* is an animal that ingests dead plants, animals, or both, all or some of the time. Vultures, termites, and hyenas are typical scavengers.

Recall from Section 4.1 that energy transfers are never 100 percent efficient. Over time, all the energy that was originally captured by producers will escape back to the environment, mostly as metabolic heat. Energy flow through an ecosystem is one-way because sunlight energy is transformed into heat energy, which cannot be captured by organisms and used to do more cellular work.

Nutrients are cycled in an ecosystem. The primary producers obtain hydrogen, oxygen, and carbon from water and carbon dioxide in the environment. They take up phosphorus, nitrogen, and other minerals as well. These are used as raw materials for biosynthesis. Later, decomposition returns the nutrients back to the environment, from which producers can take them up once again.

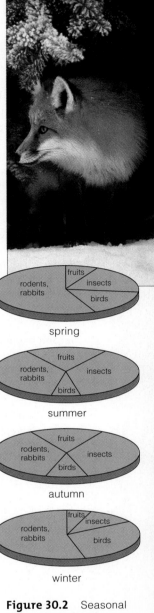

Organisms and their physical environment make up an ecosystem. A one-way energy flow and a cycling of materials interconnect them.

Primary producers capture sunlight energy and assemble organic compounds from inorganic raw materials. All other organisms are consumers; they get their carbon and energy secondhand by consuming other organisms, or their wastes or remains.

Figure 30.2 Seasonal shifts in the red fox diet.

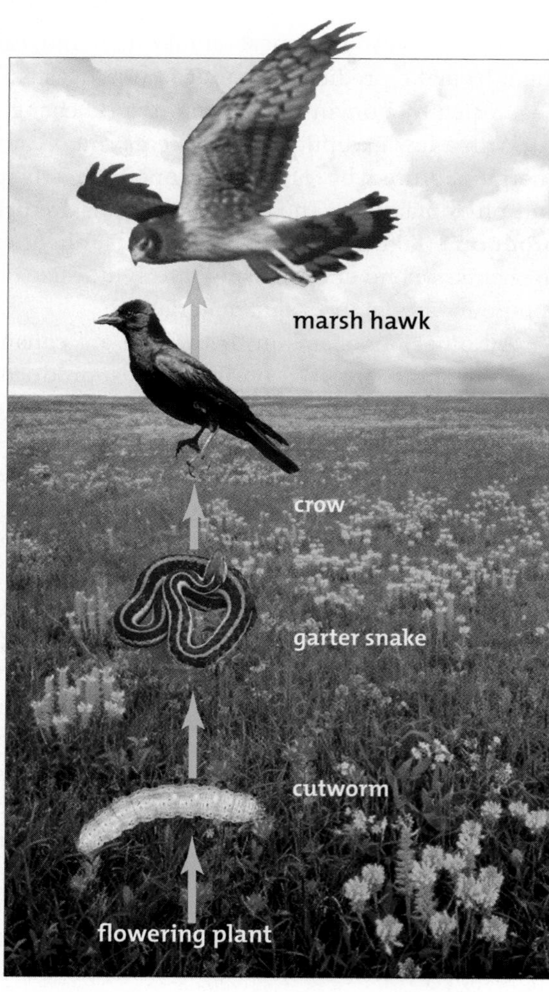

Figure 30.3 ***Animated!*** Example of a simple food chain and its corresponding trophic levels.

LINKS TO SECTIONS 4.1, 29.4

TROPHIC INTERACTIONS

We classify all organisms of an ecosystem by exactly where they fit in a hierarchy of feeding relationships called **trophic levels** (*troph*, nourishment). "Who eats whom?" we ask. If organism A is eaten by organism B, energy is transferred from A to B. All organisms at a given trophic level are the same number of transfer steps away from the energy input into the ecosystem.

As one example, think about some organisms of a tallgrass prairie ecosystem. Plants that capture energy from the sun are at the first trophic level. These plants are eaten by herbivores, such as cutworms, which are at the next trophic level. The cutworms are among the primary consumers that are eaten by animals at the third trophic level, and so on up through successively higher tiers.

At each trophic level, organisms interact with the same sets of predators, prey, or both. Omnivores feed at several levels, so we would partition them among different levels or assign them to a level of their own.

A **food chain** describes a straight-line sequence of steps by which the energy and nutrients stored in the tissues of organisms at the lowest trophic level moves up to higher levels. In one tallgrass prairie food chain, energy from a plant flows to a cutworm that eats it, then to a garter snake that eats the cutworm, to a crow that eats the snake, and finally to a marsh hawk that eats the crow. Figure 30.3 shows this food chain.

Identifying a food chain is a simple way to begin thinking about how energy and nutrients are moved among the living components of an ecosystem. Keep in mind, however, that many different organisms are usually competing for food at each trophic level. Tallgrass prairie producers (mainly flowering plants) are eaten by various mammals, birds, and insects. Many species interact in nearly all ecosystems, especially at a lower trophic level. A number of food chains *cross-connect* with one another—as **food webs**.

FOOD WEBS

By untangling many food webs, ecologists discovered unifying patterns of organization. The patterns reflect both environmental constraints and the inefficiency of energy transfers from one trophic level to the next. Plants capture only a fraction of the energy from the sun. They store some of that energy in chemical bonds of organic compounds in new plant tissues. They lose the rest as metabolic heat. Consumers make use of the energy that became stored in plant tissues, remains, and wastes. Then they, too, lose metabolic heat. Taken together, all of these heat losses represent a one-way flow of energy out of the ecosystem.

Energy transfers can never be 100 percent efficient because a bit of energy is lost at each step. This puts a limit on the length of food chains. In most food chains, energy captured by the producers passes through no more than four or five trophic levels. Even ecosystems that include complex food webs, such as the one in Figure 30.4, do not have lengthy food chains.

Field studies and computer simulations of aquatic and land food webs suggest more patterns. Chains in food webs tend to be shortest where environmental conditions often vary widely over time. Food chains tend to be longer in habitats that are more stable, such as ocean depths. The most complex webs tend to have many herbivorous species, as happens in grasslands. By comparison, the food webs with fewer connections tend to have more carnivores. Also, the number and kind of species in a food web can change over time.

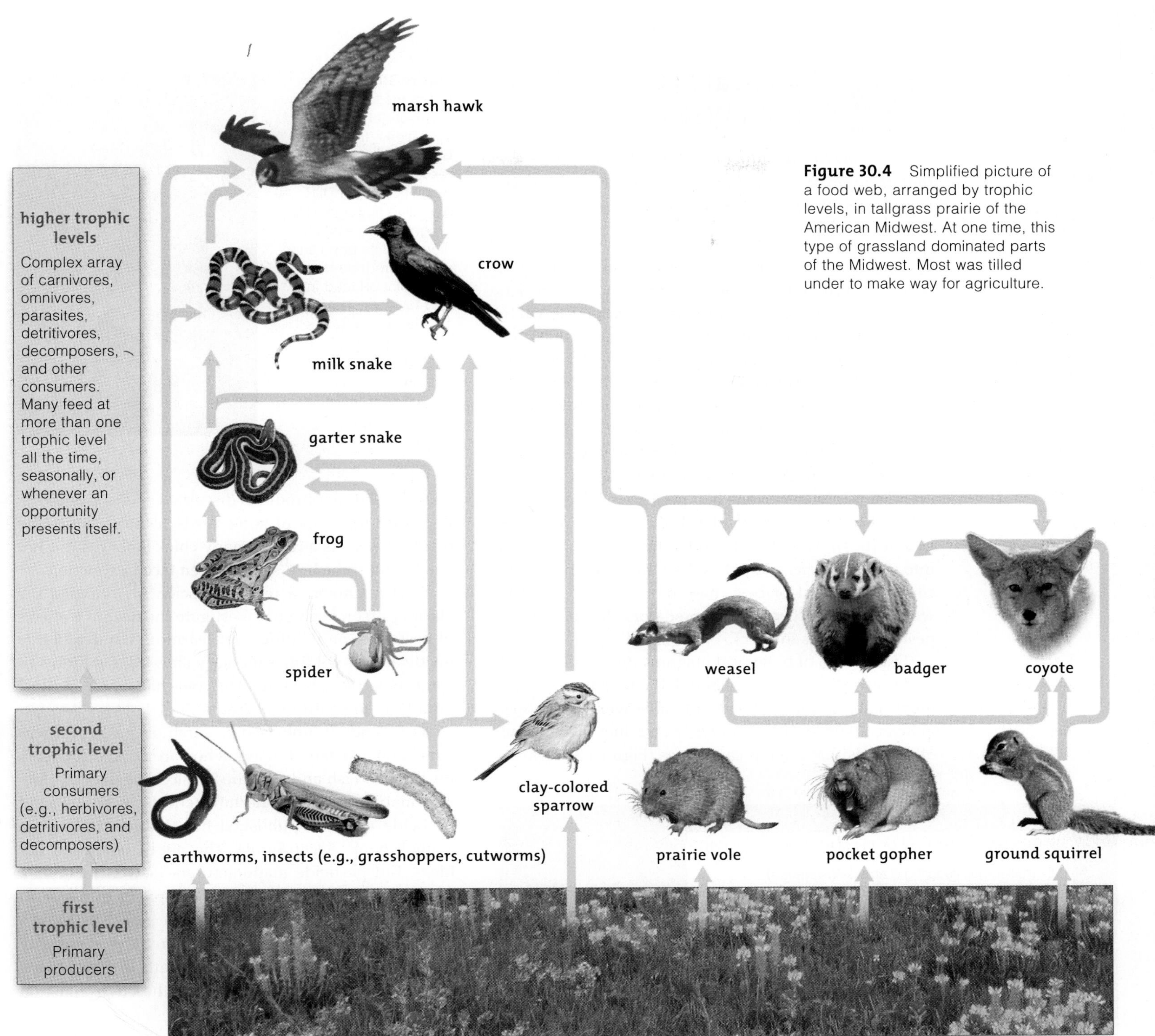

Figure 30.4 Simplified picture of a food web, arranged by trophic levels, in tallgrass prairie of the American Midwest. At one time, this type of grassland dominated parts of the Midwest. Most was tilled under to make way for agriculture.

Energy from producers flows in a single direction through two kinds of webs. In a **grazing food web**, energy stored by producers flows to herbivores, then to carnivores, and then to decomposers. In a **detrital food web**, energy from producers flows mainly into detritivores and decomposers.

In nearly all ecosystems, both kinds of webs cross-connect. For example, in a rocky intertidal ecosystem, energy captured by algae flows to snails, which are eaten by herring gulls as part of a grazing food web. The same gulls also hunt crabs, which are among the primary consumers in the detrital food web.

An organism's trophic level describes how many steps it is from the ecosystem's energy source.

With each energy transfer, some of the energy is lost as heat. These losses limit the number of trophic levels in ecosystems to four or five.

LINK TO SECTION 19.9

30.2 Biological Magnification in Food Webs

We turn now to a premise that opened this chapter—that disturbances to one part of an ecosystem often have unexpected effects on other, seemingly unrelated parts.

DDT is a synthetic organic pesticide (Section 19.9). It is nearly insoluble in water, so you may think it acts only where applied. But winds can carry DDT in vapor form; water can move fine particles of it. Being highly soluble in fats, DDT can collect in tissues, and so it can undergo **biological magnification**. By this process, a substance that degrades slowly or not at all becomes more and more concentrated in tissues of organisms at ever higher trophic levels of a food web (Figure 30.5).

Decades ago, DDT got into food webs and acted on organisms in ways no one had predicted. Where it was sprayed to control Dutch elm disease, songbirds died. In forests sprayed to kill budworm larvae, DDT got into small streams and fishes died. In fields sprayed to control one kind of pest, new pests moved in. DDT was also killing the natural predators that help keep pest populations in check.

Side effects of biological magnification also showed up far away from places where DDT was applied, and much later in time. Most vulnerable were the brown pelicans, bald eagles, ospreys, peregrine falcons, and other birds that are top carnivores (Figures 30.5, 30.6).

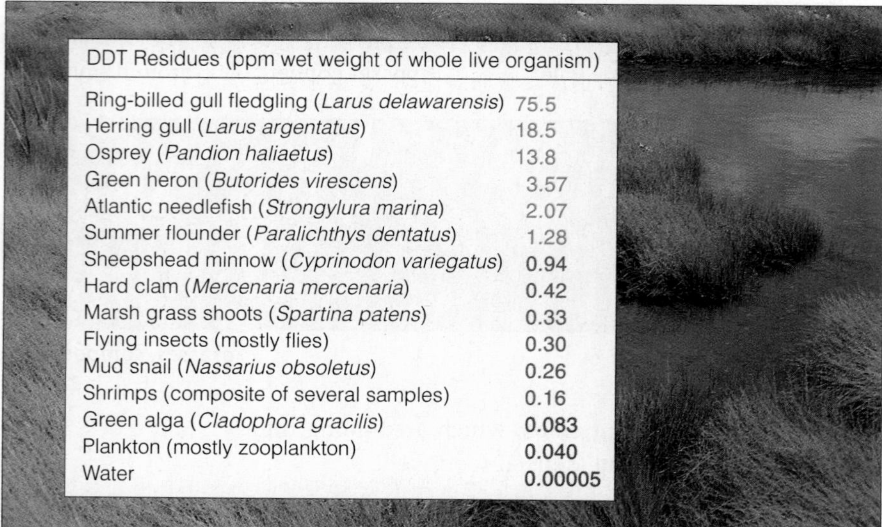

DDT Residues (ppm wet weight of whole live organism)	
Ring-billed gull fledgling (*Larus delawarensis*)	75.5
Herring gull (*Larus argentatus*)	18.5
Osprey (*Pandion haliaetus*)	13.8
Green heron (*Butorides virescens*)	3.57
Atlantic needlefish (*Strongylura marina*)	2.07
Summer flounder (*Paralichthys dentatus*)	1.28
Sheepshead minnow (*Cyprinodon variegatus*)	0.94
Hard clam (*Mercenaria mercenaria*)	0.42
Marsh grass shoots (*Spartina patens*)	0.33
Flying insects (mostly flies)	0.30
Mud snail (*Nassarius obsoletus*)	0.26
Shrimps (composite of several samples)	0.16
Green alga (*Cladophora gracilis*)	0.083
Plankton (mostly zooplankton)	0.040
Water	0.00005

Figure 30.5 Biological magnification in a Long Island, New York, estuary. In 1967, George Woodwell, Charles Wurster, and Peter Isaacson were aware that DDT exposure and wildlife deaths were connected. For instance, residues in birds that died from DDT poisoning were 41–295 parts per million (ppm) and 1–26 ppm in some fishes. In their study, some measured DDT levels were below lethal thresholds but still high enough to interfere with reproduction.

Figure 30.6 Peregrine falcon, a top carnivore of certain food webs. It almost became extinct as a result of biological magnification of DDT. A wildlife management program successfully increased its population sizes. Peregrine falcons were reintroduced into wild habitats. They also have adapted to cities. There, these raptors hunt pigeons, large populations of which are a messy nuisance.

DDT breakdown products disrupted their physiology. It caused the eggs of some birds to have unusually brittle shells. As a result, many chick embryos did not hatch, and some bird species even faced extinction.

Rachel Carson, a respected scientist, sounded the alarm. Carson began to investigate the negative effects of pesticides on wildlife after learning that dead birds inside a local wildlife sanctuary showed symptoms of some type of poisoning. The area had been sprayed with DDT to control mosquitoes.

Independent, unbiased research on pesticides was almost nonexistent. Carson, who had worked for years for the U.S. Fish and Wildlife Service, began compiling information about the harmful effects of widespread pesticide use. She published her findings in 1962 as a small book, *Silent Spring*. The public embraced her ideas. But pesticide manufacturers viewed them as a threat to sales and mounted a campaign to discredit her. At the time, Carson was in the final stages of a battle with terminal breast cancer. Yet she vigorously defended her position until her death in 1964. Later, her work became one impetus for the environmental movement in the United States.

Use of DDT has been banned in the United States since 1972. Its agricultural use is discouraged nearly everywhere, but some developing countries still use it to control disease-carrying mosquitoes. They may be trading one ill effect for another. Studies have shown that pregnant women exposed to DDT are at a higher risk for premature delivery and underweight infants.

With biological magnification, a substance that degrades slowly or not at all gets ever more concentrated in tissues of organisms at ever higher trophic levels of a food web.

30.3 Studying Energy Flow Through Ecosystems

Ecologists measure the amount of energy and nutrients that enter an ecosystem, the amounts captured, and the proportions stored in each trophic level.

ECOLOGICAL PYRAMIDS

Ecologists sometimes illustrate the trophic structure of an ecosystem in the form of an ecological pyramid. In such pyramids, the primary producers are shown as a base for successive tiers of consumers above them.

A **biomass pyramid** depicts the dry weight of all of an ecosystem's organisms at each tier (Figure 30.7). In most ecosystems, primary producers have the most biomass, and top carnivores the least. But in certain aquatic ecosystems the biomass pyramid is "upside-down." For example, in some marine ecosystems, the phytoplankton reproduce and are devoured so quickly that they never accumulate in large numbers.

An **energy pyramid** illustrates how the amount of usable energy diminishes as it is transferred through an ecosystem. Sunlight energy is captured at the base (first trophic level) and declines through successive levels to its tip (the top carnivores). Energy pyramids always have a large energy base, so they are always "right-side up."

PRIMARY PRODUCTIVITY

The rate at which producers capture and store energy in their tissues during a given interval is the **primary productivity** of an ecosystem. How much energy gets stored depends on (1) how many producers there are and (2) the balance between photosynthesis (energy trapped) and aerobic respiration (energy used). *Gross* primary production is all energy initially trapped by the producers. *Net* primary production is the fraction of trapped energy that producers funnel into growth and reproduction. **Net ecosystem production** is the gross primary production *minus* the energy used by producers and by detritivores and decomposers in soil. (In both cases, the subtracted amount of energy would not be available to organisms that eat plants.)

Many factors influence net production, its seasonal patterns, and its regional distribution (Figure 30.8). For example, energy acquisition and storage depend in part on the body size and body form of primary producers, the availability of minerals, the range of temperature, and the amount of sunlight and rain. The harsher environmental conditions are, the less new plant growth there is per season, and the lower the ecosystem's primary productivity.

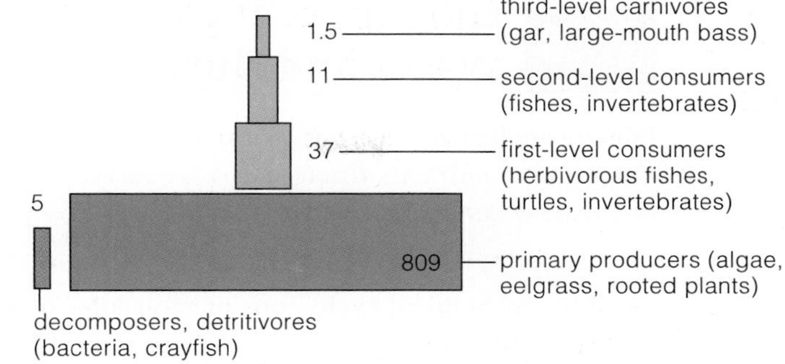

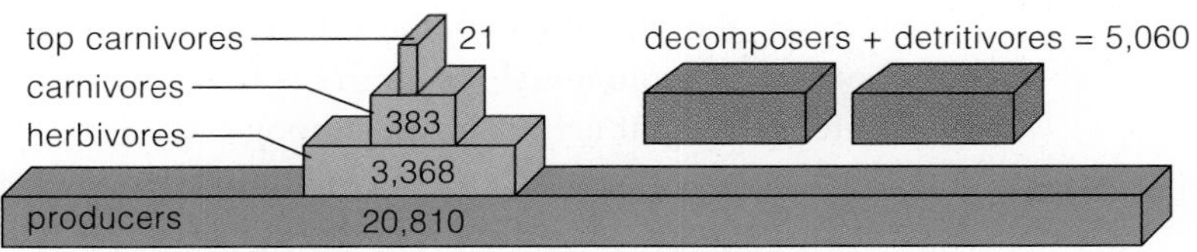

Figure 30.7 Ecological pyramids for Silver Springs, Florida, a small aquatic ecosystem. *Above*, a biomass pyramid measured in grams per square meter. In this case, biomass decreases in successively higher tiers, but some biomass pyramids are widest on the top. *Below*, pyramid of energy flow measured in kilocalories per square meter per year. All energy pyramids are widest at the bottom because energy is lost with each transfer to a higher level.

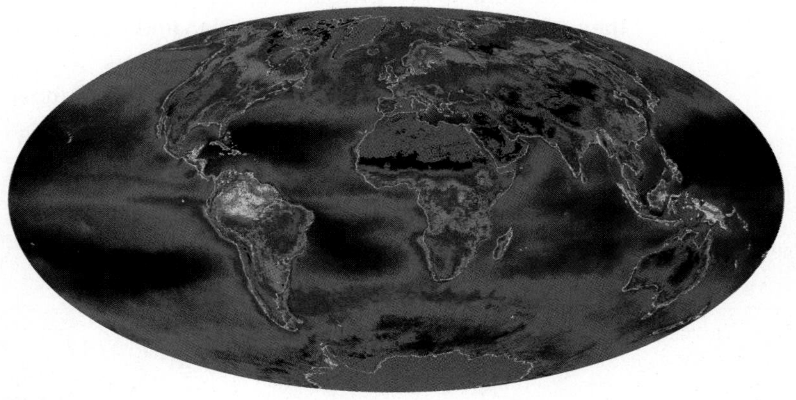

Figure 30.8 Summary of satellite data on net primary productivity during 2002. Productivity is coded as *red* (highest) down through *orange*, *yellow*, *green*, *blue*, and *purple* (lowest). Although average productivity per unit of sea surface is lower than it is on land, total productivity on land and in seas is about equal, because most of Earth's surface is covered by water.

The trophic structure of an ecosystem can be represented by an ecological pyramid. Biomass pyramids may be top- or bottom-heavy depending on the ecosystem. In contrast, an energy pyramid always has the largest tier on the bottom.

Gross primary productivity is an ecosystem's total rate of photosynthesis during a specified interval. The net amount is the rate at which primary producers store energy in tissues in excess of their rate of aerobic respiration. Heterotrophic consumption affects the rate of energy storage.

ECOSYSTEM STRUCTURE

LINKS TO SECTIONS 18.5, 19.9, 28.5

30.4 Global Cycling of Water and Nutrients

Primary productivity depends on availability of fresh water and on nutrients dissolved in it. Global cycles distribute water and nutrients.

In a **biogeochemical cycle**, ions or molecules of some essential substance flow from the environment into organisms, then back to the environment. Some part or parts of the environment function as a reservoir. In most cases, the rate at which the substance moves into and out of environmental reservoirs is far slower than the rate at which it is transferred among organisms.

Nutrients move into and out of ecosystems as a result of geologic processes. Weathering of rocks will add nutrients to an ecosystem. Erosion and runoff can remove nutrients from the system. Usually, the quantity of a nutrient that cycles through an ecosystem per year is greater than the amount that enters or leaves.

By acting as decomposers, some bacteria help keep nutrients cycling within ecosystems. Other prokaryotic species play roles in transforming solids and ions into gaseous form and then back again. By their actions, they convert some essential nutrients to and from forms that can be taken up by the primary producers.

In the *hydrologic* cycle, oxygen and hydrogen move in the form of water molecules. We call this the global water cycle. In *atmospheric* cycles, some portion of the nutrient occurs as an atmospheric gas. Nitrogen and carbon cycles are two examples. Phosphorus and other nutrients that do not commonly exist in gaseous forms move in *sedimentary cycles*. These nutrients move from land to the seafloor and eventually return to land by way of geological uplifting. Earth's crust serves as the main reservoir for nutrients with sedimentary cycles.

HYDROLOGIC CYCLE

Driven by ongoing inputs of solar energy, the Earth's waters move slowly, on a vast scale, from the ocean into the atmosphere, to land, and back to the ocean—the main reservoir. Figure 30.9 shows this **hydrologic cycle**. Water evaporating into the lower atmosphere stays aloft as vapor, clouds, and ice crystals, then falls mainly as rain and snow. Ocean circulation and wind patterns influence the cycle.

Water moves nutrients into and out of ecosystems. Long-term studies of watersheds demonstrated how. A **watershed** is any region where precipitation becomes funneled into a single stream or river. Watersheds may be an area drained by a small stream or as large as the Mississippi River watershed, which extends across one-third of the continental United States. Water entering a watershed mostly seeps into soil or joins surface runoff toward streams. Plants withdraw water and dissolved mineral ions from soil and lose water by transpiration.

You might think that surface runoff would quickly leach mineral ions. But research in New Hampshire's

Main Reservoirs	Volume (10^3 cubic kilometers)
Ocean	1,370,000
Polar ice, glaciers	29,000
Groundwater	4,000
Lakes, rivers	230
Soil moisture	67
Atmosphere (water vapor)	14

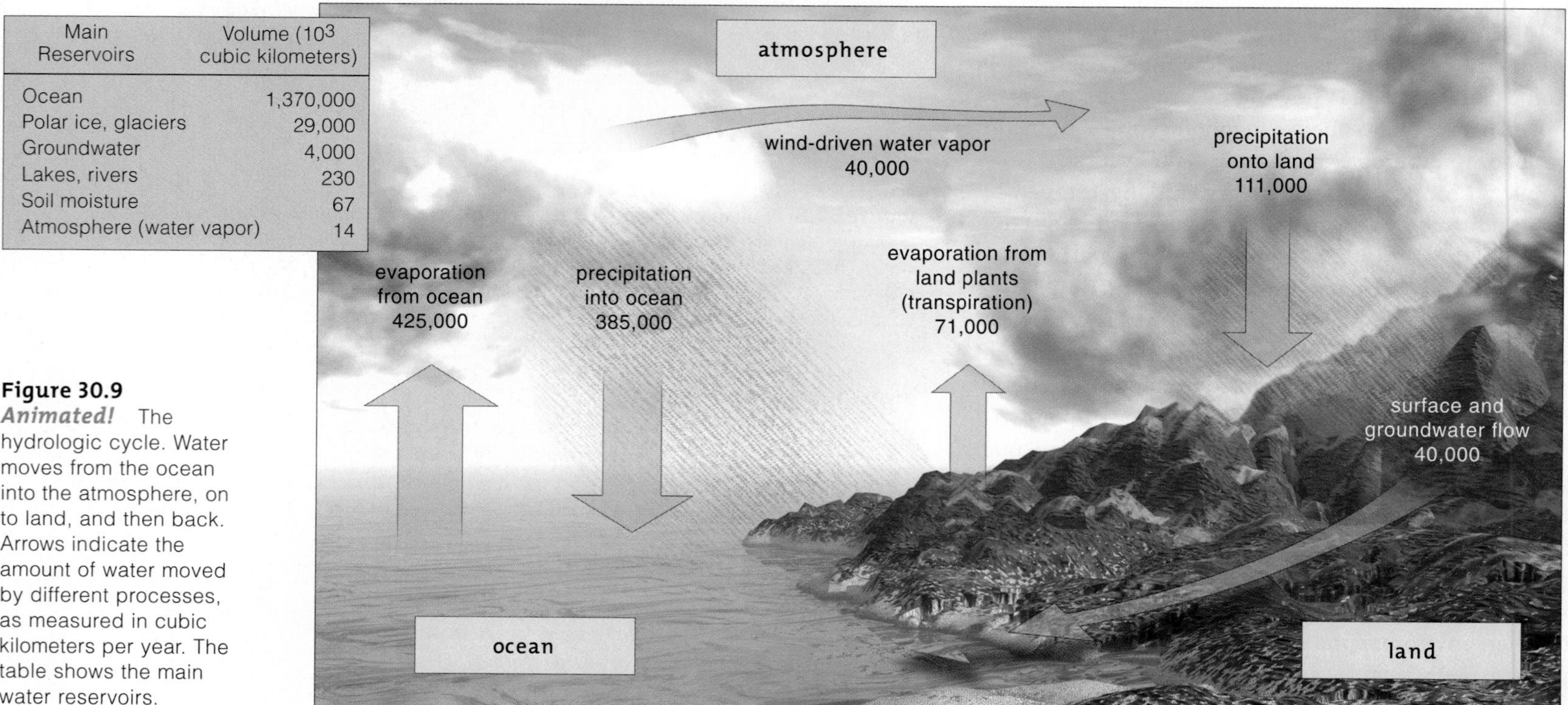

Figure 30.9 ***Animated!*** The hydrologic cycle. Water moves from the ocean into the atmosphere, on to land, and then back. Arrows indicate the amount of water moved by different processes, as measured in cubic kilometers per year. The table shows the main water reservoirs.

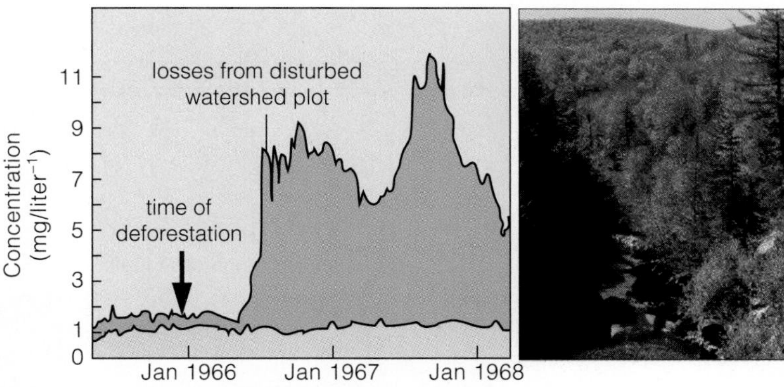

Figure 30.10 Experimental deforestation of New Hampshire's Hubbard Brook watershed. Researchers stripped vegetation from forest plots but did not disturb the soil. They applied herbicides for three years to prevent regrowth. They monitored surface runoff flowing over concrete catchments on its way to a stream below. Concentrations of calcium ions and other minerals were compared against concentrations in water passing over a control catchment, positioned in an undisturbed area.

Calcium losses were *six times* greater in deforested plots. Removing all vegetation from such forests clearly alters nutrient outputs in ways that can disrupt nutrient availability for an entire ecosystem.

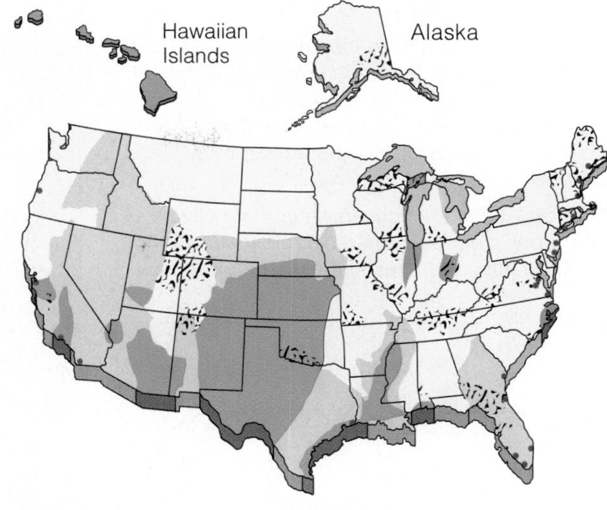

Figure 30.11 Aquifer depletion, seawater intrusion, and groundwater contamination in the United States. *Green* signifies high overdrafts, *gold*, moderate overdrafts, and *pale yellow*, insignificant withdrawals. Shaded areas are sites of major groundwater pollution. *Blue* squares indicate saltwater intrusion from nearby seas.

Hubbard Brook watershed showed this is not the case. In undisturbed forests in this watershed, each hectare lost only about eight kilograms of calcium. The lost calcium was replaced by rainfall and by weathering of rocks. Also, tree roots mine the soil, so calcium was continually being stored in the biomass of tree tissues. The experimental deforestation of these forests caused a spike in nutrient losses (Figure 30.10).

THE WATER CRISIS

Our planet has a lot of water, but most of it is too salty to be of use. If all of Earth's waters fit into a bathtub, the freshwater portion would barely fill a teaspoon.

Agriculture accounts for two-thirds of our water usage, and irrigation alters the suitability of land for agriculture. The piped-in water commonly has high concentrations of mineral salts. Where the soil drains poorly, evaporation causes **salinization**, a buildup of salt in soil that stunts crop plants and decreases yields.

Soil and permeable rock layers called aquifers hold **groundwater**. About half the United States population taps this source for drinking water. Chemicals leached from landfills, hazardous waste dumps, and underground storage tanks contaminate it. Unlike flowing rivers and streams that recover quickly, groundwater pollution is difficult and expensive to clean up.

Groundwater overdrafts, or the amount that nature does not replenish, are high in many areas. Figure 30.11 shows some regions of aquifer depletion in the United States. Overdrafts have now depleted half of the great Ogallala aquifer, which supplies irrigation water for 20 percent of the Midwest's croplands.

Contaminants, such as sewage, animal wastes, and agricultural chemicals, can make water unsuitable for drinking. These water pollutants also disrupt aquatic ecosystems, driving some species to local extinction.

If current rates of human population growth and water depletion continue, the amount of fresh water available for everyone will soon be 55 to 66 percent less than it was in 1976. In this past decade, thirty-three nations have already engaged in conflicts over reductions in water flow, pollution, and silt buildup in aquifers, rivers, and lakes.

Could **desalinization**—the removal of salt from seawater—meet our water needs? Salt can be removed by distillation or pushing water through membranes. The processes require fossil fuels, which makes them most feasible in Saudi Arabia and other countries with small populations and big fuel reserves. Most likely, desalinization will not be cost-effective for large-scale agriculture. It also produces mountains of waste salts.

In the hydrologic cycle, water slowly moves on a global scale from the world ocean (the main reservoir), through the atmosphere, onto land, then back to the ocean.

Plants absorb and store dissolved minerals. By doing so, they minimize the loss of soil nutrients in runoff.

Aquifers that supply much of the world's drinking water are becoming polluted and depleted. Regional conflicts over access to clean, drinkable water are likely to increase.

IMPACTS, ISSUES

Global warming affects the hydrologic cycle. As Earth's atmosphere is heated, its ability to hold water vapor rises. More warm water in the atmosphere has more energy to drive storms, and to raise rainfall rates. Over the past hundred years, rainfall has increased as much as 10 percent at midlatitudes and high latitudes. In the Northern Hemisphere, the number of severe storms has increased.

Figure 30.12 ***Animated!*** Global carbon cycle.

Carbon's movement through typical marine ecosystems (**a**) and through ecosystems on land (**b**). *Gold* boxes show the main carbon reservoirs. The vast majority of carbon atoms are locked in sediments and rocks, followed by increasingly lesser amounts in ocean water, soil, and the atmosphere, and finally in biomass.

The sketch below shows the loop of ocean water that delivers carbon dioxide to carbon's deep ocean reservoir. It sinks in the cold, salty North Atlantic and rises in the warmer Pacific.

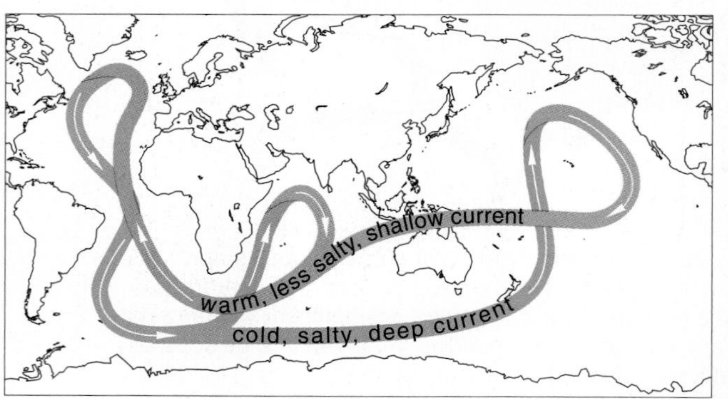

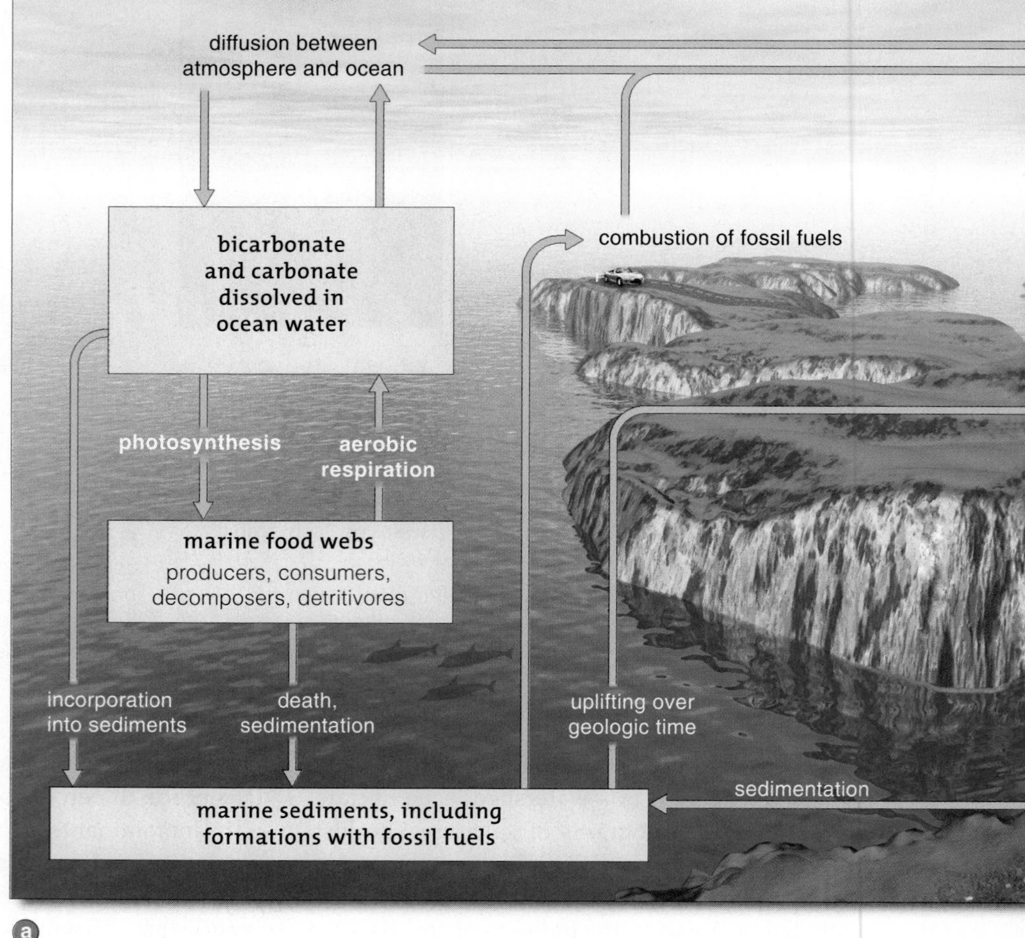

LINKS TO SECTIONS 5.3, 14.3

30.5 Carbon Cycle

Most of the world's carbon is locked in sediments and rocks. Because most carbon moves into and out of ecosystems in gaseous form, its movement is said to be an atmospheric cycle.

Carbon moves through the lower atmosphere and all food webs on its way to and from the ocean's water, sediments, and rocks. Its global movement is called the **carbon cycle** (Figure 30.12). Carbon is released as cells engage in aerobic respiration, when fossil fuels burn, and when the molten rock in Earth's crust emits carbon during volcanic eruptions.

Carbon dioxide, or CO_2, is the most abundant form of carbon in the atmosphere. In the oceans, dissolved bicarbonate (HCO_3^-) and carbonate (CO_3^-) are the main forms. All three carbon-oyxgen compounds have essential metabolic roles. For that reason, the cycle is sometimes called the *carbon–oxygen cycle*.

What keeps all the CO_2 dissolved in warm ocean surface waters from escaping into the air? Driven by winds and regional differences in water density, water makes a gigantic loop from the surface of the Pacific and Atlantic oceans down to the Atlantic and Antarctic seafloors. The CO_2 moves into deep storage reservoirs before the seawater loops back up (Figure 30.12).

Photosynthetic organisms lock billions of metric tons of carbon atoms into organic compounds each year (Chapter 5). The relative amount of time that a carbon atom spends in living and nonliving portions of an ecosystem varies. Organic wastes and remains decompose so quickly in tropical rain forests that not much carbon can accumulate at the soil surfaces. In swamps, bogs, marshes, and other anaerobic habitats, decomposers cannot degrade organic compounds into

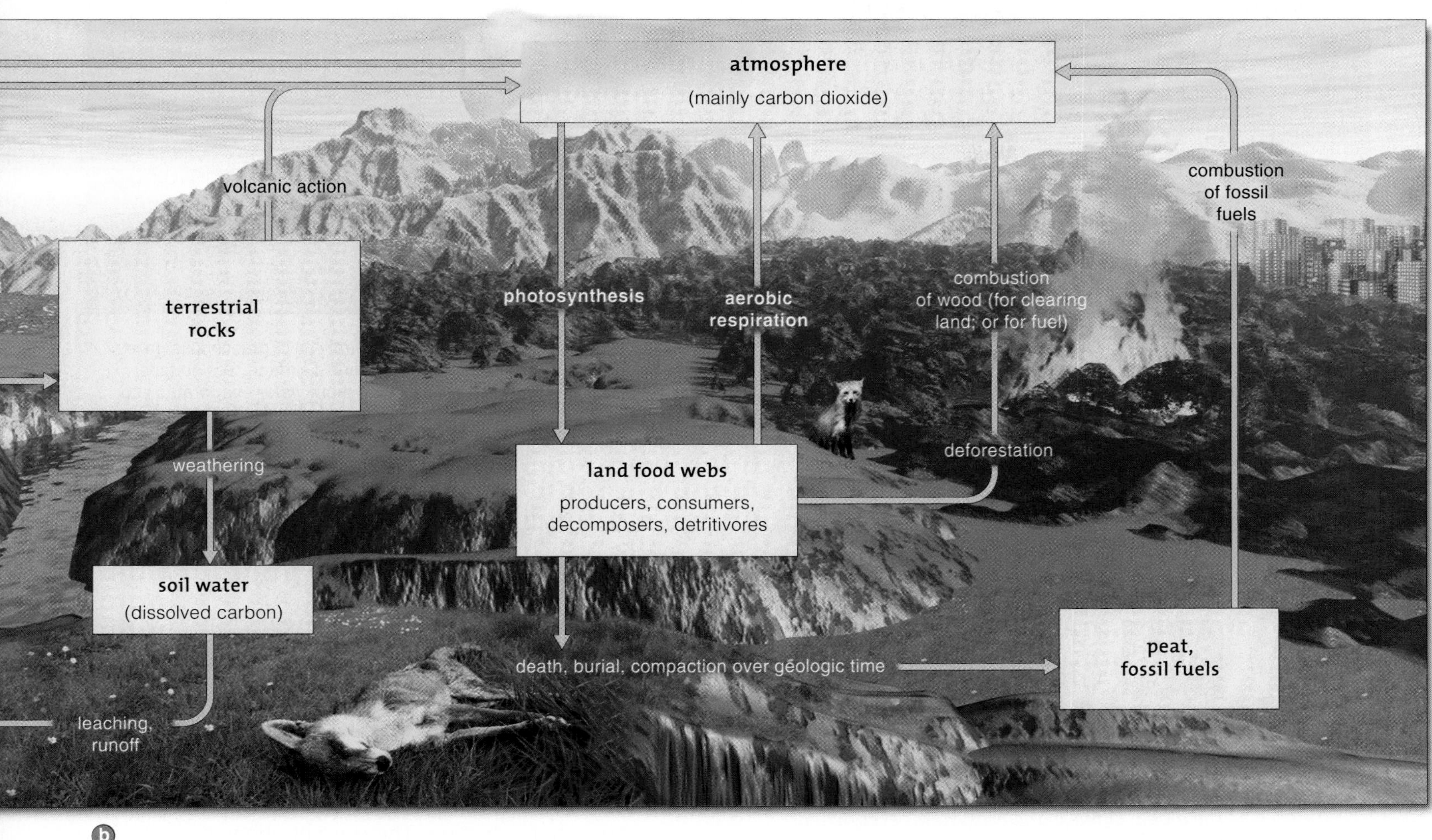

b

inorganic subunits, so carbon gradually accumulates in peat and other forms of compressed organic matter.

Also, in ancient aquatic ecosystems, carbon became incorporated in staggering numbers of shells and other hard body parts of foraminiferans and other species. Those organisms died, sank through water, then were buried in sediments. Carbon stayed buried for many millions of years until geologic forces slowly uplifted part of the seafloor. Remember the carbon-rich white cliffs of Dover, shown in Figure 14.16?

As a final example, plants of Carboniferous swamp forests incorporated carbon into organic compounds. Those compounds were gradually converted into coal, petroleum, and gas reserves—which humans now tap as fossil fuels. Over the course of a few hundred years, we have been extracting and burning fossil fuels that took millions of years to form. In other words, these fuels are nonrenewable resources. Unfortunately, they are the energy source of choice for the majority of the 6.3 billion people now on Earth—mainly in developed countries that can afford them. At present, fossil fuel burning and some other human activities are releasing more carbon into the atmosphere than can be cycled naturally to the ocean's storage reservoirs.

Only about 2 percent of the excess carbon entering the atmosphere will become dissolved in ocean water. Most researchers now think the carbon buildup in the atmosphere is amplifying the greenhouse effect, which means it may be contributing to global warming. The next section will give you a closer look at this effect.

Most carbon is trapped in Earth's rocks and in its deep sea sediments. However, carbon moves between its main reservoirs and ecosystems mostly in the form of a gas—carbon dioxide. This is why the carbon cycle is considered to be one of the Earth's atmospheric cycles.

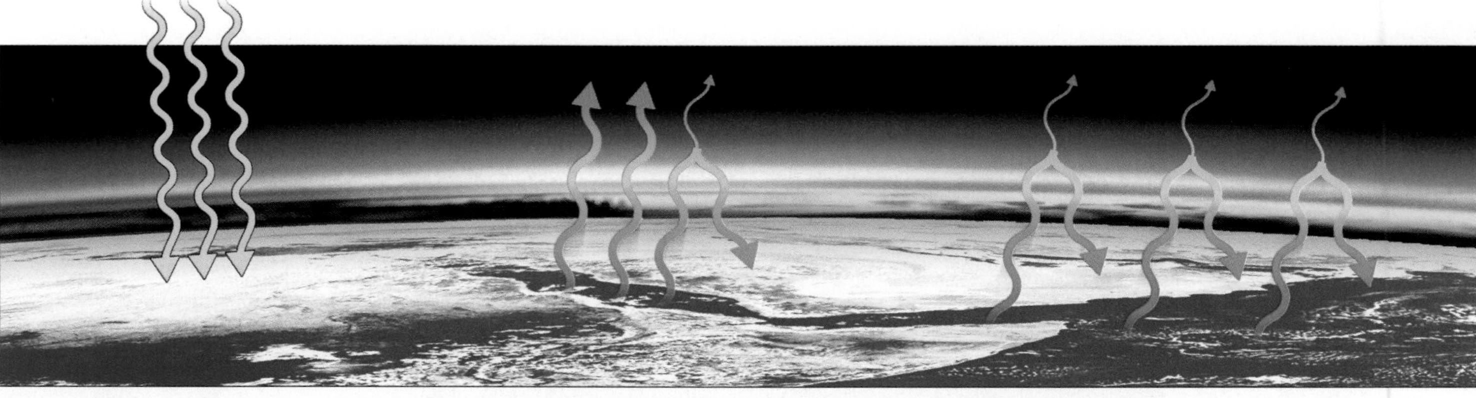

a Wavelengths in rays from the sun penetrate the lower atmosphere, and they warm the Earth's surface.

b The surface radiates heat (infrared wavelengths) to the atmosphere. Some heat escapes into space. But greenhouse gases and water vapor absorb some infrared energy and radiate a portion of it back toward Earth.

c Increased concentrations of greenhouse gases trap more heat near Earth's surface. Sea surface temperatures rise, so more water evaporates into atmosphere. The Earth's surface temperature rises.

Figure 30.13 ***Animated!*** The greenhouse effect.

LINK TO SECTION 28.5

30.6 Greenhouse Gases, Global Warming

The atmospheric concentrations of gaseous molecules help determine the average temperature near Earth's surface. That temperature has enormous impact on the global climate and ecosystems everywhere.

GREENHOUSE EFFECT

Countless molecules of carbon dioxide, water, ozone, methane, and other substances in the atmosphere affect global temperature. Collectively, these many molecules act somewhat like a pane of glass in a greenhouse—hence their name, "greenhouse gases." Incoming light energy streams past the molecules and reaches Earth's surface. This energy heats up the land, which radiates heat back into the atmosphere. Greenhouse gases slow the escape of this heat energy into space. They absorb heat and radiate it into the lower atmosphere, which warms up as a result (Figure 30.13). This warming of the lower atmosphere is called the **greenhouse effect**.

GLOBAL WARMING AND CLIMATE CHANGE

Without greenhouse gases, Earth's surface would be cold and lifeless. But there can be too much of a good thing. Primarily as the outcome of human activities, concentrations of greenhouse gases are now higher than they were in the past (Figure 30.14). Atmospheric CO_2 is at its highest level since 420,000 years ago. It may well be the highest for the past 20 million years.

In the 1950s, researchers in a laboratory on Hawaii's highest volcano set out to measure the atmospheric concentrations of greenhouse gases. That remote site is almost free of local airborne contamination. It is also representative of overall atmospheric conditions for the Northern Hemisphere. What did they find? Briefly, carbon dioxide concentrations follow annual cycles of primary production. The concentrations drop during the summer, when photosynthesis rates are high. They increase during the winter, when photosynthesis rates decline but aerobic respiration is still going on.

The alternating troughs and peaks along the graph line in Figure 30.14*a* reveal the annual lows and highs of atmospheric carbon dioxide concentrations. For the first time, scientists were able to study the integrated effects of carbon fluctuations for terrestrial and aquatic ecosystems of an entire hemisphere. Notice the overall trend. Notice especially that this line slants upward—*the concentration is steadily increasing*.

The enhanced greenhouse effect is linked to a long-term rise in the temperature near Earth's surface. We call this effect **global warming**. As an example, since direct atmospheric readings started in 1861, the lower atmosphere's temperature has risen by more than 1°F, with the biggest gains since 1946 (Figure 30.15). The trend is expected to accelerate, with temperature rising another 1°F by 2030. Polar ice is melting and glaciers are retreating. Sea level has risen an estimated twenty centimeters (eight inches) in the past century.

Greenhouse gases include carbon dioxide, methane chlorofluorocarbons (CFCs), and nitrous oxide (NO). They are among the many pollutants, substances that are new to ecosystems or are present in unnaturally

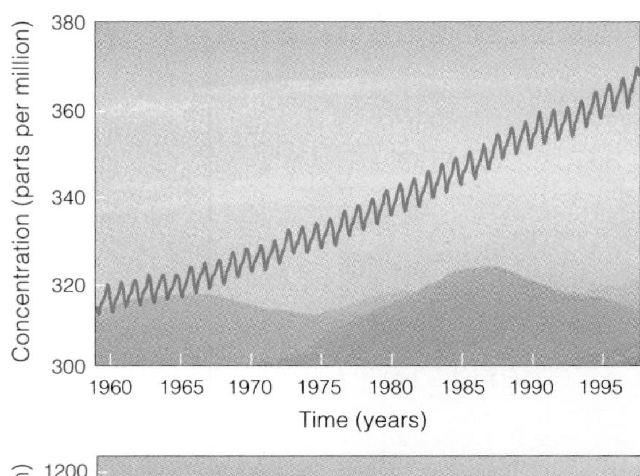

a Carbon dioxide (CO_2). Of all human activities, the burning of fossil fuels and deforestation contribute the most to increasing atmospheric levels.

c Methane (CH_4). Production and distribution of natural gas for use as fuel adds to the methane released by some bacteria that live in swamps, rice fields, landfills, and in the digestive tract of cattle and other ruminants.

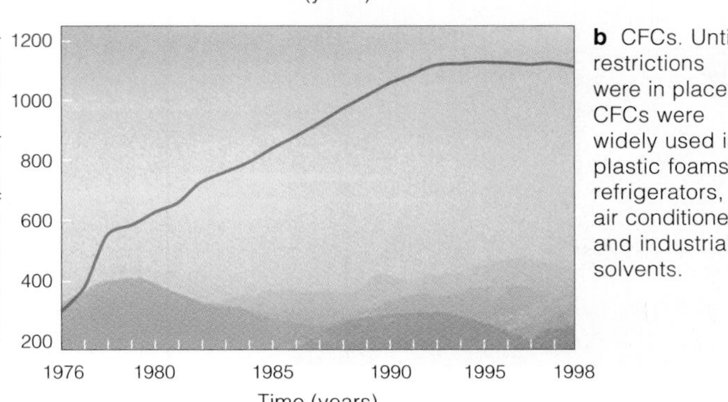

b CFCs. Until restrictions were in place, CFCs were widely used in plastic foams, refrigerators, air conditioners, and industrial solvents.

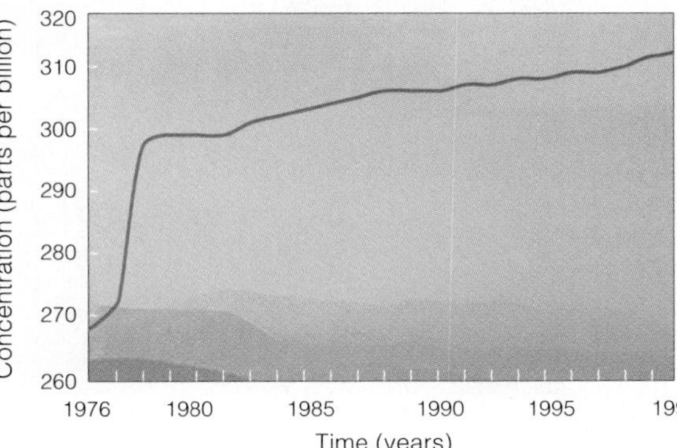

d Nitrous oxide (N_2O). Denitrifying bacteria produce N_2O in metabolism. Also, fertilizers and animal wastes release enormous amounts; this is especially so for large-scale livestock feedlots.

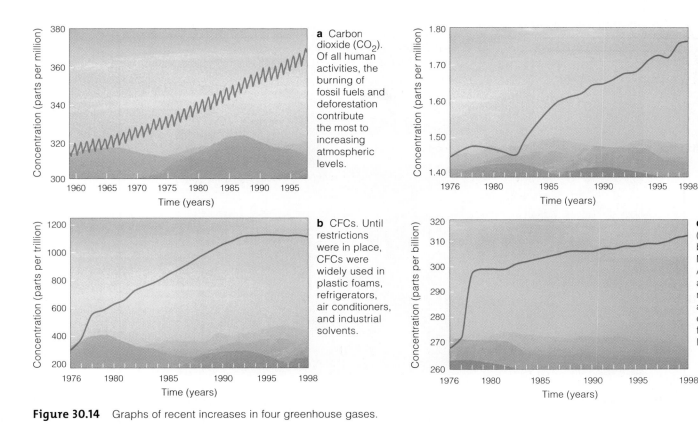

Figure 30.14 Graphs of recent increases in four greenhouse gases.

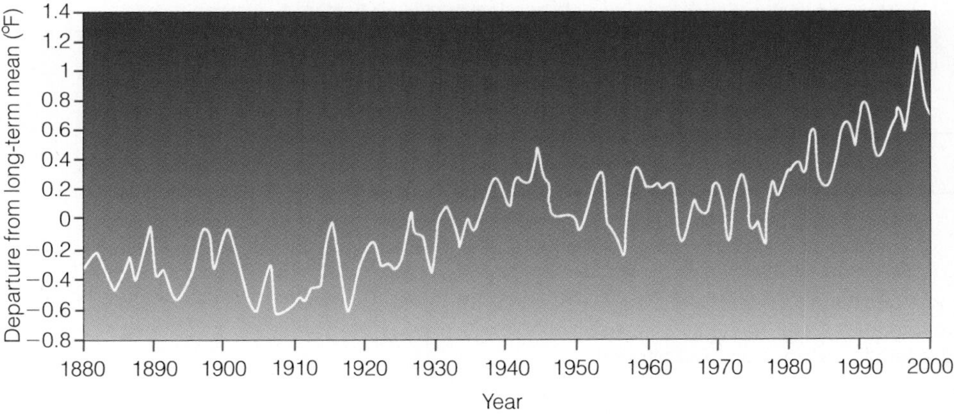

Figure 30.15 Recorded changes in global temperature between 1880 and 2000. At this writing, the hottest year on record was 2005.

Table 30.1 Per-Person CO_2 Emissions in 2002*	
Kuwait	5.97
United States	5.40
Saudi Arabia	4.77
Canada	3.87
Russian Federation	2.69
Germany	2.61
Japan	2.55
France	1.68
Mexico	1.19
China	0.60
India	0.29

* From burning fossil fuels; in metric tons.

large amounts. From a human perspective, pollutants are substances that become concentrated enough to cause adverse effects on our health or survival.

The burning of fossil fuels—including gasoline, diesel fuel, coal, and natural gas—is the main source of greenhouse gases, as well most other air pollutants. With a big population and per person CO_2 emissions exceeded only by Kuwait, the United States is the top CO_2 emitter (Table 30.1). China and India currently emit very little CO_2 on a per-person basis. Yet because of their huge populations, they are the second and fifth greatest emitters of this gas. Both are industrializing rapidly and the resulting increase in their per person fuel use is expected to accelerate global warming.

Atmospheric concentrations of greenhouse gases trap heat and help keep Earth warm enough for life.

Fossil fuel burning and other human activities add to greenhouse gases and may lead to global warming.

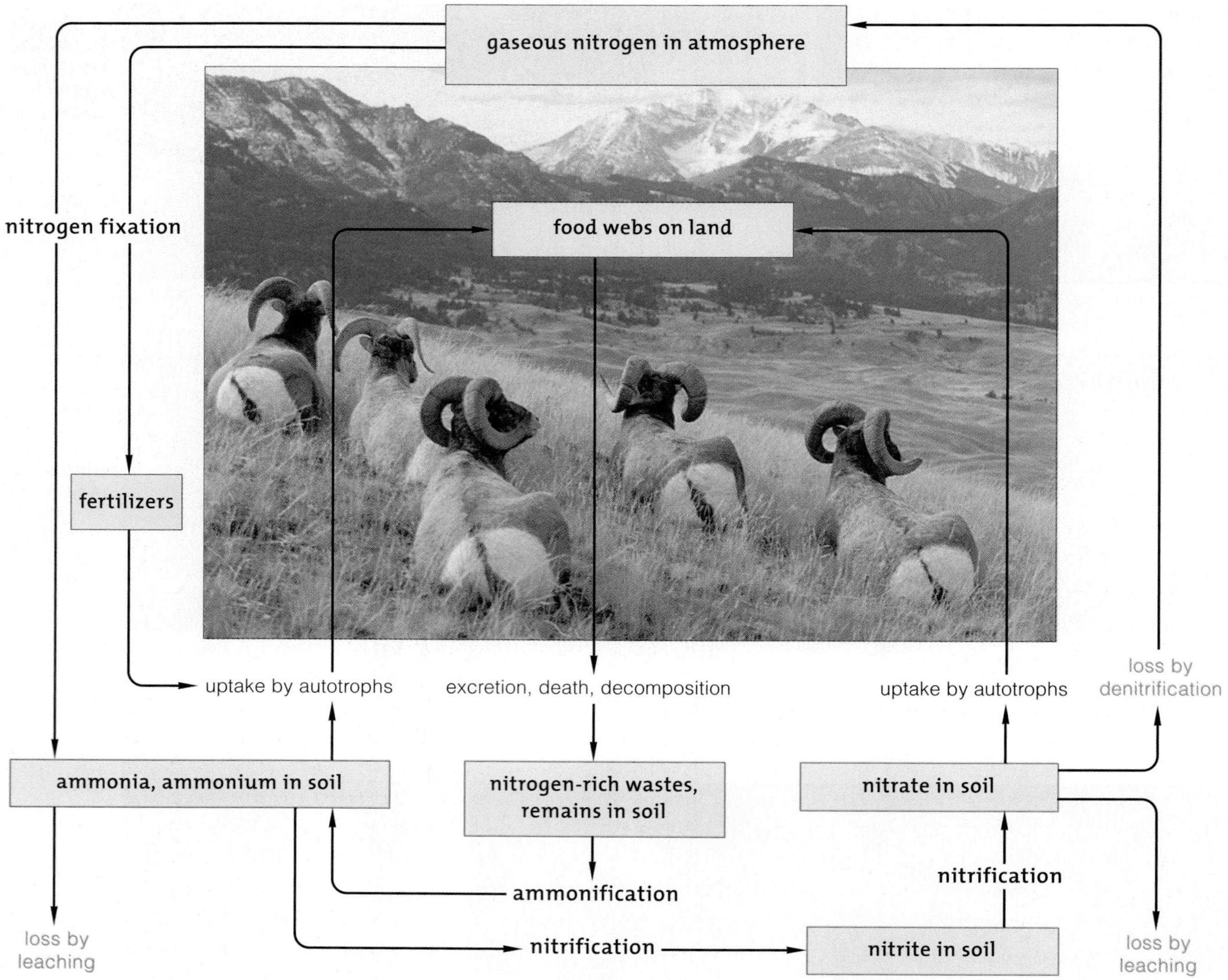

Figure 30.16 ***Animated!*** The nitrogen cycle in an ecosystem on land. Activities of nitrogen-fixing bacteria make nitrogen available to plants. Other bacterial species cycle nitrogen to plants. They break down organic wastes to ammonium and nitrates.

LINKS TO SECTIONS 14.2, 18.5

30.7 Nitrogen Cycle

Gaseous nitrogen makes up about 80 percent of the atmosphere. Successively smaller reservoirs are seafloor sediments, ocean water, soil, biomass on land, nitrous oxide in the atmosphere, and marine biomass.

Gaseous nitrogen (N_2) travels in an atmospheric cycle called the **nitrogen cycle**. Triple covalent bonds join two atoms ($N{\equiv}N$). Volcanic eruptions and lightning convert some gaseous nitrogen into forms that enter food webs. Far more enters by **nitrogen fixation**. By this metabolic process, certain bacteria in the soil split all three bonds in gaseous nitrogen, and then use the nitrogen atoms to synthesize ammonia (NH_3). Later, the ammonia is converted to ammonium (NH_4^+) and nitrate (NO_3^-), two forms of nitrogen that most plants can easily take up (Figure 30.16).

Nitrogen-fixing bacteria are found in soil, in aquatic habitats, and as the photosynthetic partners of fungi in some lichens. Others are mutualists with plant roots, as when *Rhizobium* forms nodules on the roots of peas or other legumes. Collectively, nitrogen-fixing bacteria fix about 200 million metric tons of nitrogen each year.

The nitrogen incorporated into plant tissues moves through trophic levels of ecosystems and ends up in nitrogen-rich wastes and remains. By **ammonification**, bacteria and fungi break down nitrogenous materials, forming ammonium. They use some ammonium and the rest gets into the soil. Plants take it up, as do some nitrifying bacteria. In the first step of **nitrification**, the bacteria use oxygen to strip electron from ammonium and form nitrite (NO_2^-). Other nitrifying bacteria use nitrite and oxygen in reactions that form nitrate (NO_3^-).

Ecosystems lose nitrogen through **denitrification**. Denitrifying bacteria convert nitrate or nitrite to nitrous oxide (N_2O) or nitrogen gas, which can escape into the atmosphere. Most denitrifying bacteria are anaerobic; they live in waterlogged soils and aquatic sediments.

Ammonium, nitrite, and nitrate are also lost to land ecosystems as a result of leaching and runoff. Leaching, recall, is the loss of some nutrients to water draining through the soil. Leaching removes nitrogen from land ecosystems and adds it to aquatic ones.

Figure 30.17 Dead and dying trees in Great Smoky Mountains National Park. Forests are among the casualties of nitrogen oxides and other forms of air pollution.

HUMAN IMPACT ON THE NITROGEN CYCLE

Nitrogen losses through deforestation and grassland conversion for agriculture are huge. With each forest clearing and harvest, nitrogen stored in plant biomass is removed. Soil erosion and leaching take away more.

Many farmers counter nitrogen losses by rotating crops, as by alternating wheat with a crop that has nitrogen-fixing bacterial symbionts. These symbionts add some nitrogen to the soil. In developed countries, farmers also use nitrogen-rich fertilizers. They even select new strains of crop plants that have a greater metabolic capacity to take up fertilizers from soil. By such practices, crop yields per hectare have doubled and quadrupled over the past forty years.

However, heavy fertilizer applications alter the soil water in ways that can be harmful to plants. Recall that the surface charges of soil particles attract ions in soils (Section 18.5). By a pH-dependent process called **ion exchange**, ions dissociate from soil particles, and others replace them. A change in the ion composition of soil water disrupts the number and kinds of ions that are in an exchangeable form.

Calcium and magnesium are the most important exchangeable ions. When nitrogen-rich fertilizers are utilized, ammonium also assumes importance. Heavy applications of fertilizers make soil water more acidic. Hydrogen ions from the fertilizer fill up binding sites on soil particles, displacing magnesium and calcium ions, which leave the soil in runoff.

Humans introduce excess nitrogen into ecosystems in other ways. Nitrogen-rich sewage flows into rivers, lakes, and estuaries. Fossil fuel burned in power plants and vehicles releases nitrogen oxides. Winds carry the pollutants to forests at high elevations and northern latitudes, both of which have nitrogen-poor soils.

In forests that are adapted to nitrogen-poor soils, additional nitrogen makes trees grow faster and later into the fall. The late growth is vulnerable to damage from frost. Also, nitrogen-stimulated growth outstrips the availability of phosphorus and other nutrients, so trees suffer the effects of nutrient deficiencies. Leaves yellow and fall off, photosynthesis declines, and overall growth suffers. The weakened trees are more likely to die or be damaged by diseases, drought, insects, and other stresses (Figure 30.17).

In addition, the nitrogen in acid rain has the same effect on forest soils that fertilizers have on croplands. It increases the loss of potassium, calcium, and other ions that are essential plant nutrients. Thus, nitrogen pollution of forest soils can cause a growth spurt that is followed by a decline in growth as trees become starved for other nutrients.

The nitrogen cycle is an atmospheric cycle. Nitrogen fixation by bacteria converts nitrogen gas to forms plants can use. Other bacteria make the nitrogen in wastes and remains available to plants.

Ecosystems lose nitrogen naturally by the activities of denitrifying bacteria, which convert nitrite or nitrate to forms that escape into the atmosphere.

Ecosystems often lose or gain too much nitrogen through human activities, in ways that compromise plant growth.

IMPACTS, ISSUES

Atmospheric levels of nitrous oxide are rising. Burning fossil fuels adds this greenhouse gas to the atmosphere, but agriculture is the main source. As the use of nitrogen-rich fertilizers has increased, so have the population sizes of nitrifying bacteria, and so has the rate of nitrous oxide formation. The trend is expected to continue because developing countries also are using more and more fertilizers.

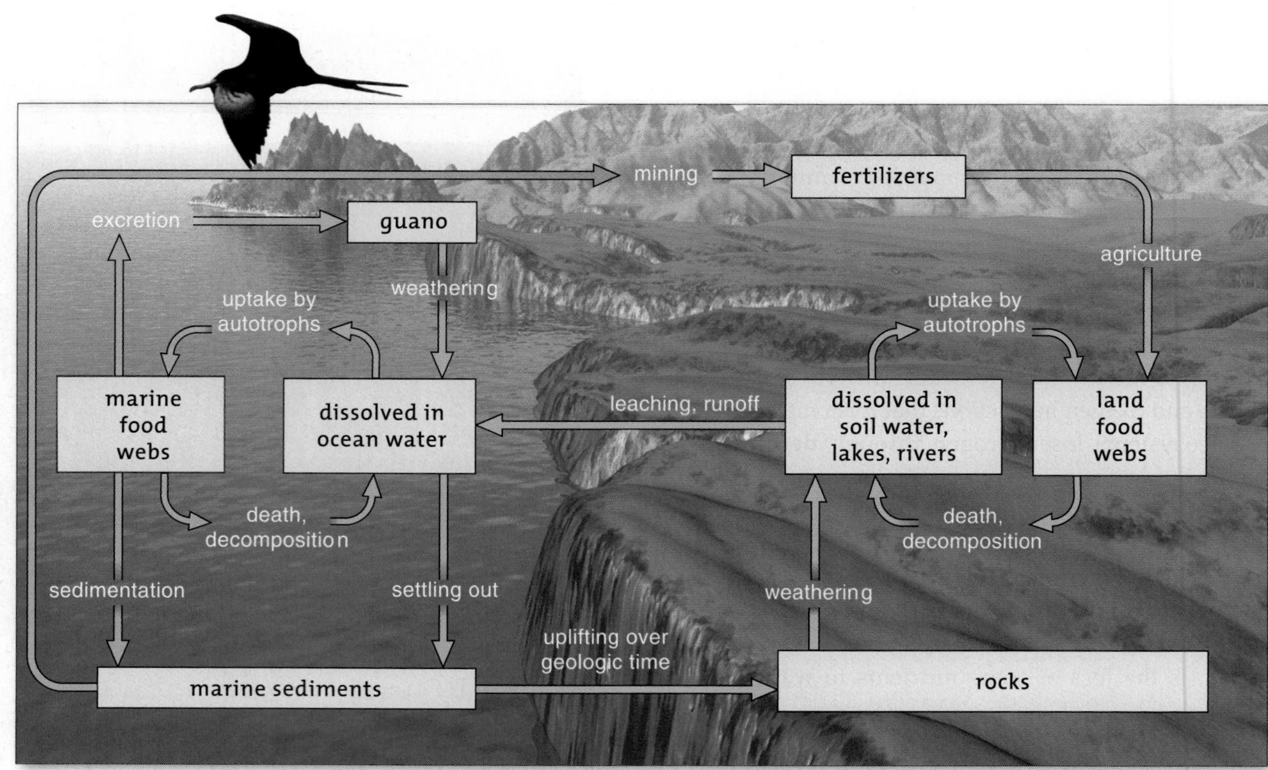

Figure 30.18 ***Animated!*** Phosphorus cycle. In this sedimentary cycle, most of the phosphorus moves in the form of phosphate ions (PO_4^{3-}) and reaches the oceans. Phytoplankton of marine food webs take up some of it, then fishes eat the plankton. Seabirds eat the fishes, and their droppings (guano) accumulate on islands. Guano is collected and used as a phosphate-rich fertilizer.

LINKS TO SECTIONS 14.3, 18.5

30.8 Phosphorus Cycle

Sedimentary cycles, again, move solid forms of nutrients into and out of all ecosystems. The phosphorus cycle is one of the most influential.

In the **phosphorus cycle**, phosphorus passes quickly through food webs as it moves from land to ocean sediments, then slowly back to land. Earth's crust is the largest phosphorus reservoir.

In rock formations, most phosphorus is in the form of phosphate ions (PO_4^{3-}). Weathering and erosion put phosphates into streams and rivers, then into ocean sediments (Figure 30.18). The ions combine with other minerals and accumulate on the submerged shelves of continents as insoluble deposits. Millions of years go by. Where movements of crustal plates uplift part of the seafloor, phosphates become exposed on drained land surfaces. In time, weathering and erosion release phosphates from exposed rocks once again.

All organisms use the ions to make phospholipids, NADPH, ATP, nucleic acids, and other compounds. Plants take up dissolved phosphates from soil water. Herbivores get needed phosphates by eating plants; carnivores get them by eating herbivores. All animals excrete phosphates as a waste in their urine and feces. Bacterial and fungal decomposers release phosphates, then plants take them up again in this rapid cycle.

The hydrologic cycle helps move minerals through ecosystems. Water evaporates from the ocean and falls on land. On its way back to the sea, it transports silt and dissolved minerals, including phosphorus.

Of all mineral nutrients, phosphorus is most often the limiting factor in natural ecosystems. Only newly weathered, young soils are high in phosphorus. Older soils have lost much of this mineral to erosion and runoff. In aquatic habitats most phosphorus is tied up in the sediments.

Some human activities are lowering the levels of phosphorus in natural ecosystems. This is especially so for tropical and subtropical regions of developing countries, which often have weathered, phosphorus-poor soils. Phosphorus stored in biomass and slowly released from decomposing organic matter sustains undisturbed forests or grasslands. But people harvest trees and clear grassland for farms, thus depleting the soil phosphorus. Crop yields start out low and become nonexistent. Fields are soon abandoned, and regrowth of natural vegetation is sparse. About 1 to 2 billion hectares are already depleted of phosphorus, mostly where impoverished farmers are not engaging in soil management practices and cannot afford fertilizers.

In developed countries, fertilizer use has left many soils with phosphorus overloads. Phosphorus becomes

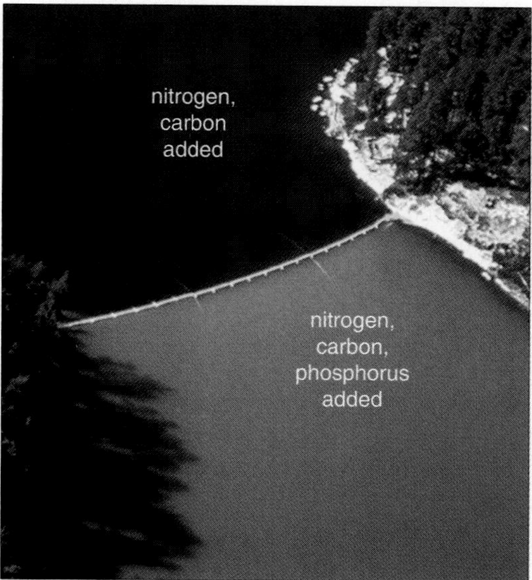

Figure 30.19 One of the eutrophication experiments. Researchers put a plastic curtain across a channel between two basins of a natural lake. They added nitrogen, carbon, and phosphorus on one side of the curtain (here, the *lower* part of the lake) and added nitrogen and carbon on the other side. Within months, the phosphorus-rich basin was eutrophic, with a dense algal bloom covering its surface.

concentrated in eroded sediments and runoff from agricultural fields. Phosphorus also is present in the outflows from sewage treatment plants and industries, in the runoff from cleared land, and even from lawns.

Dissolved phosphorus that enters streams, rivers, lakes, and estuaries can promote dense algal blooms. Like plants, photosynthetic algae cannot grow without phosphorus, nitrogen, and other mineral ions. In many freshwater ecosystems, nitrogen is not a limiting factor on algal growth because of the abundance of nitrogen-fixing bacteria. Phosphorus is. When decomposers act on the remains from algal blooms, they deplete water of oxygen, which kills fishes and other organisms.

Eutrophication refers to the nutrient enrichment of any ecosystem that is naturally low in nutrients. This is a natural process, but phosphorus inputs accelerate it, as field experiments demonstrate (Figure 30.19). The added phosphorus encourages algal growth.

Sedimentary cycles, in combination with the hydrologic cycle, move most mineral elements, such as phosphorus, through terrestrial and aquatic ecosystems.

Agriculture, deforestation, and other human activities upset the nutrient balance of an ecosystem because they accelerate the amount of minerals entering and leaving it.

http://biology.brookscole.com/BTT2

Summary

Section 30.1 An ecosystem consists of an array of organisms and their environment. There is a one-way flow of energy through an ecosystem, and a cycling of materials among the organisms. All ecosystems have inputs and outputs of energy and, to some extent, nutrients.

Sunlight is the initial energy source for almost all ecosystems. Photoautotrophs (the primary producers) convert sunlight energy to ATP and assimilate nutrients that an ecosystem's consumers require.

All organisms in an ecosystem that are at the same trophic level are the same number of steps away from the energy source for that ecosystem. Energy transfers are inefficient and so most ecosystems support no more than four or five trophic levels.

BiologyNow
Learn about trophic levels with the animation on BiologyNow.

Section 30.2 By biological magnification, a chemical substance is passed from organisms at one trophic level to those above, so that the substance becomes increasingly concentrated in body tissues.

Section 30.3 Primary productivity is the rate at which producers in an ecosystem capture and store energy. It varies with climate, season, and other factors. Energy pyramids show how usable energy lessens as it flows through an ecosystem. In contrast, a biomass pyramid is sometimes larger at the top than at the base.

Section 30.4 In a biogeochemical cycle, water or a nutrient moves through the environment, then through organisms, then back to an environmental reservoir. In the hydrologic cycle, waters move from the ocean into the atmosphere, to land, and back to the ocean—the main reservoir. Human actions are disrupting the cycle in ways that may lead to water shortages.

BiologyNow
Learn about the hydrologic cycle with the interaction on BiologyNow.

Section 30.5 The global carbon cycle moves carbon from its main reservoirs in rocks and seawater, through its gaseous form (CO_2), and through living organisms. Deforestation and burning of fossil fuels are adding more carbon to the air than the oceans can absorb.

BiologyNow
Learn about the carbon cycle with the animation and interaction on BiologyNow.

Section 30.6 The greenhouse gases trap heat in Earth's lower atmosphere and make life possible. But human activities are raising greenhouse gas levels and causing an increase in global temperatures.

BiologyNow
Learn about the greenhouse effect and greenhouse gases with the interactions on BiologyNow.

GLOBAL CYCLES

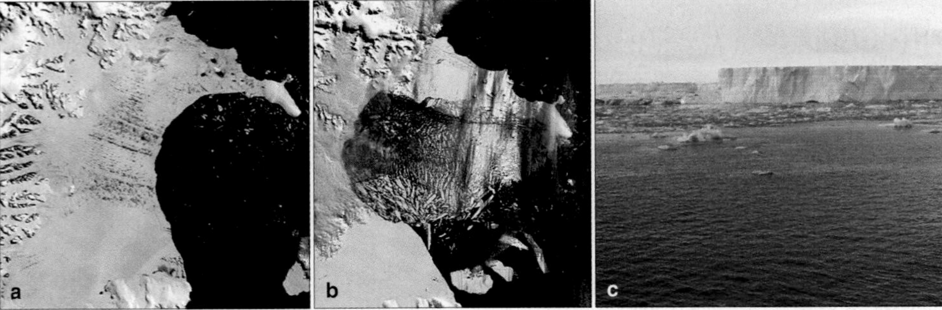

Figure 30.20 Antarctica's Larson B ice shelf in (**a**) January and (**b**) March 2002. About 720 billion tons of ice broke from the shelf, forming thousands of icebergs. (**c**) These icebergs project about twenty-five meters above the sea surface.

Section 30.7 The atmosphere is the main reservoir for nitrogen, but plants cannot use gaseous forms. Soil bacteria and the fertilizer industry fix nitrogen (convert atmospheric nitrogen to forms plants take up). Human activities are disrupting the nitrogen cycle.

BiologyNow
Learn about the nitrogen cycle with the animation on BiologyNow.

Section 30.8 Earth's crust is the largest reservoir of phosphorus. A sedimentary cycle moves the mineral from rocks and seawater into living organisms and back. Phosphorus is often a limiting factor for primary producers. Addition of phosphorus to aquatic systems contributes to eutrophication, the overloading of a system by excess nutrients. This can lead to harmful algal blooms.

BiologyNow
Learn about the phosphorus cycle with the animation and interaction on BiologyNow.

Self-Quiz

Answers in Appendix I

1. Organisms at the lowest trophic level in a tallgrass prairie are all ______ .
   a. at the first step away from the original energy input
   b. autotrophs　　d. both a and b
   c. heterotrophs　　e. both a and c
2. Which of the following is a consumer?
   a. a carnivore　　c. a detritivore
   b. an omnivore　　d. all of the above
3. As a result of biological magnification, a substance will accumulate to its highest levels in the tissues of ______ .
   a. primary producers　　c. decomposers
   b. herbivores　　d. top carnivores
4. The ______ cycle is a sedimentary cycle.
   a. hydrologic　　c. nitrogen
   b. carbon　　d. phosphorus
5. Disruption of the ______ cycle is depleting aquifers.
   a. hydrologic　　c. nitrogen
   b. carbon　　d. phosphorus
6. Earth's largest reservoir for water is ______ .
   a. lakes and streams　　c. atmospheric vapor
   b. oceans　　d. polar ice and glaciers
7. Earth's largest carbon reservoir is ______ .
   a. the atmosphere　　c. seawater
   b. sediments and rocks　　d. living organisms
8. What form(s) of nitrogen do plants easily take up?
   a. ammonium　　d. nitrous oxide
   b. nitrate　　e. both a and b
   c. nitrogen gas　　f. both c and d
9. What process converts nitrogen gas to ammonia?
   a. ammonification　　c. nitrification
   b. denitrification　　d. nitrogen fixation
10. Match the terms with suitable descriptions.
    ____ producers　　a. feed on plants
    ____ herbivores　　b. feed on small bits of organic matter
    ____ decomposers　　c. degrade organic wastes and remains to inorganic forms
    ____ detritivores　　d. capture sunlight energy

Additional questions are available on **BiologyNow™**

Critical Thinking

1. Polar ice shelves are vast, thickened sheets of ice that float on seawater. In March 2002, 3,200 square kilometers (1,300 square miles) of Antarctica's largest ice shelf broke free from the continent and shattered into thousands of icebergs (Figure 30.20). Scientists knew the ice shelf was shrinking and breaking up, but this was the single largest loss ever observed at one time. Why should this concern those of us who live in more temperate climates?

2. Tissues of predatory marine fishes, such as swordfishes, tunas, marlins, and sharks, contain high levels of mercury. Many state governments have issued health advisories to pregnant women, suggesting they limit consumption of tainted species. Other fish from the same environments have lower mercury levels and are not on the warning list. Explain how species living in the same habitat can have very different mercury levels in their tissues.

3. Marguerite is growing vegetables in her garden in Maine. What are some variables that will affect the net primary production in her garden? Which of these can she alter and which are beyond her control?

4. List some of the manufactured goods and agricultural products you purchase that contribute to amplification of the greenhouse effect and to global warming. Which might you replace with products that are environmentally friendly? Which ones might you do without entirely?

5. Lawn fertilizers are rich in nitrogen and phosphorus. Why does adding these chemicals to the soil enhance plant growth? What effect will they have if they get into nearby streams and lakes?

6. Most of the body of any living organism is made up of water. Recall that carbon, nitrogen, and phosphorus make up large proportions of the remaining body weight. The importance of these elements in ecosystems stems from their roles in building the bodies of organisms. For each of these elements, list some biological molecules that include it.

31 THE BIOSPHERE

IMPACTS, ISSUES

Surfers, Seals, and the Sea

The stormy winter of 1997–1998 was a very good time for surfers on the lookout for the biggest waves. But it was a very bad time for seals and sea lions. While Ken Bradshaw was riding the monster wave shown below, sea lions on the Galápagos Islands were dying. Half of this population disappeared, including nine out of every ten pups. That same winter in California, the number of Northern fur seals plummeted and 75 percent of the newborns survived less than a few months. The animals were among the casualties of El Niño, an event that ushers in a spectacular seesaw in the world climate.

That winter, a massive volume of warm water from the southwestern Pacific moved east. It piled into coasts from California down through Peru, displacing currents that otherwise would have churned up nutrients from the deep. Without these nutrients, primary producers for marine food webs dwindled. Warm water and scarce producers drove out the migratory big fishes. Fishes and squids that could not migrate starved to death. So did many of the seals and sea lions—which stay alive by eating squids and fishes.

Marine mammals did not dominate the headlines, because the 1997–1998 El Niño gave humans plenty of other things to worry about. It battered Pacific coasts with fierce winds and torrential rains, massive flooding, and landslides. As another rippling effect, an ice storm in New York, New England, and central and eastern Canada crippled regional electrical grids. At the same time, the global seesaw caused drought-driven crop failures and raging wildfires in Australia and Indonesia.

The **biosphere**, again, is the sum of all places where we find life on Earth. With this chapter, we turn to the physical processes that help shape the distribution of organisms through the biosphere. With few exceptions, energy from the sun fuels the processes that underlie this distribution. As you will see, it drives the circulation of air and water in ways that start weather patterns. Some are highly predictable. Others, such as El Niño, are not yet fully understood.

How Would You Vote? *We cannot prevent El Niño events, but if we knew how to predict them, we might be able to minimize damage. Should the government fund more research that might help us predict El Niño? See BiologyNow for details, then vote online.*

Key Concepts

BIOGEOGRAPHIC PATTERNS

Sunlight energy warms Earth's atmosphere, land, and seas. This creates patterns of air and water circulation that shape climate and so determine the distribution of biodiversity.

MAJOR BIOMES

Climate varies with latitude and elevation, and different types of vegetation are adapted to different climates. As a result, we find similar types of ecosystems across broad bands of latitude and at similar elevations.

THE WATER PROVINCES

Gradients of brightness, nutrient availability, and temperature help shape freshwater and saltwater communities. Seasonal changes in these factors cause corresponding changes in primary productivity.

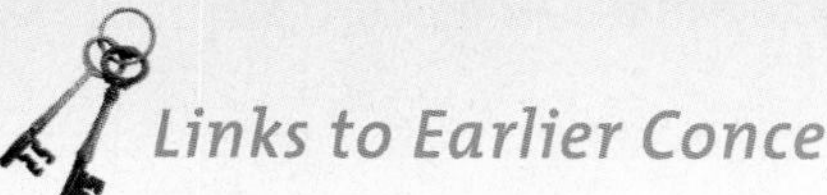

Links to Earlier Concepts

Section 2.4 introduced the concept of pH and 30.7 touched on acid rain. We revisit both those topics here, as well as the origin of life (14.1) and laws governing transfers of energy (4.1).

We also reconsider biodiversity hotspots (29.9), soil erosion (18.5), variations in plant root systems (18.3), forests and deforestation (15.7), primary productivity (5.4, 30.3), the carbon cycle (30.5), eutrophication (30.8), and the ecology of protists (14.3) and corals (16.2).

31.1 Air Circulation and Climates

Within the biosphere are ecosystems ranging from continent-straddling forests to rainwater pools in cup-shaped clusters of leaves. Except for a few ecosystems at hydrothermal vents, climate influences all of them.

Light rays from the sun strike Earth's surface most directly at the equator and in the tropics.

To reach higher latitudes, light rays pass through more atmosphere, which scatters and reflects them. Rays that do reach the surface strike at an angle, so there is less incoming energy per unit area.

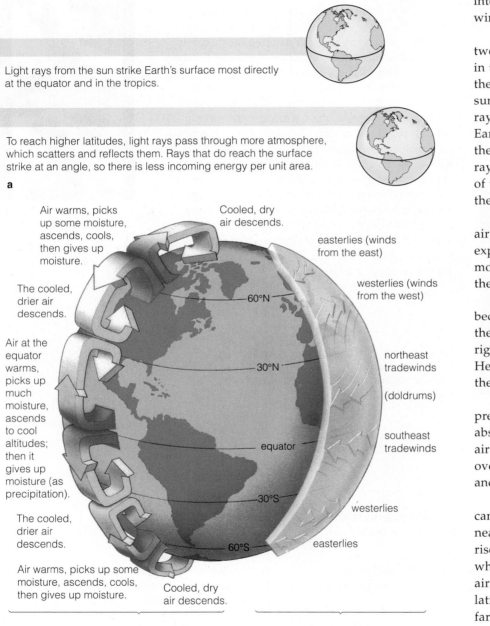

a

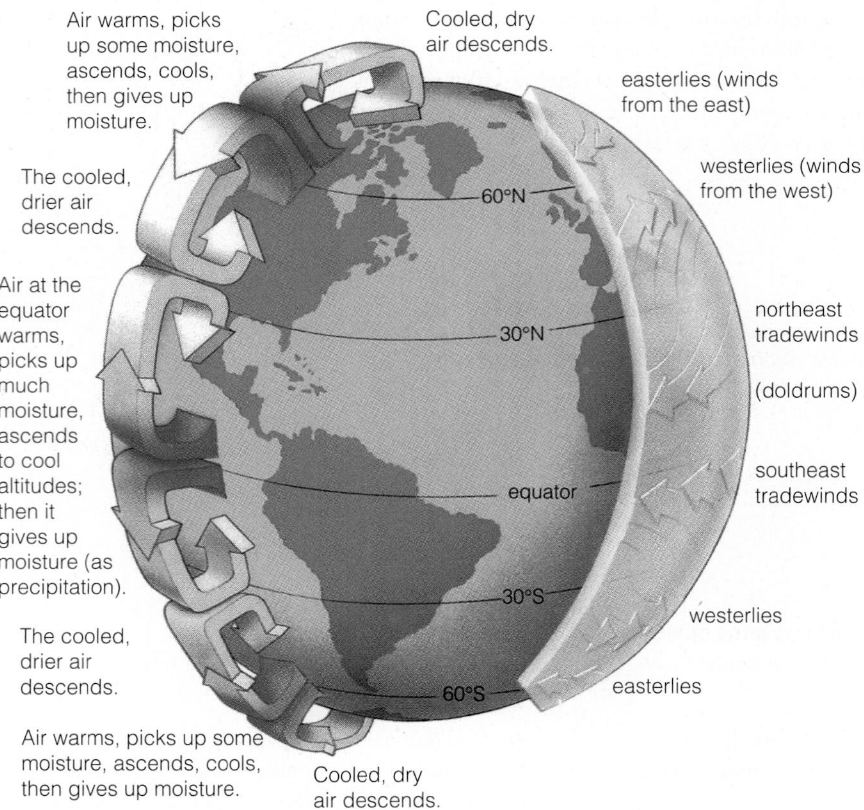

Air warms, picks up some moisture, ascends, cools, then gives up moisture.
Cooled, dry air descends.
easterlies (winds from the east)
The cooled, drier air descends.
60°N
westerlies (winds from the west)
Air at the equator warms, picks up much moisture, ascends to cool altitudes; then it gives up moisture (as precipitation).
30°N
northeast tradewinds
(doldrums)
equator
southeast tradewinds
30°S
westerlies
The cooled, drier air descends.
60°S
easterlies
Air warms, picks up some moisture, ascends, cools, then gives up moisture.
Cooled, dry air descends.

b Initial pattern of air circulation as air masses warm and rise, then cool and fall.

c Deflections in the initial pattern near Earth's surface.

Figure 31.1 ***Animated!*** Global air circulation patterns. Latitudinal differences in the amount of solar radiation reaching Earth (**a**) produce an initial north-south air circulation pattern (**b**). Earth's rotation and its curvature deflect the pattern in ways that produce prevailing east and west winds (**c**).

GLOBAL AIR CIRCULATION PATTERNS

Climate refers to average weather conditions, such as temperature, humidity, wind speed, cloud cover, and rainfall, over time. Many factors influence a region's climate. How much solar radiation a region gets varies with latitude. The global distribution of land masses and oceans affects climate. So does the topography of a region—whether it is a plain or mountainous. Complex interactions among these factors determine prevailing winds and the direction of ocean currents.

Warming effects of sunlight vary with latitude for two reasons. First, gases, bits of dust, and water vapor in the atmosphere absorb some wavelengths or reflect them back into space. Light rays that travel from the sun to the poles pass through more atmosphere than rays that travel to the equator, so fewer rays make it to Earth's polar surfaces. Second, light rays that do reach the surface at high latitudes are more spread out than rays that reach the equator (Figure 31.1*a*). As a result of these factors, Earth's surface is warmed the most at the equator and stays coolest at the poles.

Differential heating of Earth's surface sets up global air circulation patterns. Air at the equator heats up, expands, and rises. Convection causes this warm air to move north and south away from the equator, toward the cooler polar regions (Figure 31.1*b*).

Major winds do not blow directly north and south because Earth's rotation alters the initial movement. In the Northern Hemisphere, air masses become deflected rightward as Earth turns under them. In the Southern Hemisphere, this rotation deflects winds to the left of their initial course (Figure 31.1*c*).

Also, the distribution of land and water gives rise to pressure differences that produce regional winds. Land absorbs and gives up heat faster than water does, so air rises and falls more quickly over a land mass than over an ocean. Air pressure is lowest where air rises and is greatest where air sinks.

Air circulation affects rainfall patterns. Warm air can hold more moisture than cooler air. As air warms near the equator, it picks up moisture from the seas. It rises to cooler altitudes and gives up moisture as rain, which supports lush equatorial forests. Now drier, the air moves away from the equator. It descends at the latitudes of about 30°, where deserts often form. Still farther north and south, air again picks up moisture and ascends. This produces another rainy region at latitudes of about 60°. The air descends once again in polar regions. Here, frigid temperatures and minimal precipitation result in polar deserts.

Latitudinal variations in solar heating, combined with modified patterns of air circulation, help define the world's **temperature zones**, shown in Figure 31.2*a*.

The tilt of Earth's axis leads to seasonal variations in daylength and in the intensity of incoming sunlight, most notably at the poles and in temperate regions (Figure 31.2*b*). When it is summer in the Northern Hemisphere, this region is tilted *toward* the sun and sunlight strikes it most directly. In winter, this region is tilted *away* from the sun, so light rays come in at more of an angle. As described earlier, the greater the angle of incoming sunlight, the fewer light rays make it through the atmosphere to the surface, and the more widely those that do are spread out.

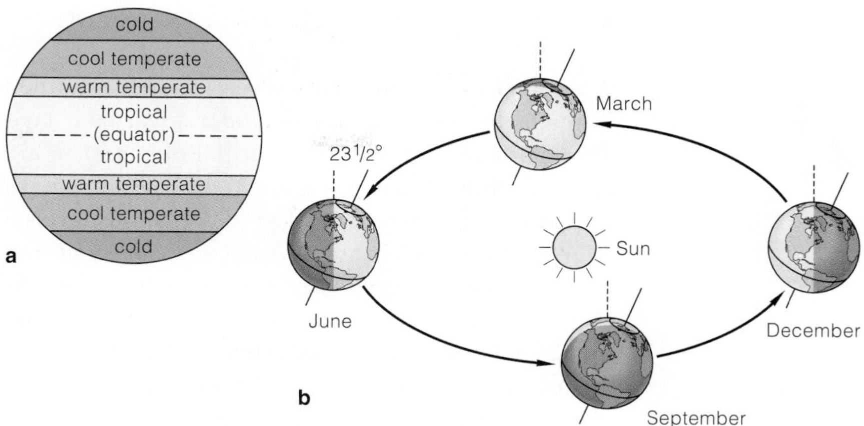

Figure 31.2 ***Animated!*** (**a**) World temperature zones. (**b**) Annual variation in incoming solar radiation. The northern end of Earth's fixed axis tilts toward the sun in June and away from it in December. This causes changes in the amount of light that reaches the surface and in daylength. Variations in sunlight intensity and daylength cause seasonal temperature variations.

A FENCE OF WIND AND OZONE THINNING

Ozone (O_3) and oxygen (O_2) molecules in the upper atmosphere absorb most ultraviolet (UV) radiation in incoming sunlight. The concentration of ozone is the highest between 17 and 27 kilometers above sea level. This region is the "ozone layer" (Figure 31.3*a*).

In the mid-1970s, scientists began to notice that the ozone layer was getting thinner. The layer's thickness had always varied a bit with the season, but this was a steady decline. By the mid-1980s, springtime ozone thinning over Antarctica had become so pronounced that some scientists described it as an "ozone hole" over the South Pole (Figure 31.3*b*).

This quickly became a matter of great concern. With less ozone in the stratosphere, people would get more UV exposure. This could increase cases of skin cancers, cataracts, and weakened immunity. High UV levels also harm plants and wildlife, which do not have the option of staying inside or slathering on a bit more sunscreen. Higher UV levels might even affect the atmosphere's composition, by slowing photosynthetic oxygen production by plants and phytoplankton.

Chlorofluorocarbons, or CFCs, are the main ozone destroyers. These odorless gases were widely used as propellants in aerosol cans, as coolants in refrigerators and air conditioners, and in solvents and plastic foam. They slowly seep into the air, and resist breakdown. In the stratosphere, CFCs release free chlorine radicals that yank oxygen away from ozone. A single chlorine radical can break apart thousands of ozone molecules.

Ozone thins the most over the poles because during the dark polar winters, swirling winds fence in CFCs. Also within this dynamic enclosure are clouds of ice crystals, upon which CFCs are converted to different molecules. In the spring, as days get longer, UV light frees chlorine radicals from these molecules.

In response to the threat posed by ozone thinning, developed countries agreed in 1992 to phase out the production of CFCs and other compounds known to destroy ozone. But climate change, discussed later in the chapter, may also be contributing to ozone loss.

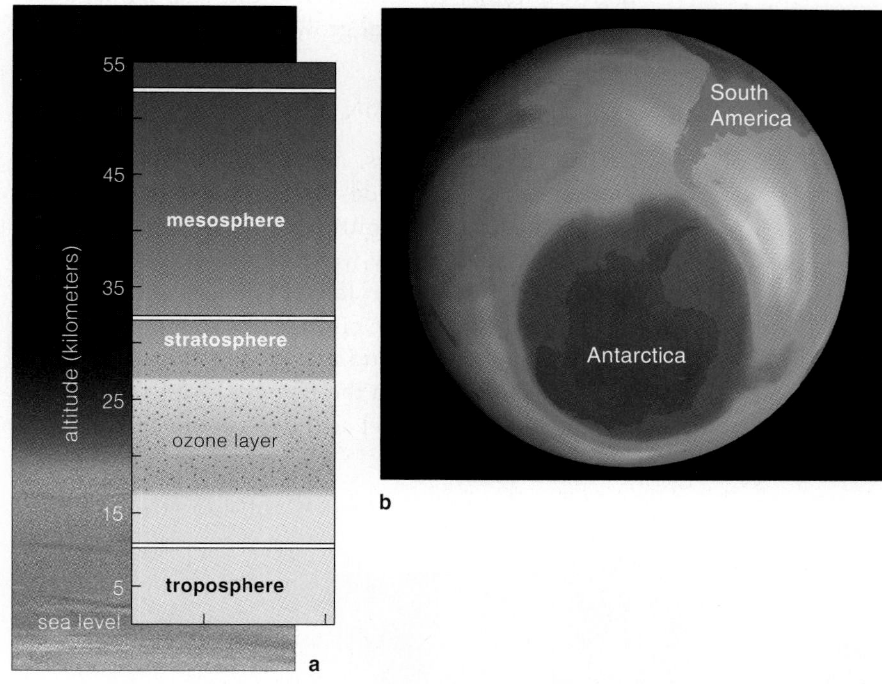

Figure 31.3 (**a**) Earth's atmosphere. Most of the air circulation occurs in the lower atmosphere (troposphere), where temperature falls rapidly with altitude. The ozone layer in the stratosphere absorbs most of the ultraviolet wavelengths in the sun's rays. (**b**) Seasonal ozone thinning above Antarctica in 2001. *Darkest blue* indicates the area with the lowest ozone level, at that time the largest recorded.

> *Heat energy from the sun warms the atmosphere and ultimately drives Earth's great weather systems.*
>
> *Ozone in the upper atmosphere is declining, exposing Earth's surface to higher levels of dangerous UV radiation.*

LINKS TO SECTIONS 2.4, 4.1, 30.7

NO WIND, LOTS OF POLLUTANTS, AND SMOG

A **thermal inversion** occurs when weather conditions trap a layer of cool, dense air under a warm air layer. Thermal inversions can keep air pollutants in an area and contribute to smog (Figure 31.4). Pollutants in the trapped layer of air accumulate to dangerous levels.

Where the winters are cold and wet, *industrial* smog may develop as a gray haze over industrialized cities that burn coal and other fossil fuels. Burning releases dust, smoke, soot, ashes, asbestos, oil, bits of lead and other heavy metals, and sulfur oxides. Most industrial smog now forms in China, India, and eastern Europe.

In large cities in warm climates, *photochemical* smog forms as a brown haze, especially where land forms a natural basin. Los Angeles and Mexico City are classic examples. Nitric oxide, the main pollutant, escapes in vehicle exhaust. It reacts with atmospheric oxygen and other pollutants to yield photochemical oxidants, such as peroxyacyl nitrates (PANs) that irritate animal eyes and respiratory systems. PANs can also damage leaf surfaces and stunt plant growth.

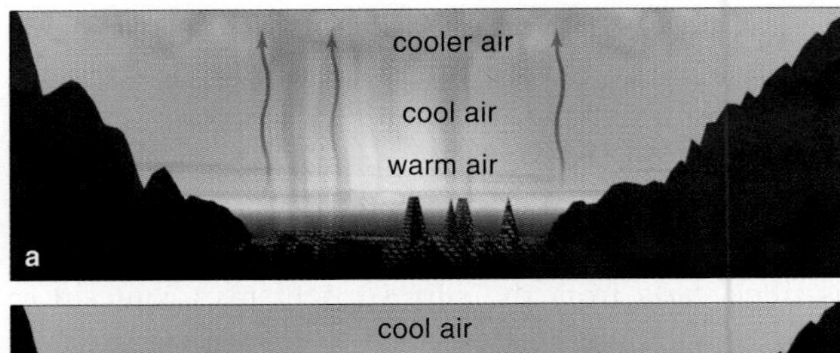

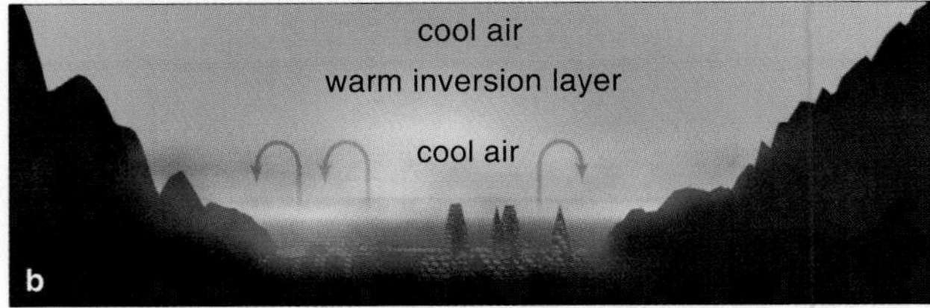

Figure 31.4 (**a**) Normal pattern of air circulation in smog-forming regions. (**b**) Air pollutants have become trapped under a thermal inversion layer.

WINDS AND ACID RAIN

Power plants, smelters, and factories that burn coal emit most sulfur dioxides. Vehicles, power plants that burn gas and oil, and nitrogen-rich fertilizers produce nitrogen oxides. In dry weather, these oxides can stay airborne for a bit, then fall as dry acid deposition. If air is moist, they form nitric acid vapor, sulfuric acid droplets, and sulfate and nitrate salts. Winds typically disperse them far from their source. They fall to Earth in rain and snow as wet acid deposition, or **acid rain**.

The pH of normal rainwater is 5 or so. Acid rain can be 0 to 100 times more acidic, as potent as lemon juice! Acid rain corrodes metals, marble on buildings, rubber, plastics, nylon, and other manmade materials. It disrupts the physiology of organisms, the chemistry of ecosystems, and so endangers biodiversity.

Depending on soil type and vegetation cover, some regions are more sensitive to acid rain (Figure 31.5). Highly alkaline soil can neutralize acids before they enter the streams and lakes of watersheds. Water that is highly alkaline also can neutralize acids. But many watersheds of northern Europe, southeastern Canada, and regions throughout the United States have only a thin soil layer covering granite bedrock. These soils have little buffering capacity and are acidified quickly.

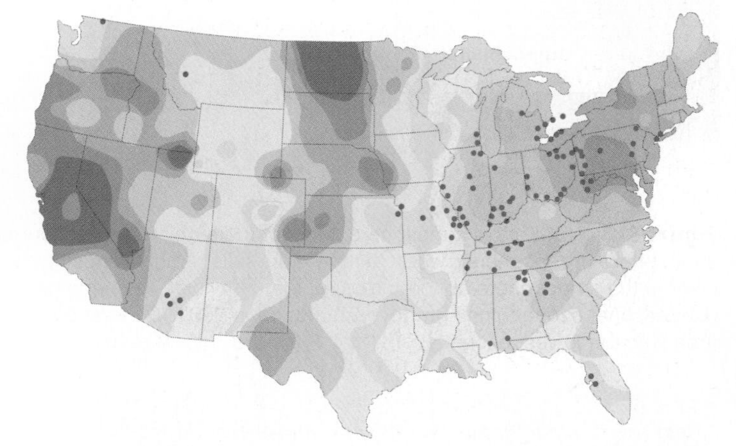

Figure 31.5 Average acidities of precipitation in the United States in 1998. *Red* dots pinpoint coal-burning power and industrial plants. *Right*, a biologist measuring the pH value of Woods Lake in New York's Adirondack Park. With a pH below 5, this lake is too acidic to support fish.

Figure 31.6 Harnessing solar energy. (**a**) A large array of electricity-producing photovoltaic cells in panels that collect sunlight energy. (**b**) A field of turbines harvesting wind energy.

Rain in much of eastern North America is thirty to forty times more acidic than it was several decades ago. Crop yields are declining. Fish populations have already disappeared from more than 200 lakes in New York's Adirondack Mountains. The air pollution from power plants and industry is increasing the acidity of rainfall. The change is contributing to the decline in forest trees and mycorrhizae that support new growth.

Coal-burning power plants are the single largest cause of acid rain. The United States produces about half of its electricity with such plants. Emissions from coal-burning plants in the midwest contribute to acid rain in Canada. Most pollutants in Scandinavia and the Netherlands, Austria, and Switzerland formed in the industrialized regions of western and eastern Europe. *Prevailing winds—and air pollutants they carry—do not stop at national boundaries.*

HARNESSING SOLAR AND WIND ENERGY

Paralleling the J-shaped curve of human population growth is a dramatic rise in total and per capita energy consumption. Some energy sources, most notably the fossil fuels, are not renewable. Others, such as direct solar energy and wind energy, are. The annual amount of incoming solar energy surpasses, by about 10 times, the energy in all the known fossil fuel reserves. That is 15,000 times more energy than human populations use on an annual basis. Maybe we should start to collect it in earnest, in something besides crops.

For example, when exposed to sunlight, electrodes in "photovoltaic cells" produce an electric current that splits water molecules into oxygen and hydrogen gas (H_2). This can be used directly as fuel to run cars, heat and cool buildings, and generate electricity. We have the technology to obtain *solar–hydrogen energy* (Figure 31.6*a*). Such energy can be stored efficiently. It costs less to distribute H_2 than electricity, and water is the only by-product. Space satellites run on this energy.

Proponents of solar–hydrogen energy say that it could end smog, oil spills, and acid rain, without the risks associated with nuclear power. However, others see potential problems. As an example, hydrogen is a tiny molecule and highly corrosive; it is likely to leak from any pipeline or container. Leakage of hydrogen gas into the air might accelerate global warming.

We already are using winds to convert solar energy to mechanical energy. Where prevailing winds travel faster than 7.5 meters per second, we find *wind farms* (Figure 31.6*b*). Wind farms now generate 1 percent of California's electricity. The winds of North and South Dakota alone might meet all but 20 percent of the energy needs of the continental United States. Winds do not blow constantly, but when they do, energy can be stored in batteries or fed into utility grids to supply local and more distant power needs.

On the downside, the rotating blades of turbines at wind farms kill birds and bats. Also wind farms have been called a form of "visual pollution" that destroys scenic views and thus lowers property values.

Accumulation of CFCs in the upper atmosphere has led to thinning of the ozone layer. A ban on these chemicals is in place and is helping to reverse this effect.

Smog formation and acid deposition are two forms of air pollution in specific regions, but prevailing winds often distribute pollutants across regional and political boundaries.

Solar and wind energy are alternatives to fossil fuels. They could help meet rising energy needs.

31.2 The Ocean, Landforms, and Climates

The ocean, a continuous body of water, covers more than 71 percent of Earth. Driven by solar heat and wind friction, its upper 10 percent moves in currents. Those currents also influence regional climates.

OCEAN CURRENTS AND THEIR EFFECTS

Latitudinal and seasonal variations in sunlight warm and cool the ocean. When heated, that vast volume of water expands; where cooled, it shrinks. So sea level is about eight centimeters, or three inches, higher at the equator than it is at either pole. The volume of water in that "slope" is enough to get surface waters moving as a response to gravity, most often toward the poles. Along the way, water warms the portion of air just above it. At midlatitudes, it transfers *10 million billion* calories of heat energy per second to the air!

Rapid bulk flows of water, or *currents*, arise from the tug of trade winds and westerlies. Their direction and properties are influenced by Earth's rotation, the distribution of land masses, and the shapes of oceanic basins. The water circulates clockwise in the Northern Hemisphere, and it circulates counterclockwise in the Southern Hemisphere (Figure 31.7).

Swift, deep, and narrow currents of nutrient-poor waters move parallel with the east coast of continents. For instance, along North America's east coast, warm water moves north as the Gulf Stream. Slower, shallow, broad currents paralleling the west coast of continents move cold water toward the equator.

Why are Pacific Northwest coasts cool, foggy, and mild in summer? Winds nearing the coast give up heat to cold coastal water, which the California Current is moving down toward the equator. Why are Baltimore and Boston so muggy in summer? Air above the Gulf Stream gains heat and moisture, which southerly and easterly winds deliver to those cities. Why are winters milder in London and Edinburgh than in Ontario and central Canada, which are all at the same latitude? The North Atlantic Current picks up warm water from the Gulf Stream. As it flows past northwestern Europe, it gives up heat energy to the prevailing winds.

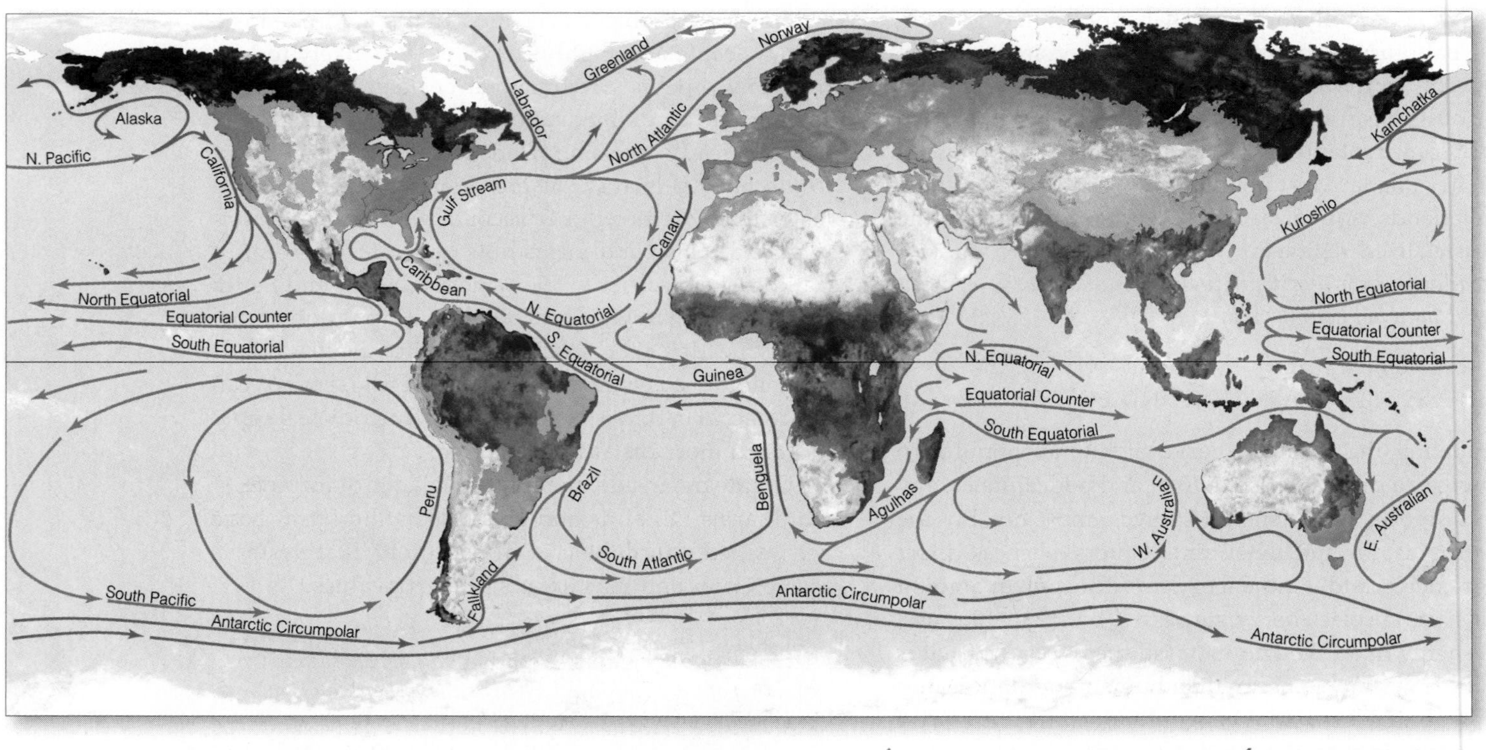

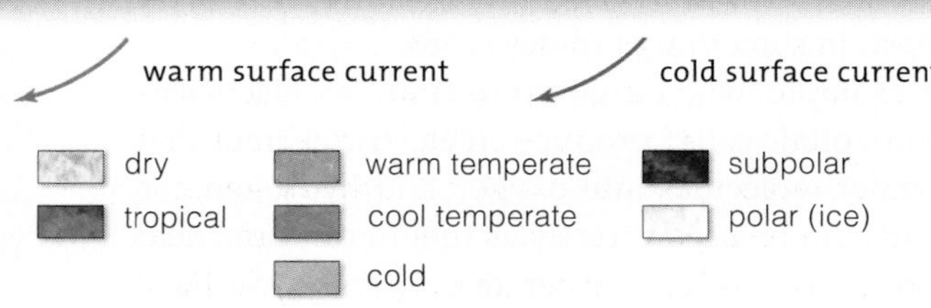

Figure 31.7 ***Animated!*** Major climate zones correlated with surface currents and drifts of the world ocean. Warm surface currents move from the equator toward the poles. Water temperatures differ with latitude and depth, which contributes to differences in air temperature and rainfall patterns. Which way a current flows depends on prevailing winds, Earth's rotation, the shapes of ocean basins, and the presence of land masses.

Figure 31.8 ***Animated!*** Rain shadow effect of the Sierra Nevada. On the side of mountains facing away from prevailing winds, rainfall is reduced. *Blue* numbers show the average yearly precipitation, in centimeters. The *white* numbers signify elevations, in meters.

RAIN SHADOWS AND MONSOONS

Mountains, valleys, and other topographical features affect regional climates. *Topography* refers to a region's variations in elevation. Consider an air mass that picks up moisture off California's coast. It moves inland to the Sierra Nevada, a mountain range paralleling the coast. The air cools as it rises to higher altitudes, then loses moisture as rain (Figure 31.8).

Belts of vegetation at different elevations reflect the temperature and moisture differences. Arid grasslands form at the western base of the range. Higher up, you see deciduous and evergreen species adapted to more moisture and cooler air. Higher still, a subalpine belt supports just a few evergreen species that withstand a rigorously cold habitat. Above the subalpine belt, only low, nonwoody plants and dwarfed shrubs can grow.

Air warms and holds more water after it flows over mountain tops. As the air descends, it retains moisture and pulls more water out of plants and soil. The result is a **rain shadow**, a semiarid or arid region of sparse rainfall on the side of mountains that faces away from the wind. The Himalayas, Andes, Rockies, and other great mountain ranges cause rain shadows. Tropical forests flourish on one side of Hawaii's high volcanic peaks, and arid conditions prevail on the other.

Monsoons are air circulation patterns with effects on continents north or south of warm oceans. The land heats intensely, so low-pressure air parcels form above it. The low pressure draws in moisture-laden air from the ocean, as from the Bay of Bengal into Bangladesh. Converging trade winds and the equatorial sun cause intense heating and heavy rain. Air moving north and south gives rise to alternating wet and dry seasons.

Recurring coastal breezes are like mini-monsoons. Water and coastal land have different heat capacities. In the morning, water does not warm as fast as the land. When warm air above land rises, cooler marine air moves in. After sunset, land loses heat faster than water. Then, land breezes flow in the reverse direction.

Surface ocean currents also influence regional climates and help distribute nutrients in marine ecosystems.

Air circulation patterns, ocean currents, and landforms interact in ways that influence regional temperatures and moisture levels. Thus they also influence the distribution and dominant features of ecosystems.

LINKS TO SECTIONS 13.3, 29.9

31.3 Realms of Biodiversity

Differences in physical and chemical properties, and in evolutionary history, help explain why grasslands, forests, deserts, and tundra form where they do, and why some water provinces are richer in species than others.

Suppose you live in the coastal hills of California and decide to tour the Mediterranean coast, the southern horn of Africa, or central Chile. There you find many-branched, tough-leafed woody plants that resemble the shrubby chaparral plants you are familiar with at home. Great geographic and evolutionary distances separate the plants. Why are they alike?

You decide to compare their locations on a global map. You discover that American and African desert plants live about the same distance from the equator. Chaparral plants and their distant look-alikes all grow along the western and southern coasts of continents positioned between the latitudes of 30° and 40°. As Alfred Wallace, Charles Darwin, and other naturalists did so long ago, you have just stumbled onto one of many biogeographic patterns.

Those naturalists divided Earth's land masses into six **biogeographic realms**—vast expanses where they could expect to find communities of specific types of plants and animals, such as palm trees and camels in the Ethiopian realm (Figure 31.9).

Biomes are finer subdivisions of the great realms, but they are still identifiable on a global scale. Their distribution has been shaped partly by evolutionary processes and events, as when evolving species were slowly rafted about on different land masses after the splitting of Pangea (Section 13.3). They also owe their distribution to topography, climate, and interactions among species. Some biomes extend across more than one continent, and their communities, although similar in many ways, each include unique species. Notice, in

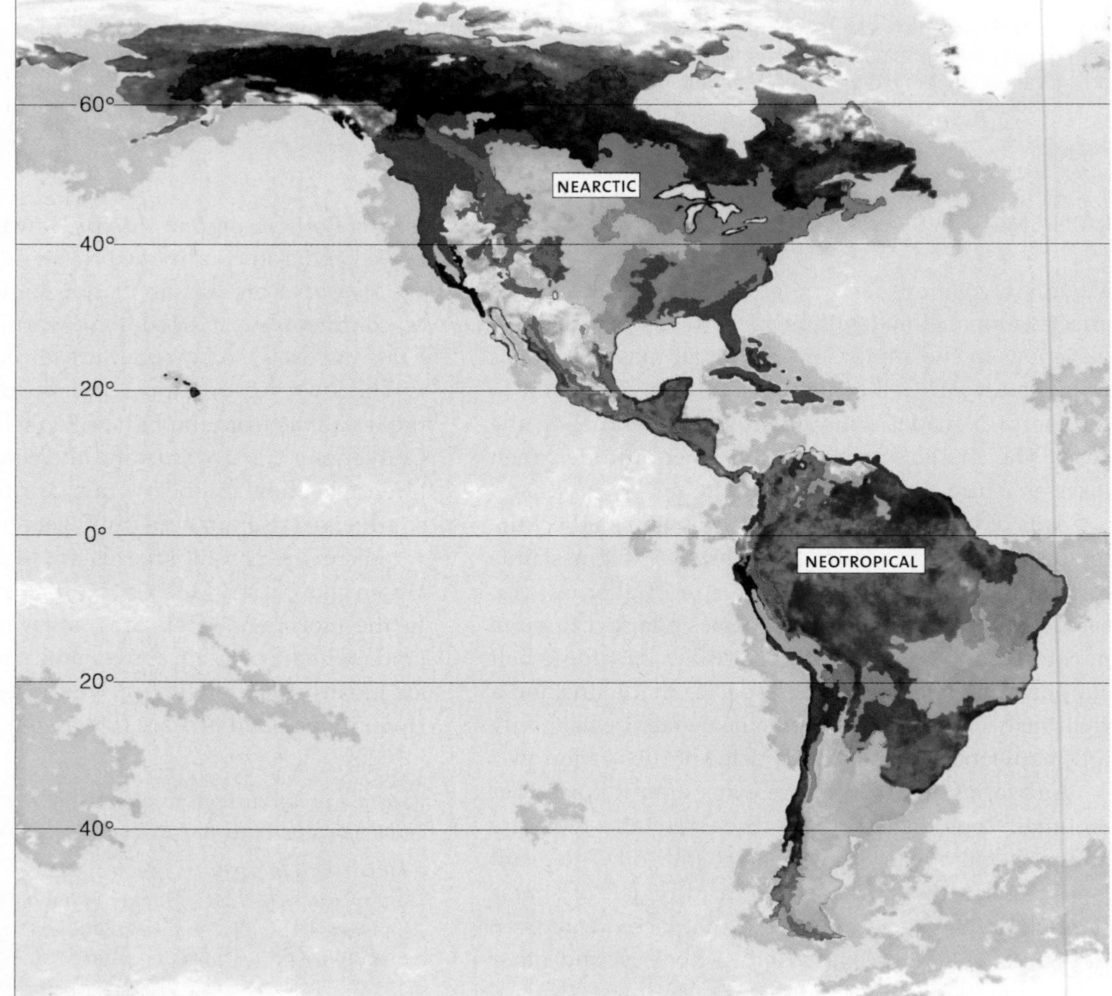

Figure 31.9 ***Animated!*** Categories of biomes and marine ecoregions among the six biogeographic realms.

Figure 31.9, the distribution of dry shrublands and woodlands (coded dark brown). Chaparral is part of that biome. Similarly, the temperate grassland biome occurs in North American prairie, South American pampa, southern African veld, and Eurasian steppe. In each case, the climax community is dominated by grasses that can tolerate strong winds and drought.

In general, communities that characterize a biome are arrays of species adapted to specific temperatures, patterns of rainfall, soil type, and to interacting with one another. Adaptations may include form, function, behavior, and life-history patterns.

Evolutionary events and environmental conditions also influence the distribution of marine ecosystems. Figure 31.9 shows the main marine ecoregions.

Conservationists are working to locate, inventory, and protect **hot spots**, those portions of biomes and ecoregions that are the richest in biodiversity and the most vulnerable to species losses. Figure 31.10 gives one example, Costa Rica's Monteverde cloud forest.

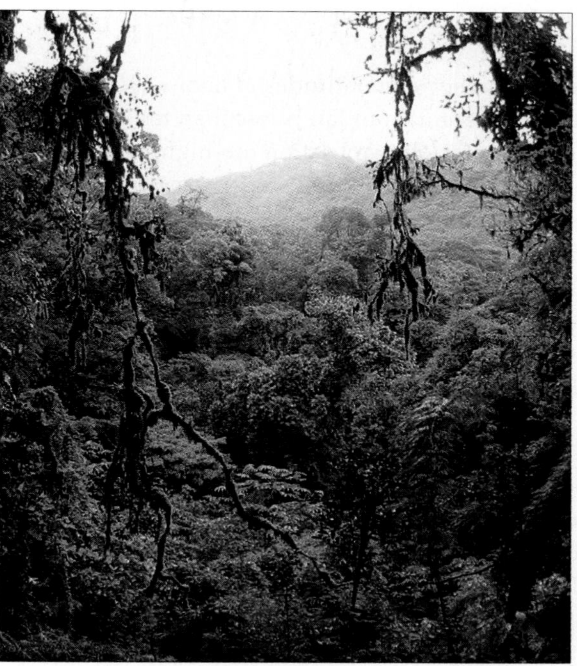

Figure 31.10 A Costa Rican hotspot. In the early 1970s, George Powell, a graduate student from the University of California at Davis, was studying this high elevation forest. When people began to move into and clear the forest, Powell got the idea to buy some land and set up a sanctuary. Powell's vision soon inspired individuals and conservation groups around the world to help protect this biological gem. Monteverde Cloud Forest Preserve now protects 5,000 hectares and is managed by a nonprofit organization. The reserve's species include 420 orchids, more than 100 species of mammals, 400 species of birds, and 120 species of amphibians and reptiles. It is one of the few habitats left for the jaguar, ocelot, and puma.

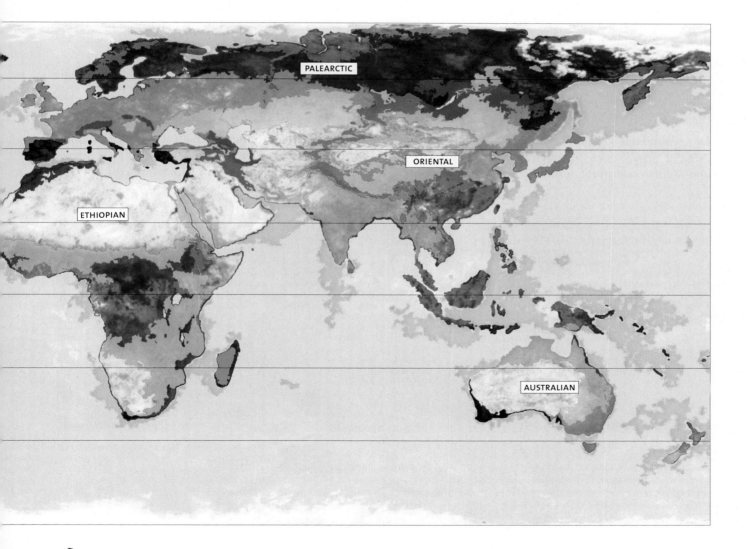

LINKS TO SECTIONS 18.3, 18.5

DESERTS, NATURAL AND MAN-MADE

Deserts form between latitudes of about 30° north and south. Here, annual rainfall is less than ten centimeters or so, and rates of evaporation are high. Heavy, brief, infrequent pulses of rainfall swiftly erode the topsoil. Humidity is so low that the sun's rays easily penetrate the air. The ground heats fast, then cools fast at night. These conditions do not favor large, leafy plants. The distribution of species within the biome is influenced by temperature and moisture gradients (Figure 31.11).

Long-term shifts in climate, and human activities, can convert grasslands to deserts. **Desertification** is the conversion of large regions of natural grasslands to desertlike conditions. Overgrazing by livestock and short-sighted agricultural practices contribute to the problem. Periodic droughts make matters worse.

Desertification has far-reaching ecological effects. For example, dust from northwestern Africa may be contributing to the decline of coral reefs far away, in the Caribbean Sea. Winds lift dust into the air and then transport it across the Atlantic. In that dust are fungal, bacterial, and viral pathogens. Since the 1980s, huge dust storms of the sort shown in Figure 31.12 have become more frequent in Africa—and so have disease outbreaks among the Caribbean corals.

DRY SHRUBLANDS, DRY WOODLANDS, AND GRASSLANDS

Dry shrublands get less than 25 to 60 centimeters of rain per year. We see them in California, South Africa, and Mediterranean regions, where they are known by such local names as fynbos and chaparral. California has 6 million acres of chaparral (Figure 31.13). In the summer, lightning-sparked fires sweep through these biomes. The shrubs have highly flammable leaves and quickly burn to the ground. But they grow back from root crowns and fire causes their seeds to germinate.

Figure 31.12 Satellite image of an immense dust storm that formed in the hot, dry Sahara Desert, then continued past the west coast of Africa, and is well on its way across the Atlantic.

Dry woodlands dominate where annual rainfall is about 40 to 100 centimeters. The dominant trees can be tall, but they do not form a dense, continuous canopy. Eucalyptus woodlands of southwestern Australia and oak woodlands of California and Oregon are like this.

Grasslands form in the interior of continents in the zones between deserts and temperate forests. Warm temperatures prevail in the summer, and winters are extremely cold. Annual rainfall of 25 to 100 centimeters prevents deserts from forming, but it is not enough to support forests. Drought-tolerant primary producers handle strong winds, sparse and infrequent rainfall, and fast evaporation. Grazing and burrowing species are the dominant animals. Their activities, combined with periodic fires, help keep shrublands and forests from encroaching on the fringes of many grasslands.

Figure 31.11 Two views of the same biome—the Sonoran Desert of Arizona. Sunlight is just as intense in the desert lowlands (**a**) as in the uplands (**b**). Vegetation differences reflect water availability, temperature, and soils. The uplands rarely frost over. Also, they receive light but prolonged rain in winter, and heavy pulses of rain in summer. Winter and summer annuals as well as perennials are not as water-stressed as they are in the lowlands, where temperatures are exceedingly high and rainfall is sparse. Widely spaced creosote bush (*Larrea*) mixed with bursage (*Ambrosia*) dominates.

North America's main grasslands are *shortgrass* and *tallgrass* prairie. Most form on flat or rolling land, as in Figure 31.14. The roots of perennials extend profusely through topsoil. In the 1930s, the shortgrass prairie of the Great Plains was overgrazed and plowed under to grow wheat, which requires more water than the area sometimes receives. Strong winds, prolonged drought, and unsuitable farming practices turned much of the prairie into the Dust Bowl. John Steinbeck's historical novel, *The Grapes of Wrath,* eloquently describes the impact of this environmental disaster on people.

Tallgrass prairie, of the sort you read about in the preceding chapter, once extended west from deciduous forests in North America. Tall grasses, legumes, and composites, such as daisies, thrived in the continent's interior, which had richer topsoil and slightly more frequent rainfall. Farmers converted nearly all of the original tallgrass prairie to agriculture, but some areas are now being restored.

Between the tropical forests and the hot deserts of Africa, South America, and Australia are broad belts of grasslands with a smattering of shrubs and trees. These are savannas (Figure 31.14). Rainfall averages 90 to 150 centimeters a year, and extended, seasonal droughts occur. Where rainfall is sparse, fast-growing grasses dominate. Acacia and other shrubs grow in regions that get a bit more moisture. Where rainfall is higher, savannas grade into tropical woodlands with low trees and shrubs and tall, coarse grasses.

Monsoon grasslands form in southern Asia where heavy rains alternate with a dry season. Dense stands of tall, coarse grasses form, then die back and often burn during the dry season.

Figure 31.13 California chaparral. The dominant plants are multibranched, woody, and typically only a meter or so tall. The leaves of many produce oils that deter herbivores, but also make the plant highly flammable. The photo at right shows a firestorm racing through chaparral in a canyon above Malibu.

Figure 31.14 *Above*, bison in a shortgrass prairie to the east of the Rocky Mountains. At one time, bison were the dominant large herbivore of North America's grasslands. Abundant grasses of this biome supported 60 million bison before Europeans arrived.

Left, the African savanna, a warm grassland with scattered stands of shrubs and trees. More varieties and greater numbers of large ungulates live here than anywhere else. They include the migratory wildebeests shown here, and giraffes, Cape buffalos, zebras, and impalas.

LINK TO SECTION 15.7

BROADLEAF FORESTS

In forest biomes, tall trees grow closely together and form a fairly continuous canopy over a broad region. In general, there are three types of trees. Which type prevails depends partly on distance from the equator. *Evergreen broadleafs* dominate between the latitudes of 20° north and south. *Deciduous broadleafs* are typical in moist, temperate latitudes where the winters are mild. *Evergreen conifers*, the third category of forest tree, are dominant in high, cold latitudes and high mountains.

Evergreen broadleaf forests sweep across tropical zones of Africa, the East Indies, Malaysia, Southeast Asia, South America, and Central America. Annual rainfall can exceed 200 centimeters and never is less than 130 centimeters. One type, tropical rain forests, flourish with regular and heavy rainfall, an annual mean temperature of 25°C, and humidity of at least 80 percent (Figure 31.15). Evergreen trees of this highly productive forest grow new leaves and shed old ones all year long. Decomposition and mineral cycling are rapid, given the hot, humid climate, so litter does not accumulate. Soils are weathered, humus-deficient, and poor nutrient reservoirs.

Deciduous broadleaf forests start in regions of mild temperatures and regular seasonal rainfall. In tropical deciduous forests, the trees drop some or all of their leaves during a pronounced dry season, a moisture-preserving adaptation. Monsoon forests of India and Southeast Asia contain such trees. Farther north, in the temperate zone, rainfall is even lower. Winters are cold and dry, and temperate deciduous forests prevail (Figure 31.16). Decomposition is not as rapid as in the humid tropics, and many nutrients are conserved in accumulated litter on the forest floor.

CONIFEROUS FORESTS

Conifers (cone-bearing trees) are primary producers of coniferous forests. Most have thickly cuticled, needle-shaped leaves and recessed stomata, adaptations that help them conserve water during droughts and winter. They dominate the boreal forests, montane coniferous forests, temperate rain forests, and pine barrens.

Boreal forests stretch across northern Europe, Asia, and North America. Such forests also are called *taigas*, which means "swamp forests." Most form in glaciated

Figure 31.15 Tropical rain forests, biomes of great biodiversity. Among the millions of known species are jaguars and *Rafflesia*, a vine with a foul-smelling fly-pollinated flower that grows three meters across. No doubt many more astonishing rain forest species are still to be discovered. Who knows how many other biological marvels we are losing as these forests disappear?

Figure 31.16 Temperate deciduous forest in North America.

Figure 31.17 *Above*, taiga in Alberta, Canada. *Right*, firefighter retreating from flames in a forest of pines, oaks, and palmettos near Daytona Beach, Florida.

regions having cold lakes and streams (Figure 31.17). It rains mostly in summer, and evaporation is low in the cool summer air. The cold, dry winters are more severe in eastern parts of these biomes than in the west, where ocean winds moderate the climate. Spruce and balsam fir dominate the boreal forests of North America. Pine, birch, and aspen forests form in areas that were burned or logged. In poorly drained soil, acidic bogs dominated by peat mosses, shrubs, and stunted trees prevail. Boreal forests are less dense to the north, where they grade into arctic tundra.

In the Northern Hemisphere, montane coniferous forests extend southward through the great mountain ranges. Spruces and firs dominate in the north and at higher elevations. They give way to firs and pines in the south and at lower elevations. Coniferous forests thrive in certain temperate lowlands. One enormous temperate rain forest parallels the coast from Alaska into northern regions of California. It contains some of the world's tallest trees—Sitka spruce to the far north and redwoods to the south. Over the past century, extensive logging has destroyed much of this biome.

Southern pine forests dominate the coastal plains of the southern Atlantic and Gulf states. Pine species are adapted to dry, sandy, nutrient-poor soil and to frequent fires. Such fires do not get hot enough to kill big trees but they burn up shrubs and accumulated debris. Forests of pine, scrub oak, and wiregrass grow in New Jersey. Palmettos grow below loblolly and other pines in the Deep South. In one recent heat wave in Florida, fires destroyed more than 320,000 acres. Counties where buildings were near the forest had not allowed any controlled burns; the resulting buildup of dry material fed the fires (Figure 31.17).

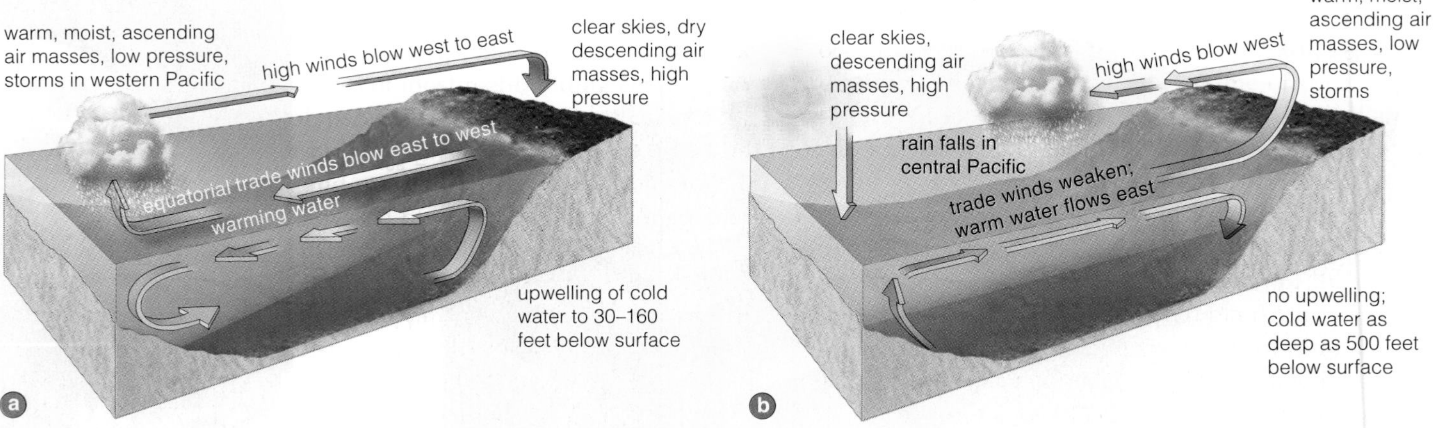

Figure 31.27 ***Animated!*** (**a**) Westward flow of cold, equatorial surface water in between ENSOs. (**b**) Eastward dislocation of warm ocean water during an ENSO.

Satellite data on primary productivity in the equatorial Pacific Ocean. The concentration of chlorophyll was used as the measure.

(**c**) During the 1997–1998 El Niño event, a massive amount of nutrient-poor water moved to the east, and photosynthetic activity was negligible.

(**d**) During the subsequent La Niña, upwelling and westward displacement of nutrient-rich water led to a huge algal bloom.

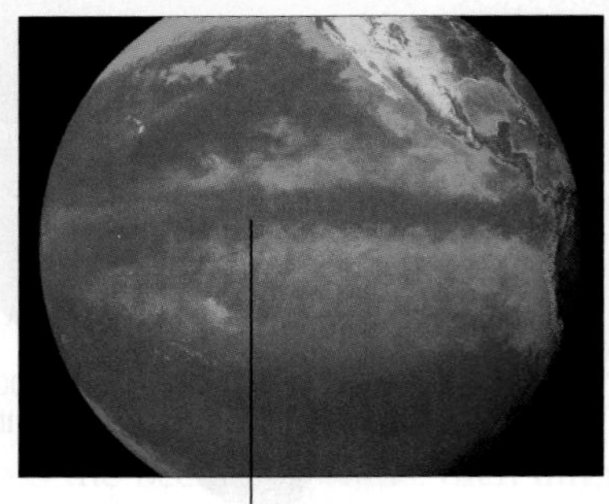

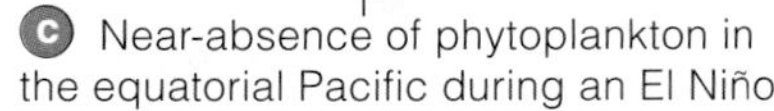

(c) Near-absence of phytoplankton in the equatorial Pacific during an El Niño.

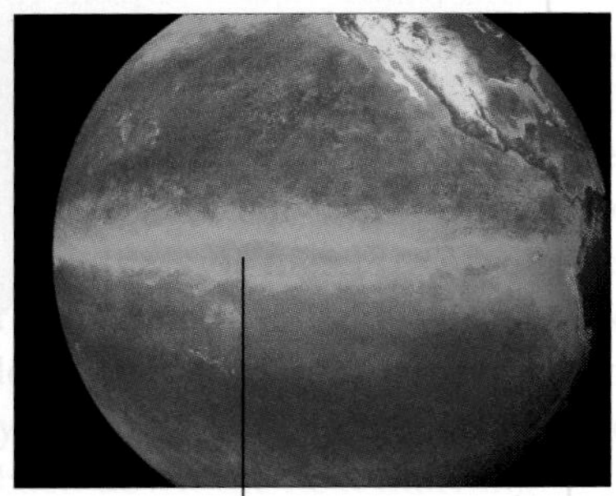

(d) Immense algal bloom in the equatorial Pacific in the La Niña rebound event.

31.5 Applying Knowledge of the Biosphere

We now return to El Niño, a global event that affects primary productivity along coasts of the Northern Hemisphere and in the equatorial Pacific Ocean. That event has rippling effects on the weather and on land ecosystems around the world.

In the Southern Hemisphere, the health of commercial fisheries depends upon wind-induced upwelling along the coasts of Peru and Chile. Prevailing coastal winds blow in from the south and southeast, tugging surface water away from shore. Cold water delivered to the continental shelf by the Peru Current moves up near the surface. Huge quantities of nitrate and phosphate are pulled up and carried north by this current. These nutrients sustain phytoplankton that are the basis for one of the world's best fisheries.

Every three to seven years, warm surface waters of the western equatorial Pacific Ocean move eastward. This massive displacement of warm water affects the prevailing wind direction. The eastward flow speeds up so much that it influences the vertical movement of water along the coasts of Central and South America.

Water driven into a coastline is forced downward and flows away from it. Near Peru's coast, prolonged **downwelling** of nutrient-poor water displaces cooler waters of the Peru Current, which prevents upwelling. The warm current usually arrives around Christmas. Fishermen in Peru named it **El Niño** ("the little one," a reference to the baby Jesus).

El Niño episodes last about six to eighteen months. An El Niño Southern Oscillation, or **ENSO**, is defined by changes in the sea's surface temperature and in the air circulation patterns. *Between* ENSOs, warm waters and heavy rains move west (Figure 31.27*a*). *During* an ENSO, the prevailing surface winds in the western equatorial Pacific pick up speed and "drag" surface waters toward the east (Figure 31.27*b*). As a result, the westward transport of water slows, the temperature of the sea's surface water rises, evaporation increases, and air pressure falls. The changes have global effects, both in the seas and on land.

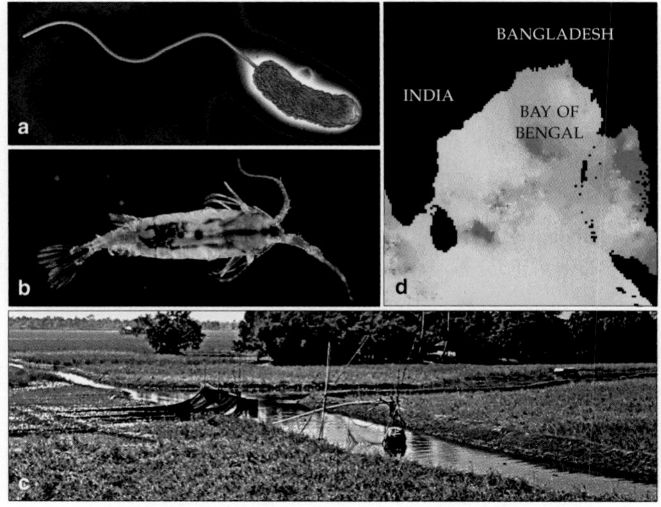

Figure 31.28 (**a**) *Vibrio cholerae*, the agent of cholera. (**b**) Copepods host a dormant stage of *V. cholerae* that waits out adverse environmental conditions that do not favor its growth and reproduction. (**c**) Typical Bangladesh waterway from which water samples were drawn for analysis. (**d**) Satellite data on rising sea surface temperatures in the Bay of Bengal correlated with cholera cases in the region's hospitals. *Red* signifies warmest summer temperatures. (**e**) Rita Colwell comparing unfiltered water with water that was filtered through folded sari fabric.

As you read in the chapter opening, 1997 ushered in an ENSO event. Average sea surface temperatures in the eastern Pacific rose 9°F (about 5°C). The unusually warm water extended 9,660 kilometers west from the coast of Peru.

The 1997–1998 El Niño event caused a tremendous change in the primary productivity of the equatorial Pacific. The massive eastward flow of nutrient-poor warm water caused phytoplankton populations to fall to almost undetectable levels (Figure 31.27*c*). During the following year, cooler, nutrient-rich water welled up and was displaced westward, along the equator in what is called a La Niña event. Satellite images taken during this period revealed a vast algal bloom that stretched across the equatorial Pacific (Figure 31.27*d*).

During the 1997–1998 El Niño event, the number of cholera cases soared. People already knew that water contaminated by *Vibrio cholerae* bacteria causes cholera epidemics (Figure 31.28*a*). The disease triggers severe diarrhea and the bacteria enter the water supply in feces. People who use the tainted water get infected.

What people did *not* know was where *V. cholerae* lurked between cholera outbreaks. It could not be found in humans or in water supplies. Then it would show up simultaneously in places that were often far apart—usually coastal cities, where the urban poor draw water from rivers flowing to the sea.

Marine biologist Rita Colwell had been thinking about cholera for some time. Colwell suspected that humans were not the host in between outbreaks. Was there an environmental reservoir for the pathogen? Maybe. But nobody had detected it in water samples that had been subjected to standard culturing.

Then Colwell had a flash of insight: Perhaps no one had found the pathogen because it sometimes enters a dormant stage that cannot be cultured.

Colwell discovered that she could use an antibody-based test to detect a protein on *V. cholerae*'s surface. Tests on samples from Bangladesh revealed a dormant form of the bacteria in fifty-one of fifty-two samples of water. Standard testing methods missed it in all but seven of these samples.

V. cholerae lives in rivers, estuaries, and the seas. As Colwell knew, plankton also thrive in these aquatic environments. Using her new method, she discovered *V. cholerae* on copepods, small marine crustaceans that graze on phytoplankton (Figure 31.28*b*).

Colwell knew that there was a seasonal pattern to cholera outbreaks in Bangladesh, and to temperatures in the Bay of Bengal (Figure 31.28*c*,*d*). She also knew that phytoplankton levels rise and fall seasonally. She hypothesized that the abundance of copepods—and of the *V. cholerae* living on them—rises and declines with phytoplankton abundance. Colwell decided to compare sea surface temperature data from 1990–1991 and 1997–1998 El Niño episodes with the incidence of cholera cases. As she expected, four to six weeks after the sea surface temperatures rose, so did the incidence of cholera cases.

Today, Colwell and Anwarul Huq, a Bangladeshi scientist, are looking into salinity and other factors that may also influence cholera outbreaks. Their goal is to design a more robust model for predicting where and when the next cholera outbreak will occur.

Colwell and Huq also devised a simple prevention method. They advised women in Bangladesh to use sari cloth as a filter to remove *V. cholerae* cells from the water (Figure 31.28*e*). Copepod hosts are too big to pass through the thin cloths, which can be rinsed in clean water, sun-dried, and used again and again. By using this simple and inexpensive method, seasonal cholera outbreaks have been reduced by 50 percent.

Summary

Section 31.1 "Climate" refers to average weather conditions over time. Climate depends upon differences in the amount of solar radiation reaching equatorial and polar regions, Earth's tilt and its annual path around the sun, the distribution of continents and seas, and the elevation of land masses.

The use of CFCs has depleted ozone in the upper atmosphere. Ozone thinning allows more damaging UV radiation to reach Earth's surface.

Smog, a form of air pollution, arises in areas where large amounts of fossil fuels are burned. Coal-burning power plants are also the main contributors to acid rain, which damages soils and kills plants and fish.

Solar energy and wind energy are renewable, clean sources of energy for the human population.

Biology Now
Learn how sunlight energy drives global patterns of air circulation with the animation on BiologyNow.

Section 31.2 Latitudinal and seasonal variations in sunlight warm ocean waters and set currents in motion. The currents distribute heat energy and affect weather patterns. Ocean currents, air currents, and topography interact to shape regional climates.

Biology Now
See the patterns of major ocean currents with the animation on BiologyNow.

Section 31.3 The world's land masses are realms of biodiversity. Each is more or less isolated from others and maintains a characteristic array of species.

Biomes are a category of major ecosystems on land. They form and are maintained by regional variations in climate, landforms, and soils.

Deserts form at latitudes about 30° north and south, where annual rainfall is sparse. Slightly wetter southern or western coastal regions support dry woodlands and dry shrublands. In midlatitude interiors of continents, vast grasslands form. Near the equator, high rainfall and mild temperatures support evergreen tropical forests. Deciduous broadleaf forests grow in regions with hot summers and cool winters. Where a cold, dry season alternates with a cool, rainy season, coniferous forests prevail. Tundra forms at high latitudes and high altitudes.

Biology Now
Examine the distribution of Earth's biomes with the animation on BiologyNow.

Section 31.4 Water provinces cover more than 71 percent of the Earth's surface. They include standing fresh water, running fresh water, and the world oceans. All aquatic ecosystems show gradients in penetration of sunlight, water temperature, salinity, and dissolved gases. These factors vary daily and seasonally, and they influence primary productivity.

Wetlands, intertidal zones, rocky and sandy shores, tropical reefs, and regions of the open ocean are major marine ecosystems. Superheated, mineral-rich waters support communities at deep-sea hydrothermal vents.

Upwelling is an upward movement of deep, cool ocean water along coasts of continents that often carries nutrients to the surface.

Biology Now
Learn about the oceanic zones with the animation on BiologyNow.

Section 31.5 An El Niño event disrupts upwelling and causes massive temporary changes in rainfall and other weather patterns with global effects.

Biology Now
Observe how an El Niño event affects ocean currents and upwelling with the animation on BiologyNow.

Self-Quiz

Answers in Appendix I

1. Solar radiation drives the distribution of weather systems and so influences _______ .
   a. temperature zones c. seasonal variations
   b. rainfall distribution d. all of the above

2. _______ shields against the sun's UV radiation.
   a. A thermal inversion c. The ozone layer
   b. Acid precipitation d. The greenhouse effect

3. Regional variations in the global patterns of rainfall and temperature depend on _______ .
   a. global air circulation c. topography
   b. ocean currents d. all of the above

4. A rain shadow is a reduction in rainfall _______ .
   a. on the inland side of a coastal mountain range
   b. along eastern coasts in the Northern Hemisphere
   c. as a result of human activities
   d. near the poles

5. The Gulf Stream flows _______ along the east coast of the United States, and the California current flows _______ along its west coast.
   a. north, south c. south, south
   b. north, north d. south, north

6. Broadleaf evergreen trees dominate in _______ .
   a. chapparal c. boreal forests
   b. tropical rain forests d. deserts

7. The major cause of acid rain is operation of _______ .
   a. nuclear power plants c. coal-burning plants
   b. automobiles d. agricultural machinery

8. Biomes are _______ .
   a. water provinces d. characterized by
   b. water and land zones dominant plants
   c. broad land regions e. both c and d

9. Biome distribution corresponds roughly with regional variations in _______ .
   a. climate c. soils
   b. topography d. all of the above

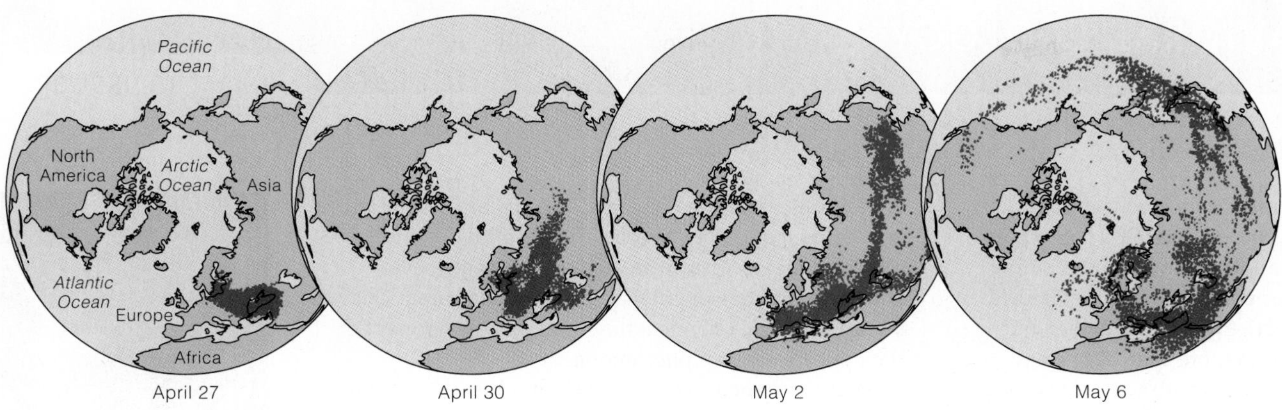

Figure 31.29 Global distribution of radioactive fallout after the 1986 meltdown of the Chernobyl nuclear power plant. The fallout put 300 million to 400 million people at risk for leukemia and other radiation-induced disorders. By 1998, the rate of thyroid abnormalities in children living immediately downwind from Chernobyl was nearly seven times as high as for those upwind; their thyroid gland concentrated the iodine radioisotopes.

10. Grasslands most often predominate ______.
   a. near the equator
   b. at high altitudes
   c. in interiors of continents
   d. b and c

11. Permafrost underlies ______.
   a. grasslands
   b. tundra
   c. temperate zone forests
   d. boreal forests

12. During ______, deeper, often nutrient-rich water moves to the surface of a body of water.
   a. spring overturn
   b. fall overturn
   c. upwelling
   d. all of the above

13. Match the terms with the most suitable description.

____ tundra	a. equatorial broadleaf forest
____ chaparral	b. area where freshwater and seawater mix
____ desert	c. type of grassland with trees
____ savanna	d. low-growing plants at high latitudes or high elevations
____ estuary	e. found at latitudes 30° north and 30° south
____ boreal forest	f. conifers dominate
____ tropical rain forest	g. dry shrubland

Additional questions are available on **Biology Now™**

Critical Thinking

1. As described in Section 14.3, the dry oak woodlands that dominate much of central California are threatened by *sudden oak death*, a fatal disease caused by a pathogenic oomycote. Trees already weakened by prolonged drought are especially vulnerable. If the oak woodlands disappear, what type of vegetation cover would you expect to replace them? Why?

2. Conifers have evergreen leaves and they maintain a higher rate of photosynthesis than broadleaf trees when temperatures are low. They are also less likely than broadleafs to form flow-interrupting bubbles in their vascular tissue as sap thaws. How do these traits help explain the distribution of boreal forests?

3. On April 26, 1986, in Ukraine, a meltdown occurred at the Chernobyl nuclear power plant. Nuclear fuel burned for nearly ten days and released up to 250 million curies of radioactive matter into the atmosphere. Inhaling as little as ten-millionths of a curie of plutonium can cause cancer. Thirty-one people died at once; others died from radiation poisoning in the following weeks. Winds carried radioactive fallout around the globe (Figure 31.29).

The Chernobyl accident stiffened opposition to nuclear power in the United States, but recent developments have some people reconsidering. Increasing the use of nuclear energy would diminish the country's dependence on oil from the politically unstable Middle East. Nuclear power does not contribute to global warming, acid rain, or smog. It does produce highly radioactive wastes. Investigate the pros and cons of nuclear power, and decide if you think the environmental benefits outweigh the risks. Would you feel differently if a nuclear power plant was going to be built ten kilometers upwind from your home?

4. Use of off-road recreational vehicles may double over the next twenty years. Many off-road enthusiasts would like increased access to government-owned desert areas. They argue that it is just the perfect place for off-roaders because "There's nothing there." How would you reply to this argument?

5. Write a short essay on the effect global warming might have on the spring overturn and thermocline formation in a Minnesota lake.

6. *Wetlands,* remember, are transitional zones between aquatic and terrestrial ecosystems in which plants are adapted to grow in periodically or permanently saturated soil. They include mangrove swamps and riparian zones. Wetlands everywhere are being converted for agriculture, home building, and other human activities. Does the great ecological value of wetlands for the nation as a whole outweigh the rights of the private citizens who own many of the remaining wetlands in the United States? Should private owners be forced to transfer title of their land to the government? If so, who should decide the value of a parcel of land *and* pay for it? Or would such seizure of private property violate laws of the United States?

32 BEHAVIORAL ECOLOGY

IMPACTS, ISSUES

My Pheromones Made Me Do It

A few years ago, as Toha Bererub (below) walked down a street near her Las Vegas home, she felt a sharp pain above her right eye. Then another. And another. A few seconds later, stinging bees encased her upper body. Firefighters wearing protective gear rescued her, but only after she had been stung more than 500 times.

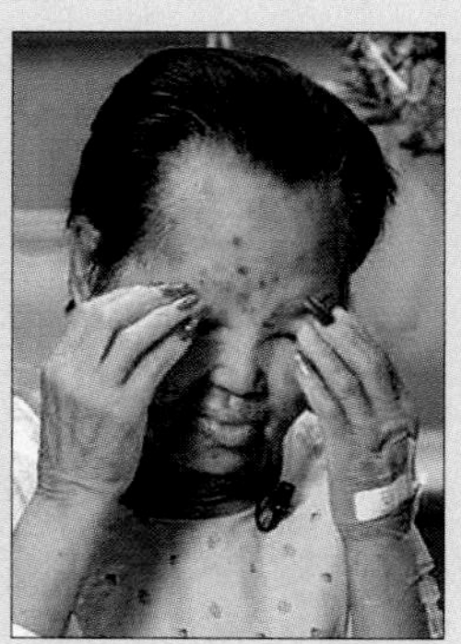

Bererub's attackers were Africanized honeybees, a hybrid between the gentle European honeybee and an aggressive African strain. African bees were imported to Brazil in the 1950s for cross-breeding experiments. Breeders were after a gentle but zippier pollinator for commercial orchards. But some of the bees from Africa escaped and started mating with the European bees that had been imported earlier. Then, in a grand example of geographic dispersal, some descendants buzzed north all the way from Brazil, into Mexico, and then into the United States.

All honeybees live in big colonies and defend them by stinging. Each bee can sting only once. Compared with their milder-mannered relatives, the Africanized bees are more easily alarmed and more persistent in their defensive behavior. Some have chased a target for a quarter of a mile. Although dubbed "killer bees" by the media, attacks are painful but generally not fatal. A healthy adult usually survives even a huge number of stings. Bererub was seventy years old and recovered fully after spending a week in the hospital.

What makes Africanized bees more aggressive? Their reaction to isopentyl acetate, for one thing. This banana-scented chemical functions as a component of the honeybee alarm pheromone.

A **pheromone** is a chemical communication signal that conveys information among members of the same species. A honeybee releases alarm pheromone when she senses a threat or as she is stinging something. Other bees follow the gradient of pheromone molecules diffusing through the air to guide them to the target.

Researchers once tested colonies of Africanized and European honeybees to compare their responses to an alarm pheromone. They positioned a tiny target in front of each colony, then released an artificial pheromone. Africanized bees flew out and attacked the target faster than European bees did. They plunged six to eight times as many stingers into it.

These differences among honeybees lead us into the world of animal behavior—coordinated responses that animal species make to stimuli. We will start with the genetic basis of behavior, then look at examples of instinctive and learned mechanisms. Along the way, we also will consider the adaptive value of behavior.

How Would You Vote? *Africanized honeybees are slowly expanding their range in North America. Should we fund more studies of their genes and behavior? See BiologyNow for details, then vote online.*

Key Concepts

THE BASIS OF BEHAVIOR
Behavior arises through interactions among gene products and the environment. Most behavior has innate components but can be modified by environmental factors.

NATURAL SELECTION AND BEHAVIOR
Like other traits, behavioral traits can evolve by natural selection. Behavior that enhances the reproductive success of an individual will be favored. For example, communication systems and mating systems have been shaped by natural selection.

GROUP LIVING
Life in social groups has reproductive benefits and costs. Self-sacrificing behavior has evolved among a few animals that live in large family groups.

Links to Earlier Concepts

An animal's sensory system (Section 25.8) influences its behavior. So do the hormones made in the brain (26.2) and neurotransmitters (25.2). Even alternative gene splicing (10.6) can affect behavior.

Much behavior is shaped by directional selection (12.5) or sexual selection (12.6). We will revisit the concept of adaptation (12.2) and consider how predation (29.4) and parasitism (29.5) influence social behavior. You may wish to review information about the primate lineage (16.13).

32.1 So Where Does Behavior Start?

LINKS TO SECTIONS 10.6, 25.8

As you learned in earlier chapters, nervous and endocrine systems govern an animal's responses to stimuli. Because genes specify the substances required for constructing those systems, they provide the heritable foundation for animal behavior.

GENES AND BEHAVIOR

An animal's nervous system is set up in a way that allows it to detect, interpret, and issue commands for responses to stimuli—specific bits of information about the world that sensory receptors have received. One way or another, diverse gene products affect each step in the formation and operation of the animal's brain, nerve pathways, and sensory receptors. Do some gene products influence *behavioral* responses to stimuli?

Stevan Arnold showed that feeding preferences of different populations of California garter snakes have a genetic basis. The garter snakes that live in the damp forests near the coast preferentially feed on large land-dwelling slugs known as banana slugs (Figure 32.1*a*,*b*). Snakes that live inland prefer tadpoles and tiny fishes. Offer them a banana slug and they ignore it.

In one set of experiments, Arnold offered newborn captive garter snakes a bit of slug as their first meal. Coastal snake offspring usually ate it and flicked their tongue at cotton swabs that smelled like slug (Figure 32.1*c*). (Tongue-flicking helps snakes to sense smells.) Inland snake offspring ignored slug-soaked swabs and rarely ate slug meat. Since both groups of snakes had no prior experience with slugs, Arnold concluded that garter snakes are genetically programmed to accept or reject slugs. He hypothesized that coastal and inland populations might have allele differences that affect development of their odor-detecting mechanisms.

To test this hypothesis, Arnold crossbred coastal and inland snakes. He predicted that hybrid offspring of these snakes would make an *intermediate* response to slug chunks and odors. The test results matched his prediction. Compared with a typical newborn inland snake, most baby snakes of mixed parentage tongue-flicked more often at slug-scented cotton swabs—but less often than newborn coastal snakes did.

Researchers are sometimes even able to pinpoint exactly which gene affects a behavior. For example, in fruit flies (*Drosophila melanogaster*) the gene known as *fruitless* (*fru*) controls male courtship behavior. Both males and females have the gene and both express it, but differences in how transcripts are spliced (Section 10.6) make their final protein products differ.

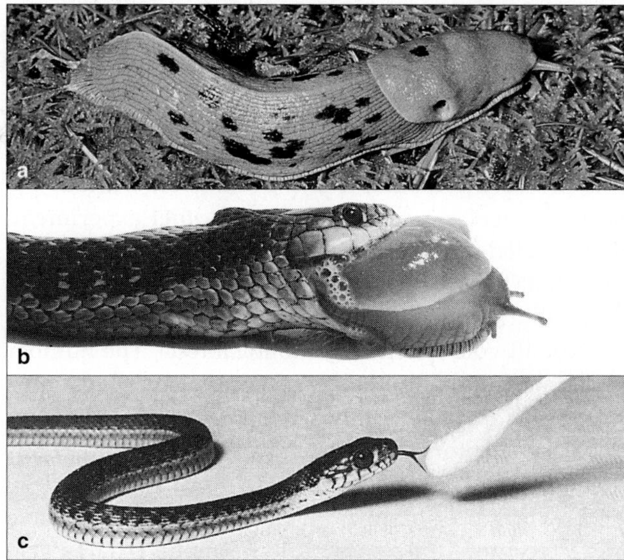

Figure 32.1 (**a**) Banana slug, food for (**b**) an adult garter snake of coastal California. (**c**) Newborn garter snake from a coastal population, tongue-flicking at a cotton swab drenched with tissue fluids from a banana slug.

As you will learn later in this chapter, specialized courtship displays precede mating in many animals, and fruit flies are no exception. A female fruit fly will mate only after her suitor performs a display in which he waves his wings and licks and taps her body. Male flies with a mutated version of the *fru* gene perform poorly at this task, so females remain unresponsive.

Perhaps even more interesting is what happened when researchers manipulated flies so that each ended up making the fruitless protein typical of the opposite sex. Males that expressed the female protein were less inclined to court females; but lavished attention on the normal males. Females with the male protein courted other females, displaying stereotypical male behaviors such as wing waving. The genetically altered females had little interest in ordinary males, but they did court males who had been made to smell like females.

The researchers concluded that the fruitless protein acts as a master switch. Which form of the protein is present affects the nervous system in ways that have far-reaching effects on complex behavior.

HORMONES AND BEHAVIOR

Gene products called hormones also guide behavior. For instance, all mammals make and secrete oxytocin, a hormone with roles in the birth process and lactation (Section 27.8). In many species, oxytocin also affects social behavior, including pair bonding.

In the prairie vole (*Microtus ochrogaster*), oxytocin is the key that unlocks a female's heart. The female of

LINKS TO SECTIONS 24.6, 26.2

these small, fluffy rodents bonds with a male after a night of repeated matings, and she mates for life. To test the effects of oxytocin, researchers placed a female vole with a male for a few hours but prevented mating. Then they injected oxytocin into the female. The result was love at first sight, a pair bond even without the normally necessary sex act. In yet another experiment, pair-bonded female prairie voles were injected with a drug that blocks oxytocin's effects. They immediately dumped their partners.

Not all vole species are monogamous. The number and distribution of oxytocin receptors may play a role. The brains of monogomous prairie voles are richer in oxytocin receptors than the brains of more promiscuous montane voles (*M. montanus*) (Figure 32.2).

The brains of monogamous vole species also have more receptors for antidiuretic hormone (ADH), also known as vasopressin. You learned how ADH affects the kidney in Section 24.6. In the brain, its effects are similar to those of oxytocin; it facilitates bonding.

Researchers isolated the gene for an ADH receptor in monogamous prairie voles and transferred copies of it into forebrain cells of male meadow voles (*M. pennsylvanicus*). After the transfer, males of this usually promiscuous species showed an increased tendency to partner with just one female.

Male meadow voles used as a control group also got copies of the gene, but in a brain region not known to be involved in pair-bonding. Unlike the experimental group, these males retained their playboy ways.

Even among prairie voles, there is variation in how much time a male devotes to his mate and offspring. Breeding studies have shown that a male's tendency to remain around his family is inherited. Compared to less devoted males, a doting dad vole has more ADH receptors in his brain.

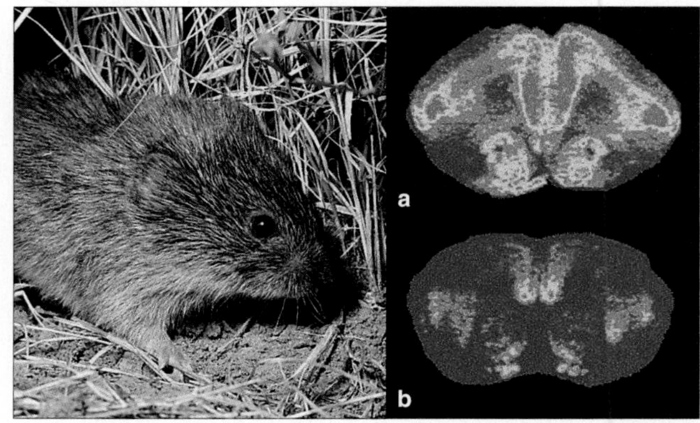

Figure 32.2 Hormones and prairie voles. Distribution of oxytocin receptors (*red*) in the brain of (**a**) a mate-for-life prairie vole and (**b**) a promiscuous montane vole.

REGARDING INSTINCT AND LEARNING

Slug-munching snakes, courting fruit flies, and pair-bonding voles all demonstrate **instinctive behavior**. This term means the behavior is performed without having first been learned by actual experience in the environment. The nervous system of a newly born or hatched animal is prewired to recognize *sign stimuli* —one or two simple, well-defined environmental cues that trigger a suitable response. The snake's response is a stereotyped motor program. When a garter snake recognizes certain sign stimuli, a **fixed action pattern**

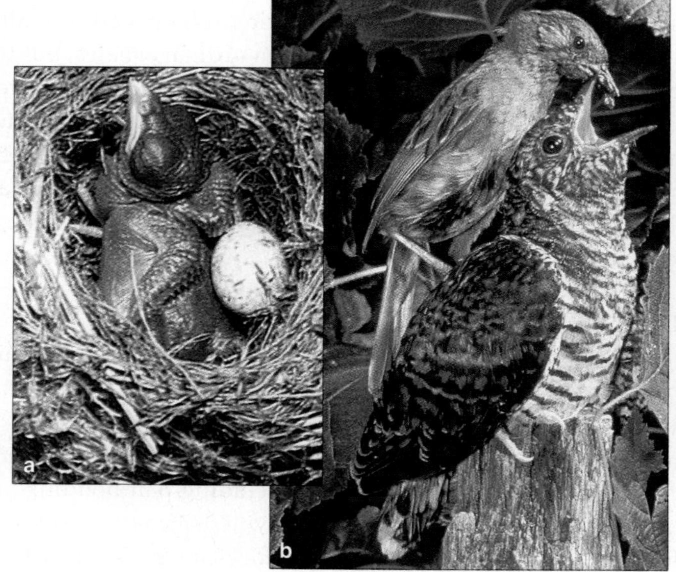

Figure 32.3 A few instinctive behaviors. (**a**) A social parasite, the European cuckoo, lays eggs in the nests of other birds. Even before this cuckoo hatchling opens its eyes, it responds instinctually to a round shape—a host's egg—and shoves it out of the nest. (**b**) The clueless foster parents show instinctual behavior as they respond to the usurper's gaping mouth, not to its size and other traits that characterize their own species. (**c**) A human baby instinctively imitates adult facial expressions.

follows. It is a program of coordinated muscle activity that will run to completion independently of feedback from the environment.

Cuckoos provide another example of instinctive behavior. Like the cowbirds described in Section 29.5, cuckoos are social parasites. The females deposit their eggs in the nests of other bird species. Cuckoos are blind when they first hatch, but contact with an egg or another round object stimulates a fixed action pattern. The cuckoo hatchling maneuvers the egg onto its back, then pushes the egg from the nest (Figure 32.3*a*). This ensures that the cuckoo receives undivided attention.

The cuckoo's "foster parents" are oblivious to the odd color and size of the usurper. They respond only to one sign stimulus—the gaping mouth of a chick—and continue to nurture the invader (Figure 32.3*b*).

Human infants display instinctive behavior. Three days after birth, a human infant already has a capacity to mimic the facial expressions of an adult who comes close (Figure 32.3*c*). This may seem easy, but the infant cannot watch its own face, nor can it feel which facial muscles the adult is using. Somehow it is able to open its mouth, protrude its tongue, or rotate its head the same way as the adult. Infants will also respond to a simplified stimulus—a flat, face-sized mask with two dark spots for eyes. One "eye" will not do the trick.

Animals also show **learned behavior**. They process information about experiences and use it to change or vary responses to stimuli. For instance, a young toad's nervous system commands it to flip its sticky tongue instinctively at any dark objects in its field of vision, which usually are edible insects. But what if the dark object is a wasp that can sting the toad's tongue? The toad will quickly learn to avoid all "black-and-yellow-banded objects that sting." For the warning coloration and mimicry described in Section 29.4 to evolve by natural selection, some predators must be able to learn to associate an unpleasant or painful experience with a particular prey coloration or pattern.

Imprinting is a classic example of learned behavior. This time-dependent form of learning is triggered by exposure to particular sign stimuli. Exposure usually takes place during a sensitive period when the animal is young. Imprinting of baby geese on their mother is a well-documented example (Figure 32.4).

Most behavior is shaped when genes interact with the "environment." Each animal's sensory experience, nutritional state, and many other variable factors that can directly or indirectly change gene activity are part of the environment.

For example, many male birds have an instinct to sing during the nesting season. Yet in different areas, songbirds may pick up variations, or dialects, of their species song. Peter Marler demonstrated that many male birds learn their full species song ten to fifty days after hatching by listening to other birds sing it. The male nervous system is prewired to recognize that song; his learning mechanism is primed to select and respond to acoustical input. What he hears during the sensitive period influences his rendition of the song.

In one study, Marler raised white-crown sparrow nestlings in soundproof chambers so they could not hear adult males. The captives sang when mature, but their songs were simpler than normal. Marler exposed some other isolated white-crown sparrow nestlings to recordings of adult males of their own species and

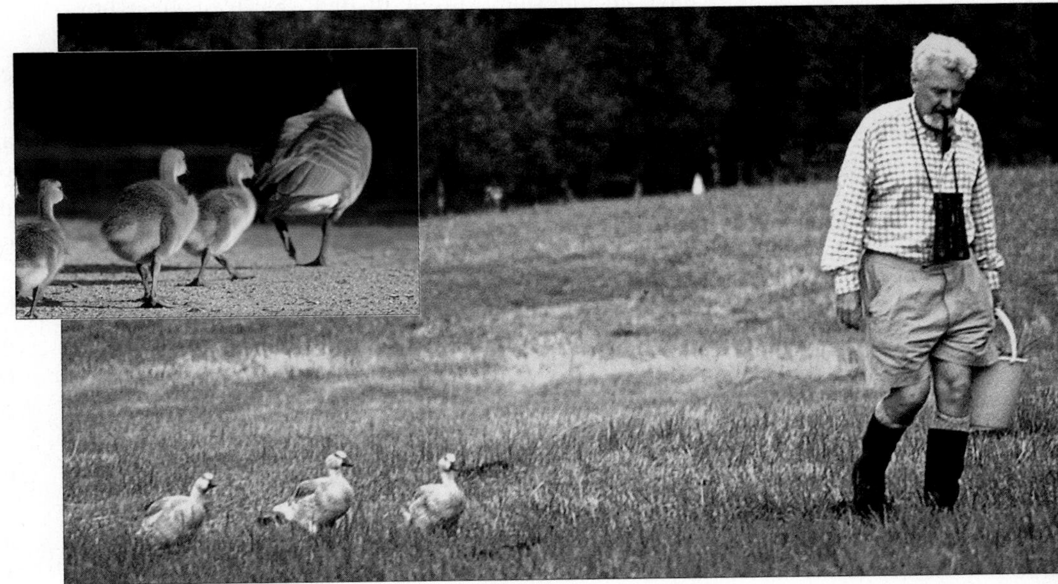

Figure 32.4 No one can tell these imprinted baby geese that Konrad Lorenz is not Mother Goose!

Imprinting is a time-dependent form of learning. It involves exposure to sign stimuli, most often early in development. In response to a moving object and probably to certain sounds, baby geese imprint on the mother and follow her during a short, sensitive period after hatching. They are neurally wired to learn crucial information —the identity of the individual that will protect them in the months ahead. Usually that individual is the mother or father. Imprinting occurs among many animals.

Konrad Lorenz, one of the early investigators of animal behavior, must have presented the baby geese in this figure with sign stimuli that made them form an attachment to him. When imprinted later in life, birds also direct sexual attention to members of the species they fixated upon when young.

LINKS TO SECTIONS 12.2, 12.5

Figure 32.5 Male marsh wren belting out a territorial song. Male birds start to imitate the species song when they are about fifteen days old, and they continue to learn songs throughout their life. The western marsh wren males learn as many as 200! Eastern marsh wrens have a more limited repertoire. The difference appears to have a genetic basis. Western birds raised in the east still mastered a far greater number of songs than native easterners.

another. At maturity, these captives sang the white-crown song only. Apparently, they had an instinctual ability to pay attention to this one song, and ignore others. These birds even picked up the dialect of their unseen same-species tutor. Marler carried out many such experiments; his results support the hypothesis that birdsong starts with *a genetically based capacity to learn* from specific acoustical cues. Many birds seem to have an inherited outline of their species song. They recognize this outline in an overheard song and then use that song as a guide to fill in their song's details.

This is not the whole story. In another experiment, Marler let young, hand-reared white-crowns interact with a "social tutor" of a different species, as opposed to listening to taped songs. The males tended to learn their tutor's song. In this case, birds were influenced by their *social experience* as well as acoustical cues.

Also, within a species there may be genetic variation in the capacity to learn different songs (Figure 32.5).

THE ADAPTIVE VALUE OF BEHAVIOR

Whenever forms of behavior vary and the variation has a genetic basis, they can be subject to evolution by natural selection. **Natural selection**, remember, is an outcome of differences in reproductive success among individuals of a species that show variation in their heritable traits. Over time, we would expect the alleles underlying the most adaptive behaviors to increase in frequency in the population, while alleles that cause less adaptive behaviors would become less frequent.

Natural selection theory can be used to develop and test possible explanations of why a behavior has endured. It helps explain how some actions bestow reproductive benefits that offset the reproductive costs (disadvantages) associated with them. If a behavior is adaptive, it must promote the *individual's* production of offspring. Here are five definitions to keep in mind:

1. *Reproductive success.* The individual produces some number of surviving offspring.
2. *Adaptive behavior.* Behavior that helps an individual promote propagation of its own genes; its frequency is maintained or increases in successive generations.
3. *Social behavior.* Interdependent interactions among individuals of the species.
4. *Selfish behavior.* Any form of behavior that improves an individual's chance to produce or protect its own offspring regardless of the impact on the population as a whole.
5. *Altruism.* Also called self-sacrificing behavior. Some individual behaves in a way that helps others in the population but lowers its chance to produce offspring.

When biologists speak of selfish or altruistic behavior, they do not mean an individual is conscious of what it is doing or of the behavior's reproductive goal. A lion does not need to know that capturing zebras will be good for its reproductive success. Its nervous system just shouts HUNTING BEHAVIOR! when the lion is hungry and sees a zebra. Hunting behavior persists in the lion population because the genes that make this behavior possible are passed along when lions reproduce.

To assess the adaptive value of behavior, biologists consider the ways that it might promote reproductive success. For example, many cavity-nesting birds, such as starlings (*Sturnus vulgaris*), decorate their nest with fresh sprigs of greenery. The birds particularly favor the leaves of strongly scented plants, such as the wild carrot (*Daucus carota*) (Figure 32.6).

Why do the birds do this? Larry Clark and Russell Mason suspected that nest-decorating behavior helps minimize the number of parasitic mites in their nests. Mites suck blood from nestlings and large numbers could affect growth and survival of the offspring.

The biologists decided to test their "nest protection hypothesis" by replacing the natural nests of starlings with experimental ones. Half of the replacement nests included wild carrot leaves and the other half had no aromatic plant matter.

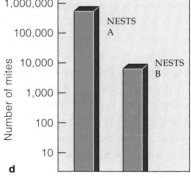

Figure 32.6 Experimental results that point to the adaptive value of starling nest-decorating behavior. (**a**) Wild carrot (*Daucus carota*) and (**b**) a nest-building starling (*Sturnus vulgaris*). Nests designated *A* were kept free of fresh sprigs of wild carrot and other plants that make aromatic compounds. The compounds prevent baby mites (*Ornithonyssus sylviarum*) from developing into (**c**) adult mites. Other nests, designated *B*, had fresh sprigs tucked in every seven days. (**d**) Counts of the number of mites infesting the nests were made at the time the starling chicks left the nest, twenty-one days after this experiment was under way.

After the starling chicks had left these nests, the researchers recorded the number of mites left behind (Figure 32.6*d*). The number of mites in sprig-free nests was far greater than the number in sprig-decorated ones. Why? Wild carrot shoots have highly aromatic compounds that repel herbivores and thus help these plants survive. By coincidence, one of the compounds also prevents mites from maturing sexually.

Mason and Clark concluded that decorating a nest with sprigs deters blood-sucking mites. They inferred that this nest-decorating behavior is adaptive because it promotes nestling survival, and thus increases the reproductive success of nest-decorating birds.

As is often the case in science, additional research suggests that the story may be a bit more complicated. Other researchers have not found the same deterrent effect of greenery on mite numbers. They suggest that nest decorating may be adaptive because the greenery deters other nest parasites, or because it enhances the immune function of the nestlings. Ongoing studies are testing these competing hypotheses.

Do not get the idea that all behavior is adaptive. A behavior may be adaptive under some circumstances but not others. Just as some morphological traits are a legacy of past selective pressures, certain behavioral characteristics persist even after the environment in which they evolved has been altered. Consider moths that exhaust themselves by fluttering against brightly lit window panes at night. Is this adaptive? No, but for millions of years, flying toward a bright, highly visible star allowed moths to navigate a straight line. Today, with the widespread use of electric lights at night, this behavior often turns out to be maladaptive. It draws the night-flying moths to windows, or even worse, to electric bug zappers.

Genes underlie animal behavior—coordinated responses to stimuli. Certain gene products are essential in constructing and operating the nervous system, which governs behavior. Genes that encode hormones or their receptors also affect many forms of behavior.

In instinctive behavior, animals make complex, stereotyped responses to specific, often simple environmental cues. In learned behavior, responses may vary or change as a result of individual experiences in the environment.

Whether instinctive or learned, behavior develops through interactions between genes and environmental inputs.

Like other genetically determined traits, behavioral traits are subject to natural selection. Adaptive behaviors increase an animal's reproductive success under current conditions.

LINK TO SECTION 26.1

32.2 Communication Signals

Competing for food, defending territory, alerting others to danger, advertising sexual readiness, forming bonds with a mate, caring for the offspring—such interactions among the members of a species require communication between two or more individuals.

THE NATURE OF COMMUNICATION SIGNALS

Intraspecific interactions involve mixes of instinctive and learned behaviors by which individuals send and respond to information-laden cues in stimuli. These **communication signals** have unambiguous meaning for a species. Specific odors, sounds, colors, patterns, postures, and movements are examples.

The communication signals sent by an individual, the signaler, are meant to change behaviors of others of the same species. A signal evolves or persists when it helps the reproductive success of both the sender *and* receiver. If a signal is usually disadvantageous to either party, then natural selection favors individuals that do not send the signal or do not respond to it.

Pheromones are chemicals released by one member of a species that influence another member of the same species. *Signaling* pheromones, such as honeybee alarm pheromone or a sex attractant, induce an immediate response. *Priming* pheromones act over a longer time frame, causing a change in the receipient's physiology. For instance, exposure to a pheromone in the urine of male mice triggers physiological changes that lead to estrus in female mice.

Acoustical signals abound in nature. Male birds, frogs, grasshoppers, whales, and other animals make sounds that attract females. Prairie dogs bark alarms. Wolves howl and kangaroo rats drum their feet on the ground when advertising possession of territory.

Some signals never vary. Zebra ears pressed flat to the head convey hostility; ears pointing up convey its absence. Other signals convey a message's intensity. A zebra that has ears laid back is not too riled up when its mouth is open a bit. When its mouth gapes, watch out. That combination is one type of *composite* signal, which has information encoded in two cues (or more).

Signals often take on different meaning in different contexts. A lion emits a spine-tingling roar to keep in touch with its pride *or* to threaten rivals. Also, a signal can convey information about signals to follow. Dogs and wolves solicit play behavior with their play bow, as in Figure 32.7*a*. Without the bow, a signal receiver may construe the behaviors that follow as aggressive, sexual, or even exploratory—but not playful.

EXAMPLES OF COMMUNICATION DISPLAYS

The play bow is a communication display, a pattern of behavior that is a social signal. The threat display is another common pattern. It unambiguously announces that a signaler is prepared to attack a signal receiver. When a rival for a receptive female confronts him, a dominant male baboon will roll his eyes upward and "yawn," which exposes his canines (Figure 32.7*b*). The signaler often benefits when the rival backs down, for he retains access to the female without having had to fight. The signal receiver benefits because he avoids a serious beating, infected wounds, and possibly death.

Figure 32.7 Communication displays. (**a**) A young male wolf using a play bow to solicit a romp. (**b**) Exposed canines are part of a male baboon's threat display. (**c**) Courtship display of king penguins.

Communication displays can evolve when behavior that occurs in one context is performed in another and becomes *ritualized*. This means that over generations, aspects of the behavior may become exaggerated and stereotyped in ways that make it easier for a receiver to recognize and interpret. Feathers or other body parts may become colored or enlarged in ways that enhance the display's effectiveness. Ritualized behavior is very commonly a part of courtship displays (Figure 32.7*c*).

In tactile displays, a signaler touches the receiver in ritualized ways. After locating a source of pollen or nectar, a foraging honeybee returns to its colony, a hive, and performs a complex dance. It moves about in a circle, jostling in the dark with a crowd of workers. Other bees may follow and maintain physical contact with the dancer. Honeybees that do so get information about the general location, distance, and direction of pollen or nectar (Figure 32.8).

ILLEGITIMATE SIGNALERS AND RECEIVERS

Sometimes the wrong parties intercept communication signals. Termites will respond with defensive behavior if they catch a whiff of the scent from an invading ant. That scent is meant to identify the ant as a member of an ant colony, and it elicits cooperative behavior from other ants. The scent does have evolved functions, but announcing the ant's presence to a termite colony is not one of them. A termite might detect the scent and kill the ant. In that case, the termite is an *illegitimate receiver* of a signal that is meant for other ants.

Male tungara frogs make two kinds of calls; one simple, the other complex. The calls say "come hither" to female frogs, but they also mean "dinner here" to fringe-lipped bats. Complex songs are more attractive to females, but make it easier for the bats to locate the caller. When bats are in the area, the male frogs call less and are more likely to make the simpler call.

There are also *illegitimate signalers*. Certain assassin bugs hook dead termite prey on their back to acquire the victim's odor. By signaling that they "belong" to a termite colony, they can more easily hunt termite prey. As another example, when the female of a predatory firefly species sees the flash from a male of a different species, she flashes back. If she lures him into attack range, she eats him—an evolutionary cost of having an otherwise useful response to a come-hither signal.

A communication signal between individuals of the same species is an action having a net beneficial effect on both the signaler and receiver. Natural selection tends to favor communication signals that promote reproductive success.

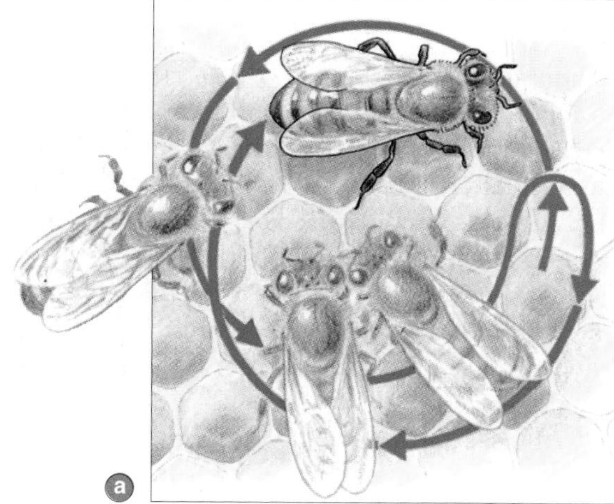

a

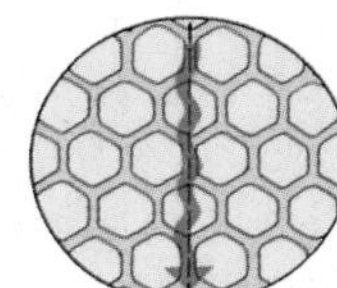

When bee moves straight down comb, recruits fly to source directly away from the sun.

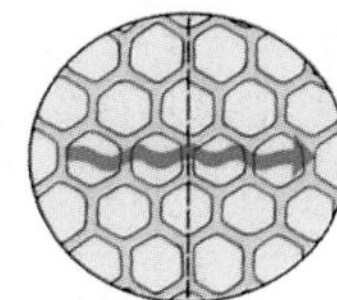

When bee moves to right of vertical, recruits fly at 90° angle to right of the sun.

b

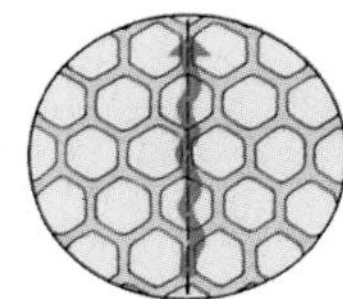

When bee moves straight up comb, recruits fly straight toward the sun.

c

Figure 32.8 ***Animated!*** Dances of the honeybees, as examples of tactile displays. (**a**) Honeybees that visit food sources close to their hive perform a *round* dance on the honeycomb. Worker bees that maintain contact with the forager through the dance go out and search for food near the hive.

(**b**) Bees trained to visit feeding stations over 100 meters from the hive do a *waggle* dance. During the dance, a bee makes a straight run and waggles its abdomen. The waggle dancers also vary the dance speed to convey more information about the distance of a food source. For example, when a site is 150 meters away, the dance is executed much faster, with more waggles per straight run, compared with a dance concerning a food source that is 500 meters away.

(**c**) As Karl von Frisch discovered, a *straight* run's orientation varies, depending on the direction in which food is located. When he put a sugar solution on a direct line between a hive and the sun, foragers that located it returned to the hive and oriented their straight runs right up the honeycomb. When he put a sugar solution at right angles to a line between the hive and the sun, foragers made straight runs 90 degrees to vertical. Thus, a honeybee "recruited" into foraging orients its flight *with respect to the sun and the hive.* Using this code, bees waste less time and energy during their food-gathering expedition.

LINK TO SECTION 12.6

32.3 Mates, Offspring, and Reproductive Success

Sexual reproduction requires a partner. An individual must attract a member of the opposite sex with which to mate. In all animals except birds, any parental care is most often provided by the female alone.

SEXUAL SELECTION AND MATING BEHAVIOR

Competition among members of one sex for access to mates is common. So is choosiness in selecting a mate. Such activities, recall, are forms of sexual selection. This microevolutionary process favors traits that give the individual a competitive edge in attracting and holding on to mates (Section 12.6).

But *whose* reproductive success is it—the male's or female's? Males, recall, produce large numbers of tiny sperm, and females produce far larger but fewer eggs. For the male, success generally depends on how many eggs he fertilizes. For the female, it depends more on how many of her eggs survive and reach adulthood. In most cases, *the key factor influencing a female's sexual preference is the quality of the mate—not the quantity of partners.* Males often employ tactics that might help them fertilize as many eggs as possible.

Female hangingflies (*Harpobittacus apicalis*) choose males that offer superior food. A male kills a moth or some other insect, then releases a sex pheromone to attract females to his "nuptial gift" (Figure 32.9*a*). A female selects a male with a large calorie-rich offering. Only *after* she has been eating for five minutes or so does she start to accept sperm from her partner, which she stores in a special internal compartment. She lets the male continue inseminating her, but only as long as the gift holds out. Before twenty minutes are up, she can break off the mating at any point. When she does, she may well mate with another male and accept his sperm. Doing so dilutes the reproductive success of her first partner.

Some females shop around for good-looking males. For example, fiddler crabs abound along many muddy shores (Figure 32.9*b*,*c*). One of the male's two claws is so enlarged that it may make up more than 50 percent of his body weight. When spring tides are favorable, the males build elaborate mating burrows next to each other and stand beside them, waving their oversized claws. Females stroll by, checking out males and their burrows. If a female likes what she sees, she enters a male's burrow and engages in sex.

Many birds show a similar pattern of mating. Male sage grouse (*Centrocercus urophasianus*) congregate in a

Figure 32.9 (**a**) Male hangingfly dangling a moth as a nuptial gift for a future mate. Females of some hangingfly species choose sexual partners on the basis of the size of prey offered to them. (**b**) A male fiddler crab waves his oversized claw to attract the interest of a female (**c**). A male sage grouse (**d**) competes for female attention at a communal display ground.

type of communal display ground called a lek. Each male stakes out a few square meters. With tail feathers erect and large neck pouches puffed out, the males emit booming calls and stamp about like wind-up toys (Figure 32.9*d*). Females tend to select and mate with one male only, then they go off to nest by themselves in the sagebrush. Many females often choose the same stand-out male, so most of the males never mate.

The females of some species cluster in defendable groups when they are sexually receptive. Where you come across such a group, you are likely to observe males competing for access to clusters. Competition for ready-made harems favors highly combative male lions, sheep, elk, elephant seals, and bison, to name a few examples (Figure 32.10).

Figure 32.10 (**a**) Sexual competition between male elephant seals, which fight for access to a cluster of females. (**b**) Male bison locked in combat during the breeding season.

PARENTAL CARE

If females are fighting over males, then males probably provide some service beyond sperm delivery, such as parental care. A midwife toad is a fine example. A male winds strings of fertilized eggs around his legs, where they stay until the eggs hatch (Figure 32.11*a*). With her eggs in the care of the male, a female is free to mate with other males—if she can find some unencumbered by eggs. As the season progresses, such males become rare, and the females fight for access to them. Females even attempt to pry mating pairs apart.

Parental behavior has reproductive costs. It drains time and energy that parents might otherwise spend on improving their own likelihood of living to reproduce another time. Yet, for many species, parental behavior improves the likelihood that offspring will live. The benefit of devoting time and energy to their offspring (immediate reproductive success) might outweigh the cost of reduced reproductive success at a later time.

Once young amphibians and reptiles have hatched, parental care is uncommon. Like birds, crocodilians are an exception. Parents build a nest where eggs are laid. The young start calling when they are ready to hatch. Parents dig them up and care for them for some time.

Many birds are monogamous and both parents often care for the young (Figure 32.11*b*). In mammals, males typically leave after mating. Females raise the young alone, and the males attempt to mate again or conserve energy for the next breeding season (Figure 32.11*c*). Mammalian species in which males do assist in care of the young tend to be monogamous.

Selection theory explains some aspects of mating behavior, for sexual selection favors behavioral traits that give the individual a competitive edge in reproductive success.

Figure 32.11 (**a**) Male midwife toad with developing eggs wrapped around his legs. (**b**) Male and female Caspian terns cooperate in the care of their chick. (**c**) A female grizzly will care for her cub for up to two years. The male takes no part in the cub's upbringing.

LINK TO SECTION 29.4

32.4 Costs and Benefits of Social Groups

Survey the animal kingdom and you'll find a range of social groups. Individuals of some species spend most of their lives alone or in small family groups. Others live in groups of thousands of related individuals. Still others live in social units with few close relatives, as humans do. Evolutionary biologists often study the costs and benefits of sociality in terms of individual reproductive success, measured by contributions to the next generation's gene pool.

Figure 32.12 Group defenses. Musk oxen adults (*Ovibos moshatus*) form a ring of horns, often around the young.

COOPERATIVE PREDATOR AVOIDANCE

By cooperatively responding to predators, some groups reduce the net risk to all. Vulnerable individuals, too, scan for predators, join in a counterattack, or engage in more effective defenses (Figures 32.12 and 32.13).

Vervet monkeys, meerkats, prairie dogs, and many other mammals give alarm calls, as in Figure 32.13*a*. A prairie dog barks different signals for eagle or coyote sightings. Others of its group then dive into burrows to escape an aerial attack or stand erect, the better to scan the horizon and pinpoint a coyote.

Ecologist Birgitta Sillén-Tullberg found evidence of group benefits for Australian sawfly caterpillars that live in clumps on branches. When something disturbs their clump, caterpillars collectively rear up from the branch and writhe about (Figure 32.13*b*). At the same time, they throw up partly digested eucalyptus leaves, which taste awful to most predators—including birds that prey on caterpillars.

As Sillén-Tullberg hypothesized, individual sawfly caterpillars benefit from their coordinated repulsion of predatory birds. She used her hypothesis to predict that birds are more likely to eat a solitary caterpillar.

To test her prediction, she gave young, hand-reared Great Tits (*Parus major*) a chance to eat the caterpillars. She offered caterpillars one at a time or in a group of twenty. When birds were offered caterpillars one at a time, they devoured an average of 5.6. Birds that were offered caterpillars as a clump only ate an average of 4.1. As predicted, each individual caterpillar was a bit safer when in the midst of a fluid-spewing group.

Figure 32.13 Sentries and slime. (**a**) Black-tailed prairie dogs bark an alarm call that warns others of predators. Does this put the caller at risk? Not much. Prairie dogs usually act as sentries only if they are done feeding and are standing beside their burrows. (**b**) Australian sawfly caterpillars form clumps and collectively regurgitate a fluid (yellow blobs) that is distasteful to most predators.

THE SELFISH HERD

Simply by their physical position in the group, some individuals form a living shield against predation on others. They belong to a **selfish herd**, a simple society formed (not consciously) by reproductive self-interest. The selfish-herd hypothesis has been tested for male bluegill sunfishes, which build adjacent nests on lake bottoms. Females deposit eggs where males have used their fins to scoop out depressions in mud.

If a colony of bluegill males is a selfish herd, then we can predict competition for the "safe" sites—at the center of a colony. Compared to eggs at the periphery, eggs in nests at the center are less likely to be eaten by snails and largemouth bass. Competition does indeed occur. The largest, most powerful males tend to claim centermost locations. Other, smaller males assemble around them and bear the brunt of predatory attacks. Even so, they are better off in the group than on their own, fending off a bass single-handedly, so to speak.

COOPERATIVE HUNTING

Many predatory mammals, including wolves, lions, and wild dogs, live in social groups and cooperate in hunts (Figure 32.14). Are group hunts more successful than solitary hunts? Surprisingly, the answer is often no. Researchers observed a solitary lion that captured prey about 15 percent of the time. Two lions hunting together captured prey twice as often, but they had to share it, so the number of successful hunts per lion balanced out. When more lions joined the hunt, the success rate per lion actually fell. Studies of wolves revealed a similar pattern. Among many cooperative hunters, group living is probably best explained by

Figure 32.14 Members of a wolf pack (*Canis lupus*) cooperate to hunt, care for young, and defend a territory. Benefits are not distributed equally. Only the highest-ranking individuals, the alpha male and alpha female, breed.

factors other than success in hunting. Individuals not only hunt together, they fend off scavengers, care for one another's young, and protect the group's territory.

DOMINANCE HIERARCHIES

In many social groups, resources are not distributed equally. Certain individuals are subordinate to others. Most wolf packs have one dominant male that breeds with a dominant female. The other pack members are nonbreeding brothers and sisters, aunts and uncles. They all hunt and bring food to individuals that stay in the den and guard the young.

Another example: Baboons live in large groups. A female spends her life within the group in which she was born, and she inherits her social standing from her mother. High-ranking females displace lower-ranked females from food, water, and grooming opportunities. As a result, offspring of higher-ranking mothers grow and mature faster than those of lower-ranking ones.

Why do subordinates bother staying in a group if they must give up resources and often do not breed? They might do so to avoid being hurt and to remain in the group. Maybe it is impossible for them to survive on their own. Maybe challenging a strong individual will invite injuries or death. Maybe a subordinate has a chance to reproduce if a predator or old age removes dominant peers. Subordinate wolves and baboons do move up the social ladder if the opportunity arises.

REGARDING THE COSTS

If social behavior can be advantageous, *then why are there relatively few social species?* In most environments, costs outweigh the benefits. For instance, when large numbers of individuals live together, they compete more. Cormorants and many other seabirds form huge breeding colonies (Figure 32.15). All must compete for a share of the same ecological pie.

Figure 32.15 The nearly uniform spacing among a densely packed cormorant colony. Imagine how easily pathogens and parasites can spread through such breeding colonies.

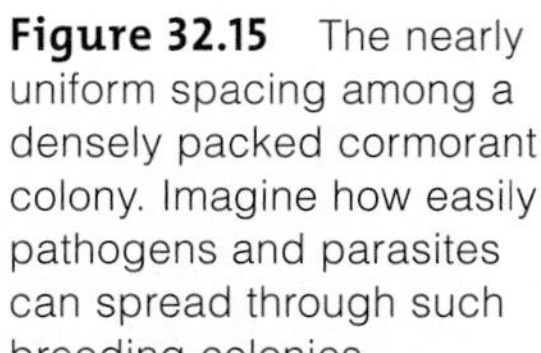

Large groups are attractive to predators. Crowded social groups also invite parasites and diseases that are readily transmitted from host to host. Still another cost is the risk of being killed or exploited by others in the group. Given the opportunity, breeding pairs of herring gulls cannibalize a neighbor's eggs and any chicks that wander away from their nest.

Living in a social group can provide benefits, as through cooperative defenses of shielding against predators.

Group living has costs, in terms of increased competition, increased vulnerability to infections, and exploitation by others of the group.

32.5 Why Sacrifice Yourself?

Extreme cases of sterility and self-sacrifice have evolved in only two insect groups and one group of mammals. How are genes of the nonreproducing individuals passed from one generation to the next?

SOCIAL INSECTS

Honeybees and fire ants are among the true social (eusocial) insects. Like termites, they stay together for generations in a group that has a reproductive division of labor. Permanently sterile individuals cooperatively care for the offspring of a few breeding individuals. Often these sterile workers are highly specialized in their body form and function (Figure 32.16).

Consider a honeybee colony, or hive. Its queen bee, the only fertile female, secretes a pheromone that other female bees distribute through the hive. This signaling molecule suppresses the development of ovaries in all other females, rendering them sterile. The queen bee is larger than the worker bees, partly because of her enlarged egg-producing ovaries (Figure 32.17*a*).

About 30,000 to 50,000 female workers feed larvae, clean and maintain the hive, and build honeycomb from waxy secretions. Adult workers live for about six weeks during the spring and summer. When foragers return to the hive after finding a rich source of nectar or pollen, they engage others in a dance. This tactile display recruits more foragers (Figure 32.8). Workers also cooperate by transferring food to one another. They guard the entrance to the hive and will sacrifice themselves to repel intruders.

Males, the stingless drones, develop only in spring and summer. They do not take part in the day-to-day work and subsist on the food gathered by their worker sisters. Drones live for sex. Each day, they fly out in search of a mate. If a drone is lucky, he will locate a virgin queen on her single flight away from her home colony. The sole function of her flight is to meet up with and mate with drones. A drone dies right after he inseminates a queen, which then starts a new colony. The queen stores and uses the drone's sperm for years, perpetuating his genes and those of the colony.

Like honeybees, termites live in huge family groups with a large queen specialized for producing eggs (Figure 32.17*b*). She mates repeatedly with a resident king. Their offspring are flightless sterile workers of both sexes and winged sexual forms that are produced and leave the colony seasonally.

SOCIAL MOLE-RATS

Vertebrates are not known for sterility and extreme self-sacrifice. The only eusocial mammals are African mole-rats. The best studied is *Heterocephalus glaber*, the naked mole-rat. Clans of this nearly hairless rodent occupy burrows in arid parts of East Africa.

One reproducing female dominates the clan and she mates with one to three males (Figure 32.17*c*). Other, nonbreeding members live to protect and care for the queen, her consorts, and their offspring. Sterile diggers excavate the underground tunnels and chambers that serve as living areas or waste-disposal sites. When a digger locates a tasty root or tuber, it hauls a bit of it back to the main chamber and emits a series of chirps. The chirps recruit others, which help carry the food back to the chamber. In this way, the queen, her retinue of males, and her offspring get fed.

Digger mole-rats also deliver food to other helpers that usually loaf about, shoulder to shoulder and belly to back, with the reproductive royals. These "loafers" actually spring to action when a snake or some other predator threatens the clan. Collectively, and at great risk, they chase away or attack and kill the predator.

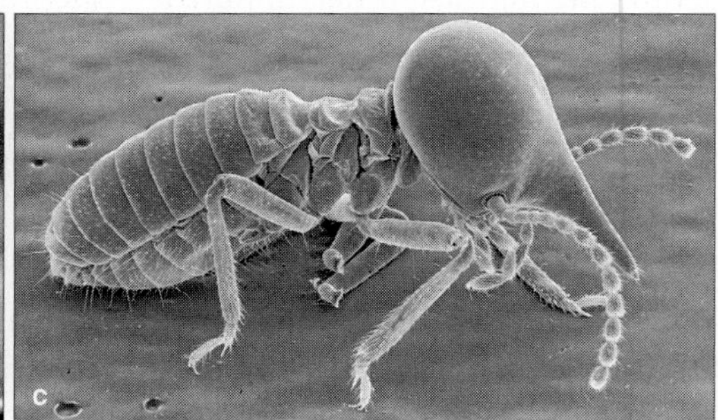

Figure 32.16 Serving and defending the colony with specializations in form and function. (**a**) An Australian honeypot ant worker. This sterile female is a living container for the colony's food reserves. (**b**) An army ant soldier (*Eciton burchelli*) displays its relatively enormous mandibles. (**c**) Eyeless soldier termites (*Nasutitermes*) bombard intruders with a stream of sticky goo from their nozzle-shaped head.

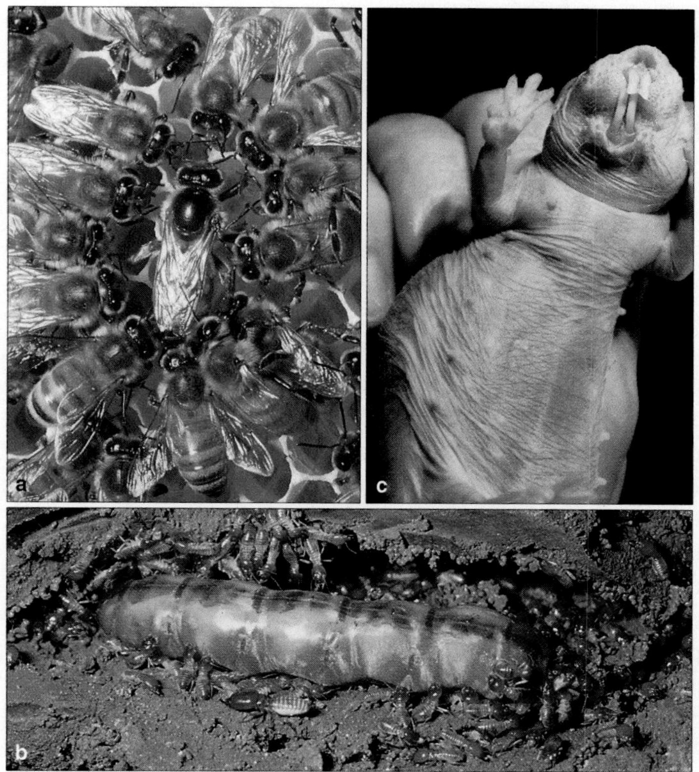

Figure 32.17 Three queens. (**a**) A queen honeybee with her court of sterile worker daughters. (**b**) A termite queen (*Macrotermes*) dwarfs her offspring and her mate. Her body pumps out thousands of eggs a day. (**c**) This queen naked mole-rat has twelve mammary glands, the better to feed her many offspring. In a laboratory colony at Cornell University, one female produced one litter of twenty-eight pups. She gave birth to more than 900 offspring during her lifetime.

Figure 32.18 Damaraland mole-rats in a burrow. Like their relatives, the naked mole-rats, they live in colonies having nonbreeding workers. But unlike the naked mole-rats, these fuzzy burrowers are not highly inbred.

INDIRECT SELECTION THEORY

None of the altruistic individuals of a honeybee hive, termite colony, or naked mole-rat clan directly passes genes to the next generation. So how are genes that underlie altruistic behavior perpetuated? According to William Hamilton's theory of **inclusive fitness**, genes associated with altruism can be favored by selection if they lead to behavior that will increase the number of offspring produced by an altruist's closest relatives.

A sexually reproducing, diploid parent caring for offspring is not helping exact genetic copies of itself. Each of its gametes, and each of its offspring, inherit one-half of its genes. Other individuals of the social group that have the same ancestors also share genes with their parents. Two siblings (brothers or sisters) are as genetically similar as a parent and its offspring. Nephews and nieces have inherited about one-fourth of their uncle's genes.

Sterile workers may be indirectly promoting genes for "self-sacrifice" through altruistic behavior that will benefit relatives. Members of a honeybee, termite, or ant colony are an extended family. The nonbreeding family members are working in support of siblings, a few of which are future kings and queens. Although a guard bee will die after driving her stinger into a bear, her death helps to ensure that relatives in the hive will live to pass along some of her genes.

Does close relatedness explain why the naked mole-rats are the only eusocial mammals? Through DNA fingerprinting studies, we know that all individuals in one naked mole-rat clan are *very* close relatives and differ genetically from individuals of other clans. Each clan is highly inbred through generations of brother–sister, mother–son, and father–daughter matings.

However, inbreeding may not be required for mole-rat eusociality. The organization of the Damaraland mole-rat (*Cryptomys damarensis*) is similar to that of *H. glaber* (Figure 32.18). Nonbreeding members of both sexes cooperatively assist one breeding pair. However, breeding pairs of wild Damaraland mole-rat colonies usually are unrelated.

Researchers are now searching for other factors that select for mole-rat eusocial behavior. According to one hypothesis, an arid habitat and patchy food supplies favor the mole-rat genes that underlie cooperation in digging burrows, searching for food, and fending off competitors and predators.

Altruistic behavior may persist when individuals pass on genes indirectly, by helping relatives survive and reproduce.

By the theory of inclusive fitness, genes associated with altruistic behavior that is directed toward relatives may spread through a population in certain situations.

LINK TO SECTION 16.13

32.6 A Look at Primate Social Behavior

Primates, especially chimpanzees and bonobos, are our closest living relatives. All live in groups, and the social environment plays an important role in determining an individual's reproductive success.

In the 1960s, a 26-year-old primatologist named Jane Goodall began what would turn out to be a life-long study of chimpanzees in Tanzania. One of her earliest discoveries was the chimpanzee's ability to make and use simple tools by stripping leaves from branches to create "fishing sticks." These long flexible sticks are inserted into a large termite mound as shown in Figure 32.19*a*. After agitated termites climb aboard the stick, it is carefully withdrawn and the chimpanzee enjoys a high-protein snack. Chimpanzees use thicker sticks to poke holes in a mound so they can insert the fishing sticks. Different chimpanzee groups exhibit somewhat different tool-shaping and termite-fishing methods. In each group, youngsters learn by imitating the adults.

Male chimpanzees spend their lives in the group in which they are born and form strong social bonds. The females are often unrelated and interact little with one another. A female's status is dictated mostly by how she gets along with the males. Before the rainy season, mature females that are entering a fertile cycle undergo hormone-driven physiological and behavioral changes. External genitalia become swollen and vivid pink as the females become sexually active. The swellings are strong visual signals to males. They are flags of sexual jamborees—of great gatherings of highly stimulated chimpanzees in which any males present may have a turn at copulating with the same female.

Male chimpanzees cooperatively hunt for monkeys, small pigs, and antelopes. They may also cooperate in attacks on neighboring groups. Males sometimes even kill infants. By one hypothesis, infanticidal behavior by males may exert selection pressure for promiscuity by females. A female who mates with many males may protect her offspring by obscuring their exact paternity. A male would be expected to avoid killing any infant that may carry his genes.

Comparative studies of the closely related bonobo reveal contrasting sexual and social behavior (Figure 32.19*b*). As in chimpanzees, adult males in a group are related and females are not. Yet bonobo females form strong bonds. Unlike chimpanzees, female bonobos are receptive to sex at any time, not only when fertile. Male bonobos interact less than male chimpanzees do. They do not hunt together, and infanticidal behavior has not been observed among bonobo males.

What explains the differences? Does a higher level of interaction help female bonobos fend off potentially infanticidal males? Does unrestricted access to sexually receptive females interfere with male–male bonding or diffuse male aggression? We do not know. Hormones that affect pair-bonding may play a role. Like prairie voles and mountain voles, chimpanzees and bonobos differ in a regulatory region near a gene that encodes one ADH receptor. In voles, a longer sequence in this region correlates with more family-oriented behavior. Interestingly, the bonobo sequence for this region is about 360 bases longer than the chimpanzee sequence. What about humans? Our sequence in this region is nearly identical to that of the bonobos. And with this in mind we turn briefly to human behavior.

Chimpanzees and bonobos are our closest living relatives. Both live in social groups but the exact details of social organization, degree of female cooperation, and extent of male aggression vary between the two species.

Figure 32.19 (**a**) A group of chimpanzees (*Pan troglodytes*) using sticks as tools to extract protein-rich termites from a nest. (**b**) A bonobo (*Pan paniscus*) mother with her offspring. Like humans, bonobos are bipedal; they often walk about on two legs. Also like humans, and unlike chimpanzees, bonobo females have sexual organs that allow them to copulate face to face with their partners and are sexually receptive at any time of year. They use sex as a means of strengthening social bonds.

32.7 Human Social Behavior

Evolutionary forces have shaped the behavior of all animals, including humans. Only humans, however, make moral choices about their behavior.

HUMAN PHEROMONES

Do humans secrete and sense pheromones? Possibly. When women live together, as in a college dormitory, their menstrual cycles typically become synchronized. Studies by Martha McClintock and Kathleen Stern showed that the length of a woman's menstrual cycle could be lengthened or shortened by exposing her to sweat secreted by a woman who was in a different part of the cycle.

Men and women secrete different chemicals and also respond differently to them. One study using PET scans found that a chemical typically released in male sweat activated certain brain areas in women but not in most men. In contrast, a chemical released in female urine activated most male brains more than it did female ones. Intriguingly, male homosexuals showed the same response to male sweat that the women did.

In most mammals, pheromones bind to receptors in a vomeronasal organ, or VNO. Neurons connect it to parts of the brain that control behavior. In humans, the VNO is a tiny ductlike structure on the septum that divides the nose into two nostrils. Many scientists contend that the human VNO is a vestigial structure (no longer functional). Others suspect that the human VNO does still connect with the brain by way of some pathway that has not yet been discovered.

HORMONES AND BONDING BEHAVIOR

The hormones oxytocin and ADH, which affect pair-bonding in other mammals, may also influence human social attachment. Consider *autism*. A person who has this behavioral disorder does not form normal social relationships. Compared to unaffected individuals, an autistic child has significantly less oxytocin circulating in the blood. Also associated with autism is a shorter than normal regulatory sequence near the gene for one ADH receptor. This is the same sequence that is shorter in chimpanzees than in bonobos and humans.

Today, researchers are studying roles of hormones in mother–infant bonding and romantic attachments. Nursing stimulates the secretion of oxytocin. So does orgasm, or even a friendly massage. Whether or not these temporary elevations in oxytocin or any other hormonal responses contribute to what we perceive as love is an interesting and still open question.

EVOLUTIONARY QUESTIONS

If we are comfortable with analyzing the evolutionary basis of the behavior of termites, naked mole-rats, and other animals, would it not be rewarding to analyze such a basis of human behavior, too? Many resist the idea. Apparently, they believe that attempts to identify the adaptive value of a human trait are attempts to define its moral or social advantage. However, there is a clear difference between trying to explain something in terms of its evolutionary history and attempting to justify it. "Adaptive" does not mean "morally right." To biologists, adaptive only means valuable in terms of transmission of the individual's genes.

One example: Infanticide is morally repugnant. Is it unnatural? No. Many animals do it and it occurs in all human cultures. Male lions often kill offspring of other males when they take over a pride. Male chimpanzees are also known to kill infants. Doing so frees up the females to breed with them, increasing the infanticidal male's reproductive success.

A biologist may predict that unrelated human males are a threat, too. In fact, the presence of an unrelated male in the home increases the risk of injury to an American child under age two by seventy times.

What about parents who kill their own offspring? In her book on maternal behavior, primatologist Sarah Blaffer Hrdy cites a study of a village in New Guinea in which about 40 percent of newborns were killed by their parents. She argues that if resources are scarce or there is little support socially, a mother might increase her fitness by killing a newborn. The mother can then allocate child-rearing energy to her other offspring or save it for children she may have in the future.

Understanding infanticide's evolutionary roots may help us prevent it. Using our knowledge of conditions that make infanticide an adaptive choice, we can strive as a society to prevent those conditions from arising.

Summary

Section 32.1 Animal behavior starts with genes that specify products required for development of the nervous, endocrine, and skeletal-muscular systems. Hormones are among the products that affect behavior.

Instinctive behavior is performed without having been learned by experience in the environment. It is a prewired response to one or two simple, well-defined environmental cues.

An animal learns when it processes and integrates information from experiences, then uses it to vary or change how it responds to stimuli. Imprinting is a type of learning that can occur only during a sensitive period.

Behavior with a genetic basis is subject to evolution by natural selection. It evolved as a result of individual differences in reproductive success in past generations. A behavior is adaptive if its reproductive benefits are greater than the reproductive costs. A behavior that has evolved under one set of circumstances may become maladaptive if circumstances change.

Section 32.2 Signalers send communication signals meant to change the behavior of receivers of the same species. Pheromones are signaling molecules that have roles in social communication.

Visual signals are key components of courtship displays and threat displays. Acoustical signals are sounds that have precise, species-specific information. Tactile signals are specific forms of physical contact between a signaler and a receiver.

Predators and parasites sometimes intercept or mimic signals of their prey species.

BiologyNow

Learn about the honeybee dance language with the animated interaction on BiologyNow.

Section 32.3 Sexual selection favors traits that give an individual a competitive edge in attracting and keeping mates. Female choice selects for males that have traits or engage in behaviors that attract females. When large numbers of females cluster in defensible areas, males may compete with one another to control these areas.

Parental care has reproductive costs in terms of future reproduction and survival. It is adaptive if the benefits to the current set of offspring offset these costs.

Section 32.4 Animals that live in social groups may benefit by cooperating in predator detection, defense, and rearing the young. The benefits of group living are often distributed unequally. Species that live in groups incur costs, including increased disease and parasitism, and increased competition for resources.

Section 32.5 Ants, termites, and some other insects as well as two species of mole-rats are eusocial. They live in colonies with overlapping generations and have a reproductive division of labor. Most colony members do not reproduce; they instead assist relatives.

According to the theory of indirect selection, such extreme altruism usually evolves because the altruistic individuals share some portion of their genes in common with their reproducing relatives. By aiding relatives, the altruists enhance the passage of shared genes—including the gene for altruism—on to the next generation.

Section 32.6 Our closest relatives, the chimpanzees and bonobos, both live in social groups, but they differ in their sexual behavior and social organization.

Section 32.7 Behavioral biologists ask questions about the mechanisms and the possible adaptive significance of human behavior. A behavior that is adaptive may still be judged by society to be morally wrong.

Self-Quiz

Answers in Appendix I

1. Genes affect the behavior of individuals by ______.
   a. influencing the development of nervous systems
   b. affecting the kinds of hormones in individuals
   c. controlling what can be learned and when
   d. all of the above

2. A behavior is considered adaptive if it ______.
   a. varies among individuals of a population
   b. has a genetic basis
   c. increases an individual's reproductive success
   d. is widespread across a species

3. The feeding preference of newly born coastal garter snakes for slugs is an example of ______.
   a. indirect selection
   b. an instinctive response
   c. imprinting
   d. a composite display signal

4. Male sage grouse gather at a lek to ______.
   a. display to and attract potential mates
   b. get protection from aerial predators
   c. share food resources with other males

5. In most mammals, the vomeronasal organ functions in ______.
   a. hormone production c. visual communication
   b. pheromone detection d. all of the above

6. Eusocial insects ______.
   a. live in extended family groups
   b. are found among almost all insect orders
   c. show a reproductive division of labor
   d. a and c
   e. all of the above

7. In ______, all sterile workers are females.
   a. honeybees c. termites
   b. naked mole-rats d. all of the above

8. Helping other individuals at a reproductive cost to oneself might be adaptive if those helped are ______.
   a. members of another species
   b. competitors for mates
   c. close relatives
   d. illegitimate signalers

Figure 32.20 Behaviorally confused rooster.

Figure 32.21 A Nazca booby tends its single surviving chick.

9. Social behavior evolves because ________ .
   a. social animals are more advanced than solitary ones
   b. under some conditions, the costs of social life to individuals are offset by benefits to the species
   c. under some conditions, the benefits of social life to an individual offset its costs to the individual
   d. under some conditions, social life has no costs to an individual.

10. Match the terms with their most suitable description.

____ fixed action pattern	a. time- and cue-dependent form of learning
____ altruism	b. interceptor of signals
____ illegitimate receiver	c. stereotyped motor program that runs to completion independently of feedback from environment
____ imprinting	d. assisting another individual at one's own expense
____ selfish herd	e. group in which individuals use others as a shield

Additional questions are available on **BiologyNow™**

Critical Thinking

1. Sexual imprinting is common in birds. During a short sensitive period in early life, the bird learns the features it will later look for when it is ready to mate. The rooster in Figure 32.20 was observed wading into the water after ducks with amorous intent. Speculate on what might have caused this behavior.

2. Nazca boobies (*Sula granti*) lay two eggs, several days apart. No matter how much food is available, only one chick survives to adulthood (Figure 32.21). The first chick to hatch pushes its younger sibling from the nest, and that sibling dies of starvation and neglect. Formulate a hypothesis on how it might be adaptive for parents to lay two eggs if one sibling generally kills the other. Design an experiment to test your hypothesis.

3. In 2002, Svante Paabo suggested how one gene may have played a pivotal role in the evolution of *language*. Humans who have a single base-pair substitution in this gene, called *FOXP2*, are unable to speak in an intelligible way, to understand complex sentences, or even to make certain movements of the mouth and face.

All mammals have the *FOXP2* gene, which has a 715 base-pair sequence. The gene has changed very little over evolutionary time. The chimpanzee gene and the mouse gene differ by only one base pair. However, two additional base-pair changes appeared and became fixed in the much briefer interval since chimpanzees and humans diverged. The rapid spread of these two substitutions suggest that these mutations may have been highly adaptive and subject to directional selection. What effect do you think mutations that favored an ability to produce and comprehend more complex auditory signals had on social behavior in the lineage leading to humans?

4. A cheetah scent-marks plants in its territory with certain exocrine gland secretions. What evidence would you require to demonstrate that the cheetah's action is an evolved communication signal?

5. Among primates, differences in sexual behavior tend to be related to the size of a male's gonads. Gorillas have relatively tiny testicles; in a male weighing 450 pounds, they may weigh only about an ounce. Gorillas live in groups consisting of a male, a few females, and offspring. This is the most typical type of social group among the primates. When a female is ready to mate, there usually is only a single adult male around to inseminate her.

In contrast, a female chimpanzee advertises her fertile period and mates with many males (Section 32.6). A 100-pound chimpanzee male has testicles about four times as weighty as a gorilla's. By producing large amounts of sperm, this male increases the chance that his sperm, rather than a rival's, will fertilize a female's egg.

Although adult men are larger than chimpanzees, their testicles are only about half the weight. What might this suggest about female promiscuity and male competition to fertilize eggs in the lineage that led to humans?

6. In moths and many other insects, potential mates locate one another with the help of species-specific pheromones. These pheromones are usually mixes of chemicals derived from fatty acids. Sometimes a gene mutation causes a change in the mix of chemicals in a moth pheromone. How might such a mutation alter mating behavior and encourage speciation?

BIOLOGICAL PRINCIPLES AND THE HUMAN IMPERATIVE

Molecules, single cells, tissues, organs, organ systems, multicelled organisms, populations, communities, then ecosystems and the biosphere. These are architectural systems of life, assembled in increasingly complex ways over the past 3.8 billion years. We are latecomers to this immense biological building program. Yet within the relatively short span of 10,000 years, our activities have been changing the character of the land, ocean, and atmosphere, even the genetic character of species.

It would be presumptuous to think we alone have had profound effects on the world of life. As long ago as the Proterozoic, photosynthetic organisms were irrevocably changing the course of biological evolution by enriching the atmosphere with oxygen. During the past as well as the present, competitive adaptations led to the rise of some groups, whose dominance assured the decline of others. Change is nothing new. What *is* new is the capacity of one species to comprehend what might be going on.

We now have the population size, technology, and cultural inclination to use up energy and modify the environment at rapid rates. Where will this end? Will feedback controls operate as they do, for example, when population growth exceeds carrying capacity? In other words, will negative feedback controls come into play and keep things from getting too far out of hand?

Feedback control will not be enough, for it operates after deviation. Our patterns of resource consumption and population growth are founded on an illusion of unlimited resources and a forgiving environment. A prolonged, global shortage of food or the passing of a critical threshold for the global climate can come too fast to be corrected; in which case the impact of the deviation may be too great to be reversed.

What about feedforward mechanisms that might serve as early warning systems? For example, when sensory receptors near the surface of skin detect a drop in outside air temperature, each sends messages to the nervous system. That system responds by triggering mechanisms that raise the body's core temperature before the body itself becomes dangerously chilled. Extrapolating from this, if we develop feedforward control mechanisms, would it not be possible to start corrective measures before we do too much harm?

Feedforward controls alone will not work, for they operate after change is under way. Think of the DEW line—the Distant Early Warning system. It is like a vast sensory receptor for detecting missiles launched against North America. By the time it does what it is supposed to, it may be too late to stop widespread destruction.

It would be naive to assume we can ever reverse who we are at this point in evolutionary time, to de-evolve ourselves culturally and biologically into becoming less complex in the hope of averting disaster. Yet there is reason to believe we can avert disaster by using a third kind of control mechanism—a capacity to anticipate events even before they happen. We are not locked into responding only after irreversible change has begun. We have the capacity to anticipate the future—it is the essence of our visions of utopia and hell. *We all have the capacity to adapt to a future that we can partly shape.*

For instance, we can stop trying to "beat nature" and learn to work with it. Individually and collectively, we can work to develop long-term policies that take into account biotic and abiotic limits on population growth. Far from being a surrender, this would be one of the most intelligent behaviors of which we are capable.

Having a capacity to adapt and using it are not the same thing. We have already put the world of life on dangerous ground because we have not yet mobilized ourselves as a species to work toward self-control.

Our survival depends on predicting possible futures. It depends on preserving, restoring, and constructing ecosystems that fit with our definition of basic human values and available biological models. Human values can change; our expectations can and must be adapted to biological reality. *For the principles of energy flow and resource utilization, which govern the survival of all systems of life, do not change.* It is our biological and cultural imperative that we come to terms with these principles, and ask ourselves what our long-term contribution will be to the world of life.

Appendix I Answers to Self-Quizzes

Italicized numbers refer to relevant section numbers

Chapter 1

1. cell	*1.1*
2. d	*1.2*
3. metabolism	*1.2*
4. homeostasis	*1.2*
5. adaptive	*1.4*
6. mutations	*1.4*
7. domains	*1.3*
8. a	*1.2*
9. d	*1.2, 1.4*
10. c	*1.6*
11. c	*1.4*
e	*1.4*
d	*1.5*
b	*1.5*
a	*1.5*

Chapter 2

1. False	*2.1*
2. c	*2.2*
3. e	*2.3*
4. d	*2.4*
5. See Table 2.1	
6. e	*2.7*
7. d	*2.5*
8. e	*2.6*
9. b	*2.7*
10. d	*2.8, 2.10*
11. d	*2.9*
12. c	*2.4*
a	*2.8*
g	*2.1*
d	*2.3*
f	*2.2*
b	*2.1*
e	*2.10*

Chapter 3

1. c	*3.3*
2. d	*3.5*
3. d	*3.8*
4. False; many have a cell wall outside the plasma membrane, and some prokaryotes have a capsule outside of that.	*3.4*
5. c	*3.1*
6. See Figures 3.12 and 3.13	
7. e	*3.5*
d	*3.5*
a	*3.1*
b	*3.5*
c	*3.5*

Chapter 4

1. c	*4.1*
2. d	*4.1*
3. d	*4.2, 4.3*
4. d	*4.3*
5. a	*4.6*
6. d	*4.4*
7. b	*4.5*
8. d	*4.7*
9. d	*4.4*
10. c	*4.3*
g	*4.5*
a	*4.2*
d	*4.2*
e	*4.2*
b	*4.3*
f	*4.1*

Chapter 5

1. carbon dioxide; sunlight	*intro essay*
2. a	*5.1, 5.2*
3. b	*5.2*
4. c	*5.2*
5. c	*5.1, 5.3*
6. b	*5.3*
7. d	*5.2*
b	*5.3*
a	*5.3*
c	*5.2*

Chapter 6

1. d	*6.2*
2. c	*6.2*
3. b	*6.3*
4. c	*6.3*
5. d	*6.4*
6. c	*6.4*
7. b	*6.4*
8. d	*6.5*
9. b	*6.2*
c	*6.4*
a	*6.3*
d	*6.3*

Chapter 7

1. d	*7.1*
2. b	*7.1*
3. a	*7.3*
4. a	*7.3*
5. b	*7.2*
6. d, b, c, a	*7.3*
7. d	*7.5*
8. d	*7.5*
9. b	*7.5*
10. d	*7.5*
11. c	*7.5*
12. d	*7.5*
13. d	*7.3*
a	*7.5*
c	*7.3*
b	*7.5*

Chapter 8

1. a	*8.1*
2. b	*8.1*
3. d	*8.1*
4. c	*8.5*
5. e	*8.8*
6. a	*8.5*
7. c	*8.4*
8. d	*8.9*
9. c	*8.9*
10. b	*8.1*
d	*8.1*
a	*8.1*
c	*8.1*
11. c	*8.5*
e	*8.8*
d	*8.9*
b	*8.8*
a	*8.4*
f	*8.5*

Chapter 9

1. c	*9.2*
2. d	*9.2*
3. c	*9.2*
4. a	*9.3*
5. d	*9.3*
6. c	*9.1*
e	*9.4*
a	*9.2*
d	*9.3*
b	*9.3*

Chapter 10

1. c	*10.2*
2. b	*10.2*
3. c	*10.1, 10.2*
4. c	*10.3*
5. a	*10.3*
6. d	*10.6*
7. d	*10.6*
8. a	*10.6*
9. b	*10.6*
10. b	*10.6*
11. c	*10.6*
12. e	*10.5*
c	*10.4*
a	*10.2*
f	*10.3*
d	*10.3*
g	*10.2*
b	*10.3*

Chapter 11

1. d	*11.5*
2. c	*11.1*
3. plasmid	*11.1*
4. b	*11.1*
5. b	*11.3*
6. d	*11.3*
7. b	*11.5, 11.7*
8. d	*11.4*
c	*11.6*
b	*11.2*
e	*11.2*
a	*11.7*

Chapter 12

1. populations	*12.3*
2. b	*intro essay*
3. a	*12.3*
4. b	*12.1*
5. b	*12.5*
6. c	*12.5*
7. b	*12.8*
8. c	*12.8*
d	*12.1*
a	*12.3*
b	*12.7*

Chapter 13

1. a	*13.2*
2. Gondwana	*13.3*
3. a.	*13.4*
4. d	*13.7*
5. a	*13.8*
6. c	*13.10*
7. c	*13.10*
8. d	*13.10*
9. e	*13.10*
f	*13.1*
g	*13.1*
c	*13.4*
b	*13.10*
d	*13.4*
a	*13.9*

Chapter 14

1. a	*14.4*
2. d	*14.2*
3. b	*14.1, 14.2*
4. b	*14.3*
5. d	*14.4*
6. c	*14.3*
7. c	*14.4*
8. d	*14.5*
9. g	*14.3*
d	*14.5*
c	*14.2*
e	*14.2*
i	*14.4*
f	*14.4*
a	*14.3*
j	*14.4*
b	*14.3*
h	*14.1*

Chapter 15

1. a	*15.1*
2. c	*15.4*
3. b	*15.2*
4. c	*15.3*
5. a	*15.3*
6. b	*15.5*
7. b	*15.4*
8. c	*15.6*
9. c	*15.5*
d	*15.1*
f	*15.3*
e	*15.2*
a	*15.1*
b	*15.1*
g	*15.6*

Chapter 16

1. c	*16.9*
2. d	*16.7*
3. f	*16.12*
4. f	*16.13*
5. a	*16.1*
6. c	*16.9*
7. a	*16.4*
8. a	*16.1*
9. b	*16.2*
g	*16.2*
l	*16.3*
j	*16.6*
e	*16.4*
c	*16.7*
d	*16.5*
f	*16.8*
k	*16.11*
a	*16.10*
h	*16.12*
i	*16.12*

Chapter 17

1 a	*17.1*
2. c	*17.1*
3. d	*17.2*
4. e	*17.2*
5. c	*17.3*
6. d	*17.5*
7. b	*17.3*
8. b	*17.4*
c	*17.3*
d	*17.3*
a	*17.5*
e	*17.3*
g	*17.3*
f	*17.3*

Chapter 18

1. a	*18.1*
2. d	*18.4*
3. c	*18.6*
4. b	*18.1*
5. b	*18.6*
6. b	*18.1*
7. c	*18.7*
8. d	*18.7*
9. b	*18.1*
d	*18.1*
e	*18.1, 18.7*
c	*18.1, 18.9*
f	*18.3*
a	*18.4*

Chapter 19

1. a	*19.1*
2. c	*19.2*
3. b	*19.1*
4. c	*19.2*
5. c	*19.3*
6. d	*19.5, 19.6*
7. b	*19.5*
8. a	*19.7*
9. f	*19.6*
g	*19.7*
b	*19.1*
d	*19.1*
e	*19.1*
a	*19.8*
c	*19.1*
h	*19.8*

Chapter 20

1. a	*20.2*
2. c	*20.2*
3. a	*20.2*
4. b	*20.2*
5. b	*20.2*
6. b	*20.2*
7. c	*20.2*
8. c	*20.2*
9. d	*20.2*
10. d	*20.2*
11. b	*20.2*
g	*20.2*
a	*20.4*
c	*20.3*
d	*20.2*
f	*20.2*
e	*20.2*

Chapter 21

1. a	*21.1*
2. d	*21.1*
3. b	*21.3*
4. b	*21.2*
5. b	*21.3*
6. d	*21.3*
7. c	*21.3*
8. d	*21.3*
9. e	*21.1*
f	*21.4*
g	*21.3, 21.4*
h	*21.1*
a	*21.3*
c	*21.1*
b	*21.1*

i	21.3
d	21.4

Chapter 22

1. b	22.1
2. b	22.3
3. a	22.4
4. c	22.3
5. c	22.4
6. d	22.2, 22.8
7. See Figure 22.6	
8. c	22.7, 22.8
9. d	22.8
10. a	22.8
11. b	22.8
12. b	22.2
e	22.7
i	22.6
h	22.4
c	22.3
g	22.7
d	22.7
f	22.3
a	22.3

Chapter 23

1. f	23.2
2. e	23.3
3. d	23.1
4. d	23.4
5. e	23.4
6. a	23.5
7. b	23.7
8. c	23.3
b	23.6
a	23.1
e	23.6
d	23.8

Chapter 24

1. b	24.1
2. c	24.2
3. b	24.2
4. d	24.2
5. f	24.2
b	24.2
a	24.2
d	24.2
e	24.2
c	24.2
6. d	24.3
7. d	24.6
8. b	24.6
9. b	24.6
10. c	24.5
e	24.6
b	24.5
a	24.5
d	24.6

Chapter 25

1. a	25.1
2. c	25.1
3. d	25.1
4. b	25.2
5. b	25.5
6. c	25.5
7. a	25.6, 25.9
8. b	25.6
9. c	25.9
10. c	25.8
11. b	25.10
12. a	25.10
13. c	25.10
14. g	25.10
a	25.10
f	25.6
b	25.6
h	25.6
e	25.10
d	25.3
c	25.6

Chapter 26

1. d	26.1
2. b	26.2
3. a	26.2
4. d	26.3
5. b	26.5
6. b	26.3
7 e	26.4
g	26.3
d	26.3
c	26.3
h	26.5
a	26.6
b	26.2
f	26.6

Chapter 27

1. d	27.1
2. c	27.2
3. c	27.2
4. b	27.2
5. a	27.3
6. c	27.4
7. c	27.4
8. c	27.7
9. a	27.7
10. e	27.6
11. d	27.7
12. e	27.8
13. c	27.3
h	27.4
a	27.7
g	27.4
d	27.4
e	27.5
b	27.3
f	27.4

Chapter 28

1. f	28.2
2. a	28.2
3. a	28.2
4. d	28.3
5. a	28.3
6. d	28.4
7. b	28.5
8. d	28.3
c	28.3
e	28.2
a	28.2
f	28.3
b	28.4

Chapter 29

1. d	29.1
2. e	29.3
3. a	29.1
4. c	29.2
5. a	29.3
6. b	29.4
7. b	29.6
8. e	29.8
9. b	29.9
10. c	29.7
d	29.8
a	29.6
e	29.1
b	29.7
g	29.9
f	29.9
h	29.1

Chapter 30

1. d	30.1
2. d	30.1
3. d	30.2
4. d	30.8
5. a	30.4
6. b	30.4
7. b	30.5
8. e	30.7
9. d	30.7
10. d	30.1
a	30.1
c	30.1
b	30.1

Chapter 31

1. d	31.1
2. c	31.1
3. d	31.2
4. a	31.2
5. a	31.2
6. b	31.3
7. c	31.1
8. e	31.3
9. d	31.3
10. c	31.3
11. b	31.3
12. d	31.4
13. d	31.3
g	31.3
e	31.3
c	31.3
b	31.3
f	31.3
a	31.3

Chapter 32

1. d	32.1
2. c	32.1
3. b	32.1
4. a	32.3
5. b	32.7
6. d	32.5
7. a	32.5
8. c	32.5
9. c	32.4
10. c	32.1
d	32.1, 32.5
b	32.2
a	32.1
e	32.4

Appendix II Answers to Genetics Problems

1. **a.** Both parents are heterozygotes (*Cc*). Their children may be albino (cc) or unaffected (CC or *Cc*).
   b. All are homozygous recessive (*cc*).
   c. Homozygous recessive (cc) father, and heterozygous (*Cc*) mother. The albino child is *cc*, the unaffected children *Cc*.
2. **a.** *AB*
   b. *AB*, *aB*
   c. *Ab*, *ab*
   d. *AB*, *Ab*, *aB*, *ab*
3. **a.** All offspring will be *AaBB*.
   b. 1/4 *AABB* (25% each genotype)
   1/4 *AABb*
   1/4 *AaBB*
   1/4 *AaBb*
   c. 1/4 *AaBb* (25% each genotype)
   1/4 *Aabb*
   1/4 *aaBb*
   1/4 *aabb*
   d. 1/16 *AABB* (6.25% of genotype)
   1/8 *AaBB* (12.5%)
   1/16 *aaBB* (6.25%)
   1/8 *AABb* (12.5%)
   1/4 *AaBb* (25%)
   1/8 *aaBb* (12.5%)
   1/16 *AAbb* (6.25%)
   1/8 *Aabb* (12.5%)
   1/16 *aabb* (6.25%)
4. A mating of two M^L cats yields 1/4 *MM*, 1/2 M^LM, and 1/4 M^LM^L. Because M^LM^L is lethal, the probability that any one kitten among the survivors will be heterozygous is 2/3.
5. **a.** Human males (XY) inherit their X chromosome from their mother.
   b. A male can produce two kinds of gametes. Half carry an X chromosome and half carry a Y chromosome. All the gametes that carry the X chromosome carry the same X-linked allele.
   c. A female homozygous for an X-linked allele produces only one kind of gamete.
   d. Fifty percent of the gametes of a female who is heterozygous for an X-linked allele carry one of the two alleles at that locus; the other fifty percent carry its partner allele for that locus.
6. Because Marfan syndrome is a case of autosomal dominant inheritance and because one parent bears the allele, the probability that any child of theirs will inherit the mutant allele is 50 percent.
7. Fred could use a testcross to find out if his pet's genotype is *WW* or *Ww*. He can let his black guinea pig mate with a white guinea pig having the genotype *ww*. If any F_1 offspring are white, then the genotype of his pet is *Ww*. If the two guinea pig parents are allowed to mate repeatedly and all the offspring of the matings are black, then there is a high probability that his pet guinea pig is *WW*. (For instance, if ten offspring are all black, then the probability that the male is *WW* is about 99.9 percent. The greater the number of offspring, the more confident Fred can be of his conclusion.)
8. **a.** Nondisjunction might occur during anaphase I or anaphase II of meiosis.
   b. One parent had a translocation in which part of chromosome 21 was exchanged with part of chromosome 14. A child receives the translocated chromosome 14 from that parent, in addition to a normal chromosome 21 from each parent. Although the somatic cells of the new individual have 46 chromosomes, the extra chromosome 21 material carried by the translocated chromosome 14 causes Down syndrome.
9. In the mother, a crossover at meiosis generates one X chromosome that carries neither mutant allele, and one that has both. A son that inherits the X chromosome with two normal alleles will be unaffected.

Appendix III Annotations to A Journal Article

This journal article reports on the movements of a female wolf during the summer of 2002 in northwestern Canada. It also reports on a scientific process of inquiry, observation and interpretation to learn where, how and why the wolf traveled as she did. In some ways, this article reflects the story of "how to do science" told in section 1.5 of this textbook. These notes are intended to help you read and understand how scientists work and how they report on their work.

1 ARCTIC

2 VOL. 57, NO. 2 (JUNE 2004) P. 196–203

3 Long Foraging Movement of a Denning Tundra Wolf

4 Paul F. Frame,[1,2] David S. Hik,[1] H. Dean Cluff,[3] and Paul C. Paquet[4]

5 *(Received 3 September 2003; accepted in revised form 16 January 2004)*

6 **ABSTRACT** Wolves (*Canis lupus*) on the Canadian barrens are intimately linked to migrating herds of barren-ground caribou (*Rangifer tarandus*). We deployed a Global Positioning System (GPS) radio collar on an adult female wolf to record her movements in response to changing caribou densities near her den during summer. This wolf and two other females were observed nursing a group of 11 pups. She traveled a minimum of 341 km during a 14-day excursion. The straight-line distance from the den to the farthest location was 103 km, and the overall minimum rate of travel was 3.1 km/h. The distance between the wolf and the radio-collared caribou decreased from 242 km one week before the excursion to 8 km four days into the excursion. We discuss several possible explanations for the long foraging bout.

7 *Key words:* wolf, GPS tracking, movements, *Canis lupus*, foraging, caribou, Northwest Territories

8 **RÉSUMÉ** Les loups (*Canis lupus*) dans la toundra canadienne sont étroitement liés aux hardes de caribous des toundras (*Rangifer tarandus*). On a équipé une louve adulte d'un collier émetteur muni d'un système de positionnement mondial (GPS) afin d'enregistrer ses déplacements en réponse au changement de densité du caribou près de sa tanière durant l'été. On a observé cette louve ainsi que deux autres en train d'allaiter un groupe de 11 louveteaux. Elle a parcouru un minimum de 341 km durant une sortie de 14 jours. La distance en ligne droite de la tanière à l'endroit le plus éloigné était de 103 km, et la vitesse minimum durant tout le voyage était de 3,1 km/h. La distance entre la louve et le caribou muni du collier émetteur a diminué de 242 km une semaine avant la sortie à 8 km quatre jours après la sortie. On commente diverses explications possibles pour ce long épisode de recherche de nourriture.

Mots clés: loup, repérage GPS, déplacements, *Canis lupus*, recherche de nourriture, caribou, Territoires du Nord-Ouest

Traduit pour la revue *Arctic* par Nésida Loyer.

9 Introduction

Wolves (*Canis lupus*) that den on the central barrens of mainland Canada follow the seasonal movements of their main prey, migratory barren-ground caribou (*Rangifer tarandus*) (Kuyt, 1962; Kelsall, 1968; Walton et al., 2001). However, most wolves do not den near caribou calving grounds, but select sites farther south, closer to the tree line (Heard and Williams, 1992). Most caribou migrate beyond primary wolf denning areas by mid-June and do not return until mid-to-late July (Heard et al., 1996; Gunn et al., 2001). Consequently, caribou density near dens is low for part of the summer.

During this period of spatial separation from the main caribou herds, wolves must either search near the homesite for scarce caribou or alternative prey (or both), travel to where prey are abundant, or use a combination of these strategies. 10

Walton et al. (2001) postulated that the travel of tundra wolves outside their normal summer ranges is a response to low caribou availability rather than a pre-dispersal exploration like that observed in terri torial wolves (Fritts and Mech, 1981; Messier, 1985). The authors postulated this because most such travel was directed toward caribou calving grounds. We report details of such a long-distance excursion by a breeding female tundra wolf wearing a GPS radio collar. We discuss the relationship of the excursion to movements of satellite-collared caribou (Gunn et al., 2001), supporting the hypothesis that tundra wolves make directional, rapid, long-distance movements in response to seasonal prey availability. 11

[1] Department of Biological Sciences, University of Alberta, Edmonton, Alberta T6G 2E9, Canada

[2] Corresponding author: pframe@ualberta.ca

[3] Department of Resources, Wildlife, and Economic Development, North Slave Region, Government of the Northwest Territories, P.O. Box 2668, 3803 Bretzlaff Dr., Yellowknife, Northwest Territories X1A 2P9, Canada; Dean_Cluff@gov.nt.ca

[4] Faculty of Environmental Design, University of Calgary, Calgary, Alberta T2N 1N4, Canada; current address: P.O. Box 150, Meacham, Saskatchewan S0K 2V0, Canada

196

1 Title of the journal, which reports on science taking place in Arctic regions.

2 Volume number, issue number and date of the journal, and page numbers of the article.

3 Title of the article: a concise but specific description of the subject of study—one episode of long-range travel by a wolf hunting for food on the Arctic tundra.

4 Authors of the article: scientists working at the institutions listed in the footnotes below. Note #2 indicates that P. F. Frame is the *corresponding author*—the person to contact with questions or comments. His email address is provided.

5 Date on which a draft of the article was received by the journal editor, followed by date one which a revised draft was accepted for publication. Between these dates, the article was reviewed and critiqued by other scientists, a process called peer review. The authors revised the article to make it clearer, according to those reviews.

6 ABSTRACT: A brief description of the study containing all basic elements of this report. First sentence summarizes the *background* material. Second sentence encapsulates the *methods* used. The rest of the paragraph sums up the *results*. Authors introduce the main *subject* of the study—a female wolf (#388) with pups in a den—and refer to later *discussion* of possible explanations for her behavior.

7 Key words are listed to help researchers using computer databases. Searching the databases using these key words will yield a list of studies related to this one.

8 RÉSUMÉ: The French translation of the abstract and key words. Many researchers in this field are French Canadian. Some journals provide such translations in French or in other languages.

9 INTRODUCTION: Gives the background for this wolf study. This paragraph tells of known or suspected wolf behavior that is important for this study. Note that (a) major species mentioned are always accompanied by scientific names, and (b) statements of fact or *postulations* (claims or assumptions about what is likely to be true) are followed by references to studies that established those facts or supported the postulations.

10 This paragraph focuses directly on the wolf behaviors that were studied here.

11 This paragraph starts with a statement of the *hypothesis* being tested, one that originated in other studies and is supported by this one. The hypothesis is restated more succinctly in the last sentence of this paragraph. This is the *inquiry* part of the scientific process—asking questions and suggesting possible answers.

12 This map shows the study area and depicts wolf and caribou locations and movements during one summer. Some of this information is explained below.

13 STUDY AREA: This section sets the stage for the study, locating it precisely with latitude and longitude coordinates and describing the area (illustrated by the map in Figure 1).

14 Here begins the story of how prey (caribou) and predators (wolves) interact on the tundra. Authors describe movements of these nomadic animals throughout the year.

15 We focus on the denning season (summer) and learn how wolves locate their dens and travel according to the movements of caribou herds.

12

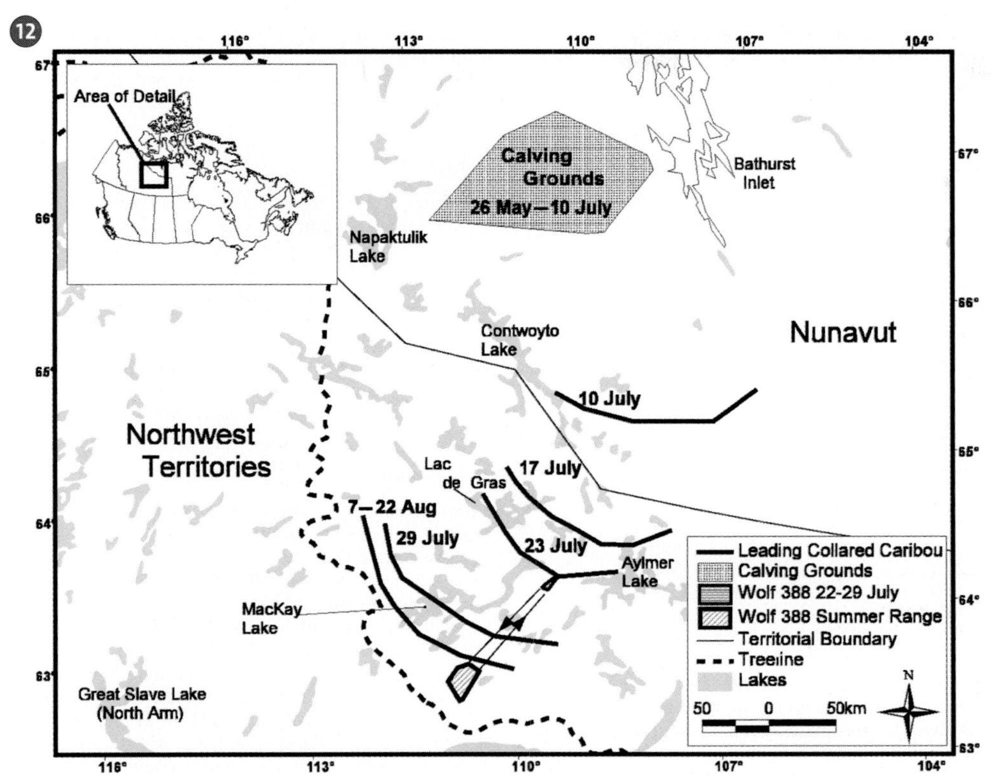

Figure 1. Map showing the movements of satellite radio-collared caribou with respect to female wolf 388's summer range and long foraging movement, in summer 2002.

13 Study Area

Our study took place in the northern boreal forest–low Arctic tundra transition zone (63° 30′ N, 110° 00′ W; Figure 1; Timoney et al., 1992). Permafrost in the area changes from discontinuous to continuous (Harris, 1986). Patches of spruce (*Picea mariana, P. glauca*) occur in the southern portion and give way to open tundra to the northeast. Eskers, kames, and other glacial deposits are scattered throughout the study area. Standing water and exposed bedrock are characteristic of the area.

14 *Details of the Caribou-Wolf System*

The Bathurst caribou herd uses this study area. Most caribou cows have begun migrating by late April, reaching calving grounds by June (Gunn et al., 2001; Figure 1). Calving peaks by 15 June (Gunn et al., 2001), and calves begin to travel with the herd by one week of age (Kelsall, 1968). The movement patterns of bulls are less known, but bulls frequent areas near calving grounds by mid-June (Heard et al., 1996; Gunn et al., 2001). In summer, Bathurst caribou cows generally travel south from their calving grounds and then, parallel to the tree line, to the northwest. The rut usually takes place at the tree line in October (Gunn et al., 2001). The winter range of the Bathurst herd varies among years, ranging through the taiga and along the tree line from south of Great Bear Lake to southeast of Great Slave Lake. Some caribou spend the winter on the tundra (Gunn et al., 2001; Thorpe et al., 2001).

In winter, wolves that prey on Bathurst caribou do not behave territorially. Instead, they follow the herd throughout its winter range (Walton et al., 2001; Musiani, 2003). However, during denning (May– 15

16 Other variables are considered—prey other than caribou and their relative abundance in 2002.

17 METHODS: There is no one scientific method. Procedures for each and every study must be explained carefully.

18 Authors explain when and how they tracked caribou and wolves, including tools used and the exact procedures followed.

19 This important subsection explains what data were calculated (average distance ...) and how, including the software used and where it came from. (The calculations are listed in Table 1.) Note that the behavior measured (traveling) is carefully defined.

20 RESULTS: The heart of the report and the *observation* part of the scientific process. This section is organized parallel to the Methods section.

21 This subsection is broken down by periods of observation. Pre-excursion period covers the time between 388's capture and the start of her long-distance travel. The investigators used visual observations as well as telemetry (measurements taken using the global positioning system (GPS)) to gather data. They looked at how 388 cared for her pups, interacted with other adults, and moved about the den area.

Table 1. Daily distances from wolf 388 and the den to the nearest radio-collared caribou during a long excursion in summer 2002.

Date (2002)	Mean distance from caribou to wolf (km)	Daily distance from closest caribou to den
12 July	242	241
13 July	210	209
14 July	200	199
15 July	186	180
16 July	163	162
17 July	151	148
18 July	144	137
19 July[1]	126	124
20 July	103	130
21 July	73	130
22 July	40	110
23 July[2]	9	104
29 July[3]	16	43
30 July	32	43
31 July	28	44
1 August	29	46
2 August[4]	54	52
3 August	53	53
4 August	74	74
5 August	75	75
6 August	74	75
7 August	72	75
8 August	76	75
9 August	79	79

[1] Excursion starts.
[2] Wolf closest to collared caribou.
[3] Previous five days' caribou locations not available.
[4] Excursion ends.

August, parturition late May to mid-June), wolf movements are limited by the need to return food to the den. To maximize access to migrating caribou, many wolves select den sites closer to the tree line than to caribou calving grounds (Heard and Williams, 1992). Because of caribou movement patterns, tundra denning wolves are separated from the main caribou herds by several hundred kilometers at some time during summer (Williams, 1990:19; Figure 1; Table 1).

16 Muskoxen do not occur in the study area (Fournier and Gunn, 1998), and there are few moose there (H.D. Cluff, pers. obs.). Therefore, alternative prey for wolves includes waterfowl, other ground-nesting birds, their eggs, rodents, and hares (Kuyt, 1972; Williams, 1990:16; H.D. Cluff and P.F. Frame, unpubl. data). During 56 hours of den observations, we saw no ground squirrels or hares, only birds. It appears that the abundance of alternative prey was relatively low in 2002.

17 Methods

Wolf Monitoring

18 We captured female wolf 388 near her den on 22 June 2002, using a helicopter net-gun (Walton et al., 2001). She was fitted with a releasable GPS radio collar (Merrill et al., 1998) programmed to acquire locations at 30-minute intervals. The collar was electronically released (e.g., Mech and Gese, 1992) on 20 August 2002. From 27 June to 3 July 2002, we observed 388's den with a 78 mm spotting scope at a distance of 390 m.

Caribou Monitoring

In spring of 2002, ten female caribou were captured by helicopter net-gun and fitted with satellite radio collars, bringing the total number of collared Bathurst cows to 19. Eight of these spent the summer of 2002 south of Queen Maud Gulf, well east of normal Bathurst caribou range. Therefore, we used 11 caribou for this analysis. The collars provided one location per day during our study, except for five days from 24 to 28 July. Locations of satellite collars were obtained from Service Argos, Inc. (Landover, Maryland).

Data Analysis

Location data were analyzed by ArcView GIS software (Environmental Systems Research Institute Inc., Redlands, California). We calculated the average distance from the nearest collared caribou to the wolf and the den for each day of the study. 19

Wolf foraging bouts were calculated from the time 388 exited a buffer zone (500 m radius around the den) until she re-entered it. We considered her to be traveling when two consecutive locations were spatially separated by more than 100 m. Minimum distance traveled was the sum of distances between each location and the next during the excursion.

We compared pre- and post-excursion data using Analysis of Variance (ANOVA; Zar, 1999). We first tested for homogeneity of variances with Levene's test (Brown and Forsythe, 1974). No transformations of these data were required.

Results 20

Wolf Monitoring

Pre-Excursion Period: Wolf 388 was lactating when captured on 22 June. We observed her and two other females nursing a group of 11 pups between 27 June and 3 July. During our observations, the pack consisted of at least four adults (3 females and 1 male) and 11 pups. On 30 June, three pups were moved to a location 310 m from the other eight and cared for by an uncollared female. The male was not seen at the den after the evening of 30 June. 21

Before the excursion, telemetry indicated 18 foraging bouts. The mean distance traveled during these bouts was 25.29 km (± 4.5 SE, range 3.1–82.5 km). Mean greatest distance from the den on foraging

22

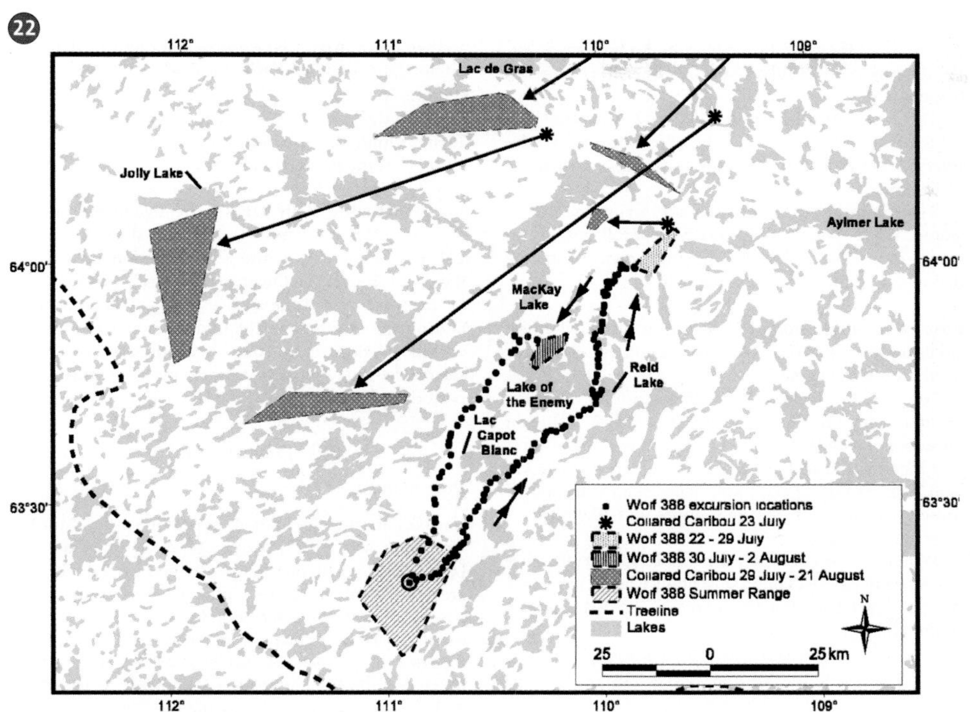

Figure 2. Details of a long foraging movement by female wolf 388 between 19 July and 2 August 2002. Also shown are locations and movements of three satellite radio-collared caribou from 23 July to 21 August 2002. On 23 July, the wolf was 8 km from a collared caribou. The farthest point from the den (103 km distant) was recorded on 27 July. Arrows indicate direction of travel.

22 The key in the lower right-hand corner of the map shows areas (shaded) within which the wolves and caribou moved, and the dotted trail of 388 during her excursion. From the results depicted on this map, the investigators tried to determine when and where 388 might have encountered caribou and how their locations affected her traveling behavior.

23 The wolf's excursion (her long trip away from the den area) is the focus of this study. These paragraphs present detailed measurements of daily movements during her two-week trip—how far she traveled, how far she was from collared caribou, her time spent traveling and resting, and her rate of speed. Authors use the phrase "minimum distance traveled" to acknowledge they couldn't track every step but were measuring samples of her movements. They knew that she went at least as far as they measured. This shows how scientists try to be exact when reporting results. Results of this study are depicted graphically in the map in Figure 2.

bouts was 7.1 km (± 0.9 SE, range 1.7–17.0 km). The average duration of foraging bouts for the period was 20.9 h (± 4.5 SE, range 1–71 h).

The average daily distance between the wolf and the nearest collared caribou decreased from 242 km on 12 July, one week before the excursion period, to 126 km on 19 July, the day the excursion began (Table 1).

23 **Excursion Period:** On 19 July at 2203, after spending 14 h at the den, 388 began moving to the northeast and did not return for 336 h (14 d; Figure 2). Whether she traveled alone or with other wolves is unknown. During the excursion, 476 (71%) of 672 possible locations were recorded. The wolf crossed the southeast end of Lac Capot Blanc on a small land bridge, where she paused for 4.5 h after traveling for 19.5 h (37.5 km). Following this rest, she traveled for 9 h (26.3 km) onto a peninsula in Reid Lake, where she spent 2 h before backtracking and stopping for 8 h just off the peninsula. Her next period of travel lasted 16.5 h (32.7 km), terminating in a pause of 9.5 h just 3.8 km from a concentration of locations at the far end of her excursion, where we presume she encountered caribou. The mean duration of these three movement periods was 15.7 h (± 2.5 SE), and that of the pauses, 7.3 h (± 1.5). The wolf required 72.5 h (3.0 d) to travel a minimum of 95 km from her den to this area near caribou (Figure 2). She remained there (35.5 km2) for 151.5 h (6.3 d) and then moved south to Lake of the Enemy, where she stayed (31.9 km^2) for 74 h (3.1 d) before returning to her den. Her greatest distance from the den, 103 km, was recorded 174.5 h (7.3 d) after the excursion

24 Post-excursion measurements of 388's movements were made to compare with those of the pre-excursion period. In order to compare, scientists often use *means*, or averages, of a series of measurements—mean distances, mean duration, etc.

25 In the comparison, authors used statistical calculations (F and df) to determine that the differences between pre- and post-excursion measurements were *statistically insignificant*, or close enough to be considered essentially the same or similar.

26 As with wolf 388, the investigators measured the movements of caribou during the study period. The areas within which the caribou moved are shown in Figure 2 by shaded polygons mentioned in the second paragraph of this subsection.

27 This subsection summarizes how distances separating predators and prey varied during the study period.

28 DISCUSSION: This section is the *interpretation* part of the scientific process.

29 This subsection reviews observations from other studies and suggests that this study fits with patterns of those observations.

30 Authors discuss a prevailing *theory* (CBFT) which might explain why a wolf would travel far to meet her own energy needs while taking food caught closer to the den back to her pups. The results of this study seem to fit that pattern.

began, at 0433 on 27 July. She was 8 km from a collared caribou on 23 July, four days after the excursion began (Table 1).

The return trip began at 0403 on 2 August, 318 h (13.2 d) after leaving the den. She followed a relatively direct path for 18 h back to the den, a distance of 75 km.

The minimum distance traveled during the excursionwas 339 km. The estimated overall minimum travel rate was 3.1 km/h, 2.6 km/h away from the den and 4.2 km/h on the return trip.

24 **Post-Excursion Period:** We saw three pups when recovering the collar on 20 August, but others may have been hiding in vegetation.

Telemetry recorded 13 foraging bouts in the post-excursion period. The mean distance traveled during these bouts was 18.3 km (+ 2.7 SE, range 1.2–47.7 km), and mean greatest distance from the den was 7.1 km (+ 0.7 SE, range 1.1–11.0 km). The mean duration of these post-excursion foraging bouts was 10.9 h (+ 2.4 SE, range 1–33 h).

When 388 reached her den on 2 August, the distance to the nearest collared caribou was 54 km. On 9 August, one week after she returned, the distance was 79 km (Table 1).

Pre- and Post-Excursion Comparison

25 We found no differences in the mean distance of foraging bouts before and after the excursion period ($F = 1.5$, $df = 1, 29$, $p = 0.24$). Likewise, the mean greatest distance from the den was similar pre- and post-excursion ($F = 0.004$, $df = 1, 29$, $p = 0.95$). However, the mean duration of 388's foraging bouts decreased by 10.0 h after her long excursion ($F = 3.1$, $df = 1, 29$, $p = 0.09$).

26 *Caribou Monitoring*

Summer Movements: On 10 July, 5 of 11 collared caribou were dispersed over a distance of 10 km, 140 km south of their calving grounds (Figure 1). On the same day, three caribou were still on the calving grounds, two were between the calving grounds and the leaders, and one was missing. One week later (17 July), the leading radio-collared cows were 100 km farther south (Figure 1). Two were within 5 km of each other in front of the rest, who were more dispersed. All radio-collared cows had left the calving grounds by this time. On 23 July, the leading radio-collared caribou had moved 35 km farther south, and all of them were more widely dispersed. The two cows closest to the leader were 26 km and 33 km away, with 37 km between them. On the next location (29 July), the most southerly caribou were 60 km farther south. All of the caribou were now in the areas where they remained for the duration of the study (Figure 2).

A Minimum Convex Polygon (Mohr and Stumpf, 1966) around all caribou locations acquired during the study encompassed 85 119 km^2.

Relative to the Wolf Den: The distance from the 27
nearest collared caribou to the den decreased from 241 km one week before the excursion to 124 km the day it began. The nearest a collared caribou came to the den was 43 km away, on 29 and 30 July. During the study, four collared caribou were located within 100 km of the den. Each of these four was closest to the wolf on at least one day during the period reported.

28 Discussion

Prey Abundance

Caribou are the single most important prey of tundra 29
wolves (Clark, 1971; Kuyt, 1972; Stephenson and James, 1982; Williams, 1990). Caribou range over vast areas, and for part of the summer, they are scarce or absent in wolf home ranges (Heard et al., 1996). Both the long distance between radio-collared caribou and the den the week before the excursion and the increased time spent foraging by wolf 388 indicate that caribou availability near the den was low. Observations of the pups' being left alone for up to 18 h, presumably while adults were searching for food, provide additional support for low caribou availability locally. Mean foraging bout duration decreased by 10.0 h after the excursion, when collared caribou were closer to the den, suggesting an increase in caribou availability nearby.

Foraging Excursion

One aspect of central place foraging theory (CPFT) 30
deals with the optimality of returning different-sized food loads from varying distances to dependents at a central place (i.e., the den) (Orians and Pearson, 1979). Carlson (1985) tested CPFT and found that the predator usually consumed prey captured far from the central place, while feeding prey captured nearby to dependants. Wolf 388 spent 7.2 days in one area near caribou before moving to a location 23 km back towards the den, where she spent an additional 3.1 days, likely hunting caribou. She began her return trip from this closer location, traveling directly to the den. While away, she may have made one or more successful kills and spent time meeting her own energetic needs before returning to the den. Alternatively, it may have taken several attempts to make a kill,

which she then fed on before beginning her return trip. We do not know if she returned food to the pups, but such behavior would be supported by CPFT.

31 Other workers have reported wolves' making long round trips and referred to them as "extraterritorial" or "pre-dispersal" forays (Fritts and Mech, 1981; Messier, 1985; Ballard et al., 1997; Merrill and Mech, 2000). These movements are most often made by young wolves (1–3 years old), in areas where annual territories are maintained and prey are relatively sedentary (Fritts and Mech, 1981; Messier, 1985). The long excursion of 388 differs in that tundra wolves do not maintain annual territories (Walton et al., 2001), and the main prey migrate over vast areas (Gunn et al., 2001).

Another difference between 388's excursion and those reported earlier is that she is a mature, breeding female. No study of territorial wolves has reported reproductive adults making extraterritorial movements in summer (Fritts and Mech, 1981; Messier, 1985; Ballard et al., 1997; Merrill and Mech, 2001). However, Walton et al. (2001) also report that breeding female tundra wolves made excursions.

Direction of Movement

32 Possible explanations for the relatively direct route 388 took to the caribou include landscape influence and experience. Considering the timing of 388's trip and the locations of caribou, had the wolf moved northwest, she might have missed the caribou entirely, or the encounter might have been delayed.

A reasonable possibility is that the land directed 388's route. The barrens are crisscrossed with trails worn into the tundra over centuries by hundreds of thousands of caribou and other animals (Kelsall, 1968; Thorpe et al., 2001). At river crossings, lakes, or narrow peninsulas, trails converge and funnel towards and away from caribou calving grounds and summer range. Wolves use trails for travel (Paquet et al., 1996; Mech and Boitani, 2003; P. Frame, pers. observation). Thus, the landscape may direct an animal's movements and lead it to where cues, such as the odor of caribou on the wind or scent marks of other wolves, may lead it to caribou.

33 Another possibility is that 388 knew where to find caribou in summer. Sexually immature tundra wolves sometimes follow caribou to calving grounds (D. Heard, unpubl. data). Possibly, 388 had made such journeys in previous years and killed caribou. If this were the case, then in times of local prey scarcity she might travel to areas where she had hunted successfully before. Continued monitoring of tundra wolves may answer questions about how their food needs are met in times of low caribou abundance near dens.

34 Caribou often form large groups while moving south to the tree line (Kelsall, 1968). After a large aggregation of caribou moves through an area, its scent can linger for weeks (Thorpe et al., 2001:104). It is conceivable that 388 detected caribou scent on the wind, which was blowing from the northeast on 19–21 July (Environment Canada, 2003), at the same time her excursion began. Many factors, such as odor strength and wind direction and strength, make systematic study of scent detection in wolves difficult under field conditions (Harrington and Asa, 2003). However, humans are able to smell odors such as forest fires or oil refineries more than 100 km away. The olfactory capabilities of dogs, which are similar to wolves, are thought to be 100 to 1 million times that of humans (Harrington and Asa, 2003). Therefore, it is reasonable to think that under the right wind conditions, the scent of many caribou traveling together could be detected by wolves from great distances, thus triggering a long foraging bout.

Rate of Travel

35 Mech (1994) reported the rate of travel of Arctic wolves on barren ground was 8.7 km/h during regular travel and 10.0 km/h when returning to the den, a difference of 1.3 km/h. These rates are based on direct observation and exclude periods when wolves moved slowly or not at all. Our calculated travel rates are assumed to include periods of slow movement or no movement. However, the pattern we report is similar to that reported by Mech (1994), in that homeward travel was faster than regular travel by 1.6 km/h. The faster rate on return may be explained by the need to return food to the den. Pup survival can increase with the number of adults in a pack available to deliver food to pups (Harrington et al., 1983). Therefore, an increased rate of travel on homeward trips could improve a wolf's reproductive fitness by getting food to pups more quickly.

Fate of 388's Pups

36 Wolf 388 was caring for pups during den observations. The pups were estimated to be six weeks old, and were seen ranging as far as 800 m from the den. They received some regurgitated food from two of the females, but were unattended for long periods. The excursion started 16 days after our observations, and it is improbable that the pups could have traveled the distance that 388 moved. If the pups died, this would have removed parental responsibility, allowing the long movement.

Our observations and the locations of radio-collared caribou indicate that prey became scarce in

31 Here our authors note other possible explanations for wolves' excursions presented by other investigators, but this study does not seem to support those ideas.

32 Authors discuss possible reasons for why 388 traveled directly to where caribou were located. They take what they learned from earlier studies and apply it to this case, suggesting that the lay of the land played a role. Note that their description paints a clear picture of the landscape.

33 Authors suggest that 388 may have learned in traveling during previous summers where the caribou were. The last two sentences suggest ideas for future studies.

34 Or maybe 388 followed the scent of the caribou. Authors acknowledge difficulties of proving this, but they suggest another area where future studies might be done.

35 Authors suggest that results of this study support previous studies about how fast wolves travel to and from the den. In the last sentence, they speculate on how these observed patterns would fit into the theory of evolution.

36 Authors also speculate on the fate of 388's pups while she was traveling. This leads to . . .

37 Discussion of cooperative rearing of pups and, in turn, to speculation on how this study and what is known about cooperative rearing might fit into the animal's strategies for survival of the species. Again, the authors approach the broader theory of evolution and how it might explain some of their results.

38 And again, they suggest that this study points to several areas where further study will shed some light.

39 In conclusion, the authors suggest that their study supports the hypothesis being tested here. And they touch on the implications of increased human activity on the tundra predicted by their results.

40 ACKNOWLEDGEMENTS: Authors note the support of institutions, companies and individuals. They thank their reviewers ad list permits under which their research was carried on.

41 REFERENCES: List of all studies cited in the report. This may seem tedious, but is a vitally important part of scientific reporting. It is a record of the sources of information on which this study is based. It provides readers with a wealth of resources for further reading on this topic. Much of it will form the foundation of future scientific studies like this one.

the area of the den as summer progressed. Wolf 388 may have abandoned her pups to seek food for herself. However, she returned to the den after the excursion, where she was seen near pups. In fact, she foraged in a similar pattern before and after the excursion, suggesting that she again was providing for pups after her return to the den.

37 A more likely possibility is that one or both of the other lactating females cared for the pups during 388's absence. The three females at this den were not seen with the pups at the same time. However, two weeks earlier, at a different den, we observed three females cooperatively caring for a group of six pups. At that den, the three lactating females were observed providing food for each other and trading places while nursing pups. Such a situation at the den of 388 could have created conditions that allowed one or more of the lactating females to range far from the den for a period, returning to her parental duties afterwards. However, the pups would have been weaned by eight weeks of age (Packard et al., 1992), so nonlactating adults could also have cared for them, as often happens in wolf packs (Packard et al., 1992; Mech et al., 1999).

Cooperative rearing of multiple litters by a pack could create opportunities for long-distance foraging movements by some reproductive wolves during summer periods of local food scarcity. We have recorded multiple lactating females at one or more tundra wolf dens per year since 1997. This reproductive strategy may be an adaptation to temporally and
38 spatially unpredictable food resources. All of these possibilities require further study, but emphasize both the adaptability of wolves living on the barrens and their dependence on caribou.

Long-range wolf movement in response to caribou
39 availability has been suggested by other researchers (Kuyt, 1972; Walton et al., 2001) and traditional ecological knowledge (Thorpe et al., 2001). Our report demonstrates the rapid and extreme response of wolves to caribou distribution and movements in summer. Increased human activity on the tundra (mining, road building, pipelines, ecotourism) may influence caribou movement patterns and change the interactions between wolves and caribou in the region. Continued monitoring of both species will help us to assess whether the association is being affected adversely by anthropogenic change.

40 Acknowledgements

This research was supported by the Department of Resources, Wildlife, and Economic Development, Government of the Northwest Territories; the Department of Biological Sciences at the University of Alberta; the Natural Sciences and Engineering Research Council of Canada; the Department of Indian and Northern Affairs Canada; the Canadian Circumpolar Institute; and DeBeers Canada, Ltd. Lorna Ruechel assisted with den observations. A. Gunn provided caribou location data. We thank Dave Mech for the use of GPS collars. M. Nelson, A. Gunn, and three anonymous reviewers made helpful comments on earlier drafts of the manuscript. This work was done under Wildlife Research Permit – WL002948 issued by the Government of the Northwest Territories, Department of Resources, Wildlife, and Economic Development.

41 References

BALLARD, W.B., AYRES, L.A., KRAUSMAN, P.R., REED, D.J., and FANCY, S.G. 1997. Ecology of wolves in relation to a migratory caribou herd in northwest Alaska. Wildlife Monographs 135. 47 p.

BROWN, M.B., and FORSYTHE, A.B. 1974. Robust tests for the equality of variances. Journal of the American Statistical Association 69:364–367.

CARLSON, A. 1985. Central place foraging in the red-backed shrike (*Lanius collurio* L.): Allocation of prey between forager and sedentary consumer. Animal Behaviour 33:664–666.

CLARK, K.R.F. 1971. Food habits and behavior of the tundra wolf on central Baffin Island. Ph.D. Thesis, University of Toronto, Ontario, Canada.

ENVIRONMENT CANADA. 2003. National climate data information archive. Available online: http://www.climate.weatheroffice.ec.gc.ca/Welcome_e.html

FOURNIER, B., and GUNN, A. 1998. Musk ox numbers and distribution in the NWT, 1997. File Report No. 121. Yellowknife: Department of Resources, Wildlife, and Economic Development, Government of the Northwest Territories. 55 p.

FRITTS, S.H., and MECH, L.D. 1981. Dynamics, movements, and feeding ecology of a newly protected wolf population in northwestern Minnesota. Wildlife Monographs 80. 79 p.

GUNN, A., DRAGON, J., and BOULANGER, J. 2001. Seasonal movements of satellite-collared caribou from the Bathurst herd. Final Report to the West Kitikmeot Slave Study Society, Yellowknife, NWT. 80 p. Available online: http://www.wkss.nt.ca/HTML/08_ProjectsReports/PDF/Seasonal MovementsFinal.pdf

HARRINGTON, F.H., and ASA, C.S. 2003. Wolf communication. In: Mech, L.D., and Boitani, L., eds. Wolves: Behavior, ecology, and conservation. Chicago: University of Chicago Press. 66–103.

HARRINGTON, F.H., MECH, L.D., and FRITTS, S.H. 1983. Pack size and wolf pup survival: Their relationship under varying ecological conditions. Behavioral Ecology and Sociobiology 13:19–26.

HARRIS, S.A. 1986. Permafrost distribution, zonation and stability along the eastern ranges of the cordillera of North America. Arctic 39(1):29–38.

HEARD, D.C., and WILLIAMS, T.M. 1992. Distribution of wolf dens on migratory caribou ranges in the Northwest

Territories, Canada. Canadian Journal of Zoology 70:1504–1510.
HEARD, D.C., WILLIAMS, T.M., and MELTON, D.A. 1996. The relationship between food intake and predation risk in migratory caribou and implication to caribou and wolf population dynamics. Rangifer Special Issue No. 2:37–44.
KELSALL, J.P. 1968. The migratory barren-ground caribou of Canada. Canadian Wildlife Service Monograph Series 3. Ottawa: Queen's Printer. 340 p.
KUYT, E. 1962. Movements of young wolves in the Northwest Territories of Canada. Journal of Mammalogy 43:270–271.
———. 1972. Food habits and ecology of wolves on barren-ground caribou range in the Northwest Territories. Canadian Wildlife Service Report Series 21. Ottawa: Information Canada. 36 p.
MECH, L.D. 1994. Regular and homeward travel speeds of Arctic wolves. Journal of Mammalogy 75:741–742.
MECH, L.D., and BOITANI, L. 2003. Wolf social ecology. In: Mech, L.D., and Boitani, L., eds. Wolves: Behavior, ecology, and conservation. Chicago: University of Chicago Press. 1–34.
MECH, L.D., and GESE, E.M. 1992. Field testing the Wildlink capture collar on wolves. Wildlife Society Bulletin 20:249–256.
MECH, L.D., WOLFE, P., and PACKARD, J.M. 1999. Regurgitative food transfer among wild wolves. Canadian Journal of Zoology 77:1192–1195.
MERRILL, S.B., and MECH, L.D. 2000. Details of extensive movements by Minnesota wolves (*Canis lupus*). American Midland Naturalist 144:428–433.
MERRILL, S.B., ADAMS, L.G., NELSON, M.E., and MECH, L.D. 1998. Testing releasable GPS radiocollars on wolves and white-tailed deer. Wildlife Society Bulletin 26:830–835.
MESSIER, F. 1985. Solitary living and extraterritorial movements of wolves in relation to social status and prey abundance. Canadian Journal of Zoology 63:239–245.
MOHR, C.O., and STUMPF, W.A. 1966. Comparison of methods for calculating areas of animal activity. Journal of Wildlife Management 30:293–304.
MUSIANI, M. 2003. Conservation biology and management of wolves and wolf-human conflicts in western North America. Ph.D. Thesis, University of Calgary, Calgary, Alberta, Canada.
ORIANS, G.H., and PEARSON, N.E. 1979. On the theory of central place foraging. In: Mitchell, R.D., and Stairs, G.F., eds. Analysis of ecological systems. Columbus: Ohio State University Press. 154–177.
PACKARD, J.M., MECH, L.D., and REAM, R.R. 1992. Weaning in an arctic wolf pack: Behavioral mechanisms. Canadian Journal of Zoology 70:1269–1275.
PAQUET, P.C., WIERZCHOWSKI, J., and CALLAGHAN, C. 1996. Summary report on the effects of human activity on gray wolves in the Bow River Valley, Banff National Park, Alberta. In: Green, J., Pacas, C., Bayley, S., and Cornwell, L., eds. A cumulative effects assessment and futures outlook for the Banff Bow Valley. Prepared for the Banff Bow Valley Study. Ottawa: Department of Canadian Heritage.
STEPHENSON, R.O., and JAMES, D. 1982. Wolf movements and food habits in northwest Alaska. In: Harrington, F.H., and Paquet, P.C., eds. Wolves of the world. New Jersey: Noyes Publications. 223–237.
THORPE, N., EYEGETOK, S., HAKONGAK, N., and QITIRMIUT ELDERS. 2001. The Tuktu and Nogak Project: A caribou chronicle. Final Report to the West Kitikmeot/Slave Study Society, Ikaluktuuttiak, NWT. 160 p.
TIMONEY, K.P., LA ROI, G.H., ZOLTAI, S.C., and ROBINSON, A.L. 1992. The high subarctic forest-tundra of northwestern Canada: Position, width, and vegetation gradients in relation to climate. Arctic 45(1):1–9.
WALTON, L.R., CLUFF, H.D., PAQUET, P.C., and RAMSAY, M.A. 2001. Movement patterns of barren-ground wolves in the central Canadian Arctic. Journal of Mammalogy 82:867–876.
WILLIAMS, T.M. 1990. Summer diet and behavior of wolves denning on barren-ground caribou range in the Northwest Territories, Canada. M.Sc. Thesis, University of Alberta, Edmonton, Alberta, Canada.
ZAR, J.H. 1999. Biostatistical analysis. 4th ed. New Jersey: Prentice Hall. 663 p.

Appendix IV A Plain English Map of the Human Chromosomes

Haploid set of human chromosomes. The banding patterns characteristic of each type of chromosome appear after staining with a reagent called Giemsa. The locations of some of the 20,065 known genes (as of November, 2005) are indicated. Also shown are locations that, when mutated, cause some of the genetic diseases discussed in the text.

For more information go to http://biology.brookscole.com/BTT2.

Glossary of Biological Terms

abscisic acid Plant hormone that stimulates stomata to close in response to water stress; induces dormancy in some buds and seeds.

abscission Process by which plant parts are shed in response to seasonal change, drought, injury, or nutrient deficiency.

accessory pigment Photosynthetic pigment other than chlorophyll *a*.

acid Any substance that releases hydrogen ions in water to lower the pH below 7.

acid rain Acidic precipitation; rain or snow with high levels of sulfur and nitrogen oxides.

actin Structural protein with roles in cell shape, cell motility, and muscle contraction.

action potential Of an excitable cell, a self-propagating, abrupt reversal in the voltage difference across the plasma membrane.

active site Chemically stable crevice in an enzyme where substrates bind and a reaction can be catalyzed repeatedly.

active transport Pumping of a solute across a cell membrane against its concentration gradient, through the interior of a transport protein. Requires energy input, as from ATP.

acute inflammation Fast, local innate immune response. Outward signs are redness, pain, swelling, warmth.

adaptation Any heritable aspect of form, function, or behavior that helps an organism survive and reproduce.

adaptive immunity Capacity of the vertebrate body to defend itself against a tremendous array of specific pathogens; includes antibody-mediated and cell-mediated immune responses.

adaptive radiation Macroevolutionary pattern. A burst of genetic divergences from a lineage; gives rise to many species able to exploit a novel resource or new habitat.

adaptive zone A set of different niches that become filled by a group of related species.

ADH Antidiuretic hormone. A hormone released by the posterior pituitary gland; induces water conservation by kidneys.

adhering junction Complex of adhesion proteins that anchors cells to each other and to extracellular matrixes.

adipose tissue Type of connective tissue specialized for storage of fats.

adrenal cortex Outer zone of the adrenal gland; secretes cortisol and aldosterone.

aerobic respiration Metabolic pathway that uses oxygen to break down carbohydrates. Typical energy yield is 36 ATP per glucose.

age structure Of a population, the way that individuals are distributed among different age categories.

aldosterone Hormone secreted by adrenal cortex; causes sodium retention in kidneys.

algal bloom Explosion in population size of phytoplankton after nutrient enrichment of an aquatic habitat; *e.g.*, a red tide.

alkylating agent Chemical that transfers small hydrocarbon groups to nucleotides in DNA. Causes mutation.

allantois One of four membranes in amniote eggs; exchanges gases and stores metabolic wastes in reptiles, birds, some mammals.

allele frequency Abundance of one allele relative to others in a population.

alleles Two or more molecular forms of a gene that encode slightly different versions of the same trait. Arise by mutation.

allergen A normally harmless substance that provokes an immune response.

allergy Sensitivity to an allergen.

allopatric speciation Speciation model in which a physical barrier that arises separates two populations, so gene flow between them ends.

alveolus, plural **alveoli** In a lung, one of the thin-walled, cup-shaped outpouchings where air exchanges gases with blood.

amino acid Monomer of polypeptide chains. Consists of a carboxyl group, an amino group, and a characteristic functional group (R).

ammonification Fungi and bacteria in soil degrade nitrogen-rich remains and wastes, then release ammonia and ammonium ions.

amnion One of the extraembryonic membranes of amniote eggs; the boundary layer of a fluid-filled sac in which an embryo develops.

amniote Animal that produces amniote eggs; a reptile, bird, or mammal.

amphibian Ectothermic (cold-blooded) tetrapod vertebrate with a three-chambered heart and thin skin that functions in respiration.

anagenesis Pattern of speciation in which one species evolves into another over time without a branching event.

analogous structures Similar body parts in lineages that are not closely related.

anatomy The study of the body parts of an organism to determine their structure and positions relative to one another.

aneuploidy Chromosome abnormality in which body (somatic) cells have one extra or one less chromosome relative to the parental number.

animal Multicelled heterotroph that moves actively during part or all of the life cycle and passes through developmental stages.

annelid Soft-bodied, bilateral invertebrate with a coelom, organs, and many segments; a leech, polychaete, or oligochaete.

anther Pollen-bearing part of a stamen.

anthocyanin Type of accessory pigment that reflects red to blue light, depending on pH.

antibiotic A product of one microorganism that inhibits or kills another. Used as drugs.

antibodies Antigen-binding proteins made and secreted only by B cells that encounter antigen; part of adaptive immune responses.

antibody-mediated immune response Part of adaptive immunity in which antibodies are produced against a specific antigen.

anticodon Series of three nucleotide bases in a tRNA that can base-pair with mRNA codons.

antigen Molecule that the body recognizes as foreign and triggers an immune response.

antioxidant Enzyme or cofactor that prevents damage to biological molecules by neutralizing free radicals and other strong oxidizers.

aorta Of vertebrates, the main artery of systemic circulation.

apical meristem *See* meristem.

apicomplexan A parasitic alveolate protist with a unique device at its anterior end that attaches to and penetrates a host cell.

apoptosis Programmed cell death.

appendix Small, narrow outpouching from the first portion of the large intestine.

Archaea Domain of prokaryotic species that share some traits with bacteria and other traits with eukaryotic species.

Archean Of the geologic time scale, the interval that extended from 3.8 billion years ago, to 2.5 billion years ago.

area effect Larger islands support more species than smaller ones the same distance from a source of colonizing species.

arterioles Blood vessels between arteries and capillaries. Constriction or dilation adjusts blood flow to different organs.

asexual reproduction Any reproductive mode by which offspring arise from one parent and inherit that parent's genes only.

atom Fundamental unit of matter; consists of protons, neutrons, and electrons.

ATP Adenosine triphosphate, a nucleotide that is the primary energy carrier between reaction sites in cells.

ATP synthase A type of membrane-bound active transport protein that catalyzes the formation of ATP.

ATP/ADP cycle How a cell regenerates ATP. ATP forms as ADP binds phosphate; ADP forms when ATP loses a phosphate.

australopith One of the early hominids; known only from fossils found in Africa.

autoimmune response Inappropriate lymphocyte attack on normal body cells.

autonomic nerves Nerves that carry signals from the central nervous system to smooth muscle, cardiac muscle, and glands of the viscera, or relay signals about the internal environment to the central nervous system.

autosome Of a sexually reproducing species, a chromosome that is the same in both males and females.

autotroph Organism that uses energy captured from sunlight or inorganic chemicals to make its own food from inorganic compounds.

auxin Major plant hormone; coordinates growth, tissue organization, and responses to environmental cues.

axon A neuron's signal-conducting zone; action potentials typically self-propagate away from the cell body on this long cellular extension.

B lymphocyte B cell. Type of white blood cell that is central to antibody-mediated immune responses; the only cell that makes antibodies.

Bacteria Domain of prokaryotic species; the most metabolically diverse organisms and the most common prokaryotes.

bacteriophage Virus that infects bacteria.

balanced polymorphism Two or more alleles for a trait persist in a population; the outcome of selection against homozygotes.

bark Of woody plants, all tissues that are external to the vascular cambium.

Barr body Inactivated X chromosome in the somatic cells of female mammals.

base Any substance that causes the pH of water to rise above 7.0. Also the nitrogen-containing component of a nucleotide.

base-pair substitution Mutation in which one nucleotide replaces another.

basophil Circulating white blood cell that releases histamine and other inflammatory mediators in response to antigen detection.

bell curve Idealized distribution of continuous variation of a trait in a population.

bilateral symmetry Body plan in which the main axis divides the body into two halves that are mirror images of one another.

bile Mix of salts, cholesterol, and pigments made by the liver and used in fat digestion.

biodiversity Biological diversity, measured by the number of genes, species, habitats.

biogeochemical cycle Slow movement of an element from environmental reservoirs, through food webs, then back.

biogeographic realm Vast expanse of land defined by a particular array of species.

biogeography Scientific study of patterns in the geographic distribution of species.

biological clock Internal time-measuring mechanism by which individuals adjust their activities seasonally, daily, or both in response to cyclic environmental cues.

biological magnification Concentration of a contaminant in body tissues increases as it is passed along a food chain.

biological species concept Defines a sexually reproducing species as one or more populations that interbreed under natural conditions, produce fertile offspring, and are reproductively isolated from other such populations.

biomass pyramid Chart in which the size of successive tiers depicts the measured biomass (dry weight) of some ecosystem's producers, consumers, and decomposers.

biome Region that supports a characteristic type of vegetation; *e.g.*, desert, tundra.

biosphere All areas of Earth's land, water, and air that can sustain some form of life.

biotic potential Theoretical maximum rate of reproduction under ideal conditions.

bipedalism Habitually walking upright on two feet, as by ostriches and hominids.

bird Endothermic (warm-blooded), feathered amniote with a four-chambered heart; the dinosaurs and crocodilians are closest relatives.

blastocyst Blastula of placental mammals.

blastula Early stage of animal embryonic development; a ball of blastomeres formed by cleavage surrounds a cavity filled with their secretions, and an inner cell mass that becomes the embryo.

blood Fluid connective tissue, a transport medium of circulatory systems. In vertebrates, it consists of plasma, cells, and platelets.

blood–brain barrier Blood capillaries that protect the brain and spinal cord by exerting some control over which solutes enter the cerebrospinal fluid.

blood pressure Fluid pressure generated by contraction of the ventricles.

bond *See* chemical bond.

bone tissue Of most vertebrate skeletons, a mineral-hardened connective tissue.

bony fish Fish that has an endoskeleton composed mostly of bone tissue.

bottleneck Severe reduction in the size of a population, brought about by intense selection pressure or a natural calamity.

brain Of most nervous systems, the major integrating center that receives, processes, and stores sensory input, and issues coordinated commands for responses.

brain stem The most ancient nerve tissue in all three divisions of a vertebrate brain.

bronchiole Finely branched airway that is part of the bronchial tree inside a lung.

bronchus, plural **bronchi** A tubular airway that extends from the trachea into a lung.

bryophyte Nonvascular land plant. The haploid stage dominates its life cycle; sperm must swim to eggs; a moss, hornwort, or liverwort.

bud Undeveloped shoot of meristem tissue; main zone of primary growth in plants.

buffer system Weak acid or base and its salt that together stabilize pH of a solution.

bulk flow Mass movement of molecules in one direction, usually in response to a pressure gradient.

C3 plant Type of plant in which three-carbon PGA is the first stable intermediate to form after carbon fixation.

C4 plant Type of plant in which four-carbon oxaloacetate is the first stable intermediate to form. Carbon is fixed twice, in two types of photosynthetic cells.

calcium pump Active transport protein; pumps calcium ions across a cell membrane against their concentration gradient.

Calvin–Benson cycle Cyclic pathway that forms sugars in the stroma of chloroplasts; also called light-independent reactions.

CAM plant Type of plant that conserves water by opening stomata only at night, when it fixes carbon by the C4 pathway.

camouflage Coloration, pattern, form, or behavior that helps an organism to blend with its surroundings.

cancer Mass of abnormally dividing cells that can invade and colonize other parts of the body. Also, malignant neoplasm.

capillary, blood Smallest diameter blood vessel; the exchanges between interstitial fluid and blood occur across its one-cell-thick wall.

capillary bed One of the diffusion zones for circulatory systems; great numbers of them exchange substances with interstitial fluid.

capture–mark–recapture method Method of estimating population size of a mobile species.

carbohydrates Organic molecules of carbon, hydrogen, and oxygen typically in a 1:2:1 ratio. Structural materials or energy reservoirs.

carbon cycle Atmospheric cycle. Carbon moves from environmental reservoirs (sediments, the ocean, rocks), through the atmosphere (mostly as CO_2), food webs, and back to the reservoirs.

carbon fixation Process by which a cell incorporates carbon atoms from CO_2 into a stable organic compound.

cardiac conduction system Specialized cardiac muscle cells that initiate and send signals that stimulate other cardiac muscle cells to contract.

cardiac cycle Sequence of muscle contraction and relaxation that occurs during one heartbeat.

cardiac muscle tissue The contractile tissue present only in the heart wall.

cardiac pacemaker The sinoatrial (SA) node; a cluster of self-excitatory cardiac muscle cells that set the normal rate of heartbeat.

carotenoid One of a class of accessory pigments in photosynthesis that reflect red, orange, or yellow light.

carpel Female reproductive part of flowers; a sticky or hairy stigma above the ovary.

carrying capacity The maximum number of individuals of a species that some particular environment can sustain indefinitely.

cartilage Connective tissue composed of collagen fibers in a rubbery matrix; resists compression.

cartilaginous fish Jawed fish having an endoskeleton of cartilage; *e.g.*, sharks.

Casparian strip Waterproof, impermeable band that seals abutting cell walls of root endodermis and exodermis; forces water and solutes to pass through the cells.

catastrophism Idea that Earth's history was shaped by repeated violent, global catastrophes unlike any we see today.

cDNA DNA synthesized from mRNA.

cell Basic structural and functional unit of all organisms; smallest entity with the properties of life.

cell cortex Supportive mesh of crosslinked cytoskeletal elements just underneath the plasma membrane and attached to it.

cell count The number of cells of a given type present in one microliter of blood.

cell cycle Of eukaryotic cells, a series of events from the time a cell forms until it reproduces.

cell differentiation In developing embryos, the process by which different cell lineages selectively express different fractions of their genome and so become specialized.

cell junction Of a tissue, a structure that connects adjoining cells, or a cell to matrix.

cell-mediated immune response Portion of adaptive immunity that targets infected or cancerous body cells. Mediated by T cells.

cell plate formation Cytoplasmic division mechanism of plant cells.

cell theory Theory that all organisms are made of one or more cells, cell is the smallest unit of organization that displays the properties of life, and all cells arise by division of another cell.

cell wall Of many cells (not animal cells), a semirigid but permeable structure that surrounds the plasma membrane.

central vacuole In many mature, living plant cells, an organelle that stores amino acids, sugars, and some wastes.

centromere Of eukaryotic chromosomes, constricted region where sister chromatids attach; binding site for spindles.

cephalization Of most bilateral animals, a concentration of sensory structures and nerve cells at the anterior end of the body.

cerebellum Hindbrain region with reflex centers that affect posture and limb movements.

cerebrum A forebrain region concerned with olfactory input and motor responses. In the mammals, it has evolved into the most complex integrating center.

chemical bond An attractive force that links two atoms.

chemical equilibrium Of reversible reactions, state in which the rate of the forward reaction equals that of the reverse.

chemical synapse A narrow space between a presynaptic neuron and postsynaptic cell that neurotransmitters diffuse across.

chemoreceptor Sensory receptor that can detect dissolved ions or molecules in fluid.

chlorophyll *a* The primary photosynthetic pigment in plants and algae; appears green.

chlorophyll *b* Yellow–green photosynthetic accessory pigment.

chloroplast Organelle of photosynthesis in plants and algae. Two outer membranes enclose a semifluid interior, the stroma. A third, much-folded inner membrane (the thylakoid membrane) forms an inner compartment and functions in ATP and NADPH formation.

cholecystokinin (CCK) Hormone that causes gallbladder contractions and the secretion of pancreatic enzymes.

chordate Animal having a notochord, a dorsal hollow nerve cord, a pharynx with gill slits, and a tail extending past the anus in the embryo.

chorion Extraembryonic membrane of amniote eggs; becomes part of the placenta.

chromatin All of the DNA molecules and the associated proteins in a nucleus.

chromosome In eukaryotic cells, one of many linear DNA molecules with attached histones and other proteins. Prokaryotic cells have one circular chromosome.

chromosome number The sum of all of the chromosomes in cells of a given type.

ciliate A heterotrophic alveolate protist that has many cilia; *e.g.*, *Paramecium*.

cilium, plural **cilia** A motile or sensory structure that projects from the plasma membrane of certain eukaryotic cells.

circadian rhythm Any biological activity repeated in cycles, each about twenty-four hours long, independently of any shifts in environmental conditions.

clade *See* monophyletic group.

cladogenesis Branching speciation pattern; genetic divergences occur after populations become reproductively isolated.

cladogram Evolutionary tree diagram that depicts relative relatedness among groups.

classification system A way to organize and retrieve information about species.

cleavage Process of animal cell division. Also, early stage of animal development in which mitosis divides a fertilized egg into smaller, nucleated cells (blastomeres).

climate Weather conditions that typically prevail in a region.

clone A genetically identical copy of DNA, a cell, or a multicelled organism.

cloning vector Any DNA molecule that can accept foreign DNA and be replicated inside a host cell.

club fungus Fungus that produces sexual spores in club-shaped cells.

cnidarian A radial invertebrate with epithelial tissues and a saclike gut; makes nematocysts.

codon Linear sequence of three nucleotides in mRNA; codes for an amino acid or a termination signal in protein synthesis.

coelom Of many invertebrates and all the vertebrates, a cavity lined with peritoneum that forms between the gut and body wall.

coenzyme An organic molecule that is a necessary participant in some enzymatic reactions; *e.g.*, a vitamin or NAD^+.

coevolution The joint evolution of two species interacting so closely that a change in one exerts selection pressure on the other.

cofactor A metal ion or a coenzyme that assists enzyme catalysis by alternately accepting and donating atoms, electrons, or functional groups.

cohesion The tendency of like molecules, particularly water molecules, to stick together.

cohesion–tension theory The theory that evaporation from leaves creates negative pressure (tension) that pulls columns of water from roots up through xylem.

cohort Group of individuals of the same age.

collenchyma A simple, flexible plant tissue that supports growing parts. Alive at maturity.

colon *See* large intestine.

commensalism Interspecific interaction; one species benefits and the other is unaffected.

communication signal Body trait, chemical, or behavior that conveys information to other members of the same species.

community All populations of all species that occupy a given habitat.

companion cell Specialized parenchyma cell that helps load sugars into sieve tubes.

comparative morphology Scientific study of comparable body parts in the embryonic and adult forms of major lineages.

compartmentalization Defensive response in some plants; an infected area is walled off by secretion of resins and toxins.

competitive exclusion A superior competitor drives a less competitive one with the same resource needs to local extinction.

complement Set of proteins that circulate in inactive form in blood; binding antigen activates it in cascading reactions. Has many roles in both innate and adaptive immune responses.

compound Molecule consisting of two or more elements in unvarying proportions.

concentration Of a mixture, the number of molecules of a substance per unit volume.

condensation reaction A type of chemical reaction in which two molecules become covalently bonded into a larger molecule; water often forms as a by-product.

cone Reproductive structure of conifers; clusters of scales with exposed ovules.

cone cell A vertebrate photoreceptor that responds to intense light and contributes to sharp daytime vision and color perception.

conifer A cone-producing gymnosperm.

conjugation Among prokaryotes, transfer of a plasmid from one cell to another.

conservation biology The field of inquiry that surveys biodiversity, researches its evolutionary history, and determines how to maintain and use it to benefit humans.

consumer Organism that feeds on the tissues of other organisms; heterotroph.

continuous variation Of individuals of a population, a range of small differences in the phenotypic expression of a particular trait.

control group In experiments, a group used as a standard for comparison with one or more experimental groups.

cork Tissue component of bark with many thickened layers; waterproofs, insulates, and protects woody stem and root surfaces.

cork cambium A lateral meristem, descendants of which replace epidermis with cork on woody plant parts.

corpus luteum A glandular structure that forms from cells of a ruptured follicle after ovulation; its progesterone secretions cause the uterine lining to thicken in preparation for pregnancy.

cotyledon Seed leaf; part of a flowering plant embryo. Eudicot seeds have two, and monocot seeds have one.

countercurrent exchange Exchange of gas or solutes between two fluids that are flowing in different directions on opposite sides of a semipermeable membrane.

covalent bond A type of chemical bond that forms when two atoms share a pair of electrons. Sharing is unequal in a *polar covalent bond*, and equal in a *nonpolar covalent bond*.

craniate A chordate with a cranium of bone or cartilage; a hagfish or vertebrate.

crossing over During prophase I of meiosis, exchange of segments between the nonsister chromatids of homologous chromosomes; puts novel combinations of alleles in gametes.

culture Sum of behavior patterns of a social group, passed between generations by learning and symbolic behavior.

cuticle Of plants, a protective surface layer secreted by epidermal cells. Of annelids, roundworms, and arthropods, a noncellular body covering. In the latter two groups it is periodically molted.

cycad A gymnosperm of subtropical or tropical habitats; has pollen-bearing and seed-bearing strobili on separate plants.

cyst Of many microbes, a resting stage with a secreted cover.

cytokines Signaling molecules of the vertebrate immune system that collectively coordinate and regulate all aspects of immunity; interleukins, interferons, and tumor necrosis factor.

cytokinins Plant hormones that promote cell division, stimulate lateral bud growth, and retard leaf aging.

cytoplasm Semifluid matrix between the plasma membrane and nucleus or nucleoid.

cytoplasmic localization The accumulation of proteins or RNAs in specific regions of egg cytoplasm. The "maternal messages" are sorted into different blastomeres during cleavage.

cytoskeleton In a eukaryotic cell, the dynamic framework of protein filaments that support, organize, and move the cell and internal structures. Prokaryotic cells have a few similar protein filaments.

decomposer Prokaryotic or fungal heterotroph; obtains carbon and energy by breaking down organic wastes or remains.

deforestation Removal of all trees from a large tract of land.

deletion Loss of a chromosome segment; often leads to genetic disorders. Also, loss of one or more nucleotides from a gene.

demographic transition model Model that correlates changes in population growth with stages of economic development.

demographics Vital statistics of a population; *e.g.*, size, age structure.

denaturation Unravelling of a molecule's three-dimensional shape caused by high temperature, shifts in pH, or detergents.

dendrite A neuron's information-receiving zone; a short extension from the cell body.

dendritic cell Phagocytic white blood cell; presents antigen to naive T cells.

denitrification Conversion of nitrate or nitrite to gaseous nitrogen by metabolic activity of certain bacteria in soil.

dense, irregular connective tissue A type of animal tissue with fibroblasts and many fibers asymmetrically arrayed in a matrix. In skin and some capsules around organs.

dense, regular connective tissue A type of animal tissue with rows of fibroblasts between parallel bundles of fibers. Occurs in tendons and elastic ligaments.

density-dependent control Any factor that comes into play in an overcrowded population; reduces birth rate or raises death and dispersal rates, as by intensified predation, parasitism, disease, competition.

density-independent factor A factor that causes fewer births or more deaths in a population regardless of its density; *e.g.*, a severe storm.

deoxyribonucleic acid *See* DNA.

derived trait A novel feature shared only by descendants of an ancestral species in which it originated.

dermis The lower layer of skin, beneath the epidermis; mostly dense connective tissue.

desalinization Removal of salt from water.

desertification Conversion of grassland or cropland to desertlike conditions.

detrital food web Cross-connecting food chains in which energy flows primarily from plants through arrays of detritivores and decomposers (rather than herbivores).

detritivore Any animal that feeds on decomposing particles of organic matter.

deuterostome A member of the bilateral, coelomate, animal lineage in which the second indentation to form on the early embryo's surface becomes a mouth.

development Of complex, multicelled species, a series of stages by which an embryo becomes an adult organism.

diaphragm A muscular partition between the thoracic and abdominal cavities. Also, a birth control device that covers the cervix and thus prevents sperm from entering uterus.

diffusion Spontaneous dispersal of ions, molecules, or energy following a gradient.

digestive system Body sac or tube, often with specialized regions where food is ingested, digested, and absorbed, and any residues are expelled. Incomplete systems have one opening; complete systems have two (mouth and anus).

dihybrid Individual that is heterozygous for two genes; *e.g.*, *AaBb*.

dihybrid experiment Cross between two identical dihybrids; *e.g.*, *AaBb* x *AaBb*.

dinoflagellate Flagellated, usually marine, protist; an alveolate with cellulose plates; photosynthetic and heterotrophic species.

dinosaur One of a group of reptiles that arose in the Triassic and became the dominant land vertebrates for 125 million years.

diploid Having two of each type of chromosome characteristic of the species.

directional selection Mechanism of natural selection by which forms at one extreme of a range of phenotypic variation are favored.

disease Illness caused by infection, diet, or environmental factors.

disruptive selection Mechanism of natural selection that favors different forms of a trait at both ends of a range of variation; intermediate forms are selected against.

distance effect Islands close to a source of colonists show greater species richness than islands farther from such a source.

DNA Deoxyribonucleic acid. Nucleic acid polymer of adenine, guanine, cytosine, and thymine bases, the sequence of which encodes primary hereditary information for all living organisms and many viruses.

DNA fingerprint The unique pattern of fragments of an individual's genomic DNA.

DNA polymerase Type of enzyme that catalyzes the addition of free nucleotides to new DNA strands during replication.

DNA replication Process by which a cell duplicates its DNA molecules before it divides into daughter cells.

DNA sequencing Method of determining the linear sequence of nucleotides in DNA.

dominant allele Of diploid cells, an allele that masks the phenotypic effect of any recessive allele paired with it.

dormancy Of many spores, cysts, seeds, perennials, and some animals, a predictable time of metabolic inactivity.

dosage compensation Gene control mechanism that balances gene expression between the sexes; one X chromosome in female mammal somatic cells is inactivated.

doubling time Time it takes for a population to double in size.

downwelling Water is forced down and away from a coast after winds shift and make a surface current pile into the coast.

drug addiction Dependence on a drug, which has assumed an "essential" biochemical role following habituation and tolerance.

duplication Base sequence in DNA that has been repeated two or more times.

echinoderm A radial, predatory invertebrate with bilateral larvae and calcified spines or plates on the body wall; *e.g.*, sea stars.

ecological succession Replacement of one array of species in a community by another.

ecology Study of the ways in which organisms interact with one another and the environment.

ecoregion Broad region of land or aquatic habitat characterized by certain physical and biological factors.

ecosystem Community interacting with its physical environment.

ectoderm Outermost primary tissue layer of animal embryos; gives rise to the vertebrate nervous tissue and outer part of skin.

Ediacarans Collection of tiny precambrian species that may have been some of the earliest multicelled life forms.

effector Muscle (or gland); helps bring about movement (or chemical change) in response to neural or endocrine signals.

egg Mature female gamete, or ovum.

El Niño Periodic warming of waters in the western equatorial Pacific; disrupts ocean currents, upwelling, and global climate.

electron Negatively charged particle that occupies orbitals around an atom's nucleus.

electron transfer chain Array of enzymes and other molecules in a cell membrane that accept and give up electrons in sequence; operation of chain releases the energy of the electrons in small, usable increments.

electron transfer phosphorylation Final stage of aerobic respiration; electron flow through electron transfer chains embedded in the inner mitochondrial membrane sets up H^+ gradients that drive ATP formation.

element Specific kind of atom defined by the number of protons in its nucleus.

embryonic induction A change in cells in an embryo after exposure to signals from cells of nearby tissues; basis of pattern formation.

emigration Permanent move of one or more individuals out of a population.

endangered species A species endemic (native) to a habitat, found nowhere else, and highly vulnerable to extinction.

endocrine gland A ductless gland that secretes hormone molecules into the blood.

endocrine system Control system of cells, tissues, and organs; secretes hormones and other signaling molecules.

endocytosis The cellular uptake of external substances by formation of vesicles from patches of plasma membrane.

endoderm Inner primary tissue layer of animal embryos; source of the inner gut lining and organs derived from it.

endodermis Cylindrical, sheetlike layer of cells around the root vascular cylinder; helps control water and solute uptake.

endomembrane system Set of organelles that includes endoplasmic reticulum, Golgi bodies, and transport vesicles.

endometrium Inner lining of the uterus.

endoplasmic reticulum (ER) Organelle that extends from the nuclear envelope through the cytoplasm. New polypeptide chains become modified in the ribosome-studded rough ER. Membrane lipids are assembled, and fatty acids and toxins are broken down in the smooth ER.

endoskeleton Of chordates, internal framework of cartilage, bone, or both; works with skeletal muscle to position, support, and move the body.

endosperm Triploid ($3n$) nutritive tissue in the seeds of flowering plants.

endosymbiosis Intimate, permanent ecological interaction in which one species lives inside and reproduces in another's body.

energy The capacity to do work.

energy pyramid Diagram that depicts the energy stored in the tissues of organisms at each trophic level in an ecosystem. The lowest tier (producers) is always the largest.

ENSO El Niño Southern Oscillation. A warming and then cooling of waters in the western Pacific Ocean.

enzyme A type of protein or RNA that speeds up (catalyzes) a specific reaction, and is also unchanged by the reaction.

eosinophil Circulating white blood cell that releases enzymes and toxins that can kill pathogens too large for phagocytosis.

epidermis Outermost tissue layer of plants and nearly all animals.

epithelial tissue *See* epithelium.

epithelium Animal tissue that covers external and internal body surfaces.

ER *See* endoplasmic reticulum.

erosion *See* soil erosion.

esophagus Muscular tube between the pharynx and stomach.

essential amino acid Any amino acid that an organism must obtain from food.

essential fatty acid Any fatty acid that an organism must obtain from food.

estrogen A sex hormone produced by the ovaries and placenta. Primes endometrium for pregnancy; affects growth, development, and female secondary sexual traits.

ethylene Gaseous plant hormone; induces fruit ripening and abscission.

eudicot Member of the most diverse group of flowering plants; has net-veined leaves, and an embryo with two cotyledons.

eukaryotic cell Type of cell that starts life with a nucleus and other organelles.

eutherian Placental mammal.

eutrophication Nutrient enrichment of water; promotes phytoplankton growth.

evaporation Process of conversion of a liquid to a gas; requires energy input.

evolution, biological Heritable change in a line of descent; occurs by microevolutionary events (mutation, natural selection, genetic drift, and gene flow).

evolutionary tree A treelike diagram in which each branch point represents a divergence from a shared ancestor.

exocrine gland A gland that releases secretions, through ducts or tubes, onto a free epithelial surface; *e.g.*, sweat gland, salivary gland.

exocytosis Fusion of a cytoplasmic vesicle with the plasma membrane and release of vesicle contents into extracellular fluid.

exodermis Cylindrical sheet of cells near root epidermis of most flowering plants; helps to control uptake of water and solutes.

exon A base sequence in eukaryotic DNA that remains in mRNA as coding sequence.

exoskeleton An external skeleton.

exotic species Species established in a new community after dispersal from its home range.

experimental group In experiments, a group of objects or individuals that display or are exposed to a variable that is being investigated. Compared to a control group.

exponential growth An increase by the same percent of the whole in each interval.

extinction Irrevocable loss of a species.

extreme halophile Prokaryote adapted to life in an extremely salty habitat.

extreme thermophile Prokaryote adapted to life in an extremely hot aquatic habitat.

FAD Flavin adenine dinucleotide; type of nucleotide coenzyme.

fall overturn Vertical mixing of a body of water in fall as upper waters cool and sink.

fat Type of lipid with one, two, or three fatty acid tails attached to a glycerol.

fatty acid Organic compound that has a long carbon backbone with only single bonds in *saturated* types, and one or more double bonds in *unsaturated* types.

feedback inhibition Mechanism by which a result of some cellular activity triggers a response that slows or stops the activity.

fermentation Type of metabolic pathway that breaks down carbohydrates without using oxygen. Net yield: 2 ATP per glucose. The end product of *alcoholic fermentation* is ethanol; that of *lactate fermentation* is lactate.

fern Seedless vascular plant that forms spores in sori on the underside of fronds.

fertilization Fusion of two gametes, the result being a single-celled zygote.

fetus In human development, the stage from nine weeks until time of birth.

fever An internally caused rise in core body temperature above the normal levels.

fibrous root system Branching roots that extend from the stem of a plant.

fin An appendage that helps stabilize, orient, and propel most fishes in water.

fitness Degree of adaptation to an environment, as measured by the relative genetic contribution to future generations.

fixation Of an allele, loss of all other alleles of a gene in a population.

fixed action pattern Instinctual program of coordinated, stereotyped movements, usually elicited by a simple stimulus.

flagellated protozoan A member of an ancient lineage of heterotrophic single-celled protists that have one or more flagella; *e.g., Giardia*.

flagellum, plural **flagella** Long, whip-like motile structure that projects from a cell.

flatworm Simple bilateral animal with no coelom, saclike gut, three primary tissue layers.

flower Reproductive structure of plants; has nonfertile parts (sepals, petals), fertile parts (stamens, carpels), and a receptacle (modified base of floral shoot).

fluid mosaic model Model that describes the organization of membranes as a mixure of lipids and proteins, the interactions and motions of which impart fluidity to it.

food chain Linear sequence of steps by which energy stored in autotroph tissues enters higher trophic levels.

food web Cross-connecting food chains.

fossil Physical evidence of an organism that lived in the distant past.

fossilization How fossils form over time. An organism or evidence of it gets buried; and metal ions replace the minerals in bones and other hardened tissues.

founder effect A form of bottlenecking. By chance, a few individuals that establish a new population differ in allele frequencies relative to the original population.

free radical Uncharged molecule or atom with an unpaired electron; highly reactive.

fruit Mature ovary, often with accessory parts; forms only in flowering plants.

functional group An atom or a group of atoms that imparts chemical properties to an organic compound.

fungus, plural **fungi** Eukaryotic consumer that obtains nutrients by extracellular digestion and absorption.

GABA Gamma amino butyric acid. A neurotransmitter with inhibitory effects; it suppresses the responses of other neurons.

gallbladder Organ that stores bile from the liver; its duct connects to the small intestine.

gametophyte Haploid, gamete-producing structure of plants and some algae.

ganglion, plural **ganglia** The clustered cell bodies of many neurons.

gap junction Cylindrical arrays of proteins that connect the cytoplasm of adjoining cells.

gastric fluid Mucus, acid, and digestive enzymes secreted by the stomach lining.

gastrula Early animal embryo with two or three primary tissue layers (germ layers).

gel electrophoresis Method of separating DNA fragments according to length. The fragments separate as they move through a gel matrix in response to an electric field.

gene Unit of heritable information; section of DNA that encodes one structural or functional component of an organism.

gene flow Microevolutionary process; alleles enter and leave a population by immigration and emigration. Counters mutation, natural selection, genetic drift.

gene pool All the alleles of all the genes in population; a pool of genetic resources.

genetic code The near-universal language of protein synthesis, the correspondence between nucleotide sequence and protein.

genetic disorder An inherited condition that causes mild to severe medical problems.

genetic divergence Accumulation of differences in the gene pools of two or more populations after gene flow stops and mutation, natural selection, and genetic drift occur in each.

genetic drift Change in allele frequencies in a population due to chance alone. Effects are most pronounced in small populations.

genetic equilibrium In theory, a state in which specific allele frequencies are not changing, so the population is not evolving with respect to that gene.

genome Complete complement of genetic information of an organism or species.

genomics The study of genes and gene function in humans and other organisms.

genus, plural **genera** Group of one or more related species that share a number of unique morphological traits.

geographic dispersal Individuals leave a home range and take root in a new community.

geologic time scale Time scale for Earth's history; major subdivisions correspond to mass extinctions.

germinate To resume metabolism and growth after a period of dormancy.

ghrelin Hormone secreted by the stomach lining; stimulates appetite control center.

gibberellin Plant hormone that induces stem elongation and helps seeds break dormancy; also has a role in flowering in some species.

gill A respiratory organ that has thin, moist, vascularized filaments for gas exchange.

gland A saclike, secretory organ that opens onto a free epithelial surface. Endocrine glands have ducts; exocrine glands do not.

global warming A long-term increase in the temperature of Earth's lower atmosphere.

glomerular filtration First step in urine formation; water and solutes from blood are forced out of glomerular capillaries and into the first portion of a kidney nephron.

glycolysis Set of metabolic reactions in which glucose is broken down to pyruvate; first stage of both aerobic respiration and fermentation. Yields 2 ATP per glucose.

gnetophyte A woody, vinelike or shrubby gymnosperm of the group most closely related to flowering plants.

Golgi body Organelle of the endomembrane system; enzymes in it modify new polypeptide chains and assemble lipids, then package both in vesicles.

Gondwana Paleozoic supercontinent that later became part of Pangea.

gradient Difference in a physical property, such as concentration, electric charge, or pressure, over distance. Drives diffusion.

gradual model, speciation Addresses the rate of speciation; morphological changes are said to accumulate over great time spans.

gravitropism Growth or movement in a direction influenced by gravity.

grazing food web Cross-connecting food chains in which energy flows from plants to an array of herbivores, then carnivores.

greenhouse effect Trapping of heat near Earth's surface by atmospheric gases. The gases absorb heat from Earth's sun-warmed surface, and then radiate some back to Earth.

groundwater Water in soil and porous rock layers called aquifers.

growth Of multicelled species, increases in the number, size, and volume of cells. Of single-celled prokaryotes, rise in the number of cells.

gut Sac or tube in which food is digested.

gymnosperm A seedbearing plant in which seeds form in exposed ovules; a ginkgo, cycad, conifer, or gnetophyte.

habitat Place where an organism or species normally lives; characterized by physical and chemical features and array of species.

habitat loss Reduction in suitable living space and closure of part of a habitat.

half-life Characteristic time it takes for half of a quantity of a radioisotope to decay.

haploid Having one of each type of chromosome characteristic of the species.

hardwood Tree with many vessels and fibers in its wood; angiosperm tree.

hearing Perception of sound.

heart Muscular pump; its contractions circulate blood through the animal body.

heartwood Dense, dry tissue at the core of aging tree stems and roots; helps trees defy gravity and store metabolic wastes.

heme Oxygen-transporting cofactor of many enzymes and pigments; one iron atom at the center of an organic ring structure.

hemoglobin A protein, produced by red blood cells, that functions in oxygen transport.

hemostasis Process that stops blood loss from a damaged blood vessel by spasm, coagulation, and other mechanisms.

heterotroph Organism that obtains energy and carbon by feeding on other organisms, their wastes, or their remains.

heterozygous Condition of having nonidentical alleles at a given locus.

higher taxon, plural **taxa** One of ever more inclusive groupings of species; *e.g.*, family, order, class, phylum, kingdom.

homeostasis State of equilibrium in which an organism maintains its internal environment within a tolerable range.

homeotic gene Gene that, when expressed in a localized region of a developing embryo, helps determine the identity of a body part.

hominid All humanlike and human species; *e.g.*, australopiths, early *Homo*.

homologous chromosome One of a pair of chromosomes in diploid cells; except for nonidentical sex chromosomes, a pair has the same size, shape, and gene sequence.

homologous structures Similar body parts in related lineages; indicate shared ancestry.

homozygous dominant State of having a pair of dominant alleles; *e.g.*, *AA*.

homozygous recessive State of having a pair of recessive alleles; *e.g.*, *Aa*.

hormone Signaling molecule secreted by one cell; alters the activities of another cell bearing receptors for it.

horsetail A seedless vascular plant with rhizomes, scale-like leaves, and hollow stems with silica-reinforced ribs.

hot spot A region where human activities are driving many species to extinction. Also, region where superplumes have ruptured Earth's crust.

human Primate of species *Homo sapiens*.

human gene therapy Transfer of normal or modified genes into a person, often with the intent to correct a genetic defect.

humus Decomposing organic matter in soil.

hybrid Heterozygous individual; also, the offspring of a cross between individuals of different species or different genera.

hydrogen bond Weak chemical bond that forms between a covalently bonded hydrogen atom and another atom taking part in a different polar covalent bond.

hydrologic cycle Driven by solar energy; water moves through atmosphere, on or through land, to the ocean and other bodies of water, and back to the atmosphere.

hydrolysis Cleavage reaction; an enzyme splits a molecule by adding —OH and —H from water to each of the two fragments.

hydrophilic Of a molecule, the property of dissolving easily in water. Also, polar.

hydrophobic Of a molecule, the property of not dissolving in water. Also, nonpolar.

hydrostatic pressure Pressure exerted by a volume of fluid against a cell wall or other enclosing structure. Also, turgor.

hydrostatic skeleton A fluid-filled cavity against which a contractile force can act.

hypertonic Of two fluids, describes the one with the higher solute concentration.

hypha, plural **hyphae** Fungal filament with chitin-reinforced walls; part of a mycelium.

hypothalamus Major homeostatic control center in forebrain; regulates hunger, thirst, core body temperature; produces hormones.

hypothesis Speculative explanation for, or generalization about, something observable.

immigration Migration of individuals into a population; adds to population size.

immunity The body's ability to resist and to combat infections.

immunization Process designed to induce immunity; *e.g.*, vaccination.

immunoglobulin One of the five classes of antibodies, each with antigen-binding and class-specific structural components.

implantation In pregnancy; a blastocyst burrows into the maternal endometrium.

imprinting A form of learning triggered by exposure to a simple stimulus during a sensitive period, usually early in life.

in vitro fertilization Conception outside the body, "in glass" petri dishes or tubes.

inbreeding Mating among close relatives.

inclusive fitness theory Theory that genes for altruism can be favored by selection if they lead to behavior that increases the reproductive success of the altruist's close relatives.

independent assortment, theory of Theory that each homologous chromosome and its partner (and the genes they carry) are assorted into different gametes independently of the other pairs. Crossing over affects the outcome.

indicator species Any species which, by its abundance or scarcity, provides a measure of the health of its habitat.

infection Successful colonization of a host species by another organism.

inhibitor Any substance that binds to a molecule and interferes with its function.

innate immunity Set of antigen-unspecific immune defenses (complement activation, phagocytosis, inflammation) triggered by pathogen-associated molecular patterns.

insertion A mutation in which one or more extra bases enter a gene.

instinctive behavior A behavioral response that does not require learning.

integrator Internal control center that receives, processes, and stores sensory input, and then coordinates the responses; *e.g.*, a brain.

intermediate Of a reaction, substance that occurs between reactants and products.

intermediate disturbance hypothesis Idea that species richness is greatest in habitats where disturbances are moderate.

intermediate filament Cytoskeletal element that mechanically strengthens some cells.

internal environment All body fluids *not* inside cells. Also, extracellular fluids.

interneuron A neuron that relays signals from one neuron to another.

interphase Of a eukaryotic cell cycle, interval between mitotic divisions.

interspecific competition Competition between members of different species.

interstitial fluid Fluid in spaces between the cells of all tissues except the blood (in which extracellular fluid is called plasma).

intervertebral disk In between vertebrae, a cartilaginous flex point and shock absorber.

intron Noncoding DNA in eukaryotic chromosomes; excised from pre-mRNA.

inversion A chromosomal alteration; part of the DNA sequence gets oriented in the reverse direction, with no molecular loss.

invertebrate Animal without a backbone.

ion exchange pH-dependent process; ions dissociate from soil particles, then other ions dissolved in soil water replace them.

ionic bond Type of chemical bond that forms between ions of opposite charges.

isotonic Describes two fluids with the same solute concentrations.

isotope One of two or more forms of the same element (same number of protons) that differ in their number of neutrons.

jaw Paired, hinged cartilaginous or bony feeding structures of most chordates.

joint Area of contact between bones.

karyotype Preparation of an individual's diploid complement of chromosomes.

key innovation Modification in body structure or function that allows a species to exploit its environment more efficiently or in a novel way.

keystone species A species that influences community structure in disproportionally large ways relative to its abundance.

kidney A vertebrate organ that removes wastes from blood and regulates solute levels.

kilocalorie 1,000 calories of heat energy; amount needed to raise the temperature of 1 kilogram of water by 1°C.

Krebs cycle The second stage of aerobic respiration in which many coenzymes and two ATP form as pyruvate from glycolysis is fully broken down to CO_2 and H_2O.

K–T asteroid impact theory Theory that a huge asteroid struck Earth 65 million years ago and caused a mass extinction.

labor A placental mammal is moved out of its mother's body by strong uterine contractions.

lac operon Set of genes and gene controls for lactose metabolism in *E. coli* bacteria.

lactation Milk production and secretion by hormone-primed mammary glands.

Langerhans cell Antigen-presenting cell in skin; engulfs viruses and bacteria.

large intestine Colon. The bacteria-rich region of the vertebrate gut that absorbs water and mineral ions and compacts undigested residues.

larva, plural **larvae** An immature stage between embryo and adult in the life cycle of animals that undergo metamorphosis.

larynx Portion of the airway that contains the vocal cords; leads into the trachea.

lateral gene transfer Movement of genes from one cell to another cell, as by conjugation.

lateral meristem *See* meristem.

leaching Removal of nutrients from soil by water passing downward through it.

leaf Chlorophyll-rich plant organ of sunlight interception and photosynthesis.

learned behavior Behavior that is modified as a result of experience.

lens Of an eye, a transparent body that bends light rays so that they converge onto the retina.

leptin Appetite-suppressing hormone produced mainly by adipose tissue.

library Collection of cells that have taken up different DNA fragments or assorted cDNAs from another species.

lichen Mutualism between a fungus and a cyanobacterium or alga.

life history pattern A description of when and how many offspring are produced in a lifetime.

ligament A strap of dense connective tissue that bridges a skeletal joint.

light-dependent reactions First stage of photosynthesis in which photon energy is converted to the chemical energy of ATP.

light-independent reactions *See* Calvin–Benson cycle.

limbic system Centers in cerebrum that govern emotions; roles in memory.

limiting factor Any essential resource that slows population growth when scarce.

lineage Line of descent.

lipid Nonpolar hydrocarbon; *e.g.*, a fat, oil, wax, sterol, or phospholipid. Cells use different kinds as storable forms of energy, structural materials, or signaling molecules.

lipid bilayer Major component of all cell membranes; mainly phospholipids arranged tail-to-tail in two layers.

liver An organ with roles in detoxification of substances in the blood, glycogen storage, and synthesis of bile and cholesterol.

loam Type of soil with roughly the same proportions of sand, silt, and clay.

local signaling molecule Chemical signal secreted into extracellular fluid. Has potent effects but is inactivated quickly.

logistic growth Of a population, a phase of nearly exponential growth; levels off as density-dependent controls slow growth.

loose connective tissue Animal tissue with fibers and fibroblasts loosely arrayed in a semifluid matrix of cell secretions.

lung Internal respiratory organ of all birds, reptiles, mammals, most amphians, some fish.

lycophyte A type of seedless vascular plant; living species are small with tiny leaves and branching stem and roots; *e.g.*, club moss.

lymphatic system Organ system; returns interstitial fluid to blood; produces, stores, and circulates lymphocytes. Includes bone marrow, thymus, lymph nodes.

lymphocyte One of a class of white blood cells that are involved in immunity. *See also* B lymphocyte, T lymphocyte, natural killer cell.

lysis Gross damage to a cell wall or plasma membrane that causes cell death.

lysosome Vesicle filled with enzymes that functions in intracellular digestion.

lysozyme Antibacterial enzyme that occurs in body secretions such as tears and mucus.

macroevolution Large-scale patterns, rates of change, and trends among lineages.

macrophage Circulating, phagocytic white blood cell that participates in both innate and adaptive immune responses.

magnoliid Least diverse of the three major flowering plant groups; *e.g.*, magnolia.

mammal Amniote with hair; feeds offspring with milk from the female's mammary glands.

marine snow Organic matter drifting down from ultraplankton to mid-oceanic water; supports food webs and marine biodiversity.

marsupial Pouched mammal, such as kangaroo.

mass extinction Catastrophic period in geologic time when major groups are lost.

mast cell White blood cell in connective tissue; releases histamines, cytokines, and some other chemicals upon antigen detection.

master gene One of the genes that encodes products that map out the body plan in the developing embryo by influencing expression of other genes.

mechanoreceptor A sensory cell that detects mechanical energy, such as a change in pressure, position, or acceleration.

medulla oblongata Hindbrain region. Controls respiration and other basic tasks; coordinates some complex reflexes; *e.g.*, coughing.

megaspore Haploid spore that forms in female parts of seed-bearing plants; gives rise to the female gametophyte.

meiosis Nuclear division mechanism that halves the parental chromosome number, to a haploid (n) number. Precedes the formation of gametes and sexual spores.

melanin A brownish-black pigment; in human skin, it helps protect against effects of damaging UV light.

menstrual cycle Approximately monthly cycle in reproductive females. Hormonal changes prompt maturation and release of an oocyte from an ovary and prime the uterine lining for pregnancy. Unless pregnancy occurs, this lining is shed and the cycle begins again.

meristem A localized region of actively dividing, undifferentiated cells that give rise to all plant tissues. *Apical meristems* lengthen shoot and root tips; *lateral meristems* increase the diameter of older stems and roots.

mesoderm Of animals with a three-layered embryo, the middle primary tissue layer that gives rise to most internal organs and part of the skin.

messenger RNA mRNA. A single strand of ribonucleotides transcribed from DNA; the only type of RNA that carries protein-building information to ribosomes.

metabolic pathway A stepwise sequence of enzyme-mediated reactions.

metabolism Sum of all chemical processes by which cells acquire and use energy to grow, maintain themselves, and reproduce.

metamorphosis Major changes in body form of some animals. Hormone-controlled growth, tissue reorganization, and body part remodeling produces the adult form.

methanogen A prokaryote that makes methane gas as a by-product of anaerobic reactions.

MHC marker Self-recognition protein on cell surfaces. Antigen fragments bound to them will trigger an adaptive immune response.

micelle formation The combining of bile salts with fatty acids into tiny droplets.

microevolution Of a population, small-scale change in allele frequencies resulting from mutation, genetic drift, gene flow, natural selection, or a combination of them.

microfilament Cytoskeletal element that functions in cell contraction, movement, and structural support; forms cell cortex.

microspore A haploid spore that gives rise to pollen grains in seedbearing plants.

microtubule Cytoskeletal element; forms a dynamic framework that organizes and moves internal parts of the cell.

microvillus Slender extension from surface of certain epithelial cells; increases surface area.

migration Of many animals, a recurring pattern of movement between two or more regions in response to seasonal change.

mimicry Resemblance between unrelated species; often a defense against predation.

mineral Element or inorganic compound required for normal cell functioning.

mitochondrion Eukaryotic organelle of ATP formation; its inner membrane is the site of the second and third stages of aerobic respiration.

mitosis Nuclear division that maintains the parental chromosome number. The basis of growth in size, tissue repair, and often asexual reproduction of eukaryotes.

mitotic spindle Of eukaryotic cells, a dynamic array of microtubules that moves chromosomes during mitosis or meiosis.

mixture Two or more types of molecules intermingled in variable proportions.

model Theoretical explanation of a partially-understood process.

molecule Atoms joined by chemical bonds.

molecular clock Model used to estimate relative times of divergences by comparing the number of neutral mutations in the DNA of different species.

mollusk Only invertebrate with a mantle; most have an external or internal shell.

molting Periodic shedding and regrowth of cuticle, feathers, or other external parts.

monocot Flowering plant characterized by embryo sporophytes having one cotyledon; floral parts usually in threes (or multiples of three); and often parallel-veined leaves.

monohybrid Individual heterozygous for one gene; *e.g.*, *Aa*.

monohybrid experiment Cross between two identical monohybrids; *e.g.*, *Aa* x *Aa*.

monophyletic group Set of all species descended from a common ancestor. Also, clade.

monotreme Egg-laying mammal.

morphogen An inducer molecule. Diffuses through embryonic tissues and the resulting gradient sequentially activates master genes.

morphogenesis Genetically programmed changes in size, proportion, and shape of body parts of an animal embryo through which specialized tissues and organs form.

morphological convergence A pattern of macroevolution in which body parts of distant lineages evolve to become similar.

morphological divergence A pattern of macroevolution in which body parts of related lineages evolve to become different.

motor neuron Neuron that relays signals from the brain or spinal cord to skeletal muscle or to a gland. Also, efferent neuron.

motor proteins Proteins that use ATP to move along a microfilament or microtubule; their concerted actions move a cell or cell structure.

motor unit One motor neuron and all the muscle cells that it can stimulate.

multiple allele system Persistence of three or more alleles of a gene in a population.

muscle fatigue Decline in muscle tension despite ongoing stimulation; occurs after a period of continuous tetanic contraction.

muscle spindle A sensory organ that detects muscle stretching.

muscle tension Mechanical force exerted by a contracting muscle.

muscle twitch Brief contraction and relaxation in response to a stimulus.

mutation Heritable change in the sequence of DNA. Original source of new alleles and life's diversity, but most are harmful. *Lethal mutations* drastically affect phenotype, and thus usually cause death.

mutualism An interspecific interaction that benefits both species.

mycelium Of a multicelled fungus, a mesh of branching hyphal filaments that absorb food.

mycorrhiza "Fungus-root." A mutualism between a fungus and young plant roots.

myofibril One of the long, thin structures divided into contractile units that parallel the long axis of a muscle fiber.

myoglobin A pigment that reversibly binds oxygen; abundant in some muscle fibers.

myosin Motor protein. Interacts with actin in muscle cells to bring about contraction.

NAD^+ Nicotinamide adenine dinucleotide. Type of nucleotide coenzyme.

natural killer cell NK cell. Cytotoxic lymphocyte active in adaptive immune responses; kills altered body cells.

natural selection A microevolutionary process; differences in survival and reproduction among individuals of a population that differ in the details of their heritable traits.

negative control Gene control mechanism; a regulatory protein slows gene expression.

negative feedback mechanism A homeostatic mechanism by which an activity changes some condition, which in turn triggers a response that reverses the change.

neoplasm Abnormal mass of cells (tumor) that lost control over the cell cycle.

nephron One of millions of tubules in a kidney that filter blood and form urine.

nerve A sheathed bundle of axons of sensory neurons, motor neurons, or both; part of the peripheral nervous system.

nerve cord A communication line that parallels the anterior–posterior axis of bilateral animals. In chordates, it develops as a hollow, neural tube and gives rise to the spinal cord and brain.

nerve net Nervous system of cnidarians and some other invertebrate groups; asymmetrical mesh of sensory and motor neurons.

nervous tissue Tissue consisting of neurons and often neuroglia.

net ecosystem production All the energy the primary producers have accumulated during growth, reproduction in a specified interval (net primary production), *minus* energy that producers, and decomposers, have used.

neuroglia Collectively, cells that structurally and metabolically support neurons; about half of the nervous tissue in vertebrates.

neuron A type of excitable cell; the basic unit of communication in nervous systems.

neurotransmitter Any of a diverse class of signaling molecules secreted by neurons. Each acts in a synaptic cleft, then is rapidly degraded or recycled.

neutron Subatomic particle with no charge. Occurs in nucleus.

neutrophil Abundant, circulating white blood cell; mainly phagocytic but also can release enzymes that kill extracellular microbes and stimulate inflammation.

niche Sum of activities and relationships in which individuals of a species engage as they secure and use resources.

nitrification Soil bacteria break down ammonia or ammonium to nitrite, then other bacteria break down nitrite to nitrate.

nitrogen cycle An atmospheric cycle in which nitrogen moves from its largest reservoir, the atmosphere, through the ocean waters and sediments, soils, and food webs, then back to the atmosphere.

nitrogen fixation Conversion of gaseous nitrogen to ammonia.

nondisjunction Failure of sister chromatids or homologous chromosomes to separate during meiosis or mitosis. Daughter cells receive too many or too few chromosomes.

nuclear envelope A double membrane that is the outer boundary of the nucleus.

nucleic acid Nucleotide polymer; used by all living cells and viruses to convey heritable information; *e.g.*, DNA, RNA.

nucleic acid hybridization Base-pairing between DNA or RNA from different sources.

nucleoid The portion of a prokaryotic cell where DNA is physically organized but not enclosed in a membrane.

nucleosome DNA wound twice around histone proteins; basic unit of structural organization in eukaryotic chromosomes.

nucleotide Small organic compound with a five-carbon sugar, a nitrogen-containing base, and phosphate. Monomers of nucleic acids are named after their component bases: adenine, guanine, cytosine, thymine, and uracil.

nucleus Defining organelle of eukaryotes; separates DNA physically and chemically from the cytoplasm.

nutrient Element with a role in metabolism that no other element can fulfill.

olfactory receptor Chemoreceptor for a water-soluble or volatile substance.

oocyte Immature egg.

operator Part of an operon; DNA binding site for a regulatory protein.

organ Two or more tissues organized in a way that allows them to perform some task.

organ system A set of organs that interact chemically, physically, or both in a task.

organelle One of many membrane-bound compartments with special functions inside eukaryotic cells; *e.g.*, a nucleus.

organic compound A carbon-based molecule that also incorporates atoms of hydrogen and, often, oxygen, nitrogen, and other elements; *e.g.*, fats, proteins.

organism Any independent, single- or multicelled form of life.

osmoreceptor Type of sensory receptor that detects shifts in water volume.

osmosis Diffusion of water across a selectively permeable membrane from a region where the water concentration is higher to a region where it is lower.

osmotic pressure Amount of pressure which, when applied to a hypertonic fluid, will stop osmosis into that fluid.

ovary Of a female animal, gonad where eggs are produced. Of flowering plants, the portion of the carpel base that contains ovules.

ovulation Release of a secondary oocyte from an ovary.

ovule Of seed plants, egg-containing structure that will, after fertilization, become a seed.

ovum Mature unfertilized egg.

oxidation–reduction (redox) reaction Type of chemical reaction in which one molecule gives up electrons to another. Primary reaction in electron transfer chains.

oxytocin Animal hormone with roles in labor, lactation, recovery of uterus after pregnancy, and often, social behavior.

pain Perception of injury to a body region.

pain receptor A nociceptor; a sensory receptor that detects tissue damage.

pancreas Glandular organ. Bulk of it makes and secretes enzymes and bicarbonate into small intestine; islets secrete the hormones glucagon and insulin into blood.

Pangea Paleozoic supercontinent; the first land plants and animals evolved on it.

parapatric speciation A speciation model in which populations in contact along a common border evolve into new species.

parasite Organism that feeds on a living host, but usually does not kill it outright.

parasitism Symbiotic interaction in which one species draws nutrients from another, usually without killing it.

parasitoid A type of insect that, in a larval stage, grows inside a host (usually another insect), feeds on its soft tissues, and kills it.

parasympathetic neuron A neuron of the autonomic nervous system. Its signals divert energy to basic housekeeping tasks.

parathyroid gland One of the four small endocrine glands embedded in the back of the thyroid gland; regulates blood calcium levels.

parenchyma Simple living tissue that composes most nonwoody parts of a plant.

passive transport Diffusion of a solute across a cell membrane, through the interior of a transport protein.

pathogen An organism that infects a host and disrupts its function, causing disease.

pattern formation In animal development, the sculpting of specialized tissues and organs from clumps of cells.

PCR Polymerase chain reaction. A method to rapidly copy specific DNA fragments.

pedigree Chart of genetic connections among individuals related by descent.

pellicle A thin, flexible, protein-rich body covering of some single-celled eukaryotes.

per capita Per individual.

permafrost An impermeable, perpetually frozen layer that underlies arctic tundra.

peroxisome Enzyme-filled vesicle that breaks down amino acids, fatty acids, and some toxic substances such as ethanol.

pH scale A scale of 0 to 14 that reflects the H^+ concentration of a solution. pH 7 is neutral.

phagocyte Cell that engulfs and digests another.

phagocytosis Process by which a cell engulfs and digests another cell.

pharynx Muscular tube used in respiration and feeding of invertebrate chordates. In humans, it leads to the esophagus and trachea.

pheromone A secreted signaling molecule that affects behavior and/or physiology of another individual of the same species.

phloem Complex, living vascular tissue of plants; interconnected sieve tubes distribute photosynthetic products through a plant.

phospholipid A lipid with a phosphate group in its hydrophilic head. The main constituent of cell membranes.

phosphorus cycle Sedimentary cycle that moves phosphates from land, through food webs, to ocean sediments, then back to land.

phosphorylation Enzyme-mediated transfer of a phosphate group to an organic compound.

photoperiodism Biological response to a change in length of day relative to night.

photoreceptor A light-sensitive sensory cell of invertebrates and vertebrates.

photosynthesis The chemical process by which photoautotrophs capture and use light energy to make sugars from CO_2 and water.

photosystem In photosynthetic cells, a cluster of pigments and other molecules; converts light energy to chemical energy.

phototropism Change in the direction of movement or growth in response to light.

phycobilin One of a class of red or blue-green photosynthetic accessory pigments that occur in cyanobacteria and red algae.

physiology Study of how the multicelled body functions in its environment; more specifically, of mechanisms by which component parts grow, develop, and are maintained and reproduced.

pigment Molecule that absorbs light of certain energy; pigments are the bridge between light and photosynthesis.

pineal gland Endocrine gland deep inside the brain; secretes melatonin. Affects biological clocks, overall activity, and reproductive cycles.

pituitary gland Vertebrate endocrine gland; interacts with the hypothalamus to control activity of many other glands.

placenta Of placental mammals, organ formed by maternal tissue and extraembryonic membranes. Allows a mother to exchange substances with a fetus without mixing their blood.

plankton Aquatic community of mostly microscopic autotrophs and heterotrophs.

plant Multicelled photoautotroph, generally with roots, stems, and leaves. The primary producers on land.

plaque Mixture of bacteria, their secretions, and saliva glycoproteins that sticks to teeth.

plasma Liquid portion of blood; water with dissolved proteins and other solutes.

plasma membrane Outer cell membrane; the structural and functional boundary between cytoplasm and extracellular fluid.

plasmid Circular bacterial DNA molecule. Some are used as cloning vectors.

plate tectonics theory Theory that great slabs or plates of Earth's outer layer float on a hot, semi-molten mantle. The plates have rafted continents to new positions over time.

pleiotropy Multiple effects of one gene.

polar body One of the four cells that form by meiotic cell division of an oocyte but that does not become the ovum.

pollen grain Of seedbearing plants, a tiny structure formed from microspores; consists of a wall around a few cells that will develop into a mature, sperm-bearing, male gametophyte.

pollination The arrival of pollen on the female parts of a plant of the same species.

pollinator Any living agent that delivers pollen to egg-containing structures of the same species.

polypeptide chain Series of many amino acids linked by peptide bonds.

polyploidy A condition in which somatic cells have three or more of each type of chromosome.

pons Hindbrain traffic center for signals between the cerebellum and forebrain.

population Group of individuals of one species living in a specified area.

population density Count of individuals of a population in a specified portion of a habitat.

population distribution The pattern in which the individuals of a population are dispersed through their habitat.

population size The total number of individuals that make up a population.

positive control Gene control mechanism by which a regulatory protein turns on or enhances gene expression.

positive feedback mechanism A homeostatic mechanism by which an activity changes some condition, which in turn triggers a response that intensifies the change.

predation Ecological interaction in which a free-living predator eats a prey organism.

predator A heterotroph that captures other living organisms (its prey), and feeds on them, usually killing them.

prediction A statement of what you would expect to observe if a hypothesis is valid.

pressure flow theory Theory that the flow of organic compounds through phloem is caused by pressure gradients between source and sink regions.

prey An organism that a predator eats.

primary producer An autotroph at the first trophic level of an ecosystem.

primary productivity The rate at which an ecosystem's primary producers secure and store energy in tissues in a given interval.

primary succession Establishment and replacement of one species array after another in an area where no community existed.

primary wall The first thin, pliable cell wall of young plant cells.

primate Placental mammal with nails rather than claws; adapted for grasping; a prosimian, tarsioid, or anthropoid.

primer Synthetic, single-stranded fragment of DNA used as an initiation site for DNA synthesis in PCR and DNA sequencing.

prion Protein particle that, in its infectious misfolded form, causes fatal brain diseases such as Creutzfeldt-Jakob disease.

probability The chance that one outcome of an event will occur; proportional to the number of possible outcomes.

probe Fragment of DNA complementary to a gene of interest, labeled with a tracer.

producer An organism that makes its own food from inorganic materials; autotroph.

product Ending substance of a reaction.

progesterone Sex hormone produced by the ovaries, corpus luteum, and placenta; stimulates thickening of the lining of the uterus during the menstrual cycle; helps maintain pregnancy; and stimulates development of mammary glands.

prokaryotic cell A single-celled organism, often walled, without a nucleus or other organelles; a bacterium or archean.

prolactin Hormone that induces synthesis of enzymes used in milk production.

promoter Short stretch of DNA to which RNA polymerase binds. Transcription then begins at the gene closest to the promoter.

protist One of the structurally simple single-celled or multicellular eukaryotes, which are now being classified into monophyletic groups.

proto-cell Membrane-bound sac enclosing metabolic machinery; presumed stage that preceded evolution of cells.

proton Positively charged particle in the nucleus of all atoms.

protostome A member of the bilateral, coelomate, animal lineage in which the first indentation to form on the early embryo's surface becomes a mouth; *e.g.*, mollusks, annelids, arthropods.

pseudopod A dynamic lobe of membrane-enclosed cytoplasm; functions in motility and phagocytosis by amoebas, amoeboid cells, and some white blood cells.

puberty Of humans, the post-embryonic stage when gametes start to mature and secondary sexual traits emerge.

pulmonary circuit Cardiovascular route in which oxygen-poor blood flows to lungs from the heart, becomes oxygenated, then flows back to the heart.

punctuation model, speciation Addresses the rate of speciation; morphological change is said to occur relatively quickly after two populations start to diverge.

Punnett-square method Method in which a matrix of genotypes is used to predict the probability of outcomes of a genetic cross.

pyruvate Three-carbon compound that forms as an end product of glycolysis.

quadrat One of a number of sampling areas of the same size and shape; data from these areas is used to estimate population density and other factors for a larger area.

radial symmetry Animal body plan with parts arrayed in a repeating pattern around a central axis, as in a sea anemone.

radiation Energy carried by a photon or subatomic particle. *Ionizing* radiation has enough energy to damage DNA and living tissues; *nonionizing* radiation does not.

radiometric dating Method of dating rocks and newer fossils in which the proportion of a radioisotope to its daughter element in a sample is used to calculate the sample's age.

rain shadow Reduction in rainfall on that side of a mountain range which faces away from prevailing winds; results in arid or semiarid conditions.

ray-finned fish A member of the most diverse group of bony fish; has fin supports derived from skin, a swim bladder, and thin flexible scales, *e.g.*, trout.

reactant Starting substance of a reaction.

reaction Process of chemical change by which a cell builds, rearranges, or breaks down organic compounds.

receptor, molecular Protein that initiates a cellular response when it binds a hormone or other signalling molecule.

receptor, sensory Sensory cell or a specialized ending of one that detects a certain stimulus.

recessive allele Allele whose expression in heterozygotes is fully or partially masked by expression of a dominant partner.

rectum Last part of the mammalian gut that briefly stores feces before their expulsion.

red blood cell Type of blood cell filled with the oxygen-transporting protein hemoglobin; also called an erythrocyte.

red marrow Site of blood cell formation in the spongy tissue of many bones.

redox reaction *See* oxidation-reduction.

reflex Simple, stereotyped movement in response to a stimulus; sensory neurons synapse on motor neurons in the simplest reflex arcs.

regulatory protein Protein that directly or indirectly adjusts gene expression.

releaser Hypothalamic signaling molecule that enhances or slows the secretion of a specific anterior pituitary hormone.

reproductive base Of some population, the number of individuals already reproducing, plus those not yet reproductively mature.

reproductive isolating mechanism A heritable feature of body form, function, or behavior that prevents interbreeding between populations; sets the stage for genetic divergences.

reptile A vertebrate with lungs, waterproof skin and lungs; amniote eggs. Living forms are all ectothermic (cold-blooded). Not a monophyletic group; *e.g.*, dinosaur, turtle, crocodile, snake.

resource partitioning Use of the same resource in different places or at different times; permits species with similar resource needs to coexist.

respiration Physiological processes that move O_2 from the surroundings to tissues of an animal body and CO_2 from these tissues to the outside.

respiratory cycle An inhalation and exhalation.

respiratory surface A thin, moist body surface that functions in gas exchange.

resting membrane potential Voltage difference across the plasma membrane of a neuron or other excitable cell that is not being stimulated.

restriction enzyme One of hundreds of proteins that recognize and cut specific base sequences in double-stranded DNA.

retina Of vertebrate and many invertebrate eyes, a tissue packed with photoreceptors.

reverse transcriptase Viral replication enzyme that uses an RNA template to assemble a strand of DNA from nucleotides.

rhizoid A rootlike part of a bryophyte.

rhizome Horizontal-growing underground stem that gives rise to roots and shoots of many vascular plants.

ribosomal RNA rRNA. The RNA component of ribosomes.

ribosome In all cells, a structure that translates genetic information in mRNAs into proteins.

riparian zone The narrow strip of vegetation along a stream or river.

RNA Ribonucleic acid. Single-stranded nucleic acid polymer of adenine, guanine, cytosine, and uracil nucleotides. Central to the processes of transcription and translation.

RNA polymerase Enzyme that catalyzes transcription of DNA into RNA.

RNA world A hypothetical period in Earth history when RNA, not DNA, stored and transmitted genetic information.

rod cell Vertebrate photoreceptor that detects dim light; contributes to the coarse perception of movement.

root Typically belowground plant part that absorbs water and dissolved minerals.

root hair Hairlike, absorptive extension of some root epidermal cells.

root nodule Localized swelling of a plant root that harbors symbiotic nitrogen-fixing bacteria.

roundworm Bilateral invertebrate with a false coelom, complete digestive system; cylindrical body is covered by a cuticle that is molted.

rubisco RuBP carboxylase. Carbon-fixing enzyme of the C3 photosynthesis pathway.

sac fungus Member of the largest fungal group; sexual spores form in sac-shaped cells.

salinization Salt buildup in soil.

salt Any compound that releases ions other than H^+ and OH^- in solution.

sampling error Difference between a statistical result derived from surveying a complete set of events or individuals, and a result derived from surveying a subset of it.

sap Sugar-rich fluid of sapwood.

saprobe Heterotroph that extracts energy and carbon from organic remains and wastes.

sapwood Of an older stem or root, the moist secondary growth in between the vascular cambium and the heartwood.

sarcomere Basic unit of contraction, along the length of a muscle fiber. Shortens by ATP-driven interactions between actin and myosin.

sarcoplasmic reticulum Specialized ER that forms around muscle fibers; stores and releases Ca^{++} that allows contraction.

scientific theory *See* Theory, scientific.

sclerenchyma Lignin-reinforced simple plant tissue that structurally supports mature parts; dead at maturity.

second messenger Molecule that relays a hormonal signal from a membrane receptor into a cell; *e.g.*, cyclic AMP.

secondary oocyte A haploid cell produced by the first meiotic division of a primary oocyte.

secondary succession Sequential changes in the array of species in a disturbed area.

secondary wall A rigid, permeable cell wall that forms inside the primary wall of many mature plant cells.

secretin A hormone secreted by cells in the lining of the small intestine in response to arrival of acidic chyme; stimulates the pancreas to secrete bicarbonate into the small intestine.

seed A mature ovule of a seed-bearing plant; contains a plant embryo.

segmentation Of animal body plans, existence of repeating units, some of which may be modified for special purposes.

segregation, theory of Mendel's theory that the two genes of a pair separate at meiosis to end up in different gametes.

selective gene expression Outcome of gene controls by which a cell uses a subset of its genome, and modulates expression of the rest. Basis of cell differentiation.

selective permeability Capacity of a membrane to prevent or allow flow of specific substances across it.

selfish herd Social group held together by reproductive self-interest.

senescence Of multicelled organisms, phase in a life cycle between maturity and death.

sensory neuron Neuron that detects a stimulus and relays information to an integrating center.

sensory receptor Cell or cell part that detects a specific form of energy.

sex chromosome One of the two chromosomes that determine gender.

sexual dimorphism A notable difference between the female and male phenotypes of a species; outcome of sexual selection.

sexual reproduction Reproductive process by which offspring arise through the combining of genetic material from two parents.

sexual selection Form of natural selection in which members of the opposite sex serve as the agents of selection.

sexually transmitted disease (STD) One of the diseases transmitted by sexual intercourse.

shell model Model that depicts electron energy levels as a nested series of shells.

shoots Plant stems and leaves.

sieve tube A conducting tube in phloem.

sister chromatids Two attached members of a duplicated eukaryotic chromosome.

six-kingdom classification system Grouping of all organisms into kingdoms Bacteria, Archaea, Protista, Fungi, Plantae, and Animalia.

skeletal muscle Organ of many muscle fibers bundled inside a connective tissue sheath and attached to bone by tendons.

skeletal muscle tissue Contractile tissue that is the functional partner of bone.

skin Integument, or outer body covering, of a vertebrate.

sliding-filament model Model for how the sarcomeres of muscle fibers contract.

small intestine Part of the vertebrate gut in which digestion is completed and from which most dietary nutrients are absorbed.

smooth muscle tissue Contractile tissue in the wall of soft internal organs.

sodium–potassium pump Protein that actively transports sodium out of a cell, and at the same time passively transports potassium into it.

softwood Type of tree with no fibers or vessels in its wood; a gymnosperm tree.

soil Mix of mineral particles of variable sizes and decomposing organic material.

soil erosion Wearing away of land surface by wind, running water, and ice.

solute Any dissolved substance.

solution Mixture of a solute and the liquid in which it is dissolved.

somatic nerve Nerve relaying information about environmental stimuli from sensory receptors to the central nervous system and/or carrying signals from the central nervous system to skeletal muscle.

somatic sensation Perception of touch, pain, pressure, temperature, motion, or positional changes of body parts.

somite One of many paired segments of mesoderm that appear on the dorsal surface of an early vertebrate embryo; they give rise to muscles, bones, and dermis of skin.

special sense Vision, hearing, olfaction, or another sensation involving receptors that are restricted to certain body parts.

speciation Macroevolutionary process by which a new species arises.

species One or more natural populations of individuals with at least one unique trait, derived from a common ancestor, that occurs in no other groups.

sperm Mature male gamete.

sphincter A ring of muscles that when contracted prevents movement of material through a tubular organ or out of the body.

spinal cord Part of a central nervous system inside the vertebral canal. Basic reflex centers, and tracts to and from the brain.

sponge Structurally simple, filter-feeding invertebrate with no body symmetry or tissues.

spore Structure of one or a few cells, formed sexually or asexually, often walled or coated, that protects and/or disperses a new generation. Some bacteria and protists, as well as fungi and plants, form spores.

sporophyte A diploid spore-producing body of a plant or multicelled alga; grows by mitotic cell divisions from a zygote.

spring overturn Of a large body of water, a downward movement of oxygenated surface water and an upward movement of nutrient-rich water from below during spring; increases primary productivity.

stabilizing selection Mode of natural selection; intermediate phenotypes are favored over extremes at both ends of the range of variation.

stamen Male reproductive part of a flower; an anther, typically atop a stalked filament.

stem cell Self-perpetuating, undifferentiated animal cell. A portion of its daughter cells becomes specialized; *e.g.*, red blood cells from stem cells in bone marrow.

sterol Lipid with no fatty acids. Important component of eukaryotic cell membranes; some kinds are hormones.

stimulus A specific form of energy that activates a sensory receptor; *e.g.*, pressure.

stoma, plural **stomata** Tiny gap that forms across the epidermis of a leaf or primary stem when paired guard cells swell; allows water vapor and gases to cross the epidermis.

stomach Muscular, stretchable sac; stores ingested food, secretes gastric juice, and mixes the two to bring about digestion.

stratification Formation of sedimentary rock layers, or the layers themselves.

strip logging A way to minimize erosion from deforestation; a narrow corridor that parallels contours of sloped land is cleared; the upper part is used as a log-hauling road, then is reseeded from intact forest above it.

strobilus Of certain nonflowering plants, a cluster of spore-producing structures.

stroma Region between the thylakoid membrane and the outer membranes of a chloroplast; where the light-independent reactions of photosynthesis occur.

stromatolite Fossilized remains of bacterial mats that formed in shallow seas.

substrate A molecule that is specifically recognized and acted upon by an enzyme.

substrate-level phosphorylation Direct, enzyme-mediated transfer of a phosphate group from a substrate to another molecule.

surface-to-volume ratio Physical relationship that constrains the size of cells; the volume of a sphere increases with the cube of the diameter, but the surface area increases with the square.

survivorship curve Plot of age-specific survival of a cohort, from the time of birth until the last individual dies.

symbiosis An interspecific interaction in which members of different species interact closely for some or all of the life cycle; *e.g.*, mutualism, commensalism, parasitism.

sympathetic neuron A neuron of the autonomic nervous system. Its activity will summon up a "fight-or-flight" response.

sympatric speciation A speciation model. Occurs inside the home range of a species in the absence of a physical barrier.

synaptic integration Summation of excitatory and inhibitory signals that arrive at an excitable cell's input zone at one time.

syndrome Set of symptoms; characterizes a genetic abnormality or disorder.

systemic acquired resistance Of many plants, a mechanism in which infection of one region induces the synthesis and distribution of compounds that protect other regions from infection.

systemic circuit Cardiovascular route in which oxygenated blood flows from the heart through the rest of the body, where it gives up oxygen and takes up carbon dioxide, then flows back to the heart.

T cell receptor TCR. Antigen receptor on the surface of T cells. Part of it recognizes MHC molecules; part recognizes antigen.

T lymphocyte T cell. White blood cell with a central role in all vertebrate adaptive immune responses.

tandem repeats On a chromosome, many identical copies of a short base sequence, the number, position, and sequence of which are unique to an individual.

taproot system A primary root and all of its lateral branchings.

taste receptor A type of chemoreceptor that detects solutes in the fluid bathing it.

TCR *See* T cell receptor.

telomere A cap of repetitive DNA sequences on the end of a chromosome that is needed for cell division. With every division it is shortened, until it becomes so short that division can no longer occur.

temperature Measure of molecular motion.

temperature zone Globe-spanning bands of temperature defined by latitude; *e.g.*, cool temperate, equatorial.

tendon A cord or strap of dense connective tissue that attaches skeletal muscle to bone.

test Systematic determination of whether a prediction based on a hypothesis is valid; also, experiment.

testis, plural **testes** A type of gonad where male gametes and sex hormones form.

tetanus A strong muscle contraction that occurs in response to repeated stimulation of a motor unit. Also, bacterial disease affecting inhibitory nervous signals; leads to continual contraction.

tetrapod A vertebrate that is a four-legged walker or a descendant of one; an amphibian, reptile, bird, or mammal.

thalamus Forebrain region; coordinates sensory input and relays signals to the cerebrum.

theory, scientific A rigorously tested, long-standing concept used to interpret a broad range of observations about some aspect of nature. Always open to revision.

thermal inversion A layer of dense, cool air trapped beneath a layer of warm air.

thermodynamics Science of the properties of energy. The *first law* states that energy cannot be either created or destroyed. The *second law* states that energy disperses spontaneously; also, every time energy becomes converted from one form to another, some is lost.

thermoreceptor Type of sensory cell that detects radiant energy (heat).

three-domain system Classification system that groups all organisms into the domains Bacteria, Archaea, and Eukarya.

thylakoid Stack of interconnected discs formed from thylakoid membrane inside chloroplasts.

thylakoid membrane Much-folded inner membrane system of a chloroplast; forms a continuous compartment. In the first stage of photosynthesis, pigments and enzymes in the membrane form ATP and NADPH.

thymus gland Lymphoid organ that lies behind sternum; secretes hormones that influence the maturation of T cells.

thyroid gland Endocrine gland at the base of the neck; secretes hormones that influence growth, development, and metabolic rate.

tight junction An array of many strands of fibrous proteins collectively joining the sides of cells that make up an epithelium; prevent fluid from leaking between the cells.

tissue Of multicelled organisms, organized group of cells and extracellular materials that interact in one or more tasks.

tissue culture propagation Inducing the vegetative growth of a plant from one cell.

topsoil Uppermost soil layer; contains the most nutrients for plant growth.

total fertility rate TFR. The average number of offspring born to a woman during her lifetime.

trachea Largest airway in vertebrates.

tracheid Type of xylem cell.

transcription Process by which an mRNA is assembled from nucleotides using a gene region in DNA as a template.

transfer RNA tRNA. One of a class of small RNA molecules that delivers amino acids to a ribosome. Its anticodon pairs with an mRNA codon during translation.

transgenic Describes an organism that has been genetically modified by experiment.

transition state The state in a chemical reaction at which the substrate bonds are at their breaking point.

translation Second stage of protein synthesis. At ribosomes, information encoded in an mRNA transcript guides the synthesis of a new polypeptide chain from amino acids.

translocation Chromosomal alteration in which a piece of chromosome becomes attached to a different chromosome. Also, process of fluid movement inside phloem.

transpiration Evaporative water loss from a plant's aboveground parts.

transposon Transposable element. Small stretch of DNA that jumps spontaneously to another, random location in a genome.

triglyceride Lipid with three flexible fatty acid tails attached to a glycerol backbone. Most abundant lipid in the body.

trophic level All organisms that are the same number of transfer steps away from the energy input into an ecosystem.

tubular reabsorption A process by which peritubular capillaries reclaim water and solutes that leak or are pumped out of a nephron's tubular regions.

tubular secretion Transport of H^+, urea, other solutes out of peritubular capillaries and into nephrons for excretion in urine.

uniformity Theory that Earth's surface has changed in uniformly repetitive ways by the typical forces of nature we see today.

upwelling Upward movement of deep, often nutrient-rich water near a coastline; replaces a mass of surface ocean water.

urea Main waste product in urine; formed in the liver when ammonia from the breakdown of proteins combines with CO_2.

urinary excretion Wastes, toxins, excess water and solutes are moved from the blood into urine, which is then eliminated from the body.

urinary system Vertebrate organ system that adjusts blood's volume and composition; helps maintain extracellular fluid.

urine Fluid consisting of excess water, wastes, and solutes; forms in kidneys.

uterus Of a female placental mammal, a pear-shaped organ in which the developing young are housed and nurtured during pregnancy.

vaccination Administration of a vaccine.

vaccine Type of antigen-containing preparation introduced into the body to prime the immune system to recognize the threat before infection.

variable Of experimental tests, a specific aspect of an object or event that may differ over time or among individuals.

vascular bundle A multistranded cluster of primary xylem and phloem that threads through the ground tissue of plants.

vascular cambium Lateral meristem that gives rise to secondary vascular tissues in older stems and roots.

vascular cylinder Core of vascular tissue in a root. Primary xylem and phloem encased by pericycle and endoderm.

vasoconstriction A decrease in the diameter of blood vessels.

vasodilation Increase in blood vessel diameter.

vector Organism that conveys a disease-causing pathogen from host to host.

vein Of a cardiovascular system, vessels that lead back to the heart. Of leaves, a vascular bundle threading through photosynthetic tissue.

vernalization Stimulation of flowering in spring by low temperature in winter.

vertebra, plural **vertebrae** One of a series of segments that make up a vertebrate backbone.

vertebrate A chordate that has a backbone.

vessel member Type of cell in xylem, dead at maturity; its wall becomes part of a water-conducting vessel.

viroid Infectious particle of short, tightly folded strands or circles of RNA.

virus A noncellular infectious agent that can be replicated only after its genetic material enters a host cell and subverts its metabolic machinery.

vision Perception of visual stimuli.

visual accommodation Adjustments in lens position or shape that focus light on the retina.

visual field The portion of the outside world that an animal sees.

vitamin An organic substance an animal needs for metabolism and must get from food.

vitamin D A fat-soluble vitamin; helps the body absorb dietary calcium.

vocal cord One of the thick, muscular folds of the larynx that help some animals produce sound waves for vocalization.

warning coloration Distinctive forms, colors, patterns, and other signals that predators learn to recognize and avoid.

watershed Region from which all water that falls as rain or snow drains into a particular stream or river.

wavelength Distance between the crests of two successive waves.

wax Type of lipid with long-chain fatty acid tails. Component of many waterproof, protective coatings on plants and animals.

white blood cell One of the many tyes of blood cells with a role in vertebrate immunity.

wood Accumulated secondary xylem.

X chromosome A type of sex chromosome that influences sex determination; *e.g.*, in mammals, an XX embryo becomes female; an XY embryo becomes male.

X chromosome inactivation *See* dosage compensation.

xanthophyll Photosynthetic accessory pigment; a type of carotenoid.

xenotransplantation Surgical transfer of an organ from one species to another.

x-ray diffraction image Film image of x-rays scattered by a crystalline sample.

xylem A complex vascular tissue of plants; its sturdy tubes formed by connected walls of dead cells conduct water and solutes.

yellow marrow Of most mature bones, a fatty tissue that produces red blood cells when blood loss from the body is severe.

yolk Protein- and lipid-rich substance that nourishes embryos in animal eggs.

yolk sac Extraembryonic membrane. In shelled eggs, it holds nutritive yolk; in humans, part becomes a blood cell formation site, and some cells give rise to forerunners of gametes.

zero population growth No net increase or decrease in population size during a interval.

zygomycetes Small fungal group in which diploid zygotes develop into zygospores, a type of thick-walled sexual spore in a thin, clear covering; e.g., black bread mold.

Art Credits and Acknowledgments

This page constitutes an extension of the book copyright page. We have made every effort to trace the ownership of all copyrighted material and secure permission from copyright holders. In the event of any question arising as to the use of any material, we will be pleased to make the necessary corrections in future printings. Thanks are due to the following authors, publishers, and agents for permission to use the material indicated.

Page i Photographer Jay Barnes

TABLE OF CONTENTS: **Page iv** From top, © Peter Turnley/Corbis; Lisa Starr with PDB file from NYU Scientific Visualization Lab; © Tony Brian, David Parker/SPL/Photo Researchers, Inc. **Page v** From top, © Paul Edmondson/Corbis; NASA; © Professors P. Motta and T. Naguro/SPL/Photo Researchers, Inc. **Page vi** From top, Micrograph, Dr. Pascal Madaule, France; © Science Photo Library/Photo Researchers, Inc.; Photo courtesy of the College of Veterinary Medicine, Texas A&M University. **Page vii** From top, PDB ID: 2AAI; Rutenber, E., Katzin, B.J., Ernst, S., Collins, E.J., Mlsna, D., Ready, M.P., Robertus, J.D.: Crystallographic refinement of ricin to 2.5 A. *Proteins* 10 pp. 240 (1991). UniProt copyright © 2002–2004 UniProt consortium; Dr. Jorge Mayer, Golden Rice Project; © Bettmann/Corbis. **Page viii** From top, NASA Galileo Imaging Team; © Philippa Uwins/ The University of Queensland. **Page ix** From top, © Clinton Webb; P. Morris/Ardea London. **Page x** From top, Star Tribune/Minneapolis-St. Paul; © Allstock/ Stone/Getty Images; © David Goodin. **Page xi** From top, © Science Photo Library/Photo Researchers, Inc.; © Steve Cole/PhotoDisc Green/Getty Images; © James Stevenson/Photo Researchers, Inc. **Page xii** From top, © NIBSC/Photo Researchers, Inc.; Courtesy of Lisa Hyche; © Cordelia Molloy/Photo Researchers, Inc. **Page xiii** From top, © Catherine Ledner Photography; © Dow W. Fawcett/Photo Researchers, Inc. **Page xiv** From top, © David Nunuk/Photo Researchers, Inc.; John Kabashima; © Harold Naideau/Corbis. **Page xv** From top, © Hank Fotos Photography; © Scott Camazine; © Joseph Sohm, Visions of America/Corbis.

INTRODUCTION: NASA Space Flight Center

CHAPTER 1: **Page 1** © Peter Turnley/Corbis. **1.1** (b, above left) PDB file courtesy of Dr. Christina A. Bailey, Department of Chemistry & Biochemistry, California Polytechnic State University, San Luis Obispo, CA; (b, above center) PDB ID: 1BBB; Silva, M.M., Rogers, P.H., Arnone, A.; A third quarternary structure of human hemoglobin A at 1.7-A resolution; *J Biol Chem* 267 pp. 17248 (1992); (b above right) PDB file from Klotho Biochemical Compounds Declarative Database; (d) © Science Photo Library/Photo Researchers; (e) © Bill Varie/Corbis; (f–h) © Jeffrey L. Rotman/Corbis; (i) © Peter Scoones; (j–k) NASA. **Page 3** Bottom right, NASA. **1.3** Left, upper, PhotoDisc, Inc.; lower, Gary Head and Lisa Starr with photograph from David Neal Parks; right, © Y. Arthrus-Bertrand/Peter Arnold, Inc. **Page 5** Bottom right, NASA. **1.6** Page 6, Upper, left, © P. Hawtin, University of Southampton/SPL/ Photo Researchers, Inc.; right, © CNRI/SPL/Photo Researchers, Inc.; lower, left, © R. Robinson/Visuals Unlimited, Inc.; right, © Dr. Harald Huber, Dr. Michael Hohn, Prof. Dr. K.O.Stetter, University of Regensburg, Germany. **1.6** Page 7, Upper left, clockwise from top, Lewis Trusty/Animals Animals; Emiliania Huxleyi photograph, Vita Pariente, scanning electron micrograph taken on a Jeol T330A instrument at Texas A&M University Electron Microscopy Center; Carolina Biological Supply Company; © Oliver Meckes/Photo Researchers, Inc.; Courtesy of James Evarts; upper right from left, © John Lotter Gurling/Tom Stack & Associates; Edward S. Ross; bottom from left, Robert C. Simpson/Nature Stock; Edward S. Ross; © Stephen Dalton/Photo Researchers, Inc. **1.7** Left, Photographs courtesy of Derrell Fowler, Tecumseh, Oklahoma; right, © Nick Brent. **Page 9** © Ed Kashi/Corbis. **Page 10** © SuperStock. **1.9** (a) © Chris D. Jiggins; (b) www.thais.it; (c) background, © Kevin Schafer/Peter Arnold, Inc.; (d) © Martin Reid. **Page 12** Gary Head. **Page 14** © Digital Vision/PictureQuest.

CHAPTER 2: **Page 15** Left, © John Collier. Great Britain, 1850–1934, "Priestess of Delphi," 1891, London, oil on canvas, 160.0 x 80.0cm. Gift of the Rt. Honourable, the Earl of Kintore, 1893; right, © Dr. John Brackenbury/ Science Photo Library/Photo Researchers, Inc.. **2.1** © David Noble/FPG. **Page 17** © Michael S. Yamashita/ Corbis. **2.3** (a) Above, Gary Head; below, Bruce Iverson. **Page 20** Left, © Dan Guravich/Corbis. **2.3** (c, above) PDB ID:1BNA; H.R. Drew, R.M. Wing, T. Takano, C. Broka, S. Tanaka, K. Itakura, R.E. Dickerson; Structure of a B-CAN Dodecamer, Conformation and Dynamics, PNAS; (c, center) PDB ID: ID:1BNA; H.R. Drew, R.M. Wing, T. Takano, C. Broka, S. Tanaka, K. Itakura, R.E. Dickerson; Structure of a B-CAN Dodecamer, Conformation and Dynamics, PNAS, V. 78 2179, 1981; (c, below) PDB ID:1BNA; H.R. Drew, R.M. Wing, T. Takano, C. Broka, S. Tanaka, K. Itakura, R.E. Dickerson; Structure of a B-CAN Dodecamer, Conformation and Dynamics, PNAS. **2.4** (a–c) PDB file from NYU Scientific Visualization Lab; (b) Right, © Steve Lissau/Rainbow. **2.6** (a) © Lester Lefkowitz/ Corbis. **2.8** Michael Grecco/Picture Group. **2.9** PDB file from NYU Scientific Visualization Lab. **2.10** (a) PDB file from NYU Scientific Visualization Lab; (b) PDB file from Klotho Biochemical Compounds Declarative Database. **2.11** Above, left, © Eric and David Hosking/Corbis; right, © Wayne Bennett/ Corbis. **2.16** © Steve Chenn/Corbis. © David Scharf/Peter Arnold, Inc. **2.18** PDB file courtesy of Dr. Christina A. Bailey, Department of Chemistry & Biochemistry, California Polytechnic State University, San Luis Obispo, CA. **Page 28** Top right, © Kevin Schafer/Corbis. **2.20** (a) PDB file courtesy of Dr. Christina A. Bailey. **2.21** (c) © Scott Camazine/Photo Researchers, Inc. **2.23** Below, PDB files from NYU Scientific Visualization Lab. **Page 32** Bottom right, © Corbis. **2.25** (a–b) PDB ID: 1BBB; Silva, M.M., Rogers, P.H., Arnone, A.; A third quarternary structure of human hemoglobin A at 1.7-A resolution; *J Biol Chem* 267 pp. 17248 (1992). **2.26** (a–b) PDB files from NYU Scientific Visualization Lab (c) Right, © Dr. Gopal Murti/SPL/Photo Researchers, Inc.; (d) Above left, Courtesy of Melba Moore. **2.29** PDB ID:1BNA; H.R. Drew, R.M. Wing, T. Takano, C. Broka, S. Tanaka, K. Itakura, R.E. Dickerson; Structure of a B-CAN Dodecamer, Conformation and Dynamics, PNAS. **2.30** © National Gallery Collection; By kind permission of the Trustees of the National Gallery, London/Corbis.

CHAPTER 3: **Page 38** © Dr. Tony Brian and David Parker/SPL/Photo Researchers, Inc. **3.2** Photographs, (a) Leica Microsystems, Inc. Deerfield, IL; (b) © George Musil/Visuals Unlimited. **3.3** (a–d) Jeremy Pickett-Heaps, School of Botany, University of Melbourne. **3.4** Art assembly, Gary Head; (hummingbird) Robert A. Tyrell; (humans) Pete Saloutos/Corbis; (redwood) Sally A. Morgan, Ecoscene/Corbis. (Left) With integrin: PDB ID:1JV2; Xiong, J.P., Stehle, T., Diefenbach, B., Zhang, R., Dunker, R., Scott, D.L., Joachimiak, A., Goodman, S.L., Arnaout, M.A.: Crystal Structure of the Extracellular Segment of Integrin; (right) Chris Keeney with Human growth hormone: PDB ID:1A22; Clackson, T., Ultsch, M. H., Wells, J. A., de Vos, A. M.: Structural and functional analysis of the 1:1 growth hormone: receptor complex reveals the molecular basis for receptor affinity. *J Mol Biol* 277 pp. 1111 (1998); HLA: PDB ID: 1AKJ; Gao, G.F., Tormo, J., Gerth, U.C., Wyer, J.R., McMichael, A.J., Stuart, D.I., Bell, J.I., Jones, E.Y., Jakobsen, B.K.; Crystal structure of the complex between human CD8alpha(alpha) and HLA-A2; *Nature* 387 pp. 630 (1997); glut1: PDB ID:1JA5; Zuniga, F.A., Shi, G., Haller, J.F., Rubashkin, A., Flynn, D.R., Iserovich, P., Fischbarg, J.: A Three-Dimensional Model of the Human Facilitative Glucose Transporter Glut1. *J.Biol.Chem.* 276 pp. 44970 (2001); calcium pump: PDB ID:1EUL; Toyoshima, C., Nakasako, M., Nomura, H., Ogawa, H.: Crystal Structure of the Calcium Pump of Sarcoplasmic Reticulum at 2.6 Angstrom Resolution. *Nature* 405 pp. 647 (2000); ATPase: PDB ID:18HE; Menz, R.I., Walker, J.E., Leslie, A.G. W.: Structure of Bovine Mitochondrial F1-ATPase with Nucleotide Bound to All Three Catalytic Sites: Implications for the Mechanism of Rotary Catalysis. *Cell* (Cambridge, Mass.) 106 pp. 331 (2001). **3.8** (a) Lisa Starr; (b) © K. G. Murti/Visuals Unlimited; (c) © R. Calentine/Visuals Unlimited; (d) Gary Gaard and Arthur Kelman. **3.9** Photographs, (a) Stephen L. Wolfe; (c) Don W. Fawcett/Visuals Unlimited, computer enhanced by Lisa Starr; page 47 (d) Don W. Fawcett/ Visuals Unlimited, computer enhanced by Lisa Starr; (e) Gary Grimes, computer enhanced by Lisa Starr. **3.10** Keith R. Porter. **3.11** © Dr. Jeremy Burgess/SPL/ Photo Researchers, Inc. **3.14** (a) Stanley M. Awramik; (b–c) Andrew H. Knoll, Harvard University; (d) P.L. Walne and J.H. Arnott, *Planta*, 77:325–354, 1967. **Page 51** © CNRI/Photo Researchers, Inc. **3.16** (a) © J.W. Shuler/Photo Researchers, Inc.; (b) Courtesy Dr. Vincenzo Cirulli, Laboratory of Developmental Biology, The Whittier Institute for Diabetes University of California, San Diego, La Jolla, California; (c) Courtesy Mary Osborn, Max Planck Institute for Biophysical Chemistry, Goettingen, FRG. **3.17** (a) John Lonsdale, www.johnlonsdale.net; (b) David C. Martin, Ph.D. **3.19** (a) © CNRI/SPL/Photo Researchers, Inc.; (b) © Don W. Fawcett/Photo Researchers, Inc.; (c) © Mike Abbey/Visuals Unlimited. **3.21** (a) George S. Ellmore; (b, left) Science Photo Library/Photo Researchers, Inc.; (b, right) Bone Clones, www .boneclones.com. **3.23** (a–b) From *Tissue and Cell*, Vol. 27, pp. 421–427, Courtesy of Bjorn Afzelius, Stockholm University.

CHAPTER 4: **Page 58** © Paul Edmondson/Corbis. **4.1** Above, © Craig Aurness/Corbis; below, © William Dow/Corbis. **4.4** (a–b) From B. Alberts et al. *Molecular Biology of the Cell*, 1983, Garland Publishing. **4.5, 4.6** Art, Lisa Starr; photograph, © Randy Ury/Corbis. **4.7** PDB ID: 1DBF; Putnam, C.D., Arvai, A.S., Bourne, Y., Tainer, J.A.: Active and Inhibited Human Catalase Structures: Lingand Nadph Binding and Catalytic Mechanism, *J Mol Biol.* 296, p. 295 (2000). **4.10** Photograph, Scott McKiernan/ZUMA Press. **4.14** After David H. MacLennan, William J. Rice and N. Michael Green, "The Mechanism of Ca2+ Transport by Sarco (Endo)plasmic Reticulum Ca2+-ATPases." *JBC* Vol. 272, No. 46, November 14, 1997, pp. 28815–28818. **Page 68** © Hubert Stadler/Corbis. **4.16** Photographs, M. Sheetz, R. Painter, and S. Singer, *Journal of Cell Biology*, 70:193 (1976) by permission, The Rockefeller University Press. **4.19** (b) © Juergen Berger/Max Planck Institute/SPL/Photo Researchers, Inc. **4.21** Frieder Sauer/Bruce Coleman.

CHAPTER 5: **Page 73** NASA. **5.2** Larry West/FPG. PDB files from NYU Scientific Visualization Lab. Harindar Keer, Thorsten Ritz Laboratory at UC Irvine, Dept. of Physics and Astronomy, using VMD proprietary software. **Page 75** Bottom right, E.R. Degginger. **Page 76** Top left, © Richard Uhlhorn Photography. **5.7** (a–d) Light Harvesting Complex PDB ID: 1RWT; Liu, Z., Yan, H., Wang, K., Kuang, T., Zhang, J., Gui, L., An, X., Chang, W.: Crystal Structure of Spinach Major Light-Harvesting Complex at 2.72 A Resolution. *Nature* 428 pp. 287 (2004). **5.8** Above left, © Bill Beatty/Visuals Unlimited; above right, Micrograph, Bruce Iverson, computer-enhanced by Lisa Starr; Art, Gary Head. **Page 78** Bottom left, NASA. **5.9** (a, above) 2001 PhotoDisc; (a, below) Micrograph, © Ken Wagner/Visuals Unlimited, computer-enhanced by Lisa Starr. **5.10** © Chris Hellier/Corbis. **5.12** (a) Herve Chaumeton/Agence Nature; (b) Douglas Faulkner/Sally Faulkner Collection.

CHAPTER 6: **Page 82** Left, © Louise Chalcraft-Frank and FARA; right, © Professors P. Motta and T. Naguro/ SPL/Photo Researchers, Inc. **6.4** (a) © Professors P. Motta and T. Naguro/SPL/Photo Researchers, Inc. **6.6** (a) Adrian Warren/Ardea, London; (b–c) © Foodpix/ Ben Fink. **6.8** Photograph, Gary Head. **Page 91** © Lois Ellen Frank/Corbis. **6.9** Lisa Starr. **6.10** Thomas D. Mangelsen/Images of Nature.

CHAPTER 7: **Page 94** Left, Courtesy of the family of Henrietta Lacks; right, Micrograph, Dr. Pascal Madaule, France. **7.1** Micrograph, C.J. Harrison et al., *Cytogenetics and Cell Genetics*, 35: 21–27 1983 S. Karger, A.G. Basel. **7.2** Above, © Tim Pannell/Corbis. **7.3** © L. Willatt, East Anglian Regional Genetics Service/SPL/Photo Researchers, Inc. **7.5** Pages 98–99, Photographs, © Jennifer W. Shuler/Science Source/Photo Researchers, Inc.; Art, Raychel Ciemma. **7.6** Micrograph, (a) © D.M. Phillips/Visuals Unlimited; (b) © R. Calentine/Visuals Unlimited. **Page 100** Bottom left, © Jennifer W. Shuler/ Science Source/Photo Researchers, Inc. **7.8** Image courtesy of Carl Zeiss MicroImaging, Thornwood, NY. **7.14** © Francis Leroy, Biocosmos/Science Photo Library/ Photo Researchers, Inc. **7.15** Lennart Nilsson from *A Child Is Born*, 1966, 1977 Dell Publishing Company, Inc. **7.16** © Phillip B. Carpenter, Department of Biochemistry and Molecular Biology. University of Texas-Houston Medical School. **7.17** © Science Photo Library/Photo Researchers, Inc. **7.19** © Ron Neumeyer, www .microimaging.ca.

CHAPTER 8: **Page 112** Left, © Simon Fraser/RVI, New Castle-Upon-Tyne/SPL/Photo Researchers, Inc.; right, © Science Photo Library/Photo Researchers, Inc. **8.2** The Moravian Museum, Brno. **8.3** (a) Jean M. Labat/ Ardea London. **8.9** David Scharf/Peter Arnold, Inc. **8.10** Upper left and right, William F. Ferguson; below, © Francesc Muntada/Corbis. **Page 118** Left, © Science Photo Library/Photo Researchers, Inc. **8.11** From left, Michael Stuckey/Comstock, Inc.; Michael Stuckey/ Comstock, Inc.; Bosco Broyer, photograph by Gary Head. **8.12** © Bettman/Corbis. **8.13** From top, Frank Cezus/FPG; Frank Cezus/FPG; 2001 PhotoDisc, Inc.; Ted Beaudin/FPG; Stan Sholik/FPG. **8.15** Courtesy of Ray Carson, University of Florida News and Public Affairs. **8.17** Left, Pamela Harper/Harper Horticultural Slide Library; (a–c) From Prof. Otto Wilhelm Thomé, Flora von Deutschland Österreich und der Schweiz. 1885, Gera, Germany. **Page 122** Right, © Gideon Mendel/Corbis. **8.18** © Omikron/Photo Researchers, Inc. **8.19** Left, upper, 2001 PhotoDisc, Inc./Getty Images; lower, 2001 EyeWire, Inc. **8.21** © Stapleton Collection/Corbis. **8.22** Dr. Victor A. McKusick. **8.23** Steve Uzzell. **8.24** © Frank Trapper/Corbis Sygma. **8.25** Left, © Lois Ellen Frank/Corbis. **8.26** Below, Photo by Gary L. Friedman, www.FriedmanArchives.com. **8.28** Right, © Bettman/Corbis. **8.29** Courtesy G.H. Valentine. **8.31** © CNRI/Photo Researchers, Inc. **8.32** (b) © Lauren Shear/Photo Researchers, Inc. **8.33** UNC Medical Illustration and Photography. **8.34** © Saturn Stills/SPL/Photo Researchers, Inc. **8.35** From Lennart Nilsson, *A Child is Born*, © 1966, 1977 Dell Publishing Company, Inc. **8.36** Left, Fran Heyl Associates © Jacques Cohen, computer-enhanced by © Pix Elation; right, © Matthew Alan/Corbis. **8.39** Upper, © Tom and Pat Leeson/Photo Researchers, Inc.; lower, Rick Guidotti, Positive Exposure. **8.40** Leslie Faltheisek. Clacritter Manx.

CHAPTER 9: **Page 137** Photo courtesy of the College of Veterinary Medicine, Texas A&M University. **9.2** Bottom, © Eye of Science/Photo Researchers, Inc. **9.3** A.C. Barrington Brown, 1968 J.D. Watson. **9.5** Right, © SPL/Photo Researchers, Inc. **9.6** PDB ID: 1BBB; Silva, M.M., Rogers, P.H., Arnone, A.; A third quarternary structure of human hemoglobin A at 1.7-A resolution; *J Biol Chem* 267 pp. 17248 (1992). **Page 143** Fran Heyl Associates © Jacques Cohen, computer-enhanced by © Pix Elation. **9.8** © James King-Holmes/SPL/Photo Researchers, Inc. **9.9** (a–b) Photos courtesy of the College of Veterinary Medicine, Texas A&M University; (c–d) Photos by Victor Fisher, courtesy Genetic Savings & Clone.

CHAPTER 10: **Page 146** Left, © Vaughan Fleming/ SPL/Photo Researchers, Inc.; (right) PDB ID:2AAI; Rutenber, E., Katzin, B.M., Ernst, S., Collins, E.J., Mlsna, D., Ready, M.P., Robertus, J.D.: Crystallographic refinement of ricin to 2.5A. *Proteins* 10 pp. 240 (1991). **10.6** Above, Model by Dr. David B. Goodin, The Scripps Research Institute. **10.7** (a) Model, courtesy of Thomas A. Steitz from *Science*. **10.10** Left, Nik Kleinberg; right, P. J. Maughan. **10.11** © John W. Gofman and Arthur R. Tamplin. From *Poisoned Power: The Case Against Nuclear Power Plants Before and After Three Mile Island*, Rodale Press, PA, 1979. **10.13** PDB ID: 1CJG; Spronk, C.A.E.M., Bonvin, A.M.J.J., Radha, P.K., Melacini, G., Boelens, R., Kaptein, R.: The Solution Structure of Lac Repressor Headpiece 62 Complexed to a Symmetrical Lac Operator. *Structure* (London) 7 pp. 1483 (1999). Also PDB ID: 1LBI; Lewis, M., Change, G., Horton, N.C., Kercher, M.A., Pace, H.C., Schumacher, M.A., Brennan, R.G., Lu, P.: Crystal structure of the lactose operon repressor and its complexes with DNA and inducer. *Science* 271 pp. 1247 (1996); lactose PDB files from the Hetero-Compound Information Centre-Uppsala (HIC-Up). **10.15** (a) © Lisa Starr; (b) © Visuals Unlimited; (c) © Gehring, P. Callaets and G. Halder; courtesy Walter J. Ghering/University of Basel, Switzerland; (d) Courtesy of Edward B. Lewis, California Institute of Technology. **10.16** (a) Dr. Karen Dyer Montgomery. **10.17** Jack Carey. **10.19** © Dr. M.A. Ansary/Science Photo Library/Photo Researchers, Inc. **Page 161** Lower right (both), Photo courtesy of the College of Veterinary Medicine, Texas A&M University.

CHAPTER 11: **Page 162** Dr. Jorge Mayer, Golden Rice Project. **11.2** (a) © Professor Stanley Cohen/SPL/ Photo Researchers, Inc.; (b) With permission of © QIAGEN, Inc. **11.7** Above, © TEK IMAGE/Photo Researchers, Inc. **11.8** Left, © David Parker/SPL/ Photo Researchers, Inc.; right, Cellmark Diagnostics, Abingdon, UK. **11.10** © Volker Steger/SPL/Photo Researchers, Inc. **11.11** (a) Courtesy Calgene LLC; (b) Dr. Vincent Chaing, School of Forestry and Wood Projects, Michigan Technology University. **11.12** Photo left, © Lowell Georgis/Corbis; right, Keith V. Wood. **11.13** (a) Transgenic goat produced using nuclear transfer at GTC Biotherapeutics. Photo used with permission; (b) © Matt Gentry/Roanoke Times. **11.14** © Adi Nes, Dvir Gallery Ltd. **11.15** © Jeans for Gene Appeal. **Page 174** © Work of Atsushi Miyawaki, Qing Xiong, Varda Lev-Ram, Paul Steinbach, and Roger Y. Tsien at the University of California, San Diego.

CHAPTER 12: **Page 176** © Bettmann/Corbis. **12.1** (a, c) © Wolfgang Kaehler/Corbis; (b) © Earl & Nazima Kowall/Corbis; (d–e) Edward S. Ross. **12.2** Left, Gary Head; right, above, © Bruce J. Mohn; below inset, Phillip Gingerich, Director, University of Michigan. Museum of Paleontology. **12.3** (a) Courtesy George P. Darwin, Darwin Museum, Down House; (b) Christopher Ralling; (c) Dieter & Mary Plage/Survival Anglia. **12.4** (a) © Karen Carr Studio/www.karencarr.com; (b) © Joe McDonald/Corbis. **12.5** (a) Heather Angel; (b) David Cavagnaro; (c) Dr. P. Evans/Bruce Coleman, Ltd.; (d) Alan Root/Bruce Coleman Ltd. **12.6** (a) Gary Head; (b) John W. Merck Jr., University of Maryland. **12.7** (a) Giorgio Gualco/Bruce Coleman, Ltd.; (b) © David Parker/SPL/Photo Researchers, Inc. **Page 182** Bottom, from left, © Peter Bowater/Photo Researchers, Inc.; © Sam Kleinman/Corbis; © Owen Franken/ Corbis; © Christopher Briscoe/Photo Researchers, Inc. **12.8** Alan Solem. **12.11** Courtesy of © Hopi Hoekstra, University of California, San Diego. **12.12** © Reuters NewMedia, Inc./Corbis. **Page 182** Bottom left (Barn owl), © Lanz Von Horsten; Gallo Images/Corbis. **12.15** Thomas Bates Smith. **12.16** Bruce Beehler. **12.17** Left, © Michael Freeman/Corbis; (a–b) After Ayala and others. **12.18** Adapted from S.S. Rich, A.E. Bell, and S.P. Wilson, "Genetic drift in small populations of Tribolium," *Evolution* 33:579–584, Fig. 1, p. 580, 1979. Used by permission of the publisher. **Page 190** Bottom left, © Steve Bronstein/The Image Bank/Getty Images. **12.19** Frans Lanting/Minden Pictures. **12.20** Left, David Neal Parks; right, W. Carter Johnson. **12.21** © Elliot Erwitt/Magnum Photos, Inc.

CHAPTER 13: **Page 194** NASA Galileo Imaging Team. **13.1** Left, H. P. Banks; right, Jonathan Blair. **13.2** © Jonathan Blair/Corbis. **Page 196** Bottom left, © David A. Kring, NASA/Univ. Arizona Space Imagery Center. **13.3** (b) © PhotoDisc/Getty Images. **13.5** © Corbis. **13.6** (a) NASA/GSFC. **13.7** Art, (a–d) After A.M. Ziegler, C.R. Scotese, and S.F. Barrett, "Mesozoic and Cenozoic Paleogeographic Maps," and J. Krohn and J. Sundermann (Eds.), *Tidal Frictions and the Earth's Rotation II*, Springer-Verlag, 1983; (e) © Martin Land/ Photo Researchers, Inc.; (f) © John Sibbick. **13.9** (a) J. Scott Altenbach, University of New Mexico, computer enhanced by Lisa Starr; (b) Frans Lanting/Minden Pictures, computer enhanced by Lisa Starr; (c) © Stephen Dalton/Photo Researchers, Inc. **13.10** (a) Above, Tait/Sunnucks Peripatus Research; below, © Jennifer Grenier, Grace Boekhoff-Falk and Sean Carroll, HMI, University of Wisconsin-Madison; (b) Above, Herve Chaumeton/Agence Nature; below, © Jennifer Grenier, Grace Boekhoff-Falk and Sean Carroll, HMI, University of Wisconsin-Madison; (c) Above, © Peter Skinner/Photo Researchers, Inc.; below, Courtesy of Dr. Giovanni Levi; (d) Dr. Chip Clark. **Page 204** © TEK IMAGE/Photo Researchers, Inc. **13.14** After F. Ayala and J. Valentine, *Evolving*, Benjamin-Cummings, 1979. **13.15** (a) John Alcock, Arizona State University; (b) © Alvin E. Staffan/Photo Researchers, Inc.; (c) G. Ziesler/ ZEFA. **13.16** Left, © Digital Vision/PictureQuest; (a) © Joe McDonald/Corbis. **13.17** (a) © Graham Neden/ Corbis; (b) © Kevin Schafer/Corbis; center, Ron Blakey, Northern Arizona University; (c) © Rick Rosen/Corbis SABA. **13.18** Po'ouli, Bill Sparklin/Ashley Dayer; All other photographs, © Jack Jeffrey Photography. **13.19** Left, above and below, © Below Water Photography/ www.belowwater.com; right, Steve Gartlan. **13.20** After W. Jensen and F.B. Salisbury, *Botany: An Ecological Approach*, Wadsworth, 1972. **13.21** Courtesy of Dr. Robert Mesibov. **13.22** Courtesy of Daniel C. Kelley, Anthony J. Arnold, and William C. Parker, Florida State University Department of Geological Science. **13.23** Carnegie Museum of Natural History. **13.24** From left, © Science Photo Library/Photo Researchers, Inc.; © Dr. Dennis Kunkel/Visuals Unlimited; © Galen Rowell/Corbis; © Kevin Schafer/Corbis; Courtesy of Department of Entomology, University of Nebraska-Lincoln; Bruce Coleman, Ltd. **13.26** From left, © Hans Reinhard/ Bruce Coleman, Inc.; © Phillip Colla Photography; © Randy Wells/Corbis; © Cousteau Society/The Image Bank/Getty Images; © Robert Dowling/Corbis. **13.28** © Gulf News, Dubai, UAE.

CHAPTER 14: **Page 218** © Philippa Uwins/The University of Queensland. **14.1** Jeff Hester and Paul Scowen, Arizona State University, and NASA. **14.2** (a) © Eiichi Kurasawa/Photo Researchers, Inc.; (b) © Dr. Ken MacDonald/SPL/Photo Researchers, Inc.; (c) © Michael J. Russell, Scottish Universities Environmental Research Centre. **14.3** (a) Sidney W. Fox; (b) From Hanczyc, Fujikawa, and Szostak, Experimental Models of Primitive Cellular Compartments: Encapsulation, Growth, and Division; www.sciencemag.org. Science, 24 October 2003; 302;529, Figure 2, page 619. Reprinted with permission of the authors and AAAS. **14.4** Above, © Chase Studios/Photo Researchers, Inc.; below left, © John Reader/SPL/Photo Researchers, Inc.; below right, © Sinclair Stammers/SPL/Photo Researchers, Inc. **14.6** (a) © P. Hawtin, University of Southampton/SPL/Photo Researchers, Inc.; (b) © Dr. Dennis Kunkel/Visuals Unlimited. **14.8** (a) © Martin Miller/Visuals Unlimited; (b) © Alan L. Detrick, Science Source/Photo Researchers, Inc.; (c) © Dr. Harald Huber, Dr. Michael Hohn, Prof. Dr. K.O. Stetter, University of Regensburg, Germany. **Page 226** Bottom left, Courtesy of Agriculture Canada. **14.9** (a) © Dr. Jeremy Burgess/SPL/Photo Researchers, Inc.; (b) P.W. Johnson and J. MeN. Sieburth, Univ. Rhode Island/ BPS. **14.10** Dr. Terry J. Beveridge, Department of Microbiology, University of Guelph, Ontario, Canada. **14.11** © Dr. Manfred Schloesser, Max Planck Institute for Marine Microbiology. **14.12** © Stem Jems/Photo Researchers, Inc.; inset, California Department of Health Services. **14.14** © Dr. Dennis Kunkel/Visuals Unlimited. **14.15** (a) © Oliver Meckes/Photo Researchers, Inc.; (b) Photo by Stephen Durr. **14.16** (a) Neal Ericsson; (b) Courtesy of Allen W.H. Bé and David A. Caron; (c, left) © Wim van Egmond; (c, right) © G. Shih and R. Kessel/Visuals Unlimited. **14.17** (a) © Andrew Syred/SPL/Photo Researchers, Inc.; (b) Courtesy © James Evarts. **14.18** (a) © Dr. David Phillips/Visuals Unlimited; (b) C.C. Lockwood; (c) © Wim van Egmond/Micropolitan Museum; (d) © Frank Borges Llosa/www.frankley.com. **14.19** From left, © Sinclair Stammers/Photo Researchers, Inc.; micrograph center, Steven L'Hernault; © London School of Hygiene & Tropical Medicine/Photo Researchers, Inc.; micrograph right, © Moredum Animal Health, Ltd./Photo Researchers, Inc. **14.20** Science Museum of Minnesota; below, Emiliania Huxleyi. Photograph by Vita Pariente. Scanning electron micrograph taken on a Jeol T330A instrument at the Texas A&M University Electron Microscopy Center. **14.21** (a) © Lewis Trusty/Animals Animals; (b) © Wim van Egmond; art (left) T. Garrison, *Oceanography: An Invitation to Marine Science*, Brooks/Cole, 1993. **14.23** (a) © Andrew J. Martinez/Photo Researchers, Inc.; (b) Courtesy of Professor Astrid Saugestad; (c) Courtesy Microbial Culture Collection, National Institute for Environmental Studies, Japan. **14.24** (a) © Astrid Hanns-Frieder Michler/SPL/Photo Researchers, Inc.; (b) M. Claviez, G. Gerish, and R. Guggenheim; (c–d) Carolina Biological Supply Company; (e) Courtesy Robert R. Kay from R.R. Kay, et al., Development, 1989 Supplement, pp. 81–90. **14.26** Left, Micrograph Garry T. Cole, University of Texas, Austin/BPS; art, After T. Rost, et al., *Botany*, Wiley, 1979. **14.27** (a–b) Robert C. Simpson/Nature Stock. **14.28** (a) Jane Burton/Bruce Coleman, Ltd.; (b) © Chris Worden. **14.29** (a) © Andrew Syred/SPL/Photo Researchers, Inc.; (a, above) Micrograph Ed Reschke; (b) John Hodgin; (c) © North Carolina State University, Department of Plant Pathology; (d) © Michael Wood/mykob.com; (e) Garry T. Cole, University of Texas, Austin/BPS. **14.30** (a) © Mark E. Gibson/Visuals Unlimited; (b) Edward S. Ross; (c) © 1977 Sherry K. Pittam; (d) Mark Mattock/Planet Earth Pictures; (e) After Raven, Evert, and Eichhorn, *Biology of Plants*, fourth edition, Worth Publishers, New York, 1986. **14.31** (a) © 1990 Gary Braasch; (b) F.B. Reeves. **14.32** (a) © Dr. P. Marazzi/SPL/Photo Researchers, Inc.; (b) © Harry Regin. **14.33** (e) © CAMR/A.B. Dowsett/SPL/Photo Researchers, Inc.; (f) © Dr. Linda Stannard, UCT/SPL/Photo Researchers, Inc. **14.34** Above left, © Science Photo Library/Photo Researchers, Inc. **14.35** Above, © NIBSC/SPL/Photo Researchers, Inc.; below, © Sercomi/Photo Researchers, Inc. **14.36** With permission of © FSIS/USDA/FDA.

CHAPTER 15: **Page 244** Left, © T. Kerasote/Photo Researchers, Inc.; right, © Clinton Webb. **15.1** (a) © Reprinted with permission from Elsevier; (b) Patricia G. Gensel; (c) Illustration by Zdenek Burian, © Jeri Hochman and Martin Hochman. **15.3** After E.O. Dodson and P. Dodson, *Evolution: Process and Product*, third edition, p. 401, PWS. **15.5** (a) © Craig Wood/Visuals Unlimited; (b) © John D. Cunningham/Visuals Unlimited; (c) National Park Services, Paul Stehr-Green; (d) National Park Services, Martin Hutten. **15.6** Jane Burton/Bruce Coleman Ltd. **Page 248** Lower left, above, © Jean Miele/Corbis; below, © Thomas D. Mangelsen/Peter Arnold, Inc. **15.7** Photo, Field Museum of Natural History, Chicago (Neg. #7500C). **15.8** (a) © Ed Reschke/Peter Arnold, Inc.; (b) © Colin Bates; (c) W.H. Hodge. **15.9** A. & E. Bomford/Ardea, London. **15.10** © Klein Hubert/Peter Arnold, Inc. **15.11** (a) © Ralph Pleasant/FPG/Getty Images; (b) © Earl Roberge/Photo Researchers, Inc.; (c) © George Loun/Visuals Unlimited; (d) Courtesy of Water Research Commission, South Africa. **15.12** (a) © Dave Cavagnaro/Peter Arnold, Inc.; (b) Robert & Linda Mitchell Photography; (c) © E. Webber/Visuals Unlimited; (d) © Michael P. Gadomski/Photo Researchers, Inc.; (e) © Sinclair Stammers/Photo Researchers, Inc.; (f) Courtesy of Wayside Gardens/www.waysidegardens.com; (g) © Gerald & Buff Corsi/Visuals Unlimited; (h) © Fletcher and Baylis/Photo Researchers, Inc. **15.13** Left, Robert Potts, California Academy of Sciences; (a) Robert & Linda Mitchell Photography; (b) © R.J. Erwin/Photo Researchers, Inc. **15.15** (a) © Michelle Garrett/Corbis; (b) © Gregory G. Dimijian/Photo Researchers, Inc.; (c) © Darrell Gulin/Corbis; (d) © Darrell Gulin/Corbis. **15.17** Gerry Ellis/The Wildlife Collection. **15.18** © William Campbell/TimePix/Getty Images.

CHAPTER 16: **Page 259** P. Morris/Ardea London. **16.5** After Eugene Kozloff. **16.6** (c) © Wim van Egmond/Micropolitan Museum; (d) © Brandon D. Cole/Corbis; (e) © Jeffrey L. Rotman/Corbis. **16.8** Right, © Andrew Syred/SPL/Photo Researchers, Inc. **16.9** (a) © Cabisco/Visuals Unlimited; (b) © Science Photo Library/Photo Researchers, Inc. **16.10** (a) Jon Kenfield/Bruce Coleman Ltd.; (b) Adapted from Rasmussen, *Ophelia*, Vol. 11, in Eugene Kozloff, Invertebrates, 1990; (c–d) J.A.L. Cooke/Oxford Scientific Films. **16.11** (b) Frank Park/ANT Photo Library; (c) Alex Kirstitch; (d) J. Grossauer/ZEFA. **16.12** (a) © Sinclair Stammers/SPL/Photo Researchers, Inc.; (b) © L. Jensen/Visuals Unlimited; (c) Dianora Niccolini. **16.13** Jane Burton/Bruce Coleman, Ltd. **16.14** (a) Dr. Chip Clark; (b) Angelo Giampiccolo/FPG/Getty Images. **16.15** (a) © Corbis; (b) John H. Gerard; (c) © D. Suzio/Photo Researchers, Inc.; (d) © Andrew Syred/Photo Researchers, Inc. **16.17** (a) © Jeff Hunter/The Image Bank/Getty Images; (b) © Peter Parks/Imagequestmarine.com; (c) © Science Photo Library/Photo Researchers, Inc. **16.18** (a–c, i–j) Edward S. Ross; (d) © Mark Moffett/Minden Pictures; (e) Courtesy of Karen Swain, North Carolina Museum of Natural Sciences; (f) Chris Anderson/Darklight Imagery; (g) Joseph L. Spencer; (h) David Maitland/Seaphot Limited. **16.19** Left, Herve Chaumeton/Agence Nature; (a) Herve Chaumeton/Agence Nature; (b) © Fred Bavendam/Minden Pictures; (c) © George Perina, www.seapix.com; (d) Jan Haaga, Kodiak Lab, AFSC/NMFS. **16.20** Right, Runk & Schoenberger/Grant Heilman, Inc. **16.21** Above, Peter Parks/Oxford Scientific Films/Animals Animals; below, © California Academy of Sciences. **16.22** (a) © Ron & Valerie Taylor/Bruce Coleman, Inc.; (b–c) Adapted from A.S. Romer and T.S. Parson, *The Vertebrate Body*, sixth edition, Saunders, 1986; (d) Left, Photo by Lisa Starr; right, Courtesy of John McNamara, www.paleodirect.com. **16.23** (a) © Jonathan Bird/Oceanic Research Group, Inc.; (b) © Gido Braase/Deep Blue Productions. **16.24** (b) © Ivor Fulcher/Corbis; (c) © Roger Archibald. **16.25** (a) © Norbert Wu/Peter Arnold, Inc.; (b) Wernher Krutein/photovault.com; (c) © Alfred Kamajian. **16.26** (a) Bill M. Campbell, MD; (b) © Stephen Dalton/Photo Researchers, Inc.; (c) © John Serraro/Visuals Unlimited. **16.27** Pieter Johnson. **16.28** (a) © Illustration by Karen Carr; (b) © Kevin Schafer/Corbis; (c) Kevin Schafer/Tom Stack & Associates; (d) Z. Leszczynski/Animals Animals; (e) © S. Blair Hedges, Pennsylvania State University. **16.29** (b) © Kevin Schafer/Corbis. **Page 279** Bottom right, above, P. Morris/Ardea London; below, James Reece, Nature Focus, Australian Museum. **16.30** (a) Sandy Roessler/FPG/Getty Images; (b) © Jean Phillipe Varin/Jacana/Photo Researchers, Inc.; (c) Corbis Images/PictureQuest; (d) © Merlin D. Tuttle/Bat Conservation International; (e) © Mike Johnson. All rights reserved, www.earthwindow.com. **16.31** (a) Larry Burrows/Aspect Photolibrary; (b) © Art Wolfe/Photo Researchers, Inc.; (c) © Dallas Zoo, Robert Cabello; (d) Allen Gathman, Biology Department, Southeast Missouri State University; right, monkey skeletons, Bone Clones®, www.boneclones.com; human, Gary Head. **16.32** Gerry Ellis/The Wildlife Collection; inset, Utah's Hogle Zoo. **16.33** (a) From left, © MPFT/Corbis Sygma; National Museum of Ethiopia, Addis Ababa. © 1985 David L. Brill; Original housed in National Museum of Ethiopia, Addis Ababa. © 1999 David L. Brill; Original housed in National Museum of Ethiopia, Addis Ababa. © 1999 David L. Brill; National Museum of Tanzania, Dar es Salaam, © 1985 David L. Brill; Transvaal Museum, Pretoria. © 1985 David L. Brill; (b) Dr. Donald Johanson, Institute of Human Origins; (c) Louise M. Robbins; (d) Kenneth Garrett/National Geographic Image Collection. **16.34** (a) National Museum of Kenya, Nairobi. © 1985 David L. Brill; (b) Jean Paul Tibbles. **16.35** Left, © Elizabeth Delaney/Visuals Unlimited; all others, John Reader © 1981. **16.37** Left, American Museum of Natural History. © 1996 David L. Brill; right, Musée de l'homme, Paris. © 1985 David L. Brill. **16.38** Left, Housed in National Museum of Ethiopia, Addis Ababa. © 2001 David L. Brill; right, NASA. **16.39** © Pete & Judy Morrin/Ardea London.

CHAPTER 17: **Page 289** Star Tribune/Minneapolis-St. Paul. **17.1** Top, Courtesy of Charles Lewallen; others, © Bruce Iverson. **17.2** Left, © 2000 PhotoDisc, Inc, with art by Lisa Starr; upper right, © CNRI/SPL/Photo Researchers, Inc.; lower right, Dr. Robert Wagner/University of Delaware, www.udel.edu/Biology/Wags. **17.3** Upper, © Pat Johnson Studios Photography; lower, © Darrell Gulin/The Image Bank/Getty Images. **17.4** Left, © Cory Gray; right, © PhotoDisc/Getty Images. **17.5** (a) Heather Angel; (b) © Biophoto Associates/Photo Researchers, Inc. **17.7** © VVG/Science Photo Library/Photo Researchers, Inc. **17.8** Top, © Niall Benvie/Corbis. **17.9** (a) © Kennan Ward/Corbis; (b) G. J. McKenzie (MGS); (c) Frank B. Salisbury. **17.10** Slim Films **17.11** (a–b) Courtesy of Dr. Kathleen K. Sulik, Bowles Center for Alcohol Studies, the University of North Carolina at Chapel Hill; (c) © John DaSiai, MD/Custom Medical Stock Photo. **17.12** (a) Courtesy of Dr. Consuelo M. De Moraes; (b–d) Andrei Sourakov and Consuelo M. De Moraes.

CHAPTER 18: **Page 300** Left, © Darrel Plowes; right, © Allstock/Stone/Getty Images. **18.2** (a) Dr. Dale M. Benham, Nebraska Wesleyan University; (b) D. E. Akin and I. L. Risgby, Richard B. Russel Agricultural Research Center, Agricultural Research Service, U.S. Dept. Agriculture, Athens, GA; (c) Kingsley R. Stern. **18.3** Below, © Andrew Syred/Photo Researchers, Inc. **18.4** (a), top to bottom, © Bruce Iverson; © Ernest Manewal/Index Stock Imagery; © Simon Fraser/Photo Researchers, Inc.; © Andrew Syred/Photo Researchers, Inc.; (b, top to bottom) Michael Clayton/University of Wisconsin, Department of Botany; © Darrell Gulin/Corbis; Gary Head; © Andrew Syred/Photo Researchers, Inc. **18.5** (a) Center, Michael Clayton/University of Wisconsin, Department of Botany; right, James W. Perry; (b) center, Carolina Biological Supply Company; right, James W. Perry. **18.6** (a) © N. Cattlin/Photo Researchers, Inc.; (c) C.E. Jeffree, et al., *Planta*, 172(1):20–37, 1987. Reprinted by permission of C.E. Jeffree and Springer-Verlag; (d) © Jeremy Burgess/SPL/Photo Researchers, Inc. **Page 305** Lower right, © 2001 PhotoDisc, Inc./Getty Images. **18.7** © Andrew Syred/Photo Researchers, Inc. **18.8** (a) © Dale M. Benham, Ph.D., Nebraska Wesleyan University. **18.10** (a) After Salisbury and Ross, *Plant Physiology*, fourth edition, Wadsworth; (b, both) Michael Clayton/University of Wisconsin, Department of Botany; (c) © Omikron/Photo Researchers, Inc. **18.12** (b) H.A. Core, W.A. Cote, and A.C. Day, *Wood Structure and Identification*, 2nd Ed., Syracuse University Press, 1979. **18.13** (a) © Peter Ryan/SPL/Photo Researchers, Inc.; (b) © Jon Pilcher; (c) © George Bernard/SPL/Photo Researchers, Inc. **18.14** (a) William Ferguson; (b) © Robert Frerch/Stone/Getty Images. **Page 311** Lower, Courtesy of NOAA. **18.15** © Andrew Syred/Photo Researchers, Inc. **18.16** Prof. DJ Read, University of Sheffield. **18.17** (a) Adrian P. Davies/Bruce Coleman; (b) NifTAL Project, Univ. of Hawaii, Maui. **18.18** (b) Mike Clayton/University of Wisconsin Department of Botany. **18.19** (a) Left, Alison W. Roberts, University of Rhode Island; (b–c) H.A. Core, W.A. Cote, and A.C. Day, *Wood Structure and Identification*, 2nd Ed., Syracuse University Press, 1979. **18.20** Left, The Ohio Historical Society, Natural History Collections. **Page 316** Lower, Courtesy of © Professor T. Mansfield. **18.21** (a–b) © Claude Nuridsany & Marie Perennou/Science Photo

Library/Photo Researchers, Inc. **18.22** George S. Ellmore. **18.24** (a) Don Hopey/Pittsburgh Post-Gazette, 2002, all rights reserved. Reprinted with permission; (b–c) © Jeremy Burgess/SPL/Photo Researchers, Inc. **18.25** (a) © J.C. Revy/ISM/Phototake; (b) Martin Zimmerman, *Science*, 1961, 133:73–79, © AAAS; (c) © James D. Mauseth, MCDB. **18.27** (a) NOAA; (b) USDA/Forestry Service; (c) David W. Stahle, Department of Geosciences, University of Arkansas.

CHAPTER 19: **Page 321** Left and right, © David Goodin. **19.1** Above, left, © John McAnulty/Corbis; right, © Robert Essel NYC/Corbis. **19.2** Left to right, David Scharf/Peter Arnold, Inc.; © David M. Phillips/Visuals Unlimited; © Dr. Jeremy Burgess/SPL/Photo Researchers, Inc. **19.3** (a) © David Goodin; (b, left) Merlin D. Tuttle, Bat Conservation International; (b, right) John Alcock, Arizona State University. **Page 323** Right, above, © Dr. John Hilty; below, © Susumu Nashinaga/Photo Researchers, Inc. **19.5** Center, Dr. Charles Good, Ohio State University-Lima; all others, Michael Clayton, University of Wisconsin, Department of Botany. **19.6** (a) R. Carr; (b) © Gregory K. Scott/Photo Researchers, Inc.; (c) © James L. Amos/Corbis; (d) Robert H. Mohlenbrock © USDA-NRCS PLANTS Database/USDA SCS. 1989. "Midwest wetland flora; field office illustrated guide to plant species." Midwest National Technical Center, Lincoln, NE; (e) left, Dr. Dan Legard, University of Florida GCREC, 2000; center, Richard H. Gross; right, © Andrew Syred/SPL/Photo Researchers, Inc. **Page 330** Left, © Corbis. **Page 326** left, above, © Dr. John Hilty; below, © Corbis. **19.7** © Darrell Gulin/Corbis. **19.8** Photo by Mike Clayton/University of Wisconsin Department of Botany. **19.9** Right, above, Barry L. Runk/Grant Heilman, Inc.; below, © James D. Mauseth, University of Texas. **19.10** Right, Herve Chaumeton/Agence Nature. **19.11** © Sylvan H. Wittwer/Visuals Unlimited. **19.12** left, © Adam Hart-Davis/Photo Researchers, Inc.; (f) © Eric B. Brennan. **19.13** (a) Michael Clayton, University of Wisconsin; (b–c) Muday, G.K. and P. Haworth (1994) "Tomato root growth, gravitropism, and lateral development: Correlations with auxin transport." *Plant Physiology and Biochemistry* 32, 193–203 with permission from Elsevier Science; (d–e) Micrographs courtesy of Randy Moore from "How Roots Respond to Gravity," M.L. Evans, R. Moore, and K. Hasenstein, *Scientific American*, December 1986. **19.14** (c) © Cathlyn Melloan/Stone/Getty Images. **19.15** Gary Head. **19.16** Cary Mitchell. **19.17** Long-day plant, © Clay Perry/Corbis; short-day plant, © Eric Chrichton/Corbis; art, Gary Head. **19.19** Left, © Roger Wilmshurst/Frank Lane Picture Agency/Corbis; right, © Dr. Jeremy Burgess/Photo Researchers, Inc. **19.20** Eric Welzel/Fox Hill Nursery, Freeport, Maine. **19.21** Dan Fairbanks. **19.22** Grant Heilman Photography, Inc.

CHAPTER 20: **Page 339** Left, © Ohlinger Jerry/Corbis Sygma; right, © Science Photo Library/Photo Researchers, Inc. **20.1** (a) Manfred Kage/Bruce Coleman, Ltd.; (b) © Science Photo Library/Photo Researchers, Inc.; right, Focus on Sports. **20.3** (a) © John Cunningham/Visuals Unlimited; (b–c) Ed Reschke; (d, g) © Science Photo Library/Photo Researchers, Inc.; (e) © Michael Abbey/Photo Researchers, Inc.; (f) © University of Cincinnati, Raymond Walters College, Biology. **Page 342** Bottom left, upper, © Science Photo Library/Photo Researchers, Inc.; lower, © Pascal Goetgheluck/Photo Researchers, Inc. **20.4** (a–b) Ed Reschke; (c) © Biophoto Associates/Photo Researchers, Inc. **20.5** © Triarch/Visuals Unlimited. **Page 343** Bottom right, upper, © Science Photo Library/Photo Researchers, Inc.; lower, Courtesy of Muscular Dystrophy Association. **Page 346** Left, upper, © Science Photo Library/Photo Researchers, Inc.; lower, © Mauro Fermariello/Photo Researchers, Inc. **20.7** (b) © John D. Cunningham/Visuals Unlimited. **20.8** Dr. Preston Maxim and Dr. Stephen Bretz, Department of Emergency Services, San Francisco General Hospital. **20.9** © CNRI/SPL/Photo Researchers, Inc.

CHAPTER 21: **Page 349** Left, Michael Neveux; right, © Steve Cole/PhotoDisc Green/Getty Images. **21.1** Above, Linda Pitkin; below, © Stephen Dalton/Photo Researchers, Inc. **21.3** Art, Raychel Ciemma; Photograph, Yokochi and J. Rohen, *Photographic Anatomy of the Human Body*, 2nd Ed., Igaku-Shoin, Ltd., 1979. **21.4** Micrograph, Ed Reschke. **21.5** © Professor P. Motta/Department of Anatomy/La Sapienza, Rome/SPL/Photo Researchers, Inc. **21.7** (a) Dance Theatre of Harlem, by Frank Capri; (b–c) © Don Fawcett/Visuals Unlimited, from D. W. Fawcett, *The Cell*, Philadelphia; W. B. Saunders Co., 1966. **21.12** Painting by Sir Charles Bell, 1809, courtesy of Royal College of Surgeons, Edinburgh. **Page 358** © Steve Cole/PhotoDisc Green/Getty Images. **21.13** (a) Paul Sponseller, MD/Johns Hopkins Medical Center; (b) Courtesy of the family of Tiffany Manning.

CHAPTER 22: **Page 360** © James Stevenson/Photo Researchers, Inc. **22.3** Above, © National Cancer Institute/Photo Researchers, Inc. **22.11** © Sheila Terry/SPL/Photo Researchers, Inc. **22.12** Left, © Biophoto Associates/Photo Researchers, Inc. **22.13** (a) Left, Lisa Starr, using © 2001 PhotoDisc, Inc.; photograph; right, © Dr. John D. Cunningham/Visuals Unlimited. **22.14** © Prof. P. Motta/Department of Anatomy/University La Sapienca, Rome/SPL/Photo Researchers, Inc. **22.15** (a) Ed Reschke; (b) © Biophoto Associates/Photo Researchers, Inc. **22.16** © Lester V. Bergman/Corbis. **22.17** © Peter Parks/Oxford Scientific Films. **22.22** Right, Photographs, Courtesy of Kay Elemetrics Corporation. **Page 375** Right, below, © Corbis. **22.23** (a–b) © SIU/Visuals Unlimited. **22.24** (a) © R. Kessel/Visuals Unlimited. **22.26** (a) Micrograph, Lennart Nilsson from *Behold Man*, © 1974 by Albert Bonniers Forlag and Little, Brown and Company, Boston; (b–c) A. Auerbach/Visuals Unlimited. **22.27** (b) Courtesy of Dr. Joe Losos.

CHAPTER 23: **Page 382** Left, © Lowell Tindell; right, © NIBSC/Photo Researchers, Inc. **23.1** © Larry Williams/Corbis. **23.2** After *Bloodline Image Atlas*, University of Nebraska-Omaha, and Sherri Wicks, *Human Physiology and Anatomy*, University of Wisconsin Web Education System, and others. **23.3** © Dr. Richard Kessel and Dr. Randy Kardon/Tissues & Organs/Visuals Unlimited. **23.4** (a) © Kwangshin Kim/Photo Researchers, Inc. **23.5** © Biology Media/Photo Researchers, Inc. **23.6** Below, Robert R. Dourmashkin, courtesy of Clinical Research Centre, Harrow, England. **23.8** © NSIBC/SPL/Photo Researchers, Inc. **Page 389** Right, both, © NIBSC/Photo Researchers, Inc. **23.9** Photograph © David Scharf/Peter Arnold, Inc. **23.17** © Dr. A. Liepins/SPL/Photo Researchers, Inc. **23.18** Left, David Scharf/Peter Arnold, Inc.; right, © Kent Wood/Photo Researchers, Inc. **23.19** © Zeva Oelbaum/Peter Arnold, Inc. **23.20** Left, © NIBSC/Photo Researchers, Inc. **23.21** Photo courtesy of MU Extension and Agricultural Infomation.

CHAPTER 24: **Page 402** Left, Courtesy of Kevin Wickenheiser, University of Michigan; right, © Gusto/Photo Researchers, Inc. **24.2** (a) © W. Perry Conway/Corbis. **24.5** After A. Vander, et al., *Human Physiology: Mechanism of Body Function*, fifth edition, McGraw-Hill, © 1990. Used by permission of McGraw-Hill. **Page 406** Left, upper, © Gusto/Photo Researchers, Inc.; lower, Courtesy of Lisa Hyche. **24.6** (a) Right, Microslide courtesy Mark Nielsen, University of Utah; (b) right, © D. W. Fawcett/Photo Researchers, Inc.; art, After Sherwood and others. **24.9** Left © Ralph Pleasant/FPG/Getty Images. **Page 411** Lower, © Elizabeth Hathon/Corbis. **Page 414** Lower left, © Gusto/Photo Researchers, Inc. **24.10** Gary Head. **24.14** Left, Evan Cerasoli. **24.15** © Air Force News/Photo by Tech. Sgt. Timothy Hoffman. **24.17** © Reuters NewsMedia/Corbis.

CHAPTER 25: **Page 422** Left, PA Photos; right, Manni Mason's Pictures. **25.1** Micrograph Manfred Kage/Peter Arnold, Inc. **25.5** (b) Micrograph, Dr. Constantino Sotelo from *International Cell Biology*, p. 83, 1977. Used by copyright permission of the Rockefeller University Press. **25.6** Micrograph by Don Fawcett, Bloom and Fawcett, 11th edition, after J. Desaki and Y. Uehara/Photo Researchers, Inc. **Page 427** Right, upper, Manni Mason's Pictures; lower, © Cordelia Molloy/Photo Researchers, Inc. **25.15** C. Yokochi and J. Rohen, *Photographic Anatomy of the Human Body*, 2nd Ed., Igaku-Shoin, Ltd., 1979. **25.19** (a–b) PET scans from E.D. London, et al., *Archives of General Psychiatry*, 47:567–574, 1990. **25.20** © Eric A. Newman. **25.21** © Colin Chumbley/Science Source/Photo Researchers, Inc.; After Penfield and Rasmussen, *The Cerebral Cortex of Man*, © 1950 Macmillan Library Reference. Renewed 1978 by Theodore Rasmussen. Reprinted by permission of the Gale Group. **25.28** Above, Micrograph Lennart Nilsson © Boehringer Ingelheim International GmbH. **25.29** www.2.gasou.edu/psychology/courses/muchinsky and www.occipita.cfa.cmu.edu. **25.30** © Ophthalmoscopic image from Webvision http://webvision.med.utah.edu/. **25.32** (a) Right, © Fabian/Corbis Sygma; (d) Medtronic Xomed; (e) Above, Micrograph by Dr. Thomas R. Van De Water, University of Miami Ear Institute. **25.33** (a) © Lily Echeverria/Miami Herald; (b) © APHIS photo by Dr. Al Jenny. **25.34** Chase Swift.

CHAPTER 26: **Page 448** © Catherine Ledner Photography. **26.4** Bottom right, Lisa Starr. **26.5** (a) Courtesy of Dr. Erica Eugster; (b–c) Courtesy of Dr. William H. Daughaday, Washington University School of Medicine, from A.I. Mendelhoff and D.E. Smith, eds., *American Journal of Medicine*, 1956, 20:133. **26.6** Left, Gary Head. **26.7** (a) © Scott Camazine/Photo Researchers, Inc.; (b) © Biophoto Associates/SPL/Photo Researchers, Inc. **Page 454** Lower left, top, © Catherine Ledner Photography; bottom, The Stover Group/D. J. Fort. **26.10** © Yoav Levy/Phototake. **26.11** John S. Dunning/Ardea, London. **Page 458** Lower left, top, © Catherine Ledner Photography; bottom, © Minnesota Pollution Control Agency. **26.7** (a) Dr. Carlos J. Bourdony; (b) Courtesy of G. Baumann, MD, Northwestern University.

CHAPTER 27: **Page 461** Left, © 1999 Dana Fineman/Corbis Sygma; right, © Dow W. Fawcett/Photo Researchers, Inc. **27.1** © Fred SaintOurs/University of Massachusetts-Boston. **27.2** (a) Rajesh Bedi; (b) © Martin Harvey/Photo Researchers, Inc. **27.3** (a) Frieder Sauer/Bruce Coleman, Ltd.; (b) Matjaz Kuntner; (c) © Ron Austing/Frank Lane Picture Agency/Corbis; (d) © Doug Perrine/seapics.com; (e) Carolina Biological Supply Company; (f) Fred McKinney/FPG/Getty Images; (g) Gary Head. **27.5** (b–j) Photographs by Carolina Biological Supply Company; art, After B. Burnside, *Developmental Biology*, 1971: 26: 416–441. Used by permission of Academic Press. **27.12** Photograph, Lennart Nilsson from *A Child is Born*, © 1966, 1977 Dell Publishing Company, Inc. **27.16** Heidi Specht, West Virginia University. **Page 474** Left, upper, © Dow W. Fawcett/Photo Researchers, Inc.; lower, © Lester Lefkowitz/Corbis. **27.17** (a) © Dr. E. Walker/Photo Researchers, Inc.; (b) © Western Ophthalmic Hospital/Photo Researchers, Inc.; (c) Kenneth Greer/Visuals Unlimited; (d) © CNRI/Photo Researchers, Inc. **Page 477** Right, upper, © Dow W. Fawcett/Photo Researchers, Inc.; lower, Gary Head. **Page 479** Right, above, © Don W. Fawcett/Photo Researchers, Inc.; center, Courtesy of © Amy Waddell/UCLA's Mattel Children's Hospital; below, © Reuters NewMedia, Inc./Corbis. **27.21** (a–d) Photographs, Lennart Nilsson, *A Child is Born*, © 1966, 1977 Dell Publishing Company, Inc. **27.22** Left, © Zeva Oelbaum/Corbis; right, James W. Hanson, MD. **27.25** Photograph, Lisa Starr; art, Adapted from L.B. Arey, *Developmental Anatomy*, Philadelphia, W.G. Saunders Co., 1965. **27.26** (a) David M. Parichy; (b–c) Dr. Sharon Amacher.

CHAPTER 28: **Page 488** © David Nunuk/Photo Researchers, Inc. **28.1** Left to right, © Amos Nachoum/Corbis; A.E. Zuckerman/Tom Stack & Associates; © Corbis. **28.2** (a) © Tom Davis; (b) © Cynthia Bateman, Bateman Photography. **28.3** © Jeff Lepore/Photo Researchers, Inc. **Page 492** Left, upper,

© David Nunuk/Photo Researchers, Inc.; lower, NASA. **28.5** (a) © Lilliam Lampas/Omni-Photo; (b) © Christine Pemberton/Omni-Photo. **28.7** © Jacques Langevin/Corbis Sygma. **28.8** Left, NASA; right, © Danny Lehman/Corbis. **28.9** (a) © Joe McDonald/Corbis; (a, inset) © Whispers Appleblossom, courtesy of Goldsmith Seeds, Inc.; (b) © Wayne Bennett/Corbis; (c) © Douglas P. Wilson/Corbis. **28.10** Helen Rodd. **28.11** Right, David Reznick/University of California-Riverside; computer enhanced by Lisa Starr. **28.12** Photo, NASA. **28.15** Left, © Adrian Arbib/Corbis; right, © Don Mason/Corbis. **28.16** © John Alcock/Arizona State University.

CHAPTER 29: **Page 503** Left, USDA; right, above, Photo by Daniel Wojak/USDA; right, below, John Kabashima. **29.1** From left, Donna Hutchins; © Martin Harvey, Gallo Images/Corbis; © Len Robinson, Frank Lane Picture Agency/Corbis; © B.G. Thomson/Photo Researchers, Inc. **29.2** Left, Harlo H. Hadow; right, © Bob and Miriam Francis/Tom Stack & Associates. **29.3** Thomas W. Doeppner. **29.4** (mountain peak) © Richard Cummins/Corbis; (a, d) Don Roberson; (b) © Kennan Ward/Corbis; (c) © D. Robert Franz/Corbis. **29.5** Left, above, © Michael Abbey/Photo Researchers, Inc.; below, © Eric V. Grave/Photo Researchers, Inc. **29.6** Art, After N. Weldan and F. Bazazz, *Ecology*, 56:681–688, © 1975 Ecological Society of America; Photographs from left, © Tony Wharton, Frank Lane Picture Agency/Corbis; © Hal Horwitz/Corbis; © Joe McDonald/Corbis. **29.7** Stephen G. Tilley. **29.8** (a) After Rickleffs & Miller, *Ecology*, fourth edition, p. 459 (Fig. 23.12a) and p. 461 (Fig. 23-14); photograph, © W. Perry Conway/Corbis. **29.9** (b) © Ed Cesar/Photo Researchers, Inc.; (c) © Kennan Ward. **29.10** (a–c) Edward S. Ross; (d) © Nigel Jones. **29.11** (a) © JH Pete Carmichael; (b) W. M. Laetsch. **29.12** (a–b) Thomas Eisner, Cornell University; (c) Douglas Faulkner/Sally Faulkner Collection; (d) © Bob Jensen Photography. **29.13** (a) Courtesy of Ken Nemuras; (b) CDC; (c) © The Samuel Roberts Noble Foundation, Inc.; (d) Courtesy of Colin Purrington, Swarthmore College. **29.14** © E.R. Degginger/Photo Researchers, Inc. **29.15** © Peter J. Bryant/Biological Photo Service. **29.16** (a) © Doug Peebles/Corbis; (b) © Pat O'Hara/Corbis; (c–d) © Tom Bean/Corbis; (e) © Duncan Murrell/Taxi/Getty Images. **29.17** (a) R. Barrick/USGS; (b) © 1980 Gary Braasch; (c) Don Johnson/Photo Nats, Inc. **29.18** (a, c) Jane Burton/Bruce Coleman, Ltd.; (b) Heather Angel; (d–e) Based on Jane Lubchenco, *American Naturalist*, 112:23–29, © 1978 by The University of Chicago Press. **29.19** Left, The University of Alabama Center for Public TV; (a) © Angelina Lax/Photo Researchers, Inc.; (b) © Pr. Alexande Meinesz, University of Nice-Sophia Antipolis. **Page 516** Left, upper, John Kabashima; lower, © Joe McDonald/Corbis. **29.20** John Carnemolla/Australian Picture Library/Westlight. **29.21** After W. Dansgaard, et al., *Nature*, 364:218–220, 15 July 1993; D. Raymond, et al., *Science*, 259:926–033, February 1993; W. Post, *American Scientist*, 78:310–326, July–August 1990. **29.22** Left, After S. Fridriksson, *Evolution of Life on a Volcanic Island*, Butterworth, London, 1975; photograph right, © Pierre Vauthey/Corbis Sygma. **29.23** (a) © Susan G. Drinker/Corbis; (b) Frans Lanting/Minden Pictures (computer modified by Lisa Starr). **29.24** © James Marshall/Corbis. **29.25** Left, © Bagla Pallava/Corbis Sygma; right, © A. Bannister/Photo Researchers, Inc. **29.28** Art, Lisa Starr with photographs © 2000 PhotoDisc, Inc. **29.29** Bureau of Land Management. **29.30** (a) © Anthony Bannister, Gallo Images/Corbis; (b) © Bob Jensen Photography; (c) Cedric Vaucher. **29.31** Courtesy of Eternal Reefs, Inc., www.eternalreefs.com.

CHAPTER 30: **Page 526** Right, above, © Harold Naideau/Corbis; below, © Chase Swift/Corbis. **30.1** Art, Gary Head and Lisa Starr; photographs, above, © PhotoDisc, Inc.; below, David Neal Parks. **30.2** Above, photograph, Alan and Sandy Carey; art, After R.L. Smith, *Ecology and Field Biology*, fifth edition. **30.3, 30.4** Lisa Starr with photographs, flowering plants, © Frank Oberle/Stone/Getty Images; hawk, © J. Lichter/Photo Researchers, Inc; crow, © Ed Reschke; milk snake, Mike Pingleton/www.pingleton.com; garter snake, Michael Jeffords; frog, John H. Gerard; yellow spider, Michael Jeffords; weasel, courtesy Biology Department, Loyola Marymount University; sparrow, © Rod Planck/Photo Researchers, Inc; cutworm, © Nigel Cattlin/Holt Studios International/Photo Researchers, Inc.; gopher and prairie vole, © Tom McHugh/Photo Researchers, Inc.; all others © 2000 PhotoDisc, Inc. **30.5** Gary Head. **30.6** © David T. Grewcock/Corbis. **30.8** NASA/GSFC. **30.10** Left, After G.E. Likens and F.H. Bormann, "An Experimental Approach to New England Landscapes," in A.D. Hasler (ed.), *Coupling of Land and Water Systems*, Chapman & Hall, 1975; right, © Dr. Elizabeth Hane. **30.11** Water Resources Council. **Page 533** Right, upper, © Harold Naideau/Corbis; lower, © Jose Luis Pelaez, Inc./Corbis. **30.12** Left, Based on NASA photograph from JSC Digital Image Collection; (a–b) After Paul Hertz, photograph, © 2000 PhotoDisc, Inc. **30.13** Based on NASA photograph from JSC Digital Image Collection. **30.14** (a) Compilation of data from Mauna Loa Observatory, Keeling and Whorf, Scripps Institute of Oceanography; (b) Compilation of data from Prinn, et al., CDIAC, Oak Ridge National Laboratory; World Resources Institute; Khalil and Rasmussen, Oregon Graduate Institute of Science and Technology, CDIAC DB-1010; Leifer and Chan, CDIAC DB-1019; (c–d) Compilation of data from World Resources Institute; Law Dome ice core samples, Etheridge, Pearman, and Fraser, Commonwealth Scientific and Industrial Research Organisation; Prinn, et al., CDIAC, Oak Ridge National Laboratory; Leifer and Chan, CDIAC DB-1019. **30.16** Photograph, © Jeff Vanuga/Corbis. **30.17** © Frederica Georgia/Photo Researchers, Inc. **Page 539** Right, upper, © Harold Naideau/Corbis; lower, © Ian Harwood; Ecoscene/Corbis. **30.18** Photograph, © 2000 PhotoDisc, Inc. **30.19** Fisheries & Oceans Canada, Experimental Lakes Area. **30.20** (a–b) Courtesy of NASA's Terra satellite, supplied by Ted Scambos, National Snow and Ice Data Center, University of Colorado, Boulder; (c) Courtesy of Keith Nicholls, British Antarctic Survey.

CHAPTER 31: **Page 543** Left, © Hank Fotos Photography; right, © Wolfgang Kaehler/Corbis. **31.3** Photographs, NASA. **31.5** Left, Adapted from *Living in the Environment* by G. Tyler Miller, Jr., p. 428. © 2002 by BrooksCole; right, © Ted Spiegel/Corbis. **31.6** Alex MacLean/Landslides. **31.7** Art based on NASA photograph. **31.8** Below left, © Sally A. Morgan, Ecoscene/Corbis; below right, © Bob Rowan, Progressive Image/Corbis. **31.9** Lisa Starr, after The World Wildlife Fund and other sources, using NASA photograph. **31.10** Hans Renner. **31.11** (a–b) Courtesy of Jim Deacon, University of Edinburgh. **31.12** © Orbimage Imagery. **31.13** Left, Jack Wilburn/Animals Animals; right, AP/Wide World Photos. **31.14** Above, © Tom Bean Photography; below, Jonathan Scott/Planet Earth Pictures. **31.15** From left, © 1991 Gary Braasch; © Adolf Schmidecker/FPG/Getty Images; Edward S. Ross. **31.16** © James Randklev/Corbis. **31.17** Left, © Raymond Gehman/Corbis; right, © Nigel Cook/Dayton Beach News Journal/Corbis Sygma. **31.18** (a) © Darrell Gulin/Corbis; (b) © Paul A. Souders/Corbis; (c) © Pat O'Hara/Corbis. **31.19** D.W. MacManiman. **31.21** (a) © Annie Griffiths Belt/Corbis; (b) © Douglas Peebles/Corbis. **31.22** © Nancy Sefton. **31.23** (a) Clockwise from top left, Douglas Faulkner/Sally Faulkner Collection; © Corbis; Alex Kirstitch; Douglas Faulkner/Sally Faulkner Collection; © Amos Nachoum/Corbis; Jeff Rotman; (b) T. Garrison, *Oceanography: An Invitation to Marine Science*, third edition, Brooks/Cole, 2000. All rights reserved. **31.24** © Greenpeace/Grace. **31.25** (b) © Peter David/FPG/Getty Images; (c) Robert Vrijenhoek, MBARI. **31.27** (c–d) NASA–Goddard Space Flight Center Scientific Visualization Studio. **31.28** (a) © Eye of Science/Photo Researchers, Inc.; (b) © Douglas P. Wilson; Frank Lane Picture Agency/Corbis; (c) Courtesy of Dr. Anwar Huq and Dr. Rita Colwell, University of Maryland; (d) CHAART, at NASA Ames Research Center; (e) © Raghu Rai/Magnum Photos. **31.29** After M.H. Dickerson, "ARAC: Modeling and Ill Wind," in *Energy and Technology Review*, August 1987. Used by permission of University of California, Lawrence Livermore National Laboratory and U.S. Dept. of Energy.

CHAPTER 32: **Page 566** Left, © Ralph Fountain; right, © Scott Camazine. **32.1** (a) Eugene Kozloff; (b–c) Stevan Arnold. **32.2** Left, © Robert M. Timm & Barbara L. Clauson, University of Kansas; (a–b) Reprinted from *Trends in Neuroscience*, Vol. 21, Issue 2, 1998, L.J. Young, W. Zuoxin, T.R. Insel, "Neuroendocrine bases of monogamy," pp. 71–75, ©1998, with permission from Elsevier Science. **32.3** (a) Eric Hosking; (b) © Stephen Dalton/Photo Researchers, Inc.; (c) © Jennie Woodcock, Reflections Photolibrary/Corbis. **32.4** Nina Leen/TimePix/Getty Images; inset, © Robert Semeniuk/Corbis. **32.5** © James Zipp/Photo Researchers, Inc. **32.6** (a) John Bova/Photo Researchers, Inc.; (b) Robert Maier/Animals Animals; (c) Jack Clark/Comstock, Inc.; (d) From L. Clark, *Parasitology Today*, 6(11), Elsevier Trends Journals, 1990, Cambridge, UK. **32.7** (a) © Monty Sloan, www.wolfphotography.com; (b) © Tom and Pat Leeson, leesonphoto.com; (c) © Kevin Schafer/Corbis. **32.8** Above, Stephen Dalton/Photo Researchers, Inc. **32.9** (a) John Alcock, Arizona State University 588; (b–c) © Pam Gardner, Frank Lane Picture Agency/Corbis; (d) © D. Robert Franz/Corbis. **32.10** (a) © Ingo Arndt/Nature Picture Library; (b) Michael Francis/The Wildlife Collection. **32.11** (a) © B. Borrell Casals; Frank Lane Picture Agency/Corbis; (b) © Frank Lane Agency/Bruce Coleman, Inc.; (c) © John Conrad/Corbis. **32.12** © Paul Nicklen/National Geographic/Getty Images. **32.13** (a) © Tom and Pat Leeson, leesonphoto.com; (b) John Alcock, Arizona State University. **32.14** © Jeff Vanuga/Corbis. **32.15** © Eric and David Hosking/Corbis. **32.16** (a) © Australian Picture Library/Corbis; (b) © Alexander Wild; (c) © Professor Louis De Vos. **32.17** (a) Kenneth Lorenzen; (b) © Peter Johnson/Corbis; (c) Nicola Kountoupes/Cornell University. **32.18** © Dr. Tim Jackson, University of Pretoria. **32.19** (a) © Steve Bloom Images/www.stevebloom.com; (b) © Gallo Images/Corbis. **Page 581** © Matthew Alan/Corbis. **32.20** F. Schutz. **32.21** © Brad Bergstrom. **Page 583** © Joseph Sohm, Visions of America/Corbis.

Index

The letter i *designates illustration;* t *designates table;* **bold** *designates defined term;* ■ *highlights the location of applications contained in text.*

D

G

H

Q

T